Supplement B

The chemistry of
acid derivatives
Part 2

THE CHEMISTRY OF FUNCTIONAL GROUPS

A series of advanced treatises under the general editorship of
Professor Saul Patai

The chemistry of alkenes (2 volumes)
The chemistry of the carbonyl group (2 volumes)
The chemistry of the ether linkage
The chemistry of the amino group
The chemistry of the nitro and nitroso groups (2 parts)
The chemistry of carboxylic acids and esters
The chemistry of the carbon—nitrogen double bond
The chemistry of amides
The chemistry of the cyano group
The chemistry of the hydroxyl group (2 parts)
The chemistry of the azido group
The chemistry of acyl halides
The chemistry of the carbon—halogen bond (2 parts)
The chemistry of the quinonoid compounds (2 parts)
The chemistry of the thiol group (2 parts)
The chemistry of the hydrazo, azo and azoxy groups (2 parts)
The chemistry of amidines and imidates
The chemistry of cyanates and their thio derivatives (2 parts)
The chemistry of diazonium and diazo groups (2 parts)
The chemistry of the carbon—carbon triple bond (2 parts)
Supplement A: The chemistry of double-bonded functional groups (2 parts)
Supplement B: The chemistry of acid derivatives (2 parts)

$$-COOH \qquad -COOR \qquad -CONH_2$$

Supplement B

The chemistry of
acid derivatives
Part 2

Edited by
SAUL PATAI

The Hebrew University, Jerusalem

1979

JOHN WILEY & SONS

CHICHESTER – NEW YORK – BRISBANE – TORONTO

An Interscience ® Publication

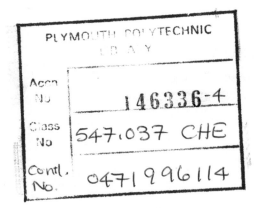
Copyright © 1979 by John Wiley & Sons Ltd.

Library of Congress Catalog Card No. 75–6913

ISBN 0 471 99610 6 (Pt. 1)
ISBN 0 471 99611 4 (Pt. 2)
ISBN 0 471 99609 2 (Set)

Typeset by Preface Ltd., Salisbury, Wiltshire.
Printed in Great Britain by Unwin Brothers Ltd.,
The Gresham Press, Old Woking, Surrey.

Contributing Authors

M. C. Baird — Department of Chemistry, Queen's University, Kingston, Canada

G. V. Boyd — Department of Chemistry, Chelsea College, London, England

C. A. Bunton — Department of Chemistry University of California, Santa Barbara, California 93106, U.S.A.

J. P. Coleman — Corporate Research Department, Monsanto Company, St. Louis, Missouri 63166, U.S.A.

I. G. Csizmadia — Department of Chemistry, University of Toronto, Toronto, Ontario, Canada

S. Detoni — University of Ljubljana, Yugoslavia

R. Foster — Chemistry Department, University of Dundee, Dundee, Scotland

O. S. Gal — Boris Kidrič Institute of Nuclear Sciences, Vinča-Belgrade, Yugoslavia

R. S. Givens — Chemistry Department, The University of Kansas, Lawrence, Kansas 66045, U.S.A.

D. Hadži — University of Ljubljana, Yugoslavia

R. Håkansson — Division of Organic Chemistry 1, Chemical Centre, University of Lund, Sweden

W. Kantlehner — Fachhochschule Aalen, Aalen, Germany

C. Kozmuta — Physical Institute, Technical University, Budapest, Hungary

N. Levi — Chemistry Department, The University of Kansas, Lawrence, Kansas 66045, U.S.A.

O. I. Mićić — Boris Kidrič Institute of Nuclear Sciences, Vinča-Belgrade, Yugoslavia

M. A. Ogliaruso — Department of Chemistry, Virginia Polytechnic Institute and State University, Blacksburg, Virginia 24061, U.S.A.

M. R. Peterson — Department of Chemistry, University of Toronto, Toronto, Ontario, Canada

W. H. Prichard — The City University, London, England

J. Ratuský — Institute of Organic Chemistry and Biochemistry, Czechoslovak Academy of Sciences, Prague, Czechoslovakia

M. A. Robb — Department of Chemistry, Queen Elizabeth College, University of London, London, England

L. S. Romsted   Department of Chemistry, University of California, Santa Barbara, California 93106, U.S.A.

R. Shaw   1162 Quince Avenue, Sunnyvale, California 94087, U.S.A.

S. W. Tam   Department of Chemistry Chung Chi College, The Chinese University of Hong Kong, Shatin, N.T., Hong Kong

R. Taylor   University of Sussex, Falmer, Brighton, U.K.

J. Voss   Institute for Organic Chemistry and Biochemistry, University of Hamburg, Hamburg, Germany

J. F. Wolfe   Department of Chemistry, Virginia Polytechnic Institute and State University, Blacksburg, Virginia 24061, U.S.A.

Foreword

Most of the originally planned volumes of the series *The Chemistry of the Functional Groups* have appeared already or are in the press. The first two books of the series, *The Chemistry of Alkenes* (1964) and *The Chemistry of the Carbonyl Group* (1966) each had a second volume published in 1970, with chapters not included in the plans of the original volumes and others which were planned but failed to materialize.

This book is the second of a set of supplementary volumes which should include material on more than a single functional group. For these volumes a division into six categories is envisaged, and supplementary volumes in each of these categories will be published as the need arises. These volumes should include 'missing chapters' as well as chapters which give a unified and comparative treatment of several related functional groups together.

The planned division is as follows:

Supplement A: The Chemistry of Double-Bonded Functional Groups (C=C; C=O; C=N; N=N etc.).

SupplementB: The Chemistry of Acid Derivatives (COOH; COOR; $CONH_2$ etc.).

Supplement C: The Chemistry of Triple-Bonded Functional Groups (C≡C; C≡N; $-\overset{+}{N}≡N$ etc.).

Supplement D: The Chemistry of Halides and Pseudohalides (−F; −Cl; −Br; −I; $-N_3$; −OCN; −NCO etc.).

Supplement E: The Chemistry of Ethers, Crown Ethers, Hydroxyl Groups and their Sulphur Analogues.

Supplement F: The Chemistry of Amines, Nitroso and Nitro Compounds and their Derivatives.

In the present volume, as usual, the authors have been asked to write chapters in the nature of essay-reviews not necessarily giving extensive or encyclopaedic coverage of the material. Once more, not all planned chapters materialized, but we hope that additional volumes of Supplement B will appear, when these gaps can be filled together with coverage of new developments in the various fields treated.

Jerusalem, May 1979   SAUL PATAI

The Chemistry of Functional Groups
Preface to the series

The series 'The Chemistry of Functional Groups' is planned to cover in each volume all aspects of the chemistry of one of the important functional groups in organic chemistry. The emphasis is laid on the functional group treated and on the effects which it exerts on the chemical and physical properties, primarily in the immediate vicinity of the group in question, and secondarily on the behaviour of the whole molecule. For instance, the volume *The Chemistry of the Ether Linkage* deals with reactions in which the C–O–C group is involved, as well as with the effects of the C–O–C group on the reactions of alkyl or aryl groups connected to the ether oxygen. It is the purpose of the volume to give a complete coverage of all properties and reactions of ethers in as far as these depend on the presence of the ether group but the primary subject matter is not the whole molecule, but the C–O–C functional group.

A further restriction in the treatment of the various functional groups in these volumes is that material included in easily and generally available secondary or tertiary sources, such as Chemical Reviews, Quarterly Reviews, Organic Reactions, various 'Advances' and 'Progress' series as well as textbooks (i.e. in books which are usually found in the chemical libraries of universities and research institutes) should not, as a rule, be repeated in detail, unless it is necessary for the balanced treatment of the subject. Therefore each of the authors is asked *not* to give an encyclopaedic coverage of his subject, but to concentrate on the most important recent developments and mainly on material that has not been adequately covered by reviews or other secondary sources by the time of writing of the chapter, and to address himself to a reader who is assumed to be at a fairly advanced post-graduate level.

With these restrictions, it is realized that no plan can be devised for a volume that would give a *complete* coverage of the subject with *no* overlap between chapters, while at the same time preserving the readability of the text. The Editor set himself the goal of attaining *reasonable* coverage with *moderate* overlap, with a minimum of cross-references between the chapters of each volume. In this manner, sufficient freedom is given to each author to produce readable quasi-monographic chapters.

The general plan of each volume includes the following main sections:

(a) An introductory chapter dealing with the general and theoretical aspects of the group.

(b) One or more chapters dealing with the formation of the functional group in question, either from groups present in the molecule, or by introducing the new group directly or indirectly.

(c) Chapters describing the characterization and characteristics of the functional groups, i.e. a chapter dealing with qualitative and quantitative methods of determination including chemical and physical methods, ultraviolet, infrared, nuclear magnetic resonance and mass spectra: a chapter dealing with activating and directive effects exerted by the group and/or a chapter on the basicity, acidity or complex-forming ability of the group (if applicable).

(d) Chapters on the reactions, transformations and rearrangements which the functional group can undergo, either alone or in conjunction with other reagents.

(e) Special topics which do not fit any of the above sections, such as photochemistry, radiation chemistry, biochemical formations and reactions. Depending on the nature of each functional group treated, these special topics may include short monographs on related functional groups on which no separate volume is planned (e.g. a chapter on 'Thioketones' is included in the volume *The Chemistry of the Carbonyl Group*, and a chapter on 'Ketenes' is included in the volume *The Chemistry of Alkenes*). In other cases certain compounds, though containing only the functional group of the title, may have special features so as to be best treated in a separate chapter, as e.g. 'Polyethers' in *The Chemistry of the Ether Linkage*, or 'Tetraaminoethylenes' in *The Chemistry of the Amino Group*.

This plan entails that the breadth, depth and thought-provoking nature of each chapter will differ with the views and inclinations of the author and the presentation will necessarily be somewhat uneven. Moreover, a serious problem is caused by authors who deliver their manuscript late or not at all. In order to overcome this problem at least to some extent, it was decided to publish certain volumes in several parts, without giving consideration to the originally planned logical order of the chapters. If after the appearance of the originally planned parts of a volume it is found that either owing to non-delivery of chapters, or to new developments in the subject, sufficient material has accumulated for publication of a supplementary volume, containing material on related functional groups, this will be done as soon as possible.

The overall plan of the volumes in the series 'The Chemistry of Functional Groups' includes the titles listed below:

The Chemistry of Alkenes (two volumes)
The Chemistry of the Carbonyl Group (two volumes)
The Chemistry of the Ether Linkage
The Chemistry of the Amino Group
The Chemistry of the Nitro and Nitroso Group (two parts)
The Chemistry of Carboxylic Acids and Esters
The Chemistry of the Carbon–Nitrogen Double Bond
The Chemistry of the Cyano Group
The Chemistry of Amides
The Chemistry of the Hydroxyl Group (two parts)
The Chemistry of the Azido Group
The Chemistry of Acyl Halides
The Chemistry of the Carbon–Halogen Bond (two parts)
The Chemistry of Quinonoid Compounds (two parts)
The Chemistry of the Thiol Group (two parts)
The Chemistry of Amidines and Imidates

The Chemistry of the Hydrazo, Azo and Azoxy Groups
The Chemistry of Cyanates and their Thio Derivatives (two parts)
The Chemistry of Diazonium and Diazo Groups (two parts)
The Chemistry of the Carbon—Carbon Triple Bond (two parts)
Supplement A: The Chemistry of Double-bonded Functional Groups (two parts)
Supplement B: The Chemistry of Acid Derivatives (two parts)

Titles in press:
The Chemistry of Ketenes, Allenes and Related Compounds
Supplement E: The Chemistry of Ethers, Crown Ethers, Hydroxyl Groups and their Sulphur Analogues
The Chemistry of the Sulphonium Group

Future volumes planned include:
The Chemistry of Organometallic Compounds
The Chemistry of Sulphur-containing Compounds
Supplement C: The Chemistry of Triple-bonded Functional Groups
Supplement D: The Chemistry of Halides and Pseudo-halides
Supplement F: The Chemistry of Amines, Nitroso and Nitro Groups and their Derivatives

Advice or criticism regarding the plan and execution of this series will be welcomed by the Editor.

The publication of this series would never have started, let alone continued, without the support of many persons. First and foremost among these is Dr Arnold Weissberger, whose reassurance and trust encouraged me to tackle this task, and who continues to help and advise me. The efficient and patient cooperation of several staff-members of the Publisher also rendered me invaluable aid (but unfortunately their code of ethics does not allow me to thank them by name). Many of my friends and colleagues in Israel and overseas helped me in the solution of various major and minor matters, and my thanks are due to all of them, especially to Professor Z. Rappoport. Carrying out such a long-range project would be quite impossible without the non-professional but none the less essential participation and partnership of my wife.

The Hebrew University   SAUL PATAI
Jerusalem, ISRAEL

Contents

1. Recent advances in the theoretical treatment of acid derivatives   1
   I. G. Csizmadia, M. R. Peterson, C. Kozmuta and M. A. Robb

2. Thermochemistry of acid derivatives   59
   R. Shaw

3. Chiroptical properties of acid derivatives   67
   R. Håkansson

4. Mass spectra of acid derivatives   121
   S. W. Tam

5. Complexes of acid anhydrides   175
   R. Foster

6. Hydrogen bonding in carboxylic acids and derivatives   213
   D. Hadži and S. Detoni

7. The synthesis of carboxylic acids and esters and their derivatives   267
   M. A. Ogliaruso and J. F. Wolfe

8. The chemistry of lactones and lactams   491
   G. V. Boyd

9. The chemistry of orthoamides of carboxylic acids and carbonic acid   533
   W. Kantlehner

10. Detection and determination of acid derivatives   601
    W. H. Prichard

11. The photochemistry of organic acids, esters, anhydrides, lactones and imides   641
    R. S. Givens and N. Levi

12. Radiation chemistry of acids, esters, anhydrides, lactones and lactams   755
    O. I. Mićić and O. S. Gal

13. The electrochemistry of carboxylic acids and derivatives: cathodic reductions   781
    J. P. Coleman

14. Decarbonylation reactions of acid halides and aldehydes by chlorotris-(triphenylphosphine)rhodium (I)   825
    M. C. Baird

15. Pyrolysis of acids and their derivatives   859
    R. Taylor

Contents

16. Transcarboxylation reactions of salts of aromatic carboxylic acids   915
     J. Ratuský

17. Micellar effects upon deacylation   945
     C. A. Bunton and L. S. Romsted

18. The chemistry of thio acid derivatives   1021
     J. Voss

19. The synthesis of lactones and lactams   1063
     J. F. Wolfe and M. A. Ogliaruso

     Author Index   1331

     Subject Index   1431

CHAPTER **12**

Radiation chemistry of acids, esters, anhydrides, lactones and lactams

O. I. MIĆIĆ and O. S. GAL

Boris Kidrič Institute of Nuclear Sciences, Vinča-Belgrade, Yugoslavia

I.	INTRODUCTION	755
II.	ACIDS	756
	A. Ionization-initiated reactions	756
	1. Ion–molecule reactions	757
	2. Neutral free radicals	757
	3. Stable products	759
	B. Radiation Chemistry of some Carboxylic Acids . . .	759
	1. Acetic acid	759
	2. Oxalic acid	760
	3. Oleic acid	762
	C. Aqueous Solutions	763
	1. Reactions with radicals of water	763
	2. Reactive intermediates	766
	a. OH and H adducts	766
	b. e_{aq}^- Adducts	768
III.	ESTERS	769
	A. Mechanism of Decomposition	769
	B. Products	770
	C. Oxidation of Methyl Oleate	771
	D. Aqueous Solutions	772
IV.	ANHYDRIDES	773
V.	LACTONES	773
	A. Aqueous Solutions of Ascorbic Acid	774
	1. Reactions with OH radical	774
	2. Reactions with e_{aq}^- radical	775
	3. Reactions with H atoms	775
VI.	LACTAMS	777
VII.	REFERENCES	778

I. INTRODUCTION

The radiolysis of carboxylic acids and their derivatives has been investigated less extensively than the radiolysis of water, hydrocarbons and alcohols. However, the general features of radiation—chemical decomposition of these compounds are known. The radiation chemistry of these compounds is essentially the chemistry of free radicals coupled with the formation of several important intermediate species. Absorption of high-energy radiation creates ionization and excitation followed by ion—molecule reactions, charge neutralization and dissociation of the resulting molecules into free radicals.

The radiolysis of carboxylic acids is simpler than that of esters, anhydrides, lactones and lactams. This is understandable, since the presence of an additional group gives rise to another combination of elementary reactions. The acids and esters have been studies for two reasons. Firstly, these compounds are widely spread as natural products and are present in almost all biochemical materials. The effect of radiation on biological macromolecules is rather complex, therefore information about radiation decomposition mechanisms of their components such as acids and esters is very important. Secondly, in the course of irradiation, a number of intermediates are produced whose nature and kinetical behaviour are of interest in free-radical chemistry.

As energy sources high-energy photons and high-energy particles are used. The pulse radiolysis techique is used to create free radicals which are then monitored by optical absorption or electron spin resonance spectroscopy, electrical conductivity, etc. Continuous irradiation, for which gamma rays from ^{60}Co are mostly used, is suitable to study the nature and the overall yield of chemical change. Radiation chemistry yields are usually expressed as G-values, the number of molecules formed or destroyed per 100 eV of energy absorbed by the systems.

Radiation studies of acids and esters are generally carried out in the liquid state, although a certain number of investigations have been devoted to ionizing radiation effects in the solid state. Most extensively studied are aqueous solutions, where modes of energy absorption differ from those in pure compounds. In very dilute aqueous solutions practically all of the energy is deposited in the water and the chemical reactions which occur are caused by interactions of reactive species originating from the water.

This review is predominantly concerned with the radiation chemistry of acids and esters. Anhydrides, lactones and lactams have been investigated to a lesser extent, lactones and lactams mainly in connection with some complex compounds containing these functional groups.

II. ACIDS

A. Ionization-initiated Reactions

Studies dealing with a number of irradiated organic compounds show that the main primary reaction induced by ionized radiation is the creation of a positive hole produced by the ejection of an electron. A molecular anion is then produced by the capture of the ejected electron[1]. Other intermediates stabilized in the solid state or detected by pulse radiolysis are believed to be secondary products originating from such positive and negative primary species.

During the past two decades carboxylic acids have been extensively investigated. The interaction of an energetic charged particle with the molecules of a carboxylic acid in the irradiated medium leads to ionization (reaction 1). The ionization is

$$RCOOH\ldots\ldots RCOOH \xrightarrow{} RCOOH\ldots\ldots RCOOH^+ + e^- \qquad (1)$$

followed by a rapid train of processes involving different intermediates which can be ions or neutral radicals. Radical intermediates have been investigated by e.s.r. at low temperatures and by optical pulse radiolysis at room temperature. Various techniques have also been applied to the identification of the stable products formed by reactions of different intermediates. These studies have subsequently improved our understanding of certain features important for the radiolysis of pure carboxylic acids.

1. Ion—molecule reactions

Irradiation of a carboxylic acid yields anion radicals produced by the capture of an ejected electron (reaction 2).[2] The capture of the electron by a COOH group is a faster process than its solvation.

$$e^- + RC \underset{OH}{\overset{O}{<}} \longrightarrow R\dot{C} \underset{OH}{\overset{O^-}{<}} \tag{2}$$

(2)

The positive hole (1) formed in reaction (1) decomposes into a carboxylic radical $R\dot{C}OO$. There is evidence that irradiated single crystals of unsaturated carboxylic acids and their salts, such as maleic acid[3] and potassium hydrogen maleate[4,5] or fumarate[6] give carboxylic radicals produced by proton tunnelling from the positive primary species (1) to the neighbouring carboxyl oxygen atom through the intermolecular hydrogen bond. A similar radical is formed in succinic acid[7]. The transfer of the proton is enhanced by the existence of hydrogen-bonded dimers. As a result of this transfer the dimer dissociates as shown in reaction (3).

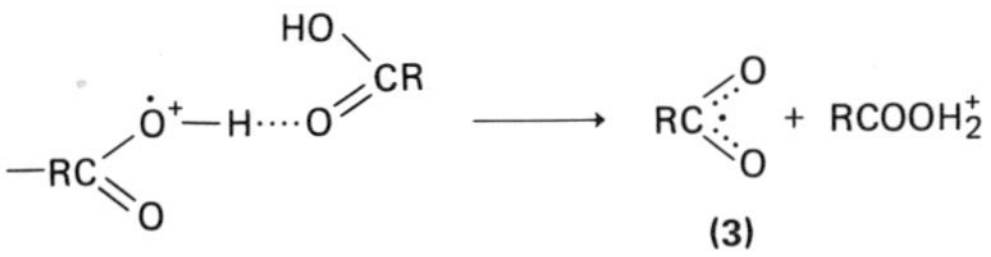

$$-R\dot{C} \underset{O}{\overset{\dot{O}^+ - H \cdots O}{<}} \overset{HO}{\underset{}{>}} CR \longrightarrow RC \underset{O}{\overset{O}{<}} + RCOOH_2^+ \tag{3}$$

(3)

The unpaired electron is localized mainly on oxygen atoms[7]. Minakata and Iwasaki[6] have obtained experimental evidence for the loss of the acidic proton and its transfer to the neighbouring molecule and propose the tunnelling model for the mechanism of the formation of the carboxyl radical 3.

2. Neutral free radicals

It has been found that the carboxyl radical formed by the proton transfer reaction (3) dissociates into an alkyl-type radical, $R\dot{C}H_2$ (4), and a CO_2 molecule[3,8] (reaction 4). When a saturated carboxylic acid is irradiated at room temperature, or when the specimen temperature rises after irradiation, a π radical

$$RCH_2C \underset{O}{\overset{O}{<}} \longrightarrow R\dot{C}H_2 + CO_2 \tag{4}$$

(4)

(5) is produced by abstraction of the hydrogen atom from the carbon atom adjacent to the carboxylic group of the neighbouring molecule[8] (reaction 5). π Radicals are also produced in unsaturated compounds by addition of H or R to a double bond[3].

$$R\dot{C}H_2 + RCH_2COOH \longrightarrow RCH_3 + R\dot{C}HOOH \qquad (5)$$

(5)

Beside the neutral radicals 3, 4 and 5 formed by decomposition of the primary positive ions, neutral acyl radicals (6) are formed by CO bond scission from the negative primary species[9,10] (reaction 6).

$$R\dot{C}\underset{OH}{\overset{O^-}{\diagdown}} \longrightarrow R\dot{C}=O + OH^- \qquad (6)$$

(6)

The radiolysis of ionic salts of carboxylic acids might produce the CO_2^- ionic radical. It seems that significant differences exist in the radiation behaviour of a neutral molecular system and an ionic salt system. Thus, there are indications that the CO_2^- ion could be formed by fragmentation from positive primaries[5]. Previously it was assumed that the CO_2^- ion radical was created in the process of decomposition of the anionic species apparently formed by trapping an electron in reaction (2)[11]. However, later observations of the ^{13}C hyperfine couplings in irradiated malonic acid[9,12] and of the ultraviolet and infrared spectra of irradiated succinic acid[13] have shown that only reaction (6) is produced by negative primary species.

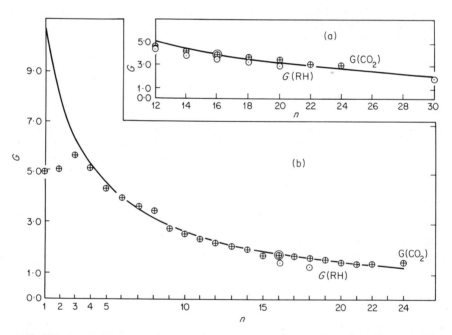

FIGURE 1. Radiolytic yield of carbon dioxide obtained by ^{60}Co- irradiation of normal carboxylic acids of chain length n (a) in the solid state and (b) in the liquid state. Taken from G.-S. Wu and D. R. Howton, *Radiation Res.*, 57, 390 (1974). Reproduced by permission of Academic Press, Inc.

3. Stable products

Decarboxylation is one of the major reactions in the radiolysis of acids[14-16], as a consequence of the instability of the RCOO radical in reaction (4). The yield of carbon dioxide is for the higher homologues of acids in fair approximation inversely proportional to the total number of electrons in the molecule. In all acids carbon dioxide is liberated on irradiation, leaving as principal organic product the saturated hydrocarbon with one carbon atom less than the organic acid, and /or forming a saturated hydrocarbon by dimerization of the radical.

Figure 1 shows the dependence of $G(CO_2)$ and $G(RH)$ of normal carboxylic acids on the chain length[17]. There is some difference between decarboxylation in the liquid and in the solid state: $G(CO_2)$ values are independent of temperature for acids in the liquid state[18], whereas $G(CO_2)$ of solid acids are temperature—sensitive[19]. Moreover, crystalline acids have generally higher $G(CO_2)$ than liquid acids[19]. In the solid state, acids of odd chain length are more sensitive towards ionizing radiation than those of even chain length[19], and, quite generally, the radiolysis of carboxylic acids can be related to their densities. The depressed decarboxylation of the dense short-chain carboxylic acids support this rule[19,20].

The hydrogen yield increases with molecular weight of the acids whereas the opposite is true for carbon dioxide. Some acids have a high radiolytic yield of water[16]. Very often a process of dimerization occurs during the course of radiolysis. There is no general rule concerning which products are formed in the decomposition of acids. In the present review only major and common products are mentioned although many other stable species are produced in small amounts, the yields depending on the structure of the acids[21]. In the following section we shall describe the radiolytic decomposition of some carboxylic acids, which, either for theoretical or practical reasons, have been extensively and thoroughly investigated.

B. Radiation Chemistry of some Carboxylic Acids

1. Acetic acid

The radiolysis of acetic acid in both the liquid[16,22-25] and solid state[26,27] has been extensively studied. Transient species have been identified by e.s.r. techniques following gamma radiolysis at low temperatures[26,27] or by pulse radiolysis combined with optical spectrophotometry at room temperature[25]. Important features of the reaction mechanism can be described as follows.

Taking into account the high yield of anion radical, $G(CH_3COOH^-) = 4.96$, found in crystalline acetic acid at 77 K[27], it can be assumed that the primary effect of the absorption of radiation is ionization:

$$CH_3COOH \xrightarrow{\hspace{1cm}} CH_3COOH^+ + e^-$$

Ionization in the condensed state is immediately followed by the ion--molecule reaction through the hydrogen bond:

$$CH_3COOH^+ \cdots \cdots CH_3COOH \longrightarrow CH_3COOH_2^+ + CH_3CO\dot{O}$$

The $CH_3CO\dot{O}$ radical is very rapidly decomposed:

$$CH_3CO\dot{O} \longrightarrow CH_3 + CO_2$$

For the $CH_3CO\dot{O}$ radical no significant corresponding singlet has been found in the e.s.r. spectrum at 77 K. Stable products are formed by reactions (7)–(9).

$$CH_3 + CH_3COOH \longrightarrow CH_4 + \dot{C}H_2COOH \tag{7}$$

$$CH_3 + CH_3 \longrightarrow C_2H_6 \tag{8}$$

$$CH_3 + CH_3\dot{C}O \longrightarrow CH_3COCH_3 \tag{9}$$

The high yield of methane ($G = 3.9$) compared to the small yield of ethane ($G = 0.48$) and acetone ($G = 0.45$) indicates that under the conditions imposed by gamma radiolysis[27], reaction (7) is more efficient than (8) and (9).

During gamma radiolysis, $\dot{C}H_2COOH$ radicals disappear by the reaction (10) giving succinic acid with a yield $G(CH_2COOH)_2 = 1.74$[28].

$$\dot{C}H_2COOH + \dot{C}H_2COOH \longrightarrow (CH_2COOH)_2 \tag{10}$$

The fate of the thermal electron is capture and formation of the unstable ahion radical CH_3COOH^- (reaction 11) which dissocciates immediately (reactions 12 and 13).

$$e^- + CH_3COOH \longrightarrow CH_3COOH^- \tag{11}$$

$$CH_3COOH^- \longrightarrow CH_3\dot{C}O + OH^- \tag{12}$$

$$CH_3COOH^- \longrightarrow CH_3COO^- + H \tag{13}$$

Reaction (13) is considerably less frequent than (12)[29,30] and this is supported by the high yield of $CH_3\dot{C}O$ radical ($G = 5.1$[25]) and the small yield of H_2 ($G = 0.5$[27]). The CH_3CO radicals disappear by recombination (reaction 14) giving biacetyl with $G(CH_3CO)_2 = 2.20$[25].

$$CH_3\dot{C}O + \dot{C}H_3CO \longrightarrow (CH_3CO)_2 \tag{14}$$

The reactions of hydrogen atoms in liquid acetic acid are the following:

$$H + H \longrightarrow H_2$$

$$H + CH_3COOH \longrightarrow H_2 + \dot{C}H_2COOH$$

$$H + CH_3COOH \longrightarrow H_2 + CH_3CO\dot{O} \tag{15}$$

Reaction (15) was proposed as an explanation for the u.v. photolysis of a solution of H_2S in CH_3COOH at 77 K when only CH_3 radicals were formed[26].

The yield of CO_2 may also indicate the yield of positive ions, i.e. $G(CH_3COOH^+) = G(CO_2) = 5.40$, although $G(CO_2)$ should be somewhat higher since the reaction (15) also indirectly produces CO_2 through $CH_3CO\dot{O}$. The sum of the yields of CO_2 and CO can be compared to the sum of the yields of CH_4, C_2H_6 and CH_3COCH_3.

The yield of negative ions can be estimated as the sum of the yields of $CH_3\dot{C}O$ radicals and H atoms, where $G(CH_3\dot{C}O)$ should be equal to the sum of the yields of CH_3COCH_3, CO_2 and $(CH_3CO)_2$. The established material balance is satisfactory for stable products formed from both positive and negative ions. Some other products present in small amounts have been detected[28], but their formation does not affect the proposed reaction mechanism.

2. Oxalic acid

Oxalic acid is decomposed radiolytically by a first-order process[31] and this fact is largely used in radiation dosimetry where the decomposition constant depends on the type of radiation. The decomposition yield for gamma rays is about $7^{3\,2-3\,4}$, this value being more certain for the anhydrous than for the dihydrated form, where the published values differ considerably[31].

The main product of decomposition is CO_2, which is formed directly in the radiolytic process, and not in subsequent side-reactions. The chemical analysis of irradiated acid has shown that CO and H_2 as well as small amounts of other compounds are formed[32].

An important factor in the decomposition of the substance is the presence of the hydrogen bonds. By analogy to the mechanism originally proposed for oleic acid[35], a scheme for dicarboxylic acid decomposition has been suggested[36]. It is assumed that the primary radiation damage might occur in any part of the molecule, but the defect precursor is in a very short time localized at the double-bonded carboxylic oxygen, i.e. near the hydrogen-bond link between two adjacent molecules in the crystal:

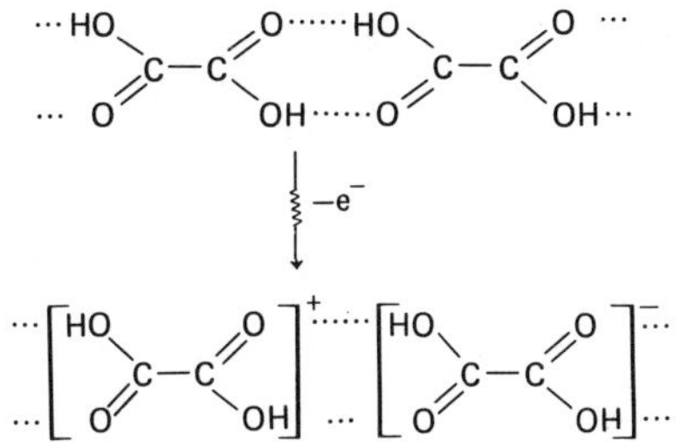

(7a)   (7b)

Instead of electron transfer, a transfer of hydrogen can occur (owing to the presence of the hydrogen bond):

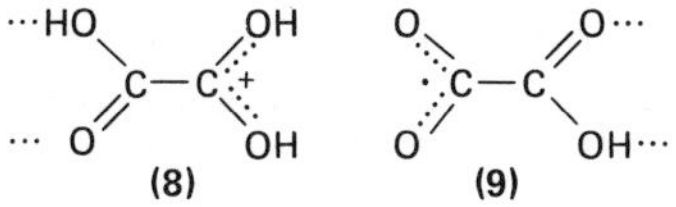

(8)   (9)

Radical 9 decomposes at room temperature by a first-order process[37] (reaction 16).

$$HOOC—COO^{\cdot} \longrightarrow CO_2 + \overset{\cdot}{C}OOH \qquad (16)$$
$$(9) \qquad\qquad\qquad (10)$$

This reaction is in accordance with the ratio $G(CO_2)/G$ (decomposition of acid) = 1 found in the radiolysis of anhydrous oxalic acid[32] and of malonic and succinic acid as well[36]. The radical 10 is relatively stable[37] and it probably disappears by the reaction (17). The ionic species 8 can yield the parent molecule by reaction with a mobile electron (reaction 18).

$$\cdot COOH \longrightarrow CO_2 + H \qquad (17)$$
$$(10)$$

$$HOOC—C(OH_2)^+ + e^- \longrightarrow (COOH)_2 + H \qquad (18)$$

Carbon monoxide, found in small amounts, could be formed from the radical $O\overset{\cdot}{C}$--COOH, whose presence is also indicated[38,39].

In the homologous series of dicarboxylic acids, the resistance toward radiation increases with the length of the aliphatic chain[36].

3. Oleic acid

The interest in the radiolysis of oleic acid is connected with its abundance in lipids of living organisms[40][-43] Oleic acid is not resistant towards radiation: its radiolytic yield of alteration in the absence of air is found to be 17[35]. Irradiation causes polymerization, *cis–trans* isomerization, decarboxylation and hydrogenation, in order of decreasing intensity. All alterations are considered to occur through reactions of highly reactive intermediates produced in primary processes of ionization or covalent-bond homolysis, or via direct *cis–trans* isomerization[35].

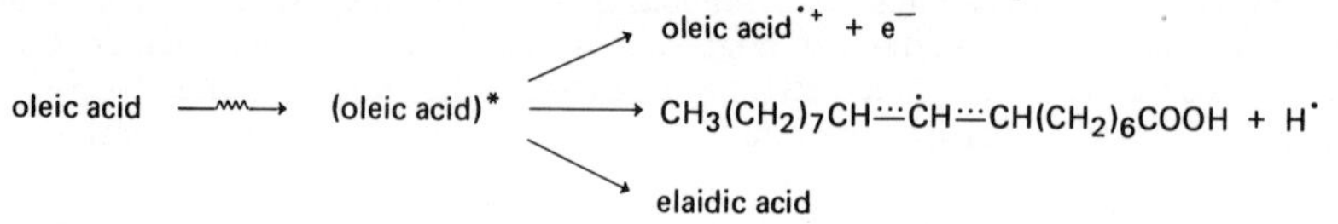

$$
\text{oleic acid} \xrightarrow{\quad\text{---}\text{www}\text{---}\quad} \text{(oleic acid)}^*
\begin{cases}
\longrightarrow \text{oleic acid}^{\cdot+} + e^- \\
\longrightarrow CH_3(CH_2)_7CH\text{---}\overset{\cdot}{C}H\text{---}CH(CH_2)_6COOH + H^\cdot \\
\longrightarrow \text{elaidic acid}
\end{cases}
$$

The allylic radicals formed by primary homolytic and hydrogen-atom abstraction processes attain a relatively high steady-state concentration (due to their resonance stabilization) and might react with a second radical of the same type to give unsaturated dimers with two double bonds, e.g. reaction (19).

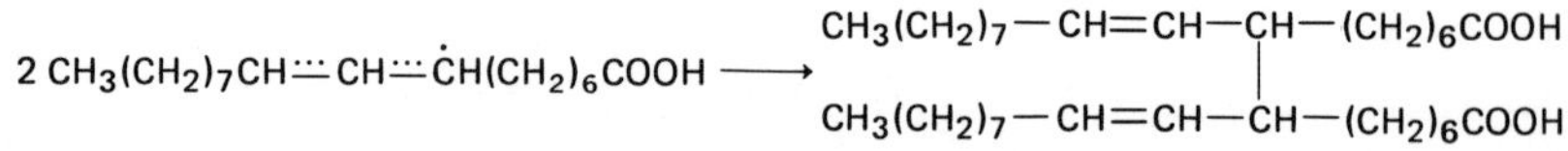

$$
2\ CH_3(CH_2)_7CH\text{---}CH\text{---}\overset{\cdot}{C}H(CH_2)_6COOH \longrightarrow
\begin{array}{l}
CH_3(CH_2)_7-CH{=}CH-CH-(CH_2)_6COOH \\
\qquad\qquad\qquad\qquad\qquad | \\
CH_3(CH_2)_7-CH{=}CH-CH-(CH_2)_6COOH
\end{array}
$$

$$(19)$$

The presence of polymer molecules having a single double bond demonstrates that polymerization of the ion–radical also takes place. In this case, a primary cationic radical. $-CH_2CH\text{---}CHCH_2-$ (formed by localization of a hole on the unsaturated group), reacts with an intact olefinic molecule, ultimately producing polymers with either vinylic or allylic branching.

Cis–trans isomerization of oleic to elaidic acid is of considerable importance for radiation yields. The mechanism of isomerization is assumed[35] to proceed through a direct route (a) (subionization excitation processes) and/or an indirect route(b), via an intermediate of ionized oleic acid:

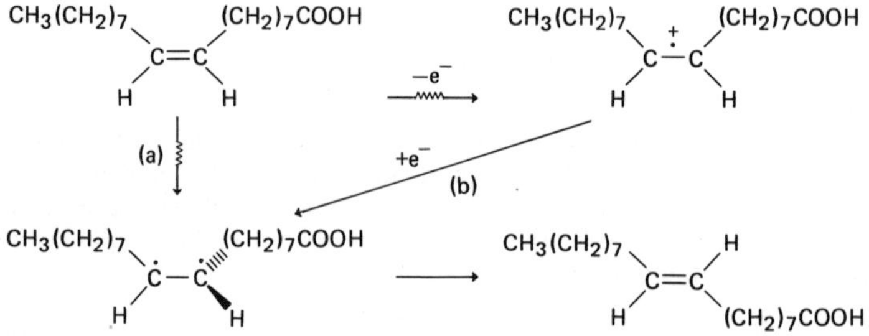

The decarboxylation process is initiated by the ionization reaction and follows the proton transfer (reaction 3). In the overall stoichiometry of the reaction three molecules of the acid are consumed, creating one monomeric hydrocarbon and a fatty acid dimer of the diallyl type:

$$3\ RCOOH \xrightarrow{\quad\text{www}\quad} RH + CO_2 + H_2 + \text{dimer}$$

Hydrogenation of oleic acid is a minor process, but it can be easily demonstrated by the presence of stearic acid and dimeric hydrocarbons. Hydrogen atoms, produced principally by CH bond homolysis may be attached to the double bond:

$$\dot{H} + -CH{=}CH- \longrightarrow -CH_2-\dot{C}H- \xrightarrow{+RH} -CH_2-CH_2- + \dot{R}$$

Experimental data obtained by GLPC and i.r. measurements as well as by gas analysis, show that the main radiolytic products are dimers, double or single unsaturated hydrocarbons, elaidic acid and gaseous CO_2 and H_2 [35,42].

There are significant differences between the radiolysis of oleic acid in the liquid and in the solid state regarding the extent of changes occurring and in the yield and chemical characteristics of the products[42]. Radiolysis in the liquid state produces a more extensive decomposition; the oligomers have a carboxy content lower than that of pure oleic acid and a different degree of unsaturation and yield of CO_2 exceeds that of C_{17} hydrocarbons by almost a factor of 3. Among the factors which contribute to these differences are: stabilization of the molecular ions by the crystalline matrix, the decrease of the overall radical yield caused by cage effects and the influence of the lattice geometry.

C. Aqueous Solutions

1. Reactions with radicals of water

The radiation chemistry of simple aliphatic carboxylic acids in aqueous solution has been extensively studied. Some of them, such as formic acid, have been among the more thoroughly investigated systems in radiation chemistry.

In fact, all products in aqueous solutions are produced by the reaction of the acids with the reactive species formed in the radiolysis of water. When an aqueous solution is irradiated by high-energy radiation most of the radiation energy is absorbed by water and various primary radicals and molecules are formed (reaction 20). The values in parentheses represent radiation-chemical yields of the primary species formed in water.

$$H_2O \xrightarrow{\sim\!\sim\!\sim} e_{aq}^-(2.6) + H(0.50) + OH(2.65) + H_2(0.45) + H_2O_2(0.7) \qquad (20)$$

TABLE 1. Rate constants for the reaction of some carboxylic acids with H, e_{aq}^- and OH radicals

Acid	$k(1\ mol^{-1}\ s^{-1})$					
	H	Ref.	e_{aq}^-	Ref.	OH	Ref.
Formic	7.4×10^5	44	1.4×10^8	46	2.5×10^8	51
Acetic	8.4×10^4	44	1.0×10^8	29	1.4×10^7	51
Propionic	6.4×10^6	44	2.2×10^8	48	4.6×10^8	51
Glycolic	1.8×10^7	44	4.3×10^8	48	4.0×10^8	51
Lactic	2.2×10^7	44	6.7×10^8	48	4.3×10^8	51
Oxalic	4.1×10^5	44	2.5×10^{10}	49	8×10^6	51
Malonic	4.2×10^5	44	1.4×10^9	48	1.7×10^8	51
Succinic	3.5×10^6	44	8.6×10^9	48	1.2×10^8	51
Fumaric	9×10^9	52	$7.5 \times 10^9\ ^a$	45	—	—
Maleic	6×10^9	52	$1.6 \times 10^9\ ^a$	45	—	—
Benzoic	8.5×10^8	44	3×10^{10}	50	2.1×10^9	51
Phenylacetic	6×10^8	47	5.1×10^7	50	$5 \times 10^9\ ^a$	51

^{a}Anion formed.

Primary radicals of water such as hydrated electrons, hydrogen atoms and OH radicals react with an acid RCOOH forming different intermediates depending on the nature of R and on the pH of solution. The rate constants for the reactions of some acids with H, OH and e_{aq}^- radicals are presented in Table 1.

The H and OH radicals abstract a hydrogen atom from the alkyl group of saturated aliphatic acids (reaction 21).

$$C_nH_{2n+1}COOH + OH(\text{or } H) \longrightarrow C_nH_{2n}COOH + H_2O(\text{or } H_2) \qquad (21)$$

The rate constants for carboxylic acids with H atoms are close to the rate constants for the hydrocarbons of the chain R[52]. Thus, methane, acetic acid and malonic acid all have rate constants about 10^5 $1\,mol^{-1}s^{-1}$ and ethane, propionic and succinic acid $(2-6) \times 10^6$ $1\,mol^{-1}s^{-1}$.

Acids containing a double or a triple bond are very reactive toward H atoms as seen from Table 1. Most unsaturated compounds add H atoms with a rate constant of about 10^9 $1\,mol^{-1}s^{-1}$. The 30% difference between maleic and fumaric acid is indeed expected and demonstrates a small steric effect on the rate of addition. Unsaturated aliphatic acids as well as aromatic acids add H and OH radicals on the double bonds forming different radicals:

$$OH + C_6H_5COOH \longrightarrow C_6H_5(OH)COOH$$

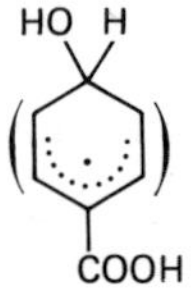

$$H + C_6H_5COOH \longrightarrow C_6H_5(H)COOH$$

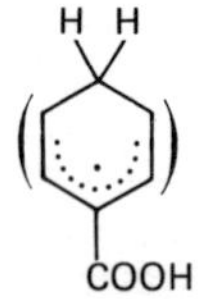

The rates of addition are several orders of magnitude faster than the rates of abstraction[51].

The hydrated electron reacts in the first step by addition of an electron to the carboxylic group, producing an anion radical (reaction 22). In the case of formic,

$$RC\overset{O}{\underset{OH}{\diagdown}} + e_{aq}^- \longrightarrow RC\overset{O^-}{\underset{OH}{\diagdown}} \qquad (22)$$

acetic and propionic acid the decomposition of this anion radical proceeds in two parallel first-order processes with comparable rates[53] (reaction 23). H atom is formed as a product in one of them.

$$RCOOH^- \begin{cases} \longrightarrow H + RCOO^- & (23a) \\ \longrightarrow OH^- + RCO & (23b) \end{cases}$$

Rate constants are quite different for the carboxylate anion and for the undissociated carboxylic acid. Reactions with OH and H are faster with the former species, while those with e_{aq}^- are slower. The reaction of e_{aq}^- with oxalic acid is a good example: the rate constants for the reactions of the acid, and its uni- and divalent anion are 2.5×10^{10} [49], 3.4×10^9 [49] and 10^7 $1\,mol^{-1}$ s^{-1} [54], respectively.

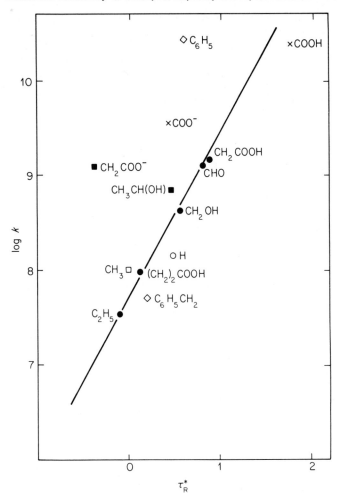

FIGURE. 2. Rate constant data for acids RCOOH with hydrated electrons according to Taft's equation. Data taken from References 46(●), 48(□), 49(×), 50(◇), and 56(■).

Owing to the localization of an electron on the carboxyl group (reaction 22), the effect of a substituent attached to that group and the degree of dissociation of the acid is important for the formation and decomposition of the $RCOOH^-$ radical. Thus the logarithm of the rate constant, log k (reaction 22), for a series of acids with various substituents increases linearly with the parameters τ_R^* of the Taft equation for those substituents, over four orders of magnitude (see Figure 2). In the case of dicarboxylic acids the statistical factor is taken into account and half the value of the rate constant is plotted on the diagram. The large positive slope, 1.8 ± 0.2, and the linear dependence of log k and τ_R^* indicate that the inductive effect of a substituent R is the most important factor influencing the rate constant of the reaction of a hydrated electron with the undissociated carboxyl group. The positive value of the slope is in accordance with the reaction mechanism by which the hydrated electron is attached to the positive centre of the carboxyl group.

Glyoxalic acid seems to belong to the same reaction series fitting well the line in Figure 2. This is not surprising since the aldehyde group should be practically fully hydrated. The behaviour of dicarboxylic acids with a negatively-charged substituent R is in accordance with the known irregularity of these substituents in other reaction series and with doubts about the universal application of τ_R^* parameters[57]. In the case of benzoic acid, resonance interaction is involved and the rate constant is higher than predicted by Taft's equation. The deviation is very small for phenylacetic acid where the resonance effect is considerably reduced since the phenyl group is separated from the reaction centre by a $-CH_2-$ group.

2. Reactive intermediates

a. OH and H adducts.   Both H atoms and OH radicals are considered to dehydrogenate carboxylic acids and their ions in a similar way, (reaction 21), forming carboxyl radicals. In a number of cases[58-60] the nature of the transient species produced in the radiolysis of these monocarboxylic acids has been established by pulse radiolysis.

There are many experimental data for radicals formed in reactions with various acids. Optical characteristics and pK values of some of these radicals are given in Table 2. For all monocarboxylic acids except formic acid the dissociated forms of radicals, $C_nH_{2n}COO^-$ (reaction 21), have absorption maxima in the region of 325–350 nm practically independent of the number of carbon atoms in the alkyl

TABLE 2. Optical characteristics and pK values of radicals produced by pulse radiolysis of aqueous solutions of some acids

Acid	Radicals	λ_{max}	$\epsilon(1 \text{ mol}^{-1} \text{ cm}^{-1})$	pK of radical	pK of acid	Reference
Formic	COOH	240	1050	1.4	3.77	63
	$\overset{\cdot}{C}O_2^-$	240	1900			
Acetic	$\overset{\cdot}{C}H_2COOH$	320	650	4.5	4.76	58
	$\overset{\cdot}{C}H_2COO^-$	350	800			
Propionic	$CH_3\overset{\cdot}{C}HCOOH$	300	700	4.9	4.88	58
	$CH_3\overset{\cdot}{C}HCOO^-$	335	950			
Trimethyl-acetic	$\overset{\cdot}{C}H_2C(CH_3)_2COOH$	<240		4.8	5.02	58
	$\overset{\cdot}{C}H_2C(CH_3)_2COO^-$	240	1800			
Glycolic	$HO\overset{\cdot}{C}HCOOH$	<245		4.6	3.83	59
	$HO\overset{\cdot}{C}HCOO^-$	245	5700	8.8		
	$^-O\overset{\cdot}{C}HCOO^-$	255	5400			
Lactic	$HO\overset{\cdot}{C}(CH_3)COOH$	<245		5.3	3.87	59
	$HO\overset{\cdot}{C}(CH_3)COO^-$	245	5100	9.8		
	$^-O\overset{\cdot}{C}(CH_3)COO^-$	275	5800			
Oxalic	$\overset{\cdot}{O}OCCOOH$	245	1800		1.25	62
	$\overset{\cdot}{O}OCCOOH^-$	255	2600		4.28	
Malonic	$HOOC\overset{\cdot}{C}HCOOH$	340	~400	5.7	2.86	59
	$^-OOC\overset{\cdot}{C}HCOO^-$	340	1000		5.7	
Benzoic	$C_6H_5(OH)COOH$	350	3800	4.4	4.2	60
	$C_6H_5(H)COOH$	350	<4200			

group, and with absorption coefficients of $800-950 \, l \, mol^{-1} cm^{-1}$ [58]. The undissociated radicals $C_n H_{2n} COOH$ show similar absorption spectra with maxima shifted to lower wavelengths. The β-carboxyalkyl radicals have absorption maxima above 240 nm with absorption coefficients considerably higher than those of the α radicals.

The carboxy-hydroxyalkyl radicals have λ_{max} at about 250 nm and the absorption coefficients are about $5000-9000 \, l \, mol^{-1} cm^{-1}$ [59], higher than those of alcohols[61] and simple carboxylic acids[58]. The absorption spectra of the polycarboxyalkyl radicals depend on the length of the alkyl group. The $\cdot OOCCOOH$ radical does not belong to the same reaction series because of the abstraction of a hydrogen atom from the hydroxyl group[62], whereas malonic and succinic acid radicals are similar to the monocarboxyalkyl radicals[59].

The pK values for the dissociation of the monocarboxyalkyl radicals are equal to those of the parent acid while for the carboxyl group in the hydroxy-carboxyalkyl radicals the pK values are higher than those of the parent acids and close to the pK values of the non-substituted acids[59]. An exception is the COOH radical which has a pK value much lower than that of formic acid[63]. This does not seem unreasonable since the electron deficiency on the carbon atom in the radical will tend to weaken the OH bond through an inductive effect and COOH should be a stronger acid than $HCO_2 H$. However, in radicals formed from higher aliphatic acids, the electron-deficient centre is separated from the carboxyl group by one or more methylene groups so that its effect is likely to be less pronounced.

Carboxyl radicals can recombine. The rate constants for the recombination of the monocarboxyl radicals are of the order of $10^9 \, l \, mol^{-1} s^{-1}$ [58,59]:

$$R_2\dot{C}COOH + R_2\dot{C}COOH \longrightarrow Product$$

The recombination of two $R_2\dot{C}COO^-$ ion-radicals is approximately half as fast since the reactants are species having the same charge. The triple-charged radicals formed from hydroxy and polycarboxylic acids decay at rates lower than $10^6 \, l \, mol^{-1} s^{-1}$ [59].

There is no rule governing which type of product is formed in the reaction of recombination. For COOH and $CH_2 COOH$ radicals the products are oxalic[54] and succinic acid[28], respectively. The formation of radicals depends on many factors, such as the type of solute, the nature of the carboxyl radicals, the concentration of the acid, the pH of the solution, the dose rate, etc.

The γ-radiolysis of acid solutions saturated with oxygen have been extensively investigated in order to obtain more information in the primary radicals formed in water[64]. Nevertheless, the observations on peroxycarboxyl radicals $\cdot OOR_2 CCOOH$, produced in oxygen-saturated solution of acids are very scarce. In the presence of oxygen, carboxyl radicals react either by forming O_2^- radicals or peroxycarboxyl radicals as in reactions (24) and (25). It was found that the transient

$$CO_2^- + O_2 \longrightarrow CO_2 + O_2^- \qquad (24)$$

$$CH_2COO^- + O_2 \longrightarrow \dot{O}OCH_2COO^- \qquad (25)$$

optical absorption spectra of the intermediates produced in the presence of oxygen are distinctly different from those observed in absence of oxygen. Peroxycarboxyl radicals from acetate, lactate and glycolate have maxima of optical absorption spectra at or below 250 nm[65]. It was also found that the ionization constants for peroxy radicals, $\cdot OOR_2 CCOOH \longrightarrow \cdot OOR_2 CCOO^- + H^+$, originating from lactate and glycolate, are much lower than those for $R_2 CCOO^-$ radicals. The transients formed in oxygenated aqueous acetic acid solutions have recently been studied extensively. It has been found that peroxy radicals, $\cdot OOCH_2COO^-$ decompose by a second-order reaction[66].

b. e_{aq}^- *Adducts.*   Little is known about the products of the reactions of e_{aq}^- with acids. Exceptions are unsaturated aliphatic and aromatic acids, whose products were determined by optical pulse radiolysis and e.s.r. techniques.

The product of the reaction of e_{aq}^- with formate ions has been investigated and is found to be the hydrated formyl radical[67] (reaction 26). The $HC(OH)_2$ radicals have also been identified as the product of the $CO + e_{aq}^-$ reaction[67].

$$HCOO^- + e_{aq}^- \xrightarrow{H_2O} HC(OH)_2 + 2OH^- \qquad (26)$$

$$k = 2.4 \times 10^4 \; l\,mol^{-1}s^{-1}$$

Three different anionic forms of e_{aq}^- adduct of oxalic acid have been identified and these are $\cdot C(O^-)OH\,COOH$, $\cdot C(O^-)OH\,COO^-$ and $\cdot C(O^-)_2COO^-$ [62]. Each of them decomposes by a first-order process and forms the $OC\overset{\cdot}{C}OOH$ radical.

CH_3CO radicals have been found in aqueous acetic acid solutions. The anionic form of the radical CH_3COOH^-, produced in the first step of the e_{aq}^- attack, is very short-lived and eliminates the OH^- group in less than 1 ns[30], forming CH_3CO radical.

In aqueous solutions radicals produced by the reaction of e_{aq}^- with a number of unsaturated carboxylate ions have been investigated. The acid–base equilibria of these radicals have been followed and protonation is found to occur on the carboxyl groups[68–72]. In an e.s.r. study of irradiated aqueous solutions of fumarate ions[70] and fumaric acid[72] the formation of uni-, di- and tervalent anions of the e_{aq}^- adduct was found. Two forms of the monovalent anion $[HO_2CCH=CHCO_2H]\cdot^-$, differing in the location of the proton on the carboxyl oxygens, were found. All forms of the e_{aq}^- adduct are in equilibrium with the pK values shown:

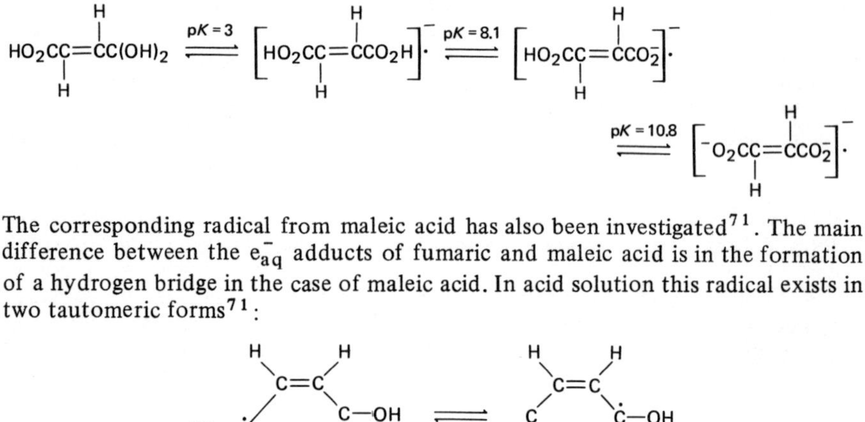

The corresponding radical from maleic acid has also been investigated[71]. The main difference between the e_{aq}^- adducts of fumaric and maleic acid is in the formation of a hydrogen bridge in the case of maleic acid. In acid solution this radical exists in two tautomeric forms[71]:

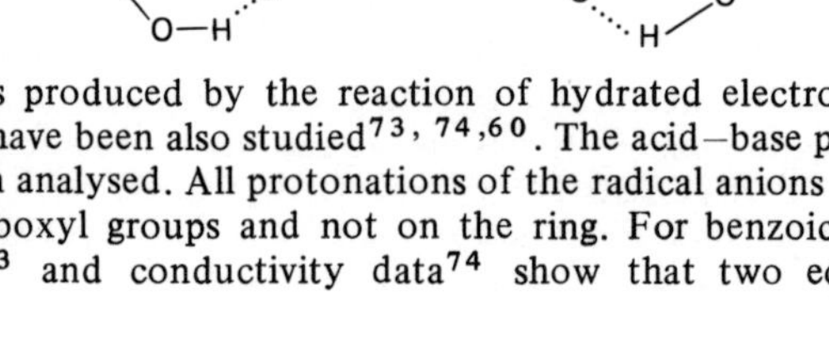

Radical anions produced by the reaction of hydrated electrons with aromatic carboxylic acids have been also studied[73, 74,60]. The acid–base properties of these radicals have been analysed. All protonations of the radical anions are found to take place on the carboxyl groups and not on the ring. For benzoic acid, the optical pulse radiolysis[73] and conductivity data[74] show that two equilibria for e_{aq}^-

adducts exist:

$$C_6H_5C(OH)_2 \overset{pK=5.3}{\rightleftharpoons} C_6H_5CO_2H^- \overset{pK=12}{\rightleftharpoons} C_6H_5\dot{C}O_2^{2-} + H^+$$

Two *ortho*-carboxyl groups attached to a benzene ring form a strong hydrogen-bonded bridge (11), which in the case of the radical produced from phthalate ions does not dissociate even at pH 14.

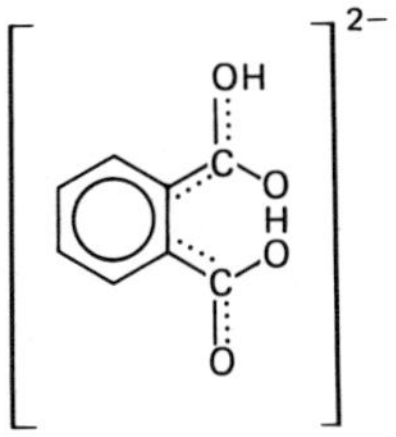

(11)

III. ESTERS

A. Mechanism of Decomposition

The radiation chemistry of esters is much more complex than that of the corresponding carboxylic acids. The investigation of intermediates in the radiolysis of a number of simple esters clearly shows this difference. Early studies of e.s.r spectra[75,76] in a number of γ-irradiated esters reveal the existence of radical anions and secondary radicals which can be formed by abstraction of hydrogen from the parent ester. Recently Ayscough and Oversby[77] investigated by using e.s.r., the intermediates formed in the radiolysis at 12 K of 12 simple aliphatic esters and the changes induced by thermal annealing. Their suggested general mechanism of decomposition is shown in reactions (27) and (28). Trapped electrons are observed

$$R^1COOR^2 \longrightarrow [R^1COOR^2]^+ + e^- \tag{27}$$

$$e^- + R^1COOR^2 \longrightarrow [R^1COOR^2]^- \tag{28}$$

in several esters (methyl acetate, *n*-propyl acetate and some others) whereas radical anions $[R^1COOR^2]^-$ are detected in all acetates and propionates.

$$[R^1COOR^2]^+ + R^1COOR^2 \longrightarrow [R^1C(OH)OR^2]^+ + (R^1CO\dot{O}R^{2'} \text{ or } \dot{R}^{1'}COOR^2) \tag{29}$$

$$[R^1COOR^2]^- \longrightarrow R^1CO_2^- + \dot{R}^2 \tag{30}$$

$$R^1CO_2^- + [R^1C(OH)OR^2]^+ \longrightarrow R^1COOH + R^1COOR^2 \tag{31}$$

The reaction sequence (29)–(31) leads to the formation of $\dot{R}^2$ and $(R^1COO\dot{R}^{2'}$ or $\dot{R}^{'}COOR^2)$ as radical products and R^1COOH as a molecular product. The nature of the secondary radicals $\dot{R}^{1'}COOR^2$ and $\dot{R}^1COO\dot{R}^{2'}$ is determined by the fact that hydrogen can be abstracted either from the R^1 or the R^2 group; thus, propionates and isobutyrates give $\dot{R}^1COOR^2$ radicals[75] whereas ethyl and isopropyl esters give $R^1COO\dot{R}^{2'}$ radicals[77]. To explain the presence of $\dot{R}^1$ radicals an alternative mode of decomposition of $[R^1COOR^2]^-$ has been suggested[77]:

$$[R^1COOR^2]^- \longrightarrow R^1\dot{C}O + [OR^2]^- \tag{32}$$

$$[R^1C(OH)OR^2]^+ + [OR^2]^- \longrightarrow R^1COOR^2 + R^2OH \tag{33}$$

$$R^1\dot{C}O \longrightarrow \dot{R}^1 + CO \tag{34}$$

It is also possible that these radicals are formed from a molecular cation by recapture of an electron followed by decomposition:

$$[R^1COOR^2]^* \longrightarrow R^1 + CO_2 + R^2 \tag{35}$$

$$(R^{1'}COOR^2 \text{ or } R^1COOR^{2'}) + H \tag{36}$$

It appears that the predominant mode of decomposition of the molecular anion is by reaction (30) in the case of acetates and propionates and by reaction (32) in the case of higher carboxylic acids. However, the mechanism given by the reactions (27)–(36) seems to be a gross oversimplification when compared with the wide range of products found in the liquid phase after irradiation. The competing decomposition reactions (30) and (32) and possibly (35) depend on the temperature and the nature of the solvent.

B. Products

The stable radiolytic products of carboxylic esters are H_2, RH, $R'H$, RCOOH, $R'OH$, CO_2, CO and hydrocarbons produced by combination of R^1 and R^2, together with smaller amounts of ethers, aldehydes and ketones. If saturated esters are irradiated in the pure state, under air-free conditions, a major product is the corresponding acid.

Methyl acetate is one of the esters which has been investigated extensively and the distribution of its radiolytic products has been examined in the liquid and gaseous phase over a wide temperature range, with and without radical scavengers[78]. The main products are H_2, CO_2, CH_4, CO, C_2H_6 and CH_3OCH_3. At room temperature, radical scavengers reduce the yield of CH_4 by 75%, that of H_2 and CO by about 20%, but have little effect on other products. These results are in agreement with the mechanism proposed above in equations (27)–(36).

The radiolysis of phenyl acetate 77 K has been investigated by analysis of stable products and by observation of reaction intermediates using optical and e.s.r spectroscopy[79]. The main products of the radiolysis are anisole, phenol and o-hydroxyacetophenone. The G-values of products depend upon the aggregate state and/or structure of the irradiated phenyl acetate. Thus the G-value of anisole in polycrystalline samples is double the value obtained in the glassy samples, while the G-value of phenol in polycrystalline phenyl acetate is half that of the value obtained in the glassy state. In the radiolysis of liquid phenyl acetate the G-value of phenol amounts to 1.5, whereas anisole is negligibly small.

It is assumed that anisole can be formed by CO elimination from an excited phenyl acetate molecule (reaction 37). Phenol may be formed by a process independent of the formation of anisole.

$$\tag{37}$$

Diethyl succinate has been irradiated with γ-rays[79] and 20 different products have been found. Among them the most abundant are ethanol, ethyl propionate, ethane, hydrogen, carbon monoxide, acetaldehyde and carbon dioxide. This abundance of products again shows that the radiolysis of esters is a rather complex process which incorporates various intermediate reactions.

The products formed in the irradiation of liquid isopropyl acetate at room temperature have also been determined. Products having G-values higher than 0.5 are acetic acid, hydrogen, carbon monoxide, methane, ethane, propylene, carbon dioxide, acetaldehyde and acetone[80]. The presence of a double bond in the case of isopropenyl acetate strongly modifies the response to irradiation: acid formation is markedly reduced and the main effect is the formation of a polymer of molecular weight 360.

C. Oxidation of Methyl Oleate

Methyl oleate is an important compound for food lipids and the oxidations of pure methyl oleate and its water emulsions, induced by γ-rays, have been examined in detail[81,82],

The high G-value for oxygen consumption indicates that the reaction has a chain mechanism. The oxidation of many organic liquids induced by ionizing radiation under suitable conditions proceeds through a peroxide chain mechanism.

The radical processes shown in reactions (38)–(40) probably occur in the radiolysis. For the sake of simplicity, the radiolysis is expressed as a process in which only R and RH_2 type radicals are formed[81].

$$RH \xrightarrow{\hspace{1cm}} R + H \tag{38}$$

$$H + RH \xrightarrow{\hspace{1cm}} RH_2 \tag{39}$$

$$H + RH \xrightarrow{\hspace{1cm}} R + H_2 \tag{40}$$

The transient H atoms either undergo addition to or take part in abstraction from methyl oleate. In both cases they produce radicals which may be subsequently scavenged by O_2.

The oxygen is quantitatively converted into the peroxide radical in the initial stage of the oxidation process, before the normal propagating sequence takes place:

$$R \text{ (or } RH_2) + O_2 \xrightarrow{\hspace{1cm}} RO_2 \text{ (or } RH_2O_2) \tag{41}$$

$$RO_2 \text{ (or } RH_2O_2) + RH \xrightarrow{\hspace{1cm}} ROOH \text{ (or } RH_2OOH) + R \tag{42}$$

The chain-termination stage is shown in reaction (43) with an unsaturated ketone as the final product.

$$2 RO_2 \xrightarrow{\hspace{1cm}} R^1CH{=}CH\dot{C}OR^2 + ROH + O_2 \tag{43}$$

It has been found that HO_2 radicals, which might be formed by scavenging hydrogen atoms, do not contribute to the initiation of the reaction since the yield of radiolytic hydrogen is $G = 1$ so that methyl oleate is unaffected by the presence of O_2[81]. The above reactions therefore present a simple mechanism of methyl oleate oxidation induced by γ-irradiation.

The kinetics of radiolitically-induced oxidation of an emulsion of methyl oleate in water[82], stabilized by sodium oleate, has been also investigated. Water dispersed in form of droplets in methyl oleate has no influence on the oxidation rate, but when the ester is dispersed in excess of water its rate of oxidation increases. The rate of oxygen consumption is proportional to the square root of the dose rate, as in pure methyl oleate, and increases with the ratio of water to methyl oleate and with the concentration of the emulsifying agent.

D. Aqueous Solutions

Data on the radiolysis of aqueous solutions of esters are rather scarce. In aqueous solution the behaviour of esters is essentially similar to that in the pure state: the radiolysis is more complex than that of the corresponding acids. The reaction with e_{aq}^- involves the addition of an electron to the carbonyl group and the formation of a radical anion (reaction 44). The latter may rapidly accept a proton producing the radical $R^1\dot{C}OHOR^2$. Another alternative is the dissociative electron capture and the formation of a radical (reaction 45).

$$e_{aq}^- + R^1COOR^2 \longrightarrow (R^1\dot{C}OOR^2)^- \tag{44}$$

$$e_{aq}^- + R^1COOR^2 \longrightarrow R^1\dot{C}=O + R^2OH + OH^- \tag{45}$$

The carbonyl group is reactive towards e_{aq}^-, but its reactivity is somewhat influenced by the nature of adjoining groups, so that various simple carboxylic esters have rate constants varying from 10^7 to several times 10^8 l mol^{-1} s^{-1}. Hart and coworkers[83] studied the effect of the substituents on the reactivity of the CO group with e_{aq}^- for a number of compounds including esters, aldehydes, ketones, carboxylic acids and oximes. They found that the electron from e_{aq}^- is placed in an orbital of the carbonyl oxygen atom and that a relationship exists between the rate constants of esters and the parameters of Taft's equation.

For reactions with OH and H radicals no similar analysis exists. Only for methyl acetate is it assumed that reactions with OH and H may involve abstraction of hydrogen atom from acyl and alkoxyl groups, leading to the formation of $\dot{C}H_2CO_2CH_3$ and $CH_3CO_2\dot{C}H_2$ radicals[84] (reactions 46 and 47). Unfortunately,

$$OH + CH_3CO_2CH_3 \longrightarrow \dot{C}H_2CO_2CH_3 \text{ (or } CH_3CO_2\dot{C}H_2) + H_2O \tag{46}$$
$$k = 7 \times 10^7 \text{ l mol}^{-1}\text{s}^{-1}$$

$$H + CH_3CO_2CH_3 \longrightarrow \dot{C}H_2CO_2CH_3 \text{ (or } CH_3CO_2\dot{C}H_2) + H_2 \tag{47}$$
$$k = 6 \times 10^4 \text{ l mol}^{-1}\text{s}^{-1}$$

there is no direct experimental evidence on the formation of intermediates in the reactions with e_{aq}^-, H and OH radicals. In the γ-radiolysis of aqueous methyl acetate solutions radical reactions lead to complex products such as ethylene diacetate, dimethyl succinate, methyl β-acetoxypropionate, to cite only those which have been identified. Acetic acid, methanol and small yields of methane and formaldehyde have been found as well[84]. Acetic acid is formed to a great extent only in the reaction with e_{aq}^-.

Elimination of the acid in the radiolysis of aqueous solutions of esters may occur through different mechanisms. Thus in the radiolysis of 2-hydroxyethyl acetate, acetic acid is eliminated from a primary radical produced by the reaction of the ester with a OH radical (reaction 48). The primary radical undergoes fast fragmentation into acetic acid and formylmethyl radical (reaction 49). The kinetics of reaction has been followed by pulse radiolysis. At higher concentration the ester is attacked by CH_2CHO and produces through a chain decomposition acetic acid and

$$OH + CH_3CO_2CH_2CH_2OH \longrightarrow H_2O + CH_3CO_2\dot{C}HCH_2OH \tag{48}$$
$$k = 8.5 \times 10^8 \text{ l mol}^{-1}\text{s}^{-1}$$

$$CH_3CO_2\dot{C}HCH_2OH \longrightarrow CH_3COOH + \dot{C}H_2CHO \tag{49}$$
$$k = 5 \times 10^5 \text{ l mol}^{-1}\text{s}^{-1}$$

acetaldehyde. This elimination of acetic acid may serve as a model for the elimination of carboxylic acids in the free-radical degradation of diglycerides, carboxylic acid esters of ceullulose and related substances.

IV. ANHYDRIDES

The radiolysis of carboxylic acid anhydrides is considerably simpler than the corresponding process for acids and esters[86]. This is attributed to the absence of intermolecular hydrogen bonds in anhydrides. Unfortunately, the overall effects of γ-radiation on anhydrides are uncertain since relevant analyses of stable products are not available. Therefore radiolysis of these compounds can only be discussed in terms of the results of e.s.r. studies related to the trapped radicals at cryogenic temperature[86].

There is evidence that carboxylic acids and esters, after γ-irradiation at 77 K, can capture thermal electrons to form radical anions which might be subsequently stabilized. In the case of four anhydrides studied, acetic, propionic, N-butyric and isobutyric, there is no evidence of such an event. Intermediates trapped at 77 K are either alkyl or acyl radicals and secondary aliphatic radicals formed by loss of a hydrogen atom from the parent anhydride.

A mechanism has been suggested for the early radiolytic stage of acetic anhydride. It follows closely the sequence postulated earlier for acetic acid[86] (reactions

$$(CH_3CO)_2O \longrightarrow\!\!\!\sim\!\!\!\longrightarrow (CH_3CO)_2O^+ + e^- \tag{50}$$

$$(CH_3CO)_2O^+ + (CH_3CO)_2O \longrightarrow (CH_3CO)_2OH^+ + \dot{C}H_2COOCOCH_3 \tag{51}$$

$$e^- + (CH_3CO)_2O \longrightarrow (CH_3CO)_2O^- \tag{52}$$

50–52). Reaction (52) is extremely rapid and seems to take place immediately after ionization. The failure to observe radical anions is partly due to the rapid

$$(CH_3CO)_2O^- \longrightarrow CH_3\dot{C}O + CH_3CO_2^- \tag{53}$$

$$CH_3\dot{C}O \longrightarrow \dot{C}H_3 + CO \tag{54}$$

$$\dot{C}H_3 + (CH_3CO)_2O \longrightarrow CH_4 + \dot{C}H_2COOCOCH_3 \tag{55}$$

dissociation (reactions 53–55) and partly to the possibility that another proton transfer can take place (reaction 56).

$$(CH_3CO)_2OH^+ + CH_3CO_2^- \longrightarrow CH_3COOH + (CH_3CO)_2O \tag{56}$$

This scheme predicts the formation of $CH_3\dot{C}O$, $\dot{C}H_3$ and $\dot{C}H_2COOCOCH_3$ in γ–irradiated acetic anhydride. These radicals have been, indeed, observed in acetic anhydride. Similar radicals have been identified in other anhydrides. Differences between the behaviour of the four anhydrides studied are attributed to the greater stability of the higher acyl radicals and to the more facile abstraction of H atoms from higher anhydrides. Thus alkyl radicals are not observed in normal and isobutyric anhydrides.

In the absence of other information about products, these e.s.r. data on acyl radicals and radicals formed by loss of a hydrogen atom are insufficient to postulate a specific mechanism for the radiolysis of carboxylic acid anhydrides.

V. LACTONES

Radiation-chemical studies of lactones have been mainly focused on aqueous solutions of ascorbic acid and related compounds. Ascorbic acid is a typical representative of a large group of lactones known to be very radiation-sensitive in aqueous solutions

A. Aqueous Solutions of Ascorbic Acid

When irradiated in aqueous solution, ascorbic acid (AH_2) is oxidized to dehydro-ascorbic acid (A) through the formation of a radical intermediate $(AH\cdot)$ (reaction 57). A G-yield of 7.8 has been found for the loss of ascorbic acid in an

$$AH_2 \xrightarrow{\text{aq}} A\overset{\cdot}{H} \longrightarrow A + H_2O \tag{57}$$

air-saturated solution[87] and owing to this high yield it has been suggested that in addition to OH radicals, HO_2 and probably also organic peroxy radicals participate in the overall oxidation. However, no indication has been obtained that oxygen takes part in a chain reaction. Analysis of the products and the decay kinetics of the free radical $AH\cdot$ suggest a disproportionation mechanism (reaction 58).

$$A\overset{\cdot}{H} + A\overset{\cdot}{H} \longrightarrow A + AH_2 \tag{58}$$

From a biological viewpoint the redox reaction is the most important chemical characteristic of ascorbic acid and therefore the nature of its free radicals and the mechanism of their formation are the main subjects of investigation. In fact, whatever radical is formed by irradiation of water, it will produce an oxidized ascorbic acid radical, $AH\cdot$, a transient intermediate of an oxidation level in between ascorbic and dehydroascorbic acid. Ample evidence of its presence is furnished by e.s.r. and optical spectroscopy[88−93].

1. Reactions with OH radical

Ascorbic acid reacts with an OH radical by electron transfer to the latter or indirectly by addition of OH to the double bond at either the $C_{(2)}$ or $C_{(3)}$ position, followed by loss of water[88,89]. Pulse radiolysis experiments show that this reaction is completed within a millisecond[89].

In the region of pH $1-13$ the radical 12 has been identified by e.s.r. and optical pulse radiolysis[88−93]. This radical is an anion with the unpaired electron spread

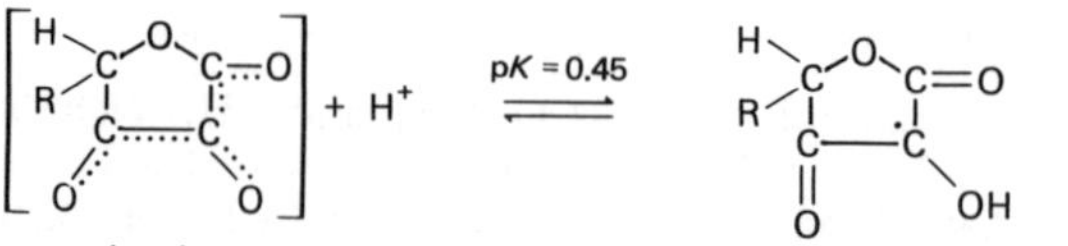

(12a)
pH = 1−13

(12b)
pH = 1

(R = CHOHCH_2OH)

over a highly conjugated tricarbonyl system. Its existence over such a large region of pH suggests an extremely acidic character. It protonates, indeed, with a pK value of 0.45[89], which means it is by four orders of magnitude more acidic than ascorbic acid itself whose first ionization corresponds to a pK_1 value of 4.1.

In the region of pH $0-6$ another radical (13) is produced in addition to radical 12, most probably by addition of OH to the $C_{(2)}=C_{(3)}$ double bond[90]. Its protonation equilibrium has a pK value of 2.0. In fact, owing to the two tautomeric forms of ascorbic acid the mechanism is more complex and several precursor radicals can be formed[93].

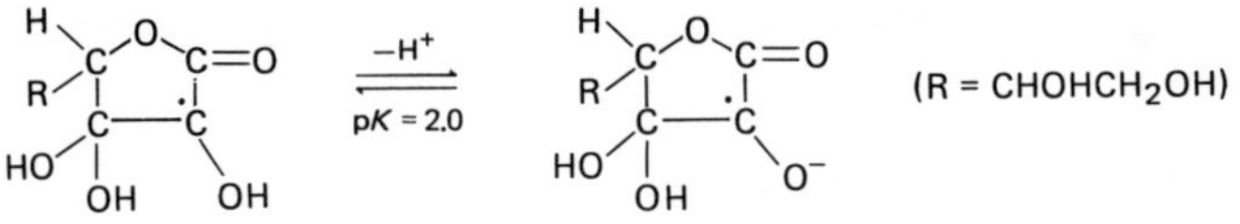

(13a)   (13b)

(R = CHOHCH$_2$OH)

The rate of reaction of OH with ascorbic acid (Table 3) suggests an almost diffusion-controlled process. This high rate is due to the fact that the acid easily undergoes reversible oxidation at the point of unsaturation, the C(OH)=C(OH) group being dehydrogenated to $-CO-CO-$[94].

2. Reactions with e_{aq}^- radical

Characteristics of the highly oxygenated parent anion AH$^-$ present in aqueous solutions of ascorbic acid and similar compounds can be deduced from their reaction with hydrated electrons.

Experiments with α-bromotetronic acid show that hydrated electrons react with this model compound producing bromide ion and tetronic acid with a high efficiency. Optical and conductometric pulse radiolysis studied[88] indicate that the radical anions resulting from bromide elimination, after electron capture, protonate very rapidly forming an intermediate, neutral entity:

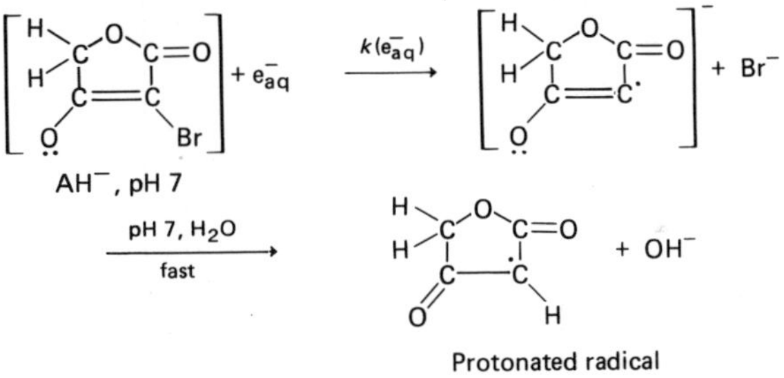

Protonated radical

It is important to note that in the above system the initial radical is an oxyanion in which the unpaired electron is coupled to oxygen through an ethylene linkage, $-C(O^-)=\dot{C}-$, so that it protonates very rapidly at the radical site. The protonated form, being a π radical, should abstract hydrogen from a hydrogen donor relatively slowly:

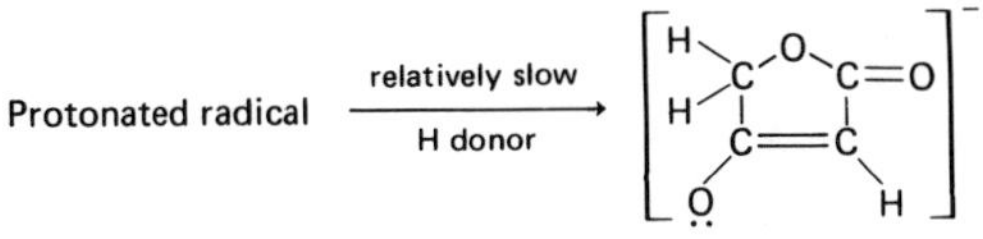

3. Reactions with H atoms

Hydrogen atoms react with ascorbic acid in aqueous solution principally by addition to the double bond and to a lesser extent, by hydrogen abstraction from the hydroxylated side-chain[95]. The rate constant depends on pH[96] (Table 3), owing to the ionic dissociation of ascorbic acid, and this is usual when acid–base equilibria are involved. Its value is only about 10^8 l mol^{-1} s^{-1}, i.e. 10 times lower

TABLE 3. Rate constants of e_{aq}^-, OH and H atom with ascorbic acid and related compounds

Compound	$k(e_{aq}^-)(l\ mol^{-1}\ s^{-1})$	pH	Ref.	$k(OH)(l\ mol^{-1}\ s^{-1})$	pH	Ref.	$k(H)(l\ mol^{-1}\ s^{-1})$	pH	Ref.
α-Bromotetronic acid	4.4×10^9 2.5×10^9	7 10	88 88	7.7×10^9	7	88			
Tetronic acid	10^8	7	88	9.2×10^9	7	88			
α-Hydroxytetronic acid				4.7×10^9	7	88			
Ascorbic acid	3×10^8	7	88	4.5×10^8 7×10^9	7	93 88	$3{-}6 \times 10^8$ 1.1×10^8	7 1	96 95
Ascorbate anion	4×10^8	7	93	7×10^9 7.2×10^9	1	93 94			

than the rate constants of other unsaturated acids, which indicates a strong deactivating effect of OH groups on the addition of H atoms to the double bond. The yield of molecular hydrogen is considerably greater than $G(H_2)$ originating directly from water. This higher yield supports the assumption that hydrogen atoms are abstracted from ascorbic acid.

VI. LACTAMS

Interest in the radiation–chemical behaviour of antibiotics has stimulated investigation into the effect of radiation on the lactam ring in various compounds. Studies of the γ-radiolytic stability of penicillins[97-100], which contain a fused β-lactam thiazolidine ring in their structure, show that the β-lactam part is most susceptible toward irradiation. Spectroscopic data of the radiolytic products isolated from irradiated penicillins show that in almost all the products no absorption bands corresponding to frequencies of the β-lactam ring can be found[98,100] The main gaseous product of radiolysis of a series of penicillins was found to be CO, which also suggests that decomposition of the β-lactam ring occurs[97]. Data obtained with γ-irradiated aminobenzylpenicillins show that the presence of water

$$C_6H_5\underset{\underset{NH_2}{|}}{C}HCONHCH-CH \overset{S}{\diagdown} C(CH_3)_2$$
$$\underset{\underset{CO-N}{|}}{} \underset{\underset{CCOOH \times nH_2O}{|}}{}$$

Aminobenzylpenicillin

of crystallization can be very important. Thus, γ-irradiated hydrated aminobenzylpenicillin exibits an e.s.r. spectrum which is assigned to the presence of an unpaired electron on the nitrogen atom of the β-lactam ring; but a similar spectrum has not been found in the dehydrated samples after irradiations. It is assumed that an interaction between the OH radicals from the water of crystallization and the β-lactam ring leads to the formation of the above mentioned radical with an unpaired electron on the nitrogen. An unpaired electron on the carbon atom formed by a C–N bond cleavage should be rapidly paired upon addition of the OH radical, so that instead of the β-lactam a COOH group appears.

An investigation dealing with the radiolysis of penicillins in aqueous solution[101] has provided more information about the radiation susceptibility of the lactam group. The radiolytic yield of degradation of benzylpenicillins after irradiation in aqueous solution is rather high. The $G(-benzylpenicillin)$ is 3.8 in 10^{-4} mol solutions, indicating that both e_{aq}^- and OH from water radiolysis participate in the degradation. The rate constants for the reactions of benzylpenicillin with e_{aq}^- and OH radicals, obtained by the pulse radiolysis technique, are 2.7×10^9 l mol^{-1} s^{-1} and 3.4×10^9 l mol^{-1} s^{-1} respectively. The radiolytic products are supposed to originate from an attack of OH and/or e_{aq}^- in the β-lactam ring. Thus, the major degradation product, benzylpenilloic acid, has similar yields in argon and N_2O saturated solutions, 1.5 and 1.2, respectively. This product could be formed from an attack of e_{aq}^- on the carbonyl group of the β-lactam ring in benzylpenicillin, followed by a loss of carbon monoxide. On the other hand, the hydrogen atom at $C_{(6)}$ is activated by the 6-amido group, so that abstraction of this hydrogen atom by OH would further weaken the already strained β-lactam ring leading to cleavage of the C–N bond. Subsequent loss of CO would again produce benzylpenilloic acid.

VII. REFERENCES

1. For a summary of work on this subject, see R. A. Holroyd in *Fundamental Processes in Radiation Chemistry* (Ed. P. Ausloos) Wiley–Interscience, New York, 1968, pp. 413–504.
2. J. E. Bennett and L. H. Gale, *Trans. Farad. Soc.*, **64**, 1174 (1968).
3. M. Iwasaki, B. Eda and K. Toriyama, *J. Amer. Chem. Soc.*, **92**, 3211 (1970).
4. K. Toriyama and M. Iwasaki, *J. Chem Phys.*, **55** 2181 (1971).
5. K. Toriyama, M. Iwasaki, S. Noda and B. Eda, *J. Amer. Chem. Soc.*, **93** 6415 (1971)
6. K. Minakata and M. Iwasaki, *J. Chem. Phys.*, **57** 4758 (1972).
7. H. C. Box, H. G. Freund, K. T. Lilga and E. E. Budzinski, *J. Phys. Chem.*, **74**, 41 (1970).
8. G. C. Moulton and B. Cernansky, *J. Chem. Phys.*, **53**, 3022 (1970).
9. R. C. McCalley and A. L. Kwiram, *J. Amer. Chem. Soc.*, **92**, 1441 (1970).
10. H. Muto, T. Inoue and M. Iwasaki, *J. Chem. Phys.*, **57**, 3220 (1972).
11. R. N. Schwartz, M. W. Hanna and B. L. Bales, *J. Chem. Phys.*, **51**, 4336 (1969).
12. R. C. McCalley and A. L. Kwiram, *J. Chem. Phys.*, **53**, 2541 (1970).
13. P. Premović, O. Gal and B. Radak, *J. Chem. Phys.*, **59**, 987 (1973).
14. A. J. Swallow, *Radiation Chemistry of Organic Compounds*, Pergamon Press, New York, 1960, pp. 111–118.
15. W. L. Whitehead, C. Goodman and I. A. Breger, *J. Chim. Phys.*, **48**, 184 (1951).
16. A. S. Newton, *J. Chem. Phys.*, **26**, 1764 (1957).
17. G.-S. Wu and D. R. Howton, *Radiation Res.*, **57**, 390 (1974).
18. A. R. Jones, *Radiation Res.*, **47**, 35 (1971).
19. A. R. Jones, *Radiation Res.*, **50**, 41 (1972).
20. A. R. Jones, *Radiation Res.*, **48**, 447 (1971).
21. G.-S. Wu and D. R. Howton, *Radiation Res.*, **61**, 374 (1975).
22. J. G. Burr, *J. Phys. Chem.*, **61**, 1481 (1957).
23. G. E. Adams, J. H. Baxendale and R. D. Segwick, *J. Phys. Chem.*, **63**, 854 (1959).
24. R. H. Johnsen, *J. Phys. Chem.*, **63**, 2041 (1959).
25. Lj. Josimović, J. Teplý and O. I. Mićić, *J. Chem. Soc., Faraday I*, **72**, 285 (1976).
26. P. B. Ayscough, K. Mach, J. P. Oversby and A. K. Roy, *Trans. Farad. Soc.*, **67**, 360 (1971).
27. S. Lukač, J. Teplý and K. Vacek, *J. Chem. Soc., Faraday I*, **68**, 1337 (1972).
28. Lj. Josimović and I. G. Draganić, *Int. J. Radiat. Phys. Chem.*, **5**, 505 (1973)
29. O. I. Mićić and V. Marković, *Int. J. Radiat. Phys. Chem.*, **7**, 541 (1975).
30. B. Cerček and O. I. Mićić, *Nature Phys. Sci.(Lond.)*, **238**, 74 (1972).
31. I. G. Draganić and O. Gal, *Radiation Res. Rev.*, **3**, 167 (1971).
32. O. Gal, Lj. Petković, Lj. Josimović and I. Draganić, *Int. J. Appl. Radiat. Isotopes*, **19**, 645 (1968).
33. O. Gal, P. J. Baugh and G. O. Phillips, *Int. J. Appl. Radiat. Isotopes*, **22**, 321 (1971).
34. Ph. L. Dougherty, *Ph.D. Dissertation*, University of Denver, 1970.
35. D. R. Howton and G.-S. Wu, *J. Amer. Chem. Soc.*, **89**, 516 (1967).
36. B. Radak, Lj. Petković and B. Bartoniček, *Int. J. Radiat. Phys. Chem.*, **1**, 77 (1969).
37. R. N. Schwartz, M. W. Hanna and B. L. Bales, *J. Chem. Phys.*, **51**, 4336 (1969).
38. Iu. H. Molin, A. T. Korickii, H. Ia. Buben and V. V. Voevodskii, *Kokl. Akad. Nauk SSSR*, **124**, 127 (1959).
39. A. Sigimori, *Bull. Chem. Soc., Japan*, **39**, 2583 (1966).
40. V. L. Burton, *J. Amer. Chem. Soc.*, **71**, 4117 (1949).
41. V. L. Burton and I. A. Breger, *Science*, **116**, 477 (1952).
42. A. Faucitano, P. Locatelli, A. Perotti and F. Faucitano Martinotti, *J. Chem. Soc., Perkin II*, 1786 (1972).
43. D. R. Howton, *Radiation Res.*, **20**, 161 (1963).
44. P. Neta, R. W. Fessenden and R. H. Schuler, *J. Phys. Chem.*, **75**, 1654 (1971).
45. E. Hayon and M. Simić, *J. Amer. Chem. Soc.*, **95**, 2433 (1973).
46. S. Gordon. E. J. Hart, M. S. Matheson, J. Rabani and J. K. Thomas, *Disc. Farad. Soc.*, **36**, 193 (1963).
47. R. A. Watter and P. Neta, *J. Org. Chem.*, **38**, 484 (1973).

48. O. I. Mićić and V. Marković, *Int. J. Radiat. Phys. Chem.*, 4, 43 (1972).
49. O. I. Mićić and I. Draganić, *Int. J. Radiat. Phys. Chem.*, 1, 287 (1969).
50. A. Szutka, J. K. Thomas, S. Gordon and E. J. Hart, *J. Phys. Chem.*, 69, 289 (1965).
51. L. M. Dorfman and G. E. Adams, *Reactivity of the Hydroxyl Radical in Aqueous Solutions*, NSRDS–NBS 46, Washington, 1973.
52. P. Neta and R. H. Schuler, *Radiation Res.*, 47, 612 (1971)
53. J. K. Thomas, *Radiation Res. Suppl.*, 4, 87 (1964).
54. E. J. Hart, J. K. Thomas and S. Gordon, *Radiation Res. Suppl.*, 4, 74 (1964).
55. R. W. Taft in *Steric Effects in Organic Chemistry*, (Ed. M. S. Neuman), John Wiley, New York, 1956, pp. 556–675.
56. F. A. Peter and P. Neta, *J. Phys. Chem.*, 76, 630 (1972).
57. V. A. Palm, *Osnovi Kolechestyenoj Teorii Organischeskih Reakcij*, Ed. Himija Leningradskoje Otdeelenije, 1967, pp. 95–132.
58. P. Neta, M Simić and E. Hayon, *J. Phys. Chem.*, 73, 4207 (1969).
59. M. Simić, P. Neta and E. Hayon, *J. Phys. Chem.*, 73, 4214 (1969).
60. M. Simić and M. Z. Hoffman, *J. Phys. Chem.*, 76, 1398 (1972).
61. M. Simić, P. Neta and E Hayon, *J. Phys. Chem.*, 73, 3794 (1969).
62. N. Getoff, F Schwörer, V. M. Marković, K. Sehested and S. O. Nielsen, *J. Phys. Chem.*, 75, 749 (1971).
63. G. V. Buxton and R. M. Sellers, *J. Chem. Soc., Faraday I*, 69, 555 (1973).
64. I. G. Draganić and Z. D. Draganić, *The Radiation Chemistry of Water*, Academic Press, New York, 1971, pp. 130–140.
65. E. Hayon and M. Simić, *J. Amer. Chem. Soc.*, 95, 6681 (1973).
66. S. Arabamovitch and J. Rabani, *J. Phys. Chem.*, 80, 1563 (1976).
67. A. J. Swallow, *Photochem. and Photobiol.*, 7, 683 (1968).
68. P. Neta and R. W. Fessenden, *J. Phys. Chem.*, 76, 1957 (1972).
69. V. Madhaven, N. N. Lichtin and E. Hayon, *J. Org. Chem.*, 41, 2320 (1976).
70. P. Neta, *J. Phys. Chem.*, 75, 2570 (1971).
71. N. H. Anderson, A. J. Dobbs, D. J. Edge, R. O. C. Norman and P. R. West, *J. Chem. Soc.(B)*, 1004 (1971).
72. O. P. Chaula and R. W. Fessenden, *J. Phys. Chem.*, 79, 76 (1975).
73. P. Neta and R. W. Fessenden, *J. Phys. Chem.*, 77, 620 (1973).
74. J. Lilie and R. W. Fessenden, *J. Phys. Chem.*, 77, 674 (1973).
75. R. S. Alger, T. H. Anderson and L. A. Webb, *J. Chem. Phys.*, 30, 695 (1959).
76. Y. Nakajima, S. Sato and S. Shids, *Bull. Chem. Soc., Japan*, 42, 2132 (1969).
77. P. B. Ayscough and J. P. Oversby, *J. Chem. Soc., Faraday I*, 7, 1153 (1972).
78. Y. Noro, M. Ochiai, T. Miyazaki, A. Torikai, K. Fueki and Z. Kuri, *J. Phys. Chem.*, 74, 63 (1970).
79. K. J. Mills and J. M. Nosworthy Peto, *J. Chem. Soc., Faraday I*, 72, 1626 (1972).
80. A. S. Newton and P. O. Storm, *J. Phys. Chem.*, 62, 24 (1958).
81. S. M. Hyde and D. Verdin, *Trans. Farad. Soc.*, 64, 144 (1968).
82. S. M. Hyde and D. Verdin, *Trans. Farad. Soc.*, 64, 155 (1968).
83. E. J. Hart, E. M. Fielden and M. Anbar, *J. Phys. Chem.*, 71, 3993 (1967).
84. T. Bernath, G. H. Parsons, and S. G. Cohen, *J. Phys. Chem.*, 97, 2413 (1975).
85. T. Matsushige, G. Koltzenburg and D. Schulte–Frohlinde, *Ber. Bun. Ges. Phys. Chem.*, 79, 657 (1975).
86. P. B. Ayscough and J. P. Oversby, *J. Chem. Soc., Faraday I*, 7, 1164 (1972).
87. N. F. Barr and C. G. King, *J. Amer. Chem. Soc.*, 78, 303 (1956).
88. M. A. Schuler, K. Bhatia and R. H. Schuler, *J. Phys. Chem.*, 78, 1063 (1974).
89. K. Y. Kirino, R. H. Schuler, *J. Amer. Chem. Soc.*, 95, 6926 (1973).
90. G. P. Laroff, R. W. Fessenden and R. H. Schuler, *J. Amer. Chem. Soc.*, 94, 9061. (1972).
91. B. H. J. Bielski, D. A. Comstock and R. O. Bowen, *J. Amer. Chem. Soc.*, 93, 5624 (1971).
92. B. H. J. Bielski, A. O. Allen, *J. Amer. Chem. Soc.*, 92, 3793 (1970).
93. M. Schöneshöfer, *Z. Naturforsch. B*, 27, 649 (1972).
94. G. E. Adams, G. W. Boag, J. Currand and B. D. Michael, in *Pulse Radiolysis*, Ed M. Ebert, Academic Press, 1965, p. 131.
95. P. Neta and R. H. Schuler, *Radiation Res.*, 47, 612 (1971).

96. P. Neta and R. H. Schuler, *J. Phys. Chem.*, **76**, 2673 (1972).
97. J. Driegielewski, B. Jeżowska-Trzebiatowska, E. Kaleciśka, I. Z. Siemion, J. Kaleciński, J. Nawojska, *Nukleonika*, **18** 513 (1973).
98. J. Driegielewski, B. Jeżowska-Trzebiatowska, I. Z. Siemion, A. Zabza, *Nukleonika*, **19**, 291 (1974).
99. J. O. Driegielewski, A. Jezierski, B. Jeżowska-Trzebiatowska, H. Koztowski, *Bull. Acad. Pol. Sci., Ser. Sci. Chim.* **22**, 635 (1974).
100. J. O. Dziegielewski, *Int. J. Radiat. Phys. Chem.*, **7**, 507 (1975).
101. G. O. Phillips, D. M. Power, C. Robinson, *J. Chem. Soc., Perkin II*, 575 (1973).

CHAPTER **13**

The electrochemistry of carboxylic acids and derivatives: cathodic reductions

JAMES P. COLEMAN

Corporate Research Department, Monsanto Company, St. Louis, Missouri 63166, U.S.A.

I.	INTRODUCTION	782
II.	CATHODIC REDUCTION REACTIONS	782
	A. Carboxylic Acids	782
	1. Aromatic acids	784
	a. Aldehyde formation	784
	b. Alcohol formation	790
	c. Other reactions	792
	2. Heterocyclic acids	794
	a. Aldehyde formation	794
	b. Alcohol formation	797
	c. Other reactions	799
	3. Aliphatic acids	800
	a. Reduction of the carboxyl group	800
	b. Other reactions	800
	B. Carboxylic Esters, Lactones and Anhydrides	801
	1. Reduction of the carboxyl group	801
	a. Anion-radical formation	801
	b. Aldehyde formation	804
	c. Alcohol and ether formation	805
	d. Other reactions	805
	2. Reductions activated by the carboxyl group	806
	a. The electrohydrodimerization (EHD) reaction	806
	b. Cathodic carboxylation of α,β-unsaturated esters	808
	c. Cleavage reactions	810
	C. Carboxylic Amides, Lactams and Imides	813
	1. Reduction of the carboxyl group	813
	a. Anion-radical formation	813
	b. Formation of aldehydes and α-amido alcohols	814
	c. Alcohol formation	817
	d. Formation of amines from amides and amides from imides	818
	2. Other reactions	820
III.	ACKNOWLEDGMENTS	821
IV.	REFERENCES	821

I. INTRODUCTION

The past fifteen years have seen a tremendous expansion in the area of organic electrochemistry. With the advent of readily available, and relatively inexpensive, electronic equipment for carrying out electrochemical experiments, synthetic and physical organic chemists have gradually adopted electrochemical techniques for both synthetic and mechanistic organic studies. From the other extreme electrochemists have turned their attention to organic problems and have begun to unravel the intricacies of organic electrode processes.

With the expansion in this area has come an abundance of literature in the form of both books[1-11] and review articles [12-15], and papers describing electrochemical synthesis are more and more frequently found in non-electrochemical journals.

Carboxylic acids and derivatives have played an important role in the development of organic electrochemistry, notably through the Kolbe oxidation of carboxylates, an extensive subject in its own right which has already been reviewed in this series[16] and in a recent book[3], and is covered annually as part of a broader review[12]. For the purposes of this review, therefore, we shall concentrate on cathodic reductions of carboxylic acids and derivatives in an attempt to redress the balance.

For each type of compound we shall consider two different types of reaction. Firstly, reactions in which the carboxyl group is electroactive and is transformed as a result of the electrochemistry and, secondly, reactions in which the carboxyl group exerts an activating or modifying influence, so that the actual transformation occurs in some other part of the molecule. In certain cases both types of reaction may occur in competition with one another. Consideration of the second type of reaction will, hopefully, accomplish the goal of giving a broader survey of reaction types.

The aim of this review is to demonstrate the wide range of electrochemical syntheses to be found in this area, but not at the expense of excluding mechanistic considerations. Wherever electroanalytical data are available, they will, of course, be included in the discussion. It is assumed that the reader is familiar with electrochemical techniques, such as controlled potential electrolysis, cyclic voltammetry, polarography, etc. These have been adequately described in the literature[4,5,16a,17].

II. CATHODIC REDUCTION REACTIONS

A. Carboxylic Acids

The cathodic reduction of carboxylic acids is an area that has been studied sporadically since the beginning of the century and is documented in Fichter's classic survey of early organic electrochemistry[18] and in more recent reviews[1,3,7]. Cathodic reduction, however, is the lesser studied of the two aspects of carboxylic acid electrochemistry and has not received as much attention as carboxylate oxidation. To some extent the carboxylic acids which undergo efficient cathodic reduction and those which undergo the Kolbe reaction fall into groups which are complementary to one another, as will become apparent.

It has been amply demonstrated that the reduction of the carboxylic acid group may take place in four different ways:

$$RCO_2H \begin{cases} \xrightarrow{1e} & RCO_2^- + \tfrac{1}{2} H_2 & (1) \\[2em] \xrightarrow[2H^+]{2e} & RCHO + H_2O & (2) \\[2em] \xrightarrow[6H^+]{4e} & RCH_2OH + H_2O & (3) \\[2em] \xrightarrow[6H^+]{6e} & RCH_3 + 2 H_2O & (4) \end{cases}$$

Successful utilization of the reactions shown in equations (2), (3) and (4) is dependent primarily on being able to avoid the trivial reaction (1). The outcome of the reaction, reduction of the proton or the carbonyl group, depends on the nature of the group R. Unless the C=O double bond is activated towards reduction, either by conjugation with an aromatic or heterocyclic ring, or by a neighbouring electron-withdrawing group, then, with few exceptions, hydrogen evolution will be the major reaction.

Even if the structure of the carboxylic acid is such that an efficient cathodic reduction may be predicted, the experimental variables must be carefully controlled. Firstly, the cathode material must be a high hydrogen overvoltage material, typically mercury or lead, so that hydrogen evolution does not occur at low potentials, masking the desired reaction.

Secondly, the pH of the solution is an important factor, Strongly alkaline solutions will lead to formation of carboxylate ion which is even more difficult to reduce than free acid. Carboxylic acid reductions are, therefore, normally carried out in slightly alkaline or acid solution where the free acid is available for reduction and, in some cases, in strongly acidic solution where the positively charged, protonated acid, $RCO_2H_2^+$, is probably the electroactive species. The acid structure, the pH of the solution, the temperature, the electrode material and the use of added complexing agents or extracting solvents may all affect the relative contributions of equations (1)–(4) to the overall reaction, and, thus, the nature of the isolated products.

The basic difficulty which has to be overcome in isolating an aldehyde (equation 2) from the reduction of a carboxylic acid is that the aldehyde is generally far more easily reducible than the acid from which it is formed. Looking at equations (2) and (3), then, one would predict that at the potential necessary to reduce the acid the reaction would go right through to the alcohol stage. However, equation (2) is an oversimplification of the reaction pathway. The first step in the reduction is actually the formation of the hydrated aldehyde (equation 5). the hydrated aldehyde is not electroactive. Loss of water (equation 6) leads to the reducible

$$RCO_2H \xrightarrow[2H^+]{2e} RCH{\begin{array}{c} \diagup OH \\ \diagdown OH \end{array}} \qquad (5)$$

$$RCH{\begin{array}{c} \diagup OH \\ \diagdown OH \end{array}} \rightleftharpoons RCHO + H_2O \qquad (6)$$

aldehyde. The hydrated aldehyde may escape from the diffusion layer around the electrode and into the bulk of the solution depending on the rate constant for equation (6). Once in the bulk of the solution, the rate and equilibrium constants for equation (6) determine whether a substantial concentration of free aldehyde builds up in the solution during the electrolysis. If the free aldehyde concentration builds up, then the aldehyde will diffuse back to the cathode and be reduced to the alcohol. In this case the aldehyde must be trapped, either by complexing or with an extracting solvent. Where the hydrated aldehyde is stable, the electrolysis may be run to completion with no alcohol formation.

Once the reduction product is formed, it normally has to be protected from reoxidation at the anode. This is accomplished by carrying out the reaction in a divided cell where the anode and cathode compartments are separated either by a porous membrane e.g. ceramic, sintered glass, or by an ion exchange membrane.

These general considerations will now be illustrated by reference to three types of acids—aromatic, heterocyclic and aliphatic.

1. Aromatic acids

a. Aldehyde formation. The cathodic reduction of benzoic acid to benzaldehyde was reported as long ago as 1908 by Mettler[19]. The electrolysis was carried out using a mercury cathode in an undivided cell and a solution of sodium benzoate, sodium sulphate and boric acid in water. Benzene was used to extract the benz-aldehyde from the electrolysis solution as it was formed and yields of 30—50% were claimed (equation 7). This method was revised by Wagenknecht[20] who studied the

$$PhCO_2H \xrightarrow[\text{benzene}]{\text{Hg, Na}_2\text{SO}_4,\ \text{H}_3\text{BO}_3} PhCHO \tag{7}$$

reduction of a series of carboxylic acids to the corresponding aldehydes. There were several differences from Mettler's procedure. Ammonium carboxylates were used in a divided cell with a mercury cathode and a constant current density of $10\ \text{mA/cm}^2$; the pH was controlled at 6 ± 0.2. Systems with boric acid, ammonium dihydrogen phosphate or no added buffer were investigated. Benzene was used to extract the aldehydes as they were formed, and analysis of the benzene solutions by gas chromatography at 5% theoretical conversion gave the results shown in Table 1.

Table 1 serves to illustrate some of the generalizations made earlier. Firstly, the effect of structure may be seen. Benzoic acid, with a conjugated carboxyl group, gives a reasonable current efficiency for aldehyde formation, whilst the non-conjugated phenylacetic and phenoxyacetic acids give no reduction at all. Electron-donating substituents in the aromatic nucleus cause a marked decrease in yield, except in the special case of salicyclic acid and its derivatives which will be discussed later. Electron-withdrawing fluoro and cyano substituents cause very little change in aldehyde yield.

Parallels may be drawn between the pK_a values of the conjugated aromatic acids and reduction efficiencies, with a maximum pK_a value of about 4.40 for reduction to occur under these conditions. A better correlation exists between yields and polarographic half-wave potentials of the methyl esters of the acids. If the ester has a reduction potential more negative than -2.20 (vs. SCE), then reduction of the acid is likely to be very inefficient.

The aldehyde hydration equilibria and efficiency of extraction of the aldehyde into benzene are such that very little alcohol was formed in any of these reactions. In an experiment taken to 50% conversion of salicylic acid an 80% chemical yield of salicylaldehyde and only 4% of the alcohol, saligenin were obtained. The type of

TABLE 1. Reduction of various carboxylic acids to the corresponding aldehydes[a]

Acid	Current efficiency[b] (%)	Buffer	$pK_a{}^c$	$E^1/_2$ vs SCE[d]
Benzoic	55	H_3BO_3	4.20	$-2.12^{d,\,e}$
Benzoic	8	None		
Benzoic	42	Phosphate		
Salicylic	73	H_3BO_2	3.00	-2.02^e
Salicylic	Trace	None		
Salicylic	Trace	Phosphate		
p-Hydroxybenzoic	3	H_3BO_3	4.48	-2.32^e
o-Methoxybenzoic	55	H_3BO_3	4.08	
o-Methoxybenzoic	7	None		
o-Methoxybenzoic	43	Phosphate		
p-Methoxybenzoic	1.75	H_3BO_3	4.47	-2.31^e
o-Toluic	3	H_3BO_3	3.91	-2.21^d
p-Toluic	15	H_3BO_3	4.37	$-2.20^{d,\,e}$
o-Acetoxybenzoic	41	H_3BO_3	3.49	
o-Acetoxybenzoic	1.5	None		
o-Acetoxybenzoic	3.5	Phosphate		
p-Cyanobenzoic	37	H_3BO_3	3.55	
o-Fluorobenzoic	41	H_3BO_3	3.27	
o-Fluorobenzoic	14	None		
o-Fluorobenzoic	46	Phosphate		
Vanillic	5	H_3BO_3	4.48	
Phenylacetic	None	H_3BO_3	4.31	
Phenoxyacetic	None	H_3BO_3	3.17	

[a] Reprinted with permission from J. H. Wagenknecht, *J. Org. Chem.*, 37, 1513 (1972). Copyright by the American Chemical Society.
[b] To the corresponding aldehyde.
[c] Values taken from *Handbook of Tables for Organic Compound Identification*, 3rd ed., Chemical Rubber Co., Cleveland Ohio, 1967.
[d] Values determined for the methyl esters in 50% ethanol containing 0.1 M $Et_4N^+ClO_4{}^-$.
[e] Values taken from T. Arai, *Nippon Kagaku Zasshi*, 89, 188 (1968), were converted to the numbers shown by addition of -0.42 V as determined from the two overlapping compounds, methyl benzoate and methyl p-toluate.

buffer system used for the reduction caused significant changes in results. For benzoic acid either a phosphate or boric acid buffer led to reasonable aldehyde yields, whilst in the solution with no added buffer the pH at the cathode surface became very high[21] and reduction inefficient, hydrogen evolution presumably accounting for most of the current. Since free carboxylic acid is not likely to exist under these very basic conditions at the cathode surface and benzoate anion shows no polarographic reduction wave[22], it was proposed that reduction of an ion pair was taking place in the unbuffered solution.

This same type of variation in yield with different buffer systems was observed for o-fluorobenzoic acid but not for salicylic acid and its methyl and acetyl derivatives. In these cases formation of a borate complex occurs and the complex is more easily reducible than the free acid.

The cathodic reduction of salicylic acid (equation 8) has probably been more intensively studied than any other carboxylic acid reduction. The desired product, salicylaldehyde, is used in the synthesis of coumarin for perfumery and other purposes.

$$\text{(salicylic acid, CO}_2\text{H, OH)} \xrightarrow[2\,\text{H}^+]{2\,\text{e}} \text{(salicylaldehyde, CHO, OH)} + H_2O \qquad (8)$$

The reduction was first carried out by Weil[23] and by Mettler[19] using Mettler's method described above for benzoic acid. Yields of 30–50% were claimed. However, Tesh and Lowy[24] were able to obtain only a 20% yield using this method. Using a modified procedure, whereby sodium bisulphite was added to the catholyte to trap salicylaldehyde as a non-reducible bisulphite compound, the yield of aldehyde was increased to 55%. Kawada and Yosida[25] improved upon this figure still further, obtaining an 80% yield of aldehyde using a catholyte containing sodium sulphate, bisulphite and salicylate, boric acid and borax.

Some twenty years later May and Kobe[26] substantiated the results of Tesh and Lowy but were not able to obtain the high yields claimed by the Japanese workers[25] and thought the attainment of higher yields improbable, even though they did not investigate Kawada and Yosida's exact conditions.

More recently, Udupa and Dey have studied this reaction using amalgamated copper cathodes and have shown that, whilst stationary cathodes give only a 10% yield of salicylaldehyde[27], rotation of the cylindrical cathodes at 1800 r.p.m. increases the yield[28] to 55–60%, possibly because of better pH control at the cathode surface.

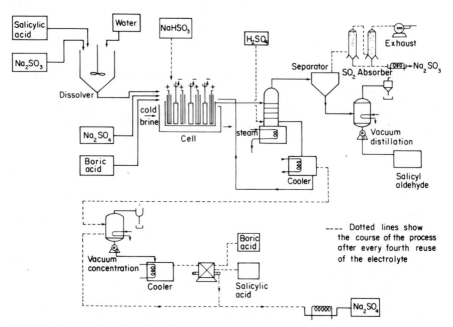

FIGURE 1.   Flowsheet for the electrolytic production of salicylaldehyde. Taken from K. S. Udupa and coworkers, *Ind. Chem.*, **39**, 238 (1963). Reproduced by permission of *Processing* (IPC Industrial Press Ltd.).

Using this technology Udupa and coworkers[29] have constructed a 300 A cell which can produce one kilogram of salicylaldehyde in about four hours. With the exception of the cathodes, the conditions are basically the same as those used by other workers. Boric acid is used to promote the reduction at pH 5.5–5.7, about 15°C and 12–15 A/dm², with sodium bisulphite to stabilize the product. Lead anodes were separated from the catholyte by microporous rubber and later[30] asbestos diaphragms. The salicylaldehyde was steam-distilled from the catholyte without acidification and the catholyte reused after makeup with salicylic acid. The average of four runs with the same batch of electrolyte gave a yield of 80% and a current efficiency of 35%.

In Figure 1 a flowsheet is given for a semicontinuous production unit capable of producing 50 kg/day of salicylaldehyde.

A pilot-scale unit has also been built by Fioshin and coworkers, in this case with a mercury cathode[31]. They have claimed that a small electrochemical unit with a capacity of 10–15 tons/year would produce salicylaldehyde at 40% of the cost of the chemically produced material. To put these figures in perspective the 1974 United States annual sales of salicylaldehyde amounted to 3.99 million pounds[32] (about 1800 tons), so that both of these units are a long way from commercial production size. Significant changes would have to be made to make either process truly commercial.

The preparative results on salicylic acid reduction are summarized in Table 2. The preparative features of this reaction are, thus, well established and more recently attention has been turned to mechanistic questions. The specific effect of boric acid on salicyclic acid reduction has been known for seventy years and complex formation postulated[26] but not pursued electrochemically until Ekel and coworkers undertook a polarographic investigation[33]. It was shown that the proton reduction wave of boric acid in aqueous tetraethylammonium iodide was replaced by a bigger wave at less negative potentials as small amounts of salicylic acid were added. The size of the wave could not be accounted for by the separate components and was, therefore, ascribed to the reduction of the hydrogen atom in a complex with a higher degree of dissociation than the original acids. However, no second wave due to aldehyde formation was observed.

Robertson and collaborators[34] went further and prepared mono- and disalicyl-borate complexes by a known method[35]. Tetrabutylammonium disalicylborate (1)

TABLE 2. Cathodic reduction of salicylic acid to salicylaldehyde

Conditions[a]	Chemical yield (%)	Current efficiency (%)	Reference
Hg cathode, benzene extraction	30–50	–	19
Hg, 6 A/dm², 15–18°C; NaHSO₃ complex	–	55	24
Hg, 2 A/dm², borax added; 8–13°C; NaHSO₃	80	–	25
Repeat of Reference 24	50	47	26
Cu–Hg cathode, 12–15 A/dm², pH 5, 4–5, 7; 15°C ; NaHSO₃	80	35	29
Hg; 10 A/dm²; 20–22°C; NaHSO₃	52	–	31
Hg; 2 A/dm²; pH 6; 10°C; benzene extraction	80	50	20

[a] All solutions contained sodium sulphate, salicylate and boric acid.

was purified and studied by a.c. and d.c. polarography and cyclic voltammetry in anhydrous DMF—tetrabutylammonium iodide. Two consecutive, reversible one-

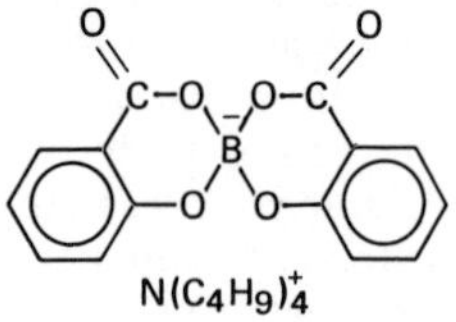

$$N(C_4H_9)_4^+$$

(1)

electron transfers were observed, at -2.07 V and -2.36 V (vs. Hg pool), respectively, corresponding to equation (9). The reducibility of the complex, relative to the non-

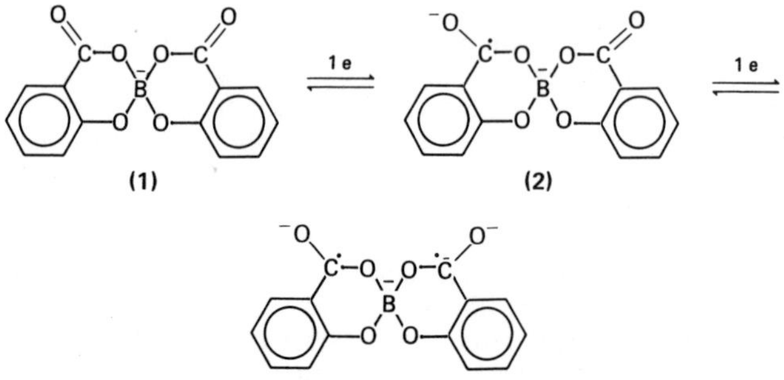

(1)   (2)

(9)

(3)

reducibility of the free acid, was attributed to a lowering of charge density on the carboxyl group resulting from complex formation. Both 2 and 3 were stable on the voltammetric time-scale (3 V/s), but, as water was added to the system, the first reduction wave increased in size and the second disappeared, indicating that 2 was being protonated and further reduced at the first wave. At concentrations of water beyond 50% the current for the first wave approached the value for a four electron transfer (two electrons/mole salicylate), presumably leading to salicylaldehyde although no product studies were made to confirm this. Mechanisms may be drawn for the decomposition of the reduced complex to salicylaldehyde.

Thus, the electrochemical behaviour of a known borate—salicylate complex has been demonstrated, albeit under conditions somewhat different from those used in synthesis, and this gives some explanation of the specific effect of boric acid on salicylic acid reduction.

Similar studies have not been made to determine the electron transfer behaviour of non complexed acids, and it is doubtful that they would prove particularly fruitful. On the basis of carbonyl compounds which form stable anion radicals[36], and the work described above, one may postulate an ECE (electrochemical, chemical, electrochemical) sequence for carboxylic acid reduction (equation 10). Detection of the carboxylic acid anion radical, even in an aprotic solvent, is unlikely, since the carboxylic acid itself is a good proton donor.

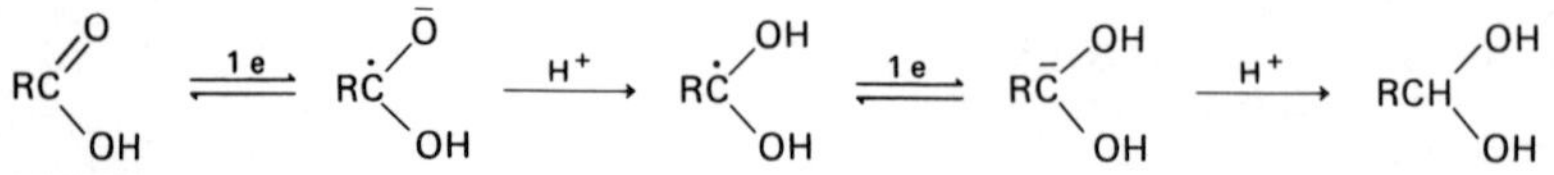

(10)

Other mechanistic studies have been made. Harrison and Shoesmith[37] carried out controlled potential reductions of benzoic and substituted benzoic acids in aqueous buffer solutions at an amalgamated copper rotating disc. Since the reaction of interest is occurring in the hydrogen evolution region, controlled potential reduction does not offer the advantage that it does with more easily reducible compounds. Analysis of current-potential curves yielded little useful information. However, analysis of product (aldehyde and alcohol) vs time curves for controlled potential electrolyses did allow calculation of approximate rate constants for the dehydration step (equation 6).

Cyclic voltammetry at a hanging mercury drop was also used. The buffer solution showed a reoxidation peak for sodium amalgam after sweeping into the background current region. This peak was decreased significantly when benzoic acid was added to the solution. The two possible explanations for this are that, firstly, the benzoic acid could be reduced indirectly by the sodium or, secondly, the benzoic acid is absorbed on the electrode and inhibits the sodium reduction. The latter explanation was favoured since tetramethylammonium electrolyte gave increased product yields vs. a sodium electrolyte and tetramethylammonium amalgam formation at the potentials used was considered unlikely. However, reduction of aromatic acids by sodium amalgam has been demonstrated[38], and it is possible that during large-scale preparative electrolyses both direct and indirect reductions are occurring.

These voltammetric studies were extended[39] and it was claimed that the voltammetry of benzoic acid in aqueous tetramethylammonium chloride at a mercury electrode showed a peak for the reduction of an adsorbed layer of tetramethylammonium benzoate to benzaldehyde hydrate. No such reaction was observed at lead and no products were obtained from a preparative-scale electrolysis at a lead cathode. Failure to observe any reaction was attributed to lack of benzoic acid adsorption at a lead cathode.

Tin, zinc and carbon cathodes, in addition to lead and mercury, were used in a study of 2-chlorobenzoic acid reduction in aqueous solution[40]. It was hoped that introducing the chloro group would be an additional guide to the mechanism of the electrode reaction. Whilst no particularly novel mechanistic conclusions were reached, some interesting observations were made. Reduction at tin and zinc cathodes occurred at much lower potentials than on lead or mercury, and, in addition to the expected reduction products, dechlorinated and decarboxylated products were observed (equations 11 and 12). These were only very small-scale

electrolyses in terms of amounts of product formed, but the use of tin and zinc cathodes would seem to offer some interesting possibilities for synthetic work if each of the different reaction types could be optimized. Reduction of the chloro group might be expected at either the acid or aldehyde stage. But cathodic decarboxylation is certainly novel although a similar chemical reaction is known[41].

In summary, the cathodic reduction of aromatic carboxylic acids to aldehydes has been studied mainly from a synthetic angle, using fairly crude electrochemical techniques. More recently, electroanalytical studies have been made but have been severely hindered by the hydrogen evolution reaction which occurs in the same potential region as carboxylic acid reduction.

b. Alcohol formation. The cathodic reduction of aromatic acids at high over-potential cathodes in strongly acidic solution generally leads to the corresponding alcohol (equation 13).

$$ArCO_2H \xrightarrow{\text{e.g. Pb, EtOH–H}_2\text{SO}_4} ArCH_2OH \tag{13}$$

Like the two-electron reduction, this reaction has been studied since early in the century and is well documented[42–46]. Mettler[43] was apparently the first to carry out this reaction, reducing benzoic acid in 30% sulphuric acid in ethanol, with a lead cathode. Benzyl alcohol was obtained in about 85% yield. Swann[44] later investigated various cathode materials for this reaction, using conditions similar to Mettler's. It was found that only lead and cadmium were suitable, lead being preferred. Surprisingly, mercury did not give any reduction.

Udupa[46] studied the reaction in aqueous sulphuric acid and confirmed that lead was the most suitable cathode material, Current efficiency was shown to increase with temperature up to 92% at about 80°C for reduction at a rotating lead cathode in 10% sulphuric acid. Current density had little effect in the range 10–40 A/dm^2 but increasing sulphuric acid concentration beyond 10% led to a decrease in current efficiency, possibly because of decreased benzyl alcohol solubility in the higher strength acid leading to electrode coating.

This reaction (equation 14) provides a fairly general synthesis of benzyl alcohols

Chemical yield = 69–78%
Current efficiency = 28–32%

$$\tag{14}$$

and anthranilic acid reduction has been described in *Organic Syntheses*[47]. Hydrogen evolution is the major cathode reaction, but the chemical yield is good.

The effect of carboxylic acid structure is less critical in these strongly acidic solutions than it was in neutral or slightly acid solutions. Thus, *o*-anisic acid gives the anisyl alcohol in 70% yield[48] and phenylacetic acid may be reduced to 2-phenylethanol in ethanol containing 50% aqueous sulphuric acid at a lead cathode[49] (equation 15). Substituting a platinized platinum cathode results in hydrogenation of the aromatic nucleus, leaving the carboxyl group intact (equation 16).

$$C_6H_5CH_2CO_2H \begin{cases} \xrightarrow{\text{Pb, EtOH–50\% aq. H}_2\text{SO}_4} \begin{array}{l} C_6H_5CH_2CH_2OH \\ \text{Chemical yield = 30–50\%} \\ \text{Current efficiency = 10–20\%} \end{array} \tag{15} \\ \\ \xrightarrow{\text{Pt–Pt, EtOH–50\% aq. H}_2\text{SO}_4} \begin{array}{l} C_6H_{11}CH_2CO_2H \\ \text{Chemical yield = 82\%} \\ \text{Current efficiency = 47\%} \end{array} \tag{16} \end{cases}$$

Mechanistic information in this area is even more scarce than it was for the two-electron reduction, most of the work having been done before controlled potential electrolysis was in common use. Once again, however, the usefulness of controlled potential reduction is reduced by hydrogen evolution.

It is quite probable that the electroactive species in these reactions are protonated carboxylic acids, $RCO_2H_2^+$, although there is no direct evidence to that effect. One would expect the reduction of the positively charged, protonated species to be less susceptible to structural electronic effects than the reduction of the neutral molecule and this seems to be confirmed by the observations. In cases where alcoholic cosolvents are used the reaction is probably further complicated by ester formation.

The dehydration reaction (6) is acid-catalysed[37]; the free aldehyde, therefore, forms rapidly, possibly in the diffusion layer, and is reduced, again possibly in the protonated form[50], to the alcohol. Aldehydes are, therefore, not isolated from reactions carried out under these conditions.

The reduction of salicylic acid to salicylaldehyde has already been described in detail, and the necessity of adding boric acid in order to carry out the reduction in an aqueous electrolyte was pointed out. However, using isopropanol as the solvent and tetraethylammonium bromide as the electrolyte this reduction has been carried out giving a good yield of the alcohol in the absence of boric acid[52] (equation 17).

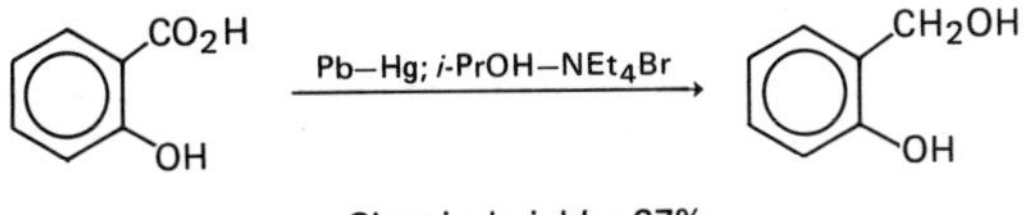

$$\text{Pb–Hg; } i\text{-PrOH–NEt}_4\text{Br} \tag{17}$$

Chemical yield = 67%
Current efficiency = 58%

Controlled potential electrolysis was used to study the reduction of pentafluoro-benzoic acid in several different aqueous solutions[51]. this work contained some interesting observations and some apparent contradictions.

The two different reduction pathways observed are shown in equations (18) and (19). The competition between these pathways was studied using different electrolytes. In neutral, aqueous tetraethylammonium tetrafluoroborate solution, fluoride

$$C_6F_5CO_2H \longrightarrow [C_6F_5CHO] \longrightarrow C_6F_5CH_2OH \tag{18}$$

$$\longrightarrow p\text{-}HC_6F_4CO_2H \longrightarrow [p\text{-}HC_6F_4CHO] \longrightarrow p\text{-}HC_6F_4CH_2OH \tag{19}$$

cleavage was the predominant reaction, as shown in Table 3. In 10% aqueous sulphuric acid both pathways were observed, fluoride cleavage predominating at low potentials. Dilute aqueous perchloric acid gave similar results to sulphuric acid, but in concentrated (50%) perchloric acid reduction of the carboxyl group was the

TABLE 3. Reduction of pentafluorobenzoic acid at a mercury cathode[a]

Conditions	Product (% yield)		
	$p\text{-}HC_6F_4CO_2H$	$p\text{-}HC_6F_4CH_2OH$	$C_6F_5CH_2OH$
−1.20 V (vs SCE), aq. H_2SO_4	73	20	6
−1.30 V, aq. H_2SO_4	−	48	24
−1.50 V, aq. H_2SO_4	−	33	45
−2.00 V, aq. NEt_4BF_4	40	55	−
−2.00 V, 50% aq. $HClO_4$	−	<1	70
−2.00 V, 7% aq. $HClO_4$	−	38	40

[a] Data taken from F. G. Drakesmith, *J. Chem. Soc., Perkin I*, 184 (1972). Reproduced by permission of the Chemical Society, London.

major reaction. These results were rationalized in terms of equations (20) and (21).

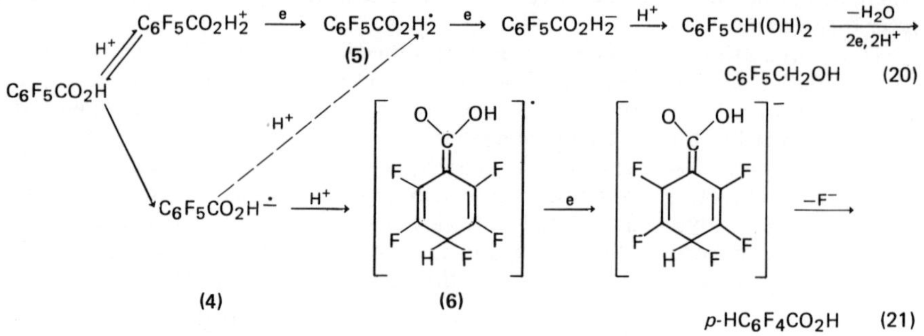

(5)

(4)   (6)

(20)

$p\text{-HC}_6\text{F}_4\text{CO}_2\text{H}$ (21)

In strongly acidic solution, reduction according to equation (20) probably occurs. For simplicity, equation (20) is condensed after the hydrated aldehyde stage.

In neutral solution, reduction to the anion radical as in equation (21) was postulated with subsequent protonation on the aromatic nucleus, electron transfer and fluoride elimination, forming the tetrafluorobenzoic acid. Reduction of the tetrafluoro acid via a combination of equations (20) and (21) would then give the tetrafluoro alcohol.

In solutions of intermediate acidity, however, the results are not easily reconcilable with equations (20) and (21). The author[51] simply says 'mechanism II (equation 21) operates at less negative potentials in a given medium than does mechanism I (equation 20)'. One would, in fact, expect just the opposite to occur. Reduction of the protonated acid should occur at lower potentials, Therefore, if the protonated acid were the electroactive species in these media, the product trend should be reversed. If the neutral carboxylic acid is the electroactive species, one has to postulate a competitive protonation of anion radical (4) at the carboxyl group or the nucleus, leading to 5 or 6, respectively. If for some reason the site of protonation changes with potential, then this would account for the results.

The diffusion layer around the electrode should become less acidic with increasing cathodic potential (current). this, again, would produce the opposite product vs. potential trend observed.

It appears, therefore, that in dilute acid media either the reduction occurs by a mechanism not accounted for in equations (20) and (21) or the site of protonation of the anion radical varies with potential.

c. *Other reactions.* Dibasic aromatic acids provide an example of a different type of reaction—selective hydrogenation of one double bond in the aromatic nucleus. Phthalic acid reduction was again originally studied by Mettler[53], but has been investigated more recently by Beck, Nohe and coworkers[54] at BASF. Mettler's original synthesis used hot (85°C) 5% sulphuric acid as the solvent, with a lead cathode. The high temperature led to isomerization of the initially formed product.

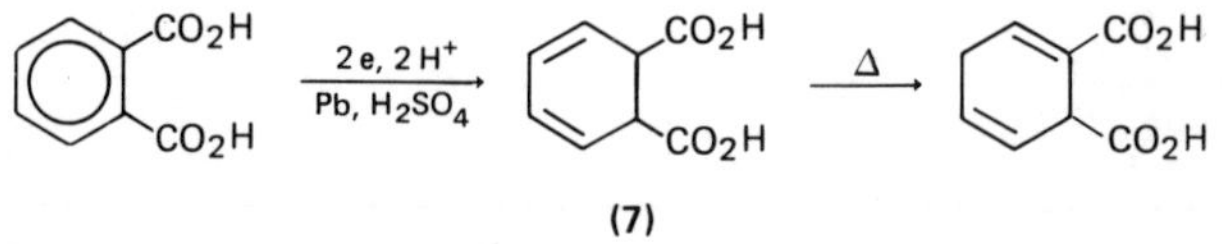

(7)

(22)

To develop a commercial synthesis of 3,5-cyclohexadiene-1,2-dicarboxylic acid

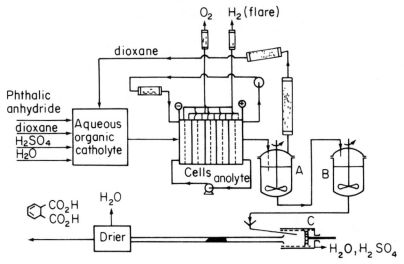

FIGURE 2.   Flowsheet for the continuous electrolytic production of 3,5-cyclo-hexadiene-1,2-dicarboxylic acid. A: stripper, B: crystallizer, C: centrifugal filter. Taken from F. Beck, *Elektro-organische Chemie*, Verlag Chemie GmbH, Weinheim, 1974. Reproduced by permission of Verlag Chemie GmbH, Weinheim.

(7) to be used in polymers, plasticisers, etc., dioxane was used as a cosolvent for this reaction, the resulting solubilization[55] of phthalic acid allowing high current densities to be used at low temperatures[55]. Using lead cathodes in a mixture of phthalic acid (20%), dioxane (55%), water (20%) and sulphuric acid (5%), a chemical yield of up to 99% with a current efficiency of 83% is attainable for the desired dihydrophthalic acid (7). This process was scaled-up to a unit producing 50 tons/month, the flowsheet for which is given in Figure 2.

The same reaction has been applied to substituted phthalic acids[56]. For instance, tetrachlorophthalic acid gives the dihydro derivative in 98% yield with 79% current efficiency.

Reduction of phthalic acid at a mercury cathode in potassium hydroxide solution, presumably at a more negative potential, gives further reduction to the tetrahydrophthalic acid[57].

$$\text{(23)}$$

78% yield

Terephthalic acid[54] behaves in the same way as phthalic acid, but with the carboxyl groups *meta* to one another in isophthalic acid, eight-electron reduction to the diol was observed[53].

$$\text{(24)}$$

82% yield

There is, thus, a delicate balance between the carboxyl groups activating the aromatic nucleus towards reduction or the nucleus activating the carboxyl groups. Changing the substitution pattern is sufficient to tip the balance one way or the other. This balance is also affected by pH. Benzoic acid which gives benzaldehyde or benzyl alcohol, in neutral or acidic solutions, respectively, gives cyclohexene-2-carboxylic acid in strongly alkaline solution[58].

The reduction of aromatic carboxylic acids to the corresponding methyl compounds (equation 4) does not seem to occur. We will, therefore, conclude the section on aromatic carboxylic acid reductions with a novel reaction which was observed with a semiconducting germanium cathode in ethanolic sulphuric acid. One-electron reduction occurred, yielding benzil[59].

$$2\,PhCO_2H \xrightarrow[\text{80\% EtOH, 0.1N } H_2SO_4]{\text{Ge, } 10^{-5}-10^{-2}\,A/cm^2} PhCOCOPh \qquad (25)$$

At the moment this has little more than curiosity value, but it does serve to show, along with Harrison's work with tin and zinc cathodes[40], that interesting results are to be found when cathodes other than the classical mercury or lead are investigated.

2. Heterocyclic acids

In discussing the cathodic reduction of heterocyclic carboxylic acids we shall be concerned mainly with N-heterocyclic acids, and most of the work we shall consider has been carried out within the last fifteen years. Electroanalytical techniques, particularly polarography, have been applied more extensively, and more profitably, in this area since certain of the heterocyclic acids are reducible at potentials less cathodic than that at which background electrolyte discharge occurs. the media used for these reactions are again aqueous solutions.

a. Aldehyde formation. Following some polarographic investigations by other workers[60], Lund, one of the major contributors to this area, undertook a study of the electrochemistry of isonicotinic acid (8) and its N-ethyl derivative (9)[61].

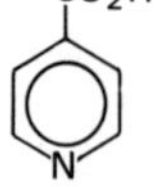

(8) 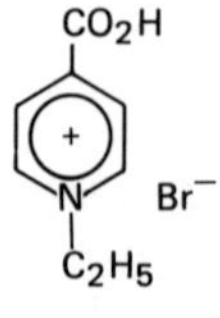   (9)

The N-ethyl compound (9) shows a polarographic reduction wave which varies in potential from -0.74 V (vs SCE) in strongly acidic solution to -1.22 V in alkaline solution. From the plot of $E_{1/2}$ vs pH, which may be approximated by three lines with changes of slope at pH 2 and pH 7, it was concluded that at pH greater than seven the reducible species was in the carboxylate anion form, whilst at Ph less than seven an undissociated carboxyl group was present. Since the pK of 9 was shown spectroscopically to be 1.75, a recombination reaction at the cathode was postulated to explain the results in the pH range 2—7.

Isonicotinic acid shows similar behaviour, explained in terms of equation (26).

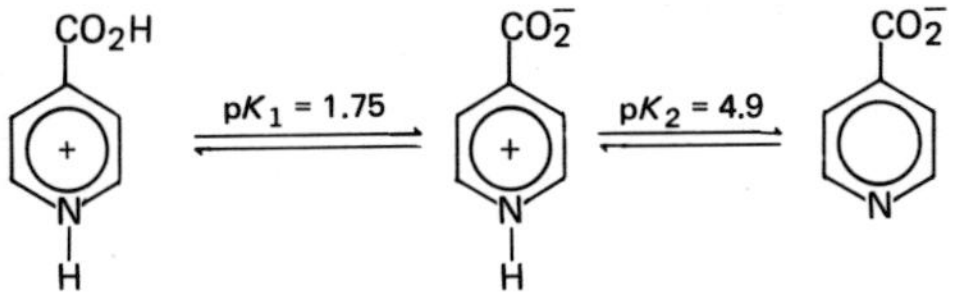

$$\text{(10)} \qquad \text{(11)} \qquad \text{(12)} \tag{26}$$

In strongly acidic solution **10** is the electroactive species, whilst the first dissociation gives the zwitterion **11**. At pH greater than 8 a splitting in the polarographic wave, not seen with **9** indicates that the anion **12** is present.

Small-scale preparative electrolyses were carried out with a mercury cathode at −1.0 V (vs SCE) in acid solution. The electrolyses were stopped after 2 F/mol of acid had been consumed and the products were analysed. The corresponding aldehyde was shown to be the major product in both cases—by isolation of the phenylhydrazone for pyridine-4-aldehyde and by polarography for the N-ethyl aldehyde.

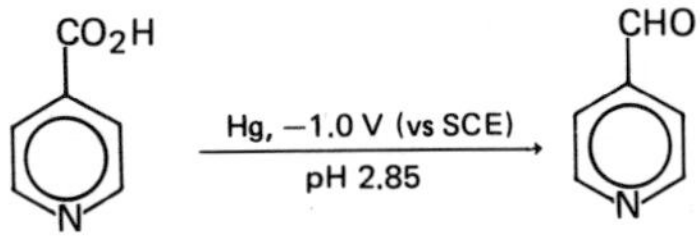

$$\tag{27}$$

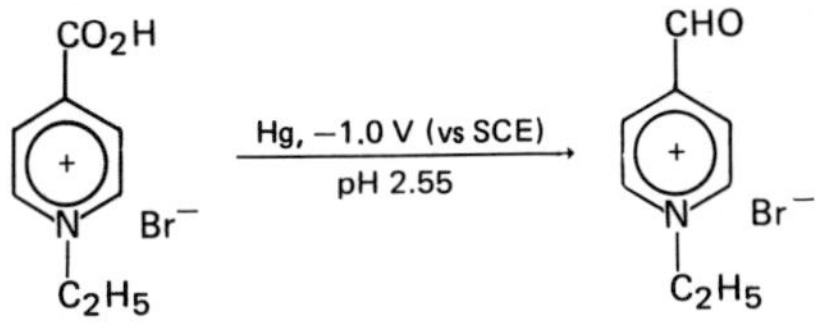

69%

$$\tag{28}$$

75%

Good yields of aldehyde were obtained even though no attempt at complexing or extracting was made. This is because these aldehydes, activated by the positive nitrogen, are strongly hydrated in acid solution, as can be shown by the abnormally low polarographic currents for these compounds in acidic media[62]. At pH greater than five, the yield of pyridine-4-aldehyde decreased dramatically, and unidentified products were formed. Possibly formation of the carboxylate anion at this pH, in contrast to the polarographic result, leads to hydrogenation of the pyridine nucleus.

A similar approach was used[63] in a study of imidazole-2-carboxylic acid (**13**) and 1-benzylimidazole-2-carboxylic acid (**14**). These imidazole acids again illustrate

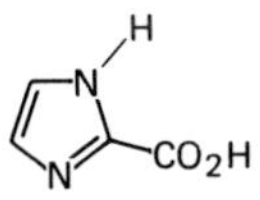

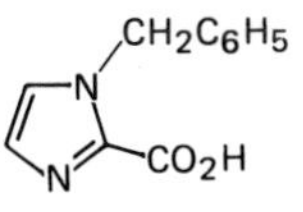

$$\text{(13)} \qquad\qquad \text{(14)}$$

the effects of structure and pH on carboxylic acid reduction. Imidazole-4-carboxylic acid shows no reduction in aqueous solution, but the 2-isomer (**13**) is sufficiently activated that it is reducible in certain pH regions. Figure 3 shows the

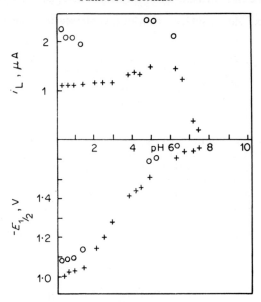

FIGURE 3.   Limiting currents (μA) and half-wave potentials vs pH for imidazole-2-carboxylic acid ($\circ$) and 1-benzylimidazole-2-carboxylic acid (+). Taken from P. E. Iversen and H. Lund, *Acta Chem. Scand.*, **21**, 279 (1967). Reproduced by permission of *Acta Chem. Scand.*

half-wave potentials and limiting currents vs. pH for **13** and **14**. In strongly acidic solution **13** shows a reduction wave which merges into the background in the pH range 2–5. The reduction is then discernible until about pH 6 when the non-reducible anion is formed. The 1-benzyl compound (**14**) shows similar behaviour except that it is reducible at slightly less negative potentials and the reduction is, therefore, observable throughout the pH range. The decreased limiting current in Figure 3 for **14** vs **13** is caused by a combination of lower concentration and lower diffusion coefficient for **14**. Similar protonation equilibria exist for these compounds as were discussed for isonicotinic acid.

     Small-scale, controlled-potential reductions were carried out, giving the aldehydes (equations 29 and 30). Strong hydration of the aldehydes in acid solution is

$$\underset{(13)}{\text{imidazole-}CO_2H} \xrightarrow[\text{0.8N HCl, 0°C}]{\text{Hg, }-1.15\text{ V}} \underset{82\%}{\text{imidazole-}CHO} \qquad (29)$$

$$\underset{(14)}{\text{1-benzylimidazole-}CO_2H} \xrightarrow[\text{1.3N HCl, 0°C}]{\text{Hg, }-1.05\text{ V}} \underset{94\%}{\text{1-benzylimidazole-}CHO} \qquad (30)$$

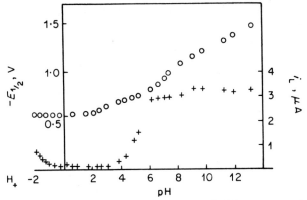

FIGURE 4.   Dependence on pH of the limiting current and half-wave potential of imidazole-2-carbaldehyde; (o) half-wave potential (vs SCE), (+) limiting current (μA). Taken from P. E. Iversen and H. Lund, *Acta Chem. Scand.*, **21**, 279 (1967). Reproduced by permission of *Acta Chem. Scand.*

again responsible for the high yields. Figure 4 shows how the limiting current for imidazole-2-carbaldehyde decreases in acid solution until only a very small, kinetically controlled wave, governed by equation (31), is seen.

$$\text{(reducible)} \quad + H_2O \rightleftharpoons \text{(non-reducible)} \tag{31}$$

The product yields in this work were determined polarographically, but, in a later paper[64], the reductions were carried out on a larger scale (0.2 mole) and the products isolated. Isolated yields of aldehydes were typically 10--20% lower than the figures obtained from polarographic analysis.

Thiazole-2-carboxylic acid (**15**) shows two polarographic reduction waves[65] in the range -0.7 to 1.7 V (vs SCE), but the reduction is more complicated than in the

(15)

imidazole series. Reduction at the first wave gave only a 25% yield of the aldehyde, and other unidentified products were formed. The nature of the second reduction was not elucidated, but reduction of the heterocyclic nucleus was mentioned as a possibility.

b. Alcohol formation.   Brown and coworkers[66] set out to synthesize the corresponding alcohols from picolinic acid (**16**) and dipicolinic acid (**17**). In less

(16)   **(17)**

acidic solutions than those used by Lund[61] four-electron reduction to the alcohol occurred (equations 32 and 33). Isolated yields of alcohols were not very high,

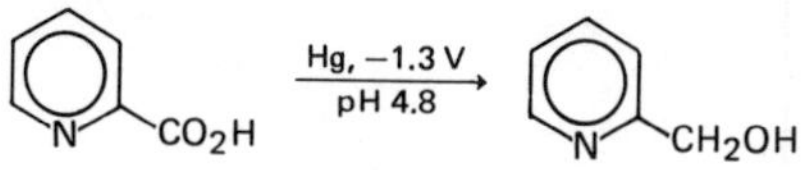

$$
\text{(16)} \quad \xrightarrow[\text{pH 4.8}]{\text{Hg, }-1.3\text{ V}} \quad \text{(32)}
$$

(16)

Chemical yield = 46%
Current efficiency = 52%

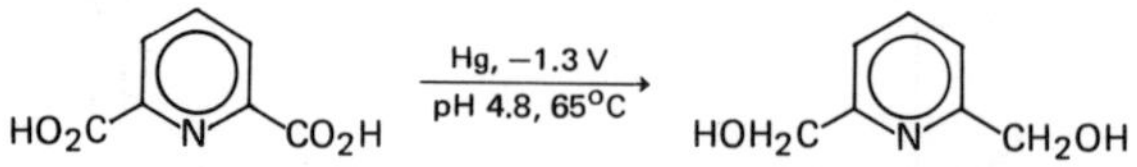

$$
\text{(17)} \quad \xrightarrow[\text{pH 4.8, 65°C}]{\text{Hg, }-1.3\text{ V}} \quad \text{(33)}
$$

(17)

Chemical yield = 26%
Current efficiency = 30%

particularly for dipicolinic acid reduction. This reaction apparently showed four-electron coulometry, not eight electrons as expected for equation (33). Whilst an aldehyde was detected during the reaction, no consideration was given to the possible intermediate carboxyaldehyde (18) or hydroxymethyl acid (19) which would, presumably, form during this reaction.

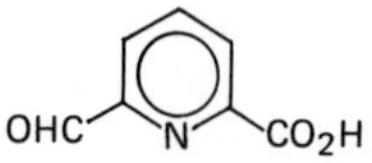

(18) 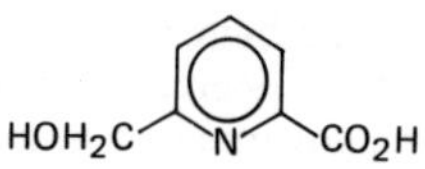   **(19)**

Carrying out the reduction of picolinic acid in two steps, the first, in acid solution, giving the aldehyde hydrate, and the second, after neutralization reducing to the alcohol, gave very low yields of alcohol (< 10%) in contrast to Lund's work[61]. It was shown that the first step was responsible for the low yield and hydrogenation of the nucleus was postulated.

The isomeric pyridine dicarboxylic acids have been studied by polarography but no product studies made[67].

In a follow-up to the picolinic acid work Brown and Bhatti used the rotating disc and rotating ring-disc electrode techniques to study the hydration kinetics of the three pyridine aldehydes[68].

$$
\text{RCHO} + \text{H}_2\text{O} \underset{k}{\overset{k'}{\rightleftharpoons}} \text{RCH(OH)}_2 \qquad (K = k/k') \qquad (34)
$$

The aldehydes were reduced at an amalgamated copper rotating disc in aqueous solution and from the plot of limiting current density (i_L) vs the square root of rotation speed ($\omega^{1/2}$) the kinetic current (i_k) at zero rotation speed was extrapolated. From the kinetic current the rate constants k and k' may be determined

$$
i_k = nFD^{1/2}Kk^{1/2}C(\text{hydrate})C(\text{H}_2\text{O})^{-1/2} \qquad (35)
$$

(equation 35), where n is the number of electrons involved, F is the Faraday, D is the diffusion coefficient and $C(i)$ are concentrations of species i. The diffusions coefficients were calculated from the slopes of the i_L vs $\omega^{1/2}$ plots and the equilibrium constants (K) were measured spectroscopically. The hydration equilibria were shown to be independent of acidity below pH 2.

Using the rotating ring-disc electrode, picolinic and isonicotinic acid were reduced at the disc and the corresponding aldehydes detected at the ring, set at the

TABLE 4. Dehydration rate constants for hydrated pyridine aldehydes in neutral (pH 6.65) solution[a]

Aldehyde	K (mol l^{-1})	k (s^{-1})	
		Rotating disc	Rotating ring-disc
Picolinic	86	0.049	0.102
Nicotinic	500	0.008	—
Isonicotinic	65	0.137	0.42

[a]Data taken from M. ud Din Bhatti and O. R. Brown, *J. Electroanal. Chem.*, **68**, 85 (1976). Reproduced by permission of Elsevier Scientific Publishing Company.

reduction potential of the aldehyde. Nicotinic acid gave only catalytic hydrogen evolution. The dehydration rate constant k determines how much free aldehyde is formed as the solution moves from the disc to the ring. From the variation in disc and ring current with rotation speed k may be calculated[69]. Table 4 shows rate constants calculated by the two methods.

Differences between the rotating disc and ring-disc methods were attributed to inaccuracies in measuring the kinetic current (i_k) and for isonicotinic acid to the non-quantitative formation of aldehyde in the ring-disc method. Three-electron coulometry was observed for isonicotinic acid, instead of four-electron coulometry which was observed for picolinic acid.

Finally, from log i vs E plots (Tafel plots) for picolinic acid reduction, it was shown that the first electron transfer is reversible (fast) and the rate-determining step is probably the subsequent electrochemical step. The following mechanism was proposed:

$$RCO_2H \rightleftharpoons RCO_2^- + H^+$$

$$RCO_2H + H^+ + e \rightleftharpoons R\dot{C}(OH)_{2\ ads.} \qquad (36)$$

$$R\dot{C}(OH)_2 + e \xrightarrow{slow} R\bar{C}(OH)_2 \xrightarrow[H^+]{fast} RCH(OH)_2$$

c. Other reactions. To complete the description of the pyridine acids six-electron reduction of picolinic (equation 37) and isonicotinic acids to the corresponding picolines was observed by Wibaut and Boer[70]. Isonicotinic acid gave a

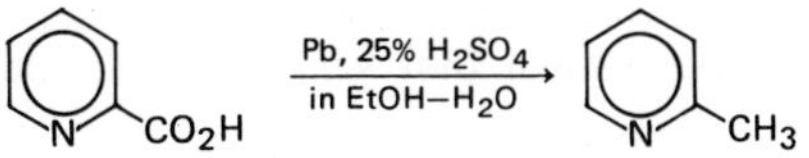

$$(37)$$

33%

31% yield of 4-picoline, but nicotinic acid gave a mixture of unidentified products, probably hydrogenated in the nucleus as claimed by Sorm[71]. Nuclear hydrogenation was observed for thiophene-2-carboxylic acid[72] (equation 38). Similarly,

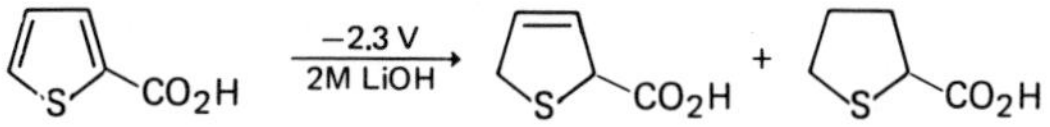

$$(38)$$

76%   15%

furan-2,5-dicarboxylic acid was reduced with sodium amalgam to give the dihydro compound[73]

3. Aliphatic acids

   a. *Reduction of the carboxyl group.*   Cathodic reductions of aliphatic acids, in contrast to anodic oxidations, are very limited in number since the unactivated carboxyl group is extremely difficult to reduce. The most widely studied acid in this category has been oxalic acid in which the carboxyl groups activate one another. There has been a lot of effort described in the old German patent literature[74], and more recently[75,76], directed at synthesizing glyoxylic acid. In one of the latest studies oxalic acid was reduced at a lead cathode in aqueous solution[76] (equation 39). This is a very efficient synthesis provided that the

$$HO_2CCO_2H \xrightarrow[<1\% \ NR_4^+ \ salts]{Pb, \ 14 \ A/dm^2} HO_2CCHO \qquad (39)$$

Chemical yield = 92%
Current efficiency = 83%

temperature is kept low enough($< \sim 15°C$) to slow down the dehydration of the hydrated aldehyde and subsequent reduction to glycolic acid. A mathematical treatment for the purpose of process evaluation has been given[77].
   None of the other aliphatic acid reductions is of comparable synthetic utility. Lactic acid was claimed to give lactic aldehyde in unspecified yield by electrolysing the acid, containing a little water, between carbon rods in an undivided cell[78].
   Butyric acid gave *n*-butanol in 6.5% yield in 80% sulphuric acid and in 17% yield in aqueous sodium hydroxide[79]. The latter result is somewhat unexpected.
   A surprising six-electron electrocatalytic reduction was observed for trifluoro-acetic acid at a platinized platinum cathode in aqueous solution[80].

$$CF_3CO_2H \xrightarrow[0.14 \ V \ vs \ SCE]{Pt-Pt, \ 50 \ \mu A/cm^2} CF_3CH_3 \qquad (40)$$

Current efficiency = 96%

   b. *Other reactions.*   Activation of the double bond by the carboxyl groups in maleic and fumaric acids leads to efficient cathodic hydrogenation, giving succinic acid. Elving[81] studied this reaction polarographically, observing reduction in the region -0.8 to -1.6 V as the pH varied from two to ten. Formation of the carboxylate anion at high pH increases the reduction potential. Elving was the first to propose an ECE mechanism for this reaction. The preparative aspects of this reaction were studied by Udupa[82], using lead cathodes and lead dioxide anodes in an undivided cell. It was shown that very high current efficiencies are possible,

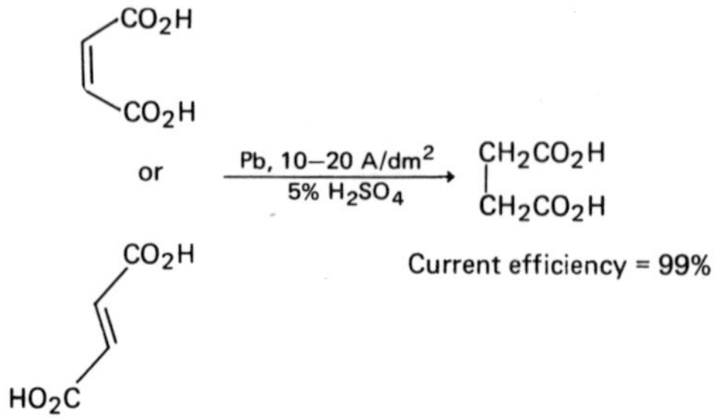

$$\xrightarrow[5\% \ H_2SO_4]{Pb, \ 10-20 \ A/dm^2} \begin{matrix} CH_2CO_2H \\ | \\ CH_2CO_2H \end{matrix} \qquad (41)$$

Current efficiency = 99%

particularly with rotating cathodes. Acrylic acid, with a singly activated double bond, has been reduced with sodium amalgam. In aqueous diglyme or dioxane propionic acid was formed, but in aqueous DMSO adipic acid was the major product[83] (equation 42). The direct electrochemical version of this reaction has

$$2\ CH_2CHCO_2H \xrightarrow[DMSO-H_2O]{Na-Hg} HO_2C(CH_2)_4CO_2H \qquad (42)$$

$$70\%$$

apparently not been pursued, although a similar reaction was observed for cinnamic acid[84]. The electrohydrodimerization (EHD) reaction will be discussed in more detail for carboxylic esters.

Numerous other examples of double-bond hydrogenation in α,β-unsaturated acids have been given in Reference 6 (p. 171)

B. Carboxylic Esters, Lactones and Anhydrides

The cathodic reduction of esters, lactones and anydrides has been studied less extensively than the reduction of carboxylic acids. Naturally there are many similarities between the two types of reductions, but there are also significant differences brought about by the absence of a labile hydrogen in this group of compounds. The most common reaction types are shown in equations (43)–(46).

$$R^1CO_2R^2 \xrightarrow{1\ e} R^1CO_2R^{2\cdot -} \qquad (43)$$

$$\xrightarrow[2\ H^+]{2\ e} R^1CHO + R^2OH \qquad (44)$$

$$\xrightarrow[4\ H^+]{4\ e} R^1CH_2OH + R^2OH \qquad (45)$$

$$\xrightarrow[4\ H^+]{4\ e} R^1CH_2OR^2 + H_2O \qquad (46)$$

Equation (43), anion radical formation, is observed for certain conjugated aromatic and aliphatic esters and anhydrides in non-aqueous solutions. Aldehyde formation, equation (44), is again observed, and the same considerations apply to aldehyde stabilization as were discussed before. Four-electron reduction leads to either an alcohol or an ether, equations (45) and (46), depending on which C–O bond is broken.

We shall also consider reductions which are activated by the carboxyl group(s) in the molecule, but in which the carboxyl group is unchanged. Since there is not an overabundance of material in this section, the results will be discussed according to reaction types and not split up further into different types of esters, etc. In the first part we shall consider reductions of the carboxyl function itself and in the second part reductions activated by the carboxyl function.

1. Reduction of the carboxyl group

a. Anion-radical formation. Certain conjugated aromatic and aliphatic carboxyl compounds are capable of accepting one electron into the lowest unoccupied molecular orbital to form anion radicals which are stable for appreciable periods of time under suitable aprotic conditions. In these stable anion radicals the unpaired electron is typically delocalized over the whole conjugated system of the molecule, so that these reductions should perhaps be considered in the second part of this section. However, since anion radicals are thought to be the initially formed intermediates, stable or not, in most of the reactions we shall consider, they will be included here.

The simplest aromatic ester, methyl benzoate, gives an anion radical which is stable on the time-scale of cyclic voltammetry[85]. Figure 5 shows the behaviour of methyl benzoate in anhydrous DMF containing tetrabutylammonium perchlorate as the electrolyte with a platinum disc cathode. The processes occurring at the three peaks are assigned in equation (47). The identity of the electrochemically generated

$$C_6H_5CO_2CH_3 \underset{\text{peak C}}{\overset{\text{peak A}}{\rightleftharpoons}} C_6H_5CO_2CH_3^{\pm} \xrightarrow{\text{peak B}} [C_6H_5CO_2CH_3^{=}] \longrightarrow \text{products} \quad (47)$$

anion radical of methyl benzoate was confirmed by e.s.r. spectroscopy by Hirayama[86a] Ethyl and isopropyl benzoates were also studied and in the e.s.r. spectra some splitting by the α-protons in the alkyl groups was observed, in addition to the expected splitting by the aromatic protons. Similar work was published by Russian workers at about the same time[86b,c].

The reduction potentials for the formation of anion radicals from some other aromatic carboxyl compounds by cyclic voltammetry, in an analogous manner to Figure 5, are given in Table 5. The e.s.r. spectra of the electrochemically generated anion radicals of phthalic anhydride and other aromatic anhydrides have also been studied[87]. The reduction of isonicotinic acid was discussed previously. Under

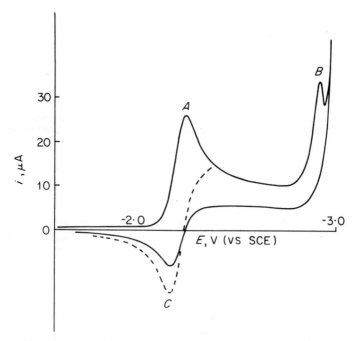

FIGURE 5.   Cyclic voltammetry of methyl benzoate[85], using DMF −0.1 M NBu$_4$ClO$_4$; Pt cathode, 50 mV/s; positive feedback iR compensation.— — — — Voltage sweep reversed after anion-radical formation.

TABLE 5. Potentials for anion radical formation[a] from some aromatic carboxyl compounds[8 5]

Compound	Peak potential, E_p (V) (vs SCE)
Methyl benzoate	−2.29
Dimethyl terephthalate	−1.68
Phthalic anhydride	−1.25
Tetrachlorophthalic anhydride	−0.80

[a]In DMF−0.1 M NBu_4ClO_4, Pt cathode, 200 mV/s.

aprotic conditions the methyl and ethyl esters of isonicotinic acid also give relatively stable anion radicals[8 6 a].

Saturated aliphatic esters do not give anion radicals which are readily accessible by electroanalytical methods, but some unsaturated esters are more amenable to study. For instance, the cyclic voltammetry of diethyl fumarate in anhydrous DMF shows stable anion radical formation[8 8]. The more sterically crowded *cis* isomer, diethyl maleate, is some 250 mV more difficult to reduce but gives the same anion radical, in the *trans* form, as may be shown by either cyclic voltammetry[8 9] or by e.s.r. spectroscopy[9 0].

With four carboxyl groups in the molecule in tetraethyl ethenetetracarboxylate (20) polarography shows two one-electron reduction waves[9 1], and cyclic voltammetry at platinum in DMF−tetraethylammonium fluoroborate shows that both the anion radical and the dianion are stable[9 2]. The structures in equation (48) are

$$(C_2H_5OOC)_2C{=}C(COOC_2H_5)_2 \xrightleftharpoons{-0.95\,V} (C_2H_5OOC)_2\dot{C}C(COOC_2H_5)_2 \xrightleftharpoons{-1.15\,V}$$
$$\text{(20)} \qquad\qquad\qquad (C_2H_5OOC)_2\overline{\dot{C}C}(COOC_2H_5)_2 \quad \text{(48)}$$

not meant to imply anything about the electron distribution but are merely simplistic localized representations of the anion radical and dianion. In liquid ammonia at −43°C, diethyl fumarate also shows a relatively stable dianion[9 3].

Diethyl oxalate, which is reduced polarographically at −1.85 V, gives an anion radical observable by e.s.r.[9 4]. However, it is not simply the anion radical of the parent molecule but is the result of the reactions shown in equation (49). The

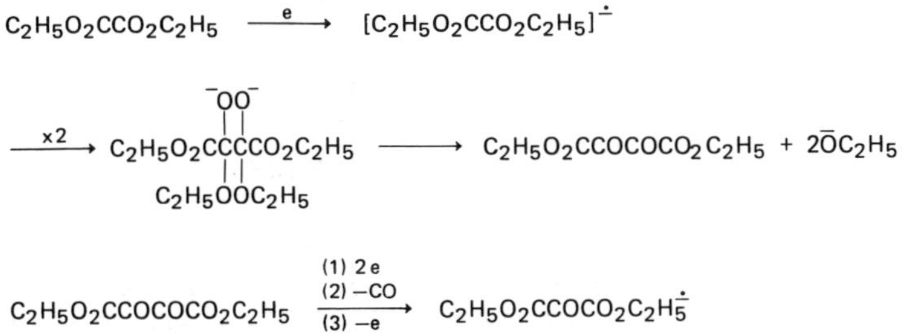

$$C_2H_5O_2CCO_2C_2H_5 \xrightarrow{e} [C_2H_5O_2CCO_2C_2H_5]^{-\cdot}$$

$$\xrightarrow{\times 2} C_2H_5O_2C\overset{\overline{O}}{\underset{\overset{|}{C_2H_5OOC_2H_5}}{\overset{|}{C}}}\overset{\overline{O}}{\underset{|}{CC}}O_2C_2H_5 \longrightarrow C_2H_5O_2CCOCOCO_2C_2H_5 + 2\overline{O}C_2H_5 \quad (49)$$

$$C_2H_5O_2CCOCOCO_2C_2H_5 \xrightarrow[\substack{(2)\,-CO\\(3)\,-e}]{(1)\,2e} C_2H_5O_2CCOCO_2C_2H_5^{-\cdot}$$
$$\text{(21)}$$

observed spectrum was identical with that of diethyl mesoxalate anion radical (21).

For certain carboxylic esters and anhydrides, then, in contrast to the corresponding acids, the initial products of the electron-transfer reaction are stable under aprotic conditions. When proton donors and other chemical reagents are added, these intermediates will react to give a variety of products which will be discussed in the following sections.

   b. Aldehyde formation.   Like the corresponding acids, activated esters will undergo two-electron reduction to give an aldehyde and an alcohol (equation 44). This reaction probably involves protonation of the anion radical, followed by a further electron transfer and protonation or elimination of alkoxide to give the hemiacetal or the free aldehyde, respectively, depending on the reaction conditions (equation 50).

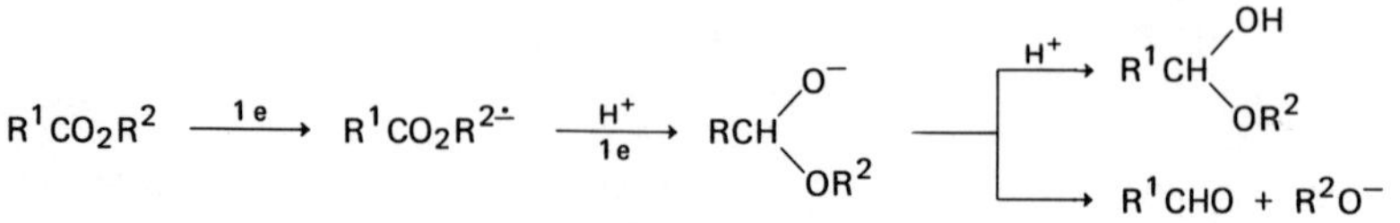

$$R^1CO_2R^2 \xrightarrow{\ 1e\ } R^1CO_2R^{2\pm} \xrightarrow[1e]{H^+} RCH\begin{smallmatrix}O^-\\ \\ OR^2\end{smallmatrix} \begin{cases} \xrightarrow{H^+} R^1CH\begin{smallmatrix}OH\\ \\ OR^2\end{smallmatrix} \\ \longrightarrow R^1CHO + R^2O^- \end{cases} \quad (50)$$

   Not too much attention has been paid to the synthesis of aldehydes from aromatic esters. However, the monoaldehyde was observed polarographically during the reduction of dimethyl terephthalate to 4-carboxymethylbenzyl alcohol[95].

   Of the heterocyclic esters ethyl 2-thiazolecarboxylate (22) was shown to give a 72% yield of the aldehyde, in contrast to the 25% obtained from the free acid[65] (equation 51). Strong hydration again prevents further reduction of the aldehyde.

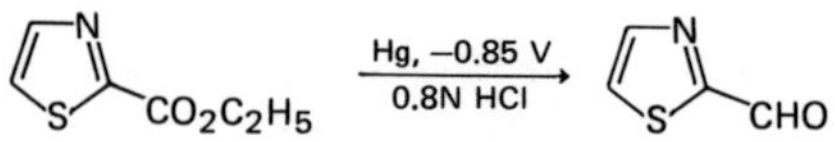

$$\underset{(22)}{\overset{\displaystyle \left[\begin{smallmatrix}N\\S\end{smallmatrix}\right]-CO_2C_2H_5}{}} \xrightarrow[0.8N\ HCl]{Hg,\ -0.85\ V} \underset{72\%}{\overset{\displaystyle \left[\begin{smallmatrix}N\\S\end{smallmatrix}\right]-CHO}{}} \quad (51)$$

Of the aliphatic esters diethyl oxalate is reduced to the hemiacetal of ethyl glyoxylate in about 50% yield in ethanolic sulphuric acid[95].

   An interesting application of electrosynthesis to carbohydrate chemistry is found in the reduction of lactones of polyhydroxy carboxylic acids to the corresponding aldoses. These have been reviewed recently[97].

   Of particular importance is the synthesis of D-ribose (24) which is used for the synthesis of riboflavin (vitamin B_2). Numerous investigators have looked at the electrochemical reduction of D-ribonolactone (23) (equation 52). It has been

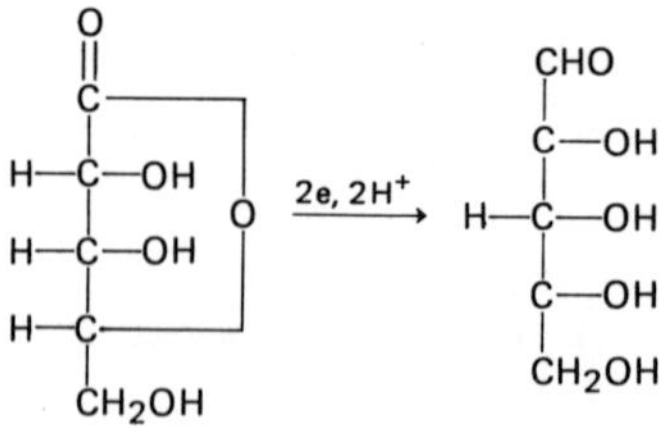

$$\underset{(23)}{\begin{array}{c} O \\ \parallel \\ C \\ | \\ H-C-OH \\ | \\ H-C-OH \\ | \\ H-C \\ | \\ CH_2OH \end{array}} \xrightarrow{2e,\ 2H^+} \underset{(24)}{\begin{array}{c} CHO \\ | \\ C-OH \\ | \\ H-C-OH \\ | \\ C-OH \\ | \\ CH_2OH \end{array}} \quad (52)$$

shown that this reaction proceeds only on a mercury or an amalgam cathode. The optimum conditions are low temperature (10–20°C) and slightly acid solution, both of which suppress hydrolysis of D-ribonolactone to non-reducible ribonic acid. Electrolytes used include ammonium salts[98] or sodium salts[99] with boric acid added as a buffer. Yields have been 75% and 87% respectively, Fioshin has shown that less expensive phosphate salts may also be used as buffers in this reaction with comparable results[100,101]. It is postulated that these reactions occur via electrochemical amalgam formation, and lack of reduction to the alcohol, D-ribotol, is ascribed to formation of a stable, non-reducible cyclic hemiacetal.

   c. *Alcohol and ether formation.*   Four-electron reduction of activated esters to alcohols is possible although there has been less interest in this reaction than in the corresponding acid reduction.

   Methyl benzoate was reduced in methanolic tetramethylammonium chloride to give a good yield of benzyl alcohol[102] (equation 53).

$$C_6H_5CO_2CH_3 \xrightarrow{\text{Hg, } CH_3OH-NMe_4Cl} C_6H_5CH_2OH + CH_3OH \qquad (53)$$

$$91\%$$

   Several dialkyl phthalates were studied polarographically in 60% ethanol and showed two reduction waves at about -1.7 and -2.1 V (vs SCE), respectively[103]. Controlled potential electrolysis at the first wave gave a 'good' yield of phthalide, resulting from four-electron reduction of one carboxyl group (equation 54).

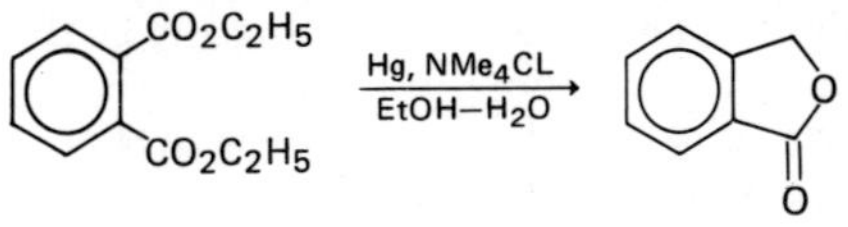

$$(54)$$

Phthalide was also obtained in 29% yield from potassium ethyl phthalate reduction[104] and in 90% yield from phthalic anhydride[105]. The last reaction was carried out in aqueous ammonia or ammonium carbonate solution, so that the electroactive species were probably ammonium phthalate and phthalamate[106]. When ester reductions are carried out in strong acid solution at low temperature ($< 25°C$) at a lead cathode, a new reaction is observed[107,108]. The C=O group in the ester is reduced to a methylene group and alkoxy group is retained, giving an ether. The alcohol is also formed at the same time in a competitive reaction. Ethyl benzoate, for instance, gives both benzyl ethyl ether and benzyl alcohol[108] (equation 55).

$$C_6H_5CO_2C_2H_5 \xrightarrow[\text{30\% } H_2SO_4 \text{ in EtOH}-H_2O]{\text{Pb, 5A/dm}^2} C_6H_5CH_2OC_2H_5 + C_6H_5CH_2OH \qquad (55)$$

$$40\% \qquad 28\%$$

   d. *Other reactions.*   The reduction of 2-alkyl-substituted acetoacetic esters in acidic solutions at a lead cathode gives rearranged products in which both the acetyl and the ester functions have been completely reduced[109,110] (equation 56). The

$$\underset{\overset{|}{R}}{CH_3COCHCO_2C_2H_5} \xrightarrow{\text{Pb, EtOH}-H_2SO_4} R(CH_2)_3CH_3 + C_2H_5OH \qquad (56)$$

$$50-60\%$$

rearrangement, known as the Tafel rearrangement, has been attributed to formation of a cyclopropane intermediate[111,112] such as 25 which may ring open to give the

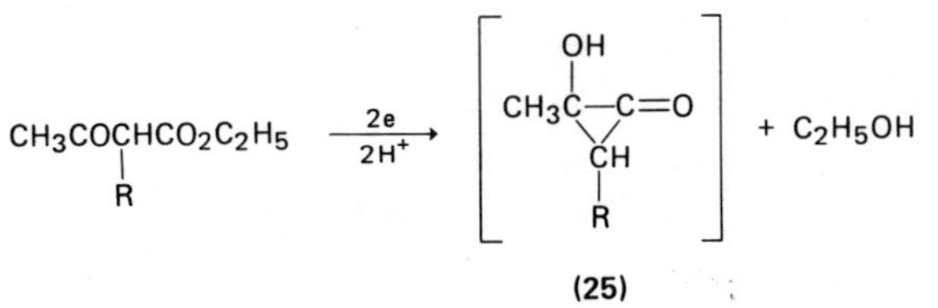

$$(57)$$

$$(25)$$

β-diketone (26); further reduction gives the hydrocarbon. Wawzonek has recently.

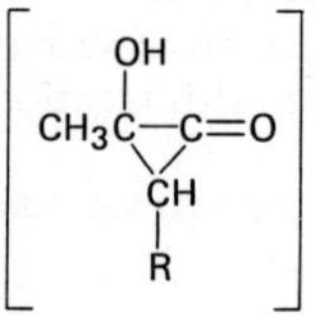

$$\left[\begin{array}{c} OH \\ CH_3C-C{=}O \\ \diagdown\!\!\diagup \\ CH \\ | \\ R \end{array} \right] \longrightarrow CH_3COCOCH_2R \xrightarrow{8e,\,8H^+} CH_3(CH_2)_3R \qquad (58)$$

$$(26)$$

confirmed the intermediacy of the β-diketone in the reduction of ethyl α-butyl-acetoacetate[112].

The mode of formation of the cyclopropane intermediate in this reaction is still the subject of conjecture. Eberson[111] favours cyclization via a carbanion formed from the acetyl group, whilst Wawzonek[112] postulates simultaneous protonation and one-electron reduction of both the acetyl and ester carbonyl groups, cycliza-tion then occurring through radical coupling. Other mechanisms may be drawn, but the evidence to date does not conclusively support any of them.

A different type of hydrocarbon formation was recently claimed by Russian workers[113]. Dimethyl adipate, when treated with electrochemically generated solvated electrons in hexamethylphosphoramide, either preformed or formed *in situ*, gave *n*-hexane (equation 59). Product identification was by g.c. retention time only, and no quantitative data were given.

$$CH_3O_2C(CH_2)_4CO_2CH_3 \xrightarrow{e_{solv},\,HMPA} n\text{-}C_6H_{14} + 2\,CH_3OH + 2\,H_2O \qquad (59)$$

It could be argued that the acyloin condensation, equation (60), should be

$$2R^1CO_2R^2 \xrightarrow{Na} R^1\overset{\overset{\displaystyle OOH}{|||}}{C}CHR^1 \qquad (60)$$

considered as an electrochemical ester reduction. However, this reaction has been reviewed in depth very recently[114] and will not be treated here, especially since the acyloin reaction via electrolysis does not seem to have been observed. Acyloin-type products were observed as minor products in the reduction of naphthoate esters[115], and electron-transfer reduction of ethyl benzoate by naphthalene anion radical to give benzil and benzoin was reported[116] (equation 61). With a 4 : 1 ratio

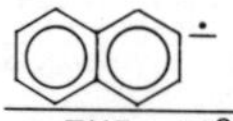

$$C_6H_5CO_2C_2H_5 \xrightarrow[THF,\,-10^\circ C]{\qquad \overset{\bar{\;}}{\qquad} Na^+} C_6H_5\overset{\overset{\displaystyle OO}{||||}}{C}CC_6H_5 + C_6H_5\overset{\displaystyle HO}{\underset{|}{C}H}\overset{\displaystyle O}{\overset{||}{C}}C_6H_5 \qquad (61)$$

of anion radical to benzoate, benzoin was obtained in 86% yield, whilst a 1 : 1 ratio gave benzil (38%) and benzoin (27%).

Bearing in mind the apparent stability of aromatic ester anion radicals in the presence of quaternary ammonium salts, it appears that ion pairing with sodium must play an important role in the coupling reaction.

2. Reductions activated by the carboxyl group

a. The electrohydrodimerization (EHD) reaction. α,β-Unsaturated nitriles, esters and amides may be reductively coupled electrochemically as shown in equation (62).

$$2\,CH_2{=}CHX \xrightarrow[2H^+]{2e} X(CH_2)_4X \quad (X = CN,\,CO_2R,\,etc.) \qquad (62)$$

This reaction had previously been carried out using amalgam reducing agents[117] but was demonstrated electrochemically for the first time by Baizer[118,119]. The reaction was developed for the hydrodimerization of acrylonitrile to adiponitrile, used to manufacture hexamethylene diamine and ultimately Nylon-6,6. It now represents the most successful commercial application of organic electrochemistry and was responsible, in part, for the recent resurgence of interest in this field. EHD reactions have been reviewed previously[1,3,6]; only the major features will, therefore, be presented here.

Amongst the compounds studied in Baizer's early work was ethyl acrylate[119]. It was shown that reduction at a mercury cathode in a concentrated solution of a salt, such as methyltriethylammonium p-toluenesulphonate, gave high yields of diethyl adipate[119] (equation 63). The reaction has been extended to include diactivated

$$2\ CH_2{=}CHCO_2C_2H_5 \xrightarrow[\text{MeNEt}_3\ \text{tosylate, DMF–H}_2\text{O}]{\text{Hg, }-1.85\ \text{V}} C_2H_5OCO(CH_2)_4CO_2C_2H_5 \qquad (63)$$

$$74\text{–}87\%$$

olefins[119], cyclization of diolefins[120] and cross-coupling with other Michael acceptor olefins[121]. Representative examples are given in Table 6 and more numerous examples are given in Reference 1.

The efficiency of the coupling reaction in aqueous, or partly aqueous, tetra-alkylammonium salt solutions has been attributed to the formation of a hydrophobic layer of adsorbed tetraalkylammonium cations on the cathode[122]. The electrogenerated intermediates may, thus, react preferentially with other organic species, rather than with water which would lead to dihydroproducts.

The exact mechanism of the EHD reaction has been, and still is, the subject of considerable discussion. In the case of monoactivated olefins, such as methyl acrylate, the chemical reactions following the initial electron transfer are so rapid that electroanalytical techniques, such as cyclic voltammetry, do not yield much

TABLE 6. EHD reactions of α, β-unsaturated esters

Ester	Product	Yield (%)	Reference
Di-n-butyl maleate	Tetra-n-butyl butane-1,2,3,4-tetracarboxylate	70	119
$CH{=}CHCO_2C_2H_5$ $\|$ $(CH_2)_3$ $\|$ $CH{=}CHCO_2C_2H_5$	$CH_2CO_2C_2H_5$ / $CH_2CO_2C_2H_5$ (cyclopentane)	ca 100	120
$OCH{=}CHCO_2C_2H_5$ $\|$ CH_2 $\|$ CH_2 $\|$ $OCH{=}CHCO_2C_2H_5$	$CH_2CO_2C_2H_5$ / $CH_2CO_2C_2H_5$ (dioxane ring)	89	120
$CH_2{=}CHCO_2C_2H_5$ + $CH_2{=}CHCN$	$CH_2 CH_2 CO_2 C_2 H_5$ $\|$ $CH_2 CH_2 CN$ + symmetrical products	–	121

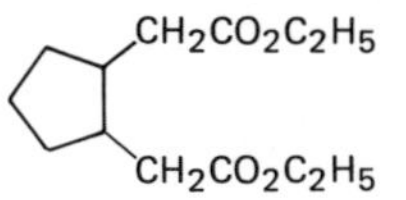

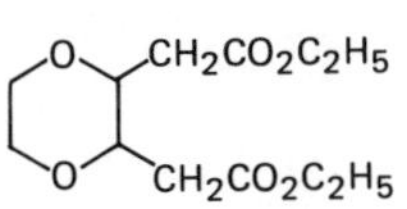

useful information. The two most likely mechanisms involve coupling of anion radicals (equation 65) or attack of an anion radical on a neutral molecule (equation 64). The experimental evidence to date does not conclusively support either mechanism.

$$CH_2{=}CHX \xrightarrow{\text{1 e}} [CH_2CHX]^{\overline{\cdot}}$$

$$CH_2{=}CHX, \xrightarrow{\quad} X\dot{C}HCH_2CH_2\bar{C}HX \xrightarrow[2\,H^+]{\text{1 e}} X(CH_2)_4X \quad (64)$$

$$\xrightarrow{\times 2} X\bar{C}HCH_2CH_2\bar{C}HX \xrightarrow{2\,H^+} X(CH_2)_4X \quad (65)$$

$$(X = CN, CO_2R, \text{etc.})$$

In the case of diactivated olefins, the initial anion radicals are more stable and the chemical reactions occur at a rate which is more amenable to study. From the results of both voltammetric and product studies, Baizer and coworkers concluded[123-125] that with tetraalkylammonium-supporting electrolyte the coupling reaction proceeds via attack of an anion radical on a neutral parent molecule (cf. equation 64). In the presence of alkali metal cations the coupling reaction was accelerated dramatically and in this case was thought to proceed via dimerization of two anion radical–metal cation ion pairs.

Bard and coworkers then carried out an extensive series of electroanalytical studies on diactivated olefins. The first technique employed was double-potential step chronoamperometry in a study of diethyl fumarate[88]. This method involves changing the cathode potential stepwise from a point where no reduction accurs to a point where anion radical is formed at a diffusion-controlled rate. At a given time, t, the cathode potential is stepped back to a point where diffusion-controlled reoxidation occurs. The cathodic and anodic currents at times t and $2t$, respectively, are recorded for varying t. Working curves for the current-time response may be obtained for various reaction mechanisms, by digital simulation, and compared with the experimental results. Using this approach it was concluded that the EHD reaction for diethyl fumarate, in DMF–tetrabutylammonium iodide, proceeds via second-order dimerization of anion radicals.

This was substantiated by later rotating ring-disc electrode studies[89,126], and the effect of alkali metal cations on the coupling, again anion-radical dimerization, was quantified[127,128].

To conclude the discussion of EHD reactions, an interesting electrohydrocyclodimerization reaction was observed recently[129]. In this reaction (equation 66),

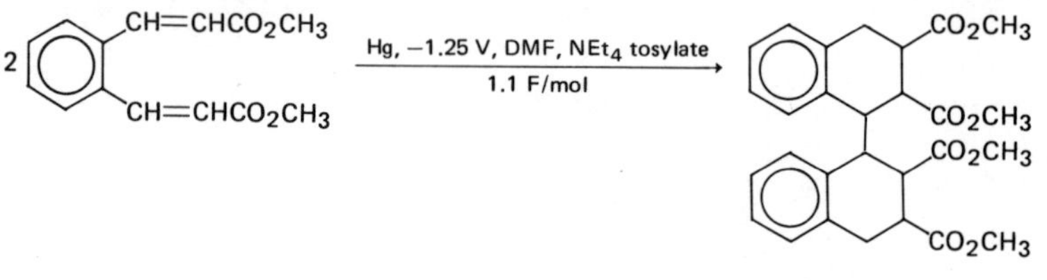

$$\text{Ca. } 47\%$$

which was carried out at a potential slightly less cathodic than the peak potential for anion-radical formation, cyclization occurred through the α-positions. This is unusual and is probably caused by a combination of conjugation through the aromatic nucleus and steric constraints on cyclization through the β-positions. Coupling of benzylic radicals and protonation gives the final product.

b. Cathodic carboxylation of α,β-unsaturated esters. The reduction of α,β-unsaturated esters in the presence of carbon dioxide gives a variety of carboxylated products. Methyl acrylate, for example, when reduced in low concentrations in acetonitrile, saturated with CO_2, gives trimethyl-1,1,2-ethanetricarboxylate, after methylation to facilitate analysis and isolation[130] (equation 67). At higher acrylate concentration, coupling occurs prior to carboxylation[131] (equation 68). The yield

$$CH_2{=}CHCO_2CH_3 \ (<0.1\,M) \xrightarrow[\text{CH}_3\text{CN--NEt}_4 \text{ tosylate--CO}_2]{\text{Hg, }-2.1\,V} CH_3O_2CCH_2CH(CO_2CH_3)_2 \quad (67)$$

(27)

61% after methylation

$$CH_2{=}CHCO_2CH_3 \ (1.32\,M) \xrightarrow[\text{CH}_3\text{CN--NEt}_4 \text{ tosylate--CO}_2]{\text{Hg, }-2.1\,V} (CH_3O_2C)_2CHCH_2CH(CO_2CH_3)_2 \quad (68)$$

(28)

47% after methylation

of **28** increases from zero to 47% on increasing the initial acrylate concentration up to 1.32 M, whilst the yield of **27** decreases from 61% to 8%. Since these reactions were carried out at constant current density, this seems to indicate that the coupling reaction probably occurs via attack of an anion radical on a neutral parent molecule. The reduction of CO_2 is not implicated since it occurs at a more negative potential than methyl acrylate reduction.

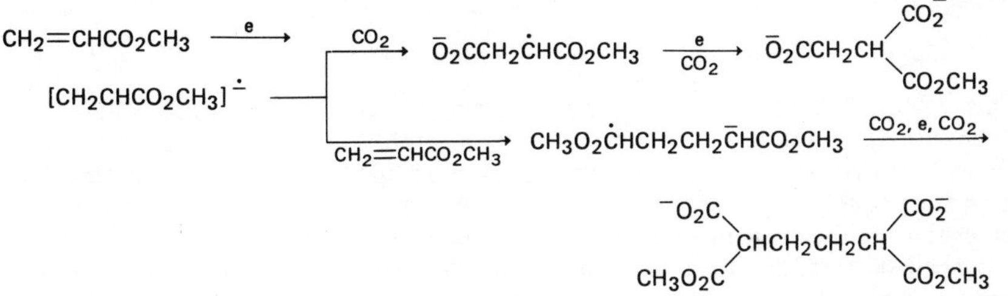

$$CH_2{=}CHCO_2CH_3 \xrightarrow{e} $$
$$[CH_2CHCO_2CH_3]^{\cdot-}$$

$$\xrightarrow{CO_2} \bar{O}_2CCH_2\dot{C}HCO_2CH_3 \xrightarrow[CO_2]{e} \bar{O}_2CCH_2CH\begin{smallmatrix}CO_2^-\\ \\CO_2CH_3\end{smallmatrix} \quad (69)$$

$$\xrightarrow{CH_2{=}CHCO_2CH_3} CH_3O_2\dot{C}HCH_2CH_2\bar{C}HCO_2CH_3 \xrightarrow{CO_2,\,e,\,CO_2}$$

$$\begin{smallmatrix}^-O_2C\\ \\CH_3O_2C\end{smallmatrix}CHCH_2CH_2CH\begin{smallmatrix}CO_2^-\\ \\CO_2CH_3\end{smallmatrix} \quad (70)$$

Dimethyl maleate shows two one-electron polarographic reduction waves in the presence of CO_2. Reduction at the potential of the first wave gives the dicarb-oxylated dimer, hexamethyl-1,1,2,3,4,4-butanehexacarboxylate[131], whilst at more negative potentials tetramethyl-1,1,2,2-ethanetetracarboxylate is formed[130]. Examples of carboxylative cyclization were also observed[131].

The cyclic voltammetry of 1,1,2,2-ethenetetracarboxylate esters is insensitive to CO_2, indicating that the anion radical and dianion do not react, or react reversibly, with CO_2[92]. This phenomenon has been used to advantage in the carboxylation of carbon acids. It was shown by cyclic voltammetry that the tetraethyl 1,1,2,2-ethenetetracarboxylate dianion is capable of deprotonating carbon acids, such as ethyl phenylacetate and 9-phenylfluorene. Preparative-scale electrolysis in the presence of CO_2 resulted in carboxylation of the carbon acid in excellent yield, according to equation (71).

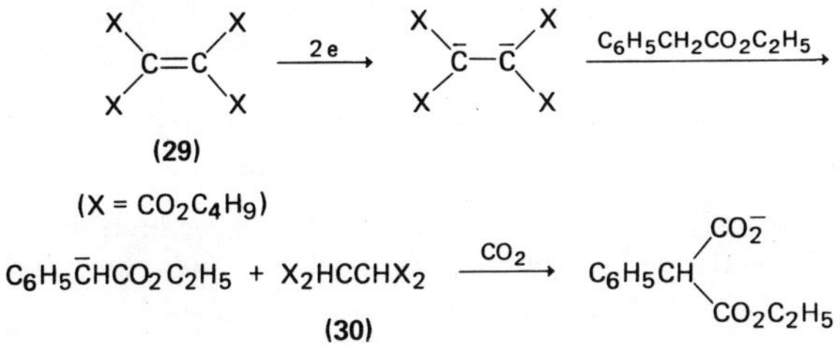

$$\begin{smallmatrix}X\\ \\X\end{smallmatrix}C{=}C\begin{smallmatrix}X\\ \\X\end{smallmatrix} \xrightarrow{2e} \begin{smallmatrix}X\\ \\X\end{smallmatrix}\bar{C}{-}\bar{C}\begin{smallmatrix}X\\ \\X\end{smallmatrix} \xrightarrow{C_6H_5CH_2CO_2C_2H_5}$$

(29)

(X = $CO_2C_4H_9$)

$$C_6H_5\bar{C}HCO_2C_2H_5 + X_2HCCHX_2 \xrightarrow{CO_2} C_6H_5CH\begin{smallmatrix}CO_2^-\\ \\CO_2C_2H_5\end{smallmatrix} \quad (71)$$

(30)

Thus, ethyl phenylacetate, with the tetrabutyl ester (29) at -1.1 V in DMF containing tetrabutylammonium iodide and CO_2, gave diethyl phenylmalonate in 78% yield, after ethylation[92]. The dihydro compound (30) was oxidized back to 29, in almost quantitative yield, with bromine.

Similarly, 9-phenylfluorene was carboxylated at the 9-position in 81% yield, and a cycle involving reoxidation and reuse of the protonated electrogenerated base was demonstrated.

c. Cleavage reactions. In this section we shall consider both reactions in which carboxylate ion acts as a leaving group and those in which the carboxyl function activates the cleavage of other groups.

Certain esters of aliphatic acids display a polarographic reduction wave at very negative potentials in anhydrous media. For example, in anhydrous DMF the benzyl ester of acetylglycine has a half-wave potential of -2.73 V (vs SCE)[132] (equation 72). It was shown that reduction at this potential resulted in cleavage of the

$$CH_3CONHCH_2CO_2CH_2C_6H_5 \xrightarrow[2\,H^+]{2\,e} CH_3CONHCH_2CO_2H + C_6H_5CH_3 \qquad (72)$$

$$70\text{--}90\%$$

carboxylate in high yield, although the preparative conditions were not described. Phenyl, diphenylmethyl and trityl esters of acetylglycine and benzyl palmitate behaved similarly.

$$RCO_2R' \xrightarrow{1\,e} RCO_2R'^{\cdot -} \longrightarrow RCO_2^- + \dot{R}' \xrightarrow[2\,H^+]{1\,e} RCO_2H + R'H \qquad (73)$$

This reaction probably occurs via cleavage of an anion radical. This is in contrast to the stability of the conjugated anion radicals discussed earlier. The fact that this reduction is observable at all must be due to the cleavage reaction shifting the reduction potential in the anodic direction. This is apparent when one observes that neither the unactivated benzene nucleus nor simple aliphatic esters are polarographically reducible, and the effect of acyloxylating the aromatic nucleus is to make it more electron-rich and, therefore, presumably, *more* difficult to reduce.

The utility of carboxylates as leaving groups is the essence of an elegant method for protecting and regenerating carboxyl, amino, hydroxyl and thiol groups[133,134]. Carboxylic acids may be protected by making 2, 2, 2-trichloroethyl esters. Cathodic reduction then results in very efficient regeneration via elimination of dichloroethylene (equation 74). For example, 2, 2, 2-trichloroethyl benzoate is reduced at -1.65 V (vs SCE) in methanol to give about 90% recovered benzoic acid[133].

$$RCO_2CH_2CCl_3 \xrightarrow{2\,e} RCO_2CH_2\dot{C}Cl_2 \longrightarrow RCO_2^- + CH_2{=}CCl_2 \qquad (74)$$

$$+ Cl^-$$

By using different haloethyl protecting groups, which are reducible at different potentials, it is possible to selectively regenerate one carboxyl group in the presence of others, using controlled-potential electrolysis.

With chloroformate derivatives it is possible to apply this reaction to other functional groups (e.g. equation 75). The effect of the ester group in activating

$$C_6H_5CH_2SCO_2CH_2CCl_3 \xrightarrow[\text{MeOH-LiClO}_4]{Hg,\ -1.5\,V} C_6H_5CH_2SH + CO_2 + CH_2{=}CCl_2 \qquad (75)$$

$$90\%$$

cathodic cleavage reactions was demonstrated in the reduction of a series of methoxycarbonyl-substituted benzyl ethers, acetates and fluorides[135]. It was mentioned earlier[132] that unsubstituted benzyl esters are reducible at ca -2.7 V (vs SCE). Introducing the carboxyl function into the aromatic nucleus makes this reduction much easier; p-methoxycarbonylbenzyl acetate was reduced at -1.8 V to the corresponding toluene (equation 76). This example, in fact, involves

$$p\text{-}CH_3O_2CC_6H_4CH_2OCOCH_3 \xrightarrow[\text{CH}_3\text{OH}-\text{NBu}_4\text{OAc}]{\text{Pb, } -1.8 \text{ V}} p\text{-}CH_3O_2CC_6H_4CH_3 \qquad (76)$$

95% yield
76% current efficiency

carboxylates as both activating and leaving groups. The corresponding benzyl methyl ether behaves similarly, giving two-electron reduction to methyl p-toluate.

In the case of methyl p-(trifluoromethyl) benzoate, six-electron reduction and loss of all three fluorine atoms, to give methyl p-toluate in 60% yield, was observed. These reactions may all be rationalized in terms of expulsion of the leaving group from an anion radical, the formation of which is assisted by the presence of the methoxycarbonyl group (equation 77). For the $para$-substituted compounds,

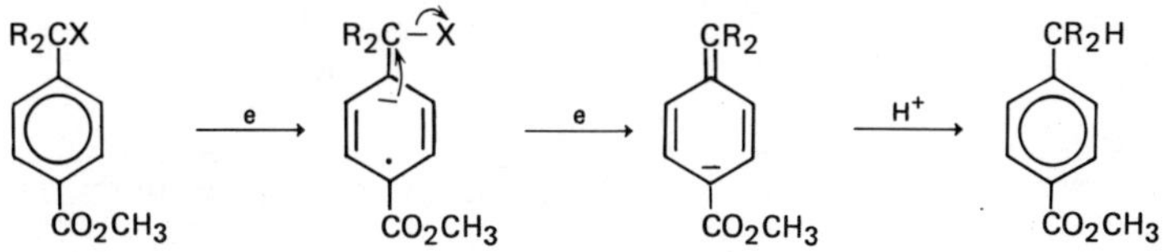

$$ (77) $$

R = H, F
X = OAc, OCH$_3$, F

expulsion of the leaving group is too rapid to observe the anion radical, but with methyl m-(trifluoromethyl)benzoate one electron, reversible cyclic voltammetry is observed[135]. Reduction of p-methoxycarboxylbenzyl methyl ether in the presence of acetic acid resulted in partial protonation of the anion radical before cleavage could occur. The major product obtained was the benzyl alcohol (equation 78).

$$p\text{-}CH_3OCH_2C_6H_4CO_2CH_3 \xrightarrow[\text{CH}_3\text{OH}-\text{AcOH}-\text{NaOAc}]{\text{Pb, } -1.5 \text{ to } -1.8 \text{ V}} p\text{-}CH_3OCH_2C_6H_4CH_2OH \qquad (78)$$

27%

The effect of leaving group on reduction potential may be seen in Table 7. The acetate cleavage reaction results in a 350 mV anodic shift in reduction potential versus the parent compound. This is significantly less for methoxide, and for

TABLE 7. Polarographic reduction potentials[a] of some α-substituted methyl toluates[b]

Substrate	$E_{1/2}$ (V vs Ag/AgI)
p-CH$_3$O$_2$CC$_6$H$_4$CH$_3$	-1.74
p-CH$_3$O$_2$CC$_6$H$_4$CH$_2$OCOCH$_3$	-1.39
p-CH$_3$O$_2$CC$_6$H$_4$CH$_2$OCH$_3$	-1.61
p-CH$_3$O$_2$CC$_6$H$_4$CF$_3$	-1.26
m-CH$_3$O$_2$CC$_6$H$_4$CF$_3$	-1.40

[a]In DMF$-$NBu$_4$I (0.1 M).
[b]Data taken from J. P. Coleman and coworkers, *J. Chem. Soc., Perkin II*, 1903 (1973). Reproduced by permission of the Chemical Society, London.

trifluoromethyl the combined synergistic and electronegativity effects cause a shift of almost 500 mV.

The reductive cleavage of fluoride from aliphatic esters has also been observed[136]. Ethyl di- and tri-fluoroacetates show reduction waves at -2.56 and -2.36 V (vs SCE), respectively, in DMF. One-electron coulometry was observed in both cases, changing to two electrons with added proton donor, and fluoride ion was detected in the electrolysis solutions; however, the organic products were not identified. Similar results were obtained with other perfluoro esters[137].

The reductive cleavage of various groups, activated by carboxyl functions, has been observed in the cathodic reduction of some cephalosporanic acids and

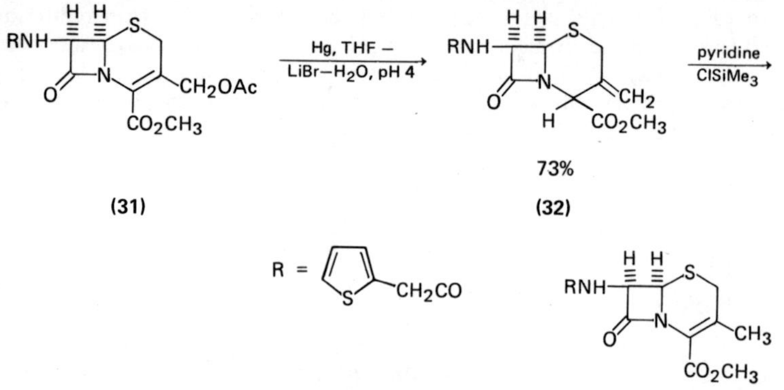

(31)

$$\xrightarrow[\text{LiBr-H}_2\text{O, pH 4}]{\text{Hg, THF} -}$$

(32)

73%

$$\xrightarrow[\text{ClSiMe}_3]{\text{pyridine}}$$

$$R = \underset{\text{S}}{\bigcirc}\text{CH}_2\text{CO}$$

(33)

(79)

derivatives[138]. For example, the cephalosporanic acid methyl ester (31), which has a half-wave potential of -1.70 V (vs SCE) in acetonitrile, gave a good yield of the 3-methylenecepham (32). This kinetically controlled product was readily isomerized to the 3-methyl-3-cephem (33).

The reaction was applied to various other cephalosporanic acids, esters and lactones with the reductive cleavage of acetate, thioacetate and pyridine and lactone ring-opening in reasonable yields. A new synthesis of the orally active cephalosporin antibiotic cephalexin (34) was, thus, demonstrated. This reaction

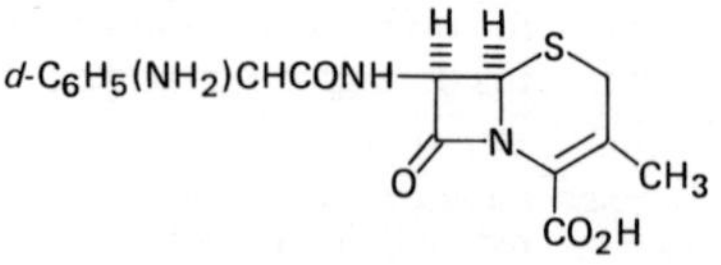

(34)

probably occurs via an ECE sequence involving anion-radical formation, expulsion of the leaving group and further reduction of the resulting allylic radical to a carbanion which is protonated in a kinetically controlled reaction.

The cathodic cleavage of allylic acetates derived from polyenes has also been observed recently[139]. Vitamin A acetate, for example, gave axerophtene (35) in good yield (equation 80). Replacement of the acetate by p-nitrobenzoate resulted in reduction of the aromatic nitro group at -0.55 V without cleavage. Electron transfer to the polyene 'electrophore' is thus necessary for the cleavage reaction to occur.

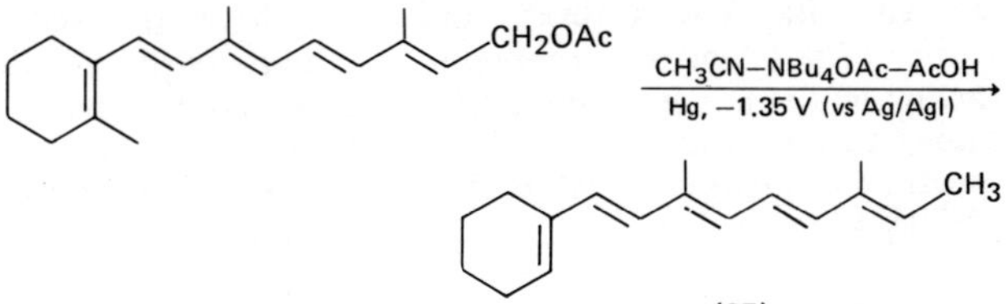

(80)

(35)

71%

C. Carboxylic Amides, Lactams and Imides

The electrochemical reactions of the nitrogen derivatives of carboxylic acids, in many respects, parallel those of the corresponding esters and anhydrides. The main reaction types are shown in equations (81)–(85). Anion radicals are again observed

$$R^1CONR^2R^3 \xrightarrow{1e} R^1CONR^2R^{3\cdot-} \qquad (81)$$

$$R^1CONR^2R^3 \xrightarrow[2H^+]{2e} R^1CH(OH)NR^2R^3 \qquad (82)$$

$$R^1CONR^2R^3 \xrightarrow[2H^+]{2e} R^1CHO + R^2R^3NH \qquad (83)$$

$$R^1CONR^2R^3 \xrightarrow[4H^+]{4e} R^1CH_2OH + R^2R^3NH \qquad (84)$$

$$R^1CONR^2R^3 \xrightarrow[4H^+]{4e} R^1CH_2NR^2R^3 + H_2O \qquad (85)$$

R^1 = alkyl, aryl; R^2 = H, acyl, alkyl, aryl; R^3 = H, alkyl, aryl

in certain cases. Two-electron reduction may lead to an aldehyde, equation (83), and with some imides the intermediate *gem*-amido alcohol is stable and may be isolated, equation (82). Equations (84) and (85) parallel the formation of alcohols and ethers from esters. The material in this section will be arranged in a similar manner to that in Section B.

1. Reduction of the carboxyl group

a. Anion-radical formation. In the same manner as esters and anhydrides, certain amides and imides are capable of forming anion radicals which are stable under anhydrous conditions. Thus, the cyclic voltammetry of *N,N*-dimethylbenzamide in anhydrous DMF is similar to that of methyl benzoate and shows a reversible one-electron reduction at –2.53 V (vs SCE)[85]. The cyclic voltammetry of *N*-methylbenzamide and benzamide, itself, shows irreversible reductions at –2.52 and –2.53 V, respectively, which become partly reversible at high voltage sweep rates (10 V/s). The difference in behaviour is probably caused by the presence of a relatively acidic hydrogen atom in the latter compounds.

Horner[140] generated the anion radicals of some *N,N*-disubstituted aromatic amides by electrolysis in acetonitrile. E.s.r. spectra were observed but no details given.

Phthalimide and its *N*-substituted derivatives have received considerable attention[141-143]. The *N*-alkyl compounds exhibit fairly simple behaviour. *N*-methylphthalimide (36), for example[141], shows two reversible, one-electron reductions leading to the anion radical (37) and dianion (38), respectively, in DMF. The e.s.r. spectra of *N*-alkylphthalimide anion radicals are fairly well documented[87,144,145].

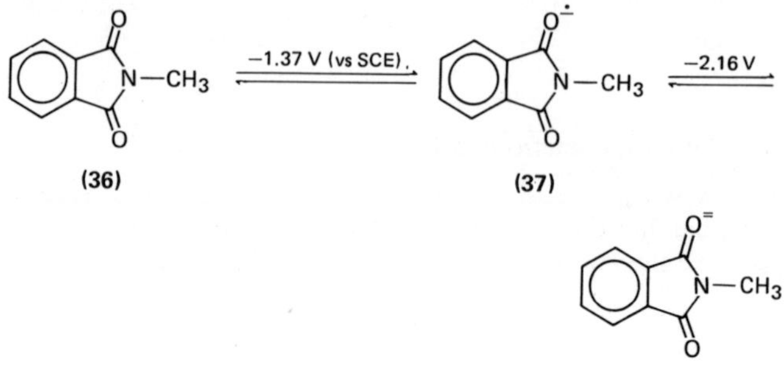

(36)   (37)

(86)

(38)

The behaviour of phthalimide, itself, is complicated by the acidic N—H proton, and there are differences of opinion as to its overall behaviour. In spite of its acidity[146] the anion radical is observable by cyclic voltammetry, and the first reduction peak in DMF (E_p = −1.49 V vs SCE) is completely reversible at a sweep rate of 20 V/s[85]. At lower sweep rates self-protonation occurs, leading to products which will be discussed later. The e.s.r. spectrum of the phthalimide anion radical has been observed on numerous occasions[87,145]. At more negative potentials several other reduction peaks are observed[85,141], together with another re-oxidation process, although Farnia and coworkers[142] reported only two reversible couples, attributed to anion-radical and dianion formation, respectively. It seems likely that the second group of reduction processes includes both the reduction of the different forms of anion radical[87,147] and the reduction of the anion (39),

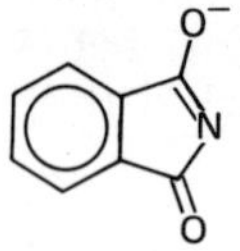

(39)

formed in an acid—base reaction at the first reduction peak. Reduction of the potassium salt of phthalimide in DMSO[145], in fact, gave an e.s.r. spectrum attributable to the dianion radical expected from the reduction of 39. The nature of the second reoxidation process cannot be assigned with any certainty.

Once again, anion radicals are thought to be the first-formed intermediates in most of the reactions we shall consider in the following sections, except those in strongly acidic solutions where protonated substrates might be the electroactive species.

    b. *Formation of aldehydes and α-amido alcohols.*   The cathodic reduction of aromatic amides to aldehydes does not appear to have been pursued, most of the work having been directed towards alcohol or amine formation. However, in view of the results on alcohol formation from these compounds, which will be discussed in the next section, it seems that, if the extraction methods applied for aromatic acid reductions (see Section II.A) were used, it should be possible to obtain aldehydes from aromatic amides.

Heterocyclic amide reductions have been studied by Lund and Iversen, and the results parallel those for the corresponding carboxylic acids. Thus, isonicotinic amide (40) shows a polarographic reduction in aqueous solution[148], the potential of which varies from -0.6 to -1.3 V as the pH varies from 0 to 13. In strongly acidic solution the aldehyde was shown to be the major product (equation 87). In less acidic solution (pH = 3.5) four-electron reduction to the alcohol was observed. this, again, reflects the stability of the hydrated aldehyde (or α-amino alcohol, equation 88). With a phenyl group on the amide nitrogen in isonicotinic anilide, aldehyde formation is almost completely suppressed, even in strongly acidic solution. The products are the alcohol and the amine, yields varing with pH (equation 89). Amides of imidazole- and thiazolecarboxylic acids[63-65] are also reducible to the corresponding aldehydes, and some of these are shown in Table 8.

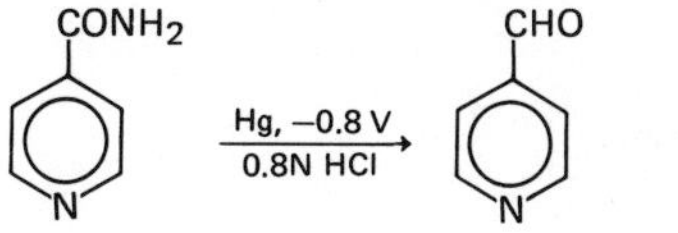

$$ (87) $$

(40)   87% by polarography

$$ RCONH_2 \xrightarrow[2\,H^+]{2\,e} RCH\underset{NH_2}{\overset{OH}{\diagup}} \underset{H_2O}{\rightleftharpoons} RCH(OH)_2 \rightleftharpoons RCHO + H_2O + NH_3 \quad (88) $$

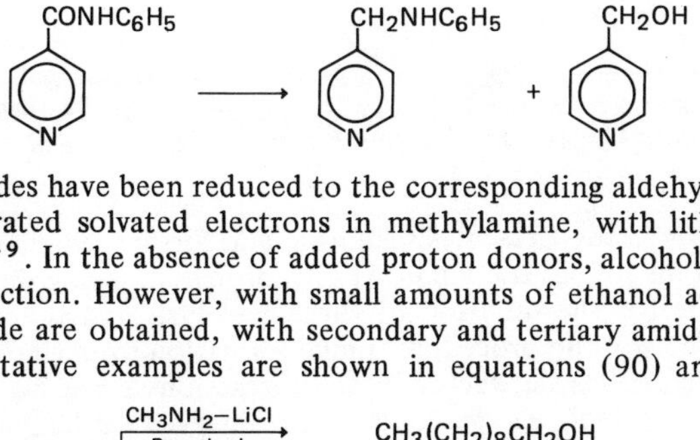

$$ (89) $$

Aliphatic amides have been reduced to the corresponding aldehydes with electrochemically generated solvated electrons in methylamine, with lithium chloride as the electrolyte[149]. In the absence of added proton donors, alcohol formation is the predominant reaction. However, with small amounts of ethanol added, reasonable yields of aldehyde are obtained, with secondary and tertiary amides giving the best yields. Representative examples are shown in equations (90) and (91). Product

$$ CH_3(CH_2)_8CONHCH_3 \xrightarrow[Pt\ cathode]{CH_3NH_2-LiCl} CH_3(CH_2)_8CH_2OH \quad (90) $$

81%

$$ \xrightarrow[Pt\ cathode]{CH_3NH_2-LiCl-EtOH} CH_3(CH_2)_8CH_2OH + CH_3(CH_2)_8CHO \quad (91) $$

24%   58%

control is rationalized in terms of equations (92) and (93). In the absence of a proton donor, the anion 41, presumably formed by an ECE sequence, decomposes

$$ R^1CONR^2R^3 \xrightarrow[H^+]{2\,e} R^1CH\underset{NR^2R^3}{\overset{O^-}{\diagup}} $$

(41)

$$ \xrightarrow{-(NR^2R^3)^-} R^1CHO \xrightarrow[2H^+]{2\,e} R^1CH_2OH \quad (92) $$

$$ \xrightarrow{EtOH} R^1CH\underset{NR^2R^3}{\overset{OH}{\diagup}} \xrightarrow[(workup)]{hydrolysis} R^1CHO \quad (93) $$

(42)

TABLE 8. Aldehyde formation via cathodic reduction of some heterocyclic amides in aqueous acid[a]

Amide	Aldehyde yield (%)		Reference
	Polarographic	Isolated	
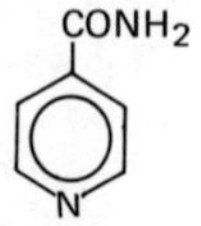 (pyridine with CONH₂)	67	49	64
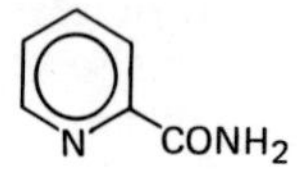 (pyridine with CONH₂)	72.5	53.5	64
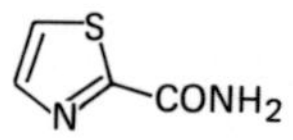 (thiazole with CONH₂)	81.5	44	64
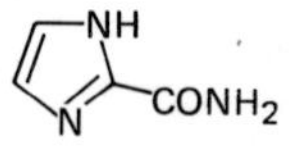 (imidazole NH with CONH₂)	85	–	63
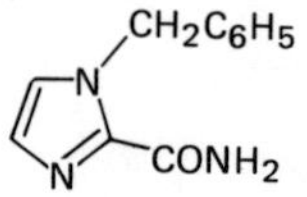 (imidazole CH₂C₆H₅ with CONH₂)	95	–	63

[a]Data taken from P. E. Iversen, *Acta Chem. Scand.*, **24**, 2459 (1970). Reproduced by permission of *Acta Chem. Scand.*

to give the aldehyde in the electrolysis solution and further reduction to the alcohol occurs. With ethanol added, the anion **41** is protonated, giving the α-aminoalcohol **42** which is stable for the duration of the electrolysis and is hydrolysed to the aldehyde during the workup procedure. Imine formation from the aldehyde and the solvent is not a complicating factor in these reactions since no N-methylamines corresponding to the starting materials were observed, although it was demonstrated that the N-methylimine of hexanal could be reduced to the amine.

It should be noted that, whilst the chemical yields from these reactions are fairly good, the current efficiencies are low. A sixfold theoretical excess of current was used, in equation (90) and in equation (91) a 7.5-fold excess.

Similar reductions in hexamethylphosphoramide[150] (HMPA) and in liquid ammonia[151] have also been studied. No aldehyde formation was observed in HMPA, but, at low temperatures ($< -70°C$) in liquid ammonia–potassium bromide, propionaldehyde was formed from propionamide. At higher temperatures, even with butanol added as a proton donor, n-propanol was formed almost exclusively.

The reduction of phthalimide and its derivatives has been studied extensively using aqueous solutions[152-158], in addition to the non-aqueous solutions described earlier[141-145]. Phthalimide undergoes a two-electron polarographic reduction at about -0.8 V in acidic aqueous ethanol solutions[155-157]. Protonated phthalimide is thought to be the electroactive species although neutral phthalimide predominates in the bulk solution. As pH increases, the reduction shifts in the cathodic direction and a second wave is observed. The first wave becomes smaller until at pH 7–9.5 two one-electron, pH-independent waves are observed. At pH greater than 10 the phthalimide anion is the bulk species, a protonation again precedes electron transfer, the neutral molecule being electroactive, and the reduction potential again becomes pH-dependent.

The product of the two-electron reduction in acidic aqueous dioxane was shown to be 3-hydroxyphthalimidine **(43)**[153,154] (equation 94). This compound is

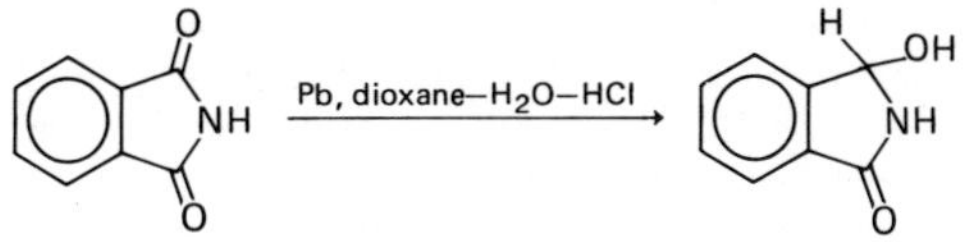

(94)

(43)

70%

analogous to the unstable intermediates proposed in the amide to aldehyde reductions but does not decompose so readily. The preparative reduction in acidic aqueous ethanol gives 3-ethoxyphthalimidine which was incorrectly identified by Sakurai[152] as the hydroxy compound. Reduction at the first wave in neutral or alkaline solution was assumed to give a pinacol-type dimer[157].

More recently, this reaction was investigated using the rotating disc electrode technique with lead and amalgamated copper in acidic aqueous acetonitrile solutions[158]. The major features described above were confirmed for phthalimide and some of its N-substituted derivatives.

Controlled potential reduction of phthalimide at the potential of the first reduction wave in DMF also gives 3-hydroxyphthalimidine[142] (equation 95). The

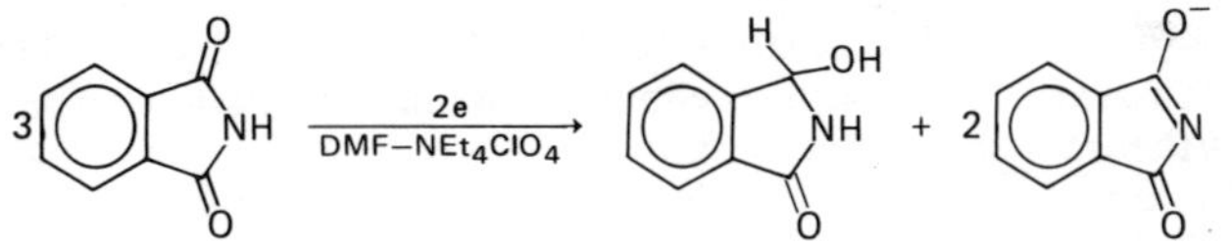

(95)

reaction only goes to 33% conversion in the absence of added proton donors. This self-protonation reaction accounts for the anomalous currents observed for phthalimide reduction by polarography and slow-sweep voltammetry[143].

c. Alcohol formation. The cathodic reduction of aromatic amides in neutral or basic alcoholic solutions generally leads to the corresponding alcohols (equation 84) as Horner and coworkers have demonstrated[102,159,160]. Polarographic reduction potentials were measured for a series of amides in ethanolic solution and some of these are shown in Table 9. The half-wave potentials for a series of substituted benzanilides were correlated fairly well with the corresponding Hammett σ values. Preparative-scale electrolysis of benzamides in alcoholic solution leads to a benzyl alcohol and ammonia, or the amine resulting from the cleavage reaction[102] (equation 96). Terephthalamides undergo eight-electron reduction

$$C_6H_5CONHCH_2C_6H_5 \xrightarrow[NMe_4Cl]{Hg, MeOH} C_6H_5CH_2OH + H_2NCH_2C_6H_5 \quad (96)$$

67%   70%

under these conditions, giving terephthalyl alcohol[160] (equation 97). The mechanism of these reductions is similar to that discussed for aliphatic amides in methyl-

TABLE 9. Polarographic half-wave potentials of some aromatic amides[a]

Amide	$E_{1/2}$ (V vs Ag/AgCl)[b]
$C_6H_5CONH_2$	−2.15
$p\text{-}CH_3OC_6H_4CONH_2$	−2.33
$C_6H_5CONHC_6H_5$	−2.02
$p\text{-}ClC_6H_4CONHC_6H_5$	−1.90
$C_6H_5CON(C_6H_5)_2$	−1.83

[a]Data taken from L. Horner and R. -J. Singer, *Annalen*, 723, 1 (1969). Reproduced by permission of Verlag Chemie GmbH, Weinheim.
[b]Measured in ethanol−NMe$_4$Cl (0.1 M).

amine, with the major difference being that the aromatic amides are reducible directly, as evidenced by the polarographic results.

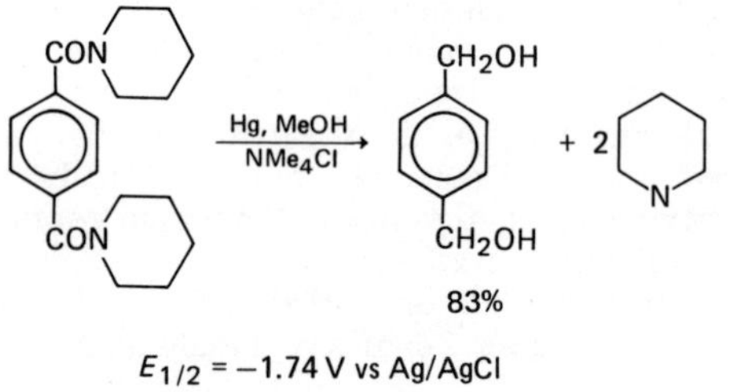

$$E_{1/2} = -1.74 \text{ V vs Ag/AgCl}$$

An ECE sequence, via the anion radical, leads to an α-amino alcohol (cf. equation 92) which, at normal temperatures, eliminates ammonia or amine. The aldehyde so formed is further reduced to the alcohol.

The reduction of ammonium phthalamate in aqueous solution gives an alcohol

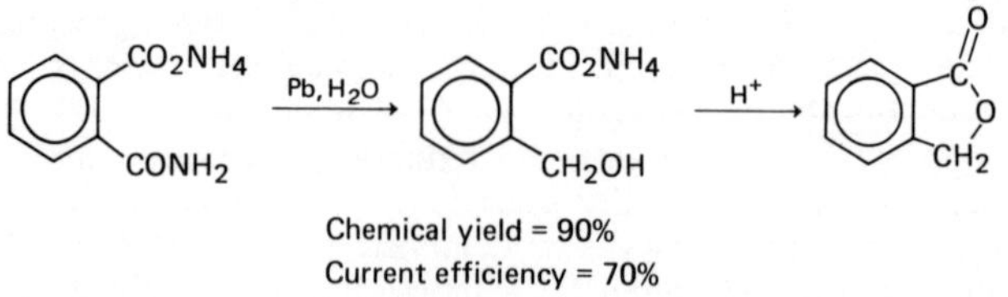

Chemical yield = 90%
Current efficiency = 70%

which cyclizes to give phthalide as the final product[161] (equation 98). Phthalide is also obtained in 60% yield by reduction of phthalimide in DMF at −2.6 V[141].

   d. *Formation of amines from amides and amides from imides.*   Reduction of the amide or imide carbonyl group to a methylene group in acidic aqueous solution seems to have been the most extensively studied cathodic reaction of these compounds and is analogous to ether formation from esters (Section B. 1. c). Since most of the work in this area is fairly old and has been summarized previously[3,5,8b], we shall do no more than outline the major features here.

These reactions have been carried out since the beginning of the century, for the most part using constant current electrolysis at lead cathodes in sulphuric acid solutions at slightly elevated temperatures. Tafel, for example, reduced

$$C_6H_5CON(CH_3)_2 \xrightarrow[35^\circ C]{Pb,\ 50\%\ H_2SO_4} C_6H_5CH_2N(CH_3)_2 \tag{99}$$

Chemical yield = 63%
Current efficiency = 11%

N,N-dimethylbenzamide to *N,N*-dimethylbenzylamine[162] (equation 99). *N,N*-dimethylphenylacetamide may be similarly reduced in aqueous hydrochloric acid solution to give *N,N*-dimethyl-2-phenylethylamine in 92% yield[163]. The monomethyl compound and phenylacetamide, itself, give 80% and < 1% yields, respectively, of the corresponding amine, thus demonstrating clearly the effect of *N*-substitution on this reaction.

Phthalimide and its derivatives have again received considerable attention in this context. Under more vigorous conditions than those described earlier for two-electron reduction, phthalimide and its *N*-alkyl derivatives are reduced in two stages to isoindolines (equation 100). Thus, isoindoline and its *N*-methyl derivative

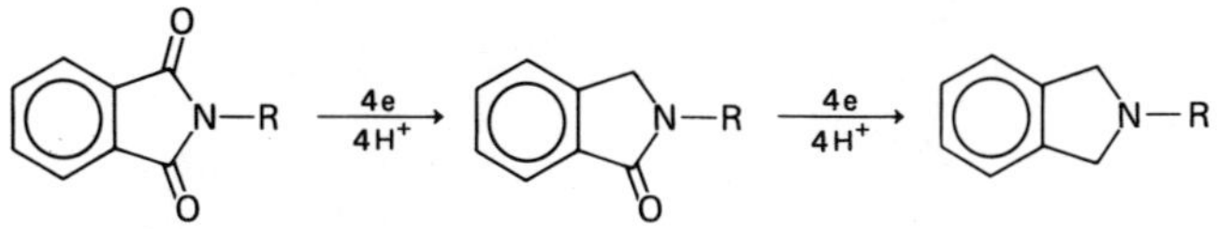

(100)

(44) (45)

R = H, alkyl

were prepared[164] in good yield from either the corresponding phthalimide or the phthalimidine (**44**, R = H, CH_3) at a lead cathode in aqueous sulphuric acid at 50°C. Current efficiencies were only 15–20%, but chemical yields were 60–70%. Allen[165] claims that current efficiency may be dramatically increased by solubilizing the starting material with acetic acid. An early demonstration of the utility of controlled potential electrolysis was given in the reduction of the tetrachlorophthalimide derivative **46**[166]. The mechanism of these reactions probably involves

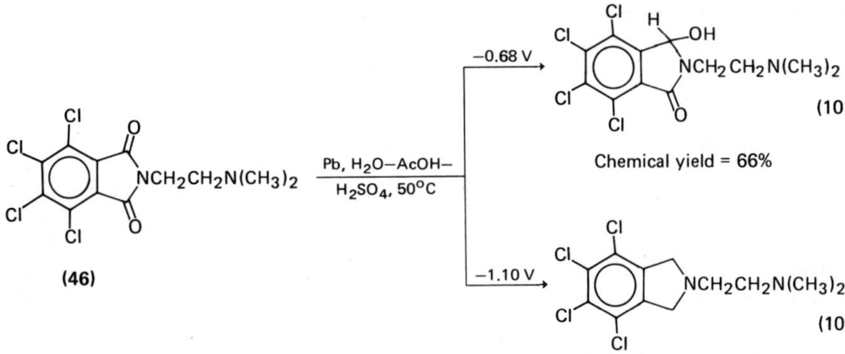

(46) (101) (102)

Chemical yield = 66%

Chemical yield = 84%
Current efficiency = 78%

reduction of the protonated amide or imide, followed by dehydration and further reduction (equation 103), but detailed mechanistic studies have not been made.

$$R^1CONR^2R^3 \xrightarrow[3H^+]{2e} R^2CH(OH)(\overset{+}{N}HR^2R^3) \xrightarrow{-H_2O} R^1CH{=}\overset{+}{N}R^2R^3 \xrightarrow[2H^+]{2e} R^1CH_2\overset{+}{N}HR^2R^3 \tag{103}$$

Certain aliphatic compounds have been reduced in a similar fashion. Swann, for example, reduced N,N-dimethylvaleramide to N,N-dimethylamylamine in 60% yield[167]. Both carbonyl groups of the diamide (47) were reduced[168]. Succinimide

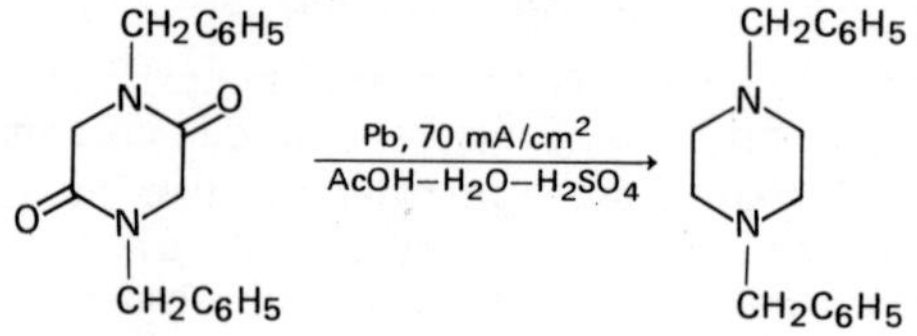

$$\text{(104)}$$

(47)   Chemical yield = 70%

and some of its derivatives have been studied. Thus, N-methylsuccinimide was reduced to the pyrrolidone in good yield[169] (equation 105). Further reduction to

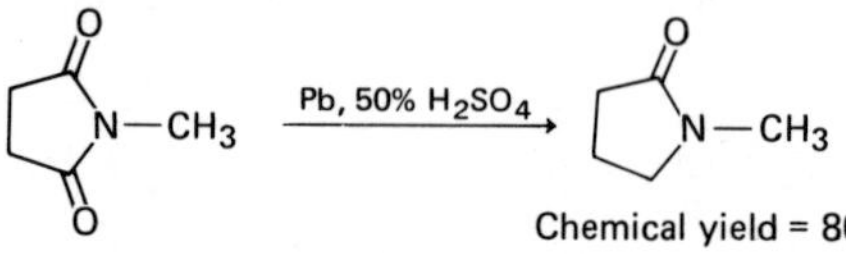

$$\text{(105)}$$

Chemical yield = 80%

pyrrolidines, in low yield, has also been reported[170]. Finally, ethylamine was reported as a major product in the reduction of acetamide with electrochemically generated solvated electrons in HMPA[150] (equations 106 and 107). Lower tem-

perature and increased acidity favour the formation of ethylamine and the results are rationalized in terms of dehydration or deamination of an intermediate amino alcohol. It is interesting to note that the use of anhydrous HCl in HMPA did not significantly affect the total current efficiency via increased hydrogen evolution.

2. Other reactions

There is not a lot of work in the literature describing carboxylic amides as activating groups in cathodic reductions. Compared with the corresponding esters, the nitrogen compounds appear to have been neglected in this context. Amide groups are capable of promoting the EHD reaction as Baizer described in his early work with acrylamide and some of its derivatives[119] (equation 108). Nicotinamides

$$2\,CH_2=CHCON(C_2H_5)_2 \xrightarrow[\text{NEt}_4 \text{ tosylate—H}_2\text{O}]{\text{Hg, } -1.95\text{ V}} (C_2H_5)_2NCO(CH_2)_4CON(C_2H_5)_2 \quad \text{(108)}$$

Current efficiency = 73%

alkylated on the pyridine nitrogen atom have been studied fairly extensively as models for the NAD$^+$—NADH system. This work has been reviewed previously[171] and, since the cathodic reduction of pyridinium compounds may be accomplished in the absence of the amide activating group[172], leading to either ring-hydrogenated or hydrodimerized products, it will not be treated here.

III. ACKNOWLEDGMENTS

I am greatly indebted to Miss Pat Merlenbach for her patience and diligence in preparing the manuscript and to Doctors Manuel Baizer, Richard Hallcher and Richard Goodin, both for helpful discussions and for the provision of unpublished results.

IV. REFERENCES

1. M. M. Baizer (Ed.), Organic Electrochemistry, Dekker, New York, 1973.
2. C. K. Mann and K. K. Barnes, *Electrochemical Reaction in Nonaqueous Systems*, Dekker, New York, 1970.
3. N. L. Weinberg (Ed.), 'Technique of Electro-organic Synthesis', *Techniques of Chemistry*, Vol. V, Parts I and II, Wiley, New York, 1974.
4. R. N. Adams (Ed.), *Electrochemistry at Solid Electrodes*, Dekker, New York, 1969.
5a. S. Swann, Jr., 'Electrolytic Reactions', *Technique of Organic Chemistry*, Vol. II, Interscience, New York, 1956, p. 385.
5b. M. R. Rifi and F. H. Covitz, *Introduction to Organic Electrochemistry*, Dekker, New York, 1974.
6. A. P. Tomilov, S. G. Mairanovskii, M. Ya. Fioshin and V. A. Smirnov, *The Electrochemistry of Organic Compounds*, Halsted Press, New York, 1972.
7. F. Beck, *Elektro-organische Chemie*, Verlag Chemie GmbH, Weinheim, 1974.
8a. A. J. Fry, *Synthetic Organic Electrochemistry*, Harper and Row, New York, 1972.
8b. M. J. Allen, *Organic Electrode Processes*, Reinhold, New York, 1958.
8c. S. D. Ross, M. Finklestein and E. J. Rudd, *Anodic Oxidation*, Academic Press, New York, 1975.
9. A. T. Kuhn (Ed.), *Industrial Electrochemical Processes*, Elsevier, London, 1971.
10. B. E. Conway, *Electrode Processes*, Ronald, New York, 1965.
11. I. Fried, *The Chemistry of Electrode Processes*, Academic Press, New York, 1973.
12. *The Chemical Society Annual Reports*, **65B–71B** (1968–74).
13. *The Chemical Society Specialist Periodical Reports, Electrochemistry*, Vol. 1–5, 1970–75.
14. A. J. Bard (Ed.), *Electroanalytical Chemistry, A Series of Advances*, Dekker, New York, 1966–.
15. P. Delahay and C. W. Tobias (Eds.), *Advances in Electrochemistry and Electrochemical Engineering*, Interscience, New York, 1961–.
16a. L. Eberson, 'Electrochemical Reactions of Carboxylic Acids and Related Processes' in *The Chemistry of Carboxylic Acids and Esters*, (Ed. Saul Patai) John Wiley and Sons., London, 1969.
16b. J. Heyrovsky and J. Kuta, *Principles of Polarography*, Academic Press, New York, 1966.
17. L. Meites, *Polarographic Techniques*, Interscience, New York, 1955.
18. Fr. Fichter, *Organische Elektrochemie*, Steinkopff, Dresden, 1942.
19. C. Mettler, *Chem. Ber.*, **41**, 448 (1908).
20. J. H. Wagenknecht, *J. Org. Chem.*, **37**, 1513 (1972).
21. M. R. Ort and M. M. Baizer, *J. Org. Chem.*, **31**, 1646 (1966).
22. J. H. Wagenknecht, *Ph.D. Dissertation*, University of Iowa, 1964.
23. H. Weil, *German Patent*, 196, 239 (1906).
24. K. S. Tesh and A. Lowy, *Trans. Electrochem. Soc.*, **45**, 37 (1924).
25. G. Kawada and J. Yosida, *Bull. Hyg. Res. Inst. Japan*, **35**, 261 (1929).
26. J. A. May and K. A. Kobe, *J. Electrochem. Soc.*, **97**, 183 (1950).
27. H. V. K. Udupa and B. B. Dey, *Bull. India Sect. Electrochem. Soc.*, **5**, 58 (1956).
28. H. V. K. Udupa, *Bull. Acad. Polon. Sci.*, *Ser. Sci. Chim.*, **9**, 51 (1961).
29. K. S. Udupa, G. S. Subramanian and H. V. K. Udupa, *Ind. Chem.*, **39**, 238 (1963).
30. T. D. Balakrishnan, K. S. Udupa, G. S. Subramanian and H. V. K. Udupa, *Chem. Ind.*, 1622 (1970).

31. V. D. Bezuglyy, V. A. Ekel, M. Ya. Fioshin, Ye. K. Nechiporenko and R. F. Ramakaeva, *The Soviet Chemical Industry* (*Khim. Prom.*) No. 11, Nov. 1970, p. 21.
32. *Annual Report on Synthetic Organic Chemicals; Cyclic Intermediates*, U.S. International Trade Commission, Washington, 1974.
33. V. A. Ekel, R. F. Ramakaeva, N. M. Przhiyalgovskaya and L. Ya. Kheifets, *Zh. Obs. Khim.*, 41, 1908 (1971).
34. J. Ch. Hofmann, P. M. Robertson and N. Ibl, *Tetrahedron Letters.*, 3433 (1972).
35. H. Schäfer, *Z. anorg. allgem. Chem.*, 250, 82 (1942).
36. N. Steinberger and G. Fraenkel, *J. Chem. Phys.*, 40, 723 (1964).
37. J. A. Harrison and D. W. Shoesmith, *J. Electroanal. Chem.*, 32, 125 (1971).
38. H. Weil, *Chem. Ber.*, 41, 4147 (1908).
39. R. G. Barradas, O. Kutowy and D. W. Shoesmith, *Electrochim. Acta,* 19, 49 (1974).
40. J. A. Harrison and K. Scoffham, *Electrochim. Acta*, 21, 585 (1976).
41. L. F. Fieser and M. Fieser, *Reagents for Organic Synthesis*, Vol. 2, Wiley, New York, 1967, p. 82.
42. Reference 3, Part II, p. 123.
43. C. Mettler, *Chem. Ber.*, 38, 1745 (1905).
44. S. Swann, Jr. and G. D. Lucker, *Trans. Electrochem. Soc.*, 75, 411 (1939).
45. A. M. Shams El Din and G. Trumpler, *Helv. Chim. Acta*, 44, 48 (1961).
46. K. Natarajan, K. S. Udupa, G. S. Subramanian and H. V. K. Udupa, *Electrochem. Technol.*, 2, 151 (1964).
47. G. H. Coleman and H. L. Johnson, *Org. Synth.*, Coll. Vol. 3, 60 (1955).
48. C. Mettler, *Chem. Ber.*, 39, 2933 (1906).
49. S. Ono and T. Hayashi, *Bull. Chem. Soc. Japan*, 26, 232 (1953).
50. Reference 1, p. 373.
51. F. G. Drakesmith, *J. Chem. Soc., Perkin I*, 184 (1972).
52. M. Rakoutz, *German Patent*, 2, 157, 560 (1972); *Chem. Abstr.*, 77 55737r (1972).
53. C. Mettler, *Chem. Ber.*, 39, 2933 (1906).
54. H. Suter, H. Nohe, F. Beck and A. Hrubesch, *U.S. Patent*, 3, 471, 381 (1969); *Chem. Abstr.*, 70, 92677x (1969).
55. H. Suter, H. Nohe, F. Beck, W. Brügel and H. Aschenbrenner, *German Patent*, 1, 618, 078, (1967); *Chem Abstr.*, 72, 74127e (1970).
56. H. Nohe and H. Suter, *German Patent*, 2, 158, 200 (1973); *Chem. Abstr.*, 79, 42045e (1973).
57. F. Fichter and C. Simon, *Helv. Chim. Acta*, 17, 1219 (1934).
58. C. Mettler, *Chem. Ber.*, 39, 2934 (1906).
59. E. A. Efimov and I. G. Erusalimchik, *Russ. J. Phys. Chem.*, 38, 1560 (1964).
60. J. Volke and V. Volkova, *Coll. Czech. Chem. Commun*, 20, 1332 (1955).
61. H. Lund, *Acta Chem. Scand.*, 17, 972 (1963).
62. J. Tirouflet and E. Laviron, *C. R. Acad. Sci.* (*Paris*), 247, 217 (1958).
63. P. E. Iversen and H. Lund, *Acta Chem. Scand.*, 21, 279 (1967).
64. P. E. Iversen, *Acta Chem. Scand.*, 24, 2459 (1970).
65. P. E. Iversen and H. Lund, *Acta Chem. Scand.*, 21, 389 (1967).
66. O. R. Brown, J. A. Harrison and K. S. Sastry, *J. Electroanal. Chem.*, 58, 387 (1975).
67. C. Tissier and M. Agoutin, *J. Electroanal. Chem.*, 47, 499 (1973).
68. M. ud Din Bhatti and O. R. Brown, *J. Electroanal Chem.*, 68, 85 (1976).
69. W. J. Albery and S. Bruckenstein, *Trans. Farad. Soc.*, 62, 1946 (1962).
70. J. P. Wibaut and H. Boer, *Rec. Trav. Chim.*, 68, 72 (1949).
71. F. Sorm, *Coll. Czech. Chem. Commun.*, 13, 57 (1948).
72. V. S. Mikhailov, V. P. Gul'tyai, S. G. Mairanovskii, S. Z. Taits, I. V. Proskurovskaya and G. Yu. Dubovik, *Izv. Akad. Nauk. SSSR, Ser. Khim.*, (4) 888 (1975).
73. D. Gagnaire and P. Monzeglio, *Bull. Soc. Chim. Fr.*, 474 (1965).
74. Reference 7, p. 157.
75. F. Beck, P. Jaeger and H. Guthke, *German Patent*, 1, 950, 282 (1971); *Chem. Abstr.*, 74, 140980 (1971).
76. D. Michelet, *German Patent*, 2, 240, 731 and 2, 240, 759 (1973); *Chem. Abstr.*, 78, 131449 and 131450 (1973).
77. D. J. Pickett and K. S. Yap, *J. Appl. Electrochem.*, 4, 17 (1974).

78. C. S. Dillon, *British Patent*, 611, 674 (1948); *Chem. Abstr.*, 43, 4589b (1949).
79. M. Masuno, T. Asahara, S. Kuroiwa, K. Shimizu and J. Nakana, *J. Chem. Soc. Jap., Ind. Chem. Sect.*, 52, 151 (1949).
80. R. Woods, *Electrochim. Acta*, 15, 815 (1970).
81. P. J. Elving and C. Teitelbaum, *J. Amer. Chem. Soc.*, 71, 3916 (1949).
82. R. Kanakam, M. S. V. Pathy and H. V. K. Udupa, *Electrochim. Acta*, 12, 329 (1967).
83. Y. Arad, M. Levy and D. Vofsi, *J. Org. Chem.*, 34, 3709 (1969).
84. C. L. Wilson and K. B. Wilson, *Trans. Electrochem. Soc.*, 84, 153 (1943).
85. R. D. Goodin, unpublished results.
86a. M. Hirayama, *Bull. Chem. Soc. Japan*, 40, 1822 (1967).
86b. A. V. Il'yasov, Yu. M. Kargin, Ya. A. Levin, I. D. Morozova and N. N. Sotnikova, *Dokl. Akad. Nauk. SSSR*, 179, 1141 (1968).
86c. A. V. Il'yasov, Yu. M. Kargin, Ya. L. Levin, I. D. Morozova and N. N. Sotnikova, *Izv. Akad. Nauk SSSR, Ser. Khim*, 5, 1030 (1968).
87. R. E. Sioda and W. S. Koski, *J. Amer. Chem. Soc.*, 89, 475 (1967).
88. W. V. Childs, J. T. Maloy, C. P. Keszthelyi and A. J. Bard, *J. Electrochem. Soc.*, 118, 874 (1971).
89. V. J. Puglisi and A. J. Bard, *J. Electrochem. Soc.*, 120, 748 (1973).
90. S. F. Nelsen, *Tetrahedron Letters*, 3795 (1967).
91. H. O. House and M. J. Umen, *J. Org. Chem.*, 38, 3893 (1973).
92. R. C. Hallcher and M. M. Baizer, *Annalen*, submitted for publication, 1976.
93. I. Vartires, W. H. Smith and A. J. Bard, *J. Electrochem. Soc.*, 122, 894 (1975).
94. G. A. Russell and S. A. Weiner, *J. Amer. Chem. Soc.*, 89, 6623 (1967).
95. S. Ono and J. Nakaya, *J. Chem. Soc. Japan, Pure Chem. Sect.*, 74, 907 (1953).
96. W. Oroshnik and P. E. Spoerri, *J. Amer. Chem. Soc.*, 63, 3338 (1941).
97. M. Fedoronko, *Adv. Carbohydrate Chem.*, 29, 107 (1974).
98. S. Sugasawa and S. Matsumoto, *U.S. Patent*, 3, 312, 608 (1967); *Chem. Abstr.*, 60, 12096g (1964).
99. T. Takahashi, T. Sekine and Y. Kikuchi, *Jap. Kokai*, 73, 28, 410 (1973).
100. I. A. Avrutskaya, M. Ya. Fioshin, E. V. Gromova and V. T. Novikov, *Elektrokhimiya*, 8, 434 (1972).
101. I. A. Avrutskaya, M. Ya. Fioshin, L. A. Muzychenko, E. V. Gromova and V. T. Novikov, *Elektrokhimiya*, 8, 609 (1972).
102. L. Horner and H. Neumann, *Chem. Ber.*, 98, 3462 (1965).
103. G. C. Whitnack, J. Reinhart and E. St. C. Ganz, *Anal. Chem.*, 27, 359, (1955).
104. W. M. Rodionow and V. C. Zvorykina, *Bull. Soc. Chim. Fr.*, 5, 840 (1938).
105. B. Sakurai, *Bull. Chem. Soc. Japan*, 7, 127, 130 (1932).
106. C. Chapman and H. Stephan, *J. Chem. Soc.*, 1791 (1925).
107. J. Tafel and G. Friedrichs, *Chem. Ber.*, 37, 3187 (1904).
108. C. Mettler, *Chem. Ber.*, 37, 3692 (1904).
109. J. Tafel and W. Jurgens, *Chem. Ber.*, 42, 2548 (1909).
110. J. Tafel, *Chem. Ber.*, 45, 437 (1912).
111. L. Eberson, Reference 1, p. 420.
112. S. Wawzonek and J. E. Durham, *J. Electrochem. Soc.*, 123, 500 (1976).
113. A. P. Tomilov, S. E. Zabusova, L. I. Krishtalik, V. F. Pavlichenko and N. M. Alpatova, *Elektrokhimiya*, 11 1132 (1975).
114. J. J. Bloomfield, D. C. Owsley and J. M. Nelke, *Organic Reactions*, 23, 259 (1976).
115. C. T. Mondodoev, N. M. Przhiyalgovskaya and V. N. Belov, *J. Org. Chem., USSR*, 1, 1257 (1965).
116. M. Vora and N. Holy, *J. Org. Chem.*, 40, 3144 (1975).
117. I. L. Knunyants and N. S. Vyazankin, *Izv. Akad. Nauk. SSSR, Otdel. Khim. Nauk*, 2238 (1957).
118. M. M. Baizer, *Tetrahedron Letters*, 973 (1963).
119. M. M. Baizer and J. D. Anderson, *J. Electrochem. Soc.*, 111, 223 (1964).
120. J. P. Petrovich, J. D. Anderson and M. M. Baizer, *J. Org. Chem.*, 31, 3897 (1966).
121. M. M. Baizer, *J. Org. Chem.*, 29, 1670 (1964).
122. I. E. Gillet, *Chem.-Ing.-Tech.*, 40, 573 (1968).
123. J. P. Petrovich, M. M. Baizer and M. R. Ort, *J. Electrochem. Soc.*, 116, 743 (1969).

124. J. P. Petrovich, M. M. Baizer and M. R. Ort, *J. Electrochem. Soc.*, **116**, 749 (1969).
125. J. P. Petrovich and M. M. Baizer, *J. Electrochem. Soc.*, **118**, 447 (1971).
126. V. J. Puglisi and A. J. Bard, *J. Electrochem. Soc.*, **119**, 829 (1972).
127. M. J. Hazelrigg, Jr. and A. J. Bard, *J. Electrochem. Soc.*, **122**, 211 (1975).
128. M. D. Ryan and D. H. Evans, *J. Electrochem. Soc.*, **121**, 881 (1974).
129. J. Anderson and L. Eberson, *J. Chem. Soc., Chem. Commun.*, 565 (1976).
130. D. A. Tysee and M. M. Baizer, *J. Org. Chem.*, **39**, 2819 (1974).
131. D. A. Tysee and M. M. Baizer, *J. Org. Chem.*, **39**, 2823 (1974).
132. V. G. Mairanovskii, N. F. Loginova and S. Ya. Mel'nik, *Elektrokhimiya*, **9**, 1174 (1973).
133. M. F. Semmelhack and G. E. Heinsohn, *J. Amer. Chem. Soc.*, **94**, 5139 (1972).
134. E. Kasafirek, *Tetrahedron Letters*, 2021 (1972).
135. J. P. Coleman, N. -ud-din, H. G. Gilde, J. H. P. Utley, B. C. L. Weedon and L. Eberson, *J. Chem. Soc., Perkin II*, 1903 (1973).
136. A. Inesi and L. Rampazzo, *J. Electroanal. Chem.*, **49**, 85 (1974).
137. A. Inesi, L. Rampazzo and A. Zeppa, *J. Electroanal. Chem.*, **69**, 203 (1976).
138. M. Ochiai, O. Aki, A. Morimoto, T. Ikada, K. Shinozaki and Y. Asahi, *J. Chem. Soc., Perkin I*, 258 (1974).
139. J. G. Gourcy, M. Hodler, B. Terem and J. H. P. Utley, *J. Chem. Soc., Chem. Commun*, 779 (1976).
140. L. Horner and R.-J. Singer, *Tetrahedron Letters*, 1545 (1969).
141. D. W. Leedy and D. L. Muck, *J. Amer. Chem. Soc.*, **93**, 4264 (1971).
142. G. Farnia, A. Romanin, G. Capobianco and F. Torzo, *J. Electroanal. Chem.*, **33**, 31 (1971).
143. A. Lasia, *J. Electroanal. Chem.*, **52**, 229 (1974).
144. M. Hirayama, *Bull. Chem. Soc. Japan*, **40**, 1557 (1967).
145. S. F. Nelsen, *J. Amer. Chem. Soc.*, **89**, 5256 (1967).
146. J. K. Wood, *J. Chem. Soc.*, **89**, 1831 (1906).
147. P. H. Rieger, *Ph.D. Thesis*, Columbia University, New York, 1961.
148. H. Lund, *Acta Chem. Scand.*, **17**, 2325 (1963).
149. R. A. Benkeser, H. Watanabe, S. J. Mels and M. A. Sabol, *J. Org. Chem.*, **35**, 1210 (1970).
150. L. A. Avaca and A. Bewick, *J. Chem. Soc., Perkin II*, 1712 (1972).
151. O. R. Brown and P. D. Stokes, *Electroanal. Chem.*, **57**, 425 (1974).
152. B. Sakurai, *Bull. Chem. Soc. Japan*, **5**, 184 (1930).
153. A. Dunet and A. Willemarte, *Compt. Rend.*, **222**, 1443 (1946).
154. A. Dunet and A. Willemarte, *Compt. Rend.*, **226**, 821 (1948).
155. J. Tirouflet, R. Robin and M. Guyard, *Bull. Soc. Chim. Fr.*, 571 (1956).
156. A. Ryvolova, *Coll. Czech. Chem. Commun.*, **25**, 420 (1960).
157. A. Ryvolova-Kejharova and P. Zuman, *Coll. Czech. Chem. Commun.*, **36**, 1019 (1971).
158. O. R. Brown, S. Fletcher and J. A. Harrison, *J. Electroanal. Chem.*, **57**, 351 (1974).
159. L. Horner and R.-J. Singer, *Annalen*, **723**, 1 (1969).
160. L. Horner and R. Weissbach, *Annalen*, **757**, 69 (1972).
161. F. Beck, *Electrochim. Acta*, **20**, 361 (1975).
162. T. B. Baillie and J. Tafel, *Chem. Ber.*, **32**, 68 (1899).
163. K. Kindler, *Chem. Ber.*, **57**, 773 (1924).
164. E. W. Cook and W. G. France, *J. Phys. Chem.*, **36**, 2383 (1932).
165. M. J. Allen, unpublished results quoted in Reference 8b, p. 75.
166. M. J. Allen and J. Ocampo, *J. Electrochem. Soc.*, **103**, 452 (1956).
167. S. Swann, Jr., *Trans. Electrochem. Soc.*, **84**, 165 (1943).
168. T. Yamazaki and M. Nagata, *Yakugaku Zasshi*, **79**, 1222 (1959).
169. L. C. Craig, *J. Amer. Chem. Soc.*, **55**, 295 (1933).
170. B. Sakurai, *Bull. Chem. Soc. Japan*, **11**, 42 (1936).
171. Reference 14, Vol. 6, p. 1.
172. Reference 1, Chap. 17.

CHAPTER **14**

Decarbonylation reactions of acid halides and aldehydes by chlorotris-(triphenylphosphine)rhodium(I)

M. C. BAIRD
Department of Chemistry, Queen's University, Kingston, Canada

I.	INTRODUCTION	825
II.	THE PREPARATION, STRUCTURE AND CHEMISTRY OF RhCl(PPh₃)₃ (1)	826
III.	REACTIONS OF 1 WITH ACYL HALIDES	827
	A. Decarbonylation Reactions of Acyl Halides	827
	1. General	827
	2. Aromatic acyl halides and cyanides	828
	3. Aliphatic and benzylic acyl halides	834
	B. Mechanisms	837
	C. Summary	842
IV.	REACTIONS OF 1 WITH ALDEHYDES	843
	A. Decarbonylation Reactions	843
	B. Mechanism of Aldehyde Decarbonylation Reactions . . .	852
	C. Summary	855
V.	REFERENCES	855

I. INTRODUCTION

The use of main-group metal organometallic compounds as reagents for effecting organic syntheses has been well known for several decades. Organolithium[1], -magnesium[2] and -zinc[3] compounds in particular have been extensively investigated and widely accepted by synthetic organic chemists, while more recently organo-thallium[4], -boron[5] and -aluminium[6] compounds have also found novel and interesting synthetic applications. Paralleling this activity to some extent is the utilization of the platinum or noble metals as heterogeneous catalysts for a wide variety of useful organic reactions, the best known, perhaps, being the hydrogenation reactions[7].

Relative newcomers to organic synthesis, being used only in the past ten to fifteen years, have been the organotransition metal compounds, used either as stoichiometric reagents[8] or as homogeneous catalysts[8,9]. Research in this area has been expanding very rapidly, with the result that a seemingly bewildering array of strange, difficult-to-handle compounds, which often exhibit a very novel organic chemistry, now lends itself to at least a degree of systematic classification[10,11].

Exploitation of this area of organic chemistry has been slow. Although a few organotransition metal compounds have been utilized, for instance, to generate otherwise inaccessible intermediates such as cyclobutadiene[12], to activate aromatic molecules to nucleophilic substitution reactions[13] and to act as protecting groups[14] or olefin hydrogenation catalysts[15], the field is as yet a rudimentary level.

Originally the intention of this review was to survey the general area of decarbonylation reactions of acid derivatives. However, an inspection of the pertinent literature revealed that, aside from a few references to decarbonylations of carboxylic acids by strong mineral acids[16], and of some acid chlorides by aluminium trichloride[17], perhaps the most synthetically useful and most intensively studied decarbonylation reactions are those of aldehydes and acyl halides by a particular compound of monovalent rhodium, chlorotris(triphenyl-phosphine)rhodium(I), $RhCl(PPh_3)_3$ (1). As a previous article in this series on the decarbonylation of aldehydes[18] was written before the investigations of the chemistry of 1 had really begun, and a previous volume in this series[19] on the chemistry of acyl halides does not appear to mention 1, it seemed appropriate to focus attention on decarbonylation reactions of acyl halides and aldehydes effected by 1. The subject is especially suitable for discussion at this point because the decarbonylation reactions of 1 are one of the very few organometallic reactions which have both been utilized to some extent in organic syntheses and which appear to be reasonably well understood mechanistically. Much of the very early work has been reviewed previously[20].

II. THE PREPARATION, STRUCTURE AND CHEMISTRY OF $RhCl(PPh_3)_3$ (1)

Compound 1 appears to have been independently prepared by several research groups during the mid 1960s[21-25]. Although it and its derivatives may be prepared by displacing olefins from rhodium(II) complexes by tertiary phosphines[26], 1 can be readily prepared in very high yield by treating commercially available hydrated rhodium trichloride(2) with an excess of triphenylphosphine in refluxing ethanol[27]. As the amount of rhodium (weight %) in 2 is three and a half times that in 1, while the ratio of costs is only a factor of two[28], the financial savings normally outweigh the small amount of time required to convert 2 to 1 (rhodium is one of the most expensive of the noble metals). Procedures are available in the literature[29] for the recovery of useable rhodium salts from rhodium residues.

A crystal-structure determination[30] shows that 1 exhibits, as expected, essentially square planar coordination about the metal. Earlier work[27] had suggested that it dissociated extensively in solution to give either the three-coordinated complex, $RhCl(PPh_3)_2$ (3) or a solvated species, $RhCl(PPh_3)_2$(solvent), but more recent[31] P n.m.r. studies show that the complex remains essentially intact[31] in solution. 1 readily undergoes both substitution and oxidation reactions; evidently one of the triphenylphosphine ligands must be very labile, as the planar compounds $RhCl(L)(PPh_3)_2$ (L = CO,C_2H_4) are rapidly formed on bubbling carbon monoxide or ethylene through solutions of 1[27,31] (equation 1).

$$1 + L \longrightarrow trans\text{-}RhCl(L)(PPh_3)_2 + PPh_3 \tag{1}$$

Addition reactions with, for instance, hydrogen[27,31], hydrogen chloride[32,33] and methyl iodide[34,35] to form five- or six-coordinated complexes of rhodium(III) are also well known (equations 2–4). [The hydrogen and methyl ligands are

$$1 + H_2 \longrightarrow \quad \begin{array}{c} Ph_3P \;\; H \;\; H \\ \backslash | / \\ Rh \\ / | \backslash \\ Ph_3P \;\; Cl \;\; PPh_3 \end{array} \tag{2}$$

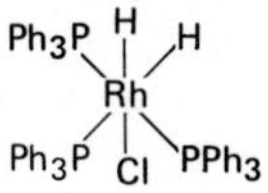

$$1 + HCl \longrightarrow RhHCl_2(PPh_3)_{2 \text{ or } 3} \tag{3}$$

$$1 + MeI \longrightarrow \quad \begin{array}{c} Ph_3P \;\; Me \;\; I \\ \backslash | / \\ Rh \\ / \backslash \\ I \qquad PPh_3 \end{array} \tag{4}$$

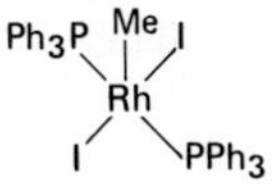

(4)

generally regarded as being anionic, two-electron donors, i.e. hydrido and carbanionic groups; thus addition of molecular hydrogen is considered as being an oxidation process, contrary to the conventions of organic chemistry. As reactions (2), (3) and (4) involve increases in both the oxidation state and the coordination number of the central metal atom, they are often referred to as *oxidative addition* reactions].

Investigations of the reactions of 1 with aldehydes and acyl halides were initiated independently in about 1965 by research groups in Japan[36], England[37] and Israel[38]. Subsequent research by these and other groups has shown that while both aldehydes and acyl halides react with 1 by similar mechanisms, reaction intermediates can only be detected during reactions of the latter. Thus, in the following sections, decarbonylation reactions of acyl halides by 1 will be treated first, followed by decarbonylation reactions of aldehydes.

III. REACTIONS OF 1 WITH ACYL HALIDES

A. Decarbonylation Reactions of Acyl Halides

1. General

Reactions of acyl halides with 1 fall into two main categories, those which proceed with elimination of carbon monoxide only to give the aryl (equation 5) or alkyl (equation 6) halide, and those which result in elimination of both carbon monoxide and hydrogen halide to give olefin. The latter situation occurs only when $C_{(3)}$ of aliphatic acyl halides is bonded to a hydrogen atom (equation 7). The

$$ArCOCl + 1 \longrightarrow ArCl + 5 \tag{5}$$

$$MeCOCl + 1 \longrightarrow MeCl + 5 \tag{6}$$

$$RCH_2CH_2COCl + 1 \longrightarrow RCH=CH_2 + HCl + 5 \tag{7}$$

rhodium-containing product in all cases is the planar rhodium(II) complex, *trans*-RhClCO(PPh$_3$)$_2$ (5), the same compound obtained on treatment of 1 with carbon monoxide (equation 1). In addition, decarbonylation reactions may be stoichiometric (temperature range $80-100°C$) or catalytic (temperatures above $\sim180°C$).

2. Aromatic acyl halides and cyanides

Experimental conditions and results for a number of aromatic acyl halides (and cyanides) are presented in Table 1; most studies were carried out using solutions of 1 in the neat acyl halides. All yields in this and subsequent tables are based on the amount of acyl halide used. Although the homogeneous decarbonylation reaction proceeds stoichiometrically under relatively mild conditions $(80°C)$[37,38], the reaction is generally catalytic at about 200°C. Indeed, a recent kinetic study of the decarbonylation of benzoyl bromide by 1 shows that maximum catalytic activity is achieved only in the temperature range $190–210°C$[44]. The reaction is sluggish at lower temperatures, while the catalyst is deactivated at higher temperatures. However, the 1-catalysed reaction appears to be much more efficient and useful than other procedures for the decarbonylation of aromatic acyl halides[38-40].

The catalytic decarbonylations of aroyl fluorides have been attempted using both neat substrates and solutions in a variety of inert solvents; o-xylene appeared to be the best solvent[39]. Ideal conditions for decarbonylation of aroyl fluorides have unfortunately not yet been achieved, however, since the average number of reaction cycles per catalyst molecule ranges only between three and five. Inactive fluororhodium species may form as the reaction proceeds[39].

Catalyst efficiency is generally much higher with aroyl chlorides, bromides and iodides, although the above-mentioned kinetic study[44] suggests that reactions may not always have been carried out at the optimum temperature. Work by Blum and coworkers[38,42] has amply demonstrated that consideration of the thermal stabilities of the products is necessary if efficiency is generally to be achieved. It was found best in many cases that the reaction mixture be heated only to the reflux temperature of the aryl halide product. If the latter were not distilled off, the catalyst became deactivated in a manner not yet elucidated[42]. (An exception to this generalization is 4-acetamidobenzoyl chloride (Table 1, no,22), which decomposed by other routes[42]). The 'distillation technique' did not apply to the high-boiling polycyclic acyl compounds (nos. 29–38, 43, 44), however, as the products seemed to be unstable with respect to further reaction with one or more rhodium compounds in solution at high temperatures. These were best decarbonylated by brief heating with 1 at 235–300°C followed by chromatographic purification[42].

Although 1,2-phthaloyl chloride is not decarbonylated, perhaps forming a stable rhodium complex, the 1,3- and 1,4-isomers react in two steps[41,42]. Thus the two acyl chloride groups in these cases react independently with the rhodium catalyst.

Little effort appears to have been made to study decarbonylations of aroyl bromides; it seems generally to have been assumed (and found) that they behave similarly to aroyl chlorides. However, the possible utilization of 1 to synthesize rather labile, difficult to prepare aryl iodides has prompted one study of aroyl iodides (nos. 45–54)[43]. Good yields were obtained at 200°C in all examples studies when the products were distilled off at reduced pressure as they formed. Indeed, the yields of 1,2-diiodobenzene from 2-iodobenzoyl iodide (no. 50) was much higher than the yields from the corresponding reaction of 2-iodobenzoyl chloride (no. 18), prompting the suggestion of a special benzyne-type intermediate in this case. No rhodium intermediates were isolated, however, and in view of the facts that not only can the temperature determine the distribution of products (see above), but also that the chemistry of iodorhodium complexes is very incompletely understood[45], suggestions regarding mechanisms seem premature in this case.

The reactions of the aroyl cyanides (nos. 55–57) are included for reasons of completeness. Benzoyl cyanide was not decarbonylated, probably because the

TABLE 1. Decarbonylation reactions of aroyl halides and cyanides

No.	Acyl Halide	Product(s)	Temperature (°C)	Yield (%)	References
Fluorides					
1	PhCOF	PhF	80–120	Essentially quantitative	39
2	4-MeC$_6$H$_4$COF	4-MeC$_6$H$_4$F			39
3	3-MeC$_6$H$_4$COF	3-MeC$_6$H$_4$F			39
4	4-ClC$_6$H$_4$COF	4-ClC$_6$H$_4$F			39
5	3-ClC$_6$H$_4$COF	3-ClC$_6$H$_4$F			39
6	4-FC$_6$H$_4$COF	1,4-C$_6$H$_4$F$_2$			39
Chlorides					
7	PhCOCl	PhCl	197	60–90	38, 40, 41
8	4-MeC$_6$H$_4$COCl	4-MeC$_6$H$_4$Cl	230	66	41
9	4-EtC$_6$H$_4$COCl	4-EtC$_6$H$_4$Cl	180[a]	82	42
10	4-MeOC$_6$H$_4$COCl	4-MeOC$_6$H$_4$Cl	240–250, 200[a]	50	41, 42
11	4-EtOC$_6$H$_4$COCl	4-EtOC$_6$H$_4$Cl	212[a]	59	42
12	3-MeC$_6$H$_4$COCl	3-MeC$_6$H$_4$Cl	162[a]	74	42
13	2,4,6-Me$_3$C$_6$H$_2$COCl	2,4,6-Me$_3$C$_6$H$_2$Cl	205[a]	55	42
14	4-ClC$_6$H$_4$COCl	1,4-C$_6$H$_4$Cl$_2$	222[b]	79	38
15	2,4-Cl$_2$C$_6$H$_3$COCl	1,2,4-C$_6$H$_3$Cl$_3$	b	98	38
16	3,4-Cl$_2$C$_6$H$_3$COCl	1,3,4-C$_6$H$_3$Cl$_3$	250[a]	96, 76	41, 42
17	2-BrC$_6$H$_4$COCl	2-BrC$_6$H$_4$Cl	245[b]	78	38
18	2-IC$_6$H$_4$COCl	2-IC$_6$H$_4$Cl	234[a]	38	38
19	4-IC$_6$H$_4$COCl	4-IC$_6$H$_4$Cl	227[a]	78	38
20	4-NCC$_6$H$_4$COCl	4-NCC$_6$H$_4$Cl	223[a]	90	42
21	4-O$_2$NC$_6$H$_4$COCl	4-O$_2$NC$_6$H$_4$Cl	a	79	42
22	4-(MeCONH)C$_6$H$_4$COCl	4-(MeCONH)C$_6$H$_4$Cl	a	'Low'	42
23	1,2-C$_6$H$_4$(COCl)$_2$	—	—	0	42
24	1,3-C$_6$H$_4$(COCl)$_2$	3-ClC$_6$H$_4$COCl	240–250, 225[a]	75	41, 42
		1,3-C$_6$H$_4$Cl$_2$	240–250, 162[a]	20	41, 42
25	1,4-C$_6$H$_4$(COCl)$_2$	4-ClC$_6$H$_4$COCl	222[a]	79	42
		1,4-C$_6$H$_4$Cl$_2$	173[a]	5	42
26	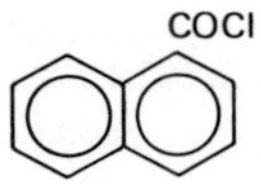 [naphthalene]COCl	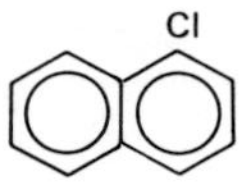 [naphthalene]Cl	260[a], 240	96, 97	38, 41

TABLE 1. (Continued)

No.	Acyl Halide	Product(s)	Temperature (°C)	Yield (%)	References
27	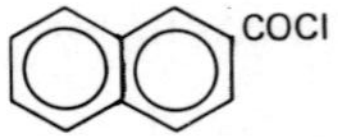	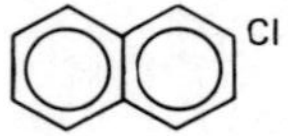	265[a]	94	38
28	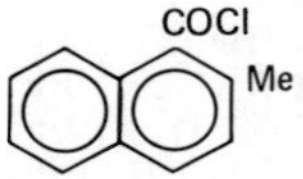	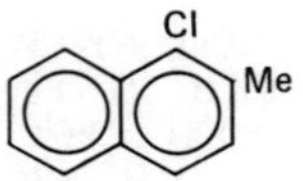	a	93	38
29	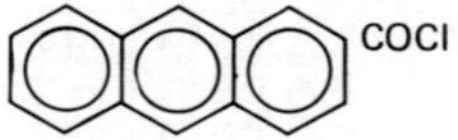	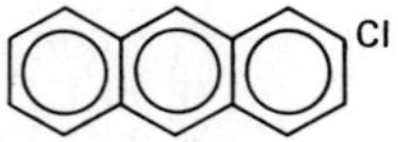	310	81	42
30	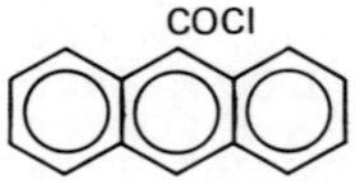	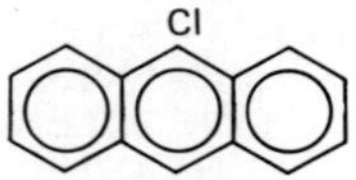	310	24	42
31	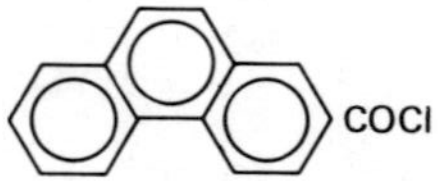	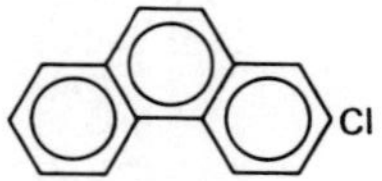	235	95	42
32	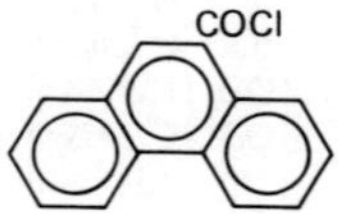	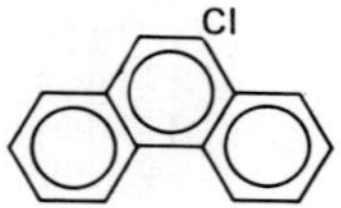	260	60	42
33	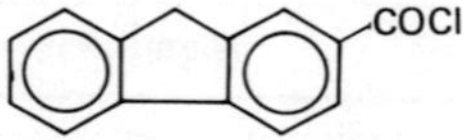	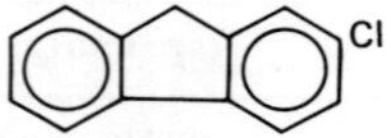	280	60	42

	Acid halide	Product			Ref.
34	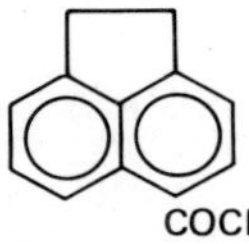 COCl	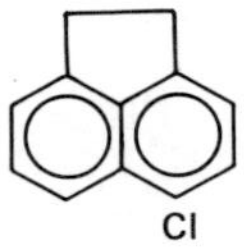 Cl	300	81	42
35	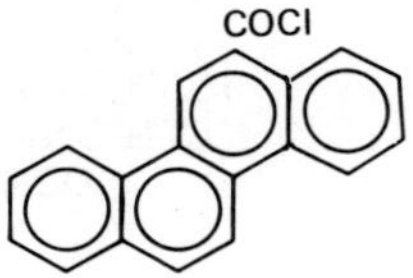 COCl	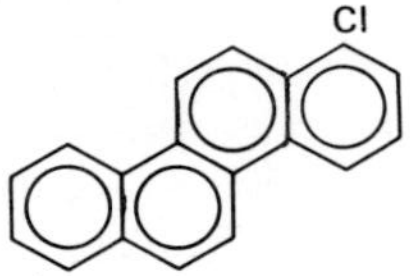 Cl	280	48	42
36	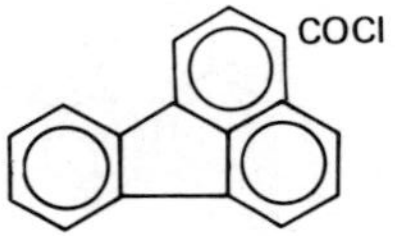 COCl	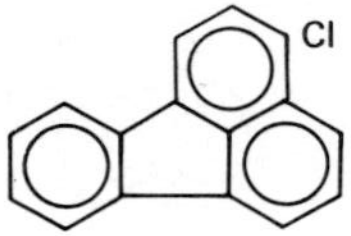 Cl	235	84	42
37	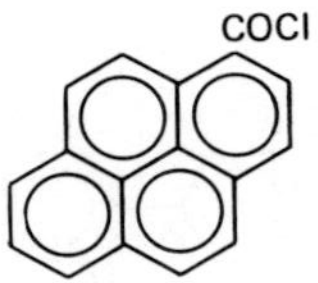 COCl	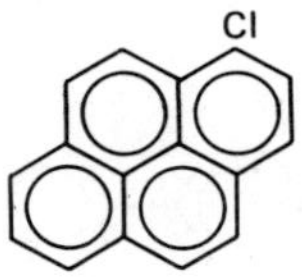 Cl	250	94	42
38	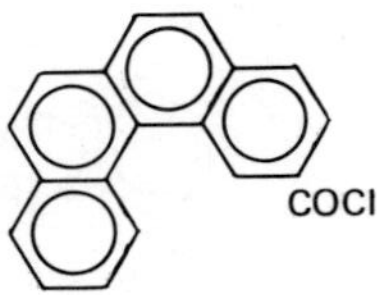 COCl	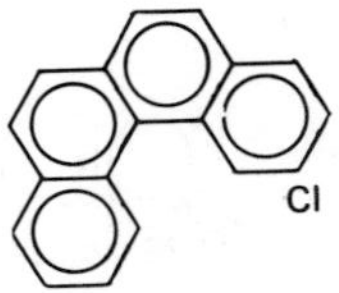 Cl	260	89	42
Bromides					
39	PhCOBr	PhBr	220, 156[a]	80–87	40, 42
40	3-BrC$_6$H$_4$COBr	1,3-C$_6$H$_4$Br$_2$	219[a]	85	42
41	4-BrC$_6$H$_4$COBr	1,4-C$_6$H$_4$Br$_2$	219[a]	95	42

TABLE 1. (Continued)

No.	Acyl Halide	Product(s)	Temperature (°C)	Yield (%)	References
42	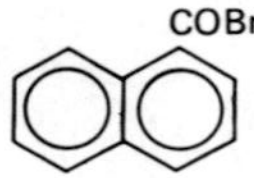 COBr (structure)	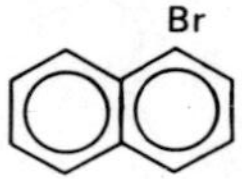 Br (structure)	281[a]	84	42
43	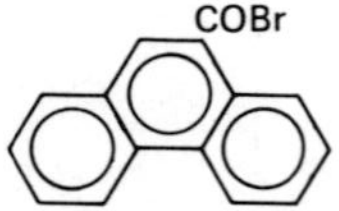 COBr (structure)	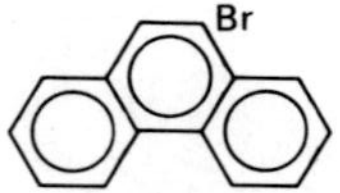 Br (structure)	300	95	42
Bromides					
44	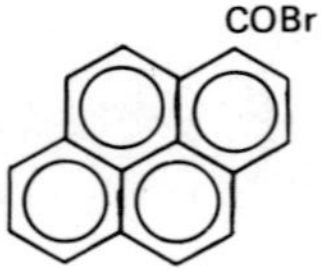 COBr (structure)	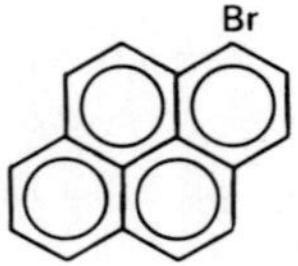 Br (structure)	290	89	42
Iodides					
45	PhCOI	PhI	c	62	43
46	4-MeC$_6$H$_4$COI	4-MeC$_6$H$_4$I	c	63	43
47	2-ClC$_6$H$_4$COI	2-ClC$_6$H$_4$I	c	53	43
48	4-ClC$_6$H$_4$COI	4-ClC$_6$H$_4$I	c	98	43
49	4-BrC$_6$H$_4$COI	4-BrC$_6$H$_4$I	c	82	43
50	2-IC$_6$H$_4$COI	1,2-C$_6$H$_4$I$_2$	c	72	43
51	3-IC$_6$H$_4$COI	1,3-C$_6$H$_4$I$_2$	c	64	43
52	3,4-Cl$_2$C$_6$H$_3$COI	3,4-Cl$_2$C$_6$H$_3$I	c	98	43
53	1,3-C$_6$H$_4$(COI)$_2$	1,3-C$_6$H$_4$I$_2$	c	65	43

54	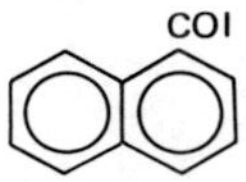 COI	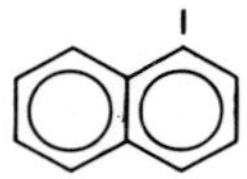 I	146 (0.1 mm)	60	43
55	PhCOCN	PhCN	205^b	8	42
56	4-ClC$_6$H$_4$COCN	4-ClC$_6$H$_4$CN	223^a	95	42
57	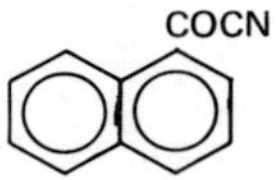 COCN	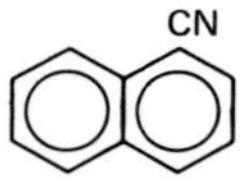 CN	294^a	87	42

[a] Temperature at which reaction mixture was heated and at which product distilled.
[b] Reflux temperature of acyl halide.
[c] Reaction mixture heated between 175° and 250° at sufficiently reduced pressure that aryl iodide product distilled.

reflux temperature was not sufficiently high to decompose the intermediate cyanorhodium complexes. The higher boiling substrates, on the other hand, were decarbonylated smoothly.

3. Aliphatic and benzylic acyl halides

Experimental conditions and results for decarbonylation reactions of alkyl and benzylic acyl halides are listed in Table 2. In contrast to the situation with aroyl halides (Table 1), most studies have been done on solutions in a variety of inert solvents, nitriles being preferred[50,51]. In general, therefore, decarbonylation reactions carried out at temperatures much below 200°C are only stoichiometric (equations 6 and 7) while those at about 200°C or higher are catalytic. As 5 is the observed product in stoichiometric decarbonylations, it is often assumed to be an intermediate in catalytic reactions, and therefore a catalyst also. Both 1 and 5 have therefore been used in catalytic decarbonylations, although reservations concerning their interchangeability have been expressed[42]. This issue will be discussed in Section III.B. Yields are based on the amount of acyl halide used.

Experiment nos. 1—4 and 6 of Table 2 show that methyl and benzylic derivatives are decarbonylated very smoothly, although no. 4 suggests that optically active acyl halides are unfortunately not decarbonylated stereospecifically. Bulky substituents on $C_{(2)}$ of the acyl halide can affect both the ease of the reaction and the nature of the products[50]; a π-benzylrhodium complex (6) was formed in no. 5, effectively destroying the catalyst.

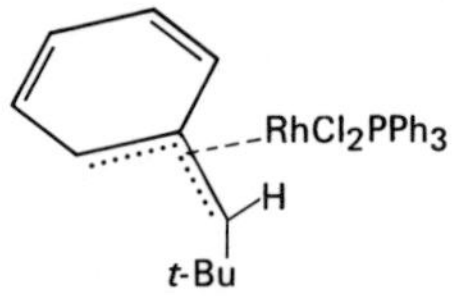

(6)

The remaining entries in Table 2 show that 1 very effectively catalyses simultaneous decarbonylation and dehydrohalogenation of aliphatic acyl chlorides; data on bromides and iodides are very sparse. Whereas similar reactions using palladium metal as catalyst give predominantly the thermodynamically more stable internal olefins[40,41,55], the results of both double-bond migration and cis—trans isomerization of the products, the rhodium catalysts 1 and 5 give mixtures of terminal and internal olefins. Not mentioned in Table 2 are 3-(methoxycarbonyl) propionyl chloride, which gives succinic anhydride by an unknown mechanism[41], and adipic and suberic acid dichlorides, which give resinous materials[41].

Of great synthetic importance is the fact that addition of excess triphenyl phosphine to solutions of the rhodium catalysts[51,52] accompanied by immediate distillation of olefins as formed[41] minimizes olefin isomerization (nos. 7, 11, 18, 19, 20, 21; ratios indicated are mole ratios). Although both reaction rates and yields are lowered in the presence of excess triphenylphosphine, the procedure demonstrates that conversion of a variety of acy halides to olefins without isomerization of the latter may be possible.

Experiment nos. 15—21 further clarify the possible synthetic utility of 1 and 5 in this respect. Although catalytic decarbonylation of neat threo-PhCHDCHDCOCl (no. 15) results in very extensive scrambling of deuterium in the styrene produced, the much bulkier threo- and erythro-PhMeCHPhCHCOCl (nos. 16 and 17) are

TABLE 2. Decarbonylation reactions of alkyl and benzylic acyl halides

No.	Acyl Halide	Product(s)	Solvent temperature (°C)	Yield (%)	Reference
Acyl halides without a hydrogen on $C_{(3)}$					
1	MeCOCl	MeCl	CHCl$_3$, 20; C$_6$H$_6$, 80	100	37, 40, 46–48
2	PhCH$_2$COCl	PhCH$_2$Cl	C$_6$H$_6$, 80 Neat, 200–220 Neat, 179[a]	81 60 97	41, 49 40, 41 42
3	Ph$_2$CHCOCl	Ph$_2$CHCl	Neat, 210–230 reduced pressure[a]	94	42
4	(S)-Ph(CF$_3$)CHCOCl	(R,S)-PhCF$_3$CHCl	C$_6$H$_6$, 80; PhMe, 111; MeCN, 81	71	50
5	t-Bu(Ph)CHCOCl	—	—	0	50
6	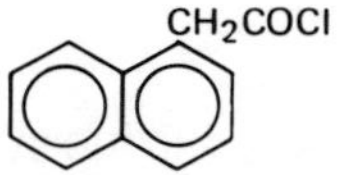 (naphthalene)–CH$_2$COCl / CH$_2$COCl	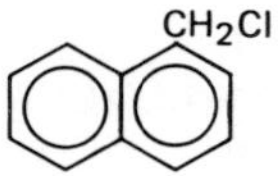 (naphthalene)–CH$_2$Cl	Neat, reduced pressure[a]	87	38
Acyl halides with a hydrogen on $C_{(3)}$					
7	Me(CH$_2$)$_4$COCl	1-Pentene(5),2-pentene(95) 1-Pentene(66),2-pentene(34) 1-Pentene(95),2-pentene(5)	Xylene, 140 PhCN, 190 PhCN, 190 +5 PPh$_3$/Rh	— 62 57	51 51 51
8	Me(CH$_2$)$_5$COBr	1-Hexene(61),2,3-hexenes(39)	Neat, 200	85	41
9	Me(CH$_2$)$_6$COCl	'Heptenes'	Neat, 190–200	91	41
10	Me(CH$_2$)$_6$COBr	1-Heptene(71),trans-2-heptene(24) cis-2-heptene(5)	Neat, 200	90	41
11	Me(CH$_2$)$_8$COCl	Decenes, ~50% 1-decene 1 decene	Neat, >200 Neat, >200, +100 PPh$_3$/Rh	70 50	52 52
12	Me(CH$_2$)$_{14}$COCl	Pentadecene	Neat, 230	54	41
13	(CH$_2$)$_8$(COCl)$_2$	1,7-octadiene(25),1,6,-octadiene(34), 1,5-octadiene(13), other octadienes(28)	210–220	85	41
14	PhCH$_2$CH$_2$COCl	PhCH=CH$_2$	PhMe, 111	71	41, 49
15	threo-PhCHDCHDCOCl	All possible d$_1$-, d$_2$-styrenes deuterated in the vinyl positions	Neat, 145–200	—	53, 54

TABLE 2. (Continued)

No.	Acyl Halide	Product(s)	Solvent, temperature (°C)	Yield (%)	References
16	erythro-PhMeCHPhCHCOCl	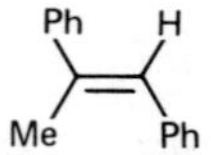 Ph, H / Me, Ph (alkene, $Ph(Me)C{=}C(Ph)H$)	C_6H_6, 30	90	51
17	threo-PhMeCHPhCHCOCl	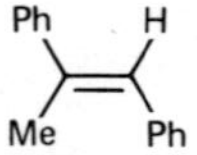 Ph, H / Me, Ph and Ph, Ph / Me, H (11) 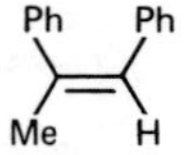	C_6H_6, 30	90	51
18	$MeCH_2CH_2\overset{\mid}{C}HCOCl$ (Me)	$MeCH_2CH_2CH{=}CH_2(30)(1)^b$, $MeCH_2CH{=}CHCH_3(70)(3.5)^b$ (cis and trans)	PhCN, 190 +10 PPh_3/Rh	52	51
19	$MeCH_2CH_2\overset{\mid}{\underset{\mid}{C}}COCl$ (Me, Me)	$MeCH_2CH_2\overset{\mid}{C}{=}CH_2(49)(1)^b$ (Me), $MeCH_2CH{=}\overset{\mid}{C}CH_3(47)(3)^b$ (Me), $MeCH{=}CH\overset{\mid}{C}HMe(4)$ (Me) (cis and trans)	PhCN, 190 +10 PPh_3/Rh	36	51
20	$Me\overset{\mid}{\underset{\mid}{C}}HCH_2CH_3$ (Me, $COCl$)	$MeCH_2CH_2\overset{\mid}{C}{=}CH_2(10)$ (Me), $MeCH_2CH{=}\overset{\mid}{C}Me(44)(2)^b$ (Me), $MeCH{=}CH\overset{\mid}{C}HMe(46)(1)^b$ (Me) (cis and trans)	PhCN, 190 +10 PPh_3/Rh	80	51
21	$Me\overset{\mid}{C}H\overset{\mid}{C}HCOCl$ (Me Me)	$Me\overset{\mid}{C}HCH{=}CH_2(12.4)(1)^b$ (Me), $MeCH_2CH{=}CH_2(9.6)$, $MeCH{=}\overset{\mid}{C}CH_3(78)(19)^b$ (Me)	PhCN, 190 +10 PPh_3/Rh	75	51

^{a}Temperature at which the reaction mixture was heated and at which product distilled.
^{b}Statistically corrected ratios.

decarbonylated and dehydrohalogenated stereospecifically. A rationale of the difference[54] will be developed in Section III.C. These observations, together with those of experiments 18–21, which demonstrate that ease of removal of a hydrogen from $C_{(3)}$ of the acyl halide is tertiary > secondary > primary, suggest that regioselective decarbonylation and dehydrohalogenation of complex acyl halides may prove to be of great synthetic utility.

Although little research in this area with other metal compounds has been carried out, a patent[56] suggests that the compounds $RhBrCO(PBu_3)_2$ and $RhClCO(PBuPh_2)_2$, at least, will decarbonylate acyl halides.

B. Mechanisms

As mentioned in Section II, 1 undergoes oxidative addition with methyl iodide to give the five-coordinated rhodium(III) complex, $RhMeI_2(PPh_3)_2$ (4). Numerous studies have shown that, at room temperature or slightly above, 1 undergoes similar reactions with acyl chlorides[32,41,42,44,47,48,50,51,54,57] (equation 8).

$$1 + RCOCl \longrightarrow RhCl_2(COR)(PPh_3)_2 + PPh_3$$

$$(7) \hspace{4cm} (8)$$

$$R = alkyl, aryl$$

The mechanism of the oxidative addition step is not clear, but probably involves a solvated or three-coordinated intermediate, $RhCl(PPh_3)_2$[3,48], resulting from slight dissociation of one of the triphenylphosphines[31,58]. Low-valent, electron-rich planar molecules such as 1, 3 and 5 have filled $4d_z^2$ orbitals of essentially σ symmetry (Figure 1) and are believed to be reasonably good nucleophiles; they protonate readily on the metal[32], while their oxidative addition reactions with methyl iodide are believed to involve S_N2 displacement of iodide from the saturated carbon by the metal nucleophile[59]. The oxidative addition of acyl chlorides to 3 may well involve interaction of the high-energy, filled $4d_z^2$ orbital with the vacant π^* orbital of the carbonyl group (probably the lowest unoccupied molecular orbital[60]), in other words a nucleophilic attack on the carbonyl carbon atom (Figure 2).

A wide variety of acyl compounds (7) has been reported; a few are listed in Table 3. In the case of the addition of acetyl chloride, low-temperature[31] P n.m.r. spectroscopy has shown[48] that the acylrhodium compound initially formed in chloroform solution is the monomeric species, 8. Compound 8 can be obtained from solution as a reasonably stable, pale yellow solid, and has been characterized

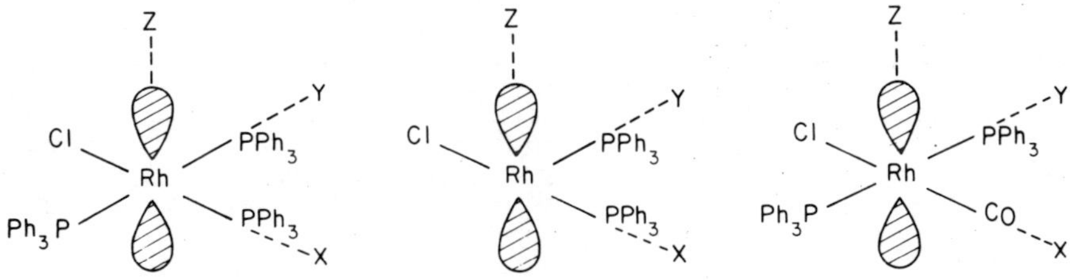

FIGURE 1. Structures of $RhCl(PPh_3)_3$ (1) and $RhClCO(PPh_3)_3$ (5), and probable structure of $RhCl(PPh_3)_2$ (3), illustrating the orientation of the filled $4d_{z^2}$ orbital (omitting the 'donut').

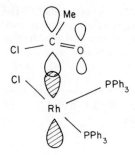

FIGURE 2. Possible interaction between the highest, occupied molecular orbital of RhCl(PPh$_3$)$_2$, predominantly of metal $4d_{z^2}$ character, with the lowest unoccupied molecular orbital of acetyl chloride. Hatching denotes a filled orbital.

by i.r., far i.r. and ^{1}H and ^{31}P n.m.r. spectroscopy[48]. It has the structure expected for a *cis* oxidative addition of acetyl chloride to 3 via the transition state illustrated in Figure 2.

$$\underset{(\textbf{8})}{\overset{\displaystyle COMe}{\underset{\displaystyle Cl}{\overset{\displaystyle Cl}{|}}\!\!\!\!\!\!\underset{\displaystyle Rh}{\diagdown}\!\!\!\!\!\!\overset{\displaystyle PPh_3}{\diagup}}} \qquad \underset{(\textbf{9})}{\overset{\displaystyle COMe}{\underset{\displaystyle Ph_3P}{\overset{\displaystyle Cl}{|}}\!\!\!\!\!\!\underset{\displaystyle Rh}{\diagup}\!\!\!\!\!\!\overset{\displaystyle PPh_3}{\diagdown}Cl}}$$

(8)   (9)

In chloroform solution at room temperature, **8** isomerizes irreversibly to an orange acylrhodium compound, **9**[48]. The structure of the latter, inferred from spectroscopic data, has been verified by the X-ray crystal structure determination of the orange RhCl$_2$(COCH$_2$CH$_2$Ph)(PPh$_3$)$_2$[48]. It is square pyramidal, with *trans* basal chlorines, *trans* basal phosphines and an apical acyl group, and has very similar i.r. and n.m.r. spectral parameters to **9**. As most of the acylrhodium compounds listed in Table 3 are also orange, it seems likely that they all have structures similar to that of **9**.

$$\underset{(\textbf{9})}{\overset{\displaystyle COR}{\underset{\displaystyle Ph_3P}{\overset{\displaystyle Cl}{|}}\!\!\!\!\!\!\underset{\displaystyle Rh}{\diagdown}\!\!\!\!\!\!\overset{\displaystyle PPh_3}{\diagup}Cl}} \quad \rightleftharpoons \quad \underset{(\textbf{10})}{\overset{\displaystyle CO}{\underset{\displaystyle Ph_3P}{\overset{\displaystyle Cl}{|}}\!\!\!\!\!\!\underset{\displaystyle Cl}{\overset{\displaystyle Rh}{\diagdown}}\!\!\!\!\!\!\overset{\displaystyle PPh_3}{\diagup}\;R}} \qquad\qquad \textbf{(9)}$$

(9)   (10)

TABLE 3. Representative compounds of the type RhCl$_2$(COR)(PPh$_3$)$_2$

No.	R	$\nu_{C=O}(cm^{-1})$	N.m.r. data[a]	References
1	Me (9)	1700	$\delta(CH_3) = 2.49$ ($J_{PH} = 1$ Hz) $\delta(P) = 23.6$ ($J_{RhP} = 108$ Hz)	48
2	Me(CH$_2$)$_5$	–	$\delta(CH_3) = 0.75$, $\delta(CH_2)_4 = 0.9-1.5$ $\delta(CH_2CO) = 2.93$	41
3	PhCH$_2$CH$_2$	1710	$\delta(CH_2Ph) = 2.60$, $\delta(CH_2CO) = 3.11$ $\delta(P) = 23.2$ ($J_{RhP} = 108$ Hz)	41, 48, 51
4	PhCH$_2$	1708	--	57
5	Ph	1666	--	57

[a] ^{1}H chemical shifts in p.p.m. from TMS, ^{31}P chemical shifts in p.p.m. downfield from external H$_3$PO$_4$.

TABLE 4. Representative compounds of the type $RhRCl_2CO(PPh_3)_2$

No.	R	$\nu_{C=O}(cm^{-1})$	N.m.r. data[a]	References
1	Me	2045	$\delta(CH_3$ = 0.84 (J_{PH} = 5 Hz, J_{RhH} = 2 Hz) $\delta(P)$ = 18.7 (J_{RhP} = 90 Hz)	48
2	Ph	2074	--	57
3	$PhCH_2$	2069	—	57
4	$PhCH_2CH_2$	2096[b]	$\delta(P)$ = 15.9 (J_{RhP} = 89 Hz)	48, 54

[a] ^{1}H chemical shifts in p.p.m. from TMS, ^{31}P chemical shifts in p.p.m. downfield from external H_3PO_4.
[b] Incorrectly stated to be 1996 cm^{-1} in Reference 54.

Simultaneously with isomerization of 8 to 9, 9 isomerizes to the six-coordinated, methylrhodium compound, 10 (R = Me). A number of six-coordinated compounds of the type $RhRCl_2CO(PPh_3)_2$ have been reported (Table 4) and are generally believed to have structures similar to that of 10.

The isomerization reaction is of a type which is well known in organotransition metal chemistry[61]. Its mechanism is believed to involve a 1,2-shift of the methyl from the carbonyl group to a *cis* position on the metal[47,48,54]. Such migration reactions of primary and secondary groups generally appear to proceed with retention of configuration at $C_{(2)}$ of the acyl group[54,62−65], while the ρ values for the same isomerization reactions of *para*-substituted benzyl and phenylacetyl complexes suggest little charge imbalance in the transition state[57]. These data are consistent with a concerted mechanism involving a three-centred transition state, as in equation (10).

$$(10)$$

The isomerization of 9 to 10 is reversible, the equilibrium constant being about 0.3 in chloroform and varying little with *para* substitution on the triarylphosphine. Both the enthalpy and the entropy changes in the 9 → 10 conversion are small and negative[48]. Electron-withdrawing *para* substituents on the migrating group of a number of benzyl and phenylacetyl compounds enhance the rates of aryl migration, but decrease the rates of benzyl migration[57], observations not readily understood at present but which are consistent with the observed substituent effects on the rates of catalytic decarbonylation of aroyl halides[41]. Added triphenylphosphine has negligible effect on either the position of equilibrium[48] or the rate of migration[57] in the isomerization of acylrhodium complexes 7 to their alkyl or aryl isomers.

Interestingly, however, the positions of equilibrium in the acyl ⇌ alkyl (aryl) isomerizations are strongly dependent on the nature of the migrating group. Representative equilibrium constants are listed in Table 5. Both steric and electronic factors are believed to be responsible for the large differences[48].

Most compounds of the type $RhRCl_2CO(PPh_3)_2$ are unstable in solution at room temperature, slowly decomposing to 5 by either elimination of RCl (R = methyl, aryl, benzyl) or of olefin and hydrogen chloride (when $C_{(2)}$ of R is bonded to hydrogen).

$$RhRCl_2CO(PPh_3)_2 \longrightarrow 5 + RCl \qquad (11)$$

$$Rh(CH_2CH_2R')Cl_2CO(PPh_3)_2 \longrightarrow 5 + R'CH=CH_2 + HCl \qquad (12)$$

TABLE 5. Equilibrium constants for the
reactions $RhCl_2(COR)(PPh_3)_2 \rightleftharpoons RhRCl_2CO$-$(PPh_3)_2$

R	K	References
Me	0.29	47, 48
Ph	>20	47, 48, 57
$R'CH_2CH_2$	<0.1	47, 48, 54
$p\text{-}ClC_6H_4CH_2$	0.07	57
$ClCH_2$	>20	48

The reactions exemplified by equation (10) are the reverse of the oxidative addition reactions discussed in Section II, and are referred to as reductive elimination reactions. Unfortunately little is known at present of the mechanisms of reductive elimination reactions although, invoking the principle of microscopic reversibility, they presumably proceed via the same reaction paths as do oxidative addition reactions. The latter have been shown to involve a number of different mechanisms, depending in large part on the nature of the alkyl (aryl) group[59], and thus it seems unlikely that a single process is involved in the reactions under consideration here.

In agreement with this suggestion, Stille and Regan[57] have shown that the elimination of *para*-substituted aryl and benzyl halides from **11** and **12**, respectively, must proceed by quite different mechanisms. The rates of the former are

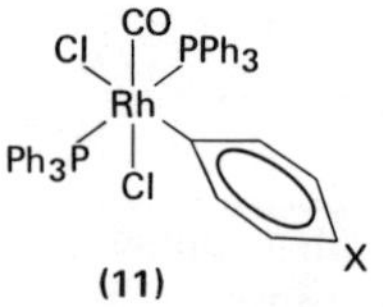

(11) 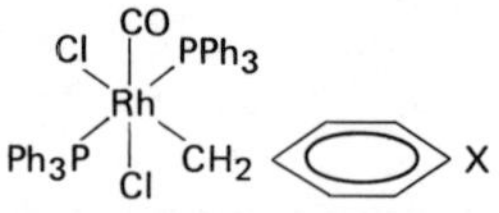   **(12)**

inhibited by electron-withdrawing substituents and free triphenylphosphine, while the rates of the latter are enhanced by electron-withdrawing substituents but are unaffected by free triphenylphosphine. It was suggested that elimination of aryl halides, at least, involves preliminary dissociation of one triphenylphosphine, a hypothesis supported by a positive entropy of activation[57].

The mechanism of equation (13) is probably better understood[54]. The reaction most likely proceeds in steps involving dissociation of a phosphine, migration of a hydrogen atom from $C_{(2)}$ of the alkyl group to the metal to give a rhodium complex containing coordinated hydrogen and olefin ligands (**13**), and then dissociation of olefin and hydrogen halide to give **5**.

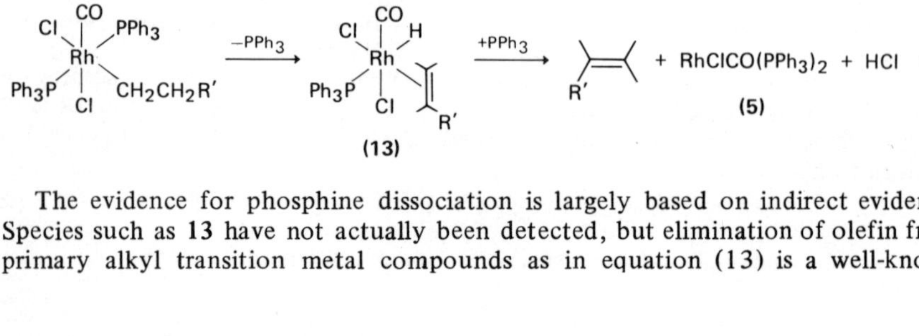

(13)

The evidence for phosphine dissociation is largely based on indirect evidence. Species such as **13** have not actually been detected, but elimination of olefin from primary alkyl transition metal compounds as in equation (13) is a well-known reaction[66]. Mechanistically, it probably involves a concerted, *cis*-1,3 or β-elimination *via* a four-centred intermediate (equation 14; see reference 54 for a detailed discussion of the evidence for this mechanism). The order in which 13 dissociates olefin, eliminates hydrogen chloride and recoordinates the triphenylphosphine to form 5 is not known.

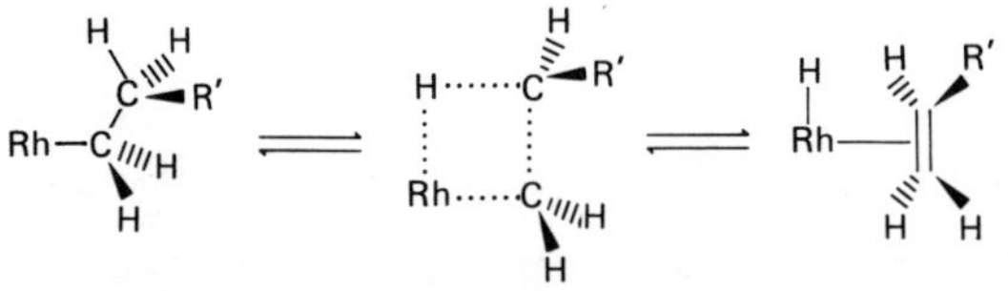

$$\tag{14}$$

The sequence of reactions shown in equations (8), (9) and either (11) or (12) adequately describes the important steps in the *stoichiometric* decarbonylation of acyl halides by 1. Although all the steps generally proceed at measurable rates in solution at room temperature, decarbonylation is not catalytic because the product (5) is inert under these conditions.

At higher temperatures, however, the decarbonylation reaction becomes catalytic. Either 5 becomes reactive, or it is not formed. Opinions on this matter vary in the literature, some authors[41] suggesting that oxidative addition of acyl halides to 5 becomes important over 200°C. Species such as 14 have not been detected

$$RCOCl + 5 \longrightarrow RhCl_2(COR)CO(PPh_3)_2 \tag{15}$$
$$(14)$$

$$RhCl_2(COR)CO(PPh_3)_2 \xrightarrow{-CO} RhCl_2(COR)(PPh_3)_2 \longrightarrow etc. \tag{16}$$
$$(7)$$

in the rhodium triphenylphosphine catalytic system, but are reasonable intermediates. Loss of CO from the formally rhodium(III) compounds to give the five-coordinated acyl compounds might also be expected, as π-bonding between the metal and the carbonyl group would be relatively weak[67]. Loss of carbon monoxide from probably very similar rhodium(III) compounds has been observed[32].

Other authors have suggested[42] that the five-coordinated acyl complexes decompose in a different manner at higher temperatures (equation 17). Evidence for

$$RhCl_2(COR)(PPh_3)_2 \xrightarrow{-CO} RhRCl_2(PPh_3)_2 \tag{17}$$
$$(7) \qquad\qquad (15)$$

$$15 \longrightarrow RCl + RhCl(PPh_3)_2 \longrightarrow etc. \tag{18}$$
$$(3)$$

such a path comes from the observations that 1 has been observed to apparently be more catalytically efficient than 5 on some occasions[42], less so on others.

Comparative studies of the decarbonylation of benzoyl bromide at 200°C by 1, 5 and 5 prepared *in situ* show definite rate differences, which are attributable to the catalyst and not to the organic substrate[44]. Interestingly, while compounds such as 15 have not been satisfactorily characterized, there have been two reports that 1 reacts on occasion with aroyl chlorides to give compounds containing no carbonyl stretching bands in their infrared spectra[41,42]. In addition, there are suggestions[42] that the *cis* isomer of 5 may be involved.

It is this author's opinion that the identity of the true catalyst is as yet not known, and that it may actually be different for different acyl halides. The work of Ströhmeier and Pföhler[44], which shows that there is a definite, narrow optimum temperature range for the decarbonylation of benzoyl bromide, has not been repeated for any other acyl halide. Thus published comparative studies may be meaningless in terms of elucidating mechanisms.

C. Summary

With respect to the utility of 1 and 5 for the decarbonylation of acyl halides, it is quite clear that stoichiometric reactions of aroyl and arylacetyl halides with 1 to form the corresponding aryl and benzyl halides can be induced under very mild conditions. In addition, the many examples listed in Tables 1 and 2 show that catalytic decarbonylations are also normally possible, a major limitation being the necessity for rapid separation of products as they form. Notable exceptions to this generalization are experiments no. 30 and 55 of Table 1. A major product in the former was the ketone 16, formed by an unknown route[42].

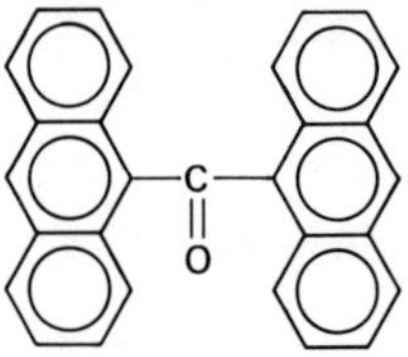

(16)

In the latter case (benzoyl cyanide) an intermediate such as 17 may be thermally stable at the temperature used. Thus the reaction would not be catalytic. Aroyl cyanides which are successfully decarbonylated by 1 were reacted at higher temperatures[42].

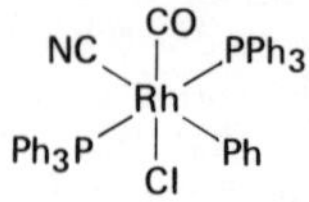

(17)

Decarbonylation reactions of aliphatic acyl halides (Table 2) generally give mixtures of terminal and internal olefins, and thus are not of general synthetic value unless the isomerization of the products can be inhibited. Fortunately, in some cases at least, the addition of at least a tenfold excess of free triphenylphosphine does inhibit product isomerization[51,52]. Possibly the terminal olefin is displaced from a species such as 13 before it can isomerize. In cases where the alkylrhodium compound such as that in equation (12) can eliminate olefin by the migration of more than one type hydrogen, some measure of selectivity is obtained by the fact that the decreasing order of rates of migration is tertiary > secondary > primary[51] (experiments no. 18–21). In addition, there are indications that highly substituted olefins can also be obtained without significant isomerization[51] (experiments no. 16 and 17), probably because of the greater tendency of such olefins to dissociate from the rhodium[54].

Solvent effects on decarbonylation reactions have not been reported, but a preference has been stated for benzonitrile[51].

IV. REACTIONS OF 1 WITH ALDEHYDES

A. Decarbonylation Reactions

Table 6 lists the products of the reactions of 1 with a variety of aldehydes. A recent patent[82] also describes the use of 1 to prepare furan, nonane, ethyl octanoate and 20-deoxyprogesterone 'from the corresponding aldehydes'. In most cases, the reactions below about 200°C are stoichiometric in accord with equation (19). Indeed, experiment no. 1 of Table 6 represents the method of choice for the preparation of 5[68].

$$1 + RCHO \longrightarrow RH + 5 \qquad (19)$$

In contrast to the decarbonylation of acyl halides, aliphatic aldehydes generally give no olefinic products, although small amounts of olefins have been reported in some cases[41,69,70] (experiment no. 5). In the decarbonylation of n-heptanal by the compounds $RhX(PPh_3)_3$ the ratio of hexene:hexane produced decreased in the order $X = Cl > Br > SnCl_3$[70].

In general, aryl and primary alkyl aldehydes are smoothly decarbonylated by 1 under relatively mild conditions, i.e. in solution at room temperature or in refluxing benzene. Decarbonylation of secondary aldehydes requires more forcing conditions, presumably because of steric hindrance, (experiments 14 and 30 do not proceed in benzene and methylene chloride), and both toluene and xylene have been utilized[41,74]. In some cases, however, the rates of decarbonylation are much slower in aromatic solvents than the rate of conversion of 1 to the rather insoluble bisphosphine dimer, 18 (equation 20). The latter does not react with aldehydes

$$2 \; 1 \xrightarrow[\substack{\text{aromatic} \\ \text{solvent}}]{\text{refluxing}} \underset{Ph_3P}{\overset{Ph_3P}{\diagdown}} \underset{Cl}{\overset{Cl}{Rh}} \underset{Cl}{\overset{}{Rh}} \underset{PPh_3}{\overset{PPh_3}{\diagup}} + 2 \; PPh_3 \qquad (20)$$

(18)

under these conditions, and thus the decarbonylation reaction stops. Tsuji and Ohno[41,74] have therefore recommended nitriles as solvents, especially benzonitrile. The latter can presumably coordinate to the rhodium as in 19 to prevent the formation of 18; the boiling point of benzonitrile is also suitably high, and an

$$\underset{Ph_3P}{\overset{Cl}{\diagdown}} \underset{}{\overset{}{Rh}} \underset{PPh_3}{\overset{NCPh}{\diagup}}$$

(19)

inspection of Table 6 reveals that benzonitrile has indeed been used by a number of workers. In some cases, where the liberated triphenylphosphine from 1 would be difficult to separate from the hydrocarbon products, it is recommended that 18, which may readily be obtained as somewhat air-sensitive, reddish crystals by refluxing 1 in aromatic solvents, be dissolved in benzonitrile to give the reactive 19[41,74].

Experiments 8–13 show that the stoichiometric decarbonylation of tertiary aldehydes is often highly stereospecific (compare these results with those of experiment no. 4 in Table 2), in contrast with aldehyde decarbonylations by other methods[72,73]. Experiment no. 10 indicates the value of 1 for introducing deuterium into an alkane stereospecifically; similar experiments with $PdCl_2$ as decarbonylation catalyst gave a 90% yield of 1,1-diphenyl-1-butene, a product of ring-opening[73]. The cyclopropyl products are reasonably configurationally stable under the reaction conditions, although they react further with 5 above 230°C[73].

TABLE 6. Decarbonylation reactions of aldehydes

No.	Aldehyde	Product(s)	Solvent, temperature (°C)	Yield	References
1	HCHO	H$_2$ (?)	Aqueous ethanol	—	68
2	CH$_3$CH$_2$CH$_2$CHO	C$_3$H$_8$	C$_6$H$_6$, 20	—	36, 41
3	Me$_2$CHCHO	C$_3$H$_8$	C$_6$H$_6$, 20, 80	—	36, 41
4	CH$_3$(CH$_2$)$_3$CHO	n-C$_4$H$_{10}$ (6)	CH$_2$Cl$_2$, 20	—	37, 69
5	CH$_3$(CH$_2$)$_5$CHO	n-C$_6$H$_{14}$ (6) 1-hexene (1)	CH$_2$Cl$_2$, 20	—	41, 70
6	PhCH$_2$CH$_2$CHO	PhCH$_2$CH$_3$	C$_6$H$_6$, 20, 80	67	36, 41
7	PhMe$_2$CCH$_2$CHO	Me$_3$CPh	—	—	71
8	(−)-(R)-PhMeEtCCHO	(+)-(S)-PhMeEtCH (81% retention)	PhCN, 160	51	72, 73
9	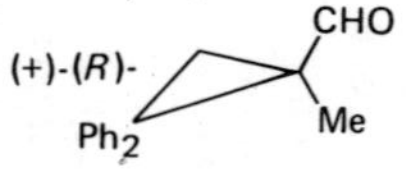 (+)-(R)-	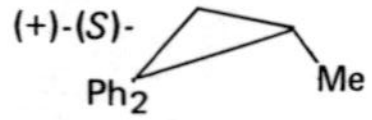 (+)-(S)- (94% retention)	Xylene, 140	70	72, 73
10	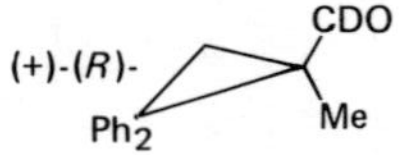 (+)-(R)-	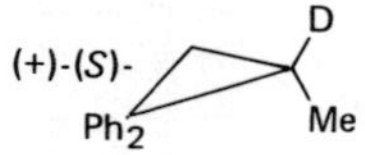 (+)-(S)-	Xylene, 140	—	72, 73
11	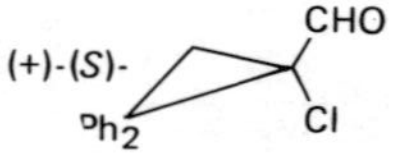 (+)-(S)-	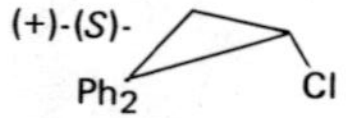 (+)-(S)- (83% retention)	PhCN, 160	40	73
12	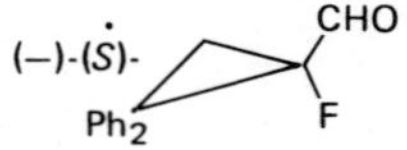 (−)-(S)-	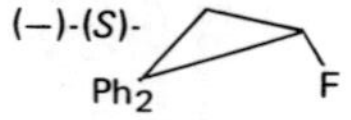 (−)-(S)- (73% retention)	PhCN, 160	62	73
13	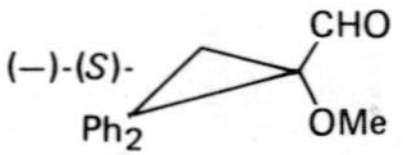 (−)-(S)-	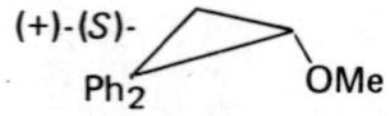 (+)-(S)- (6% retention)	PhCN, 160	84	73

No.	Aldehyde	Product	Conditions	Yield (%)	References
14	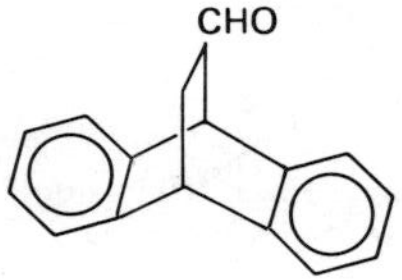 (dibenzobicyclic)–CHO	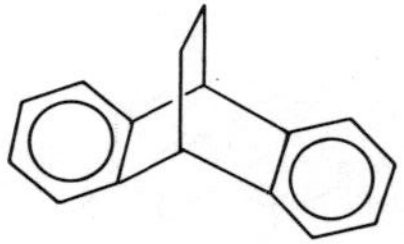 (dibenzobicyclic)	PhCN, 160	67	41, 74
15	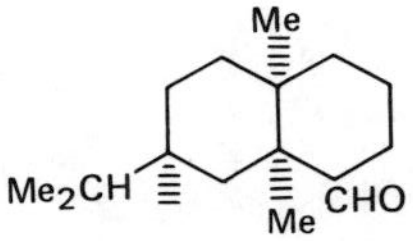 Me₂CH–, Me, Me–CHO decalin	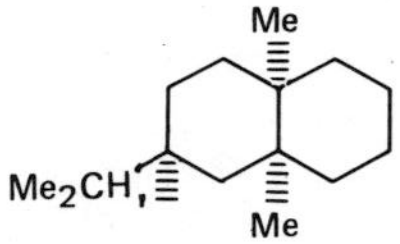 Me₂CH–, Me, Me decalin	C₆H₆, 80	43	75
16	PhCHO	C₆H₆	Neat, 20, 120, 179	—	36, 41
		C₆H₆	Toluene, 111	83	41
17	4-ClC₆H₄CHO	PhCl	C₆H₆, 80	85	36, 41
		PhCl	Neat, 220	71	74
18	2-HOC₆H₄CHO	PhOH	Toluene, 110	70	36, 41
		PhOH	Neat, 210	80	74
19	C₆F₅CHO	C₆F₅H	CH₂Cl₂, 20	—	69
20	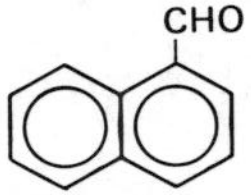 (naphthyl)–CHO	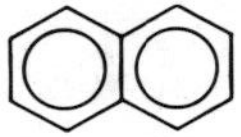 (naphthalene)	Neat (?)	88	42
21	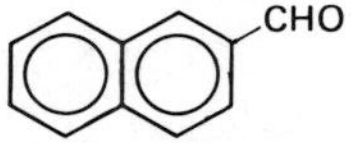 (naphthyl)–CHO	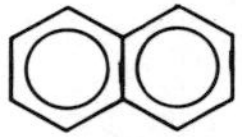 PhCH=CH₂	Neat (?)	77	42
22	trans-PhCH=CHCHO	PhCH=CH₂	C₆H₆, 20, 80	77	36, 41
			CH₂Cl₂, 20	60	36, 41
			Neat, 230–240	76	74
	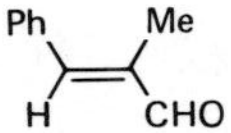	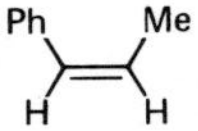	Toluene, 110	88	41, 74
			PhCN, 160	86	41, 74
23	Ph, Me / CHO alkene	Ph, Me / H alkene	PhCN, 160	82	74
24	Ph, Et / CHO alkene	Ph, Et / H alkene	C₆H₆, 80	—	72, 73
			Neat, 128/30 mm	—	72, 73

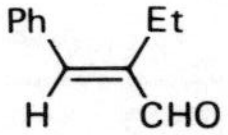

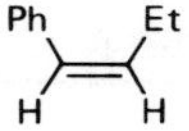

TABLE 6. (Continued)

No.	Aldehyde	Product(s)	Solvent, temperature (°C)	Yield	References
25	Me$_2$C=CHCH$_2$CH$_2$CMe=CHCHO	—	C$_6$H$_6$, 20	—	76
26	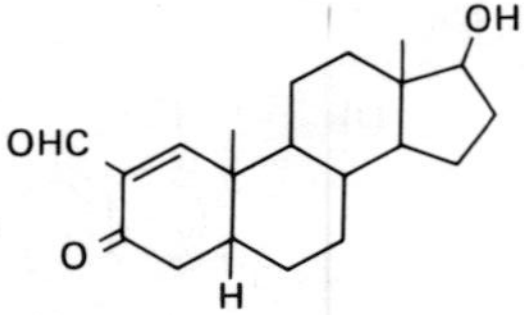 (steroid, OHC)	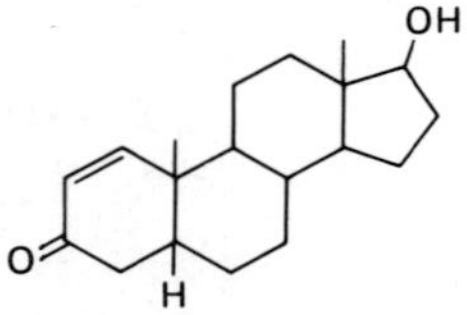 (steroid, OH)	C$_6$H$_6$, 80	68	77
27	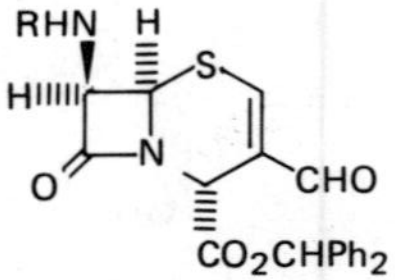 (β-lactam, RHN, CHO, CO$_2$CHPh$_2$) R = PhCH$_2$CO—, [D]-PhCHNH$_2$CO—	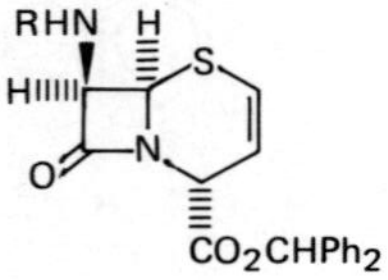 (penem, RHN, CO$_2$CHPh$_2$) R = PhCH$_2$CO—, [D]-PhCHNH$_2$CO—	C$_6$H$_6$, 80	19	78
28	(β-lactam, RHN, CHO, CO$_2$CHPh$_2$) R = PhCH$_2$CO—, [D]-PhCHNH$_2$CO—	(penem, RHN, CO$_2$CHPh$_2$)	C$_6$H$_6$, 80	84	78
29	(bicyclic, Ph, H, OTAF^a, CHO)	(bicyclic, Ph, H, OTAF, H)	—	79	79

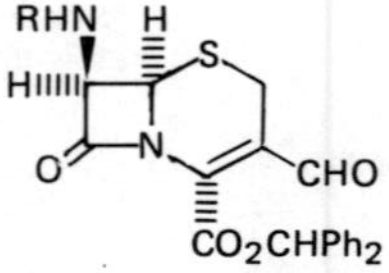

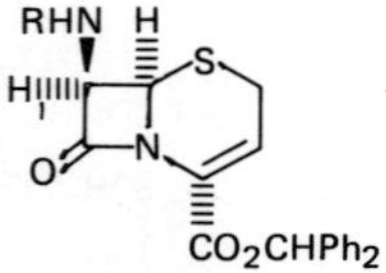

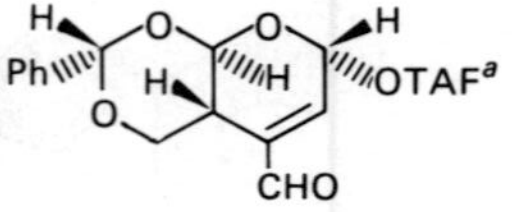

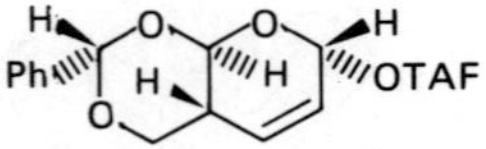

30   PhCN, 160   71   41, 74

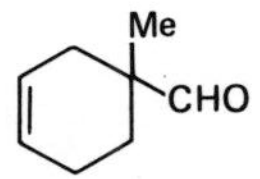

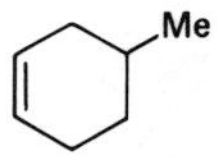

31   C_6H_6, 80   80   81

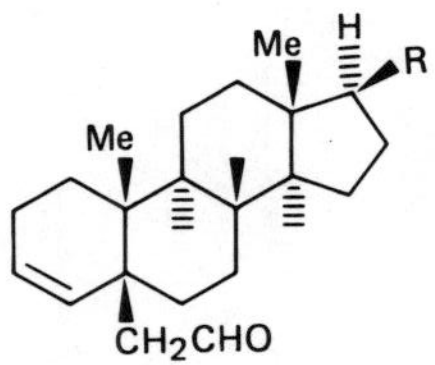

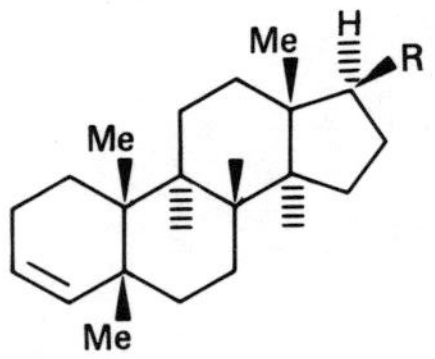

R = CHMe(CH$_2$)$_3$CHMe$_2$

32   C_6H_6 —   80

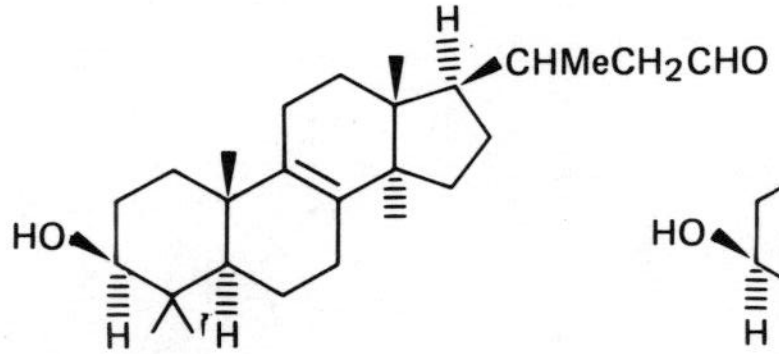

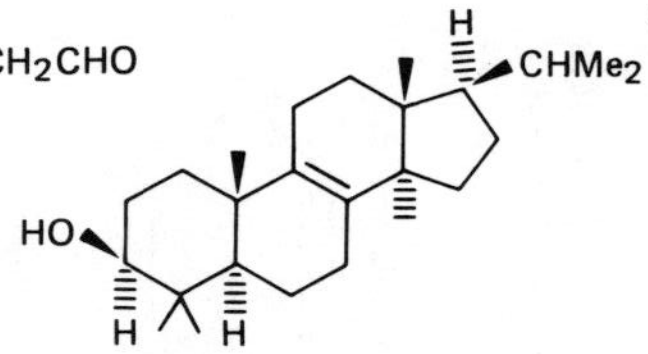

^{a}TAF = tetraacetylfructofuranose.

Experiments 22–24 and 26–29 show that vinylic aldehydes are also decarbonylated stereospecifically. Although the earlier work[41,74] suggested that some isomerization of the products to the *trans* isomers occurred, more recent work[72,73] has shown that the decarbonylation is stereospecific but that the products are isomerized by **5** if the reaction conditions are too drastic[73]. Again, decarbonylation by $PdCl_2$ is much less stereospecific and can lead to undesirable side-reactions[74].

As with acyl halides (Section III.A), decarbonylation reactions of aldehydes become catalytic above about 200°C, leading to the suggestion that **5** can also be used as a catalyst[41,74]. Offsetting the advantages of a catalytic system, however, are the above-mentioned possible losses of stereospecificity and, in the case of aliphatic aldehydes, some undefined aldol condensation side-reactions[41]. The utilization of **1** and **5** as decarbonylation catalysts has really been only cursorily investigated.

Experiments 15 and 26–32 provide interesting examples of the value of **1** in organic syntheses. Experiment no. 15 provides a novel method for the introduction of an angular methyl group into bicyclic compounds, while experiments 26, 31 and 32 illustrate uses of **1** in steroid syntheses. The reasons for the low yield in the decarbonylation of the 3-formylcephem compound in no. 27 are not known; unreacted starting materials were isolated. More forcing conditions (refluxing toluene) resulted in partial isomerization of the product to the product of no. 28, possibly catalysed by **5**[28].

Experiment no. 29 illustrates a useful application of **1** to carbohydrate chemistry[79]. The sugars **20** and **21** are unaffected by **1** in a refluxing mixture of ethanol–benzene–water $(7:3:1)$[83] suggesting that the linear isomers of these

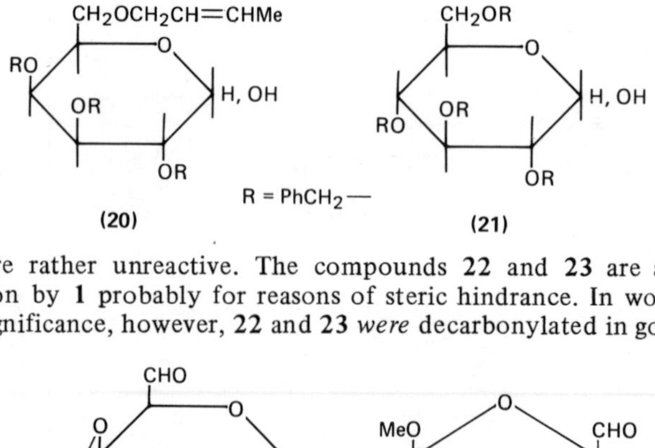

$R = PhCH_2—$

(20)   (21)

compounds are rather unreactive. The compounds **22** and **23** are also inert to decarbonylation by **1** probably for reasons of steric hindrance. In work of potentially major significance, however, **22** and **23** *were* decarbonylated in good yields by

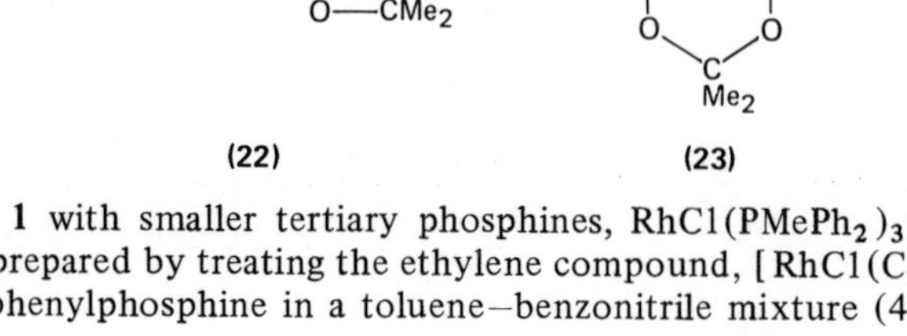

(22)   (23)

a derivative of **1** with smaller tertiary phosphines, $RhCl(PMePh_2)_3$ **(24)**[84]. Compound **24** was prepared by treating the ethylene compound, $[RhCl(C_2H_4)_2]_2$ **(25)** with methyldiphenylphosphine in a toluene–benzonitrile mixture (49:1), a stand-

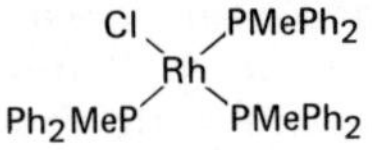

(24)

ard procedure for making such compounds involving substitution of both ethylenes and one bridging chloride on each rhodium atom. Compound **24** was not actually

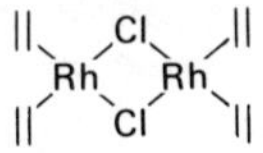

(25)

isolated, but was reacted *in situ* with **22** and **23**. The products, **26** and **27**, were obtained in about 35% yield.

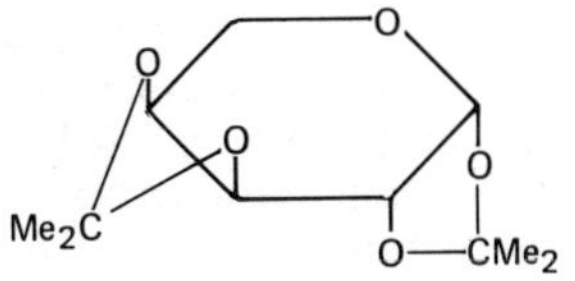

(26) 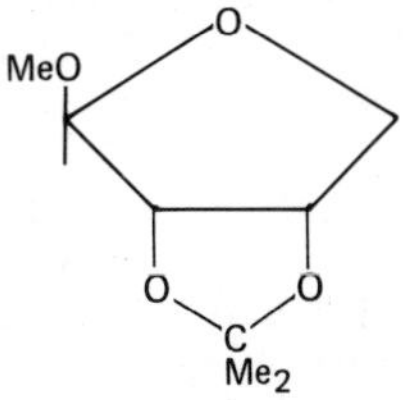   (27)

This work suggests very strongly that other, similar rhodium compounds should be investigated as decarbonylation reagents. Little has been done in this area, with the exception of very brief mention of the compounds $RhX(PPh_3)_3(X = Br^{70}$, $SnCl_3^{70}$, OAc^{85}), $RhCl(PF_2NMe_2)_3^{86}$ and $[RhClL_2]_2$ $(L = PF_2NMe_2, CO)^{86}$.

TABLE 7. Decarbonylation reactions of allylic alcohols

No.	Allylic alcohol	Product(s) (yields %)	Solvent, temperature	Reference
1	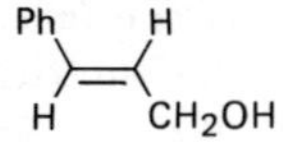	$PhCH_2CH_3$ (75) $PhCH=CH_2$ (4)	MeCN, PhCN, 150	88
2	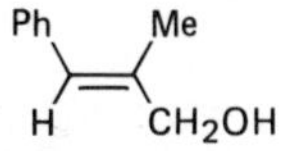	$PhCH_2CH_2Me$ (67) $PhCH=CHMe$ (20)	MeCN, PhCN, 150	88
3	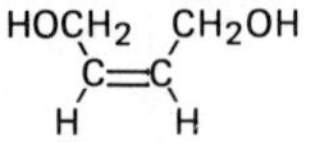	$CH_3CH_2CH_2OH$ (89)	MeCN, PhCN, 150	88
4	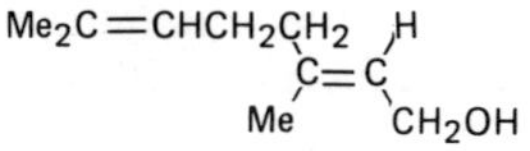	?	C_6H_6, 20	76
5	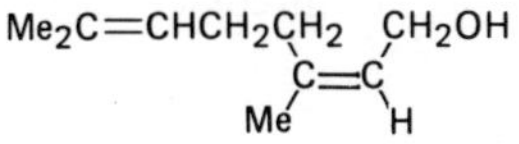	?	C_6H_6, 20	76

A ruthenium compound has also been shown to decarbonylate aldehydes[87], although its mode of action is not known. It appears to be much less specific than 1, and is thus unlikely to prove as useful as the rhodium system.

Besides the work, already discussed, on the decarbonylation of vinylic aldehydes, Table 6 contains four entries (nos. 26 and 30–32) in which the organic substrate contains an olefinic linkage sufficiently remote from the aldehyde group that it appears to play no role in the decarbonylation reaction. Not included, however, are a number of allylic alcohols, which are also decarbonylated by 1, probably after isomerization, catalysed by 1 (equation 21), to their aldehyde tautomer (Table 7).

$$RCH = CHCH_2OH \longrightarrow RCH_2CH = CHOH \longrightarrow RCH_2CH_2CHO \qquad (21)$$

A possible mechanism for the isomerization step will be discussed in Section IV.B. Decarbonylation of the aldehyde tautomer should be unexceptional. Similar experiments with (3-cyclohexenyl)methanol and 3-phenylpropyn-1-ol gave very low yields[88]. Entries 4 and 5 of Table 7 were carried out during attempts to hydrogenate the substrates using 1 as a hydrogenation catalyst. Failure of the hydrogenation was indicated by the formation of 5, the expected rhodium-containing product of decarbonylation, although the organic products do not appear to have been isolated or characterized. In the same study, it was shown that compound 28 can be hydrogenated, while it has been shown that olefinic aldehydes can generally be hydrogenated if suitable reaction conditions are employed[89].

$$Me_2C = CHCH_2CH_2CMe(OH)CH = CH_2$$

(28)

Normally aldehyde decarbonylation reactions by 1 proceed quite cleanly and, unless steric factors are important, in good yields. Recent work has shown that an important exception to this generalization is long-chain *flexible* aldehydes containing a distant olefinic linkage. In contrast to, for instance, experiments 30–32 of Table 6, where the olefinic groups of the organic molecules are held in positions remote from the aldehyde groups, the olefinic aldehydes of Table 8 are not decarbonylated in the presence of 1, but are rather cyclized to form cyclic ketones or similar compounds. Experiment 1 of Table 8 suggests that an aldehyde can condense with an olefin to some extent to form a ketone[90], but no reaction conditions are given and similar experiments with heptanal and ethylene failed to give a ketone[91]. Experiment 2, however, shows that 4-pentenal and substituted 1-al-4-enes can cyclize to form a series of cyclopentanones and, in some cases, substituted cyclopropanes. In all cases, decarbonylation was a minor process; yields of the ketones could be increased by carrying out the reactions in the presence of ethylene[91]. Experiment 3, with (+)-citronellal, shows that a 1-al-6-ene system will also cyclize, yielding in this case (+)-neoisopulegol and (−)-isopulegol[93]. The significance of the formation of cyclohexanol rather than cyclohexanone derivatives is not known, but the reactions represented by Table 8 probably represent a major exception to the type of chemistry under consideration in this article. Such cyclization reactions may, however, prove to be very synthetically useful in themselves; the work discussed in Reference 92. for instance, is concerned with synthese of prostaglandin derivatives.

Possible mechanisms for the cyclization reactions will be discussed in Section IV.B.

TABLE 8. Cyclization reactions of olefinic aldehydes

No.	Reactant(s)	Product(s)	Solvent, Temperature	Yield (%)	Reference
1	MeCHO + n-C$_5$H$_{11}$CH=CH$_2$	n-C$_7$H$_{15}$COMe		Low	90
2 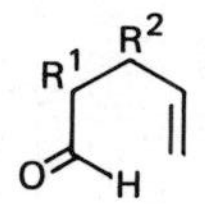 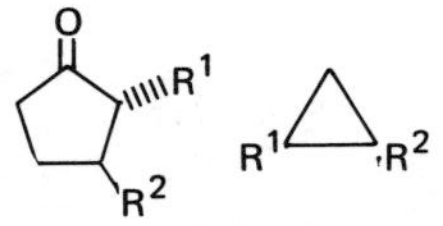					

	R^1	R^2			
(a)	H	H	CHCl$_3$, 20	72, 0	91
(b)	(CH$_2$)$_6$CO$_2$Me	(CH$_2$)$_7$Me	CHCl$_3$; C$_6$H$_6$, MeCN, 20	30, 30	92
(c)	(CH$_2$)$_3$Me	(CH$_2$)$_2$Me		29, 35	92
(d)	(CH$_2$)$_6$CO$_2$Me	Me		— —	92
(e)	(CH$_2$)$_6$CO$_2$Me	H		26, 23	92
(f)	(CH$_2$)$_6$CO$_2$Me	(CH$_2$)$_2$Me		34, 32	92
(g)	H	(CH$_2$)$_7$Me		30, 32	92
(h)	(CH$_2$)$_6$CO$_2$Me	CH$_2$OMe		17, 20	92
(i)	(CH$_2$)$_3$Me	(CH$_2$)$_7$Me		28, 33	92

| 3 | 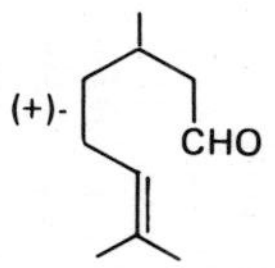 (+)- | 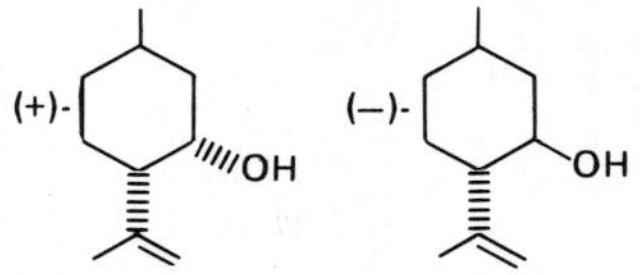 (+)-   (−)- | CHCl$_3$, 20 | 41, 14 | 93 |

B. Mechanism of Aldehyde Decarbonylation Reactions

As mentioned in Section II, unstable intermediates are not as readily detected in aldehyde decarbonylation reactions as in acyl halide decarbonylation reactions and thus much less is known of the former. In general, however, opinion[41,69] favours a very similar sequence of steps in both cases, i.e.:

$$
1 + RCHO \longrightarrow
\begin{array}{c}
COR \\
H \diagdown \;\;|\;\; \diagup PPh_3 \\
Rh \\
Ph_3P \diagup \;\; \diagdown Cl
\end{array}
\qquad (22)
$$

$$(29)$$

$$
29 \longrightarrow
\begin{array}{c}
CO \\
R \diagdown \;\;|\;\; \diagup PPh_3 \\
Rh \\
Ph_3P \diagup \;\; | \;\; \diagdown Cl \\
H
\end{array}
\longrightarrow 5 + RH
\qquad (23)
$$

$$(30)$$

The reaction represented in equation (22) would involve oxidative addition of the RCO—H group to rhodium (I), a reaction with few precedents, but which is regarded as reasonable in view of the apparently great similarity in acyl halide and aldehyde decarbonylation reactions by 1. Species such as 29 and 30 have not been detected spectroscopically[69], but kinetic studies are consistent with equation (22) being the rate-determining step[69], and the complexes $FeH(COR)(Me_2PCH_2CH_2P-Me_2)_2$ (R = Ph, Et)[94] are presumably formed by oxidative addition of aldehyde to an iron(O) complex generated by dissociation of naphthalene from $FeH(2-naphthyl)(Me_2PCH_2CH_2PMe_2)_2$[95].

Isomerization of 29 to 30 would be unexceptional and, by analogy with many other such alkyl migration reactions, should proceed with retention of configuration of the group R (Section III.B). Alklhydrido compounds such as 30 are generally thermally quite unstable, and reductive elimination of alkane from 30 to form 5 would be expected[66]. While the stereochemistry of such a step has not yet been investigated, it is the author's opinion that a concerted process occurring with retention of configuration would not overly disturb or surprise most organometallic chemists.

On this basis, the observation of retention of configuration during the decarbonylation of a number of aliphatic and vinylic aldehydes (Table 6) is quite consistent with a series of concerted steps as in equations (22) and (23), and the proposal[73] of a radical cage mechanism does not seem necessary.

The elimination of olefins, reported to be a side-reaction in several of the examples in Table 6, could well occur from an intermediate such as 30. Indeed, the analogy with acyl chlorides suggests that olefin elimination should be an important process (Section III.A.3). The difference may lie in the relative rates of reductive elimination and olefin β-elimination from 10 (R = alkyl group containing a hydrogen on $C_{(2)}$) and 30. In the case of 10, elimination of RCl would appear to be relatively slow compared with elimination of olefin. In the case of 30, in contrast, elimination of RH must be the kinetically preferred process.

The mechanism described in equations (22) and (23) need only be modified slightly to accommodate the results in Table 8. As discussed in Section II, substitution of one of the labile phosphines of 1 by ethylene (equation 1, L = C_2H_4) or terminal olefin (equation 24) occurs readily. Compound **32** could then undergo

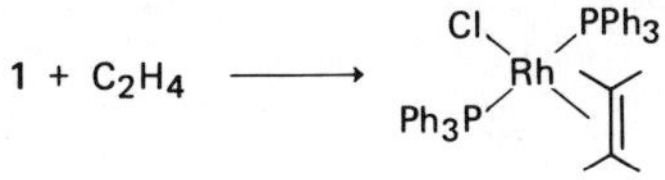

$$1 + C_2H_4 \longrightarrow \underset{(31)}{\text{(structure 31)}} \qquad (1)$$

$$1 + \underset{(32)}{\text{(structure)}} \longrightarrow \underset{(32)}{\text{(structure 32)}} \qquad (24)$$

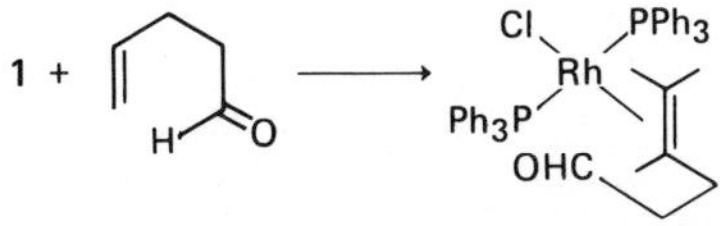

intramolecular nucleophilic attack by the rhodium(I) on the aldehyde carbonyl group (Figure 2) to forms a species **33** related to **29** (equation 25). Hydrogen

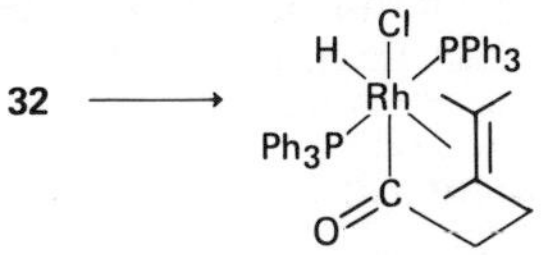

$$32 \longrightarrow \underset{(33)}{\text{(structure 33)}} \qquad (25)$$

migration from rhodium to $C_{(2)}$ of the olefin would generate a cyclic intermediate, **34**, which could in turn undergo reductive elimination of a cyclic ketone (equation 26).

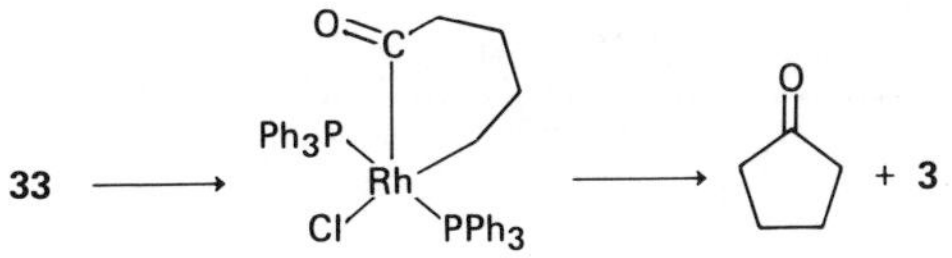

$$33 \longrightarrow \underset{(34)}{\text{(structure 34)}} \longrightarrow \text{(cyclopentanone)} + 3 \qquad (26)$$

This mechanism, proposed by Lochow and Miller[91], suggests that cyclization should actually be catalytic, as the three-coordinated $RhCl(PPh_3)_2$ (3) could readily interact further with another molecule of organic substrate. Compound **31** has been reported[27], and oxidative addition of an aldehyde to a species such as **31** could explain the results of experiment 1 of Table 8. The fact that free ethylene appears to increase the yield of cyclopentanone in experiment 2 suggests that free aldehyde does not readily add to **31**. As formation of **31** would be rapid[27], it seems much more likely that the 4-pentenal interacts initially with the rhodium by displacement of the ethylene and coordination of the olefinic group. The aldehyde group would then be in a position to interact intramolecularly with the rhodium to form **33**, an hypothesis which suggests that the formation of cyclic ketones should be very sensitive to the length and flexibility of the hydrocarbon chain of the enal substrate. Experiment 3 of Table 8 may proceed similarly, the cyclohexanols being produced by isomerization of a cyclohexanone intermediate.

The formation of cyclopropane derivatives (experiment 2) would be consistent with collapse of **33** by hydrogen migration to $C_{(1)}$ of the olefin to yield **35**, followed by migration of $C_{(4)}$ from the acyl carbon to rhodium, as in equations (9) and (27). The final step would be reductive elimination of cyclopropane, as in equation (28).

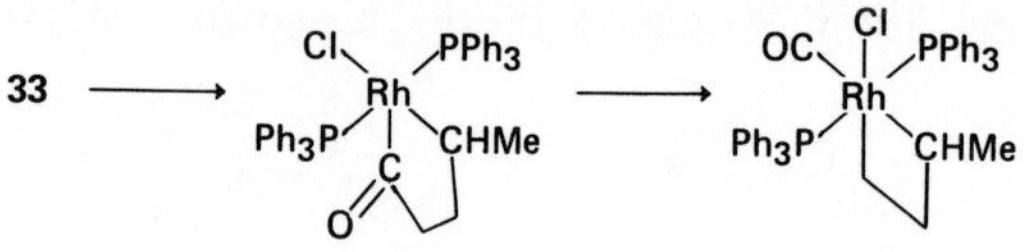

$$33 \longrightarrow \underset{(35)}{\text{[structure: Cl, PPh}_3\text{, Rh, Ph}_3\text{P, C=O, CHMe]}} \longrightarrow \underset{(36)}{\text{[structure: OC, Cl, PPh}_3\text{, Rh, Ph}_3\text{P, CHMe]}} \quad (27)$$

$$36 \longrightarrow \triangleright\!\!-\!\text{Me} + 5 \qquad\qquad (28)$$

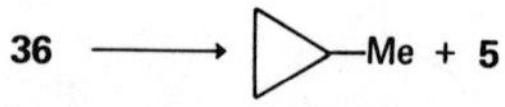

As mentioned in Section III.A, decarbonylation of the allylic alcohols in Table 7 probably involves isomerization of the substrates to their corresponding aldehyde tautomers, followed by a normal decarbonylation sequence. Olefin isomerization by transition metal compounds is well known; a reasonable mechanism in the present situation would involve coordination of the olefin to the metal followed by hydrogen transfer to give a π-allylic intermediate[96].

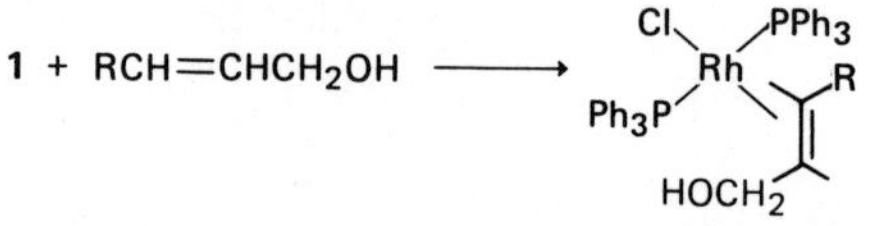

$$1 + \text{RCH}=\text{CHCH}_2\text{OH} \longrightarrow \underset{(37)}{\text{[structure: Cl, PPh}_3\text{, Rh, Ph}_3\text{P, R, HOCH}_2\text{]}} \quad (29)$$

$$37 \longrightarrow \underset{(38)}{\text{[structure: Cl, PPh}_3\text{, Rh, H, Ph}_3\text{P, R, H, OH]}} \longrightarrow \underset{(39)}{\text{[structure: Cl, PPh}_3\text{, Rh, Ph}_3\text{P, CH}_2\text{R, OH]}} \quad (30)$$

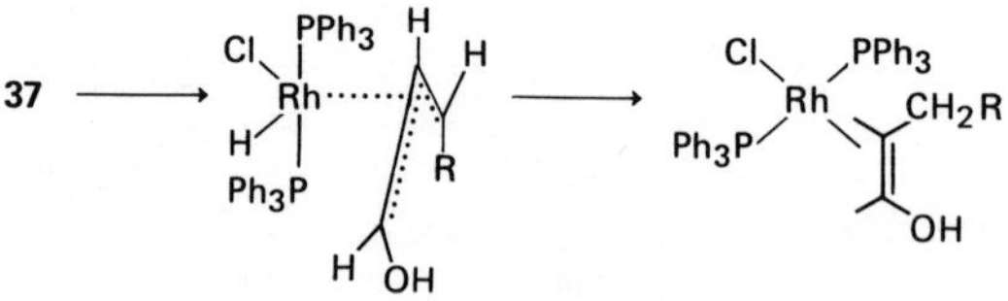

$$39 \longrightarrow \text{RhCl(PPh}_3)_2 + \text{RCH}_2\text{CH}=\text{CHOH} \longrightarrow \text{RCH}_2\text{CH}_2\text{CHO} \longrightarrow \text{etc.} \quad (31)$$

Vinyl alcohol complexes such as 39 have been previously characterized[97]. Dissociation of the vinyl alcohol would lead to its tautomerization to the aldehyde, which would undergo decarbonylation. The scheme outlined in equations (29)–(31) is also consistent with a deuteration study[88] in which 40 was decarbonylated to 41.

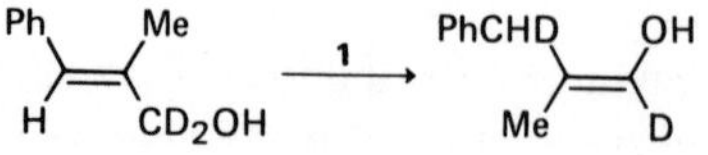

$$\underset{(40)}{\text{[structure: Ph, Me, H, CD}_2\text{OH]}} \xrightarrow{\;1\;} \text{[structure: PhCHD, OH, Me, D]} \quad (32)$$

$$\underset{}{\text{[structure: PhCHD, OH, Me, D]}} \longrightarrow \text{PhCHDCHMeCDO} \xrightarrow{+5} \underset{(41)}{\text{PhCHDCHDMe}} \quad (33)$$

C. Summary

The compound $RhCl(PPh_3)_3$ (1) provides a very convenient and mild reagent for the decarbonylation of aldehydes. Reactions are often stereospecific if conditions are not too forcing, and catalytic if the products are sufficiently robust to withstand temperatures in excess of $200°C$. The use of compounds similar to 1 but with smaller phosphines can be expected to increase the number of organic compounds susceptible to decarbonylation, but substrates containing olefinic groups will undoubtedly lead to interesting side-reactions.

V. REFERENCES

1. B. J. Wakefield, *The Chemistry of Organolithium Compounds*, Pergamon Press, Oxford, 1974; U. Schölkopf in *Methoden der organischen Chemie* (Ed. O. Bayer, E. Müller and K. Ziegler) Vol. XIII/1, 1970, p. 86.
2. K. Nützel in *Methoden der organischen Chemie* (Ed. O. Bayer, E. Müller and K. Ziegler) Vol. XIII/2a, 1973, p. 46.
3. K. Nützel in *Methoden der organischen Chemie* (Ed. O. Bayer, E. Müller and K. Ziegler), Vol. XIII/2a 1973, p. 552; J. Furukawa and N. Kawabata, *Adv. Organometal. Chem.*, 12, 83 (1974).
4. E. C. Taylor and A. McKillop, *Acc. Chem. Res.*, 3, 338 (1970); *Chemistry in Britain*, 9, 4 (1973).
5. H. C. Brown, *Organic Syntheses via Boranes*, John Wiley and Sons, New York, 1975; *Boranes in Organic Chemistry*, Cornell University Press, Ithaca, New York, 1972.
6. H. Reinheckel, K. Haage and D. Jahnke, *Organometal. Chem Rev. A*, 4, 47 (1969); H. Lehmkuhl, K. Ziegler and H. G. Gellert in *Methoden der organischen Chemie* (Ed. E. Müller, O. Bayer and K. Ziegler), Vol XIII/4, 1970, p. 1.
7. See, for instance, P. N. Rylander, *Catalytic Hydrogenation over Platinum Metals*, Academic Press, New York, 1967.
8. In-depth studies of the possible uses of nickel and palladium compounds, respectively, are presented by: P. W. Jolly and G. Wilke, *The Organic Chemistry of Nickel*, Vols I and II, Academic Press, New York, 1975; P. M. Maitlis, *The Organic Chemistry Of Palladium*, Vols. I and II, Academic Press, New York, 1971.
9. G. N. Schrauzer, (Ed.), *Transition Metals in Homogeneous Catalysis*, Marcel Dekker, New York, 1971; R. Ugo (Ed.), *Aspects of Homogeneous Catalysis*, Vol. I, Carlo Manfredi, Milan, 1970; Vol. II, Reidel Publishing Co., Dordrecht, Holland, 1974.
10. M. L. H. Green, *Organometallic Compounds, The Transition Elements*, Chapman and Hall, London, 1968.
11. R. F. Heck, *Organotransition Metal Chemistry*, Academic Press, New York, 1974.
12. P. M. Maitlis and K. W. Eberius in *Nonbenzenoid Aromatics*, Vol. II, (Ed. J. P. Snyder), Academic Press, New York, 1971, p. 360.
13. M. F. Semmelhack and H. T. Hall, *J. Amer. Chem. Soc.*, 96, 7091 (1974); M. F. Semmelhack, M. Yoshifuji, H. T. Hall and G. Clark, *J. Amer. Chem. Soc.*, 97, 1247 (1975); M. F. Semmelhack, H. T. Hall and M. Yoshifuji, *J. Amer. Chem. Soc.*, 98, 6387 (1976).
14. D. H. R. Barton and H. Patin, *J. Chem. Soc., Perkin I*, 829 (1976).
15. B. R. James, *Homogeneous Hydrogenation*, John Wiley and Sons, New York, 1973.
16. See, for instance, J. T. D. Cross and V. R. Stimson, *J. Chem. Soc.(B)*, 88 (1967). *Australian J. Chem.*, 21, 687, 701, 713, 725, 1711 (1968); D. A. Kairaitis and V. R. Stimson, *Australian J. Chem.*, 23, 1149 (1970).
17. See M. H. Palmer and G. J. McVie, *Tetrahedron Letters*, 6405 (1966); E. Rothstein and F. Vallely, *J. Chem. Soc. Perkin I*, 443 (1974), and references therein.
18. W. M. Schubert and R. R. Kintner in *The Chemistry of the Carbonyl Group* (Ed. S. Patai), John Wiley and Sons, London, 1966, p. 695.
19. S. Patai (Ed.), *The Chemistry of Acyl Halides*, John Wiley and Sons, London, 1972.
20. J. Tsuji and K. Ohno, *Synthesis*, 1, 157 (1969).

21. J. F. Young, J. A. Osborn, F. H. Jardine and G. Wilkinson, *Chem. Commun.*, 131 (1965).
22. M. A. Bennett and P. A. Longstaff, *Chem. Ind. (Lond.)*, 846 (1965).
23. A. Sacco, R. Ugo and A. Moles, *J. Chem. Soc. (A)*, 1670 (1966).
24. L. Vaska and R. E. Rhodes, *J. Amer. Chem. Soc.*, 87, 4970 (1965).
25. J. P. Candlin and A. R. Oldham, *Disc. Faraday Soc.*, 46, 60 (1968).
26. Reference 15, pp. 229 and 230.
27. J. A. Osborn, F. H. Jardine, J. F. Young and G. Wilkinson, *J. Chem. Soc.* (*A*), 1711 (1966).
28. Estimates based on 1976 prices.
29. *Inorganic Syntheses*, 7, 214 (1965); 8, 220 (1966).
30. P. B. Hitchcock, M. McPartlin and R. Mason, *Chem. Commun.*, 1367 (1969).
31. C. A. Tolman, P. Z. Meakin, D. L. Lindner and J. P. Jesson, *J. Amer. Chem. Soc.*, 96, 2762 (1974), and references therein.
32. M. C. Baird, J. T. Mague, J. A. Osborn and G. Wilkinson, *J. Chem. Soc.* (*A*), 1347 (1967).
33. A. Sacco, R. Ugo and A. Moles, *J. Chem. Soc.* (*A*), 1670 (1966).
34. D. N. Lawson, J. A. Osborn and G. Wilkinson, *J. Chem. Soc. (A)*, 1733 (1966).
35. P. G. H. Troughton and A. C. Skapski, *Chem. Commun.*, 575 (1968).
36. J. Tsuji and K. Ohno, *Tetrahedron Letters*, 3969 (1965).
37. M. C. Baird, D. N. Lawson, J. T. Mague, J. A. Osborn and G. Wilkinson, *Chem. Commun.*, 129 (1966).
38. J. Blum, *Tetrahedron Letters*, 1605 (1966).
39. G. A. Olah and P. Kreienbühl, *J. Org. Chem.*, 32, 1614 (1967).
40. J. Tsuji and K. Ohno, *Tetrahedron Letters*, 4713 (1966).
41. K. Ohno and J. Tsuji, *J. Amer. Chem. Soc.*, 90, 99 (1968).
42. J. Blum, E. Oppenheimer and E. D. Bergmann, *J. Amer. Chem. Soc.*, 89, 2338 (1967).
43. J. Blum, H. Rosenman and E. D. Bergmann, *J. Org. Chem.*, 33, 1928 (1968).
44. W. Ströhmeier and P. Pföhler, *J. Organometal. Chem.*, 108, 393 (1976).
45. W. P. Griffith, *The Chemistry of the Rarer Platinum Metals*, Interscience, New York, 1967, Chap. 6.
46. M. C. Baird, D. N. Lawson, J. T. Mague, J. A. Osborn and G. Wilkinson, *Chem. Commun.*, 129 (1966).
47. D. Egglestone and M. C. Baird, *J. Organometal. Chem.*, 113, C25 (1976).
48. D. Egglestone, M. C. Baird, C. J. L. Lock and G. Turner, *J. Chem. Soc., Dalton.*
49. J. Tsuji and K. Ohno, *J. Amer. Chem. Soc.*, 88, 3452 (1966).
50. J. K. Stille and R. W. Fries, *J. Amer. Chem. Soc.*, 96, 1514 (1974).
51. J. K. Stille, F. Huang and T. R. Regan, *J. Amer. Chem. Soc.*, 96, 1518 (1974).
52. J. Blum, S. Kraus and Y. Pickholtz, *J. Organometal. Chem.*, 33, 227 (1971).
53. M. C. Baird, *J. Mag. Res.*, 14, 117 (1974).
54. N. A. Dunham and M. C. Baird, *J. Chem. Soc. Dalton.* 774 (1975).
55. J. Tsuji and K. Ohno, *J. Amer. Chem. Soc.*, 90, 94 (1968).
56. J. Tsuji and K. Ohno, *Japanese Patent*, 6808, 442 (1968); *Chem. Abstr.*, 70, 11368 (1969).
57. J. K. Stille and M. T. Regan, *J. Amer. Chem. Soc.*, 96, 1508 (1974).
58. J. Halpern and C. S. Wong, *Chem. Commun.*, 629 (1973).
59. J. A. Osborn in *Organotransition-metal Chemistry*, (Ed. Y. Ishi and M. Tsutsui), Plenum Press. New York, 1975, p. 65.
60. For examples of orbital shapes of similar molecules, see W. L. Jorgensen and L. Salem, *The Organic Chemists Book of Orbitals*, Academic Press, New York, 1973.
61. A. Wojciki, *Adv. Organometal. Chem.*, 11, 87 (1973).
62. K. M. Nicholas and M. Rosenblum, *J. Amer. Chem. Soc.*, 95, 4449 (1973).
63. P. L. Bock, D. J. Boschetto, J. R. Rasmussen, J. P. Demers and G. M. Whitesides, *J. Amer. Chem. Soc.*, 96, 2814 (1974).
64. L. F. Hines and J. K. Stille, *J. Amer. Chem. Soc.*, 94, 485 (1972).
65. J. A. Labinger, D. W. Hart, W. E. Seibert and J. Schwartz, *J. Amer. Chem. Soc.*, 97, 3851 (1975).
66. M. C. Baird, *J. Organometal. Chem.*, 64, 289 (1974).

67. F. A. Cotton and G. Wilkinson, *Advanced Inorganic Chemistry*, 3rd ed. Wiley–Interscience, New York, 1972, p. 684.
68. D. Evans, J. A. Osborn and G. Wilkinson, *Inorg. Synth.*, **XI**, 99 (1968).
69. M. C. Baird, C. J. Nyman and G. Wilkinson, *J. Chem. Soc. (A)*, 348 (1968).
70. F. J. Huang, *Diss. Abstr. Int. B.*, 35, 3246 (1975).
71. R. W. Fries, *Diss. Abstr. Int. B.*, 31, 4285 (1974).
72. H. M. Walborsky and L. E. Allen, *Tetrahedron Letters*, 823 (1970).
73. H. M. Walborsky and L. E. Allen, *J. Amer. Chem. Soc.*, 93, 5465 (1971).
74. J. Tsuji and K. Ohno, *Tetrahedron Letters*, 2173 (1967).
75. D. J. Dawson and R. E. Ireland, *Tetrahedron Letters*, 1899 (1968).
76. A. J. Birch and K. A. M. Walker, *J. Chem. Soc. (C)*, 1894 (1966).
77. Y. Shimizu, H. Mitsuhashi and E. Caspi, *Tetrahedron Letters*, 4113 (1966).
78. H. Peter and H. Bickel, *Helv. Chim. Acta.*, 57, 2044 (1974).
79. D. E. Iley and B. Fraser-Reid, *J. Amer. Chem. Soc.*, 97, 2563 (1975).
80. R. J. Anderson, R. P. Hanzlik, K. B. Sharpless, E. E. vanTamelen and R. B. Clayton, *Chem. Commun.*, 53 (1969).
81. R. E. Ireland and G. Pfister, *Tetrahedron Letters*, 2145 (1969).
82. J. Tsuji and K. Ohno, *Japanese Patent*, 7032, 402 (1970); *Chem. Abstr.*, 74, 63593 (1971).
83. P. A. Gent and R. Gigg, *Chem. Commun*, 277 (1974).
84. D. J. Ward, W. A. Szarek and J. K. N. Jones, *Chem. Ind. (Lond.)*, 162 (1976).
85. G. Wilkinson, *British Patent*, 1,368,432 (1975); *Chem. Abstr.*, 82, 30930 (1975).
86. D. A. Clement and J. F. Nixon, *J. Chem. Soc., Dalton*, 195 (1973).
87. R. H. Prince and K. A. Raspin, *J. Chem. Soc. (A)*, 612 (1969).
88. A. Emergy, A. C. Oehlschlager and A. M. Unrau, *Tetrahedron Letters*, 4401 (1970).
89. F. H. Jardine and G. Wilkinson, *J. Chem. Soc. (C)*, 270 (1967).
90. I. S. Kolomnikov, M. B. Erman, V. P. Kukolev and M. E. Vol'pin, *Kinetics and Catalysis*, 13, 227 (1972).
91. C. F. Lochow and R. G. Miller, *J. Amer. Chem. Soc.*, 98, 1281 (1976).
92. K. Sakai, J. Ide, O. Oda and N. Nakamura, *Tetrahedron Letters*, 1287 (1972).
93. K. Sakai and O. Oda, *Tetrahedron Letters*, 4375 (1972).
94. C. A. Tolman, *Organometallic Gordon Conference*, Proctor Academy, August, 1976. The author thanks Dr. Tolman for permission to cite this work.
95. S. D. Ittel, C. A. Tolman, A. D. English and J. P. Jesson, *J. Amer. Chem. Soc.*, 98, 6073 (1976).
96. Reference 10, p. 317.
97. J. Hillis, J. Francis, M. Ori and M. Tsutsui, *J. Amer. Chem. Soc.*, 96, 4800 (1974).

CHAPTER **15**

Pyrolysis of acids and their derivatives

R. TAYLOR
University of Sussex, Falmer, Brighton, U.K.

I.	INTRODUCTION	860
II.	PYROLYSIS OF CARBOXYLIC ACIDS	860
	A. Acids with Unsaturation at the β-Carbon	860
	B. Saturated Acids	864
III.	PYROLYSIS OF ACID HALIDES	867
IV.	PYROLYSIS OF ACID AMIDES	867
	A. Primary Amides and Higher Amides lacking β-Hydrogen in the *N*-alkyl group	867
	B. Amides with *N*-alkyl Groups containing β-Hydrogen Atoms	870
V.	PYROLYSIS OF ACID ANHYDRIDES	871
	A. Anhydrides which possess α-Hydrogen Atoms	871
	B. Anhydrides which either lack β-Hydrogen Atoms and are α,β-Unsaturated or are Cyclic	873
VI.	PYROLYSIS OF ESTERS	874
	A. Esters without β-Hydrogen Atoms in the Alkyl Group	874
	B. Vinyl Esters	876
	C. Allyl Esters	877
	D. Esters containing Non-vinylic β-Hydrogen Atoms	880
	1. The cyclic nature of the elimination	881
	2. The *cis* nature of the elimination	881
	3. Direction of the elimination	884
	a. Statistical effects	884
	b. Thermodynamic stability of the products	885
	c. Electronic effects	890
	4. Isotope effects	898
	5. Neighbouring-group effects	900
	a. Steric acceleration	900
	b. Anchimeric assistance	901
	6. Rearrangements	901
	7. Summary of the mechanism	904
	8. Use of the reaction as a model for electrophilic aromatic substitution	904
	E. β-Hydroxy Esters	908
VII.	PYROLYSIS OF LACTONES	908
VIII.	PYROLYSIS OF LACTAMS	909
IX.	REFERENCES	909

I. INTRODUCTION

A review of the pyrolysis of carboxylic acids and their derivatives is timely for two reasons. Firstly, the advances in gas-phase kinetic technology have permitted accurate measurements of rates of elimination of high-boiling compounds. Such compounds are needed for analysis of kinetic data in terms of linear free-energy relationships, vital for unambiguous interpretation of the electronic effects of substituents and hence of charge distribution in transition states which, in turn, leads to the reaction mechanism. Not only has the information gained in this way now produced a very detailed picture of the mechanisms of some eliminations, but extension of the technique has produced a very important tool for determining the true electronic effects of groups, especially heterocyclic molecules, and free of solvent complications. One may anticipate considerable expansion in the use of gas-phase studies directed towards the latter general direction, thereby reducing the bias hitherto in favour of solution chemistry in physical-organic research.

Secondly, previous attempts to correlate the mechanisms of the eliminations of the various acid derivatives with each other have been either scant or (more often) non-existent. As a result, authors have tended to consider the data in isolation and overlook important analogies in mechanism. Attention is drawn to these in this critical review and also to reasons for the differences in reactivity of derivatives, possible new reactions, and areas where further research is needed; mechanisms are proposed, alternative to those in the literature which are improbable or impossible.

Two features dominate these pyrolyses. Firstly, fragmentation of the molecule takes place, two smaller molecules being most commonly produced. Secondly, the majority of these fragmentations involve concerted cyclic processes which are now recognized as being non-synchronous, i.e. in the transition state some polarization is produced within the cyclic structure. The most common cyclic transition state is six-membered, this being favoured by orbital symmetry considerations. At higher temperatures 'symmetry-forbidden' four-centre processes are also involved; these are in the event permitted because of the semiheterolytic nature of the transition state.

II. PYROLYSIS OF CARBOXYLIC ACIDS

A. Acids with Unsaturation at the β-Carbon

This subgroup includes β,γ-unsaturated acids and β-keto acids and they decompose by a unimolecular process to give carbon dioxide and an alkene, or aldehyde (or ketone), respectively. Arnold and coworkers proposed the mechanism (Equation 1) for alkenoic acids as a result of studies on 2,2-dimethylbut-3-enoic acid[1].

$$Me_2C \xrightarrow{\Delta} Me_2C{=}CHCH_3 + CO_2 \qquad (1)$$

This was much more reactive than 4,4-dimethylbut-2-enoic acid so that isomerization of α,β-unsaturated acids to β,γ-unsaturated acids (suggested as a mechanism for elimination from the former) is evidently slower than the elimination. This mechanism is supported by the unreactivity of styrylacetic acid towards elimination since the latter destroys the conjugation between the phenyl ring and the double bond. The conjugation effect causes the equilibrium between 4-phenylcrotonic acid and styrylacetic acid to strongly favour the latter[2].

Bigley confirmed mechanism (1) by showing that in pyrolysis of the acids (1): (i) the terminal alkene is obtained when R^1 = H even though these are not the most thermodynamically stable, (ii) when R^1 = alkyl the *trans* alkene is obtained and (iii) deuterium is transferred from the carbonyl group to the terminal carbon (in 2,2-dimethylpent-3-enoic acid[3]. Likewise Smith and Blau showed that the entropy of activation for pyrolysis of but-3-enoic acid ($E_{act.}$ = 39.3 kcal/mol, $\Delta S\ddagger$ = −10.2 cal/mol/K) was consistent with a concerted cyclic mechanism[4]; they noted that this mechanism also accounts for the formation of pent-4-enoic acid from hex-3-endioic acid (2)[5].

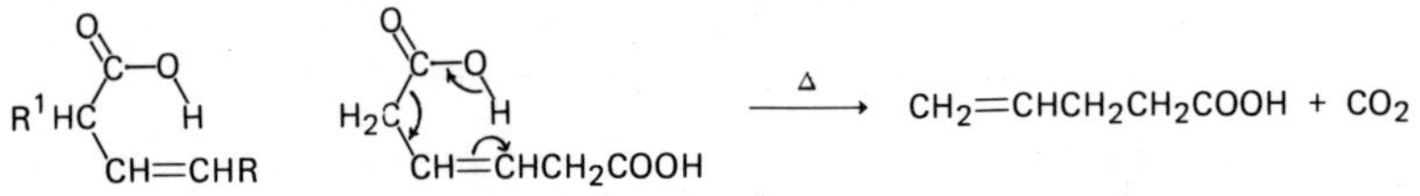

$$\xrightarrow{\Delta} CH_2{=}CHCH_2CH_2COOH + CO_2$$

(1)   (2)

The nucleophilicity of the carbon—carbon double bond provides the driving force for the reaction and it logically follows that the electron pair (i) in equation (1) will move prior to the other pairs, i.e. this pair will have been displaced further in the transition state. (Each pair cannot move precisely at the same time because this would produce no charge separation in the transition state, a possibility disproved by the data below). Two points follow from this. Firstly, β-keto acids, having a more nucleophilic double bond (at the end to which the hydrogen is transferred) should eliminate more rapidly than comparable β,γ-unsaturated alkenoic acids, and this is certainly true. Secondly, the electrons do not move in the clockwise direction depicted in some papers[6,7], as this would require attack of nucleophilic hydrogen upon the double bond; such a process would also require unacceptable polarization of the H—O and C—CO bonds. The transfer of a hydride ion should not lead to a substantial isotope effect, whereas transfer of a proton (mechanism 1) should, and this isotope effect should show substantial variation according to electron demand in the cyclic structure[8]; both these predictions are fulfilled[9,10]. The carbonyl carbon also shows a kinetic isotope effect (1.035 ± 0.01 for 2,2-dimethylpent-3-enoic acid) confirming that the C—CO bond is broken in the rate-determining step.

The non-synchronous electron movements in (1) means that charges will tend to appear in the transition state as shown in (3). This is necessarily an approximation and the delta charges of like and unlike charge will not be of equal and opposite magnitude respectively. It follows that electron-supplying substituents at the α- and

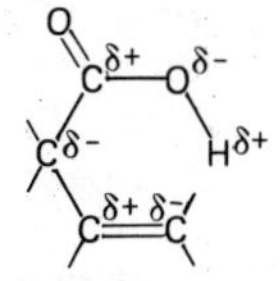

(3)

γ-carbons should decrease the rate whereas at the β-carbon they should increase the rate. However, the situation is not quite as clear as this, because substituents at the γ-carbon lose conjugation with the double bond whereas those at the α-carbon gain it. The results of a number of studies of substituent effects (Table 1)[6,11] show the following features:

(a) Comparison of data for acids 1,2 and 5 show that whereas a methyl on C_α increases the rate 2.8-fold per methyl, on C_γ the rate is decreased 6.7-fold per

TABLE 1. Relative rates of pyrolysis of
$RCH=CR^1CR_2^2COOH$ at $500°C$

Acid no.	Acid	k_{rel}
1	$CH_2=CHCH_2COOH$	1
2	$CH_2=CHCMe_2COOH$	5.6
3	$PhCH=CHCMe_2COOH$	0.75
4	$CH_2=CPhCMe_2COOH$	580
5	$MeCH=CHCMe_2COOH$	0.84
6	$CH_2=CMeCMe_2COOH$	168
7	$MeCH=CEtCMe_2COOH$	54.1

methyl. These results show clearly the effect of a gain and loss of conjugation respectively, superimposed upon the expected substituent effect. Comparison of data for acids 3 and 5 shows that the loss of conjugation at C_γ causes the γ-phenyl substituent to be slightly more rate-retarding than γ-methyl.

(b) Comparison of data for acids 2 and 4, 5 and 7, and 2 and 6, show that β-phenyl, β-ethyl and β-methyl substituents produce large rate accelerations of 100-fold, 64-fold and 30-fold respectively. These are predicted by mechanism (1) and it should be noted that the conjugation at the β-carbon is unchanged as a result of the elimination. It has been argued however that these large factors are produced because conjugation is inhibited in the ground state and relieved in the product; u.v. evidence was believed to support this view[6]. On the other hand, examination of molecular models gives no evidence in support of this. In order to obtain a more meaningful measure of the substituent effect at the β-carbon, Bigley and Thurman examined the effects of p-substituted phenyl groups and argued that the results would be free of any complication from steric inhibition of conjugation[9]. However, the substituent effects acting through the benzene ring will depend upon the ability of the ring to conjugate with the double bond, if this were an important factor. In the writer's view the β-substituent effects are purely electronic and this is supported by the data in Table 2 which compares them with the effects upon the rate of elimination of ethyl acetate when substituted at the α-position (see section VI.D) and for which the steric considerations do not apply. The ratios of the ρ-factors for the ring substituents are certainly not greater than the ratios of the logarithms of the direct substituent effects. Indeed the direct substituent effect in the acids appears to be proportionally *less* than the effect acting through the phenyl ring.

The relative rates of decarboxylation[6] of the acids 4−6 demonstrate the rate enhancement resulting from removal of a double bond and hence strain from a five-membered ring, though in these terms the higher reactivity of compound 4 cannot be explained. The explanation for the high reactivity of 4 comes from studies of the rates of acid-catalysed hydrogen exchange of benzocycloalkenes[12].

TABLE 2. Substituent effects in pyrolysis of $CH_2=CRCMe_2COOH$ and $AcOCHR(CH_3)$

R	Acids (500°C)	Esters (327°C)	$\log f_{acids}/\log f_{esters}$
Ph	104	66	0.90
Me	30	14.4	0.80
p-XC_6H_4	ρ = ca −1.1	ρ = −0.66	$(0.60)^a$

^{a} Ratio of ρ-factors.

In the seven-membered ring of **4** the hydrogens on the carbons adjacent to the double bond are largely precluded from hyperconjugating with it. When the double bond is transferred to the side-chain, hyperconjugation becomes possible and so there is a gain in conjugation on going from reactants to products.

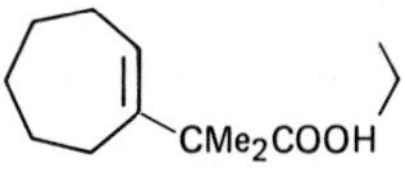

The gas-phase elimination of carbon dioxide from β-keto acids has been less well studied since the high reactivity means that reaction can occur at solution temperatures; this latter aspect has been reviewed[13]. The zwitterionic intermediate proposed to account for the elimination in solution[14]. may be ruled out in the gas phase on energy grounds. (Even in solution this mechanism is unlikely in view of the absence of any significant effect of polar solvents upon the rate[15].) The mechanism of the elimination was proposed as equation (2), i.e. the analogue of

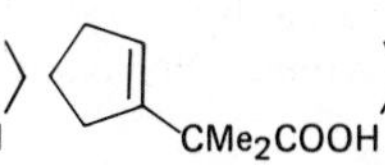

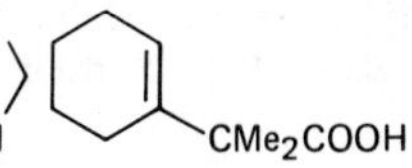

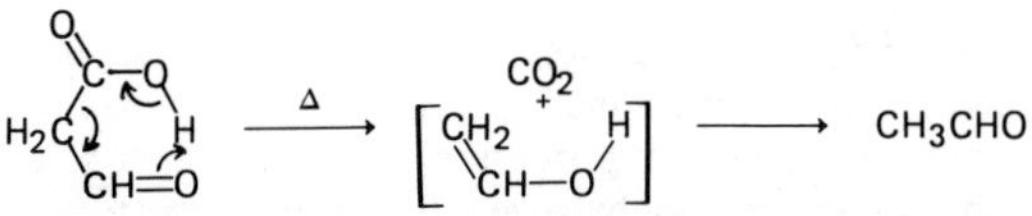

$$ (2) $$

equation (1)[15], an enolic intermediate having been shown to be involved, since in the presence of bromine, bromoacetone is formed from acetoacetic acid even though acetone does not react with bromine under the reaction conditions[14]. Electron-supplying substituents attached to the β-carbon should, as in the case of the alkenoic acids, increase the rate and vice versa; this is found, the ρ-factor of ca -1.0^8 being similar to that for the alkenoic acids. (This factor makes trifluoroacetylacetic acid a particularly stable β-keto acid). Similarly a large kinetic isotope effect is obtained consistent only with proton transfer, and there is no significant effect of polar solvents upon the rate of decarboxylation (of 2-ethyl-3-ketohexanoic acid)[8]. Brouwer and coworkers have argued that dipolar structures are involved in the transition state because the decarboxylation has a positive volume of activation, this parameter being in their view a more reliable indicator of transition state polarity than the effect of solvent upon the rate[16]. While these structures may be involved in the decarboxylation of β-keto acids in solution, they are unlikely to be involved in the gas phase.

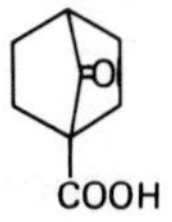

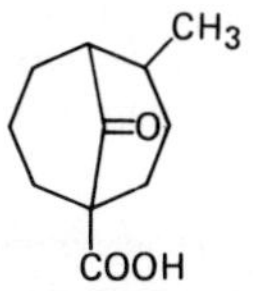

The intermediacy of enols in the reaction pathway means that acids such as **7** and **8** do not eliminate because of the difficulty of forming a double bond at a bridgehead. Acid **9**, by contrast, eliminates easily and it has been suggested that other aspects of the geometry in the 6-membered transition state may be more important, and that such factors account for the markedly different stabilities of naturally occurring β-keto acids, e.g. lycoctonamic acid[17].

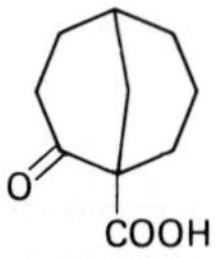

B. Saturated Acids

Three mechanisms dominate the decomposition of alkanoic acids, the importance of each depending upon the pyrolysis conditions. Blake and Hinshelwood observed that formic acid is decarboxylated in a static system by a first-order process (equation 3) and dehydrated by a second-order process[18], probably as shown in equation (4). Similar kinetic behaviour was observed (using a static

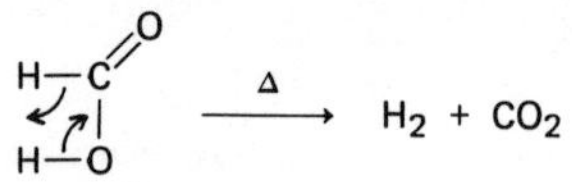

$$H-C\overset{O}{\underset{O-H}{\Big|}} \quad \xrightarrow{\Delta} \quad H_2 + CO_2 \tag{3}$$

$$\xrightarrow{slow} \quad H_2O\,[+(HCO)_2O] \quad \xrightarrow{fast} \quad HCOOH + CO \tag{4}$$

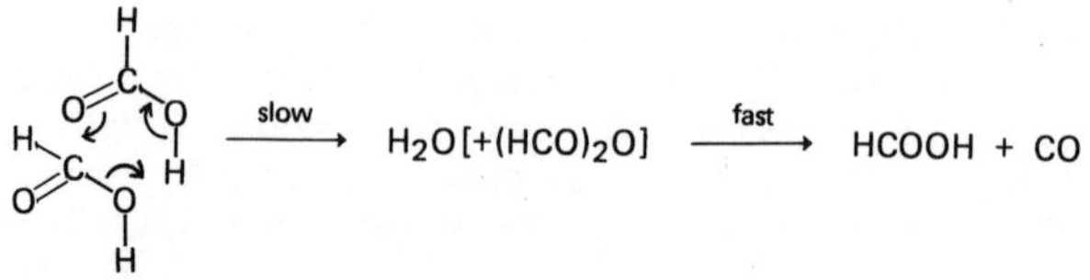

system) for acetic acid[19] and propanoic acid[20], the former giving methane with ketene, and the latter ethane with methylketene, along with water, carbon monoxide and carbon dioxide. Radical processes were shown not to be involved, and the similarity to the reaction mechanisms for dehydration is illustrated by the Arrhenius parameters in Table 3. These also show the lower activation energy for the more favourable 6-centre process involved in the dehydration.

The kinetic form for decarboxylation remains the same over a wide range of temperatures (530–760°C) with, for acetic acid, $E_{act.} = 62.0^{21} - 69.8^{22}$ kcal/mol and $\log A = 11.9^{21} - 13.6^{22}$ s^{-1}. However, the kinetics of dehydration change to first order at higher temperatures ($E_{act.} = 67.5^{21} - 64.9^{22}$ kcal/mol, $\log A = 12.95^{21} - 12.45^{22}$ s^{-1}), probably because the higher energy input facilitates a change to the simpler 4-centre process (equation 5). Although both flow and static systems show the same kinetic behaviour, the rates are lower at a given temperature under flow conditions, probably because of failure to attain true thermal equilibrium.

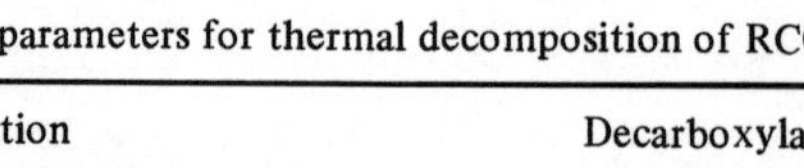

$$CH_2-C\overset{O}{\underset{O-H}{\Big|}} \quad \xrightarrow{\Delta} \quad H_2O + CH_2=C=O \tag{5}$$

The decarboxylation mechanism (equation 3) should be aided by electron withdrawal by the α-substituent and this is certainly the general observation for saturated carboxylic acids. This mechanism appears to have been overlooked in

TABLE 3. Arrhenius parameters for thermal decomposition of RCOOH

R	Dehydration		Decarboxylation	
	$E_{act.}$(kcal/mol)	$\log A$(s^{-1})	$E_{act.}$(kcal/mol)	$\log A$(s^{-1})
H	28.5	7.46		
CH$_3$	34.2	8.45	58.5	11.1
C$_2$H$_5$	35.15	8.76	>49.3[a]	>9.8[a]

[a] This included a component from the dehydration; the true value for pure decarboxylation would therefore be higher.

discussion of conclusions based upon a very extensive study of heterocyclic substituted acetic acids[23]. The mechanism corresponding to equation (1), i.e. **10**, (R = H) was also considered invalid because 4-pyridylacetic acid eliminates at a similar

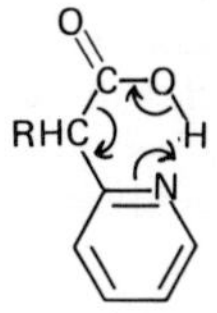

(10)

(11)

R = H, p-ClC$_6$H$_4$

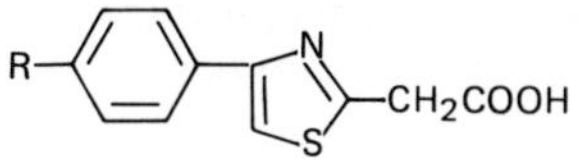

rate to 2-pyridylacetic acid, both being faster than 3-pyridylacetic acid. Moreover, an analogue of the mechanism shown in equation (1) is very unlikely simply because of the insufficient nucleophilicity of the aryl ring bonds; the observed order for the pyridyl substituents corresponds to their known abilities to withdraw electrons. Indeed, substitution of an electron-withdrawing p-chloro substituent at the α-position of 2-pyridylacetic acid (**10**, R = p-ClC$_6$H$_4$) caused a 10^3 increase in the rate of decarboxylation[23]. Consequently the zwitterion mechanism proposed for the elimination seems less likely than indicated by the authors' analysis. Support for the zwitterion mechanism in solution was thought to be provided by the fact that *para* substituents R in (**11**) gave a Hammett correlation with a small negative ρ-factor, *i.e.* these substituents altered the electron density on the nitrogen and hence the ease of zwitterion formation[23]. However, it is by no means certain that the ability of sulphur to conjugate with the C=N double bond does not in fact cause these acids to decarboxylate via the mechanism shown in **10** and this would be facilitated by electron supply from the group R as observed.

Cyclopropanecarboxylic acid decarboxylates to propene via ring-opening to crotonic acid which isomerizes to but-3-enoic acid and thence as described in Section II.A; the isomerization is high order and hence more rapid at high pressure[24]. Likewise cyclopropylacetic acid ring-opens in a rate-determining step to pent-4-enoic acid[24]; the cyclopropyl ring is insufficiently nucleophilic to permit a mechanism analogous to equation (1).

Comprehensive studies of the decarboxylation of mono-[25], di-[26], and tri-[27] fluoroacetic acids (by passage over silica) have shown that the elimination products include hydrogen fluoride, carbon monoxide, carbon dioxide and acyl fluorides. In each case the elimination of hydrogen fluoride is the first step e.g. for fluoroacetic acid the main reactions are (6)–(8). For difluoroacetic acid the fluoroformaldehyde product from (7) decomposes further to hydrogen fluoride and carbon monoxide, and for trifluoroacetic acid, decomposition of the intermediate CF_2CO_2 gives carbon dioxide and difluorocarbene which inserts into the starting acid to give difluoromethyl trifluoroacetate.

$$FCH_2COOH \longrightarrow HF + (CH_2CO_2) \qquad (6)$$

$$(CH_2CO_2) \longrightarrow HCHO + CO \qquad (7)$$

$$FCH_2COOH + HF \longrightarrow FCH_2COF + H_2O \qquad (8)$$

The pyrolysis of iodoacetic acid proceeds via the expected homolysis of the C–I bond, the carboxymethyl radical being detected at low temperature[28]. Decomposition of phenylmercaptoic acid gives thiophenol, carbon monoxide, carbon dioxide, acetic acid, methyl phenyl thioether and dibenzyl, evidently via a combination of 4-centre and radical mechanisms, the latter involving $S-CH_2$ bond cleavage[29]. Radical mechanisms were proposed to account for the products obtained in the decomposition of phenyl- and diphenylacetic acids[30], though the use of toluene as a carrier makes the origin of products such as dibenzyl rather unclear (at least for the former acid), and 4-centre processes cannot therefore be ruled out. For example the formation of phenylketene must have involved a 4-centre process (12) and a similar process (13) leading to toluene could not have been detected.

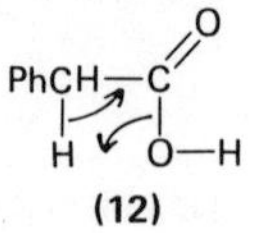

(12) 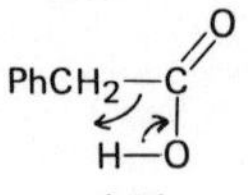   (13) 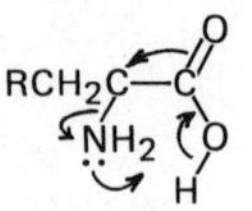   (14)

Zwitterionic intermediates have been proposed to account for the formation of α-lactones (and hence ketones) in pyrolysis of α-amino acids[31], but since the pyrolysis temperature was 500°C this must be regarded as very improbable; a concerted process such as (14) would be more likely. Thermodynamic parameters have been quoted for pyrolysis of malonic, oxanilic, picolinic, anthranilic, p-amino-benzoic and benzylmalonic acids[32] but are unlikely to be meaningful in view of the mere 10°C range used in their derivation.

An interesting report of the pyrolysis of acrylic, methacrylic and crotonic acids show these to produce respectively, acetaldehyde, acetone and propionaldehyde[33]. Decarbonylation is involved and this has previously been reported only for benzoic acid (as a minor reaction accompanying decarboxylation[34]). The proximity of a double bond seems to be important, so that a process, involving electron acceptance by this bond, may be involved, i.e. attack of OH on the adjacent carbon is the initial step (equation 9).

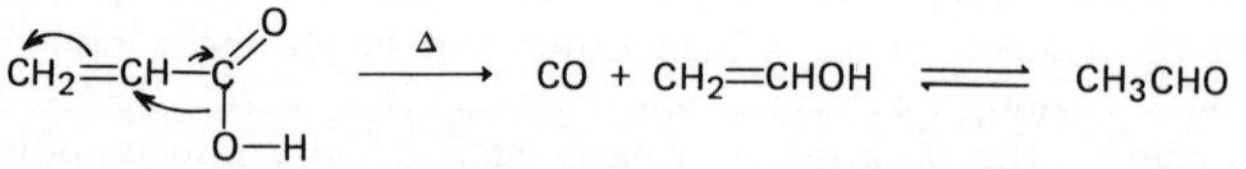

$$CH_2=CH-C(O)(O-H) \xrightarrow{\Delta} CO + CH_2=CHOH \rightleftharpoons CH_3CHO \qquad (9)$$

In the decomposition of propanoic acid noted above, ethylene is also a primary product and is probably formed via a 5-membered transition state (15) which is not

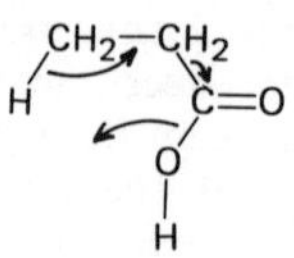

(15) 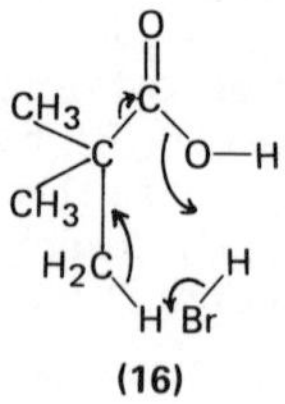   (16)

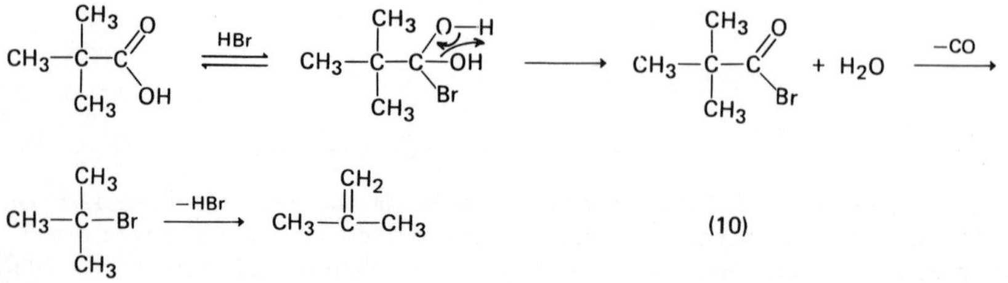

$$(10)$$

particularly favourable. Likewise trimethylacetic acid gives isobutylene, but the rate of formation of the latter is increased dramatically in the presence of hydrogen bromide as a catalyst[35]. The intervention of a 7-membered transition state (16) was postulated though this would also not be expected to be very favourable. An alternative scheme which was not considered is that shown in equation (10) in which the intermediate acyl bromide decarbonylates (see Section III) to give *t*-butyl bromide which would rapidly lose hydrogen bromide.

III. PYROLYSIS OF ACID HALIDES

These compounds are unable to undergo the general β-elimination process, so the only pyrolytic decompositions possible are the extrusion of carbon monoxide or hydrogen halide. This appears to have been the subject of only one study[36], in which acetyl bromide at 600–800°C was found to give carbon monoxide and hydrogen bromide in the ratio of 1:5 with no free-radical products. Only the processes (11) and (12) seem therefore to be involved.

Reaction (12) would be expected to be much faster if methyl were replaced by hydrogen and this undoubtedly accounts for the instability at room temperatures of the formyl halides[37]. Reaction (11) is analogous to reaction (5) for carboxylic acids and it may be noted that with acyl halides there can be no analogue of the alternative mechanism (4) for the carboxylic acids.

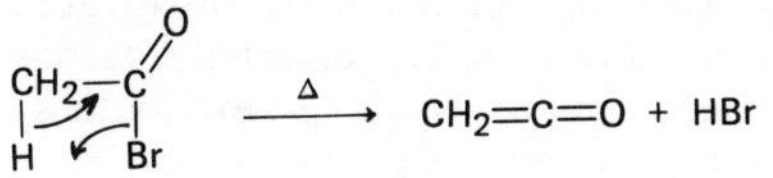

$$CH_2\!\!-\!\!C\overset{O}{\underset{Br}{\diagdown}} \quad \xrightarrow{\Delta} \quad CH_2\!\!=\!\!C\!\!=\!\!O \; + \; HBr \tag{11}$$

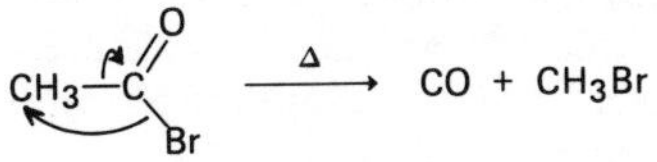

$$CH_3\!\!-\!\!C\overset{O}{\underset{Br}{\diagdown}} \quad \xrightarrow{\Delta} \quad CO \; + \; CH_3Br \tag{12}$$

IV. PYROLYSIS OF ACID AMIDES

A. Primary Amides and Higher Amides lacking β-Hydrogen in the *N*-Alkyl Group

Most studies in this subgroup have concerned acetamide and its *C*-substituted derivatives. Although Boehner and Andrews reported some sixty years ago that nitriles were formed in thermal decomposition of amides[38], forty years elapsed before Davidson and Karten established the main features of the reaction (of

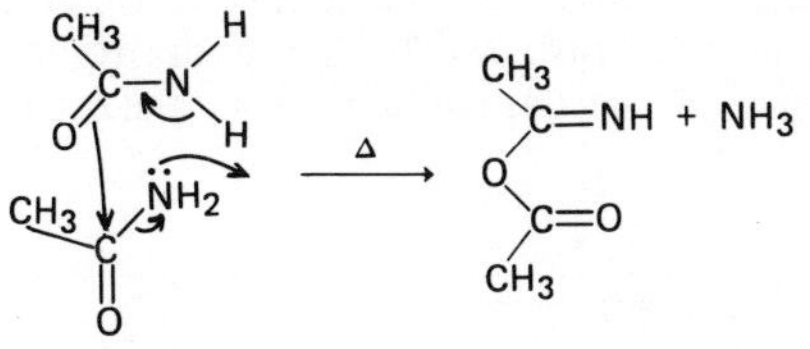

$$\xrightarrow{\Delta} \quad \begin{array}{c} CH_3 \\ C\!\!=\!\!NH \; + \; NH_3 \\ O \\ C\!\!=\!\!O \\ CH_3 \end{array} \tag{13}$$

$$\xrightarrow{\Delta} \quad CH_3C\!\!\equiv\!\!N \; + \; CH_3COOH \tag{14}$$

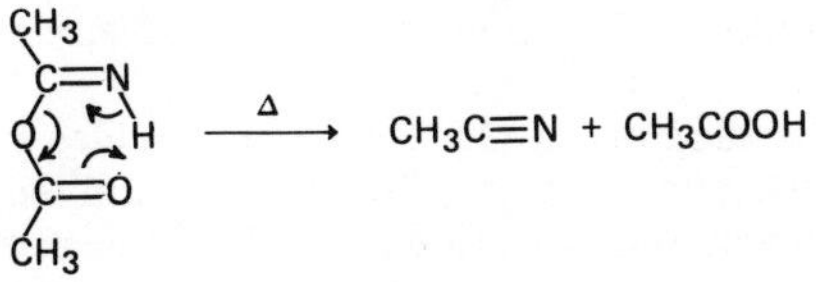

acetamide)[39]. They suggested that the primary step of the reaction is formation of the isoimide in a 6-centre process (13). The isoimide may decompose by a number of mechanisms[40], the most likely of which is the 6-centre process (14). The acetic acid produced may then combine with the ammonia to form ammonium acetate which dehydrates to regenerate acetamide[39]. A more recent study suggests that the intermediate is the imide[41] but this is in any case in equilibrium with the isomide[42].

It will be immediately apparent that equations (13) and (14) are the nitrogen analogues of the bimolecular reaction (4) leading to dehydration of carboxylic acids. The analogy goes further, for pyrolysis of acetamide goes over to a unimolecular decomposition mechanism at temperatures high enough to permit the 4-centre process (15), which produces ammonia and ketene (cf. reaction 5 for the acids)[43]. Comparison of the rates of this decomposition of acetic acid[22] and acetamide[43] indicates that the latter decomposes ca 30% faster at 700°C and this is consistent with the greater nucleophilicity of NH_2 compared to OH and as required by mechanism (15). Mechanism (15) is also analogous to (11) for reaction of acyl halides.

$$\text{CH}_2\text{—C}\overset{\text{O}}{\underset{\text{NH}_2}{\diagup}} \quad\overset{\Delta}{\longrightarrow}\quad \text{NH}_3 + \text{CH}_2\text{=C=O} \tag{15}$$

The pyrolyses of a range of amides with substituted acyl groups have been investigated. Just as β-keto acids readily eliminate carbon dioxide via a 6-centre process (2) so their nitrogen analogues, the β-keto amides, eliminate isocyanates to give acetone via the 6-centre process (16). The pyrolysis of N-t-butylacetoacetamide gave isobutene as an additional product, and an 8-membered transition state was, incorrectly, proposed for this [44]. At the temperature of the study (up to 740°C) the isobutene would be very readily eliminated (see Section IV.B) to give aceto-acetamide which would then eliminate as in (16).

$$\text{H}_2\text{C}\overset{\text{O}}{\underset{\overset{|}{\text{C=O}}}{\diagdown}}\overset{R}{\underset{H}{\diagdown}}\quad\overset{\Delta}{\longrightarrow}\quad \text{RNCO} + \text{CH}_2\overset{\text{OH}}{=}\text{CCH}_3 \tag{16}$$
$$\underset{\text{CH}_3\text{COCH}_3}{\|}$$

In view of the enhanced acidity of the hydrogen on the methylene between the carbonyl groups it would be expected that an alternative elimination (17) analogous to (15) would take place and this is so[44]. The failure to observe acetylketene in an analogous decomposition of acetoacetic acid is surprising in view of the similarity

$$\text{CH}_3\text{COCH—C}\overset{\text{O}}{\underset{\text{NH}_2}{\diagup}} \quad\overset{\Delta}{\longrightarrow}\quad \text{NH}_3 + \text{CH}_3\text{COCH=C=O} \tag{17}$$

between (15) and (5). Elimination of water is also observed giving 5-carbamoyl-4-dimethyl-2(1H)-pyridone (17)[45]; again the lack of any report of the oxygen analogue is surprising.

By a mechanism analogous to (16), cyanoacetamides give isocynates and methyl cyanide[46]. The decomposition of 2-phenyl-2,2-diphenyl- and 2,2,2-triphenyl-acetamides gives isocyanates and toluene, diphenylmethane and triphenylmethane,

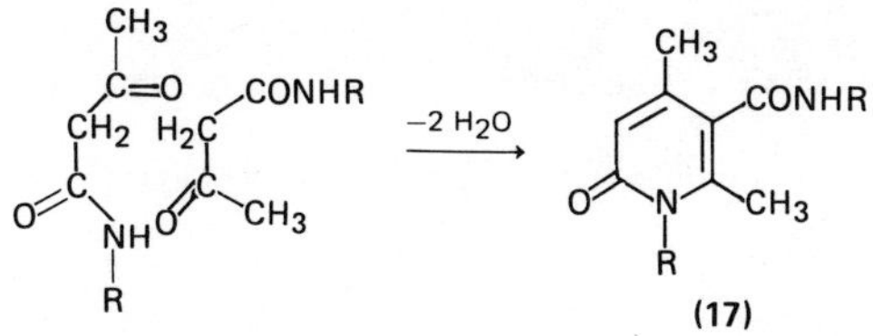

(17)

respectively[46], though with poor reproducibility. Ionic intermediates were incorrectly proposed by Mukaiyama and coworkers to account for these results, but a more reasonable interpretation would be in terms of 4-centre transition states such as **18**, and radical processes may also be involved. The isolation of toluene from the

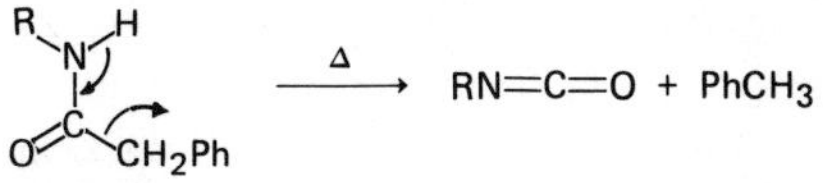

(18)

2-phenyl compounds suggests that toluene would have been obtained in pyrolysis of the analogous phenylacetic acid (see Section II.B) had it not been used as a carrier gas. Ionic intermediates were also proposed, incorrectly, to account for the formation of nitriles, phosgene and hydrogen chloride together with isocyanates, from the pyrolysis of 2,2,2-trichloroacetamides at $500-600°C$[47]. A more detailed study would probably reveal a mechanism somewhat analogous to that for decomposition of halogenoacetic acids (see Section II.B). Indeed, such a mechanism (equation 18) has been proposed to take account of the formation of formaldehyde in pyrolysis of 2-chloroacetamide at high temperatures (ca $800°C$)[48].

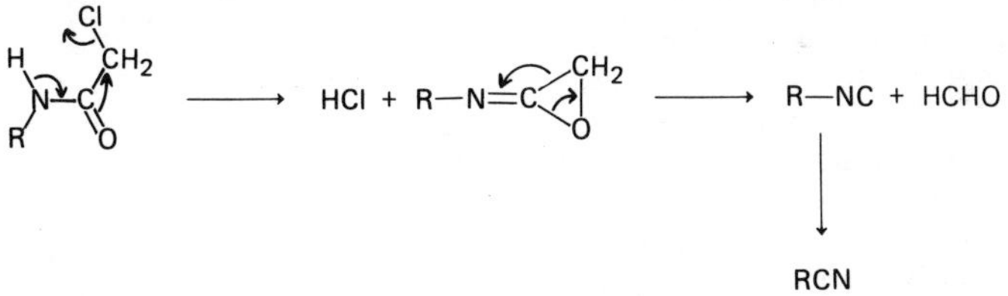

(18)

Formamide pyrolyses to ammonia and carbon monoxide (and also to hydrogen cyanide through dehydration[49]. The mechanism is likely to be analogous to (12) for acyl halides. In the presence of hydrogen chloride the elimination is accelerated and the 5-centre process (**19**) was proposed to account for this[50]. An alternative

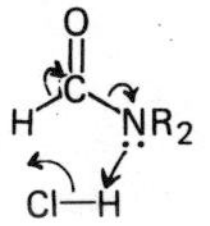

(19)

which must be considered however is equation (19), involving formyl chloride as a highly unstable intermediate.

There have been reports of the acid-catalysed pyrolysis of β-alkoxyamides to water, alkyl cyanide and alkanol[51], and of the decomposition of diazoamides in the presence of oxygen[52]. The pyrolysis of polyamides to amines and carbon dioxide[53] is of commercial interest.

$$\text{HCONR}_2 \;\; \underset{\text{HCl}}{\rightleftharpoons} \;\; \text{H}-\overset{\overset{\displaystyle O-H}{|}}{\underset{\underset{\displaystyle Cl}{|}}{C}}-\text{NR}_2 \;\; \longrightarrow \;\; \text{HCOCl} + \text{NR}_2\text{H} \tag{19}$$

$$\text{CO} + \text{HCl}$$

B. Amides with *N*-Alkyl Groups containing β-Hydrogen Atoms

Amides with *N*-alkyl groups containing β-hydrogen atoms are able to eliminate alkenes as do esters, their oxygen analogues (see Section VI.B), and give the nitrogen analogue of carboxylic acids, namely a primary amide. Since nitrogen is less electron-withdrawing than oxygen, polarization of the C—N bond, the principal driving force for the reaction (20), is more difficult. Consequently the temperature of elimination is approximately $100°$ higher than that needed for esters[54]. Bailey

$$\overset{\displaystyle |\quad|}{\underset{\displaystyle \underset{R^1}{\overset{|}{C=O}}}{\underset{NR}{C-C}}}\text{H} \;\; \xrightarrow{\;\Delta\;} \;\; \Big\rangle C=C\Big\langle \; + \; R^1\text{CONHR} \tag{20}$$

and Bird, who proposed the 6-centre mechanism (20), showed that as in the case of esters, tertiary amides eliminate more readily than secondary amides, and also the *N*-phenyl amide (R = Ph) eliminates faster than the corresponding *N*-methyl amide (R = Me)[54]; this would follow from the greater electron withdrawal of phenyl relative to methyl, thereby aiding C—N cleavage.

The similarity of the reaction to ester elimination was shown by the fact that *N*-(1-methylcyclohexyl)acetamide gave methylenecyclohexane and 1-methylcyclo-hexene in the ratio 28:72, as do corresponding esters[55]. Similar studies by Baumgarten and coworkers confirmed this and also indicated that the reaction is less selective than is ester elimination[56]. This would accord with the lower charge separation expected, in the writers view, in the transition state for the reaction, though there have not been any studies of Hammett correlations in the reaction to confirm this. These in fact might be difficult because of the accompanying side-re-actions which are compounded by the higher temperatures needed. For example, the pyrolysis of *N*-*t*-butylacetamide gave in addition to isobutene and acetamide ($\log A = 12.4$ s^{-1}, $E_{act.} = 51.4$ kcal/mol) *t*-butylamine and ketene[57], evidently via a 4-centre analogue of equation (15). Decomposition of the acetamide gave acetic acid which catalysed the formation of *t*-butylamine and ketene, with $\log A = 13.65$ s^{-1} and $E_{act.} = 34.9$ kcal/mol for this catalysed reaction.

The decomposition of the 2-halogen-substituted derivatives of *N*-*t*-butyl-acetamide showed carbon monoxide to be produced in a first-order reaction which increases in rate with increasing substitution of the acyl group[58]. A homolytic process was suggested, though not confirmed by the detection of free-radical products, and it is probable that the carbon monoxide is formed by (21) which is analogous to (12) for acyl halides, and could be expected to be faster because of

$$\text{Cl}_3\text{C}-\overset{\overset{\displaystyle O}{\diagup\diagup}}{\underset{\underset{\displaystyle \cdot \text{NHR}}{\diagdown}}{C}} \;\; \xrightarrow{\;\Delta\;} \;\; \text{Cl}_3\text{CNHR} + \text{CO} \tag{21}$$

the greater nucleophilicity of nitrogen and the greater positive charge on the α-carbon. Isobutene was also obtained (by process 20) and this would be faster than with the non-halogenated amides since withdrawal of electrons from the carbonyl group will aid polarization of the C—N bond.

V. PYROLYSIS OF ACID ANHYDRIDES

A. Anhydrides which possess α-Hydrogen Atoms

If we exclude from this group anhydrides which are cyclic or are α β-unsaturated then a common mechanism (equation 22) applies[59] which leads to a carboxylic acid and ketene as first observed by Wilsmore[60] and by a number of subsequent workers (e.g. Reference 61). At 355°C, acetic anhydride eliminates acetic acid 13,000 times faster than ethyl acetate, i.e. 6,500 times faster per β-hydrogen atom.

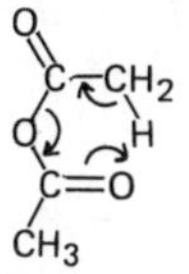

$$\begin{array}{c}
O \\
\| \\
C \!\!-\!\! CH_2 \\
/ \quad | \\
O \quad H \qquad \xrightarrow{\Delta} \quad CH_3COOH + CH_2CO \qquad\qquad (22) \\
\backslash \\
C{=}O \\
| \\
CH_3
\end{array}$$

The reason for this is not entirely clear, for although a β-acetyl group speeds up the rate of elimination of ethyl acetate by a factor of 240 times per β-hydrogen atom, due to the electron-withdrawing effect of the carbonyl group upon the acidity of the adjacent β-hydrogen (see Section VI.B), this still leaves a factor of ca 25 unaccounted for. One reason could lie in the fact that the most stable conformation of acetic anydride (taking into account the conjugation between the carbonyl groups and the central oxygen) is indeed that shown in (22). Little reorganization of structure is therefore needed to achieve the transition state, though this conformational advantage does not show up in an enhanced log A factor for the reaction.

Anhydrides of carbonic acid and a carboxylic acid pyrolyse to give an ester with elimination of carbon dioxide, e.g. benzoic carbonic anhydrides give alkyl benzoate and carbon dioxide (equation 23). An accompanying reaction gives carbon dioxide, benzoic anhydride and diethyl carbonate (equation 24)[62]; the former takes place

$$PhCOOEt + CO_2 \qquad\qquad (23)$$

$$\begin{array}{c}
O \ O \\
\| \ \| \\
PhCOCOEt \quad \Delta
\end{array}$$

$$PhCOOCOPh + EtOCOOEt + CO_2 \qquad\qquad (24)$$

more readily the more electron-supplying the alkyl group. This reaction does not proceed via the expected 6-centre process (20) but via the less favourable 4-centre process (21), since labelling experiments (with benzoic s-butylcarbonic anhydride) have shown that alkyl-oxygen cleavage does not take place[63]; free-radical mechanisms were also ruled out[64]. The failure to observe 20 must stem from the fact that carbon would need to be attacked, rather than the more electropositive hydrogen which is normally involved in these 6-centre processes. Also the carbon centre is more sterically hindered particularly in the anhydride chosen for this study and it is

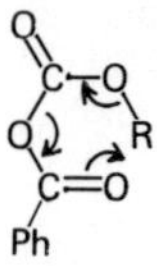

(20) 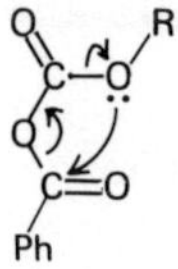   (21) 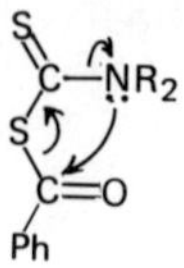   (21a)

possible that alkyl-oxygen cleavage might be observed with benzoic methylcarbonic anhydride. The generality of (21) is shown by the fact that carboxylic dithoicarbamic anhydrides eliminate carbon disulphide to give amides (21a)[64a].

Process 21 is an S_Ni reaction and should be aided by electron supply in R and by electron withdrawal in the phenyl group though these aspects have not been investigated. Interestingly, the S_Ni reaction has very recently been discovered in pyrolysis of carbonate esters (and related esters)[65] but takes place less readily with these because there is far less electron withdrawal from the carbon being attacked (see Section VI.B). We may predict on the basis of 21 that mixed anhydrides of carbamic acid and, for example, benzoic acid will also eliminate carbon dioxide (even more readily than the above) to give amides.

The mechanism of reaction (24) is not known, but the 6-centre process (22) must be a strong possibility, and such a process should be largely unaffected by the

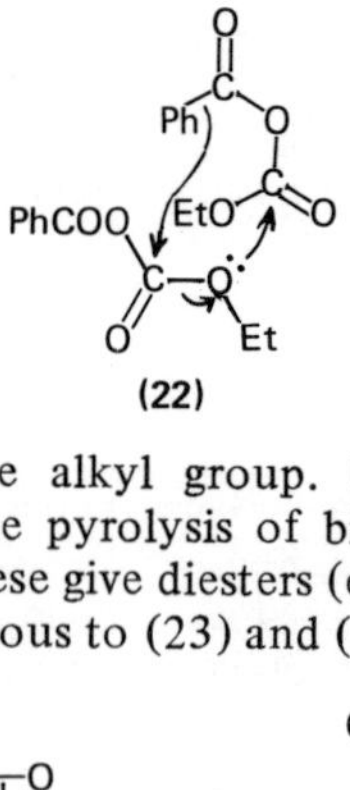

(22)

electron-supplying nature of the alkyl group. In order to provide further information on these reactions, the pyrolysis of bis(ethylcarbonic)dicarboxylic anhydrides has been studied[66]. These give diesters (equation 25) or cyclic anhydrides (equation 26) in reactions analogous to (23) and (24) respectively. Mechanism (26)

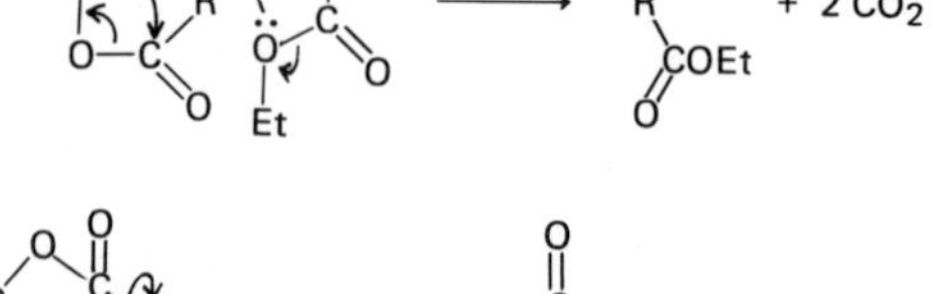

$$R\begin{matrix} COEt \\ COEt \end{matrix} + 2\,CO_2 \qquad (25)$$

$$R\begin{matrix} C \\ O \\ C \end{matrix}O + EtOCOOEt + CO_2 \qquad (26)$$

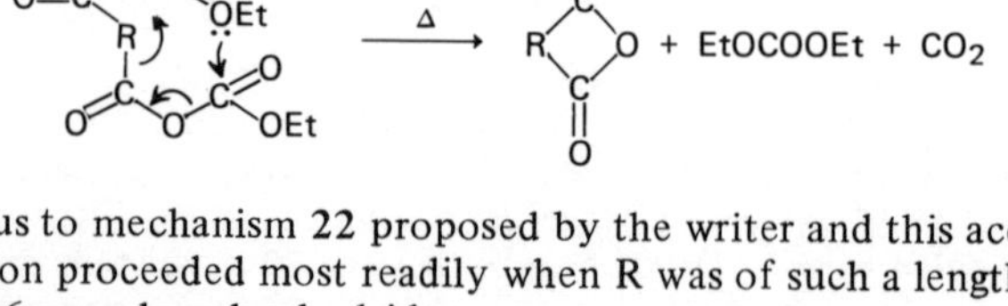

is thus analogous to mechanism 22 proposed by the writer and this accords with the fact that reaction proceeded most readily when R was of such a length as to be able to form a 5- or 6- membered anhydride.

Three mechanisms operate in the pyrolysis of the anhydrides of crotonic acid (and methyl derivatives) and ethyl carbonic acids[67] The *trans* acid gives crotonic anhydride, ethyl crotonate, diethyl carbonate and carbon dioxide via mechanisms (23) and (24). By contrast the *cis* acid gives ethyl but-3-enate via equation (27), in which the reaction products ethanol and vinylketene combine.

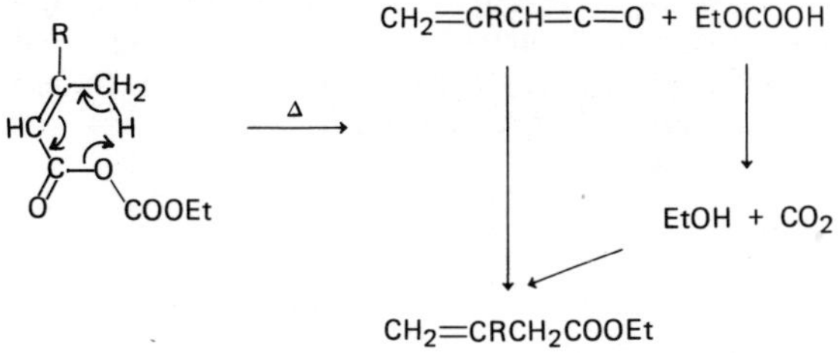
R
CH2
HC
H
C—O
O
COOEt
Δ
CH2=CRCH=C=O + EtOCOOH
EtOH + CO2
CH2=CRCH2COOEt

(27)

B. Anhydrides which either lack β-Hydrogen Atoms and are αβ-Unsaturated, or are Cyclic

The first category of anhydrides undergo a different type of pyrolytic decomposition which involves carbon-acyl scission and requires much higher temperatures than those described in Section V.A. This is to be expected in view of the postulated 3- and 4-centre mechanisms (e.g. equation 28) for methacrylic anhydride.

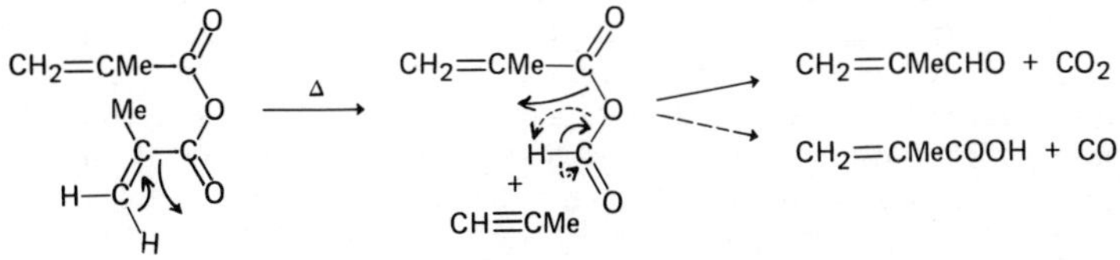
CH2=CMe—C
Me
H—C
H
Δ
CH2=CMe—C
H—C
+
CH≡CMe
CH2=CMeCHO + CO2
CH2=CMeCOOH + CO

(28)

Mechanisms involving oxygen-acyl scission have been proposed to account for the loss of carbon monoxide and carbon dioxide and the formation of a more unsaturated hydrocarbon residue in the pyrolysis of cyclic anhydrides[68]. The early reports of the formation of ethylene from succinic anhydride, of alkene and propene from aconitic anhydride[69], and of acetylene and fluoroacetylene from maleic and fluoromaleic anhydrides, respectively[70], are all examples of this reaction. Detailed studies of the pyrolysis of succinic, methylsuccinic, adipic, and maleic anhydrides have indicated a process such as (29)[68], the intermediate

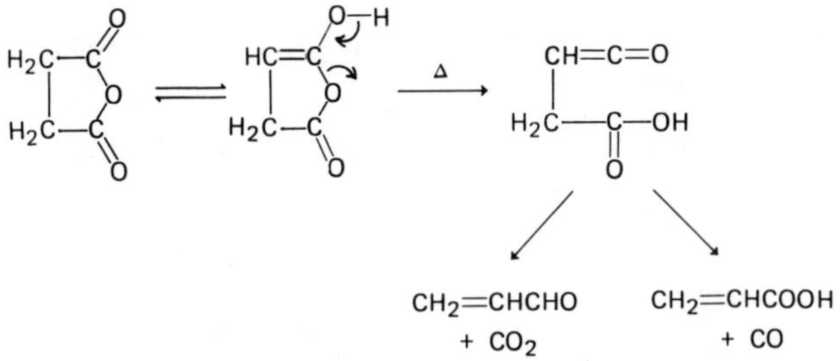
H2C—C
H2C—C
HC=C
O—H
H2C—C
Δ
CH=C=O
H2C—C—OH
CH2=CHCHO
+ CO2
CH2=CHCOOH
+ CO

(29)

propenal and propenoic acid undergoing further elimination to give ethylene, though the manner in which this takes place is by no means clear.

When anhydrides such as this form part of an aromatic ring system, the reaction is of preparative importance and here the present balance of evidence indicates that a concerted reaction is not involved[71]. For example, phthalic anhydride eliminates carbon monoxide and carbon dioxide to give benzyne which via subsequent insertion, 1,2- and 1,4-addition reactions (the latter confirmed by labelling experiments)[72], gives biphenylene, biphenyl and naphthalene (as well as acetylene). Zwitterionic intermediates (e.g. 23)[71] have been proposed, though other workers favour diradical intermediates[73]; the reaction is particularly valuable as a source of biphenylene and derivatives[74].

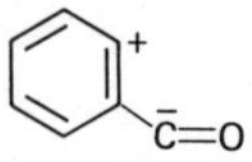

(23)

The thermal decomposition of benzoic anhydride has also been studied and at $500°C$ it gives mainly benzene together with benzoic acid, benzophenone, biphenyl, benzaldehyde, carbon monoxide and carbon dioxide[75].

VI. PYROLYSIS OF ESTERS

A. Esters without β-Hydrogen Atoms in the Alkyl Group

Methyl esters do not have β-hydrogen atoms in the alkyl group and are therefore unable to undergo the *cis* β-elimination described in Section VI.D; they are therefore stable to ca $550°C$. However, elimination does take place under the conducive conditions of either high temperature or a suitable molecular structure. For example, dicyanophenylmethyl benzoate eliminates carbon dioxide and also rearranges to benzoyl cyanide, (though 70% of the products are tars)[76]. The reactions may be formulated as the 4-centre processes (30) and (31) respectively;

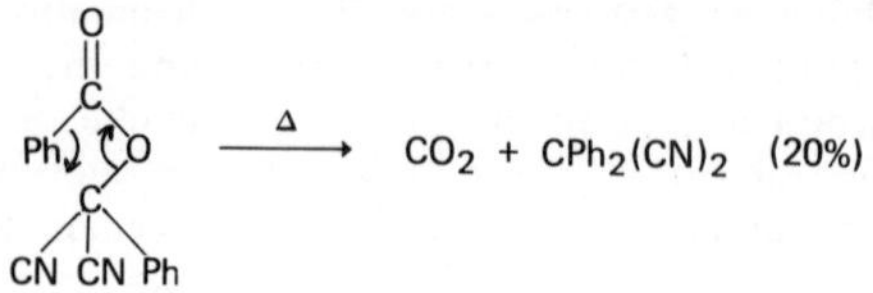

$$CO_2 + CPh_2(CN)_2 \quad (20\%) \qquad\qquad (30)$$

$$2\ PhCOCN\ (\rightarrow PhCN + CO)\ (10\%) \qquad\qquad (31)$$

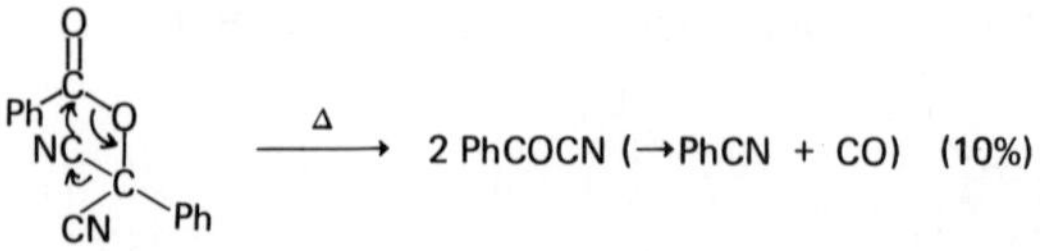

the former appears to be an intramolecular electrophilic aromatic substitution and examination of substituent effects in the aryl rings would easily pinpoint the mechanism. The formation of acyl cyanide (reaction 31) was also observed to accompany, to the extent of 3%, the normal elimination of acetic acid from 1,1-dicyanoethyl acetate[77]; (the presence of the electron-withdrawing cyano groups on the α-carbon necessitated an elimination temperature, for the normal *cis* β-reaction, of $>600°C$).

Other examples of the behaviour shown in equation (31) are known. Newallis and Lombardo[78] found that the ester **24** decomposes to ethyl acetate and bis-(chlorodifluoromethyl) acetone by, in the writer's view, the mechanism shown,

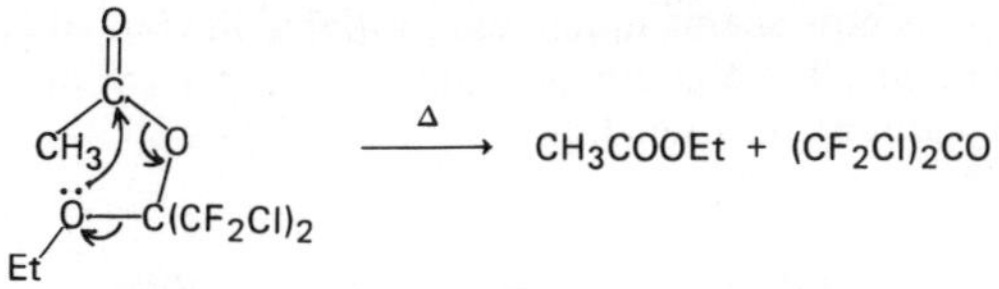

(24)

which also accounts for the higher rate of elimination found for the carbonate analogue. By contrast this is not explained by the 6-centre mechanism (involving less probable attack of carbonyl oxygen upon carbon) given in the literature[78]. The t-butyl derivative was said not to undergo this reaction since isobutylene and acetic acid were obtained, though this makes it certain that the mechanism shown in **24** was followed, since the t-butyl acetate produced would undergo very rapid elimination by the normal route (Section VI.D).

Likewise the dialkoxyalkyl esters (**25**) decompose to diesters[79], almost certainly by the 4-centre mechanism shown and again involving nucleophilic attack upon the

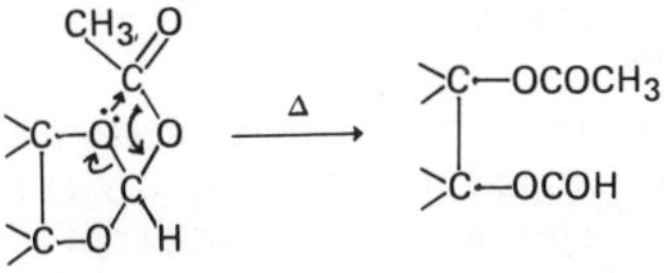

(25)

carbonyl carbon. S-Methoxymethyl thioacetates and acetates decompose to this ester and (thio)aldehyde in a similar fashion (equation 32)[80]. By contrast the methoxy analogue does not undergo this reaction (in fact it participates in an alternative elimination described below), and this follows from the greater nucleophilicity of sulphur compared to oxygen. Likewise the high nucleophilicity of the dialkylamino group causes the α-dimethylamino analogues to undergo reaction (32)[80a] (to give an aldehyde and amide).

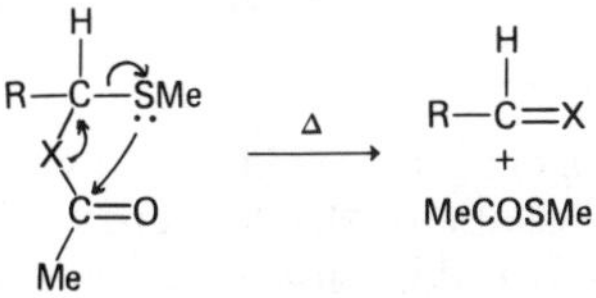

$$(X = O, S)$$

Extrusion of carbon dioxide from esters lacking β-hydrogens was first noted by Anschütz who found that diphenyl maleate gave stilbene[81]. However, a more detailed study of the decomposition of phenyl acrylate (or α-methylacrylate) showed two processes to occur. One is a molecular reaction which gives styrene (or methylstyrene) and carbon dioxide and is therefore similar to (30). The other is a free-radical reaction which gives acetylene (or methylacetylene) and phenyl formate which subsequently decomposes to phenol and carbon monoxide[82].

Methyl esters which have electron-withdrawing substituents (e.g. OR,SR) in the methyl group are reported to be able to eliminate acetic acid in a molecular 5-centre process (33). The stability of the resultant carbene rather than the acidity of the α-hydrogen is apparently the most important factor. Decomposition of the resultant carbenes gave a variety of products, e.g. dimethoxycarbene gave methyl acetate[83]. When the two methoxy groups are bound into a cyclic structure, the products are acetic acid, alkene and carbon dioxide[84]. A complex mechanism was proposed for this, but now seems much less probable in view of equation (33) since the carbon dioxide and alkene could be obtained simply by subsequent decomposition of the cyclic carbene produced.

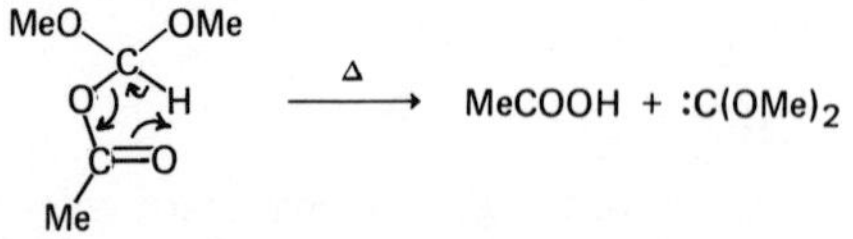

$$\text{(structure)} \xrightarrow{\Delta} \text{MeCOOH} + :\text{C(OMe)}_2 \qquad (33)$$

In pyrolysis of methylene dibenzoate, formaldehyde and benzoic anhydride are produced along with many minor products[75]. The mechanism here must almost certainly involve nucleophilic attack of one oxygen upon the remote acyl carbon in a 4-centre process (34, where $n = 0$). Process (34) thus differs from 21 only in that

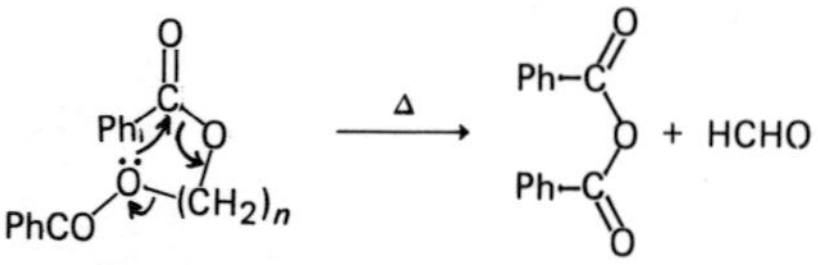

$$\text{(structure)} \xrightarrow{\Delta} \text{(structure)} + \text{HCHO} \qquad (34)$$

the methylene and carbonyl groups have interchanged positions. Given this mechanism it is not difficult to understand why trimethylene dibenzoate (containing a saturated chain, i.e. $n = 3$) is reported to fail to undergo the same reaction[75]. Instead alkyl-oxygen scission takes place, most probably as in equation (35), giving benzoic acid and alkyl benzoate. Likewise this type of process is the only one open to propylene dibenzoate (and diacetate) which give the corresponding 2-methyl vinyl ester and carboxylic acid. Likewise ethylene dibenzoate gives benzoic acid and vinyl benzoate[85], though here acetaldehyde is also produced, presumably via reaction (34) with subsequent rearrangement.

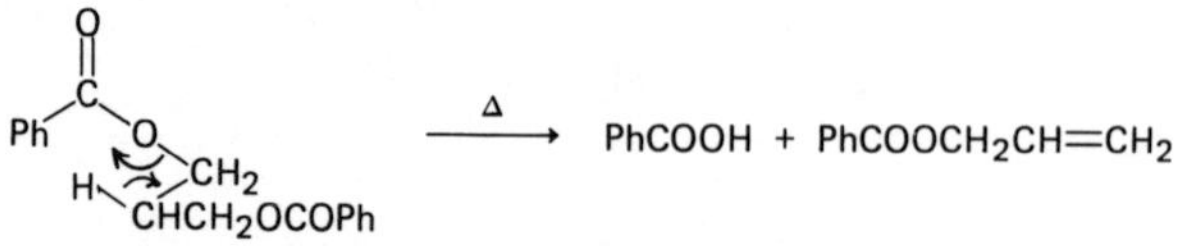

$$\text{(structure)} \xrightarrow{\Delta} \text{PhCOOH} + \text{PhCOOCH}_2\text{CH}=\text{CH}_2 \qquad (35)$$

Yet another rearrangement is possible in pyrolysis of methyl N-methylcarbamates and is unique because of the N-hydrogen. The products methanol and methyl isocyanate are most probably formed via the 4-centre process (36)[86].

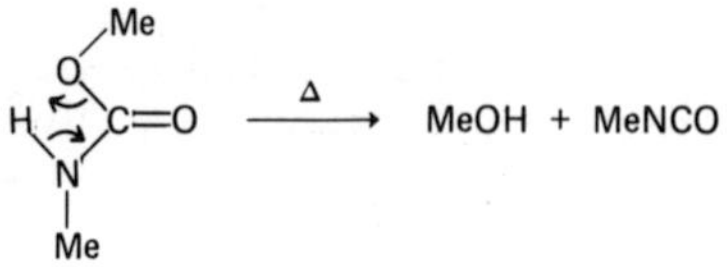

$$\text{(structure)} \xrightarrow{\Delta} \text{MeOH} + \text{MeNCO} \qquad (36)$$

B. Vinyl Esters

A study of the thermal decomposition of vinyl benzoate[85] showed that although normal cis β-elimination (37) will take place, this reaction is a minor one accompanying (38) and the major reaction (39); the lesser importance of (32) may be due to the instability of the acetylene product. The direction of the electron movements in (38) and the second step of (39) are not at all certain.

$$\text{(structure)} \xrightarrow{\Delta} PhCOOH + HC\equiv CH \tag{37}$$

$$\text{(structure)} \xrightarrow{\Delta} PhCH=CH_2 + CO_2 \tag{38}$$

$$\text{(structure)} \xrightarrow{\Delta} Ph\text{(structure)} \longrightarrow PhCOCH_3 + CO \tag{39}$$

$$\text{(structure)} \xrightarrow{\Delta} RR^1C=CO + CH_3CHO \tag{40}$$

For non-aromatic vinyl esters an additional reaction (40) takes place, and is a major reaction as expected since a 6-centre process must be involved. It should be noted that (40) is related to the decomposition of vinyl ethers in just the same way that esters are related to anhydrides. Although there are no kinetic studies available one may predict that reaction (40) takes place more readily than would the decomposition of the analogous vinyl ether.

C. Allyl Esters

Like the esters described above, the allyl esters possess a vinylic β-hydrogen atom, elimination of which should give an alkene (cf. 37); this reaction is not observed.

The allyl esters of formic acid are able to undergo a reaction not available to the allyl esters of other acids. This process (equation 41) is a 6-centre one and differs from (40) only in that the alkyl carbon chain is one unit longer and the acyl carbon

$$\text{(structure)} \xrightarrow{\Delta} R^1CH_2CR=CH_2 + CO_2 \tag{41}$$
$$(R = H, Me)$$

chain is one unit shorter. The Arrhenius parameters for allyl formate and 2-methylallyl formate ($E_{act.}$ = 43.0, 42.1 kcal/mol; log A = 10.1, 9.8 s^{-1}, respectively)[87,88] are consistent with the cyclic process; the latter compound decomposes 2.1 times faster than the former. A non-radical pathway was confirmed by the fact that propene derived from simultaneous decomposition of allyl formate and tritiated 2-methylallyl formate contained only 1% of the initial tritium content of the latter, i.e. there were no significant hydrogen-atom extraction processes indicative of the presence of free radicals.

It should be noted that the elimination shown in equation (41) is closely similar to the mechanism for the pyrolysis of β,γ-alkenoic acids (equation 1) and, like that reaction, is accelerated by electron-supplying substituents on the 2-carbon atom (the -carbon atom in the case of the alkenoic acids). This provides further support for the mechanism given in (equation 41), i.e. nucleophilic attack of the double bond upon the hydrogen is a driving force for the reaction. Moreover, the O—H bond in but-3-enoic acid should be more easily polarized in the correct direction than should the H—CO bond in allyl formate and the significantly lower activation energy for elimination from the former is consistent with this. Evidence for the *cis* nature of the elimination is also provided by the elimination of *trans*-cinnamyl formate (equation 41; R^1 = Ph, R = H) to carbon dioxide and allylbenzene[88].

It follows that a 6-membered transition state analogous to that in equation (41) cannot be obtained with allyl acetate. Neither elimination nor molecular rearrangement (see below) are favourable, consequently decomposition takes place via a radical pathway. This is indicated by the variety of products (in decreasing order of yield): carbon dioxide, methane, carbon monoxide, but-1-ene, propene, acrolein, ethene and ethane. Likewise the allyl esters of benzoic, phenylacetic, trifluoroacetic and oxalic acids similarly decompose by radical pathways[89].

Allyl esters readily undergo rearrangement of the type shown generally in (equation 42). The reaction will tend to occur when *cis* β-elimination is possible in

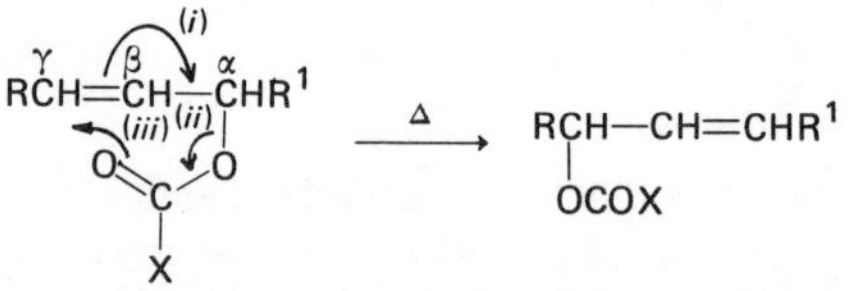

$$\text{(42)}$$

R but not in R^1; where elimination is possible in both R and R^1 the extent of the rearrangement will be governed by the relative ease of the elimination in each. The product of rearrangement and elimination is a conjugated diene and this provides a further driving force for the reaction. Some examples illustrating the reaction are given in Table 4. Note that esters nos. 4 and 5 both give a 1,3-butadiene, the former via the normal *cis* β-elimination (see Section VI.D below) and the latter via rearrangement and elimination. In ester no. 6 the strong electron withdrawal from the α-carbon makes *cis* β-elimination so unfavourable that rearrangement to the internal alkene is preferred. The products from esters nos. 7 and 8 confirm the proposed mechanism since, in addition to the 1,3-diene, the rearranged but un-eliminated ester is obtained in each case.

A detailed study of the effects of substituents upon the rate of rearrangement[99] has shown the following:

(*a*) Increased electron withdrawal in the group X increases the rate of rearrangement. This follows if breaking of the C—O bond is a primary step (as it is in *cis* -elimination of esters).

(*b*) Electron supply to the α-carbon increases the rate by a large factor. This will aid C—O bond breaking and also increase conjugation with the forming double bond.

(*c*) Electron supply to the γ-carbon also increases the rate by a large factor. This follows if breaking of the double bond is a primary step in the rearrangement. It would *not* follow if attack of the carbonyl oxygen upon the γ-carbon were kinetically important. We may therefore assert that the extent to which the electrons have moved in the transition state of equation (42) follows the order (*i*) > (*ii*) > (*iii*) or (*ii*) > (*i*) > (*iii*).

(*d*) Replacement of hydrogen on the α-carbon by deuterium produced a small rate retardation.

TABLE 4. Products of pyrolysis of allylic esters

Ester	No.	Major product	Minor product	Reference
$C_6H_{11}-CH=CHCHCH_3$, $\overset{\mid}{O}Ac$	1	$C_6H_{11}-CH=CHCH=CH_2$	$C_6H_{11}-CHCH=CHCH_3$	90
$CH_3CH_2CH=CHCHCH_3$, $\overset{\mid}{O}Ac$	2	$CH_3CH_2CH=CHCH=CH_2$	$CH_3CH=CHCH=CHCH_3$	91
$CH_3(CH_2)_2CH=CHCHCH_3$, $\overset{\mid}{O}Ac$	3	$CH_3(CH_2)_2CH=CHCH=CH_2$	$CH_3CH_2CH=CHCH=CHCH_3$	92
$CH_2=\overset{R}{\underset{\mid}{C}}CHCH_3$, $\overset{\mid}{O}Ac$	4	$CH_2=\overset{R}{\underset{\mid}{C}}CH=CH_2$		93
$CH_3CH=CHCH_2$, $\overset{\mid}{O}Ac$, CN	5	$CH_2=CHCH=CH_2$, CN		94, 95
$CH_2=\overset{CN}{\underset{\mid}{C}}CH_3$, $\overset{\mid}{O}Ac$	6	$AcOCH_2CH=\overset{CN}{\underset{\mid}{C}}CH_3$		96
cis-$CH_2CH=CHCH_2$, $\overset{\mid}{O}Ac$ $\overset{\mid}{O}Ac$	7	$CH_2=CHCH=CH$, $\overset{\mid}{O}Ac$	$CH_2=CHCHCH_2$, AcO OAc	97
$CH_2=CHCHCH_2$, AcO OAc	8	$CH_2=CHCH=CH$, $\overset{\mid}{O}Ac$	cis-$CH_2CH=CHCH_2$, OAc OAc	97
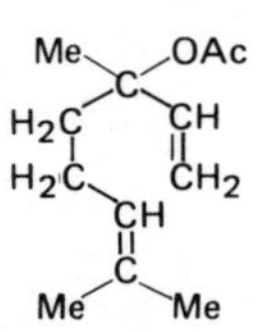 Linalyl acetate	9	Myrcene cis and $trans$-Ocimene	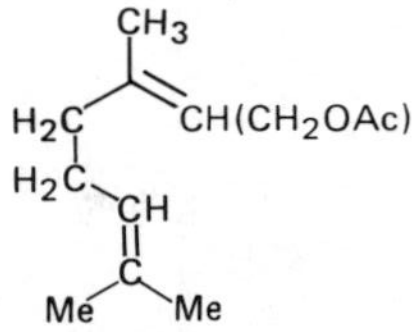 Neryl acetate and geranyl acetate	98

(*e*) Electron supply to the β-carbon atoms has a trivial (and indeed inconsistent) effect upon the rate. This confirms that the electron movements are not in the opposite direction to that shown in equation (42) since this would cause the -carbon to be significantly electron-deficient.

(*f*) ^{18}O-Labelled ether oxygen in the starting material largely becomes labelled carbonyl oxygen in the product confirming the involvement of the 6-centre process (42).

D. Esters containing Non-vinylic β-Hydrogen Atoms

Esters containing non-vinylic β-hydrogen atoms in the alkyl group undergo elimination to an alkene and a carboxylic acid, and this takes place most readily if the hydrogen and the acyloxy group are *cis* to each other. This is one of the oldest known organic reactions[100] and was even studied in the gas phase more than a century ago[101]. The reaction is remarkably straightforward and relatively unaffected by surface conditions, so much so that activation energies measured at that time[102] do not differ significantly from present values. Of all the concerted β-elimination reactions, ester pyrolysis has been the most intensively studied and our knowledge and understanding of the transition states for these reactions derives very considerably from these studies.

The experimental features which have lead to the current view of the mechanism are described in detail below. It is, however, helpful to have to begin with a general view of the transition state **26** which applies not only to elimination from carboxylates but from a whole class of closely related compounds. For carboxylic acid

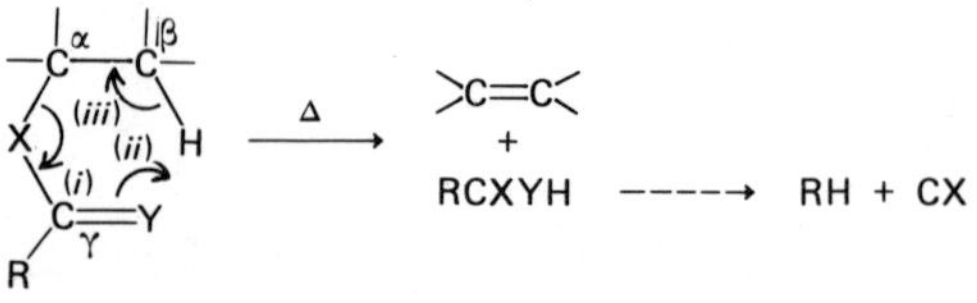

(26)

esters X and Y = O, and R = alkyl; the three carbon atoms in **26** are designated α,β and γ. Notable features are the following:

(1) The reaction pathway is a semiconcerted process involving 6 atoms in a cyclic array, which is not, however, necessarily planar.

(2) The reaction is a *cis* elimination.

(3) Where there are *cis* β-hydrogens in different environments, elimination tends to take place so as to produce the most sterically favourable product.

(4) Where elimination can produce either *cis* or *trans* products the latter is favoured.

(5) The reaction is aided by electron supply at C_α so that the order of reactivity of alkyl esters is $3^0 > 2^0 > 1^0$ and this is true even when a statistical correction is made for the different number of β-hydrogen atoms. The transition state becomes more polar along the series $1^0 < 2^0 < 3^0$, the biggest difference in polarity coming between the secondary and tertiary esters.

(6) The reaction is aided by greater electron withdrawal by R. Reactivity series are therefore:

(*a*) formates > acetates > propanoates (R = H,CH_3, C_2H_5),
(*b*) chloroacetates > acetates (R = $ClCH_2$, CH_3)
(*c*) chloroformates > formates (R = Cl, H),
(*d*) carbonates > carbamates > acetates (R = R^1O, R^1NH, CH_3)

The polarity of the transition state decreases along each of these series. This variation, coupled with that brought about by C_α substitution means that a spectrum of transition-state structures are obtained. Thus the most E_i-like transition state will be obtained in the pyrolysis of ethyl acetate and the most E1-like transition state will be obtained in the pyrolysis of t-butyl carbonates or chloroformates.

(7) The reaction is aided by increased electronegativity of X since this aids polarization of the C—X bond. A reactivity series is therefore acetates > thioacetates > amides; the polarity of the transition state decreases along this series.

(8) The reaction is slightly aided by greater nucleophilicity in Y. It therefore takes place more readily with thionacetates (X = O, Y = S) than with acetates.

(9) The reaction is aided by electron-withdrawal at C_β but superimposed upon this is the effect of steric acceleration, so that bulky groups on C_β produce a rate increase even if they are inductively electron-supplying, and this is greater the bulkier the groups on C_α. The effect of β-substituents diminishes along the series of esters: $1^0 > 2^0 > 3^0$.

(10) The reaction shows a β-deuterium kinetic isotope effect which is close to the theoretical maximum for primary esters, but may diminish along the series $1^0 > 2^0 > 3^0$ and as R is more electron-withdrawing.

(11) The reaction is aided by steric acceleration.

(12) The decomposition is a first-order (and hence unimolecular) process and gives a stoichiometry of 2.0 if R is electron-supplying, so that the decomposition of the acid is relatively slow (e.g. for acetates). If R is strongly electron-withdrawing, the subsequent decomposition of the acid is instantaneous and the stoichiometry becomes 3.0 (e.g. for carbonates). For esters of acids of intermediate strength (e.g. benzoates) under static conditions, a stoichiometry of 2.0 is rapidly established and the pressure continues to rise slowly to give a final value of 3.0.

Evidence upon which these conclusions are based are the following:

1. The cyclic nature of the elimination

This mechanism, first proposed by Hurd and Blunck[103]. is now recognized as a symmetry-allowed 1,5-hydrogen shift of which many examples are known. This in itself constitutes an important piece of evidence, as does the unimolecularity of the reaction and the negative entropy of activation (most $\log A/s^{-1}$ values fall within the range 12.5—13.5).

2. The cis nature of the elimination

At its simplest, this is demonstrated by the fact that esters with $trans$ β-hydrogens only undergo elimination at temperatures which are very much higher than those at which esters with cis β-hydrogens will eliminate. However, there are very few esters with sufficiently locked conformations for this aspect to be demonstrated. The only clear example concerns the $trans$ isomer of 2-methyl-1-indanyl acetate (27) which gives 2-methylindene (28). By contrast the cis isomer (29) requires a 150°C temperature increase to bring about the same reaction[104] and this corresponds to a reactivity difference of > 10^4.

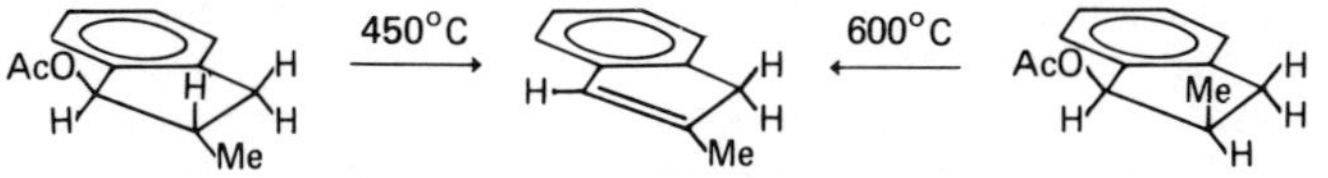

(27)   (28)   (29)

Esters with a choice of both types of β-hydrogen preferentially eliminate the *cis* hydrogen. Thus *cis*-2-substituted cyclohexyl acetates (**30**) give predominantly the corresponding 3-substituted cyclohexene; by contrast the *trans* isomers give a

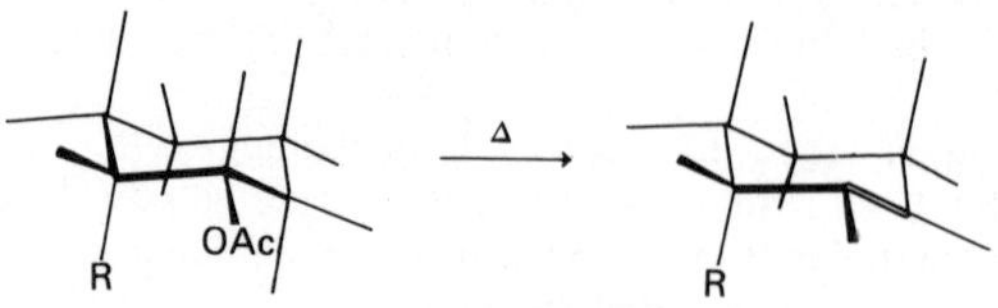

(**30**)   (R = Me, Ph, COOMe)

mixture of the 1- and 3-substituted cyclohexenes[105-107]. The fact that the 1-substituted cyclohexene is obtained to some extent from the *cis* isomers does not mean, as commonly suggested, that a different mechanism applies. There are two conformations for the *trans* isomer and one for the *cis* isomer in which the acetoxy group and β-hydrogen lie *gauche* to each other, permitting *cis* elimination. The *trans* isomer gives proportionally more of the 1-substituted cycloalkene, probably because one of its conformers has both bulky groups axial; elimination from this conformer will be *sterically accelerated*.

Curtin and Kellom produced the most elegant demonstration of the *cis* nature of the elimination by pyrolysing the *dl-erythro* and *-threo*-2-deuterio-1,2-diphenylethyl acetates **31** and **32** respectively) to *trans*-stilbene (**33**)[108]. The

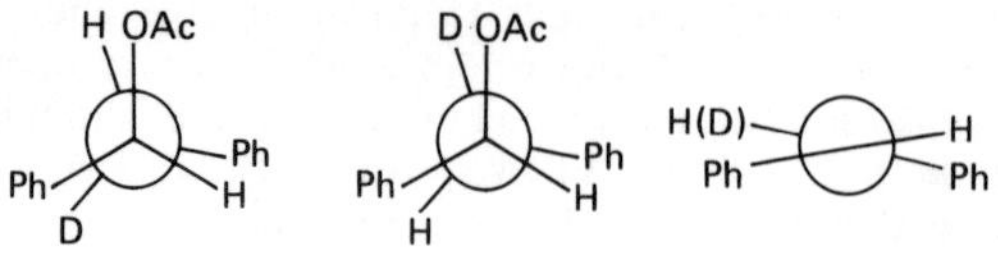

(**31**)   (**32**)   (**33**)

former compound retained 97% of the initial deuterium content whereas the latter retained only 26% and this clearly arises from *cis* elimination. However, the *threo* isomer (**32**) would, by comparison with the result for the *erythro* isomer, be expected to retain 3% of the initial deuterium and the discrepancy was assumed by later workers[109] to have been caused by isomerization during elimination. This view was considered to be supported by the fact that pyrolysis of *dl-erythro*- and *threo*-3-deuterio-2-butyl acetates (i.e. the analogues of **31** and **32** with Ph replaced by Me) produced but-2-ene with 97% retention and loss, respectively, of deuterium[109]. However, a more recent and accurate kinetic study of the rate of elimination from **31** and **32** showed that in fact 94% of deuterium is retained and lost, respectively[110]. The anomaly in Curtin and Kellom's work most probably stems from the fact that *cis*-stilbene oxide (the precursor of **32**) readily isomerizes to the *trans* isomer[109], which if it was not reduced immediately after preparation, would result in a *erythro*-contaminated *threo* product.

The *cis* nature of the elimination was used by Barton and Rosenfelder in analysis of the conformation of natural products[111]. A typical example concerned the configuration at the 7-carbon in allocholene steroids. Whereas the 7-benzoate-3-acetate of the supposed 7-'β'-epimer (**34**) eliminated benzoic acid to give cholest-6-en-3(β)yl acetate, the 7-benzoate-3-acetate of the 7-'α'-epimer (**35**) gave cholest-7-en-3(β)yl acetate. Consequently **34** was in fact the α-epimer and **35** the β-epimer. Similarly pyrolysis of cholestan-4-yl benzoates gave entirely cholest-3-ene from one

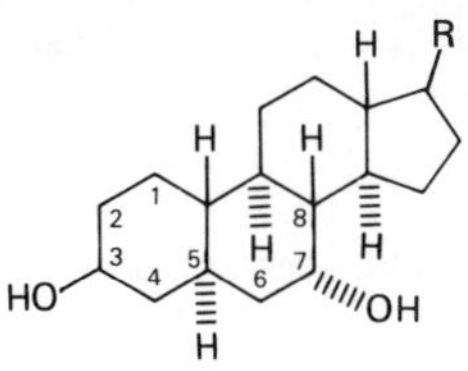

(34, R = $C_{18}H_{17}$) 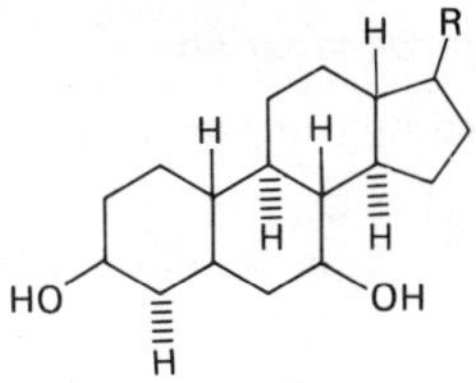   (35, R = $C_{18}H_{17}$)

isomer and a mixture of cholest-3- and -4-enes (in the ratio 1:1.5) from the other; these were therefore the β- and α-isomers respectively i.e. 36 and 37.

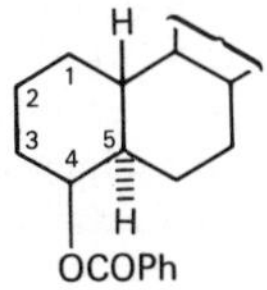

(36) 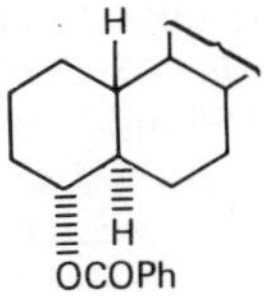   (37)

Although the above indicates that elimination is uniquely *cis*, it is possible that this depends upon ester type. For example, xanthates (which pyrolyse so much faster than they eliminate before gas-phase temperatures can be attained) are reported to be able to undergo *trans* elimination if the *trans* hydrogen is activated by a strongly electron-withdrawing group[112]. This may not be as anomalous as it seems. The transition state for ester pyrolysis is more polar the faster the elimination takes place, and consequently may be more ionic for xanthates. (This is indicated by the Hammett ρ-factor of +0.8 for pyrolysis of cholesteryl-S-aryl xanthates at 176°C[113] which is equivalent to 0.65 at 600 K and larger than for comparable secondary carboxylates — see below). The transition state for xanthate pyrolysis may therefore be more E1- and less E_i-like and hence less stereospecific. More work is needed to evaluate this aspect, especially since the xanthate derived from *cis*-2-phenyl-cyclohexanol was reported to give *less trans* elimination than did the acetate[105]. Reinvestigation of this aspect using modern physical analytical techniques might be valuable.

It is by no means certain that the 6-centre transition state is planar. Indeed this would require eclipsing of neighbouring groups so that a partly staggered conformation is more likely. Evidence to support this is twofold. Firstly, if eclipsing was required elimination in the cyclohexyl system would require intervention of the high-energy boat form with significantly slower elimination[114]; this is not observed. Secondly, elimination of the acetate derived from 1-methylcyclohexanol gives quite different *endo:exo* product yields [75% of 1-methylcyclohexene (38) and 25% of methylenecyclohexane (39)[115-119] than does elimination of the amine oxide derivative (3% and 97% respectively[120]). Now if elimination takes place

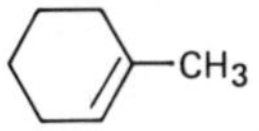

(38) 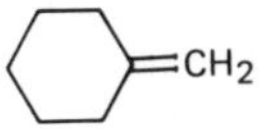   (39)

through the staggered conformation, the extra distance between the oxygen and the hydrogen in the amine oxide compared to the acetate (or carboxylate in general) makes elimination in the ring very difficult for the former compounds, but not at all disadvantageous for the esters.

3. Direction of the elimination

This is governed by three main factors:

    (*a*) statistical effects,
    (*b*) thermodynamic stability of the products, arising from steric effects,
    (*c*) electronic effects.

In any given situation all three effects may be in operation and they are, in any case, not independent of each other, e.g. (*b*) is a function of (*c*).

    a. Statistical effects. These are evident from the data gathered in Table 5. For example, consider elimination from 2-butyl acetate, ester no. 1 (**40**). On statistical grounds this should give 1- and 2-butenes in the ratio of 60% : 40% and the observed ratio of 57% : 43%[119,121-126] is close to this. Esters nos. 2−4 all show a similar result, and it should be noted that these results are not subject to complications arising from isomerization which does not take place under homogeneous gas-phase conditions[130]. Esters nos. 5 and 6 have one less 'internal' hydrogen so

TABLE 5. Product distribution in pyrolysis of aliphatic acetates

No.	Ester	Products	Reference
1	$MeCH_2CH(OAc)CH_3$	$MeCH_2CH=CH_2$ (57%) $MeCH=CHCH_3$ (15% *cis*, 28% *trans*)	119, 121−126
2	$EtCH_2CH(OAc)CH_3$	$EtCH_2CH=CH_2$ (55%) $EtCH=CHCH_3$ (45% *cis* and *trans*)	126
3	$i\text{-}PrCH_2CH(OAc)CH_3$	$i\text{-}PrCH_2CH=CH_2$ (46%) $i\text{-}PrCH=CHCH_3$ (54% *cis* and *trans*)	119, 127
4	$n\text{-}BuCH_2CH(OAc)CH_2$	$n\text{-}BuCH_2CH=CH_2$ (54%) $n\text{-}BuCH=CHCH_3$ (17% *cis*, 29% *trans*)	126
5	$Me_2CHCH(OAc)CH_3$	$Me_2CHCH=CH_2$ (80%) $Me_2C=CHCH_3$ (20%)	94, 119
6	$Et(Me)CHCH(OAc)CH_3$	$EtCH(Me)CH=CH_2$ (76%) $EtC(Me)=CHCH_3$ (24% *cis* and *trans*)	126
7	$(CH_3CH_2)_2C(OAc)CH_3$	$(CH_3CH_2)_2C=CH_2$ (35%) $CH_3CH_2C(CH_3)=CHCH_3$ (22% *cis*, 43% *trans*)	126
8	$MeCH_2CH(OAc)CH_2CH_3$	$MeCH_2CH=CHCH_3$ (40% *cis*, 60% *trans*)	128
9	$EtCH_2CH(OAc)CH_2CH_3$	$EtCH_2CH=CHCH_3$ (17% *cis*, 35% *trans*) $EtCH=CHCH_2CH_3$ (15% *cis*, 33% *trans*)	128
10	$n\text{-}PrCH_2CH(OAc)CH_2CH_3$	$n\text{-}PrCH_2CH=CHCH_3$ (35% *cis*, 12% *trans*) $n\text{-}PrCH=CHCH_2CH_3$ (53% *cis* and *trans*)	127
11	$i\text{-}PrCH_2CH(OAc)CH_2CH_3$	$i\text{-}PrCH_2CH=CHCH_3$ (12% *cis*, 33% *trans*) $i\text{-}PrCH=CHCH_2CH_3$ (5% *cis*, 50% *trans*)	128
12	$t\text{-}BuCH_2CH(OAc)CH_2CH_3$	$t\text{-}BuCH_2CH=CHCH_3$ (9% *cis*, 21% *trans*) $t\text{-}BuCH=CHCH_2CH_3$ (5% *cis*, 65% *trans*)	128
13	$MeCH_2C(OAc)(CH_3)_2$	$MeCH_2C(CH_3)=CH_2$ (76%) $MeCH=C(CH_3)_2$ (24%)	115, 119, 129
14	$EtCH_2C(OAc)(CH_3)_2$	$EtCH_2C(CH_3)=CH_2$ (72%) $EtCH=C(CH_3)_2$ (28%)	115, 129
15	$Me_2CHC(OAc)(CH_3)_2$	$Me_2CHC(CH_3)=CH_2$ (89%) $Me_2C=CHMe$ (11%)	115, 129
16	$(CH_3)_2CHCH(OAc)CH_2Et$	$(CH_3)_2CHCH=CHEt$ (73% *cis* and *trans*) $(CH_3)_2C=CHCH_2Et$ (27%)	119
17	$CH_2=CHCH_2CH(OAc)CH_3$	$CH_2=CHCH_2CH=CH_2$ (26% *cis* and *trans*) $CH_2=CHCH=CHCH_3$ (74%)	94
18	$CH_2=CHCH_2C(OAc)(CH_3)_2$	$CH_2=CHCH_2C(CH_3)=CH_2$ (50%) $CH_2=CHCH=C(CH_3)_2$ (50%)	94

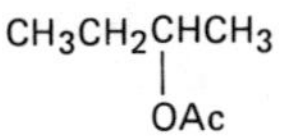

CH$_3$CH$_2$CHCH$_3$
|
OAc

(40)

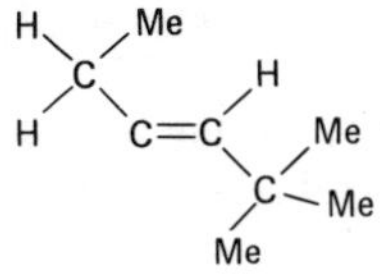

(41)

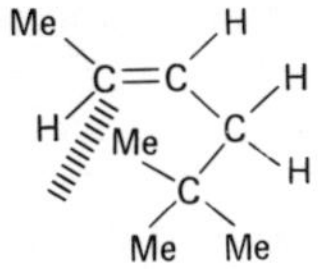

(42)

the amount of terminal alkene is increased to approximately the 75% expected. Ester no. 13 has the same ratio of terminal to 'internal' hydrogen as do esters nos. 5 and 6, and the product distribution is therefore similar.

b. *Thermodynamic stability of the products.*   An alkene has greater thermodynamic stability if it has maximum conjugation with the double bond, and minimum steric interactions. The effect of thermodynamic stability in governing the direction of elimination (and the former contributor in particular) has hitherto been considered to be very important. This follows from the work of DePuy and Leary[131] who pyrolysed ester 43 and obtained the alkenes 44 and 45 in 74% and 26% yields respectively. Very recent work has however shown that 45 is in fact the *major* product, the error in the original work arising most probably from 43 being contaminated with isomers[127a]. Conjugative stabilization of the product is therefore *unimportant* in governing the direction of elimination, and this is confirmed by other recent work (Section VI.D.6). It is now evident that esters eliminate so as to produce the minimum steric interaction in the product. However, this may not necessarily be a question of thermodynamic stability of the product, but due rather to the need to minimize steric interactions in the transition state, or to relieve steric interactions in the ground state.

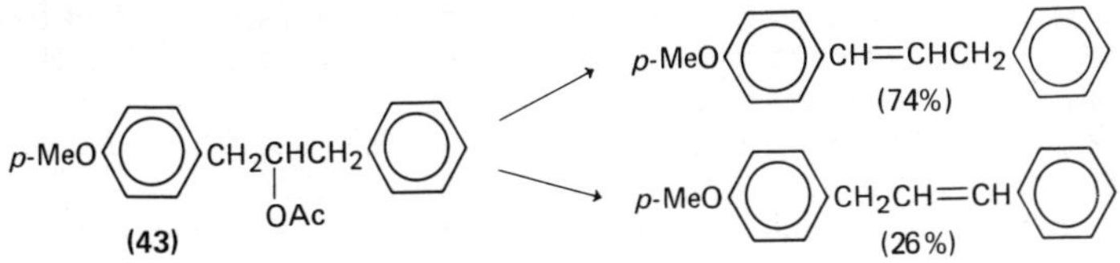

p-MeO⟨○⟩CH=CHCH$_2$⟨○⟩   (44)
(74%)

p-MeO⟨○⟩CH$_2$CHCH$_2$⟨○⟩
|
OAc
(43)

p-MeO⟨○⟩CH$_2$CH=CH⟨○⟩   (45)
(26%)

The effect of steric interactions is demonstrated by considering the product ratios given in Table 5. Esters nos. 1–4 show that the proportion of terminal alkene diminishes with increasing bulk in the terminal group; models indicate that the transition state for 2-alkene formation has the eclipsing interactions of lowest energy. The same argument accounts for the change in the 2-:3-alkene ratio for esters nos. 8–12, and here the marked change in the ratio with bulk of the terminal group demonstrates clearly that a steric rather than a conjugative effect is involved. It has been argued that for ester no. 12 this bulk produces *steric acceleration* towards formation of the 3-alkene[128]. This is not easily visualized without models, but is clearly evident in the products 41 and 42, and this is a permissible approach in view of the product-like nature of the elimination transition state. The 2-alkene (42) is severely hindered in one conformation in contrast to the 3-alkene. This does not of itself provide proof of *acceleration* of the formation of the 3-alkene, but such evidence is unambiguously provided by rate studies, described below (Section VI.D.3.c.iv).

By contrast, esters nos. 5–7 and 13–15 give *more* terminal alkene than statistically predicted, i.e. more of the least conjugatively stabilized product, and models indicate that the transition states for formation of the terminal alkene now has the eclipsing interactions of lowest energy. Ester no. 16 also gives the product which is least conjugatively stabilized.

Only esters nos. 17 and 18 give products which could be governed by conjugative stability, but even here it is not possible to rule out steric strain in the ground state as the important factor.

(*ii*) The need to minimize steric interactions also shows up in the tendency to form the *trans* alkene. Nevertheless, considerably less of this is produced than would be the case if a carbocationic intermediate were formed, permitting free rotation before loss of the proton. Thus the formation of a substantial amount of *cis* alkene again confirms the concerted nature of the transition state. Esters nos. 8–12 show the increasing amount of *trans* product formed with increasing size of the alkyl group (the anomalous result for 2-alkene formation from ester No. 10 is almost certainly the result of incorrect assignment of the isomers) and this reflects adoption of the least hindered conformer in the transition state.

It follows from the above that as the transition state becomes more polar i.e. move E1- and less E_i-like, two changes should be observed. First more *trans* product should be obtained and secondly, since loss of the hydrogen is not rate-determining for the E1 reaction, the importance of the statistical factor should disappear, i.e. the proportion of the thermodynamically less stable terminal alkene should diminish. A more polar transition state is produced on making R in **26** more electron-withdrawing, and examination of data for elimination from esters derived from butan-2-ol confirms both these predictions. Along the series acetates, chloro-, dichloro-, and trifluoroacetates, a small but definite increase in the amount of 2-butene, and of the *trans* isomer, is observed[124]. Likewise *bis*-but-2-yl carbonate pyrolyses to give more but-2-ene and more of the *trans* isomer than does but-2-yl acetate. One may predict that pyrolysis of 2-butyl 2,4,6-trinitrophenyl carbonate (which would have an even more polar transition state) would show further trends in this direction. Halide pyrolysis has a more polar transition state than does ester pyrolysis, and the polarity increases along the series chlorides < bromides < iodides. In agreement with the above analysis, the proportion of 2-butene and of *trans*-butene in pyrolysis of 2-butyl halides increases along this series[132].

The above features, and some interesting consequences, are evident from the

TABLE 6. Product distribution in pyrolysis of cycloalkyl acetates

No.	Ester	Products			Reference
19	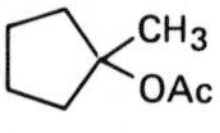	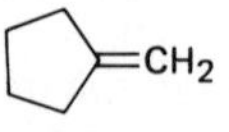 0–16%	+	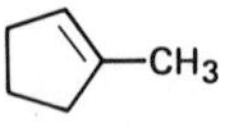 84–100%	119, 133
20	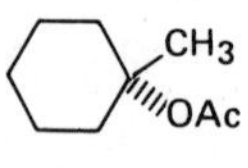	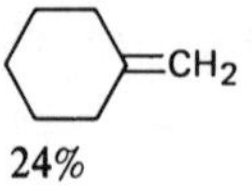 24%	+	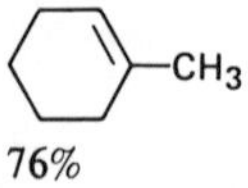 76%	115, 116, 119, 124, 126
21	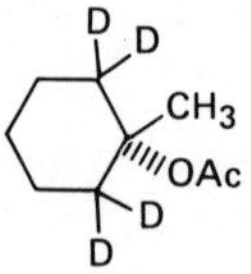	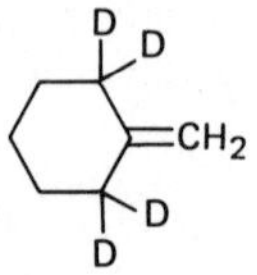 40%	+	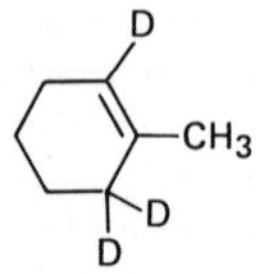 60%	117, 133

TABLE 6. (Continued).

No.	Ester	Products	Reference

22 · · 134

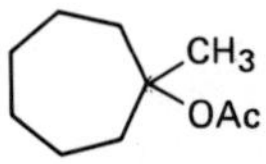

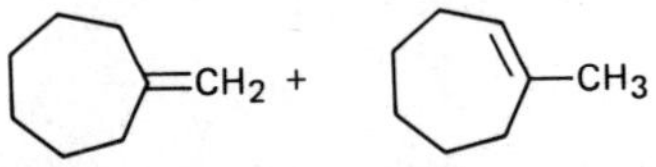

24% · 76%

23 · $\boxed{C_8-C_{10}}$ · Mainly ring-open α,ω-dienes · 134

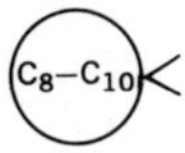

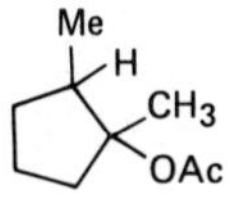

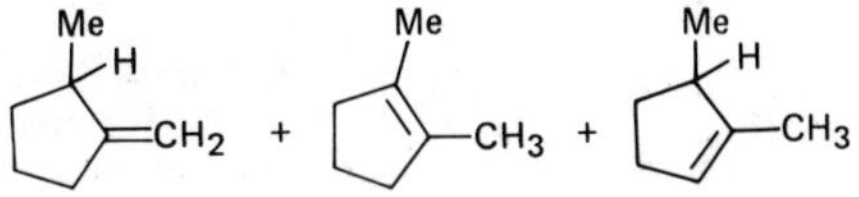

· 119

| 24 | cis | 20–30% | 40–50% | 25–30% |
| 25 | trans | 1–4% | 5–10% | 85–90% |

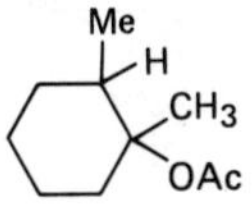

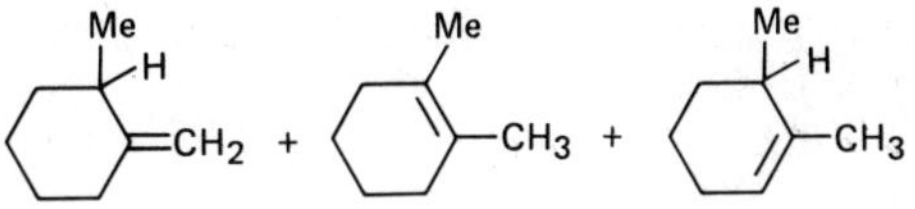

· 119

| 26 | cis | 46% | 28% | 26% |
| 27 | trans | 55% | 0% | 45% |

28 · · 135

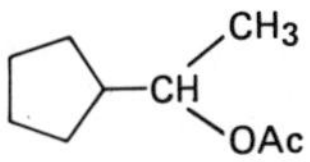

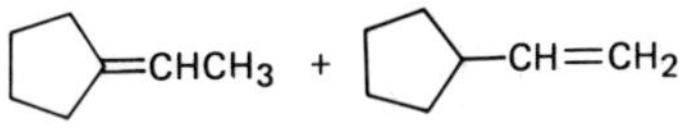

12.5% · 87.5%

29 · · 135

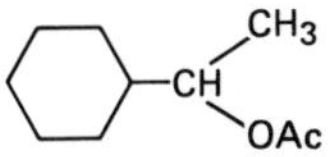

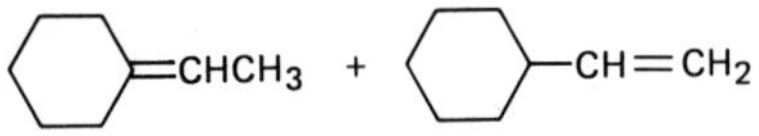

10% · 90%

30 · · 136

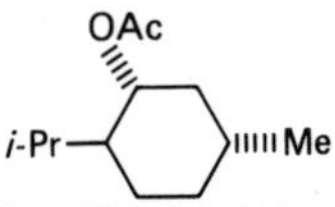

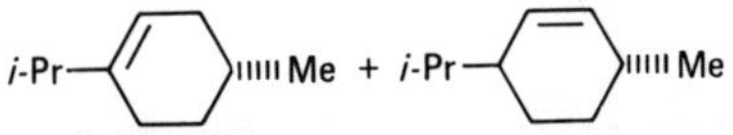

Menthyl acetate[a] (+)-p-menth-3-ene   trans-p-menth-2-ene
65%   35%

31 · · 137

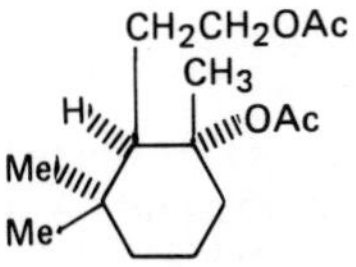

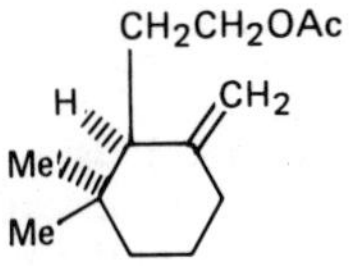

84%

TABLE 6. (Continued).

No.	Ester	Products	Reference
32	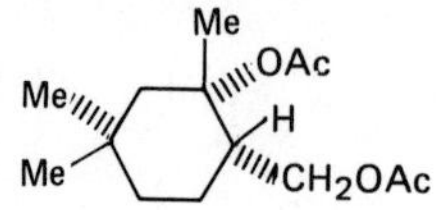	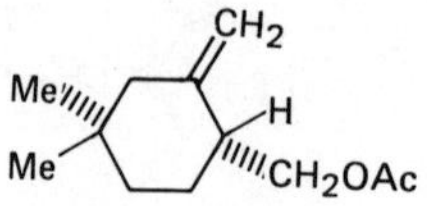 52%	138
33	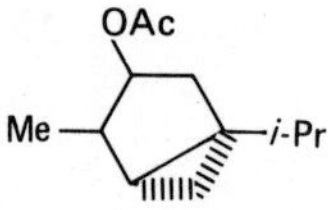	Thujene, 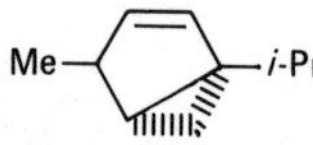 Me—⟨⟩—i-Pr, is *not* obtained, only ring-opened products (menthadienes and p-cymenes)	139
34	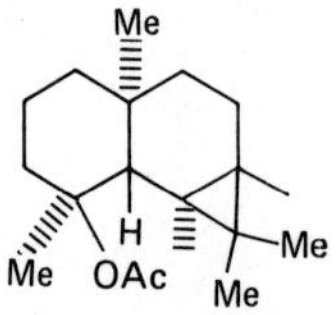	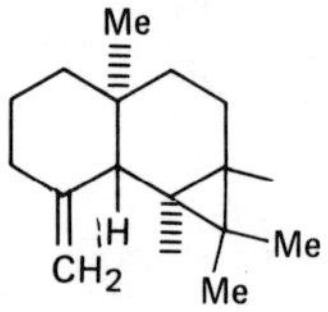	140
35	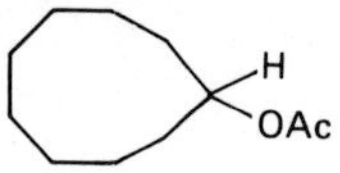	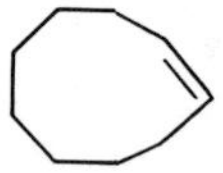 + 1,8-nonadiene 70% 27% *cis* + 1.5% *trans*	141
36	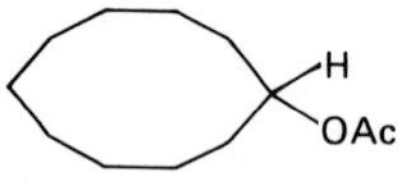	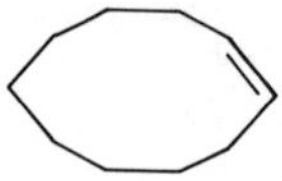 19% *cis* + 69% *trans*	142
143	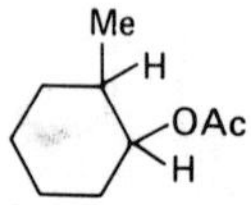	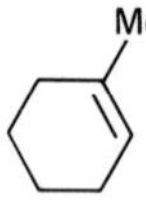 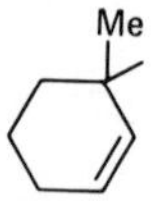	143
37	*cis*	25%[b] 75%	
38	*trans*	55%[b] 45%	

[a] The yield of menth-3-ene increased, and that of menth-2-ene decreased along the series acetate, (stearate), benzoate, carbonate and 5-methylthiolcarbonate.
[b] Under the reaction conditions, isomerization to 3-methylcyclohexene tended to occur.

data for pyrolysis of cycloalkyl esters (Table 6), discussion of which in the literature has tended to be either misleading, incorrect or non-existent. A detailed analysis is therefore given here, and the following features are notable:

(i) Pyrolysis of (−)-menthyl acetate gives the product with the fewest eclipsing interactions between the i-propyl group and the cyclohexane ring.

(*ii*) Pyrolysis of thujyl acetate produces ring-opening rather than forming thujene which is highly strained.

(*iii*) The products from pyrolysis of esters nos. 19–22 and 24–29 show that formation of a double bond *exo* to a ring is generally unfavourable. The *endo*cyclic alkene is favoured not only statistically, but more importantly because its formation produces the greater reduction in eclipsing interactions (within the ring or between the side-chain and ring) present in the initial ester.

(*iv*) The results for esters nos. 31, 32 and 34 appear to contradict the above observation. However, a conformational effect operates here, because in each ester there are methyl groups on the α- and γ-carbon atoms. For ester no. 34 these are fixed in an axial position, and they must adopt this conformation in esters nos. 31 and 32 so that the bulky acetoxy group and substituted alkyl groups may be equatorial. There are thus strong methyl—methyl steric interactions in each ester which are relieved on formation of the *exo*cyclic alkenes; these latter are therefore formed as a result of *steric acceleration*.

(*v*) Comparison of the products from esters nos. 19 and 20 show that formation of the *endo*cyclic alkene is preferred to a greater extent for the cyclopentyl ester than for the cyclohexyl ester. Since the acetoxy group and the β-hydrogens are held in a coplanar *cis* configuration in the cyclopentyl ester, the result follows simply from a more favourable entropy of activation. Since the methyl group and the adjacent C—H bond are eclipsed in ester no. 19, elimination from this ester should be faster than from ester no. 20, and this is found, a factor of 3.4–5.2 being obtained[144].

(*vi*) The tendency to avoid formation of the *exo*cyclic alkene is also evident from the products from esters nos. 28 and 29. These indicate that here, formation of the *exo*cyclic alkene from the cyclopentyl ester is not so unfavourable as from the cyclohexyl ester. This does not contradict, as might first appear, the results for esters nos. 19 and 20. Eclipsing interaction between the methyl group attached to the double bond, and the C—H bonds on $C_{(2)}$ and $C_{(6)}$ is worse for the 6-membered ring than for the 5-membered ring because the distance between them is shorter.

(*vii*) The 1-methylcycloalkyl acetates with 8–10 carbons in the ring undergo ring-opening on pyrolysis. This is somewhat surprising because models indicate that not all of the possible alkene products should be sterically precluded from being formed, and moreover cyclodecyl acetate (ester no. 36) itself does *not* undergo mainly ring-opening.

(*viii*) The pyrolysis of cyclodecyl acetate is particularly interesting because it yields mainly a *trans* product in what *appears* therefore to be a *trans* elimination. In fact however, because of the need to avoid steric interactions between the hydrogens on $C_{(3)}$ and $C_{(8)}$ in the transition state, the hydrogen which is streochemically *trans* to the acetoxy group is forced to lie *gauche* to it, so that a normal *cis* elimination occurs. The same applies to elimination from *cis*-2-methylcyclohexyl acetate, ester no. 37, in which formation of 1-methylcyclohexene implies a *trans* elimination, and it has been assumed that a different mechanism applies here[145]; this is not so. The bulky acetoxy group must adopt an equatorial position (with the methyl groups axial), and the *trans* β-hydrogen then becomes equatorial also, and therefore *gauche* to the acetoxy group; a normal *cis* elimination can therefore occur.

(*ix*) The results for esters nos. 24–27 appear at first sight to be irrational, yet in fact they can be simply explained. For both *cis* compounds the adjacent methyl groups sterically interact, and removal of this interaction is favourable. This is best achieved by converting one methyl to methylene, consequently the yields of *exo* alkene are increased relative to the monomethyl analogues, esters nos. 19 and 20. In the *trans* cyclopentyl compound there is no such methyl—methyl interaction so that steric acceleration towards formation of the *exo* alkene no longer applies and the yield of this is very small. Formation of 1,2-dimethyl-cyclopentene appears to involve a *trans* elimination. However, the eclipsing interactions in the cyclopentane ring can cause sufficient puckering to place the *trans* β-hydrogen *gauche* to the acetoxy group and a normal *cis* β-elimination therefore takes place. In the *trans* cyclohexyl ester, there is a choice between having either the acetoxy group, or both methyl groups, in an axial position. The latter are evidently bulkier and this places the acetoxy group *trans* to the β-hydrogen on $C_{(2)}$; consequently no elimination takes place in this direction. A second consequence is that the equatorial methyl groups are *gauche* to each other, so conversion of one of them to methylene is sterically favourable, a high yield of the *exo*cyclic alkene is therefore produced.

 c. Electronic effects. Electronic effects are now recognized as the dominant factor in governing the direction and rate of ester elimination. The effects diminish in magnitude when a given group is substituted at the carbons along the series $\alpha > \gamma > \beta$. One of the difficulties which has delayed evaluation of the true electronic effects of substituents has been that a given substituent can affect more than one site, produce steric effects and alter the number of β-hydrogens available for elimination. For example, comparison of the rates of elimination of 2-propyl acetate (46) and 2-butyl acetate (47) does not simply give the electronic effect of a methyl vs an ethyl substituent. In 47 there are now two types of β-hydrogen which

CH₃CHCH₃   CH₃CH₂CHCH₃
| |
OAc   OAc

(46)   (47)

will eliminate at different rates because one of them is affected by the adjacent methyl group. In addition the increased steric interaction in 47 causes elimination to be *sterically accelerated*. Consequently the most recent and meaningful studies have employed linear free-energy analysis of the rates of elimination of aryl esters in which the statistical and steric effects can be kept constant.

 (*i*) *Substituents at the α-carbon.* Esters pyrolyse through partial formation of a carbocation at the α-carbon and this was first shown by Taylor, Smith and Wetzel, who found that the logarithms of the relative rates of elimination of 1-arylethyl acetates (48) gave a linear correlation with σ^+-constants with $\rho = -0.66$ at 600 K[146,147]. (Linear free-energy correlations against σ^+-constants are diagnostic of transition states with electron-deficient centres at aromatic side-chain α-positions). Thus a positive charge develops where shown in 48. This was an important discovery not only with regard to elucidating the mechanism of ester pyrolysis, but because it provides a unique tool for measuring electrophilic substituent effects in the gas phase and this is discussed in detail below.

δ+
ArCHCH₃   ArCHCH₂Ph
| |
OAc   OAc

(48)   (49)

 Pyrolysis of 1,2-diaryl ethyl acetates (49) gave a similar correlation with $\rho = -0.62$ at 600 K[146,148]. These values are equivalent to ca -1.3 at 25°C, and the extent of carbocation formation may be judged by the fact that formation of a full carbocation at a side-chain α-position is estimated to give a ρ-factor of ca -20[149].

TABLE 7. Hammett ρ-factors for pyrolysis of 1-arylethyl esters at 600 K

Ester	ρ	Reference
Acetates	−0.66	146, 147
Benzoates	−0.72[a]	150
Methyl carbonates	−0.825	151a
Phenyl carbonates	−0.84	152

[a]The value of −0.80 given in the literature[151] is incorrect[150].

It should not be thought however that the extent of carbocationic formation is constant for all esters for this is certainly not the case. The above result logically indicates that the carbon–oxygen bond is polarized thus: $C^{\delta+}$—$O^{\delta-}$. It then follows that increased electron withdrawal towards oxygen (e.g. by making the R groups in **26** more electron-withdrawing) should increase the positive charge on C_α and hence increase the ρ-factor. This has recently been shown to be so from pyrolysis of the esters shown in Table 7.

A consequence of the formation of a partial carbocation at the α-carbon is that the ease of pyrolysis of alkyl acetates follows the order $3^0 > 2^0 > 1^0$ and indeed this was used as evidence to confirm the intermediacy of a carbocation[126]. There have been a large number of studies of the rates of elimination of various alkyl acetates (and formates)[153]. However, these have been obtained by a variety of workers using different techniques and widely different temperature ranges. (One of the problems associated with high-temperature kinetic studies is that although differences in temperature can be measured very accurately with thermocouples, the *absolute* value of the temperature is generally not known to better than ±0.5⁰. Since the rate spread is small, comparison of two sets of work can give misleading conclusions.) The results gathered in Reference 153 therefore show poor agreement and little quantitative information can be gained from them. The relative rates of pyrolysis of the primary, secondary and tertiary acetates have therefore been reexamined *under one condition* along with the corresponding esters of phenyl-acetic, benzoic, N-phenylcarbamic and phenylcarbonic acids; the data are gathered in Table 8[154]. These show the following important features:

(1) The rate spread increases regularly as inductive electron withdrawal in R is increased, i.e. *the transition state becomes more polar* on going from acetates through to phenylcarbonates. This confirms the evidence given in Table 7, and

TABLE 8. Relative rates of pyrolysis of esters RCOOAlkyl at 600 K[153]

R	$k_{rel.}$(t-Bu ester)	Alkyl-group rate ratios		
		$k(i\text{-Pr})/k(\text{Et})$	$k(t\text{-Bu})/k(i\text{-Pr})$	$k(t\text{-Bu})/k(\text{Et})$
CH₃	1	28.8[a]	115	3,315[a]
PhCH₂	1.55	32.3	121	3,910
Ph	2.22	36.3	125	4,540
PhNH	7.00	—	—	—
PhO	17.8	39.8	126	5,020

[a]These have been determined under the same conditions and supersede the extrapolated values given in the literature[126,155].

again shows that C—O bond polarization is increased by increased electron withdrawal in R.

It follows that if O is replaced by S then the polarization should decrease and thus the rate spread for S-alkyl thioacetates is smaller than for acetates[156,157], e.g. the relative ethyl: i-propyl: t-butyl rates at 600 K may be calculated from the data in Reference. 155 to be 1:17:1300.

(2) The rate spread increases the more reactive the esters. Thus the more reactive an ester type, the more polar is the transition state. (Again thioacetates are less reactive than acetates[156–158], e.g. t-butyl thioacetate is ca 30 times less reactive than t-butyl acetate at 600 K.) Likewise amides, where X = NH, are much less reactive than acetates because of the lower electronegativity of NH relative to O; the spread of rates for different N-alkyl groups has not however been measured.

(3) From (2) one can infer that the polarity of the transition state increases along the series $1^0 < 2^0 < 3^0$ and definite proof of this is given below. It is also indicated by the activating effects of the phenyl and cyclopropyl groups (X) in the secondary and tertiary esters **50** and **51**, respectively. In **50** the statistically

$$CH_3-\underset{\underset{OAc}{|}}{\overset{\overset{X}{|}}{C}}-H \qquad CH_3-\underset{\underset{OAc}{|}}{\overset{\overset{X}{|}}{C}}-CH_3 \qquad CH_3-\underset{\underset{OAc}{|}}{\overset{\overset{X}{|}}{C}}-H \qquad CH_3-\underset{\underset{OAc}{|}}{\overset{\overset{X}{|}}{C}}-CH_3$$

(50)   **(51)**   **(52)**   **(53)**

X = Me, Ph, cyclopropyl   X = H, Ph

corrected factors for acceleration relative to a methyl group are 3.45 (phenyl) and 3.6 (cyclopropyl) at 650 K, whereas in **51** these values become 4.5 and 8.6, respectively at 570 K[94,159], i.e. there is greater activation in the tertiary series because a larger charge is created at the α-carbon. Likewise the effect of a phenyl group relative to hydrogen in the primary and secondary acetates **52** and **53** are 50 and 122, respectively, at 650 K[152,153].

(4) Since the $k(t\text{-Bu})/k(i\text{-Pr})$ ratios are greater than the $k(i\text{-Pr})/k(\text{Et})$ ratios it follows that the polarity difference between the transition states for the 3^0 and 2^0 esters is greater than that between the 2^0 and 1^0 esters. It also follows that since the transition states for the t-butyl esters are already fairly polar, then these will be subject to smaller changes in polarity as the nature of R is varied, than will the transition states for the i-propyl esters. Consequently we see that on going from (thioacetates) through acetates to phenylcarbonates, the $k(i\text{-Pr})/k(\text{Et})$ ratio changes much more than does the $k(t\text{-Bu})/k(i\text{-Pr})$ ratio. This trend is maintained through to alkyl halide pyrolysis (which has a much more polar transition state) to the extent that these two ratios now become comparable, e.g. for alkyl chlorides these are 150 and 167, respectively at 633 K[160]. For chloroformates the $k(i\text{-Pr})/k(\text{Et})$ ratio is 220 at 513 K[161] $\equiv$ 100 at 600 K and from this one can calculate that t-butyl chloroformate will be thermally unstable at room temperature.

(ii) *Substituents at the* γ-*carbon.* Increasing electronegativity in the group R in **26** should increase the rate of elimination provided that in the transition state the electron pair (ii) has not moved as far as the pair (i). This is in fact observed. Earlier work indicated that the elimination rates were proportional to the pK_a of the acid[162,163], but this relationship is not general. For example, carbonates are very reactive towards elimination yet carbonic acid is very weak. This has previously been considered inexplicable[164], but can be rationalized as follows: Carbonic acid is weak because of the contribution of **54** to the resonance hybrid and the negative

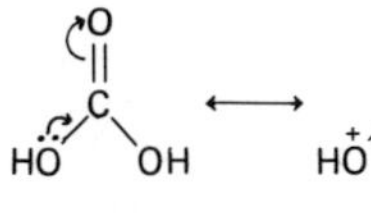

(54)

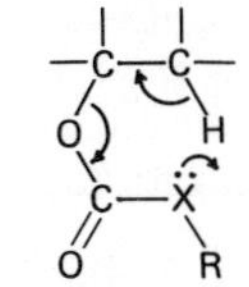

(55; X = O, NH)

charge on oxygen retards polarization of the adjacent OH bond. However, in ester pyrolysis the nucleophilicity of the carbonyl oxygen is a relatively unimportant factor in determining elimination rates, compared to the inductive effect of the group R. Consequently carbonates (R = O-alkyl) are much more reactive than acetates (R = alkyl). Moreover, the reactivity order carbonates > carbamates rules out the alternative elimination mechanism (55) because O is less nucleophilic than NH. On the other hand the greater −I effect for O relative to NH predicts the observed reactivity order provided the mechanism for these esters is as shown in 26; this conclusion has been reached on other grounds for carbonates[165] and for carbamates[166-168].

The effect of substituents at the γ-carbon was first clearly defined by use of the Hammett equation and the results of a number of studies which have utilized this are gathered in Table 9. The kinetic data correlated with σ^0- rather than with σ-values and this would be expected for the phenylacetates, carbamates and carbonates because of the atom which intervenes between the ring and the γ-carbon. But this correlation is true also for the benzoates and has hitherto remained unexplained. However, it can now be seen to arise from the fact that nucleophilic attack of the carbonyl oxygen upon the β-hydrogen is relatively unimportant as therefore is conjugation between the carbonyl group and the substituents. This therefore reinforces the above conclusion based upon the relative reactivities of carbonates and acetates.

The results in Table 9 show clearly that electron-withdrawal at the γ-carbon increases the reactivity of the esters, that the transition state polarity increases along the series $1^0 < 2^0 < 3^0$, and that the biggest increase in polarity comes between the 2^0 and 3^0 esters; these latter two conclusions confirm those deduced from analysis of the substituent effects at the α-carbon, noted above Section VI. D. 3. c. i. Although the data in Table 9 are as yet incomplete, the indication is that the charge developed at the γ-carbon increases on going from phenylacetates through to phenylcarbonates. Note that the apparent anomaly for the benzoates derives simply from the fact that the ρ-factors are larger because the phenyl group is attached directly to the γ-carbon. This trend therefore confirms the indications

TABLE 9. Hammett ρ-factors for pyrolysis of esters R¹ COOR at 600 K

| R | R^1 | | | |
	PhCH₂ (phenylacetates)	Ph (benzoates)	NHPh (N-phenylcarbamates)	OPh (phenylcarbonates)
Et	—	⎡0.26[169] ⎤^{a}	—	0.19[165]
i-Pr	—	⎢0.335[170] ⎥	—	—
t-Bu	0.39[150]	⎣0.59[171] ⎦	0.48[172]	—

^{a}When considering the ρ-factor for benzoates, the absence of any group between the aryl ring and the γ-carbon must be taken into account.

relating to transition-state polarity given above (Section VI. D. 3. c. i). The data have also been used to show that NH is a poorer transmitter of conjugative effects than are O or CH_2[172], and they have also been used to determine σ^0-values, especially for *ortho* substituents[165,169,170]. For the latter purpose the pyrolysis of the *t*-butyl esters is preferable since they give a larger rate spread, and carbonates and carbamates are the most suitable since the subsequent decomposition of the acid by-product is instantaneous and leads therefore to excellent first-order kinetics.

There have been a number of subsequent reports showing that electron withdrawal in R increases the rate of elimination. For example, the relative rates of pyrolysis of *t*-butyl acetate, chloroacetate and dichloroacetate may be calculated to be 1 : 3.7 : 8.5 at 650 K[163]. Cyclohexyl trifluoroacetate is 19 times more reactive than cyclohexyl acetate[144]. Tertiary hydrogen phthalate esters (where R = *o*-$HOOCC_6H_4-$) contain a very strongly electron-withdrawing group and therefore readily undergo liquid-phase pyrolysis[173]. Ethyl *trans*-crotonate is more reactive than ethyl acetate[174]. Diethyl carbonate is more reactive than ethyl methyl carbonate by a factor of 1,6[175] which becomes 0.8 after statistical correction, i.e. the greater electron supply at the γ-carbon by EtO relative to MeO reduces the rate. There have been other reports for the effects of altering the γ-substituents in pyrolysis of carbonates[125,176] and carbamates[168], but since these relate to symmetrical esters, the effect at the γ-position is compensated by the change in the nature of the α-substituent and no conclusions can be reached.

The general importance of electron withdrawal at the γ-position may also be judged by the fact that replacement of carbon by the more electronegative phosphorus gives esters such as tributyl phosphate (**56a**), which are much more reactive than the corresponding carboxylate[177]. Moreover, decreasing the electron withdrawal from the α-position, i.e. on going to **56b** and **56c**, produced a rate decrease which was greater than the statistical reduction in the number of β-hydrogens[177].

(56)   **(57)**

(a) R = R^1 = BuO
(b) R = BuO, R^1 = Bu
(c) R = R^1 = Bu

For the alkyl diphenylphosphinates (**57** the rate spread for the 1^0, 2^0 and 3^0 esters was $1 : 400 : 10^6$ at $126°C$[178], whereas at this temperature the values for the corresponding acetates may be calculated to be $1:70:81,000$[154]. Thus increasing electron withdrawal at the γ-position produces a more polar transition state. (Note also that the more polar reaction again produces a $1^0:2^0:3^0$ rate spread which is more nearly statistical — see Section IV. D. 3. c.i above.) On this basis germaacetates should be less reactive than acetates. The only report so far concerns ethyl germaacetate (GeH_3COOEt)[179], but with this the temperature of elimination produces secondary decomposition of germaacetic acid; this decomposition could probably be avoided by using the *t*-butyl derivative.

For chloroformates and cyanoformates, which have strong electron withdrawal

from the γ-carbon, two further aspects of the elimination have to be considered:
Firstly, the mechanism as shown in **58** has been proposed[180,181] as an alternative

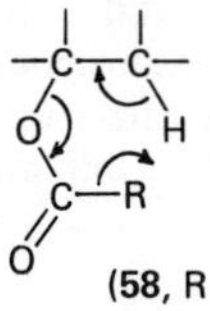

(58, R = Cl, CN)

to that in **26** and is not easily distinguished from it because the products and
stoichiometry will be the same. (Elimination via **26** will produce chloroformic and
cyanoformic acids which decompose instantaneously.) However on the basis of
analogy, the mechanism in **58** must appear less likely and it may be noted that even
xanthates, which possess a strongly nucleophilic SR group on the γ-carbon, eliminate via the mechanism in **26**[182] Also against **58** is the fact that the activation
energies for a range of ethyl esters including the chloroformate and cyanoformate
correlate with Taft σ^*-values for the group R indicating a common mechanism for
all of them[183]. Moreover the log A/s^{-1} value for the chloroformate is the same as
for the other esters[183]. (One way of distinguishing the mechanism for chloroformates might be to measure the rate vs bromoformates; these latter will pyrolyse
faster if **58** applies and slower if (**26**) applies). The log A/s^{-1} value for cyanoformates, however, is less than for the other esters, and while this may be due to
experimental error[183], it is also significant that the rate of elimination is lower
than for chloroformates[181,183], which is anomalous in terms of the electron-withdrawing abilities of Cl and CN. It may be that **58** therefore applies for cyanoformates and given this it is possible that a 7-membered transition state applies
(with attack of nitrogen on the β-hydrogen[183].

Secondly the reactions are accompanied by an $S_N i$ reaction[65,180,181], and this
is general for ester pyrolysis. Characteristics of this reaction (43) are[65]:

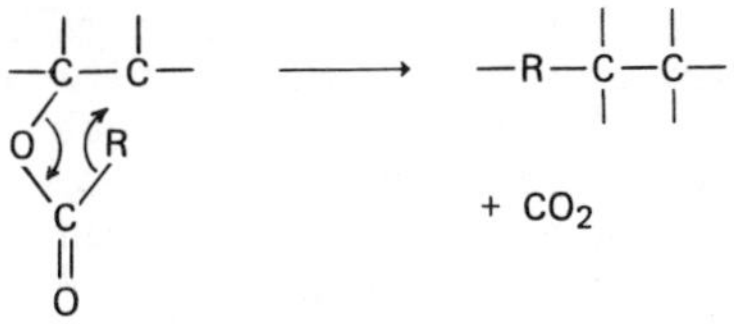

$$+ CO_2 \tag{43}$$

(1) It becomes more significant as R is more electron-withdrawing,
(2) it is surface catalysed,
(3) it is sterically hindered,
(4) it becomes less important along the ester series $1^0 > 2^0 > 3^0$,
(5) it has a lower activation energy than the elimination.

Thus just as in solution, elimination is accompanied by nucleophilic substitution
which is sterically hindered, so the exact parallel is found in the gas phase.
Moreover, as in solution, the nucleophilic substitution has the lower activation
energy and this may stem from the fact that a strong C–H bond has to be broken in
the elimination.

It follows therefore that in order to obtain the maximum alkene yield, the
surface should be inactive and the temperature as high as possible and this is more
important for primary than for tertiary esters. The identification of the $S_N i$ criteria
has led to the discovery that carbonates can be decomposed to ethers[65], and since
then the analogous formation of thioethers from thiocarbonates has been

reported[184]. Nucleophilic attack on carbon in the gas phase is of course not unique, and reported examples include the rearrangement of xanthates to dithiol-carbonates[185] and of thioncarbonates and -carbamates to thiolcarbamates and -carbonates (equation 44)[186]; the former process may account for the 'stable xanthates' produced on pyrolysis of xanthates[187], though the removal of free radicals has been suggested as an alternative explanation[188].

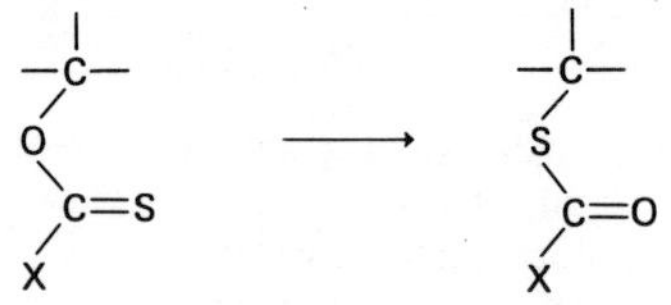

$$
\begin{array}{ccc}
\begin{array}{c} | \\ -C- \\ | \\ O \\ \diagdown \\ C=S \\ \diagup \\ X \end{array}
& \longrightarrow &
\begin{array}{c} | \\ -C- \\ | \\ S \\ \diagdown \\ C=O \\ \diagup \\ X \end{array}
\end{array} \qquad (44)
$$

(X = SR, OR, NR$_2$)

The pyrolytic cyclization of β-keto amides has been noted above (Section IV.A); the oxygen analogue, acetoacetic ester also undergoes a pyrolytic cyclization in the presence of quinoline to give **59** and **60** as a result of elimination of ethanol or ethanol and carbon dioxide, respectively[189]; nucleophilic attack on carbon must be involved in these reactions.

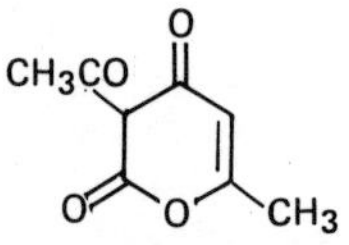

(59) 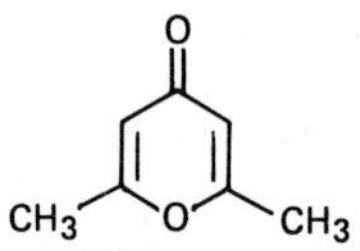   **(60)**

(*iii*) *The nature of Y.* The lower reactivity of carboxylates relative to xanthates and thionacetates[157] (Y = S) could stem from three possibilities. First is the greater nucleophilicity of thion sulphur relative to carbonyl oxygen, though the indication in Section VI. D. 3. c. ii is that the nucleophilicity of Y is not very important. Second, and not previously considered, is the larger size of sulphur which brings it closer to the β-hydrogen. If this is important it should show a more favourable entropy of activation, though the available evidence does not indicate this[157,190]. Third, and probably most important, is the energetically favourable conversion in the transition state of the system −O−C=S into −S−C=O[190].

(*iv*) *Substituents at the β-carbon.* The true nature of the effects of β-substituents has been elucidated only recently[192]; discussions published prior to this are misleading in several respects. Esters of types **61**−**63** have been studied. For

XC$_6$H$_4$CH$_2$CH$_2$OAc   XCH$_2$CH$_2$OAc   XCH$_2$CH$_2$CH$_2$OAc

(61)   **(62)**   **(63)**

substituents in **61** a Hammett correlation was obtained with $\rho = 0.2$ at 650 K (this supersedes an earlier approximate value[146]) showing that electron supply from a substituent retards the reaction and vice versa. The fact that this value is smaller than the values produced by aryl-group substitution at the α- and γ-carbons shows that C−H bond-breaking is not so important kinetically as C−O bond-breaking. In general substituents X in **61** and **62** produce the same qualitative effect upon the rate, with the exception of the alkyl substituents. These *accelerate* the reaction in **62** and this becomes more marked the bulkier the alkyl group, and the bulkier the substituents on the α-carbon, as the data in Table 10 show. (Note that β-methy-lation increases the rate per β-hydrogen for 1°, 2° and 3° esters; these data have

TABLE 10. Rate per -hydrogen due, and adjacent to, the group R[126,192]

	R		
	Me	Et	i-Pr
HCHR$\mid$OAc	1	1.37	1.32
CH$_3$CHR$\mid$OAc	1	2.6	3.8[a,b]
(CH$_3$)$_2$CR$\mid$OAc	1	3.5	3.05[c]
t-BuCHR$\mid$OAc	1	2.6	1.4

[a]For CHMe(Et). [b]For i-Pr a value of 4.5 is reported[94].
[c]Reference 94 gives 4.1.

been incorrectly analysed in the literature[155].) The anomalous effect of alkyl groups is therefore due to *steric acceleration*[191], and the data in Table 10 arise from a combination of rate retardation due to the electronic effects of the alkyl groups, and rate acceleration due to their bulk[191]. The steric acceleration by β-alkyl groups is most dramatically demonstrated by the rates of elimination of *trans*-2-alkylcyclohexyl acetates which are 1.1(Me), 1.8(i-Pr) and 13(t-Bu), relative to the unsubstituted ester[144]. The explanations of these effects in terms of long-range inductive stabilization of the remote α-carbocation has been shown to be incorrect on a number of grounds[191]. The combination of electronic and steric effects account nicely for the differing rates produced by β-phenyl, vinyl and ethynyl substituents which have qualitatively similar electronic effects but quite different bulks[191].

The best correlation of the data for the 2-arylethyl acetates (61) required the Yukawa—Tsuno version of the Hammett equation suggesting that conjugative effects of substituents at the β-carbon are the more important. This was confirmed by comparison of rates for esters 62 and 63, e.g. MeO and PhO ($-I$, $+M$) deactivate in 62 but activate in 63 where only the $-I$ effect can operate. Likewise the 240-fold acceleration per β-hydrogen for the effect of acetyl ($-I$, $-M$) in 62[193] becomes a mere 3.14-fold in 63[191]. The conjugative effect stems from the favourability and unfavourability of 64 and 65, respectively.

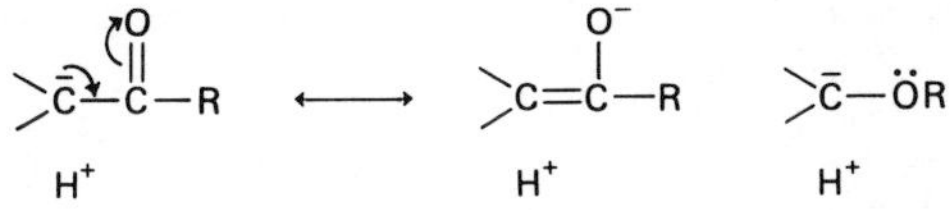

(64)   (65)

Throughout this account the variability of the transition-state structure with reactivity of the ester type has been stressed, and it has been argued that the transition state will tend from E$_i$ to E1 on going from 1⁰ to 3⁰ esters and from acetates to carbonates. This being so the importance of C—H bond-breaking, and

TABLE 11. Hammett ρ-factors for 2-arylethyl esters

Ester	ρ	Reference
CH$_2$CH$_2$Ar | OAc	0.2	191
PhCHCH$_2$Ar | OAc	0.08	148
CH$_2$CH$_2$Ar | OCOOMe	0.1	194

therefore of the β-substituent effect should diminish along this series. Evidence to suggest this is given in Table 11; it is also confirmed by the fact that a 2-methoxy substituent is less deactivating in a secondary ester than in a primary one[126,191], and in particular by the acceleration per β-hydrogen atom due to an adjacent acetyl and phenyl group (Table 12). The decreasing effect of the β-substituents is maintained through to halide pyrolysis which has a more E1-like transition state than ester pyrolysis. An interesting feature is that the results again show (cf. Section VI. D. 3. c. *i, ii*) that there is a bigger change in transition-state polarity between the 3^0 and 2^0 esters than between the 2^0 and 1^0 esters.

TABLE 12. Rate per β-hydrogen due, and adjacent to, the group X[144,191]

Ester	X	
	Ph	COCH$_3$
CH$_2$CH$_2$X | OAc	6.9	257
MeCHCH$_2$X | OAc	–	241
PhCHCH$_2$X | OAc	3.95	–
Me$_2$CCH$_2$X | OAc	–	17.3
Me$_2$CCH$_2$X | OCOCF$_3$	–	3.5
[CH$_2$CH$_2$X | Cl]	1.0	2.4

4. Isotope effects

Both the ether oxygen and α- and β-hydrogen isotope effects have been examined. Since in the transition state partial ionization of the C—O bond occurs, the extent to which an ion pair might be formed has been evaluated using ^{18}O-enriched ethyl acetate[195]. No scrambling of the label occurred on partial pyrolysis, but this is to be expected since of all esters, ethyl acetate has almost the least polar transition state. However, the same result was found using t-butyl N,N-dimethyl-carbamate which has a more polar transition state[196]. The greatest chance of observing scrambling would be to examine an ester such as t-butyl 2, 4, 6-trinitro-phenyl carbonate as this will have just about the most polar transition state for an ester.

Since breaking of the β-C−H bond is believed to be partially rate-determining in ester pyrolysis, a rate reduction should be obtained on replacing the β-hydrogen by deuterium, and this has been confirmed by a number of studies[108-110,117, 125,143,172,197,198]. Notable features of the data obtained are the range of k_H/k_D values (1.79[143]−2.8[108]) and the fact that the values are close to (or even exceed) the theoretical maximum (calculated from the i.r. stretching frequencies of the C−H and C−D bonds). Part of the variation derives from the different temperatures employed, and part is experimental error since most values have been obtained from product studies only, and this accounts for example for the higher than theoretical value of 2.8[110]. Additional complications are that α-hydrogens were deuteriated in some studies[197] and the effect of this was not separately deter-mined, the observed β-deuterium isotope effect might contain a contribution from hyperconjugative stabilization of the α-carbocation, and, because the isotope effect is so large, an error of $x\%$ in the isotopic purity of the starting material produces an error of $> 2x\%$ in observed value.

The importance of these factors has been evaluated by the reviewer using compounds **66−70**. Ester **66** pyrolysed 1.025 times slower than its non-deuterated

$$PhCDCH_3 \quad PhCHCD_3$$
$$| \qquad\qquad |$$
$$OAc \qquad\quad OAc$$

(66)   **(67)**   **(68)**   **(69)**   **(70)**

analogue and this secondary effect can be attributed to rehybridization changes at the α-carbon in the transition state[110] This result means that the values of Blades and Gilderson[197] (who used d_5-ethyl acetate) should have been up to 5% lower. Ester **69**, after corrections for the 4.5% of 'trans' elimination that takes place, gave an identical isotope effect to that (2.14 at 632 K) obtained with ester **70**. Likewise ester **68** after the same correction gave no isotope effect, both these results thereby showing that the hyperconjugative effect is trivial (at least for 2^0 esters). This was further confirmed by the fact that ester **67** (which has an additional β-deuterium available for hyperconjugation) gave almost the same isotope effect (2.15 at 656 K). In this work all of the esters were shown by n.m.r. to be $> 99\%$ iso-topically pure.

Since the transition state becomes more E1-like as the ester becomes more reactive, a smaller β-deuterium isotope effect might be expected in this direction. On the other hand, the hyperconjugative effect might be more important for tertiary esters so that the two effects might tend to cancel out. As yet insufficient data have been obtained under a given set of conditions to permit a definite conclusion.* However, t-butyl N-p-tolylcarbamate gives a value of 2.56 at 469 K

Added in proof: A comparison of the isotope effects in pyrolysis of acetates and carbonates has now produced evidence for these two phenomena[152].

which is smaller than obtained with the above (and less reactive) secondary acetates ($\equiv$ 2.92 at 469 K); it is also considerably smaller than the theoretical maximum of 3.12 at this temperature[172]. On the other hand, Kwart and Slutsky have reported a much higher value of 2.6 at 550 K for t-butyl N,N-dimethylcarbamate which is only slightly less reactive[198].

It is evident that accurately determined β-deuterium isotope effects are ca 80% of the theoretical maximum, but this by no means implies that the C—H bond is largely broken in the transition state. Since the maximum effect should be obtained at half-transfer from C to C and less from C to O, the results argue for as little as 25% C—H bond breaking in the transition state[110]. It has been argued that the large effect requires linearity of the OHC angle[198]; this can only be accommodated satisfactorily by having the C_α—O bond very stretched (71) and it should be noted that all of the other kinetic data indicate that stretching of this bond is the most important feature of the transition state.

R (71)

One report has indicated that the elimination of *trans* β-hydrogen from 2-methylcyclohexyl acetates (and xanthates) proceeds with a much reduced isotope effect implying a different (and ionic) mechanism for the *trans* elimination[143]. Against this conclusion must be set the following: (1) a *cis* elimination is feasible merely by a change in ring conformation (and which would place the bulky acetoxy group in the more favourable equatorial position) and (2), for the normal *cis* elimination, the observed isotope effects (1.79−1.89) are very much less than in any other study of acetates.

a. Solvent effects.   The high reactivity of carbamates has resulted in their pyrolytic rates of elimination being studied *in solution*[167,199] as well as in the gas phase[172] and this is one of the very few reactions which have been studied under both conditions. the Hammett ρ-factors (for t-butyl N-arylcarbamates) under both conditions are (after correction for the temperature difference) almost identical, which is consistent with the relatively non-polar nature of the transition state. The solvent does however increase the overall rates by the following approximate factors: 3.6 (dodecane), 8.0 (diphenyl ether), 9.3 (acetophenone), 15.4 (nitrobenzene) and 25 (decanol)[172].

The elimination of acetates in solution is catalysed by alumina especially if this is acidic, to the extent that it takes place readily *at room temperature*[200]. This is a remarkably easy way of preparing alkenes, passage of a solution of the ester in carbon tetrachloride down an alumina column being all that is required. The relative rates of elimination of 1-arylethyl acetates indicate a carbocationic mechanism[200].

5. Neighbouring-group effects

a. Steric acceleration.   A number of examples of this have already been revealed in this review. Steric acceleration also probably accounts for the anomalous effects of alkyl groups at the γ-position in elimination. Just as at the β-carbon we find that the effect when attached directly to this carbon is opposite to that when acting through a phenyl ring, so this is true also at the γ-carbon, i.e. the larger alkyl groups cause rate acceleration instead of the expected diminution. This is most clearly shown by the fact that secondary pivalates eliminate more readily than secondary acetates[144,201] and the accelerating interaction may be envisaged as being between the alkyl groups on the α- and γ-carbon atoms. This view is reinforced by the report that ethyl pivalate is not more reactive than ethyl acetate[202].

b. *Anchimeric assistance.* the relative rates of pyrolysis of *anti*- and *syn*-7-acetoxy-7-methylnorbornene (72, 73) and of 7-acetoxy-7-methylnorbornane (74)

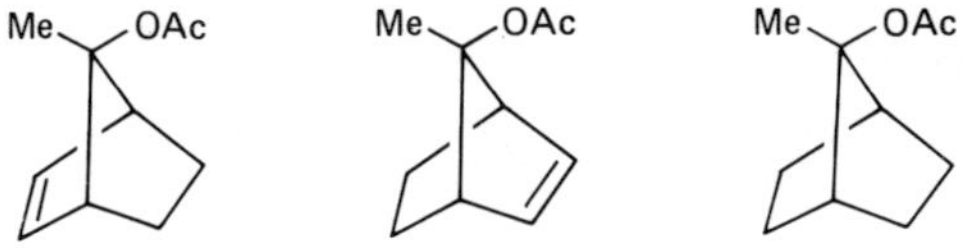

(72)   (73)   (74)

are $1.9 : 0.87 : 1.0$[203]. The relative rates of 72 to 74 can be explained by anchimeric assistance, i.e. through space stabilization of the incipient carbocation by the double bond which is on the opposite side of the α-carbon to the departing acetoxy group. However, since the allyl group is rate-enhancing[94], both 72 *and* 73 should be more reactive than 74. Since this is not observed this may be a further example of *steric acceleration* (in 74) since it is more crowded in the vicinity of the γ-carbon than is 73.

6. Rearrangements

Since a free carbocation is not formed in ester pyrolysis, rearrangements indicative of such intermediates are generally insignificant. More common are rearrangements which can be ascribed to favourable juxtaposition of the acyloxy group and a hydrogen which is not in the β-position, and their consequent elimination. This accounts for example for the formation of 1.8-nonadiene from pyrolysis of cyclononyl acetate (equation 44)[141], (see Table 6). Similar results have been noted by

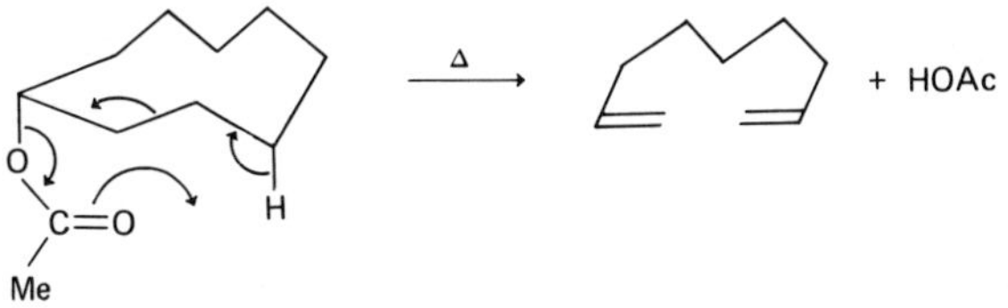

$$+ \text{ HOAc} \qquad (44)$$

Cope and Youngquist[204]. In some instances such rearrangements arise from the very enhanced acidity of a suitably situated hydrogen, such as is found in α-acetoxy ketones[205]. Elimination from these results in extrusion of CO, as for example shown in equation (45)[206]. Rearranged products have also been obtained in

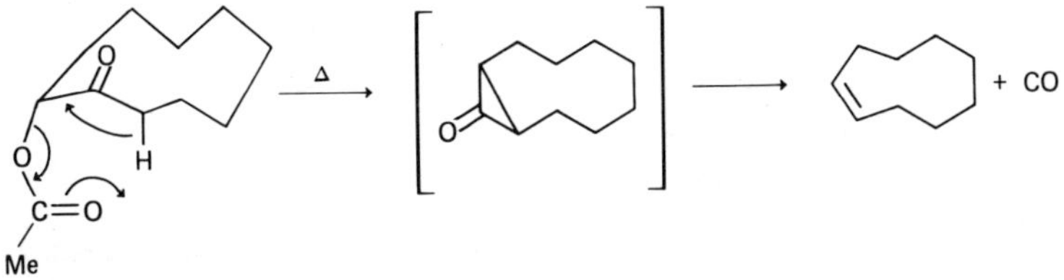

$$+ \text{ CO} \qquad (45)$$

pyrolysis of esters derived from strained-ring compounds[207], but it is not certain that these are the result of the elimination process itself. The extent to which rearranged products are the result of the elimination alone is also not clear in pyrolysis of bornyl acetate (75) and isobornyl acetates (76)[208] (cf. References 209 and 210), in which camphene (78) and tricyclene (79) accompany the normal *cis* elimination product bornylene (77). The latter acetate eliminated 6.8 times

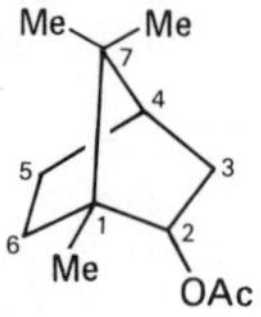

(75) 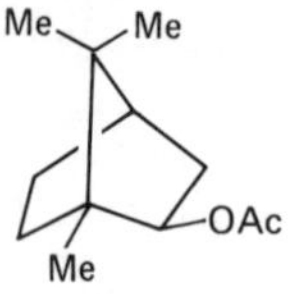   (76) 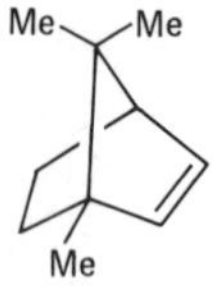   (77) 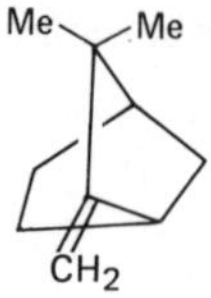   (78) 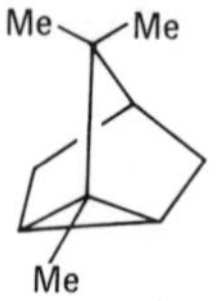   (79)

faster than the former at 638.4 K; this result qualitatively parallels those obtained for related reactions in solution and which have been interpreted in terms either of steric acceleration from the methylated bridge, or from synartetic assistance by the electrons of the 1,6–bond. Although steric acceleration has been identified in a number of instances in this review, the large factor noted above does seem to rule this out as a primary explanation. Synartetic assistance on the other hand could lead to rearranged products such as those noted above. For example camphene could be produced from the initial rearrangement (equation 46). Such 1,2-migrations of ester function have been reported for xanthates[185], and should for

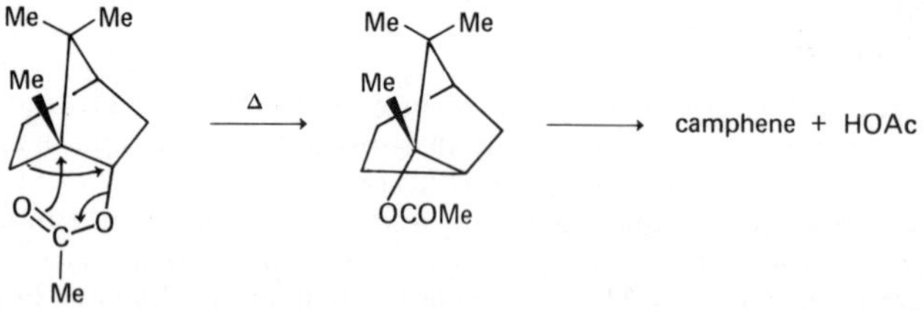

$$\xrightarrow{\Delta} \qquad \longrightarrow \quad \text{camphene} + \text{HOAc} \qquad (46)$$

these be easier because of the greater nucleophilicity of the thion sulphur. Consequently, it is consistent with this view that xanthates give a higher yield of camphene in the above reaction than do acetates and also that the isobornyl esters give a considerably higher yield than do the bornyl esters; in the former the electrons of the 1,6–bond. Although steric acceleration has been identified in sides of the 2-carbon atom.

An alternative explanation of these results is that camphene is formed via a 7-membered transition state. However this must be considered improbable as is the 7-membered transition state and α-elimination proposed by Kwart and co-workers[211] to account for the formation of 2-benzylpropene (81) and 2,2-dimethylstyrene (82) from neophyl acetate (80) (and the corresponding methyl

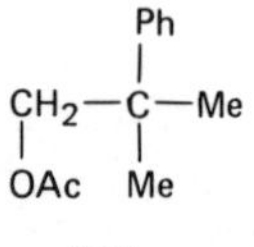

(80) 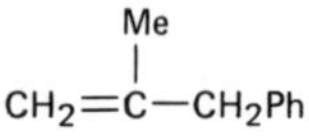   (81) 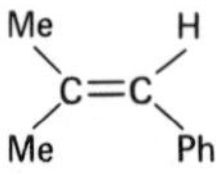   (82)

carbonate). The simplest and most attractive explanation of these results is that synchronous phenyl- and acetoxy-1,2-migration occurs to give benzyldimethyl-carbinyl acetate (equation 47), *and this migration is exactly analogous to reaction (46)*. Elimination following reaction (47) would then give **81** and **82**. This mechanism was ruled out by Kwart and coworkers because **81** and **82** were formed in the

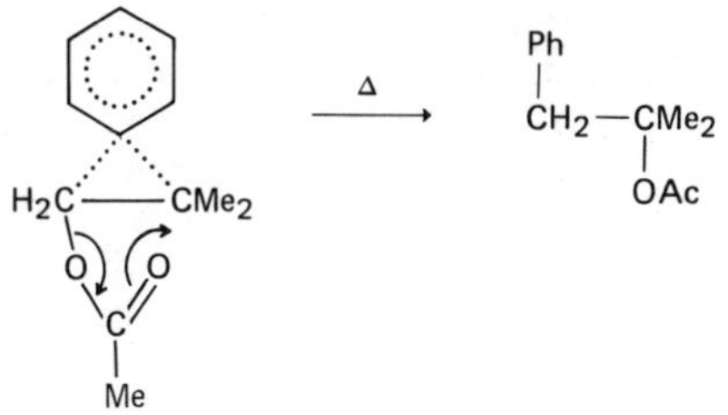

$$\text{(47)}$$

Ph
|
CH$_2$—CMe$_2$
|
OAc

H$_2$C——CMe$_2$

Me

ratio of 1.6 : 1.0, and greater than they considered should be the case based upon *liquid-phase* pyrolysis data. However, it should be noted that on statistical grounds alone the ratio should be 3 : 1 (see Table 5 for data on the related alkyl esters) and halving this to allow for the conjugative effect does not seem unreasonable. Moreover, if the activation energy for removal of the more acidic β-hydrogen to give the conjugated alkene is lower, as is certain to be the case, then at the high temperature used in this study, elimination from the terminal methyl groups could be more significant than indicated by data obtained at lower temperatures*. Related migrations may also account for the formation of 4-methyleneprotoadamantane and 3-vinylnoradamantane from pyrolysis of 3-homoadamantyl acetate[212].

An unusual, and as yet unexplained, rearrangement takes place in the pyrolysis of 2-(1-acetoxyethyl)pyridine *N*-oxide (**83**) which gives 2-acetylpyridine instead of the 2-vinylpyridine *N*-oxide. The latter does not appear to rearrange to the former under the reaction conditions, nor can the product be derived from nucleophilic attack of the oxide upon the carbonyl group of the ester since it is also obtained on pyrolysis of 2-(1-hydroxyethyl)pyridine *N*-oxide. A possible explanation is that the oxygen on nitrogen inserts into the side-chain C$_\alpha$—H bond (a well-known reaction under certain conditions) to give **84** which would then rapidly lose acetic acid (or water in the case of the alcohol) because it is a 1,1-diol or derivative[213].

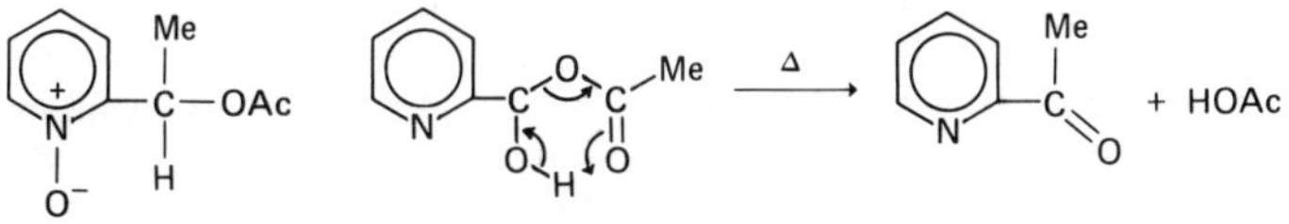

Me
|
C—OAc
|
H

N
|
O⁻

(83)

O
‖
C—Me
N

H

(84)

Me
|
C
‖
O

N   + HOAc

Rearrangement due to ring strain in the product takes place in the pyrolysis of 1,2-diacetoxymethylcyclobutane. Although this gives 1-methylene-2-acetoxy-methylcyclobutane, the expected 1,2-dimethylenecyclobutane is not obtained. This latter contains two *sp*^{2} carbon atoms in the 4-membered ring so the strain is, at the temperature of pyrolysis, sufficient to rupture the ring; in solution eliminations at much lower temperatures, this molecule can be formed. Rupture of the ring prior to pyrolysis also occurs, giving allyl acetate[213a].

Added in proof: All of these proposals have now been confirmed. Benzyldimethylcarbinyl acetate gives precisely the same isomer ratios as does neophyl acetates[127a].

7. Summary of the mechanism

All of the foregoing evidence relating to the mechanism of ester pyrolysis fits into one coherent picture. The process is an E_i elimination, in the transition state of which (26) the electron pair (*i*) has moved further than pair (*ii*) which in turn has moved further than pair (*iii*); this accounts for the signs and magnitudes of all of the ρ-factors observed. On going from 1^0 to 3^0 esters, and to esters with more electron-withdrawal in the γ-position (or containing a more electronegative element X) electron pair (*i*) will have moved further, and further also in relation to pairs (*ii*) and (*iii*), i.e. the process becomes more E1-like. Thus for the more reactive esters with the more polar transition states, the α- and γ-substituent effects become larger, whilst the β-substituent effects and the β-deuterium kinetic isotope effect should become smaller; full confirmation of the latter point is still needed. Attention has been drawn to the fact that interchange of the groups X and Y (26) produces a large alteration in the rate of the elimination[214] (and this has a generality beyond elimination from esters[215]). This follows from the fact that the factors which aid elimination (electronegativity in X, basicity in Y) tend to oppose one another; this provides an alternative, though related, explanation to the one based upon bond energies[214].

8. Use of the reaction as a model for electrophilic aromatic substitution

Since ester elimination proceeds via partial formation of a carbocation at the side-chain α-position, the effects of aryl substituents on stabilizing this cation will be the same as on stabilizing an incoming electrophile. The general technique of using reactions with wide-chain α-carbocations as models was innovated by H. C. Brown and coworkers[216], who drew attention to the formal similarity of 85 and

(85)   (86)

86 (E = electrophile). The S_N1 solvolysis of *t*-cumyl chlorides was therefore introduced as the standard reaction for determination of electrophilic substituent constants σ^+. However this reaction suffers from a number of disadvantages, which are: (*i*) it is a solution reaction and differential solvation factors can, and indeed in this reaction do, have a large effect upon the solvolysis rate, (*ii*) there is a large rate spread so that different solvent systems have to be used to measure pairs of rates (the overlap technique) and this can lead to systematic errors, (*iii*) the compounds eliminate HCl so readily that it is difficult or impossible to isolate them pure to begin with. All of these difficulties are overcome by using the pyrolysis of 1-arylethyl acetates (for which $\rho = -0.66$ at 600 K, -0.63 at 625 K) and this technique was introduced by Taylor, Smith and Wetzel[146,147]. Subsequently it has been extended by the writer to the determination of the electrophilic reactivities of heterocycles[217] and here it has been of major importance because of the freedom from protonation and hydrogen-bonding effects. The method has been used to provide the first quantitative reactivities of pyridine, quinoline, isoquinoline, pyridine *N*-oxide, furan and thiophene, and all under the same conditions. This general method has now been adopted by others to determination of heterocyclic reactivities in solution reactions[218] though some of the results at least are affected by solvent factors.

The data obtained by this method are gathered in Table 13. The implications are

TABLE 13. Electrophilic aromatic substituent constants determined from pyrolysis of 1-arylethyl acetates

Aromatic compound	Position	$\log k_{rel.}$(600 K)	$\log k_{rel.}$(625 K)	σ^+	Reference
Anisole	4	0.500		−0.76	146
	2	0.260			146
Thioanisole	4	0.17		−0.26	219
	2	0.01			219
Diphenyl ether	4	0.35		−0.53	219
	2	0.08			219
Diphenyl sulphide	4	0.12		−0.18	219
	3	−0.095		−0.145	219
	2	−0.05			219
Toluene	4	0.19		−0.29	146
	3	0.065		−0.098	220
	2	0.175			219, 221
Diphenylmethane	4	0.18		−0.27	219
	2	0.18			219
t-Butylbenzene	4		0.23	−0.365	222
	3		0.12	−0.19	222
Cyclohexylbenzene	4		0.24	−0.38	222
o-Xylene	4	0.27			221
	3	0.27			221
m-Xylene	4	0.425			221
p-Xylene	2	0.27			221
Mesitylene	2	0.45			221
1,2,3-Trimethylbenzene	2	0.50			221
1,2,4-Trimethylbenzene	5	0.51			221
Biphenyl	4	0.14		−0.21	146
	3	0.0		0	146
	2	0.235			146
Naphthalene	1	0.14		−0.21	146
	2	0.115		−0.175	146
Biphenylene	1		0.081	−0.13	223
	2		0.391	−0.625	224
Fluorene	2	0.395		−0.60	146
Trimethylsilylbenzene	4	0.058		−0.09	225
	3	0.105		−0.16	225
	2	0.330			225
Benzene	1	0		0	146
Fluorobenzene	4	0.035		−0.05	146
	3	−0.245		+0.389	146
	2	−0.245			151a
Chlorobenzene	4	−0.070		+0.106	146
	3	−0.245		+0.389	146
	2	−0.325			151a
Bromobenzene	4	−0.110		+0.167	146
	2	−0.340			151a
Iodobenzene	4	−0.08		+0.121	140
	3	−0.230		+0.348	146
	2	−0.312			151a
Benzotrifluoride	4	−0.405		+0.513	151a
	3	−0.373		+0.565	151a
	2	−0.370			151a
Nitrobenzene	4	−0.625[a]		+0.745[a]	152
	3	−0.590[a]		+0.71[a]	152
	2	−0.555			151a

TABLE 13. (Continued)

Aromatic compound	Position	$\log k_{rel.}$(600 K)	$\log k_{rel.}$(625 K)	σ^+	Reference
1,2-Dichlorobenzene	3		−0.428		226
	4		−0.278	+0.44	226
1,3-Dichlorobenzene	4		−0.356		226
1,4-Dichlorobenzene	2		−0.511		226
1,2,3-Trichlorobenzene	4		−0.45		226
1,2,4-Trichlorobenzene	3		−0.639		226
	5		−0.472		226
1,3,5-Trichlorobenzene	2		−0.561		226
1,2,4,5-Tetrachlorobenzene	3		−0.717		226
Pentachlorobenzene	6		−0.70		226
Pentafluorobenzene	6		−0.663		227
Pyridine	2		−0.19	+0.30	228
	3		−0.505	+0.80	228
	4		−0.55	+0.87	228
Quinoline	2		−0.46	+0.730	228
	3		−0.05	+0.079	228
	4		−0.47	+0.747	228
	5		+0.068	−0.108	228
	6		−0.040	+0.063	228
	7		−0.096	+0.152	228
	8		−0.040	+0.063	228
i-Quinoline	1		−0.32	+0.51	159
	3		−0.26	+0.41	159
	4		+0.015	−0.025	159
	5		−0.045	+0.07	159
	6		−0.195	+0.31	159
	7		−0.045	+0.07	159
	8		−0.16	+0.255	159
Pyridine N-oxide	2		1.73		213
	3		−0.51	+0.81	213
	4		−0.01	+0.016	213
Furan	2	0.588		−0.885	229
	3	0.274		−0.415	229
Thiophen	2	0.524		−0.79	229
	3	0.251		−0.38	229
Selenophene	2	0.555		−0.855	230

[a]These values were obtained in the pyrolysis of 1-arylethyl phenyl carbonates for which $\rho = -0.84$ at 600 K[152].

outside the scope of this review, so attention is drawn here only to the most significant features.

(1) The σ^+-values obtained for the m-Ph, m-Me, m-CF$_3$, m-NO$_2$, m-SiMe$_3$ and m-t-Bu substituents all give better correlation with electrophilic aromatic substitution data than do the values derived from the solvolysis of t-cumyl chlorides. This is because the latter reaction is extremely susceptible to steric hindrance to solvation[222].

(2) The activating effect of the p-t-butyl substituent in the elimination is proportionally greater, and greater relative to p-methyl than for any other known reaction in which this substituent can conjugate with the reaction site. This proves

that C—C hyperconjugation is *greater* than C—H hyperconjugation and that the Baker—Nathan electron-releasing order of these substituents is a result of steric hindrance to solvation being superimposed upon the above conjugative electron-releasing order[222]. Steric hindrance to solvation is very marked in the solvolysis of *t*-cumyl chlorides, so this reaction produces very attenuated sigma values for both the *p-t*-butyl and *p*-trimethylsilyl substituents, and indeed for any other bulky substituent[222].

(3) In contrast to electrophilic aromatic substitutions (where the transition state is approached through a bond-making process rather than through bond-breaking processes as in the elimination) the activating effects of multiple methyl substituents is *greater* than calculated on the basis of additivity. This confirms that the polarity of the transition state varies with each ester, and that great electron supply to the α-carbon produces a more polar transition state. Consequently each additional methyl substituent increases the polarity at the α-carbon and produces a rate enhancement which is greater than calculated[221]. Likewise multiple chloro substituents deactivate *less* than predicted because a given substituent decreases the charge at the α-carbon so that the deactivating effect of an additional chloro substituent becomes less[226].

(4) The positional reactivities in pyridine, quinoline, isoquinoline and pyridine N-oxide agree very well with π-electron densities calculated by the Hückel method. Positions conjugated with the nitrogen in the former three molecules are the least reactive, and the positional reactivities in quinoline and isoquinoline are approximately the products of the reactivities of the corresponding positions in pyridine and naphthalene.

(5) The positional reactivity order for quinoline is precisely followed in nitration (even though the quinolinium ion is the nitrated species). The gas-phase σ^+-values confirm that the protonated species is being nitrated[228]. The 4-position of isoquinoline is activated towards electrophilic aromatic substitution and the predominance of substitution at this site under neutral conditions is to be expected[159] special mechanisms previously proposed to account for this substitution pattern are unnecessary.

(6) The data show the great importance of bond-fixation effects which have been highlighted by hydrogen exchange of substituted naphthalenes. Thus for example the deactivating effect of the nitrogen in isoquinoline is greater across the 1,2-bond than across the 2,3-bond[159].

(7) The reactivity order of the 2- and 3- positions in furan and thiophene (first quantitatively evaluated in this elimination) has been confirmed by more recent data obtained in solution[231].

(8) The high 2 : 1 positional reactivity order for biphenylene provides strong confirmation that strain rather than electronegativity effects produce this order in electrophilic substitution of this and related molecules[232].

(9) No values of σ^+_{ortho} are derivable from this work, and attempts to do this[151] have been shown to be invalid[219]. The reason is that direct field effects operate between the side-chain α-carbocation and the substituent. Indeed, this effect, first identified in the elimination, has been shown to be a factor governing the rates of *all ortho*-substituted compounds in side-chain reactions which are analogues of electrophilic aromatic substitution[219].

(10) The reactivity of sulphur-containing compounds emphasizes the importance of *d*-orbital conjugation of sulphur. Substituents containing sulphur thereby appear to be able to show both $-M$ and $+M$ effects[219].

E. β-Hydroxy Esters

These are esters which contain a β-hydroxy substituent in the alkyl group attached to the carbonyl group. On pyrolysis they give an ester and an aldehyde or ketone[233]. The mechanism is the 6-centre process shown in **87** which differs from the analogous **26** in that the polarization of the C_β—C bond in **87** will be less than

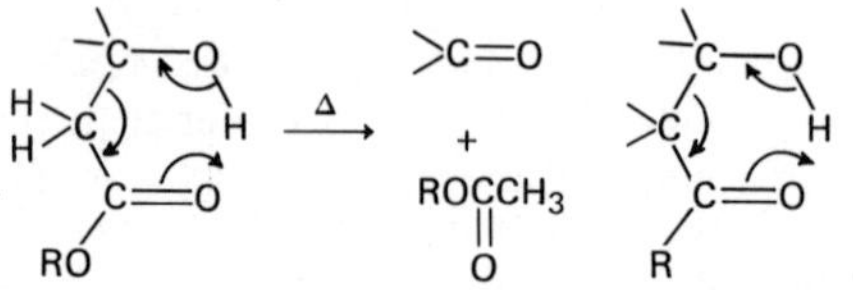

(87)   (88)

that of the corresponding C_α—O bond in **26**. On the other hand the polarization of the O—H bond will be greater than that of the C—H bond in **26**. However, since the former factor is the most important in ester elimination, the overall polarity of the transition state should be less than for normal esters and this is found to be so[233]. The similarity of the two processes is indicated by the following facts: (1) Electron withdrawal at the acyl carbon increases the rate, i.e. β-hydroxy esters are more reactive than β-hydroxy ketones (**88**). (2) The reaction is aided by electron supply to the β-carbon which corresponds to the α-carbon in normal ester elimination. (3) The reaction may be subject to steric acceleration by bulkier groups at the acyl carbon since ethyl β-hydroxy esters are more reactive than methyl β-hydroxy esters[233].

For pyrolysis of a series of related compounds viz. β-hydroxy-esters, -ketones, -alkenes and -alkynes, the transition state for the former is the most polar[234], and this is mechanistically reasonable.

VII. PYROLYSIS OF LACTONES

As long ago as 1883, Einhorn observed that the β-lactone **89**, derived from β-hydroxy-β(o-nitrophenyl)propionic acid, readily eliminated carbon dioxide to give the o-nitrostyrene[235]. Despite the length of time for which this reaction has been known, the ease with which the reaction takes place has been such as to preclude

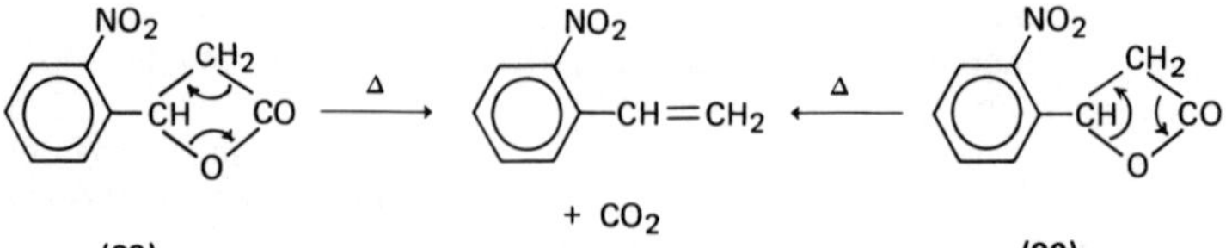

+ CO_2

(89)   (90)

any mechanistic studies, so that it is not known if the electron movements are as in **89** or **103**. Attempts to prepare the corresponding lactone without the o-nitro substituent have been unsuccessful, only the elimination product styrene being obtained[236]. This suggests that the reaction is retarded by electron withdrawal which is consistent with the mechanistically more reasonable **89** rather than with **90**.

The cis (and also the concerted) nature of the elimination was confirmed by the fact that the β-lactone of cis-α-methyl-β-(p-chlorophenyl)propionic acid (**91**) gave cis-1-(p-chlorophenyl) propene (**92**)[237]. This feature has been developed as a method for formation of alkenes with retention of the geometry present in the original lactone, and yields of up to 100% of the desired isomer have been recorded[238]. So easily does the reaction take place that many eliminations will take place merely on heating in water, e.g. a 68% yield of isobutylene results from heating the β-lactone of β-methyl-β-hydroxybutyric acid[239]; indeed this method for alkene formation has been the subject of at least two patents[240].

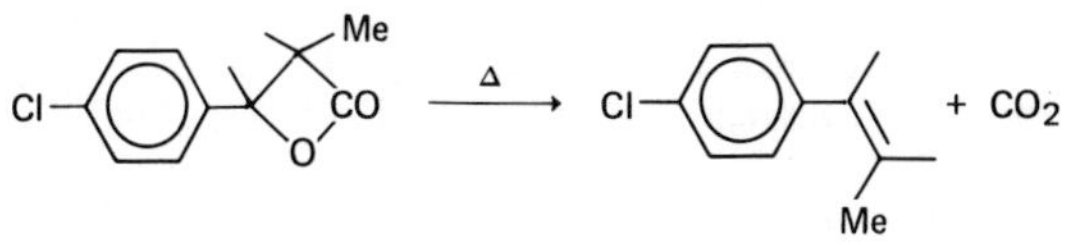

(91)   (92)

Pyrolysis of 2-pyrone and coumarin (93) produces CO extrusion in each case to give furan and benzo[b]furan, respectively; the former was also accompanied by formation of propyne and alkene[241]. The pyrolysis of α-coumaranone (94) also produces CO extrusion to give (after decomposition of the intermediate), fulvene and benzene. By contrast the isomer phthalide (95) predominantly undergoes extrusion of CO_2 to give a high yield of fulvene and ethynylcyclopentadiene, toluene and benzene; various radical pathways have been proposed to account for these results[242].

γ-Lactones are stable towards elimination as expected as no suitable transition state can be written. The same is true of δ-lactones except when these are suitably unsaturated as in the case of 96 which undergoes the elimination shown; by contrast 97 is thermally stable as expected. Kinetic evidence suggests that in the gas phase the concerted mechanism shown in (96) applies[243].

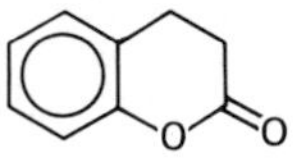

(93) 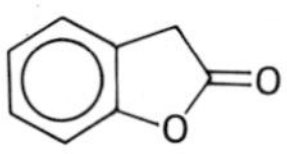   (94) 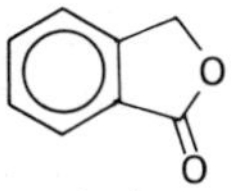   (95)

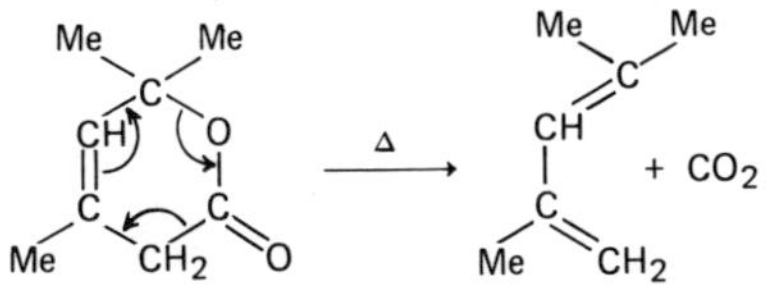

(96) 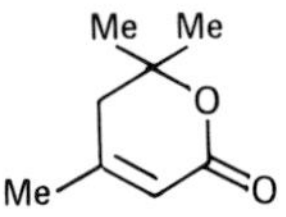   (97)

VIII. PYROLYSIS OF LACTAMS

Since the species O=C=NH is unstable, it follows that β-lactams do not undergo the elimination reactions analogous to those for β-lactones. The only reported decomposition is of 2-pyridone which extrudes carbon monoxide to give pyrrole[242].

IX. REFERENCES

1. R. T. Arnold, O. C. Elmer and R. M. Dodson. *J. Amer. Chem. Soc.,*, 72, 4359 (1950).
2. R. P. Linstead, *J. Chem. Soc.*, 1603 (1930).
3. D. B. Bigley, *J. Chem. Soc.*, 3897 (1964).
4. G. G. Smith and S. E. Blau, *J. Phys. Chem.*, 68, 1231 (1964).
5. F. Bennington and R. D. Morin, *J. Org. Chem.*, 26, 5210 (1961).

910 R. Taylor

6. D. B. Bigley and R. W. May, *J. Chem. Soc. (B)*, 557 (1967).
7. G. G. Smith and F. W. Kelly, *Progr. Phys. Org. Chem.*, **8**, 150 (1971).
8. C. G. Swain, R. F. W. Baker, R. M. Esteve and R. N. Griffin, *J. Amer. Chem. Soc.*, **83**, 1951 (1961).
9. D. B. Bigley and J. C. Thurman, *Tetrahedron Letters.*, 2377 (1967); *J. Chem. Soc. (B)*, 436 (1968).
10. D. B. Bigley and J. C. Thurman, *J. Chem. Soc. (B)*, 94 (1967).
11. D. B. Bigley and J. C. Thurman, *J. Chem. Soc. (B)*, 1076 (1966); *J. Chem. Soc.*, 6202 (1965).
12. M. M. A. Stroud and R. Taylor, *J. Chem. Research (S)*, 425 (1978).
13. B. R. Brown, *Quart. Rev.*, **5**, 131 (1951).
14. K. J. Pederson, *J. Amer. Chem. Soc.*, **51**, 2098 (1929); *J. Phys. Chem.*, **38**, 559 (1934).
15. F. H. Westheimer and W. A. Jones, *J. Amer. Chem. Soc.*, **63**, 3283 (1941).
16. K. B. Brouwer, B. Gay and T. K. Konkol, *J. Amer. Chem. Soc.*, **88**, 1681 (1966).
17. J. P. Ferris and N. C. Miller, *J. Amer. Chem. Soc.*, **85**, 1325 (1963); **88**, 3522 (1966).
18. P. G. Blake and C. Hinshelwood, *Proc. Roy. Soc. (A)*, **255**, 444 (1960)
19. P. G. Blake and G. E. Jackson, *J. Chem. Soc. (B)*, 1153 (1968).
20. P. G. Blake and K. J. Hole, *J. Chem. Soc. (B)*, 577 (1966).
21. C. H. Bamford and M. J. S. Dewar, *J. Chem. Soc.*, 2873 (1949).
22. P. G. Blake and G. E. Jackson, *J. Chem. Soc. (B)*, 94 (1969).
23. P. J. Taylor, *J. Chem. Soc., Perkin II*, 1077 (1972).
24. D. B. Bigley and J. C. Thurman, *Tetrahedron Letters*, 4687 (1965).
25. P. G. Blake and A. D. Tomlinson, *J. Che,. Soc. (B)*, 1593 (1971).
26. P. G. Blake, H. Pritchard and A. D. Tomlinson, *J. Chem. Soc. (B)*, 607 (1971).
27. P. G. Blake and A. D. Tomlinson, *J. Chem. Soc. (B)*, 282 (1967).
28. P. H. Kasai and D. McLeod, *J. Amer. Chem. Soc.*, **94**, 7975 (1972).
29. A. T. C. H. Tan and A. H. Sehon, *Can. J. Chem.*, **46**, 191 (1968).
30. M. H. Black and A. H. Sehon, *Can. J. Chem.*, **38**, 1261, 1271 (1960).
31. M. A. Ratcliff, E. E. Medley and P. G. Simmonds, *J. Org. Chem.*, **39**, 1481 (1974).
32. L. W. Clark, *J. Amer. Chem. Soc.*, **77**, 3130 (1955); *J. Phys. Chem.* **66**, 125, 1545 (1962); **67**, 138 (1963).
33. R. L. Forman, H. M. Mackinnon and P. D. Ritchie, *J. Chem. Soc. (C)*, 2013 (1968).
34. W. Moser, *Helv. Chim. Acta*, **14**, 971 (1931).
35. J. T. D. Cross and V. R. Stimson, *J. Chem. Soc. (B)*, 88 (1967).
36. M. Szwarc and J. Murawski, *Trans. Faraday Soc.*, **47**, 269 (1951).
37. H. A. Staab and A. P. Datta, *Angew. Chem.*, **75**, 1203 (1963).
38. R. S. Boehner and C. E. Andrews, *J. Amer. Chem. Soc.*, **38**, 2503, 2505 (1916).
39. D. Davidson and M. Karten, *J. Amer. Chem. Soc.*, **78**, 1060 (1956).
40. J. Aspden, A. Maccoll and R. A. Ross, *Trans. Faraday Soc.*, **64**, 965 (1968).
41. E. N. Zilberman, R. I. Spasskaya, B. N. Milyakov and G. G. Shchukina, *Zh. Prikl. Khim.*, **47**, 2293 (1974).
42. D. J. Hoy and E. J. Poziomek, *J. Org. Chem.*, **33**, 4050 (1968).
43. M. Hunt, J. A. Kerr and A. F. Trotman-Dickenson, *J. Chem. Soc.*, 5074 (1965).
44. T. Mukaiyama, M. Tokizawa, H. Nohira and H. Takei, *J. Org. Chem.*, **26**, 4381 (1961).
45. Z. Bukač and J. Sebenda, *Coll. Czech. Chem. Commun.*, **32**, 3537 (1967).
46. T. Mukaiyama, H. Takei and Y. Koma, *Bull Chem. Soc. Japan*, **36**, 95 (1963).
47. T. Mukaiyama, M. Tokizawa and H. Takei, *J. Org. Chem.*, **27**, 803 (1962).
48. R. F. C. Brown, M. Butcher and R. A. Fergie, *Australian J. Chem.*, **26**, 1319 (1973).
49. G. M. Schwab, *Z. Anorg. Chem.*, **262**, 41 (1950).
50. A. Maccoll and S. S. Nagra, *J. Chem. Soc. (B)*, 1867 (1971).
51. G. P. Armstrong, *British Patent*, No. 703, 398 (1954).
52. N. T. Buu, *Dissertation Abstracts (B)*, **32**, 3817 (1972).
53. I. Goodman, *J. Polymer Sci.*, **13**, 175 (1954): B. Leclerq, *Bull. Inst. Text. Fr.*, **22**, 915 (1968) E. P. Krasnov, V. P. Aksenova and S. N. Kharkov, *Vysokomol Soedin Ser. A.*, **11**, 1930 (1969); *Chem. Abstr.*, **72**, 22116 (1970).
54. W. J. Bailey and C. N. Bird, *J. Org. Chem.*, **23**, 996 (1958).
55. W. J. Bailey, J. J. Hewitt and C. King, *J. Amer. Chem. Soc.*, **77**, 357 (1955).

56. H. E. Baumgarten, F. A. Bower, R. A. Setterquist and R. E. Allen, *J. Amer. Chem. Soc.*, **80**, 4588 (1958).
57. A. Maccoll and S. S. Nagra, *J. Chem. Soc. Faraday I*, **69**, 1108 (1973).
58. T. E. Peel and R. T. B. Rye, *J. Chem. Soc. Chem. Commun.*, 184 (1967).
59. M. Szwarc and J. Murawski, *Trans. Faraday Soc.*, **47**, 267 (1951).
60. N. T. M. Wilsmore, *J. Chem. Soc.*, 1938 (1907).
61. G. J. Fisher. A. F. McLean and A. W. Schuizer, *J. Org. Chem.*, **18**, 1055 (1953).
62. D. S. Tarbell and E. J. Longosz, *J. Org. Chem.*, **24**, 774 (1959).
63. C. J. Michejda, D. S. Tarbell and W. H. Saunders, *J. Amer. Chem. Soc.*, **84**, 4113 (1962).
64. E. J. Longosz and D. S. Tarbell, *J. Org. Chem.*, **26**, 2161 (1961).
64a. D. S. Tarbell and R. P. F. Scharrer, *J. Org. Chem.*, **27**, 1972 (1962).
65. R. Taylor, *Tetrahedron Letters.* 593 (1975).
66. T. B. Windholz and J. B. Clements, *J. Org. Chem.*, **29**, 3021 (1964).
67. E. V. Dehmlow, *Z. Naturforsch. (B)*, **28**, 693 (1973).
68. A. L. Brown and P. D. Ritchie, *J. Chem. Soc. (C)*, 2007 (1968).
69. F. O. Rice and M. T. Murphy, *J. Amer. Chem. Soc.*, **64**, 896 (1942).
70. W. J. Middleton and W. H. Sharkey, *J. Amer. Chem. Soc.*, **81**, 803 (1959).
71. E. K. Fields and S. Meyerson, *J. Chem. Soc., Chem. Commun.*, 474 (1965); *J. Org. Chem.*, **31**, 3307 (1966).
72. L. Friedman and D. F. Lindow, *J. Amer. Chem. Soc.*, **90**, 2329 (1968).
73. M. P. Cava, M. J. Mitchell, D. C. de Jongh and R. Y. van Fossen, *Tetrahedron Letters*, 2947 (1966).
74. R. F. C. Brown, D. Gardner, J. F. W. McOmie and R. Solly, *J. Chem. Soc., Chem. Commun*, 408 (1966); *Australian J. Chem.*, **20**, 139 (1967).
75. R. J. P. Allen, E. Jones and P. D. Ritchie, *J. Chem. Soc.*, 524 (1957).
76. R. N. Bennett, E. Jones and P. D. Ritchie, *J. Chem. Soc.*, 2628 (1956).
77. A. E. Ardis, S. J. Averill, H. Gilbert, F. F. Miller, R. F. Schmidt, F. D. Stewart and H. C. Turnbull, *J. Amer. Chem. Soc.*, **72**, 1305 (1950).
78. P. E. Newallis and P. Lombardo, *J. Org. Chem.*, **30**, 3834 (1965).
79. A. P. M. van der Week and F. H. van Putten, *Tetrahedron Letters*, 3951 (1970).
80. P. C. Oele and R. Louw, *J. Chem. Soc., Chem. Commun.*, 849 (1972).
80a. S. Searles and S. Nakina, *J. Amer. Chem. Soc.*, **78**, 5656 (1956).
81. R. Auschütz, *Chem. Ber.*, **60**, 1320 (1927).
82. H. M. Mackinnon and P. D. Ritchie, *J. Chem. Soc.*, 2564 (1957).
83. P. C. Oele and R. Louw, *Tetrahedron Letters*, 4941 (1972).
84. J. W. Scheeren, A. P. M. van der Week and W. Stevens, *Rec. Trav. Chim.*, **88**, 195 (1969).
85. R. J. P. Allan, R. L. Forman and P. D. Ritchie, *J. Chem. Soc.*, 2717 (1955).
86. N. J. Daly and F. Ziolkowski, *Australian J. Chem.*, **35**, 1453 (1972).
87. J. M. Vernon and D. J. Waddington, *J. Chem. Soc., Chem. Commun.*, 623 (1969).
88. K. H. Leavall and E. S. Lewis, *Tetrahedron*, **28**, 1167 (1972).
89. R. Louw and E. C. Kooyman, *Rec. Trav. Chim.*, **84**. 1151 (1965); **86**, 147 (1967)); R. Louw, M. van den Brink and H. P. W. Vermeeren, *J. Chem. Soc., Perkin II*, 1327 (1973).
90. O. Grummitt and J. Splitter, *J. Amer. Chem. Soc.*, **74**, 3924 (1952).
91. F. L. Greenwood, *J. Org. Chem.*, **27**, 2308 (1962).
92. F. L. Greenwood, *J. Org. Chem.*, **24**, 1735 (1959).
93. C. S. Marvel and J. L. R. Williams, *J. Amer. Chem. Soc.*, **70**, 3842 (1948).
94. K. K. Lum and G. G. Smith, *Int. J. Chem. Kinetics*, **1**, 401 (1969).
95. I. I. Ostromuisslenskii, *J. Russ. Phys. Chem. Soc.*, **47**, 1472 (1915); *Chem. Abstr.*, **10**, 3178 (1916).
96. C. S. Marvel and N. O. Brace, *J. Amer. Chem. Soc.*, **70**, 1775 (1948).
97. W. J. Bailey and R. Barclay, *J. Org. Chem.*, **21**, 328 (1956).
98. F. H. A. Rummens, *Rec. Trav. Chim.*, **83**, 901 (1964).
99. E. S. Lewis, J. T. Hill and E. R. Newman, *J. Amer. Chem. Soc.*, **90**, 662 (1968); E. S. Lewis and J. T. Hill, *J. Amer. Chem. Soc.*, **91**, 7458 (1969).
100. E. Smith and J. Laurence, *Ann. Chim. Phys.*, **6**, 40 (1942); *Amer. J. Sci.*, **43**, 301 (1942).
101. A. Oppenheim and H. Precht, *Chem. Ber.*, **9**, 325 (1876).

102. N. Menschutkin, *Chem. Ber.*, **15**, 2512 (1882).
103. C. D. Hurd and F. H. Blunck, *J. Amer. Chem. Soc.*, **60**, 2419 (1938).
104. E. R. Alexander and A. Mudrak, *J. Amer. Chem. Soc.*, **72**, 3194 (1950); **73**, 59 (1951).
105. E. R. Alexander and A. Mudrak, *J. Amer. Chem. Soc.*, **72**, 1810 (1950).
106. R. T. Arnold, G. G. Smith and R. M. Dodson, *J. Org. Chem.*, **15**, 1256 (1950).
107. W. J. Bailey and L. Nicholas, *J. Org. Chem.*, **21**, 854 (1956).
108. D. Y. Curtin and D. B. Kellom, *J. Amer. Chem. Soc.*, **75**, 6011 (1953).
109. P. S. Skell and W. L. Hall, *J. Amer. Chem. Soc.* **86**, 1557 (1964).
110. R. Taylor, *J. Chem. Soc., Perkin II*, 165 (1972).
111. D. H. R. Barton, *J. Chem. Soc.*, 2174 (1949); D. H. R. Barton and W. J. Rosenfelder, *J. Chem. Soc.*, 2457, 2459 (1949); 1048 (1951).
112. F. G. Bordwell and P. S. Landis, *J. Amer. Chem. Soc.*, **80**, 2450 (1958).
113. G. L. O'Connor and H. R. Nace, *J. Amer. Chem. Soc.*, **75**, 2118 (1953).
114. C. H. De Puy and R. W. King, *Chem. Rev.*, **60**, 436 (1960).
115. W. J. Bailey and W. F. Hale, *J. Amer. Chem. Soc.*, **81**, 647 (1959).
116. R. A. Benkeser and J. J. Hazdra, *J. Amer. Chem. Soc.*, **81**, 228 (1959).
117. C. H. De Puy, R. W. King and D. H. Froemsdorf, *Tetrahedron*, **7**, 123 (1959).
118. G. Eglinton and M. N. Rodger, *Chem. Ind. (London)*, 256 (1959).
119. D. H. Froemsdorf, C. H. Collins, G. S. Hammond and C. H. De Puy, *J. Amer. Chem. Soc.*, **81**, 643 (1959).
120. A. C. Cope, C. L. Bumgardner and E. E. Schweizer, *J. Amer. Chem. Soc.*, **79**, 4729 (1957).
121. J. P. W. Houtman, J. Van Steenis and P. M. Heertjes, *Rec. Trav. Chim.*, **65**, 781 (1946).
122. W. O. Haag and H. Pines, *J. Org. Chem.*, **24**, 877 (1959).
123. R. Borkowski and P. Ausloos, *J. Amer. Chem. Soc.*, **83**, 1053 (1961).
124. C. H. De Puy, C. A. Bishop and C. N. Goeders, *J. Amer. Chem. Soc.*, **83**, 2151 (1961).
125. D. B. Bigley and C. M. Wren, *J. Chem. Soc., Perkin II*, 1744 (1972).
126. J. C. Scheer, E. C. Kooyman and F. L. J. Sixina, *Rec. Trav. Chim.*, **82**, 1123 (1963).
127. E. E. Royals, *J. Org. Chem.*, **23**, 1822 (1958).
127a. R. Taylor, *J. Chem. Research (S)*, 267 (1978), and unpublished work.
128. R. A. Benkeser, J. J. Hazdra and M. L. Burrous, *J. Amer. Chem. Soc.*, **81**, 5374 (1959).
129. R. Onesta and G. Castelfranchi, *Chimi. Ind. (Milan)*, **41**, 222 91959).
130. Reference 7, pp. 89–91.
131. C. H. De Puy and R. E. Leary, *J. Amer. Chem. Soc.*, **79**, 3705 (1957).
132. A. Maccoll and R. H. Stone, *J. Chem. Soc.*, 2756 (1961).
133. W. J. Bailey and W. F. Hale, *J. Amer. Chem. Soc.*, **81**, 651 (1959).
134. A. T. Blomquist and G. Denning, *Abstracts, 139th National Meeting of the American Chemical Society*, St. Louis, March, 1961.
135. J. R. van der Biz and E. C. Kooyman, *Rec. Trav. Chim.*, **71**, 837 (1952).
136. N. L. McNiven and J. Read, *J. Chem. Soc.*, 2067 (1953); J. P. Wibaut, H. C. Beyerman and H. B. Leeuwen, *Rec. Trav. Chim.*, **71**, 1027 (1952).
137. G. Ohloff and G. Schade, *Chem. Ber.*, **91**, 2017 (1958).
138. A. Brenner and H. Schinz, *Helv. Chim. Acta*, **35**, 1333 (1952).
139. D. V. Banthorpe and H. S. Davies, *J. Chem. Soc. (B)*, 1339 (1968).
140. G. Büchi, M. Schach von Wittenau and H. Schechter, *J. Amer. Chem. Soc.*, **81**, 1968 (1959).
141. A. T. Blomquist and P. R. Taussig, *J. Amer. Chem. Soc.*, **79**, 3505 (1957).
142. A. T. Blomquist and A. Goldstein, *J. Amer. Chem. Soc.*, **77**, 1001 (1955).
143. W. S. Briggs and C. Djerrassi, *J. Org. Chem.*, **33**, 1625 (1968).
144. T. O. Bamkole and E. U. Emovon, *J. Chem. Soc. (B)*, 187 (1969); A. Tinkelberg, E. C. Kooyman and R. Louw, *Rec. Trav. Chim.*, **91**, 3 (1972).
145. Reference 7, p. 113.
146. R. Taylor, G. G. Smith and W. H. Wetzel, *J. Amer. Chem. Soc.*, **84**, 4817 (1962).
147. R. Taylor and G. G. Smith, *Tetrahedron*, **19**, 937 (1963).
148. G. G. Smith, F. D. Bagley and R. Taylor, *J. Amer. Chem. Soc.*, **83**, 3647 (1961).
149. A. G. Harrison, P. Kebarle and F. P. Lossing, *J. Amer. Chem. Soc.*, **83**, 772 (1961).
150. H. B. Amin and R. Taylor, *J. Chem. Soc., Perkin II*, 1095 (1978).
151. G. G. Smith, K. K. Lum, J. A. Kirby and J. A. Posposil, *J. Org. Chem.*, **34**, 2090 (1969).

151a. R. Taylor, *J. Chem. Soc. (B)*, 622 (1971).
152. H. B. Amin and R. Taylor, *J. Chem. Soc. Perkin II*, 1090 (1978).
153. Reference 7, pp. 94–96.
154. R. Taylor, *J. Chem. Soc., Perkin II*, 1025 (1975).
155. A. Maccoll and P. J. Thomas, *Prog. Reaction Kinetics*, 4, 119 (1967).
156. D. B. Bigley and R. E. Gabbott, *J. Chem. Soc., Perkin II*, 1293 (1973).
157. P. C. Oele, A. Tinkelberg and R. Louw, *Tetrahedron Letters*, 2375 (1972).
158. M. M. Brubaker, *British Patent*, No. 568,565 (1945).
159. E. Glyde and R. Taylor, *J. Chem. Soc., Perkin II*, 1783 (1975).
160. A. Maccoll, *Chem. Rev.*, 69, 33 (1969).
161. E. S. Lewis and W. S. Herndon, *J. Amer. Chem. Soc.*, 83, 1955 (1961).
162. W. J. Bailey and J. J. Hewitt, *J. Org. Chem.*, 21, 543 (1956).
163. E. U. Emovon, *J. Chem. Soc.*, 1246 (1963).
164. Reference 7, p. 102.
165. G. G. Smith, D. A. K. Jones and R. Taylor, *J. Org. Chem.*, 28, 3547 (1963).
166. E. Dyer and G. C. Wright, *J. Amer. Chem. Soc.*, 81, 2138 (1959).
167. M. P. Thorne, *Canad. J. Chem.*, 45, 2537 (1967).
168. N. J. Daly, G. M. Heweston and F. Ziolkowski, *Australian J. Chem.*, 26, 1259 (1973).
169. G. G. Smith and D. A. K. Jones, *J. Org. Che,.*, 28, 3496 (1963).
170. G. G. Smith, D. A. K. Jones and D. F. Brown, *J. Org. Chem.*, 28, 403 (1963).
171. H. B. Amin and R. Taylor, *J. Chem. Soc., Perkin II*, 1802 (1975).
172. R. Taylor and M. P. Thorne, *J. Chem. Soc., Perkin II*, 799 (1976).
173. K. C. Rutherford and D. P. C. Fung, *Canad. J. Chem.*, 42, 2657 (1964).
174. D. A. Kairaitis and V. R. Stimson, *Australian J. Chem.*, 21, 1349 (1968).
175. A. S. Gordon and W. P. Norris, *J. Phys. Chem.*, 69, 3013 (1965).
176. D. B. Bigley and C. M. Wren. *J. Chem. Soc., Perkin II*, 926 (1972).
177. C. E. Higgins and W. H. Baldwin, *J. Org. Chem.*, 26, 846 (1961).
178. P. Haake and C. E. Diebert, *J. Amer. Chem. Soc.*, 93, 6931 (1971).
179. K. A. Strom and W. L. Jolly, *J. Organometallic Chem.*, 69, 201 (1974).
180. E. S. Lewis and W. C. Herndon, *J. Amer. Chem Soc.*, 83, 1955, (1961); E. S. Lewis, W. C. Herndon and D. C. Duffey, *J. Amer. Chem. Soc.*, 83, 1959 (1961). E. S. Lewis and K. Witte, *J. Chem. Soc. (B)*, 1198 (1968).
181. W. A. Sheppard, *J. Org. Chem.*, 27, 3756 (1962).
182. R. F. W. Bader and A. N. Bourns, *Canad. J. Chem.*, 39, 348 (1961).
183. N. Barroeta, V. de Santis and M. Rincon, *J. Chem. Soc., Perkin II*, 911 (1974).
184. G. H. Hanson, *Dissertation Abstr. (B)*, 33, 4192 (1973).
185. T. Taguchi and M. Nakao, *Tetrahedron*, 18, 245 (1962). T. Taguchi, Y. Kawazoe and M. Nakao, *Tetrahedron Letters*, 131 (1963). C. G. Overberger and A. E. Borchert, *J. Amer. Chem, Soc.*, 82, 4896 (1960). P. V. Laakso, *Suomen Kemi*, 13B, 8 (1940); *Chem. Abstr.*, 34, 5059 (1940); *Suomen Kemi*, 16B, 19 (1943); *Chem. Abstr.*, 40, 4687 (1946).
186. H. Kwart and E. R. Evans, *J. Org. Chem.*, 31, 410 (1966); K. Miyazaki, *Tetrahedron Letters*, 2793 (1968).
187. I. M. McAlpine, *J. Chem. Soc.*, 1114 (1931); 906 (1932).
188. H. R. Nace, D. Manly and S. Fusco, *J. Org. Chem.*, 23, 687 (1958).
189. T. Jaworski, *Rocz. Chim.*, 48, 263 (1974).
190. Reference 114, p. 447.
191. S. de Burgh Norfolk and R. Taylor, *J. Chem. Soc., Perkin II*, 280 (1976).
192. G. Chuchani, S. P. de Chang and L. Lombana, *J. Chem. Soc., Perkin II*, 1961 (1973).
193. E. U. Emovon and A. Maccoll, *J. Chem. Soc.*, 227 (1964).
194. G. G. Smith and B. L. Yates, *J. Org. Chem.*, 30, 434 (1965).
195. G. G. Smith, K. J. Vorhees and F. M. Kelly, *J. Chem. Soc., Chem. Commun.*, 789 (1971).
196. H. Kwart and J. Slutsky, *J. Chem. Soc., Chem. Commun.*, 552 (1972).
197. A. T. Blades and P. W. Gilderson, *Canad. J. Chem.*, 38, 1401, 1407 (1960).
198. H. Kwart and J. Slutsky, *J. Chem. Soc., Chem. Commun.*, 1182 (1972).
199. S. J. Ashcroft and M. P. Thorne, *Canad. J. Chem.*, 50, 3478 (1972).
200. R. Taylor, *Chem. Ind. (London)*, 1684 (1962).
201. B. S. Lennon and V. R. Stimson, *Australian J. Chem.*, 21, 1659 (1968).

202. J. T. D. Cross and V. R. Stimson, *Australian J. Chem.*, **20**, 177 (1967).
203. D. J. Kramer and G. G. Smith, *Int. J. Chem. Kinetics*, **6**, 849 (1974).
204. A. C. Cope and M. J. Youngquist, *J. Amer. Chem. Soc.*, **84**, 2411 (1962).
205. T. A. Spencer, S. W. Baldwin and K. K. Schmiegel, *J. Org. Chem.*, **30**, 1294 (1965). J. Cologne and J. C. Dubin, *Compt. Rend.*, **250**, 553 (1960); *Bull. Soc. Chim. Fr.*, 1180 (1960). K. L. Williamson, R. T. Keller, G. S. Fonken, J. Szmuszkovicz and W. S. Johnson, *J. Org. Chem.*, **27**, 1612 (1962).
206. R. G. Carlson and J. H. Bateman, *J. Org. Chem.*, **32**, 1608 (1967).
207. M. Hanack, H.-J. Schneider and H. Schneider-Berlöhr, *Tetrahedron*, **23**, 2195 (1967).
208. E. U. Emovon, *J. Chem. Soc. (B)*, 588 (1966).
209. T. Sato, K. Murata, A. Nishimura, T. Tsuchiya and N. Wasada, *Tetrahedron*, **23**, 1791 (1967).
210. C. A. Bunton, K. Khaleeluddin and D. Whittaker, *Nature*, **190**, 715 (1961).
211. H. Kwart and D. P. Hoster, *J. Chem. Soc., Chem. Commun.*, 1155 (1967); H. Kwart and H. G. Ling, *J. Chem. Soc., Chem. Commun.*, 302 (1969).
212 B. L. Adams and P. Kovacic, *J. Chem. Soc., Chem. Commun.*, 1310 (1972).
213. R. Taylor, *J. Chem. Soc., Perkin II*, 277 (1975).
213a. A. T. Blomquist and J. A. Verdol, *J. Amer. Chem. Soc.*, **77**, 1806 (1955); W. J. Bailey, C. H. Cunor and L. Nicholas, *J. Amer. Chem. Soc.*, **77**, 2787 (1955).
214. C. H. De Puy and C. A. Bishop, *Tetrahedron Letters*, 239 (1963).
215. R. C. Cookson and S. R. Wallis, *J. Chem. Soc. (B)*, 1245 (1966).
216. Summarized in L. M. Stock and H. C. Brown, *Adv. Phys. Org. Chem.*, **1**, 35 (1963).
217. R. Taylor, *J. Chem. Soc.*, 4881 (1962).
218. E. A. Hill, M. L. Gross, M. Stasiewicz and M. Manion, *J. Amer. Chem. Soc.*, **91**, 7381 (1969). T. J. Broxton, G. L. Butt, L. W. Deady, S. H. Toh and R. D. Topsom, *Canad. J. Chem.*, **51**, 1620 (1973). G. T. Bruce, A. R. Cooksey and K. J. Morgan *J. Chem. Soc., Perkin II*, 551 (1975); D. S. Noyce and D. A. Forsyth, *J. Org. Chem.*, **39**, 2828 (1974) and earlier papers in this series.
219. R. Taylor, *J. Chem. Soc. (B)*, 1450 (1971).
220. E. Glyde and R. Taylor, *J. Chem. Soc., Perkin II*, 1463 (1975).
221. E. Glyde and R. Taylor, *J. Chem. Soc., Perkin II*, 1537 (1977).
222. E. Glyde and R. Taylor, *J. Chem. Soc., Perkin II*, 678 (1977).
223. R. Taylor, M. P. David and J. F. W. McOmie, *J. Chem. Soc., Perkin II*, 162 (1972).
224. J. M. Blatchly and R. Taylor, *J. Chem. Soc. (B)*, 1402 (1968).
225. E. Glyde and R. Taylor, *J. Chem. Soc., Perkin II*, 1632 (1973).
226. E. Glyde and R. Taylor, *J. Chem. Soc., Perkin II*, 1541 (1977).
227. R. Taylor, *J. Chem. Soc. (B)*, 255 (1971).
228. R. Taylor, *J. Chem. Soc. (B)*, 2382 (1971).
229. R. Taylor, *J. Chem. Soc. (B)*, 1397 (1968).
230. S. Clementi, G. Marino and R. Taylor, unpublished work.
231. R. Taylor, *J. Chem. Soc. (B)*, 1364 (1970).
232. R. Taylor, *J. Chem. Soc. (B)*, 1559 (1968).
233. B. L. Yates, O. Ramirez and O. Velasquez, *J. Org. Chem.*, **36**, 3579 (1971).
234. G. G. Smith and B. L. Yates, *J. Org. Chem.*, **32**, 2067 (1967); A. Viola, J. H. MacMillan, R. J. Proverb and B. L. Yates, *J. Chem. Soc., Chem. Commun.*, 936 (1971); *J. Amer. Chem. Soc.*, **93**, 6967 (1971).
235. A. Einhorn, *Chem. Ber.*, **16**, 2208 (1883).
236. E. Erlenmeyer, *Chem. Ber.*, **13**, 303 (1880); G. Senter and A. M. Ward, *J. Chem. Soc.*, 1847 (1925).
237. D. S. Noyce and E. H. Banitt, *J. Org. Chem.*, **31**, 4043 (1966).
238. W. Adam, J. Baeza and J. C. Liu, *J. Amer. Chem. Soc.*, **94**, 2000 (1972).
239. P. D. Bartlett and H. T. Liang, *J. Amer. Chem. Soc.*, **80**, 3585 (1958).
240. D. J. Loder, *U. S. Patent*, No. 2,480,586 (1949); *Chem. Abstr.*, **44**, 1133c (1950). J. R. Caldwell, *U. S. Patent*, No. 2,585,223 (1952); *Chem. Abstr.*, **46**, 8672i (1952).
241. D. A. Brent, J. D. Hribar and D. C. De Jongh, *J. Org. Chem.*, **35**, 135 (1970).
242. C. Wentrup, *Tetrahedron Letters*, 2915, 2919 (1973).
243. R Louw, *Doctoral thesis*, University of Leiden, 1964.

CHAPTER **16**

Transcarboxylation reactions of salts of aromatic carboxylic acids

JOSEF RATUSKÝ
Institute of Organic Chemistry and Biochemistry, Czechoslovak Academy of Sciences, Prague, Czechoslovakia

I.   INTRODUCTION 915
II. THE INTERMOLECULAR IONIC MECHANISM OF
     TRANSCARBOXYLATION REACTIONS 916
     A. Benzenecarboxylic Acids 916
        1. Ionic character of the liberation of the carboxylate group from the
           benzene ring in transcarboxylations 917
        2. Time dependence of the content of reaction mixtures in
           transcarboxylations of potassium benzenecarboxylates . . . 918
           a. Transcarboxylation of potassium benzenedicarboxylates . . . 919
           b. Transcarboxylation of potassium benzoate 921
           c. Transcarboxylation of mixtures containing potassium benzoate and
              potassium benzenepolycarboxylates 923
        3. Transcarboxylation of benzenecarboxylates in molten KCNO or KCNS . 924
        4. Transcarboxylation of benzenecarboxylates in an atmosphere of
           ^{14}C-labelled carbon dioxide 926
        5. Exchange of deuterium by protium in transcarboxylations of mixtures
           of benzenecarboxylates labelled and non-labelled by deuterium on the
           benzene ring 929
        6. Mechanism of transcarboxylation of benzenecarboxylates . . . 930
        7. Effect of catalysts on the transcarboxylation of benzenecarboxylates . 932
        8. Effect of various cations on the transcarboxylation of benzenecarboxylates. 933
     B. Naphthalenecarboxylic Acids 935
     C. Biphenylcarboxylic Acids 938
     D. Substituted Benzoic Acids 938
     E. Heterocyclic Carboxylic Acids 939
     F. Polynuclear Carboxylic Acids 939
III. MIXED TRANSCARBOXYLATIONS OF SALTS OF AROMATIC
     CARBOXYLIC ACIDS 940
IV. CONCLUSION 942
V.   REFERENCES 942

I. INTRODUCTION

The thermal transformation of salts of benzenecarboxylic acids was first attempted in the nineteenth century by von Richter[1] by fusing potassium benzoate with sodium formate and by Wislicenus and Conrad[2] by fusing sodium benzoate. After World War I, the experiments were continued by Schrader and Wolter[3]. The reactions were performed very primitively and the yields were rather discouraging.

The need for terephthalic acid in producing polyester fibre or film prompted Raecke and coworkers[4] to develop technological processes for the production of terephthalic acid by a catalysed 'rearrangement' of potassium salts of phthalic acid and isophthalic acid or by disproportionation of potassium benzoate[5-8].

Regarding the course of these reactions, some ideas were presented by Raecke but neither he nor other authors of numerous patents[4-9] could propose a detailed reaction mechanism[4]. Later on, when the reaction conditions and interpretation of results were improved, some authors became interested in the mechanism of the thermal transcarboxylation. Thus, on the basis of a very low incorporation of the ^{14}C radionuclide from radioactive K_2CO_3-^{14}C into the molecules of potassium terephthalate[10,11] or by analogy to the rearrangement of potassium salicylate[12], an intramolecular mechanism was first favoured. However, improved experiments indicated a considerable incorporation of ^{14}C from the atmosphere of ^{14}C-labelled carbon dioxide and removal of the carboxylate group; these observations were in accordance with an internuclear reaction course, though the proposed mechanism was not correct[13-15]. Of great importance was the finding that the isomerization of potassium phthalate and the disproportionation of potassium benzoate take place with the formation of the other benzenecarboxylates as intermediates[16-18] and that the protons and all carboxylate groups on the benzene ring are removed by an ionic mechanism[18]. The evidence favouring the intermolecular, ionic mechanism gradually accumulated and resulted in a general acceptance of this mechanism[19-26] which governs transcarboxylation reactions of all carboxylic acids derived from aromatic systems[27].

II. THE INTERMOLECULAR IONIC MECHANISM OF TRANSCARBOXYLATION REACTIONS

A. Benzenecarboxylic Acids

The course and mechanism of transcarboxylations may be elucidated using the simplest type of reaction, the transcarboxylation of potassium benzenecarboxylates. These transformations have been most intensively investigated because of the great industrial importance of benzenecarboxylic acids, especially of terephthalic acid. The reactions may be expressed by equations (1) and (2).

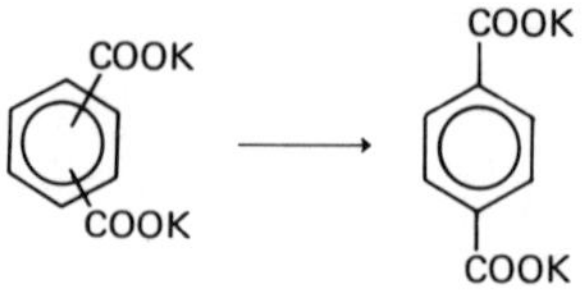

$$(1)$$

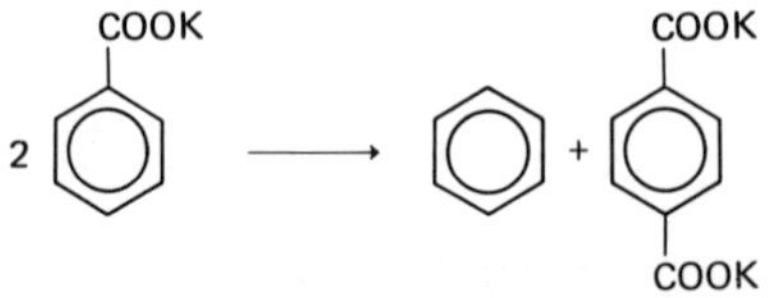

$$(2)$$

Optimum reaction conditions in transcarboxylations of potassium salts of benzenecarboxylic acids are as follows: reaction temperature above 400°C, the presence of suitable catalysts, a protective reaction atmosphere and the absence of compounds releasing protons.

The course and the mechanism of transcarboxylations was examined as follows: (i) The ionic character of transcarboxylations was established. (ii) The time dependence of the ratio of components in reaction mixtures containing potassium benzenecarboxylates was determined. (iii) The analogous time dependence was determined for transcarboxylations performed in fused KCNO or KCNS. (iv) Transcarboxylations of the above potassium salts were performed in an atmosphere of ^{14}C-labelled carbon dioxide. (v) Deuterium was exchanged by protium in transcarboxylations of mixtures of potassium benzenecarboxylates labelled and non-labelled by deuterium in the benzene ring. (vi) Potassium benzoate was carboxylated using various carboxylating agents. (vii) The influence of various cations in benzenecarboxylates on the course of the transcarboxylation was determined. (viii) The influence of catalysts on the course of the transcarboxylation was determined.

1. Ionic character of the liberation of the carboxylate group from the benzene ring in transcarboxylations

The ionic character of the cleavage of the C—C bond between the benzene ring and the carboxylate group in transcarboxylations of salts of benzenecarboxylic acids is indicated by the following experimental observations.

(a) The reaction mixture was free of diphenyl, diphenylcarboxylic acids or oxalic acid which would be formed in the radical cleavage of the above C—C bond[18,28].

(b) The reaction is not accelerated by irradiation with u.v. light or by initiators (catalysts) of radical reactions.

(c) The presence of compounds releasing protons in the reaction mixture, e.g. water, acids, hydrogen benzenedicarboxylates and the like or the presence of an incompletely neutralized starting material results in a smooth decarboxylation of an aliquot of the starting transcarboxylated salt[4,15,18].

(d) On the other hand, the presence of compounds capable of binding the protons formed in transcarboxylations can make possible a direct carboxylation of salts of aromatic monocarboxylic acids when substances are present that can supply cations for the carboxylates formed. Thus for example, potassium benzoate can be carboxylated to potassium terephthalate by the action of potassium carbonate and carbon dioxide in the presence of calcium carbide or aluminium carbide[8] with the simultaneous formation of acetylene and the corresponding carbonate (equation 3).

$$2 \; \text{C}_6\text{H}_5\text{COOK} + \text{K}_2\text{CO}_3 + 2\,\text{CO}_2 + \text{CaC}_2 \longrightarrow 2 \; \text{C}_6\text{H}_4(\text{COOK})_2 + \text{CaCO}_3 + \text{C}_2\text{H}_2 \quad (3)$$

(e) The ionic character of transcarboxylations is also supported by the unusually easy and rapid incorporation of the ^{14}C radionuclide from the reaction atmosphere of ^{14}C-labelled carbon dioxide into the carboxylate groups of salts of carboxylic acids, proceeding through the highly labile COOK^+ carboxylate cation[18,23,24].

(f) Finally, the exchange of deuterium by protium during the simultaneous decarboxylation—recarboxylation in mixtures of deuterium-labelled and non-labelled salts of carboxylic acids may be most readily explained by the ionic character of the reaction[18,25,30].

2. Time dependence of the content of reaction mixtures in transcarboxylations of potassium benzenecarboxylates

The course of transcarboxylation reactions of salts of aromatic carboxylic acids is highly complex not only from the chemical but also from the physical standpoint. Thus for example in the transcarboxylation of potassium phthalate, the reaction mixture passes from the solid phase into the liquid phase and then a compact solid phase is recovered through the stage of a pasty phase[23]. The reaction is endothermic in the fusing stage and exothermic in the subsequent solidification of the reaction mixture[4,29,31]. The heat transfer into such a reaction mixture and the stirring is very difficult, especially in a thick layer. Furthermore, the reaction must be conducted at a high temperature and under pressure of carbon dioxide. It

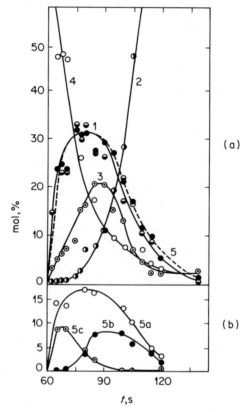

FIGURE 1.   Time dependence of the ratio of reaction products in the transcarboxylation of potassium phthalate. (a) Components of the reaction mixture: (1) benzoic acid, (2) terephthalic acid, (3) isophthalic acid, (4) phthalic acid and (5) a mixture of isomeric benzenetricarboxylic acids. (b) Detailed content of isomeric benzenetricarboxylic acids: (5a) trimellitic acid, (5b) trimesic acid and (5c) hemimellitic acid.

is therefore very difficult to determine the reaction conditions which vary from author to author[10,11,16,17,22,28]. The work on a micro scale in sealed glass ampoules, which allows a visual check to be made on the reaction course, appears to be the most advantageous technique for the determination of components of mixtures resulting from transcarboxylations[18,22-24].

   a. Transcarboxylation of potassium benzenedicarboxylates.   As indicated by the relationship between the composition of reaction mixtures and the reaction time in transcarboxylations of potassium phthalate (Figure 1), the first stage of the transcarboxylation consists in disproportionation of the phthalate (curve 4) into an equimolecular mixture of potassium benzoate (curve 1) on the one hand and a pair of benzenetricarboxylates (curve 5) on the other, namely, potassium hemimellitate (curve 5c) and potassium trimellitate (curve 5a). The third positional isomer, the symmetric potassium trimesate (curve 5b) with mutual *meta* positions of the three carboxylic groups, cannot be formed in this stage. In the next stage, the resulting mixture of potassium benzoate and the two potassium benzenetricarboxylates affords a mixture of all three isomeric potassium benzenedicarboxylates in which potassium isophthalate predominates (curve 3). In the third phase, the benzenedicarboxylates disproportionate into potassium benzoate and the benzenetricarboxylates. At this stage, the hemimellitate (curve 5c) rapidly disappears and is replaced in the mixture by the trimesate (curve 5b) while the amount of the trimellitate (curve 5a) remains almost unchanged. This effect can be readily explained since the symmetric trimesate can be exclusively formed from the isophthalate, i.e. from the most abundant benzenedicarboxylate in the second stage of the transcarboxylation. Potassium trimellitate (curve 5a) can be formed not only from the isophthalate (curve 3) but also from the remaining two positional isomers, i.e. from the phthalate (curve 4) and the terephthalate (curve 2), which are however less abundant in this stage of the reaction. The formation of the hemimellitate (curve 5c) from the isophthalate (curve 3) is sterically hindered. For this reason, the content of the hemimellitate (curve 5c) strikingly decreases in the third stage whereas the amount of the trimesate (curve 5b) and the trimellitate (curve 5a) increases. In this complex transcarboxylation the respective reaction stages can proceed almost simultaneously and are many times repeated as it will be shown below. Finally, potassium terephthalate (curve 2) crystallizes from the reaction mixture because of the great thermal stability and crystallizing ability of this salt. The transcarboxylation of potassium phthalate can be expressed by equation (4)[18,23].

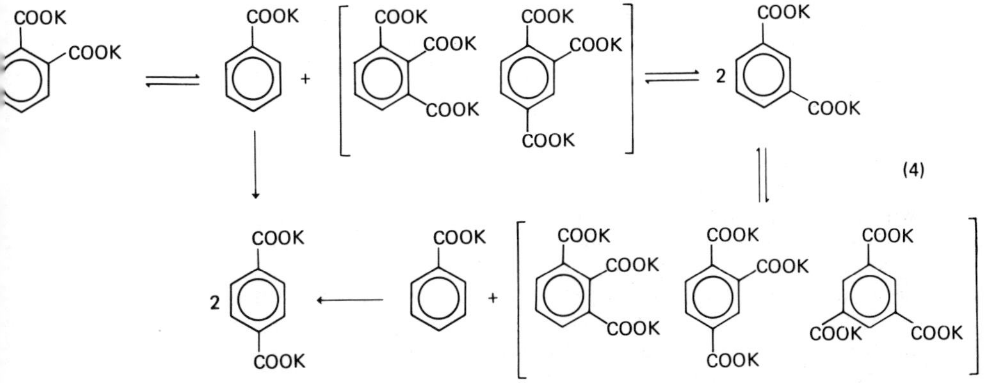

(4)

This reaction course is the same under various reaction conditions e.g. in the presence or absence of catalysts, at various pressures of carbon dioxide, and at various temperatures from $330-500°C$.

It is of interest that no benzene is formed by the irreversible disproportionation of potassium benzoate (during the initial stages up to 30% of the benzoate and the benzenetricarboxylates is formed, see curves 1 and 5). Most probably, the molecules of benzenecarboxylates react in 'dynamically' associated pairs, the protons and carboxylate groups being exchanged between the two molecules. However, this association is very labile as indicated by the statistic exchange of deuterium by protium on benzene rings of benzenecarboxylates in the transcarboxylation of a mixture of salts labelled and non-labelled with deuterium. The associated pairs of salts also operate in transcarboxylations in the absence of catalysts[34]. This observation is not at variance with the interesting hypothesis of Ogata and coworkers[19-21] on the sandwich-like π complexes of salts of benzenecarboxylic acids with the catalyst in transcarboxylations, but also indicates that the transcarboxylation can proceed without these complexes.

The transcarboxylation of potassium isophthalate resembles that of potassium phthalate except for the forced reaction conditions[32-35]. Since the carboxylate groups are placed in *meta* positions, the disproportionation of the isophthalate in the first stage of the reaction (Figure 2, curve 3) to the benzoate (curve 1) and the benzenetricarboxylates (curve 5) mainly affords the trimellitate (curve 5a) and the trimesate (curve 5b), whereas the hemimellitate (curve 5c) is almost absent. The first stage of the transcarboxylation of potassium isophthalate thus corresponds to the third phase in the transcarboxylation of potassium phthalate. The second stage, i.e. the formation of a mixture of benzenedicarboxylates from potassium benzoate (curve 1) and potassium benzenetricarboxylates (curve 5), gives rise to the terephthalate (curve 2) and isophthalate (curve 3) while the content of the phthalate (curve 4) is very low. The content of the phthalate and hemimellitate in the reaction mixture in the course of the whole reaction is thus maintained at a very low level (curves 4 and 5c). Consequently, when these small amounts of the phthalate and hemimellitate are not taken into account, the transcarboxylation scheme of potassium isophthalate, when compared with equation (4) for the phthalate, may be simplified to equation (5)[18,23].

$$\tag{5}$$

As has been shown earlier, potassium terephthalate is stable under reaction conditions that bring about transcarboxylation of the other benzenecarboxylates. With the use of considerably elevated temperatures (above $450°C$) and long reaction times, a negligible amount of the terephthalate is converted (under partial carbonization) to the other benzenecarboxylates. The resistance of potassium terephthalate to the transcarboxylation is of a physical nature (thermal stability and crystallizing ability) and does not result from the thermodynamic equilibrium in the reaction mixture as indicated by transcarboxylations of potassium terephthalate in the melt of potassium cyanate[36,37]. The low transcarboxylation activity of

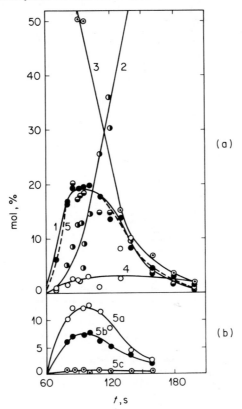

FIGURE 2.   Time dependence of the ratio of reaction products in the transcarboxylation of potassium isophthalate. Designations as in Figure 1.

potassium terephthalate (present in a stable crystalline lattice in the course of the reaction) is also indicated by the zero incorporation of the ^{14}C radionuclide from the ^{14}C-labelled carbon dioxide into the carboxylate groups of potassium terephthalate[14] and by the negligible exchange of deuterium by protium in experiments with a mixture of deuterium-labelled and non-labelled potassium terephthalate conducted under reaction conditions that favour the transcarboxylation[30].

    b. *Transcarboxylation of potassium benzoate.*   In contrast to the transcarboxylation of potassium benzenedicarboxylates, potassium benzoate undergoes transcarboxylation in two stages. In the first irreversible stage, two molecules of the benzoate disproportionate, with the formation of benzene and a mixture of potassium benzenedicarboxylates (Figure 3); in this mixture*, the phthalate (curve 4), isophthalate (curve 3) and terephthalate (curve 2) are present in the ratio of 25 : 5 : 1. This stage can be illustrated by equation (6). It may be seen from this equation that the amount of the resulting benzene is the measure of completion of the first stage[18,24].

_____
*At the beginning of the reaction.

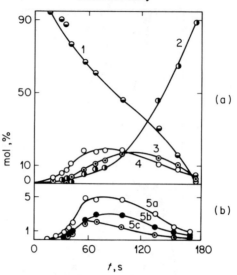

FIGURE 3. Time dependence of the ratio of reaction products in the transcarboxylation of potassium benzoate. (a) Components of the solid reaction mixture [without curve (5) for a mixture of benzenetricarboxylic acids]: (1) benzoic acid, (2) terephthalic acid, (3) isophthalic acid and (4) phthalic acid. (b) Detailed content of the isomeric benzenetricarboxylic acids: (5a) trimellitic acid, (5b) trimesic acid and (5c) hemimellitic acid.

In the second stage of the transcarboxylation of potassium benzoate, the benzenedicarboxylates resulting from the first stage react according to equation (7),

$$2 \quad \text{(COOK)} \quad \longrightarrow \quad \text{( )} \quad + \quad \text{(COOK, COOK)} \tag{6}$$

$$2 \quad \text{(COOK, COOK)} \quad \rightleftharpoons \quad \text{( )} \quad + \quad \text{(COOK, COOK, COOK)} \quad \longrightarrow \quad 2 \quad \text{(COOK, COOK)} \tag{7}$$

i.e. analogously to the above transcarboxylation of potassium phthalate and potassium isophthalate. The two stages obviously overlap to a certain degree. The transformation of benzenedicarboxylates into potassium benzoate and potassium benzenetricarboxylates is less marked because in the first stage of the reaction the mixture contains mainly the starting potassium benzoate and only very little potassium benzenedicarboxylate. Consequently, the content of benzenetricarboxylates in the whole reaction mixture appears as relatively low (curves 5a—5c).

The irreversibility of equation (6) was unambiguously established by the reaction of non-radioactive potassium benzoate in the presence of ^{14}C-labelled benzene. From the reaction mixture, non-radioactive terephthalic acid was isolated and the whole radioactivity remained in the benzene[38].

c. *Transcarboxylation of mixtures containing potassium benzoate and potassium benzenepolycarboxylates.*   The transcarboxylation of an equimolecular mixture of potassium benzoate and potassium benzenetricarboxylates (equation 8)

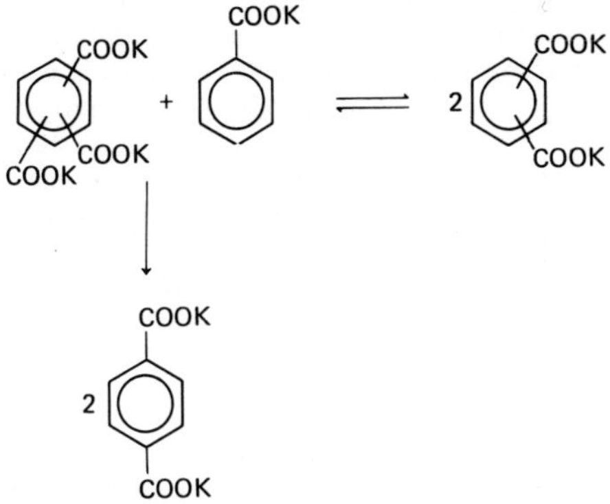

(8)

shows the reverse of equation (7). In the present case, however, the stirring and homogenization of the reaction mixture cannot be as thorough as with the benzenedicarboxylates in equation (7). For this reason, a proportion of the benzoate

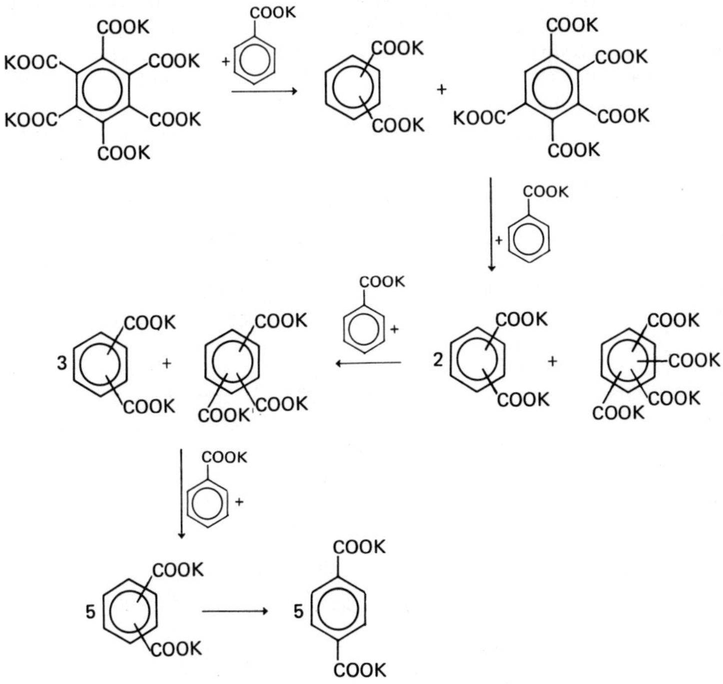

(9)

undergoes an irreversible disproportion to benzene and benzenedicarboxylates according to equation (6) and the excess benzenetricarboxylates are subjected to a complex reaction, with the formation of potassium terephthalate and potassium benzenetetracarboxylates; the latter salts undergo a partial decarboxylation and carbonization. The amount of benzene produced by the irreversible disproportionation of potassium benzoate depends on the homogeneity of the starting reaction mixture[24].

Benzenetetra-, benzenepenta- and benzenehexacarboxylates react analogously to benzenetricarboxylates with two, three, and four equivalents of potassium benzoate[24]. However, the homogenization of such mixtures is very difficult. Moreover, the successive reaction of one molecule of the benzenepolycarboxylate (e.g. mellitate, see equation 9) with the benzoate molecules in 'associated' pairs is less probable than the interaction of two benzoate molecules. For this reason, the successive reaction of one equivalent of potassium mellitate with four equivalents of potassium benzoate does not afford the theoretical five equivalents of potassium terephthalate (equation 9), but a proportion of the benzoate undergoes an irreversible transformation to benzene and benzenedicarboxylates according to equation (6).

3. Transcarboxylation of benzenecarboxylates in molten KCNO or KCNS

In order to throw some light on the role of reaction equilibria in the complex course of transcarboxylations of potassium benzenecarboxylates, these reactions were examined with fused inert compounds, such as fused potassium cyanate or fused potassium thiocyanate, as solvents. Under the transcarboxylation conditions, both these potassium salts are chemically inert towards potassium benzenecarboxylates and no competitive reactions take place. The transcarboxylation can thus be conducted under the same reaction conditions as in the absence of fused KCNO or KCNS. On the other hand, these salts interfere with the crystallization of some reaction products (mainly the terephthalate) and with the shift of equilibrium The ratio of fused KCNO or KCNS to the reactants should be high enough to ensure a homogeneous melt.

Thus, under the above stated reaction conditions, the transcarboxylation affords[36,37] an equilibrium mixture containing (approximately) 20% of the benzoate, 18% of the terephthalate, 35% of the isophthalate, 5% of the phthalate, 9% of the trimesate, 10% of the trimellitate and 1% of potassium hemimellitate, irrespective of the starting benzenecarboxylate (i.e. the same product ratio is also obtained from potassium terephthalate) (see Figures 4–7). When an insufficient amount of fused KCNO is used[19,20,39], or when an improper mixture (e.g. a mixture of potassium cyanate and potassium carbonate[40]) is applied that does not entirely suppress the crystallization of potassium terephthalate, the equilibrium is shifted in favour of the formation of potassium terephthalate as the final reaction product. The transcarboxylation of potassium benzenedicarboxylates or of a mixture of potassium benzoate with potassium benzenetricarboxylates in fused KCNO as solvent can be thus expressed by equation (10).

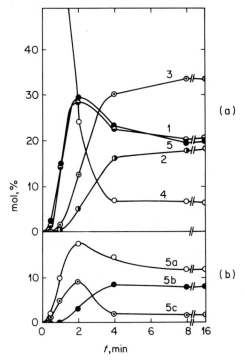

FIGURE 4. Time dependence of the ratio of reaction products in the transcarboxylation of potassium phthalate in fused KCNO. (a) Components of the reaction mixture: (1) benzoic acid, (2) terephthalic acid, (3) isophthalic acid, (4) phthalic acid and (5) a mixture of isomeric benzenetricarboxylic acids. (b) Detailed content of isomeric benzenetricarboxylic acids: (5a) trimellitic acid, (5b) trimesic acid and (5c) hemimellitic acid.

The analogous transcarboxylation of potassium benzoate in fused KCNO proceeds similarly to that performed in the absence of fused KCNO, i.e. in two stages. The first irreversible stage can be expressed by equation (11) and the equilibration in the second stage is in accordance with equation (10).

$$2 \underset{}{\overset{COOK}{\bigcirc}} \xrightarrow{(KCNO)} \bigcirc + \underset{COOK}{\overset{COOK}{\bigcirc}} \tag{11}$$

The examination of the transcarboxylation of potassium benzenetricarboxylates in fused KCNO was very helpful in the elucidation of the reaction mechanism of all transcarboxylations of benzenecarboxylates. In the absence of fused KCNO, a considerable carbonized reaction mixture is obtained containing potassium terephthalate along with potassium salts of the other benzenecarboxylic acids including

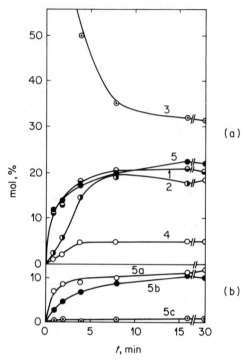

FIGURE 5.   Time dependence of the ratio of
reaction products in the transcarboxylation of
potassium   isophthalate   in   fused   KCNO.
Designations as in Figure 4.

the benzenetetracarboxylic acids. In the presence of fused KCNO, potassium
benzenetricarboxylates undergo disproportionation to a mixture of benzenedicarboxylates and benzenetetracarboxylates (equation 12), as has been determined by

$$2 \; \text{(COOK benzenetricarboxylate)} \; \underset{\text{(KCNO)}}{\rightleftarrows} \; \text{(benzenedicarboxylate)} \; + \; \text{(benzenetetracarboxylate)} \tag{12}$$

analysis of the final reaction mixture[41]. Disproportionation of the benzenedicarboxylates to the benzoate and benzenetricarboxylates (according to equation 10) is
substantially suppressed by the presence of the starting potassium benzenetricarboxylates.

4. Transcarboxylation of benzenecarboxylates in an atmosphere of ^{14}C-labelled carbon dioxide

Successful incorporation of the ^{14}C radionuclide into the carboxylate groups
during the transcarboxylation of salts of aromatic carboxylic acids requires the
following reaction conditions. (*i*) The ^{14}C-labelled carbon dioxide cannot be replaced

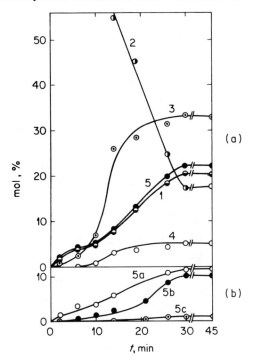

FIGURE 6.   Time dependence of the ratio of
reaction products in the transcarboxylation of
potassium   terephthalate   in   fused   KCNO.
Designations as in Figure 4.

by ^{14}C-labelled potassium carbonate, the decomposition temperature (above 800°C) of which is too high. (*ii*) The reactants (salts of benzenecarboxylic acids) should be applied to the support in thin layers to ensure the highest contact with the atmosphere of ^{14}C-labelled carbon dioxide[18,23,24]. (*iii*) Efficient stirring is indispensable[24,42]. Under such reaction conditions and in the initial stage of the transcarboxylation (when the reaction mixture contains intermediates along with some potassium terephthalate), the incorporation of the ^{14}C radionuclide into the salts corresponds to a complete equilibrium between the carboxylate groups of benzenecarboxylates and the [14]CO_2 molecules in the reaction atmosphere.

When the transcarboxylation does not proceed or when it proceeds to a low extent only (in the absence of catalysts, at low temperatures or when the reaction time is too short), no incorporation of ^{14}C into the carboxylate groups of salts takes place, or it occurs to a lesser extent corresponding to the degree of the transcarboxylation. Consequently, no ^{14}C is incorporated from the [14]CO_2-containing reaction atmosphere into the carboxylate groups of potassium terephthalate in spite of reaction conditions that favour transcarboxylations of the other benzene-carboxylates[14,23].

In conclusion (*a*) the incorporation of ^{14}C from the [14]CO_2-containing reaction atmosphere depends on the removal and readdition of carboxylate groups (the ionic mechanism of this process is claimed above) and (*b*) the transcarboxylation of

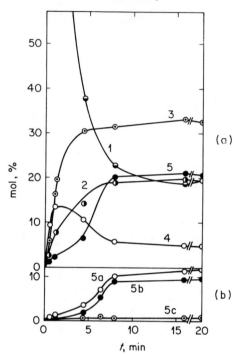

FIGURE 7. Time dependence of the ratio of the solid reaction products in the transcarboxylation of potassium benzoate in fused KCNO. Designations as in Figure 4.

potassium benzenecarboxylates consists in repeated decarboxylations—recarboxylations of all carboxylate groups.

Concerning the mechanism of the ^{14}C incorporation, the $^{14}CO_2$ molecules from the reaction atmosphere most probably replace the carbon dioxide liberated from the removed carboxylate cation (equation 13). The equilibrium in this equation is

$$COOK^+ \rightleftharpoons CO_2 + K^+ \tag{13}$$

strongly shifted to the left in favour of the carboxylate cation which is simultaneously bound by the strongly basic carbanion (arisen by ionic removal of the carboxylate group); thus the free carbon dioxide does not escape into the reaction atmosphere[43].

As mentioned above, the reaction conditions strongly affect the ^{14}C incorporation into the carboxylate group of salts. In the case of insufficient contact of the labelled carbon dioxide with salts to be transcarboxylated, the incorporation of the ^{14}C radionuclide proceeds to a low extent only[13-15]. A similar low incorporation is observed when ^{14}C-labelled potassium carbonate[10,11,19,44] is used instead of $^{14}CO_2$ since decomposition of $K_2{}^{14}CO_3$ takes place at $> 800°C$ while the transcarboxylation temperature is about $400°C$. At this temperature the partial pressure of $^{14}CO_2$ in the reaction atmosphere is very low[23,24].

A higher incorporation of the ^{14}C radionuclide from $K_2^{14}CO_3$ into the carboxylate group of salts would require a very long reaction time. However, the formation time of the final terephthalate by transcarboxylation is relatively short and the terephthalate is known not to incorporate any ^{14}C radionuclide from the $^{14}CO_2$-containing reaction atmosphere[14].

5. Exchange of deuterium by protium in transcarboxylations of mixtures of benzenecarboxylates labelled and non-labelled by deuterium on the benzene ring[18,25,30]

The assumed formation of carbanions in transcarboxylations of salts of benzenecarboxylic acids may occur either by rupture of the C–C bond between the benzene ring and the carboxylate group, or by liberation of protons from the benzene ring as inferred from the statistical exchange of deuterium by protium in transcarboxylations of mixtures of benzenecarboxylates labelled and non-labelled by deuterium in the benzene ring. Benzene itself does not undergo this exchange reaction or the transcarboxylation reaction[30]. Furthermore, the exchange of deuterium by protium in benzenecarboxylates is faster than the transcarboxylation, as exemplified by transcarboxylation of an equimolar mixture of potassium benzoate-5d and non-deuterated benzoate: an exchange of deuterium by protium was already accomplished in molecules of benzoate in the initial stage of the transcarboxylation[30].

When compared with the rate of the corresponding decarboxylation–recarboxylation process, the rate of exchange of deuterium by protium is so high that the exchange may also be effected (under suitable conditions) with deuterium oxide without any substantial occurrence of the transcarboxylation reaction[45]. This observation was made in two experiments on the transcarboxylation of potassium phthalate in the presence of deuterium oxide under different reaction conditions. Thus, when the transcarboxylation is performed in excess D_2O at an elevated temperature and with a sufficiently long reaction time, complete exchange of deuterium by protium and a simultaneous decarboxylation take place with the formation of deuterated benzene. On the other hand, when the mixture is briefly heated at a low temperature, deuterated potassium phthalate is obtained along with a small amount of transcarboxylation products[45]. Fast exchange of deuterium by protium takes place with all benzenecarboxylates susceptible to transcarboxylations. When the transcarboxylation for some reason does not take place or occurs to a negligible extent only, no deuterium or only a trace amount of deuterium is exchanged by protium. Thus for example in the absence of catalysts or at temperatures lower than needed for transcarboxylations, no D/H exchange takes place in the interaction between deuterated and non-deuterated benzoate. Similarly, heating a mixture of deuterated and non-deuterated potassium terephthalate at the transcarboxylation temperature does not result in any transcarboxylation or exchange since the salt of terephthalic acid is unreactive even in the presence of catalysts under these conditions[30].

When the transcarboxylation is effected with a mixture of a deuterated phthalate and a non-deuterated terephthalate or with a mixture of a benzoate-5d and a terephthalate, a negligible exchange may be observed. As suggested by this low exchange, potassium terephthalate in the fused reaction mixture is not completely resistant to the transcarboxylation reaction[30].

The above results can be explained by involvement of a relatively stable 'dynamic' carbanion species which is formed in the initial stages of the transcarboxylation reaction, since the liberation and addition of protons is much faster than the decarboxylation–recarboxylation process of the carboxylate group[30] (for the case of potassium benzoate see equation 14).

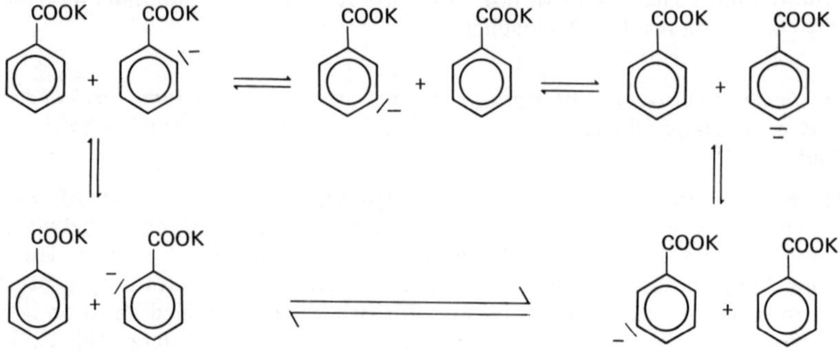

(14)

6. Mechanism of transcarboxylation of benzenecarboxylates

As may be inferred from the above observations, the transcarboxylation of benzenecarboxylates can be interpreted by an intermolecular, ionic mechanism. It may also be seen that the transcarboxylation of all benzenecarboxylates proceeds by a few fundamental pathways, though a very complex and time-dependent mixture of intermediary products is involved in the course of the reaction. Thus for example, the transcarboxylation of potassium phthalate can be interpreted by a few simple consecutive reactions[23]:

(a) Removal of the carboxylate groups with the formation of a carbanion and a carboxylate cation (equation 15).

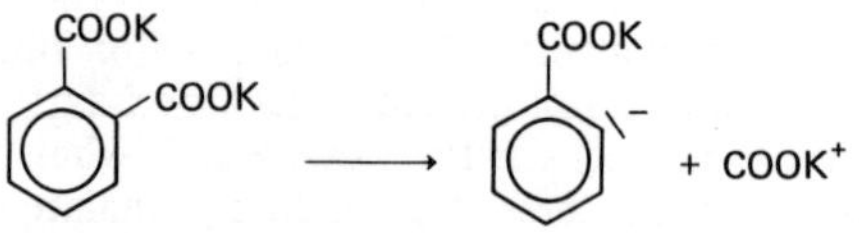

(15)

(b) Liberation of a proton from the molecule of the original phthalate by the action of a carbanion (as a very strong base) and addition of the liberated proton to this carbanion with the formation of the benzoate and two new carbanions which are precursors of potassium trimellitate and potassium hemimellitate (equation 16).

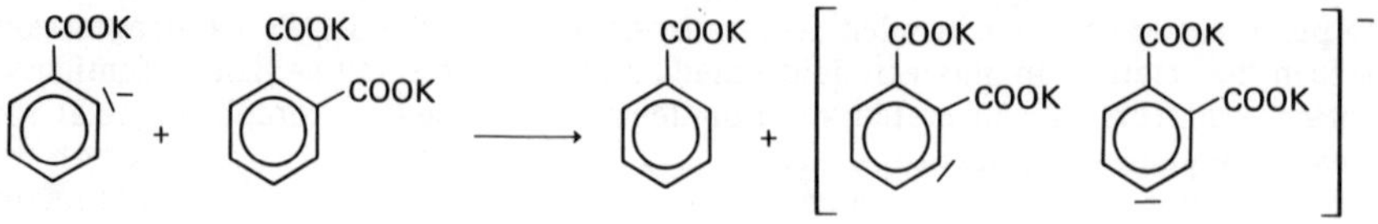

(16)

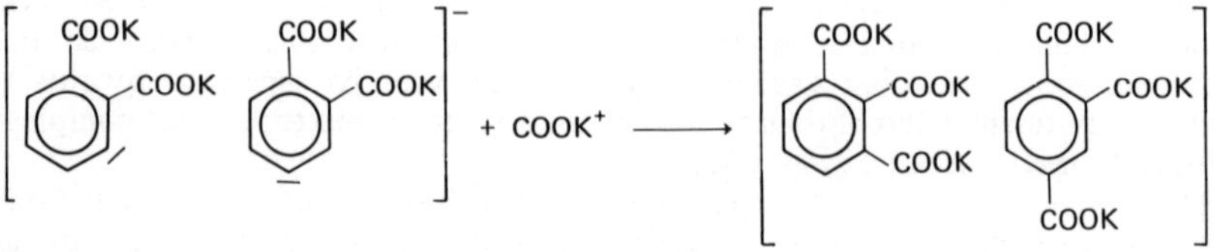

(17)

(*c*) Addition of the carboxylate cation to the resulting carbanions with the formation of the corresponding salts of benzenetricarboxylic acids (equation 17).

The liberation and addition of protons (equation 16) is faster than the decarboxylation—recarboxylation process (equations 15 and 17). This transprotonization takes place on all carbon atoms of the benzene ring to which hydrogen atoms are attached, with the formation of a 'dynamic' carbanion species bearing an unshared electron pair at various positions of the benzene ring[30] (see for example equation 14 for the case of a benzoate).

These processes occur during the transcarboxylation in repeating cycles and yield a complex mixture of intermediates which finally deposits the potassium terephthalate.

The carboxylate group is split off and added again, more likely in the form of the carboxylate cation $COOK^+$ than in the form of CO_2 and then the potassium cation K^+, since no exchange of the ^{14}C radionuclide is observed in the transcarboxylation of potassium benzoate-7-^{14}C and non-radioactive potassium benzoate placed in the same nitrogen atmosphere but in separate vessels[43].

As indicated by the fast (faster than the C—C bond rupture in the decarboxylation of the carboxylate group) exchange of deuterium by protium with the formation of a 'dynamic' carbanion[30,43] (equation 14), the liberation of protons from the benzene-ring carbon atoms can be more likely ascribed to the action of the unshared electron pair of the primarily arisen carbanion than to the action of the negative charge of the anion of the dissociated salt[43].

The formation of primary carbanions is initiated by catalysts[47], which accelerate the transcarboxylation of potassium salts of aromatic carboxylic acids. This transcarboxylation can also take place in the absence of catalysts, but the reaction course is very slow. In the presence of an inorganic cadmium salt, the transcarboxylation of aromatic potassium salts proceeds as follows. A mixed cadmium salt of the benzenecarboxylate is first formed and then readily decarboxylated to the primary carbanion. By liberation of a proton from the benzene ring of another benzenecarboxylate, this primary carbanion gives rise to a new carbanion (the 'dynamic' carbanion, equation 14). By the addition of a carboxylate cation to this carbanion, the salt of the corresponding benzenecarboxylic acid is formed (e.g. equation 17).

The ionic charge of the carbanion also makes possible the formation of the above mentioned associated pairs of the reacting salts. The existence of these pairs explains the absence of benzene in the course of the transcarboxylation of benzenedicarboxylates, though a disproportionation of benzenedicarboxylates to the benzoate and benzenetricarboxylates takes place in the initial stage of the reaction (e.g. equation 4). These associated pairs (the formation of which does not depend on the presence of catalysts) are unusually labile as indicated by the rapid random exchange of deuterium by protium in the above mentioned transcarboxylation of benzenecarboxylates labelled and non-labelled by deuterium. The lability of associated pairs is also due to the repeated disappearance of the negative charge of the carbanion by additions of the carboxylate cation or proton to the unshared electron pair of the carbanion.

The mechanism of the repeated formation of carbanions (by decarboxylation of carboxylate groups or by deprotonization) and their repeated disappearance (by additions of carboxylate cations or protons) expresses better the course of the transcarboxylation than the mechanism of the 'sandwich' complex of the reacting salts with the catalyst[19-21]. The 'sandwich' theory could hardly explain the course of the transcarboxylation even in the absence of catalysts, yet alone the high rate of transcarboxylations in the solid or semisolid state in the presence of a small amount of the catalyst. The sandwich formation from two molecules of the salt to be transcarboxylated and one molecule of the catalyst would require a high mobility of salts in the reaction mixture. Remembering the solid state of the reaction mixture, such a mobility can be hardly assumed.

In conclusion, the transcarboxylation of salts of benzenecarboxylic acids can be characterized as an intermolecular, ionic decarboxylation—recarboxylation process in combination with an intermolecular transprotonation proceeding in repeating cycles.

7. Effect of catalysts on the transcarboxylation of benzenecarboxylates

Transcarboxylations are markedly accelerated by the presence of catalysts[4,46], though they are known[4,34,47] to proceed even in the absence of catalysts. Salts of metals of the subgroup IIb of the Periodic Table, especially the salts of cadmium and zinc[4,48], have proved to be the most efficient catalysts. The catalytic effect in transcarboxylations largely depends on the chemical and physical form of the catalyst. Among the most frequently used cadmium catalysts, the highest potency is exhibited by inorganic acdmium salts, particularly cadmium iodide. Somewhat less potent are cadmium salts of the acids to be transcarboxylated, cadmium carbonate and cadmium oxide; the least potent catalyst of the cadmium series is cadmium metal. The salts of other metals, for example zinc, show a similar decrease in catalytic effects.

The high effectiveness of inorganic salts as catalysts of transcarboxylations might be explained by a high disperion of the catalyst in the reaction mixture by means of the primary reaction with the potassium salt of the aromatic acid to be transcarboxylated[47] (e.g. potassium phthalate in equation 18).

$$\underset{\text{COOK}}{\overset{\text{COOK}}{\bigcirc}} + CdX_2 \longrightarrow \underset{\text{COO}\frac{Cd}{2}}{\overset{\text{COO}\frac{Cd}{2}}{\bigcirc}} + 2\,KX \qquad (18)$$

This dispersion cannot be as fast and efficient when previously prepared cadmium phthalate is used as catalyst. Furthermore, the dispersion of the catalyst is obviously affected by the nature of the anions. Thus for example, the higher activity of cadmium iodide in comparison to the other cadmium halides is due to its low melting point ($388°C$) which facilitates the reaction according to equation (18)[47]. Since the anion of the salt must be chemically inert in the transcarboxylation reaction, most organic salts are thus excluded as catalysts.

The low activity of powdered metals as catalysts is not due to the metal itself but to a very small amount of oxides or salts on the surface of the metal powder from atmospheric corrosion during the storage or from the action of reactants or traces of moisture directly in the reaction mixture. The powdered metal itself is not engaged in the catalysis and does not suffer any change during the reaction.

The catalytically active inorganic salts of metals are transformed to the corresponding aromatic carboxylates (equation 18), which undergo a transcarboxylation analogous to that of the potassium salts except for the formation of a mixture of benzenecarboxylates instead of the exclusive formation of terephthalate (equations 19 and 20). The aromatic salts of cadmium being less stable than aromatic potassium carboxylates, undergo the transcarboxylation more readily and at lower temperatures, thus initiating the primary formation of carbanions. On the other hand, the low thermal stability of aromatic cadmium salts results in considerable

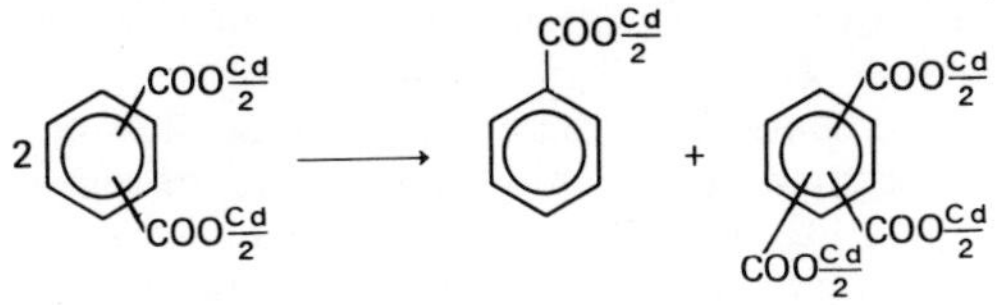

$$2 \; C_6H_4(COO\tfrac{Cd}{2})_2 \longrightarrow C_6H_6 + C_6H_3(COO\tfrac{Cd}{2})_3 \tag{19}$$

$$2 \; C_6H_5(COO\tfrac{Cd}{2}) \longrightarrow C_6H_6 + C_6H_4(COO\tfrac{Cd}{2})_2 \tag{20}$$

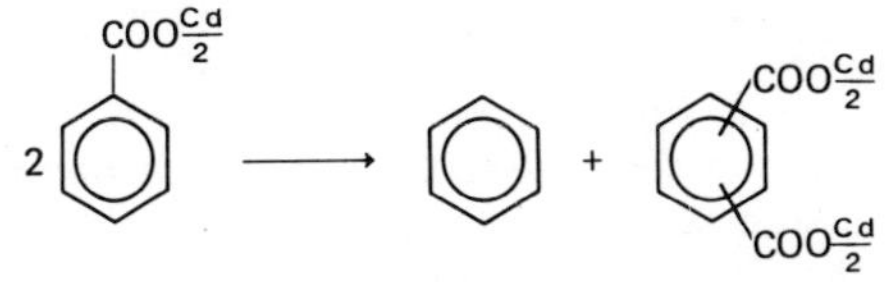

destruction of these salts at the higher transcarboxylation temperature of potassium salts. Consequently, an equimolar portion (corresponding to the amount of the catalyst) of the potassium benzenecarboxylates to be transcarboxylated is carbonized[47].

It is useful to compare the melting point and the transcarboxylation temperature of potassium and cadmium salts of benzoic acid. Thus, potassium benzoate melts at $425-430°C$, its transcarboxylation rate in the absence of catalysts being negligible at this temperature. On the other hand, cadmium benzoate melts at $235°C$ and undergoes a fast transcarboxylation at $290°C$. At $400°C$ (i.e. the temperature applied to transcarboxylations of potassium benzoate in the presence of catalysts), the use of cadmium benzoate as the reactant results in destruction of a considerable portion of the reaction mixture[47]. Transcarboxylations of the other cadmium benzenecarboxylates show a similar reaction course.

The catalytic activity largely depends on the dispersion of the catalyst in the reaction mixture and on the relatively low melting point of the catalyst since the lower temperature makes possible its dispersion and the formation of primary carbanions. The rapid initiation of the transcarboxylation reaction by the action of catalysts has a further intensified effect on the transcarboxylation rate of aromatic potassium salts since the initial decrease of the melting point of the reaction mass due to the formation of a complex mixture of intermediates results in an easier agitation and thus homogenization of the reaction mixture. In an atmosphere of carbon dioxide at 3 atm, with potassium phthalate melting at $385-390°C$ the rate of transcarboxylation is very low. However, in the presence of 5% of cadmium iodide, the mixture melts at $330-335°C$ and the transcarboxylation process is very fast. In addition to the unreacted phthalate, the melt contains an appreciable amount of the other benzenecarboxylates (Figure 1). The transcarboxylation rate of potassium isophthalate is much slower. This can be ascribed to the higher melting point of potassium isophthalate: the melt is less easily formed. Consequently, the primary reaction with the catalyst, its dispersion in the mixture and the formation of primary carbanions are all slower.

In conclusion, the effect of catalysts in transcarboxylations of salts of aromatic acids is of a complex nature since the chemical effect which initiates the formation of primary carbanions is accompanied by the physical effect which easily converts the reaction mixture to the melt.

8. Effect of various cations on the transcarboxylation of benzenecarboxylates

The transcarboxylation of potassium benzenecarboxylates to potassium terephthalate is a special case of a general reaction since instead of the potassium cation other cations can be used[47]. As shown in the preceding section, the use of other cations such as Cd^{2+} or Zn^{2+} is of great importance for the catalysis of transcarboxylation of potassium salts. The common feature of all transcarboxylations is the formation of the corresponding carbanions by removal of carboxylate groups in the form of cations or the liberation of protons by the action of carbanions. On the other hand, the formation of the final product depends on the thermal stability and crystallizing ability of some of the intermediary products; the final product is usually deposited in crystalline form by the reaction mixture. The structure of these final products depends on the nature of both the aromatic acid as anion and the metal as cation. In the field of salts of benzenecarboxylic acids, the transcarboxylation of potassium salts has been examined in detail; in this case, potassium terephthalate is the final product of the complex reaction.

Using salts of other alkali metals, the final product differs from that resulting from the transcarboxylation of potassium salts. The rubidium and caesium salts are easy to melt and decarboxylate with the formation of carbanions, but their thermal stability is relatively low and the crystallizing ability of the corresponding rubidium or caesium terephthalate is rather low. The final complex mixture thus contains a lower amount of the terephthalate than in the case of the potassium cation[47,49]. Furthermore, the amount of the carbonized material is considerably high.

On the other hand, the lithium and sodium benzenecarboxylates are more difficult to melt than the potassium salts and their crystallizing ability differs from that of potassium salts. Particularly in the case of sodium salts, the final reaction mixtures exhibit a higher content of sodium trimesate at the expense of sodium terephthalate[47,49]. By the addition of suitable fusing agents, the transcarboxylation is accelerated and the destructive side-reactions suppressed; in such a case, sodium trimesate crystallizes as the main product from the reaction mixture[47,49,50] (equations 21 and 22).

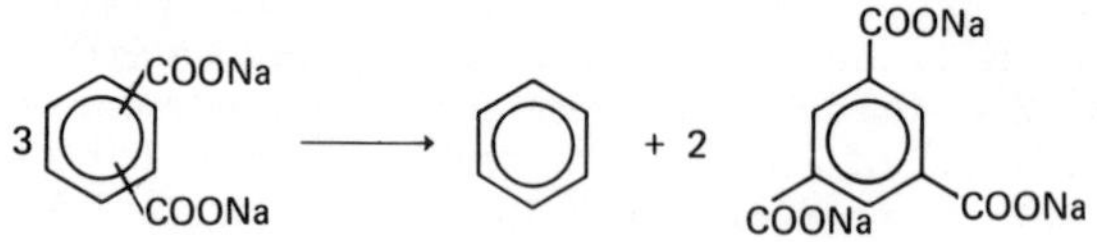

$$\text{(21)}$$

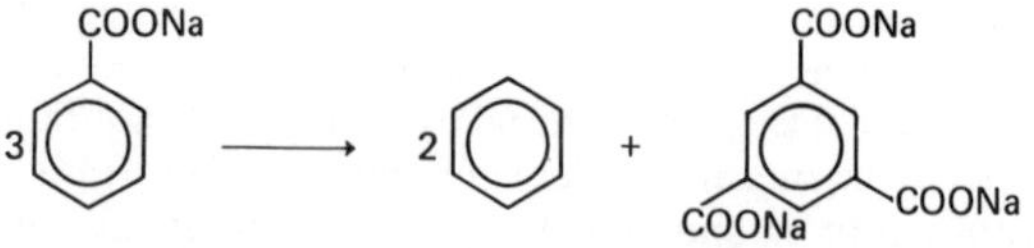

$$\text{(22)}$$

The transcarboxylation mechanism of Li, Na, Rb and Cs benzenecarboxylates is analogous to that of potassium salts as indicated by H/D exchange on the benzene rings and by incorporation of the ^{14}C radionuclide into the carboxylate groups of salts[47,51,52].

In the transcarboxylation of salts with bivalent cations of calcium, strontium and barium, the reaction course differs markedly from that with alkali metal salts. In the case of alkaline-earth metal salts, the reaction stops at the stage of *ortho* isomers, which are less fusible than the alkali metal salts and crystallize from the reaction mixture[47,53] (cf. equations 23 and 24 for the transcarboxylation of barium salts of benzoic acid and isophthalic acid).

The content of the *ortho* isomers of alkaline-earth metal benzenetricarboxylates in the final transcarboxylation mixture can be markedly increased by the use of a

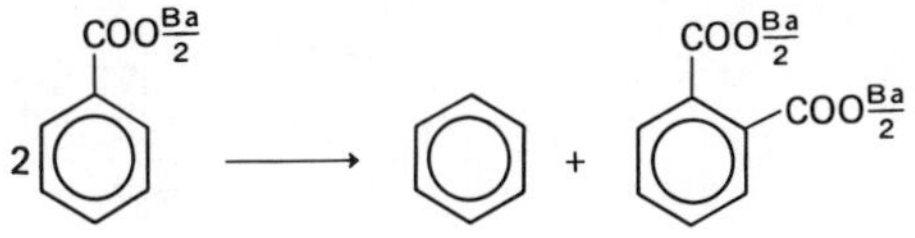

$$2 \bigcirc\text{--}COO\tfrac{Ba}{2} \longrightarrow \bigcirc + \bigcirc\genfrac{}{}{0pt}{}{COO\tfrac{Ba}{2}}{COO\tfrac{Ba}{2}} \tag{23}$$

$$2 \bigcirc\genfrac{}{}{0pt}{}{COO\tfrac{Ba}{2}}{COO\tfrac{Ba}{2}} \longrightarrow \bigcirc + \left[\bigcirc\genfrac{}{}{0pt}{}{COO\tfrac{Ba}{2}}{COO\tfrac{Ba}{2},\,COO\tfrac{Ba}{2}} \quad \bigcirc\genfrac{}{}{0pt}{}{COO\tfrac{Ba}{2},\,COO\tfrac{Ba}{2}}{COO\tfrac{Ba}{2}} \right] \tag{24}$$

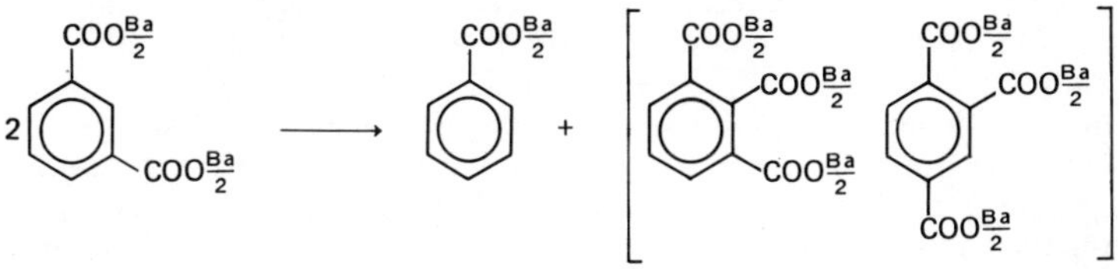

suitable fusing agent[53]; this increase is particularly surprising in the case of the sterically hindered hemimellitic acid. The complex course of the transcarboxylation of alkaline-earth metal benzenecarboxylates requires an efficient analytical method to detect all the occurring isomers in the final reaction mixture, since otherwise incomplete results are obtained[54,55]. Thus, the transcarboxylation of barium phthalate or barium isophthalate has been claimed to give a low yield (up to 10%) of a single product, namely, barium hemimellitate, while barium phthalate and an equimolar amount of benzene should result from the transcarboxylation of barium benzoate. However, as indicated by a detailed analysis, the above transcarboxylations afford a similar mixture of products, namely, the barium salts of hemimellitic and trimellitic acid (the *ortho* isomers) and of benzoic and phthalic acid, and a lesser amount of barium trimesate, barium isophthalate and barium terephthalate (the *meta* and *para* isomers)[47,53]. The calcium and strontium salts of b enzenecarboxylates are transcarboxylated similarly to the barium salts.

The transcarboxylation of bivalent alkaline earth metal benzenecarboxylates is again based on the formation of carbanions arising from decarboxylation of carboxylate groups. This is confirmed by H/D exchange on the benzene rings and by incorporation of the ^{14}C from the ^{14}C-labelled carbon dioxide into the carboxylate groups of the corresponding salts[56].

This group of salts with bivalent cations also contains the earlier mentioned cadmium and zinc salts, which are important as catalysts in the transcarboxylation of the other salts. The course of their transcarboxylation (see equations 19 and 20) is similar to that of the alkaline-earth metal salts; moreover, the cadmium and zinc benzenecarboxylates melt and undergo decarboxylation to carbanions at much lower temperatures[47]. The behaviour of lead (II) and cupric benzenecarboxylates in transcarboxylations is similar to that of the cadmium and zinc salts except for a lower catalytic activity.

B. Napthalenecarboxylic Acids

Naphthalenecarboxylic acid salts also undergo transcarboxylation reactions[4]. In this series, too the potassium salts give the best results. As the final product of the complex reaction, potassium 2,6-naphthalenedicarboxylate is isolated (equations 25 and 26). When compared with benzenecarboxylates, the trans-

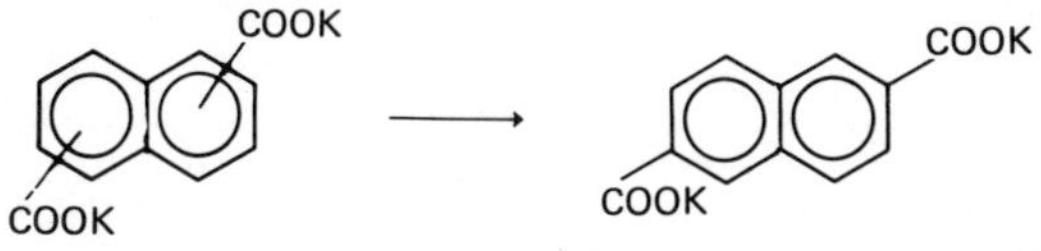

$$\bigcirc\!\bigcirc\genfrac{}{}{0pt}{}{COOK}{COOK} \longrightarrow \bigcirc\!\bigcirc\genfrac{}{}{0pt}{}{COOK}{COOK} \tag{25}$$

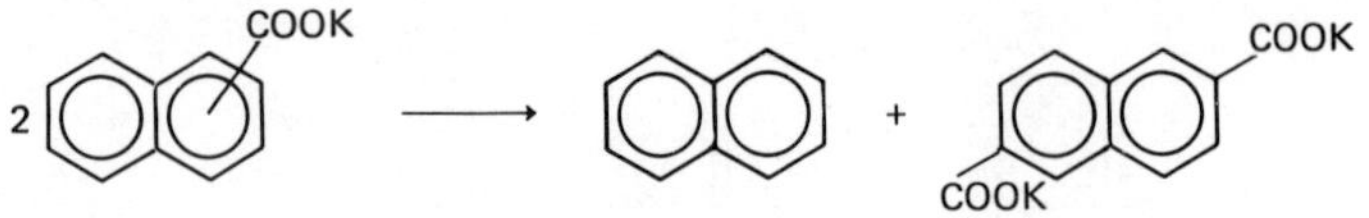

carboxylation of naphthalenecarboxylates is more complex and so is the final reaction mixture too, obviously because of the more complex structure of the naphthalene ring system[57−62]. Nevertheless, the removal of carboxylate groups from the naphthalene ring follows the same ionic mechanism as established with benzenecarboxylates. Thus, smooth decarboxylation takes place in the presence of protons while potassium 2,6-naphthalenedicarboxylate is obtained from potassium α- or β-naphthoate in the presence of potassium carbonate, carbon dioxide and calcium carbide.

Similarly, both the incorporation of ^{14}C from ^{14}C-labelled carbon dioxide in the carboxylate groups of naphthalenecarboxylates[58,62] and the exchange of deuterium by protium on the carbon atoms of the naphthalene ring system[63] take place in the initial stage of the reaction with involvement of the original derivatives that have not yet undergone the transcarboxylation.

The course of the transcarboxylation as inferred from the time dependence of the components of reaction mixtures is similar to that of the benzenecarb-oxylates[57]. In the first stage of the reaction, the salts of naphthalenedicarboxylic acids disproportionate into a naphthalenemonocarboxylate and naphthalenetricarb-oxylates, the mutual interaction of which affords a mixture of naphthalenedicarb-oxylates. From this complex, the crystalline potassium 2,6-naphthalenedicarb-oxylate is finally deposited (equation 27).

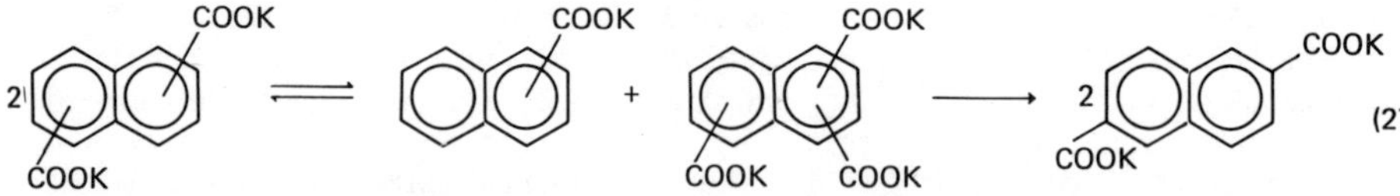

Potassium α- and β-naphthoate (similarly to potassium benzoate) undergoes an irreversible disproportionation in the initial step with the formation of naphthalene and naphthalenedicarboxylates (equation 28). Contrary to the view of McNelis[57],

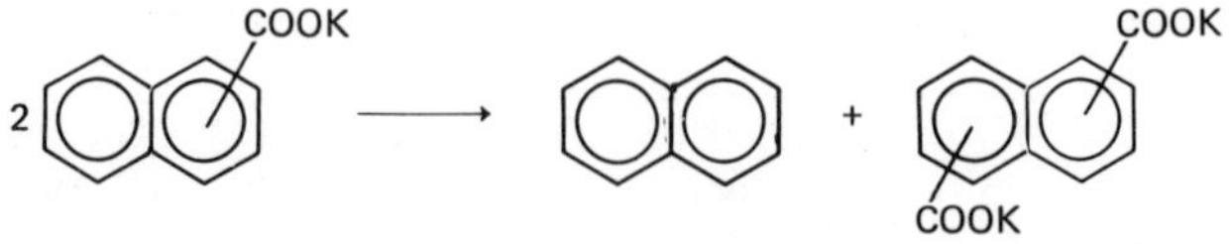

the irreversibility of this initial step was unambiguously established[38] by transcarb-oxylation of potassium α-naphthoate in the presence of ^{14}C-labelled naphthalene. This experiment yielded non-radioactive potassium 2,6-naphthalenedicarboxylate while the whole radioactivity remained in the naphthalene.

When the transcarboxylation of naphthalenecarboxylates is performed in the melt of an inert diluent such as potassium cyanate, a complex equilibrium mixture is obtained even when using potassium 2,6-naphthalenedicarboxylate as the starting material (equations 29 and 30). The initial step in the transcarboxylation of

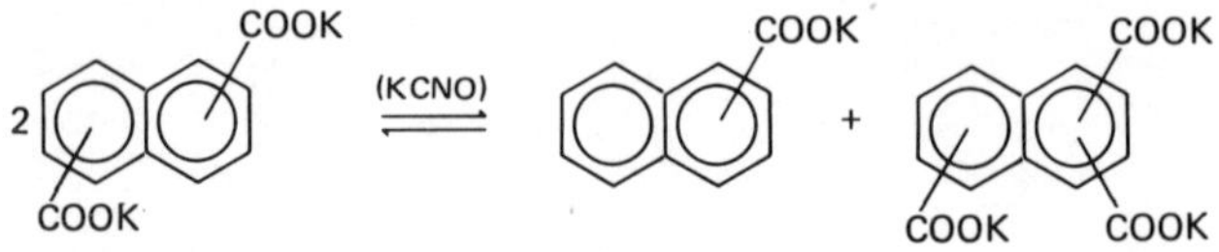

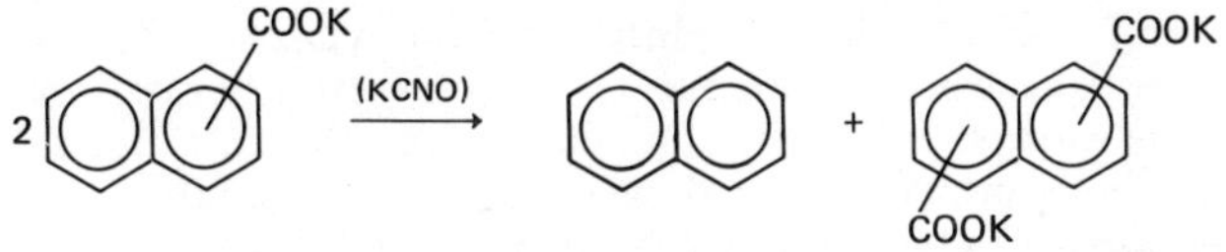

$$\text{(30)}$$

potassium α- or β-naphthoate is irreversible; this is true also in the presence (equation 30) of fused potassium cyanate. The above results show that the transcarboxylation of naphthalenedicarboxylates also occurs by repeated ionic, intermolecular decarboxylation–recarboxylation processes with simultaneous liberation and addition of protons. The transcarboxylation is again based on the formation of carbanions by removal of the carboxylate cation as exemplified by equation (31)

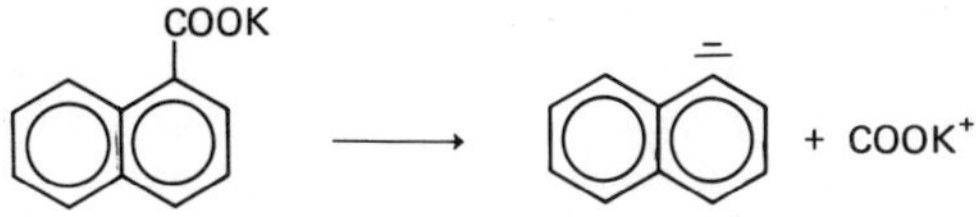

$$+ \text{COOK}^+ \qquad \text{(31)}$$

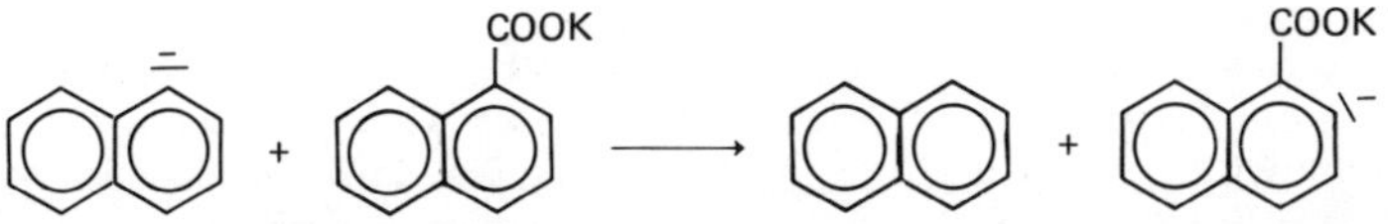

$$\text{(32)}$$

for the transcarboxylation of potassium α-naphthoate. Protons are then liberated from a further molecule of the α-naphthoate with the formation of a new carbanion and naphthalene (e.g., equation 32). Finally, potassium naphthalenedicarboxylate is formed by the addition of the carboxylate cation to the carbanion (equation 33).

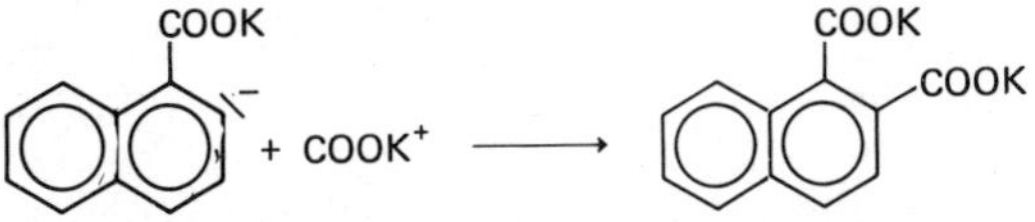

$$+ \text{COOK}^+ \longrightarrow \qquad \text{(33)}$$

Seven new carbanions can theoretically arise by reaction of the naphthalene carbanion with the naphthoate according to equation (32), leading to the production of seven isomeric naphthalenedicarboxylates according to equation (33). It is of interest that the initial step consists of the preferential formation of the *ortho* isomer according to equations (32) and (33). The isolation of this isomer from the complex reaction mixture of potassium salts is very difficult. However, when barium α- or β-naphthoate is used as the starting material, the reaction is interrupted at the stage of the poorly fusible *ortho* isomers. Thus, a mixture of barium 1,2-(major product) and 1,8-naphthalenedicarboxylate is obtained from barium α-naphthoate and a mixture of barium 2,3-(major product) and 1,2-naphthalenedicarboxylate results from barium β-naphthoate[57]. A similar formation of the *ortho* derivatives was encountered in the transcarboxylation of sodium salts of α- and β-naphthoic acids[59].

Analogously to the series of benzenecarboxylic acids, the exchange of deuterium by protium in transcarboxylations of mixtures of deuterium-labelled and deuterium-free naphthalenecarboxylates is faster than the decarboxylation, irrespective of the initiation of the reaction by the formation of primary carbanions which is accomplished by elimination of the carboxylate cation. In the napthalenecarboxylate series, too, the existence of a 'dynamic' carbanion with an unshared pair of electrons on different carbon atoms of the naphthalene ring can thus be assumed[63].

The mechanism of the transcarboxylation of potassium naphthalenecarboxylates can also be applied to transcarboxylations of salts with other cations (e.g. sodium, rubidium, caesium and barium naphthalenecarboxylates[57,59]), when the final products can be considerably different. Of great importance again are the cadmium and zinc salts as catalysts in transcarboxylations of other salts due to the initiation of the formation of primary carbanions in the reaction mixtures. Analogously to the benzenecarboxylate series the low thermal stability of cadmium and zinc naphthalenecarboxylates results in various side-reactions and destruction of an aliquot portion of the reaction product[47].

C. Biphenylcarboxylic Acids

In the group of biphenylcarboxylic acids[64–66] the mechanism of the transcarboxylation is again governed by an intermolecular, ionic decarboxylation–recarboxylation process combined with an ionic intermolecular transprotonation. This was determined by analogous methods with the use of [14]C- and [2]H-labelled specimens and by the smooth migration of the carboxylate group from the biphenyl to the benzene ring in the mixed transcarboxylation of potassium benzoate with potassium 2-biphenylcarboxylate[66].

The transcarboxylation of biphenylcarboxylates is a very complex process affording a complex mixture of final products, even when potassium salts are used which are known to yield a uniform product in the series of benzenecarboxylates or naphthalenecarboxylates. The reaction course varies according to whether the carboxylate groups are attached to a single phenyl group only or to both phenyl groups. In the former case, the unsubstituted benzene ring is highly unreactive and the migration of carboxylate groups takes place on the substituted benzene ring only, as shown by detailed analyses of the products and by D/H exchange experiments. Thus, when an equimolar mixture of potassium benzoate-5d and non-deuterated potassium 2-biphenylcarboxylate is subjected to transcarboxylation, the exchange proceeds on the substituted benzene ring only[66]. When an equimolar mixture of potassium benzoate-5d and non-deuterated potassium 2,2'-biphenyldicarboxylate is used as the starting material, the exchange proceeds on both rings of the biphenyl system[66].

Compared to benzenecarboxylates (cf. equations 15–17) the transcarboxylation of biphenyldicarboxylates is more complex and the number of theoretically possible products is much greater. The amount of the thermally very stable potassium 4,4'-biphenyldicarboxylate (resistant to transcarboxylation) is low. A very small amount only of this compound is obtained in the transcarboxylation of potassium phenylbenzenedicarboxylates (i.e. with carboxyl groups on the same benzene ring of the biphenyl ring system), probably due to the low reactivity of the unsubstituted phenyl group.

D. Substituted Benzoic Acids

When an additional substituent (inert under conditions of the transcarboxylation reaction) interferes with the symmetry of the benzene ring in benzenecarboxylates, e.g. in the transcarboxylation of potassium toluates (the mechanism of which is similar to that of benzenecarboxylates and naphthalenecarboxylates), a complex final reaction mixture is obtained since none of the products has the ability to crystallize from the reaction mixture[64].

The transcarboxylations of halogenated benzenecarboxylates also afford complex mixtures of products along with a substantial amount of tar, due to side-reactions with the formation of benzyne derivatives as intermediates[67,68].

Of special interest are the mixed transcarboxylations of salts of aromatic carboxylic acids and aromatic sulphonic acids. Thus, for example, sodium benzenesulphonate alone thermally disproportionates into benzene and sodium 1,4-benzenedisulphonate, while the thermal disproportionation of sodium naphthalenesulphonate alone affords naphthalene and 2,6-naphthalenedisulphonate[69]. When the benzenesulphonate is heated with potassium benzoate, a mixture of benzene and potassium 4-sulphobenzoate results; when potassium benzoate is replace by potassium α- or β-naphthoate, a mixture of naphthalene and potassium 4-sulphobenzoate is obtained. When these reactions are effected in an atmosphere of ^{14}C-labelled carbon dioxide, the whole equivalent of radioactivity is concentrated in the carboxyl group of the resulting 4-sulphobenzoate. When a mixture of the benzenesulphonate and benzoate-5d is subjected to the transcarboxylation, the benzene rings of both salts are involved in the statistic exchange of deuterium by protium. The above transcarboxylations—transsulphonations thus appear to obey a similar mechanism.

E. Heterocyclic Carboxylic Acids[4,70]

As shown with salts of toluic acids, the asymmetry introduced into the molecule by substitution of the benzene ring with a methyl group, highly affects the course of the transcarboxylation since complex mixtures of products are formed. It was, therefore, of interest to examine the influence of the introduction of a heteroatom into the ring of aromatic carboxylic acids on the course of the transcarboxylation since transcarboxylations of heterocyclic carboxylates were reported to afford the expected products in a low yield in addition to numerous isomers[71,72]. Using the earlier described methods such as incorporation of [14]CO_2 into the carboxylate groups[62], exchange D/H, time dependence of the composition of reaction mixtures and the like, the transcarboxylation of salts of heterocyclic carboxylic acids (i.e. pyridine-, furan-, thiophene-, pyrrole-, quinoline- and pyrazinecarboxylic acids, etc.) was again shown to be governed by an intermolecular, ionic decarboxylation—recarboxylation process combined with ionic transprotonation[70]. Evidence in favour of such a mechanism was also provided by mixed transcarboxylations of salts of heterocyclic carboxylic acids with potassium benzoate or potassium naphthalenemonocarboxylates[27,73] (see Table 1).

F. Polynuclear Carboxylic Acids

As shown in the preceding chapters, the aromatic character of aromatic carboxylic acids is an important proviso in transcarboxylations of their salts. It was therefore of interest to examine the transcarboxylation of salts of carboxylic acids derived from anthracene, phenanthrene and the like. Salts of this type were shown to undergo a smooth disproportionation to the hydrocarbon and the corresponding salt. Further heating led to complex reaction mixtures contaminated with decomposition products. Nevertheless, the incorporation of ^{14}C from the [14]CO_2 atmosphere, exchange D/H, and mixed transcarboxylations with benzenecarboxylates or naphthalene monocarboxylates (see Table 1) again indicated that the transcarboxylation of salts of polynuclear carboxylic acids obeys the mechanism of an intermolecular, ionic decarboxylation—recarboxylation process combined with an ionic transprotonation[27,74].

TABLE 1. Ratio of products in mixed transcarboxylations of salts of aromatic carboxylic acids[a]

Experiment No.	Reactants[b]	Products			
		Hydrocarbons	Molar ratio (%)	Acids	Molar ratio (%)
1	BK + α-NK	B:N	5:95	BH_x:NH_x	95:5
2	BK + β-NK	B:N	10:90	BH_x:NH_x	90:10
3	2 BK + 1,8-NK$_2$	B:N	21:79	BH_x:NH_x	77:23
4	BK + PicK	B:Pyr	53:47	BH_x:$PyrH_x$	50:50
5	BK + NicK	B:Pyr	27.5:72.5	BH_x:$PyrH_x$	70:30
6	BK + IsoNicK	B:Pyr	25.5:74.5	BH_x:$PyrH_x$	75:25
7	2 BK + 3,4-PyrK$_2$	B:Pyr	38:62	BH_x:$PyrH_x$	69:31
8	BK + 2-DiphK	B:Diph	20:80	BH_x:$DiphH_x$	84:16
9	2 BK + 2,2'-DiphK$_2$	B:Diph	25:75	BH_x:$DiphH_x$	78:22
10	BK + α-FurK	B:Fur	10:90	BH_x:$FurH_x$	92:8
11	BK + α-PyrrolK	B:Pyrrol	15:85	BH_x:$PyrrolH_x$	87:13
12	BK + α-ThiophK	B:Thioph	10:90	BH_x:$ThiophH_x$	90:10
13	BK + PyrazinK	B:Pyrazin	19:81		
14	BK + AnthK	B:Anth	14:86		
15	BK + PhenK	B:Phen	16:84		
16	α-NK + PicK	N:Pyr	90:10	NH_x:$PyrH_x$	15:85
17	α-NK + NicK	N:Pyr	70:30	NH_x:$PyrH_x$	33:67
18	α-NK + isoNicK	N:Pyr	77:23	NH_x:$PyrH_x$	28:72
19	α-NK + 2-DiphK	N:Diph	40:60	NH_x:$DiphH_x$	55:45
20	α-NK + α-FurK	N:Fur	12:88	NH_x:$FurH_x$	85:15
21	α-NK + α-PyrrolK	N:Pyrrol	31:69	NH_x:$PyrrolH_x$	70:30
22	α-NK + α-ThiophK	N:Thioph	32:68	NH_x:$ThiophH_x$	74:26
23	β-NK + PicK	N:Pyr	81:19	NH_x:$PyrH_x$	24:76
24	β-NK + NicK	N:Pyr	71:29	NH_x:$PyrH_x$	30:70
25	β-NK + isoNicK	N:Pyr	72.5:27.5	NH_x:$PyrH_x$	29:71

[a] Abbreviations: B, benzene; N, naphthalene; BK, potassium benzoate; α-NK, potassium α-naphthoate; β-NK, potassium β-naphthoate; PicK, potassium picolinate; NicK, potassium nicotinate; isoNicK, potassium isonicotinate; 3,4-PyrK$_2$, potassium 3,4-pyridinedicarboxylate; 2-DiphK, potassium 2-diphenylcarboxylate; 2.2'-DiphK, potassium 2.2'-diphenyldicarboxylate; α-FurK, potassium 2-furoate; α-PyrrolK, potassium α-pyrrolecarboxylate; α-ThiophK, potassium α-thiophenecarboxylate; PyrazinK, potassium pyrazinecarboxylate; AnthK, potassium 9-anthracenecarboxylate; PhenK, potassium 9-phenanthrenecarboxylate. The symbols BH_x, NH_x, $PyrH_x$, $DiphH_x$, $FurH_x$, $PyrrolH_x$ and $ThiophH_x$ designate a mixture of mono-, di-, tri- to polycarboxylic acids derived from benzene, naphthalene, pyridine, diphenyl, furan, pyrrole and thiophene, respectively. Reaction conditions: reactions were carried out in a sealed ampoule on a mMole scale; catalyst, CdI$_2$ (3% Cd^{++} w/w); reaction atmosphere, 3 atm CO$_2$; reaction temperature, 400–410°C (expt. 1–15), 380–390°C (expt. 16–22), 390–400°C (expt. 23–25); reaction time, 10 min (expt. 1–15), 12 min (expt. 16–25). Analyses were performed by gas chromatography; the acids were analyzed in the form of methyl esters; in experiments 13–15, the ratio of acids was not determined.
[b] The ratio of reactants is indicated (1:1 or 2:1).

III. MIXED TRANSCARBOXYLATIONS OF SALTS OF AROMATIC CARBOXYLIC ACIDS

As briefly mentioned in the preceding sections, the mixed transcarboxylations of salts of two different aromatic acids result in a more or less predominant carboxylation of the salt of one acid by the action of the salt of the other acid. Thus for example in the mixed transcarboxylation of potassium benzoate and salts of naphthalenecarboxylic acids, a mixture of potassium terephthalate and naphthalene is almost quantitatively formed.

A similar reaction course in which the salt of one acid acts as a donor of carboxylate groups and the salt of the other acid as their acceptor, can be observed in transcarboxylations of salts of numerous aromatic acids[27,73]. The salt which in one pair of carboxylates acts as a donor of carboxylate groups can be their acceptor in another pair of salts.

The course of the mixed transcarboxylations can be explained using a simple transcarboxylation of an equimolar mixture of potassium benzoate and potassium α-naphthoate. The separate transcarboxylations of the two components are shown to be governed by an intermolecular, ionic decarboxylation–recarboxylation mechanism (equations 15–17 and 31–33, respectively). In both these separate transcarboxylations, the final reaction mixtures deposit a crystalline product (equations 6, 7, 27 and 28). The identical transcarboxylation mechanism of the two separate salts suggests the possibility of a mixed transcarboxylation. Virtually 95% of the carboxylate groups is transferred from potassium α-naphthoate to potassium benzoate. The hydrocarbon mixture then contains about 5% of benzene and 95% of naphthalene and the mixture of salts contains 95% of potassium terephthalate (along with a small amount of potassium salts of some other benzenecarboxylic acids) and 5% of potassium naphthalenecarboxylates (mainly the 2,6-isomer).

The ratio of the two hydrocarbons is almost constant during the whole reaction until the reaction is completed. Concerning the salts, a considerable amount of potassium α-naphthoate disproportionates in the initial stages of the transcarboxylation into naphthalene and a mixture of naphthalenedicarboxylates, while potassium benzoate remains almost unchanged at this stage of the reaction. In transcarboxylations, the hydrocarbons arising by decarboxylation are known to not undergo a recarboxylation. In summary, in the mixed transcarboxylation of potassium benzoate and potassium α-naphthoate, the latter salt is first transcarboxylated to naphthalene and a mixture of naphthalenedicarboxylates (equation 28) and then potassium benzoate is carboxylated by the action of these naphthalenedicarboxylates with the formation of a mixture of benzenecarboxylates (equation 34)

$$(34)$$

This mixture finally deposits crystalline potassium terephthalate (equation 7)[27]. The small amount of benzene produced in the initial stages of the reaction is due to the transcarboxylation of cadmium benzoate as catalyst which initiates the formation of primary carbanions.

In view of the easier rupture of the C–C bond between the naphthalene ring and the carboxylate group, the transcarboxylation of naphthalenecarboxylates is faster than that of benzenecarboxylates. In the next stage of transcarboxylation, the resulting mixture of naphthalenedicarboxylates acts as a donor of carboxylate groups.

Potassium benzoate acts as an acceptor of carboxylate groups in mixed transcarboxylations with potassium salts of numerous additional carboxylic acids, though the amount of the accepted carboxylate groups may be different depending on the particular acid. The reactivity of the appropriate aromatic carboxylates with respect to potassium benzoate can be expressed by the ratio of the resulting benzene to the other hydrocarbon or heterocycle, or by the ratio of the resulting potassium terephthalate (along with a small amount of admixed salts of the other benzenecarboxylic acids) to the other aromatic carboxylates.

Thus, in the above mixed transcarboxylation of potassium benzoate and potassium α-naphthoate, the ratio of benzene to naphthalene is 5:95 and the ratio of terephthalic acid (with a small admixture of the other benzenecarboxylic acids) to naphthalenecarboxylic acids is 95:5. For the ratios in mixed transcarboxylations of potassium benzoate with potassium salts of some other aromatic carboxylic acids see Table 1[27].

Furthermore, the mixed transcarboxylation can be used as a method for the determination of the unknown transcarboxylation mechanism of the other reaction partner[27]. The first information is supplied by the ratio of the hydrocarbons formed. In the mixed transcarboxylation of a salt with an unknown transcarboxylation mechanism, the other salt (of a known transcarboxylation mechanism) must be selected in such a manner that easy separation of the resulting two hydrocarbons and of the resulting carboxylic acids is possible, i.e. the transfer of carboxylate groups from the salt of one acid (donor) to the salt of the other acid (acceptor) must be as complete as possible. Since the number of aromatic carboxylic acids, the salts of which are susceptible to transcarboxylations, is great, such a selection can be easily accomplished.

IV. CONCLUSION

The knowledge of the mechanism of transcarboxylation reactions of salts of aromatic carboxylic acids is very useful in many respects, for example, in the determination of reaction conditions in the production of some important aromatic carboxylic acids, especially terephthalic acid, and in the preparation of some otherwise almost inaccessible aromatic carboxylic acids labelled with ^{14}C or with deuterium. Furthermore, the mixed transcarboxylation can be used as a simple method for the determination of the course and mechanism of transcarboxylations if the mechanism is unknown for the respective acid.

V. REFERENCES

1. W. von Richter, *Chem. Ber.*, 6, 876 (1873).
2. W. Wislicenus and Conrad, *Chem. Ber.*, 6, 1395 (1873).
3. H. Schrader and H. Wolter, *Gesammelte Abh. Kenntn. Kohle*, 6, 81 (1921).
4. B. Raecke, *Angew. Chem.*, 70, 1 (1958).
5. B. Raecke (Henkel & Cie), *German Patent*, 936 036 (1955); *Chem. Abstr.*, 52, 20 065 (1958).
6. B. Raecke, W. Stein and H. Schirp (Henkel & Cie), *German Patent*, 965 399 (1957); *Chem. Abstr.*, 53, 10 132 (1959).
7. B. Raecke (Henkel & Cie), *German Patent*, 958 920 (1957); *Chem. Abstr.*, 53, 3 155 (1959).
8. Henkel & Cie, *French Patent*, 1 184 005 (1959).
9. J. Ratuský, *Chemie*, 9, 422 (1957).
10. Y. Ogata, M. Tsuchida and A. Muramoto, *J. Amer. Chem. Soc.*, 79, 6005 (1957).
11. Y. Ogata, M. Hojo, M. Morikawa *J. Org. Chem.*, 25, 2082 (1960).
12. J. I. Jones, A. S. Lindsey and H. S. Turner, *Chem. Ind. (Lond.)*, 659 (1958).
13. F. Šorm and J. Ratuský, *Chem. Ind. (Lond.)*, 294 (1958).
14. J. Ratuský and F. Šorm, *Coll. Czech. Chem. Commun.*, 24, 2553 (1959). .
15. O. Riedel and H. Kienitz, *Angew. Chem.*, 72, 738 (1960).

16. M. Kraus, K. Kochloefl, L. Beránek, V. Bažant and F. Šorm, *Chem. Ind.* (*Lond.*), 1160 (1961).
17. M. Kraus, K. Kochloefl, K. Setínek, L. Beránek, M. Houda and V. Bažant, *Chem. Prům.*, 12, 529 (1962).
18. J. Ratuský, *Final Report on Preparation of Dimethyl Terephthalate*, Part III, Czechoslovak Academy of Sciences, Prague, 1961, pp. 20–23.
19. Y. Ogata and K. Sakamoto, *Chem. Ind.* (*Lond.*), 749 (1964).
20. Y. Ogata and K. Sakamoto, *Chem. Ind.* (*Lond.*), 2012 (1964).
21. Y. Ogata and K. Nakajima, *Tetrahedron*, 21, 2393 (1965).
22. J. Ratuský and F. Šorm, *Chem. Ind.* (*Lond.*), 1798 (1966).
23. J. Ratuský, R. Tykva and F. Šorm. *Coll. Czech. Chem. Commun.*, 32, 1719 (1967).
24. J. Ratuský, *Coll. Czech. Chem. Commun.*, 32, 2504 (1967).
25. S. Furuyama, *Bull. Chem. Soc. Japan*, 40, 1212 (1967).
26. S. Furuyama, *Sci. Papers Gen. Educ.* (*University Tokyo*), 16, 203 (1966).
27. J. Ratuský, *Coll. Czech. Chem. Commun.*, 38, 87 (1973).
28. J. Nelles and R. Streicher, *Zeit. Chem.*, 3, 188 (1963).
29. Málek, Z. Novosad, V. Šváb, J. Siládi, K. Kočí and J. Smutaý, *Chem. Prům.*, 15, 479 (1965).
30. J. Ratuský, *J. Lab. Comp.*, 6, 124 (1970).
31. K. Setínek and V. Bažant, *Chem. Prům.*, 13, 509 (1963).
32. J. Ratuský, *Czechoslovak Patent*, 96 503 (1959); *Chem. Abstr.*, 55, 19 867 (1961).
33. B. Raecke (Henkel & Cie), *German Patent*, 951 566 (1956); *Chem. Abstr.*, 53, 3 155 (1959).
34. B. D. Kruzhalov, Kh. E. Khcheyan and A. F. Pavlichev, *Khim. Prom.*, 10 (1959).
35. B. Raecke, B. Blaser, W. Stein and H. Schirp (Henkel & Cie), *Australian Patent*, 202 893 (1956).
36. J. Ratuský, *Chem. Ind.* (*Lond.*), 1093 (1967).
37. J. Ratuský, *Coll. Czech. Chem. Commun.*, 33, 2346 (1968).
38. J. Ratuský and R. Tykva, *Radiochem. Radioanal. Lett.*, 1, 325 (1969).
39. Y. Dozen and S. Fujishima, *J. Soc. Org. Synth. Chem., Japan*, 25, 155 (1967).
40. Henkel & Cie, *French Patent*, 1 201 434 (1959).
41. J. Ratuský, *D.Sc. Thesis*, 1970, p. 38.
42. J. Ratuský, R. Tykva and F. Šorm: *Czechoslovak Patent*, 121 513 (1965); *Chem. Abstr.*, 68, 77 982 (1968).
43. J. Ratuský, *D.Sc. Thesis*, 1970, p. 49.
44. Y. Ogata, Y. Furuya and T. Go, *Chem. Ind.* (*Lond.*), 2011 (1964).
45. J. Ratuský, *D.Sc. Thesis*, 1970, p. 62.
46. K. Chiba and T. Murakami, *J. Chem. Soc. Japan, Ind. Chem. Sec.*, 69, 1285, 1289 (1966).
47. J. Ratuský, *Coll. Czech. Chem. Commun.*, 38, 74 (1973).
48. F. Šorm, J. Ratuský and J. Novosad, *Final Report on Preparation of Dimethyl Terephthalate*, Part I, Czechoslovak Academy of Sciences, Prague, 1958, pp. 37–39.
49. S. Furuyama and N. Ebara, *Sci. Papers Gen. Educ.* (*University Tokyo*), 17, 81 (1967).
50. Mitsubishi Chemical Co., Ltd., *Japanese Patent*, 61, 19 264; 61, 19 265; 61, 19 266 (1961).
51. J. Ratuský, *Czechoslovak Patent*, 159 608 (1975); *Chem. Abstr.*, 84, 150 357 (1976).
52. J. Ratuský, *Czechoslovak Patent*, 159 614 (1975).
53. J. Ratuský, *Czechoslovak Patent*, 161 164 (1975).
54. Henkel & Cie, *British Patent*, 806 569 (1958); *Chem. Abstr.*, 53, 12 246 (1959).
55. B. Raecke, H. Schirp and B. Blaser (Henkel & Cie), *German Patent*, 1 028 984 (1958); *Chem. Abstr.*, 54, 10 963 (1960).
56. J. Ratuský, *Czechoslovak Patent*, 169 541 (1975).
57. E. McNelis, *J. Org. Chem.*, 30, 1209 (1965).
58. J. Ratuský, R. Tykva and F. Šorm, *Proceedings Sec. Int. Conf. Met. Prep. Stor. Lab. Comp., Brussels*, 897 (1966).
59. Y. Dozen, *Bull. Chem. Soc. Japan*, 41, 664 (1968).
60. J. W. Patton and M. O. Son, *J. Org. Chem.*, 30, 2869 (1965).

61. J. Yamashita, K. Enomoto, H. Ebisawa and S. Kato, *J. Soc. Synth. Chem. Japan*, **20**, 501 (1962).
62. J. Ratuský and R. Tykva, *J. Lab. Comp.*, **5**, 211 (1965).
63. J. Ratuský, *J. Radioanal. Chem.*, **8**, 107 (1971).
64. Y. Ogata, M. Hojo, M. Morikawa and J. Maekawa, *J. Org. Chem.*, **27**, 3373 (1962).
65. Y. Dozen, *Bull. Chem. Soc. Japan*, **40**, 1218 (1967).
66. J. Ratuský, *Coll. Czech. Chem. Commun.*, **37**, 2436 (1971).
67. Y. Dozen and H. Shingu, *J. Chem. Soc. Japan, Ind. Chem. Soc.*, **67**, 581 (1964).
68. E. McNelis, *J. Org. Chem.*, **28**, 3188 (1963).
69. I. Goodman and R. A. Edington (JCJ., Ltd.), *German Patent*, 1 083 258 (1960); *see British Patent*, 834 250—1 (1959), *Chem. Abstr.*, **54**, 20 985—6 (1960).
70. J. Ratuský, *Coll. Czech. Chem. Commun.*, **36**, 2843 (1971).
71. Henkel & Cie, *British Patent*, 816 531 (1959); *Chem. Abstr.*, **54**, 1 552 (1960).
72. B. Raecke, B. Blaser, W. Stein, H. Schirp and H. Schütt (Henkel & Cie), *German Patent*, 1 095 281 (1960); *Chem. Abstr.*, **56**, 2 425 (1962).
73. J. Ratuský, *Chem. Ind. (Lond.)*, 1347 (1970).
74. J. Ratuský, *D.Sc. Thesis*, 1970, p. 148.

CHAPTER **17**

Micellar effects upon deacylation

C. A. BUNTON and L. S. ROMSTED
Department of Chemistry, University of California, Santa Barbara, Ca. 93106, U.S.A.

I. SYMBOLS USED IN TEXT 946

II. INTRODUCTION 947

III. CARBOXYLIC ESTERS AS PROBES FOR MICELLAR CATALYSED
REACTIONS 948

IV. MICELLAR STRUCTURE 951
    A. Distribution of Hydrophilic Ions in Micellar Solutions . . . 955

V. THE SIMPLE KINETIC MODEL 957
    A. Effect of Micelles on Unimolecular, Spontaneous Reactions . . 958
    B. Effect of Micelles on Bimolecular Reactions 959

VI. KINETIC MODELS FOR MICELLAR CATALYSED REACTIONS . . . 960
    A. Quantitative Treatment of Micellar Catalysed Bimolecular Reactions . . 961
    B. Second-order Reactions Between Hydrophobic Substrates 964
    C. Second-order Reactions Between Hydrophilic Ions and Organic Substrates . 969
    D. The Problem of Buffers – an Aside 971
    E. Qualitative Comparison between Theory and Experiment 972
    F. Unimolecular and Third-order Reactions 975
    G. Micellar Inhibition of Bimolecular Reactions 976
    H. Effect of Added Electrolytes 977
    J. Effect of Changing the Concentration of the Hydrophilic Nucleophile . . 980
    K. Summary and Implications of the Current Models 982
    L. Other Kinetic Models 983

VII. MISCELLANEOUS 984
    A. Non-ionic Micelles 984
    B. The Effect of Non-ionic Additives on Reaction Rates . . . 984
    C Zwitterionic Surfactants 985
    D. Self-aggregating Systems 985
    E. Reactions in Reversed Micelles 985

VIII. EXPERIMENTAL PROBLEMS 986

IX. PREPARATIVE ASPECTS 988

X. DEACYLATIONS IN NON-FUNCTIONAL MICELLES . . . 989
    A. Carboxylic Esters 989
    B. Amides 991
    C. Acyl Anhydrides 993

XI.   FUNCTIONAL MICELLES AND COMICELLES   995
      A. Factors which Contribute to Catalysis by Functional Surfactants .   .   . 996
      B. Thiols   1001
      C. Amines and Imidazoles 1002
      D. Oximes and Hydroxamates   1007
      E. Hydroxyalkyl Derivatives   1008
XII. THE QUESTION OF BIFUNCTIONAL CATALYSIS 1011
XIII. CONCLUSIONS, CONNECTIONS AND CONJECTURES   .   .   . 1013
XIV. ACKNOWLEDGMENTS   1015
XV.   REFERENCES   1016

I. SYMBOLS USED IN TEXT*

α	= degree of ionization of the micelle
β	= fraction of counterions bound to micelle, $\beta = 1 - \alpha$
cmc	= critical micelle concentration
CTABr(Cl)	= n-hexadecyltrimethylammonium bromide (chloride) or cetyltrimethylammonium bromide (chloride)
[D]	= stoichiometric concentration of surfactant
$[D_n]$	= micelle concentration
f	= fraction of nucleophile incorporated into micellar phase
K	= ion-exchange constant for two counterions between micellar surface and the aqueous phase
K_X	= ion exchange constant for a hydrophilic anionic nucleophile and any other non-reactive counterion
K_F	= ion-exchange constant for a hydrophilic anionic nucleophile and fluoride ion
K_a	= ionization constant of a weak acid in water
K_{app}	= observed or apparent ionization constant of a weak acid in micellar solution
K_s	= substrate binding constant
K_N, K_N', K_N''	= nucleophile binding constants
$K_{\overline{N}}, K_{\overline{N}}', K_{\overline{N}}''$	= binding constants of the anion of the nucleophile
k_ψ	= observed first-order rate constant
k_w'	= first-order rate constant for a reaction in water
k_m	= first-order rate constant for reaction in micelles
k_{max}	= maximum observed first-order rate constant for reaction in micellar solution
k_2	= observed second order rate constant
k_{rel}	= k_2/k_w or k_ψ/k_w' ratio of the rate of reactions in the micellar phase with the rate in water
k_w	= second-order rate constant for the reaction in water
k_M	= second-order rate constant for the reaction in the micellar phase using concentration in mole ratio

*The symbols [ ] denote concentrations in mole 1^{-1}.

The rate constants are generally expressed in terms of seconds, except for those figures from the literature in which the authors expressed time in minutes. In some captions the observed first-order rate constants are denoted as k_{obsd}, min^{-1}.

k_2^m, $k_2^{m'}$	= second-order rate constants in the micellar phase in terms of the Stern layer volume and the total micellar volume respectively
m_N^s	= moles of micellar-bound nucleophile per mole of micellized surfactant
m_X^s	= moles of micellar-bound non-reactive counterion per mole of micellized surfactant
[N]	= concentration of nucleophile or second reactant
NaLS	= sodium lauryl sulphate, often designated as SDS in the literature
PNPA	= p-nitrophenyl acetate
PNPH	= p-nitrophenyl hexanoate
PNPL	= p-nitrophenyl laurate
Subscript m	= micellar phase
Subscript w	= aqueous phase
Subscript t	= total concentration of a species in both phases
TTABr	= n-tetradecyltrimethylammonium bromide
TTACl	= n tetradecyltrimethylammonium chloride
[X]	= concentration of non reactive counterion
AcHis	= N^α-acetylhistidine
MirHis	= N^α myristoylhistidine

II. INTRODUCTION

Mechanistic organic chemists generally choose to follow reactions in the condensed phases under homogeneous conditions. Although finely divided solids can often strongly catalyse organic reactions it is generally difficult to obtain reproducible results, in part because interactions between the solid surface and the reactants change as the reaction proceeds. However many preparative reactions are carried out under heterogeneous conditions, and most biological reactions occur at interfaces.[1] Micelles are submicroscopic aggregates and reactions can occur at their surface. Micelles, therefore, provide simple models for the study of interfacial effects on chemical reactivity, and recent studies of their physical properties have been very helpful.

Early experimental work on micellar catalysed reactions delineated a number of factors which contribute to rate enhancement or inhibition in micellar solutions. The result of this work, covering a wide range of chemical reactions under a variety of experimental conditions, is the subject of a number of recent monographs and reviews[2-11]. A number of models have been developed over the past decade to account for rate changes as a function of a number of experimental variables: for example, surfactant concentration, head-group charge and structure, surfactant chain length and the effect of added electrolytes and non-ionic solutes[3,8,9, 12-15]. However, none of the models currently in use are sufficiently precise or general to provide consistently reliable interpretations. One of the most difficult problems has been the clear separation of the rate effects of the micelle as a medium, isolated from the surrounding water, from its ability to alter the distribution of reactants (and transition states) between the micellar and aqueous phases for bimolecular reactions.

The purpose of this and the following sections is threefold. First, to introduce the current models used to interpret micellar catalysis and inhibition. Second, to consider the factors that contribute to observed rate effects and to test the explanatory and predictive power of the models. Third, to use this information to outline the problems in the design of experiments and interpretation of data. In the generalizations listed below and in the discussion to follow, attention will be focused largely on reactions in water catalysed by ionic surfactants composed of a long hydrocarbon chain, a cationic or anionic head group and a small and usually hydrophilic counterion. Much of the experimental work has been done in solutions of these types of surfactants. A small amount of work has been done on non-ionic and zwitterionic micelles and even less in mixed, or co-micelles, composed of two different surface active solutes. Finally, all of the kinetic treatments have been developed specifically for aqueous solutions and no attempt has been made to extend the treatments to inverse micellar solutions composed of surfactant aggregates dissolved in non-aqueous solvents.

III. CARBOXYLIC ESTERS AS PROBES FOR MICELLAR CATALYSED REACTIONS

Much of the early work in micellar catalysis and inhibition was done on the deacylation of carboxylic esters. There are several reasons for the selection of esters, especially p-nitrophenyl esters, as substrates. First, micellar catalysis was viewed as a potentially simple model for enzymic catalysis, so it was natural to select a substrate that was commonly used in enzyme mechanism studies[16,17]. This hope has not been realized since micellar catalysis has also proved to be complicated, but the attempt has produced results that will provide new insight into the factors contributing to enzyme-catalysed reactions; and micellar catalysis has proved interesting in its own right.

Second, deacylation of p-nitrophenyl esters is an excellent reaction for mechanistic study because the formation of the nitrophenoxide ion product can easily be followed spectrophotometrically in the visible region at very low concentrations (on the order of 10^{-5}M). Also, the mechanism of the reaction in aqueous solution is well understood. Deacylation is a multi-step reaction, and the initially formed tetrahedral intermediate may go forward to products or back to starting material[16].

$$N^- + R-\overset{\overset{\displaystyle O}{\|}}{C}-OAr \rightleftharpoons R-\overset{\overset{\displaystyle O^-}{|}}{\underset{\underset{\displaystyle N}{|}}{C}}-OAr \longrightarrow R-\overset{\overset{\displaystyle O}{\|}}{C}-N + {}^-OAr$$

Often the rate-limiting step is nucleophilic attack on the ester and the back-reaction is of little importance. This is true for example, of the most commonly studied reaction, the attack of hydroxide ion on p-nitrophenyl esters[18,19].

The early results of the work on micellar catalysed deacylations provided the initial experimental observations that must be accounted for by any model of micellar catalysis in dilute aqueous solution.

(1) Almost all reactions between an organic substrate and a charged nucleophile that are catalysed by micelles of one charge are inhibited or unaffected by micelles of the opposite charge (Sections VI. A. C. and G), whereas some reactions between neutral species are catalysed by micelles of either charge (Section VI. B). For example cationic micelles speed, while anionic micelles inhibit, the attack of the hydroxide ion on p-nitrophenyl esters, while non-ionic and zwitterionic micelles have little effect[11], and the benzimidazole-induced deacylation of p-nitrophenyl esters is catalysed by both anionic and cationic micelles[20].

(2) The shape of the rate–surfactant conentration profile changes dramatically when the molecularity of the reaction changes. The profiles for ester hydrolysis

with either anionic or non-ionic nucleophiles are typical of bimolecular reactions (Sections VI. B and E). Once micelle formation begins the rate increases rapidly to a maximum followed by a gradual but steady decrease in rate. Unimolecular reactions, on the other hand, have a plateau region which may extend to very high surfactant concentrations and rate maxima are not observed (Section V.A), Third-order reactions have the same basic shape as bimolecular second-order reactions but with much higher and sharper maxima (Section VI. F). There are no examples of micellar catalysed unimolecular deacylations, but micellar catalysed ester amino-lysis has a third-order component of reaction in which the amine also acts as a general base[21].

Anionic surfactants containing either sulphate of carboxylate head groups are effective inhibitors of ester deacylation[11], and the carboxylate ion is not sufficiently nucleophilic to displace the p-nitrophenoxide ion. This accords with the observation that in homogeneous solution carboxylate ions are generally poor catalysts of ester deacylation[16].

(3) Increasing either substrate or nucleophile hydrophobicity increases the magnitude of the rate effects for both micellar-catalysed and inhibited reactions (Sections VI. B, E and G). The hydrolyses of p-nitrophenyl alkanoates provide excellent examples of this behaviour and are good probes. The hydrophobicity of an ester can be increased by extending the length of the alkyl chain with little or no effect on the chemical reactivity, and hydrolysis of the more hydrophobic esters is catalysed more strongly by cationic surfactants and is inhibited more strongly by anionic surfactants than that of the shorter chain-length esters.

(4) Bimolecular reactions between an organic substrate and an anionic nucleophile are inhibited by inert electrolytes (Section VI. G). The saponification of p-nitrophenyl esters in the presence of cationic surfactants is inhibited by added salts and the inhibition increases with the increasing size of the anion ($F^- < Cl^- < Br^- \approx NO_3^-$). Although changing the cation type has little effect on micellar-catalysed deacylation, it will change the inhibition of reactions catalysed by anionic surfactants, e.g. of acid-catalysed acetal hydrolysis. Added non-reactive electrolytes have complex effects on unimolecular reactions.

(5) In the few examples studied, addition of such hydrophobic non electrolytes as long-chain alcohols always produces a marked rate decrease (Section VII. B), but the effect on deacylation has not been studied.

(6) Cationic surfactants containing functional groups such as hydroxyl or imidazole are much better catalysts of deacylation than are simple surfactants (Section XI). The catalytic action of these surfactants is complex. As might be expected, the rate of the reaction of in the functional micelle is strongly dependent upon pH because the anionic form of the nucleophile is generally the active species. However, cationic micelles lower the apparent pK_a of these functional groups by as much as $1-2$ pK_a units (micelles have similar effects on the ionization of micellar bound weak acids) and the contribution of this change of acid dissociation to the overall effect has not been established.

Some examples of micellar effects on ester deacylations are shown in Table 1. A more comprehensive list is available in Reference 11*. In using compilations here and in the literature the reader should be aware that the extent of micellar catalysis or inhibition can be very sensitive to the reaction conditions. For example, most of

*This reference is of special importance because the authors have compiled all the kinetic and equilibrium studies done in micellar solutions and in such related systems as inverse micelles, polyelectrolytes and cyclodextrins to date (through 1974).

TABLE 1. Micellar effects on rates of ester saponification for various types of surfactants and substrates

(a) *Effect of surfactant type on the rate of saponification of* p-*nitrophenyl hexanoate*

Surfactant	Effect
n-$C_{16}H_{33}\overset{+}{N}(CH_3)_3 Br^-$ [a]	Catalysis
n-$C_{12}H_{25}SO_4^- Na^+$ [b]	Inhibition
Polyoxyethylene(18)dodecylphenol [b]	Inhibition
n-$C_{10}H_{21}\overset{+}{N}(Me)_2 CH_2 COOH Br^-$ [c]	Inhibition
Sodium deoxycholate [d]	Inhibition

(b) *Effect of surfactant chain length on the maximum rate enhancements for saponification of* p-*nitrophenyl-hexanoate in* n-*alkyltrimethylammonium bromide micelles* $(R\overset{+}{N}(CH_3)_3 Br^-)$[a]

R	k_{max}/k_w' [e]
n-$C_{10}H_{21}$	≈ 1
n-$C_{12}H_{25}$	≈ 2
n-$C_{14}H_{29}$	≈ 5
n-$C_{16}H_{33}$	5.5
n-$C_{18}H_{37}$	> 6

(c) *Effect of substrate hydrophobicity on the maximum rate enhancements for saponification of* p-*nitrophenyl alkanoates in* n-*tetradecyltrimethylammonium chloride*[a]

Substrate	k_{max}/k_w' [e]
p-Nitrophenyl acetate	1.8
p-Nitrophenyl hexanoate	5
p-Nitrophenyl laurate	8

[a]Reference 22.
[b]Reference 23.
[c]Reference 24.
[d]Reference 25.
[e]k_{max} = maximum observed first-order rate constant for the reaction in micellar solution; k_w' = first-order rate constant in water.

the experiments relied on buffers for pH control and added salt to control the ionic strength, ignoring micellar effects on pH and the question of the distribution of ions between micellar and solvent solution and the effect of such changes on reaction in the micelle. Consequently, experimental observations of different workers are generally not comparable, and the extents of micellar catalysis and inhibition are very dependent on the actual experimental conditions.

Finally, much of the data in the literature is too fragmentary to allow a complete interpretation of the results. For example, in many reactions only a few surfactant concentrations were used, leaving the shape of the rate—surfactant concentration profile undefined, and sometimes the rate constants were not measured at sufficiently high concentrations of surfactant to determine whether or not a maximum was present in the profile.

IV. MICELLAR STRUCTURE

Micelles have gross structural features in common with enzymes: e.g. molecular weights within the same order of magnitude, hydrocarbon-like interiors, with ionic or polar surfaces, and the capacity to bind ionic and non-ionic solutes[26]. In addition, both enzymes and micelles undergo strong hydrophobic interactions with, and are structurally affected by, solutes which disrupt water structure[6,27,28]. Finally, the polarities of the surfaces of ionic micelles are similar to those of proteins, at least in terms of spectrally measured polarity scales[29]. However, here the similarity ends, for there are enormous differences in structural detail and consequently in catalytic efficiency and especially selectivity.

The general structural features of micelles and micellar solutions are well established and are the subject of numerous reviews and monographs[6,10,11,26-28,30-33]. Micelles are stable but dynamic aggregates composed of long-chain surfactant molecules. Surfactants are amphiphatic species, i.e. they have both hydrophilic and hydrophobic properties. Large structural variation is possible in both parts of the molecule: the head groups can be of varying charge (e.g. alkyl sulphate, alkyl phosphate or alkylammonium ion), and size (e.g. ammonium, trimethylammonium or triethylammonium ion), with accompanying counterions, and they can be attached to alkyl groups of various lengths (8—18 carbons) or to other hydrophobic moieties (Table 2).

In very dilute solutions, of the order of 10^{-2} M to 10^{-4} M and below, surfactants exist as monomers. When their concentration exceeds a certain minimum, the so-called critical micelle concentration (cmc), approximately spherical aggregates form. These aggregates generally contain at least 50 monomers when the surfactant is ionic, but are usually much larger when the surfactant is non-ionic.

In water, micellization is generally detected by a sharp change in some physical property of the solution, e.g. conductivity, surface tension, refractive index or light scattering[34]. Spectral changes that result from the micellar incorporation of dyes is often used to detect micellization, although the interpretation of results and comparison with other methods are complicated by the fact that such sparingly soluble solutes may artificially induce micellization. In any event these abrupt changes suggest that micelles exist in equilibrium with monomers and that sub-micellar aggregates are relatively unimportant, at least in water[33]. However, even though it is difficult to provide a theoretical basis for the concept of the critical micelle concentration, it is still an experimentally useful parameter.

As the surfactant concentration continues to increase above the cmc, additional monomers form new micelles, and the monomer concentration changes slowly, if at all. This fact leads to the commonly used approximation that the monomer concentration, at any surfactant concentration above the cmc, is equal to the cmc[26,33].

Micelle formation is a non-specific phenomenon, and requires that the hydrophobic effect, i.e. the tendency for the hydrocarbon chains to aggregate and reduce their contact with water, overcomes a net repulsive force between head groups which is reduced by absorption of counterions at the micellar surface[34]. Consequently, both micelle formation (as measured by changes in the cmc) and micellar size and shape (as measured by the aggregation number) are sensitive functions of many experimental variables, including the surfactant concentration and chain length, head-group type and structure, counterion type and concentration, temperature and the concentration and type of added non-electrolytes[26,28,35].

The actual size and shape of micelles is still in dispute. Most workers in the field assume that a micelle in dilute solution is approximately spherical and monodispersed, having a narrow range of aggregation numbers[36] (Figure 1). Tanford, however, has recently concluded that experimental data from a variety of studies support a disk-like shape for NaLS micelles at moderate ionic strength[37-40], while Israelachvili and coworkers suggest that micelles in dilute solution can be treated as spheres without significant error[41].

In more concentrated surfactant solutions, or in the presence of added electrolyte, the solution viscosity often increases enormously, although the concentration range at which the transition occurs is variable and depends on the natures of the surfactant and salt[42]. This increase in viscosity is generally assumed to be caused by formation of long flexible rods of high molecular weight having a wide range of aggregation numbers (Figure 1)[43,44]. At very high surfactant concentrations new phases appear[28]. The treatment here is restricted to relatively dilute surfactant solutions below the concentration required for any major phase change.

Hydrophobic interactions between solutes are the consequence of the three-dimensional structure of water and therefore micelles do not form in associated

TABLE 2. Examples of amphiphatic molecules which form micelles

Anionic surfactants

$$CH_3(CH_2)_nOSO_3^- M \qquad CH_3(CH_2)_nSO_3^- M$$
$$CH_3(CH_2)_nOPO_3^{2-} M \qquad CH_3(CH_2)_nCO_2^- M$$

$$M = Li^+, Na^+, K^+, Ca^{2+}, Mg^{2+}, \text{etc.}$$

Cationic surfactants

$$CH_3(CH_2)_n\overset{+}{N}(CH_3)_3 X \qquad CH_3(CH_2)_n\overset{+}{N}H_3 X$$

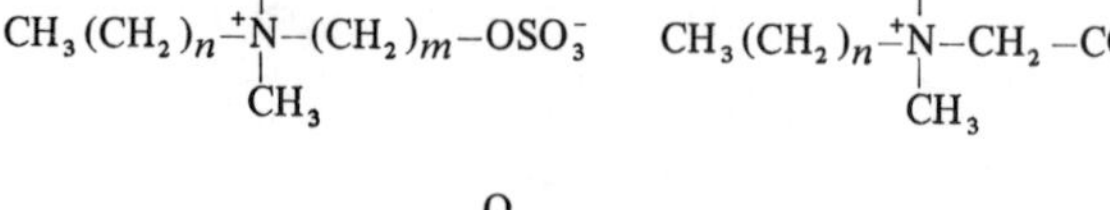

$$X = F^-, Cl^-, Br^-, I^-, NO_3^-, \text{etc.}$$

Zwitterionic surfactants

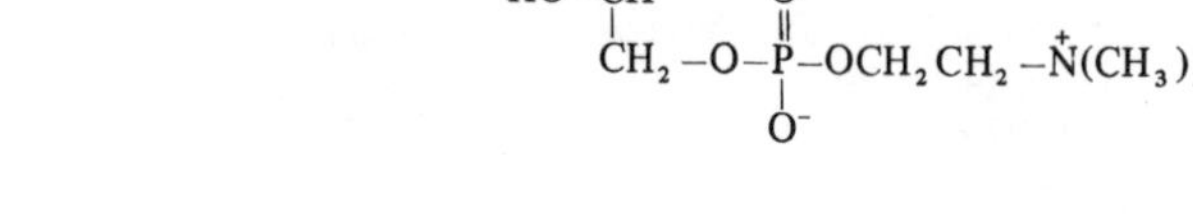

$$CH_3(CH_2)_n-\overset{O}{\overset{\|}{C}}-O-CH_2$$
$$HO-CH$$
$$CH_2-O-\overset{O}{\overset{\|}{P}}-OCH_2CH_2-\overset{+}{N}(CH_3)_3$$
$$O^-$$

TABLE 2   (*continued*)

Non-ionic surfactants

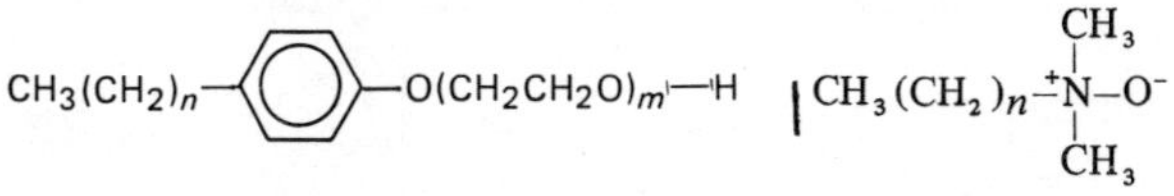

Bile salts

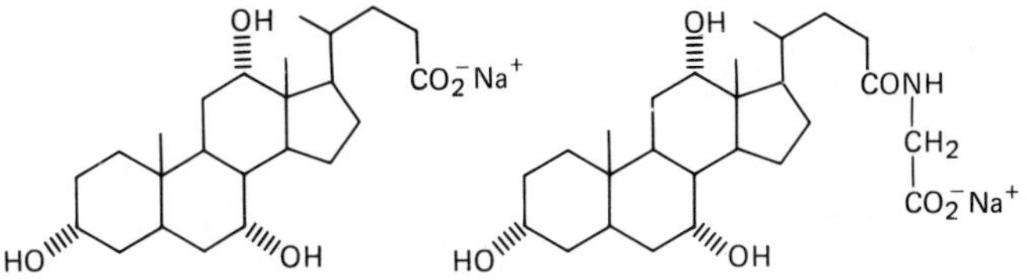

Polysoaps[a]

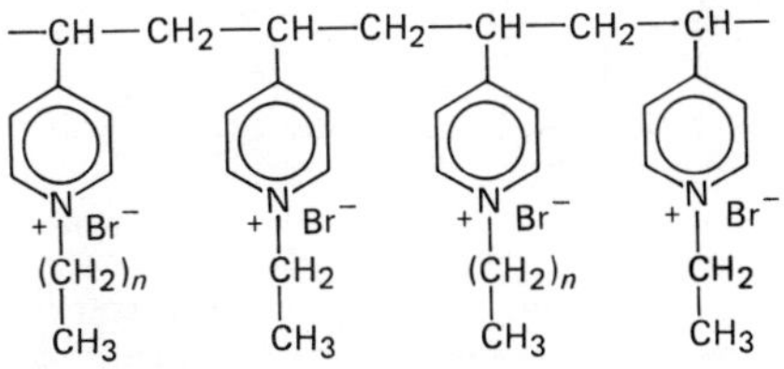

[a]Polysoaps have interesting but poorly understood structures, which combine the properties of both micelles and polyelectrolytes. When polyelectrolytes contain a sufficient percentage of long-chain alkyl side-chains they undergo a transition from random-coiled to compact structures with micellar properties[49,50]. For example, poly-4-vinylpyridine alkylated with dodecyl bromide will catalyse the saponification of *p* nitrophenyl hexanoate[51]. The rate enhancements are quite similar to those of cationic surfactants, and the polysoap-catalysed reaction is also inhibited by added salt.

solvents which have only a two-dimensional structure, e.g. in monohydric alcohols or dipolar aprotic solvents. Many of these solvents, e.g. ethanol, tend to break up micelles by disrupting water structure, as do solutes such as urea and guanidine in relatively high concentration[27,28]. However, micelles form in some associated solvents such as dihydroxy alcohols and primary amides in which there is a degree of three-dimensional structure[45-48].

Reverse micelles can form in non-polar aprotic solvents, especially if a small amount of water is present[11]. The properties of these micelles depend critically upon added solutes, and in their absence the aggregation number of the micelle is very small, but the size can increase sharply as solutes are incorporated. When water is present these micelles have an aqueous interior, the so-called water pools, and an apolar exterior. The structures of these reverse micelles are very different from those of normal micelles where a hydrocarbon-like interior is surrounded by polar

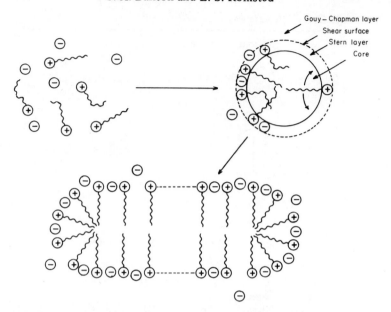

FIGURE 1. Representations of cationic surfactant monomers with accompanying counterions and cross-sections of spherical and rod-like micelles. Curved arrows illustrate liquid-like nature of the micelle core.

or ionic head groups which are in contact with the solvent. However, both normal and reverse micelles can attract solutes out of bulk solvent and provide a submicroscopic reaction medium and so control reactivity.

Many surfactants are synthetic materials, but many natural products have surfactant properties and examples are shown in Table 2. The properties of both synthetic and natural surfactants depend critically upon the polar and apolar groups, and both classes of surfactants have structures appropriate for their functions. To date, most physical studies have been made with the simple synthetic surfactants or with such biologically important surfactants as the bile salts or the phospholipids. Kineticists have generally used commercially available surfactants, although there is an increasing use of surfactants with structures designed for kinetic work.

Two widely used approaches describe the thermodynamics of micelle formation, the pseudo-phase and mass-action models. The mass-action model treats micelles as individual aggregates in dynamic equilibrium with monomers[27,33]. This approach is probably more formally correct because it provides an explanation of the increase in micelle size as a function of increasing surfactant or salt concentration. However, its application to micellar catalysed reactions is difficult because aggregation numbers are seldom accurately known, and even less information is available on the change in aggregation number as a function of the structural variables listed at the beginning of this section.

The pseudo-phase model treats micelles as a homogeneously distributed separate phase[26,52,53]. This model is conceptually simpler to use because: (a) it focuses on the change in volume or mass of the whole micellar phase as a function of surfactant concentration; (b) it assumes the monomer concentration to be constant above the cmc; and (c) consequently, its mathematical description for use in kinetics requires information only on the cmc, and the density of the micelle if rate constants are measured using the conventional concentration units of moles per litre.

A. Distribution of Hydrophilic Ions in Micellar Solutions

While the basic structure of the micelle in dilute aqueous solutions is not in doubt, there is no general consensus on how best to describe the distribution of counterions (and coions) from the head groups at the surface out into the aqueous phase. Another problem is that most of the theoretical and experimental work on micellar solutions has been done on systems with only one type of counterion present, while kinetic studies are usually done on solutions containing two or more counterions, and often in the presence of buffers which contribute additional ions to the system. Consequently, none of the theoretical treatments of micelle structure can be applied unambiguously to kinetic systems.

Two basic approaches are currently used to describe the distribution of counterions: the Gouy–Chapman theory[54] originally developed to account for the effect of added salts on the properties and surface potentials of electrodes[55] and hydrophobic colloids[56], and its successor, the Stern theory, which modified the Gouy–Chapman theory to account for specific ion effects[57,58].

The Gouy–Chapman treatment is inadequate because it assumes that ions can be treated as point charges whose distribution is governed only by coulombic interactions with charges on a smooth surface and with each other, neglecting the dimensions of the small ions[59a]. This approach is inconsistent with the dependence of such properties as the cmc and the aggregation number on the charge density of the counterion. In addition, the theory predicts a much higher electrophoretic mobility for micelles than is generally observed. To overcome some of these discrepancies Stigter[59a], in 1964, developed the concept of the rough- rather than smooth-surfaced micelle first proposed in 1955 (see Figure 1)[60].

The innovative part of Stigter's treatment is contained in his assumptions about the structure of the Stern layer. A spherical micelle is assumed to have a liquid-like hydrocarbon core surrounded by the Stern layer composed of n fully hydrated surfactant head groups, $(1 - \alpha)n$ fully hydrated counterions which are located between the surfactant head groups, and free water. The remaining αn counterions are in the Gouy–Chapman layer. The degree of dissociation, α, defines the fraction of counterions contributed to the aqueous phase. The counterions are considered to be 'bound' only in the sense that they are part of the kinetic micelle, that is the hydrocarbon core plus head groups plus counterions, whose properties are amenable to experimental measurement. Thus counterions neutralize head groups only in the sense that they are associated with the micelle, and they move freely within the Stern layer and between it and bulk water.

Using this model, Stigter calculated both the specific adsorption potential of hydrated counterions to the Stern layer and the surface potential at the Stern layer as a function of increasing counterion concentration. He found that the specific adsorption potential of counterions to micelles of both sodium lauryl sulphate (NaLS) and dodecylammonium chloride showed no specific trends with increasing counterion concentration. Also, the *difference* between the calculated surface potential at the Stern layer and the experimental values for the zeta potential at the shear surface between the Stern and Gouy–Chapman layer (see Figure 1) changed little for either surfactant with increasing ionic strength, although the absolute values of both potentials decreased steadily*.

These results suggest that the degree of ionization of micelles, α, is not a sensitive function of changes in counterion or surfactant concentration. Significant changes in the Stern layer counterion concentration should be reflected by changes in α values. A recent review[6] of experimentally estimated α values as a function of a number of different variables showed that α generally increases with temperature, non-electrolyte concentration, increasing surfactant head group size, decreasing surfactant chain length, and increasing ionic radius of the hydrophilic counterion (following a Hofmeister series). However, *no* consistent trends were found for α values with increasing hydrophilic counterion concentration *or* surfactant concentration. The validity of these generalizations must be tempered by the fact that the numerical agreement between α values determined by different experimental methods is seldom good, but these trends showed up repeatedly using several methods. The few exceptions were attributed to the presence of an additional binding force, e.g. to charge transfer interactions between the surfactant head group and its counterion.

The concept that α values are essentially independent of ionic strength has received recent theoretical support[61]. Using the Langmuir adsorption isotherm combined with either the Gouy–Chapman or Stern layer model for monolayers, spherical micelles and rod-shaped micelles and polyelectrolytes, Stigter found that calculated values of α as a function of salt concentration were roughly constant for reasonable values of intrinsic binding constants between the counterion and the interface.

Typical α values are usually within the range $\alpha = 0.3$ to $\alpha = 0.1$. This means that 70–90% of the counterions contributed to the solution by micelles are contained within the Stern layer volume of the micellar phase. Using Stigter's molecular dimensions for the NaLS micelle[59a], an aggregation number of 62, and a density of $\lg ml^{-1}$ for the micellar phase, the estimated molar concentration of sodium ions in the Stern layer in moles of sodium ion per litre of Stern layer volume varies from 4.5 M ($\alpha = 0.1$) to 3.5 M ($\alpha = 0.3$)[6]. These differences in ionic concentration are small compared with the 10–100-fold changes in added counterion concentration used in experiments designed to measure α values as a function of salt concentration. More importantly, this calculation supports the concept that the Stern layer is saturated with respect to its counterion because small changes in α may result in insignificant changes in the Stern layer concentration of counterions.

The physical model used by Stigter can be adapted to micellar systems containing a mixture of ions. The Stern layer can be viewed as a loosely cross-linked ion exchange resin with the concentration of an ion at the surface being determined by the selectivity of the surface for the ions present and their relative concentrations[6].

This analysis, if correct, has important implications for the analysis of the factors that contribute to the observed catalysis. The operational assumption, now supported by a large body of experimental evidence, is that the Stern layer is the reactive 'site' for bimolecular reactions between polar organic molecules and hydrophilic ions[4,5,7]. This means that both the ground state (hydrophilic ion and organic reactant) and the transition state for the reaction experience an essentially invariant electrostatic potential when bound to the micelle even though the counterion or surfactant concentration is increased. Consequently, any changes in the observed

*In his more recent papers Stigter has refined his original treatment and delineated more clearly the forces contributing to micelle formation and growth[59b–e].

reaction rate which occur at surfactant concentrations sufficient to bind all the organic substrate can only be caused by changes in the concentration of the reactive hydrophilic ion in the Stern layer. This approach has recently been used with some success to interpret the effect of increasing concentation and type of counterion on micelle-catalysed reactions (see Section VI. E). Finally, although similar kinetic studies have been carried out using soluble polyelectrolytes[10,62,63], and ion-exchange resins as catalysts[64], little work has been done on the effects of added inert electrolyte. Perhaps the models developed here will also serve these systems.

V. THE SIMPLE KINETIC MODEL

Because early work on micellar catalysed and inhibited reactions was pursued with the hope that micelles would function as simple models for enzyme catalysis, it became the paradigm directing both the design of experiments and the interpretation of results. The enzyme-like model was developed for, and successfully applied to, micellar catalysis of unimolecular, spontaneous reactions and micellar inhibited reactions involving hydrophobic reactants and hydrophilic ions: for example, ester saponification. The basic kinetic equations were developed by Menger and Portnoy for the inhibition of ester saponification[65] and Bunton extended the treatment to catalysed unimolecular reactions[2]. The primary assumptions involved in the kinetic treatment illustrated in Scheme 1 are: (a) below the cmc the substrate reacts at a rate characteristic of its concentration in water and the appropriate rate

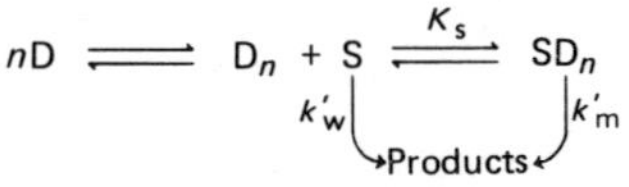

$$nD \;\rightleftharpoons\; D_n + S \;\overset{K_s}{\rightleftharpoons}\; SD_n$$
$$k'_w \Big\downarrow_{\text{Products}} \Big\downarrow k'_m$$

SCHEME 1.

constant, k'_w; (b) above the cmc, increasing surfactant concentration increases the binding of the organic reactant by micelles, and reaction in the micellar pseudophase has the rate constant, k'_m; (c) the concentration of organic substrate [S] is always sufficiently small compared to the surfactant concentration [D] that binding of the organic substrate to the micelle does not significantly alter the catalytic properties of the micelle; (d) changes in aggregation number or micellar structure do not alter the reactivity of the reactants in the micelle; and (e) the concentration of surfactant monomer in solution is constant and equal in value to the cmc once micelle formation begins.

The observed first-order rate constant, k_ψ, is given by:

$$k_\psi = \frac{k'_w + k'_m K_s [D_n]}{1 + K_s [D_n]} \tag{1}$$

and

$$[D_n] = [D] - \text{cmc} \tag{2}$$

$$K_s = \frac{[SD_n]}{[S][D_n]} \tag{3}$$

The binding constant, K_s, is written in terms of the concentration of micellized surfactant. It can also be written in terms of the concentration of micelles, by including the aggregation number of the micelle in equation (2). This equation is formally identical to the classical Michaelis–Menten equation for enzyme-catalysed reactions, except that significantly, in micellar systems the substrate is being saturated with catalyst, whereas in enzyme kinetic studies the substrate is generally in excess over the enzyme.

A. Effect of Micelles on Unimolecular, Spontaneous Reactions

As mentioned above the effects of micelles on the rates of unimolecular reactions can be successfully described by the model shown in Scheme 1, and they illustrate the ability of the micelle to provide a submicroscopic reaction medium.

Only two types of micellar catalysed unimolecular reactions have been examined to date: decarboxylations, and hydrolyses of aryl phosphate and aryl sulphate esters[7,11]. These systems have been considered in detail and we shall only summarize the results here. The reactions are catalysed by cationic surfactants but are unaffected by anionic surfactants. Non-ionic surfactants usually give small rate increases, but zwitterionic surfactants either speed reaction or have little effect, for example, the decarboxylation of 6-nitrobenzisoxazole-3-carboxylate ions (1) is catalysed not only by cationic surfactants but also by the long-chain derivatives of glycine, N,N-dimethyl-N-dodecylglycine, but is only weakly catalysed by lyso-lecithin[66].

(1)

The rate profiles are characteristic of all unimolecular reactions: once the cmc is reached there is a rapid increase in the rate constant which rises sigmoidally to a plateau, extending in some cases to high surfactant concentrations. Equation (1) successfully describes these profiles where the onset of catalysis marks the kinetic cmc, the steepness of the increase in rate constant is determined by the substrate binding constant (K_s) and the extent of catalysis is determined by the value of the micellar rate constant (k'_m).

However, even though equation (1) provides the correct qualitative picture for the effect of micelles on unimolecular reactions, more precise quantitative results are nearly impossible at this time. First, added reactant may lower the cmc of the pure surfactant to a new value (the kinetic cmc) depending upon the type and concentration of the reactant. This change has several implications, including induced micellization by the organic reactant or possibly the formation of small catalytic premicellar aggregates. This observation is consistent with the known dependence of the cmc on the concentration and type of added hydrophobic solutes. Second, the binding constant for these substrates cannot be measured independently under the experimental conditions because of the substrate's reactivity; Independent estimates of the binding constant, which have not yet been made, will require the use of non-reactive model compounds.

Micellar catalysis of a unimolecular reaction demands that k'_m be greater than k'_w, and this 'solvent' effect can be ascribed at least in part to initial state destabilization due to reduced solvation of the anionic moiety of the substrate, and to the lower polarity of the micelle–water interface as compared with water. In addition the transition state could be stabilized by transfer of charge towards the organic portion of the substrate and interaction of this charge with the cationic head groups of the micelle. For example, for decomposition of an aryl phosphate dianion:

$$ArO-PO_3^{2-} \longrightarrow ArO^- + PO_3^- \xrightarrow{H_2O} H_2PO_4^-$$

the forming phenoxide ion could interact beneficially with cationic head groups.

The effective catalysis of decarboxylation by zwitterionic micelles can be rationalized on the assumption that there are unfavourable initial-state coulombic inter-

$$R\overset{+}{N}Me_2CH_2CO_2^-$$

actions between the carboxylate groups of the substrate and the micelle which are relieved in the carbanion like transition state[66].

$$ArCH(CN)CO_2^- \longrightarrow Ar\overset{-}{C}HCN + CO_2$$
$$ \longrightarrow Products$$

Both cationic and anionic micelles typically inhibit S_N1 reactions, probably because of the lower polarity of the micelle–water interface relative to water[67,68].

B. Effect of Micelles on Bimolecular Reactions

The application of the simple kinetic model to bimolecular reactions in micellar solutions creates difficulties in interpretation for several reasons. Direct application of the experimental conditions often used in enzymatic studies to micellar systems, for example the use of buffers and high salt concentrations to control the ionic strength, complicates the interpretations, because micelles affect buffer equilibria and added salts affect micellar catalysis (see Section VI. D).

More seriously, early workers also adopted a second operational assumption used in studies of enzymic catalysis which created fundamental difficulties in interpretation. They assumed that a small reactive hydrophilic ion (e.g. the hydroxide ion in ester saponification) was located almost wholly in the aqueous phase and that therefore an increase in the micelle concentration had an insignificant effect on the distribution of the reactive hydrophilic ion between the aqueous and micellar phases[3]. While it is true that both micelles and enzymes have charged surfaces, protein surfaces generally have a low charge density and bind only a small number of counterions[69], although their effective concentrations may be high at the active site. Micelles, however, bind a large number of counterions. In addition, typical reaction conditions for enzyme-catalysed reactions hold the reactive ion concentration constant with a buffer, or maintain it much larger than the enzyme concentration, so that a large change in enzyme concentration will usually not affect distribution of the reactive ion[17]. In unbuffered and especially in buffered surfactant solutions, however, the micellized monomer concentration, even near the cmc, may be similar to that of the reactive ion and generally exceeds it at high surfactant concentration. Consequently, just as changing the micelle concentration has a profound effect on the distribution of the organic substrate so it may have a large effect on the distribution of hydrophilic ions between the aqueous and micellar pseudo-phases. For example, approximately 75% of the hydrogen ions were bound to NaLS micelles when the surfactant concentration was approximately 10 times the cmc (8mM) and the acid concentration was in the millimolar range[70]. The presence of a buffer does not guarantee a constant distribution of ions between the micellar and aqueous phases. Micelles alter the apparent pK_as of hydrophobic weak acids such as phenols and amines by as much as $1-2$ pK_a units[11], and they may also change the buffering capacity of hydrophilic organic or inorganic buffers in the micellar phase. If this occurs, the concentration of reactive ions, for example hydrogen or hydroxide, in the micellar phase will be uncontrolled and impossible to estimate. Unfortunately many kinetic studies of acid- and base-catalysed reactions in micellar systems suffer from this defect.

The original assumption that micelles do not significantly alter the distribution of reactive ions in buffered solutions suggested that rate—surfactant concentration profiles for bimolecular reactions between hydrophobic substrates and hydrophilic ions should reach a plateau at high surfactant concentrations just like unimolecular reactions (Scheme 1). This assumption is completely wrong.

VI. KINETIC MODELS FOR MICELLAR CATALYSED REACTIONS

The simple distribution model, Scheme 1 and equation (1), fails for bimolecular micellar catalysed reactions which generally show maxima in the rate—surfactant profiles. For a unimolecular reaction the rate will be the sum of the reaction rates in the water and the micellar pseudo-phase, and each will depend only upon the concentration of the reactant in each pseudo-phase, and the first-order rate constant in that pseudo-phase. For bimolecular micellar inhibited reactions the micelle merely keeps the reactants apart by incorporating a hydrophobic substrate and repelling the ionic reagent, so that again the overall rate depends only upon the distribution of substrate between water and micelle.

For bimolecular micellar catalysed reactions we must consider, for the reasons discussed above, the distribution of *both* reactants between water and micelles.

There are two general and complementary ways of solving this problem. In the first the distributions of the reactants between water and the micelles are determined by physical measurement. For example, for hydrophobic ionic or non-ionic solutes we can estimate the concentration of solute in the micelle by assuming that any increase in solubility is caused by incorporation of the solute into the micelle. For a hydrophilic ion, e.g. the hydrogen or bromide ion, we can use a specific ion electrode to determine the ionic concentration in water, and therefore estimate the concentration in the micellar pseudo-phase by difference.

The alternative approach is to use a theoretical model for the distribution of the reactants between water and the surfactant and to estimate their concentrations in each phase from binding constants calculated from kinetic data. Berezin and his coworkers have used both of these approaches and have obtained reasonable agreement for reactions involving hydrophobic reactants (Section VI. B).

In treating reactions of hydrophilic ions one can assume that the distribution of ionic reactants depends upon the competition between all types of counterions present in the solution for a limited number of 'sites' on the micelle surface. A direct analogy can be drawn between the micelle surface and an ion exchange resin, allowing ion exchange constants to be estimated for mixtures of two (and possibly more) ions (Section VI. C).

All the treatments contain assumptions that limit their generality, as well as a number of experimental limitations. In fact, because of the large number of special interactions that are possible between reactants and surfactant molecules, both individually and in aggregates of variable size, no general method for treating the kinetic form of all bimolecular catalysed reactions may be feasible.

A. Quantitative Treatment of Micellar Catalysed Bimolecular Reactions

For reaction of substrate S and nucleophile N (or any second reactant):

$$S + N \longrightarrow products$$

we follow Scheme 1 and write the overall first-order rate constant with respect to S as:

$$k_\psi = \frac{k'_w + k'_m K_s[D_n]}{1 + K_s[D_n]} \tag{4}$$

$$[D_n] = [D] - cmc$$

where k'_w and k'_m are first-order constants in water and the micelle respectively, and K_s is the binding constant of S, written in terms of the molar concentration of micellized surfactant.

The rate constants k'_w and k'_m will depend on the concentrations of N in water and the micelle, and it is convenient to write them in terms of the second-order rate constants k_w and k_M, so that:

$$k'_w = k_w[N_w] \tag{5}$$
$$k'_m = k_M m_N^s \tag{6}$$

and equation (4) becomes:

$$k_\psi = \frac{k_w[N_w] + k_M K_s m_N^s[D_n]}{1 + K_s[D_n]} \tag{7}$$

For dilute surfactant solutions, in which the volume of the micellar pseudophase makes a negligible contribution to the total solution volume, the concentration of N in the aqueous phase is conveniently expressed as moles of N per litre of solution. However, the concentration of N in the micellar phase is expressed in mole ratio units, i.e. as the moles of N per mole of micellized surfactant, D_n, so that:

$$m_N^s = [N_m]/[D_n] \tag{8}$$

where $[N_m]$ is moles of micellized N per litre of solution*.

The advantage of writing the concentration of N in the micelles in mole ratio units is that we need consider only the concentration of micellized surfactant, eliminating the need to estimate the total volume of the micelles in the solution, or to decide whether the reaction site is restricted to the micellar Stern layer.

If we express the concentration of N in these terms, equation (7) takes the very simple form:

$$k_\psi = \frac{k_w[N_w] + k_M K_s[N_m]}{1 + K_s[D_n]} \tag{9}$$

where as before $[D_n] = [D] - cmc$, and $[D]$ is the total concentration of surfactant.

Equation (9) is written in terms of the observed first-order rate constant, but can easily be converted into the second-order form by writing the fraction, f, of

*The concentration of micellized nucleophile, m_N^s, is expressed as a mole ratio instead of a mole fraction because its concentration may not be small relative to that of micellized surfactant.

micellar incorporated N as:

$$f = \frac{[N_m]}{[N_m] + [N_w]} \tag{10}$$

Then, if the volume of the micellar pseudo-phase remains small, the second-order rate constant, k_2, is given by:

$$k_2 = \frac{k_w(1 - f) + k_M K_s f}{1 + K_s[D_n]} \tag{11}$$

The values of f can be measured directly by methods such as gel filtration[71], ultrafiltration[72], solubility[73] or spectral changes for hydrophobic reactants[74], or for hydrophilic ions such as hydrogen or bromide by the use of specific ion electrodes[75]. When the concentration of reactant is small compared to the concentration of micellized surfactant, the fraction of bound reactant can also be expressed as a binding constant, K_N:

$$N_w + D_n \; \underset{}{\overset{K_N}{\rightleftharpoons}} \; N_m$$

$$K_N = \frac{[N_m]}{[N_w][D_n]} = \frac{f}{(1-f)[D_n]} \tag{12}$$

and equation (11) becomes:

$$k_2 = \frac{k_w + k_M K_s K_N[D_n]}{(1 + K_s[D_n])(1 + K_N[D_n])} \tag{13}$$

This form of the equation is identical to one derived by Berezin and coworkers[14], except that theirs involves the volume of the micellar pseudo-phase, and either one can be used for reactions in which the binding constants K_s and K_N are measureable. Equation (13) predicts a maximum in the value of k_2 as the surfactant concentration increases, conforming to the common observation for bimolecular micellar catalysed reactions (see Figures 2 and 3).

For bimolecular reactions where one of the reactants is a hydrophilic ion, either equation (11) or (13) can be used, provided that f can be measured directly or estimated from kinetic data using the ion exchange model as described in Section VI. C.

Most of the approximations involved in the derivation of equations (11) and (13) have been discussed. Several other untested assumptions have been made both in the theoretical and experimental work. Most workers have assumed that the binding of one reactant does not affect the other, i.e., that K_s and K_N are independent parameters[14]. The error introduced by this assumption may be especially serious when the reactants are hydrophobic and the surfactant concentration is near the cmc. For example the incorporation of phenol and p-cresol, and their aryloxide ions into CTABr has been determined spectrally, by working at low and high pH[76]. In principle it should be possible to use these results and the values of the apparent pK_a to estimate the amount of micellar incorporated aryloxide ions in solutions buffered at pH 10 where both phenol and the aryloxide ion are present. However, the extents of incorporation so estimated do not agree with those determined by direct measurement at pH 10. Also, the extent of incorporation of the aryloxide ion at pH 12 decreases as the aryloxide ion concentration increases. Both of these discrepancies suggest that in general the micellar binding of organic molecules may not be independent parameters, so that the binding of a reactant to a micelle should be measured whenever possible under the actual reaction conditions and not inferred from indirect measurements. Changes in counterion type and concentration may also alter the extent of incorporation of reactants. For example, the addition of tosylate ions to CTABr solutions displaces the bromide ions from the micelle surface[77], and the exchange of hydrogen ion for sodium ions on micelles of NaLS has been observed over a range of surfactant concentrations[78].

The method of estimating the amount of micellized surfactant using the cmc also has limitations. Solutes, especially hydrophobic ones, may reduce the cmc significantly[27], so that we customarily use the cmc determined under kinetic conditions, by assuming that the onset of catalysis marks the beginning of micelle formation. However, as the surfactant concentration increases and the extent of incorporation of the solute increases, the solute's ability to alter the cmc will be reduced, and the value of the cmc will increase steadily towards the value in pure surfactant solution. We see no simple way of treating this problem.

The treatment described in this section has not been applied to deacylations, but it has been used with similar nucleophilic displacements and some examples are shown in Table 3.

It should be noted that the dimensions of k_M are s^{-1} because mole ratios are dimensionless quantities. However, k_M can be converted into a second-order rate constant expressed in terms of moles per litre of micelles, or moles per litre of Stern layer, by estimating the volume of one mole of micellized surfactant. We estimated this volume assuming the density of the micelles to be 1 g ml^{-1} and using Stigter's model of the Stern layer of an ionic micelle. Multiplying k_M by the factors shown below gives k_2^m, $M^{-1}s^{-1}$, which is the second-order rate constant in terms of one litre of Stern layer*. (Second-order rate constants in water can also be written in terms of mole fractions, although solution kineticists are probably too wedded to molarities to indulge in such a transformation.)

$$\left.\begin{array}{c} CTABr \\ NaLS \end{array}\right\} : k_2^m = 0.14 k_M$$

$$H_2O \; : \; k_2 = 0.018k$$

TABLE 3. Second-order rate constants in micelles and in water

Reaction	Surfactant	K_s	k_{rel}	$k_M(s^{-1})^e$	$k_2^m(M^{-1} s^{-1})^e$
$NC_6H_4OPO(OPh)_2 + OPh^-$ [a]	CTABr	$\approx 2 \times 10^4$	2800	0.11 (1.8)	0.015 (0.032)
$NC_6H_4OPO(OPh)_2 + {}^-OC_6H_4Me$ [a]	CTABr	$\approx 2 \times 10^4$	3700	0.12 (1.9)	0.017 (0.034)
$4\text{-}(O_2N)_2C_6H_3F + {}^-OPh$ [b]	CTABr	55	230	3.4 (38)	0.5 (0.7)
$4\text{-}(O_2N)_2C_6H_3F + PhNH_2$ [c]	CTABr	55	8	0.035 (1.7)	0.005 (0.03)
$4\text{-}(O_2N)_2C_6H_3CF + PhNH_2$ [c]	NaLS	23	4	0.026 (1.7)	0.004 (0.03)
$O_2NC_6H_4CH(OEt)_2 + H^+$ [d]	NaLS		8–28	0.1 (16)	0.014 (0.3)

Reference 79.
Reference 80.
Reference 81.
Reference 70.
The values in parentheses are for reaction in water.

*Estimates of kinetic solvent effects upon bimolecular reactions change markedly if second-order rate constants are written in terms of mole fractions instead of molarities.

Values of k_m^2 are included in Table 3. They are generally smaller than the rate constants in water, especially for the reaction of the hydrophilic hydrogen ion with p-nitrobenzaldehyde diethyl acetal and of aniline with 2,4-dinitrofluorobenzene. In the latter case we would expect the lower polarity of the micellar surface, as compared with water, to reduce the rate constant. The value of k_2^m for acetal hydrolysis in NaLS may be less than in water because the sulphate head group interacts strongly with a hydrogen ion, by hydrogen bonding or covalent interaction, thus decreasing its acidity, and this effect would decrease the second-order rate constant.

For the other reactions the values of k_2^m are similar to the second-order rate constants in water, expressed in $M^{-1}s^{-1}$, as has been noted for a number of deacylations (Sections VI. A. and E). If this similarity of the second-order rate constants in the two phases is general, then it is relatively easy to estimate the contribution of the proximity effect to catalysis in a functional micelle or other submicroscopic aggregate such as a polyelectrolyte or enzyme.

The maximum reasonable value of m_N^s is one, assuming one ionic reagent per head group of a functional micelle. If the substrate is fully bound to the micelle, $K_s[D_n] \gg 1$ and there is no reaction in water, then from equation (4):

$$k_\psi = k_m' \qquad (14)$$

But $0.1 k_m' \sim k_2^m$, and $k_2^m \sim k_w$ (Table 3), so that if the concentration of reactants into a small volume in the Stern layer is the sole source of micellar catalysis, the maximum value of the first-order rate constant, k_ψ, with respect to substrate for catalysis by a functional micelle should be given by:

$$k_\psi \sim 10 k_w$$

where k_w is the second-order rate constant in water*. These conclusions, however, do not apply to spontaneous, unimolecular, reactions, where the maximum rate effects are due solely to the differences between water and the micellar surface as reaction media.

B. Second-order Reactions between Hydrophobic Substrates

The most reliable interpretations of reaction rate—surfactant concentration profiles are for reactions between a neutral organic substrate and a neutral organic nucleophile. The kinetic analysis for this type of reaction was first developed by Berezin and coworkers and the results are available in a series of papers[14,20,73,82–88], including a major review[8]. Although the approach used in the derivation of their equations is different from the one used here, the basic assumptions are the same, so that the final form of their equation is very similar to equation (13). From the kinetic data, they used linearized forms of their equation to calculate binding constants for some reactants. The calculated binding constants agreed reasonably well with ones determined independently, usually by solubility measurements (Table 4, and Reference 14).

The treatment was extended to reactions between neutral hydrophobic substrates and nucleophiles which were either fully charged or partially ionized under the reaction conditions. In these cases, verifiable values for the extent of binding of the reactants were not obtained. We believe that this may create a problem at least in

*Note that k_w is the numerical value of the first-order rate constant in water when the concentration of the nucleophile is 1 M.

TABLE 4. Binding constants, K_S (M^{-1}) characterizing the incorporation of the reagents into the CTAB and NaLS micelles[20]. Conditions: 30°C, 0.02 M borate buffer, 1 vol. % of dimethylsulphoxide

Surfactants	Reagents						
	N-methyl-benzimidazole	Benzimidazole cation	Benzimidazole electroneutral form	Benzimidazole anion	p-Nitrophenyl acetate	p-Nitrophenyl butyrate	p-Nitrophenyl heptanoate
CTAB	34[a]	<1[b]	33[a] 37[b]	4000–5000[c] 50[d,e]	27[f]	530[g]	3800[d,e] 3600[g,e] 3000[g]
NaLS	30[a]	2400[b]	28[a] 30[b] 30[h]	—	—	—	1500[g] 2000[h]

[a] From the dependence of the difference spectrum of the reagent on the surfactant concentration[73].
[b] From the dependence of the apparent pK_a value on the surfactant concentration[73].
[c] At [CTAB] → cmc. Found from the dependence of the apparent pK_a value on the CTAB concentration[73]. The dependence of $K_{\bar{N}}$ on CTAB concentration is given in Reference 73.
[d] From the dependence of the apparent rate constant on the CTAB concentration with the use of an equation similar to equation (15).
[e] In the presence of 0.12 M KNO$_3$.
[f] Found by gel filtration[22].
[g] From the dependence of the solubility of the reagent on the surfactant concentration[83].
[h] From the dependence of the observed second-order rate constant, k_2, on the NaLS concentration with the use of an equation similar to equation (13).

part, because of the assumptions about the effect of micelles on the distribution of hydrophilic ions and moderately hydrophobic ions, especially when a mixture of counterions is present in solution. All the work of Berezin and his group was done in the presence of buffers, and they assumed, as one normally would for enzyme-catalysed reactions, that the presence of a buffer controls the pH at the micelle surface, even though the surfactant concentration or salt concentration may change. We believe that this assumption may be invalid. As noted earlier, it has been known for some time that micelles strongly affect acid–base equilibria[89]; in fact Hartley developed a series of rules which successfully predict the direction, but not the extent of the change, as a function of the charge on the acid and base forms of indicators and the surfactant charge[90,91]. While it is easy to measure the change in the overall acid dissociation, i.e. the apparent pK_a, it is difficult to separate the contributions due to the (different) distributions of the acidic and basic forms between water and the micellar pseudo-phase *and* the actual acidity at the micellar surface, i.e. the concentration of hydrogen or hydroxide ion in the Stern layer.

It is simplest, therefore, to consider first deacylation by non-ionic nucleophiles. The rates of reaction of *p*-nitrophenyl alkanoates with *N*-alkylimidazoles are affected in a parallel fashion by both CTABr and NaLS[88] (Scheme 2). The results

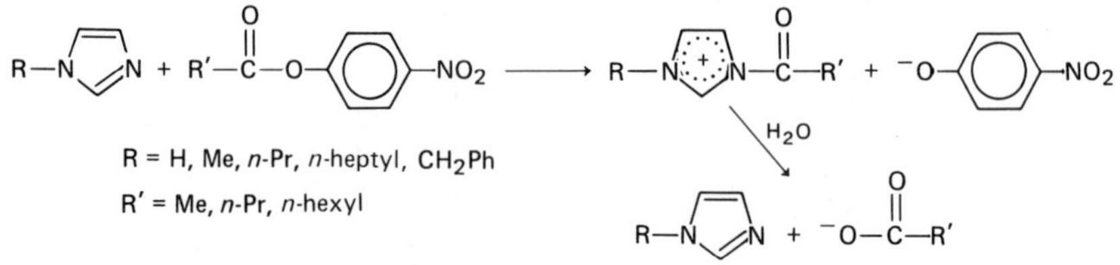

R = H, Me, *n*-Pr, *n*-heptyl, CH$_2$Ph

R′ = Me, *n*-Pr, *n*-hexyl

SCHEME 2.

are: (*a*) the maximum catalysis observed is small and little affected by micellar charge; (*b*) increasing the hydrophobicity of the alkylimidazole (increasing the chain-length) increases the catalysis, but increasing the hydrophobicity of the ester inhibits the reaction. Data for the acylation of *N*-heptylimidazole are shown in Table 5.

Several factors must be considered in the interpretation of the results. First, as the authors point out, the Stern layer of the micelle is a medium of lower polarity than water, making it more difficult to form the partially charged transition state from the neutral starting materials, c.f. Reference 86, and the calculated values of the micellar rate constants (using an equation that is formally identical with equation 13) are considerably smaller than the rate constants in water. Second, the rate increase found with increasingly hydrophobic alkylimidazoles attacking *p*-nitrophenyl acetate no doubt reflects an increase in the extent of binding of the nucleophile. The decreasing values of k_2^m with increasing hydrophobicity of the ester is more difficult to understand. It may be that the more hydrophobic esters are drawn into the micellar core away from the hydrophilic imadazole moiety[14].

The reactions of benzimidazole with *p*-nitrophenyl esters are more difficult to interpret. The rate enhancements by CTABr micelles are very large, in part because of increased formation of the highly nucleophilic benzimidazole anion in CTABr[20] (Scheme 3). The problem lies in the treatment of the micellar effect upon the pK_a of benzimidazole. To describe the effect of the shift in pK_a of benzimidazole as a function of CTABr concentration mathematically on the observed second-order rate constant, the authors added an additional term to their basic equation. We have followed their approach and modified the equivalent equation (13) in the same way

TABLE 5. Acylation of N-heptylimidazole by p-nitrophenyl alkanoates in the presence of CTABr micelles[88]

| $\begin{matrix}O\\\parallel\\R{-}C{-}OC_6H_4NO_2\end{matrix}$ R | Binding constants | | $k_M(s^{-1})$ | $k_2^{m'}/k_W{}^c$ | $k_2^m/k_W{}^d$ | k_{max}/k_ψ |
	N-Heptyl imidazole $K_N{}^a$	Ester $K_s{}^b$				
$n{-}C_6H_{13}$	110	4500	0.01	0.0053	0.0021	1.4
$n{-}C_3H_7$	130	530	0.015	0.0070	0.0028	1.6
CH_3	120	27	0.08–0.17	0.03–0.06	0.012–0.024	1.9

[a]From the dependence of the observed second-order rate constant on the CTABr concentration.
[b]Values for K_s from Reference 83.
[c]Value of k_2^m calculated from k_M with $k_2^{m'} = \bar{V}k_M$. $\bar{V}$ is the partial molar volume of CTABr in water; $\bar{V} = 0.35$ litre/mole (Reference 83).
[d]Value of k_2^m calculated from k_M with $k_2^m = 0.14k_M$. k_2^m is the second order rate constant in terms of one litre of Stern layer (see p. 963).

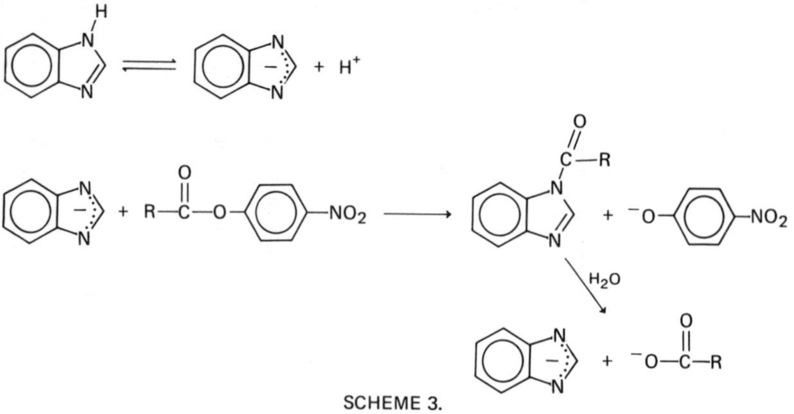

SCHEME 3.

to give:

$$k_2 = \frac{k_W + k_m K_s K_N^- [D_n]}{(1 + K_s[D_n])(1 + K_N^-[D_n])(1 + [H_b^+]/K_{app})} \tag{15}$$

where K_N^- is the binding constant of the benzimidazole anion, $[H_b^+]$ is the hydrogen ion concentration in in the aqueous phase, which is assumed to be held constant by a buffer, and the apparent ionization constant, K_{app}, of benzimidazole is defined by equation (16):

$$K_{app} = K_a \left\{ \frac{1 + K_N^-[D_n]}{1 + K_N[D_n]} \right\} \tag{16}$$

where K_a is the ionization constant of benzimidazole in water.

A serious problem appears at this point. While the binding constants of many other solutes appear to be essentially constant over a range of surfactant concentration, the calculated binding constant for the benzimidazole anion, K_N^-, decreases sharply as a function of surfactant concentration, when K_{app} is measured spectrophotometrically at several surfactant concentrations. This effect is ascribed to 'an increase in counterion concentration, and hence, a decrease in the surface potential of the micelles which must cause a weaker bonding of the anionic species, in this case the [benzimidazole anion]'[73b]. In support of this argument it was shown that added 0.12 M KNO$_3$ not only decreases the rate in CTABr but also dramatically lowers the apparent binding of the benzimidazole anion (Table 4).

This apparent dramatic decrease in the binding of the benzimidazole anion with increasing surfactant concentration or added potassium nitrate is a singular observation that conflicts with observations in other systems. For example, the relatively hydrophobic benzimidazole anion should not be displaced significantly from the micelle surface by either bromide or nitrate ion and neither increasing surfactant concentration nor added electrolytes are expected to have a significant effect on the binding of neutral substrates in this reaction. Also, the apparent change of binding requires that the very favourable distribution of the benzimidazole anion towards the micellar phase decreases as the volume of the micellar phase increases, which if true is a novel observation. Direct measurement on the binding of phenoxide ions to CTABr shows that the incorporation of phenoxide into the cationic micelles increases steadily with increasing surfactant concentration, and we see no reason why this pattern should not be followed by the benzimidazole anion[76]. However, the greatest contradiction appears when this conclusion is compared with results on unimolecular reactions. The rates of unimolecular reactions are constant once all the substrate is bound, indicating that the binding constant of these substrates, some of which are structurally similar to the benzimidazole anion, are independent of changes in surfactant concentration (see Section V. A.). Also, added hydrophilic salts have produced both modest rate increases and decreases in the observed rate of unimolecular reactions[7,92], but never the approximately 100-fold rate decrease the authors attribute to decreased binding of the benzimidazole anion. Finally, the authors have generally assumed that the Gouy–Chapman model provides an adequate description of the distribution of counterions between micelles and the aqueous phase. However, the validity of this model has been severely criticized and an alternative model and experimental approach has been proposed (Section VI. C).

C. Second-order Reactions Between Hydrophilic Ions and Organic Substrates

Recently Romsted modified the approach developed by Berezin and coworkers, expanding it to include the effect of micelles on reactions of hydrophobic substrates with hydrophilic ions[6,15]. The fundamental change is to assume that the distribution of the reactive ion is controlled by competition with other counterions present in the solution for a limited but large number of 'sites' available in the Stern layer of the micelle, instead of being controlled by changes in surface potential. Based on the structural model of the micelle developed by Stigter (Section IV. A), the micellar surface is treated as very concentrated and saturated salt solution composed of a constant ratio of surfactant head groups to counterions. Furthermore, this surface is assumed to act like the surface of a slightly cross-linked ion exchange resin or soluble polyelectrolyte to changes in the concentration of ions and mixtures of ions in the aqueous phase[93].

The development of the original equation followed the approach used by Berezin and coworkers. All the assumptions were the same except for the factors controlling the distribution of the hydrophilic ions. Consequently, micelles were treated as a separate uniformly distributed pseudo-phase in water unaffected by changes in the concentration or type of counterions or the presence of organic substrates, making the kinetic equation once again independent of micellar shape and size.

While the original derivation is different from the approach used here, the final expression differs from equation (13) only in terms of the factors controlling the fraction of the nucleophilic (or any reactive) ion bound to the surface, e.g. hydrogen ions in anionic surfactant solutions or hydroxide ions in cationic systems. Therefore, only the portion of the derivation concerned with distribution of hydrophilic counterions will be presented here.

If the interaction of a single counterion with the micelle surface is independent of increasing surfactant concentration or increasing counterion concentration, then the observed changes in rate under these conditions (see, for example, Figure 2) must come from the fact that the micellar solutions contain a mixture of reactive and non-reactive counterions competing for 'sites' on the micelle surface. In a micellar system containing a mixture of two counterions, the distribution of a reactive hydrophilic ion, N, must contain a term for the distribution of non-reactive ion, X, whose concentration is equal to the total surfactant concentration, [D] (remembering that the addition of each ionic surfactant also adds one counterion), plus the concentration of added salt ($[X_t = [D] + [MX]$).

The concept that the Stern layer is saturated with respect to a mixture of counterions is assumed to be expressed by:

$$m_N^s + m_X^s = \beta \tag{17}$$

where m_N^s and m_X^s are the concentrations of the two ions in the Stern layer expressed in mole ratio units $m_N^s = [N_m]/[D_n]$ and $m_X^s = [X_m]/[D_n]$ and $\beta = (1 - \alpha)$. Equation (17) contains the implicit assumption that the rate of reaction in the micellar phase is not significantly dependent on the nature of the counterion in the Stern layer and the explicit assumption that the rate does not depend upon the difference in β values for different ions, but only on the ionic concentration in the Stern layer. While neither of these assumptions can be completely valid, for example they fail for some unimolecular reactions[92], they are reasonable, primarily because β values are in the range 0.7–0.9, and reaction rates do not appear to be strongly dependent upon changes in electrostatic interactions at the Stern layer, at least for hydrophilic counterions[6].

The two counterions are assumed to exchange rapidly between the aqueous and micellar phases:

$$m_N^s + [X_w] \quad \rightleftharpoons \quad m_X^s + [N_w]$$

so that their distribution will be expressed by a simple ion exchange constant, K, where:

$$K = \frac{[N_w]\, m_X^s}{m_N^s\, [X_w]} \tag{18}$$

One important consequence of this approach is that even when the non-reactive counterion concentration is much greater than the reactive ion concentration ($[X] \gg [N]$), the concentration of the reactive ion in the micelle can still be greater than in the aqueous phase. Also, because both the reactive and non-reactive ions are assumed to affect the properties of the Stern layer approximately equally relative to the reaction being studied, the concentration of N need not be kept small relative to the surfactant concentration to make the assumptions of the treatment valid, although it often is.

If the surfactant volume is small relative to the total solution volume, the material balance equations for the two ions are:

$$[N_t] = m_N^s[D_n] + [N_n] \tag{19}$$

$$[X_t] = m_X^s[D_n] + [X_w] \tag{20}$$

Equations (17) through (20) are combined, rearranged and solved for the concentration of reactive ions in the micellar phase, m_N^s.

$$m_N^s = \frac{\beta[N_t]}{([N_t] + [X_t]\,K)} \tag{21}$$

The major simplifying assumption required in the derivation (to eliminate higher order terms), that $[N_w] \gg [D_n]K m_N^s$, limits the applicability of the final expression to relatively dilute micellar solutions. Substituting equation (21) into equation (7) and setting $[N_t] = [N_w]$ gives the observed second-order rate constant k_2:

$$k_2 = \frac{k_w}{(1 + K_s[D_n])} + \frac{k_M \beta K_s[D_n]}{([N_t] + [X_t]K)(1 + K_s[D_n])} \tag{22}$$

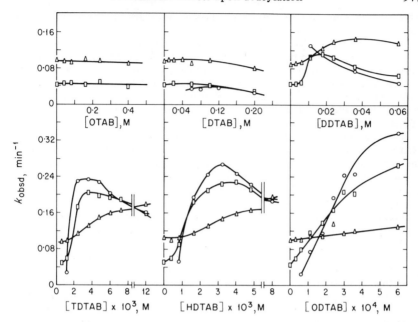

FIGURE 2. First-order rate constants for the basic hydrolysis of *p*-nitrophenyl acetate (△), *p*-nitrophenyl hexanoate (□), and *p*-nitrophenyl laurate (○) at 25°C and pH 10.07 plotted as a function of the concentration of a series of *n*-alkyltrimethyl-ammonium bromides: From left to right starting at the top, the *n*-alkyl groups are octyl, decyl, dodecyl, tetradecyl, hexadecyl and octadecyl. Taken from L. R. Romsted and E. H. Cordes, *J. Amer. Chem. Soc.*, **90**, 4404 (1968). Reprinted with permission of the American Chemical Society.

The complete derivation of equation (22) in slightly different form is published elsewhere[6].

D. The Problem of Buffers — an Aside

Figure 2 shows the observed first-order rate constants for the hydrolysis of *p*-nitrophenyl acetate (PNPA), hexanoate (PNPH) and laurate (PNPL) esters as a function of surfactant concentrations for a series of *n*-alkyltrimethylammonium bromide surfactants (*n* = 10,12,14,16 and 18)[22]. A complete quantitative analysis of the curves in Figure 2 is impossible at this time (as in all ester saponifications studied to date), because a buffer was used to control the pH (0.01 M carbonate, 50% base, pH = 10.1). The use of equation (22) requires knowledge of the total hydroxide ion concentration, $[N_t]$, which is impossible in the presence of the buffer since the amount of hydroxide ion bound to the micellar phase is an unknown, and at present an unmeasureable, variable. It might be tempting to discard the analysis leading to equation (22) because it does not apply directly to buffered solutions, however, the profiles in Figure 2 fit the requirements of the equation, in that they are identical in form to those of many other micellar catalysed reactions between a hydrophobic reactant and a hydrophilic ion when no buffer is present[70,94,95], and they have the same basic profile as reactions between two hydrophobic substrates.

This striking similarity to other second-order reactions implies a similar catalytic mechanism.

As noted earlier (Section VI) the parallel nature of these results indicates that the buffer cannot be operating efficiently, if at all, in the micellar phase. If the buffer was acting effectively in both phases, then it would be holding the pH constant in both phases. Consequently, once the micelle concentration was sufficient to bind all the hydrophobic reactant, further increase in micelle concentration would produce no change in the observed rate, because the pH of the micellar phase would remain constant.

There is support for this analysis. Spectrophotometric measurements of concentration ratios of the acid and base forms of indicators show that micelles alter the apparent pK_a of organic acids by $1-2$ pK_a units[89,96,97]. Measurements of the apparent pK_as of hydrophobic buffers using the glass electrode show shifts of around 0.5 pK_a units[98]. These results suggest that organic buffers which bind strongly to micelles may have their ratio of acid/base forms altered as much as 100-fold, destroying their capacity to buffer effectively in the micellar phase. Alternatively, if the buffer is extremely hydrophilic, little or any of the buffer may be bound. Or, if one form of the buffer has the same charge as the surfactant monomer, it will be totally excluded from the Stern layer. Either condition will prevent the buffer from operating in the micellar phase.

If this latter supposition is correct, the conceptual approach leading to equation (22) can be applied to buffered solutions even through the total reactive ion concentration, N_t, cannot be estimated. For example, in the CTABr-catalysed reactions in Figure 2, the hydroxide ion concentration on the micelle surface will be controlled by its concentration relative to the concentration of the other ions. in solution (Cl^-, Br^-, CO_3^{2-} and HCO_3^-) and their relative ion exchange constants with hydroxide ion.

Another indirect piece of evidence for the lack of buffering action in the micellar phase comes from an experiment that preceded the development of equation (22). The concept of ion exchange implies, and equation (22) predicts, that if the ratio of reactive to non-reactive stoichiometric counterion concentration is held constant while the surfactant concentration is increased, then the apparent first-order rate constant will plateau once all the substrate is bound (equation 23).

$$k_\psi = k_2[N_t] = \frac{k_M \beta [N_t]}{([N_t] + [X_t]K)} \tag{23}$$

when $[D] \gg cmc$ and $\dfrac{k_M \beta}{([N_t] + [X_t]K)} \gg \dfrac{k_w}{K_s[D_n]}$

This result has been observed in several systems. One example is the base-catalysed hydrolysis of PNPH in TTACl solutions with the total chloride ion concentration held constant and the hydroxide ion concentration controlled with a trimethyl-amine/ammonium chloride buffer[22]. Another example is the acid-catalysed hydrolysis of methyl orthobenzoate in NaLS solutions with the total sodium ion concentration held constant and the pH buffered at about 5[3].

E. Qualitative Comparison between Theory and Experiment

On the assumption that the buffer holds the hydroxide ion concentration constant only in the aqueous phase, the difference between the curves in Figure 2

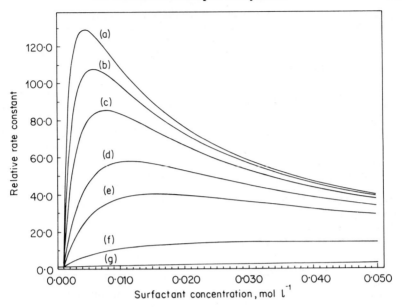

FIGURE 3. Computer-generated plots of the change in the relative rate constant for a second-order reaction, k_2, as a function of the surfactant concentration, $[D]$. The substrate binding constant, K_s is the second independent variable. K_s = (a) 1000, (b) 500, (c) 250, (d) 100, (e) 50, (f) 10, (g) 1; cmc = 0.001; $k_2^m = k_w = 1$; $\beta = 0.8$; $K = 1.0$; $[N_t] = 0.01$; $[X_t] = [D]$. Taken from L. S. Romsted, *Ph.D. Thesis*, Indiana University, 1975.

can be interpreted almost completly in terms of the concentration and ion-exchange effects. To illustrate the relation between the model and equation (22) a FORTRAN progam was written to produce relative rate constants ($k_{rel} = k_2 / k_w$) as a function of two independent variables[6]. In each example the rate constants in the micellar phase and the aqueous phase are assumed to be equal ($k_m = k_w = 1$). This assumption gives full play to the concentration effect. Values for the constants used to produce the curves are specified in the figure legends. the families of curves produced in these plots will be compared with published results for changes in equivalent experimental variables.

Figure 3 shows a family of curves produced to illustrate the effect of increasing substrate hydrophobicity, which is mathematically equivalent to increasing the binding constant K_s. The general shapes of the curves in Figure 3 can be interpreted as the summation of two oppositing effects. Once the surfactant concentration has exceeded the cmc, the relative concentrations of the organic substrate and the hydrophilic ion increases rapidly in the Stern layer of the micellar phase. the larger the binding constant, K_s, the greater the concentration increase in the micelle, the faster the rate increase, and the greater the rate attainable at a lower surfactant concentration. In addition, the rate maximum shifts to lower surfactant concentration with increasing K_s because the concentration effect is opposed by a continuous decrease in the Stern layer concentration of the reactive counterion. m_N^s. The non-reactive counterion concentration increases continuously

$([X_t] = [D])$, while the total reactive counterion concentration is constant. Consequently, because there is a limited number of binding 'sites' available, the ratio m_N^s/m_X^s decreases continuously. This effect will predominate at higher surfactant concentration producing the observed maximum followed by the gradual decrease in the observed rate of reaction.

In Figure 2, the experimental rate profiles for the micellar catalysed reactions generally agree with the shapes of the curves shown in Figure 3. For example, for CTABr in Figure 2, the rate maximum clearly increases and shifts to lower surfactant concentration with increasing substrate hydrophobicity.

The differences between the plots in Figures 2 and 3 can be rationalized by assuming that $k_w > k_2^m{}'(PNPA) > k_2^m{}'(PNPH) \approx k_2^m{}'(PNPL)$. This assumption correlates the following facts. (a) The differences in maximum attainable rates are small compared to those shown in Figure 3, even though the binding constants (as measured independently by gel filtration) are very different, 1.6×10^4 M^{-1} for PNPH and 33 M^{-1} for PNPA in $C_{14}H_{29}NMe_3Cl^{22}$. (b) At high surfactant concentrations the rate constants for PNPL and PNPH are below that of PNPA for all surfactants except $C_{18}H_{37}NMe_3Br$. (c) The order for maximum rate enhancement with substrate hydrophobicity is reversed in octyl-, decyl-, and dodecyl-trimethylammonium bromides. Also, because the cmc values of these surfactants are quite high, the m_N^s/m_X^s ratio is already small at the cmc, so that little if any catalysis can be produced by the concentration effect. The results for PNPL are slightly different because the substrate is insoluble in water and no catalysis can be

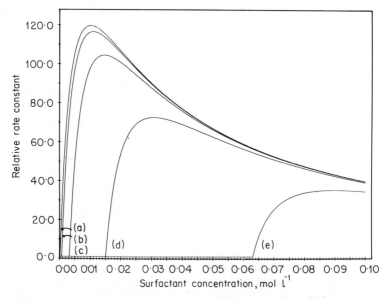

FIGURE 4.   Computer-generated plots of the change in the relative rate constant for a second-order reaction, k_2, as a function of the surfactant concentration, [D]; The cmc is the second independent variable; cmc = (a) 0.00016, (b) 0.00071, (c) 0.0032, (d) 0.015, (e) 0.063; $k_w = k_2^m = 1$; $K_S = 100$; $\beta = 0.8$; $K = 1.0$; $[N_t] = 0.01$; $[X_t] = [D]$. Taken from L. S. Romsted, *Ph.D. Thesis*, Indiana University, 1975.

observed until the micelle concentration is sufficient to solubilize it. The rate profiles for the reactions in $C_{18}H_{37}NMe_3Br$ are incomplete because the surfactant precipitates at high concentrations.

Figure 4 is a computer plot based on an equation with the same form as equation (22) and Figure 5 gives the experimental data from Figure 2 for PNPH which show the effect of increasing surfactant chain length (or its mathematical equivalent which is a decreasing cmc) on the reaction rate–surfactant concentration profiles. The experimental profiles for the saponification of PNPH in Figure 5 look remarkably like the computer plots. Even more important, however, the increase in the rate maximum can be accounted for without invoking a change in the micellar rate constant. A smaller value for the cmc simply means that micelles form at lower surfactant concentrations. At lower surfactant concentrations, the m_N^s/m_X^s ratio in the Stern layer is higher, making the concentration effect more important and the potential maximum rate increase higher.

The analysis outlined above is supported by the experimental results from other micelle-catalysed reactions which show similar trends for both increasing substrate hydrophobicity and increasing surfactant chain length[6].

F. Unimolecular and Third-order Reactions

Since the underlying assumptions of all the kinetic models presented are essentially the same, these various equations reduce to the same form as equation (4) for

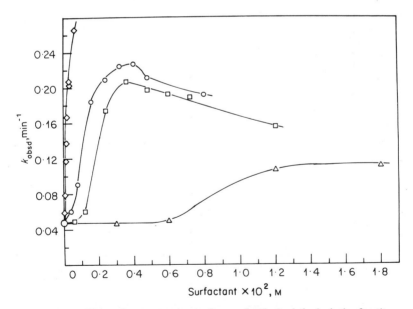

FIGURE 5.   First-order rate constants, k_{obsd}, for the basic hydrolysis of p-nitrophenyl hexanoate at 25°C and pH 10.07 as a function of the concentration of a series of n-alkyltrimethylammonium bromides: n-dodecyl- (△), n-tetradecyl- (□), n-hexadecyl- (○), and n-octadecyl- (◇). Taken from L. S. Romsted, *Ph.D. Thesis*, Indiana University, 1975.

unimolecular reactions, predicting the often observed plateau in the first-order rate constant at high surfactant concentration.

Two examples of third-order reactions catalysed by micelles have been studied. Bunton and Rubin measured the specific acid-catalysed hydrolysis of the benzidine rearrangement in NaLS[99,100], which is first order in hydrazobenzene and second order in hydrogen ions, and Oakenfull studied the reactions between long-chain alkylamines and long-chain carboxylic esters of p-nitrophenol[21].

Both reactions show dramatic rate enhancements. The rate of the benzidine rearrangement reaction increased more than 1000-fold at surfactant concentrations slightly above the cmc with a very sharp maximum in the rate—surfactant concentration profile. This result is in accord with the predictions of an equation for a third-order reaction derived in the same way as equation (22), but containing a second-order term in hydrogen ion concentration[6,15].

Oakenfull found that the reaction of the ester with amine is base-catalysed and is therefore second order in amine[21]. The mixed micelles formed from the reactants produced up to 10^7-fold rate enhancements. While the data were not gathered in a form amenable to quantitative interpretation, the rate enhancements are much larger than those found for otherwise similar second-order reactions.

G. Micellar Inhibition of Bimolecular Reactions

Although the simple kinetic Scheme 1 fails for bimolecular micellar catalysed reactions, it is quite successful for bimolecular micellar inhibited reactions, simply

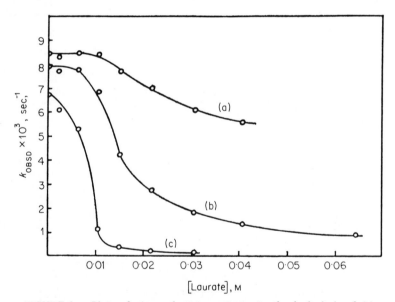

FIGURE 6.   Plots of observed rate constants for the hydrolysis of (a) p-nitrophenyl acetate (b) a mono-p-nitrophenyl dodecanedioate and (c) p-nitrophenyl octanoate at pH 9.59, ionic strength 0.1, 50.0°C vs concentration of laurate. The rate constants of (a) have been divided by 2.00 to bring the curve on scale. Taken from F. M. Menger and C. E. Portnoy, *J. Amer. Chem. Soc.*, **89**, 4698 (1967). Reprinted with permission of the authors and the American Chemical Society.

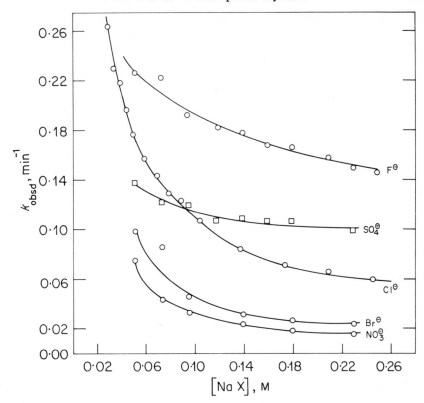

FIGURE 7. First-order rate constants for hydrolysis of *p*-nitrophenyl hexanoate in the presence of 0.009 M tetradecyltrimethylammonium chloride, pH 10.15, as a function of the concentration of several anions. Total anion concentration is plotted on the abscissa: 0.029 M Cl$^-$ (0.009 M TTACl and 0.02 M triethylamine–ammonium chloride buffer) + [NaX], M. Taken from L. R. Romsted and E. H. Cordes, *J. Amer. Chem. Soc.*, 90, 4404 (1968). Reprinted with permission of the American Chemical Society.

because the sole function of the micelle is to keep the reactants apart[6]. For example, in the reaction of several different carboxylic esters with hydroxide ion in the presence of anionic micelles of sodium laurate, the ester enters the micelle which excludes the hydroxide ion, so that the reaction occurs only in the water and the reaction rate decreases steadily as ester is transferred from water to the micelle (Figure 6)[65]. Also, as the hydrophobicity of the ester increases the inhibition occurs more rapidly, in accord with the increased binding of the ester.

H. Effect of Added Electrolytes

A consistent observation in micellar catalysed reactions between hydrophobic substrates and hydrophilic ions is the sharp and apparently hyperbolic decrease in the observed rate with added inert counterion at constant surfactant concentration well above the cmc (Figure 7)[2,3,6,7,22,92]. These salt effects are different from those on unimolecular reactions because they seem to depend only on the concentration and type of counterion and not on the reaction studied[15]. This behaviour is interpreted in terms of the ion-exchange model developed above (Section VI.C).

As noted earlier, all the effects of added inert hydrophilic salts are attributed in the ion-exchange model to competition between the added inert ion and the reactive ion for the micelle surface, i.e. the binding of counterions to the surface is assumed to be dominated by the individual specific adsorption potentials for each counterion. Equation (22) expresses these assumptions mathematically, with the reaction rate being inversely proportional to the reactive ion, N, and non-reactive counterion, X, concentrations, and the magnitude of the ion-exchange constant, $K*$.

If these assumptions are correct, then at surfactant concentrations well above the cmc and sufficient to incorporate all the organic substrate, equation (22) can be simplified and rearranged into an expression that predicts a linear relation between

$$\frac{1}{k_2} = \frac{[N_t]}{k_M \beta} + \frac{K}{k_M \beta} [X_t] \tag{25}$$

the reciprocal of the second-order rate constant and the concentration of added counterion (equation 25). If values for N_t and β are known, or assumed, values for both the micellar rate constant, k_M, and the ion-exchange constant, K, can be calculated.

Figure 7 shows the effect of increasing the concentration and changing the type of non-reactive counterion on the saponification of p-nitrophenyl hexanoate in n-tetradecyltrimethylammonium chloride (TTACl)[22]. These curves are typical of the effect of anions on the observed rates. The effectiveness of an anion in decreasing the rate generally follows the Hofmeister series, so that the rate decreases with the increasing ionic radius of the ion[101,102]†. Although electrolyte effects on anion—molecule reactions do not depend upon the added cation, it is the added cation which determines the electrolyte effects in reactions catalysed by anionic micelles[103]. Finally, the importance of interactions between an ionic micelle and added counterions is also exemplified by the decrease of the cmc and increase in the size of the micelle on addition of electrolytes. These structural effects, like the rate effects, increase with decreasing charge density of the added counterion[3,4,6].

Subject to the assumptions which we have discussed, the salt-effect data can be used to calculate the ratio of ion-exchange constants from the ratios of the slopes

---

*A second model, adopted by Berezin and coworkers assumes that the Gouy—Chapman model, criticized earlier, describes the effect of an added inert counterion on the distribution of reactive counterion[14]. They write the distribution coefficient for the reactive ion P_I, as:

$$P_I = P_{I_0} \exp(-4e/kT) \approx P_{I_0} A/C_i \tag{24}$$

At high surface potentials, typical of micellar systems, the exponential dependence of the surfact potential on the ionic strength can be approximated by a simple linear dependence on the reciprocal of the counterion concentration, C_i, with P_{I_0} being a factor which represents the non-electrostatic contribution to ion binding, and A is a constant. The distribution coefficient, P_I should therefore decrease hyperbolically with increasing salt concentration and so therefore should k_{app} because it is assumed to be directly proportional to P_I. While Equation (24) agrees qualitatively with the observed effects of added salts, we do not believe that it will prove to be generally applicable, in part because it does not predict the dependence of the salt effects on the type of counterion added.

†The effect of the sulphate ion appears to be anomalous in these anion-molecule reactions.

TABLE 6. Calculated values for the micellar rate constant, $k_2^{m'}$, the ion-exchange constant, K, and the ion-exchange constant ratio, K_X/K_F

	Counterion				
	Br(TTABr)[a]	Br	NO$_3$	Cl	F
$k_2^{m'}$ (M^{-1}s^{-1})[b]	0.004	0.005	0.0048	0.0047	0.0045
k_w(M^{-1}s^{-1})[c]			(0.014)		
K^b	0.019	0.037	0.024	0.019	0.0026
$K_X/K_F{}^b$		14.5	9.4	7.6	1.0
$K_X/K_F{}^d$		18.5	25.1	5.9	1.0
Dowex 1 [e]	K_X/K_F	31.1	42.2	11.1	1.0
PVCP-8 [f]	K_X/K_F	5.8	40.2	2.7	1.0

[a]Addition of cyanide ion (0.004 M) to n-dodecyl-3-carbamoylpyridinium bromide in TTABr at 25 °C, with 0.001 M hydroxide added to ensure that all cyanide is present as the anion[104].
[b]The same reaction as in a, but with increasing counterion concentration (to 0.5 M) in 0.02 M TTABr and 0.001 M cyanide ion at 30°C with the pH maintained at 10.4 by 0.01 M triethylamine—ammonium buffer[104]. The value of $k_2^{m'}$ was calculated from k_2^m β assuming that β = 0.8 and that the partial molar volume of the micelles is 0.33 litres per mole.
[c]Second-order rate constant for the addition of cyanide ion to N-propyl-3-carbomoyl-pyridinium iodide measured at 25°C with an ionic strength of 0.5[105].
[d]Base-catalysed hydrolysis of PNPH in 0.009 M TTACl and 0.02 M trimethylamine—ammonium chloride buffer at 25°C. Counterion concentration was increased up to 0.2 M [22].
[e]Dowex 1 has a trimethylammonium head group on a polystyrene backbone[93].
[f]The ionic strength is about 10^{-3} at 25 ± 0.5°C[106].

of the lines from the reciprocal plots of the data in Figure 7[6]. The ion-exchange ratios k_X/k_F (X = F$^-$, Cl$^-$, Br$^-$ and NO$_3^-$) are listed in Table 6 together with the data of Baumrucker and coworkers for the effect of increasing concentration of the same ions on the rate of addition of cyanide ion to N-dodecyl-3-carbamoylpyridinium bromide in n-tetradecyltrimethylammonium bromide (TTABr)[104]. Because no buffer was used in this system, the values of the micellar rate constant and the ion-exchange constant could be calculated from the data, and the results and the ion-exchange ratios are included in Table 6. For comparison, K_X/K_F values for ion exchange in two other systems are included in the table; they are Dowex 1, a strongly basic anion-exchange resin[93], and water-insoluble polyvinylpyridinium chloride spread as a thin film on aqueous salt solutions[106].

The results are important for several reasons. First, when the calculated values of $k_2^{m'}$ are compared with the second-order rate constant in water for the addition cyanide ion to the water-soluble substrate N-propyl-3-carbamoylpyridinium iodide (see Table 6) $k_2^{m'} < k_w$. Second, the assumptions used to derive equation (25) are tentatively confirmed since the values of k_2^m and K, determined from added bromide ion and increasing TTABr, are similar. Third, the absolute values of K for the cyanide ion show that, as expected, it binds more tightly to micelles than any of the added counterions. Fourth, for the halide ions, the values for the K_X/K_F ratios for the two reactions are similar, providing additional support for the assumption of a Stern layer saturated with counterions. A complete discussion of these calculations is published elsewhere[6]. The K_X/K_F ratios for the ion-exchange resin and the monolayer show the same trends, indicating the generality of the approach as well as its limitations. While little is known about the ion-exchange properties of monolayers, ion-exchange resins have been carefully studied[107-110]. Ion-exchange constants for resins are empirical constants whose values depend upon a large number of variables including resin cross-linking, composition, capacity and functional group, solution ionic strength and composition, and other variables including temperature and pressure[109]. Ion-exchange constants for micelles no doubt will depend upon a similar set of variables.

The validity of the kinetic model for bimolecular reactions between neutral organic substrates has been substantiated by the reasonable agreement of the substrate binding constants determined both kinetically and by other methods (see Section VI. B). However, only a few independent measurements of the reactive ion distribution for other hydrophilic or hydrophobic nucleophiles have been completed in micellar solutions containing mixtures of counterions, so that independent varification of the ion-exchange model is incomplete. Two experimental methods are immediately available, ion-selective electrodes[75,77] and ultrafiltration[72].

Ion-selective electrodes are believed to be sensitive only to the concentration of ions in the aqueous phase, allowing the concentration of the ion in the micellar phase to be estimated by difference. One potential problem is that the interaction of the surfactant itself with the electrode may complicate the results. For example, in concentrated NaLS solutions, insoluble potassium lauryl sulphate may precipitate on a KCl salt bridge. But if such complications can be avoided, the results can be used in conjunction with the binding constant for the organic substrate, the experimental kinetic data and equation (7) to calculate the micellar rate constant at various surfactant concentrations. This approach has already been applied with some success to the acid-catalysed hydrolysis of p-nitrobenzaldehyde diethyl acetal in NaLS (Section VI. J)[70]. Also, Larsen and others have used ion-selective electrodes to show that one ion can displace another from the micelle surface, for example that relatively hydrophobic ions like tosylate can displace bromide ions from the Stern layer of CTABr[77]. Alternatively, two ion-selective electrodes could be used to measure the distribution of both the reactive and non-reactive ions simultaneously at several surfactant concentrations and ionic strength to see if the ion-exchange constant, K, is a true constant.

Ultrafiltration techniques have already been successfully used to measure the binding of neutral hydrophobic molecules to micelles[72] and can easily be adapted to estimate the binding of the reactive ion and ion-exchange constants. All that is required is an analytical method to measure the concentration of one or both ions in the filtrate.

J. Effect of Changing the Concentration of the Hydrophilic Nucleophile

Equation (22) predicts an inverse relationship between the concentration of a hydrophilic anionic nucleophile, N, and the apparent second-order rate constant, and shows that the form of the rate—surfactant concentration profiles depends upon the relative concentrations of reactive and non-reactive counterions in solution.

Figure 8 is a computer plot showing the predicted effect of increasing the reactive counterion concentration on the shape of the rate—surfactant concentration profiles. Increasing the concentration of reactive ion generally decreases the apparent second-order rate constant. A similar set of profiles is obtained if the concentration of the non-reactive counterion is increased while the reactive ion concentration is held constant[6].

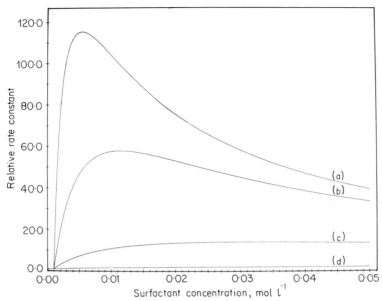

FIGURE 8. Computer-generated plots of the change in the relative rate constant for a second-order reaction, k_2, as a function of the surfactant concentration, [D]. The concentration of the reactive counterion, [N_t], is the second independent variable. [N_t] = (a) 0.001, (b) 0.01, (c) 0.1, (d) 1.0; cmc = 0.001; $k_w = k_2^m = 1$; $K = 100$; $\beta = 0.8$; $K = 1.0$; [X_t] = [D]. Taken from L. S. Romsted, *Ph.D. Thesis*, Indiana University, 1975.

Most micellar catalysed ester saponifications are studied in buffered solutions at a pH which corresponds to relatively low hydroxide ion concentration, simply because the reactions become too fast for convenient rate measurement at high pH. However, for a number of other reactions between hydrophilic ions and organic substrates the relations between rate, surfactant and reactive ion concentration are consistent with the predictions of equation (22). For example, the rate–surfactant concentration profiles for the acid-catalysed hydrolysis of *p*-nitrobenzaldehyde diethyl acetal in NaLS with increasing amounts of added HCl (Figure 9) show changes in the shape of the profiles which are essentially the same as in Figure 8[70]. In addition, the overall rate constants were correlated directly with the concentration of hydrogen ions in the micellar phase, which was estimated independently from the difference between the stoichiometric proton concentration and the proton concentration in the aqueous phase determined by the glass electrode.

The analysis leading to equation (22) resulted in the prediction that for the special case when the counterion is also the reactive ion (X=O), the observed *first*-order rate will also rise to a plateau instead of passing through a maximum once all the substrate is bound. This result has recently been observed several times under very different experimental conditions. The rate constants for the acid-catalysed hydrolysis of *p*-nitrobenzaldehyde diethyl acetal approach a plateau at high concentration of *p*-dodecyloxybenzenesulphonic acid[110], as do those for hydroxide ion attack on *p*-nitrophenyldiphenyl phosphate at high concentrations of *p*-octyloxybenzyltrimethylammonium hydroxide[111].

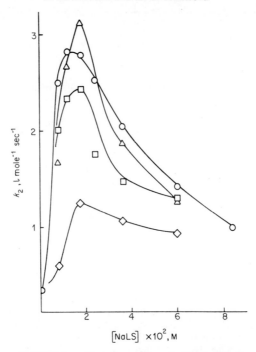

FIGURE 9. Variation of $k_2 = k_\psi/[H_t^+]$ with the concentration of NaLS in dilute HCl at $25.0°$C for the acid-catalysed hydrolysis of p-nitrobenzaldehyde diethyl acetal: ($\triangle$) 10^{-3} M HCl; ($\circ$) 3.16×10^{-3} M HCl; ($\square$) 10^{-2} M HCl; ($\diamond$) 3×10^{-2} M HCl. Taken from C. A. Bunton and B. Wolfe, *J. Amer. Chem. Soc.*, **95**, 3742 (1973). Reprinted with permission of the American Chemical Society.

K. Summary and Implications of the Current Models

The ion-exchange model successfully accounts for a number of properties of reactions between hydrophobic reactants and hydrophilic ions and includes the effect of changing the concentration and type of non-reactive counterion, making the approach an extension of the work of Berezin and coworkers. However, the relationship between theory and experiment remains qualitative, because of the lack of independent determinations of the substrate and reagent binding constants in many reactions. Also, none of these current theories account for the factors contributing to salt effects on unimolecular reactions, and we cannot exclude the possibility that some of these factors affect the second-order rate constants in the micellar pseudo-phase.

However, if the theoretical difficulties can be removed the results discussed above indicate that it will be possible to determine accurately micellar rate constants which will in turn provide information on the properties of the micelles themselves.

Unfortunately, meaningful study of the effects of micelles on the energies of activation of bimolecular reactions is the task of future research. Most of the past work on these systems is not amenable to the types of calculations outlined in the preceding pages. Often insufficient data are available for a particular system, for example: only a few data points were collected or only a narrow range of surfactant concentration was used, substrate binding constants were seldom measured, and unnecessarily high concentration of counterions were used to 'control' the ionic strength, or the solutions contained mixtures of counterion type, or the type of buffer was changed when a different pH was needed. As noted in the introduction, this makes comparison of maximum rate enhancements and activation parameters almost meaningless, because the experimental conditions at that surfactant concentration may vary considerably from system to system. In addition, a rate maximum does not necessarily indicate complete binding of the substrate, because it depends on a balance between the rate enhancement caused by the concentration of reactants in the micelle and the inhibition resulting from the continuous dilution of the reactants in the increasing volume of the total micellar pseudo-phase.

L. Other Kinetic Models

Several other models for micellar catalysed reactions have been proposed. Dougherty and Berg's model[13] is similar to that of Berezin and coworkers. It was applied successfully to the reaction between the neutral substrates, 2,4-dinitro-fluorobenzene and aniline in NaLS. The calculated micellar rate constants, in accord with the results presented here, were less than the rate constant for the reaction in water, but were about three times larger than the value listed in Table 3, probably in part because of differences in the assumed volume elements.

Shirahama fitted the rate maximum for the rate–surfactant concentration profile for reaction of methyl orthobenzoate with hydrogen ions using a distribution model similar to the ones presented above, but assuming that the hydrogen-ion concentration at the surface was controlled by the micelle surface potential which decreased continuously with increasing surfactant concentration[112]. We have already pointed out the limitations of this approach (see Section IV.A). However, like all the models discussed, the calculated micellar rate constant required to fit the data was less than the rate constant in water, again indicating that observed rate enhancement is due largely to the concentration effect.

Finally, Piszkiewicz has modified the theoretical treatment of cooperative catalysis which has been applied to a number of regulatory enzymes to micellar systems[113]. Binding of additional substrate to oligomeric enzymes may increase or decrease the observed reaction rate, and by analogy cooperativity between surfactant molecules and substrate may account for micellar catalysis. The author accounts for inhibition at high surfactant concentration and thus the maximum in the rate-surfactant profile by assuming that excess surfactant forms non-productive aggregates with the substrate[114]. The treatment predicts that the number of surfactant monomers in each aggregate is of the order 1−5, making the aggregate considerably smaller than typical micelles. We feel that the linearities obtained in the empirical rate versus log surfactant plots used to interpret the data do not give accurate information about the substrate–micelle aggregate, although the approach may be helpful in indicating the way in which substrates may assist micellization or comicellization. A major problem with this treatment is that it ignores existing evidence on the incorporation of reactants into the micellar pseudo-phase.

VII. MISCELLANEOUS

A. Non-ionic Micelles

Non-ionic micelles can also incorporate substrate and there is evidence that different substrates may reside at different locations in the micelle. Non-ionic micelles have little effect upon the rates of ester saponification, or aromatic nucleophilic substitution, probably because the substrates are located in the poly-oxyethylene portion of the micelle, which is extensively hydrated and accessible to the hydroxide ion[35]. However, the reaction of p-nitrophenyl diphenyl phosphate with hydroxide ion is strongly inhibited by non-ionic micelles, suggesting that very hydrophobic substrates enter the hydrocarbon core of the micelle which protects them from the ionic reagent[115,116].

B. Effect of Non-ionic Additives on Reaction Rates

While the effect of added non-electrolytes on micellar properties has been studied in detail for some time, as part of the general phenomenon called solubi-lization[35,117-119], only a little work has been done on the effect of uncharged additives on reaction rates in micellar solutions. Dunlap and Cordes studied the effect of alcohols of increasing chain length on the rate of acid-catalysed hydrolysis of methyl orthobenzoate in 0.01 M NaLS[120]. All the alcohols inhibit the reaction to some extent with ethanol decreasing the rate only slightly, about 10% at 0.1 M, while decanol has a very powerful effect, lowering the rate two-fold at 7.9×10^{-5} M.

Ethanol and decanol represent two commonly observed extremes in the effects of non-ionic additives on micellar properties. Ethanol, like urea, is a non-pene-trating additive which is believed to alter micellar properties indirectly by breaking up water structure[119]. Thus ethanol would inhibit the reaction by reducing the driving force for micelle formation (the hydrophobic effect), increasing the cmc and decreasing the concentration of micelles[26]. When the ethanol concentration is large the micelles are completly disrupted. Decanol is a penetrating additive which probably orients itself like a surfactant molecule within the micelle, forming a comicelle. It may inhibit the reaction in several ways: (a) by displacing the substrate from the micelle surface, (b) by decreasing the activity of the substrate in the micellar phase, or (c) by decreasing ionic concentration at the surface. There is some indirect evidence for this last possibility. The addition of long-chain alcohols and non-ionic surfactants increases the degree of ionization of micelles[121,122]. Lawrence has rationalized this observation by assuming that the comicellized alcohol 'wedges' itself between the surfactant molecules, spreading out the head groups, reducing the charge density of the surface and freeing some of the bound counterions[121]. These explanations will certainly not exhaust the possible effects of non-ionic additives on micelle—catalysed reactions. Addition of short-chain alcohols[123,124] and urea[125] will also increase the degree of ionization of micelles, although much higher additive concentrations are required. Also, the addition of hexadecanol increases the rate of acid-catalysed hydrolysis of NaLS micelles at low alcohol concentrations, but inhibits the reaction at higher concentrations, and this rate discontinuity is associated with formation of a new, visible phase[126].

C. Zwitterionic Surfactants

Although many of the functional surfactants used in deacylation studies are zwitterionic, or become so on deprotonation (see Section XI), non-functional zwitterionic micelles are seldom used. However, Katzhendler, Sarel and their coworkers have shown that the effect of micellized 2 (R = $C_{10}H_{21}$) upon the

$$R-\overset{+}{N}Me_2CH_2CO_2^-$$

(2)

saponification of p-nitrophenyl hexanoate and decanoate is similar to that of an anionic rather than a cationic micelle[127a,b].

Anionic micelles of n-alkane carboxylate ions inhibit saponification[65]. Micelles of 2 (R = $C_{12}H_{25}$) are ineffective catalysts of the spontaneous hydrolysis of 2,4-dinitrophenyl phosphate dianion, but they are very good catalysts of spontaneous anionic decarboxylation[128]. The catalysis can be rationalized in terms of a decrease in coulombic repulsions in going from the initial to the transition state[129].

D. Self-aggregating Systems

In the reactions already considered substrates have been incorporated into a non-functional micelle, but the effect of self-micellization of a hydrophobic substrate has also been examined.

One of the earliest kinetic investigations of micellar catalysis and inhibition was on the hydrolysis of monomeric and micellized sodium salts of monoalkyl sulphate[130,131]. Micellization inhibited the base-catalysed reaction, presumably because the hydroxide ion was excluded from the micellar surface and prevented from attacking the alkyl group. However, the acid-catalysed reaction was strongly assisted by micelle formation which increased the local hydrogen ion concentration. Furthermore, the rate enhancement increased with increasing chain length of the surfactant as expected from the forms of the kinetic models discussed above.

Long-chain p-nitrophenyl esters aggregate in water, although the aggregates are probably small dimers or trimers rather than micelles. These aggregates are less reactive towards hydrolysis than the monomeric ester, which is understandable because aggregation would, if nothing else, lower the concentration of ester in solution[132], and factors which disrupt the aggregates speed saponification. Indeed, urea, which has little affect upon the rate of hydrolysis of non-associated p-nitrophenyl acetate, speeds the saponification of aggregated p-nitrophenyl dodecanoate, probably by disrupting water structure and thus reducing the tendency for the ester to aggregate in solution.

The saponification of the micellized cationic ester 3 is considerably faster than that of the similar monomeric ester 4, and addition of dioxane disrupts the micelles and reduces the reaction rate[133].

$$p\text{-}O_2NC_6H_4COOCH_2CH_2\overset{+}{N}Me_2C_{16}H_{33} \qquad p\text{-}O_2NC_6H_4COOCH_2CH_2\overset{+}{N}Me_3$$

(3)   (4)

E. Reactions in Reversed Micelles

The normal micelles which form in aqueous solution have the apolar organic groups in the interior and the polar or ionic head groups at the micelle–water interface. However, completely different aggregates form in aprotic, non-polar solvents such as benzene and hexane, especially if small amounts of water or other polar solutes are present. In these reversed micelles the apolar groups are at the exterior, and the ionic or polar head groups are in the interior, which also contains the water molecules. These aqueous regions of the micelles have been described as 'water pools'[134].

Electrolytes are extensively ion-paired in these solvents, and the aggregates increase in size around ionic or polar solutes. These aggregates are generally assumed to be micelle-like, although we should be cautious in drawing too close analogies between their formation and micellization in water. For example, it is not certain that the concept of a critical micelle concentration is applicable to these systems*.

There are now a striking number of reactions for which reversed micelles are powerful catalysts, but most of these do not involve deacylation.

The imadazole-promoted deacylations of the non-ionic p-nitrophenyl acetate and the cationic p-nitrophenyl p-guanidinobenzoate hydrochloride have been examined in reverse micelles of Aerosol O. T. [sodium di(2-ethylhexyl)sulphosuccinate in octane][135]. These micelles apparently take up water in their interior, and because the rate constant at constant imidazole concentration increases with increasing $[H_2O] / [surfactant]$ it was suggested that reaction occurs in the 'water pools' in the micellar interior which take up both imidazole and substrate, and the reaction rate decreases with increasing hydrophobicity of the substrate. However the reaction in this system is slower than in pure water. This is contrary to the evidence for rate enhancements in normal micelles which can be related directly to the concentration of reagents into a small volume at the micelle water interface.

There is a serious question as to the structure of the reversed micelles which catalyse these reactions. For example, in toluene, tetra-n-hexylammonium benzoate is an effective catalyst for the deacylation of p-nitrophenyl acetate by piperidine or imidazole, whereas octadecylammonium benzoate is relatively ineffective[136]. In water an octadecylammonium salt forms a normal micelle, whereas salts of the highly symmetrical tetrahexylammonium ion would not, although they are very effective phase-transfer catalysts. It may be that the aggregates which form in apolar solvents are so different from normal micelles in water that comparisons of their structures to 'normal' micelles may not be very instructive. Nor is the structure of the catalytic aggregate in aqueous solutions always clear. Kunitake and his coworkers have shown that aggregates of tetraalkylammonium ions in water can be very effective catalysts of deacylation under conditions discussed in Section XI. D

VIII. EXPERIMENTAL PROBLEMS

Experimental work on micellar catalysed reactions contains traps which if not recognized can confound quantitative and sometimes even qualitative interpretation of results. one of the most common problems is ensuring surfactant purity, but probably the most important is the assumption that experimental precautions and controls used in enzyme mechanism studies are adequate for micellar systems.

A number of problems have already been discussed, but they deserve review. First, the assumption that buffers will effectively control the Ph is misleading. For example, consider the micellar catalysed hydrolysis of p-nitrophenyl esters shown

*For a comprehensive discussion of reversed micelles see Reference 11, Chapter 10.

in Figure 2. The experimental conditions were, $25°C$, $0.01 M$ carbonate, 50% base (pH = 10.1), and $0.01 M$ KCl added to control the ionic strength. The fact that the rate—surfactant concentration profiles exhibit maxima shows that the relative concentrations of ester and hydroxide ion are changing in the micellar phase, i.e. that the buffer is not holding the hydroxide-ion concentration constant on the micellar surface, so that quantitative comparisons are impossible in this system. Every solution contains Br^-, Cl^-, HCO_3^-, $CO_3^=$ and OH^- ions, all of which are potential counterions. In fact at low surfactant concentrations the predominant counterion at the Stern layer may not be the bromide ion. Also, as the surfactant concentration increases, the ratio of the ions at the surface changes continuously. Thus, while the buffer should control the pH of the aqueous phase, nothing quantitative can be said about either the total concentration of hydroxide ion in solution, or the actual quantity of hydroxide ion at the micelle surface, because of uncertainties about the relative affinities of the various ions for the micelle surface. Addition of $0.01 M$ KCl is an additional complication. Added salts inhibit the reaction, and the ionic strength may be altered significantly at high surfactant concentration by the ions contributed by the surfactant; moreover nothing is known about the significance of 'constant ionic strength' as applied to micellar systems. Finally, the addition of potassium chloride inevitably introduces a new counterion, and we know that different counterions affect the rates of micelle-catalysed reactions to different extents.

One should not assume that different buffer systems will behave identically in the presence of added surfactant; indeed it may be wise to avoid the use of buffers wherever possible in order to avoid the concomitant salt effects. Also, because the pH may change on addition of surfactant it is best to measure the pH under the actual reaction conditions whenever possible. When studying micellar catalysed reactions, the effectiveness of the experimental controls cannot be assumed, but must be established and specified for each experimental variable.

The other factor which must be carefully controlled is surfactant purity. Commercial surfactants are often very impure and need extensive purification before use. Some surfactants are notoriously difficult to purify completely (NaLS, for example)[137], and many of the usual analytical techniques are not very useful because significant quantities of impurities may be only 1%, and often less. Checking the accuracy of the cmc by independent physical methods provides an important test for the presence of small amounts of hydrophobic impurities, which are often powerful inhibitors of micellar catalysed reactions[121]. Surface tension measurements for determination of the cmc provide additional evidence, because traces of highly surface-active impurities produce noticeable minima in the surface tension—surfactant concentration plots[138]. Mukerjee and Mysels have compiled cmc data for a large number of surfactants from the literature and have included the methods of measurement, experimental conditions (temperature, concentration and type of additives) when available, and critical evaluations of the reliability of the measurements[34].

If the reaction followed is to be used as a probe of miceller structure, then effects of reactants and products on the structure must be considered. Selection of a suitable substrate is often difficult. Surfactants solubilize solutes, so that it is often difficult to isolate products, and if reactant concentrations are increased the micellar structure will surely be significantly perturbed. Product determination therefore requires a sensitive analytical method if kinetic probes are to be used.

The desirability of using low substrate concentrations complicates rate measurements, because chemical analysis is generally too insensitive, and conductivity is useless in the presence of ionic surfactants. The best methods are spectrophoto-metric, and most investigators choose reactions which can be followed in the ultraviolet or visible spectral regions, although fluorescence, phosphorescence and electron paramagnetic resonance spectra can also be useful[11,32]. Nitrophenyl moieties are probably the most useful chromophores as seen from our discussion, since they can be easily used at 10^{-5} M concentration. However, even these low substrate concentrations may significantly lower the cmc of the surfactant, and the more pronounced the effect, the greater the possibility that the initial change in rate is due to the formation of small aggregates of substrate and surfactant rather than micelles. Only independent physical measurements, such as light scattering or conductivity will provide information on the nature of such solutions.

Some of these problems are relatively unimportant if one is concerned only with qualitative information on micellar effects. However, even here, there are potential problems. For example, consider hydrolyses by functional micelles which contain hydroxyl groups as potential models for serine hydroxyl groups in enzyme-catalysed ester hydrolysis (see Section XI. E). In strongly basic solution the attack-ing species is probably the alkoxide ion, whose concentration at the micelle surface will be controlled by the hydroxide ion concentration in the Stern layer. In turn, the hydroxide ion concentration will be determined by the relative amount of other anions in solution, primarily by the counterion contributed by the surfactant. If the stoichiometric amount of hydroxide ion is held constant, then the surface concen-tration of hydroxide ion will decrease continuously with increasing surfactant concentration because of the increasing concentration of unreactive counterion. Consequently, the concentration of alkoxide ion will decrease simultaneously, making it difficult to make meaningful comparisons of the reaction rates at low and high surfactant concentrations.

There are several other kinds of problems that can prove frustrating. Surfactants have widely different solubilities that depend on both chain length and counterion type. For example it is possible to make a solution of almost 1 M n-tetradecyltri-methylammonium bromide at room temperature while a 0.1 M CTABr solution tends to precipitate, especially in the presence of hydrophobic substrates. Concen-trated surfactant solutions may be extremely viscous, making rapid mixing of the reaction solution impossible. Some added salts markedly affect viscosity and surfactant solubility, for example, CTABr becomes viscous and precipitates readily when only a small amount of NaBr is added, whereas CTACl remains quite fluid and does not precipitate even in the presence of a large quantity of NaCl; but the addition of other ions such as tosylate, NO_3^- or CN^- markedly increases the viscosity of CTABr solutions, but the surfactant does not precipitate. These different effects upon viscosity and solubility are not surprising, because viscosity will depend upon micellar structure, and will increase markedly as the micelles go from spherical to rod-like[92], whereas solubility depends not only upon the proper-ties of the solutes, but also upon the packing of the solids into the crystal.

IX. PREPARATIVE ASPECTS

Although most work in kinetic micellar effects has relied wholly on rate measurement there are a few studies of reactions in which micelles control product composition. Examples come from the decomposition of β-bromocarboxylate ions, which react by the S_N1 mechanism in dilute aqueous alkali and the E2 mechanism in CTABr[139], and reactions of 2,4-dinitrophenylsulphate where spontaneous de-composition and amine attack are catalysed to different extents by cationic micelles[140]. There are also a number of S_N1 reactions and deaminations of amines in which the stereochemistry is changed when the substrate is micellized[141-144].

Large rate enhancements by micelles are commonplace for reactions carried out in the presence of normal functional and non-functional micelles in aqueous solutions, and reverse micelles in organic solvents. But it is difficult to make use of these large rate enhancements in preparative chemistry. Micelles are dynamic aggregates whose structures are easily perturbed by even moderately high concentrations of reactants, and in addition surfactants have molecular weights in the range 300–500, so that surfactant solutions of low molarity may contain large amounts of surfactants on a weight basis. These disadvantages are increased by the ability of surfactants to solubilize organic solutes in water, and thereby hinder the isolation of reaction products. These considerations are discussed by Menger and his coworkers who have evaluated the use of surfactants in preparative chemistry[145].

Although we see considerable difficulties in the use of aqueous micelles as catalysts in laboratory-scale preparative chemistry we believe that they may be very useful as inhibitors of undesired reactions, and in stabilizing relatively unstable solutes. For example, a large number of drugs can be successfully stabilized in surfactant solutions, and in some cases can be used in place of aqueous alcohol solutions for administering a drug[35].

Phase-transfer catalysis neatly avoids some of the problems discussed above, and it is worth noting that in some reactions there is a blurring of the distinctions between catalysis by micelles and by monomers or small aggregates of quaternary ammonium ions akin to phase-transfer catalysts. For example, there is the continuing question of the importance of micellar aggregates as compared with 1 : 1 complexes in ester deacylations in the presence of hydrophobic amines (Section XI).

Probably the most promising approach in the preparative use of micelles or micelle-like catalysts is immobilization. For example, nucleophilic or basic groups bonded to a polymeric backbone are effective reagents or catalysts of deacylation[146,147], and their efficacy is increased by attaching hydrophobic residues which assist substrate incorporation.

Another approach which has only been used for dephosphorylation is to bind a cationic functional surfactant (e.g. structure 19, Section XI. C, p. 1005) coulombically to an anionic ion-exchange resin and to use the beads as a reusable catalyst[148].

A very interesting discussion of some of the practical applications of micelles is given in Reference 149.

X. DEACYLATIONS IN NON-FUNCTIONAL MICELLES

A. Carboxylic Esters

Much of the initial work on micellar catalysis and inhibition was on the deacylation of p-nitrophenyl esters. In most of the experiments buffers were used to control pH and salts were often added to maintain the ionic strength of the solution. For the reasons mentioned in Section VI. D, we believe that it is impossible to treat these data quantitatively. However, these early experiments were extremely valuable in that they demonstrated conclusively that cationic micelles catalysed deacylation by hydroxide ion whereas anionic micelles inhibited the reaction by keeping the reagents apart. In addition it was shown that monomeric surfactant had little or no effect on the reaction rate, but that the micellar rate effects increased markedly with increasing substrate hydrophobicity, and that micellar incorporation of the substrate was therefore of key importance.

For the inhibited reactions in anionic micelles reliable substrate binding constants were estimated (see Reference 132 and Section VI.G). The binding constants between hydrophobic esters and micelles were of the same order of magnitude as those found in enzymic reactions, and these observations suggested that there were similarities between the surfaces of micelles and enzymes. Spectrophotometric measurements of surface polarities support this view[29].

As we have mentioned before, these results have been compiled in an important book by Fendler and Fendler which includes an exhaustive compilation of rate data up to 1974[11].

Within recent years studies of micellar effects upon ester deacylation have shifted largely to the use of functional surfactants and this work is discussed in Section XI, and except for brief comparisons, only recent work on non-functional micelles is considered here.

Deacylations of p-nitrophenyl alkanoates usually follow the $B_{Ac}2$ mechanism, but an ElcB mechanism of decomposition[150] is found in the presence of cationic micelles if an electron-withdrawing α-substituent can stabilize a carbanion (Scheme 4)[151].

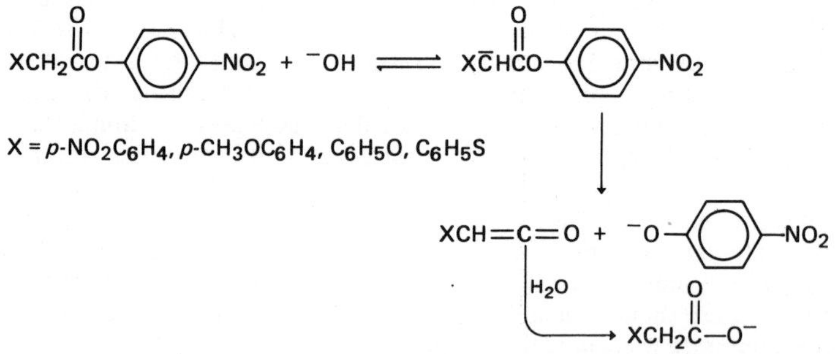

$$X = p\text{-}NO_2C_6H_4, p\text{-}CH_3OC_6H_4, C_6H_5O, C_6H_5S$$

SCHEME 4.

It appears that in water all these substrates react by the $B_{Ac}2$ mechanism, as indicated by a ρ-value of ca 0.7, but that CTABr catalyses proton loss so strongly that the ElcB mechanism is followed in the micelles, as shown by the value of $\rho \approx 2$. These conclusions are consistent with evidence for strong micellar catalysis of other proton eliminations[139,152].

Oximate ions are very effective deacylating agents and the reactions are catalysed very strongly by cationic micelles. Berezin and his coworkers measured the catalysis by CTABr of the acylation of m-bromobenzaldoxime by a series of

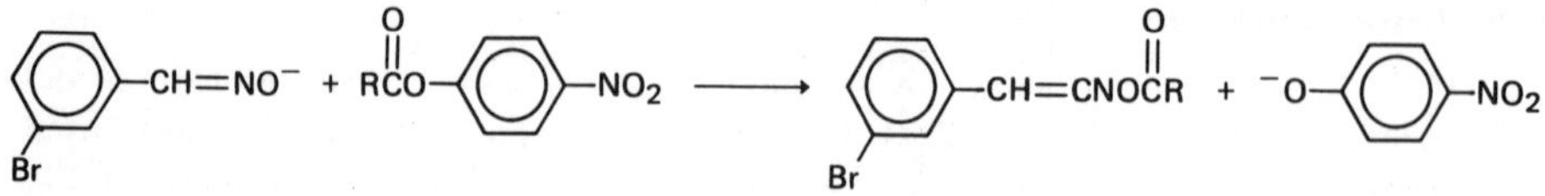

$$R = Me, t\text{-}Bu, n\text{-}Pr, n\text{-}hexyl, PhCH_2, o\text{-}HOC_6H_4$$

p-nitrophenyl alkanoates[83]. The hydrophobicity of the ester was increased from acetate to heptanoate and the extent of catalysis was related directly to the extent of micellar incorporation of the substrate as determined by solubility measurements. It was assumed that the nucleophilic oximate ion was preferentially incorporated into the micelle and the amount of incorporated oximate ion was then estimated from the apparent pK_a of the oxime under the reaction conditions. The overall rate—surfactant profiles were interpreted in terms of the amount of each reactant in the micellar pseudo-phase following the general approaches outlined in Sections VI. A and B.

The deacylation of p-nitrophenyl 3-phenylpropionate by oximate ions of pyridine and quinoline aldehydes is also effectively catalysed by micelles of CTABr (Scheme 5), as is the phosphorylation of these nucleophiles by p-nitrophenyldiphenylphosphate[153].

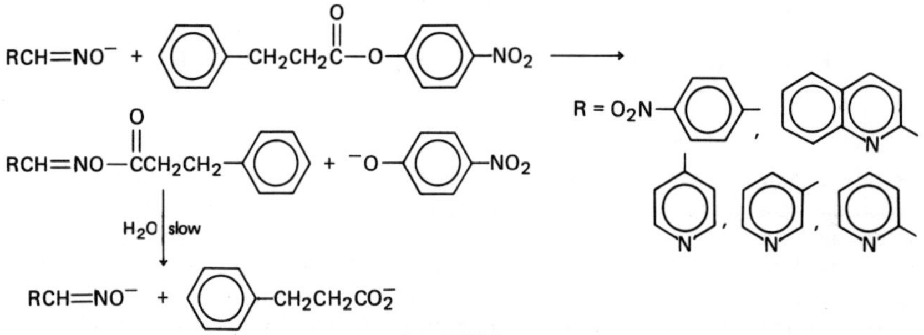

SCHEME 5.

The reaction was carried out in the usual way with oxime in excess over ester, and also with ester in excess over oxime. Under the latter conditions there is an initial acylation of the oxime which is followed by its slow regeneration by hydrolysis. The rate constants for the initial acylation determined under these different conditions did not agree, and the differences between them are not unexpected because of the change in the extent of micellar incorporation of the reactants. As expected the maximum micellar catalysis depended upon the hydrophobicity of the oxime.

B. Amides

There have been only two studies of micellar effects upon the hydrolysis of simple amides*. The first was on the alkaline hydrolysis of $para$-substituted N-methyl acetanilides (5) in the presence of CTABr and the functional surfactant (6) derived from choline[154,155].

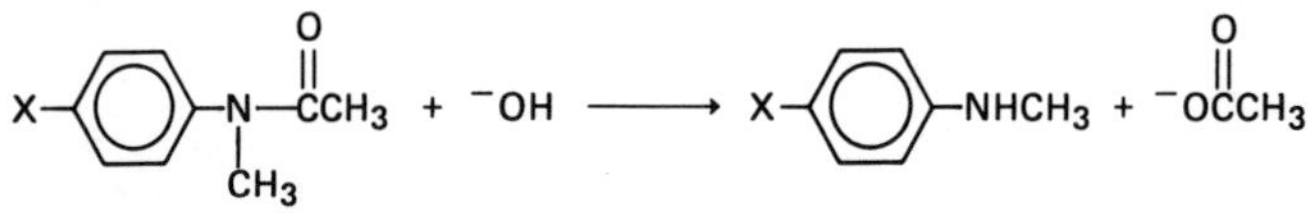

(5)

X = NO$_2$, OCH$_3$, H

$$C_{16}H_{33}\overset{+}{N}Me_2CH_2CH_2OH \quad Br^-$$

(6)

---

*A recent paper describes modest rate enhancements of the acid hydrolysis of hydroxamic acids by NaLS – D. C. Berndt and L. E. Sendelbach, *J. Org. Chem.*, 42, 3305 (1974).

The rate enhancements are small for all the substrates ($k_{max} / k_w < 4$). When X = H and OCH_3 the maximum rate enhancements are only slightly greater than one, but when X = NO_2, the functional surfactant, 6 is about twice as effective as CTABr. It is difficult to draw definitive conclusions from these data.

A more comprehsive study is being made by Broxton and his coworkers on the reactions of a series of anilides in 0.00053 M NaOH[156]. The rate constants go through maxima with increasing concentration of CTABr, as is typical of bimolecular deacylations.

For the reactions:

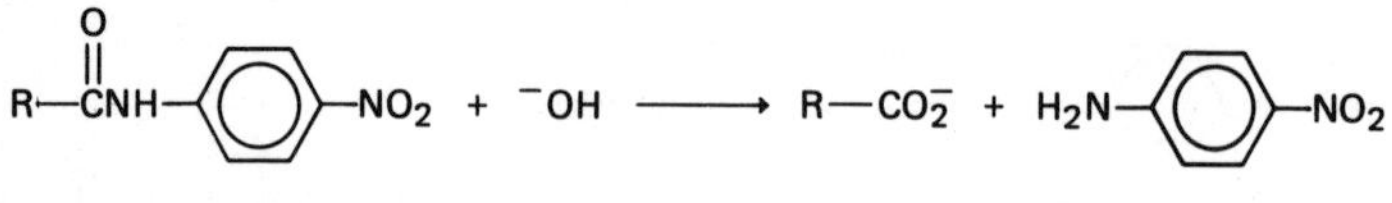

$$R-\overset{\overset{\displaystyle O}{\|}}{C}NH-\!\!\!\left\langle\bigcirc\right\rangle\!\!\!-NO_2 \ + \ ^-OH \ \longrightarrow \ R-CO_2^- \ + \ H_2N-\!\!\!\left\langle\bigcirc\right\rangle\!\!\!-NO_2$$

R = alkyl

the rate enhancements range from ca 6 for R = Me to ca 80 for R = C_8H_{17}, and the surfactant concentration required for maximum catalysis decreases steadily with increasing length of the alkyl group, R. This dependence on substrate hydrophobicity is very similar to that found for other deacylations (see Section VI. E and Figure 2), although it is not completely clear at present whether this shift reflects only greater substrate binding or also a contribution from an increase in transition-state stabilization.

Activated amides, e.g. acylimidazoles, are formed in deacylations catalysed by functional micelles which contain the imidazole moiety, and a number of investigators have examined the turnover step which regenerates the imidazole catalyst by hydrolysis of the acyl intermediate. Of particular interest is the work of Tonellato and Moss and coworkers on the transfer of the acyl group from an imidazole to a hydroxyl moiety in micelles which contain both imidazole and hydroxyalkyl head groups. These systems are discussed in Section XI. C.

Micellar effects upon hydrolysis of 1-trifluoroacetylindole (7) have been examined (Scheme 6)[157]. This very reactive amide is strongly hydrated in aqueous

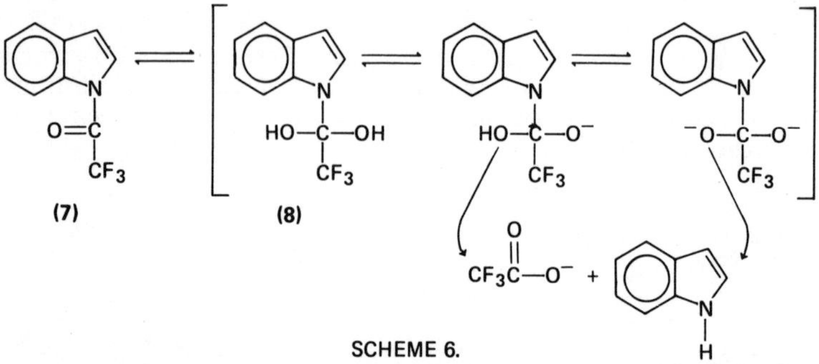

SCHEME 6.

solution, and except at low pH its hydrolysis probably involves the spontaneous decomposition of the hydrate (8) or its anions.

The decomposition at pH > 4 is inhibited by NaLS and speeded by CTABr micelles, which may simply be due to changes in the extent of ionization of the hydrate, with a decrease by the anionic and an increase by the cationic micelle.

The acid hydrolysis of ureas is akin to that of amides. In water the hydrolysis of 4-tolyl and 4-nitrophenyl urea (9) are subject to both general acid and base

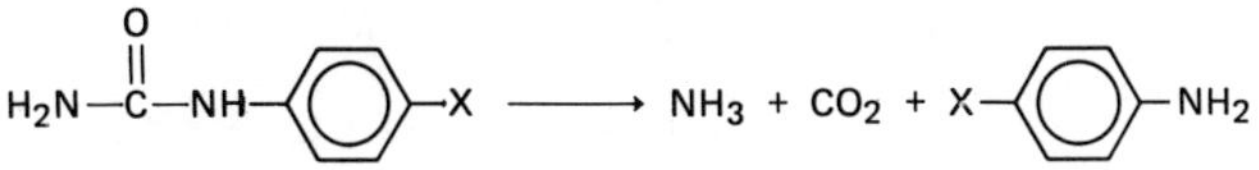

$$H_2N-\overset{\overset{\displaystyle O}{\|}}{C}-NH-\underset{\text{(9)}}{\bigcirc}-X \longrightarrow NH_3 + CO_2 + X-\bigcirc-NH_2$$

X = Me, NO$_2$

catalysis, but only small rate effects are observed on the addition of cationic, non-ionic or anionic micelles[158]. However, the hydrolyses are strongly catalysed by reversed micelles of dodecylammonium propionate, with rate enhancements of ca 3×10^3 relative to aqueous solution. The authors attribute these large catalyses to incorporation of the substrate into the interior of a reverse micelle which brings the acidic and basic groups of the catalyst into close proximity with the substrate.

C. Acyl Anhydrides

The catalysis by cationic micelles of the reactions of p-nitrobenzoyl phosphate and its salts has been examined in detail. The micellar effects show common features with the catalysis of bimolecular reactions of carboxylic esters and of unimolecular reactions of aryl phosphate dianions[159a].

The unimolecular reaction is a spontaneous elimination of a metaphosphate ion from either the mono- (10) or dianion (11).

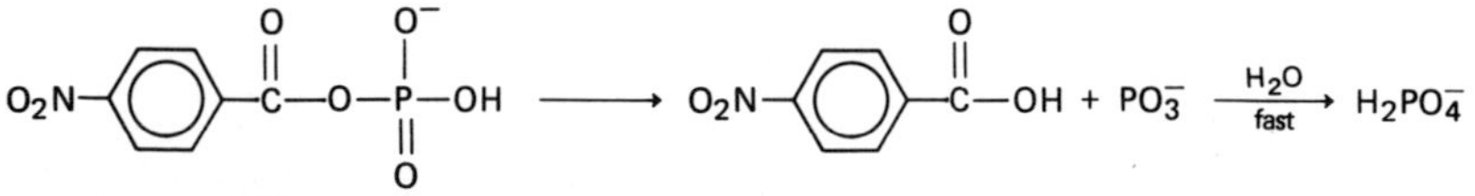

$$O_2N-\bigcirc-\overset{\overset{\displaystyle O}{\|}}{C}-O-\overset{\overset{\displaystyle O^-}{|}}{\underset{\underset{\displaystyle O}{\|}}{P}}-OH \longrightarrow O_2N-\bigcirc-\overset{\overset{\displaystyle O}{\|}}{C}-OH + PO_3^- \xrightarrow[\text{fast}]{H_2O} H_2PO_4^-$$

(10)

$$O_2N-\bigcirc-\overset{\overset{\displaystyle O}{\|}}{C}-O-\overset{\overset{\displaystyle O^-}{|}}{\underset{\underset{\displaystyle O}{\|}}{P}}-O^- \longrightarrow O_2N-\bigcirc-\overset{\overset{\displaystyle O}{\|}}{C}-O^- + PO_3^- \xrightarrow[\text{fast}]{H_2O} H_2PO_4^-$$

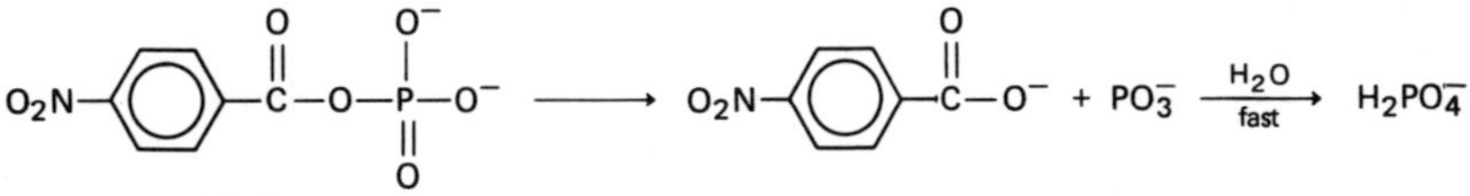

(11)

The reaction of the monoanion 10 is only slightly affected by incorporation into micelles, while the reaction of the dianion 11 is catalysed by cationic micelles of CTABr, and as is typical of unimolecular reactions, the first-order rate constants increase to a plateau as the substrate becomes completely incorporated into the micelles. This kinetic form is very similar to those observed in hydrolyses of dinitrophenyl phosphate dianion[159b] but the overall rate enhancements are much smaller, being ca fivefold for the acyl phosphate as compared with ca twentyfold for the dinitrophenyl phosphates.

The situation is more complex when deacylations of the mono or dianion of the mixed anhydride are catalysed by nucleophiles. At pH > 8, the bulk species is the dianion 11, which is deacylated by both hydroxide ion and dodecylamine (Scheme 7).

Whereas the maximum rate enhancement of the CATBr catalysis of the reaction with OH$^-$ is only about eightfold, it is thirtyfold for the reaction with dodeclamine. These differences, as before, are similar to those generally found for micellar catalysis, with the rate enhancements increasing with increasing hydrophobicity of the nucleophile. For example, in water, hydroxide ion is much more nucleophilic

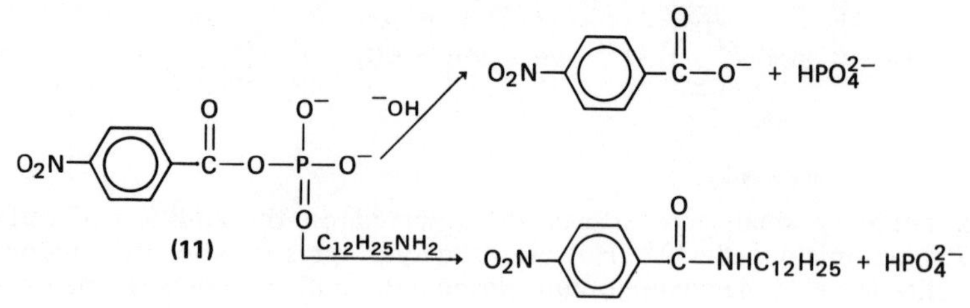

SCHEME 7.

than an alkylamine toward an acyl phosphate dianion, but when the reaction is carried out in a cationic micelle aminolysis competes effectively with hydrolysis. A somewhat similar pattern is found for reaction of the monoanion **10** (Scheme 8).

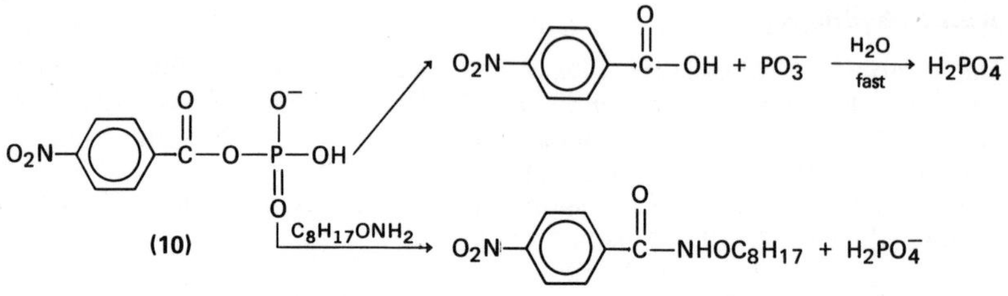

SCHEME 8.

While there is a little micellar catalysis of the spontaneous hydrolysis of the monoanion **10**, there is extensive catalysis of the bimolecular deacylation by octyloxyamine. In this experiment it was necessary to use a weakly basic amine, because the monoanionic substrate is extensively deprotonated at pH > 5.

In these reactions the micelles of CTABr bring reagents together into the Stern layer and so speed bimolecular reactions, and they also provide a reaction environment which assists the unimolecular decomposition of the dianion. The rate attack of 0.01 M hydroxide ion upon the dianion in CTABr is approximately doubled when n-decylguanidinium ion is added to the micelle, possibly because the guanidinium ion hydrogen bonds to the leaving phosphate ion and provides electrophilic assistance to the reaction.

Finally, the functional surfactant **6**, derived from choline, is an effective reagent for the deacylation at high pH (Scheme 9), and the reaction almost certainly involves nucleophilic attack by the alkoxide moiety. Evidence for nucleophilic

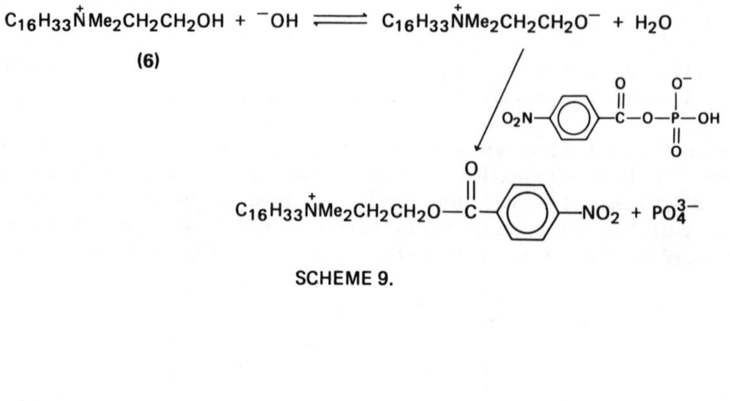

$$C_{16}H_{33}\overset{+}{N}Me_2CH_2CH_2OH + {}^-OH \rightleftharpoons C_{16}H_{33}\overset{+}{N}Me_2CH_2CH_2O^- + H_2O$$

(6)

SCHEME 9.

attack as opposed to general acid or general base catalysis by the hydroxyl group of **6** is presented in Section XI. E.

XI. FUNCTIONAL MICELLES AND COMICELLES

Elucidation of the factors which govern catalysis by micelles of non-functional surfactants was soon followed by the design of functional micelles in which a nucleophilic or basic group was covalently bound to the surfactant. So far as we are aware there is no compelling evidence for electrophilic catalysis in these systems.

In many reactions it is evident that the reaction is occurring in a micellar pseudo-phase, for example the reaction–surfactant relationships are similar to those observed with non-functional micelles in that the reaction is very slow at surfactant concentrations below the cmc, and the rate increases as substrate is incorporated into the micelle; but sometimes it is difficult to decide whether micelles or some other aggregate are involved.

Bruice and his coworkers examined deacylation of *p*-nitrophenyl esters of long-chain alkane carboxylic acids by long-chain alkylamines[160]. By introducing charged centres with the reactants they hoped to observe rate enhancements by 'twinning' of the reactants when the structures of the reactants permitted a maximum in the coulombic and hydrophobic interactions.

While these results were best interpreted simply in terms of micelle formation, in a slightly different system large rate enhancements were found for reactions of long-chain *n*-alkylamines with *p*-nitrophenyl *n*-alkane carboxylates[161]. Low ester concentrations were used, and rate enhancements were found at amine concentrations well below the cmc, suggesting that micelles were not involved in these reactions and that the rate enhancements were probably due to formation of 1 : 1 complexes. A major problem with this explanation is that the long-chain non-ionic esters are themselves surface-active, and may also induce micellization of the *n*-alkylamine, and therefore the rate increases with increasing hydrophobicity of the amines could be interpreted in terms of micellization[162].

Although the examples cited above may involve micellar catalysis, the catalysis of deacylation of a series of aryl alkanoates by the imidazole derivative **12** appears to involve formation of 1 : 1 complexes[163].

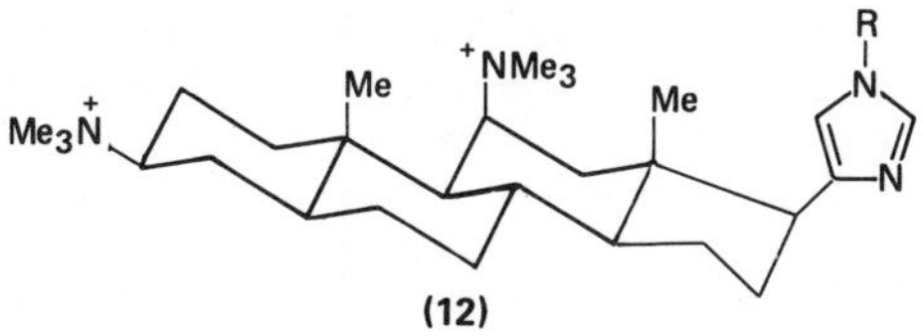

(12)

(a) R = H
(b) R = *i*-Pr

The rate enhancements were generally only of one order of magnitude, and the isopropyl derivative (**12b**) was, as expected, a poorer catalyst than the unsubstituted compound (**12a**).

Recent work by Kunitake and his coworkers raises questions about the structures of the aggregates involved in reactions of functionalized surfactants and related hydrophobic solutes. One striking example of this problem is deacylation by alkyl hydroxamates[164].

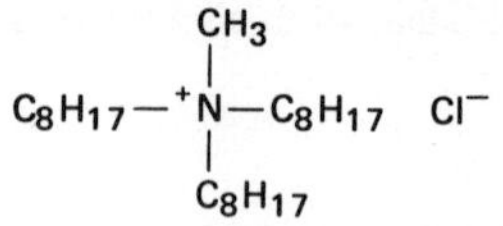

$$C_8H_{17}-\overset{\overset{\displaystyle CH_3}{|}}{\underset{\underset{\displaystyle C_8H_{17}}{|}}{\overset{+}{N}}}-C_8H_{17} \quad Cl^-$$

$$C_{12}H_{25}-\underset{\underset{\displaystyle OH}{|}}{N}-\overset{\overset{\displaystyle O}{\|}}{C}-\bigcirc$$

(14) 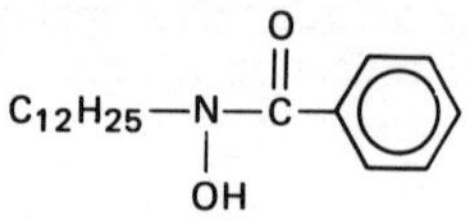   **(13)**

The reaction of p-nitrophenyl acetate with the hydroxamate **13**, in water, is very strongly caralysed by tri-n-octylmethylammonium chloride (**14**), and the catalysis is much greater than that given by cetyltrimethylammonium bromide, which forms normal cationic micelles. Salts such as **14** are typically phase-transfer catalysts and are generally not regarded as micelle-forming agents, so it seems that in the systems studied by the Kunitake group we are dealing with aggregates whose structures are different from those of the typical micelle. However, there are many examples in which all the evidence points to the existence of functional micelles which incorporate the substrate, and these cases are discussed in terms of the specific functional group.

In many deacylations catalysed by functional micelles the initial step is acylation of a nucleophilic group, e.g. imidazole, which is followed by decomposition of the acyl intermediate with regeneration of the catalyst (Scheme 10).

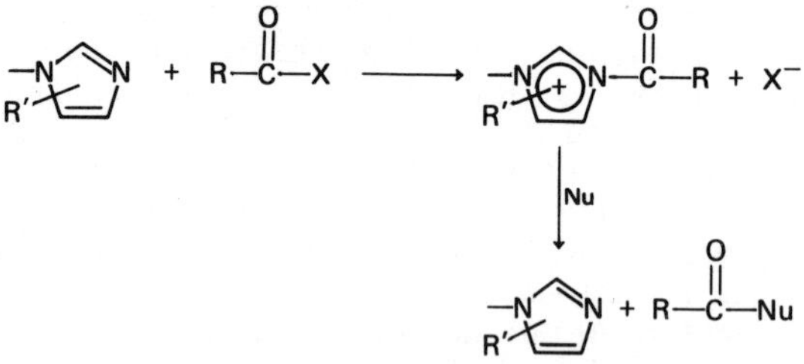

SCHEME 10.

One key test for formation of a covalent intermediate is the kinetic form of reaction when the electrophile, e.g. RCOX, is in excess over the nucleophile, which in this case is a functional surfactant. There is an initial 'burst' of X^-, e.g. p-nitrophenoxide ion, with consumption of the nucleophile which is then regenerated in a slower second step. This test has been applied to catalysis by a variety of functional micelles which will be considered in this chapter. But in these experiments with low concentrations of nucleophilic surfactant and relatively high substrate concentration the aggregates present may be different from normal micelles which are formed when the surfactant is in high concentration and the substrate is low.

A. Factors which Contribute to Catalysis by Functional Surfactants

Before discussing particular reactive head groups it will be useful to analyse the factors which control catalysis by functional surfactants. Like the early work on non-functional surfactants, the impetus for work in this area comes from a search for models of enzymic catalysis.

One of the goals in studying the mechanism of enzyme-catalysed reactions is the dissection of the overall reaction into its individual steps and determination of the

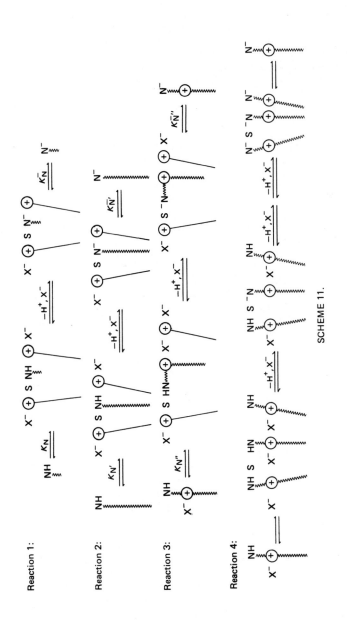

SCHEME 11.

rate of each step. This work should be facilitated by the study of simpler systems and micelles, and also soluble polyelectrolytes and resins represent potentially useful models. However, Herries, Bishop and Richards, who made one of the earliest studies of micellar catalysis, concluded that 'micelles appear just as complicated as the proteins for which we had hoped to use them as models'[71]. In addition, micelles composed of simple non-functional surfactants generally show poor selectivity and low catalytic activity.

A more promising approach is the construction of micelle models that contain part or all of the active sites of the enzymes under consideration. A number of studies of this type will be discussed briefly in later parts of this section, with much of the work focused on duplicating the acylation–deacylation activity of chymotrypsin[16,17]. Functional micelles, comicelles and mixed micelles have been constructed containing groups which serve as models for the amino-acid side-chains at the active site of chymotrypsin, e.g. the serine-195 hydroxyl group, the histidine-57 group, and carboxylate group of asparate-102. Other groups whose catalytic properties have been studied in micellar systems include oxime, hydroxamic acid, thiol and their respective anions, and the amino group.

If these models are to be fully exploited the experiments must be in sufficient detail to isolate the individual steps and their respective rate constants for the whole mechanism. This means that the reaction conditions must be standardized and experimental controls established so that meaningful comparisons between various systems can be made. Inspection of the literature indicates that neither condition has yet been met. Most of the studies are preliminary, having demonstrated only the presence of, or catalysis (albeit, sometimes very large) by, a particular functional group. Most workers use consistent experimental conditions for their own work, but these often differ considerably from group to group. For example, there may be differences in buffer type and concentration, counterion type and concentration, surfactant chain length or head-group structure, substrate and nucleophile structure, making quantitative comparisons almost impossible. Meaningful comparisons can be made even more difficult by the formation of complexes between the surfactant monomer and a hydrophobic substrate or nucleophile at low surfactant concentrations, by different extents of binding of chemically similar reactants differing only in their hydrophobicity, or simply by there being insufficient data for separation of the concentration and medium effects of the micelle.

Scheme 11 illustrates the importance of controlling the variables listed above. Any of these micellar reaction systems can be used as potential models for the various steps involved in enzyme catalysis. Reaction 1 shows the binding of a hydrophilic organic nucleophile to a non-functional cationic surfactant. Reaction 2 shows the binding of the chemically similar nucleophile whose hydrophobicity has been increased by the attachment of a long hydrocarbon chain, which produces comicelles. Reaction 3 illustrates the formation of mixed micelles when the nucleophile is attached directly to the surfactant monomer, which is combined in varying ratios with a non-functional surfactant. Reaction 4 shows functional micelles composed only of functional surfactant monomers. The non-ionic substrate, S, is assumed to be the same in all cases, and in low concentration so that it neither perturbs the micelle structure nor depletes the nucleophile concentration significantly during the course of the reaction, thus maintaining first-order conditions. Scheme 11 also illustrates the important point that many of the nucleophiles studied can ionize forming highly nucleophilic anions at an experimental pH well below the pK_a of the nucleophile in water.

Comparison of the catalytic properties of nucleophiles (or substrates) of different hydrophobicity in micellar solutions (Reaction 1 with 2 or 3) requires that the surfactant concentration is high enough to ensure complete binding of the nucleophile (or substrate) or that the binding constant is known under the experimental conditions. If the extent of binding is unknown, it will be impossible to decide whether rate increases with increasing reactant hydrophobicity are related to an increase in the incorporation of the reactant or to changes in its location within the micelle.

One of the earliest studies of the importance of nucleophile and substrate hydrophobicity illustrates many of the potential problems in the interpretation of data. Gitler and Ochoa-Solano studied the deacylation of p-nitrophenyl esters by N^α-acetyl (AcHis) and myristoyl histidines (MirHis) in CTABr at pH 7.02 using Tris buffer (Scheme 12)[165]. They found that CTABr catalysed deacylation by MirHis

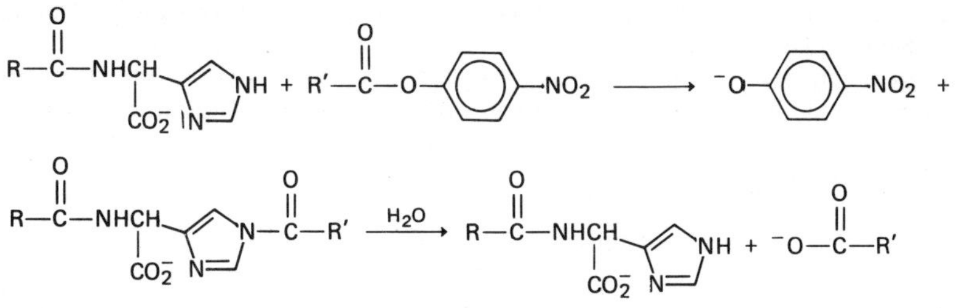

$$R = Me, C_{13}H_{27}$$
$$R' = Me, Et, n\text{-Pr}, n\text{-Bu}, n\text{-amyl}$$

SCHEME 12.

but inhibited that by AcHis, and that the extent of catalysis and inhibition increased with increasing length of the alkyl group (hydrophobicity) of the ester. The simplest interpretation of these results is that MirHis is incorporated into micelles, while AcHis is not. There is a linear relation between the logarithm of the second-order rate constant at constant MirHis and the length of the n-alkyl group of the ester at constant surfactant and MirHis concentration. However, the authors did not measure the extent of incorporation of the esters, making it impossible to decide whether this free-energy relationship depends on increased binding of the ester or on a change in the rate of the reaction in the micellar phase.

This study also illustrates some of the other important controls required for the interpretation of the kinetics of micelle-catalysed reactions. Turbidity measurements provided strong evidence for formation of a 1 : 1 complex between MirHis and CTABr below the cmc of pure CTABr. The turbidity disappeared at higher CTABr concentrations, and the complexes were apparently not catalysts. At very low CTABr concentrations, MirHis was insoluble, making it impossible to compare the effect of increasing nucleophile hydrophobicity over a wide concentration range of surfactant. In addition a soluble nucleophile, like other added solutes, may lower the cmc, so that the cmc under kinetic conditions is generally lower than that of the pure surfactant in water.

Gitler and Ochoa-Solano showed that the initial step of the reaction was acylation of the MirHis with release of p-nitrophenoxide ion. The pH–rate profile gave an apparent pK_a of 6.2 for the micellar-bound MirHis, compared to a pK_a of

7.0 for imidazole in water, and was consistent with the neutral imidazole moiety being the nucleophile. However, the pH was not increased sufficiently to exclude the possibility that the anion of the nucleophile was the reactive species, and in many reactions involving imidazole-derived functional surfactants the anion is the reactive species. There are, however, examples of nucleophilic participation by both the non-ionic imidazole moiety and its anion (see Section XI. C and Figure 10).

While reactions of non-ionic nucleophiles and substrates are amenable to the analysis discussed earlier (see Section VI. A and B), the complete interpretation becomes much more complex as the pH is increased and the nucleophile ionizes. Some of the enhanced catalytic activity of functional surfactants and comicelles can be attributed to the presence of the nucleophilic anion at operational pH's well below the pK_a in water, a conclusion supported by the fact that cationic micelles lower the pK_a of weak acids while anionic surfactants increase it[11]. However, a complete interpretation of the effect of micelles on ionic nucleophiles is impossible at this time because the data are incomplete and more importantly, the current experimental controls are inadequate.

The complexity of this problem is greatest for functional micelles (Scheme 11, Reaction 4), as is demonstrated by the effect of micellization and added salts on the apparent pK_a of a series of acylcarnitines determined by potentiometric titration (Scheme 13)[166,167].

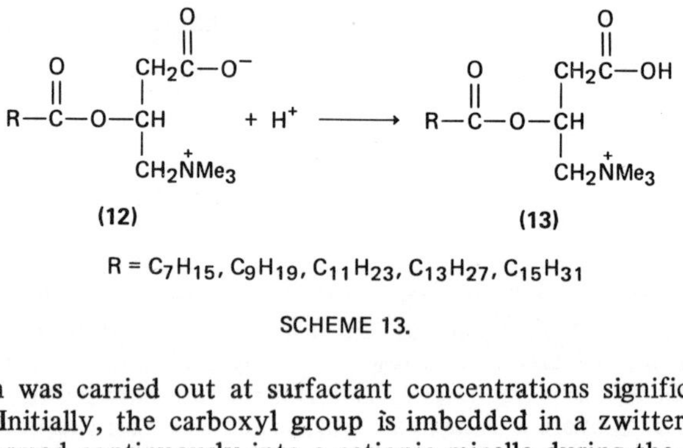

$$
\begin{array}{ccc}
& \text{O} & \text{O} \\
& \| & \| \\
\text{O} & \text{CH}_2\text{C}-\text{O}^- & \text{O} \quad \text{CH}_2\text{C}-\text{OH} \\
\| & | & \| \quad | \\
\text{R}-\text{C}-\text{O}-\text{CH} \quad + \text{H}^+ \longrightarrow & \text{R}-\text{C}-\text{O}-\text{CH} \\
| & | \\
\text{CH}_2\overset{+}{\text{N}}\text{Me}_3 & \text{CH}_2\overset{+}{\text{N}}\text{Me}_3
\end{array}
$$

(12)   (13)

$R = C_7H_{15}, C_9H_{19}, C_{11}H_{23}, C_{13}H_{27}, C_{15}H_{31}$

SCHEME 13.

The titration was carried out at surfactant concentrations significantly higher than the cmc. Initially, the carboxyl group is imbedded in a zwitterionic micelle which is transformed continuously into a cationic micelle during the course of the titration. This change in surface structure produces a dramatic change in the apparent pK_a of the carboxyl group.

When the micelle is initially zwitterionic, the apparent pK_a is 4.85 for the carboxyl groups in the lauryl carnitine micelles 12, $R = C_{11}H_{23}$, but it decreases 2 pK_a units on complete protonation. The apparent pK_a of the carboxyl group in the cationic form of the micelle 13 is sensitive to changes in the concentration of added salt, with a pK_a of 3.4 for the protonated form of lauryl carnitine in 0.2 M potassium chloride compared to a pK_a of 2.9 in the absence of salt. However, at any pH the apparent pK_a of micellized myristyl carnitine, $R = C_{13}H_{27}$, is sensitive to the type of anion present, but not to the cation. In 0.2 M added salt the salt order on pK_a of myristyl carnitine is $I^- > Br^- > Cl^-$. Similar trends are observed in pK_a determinations of micellized alkylamine oxides[168].

These results indicate that the rates of reactions catalysed by functional micelles may be sensitive to changes in surface charge and the type and concentration of buffer and counterions present. Changes in the extent of acid dissociation of the functional group at the surface will alter the concentration of counterions at the surface, and perhaps also the pH at the surface and the intrinsic pK_a of the functional group, and conversely, changes in the salt concentration and counterion type may alter the surface pH and the intrinsic pK_a. The problem may be even more complex when one buffer component is also a counterion to the micelle. Some of these problems can be avoided by measuring the apparent pK_a of the functional surfactant under kinetic conditions, or by using a mixed micelle system with a very low concentration of the functional surfactant to minimize perturbation of micellar structure.

Ionization of the nucleophile and the change in micellar charge may also affect the binding of hydrophilic nucleophiles, e.g. of acetyl histidine, whose anion will bind more strongly than the neutral precursor (Scheme 11, Reaction 1), i.e. $K_{\bar{N}} \gg K_N$. Addition of non-reactive counterions may alter the pH at the micellar surface by displacing hydroxide ions, and possibly the anion of the nucleophile, thus further complicating the interpretation of the changes in apparent pK_a of the functional group, the role of hydrophobicity of added nucleophile, and the significance of the extent of catalysis by a functional micelle.

Although the various factors have not been separated, in combination they can account for rate enhancements of several orders of magnitude. In Section VI. A we predicted the magnitude of the catalysis to be expected from the concentration of the reacting entities at the micellar surface. There are currently insufficient data to permit a test of this prediction because the conditions of the micellar and non-micellar experiments are generally very different.

B. Thiols

The first example of the use of a functional micelle containing the thiol group as a catalyst of ester deacylation was provided by Heitmann[169], who used N-dodecanoyl-D L-cysteinate in aqueous solutions of anionic and cationic surfactants in the deacylation of p-nitrophenyl acetate and nucleophilic substitutions as models for —SH enzymes. Incorporation of the functional surfactant into anionic micelles of N-dodecanoylglycinate inhibits reaction while incorporation of the nucleophile into cationic micelles of CTABr strongly catalyses deacylation. The results are consistent with RS$^-$ being the reactive species. The rate enhancement in CTABr is due in part to increased ionization of the —SH group, while incorporation in the anionic micelles decreases ionization producing net inhibition.

More recently Tagaki and his coworkers have shown that comicelles of alkanethiols[170] and octadecyltrimethylammonium bromide are effective deacylating reagents towards p-nitrophenyl acetate (Scheme 14). The rate enhancements over reaction in aqueous solution range from a factor of 110 for ethane thiol to ca.

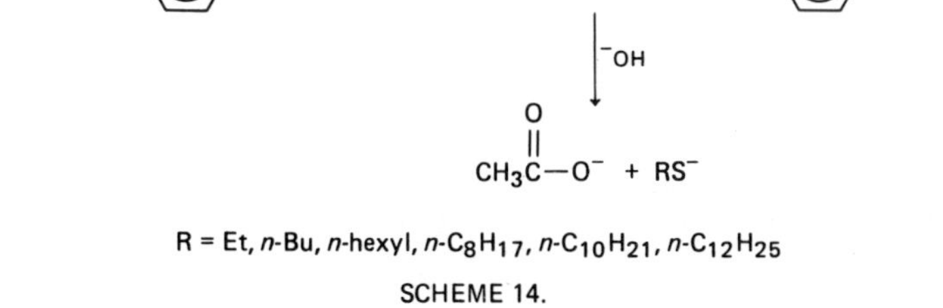

$R = Et, n\text{-}Bu, n\text{-hexyl}, n\text{-}C_8H_{17}, n\text{-}C_{10}H_{21}, n\text{-}C_{12}H_{25}$

SCHEME 14.

5×10^4 for n-dodecane thiol. This dependence of reaction rate upon reactant hydrophobicity is typical of micellar catalysis, and can be related to the extent of partitioning of reagents between water and micellar pseudo-phase. Long-chain alkane thiols are hydrophobic and therefore should be incorporated largely into the micelle, so that the rate constants in the micelle increase with increasing length of the alkyl group of the thiol. The reactive species is the thiolate ion and incorporation should increase the extent of ionization of the thiol, as well as the reactant concentration in the micellar pseudo-phase. These factors have not been separated.

It is reported that the thiol ester formed by initial deacylation is hydrolysed under the reaction conditions, with turnover of the nucleophile. However, thiol esters are not especially reactive at the pH of these experiments[171] and it would be useful to have independent evidence for this turnover step.

C. Amines and Imidazoles

We consider first reactions in which the nucleophile is a primary amine. Both Knowles and coworkers[161] and Oakenfull[21] have used primary n-alkylamines as deacylating agents. The reaction rates increase sharply with increasing hydrophobicity of both the n-alkylamine and the substrate, which was a p-nitrophenyl derivative of an n-alkane carboxylic acid. Oakenfull used aqueous ethanol as a solvent, with the aim of working at reagent concentrations below the critical micelle concentrations of the reactants, but it is not easy to decide whether comicelles or 1 : 1 amine—ester complexes are present in these systems, because small amounts of hydrophobic solutes can induce micellization*.

Knowles and coworkers obtained similar results using n-alkylamines, and showed that both aminolysis and hydrolysis occurred more readily when the reagents could form aggregates. However, much of the work of both groups was done using n-alkylimidazoles, and again very large rate enhancements were found with the more hydrophobic reactants.

The major unsolved problem in these systems is distinguishing between the formation of reactive 1 : 1 complexes, by 'twinning' of the reagents, or the formation of comicelles as was suggested by Guthrie[162]. Unfortunately, an increase in reaction rate with increasing reactant hydrophobicity may not distinguish between these possibilities. Blyth and Knowles[161] for example, cited a parallel between the rate enhancements of these reactions and the increase in hydrophobic bonding between the long-chain esters and amines as evidence for the formation of 1 : 1 complexes. However, this trend is equally consistent with comicellization, because the driving force for the formation of both types of aggregates is the release of water molecules from the surface of hydrophobic groups in aggregation. Elucidation of this problem will probably require physical study of the aggregates using chemically inert model compounds.

Imidazole derivatives provide many examples of intramolecular catalysis in homogeneous systems, and functional surfactants which contain imidazole in the head groups have been used extensively in deacylations catalysed by normal micelles in aqueous solutions. The pattern of reactivity is similar to that discussed earlier for deacylation by n-alkylamines, in that the rate constants increase with increasing hydrophobicity of both reagents.

*Another example of powerful catalysis of deacylation by aggregates which are apparently not typical micelles is provided by the work of Kunitake and coworkers on the deacylation of p-nitrophenyl acetate by hydrophobic hydroxamate ions in the presence of trioctylmethylammonium chloride (see Section XI. D, Reference 164).

The ambiguities regarding the nature of the association between reactants are absent in some, but not all, or the systems composed of mixed micelles and functional micelles. Mixed micelles of CTABr and N^α-acyl histidines discussed earlier (Section XI. A) provide an excellent example of the properties of micelles composed of two types of surfactants.

An early example of catalysis by functional micelles is the decomposition of the cationic carbonate ester **14** by micelles of N^α-stearoylhistidine (**15**) (Scheme 15)[172].

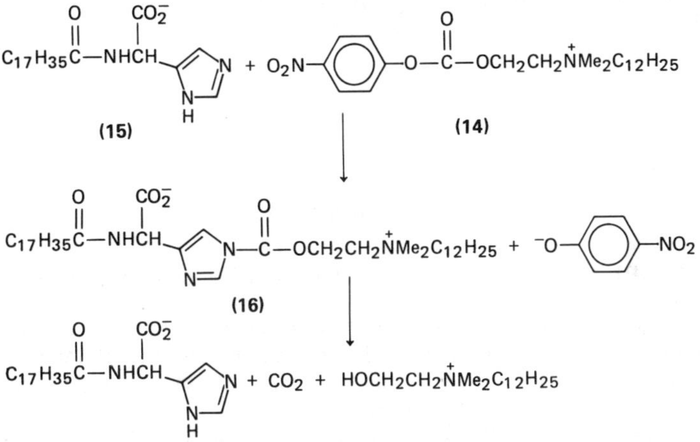

SCHEME 15.

In this system the rate constant increases sigmoidally to a plateau with increasing concentration of N^α-stearoylhistidine, which is typical of micellar incorporation of a substrate, suggesting that this catalysis almost certainly involves micelle formation rather than 1 : 1 complexation. However, the authors interpreted the catalysis in terms of the formation of 1 : 1 complexes because it was observed at surfactant concentrations well below the cmc. This system, like others discussed in this section, illustrates the difficulty in distinguishing between the formation of micelles and small substrate–surfactant aggregates.

The pH–rate profile shows that the neutral form of the imidazole moiety was reactive under the experimental conditions (pH < 8.5). This result is understandable because the imidazole group is physically close to the carboxylate anion and kinetic evidence for the presence of a reactive imidazole anion has been found only in cationic functional micelles at pH > 8 (see below).

The acyl intermediate (**16**) breaks down rapidly with loss of CO_2, but there are several deacylations in which the formation of acyl imidazoles has been demonstrated kinetically or by trapping (cf. Section XII).

Tagaki and his coworkers have used the functional surfactant **17** (Scheme 16) to catalyse the deacylation of p-nitrophenyl acetate and they have also followed the decomposition of the acyl intermediate[173]. They suggest that the initial acylation step involves the anion of **17**, and ionization should be assisted by the cationic micelle, as is deacylation of the intermediate by attack of hydroxide ion. This explanation is supported by the observation that comicelles of **17** and sodium tetradecyl sulphate are poor catalysts. The effect of pH upon the reaction of p-nitrophenyl acetate in the presence of micelles of the functional surfactant N^α-dodecanoly-L-histidine (DoHis) and in mixed micelles of DoHis with CTABr

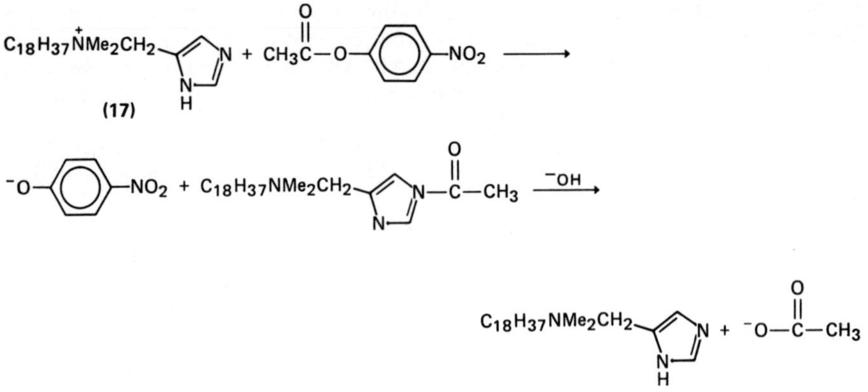

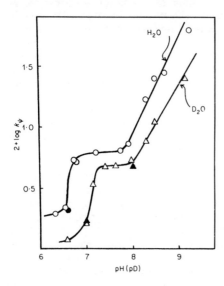

SCHEME 16.

and DoHis with N^{α}-dodecylglycine was investigated in detail by Heitmann and his coworkers who concluded that here too the active nucleophile is the imidazole anion[174]. However, they did not attempt to estimate the rate constant in the micellar phase with respect to the imidazole anion. The uncertaintaies as to the second pK_a of imidazole and its derivatives, both in water and in micelles, complicate quantitative interpretations.

Berezin and coworkers studied the deacylation of p-nitrophenyl heptanoate by a variety of imidazole derivatives in CTABr in 0.02 M borate and phosphate buffers[87b]. Some of these derivatives were sufficiently hydrophobic that the aggregates could be regarded as comicelles with CTABr rather than as CTABr micelles which contained small quantities of bound nucleophile. These results indicate quite clearly that in moderately basic solutions the anionic form of the imidazole derivative is the active species.

FIGURE 10.   Log k vs pH (pD) profile for the hydrolysis of (R)-p-nitrophenyl-N-acetylphenylalanine in micelles of the chiral surfactant, 19. Solid symbols denote different samples of surfactant. Taken from S. Diaz, *Ph.D. Thesis*, University of California, Santa Barbara, 1977.

Taken together, these observations suggest that at pH > 8 in cationic surfactant solutions the active nucleophile is the imidazole anion, but that in anionic or zwitterionic micelles, or in 1 : 1 complexes, the neutral imidazole group may be the active species. However, the evidence is unambiguous for the deacylation of 18 by the histidine derivative 19, examined over a wide range of pH[148]. In these reactions

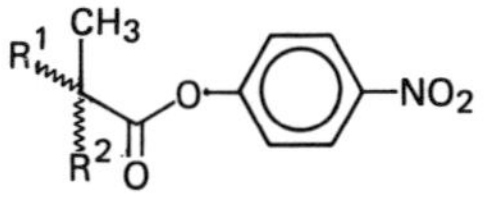

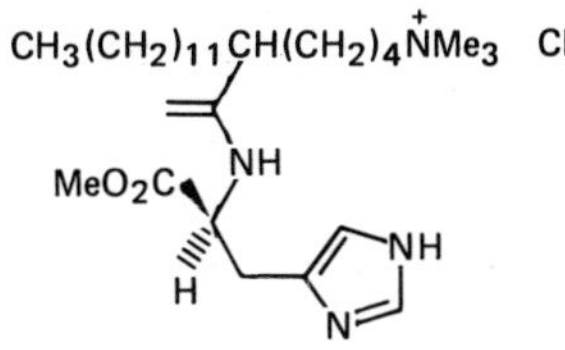

CH$_3$(CH$_2$)$_{11}$CH(CH$_2$)$_4$N$^+$Me$_3$ Cl$^-$

(18)

(a) R^1 = Ph, R^2 = Me   (19)

(b) R^1 = PhCH$_2$, R^2 = MeCONH

the surfactant was in large excess over the substrate and the rate constants increase sigmoidally as the micelle incorporates the substrate. Under conditions in which the substrate is largely micellar incorporated the rate constants increase in going from pH 6.5 to 6.9, then become approximately constant, and then increase again at pH > 8 (Figure 10). This general form of the rate–pH (pD) profile has been observed with p-nitrophenyl-3-phenylpropionate (18a) and (R)-p-nitrophenyl-N-acetylphenylalanine (18b). The pH–rate profile for this system can be rationalized using the acid–base equilibria in Scheme 17, on the assumption that the cation (20)

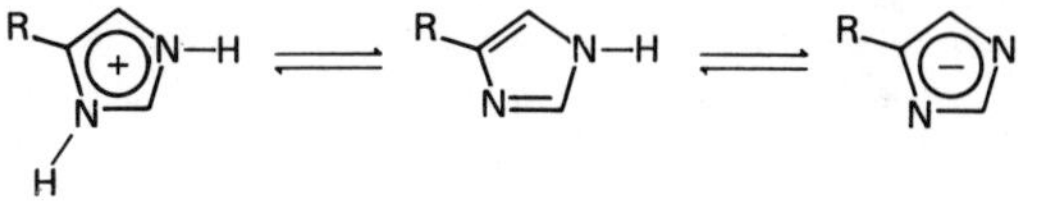

(20)   (19)   (21)

SCHEME 17.

is unreactive and that the anion (21) is a better nucleophile than the undissociated form (19). The apparent pK_a of the surfactant is consistent with the hypothesis, and the deuterium solvent isotope effect, $k_{H_2O}/k_{D_2O} \approx 1.3$, for the hydrolysis of 18b catalysed by 19, is in the range expected for nucleophilic attack. As expected the kinetic solvent deuterium isotope effect increases with increasing pH as the imidazole group becomes deprotonated, because of the normal deuterium solvent isotope effect on acid dissociation.

The surfactant 19 has two chiral centres and a single diastereoisomer could be isolated. It was a stereospecific catalyst for the hydrolysis of 18b, with the rate of reaction of the S-substrate being approximately three times that of the R[175]. This stereospecificity was not derived from differences in substrate binding, because the binding constants estimated from the rate–surfactant profiles in Figure 11 were equal within experimental error, but it stemmed from differences in the transition-state interactions. These interactions probably involve hydrogen bonding between the amido group of the surfactant and the acetamido group of the substrate 22, a conclusion supported by the fact that there was little stereospecificity in the catalysed hydrolysis of R- and S-p-nitrophenyl-2-phenylpropionate where there can be no such interaction.

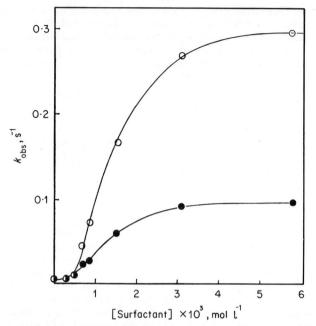

FIGURE 11.   Variation of k_{obs} with concentration of the
chiral surfactant, **19**. Solid circles are the R points; open circles
are the S. Taken from J. M. Brown and C. A. Bunton, *J. Chem.
Soc., Chem. Commun.*, 969 (1974). Reprinted with permission
of the Chemical Society, London.

(22)

Chiral recognition has not been observed in deacylations catalysed by non-
functional cationic micelles which contain chiral head groups[176]. These observations are consistent with micelles having structures which are mobile and easily
deformable by added substrates. Thus, chiral non-functional micelles do not distinguish between chiral substrates or the transition states derived from them.

(Non-functional micelles can change the stereochemical course of S_N1 reactions
of chiral substrates bound to them[141,142,177-179]. For example, the deamination
of non-micellized chiral primary amines involves predominant inversion of configuration, while deamination of the micellized amines leads to predominant retention
of configuration[179]. These observations can be rationalized in terms of preferred
front-side nucleophilic attach upon the carbocationic intermediates in these reactions.)

D. Oximes and Hydroxamates

Oximate ions are excellent nucleophiles and recently functional surfactants containing oxime or hydroxamic acid head groups have been shown to be excellent micellar catalysts of deacylation. For example, mixed micelles of CTABr and *N*-methyl-*N*-laurylhydroxamate ion are very effective deacylating agents of *p*-nitrophenyl acetate and their activity is reported to be similar to that of chymotrypsin[180].

In another example, the deacylation of *p*-nitrophenyl acetate by the cationic oxime derivative **23** or the hydroxamic acid **24** is strongly accelerated by cationic

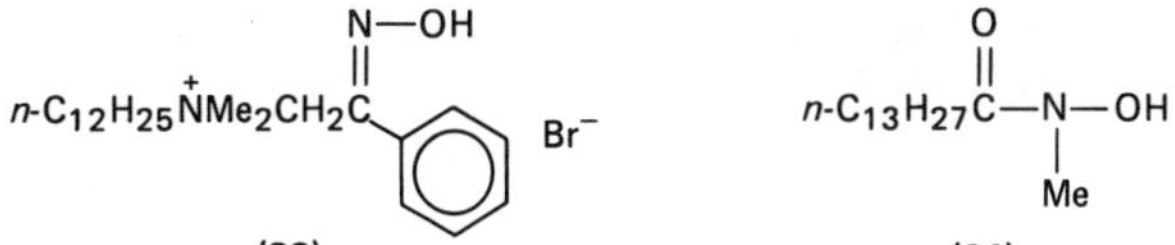

(23)   (24)

micelles of CTABr (by factors of ca 10^2), and to a smaller extent by non-ionic micelles, but anionic micelles of NaLS retard the reaction[181]. These reactions almost certainly involve nucleophilic attack by the oximate or hydroxamate derivatives of **23** and **24** because the reaction rates increase with increasing pH and begin to level off when acid dissociation is complete at pH > 9.

As in other studies of reactions with functional surfactants, it is difficult to separate the various factors which contribute to the rate enhancements because we do not know the extent of micellar incorporation of *p*-nitrophenyl acetate, and in addition cationic micelles will assist formation of the nucleophilic anion.

Another example of the importance of reactant hydrophobicity in a hydroxamate-induced deacylation is a study of the reaction of a series of anionic benzoates, **25**[182]. The reactions are catalysed by micelles of CTABr provided that the

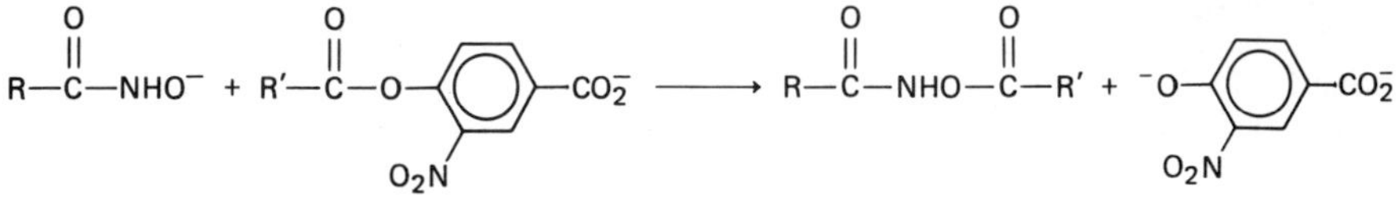

(25)

R = *n*-Pr, *n*-C$_5$H$_{11}$, *n*-C$_7$H$_{15}$, *n*-C$_9$H$_{19}$

R' = Me, *n*-Pr, *n*-C$_5$H$_{11}$, *n*-C$_7$H$_{15}$, *n*-C$_9$H$_{19}$, *n*-C$_{11}$H$_{23}$, *n*-C$_{13}$H$_{27}$

hydrophobic residues R and R' in either the nucleophile or the electrophile can bring reactants together in a comicelle. The rate enhancements are up to 500-fold on the addition of CTABr, and as expected they depend critically upon the lengths of the alkyl groups R and R'. An interesting observation is that the phase-transfer catalyst trioctylmethylammonium chloride strongly enhances the reactivity of the hydroxamate derivative in the presence of non-ionic micelles[164,181]. Trioctylmethylammonium chloride does not micellize, but it apparently aggregates with non-ionic surfactants. the rate enhancement is ascribed to the formation of dehydrated hydrophibic ion pairs, making the anions much more reactive than in bulk water. This suggestion appears to be contrary to evidence that hydrophilic ions are fully hydrated in the Stern layer in ionic micelles, and resolution of this question is needed. Alternatively, the rate enhancement can be attributed to a proximity effect, a form of forced ion pairing brought about by binding to the non-ionic surfactant which bring the reagents together.

A functional surfactant which contains both hydroxamate and imidazole groups has been synthesized by Kunitake and his coworkers and is discussed in Section XII.

E. Hydroxylalkyl Derivatives

Cationic surfactants of the general structure **26** are effective catalysts of deacylation[127,183] and dephosphorylation[94,184] and the hydroxy moiety reacts nucleophilically with alkyl[185] or aryl halides[186] and with carbocations[187]. The surfactants are generally choline derivatives, R = H, $n = 1$, but ephedrine derivatives have been used in some experiments[185], and homocholine derivatives, R = H, $n = 2$, have also been studied[127].

$$R'\overset{+}{N}Me_2CR_2(CR_2)_nOH \quad Br^-$$

(26)

There is compelling evidence that at high pH the hydroxy group generates the alkoxide moiety which is an effective nucleophile. Evidence which excludes general acid or base comes not only from deacylation, but also from reactions with phosphate esters and aryl halides, and these will be considered together.

$$n\text{-}C_{16}H_{33}\overset{+}{N}Me_2CH_2CH_2OH + {}^-OH \rightleftharpoons n\text{-}C_{16}H_{33}\overset{+}{N}Me_2CH_2CH_2O^- + H_2O$$

Covalent intermediates have been detected kinetically and spectrophotometrically. For example, the deacylation of p-nitrophenyl acetate is catalysed by micelles of $n\text{-}C_{18}H_{37}\overset{+}{N}Me_2CH_2CH_2OH$, and with substrate in large excess over surfactant an initial burst of p-nitrophenoxide ion is observed followed by a slow 'turnover' step which is deacylation of the catalyst[87a]. This approach is similar to the typical 'burst' experiment in enzyme kinetics.

Formation of the aryl ether **27** was demonstrated spectrophotometrically and kinetically in reactions of 2,4-dinitrofluoro- and chlorobenzene in dilute aqueous alkali (Scheme 18)[186]. The rate constants for the formation and decomposition of

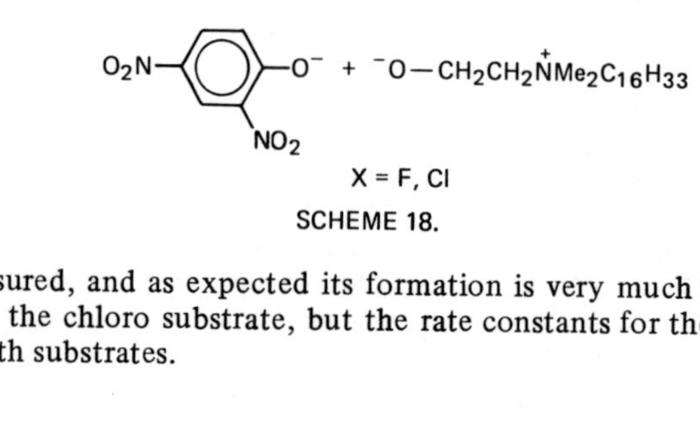

X = F, Cl

SCHEME 18.

27 can be measured, and as expected its formation is very much more rapid with the fluoro than the chloro substrate, but the rate constants for the second step are the same for both substrates.

Other evidence for nucleophilic attack by the hydroxy alkyl group comes from reactions of the cationic substrates **28** and **30** (Schemes 19 and 20). The hydroxy-

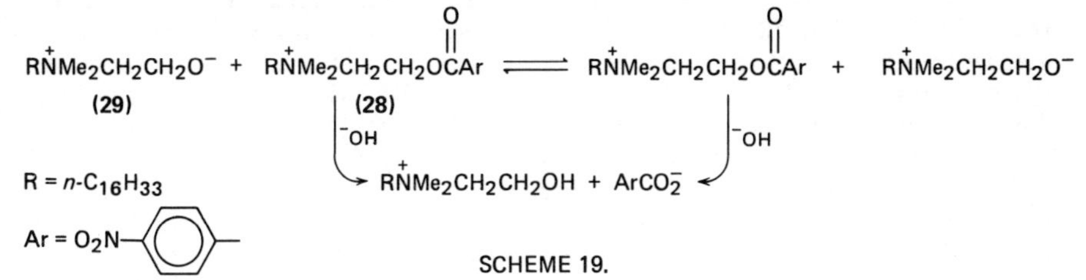

$$R\overset{+}{N}Me_2CH_2CH_2O^- + R\overset{+}{N}Me_2CH_2CH_2O\overset{O}{\overset{\|}{C}}Ar \rightleftharpoons R\overset{+}{N}Me_2CH_2CH_2O\overset{O}{\overset{\|}{C}}Ar + R\overset{+}{N}Me_2CH_2CH_2O^-$$

(29)   (28)   (29)

R = n-C$_{16}$H$_{33}$

$^-$OH ↓   ↓ $^-$OH

R$\overset{+}{N}$Me$_2$CH$_2$CH$_2$OH + ArCO$_2^-$

Ar = O$_2$N—⟨benzene⟩—

SCHEME 19.

$$R\overset{+}{N}Me_2CH_2CH_2O^- + R\overset{+}{N}Me_2CH_2CH_2CH_2O\overset{O}{\overset{\|}{C}}Ar \rightleftharpoons R\overset{+}{N}Me_2CH_2CH_2O\overset{O}{\overset{\|}{C}}Ar + R\overset{+}{N}Me_2CH_2CH_2CH_2O^-$$

(30)   (31)

↓ $^-$OH   ↓ $^-$OH

ArCO$_2^-$ + R$\overset{+}{N}$Me$_2$CH$_2$CH$_2$CH$_2$O$^-$ R$\overset{+}{N}$Me$_2$CH$_2$CH$_2$O$^-$ + ArCO$_2^-$

R = n-C$_{10}$H$_{21}$

Ar = O$_2$N—⟨benzene⟩—
                    NO$_2$

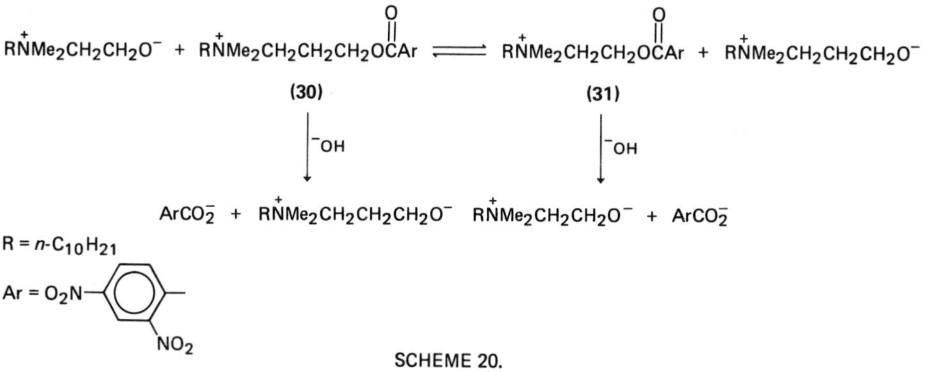

SCHEME 20.

ethyl surfactant **29** is a relatively ineffective catalyst for deacylation of **28**, simply because nucleophilic attack by **29** gives no chemical change, and the products arise only by attack of hydroxide ion[159a]. This evidence appears to exclude general acid or base catalysis by the hydroxyethyl surfactant. However, in a rather similar system, using the homocholine derivative **30** as a substrate, transesterification was observed kinetically and the final products were shown to be derived from attack of hydroxide ion on **30** and **31**[127a].

Additional evidence for formation of the reactive zwitterion **33** by acid dissociation of **32** is provided by the values of the solvent deuterium isotope effect for dephosphorylation (Scheme 21)[188]. Deuterium solvent isotope effects are gener-

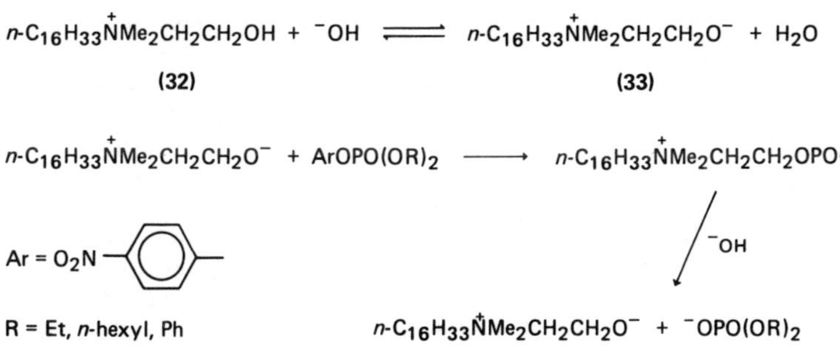

$$n\text{-}C_{16}H_{33}\overset{+}{N}Me_2CH_2CH_2OH + {}^-OH \rightleftharpoons n\text{-}C_{16}H_{33}\overset{+}{N}Me_2CH_2CH_2O^- + H_2O$$

(32)   (33)

$$n\text{-}C_{16}H_{33}\overset{+}{N}Me_2CH_2CH_2O^- + ArOPO(OR)_2 \longrightarrow n\text{-}C_{16}H_{33}\overset{+}{N}Me_2CH_2CH_2OPO(OR)_2 + ArO^-$$

Ar = O$_2$N—⟨benzene⟩—

R = Et, n-hexyl, Ph

↓ $^-$OH

$$n\text{-}C_{16}H_{33}\overset{+}{N}Me_2CH_2CH_2O^- + {}^-OPO(OR)_2$$

SCHEME 21.

ally large, with $k_{H_2O}/k_{D_2O} > 2.5$, for general acid- or base-catalysed reactions in which there is proton transfer in the rate-limiting step. However, the isotope effects are small and inverse for dephosphorylation of micellar bound **34** in dilute alkali. There should be a small inverse isotope effect on the formation of the zwitterion, and with the n-hexyl derivative $k_{H_2O}/k_{D_2O} = 0.75$ in 0.01 M OH⁻ where the hydroxyl group is only partially ionized. In 0.2 M OH⁻ where this group should be almost completely ionized, this inverse istope effect should have disappeared and in the reaction with the n-hexyl derivative $k_{H_2O}/k_{D_2O} \approx 1$

In the foregoing discussion it has been assumed implicitly that zwitterions are present in these micellized hydroxyalkyl surfactants in mildly alkaline solution. The pK_a of choline is 13.9 and micellization should increase the acidity of the hydroxy group. A γ-hydroxy surfactant should be less acidic than the corresponding β-derivative, but these surfactants, e.g. n-$C_{10}H_{21}NMe_2(CH_2)_3OH$, are also effective nucleophiles[127], suggesting that here too there is some acid dissociation of the hydroxy group, although presumably less than with the corresponding β-derivatives*.

It is difficult to isolate the general factors which contribute to catalysis by these functional micelles. The reaction in most cases involves nucleophilic attack by an anionic moiety, e.g. oximate or alkoxide ion, generated by acid dissociation, and this dissociation should be increased by incorporation into a cationic micelle.

For example, the pK_a of the octadecyl surfactant **34** has been estimated as

$$C_{18}H_{37}\overset{+}{N}Et_2CH_2CH_2OH \quad Br^-$$

(34)

10.5[87a]. This value is much lower than that of ca 12.3 estimated for the hexadecyl derivative[94,184], suggesting that increasing the length of the n-alkyl group, and the hydrophobicity of the groups at the cationic centre, increases the acid dissociation of the head group, resulting in greater rate enhancements. Also, as we have mentioned before, comparison of the reactivity of a micellized nucleophilic anion such as **33** with that of hydroxide ion is very difficult. Allowance must be made for the differences in distribution of substrate and nucleophile between the micellar and aqueous phases in the two systems. Thus, comparisons of raw rate data are not particularly helpful.

An alternative approach, used with the C_{16} hydroxyethyl surfactant, was to work under conditions in which the substrate should be wholly micellar-bound and to estimate the fraction of the micelle which is present as the reactive zwitterion[94,184] using the apparent pK_a. The observed first-order rate constant in a (hypothetical) zwitterionic micelle can then be compared with the second-order rate constant for reactions of hydroxide ion in water. In comparing reactivities for intra- and inter-molecular reactions one usually divides the first- by the second-order rate constants and expresses the reactivity difference in terms of concentration of the nucleophile or base, which is in this case the hydroxide ion[16]. The rate enhancements calculated in this way range from 4 M for deacylation of the p-nitrobenzoyl phosphate dianion (Section X.C) to 410 M for the decomposition of

---

*There are problems in defining pK_a in micelles as we have noted several times earlier, in part because of the uncertainty in the distribution of ions between micelles and bulk water, but also because of the change in composition of the surface. In this case the surface of micellized β-hydroxyethyl surfactant is transformed from cationic to zwitterionic (see also discussion on Scheme 11, Reaction 4, Section XI. A).

$$(R\overset{+}{N}Me_2CH_2CH_2OH)_n \;\rightleftharpoons\; (R\overset{+}{N}Me_2CH_2CH_2OH)_{n-m} \quad (R\overset{+}{N}Me_2CH_2CH_2O^-)_m + H^+$$

2,4-dinitrochlorobenzene[189]. These comparisons show that micellized hydroxy-alkyl surfactants are much more effective reagents than hydroxide ion in the presence of micellized non-functional surfactants, e.g. of the n-alkyltrimethyl-ammonium halides.

XII. THE QUESTION OF BIFUNCTIONAL CATALYSIS

The possibility of bifunctional catalysis has intrigued solution kineticists for many years, and although many possible systems have been studied there appears to be no certainty regarding bifunctional catalysis of reactions in homogeneous solutions (cf. Reference 190). Nevertheless, this mode of catalysis is often postulated for enzymic reactions[191].

Many reactions, especially those catalysed by acids or bases, involve the creation of both acidic and basic centres during transition-state formation, and it seems not unreasonable to expect that a general base might abstract a proton from an acidic centre to generate a nucleophile which adds to an electrophilic centre. For example, Scheme 22 shows a proposed model for bifunctional catalysis of ester deacylation

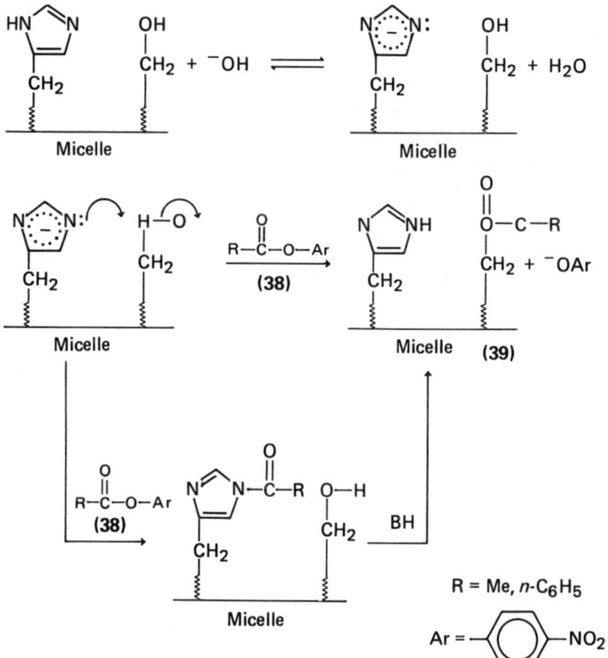

SCHEME 22.

at the micellar surface[192,193]. However, there is considerable evidence which suggests that these transfers are generally consecutive rather than concerted, probably because the unfavourable entropy change required to incorporate a third reactant into the transition state offsets any benefits from increased covalent participation*.

*It has been suggested that even S_N2 and E2 reactions of alkyl halides and related compounds are in fact stepwise and the arguments against concertedness have been cogently developed[194].

A major source of the micellar catalysis is the partial avoidance of the entropy loss in bringing reactants together in the transition state; thus it should be easy to obtain concerted bifunctional catalysis in a micelle. For example, the powerful catalysis of mutarotation in reverse micelles of alkylammonium carboxylates has been interpreted in terms of concerted proton transfers[195].

The question of bifunctional catalysis of deacylation was answered in part using normal micelles of surfactants which contain imidazole and β-hydroxyethyl groups, and comparing the reactivity of the bifunctional surfactant 37 with mixed micelles of the monofunctional surfactants, 35 and 36[192,193]. All the functional surfactants

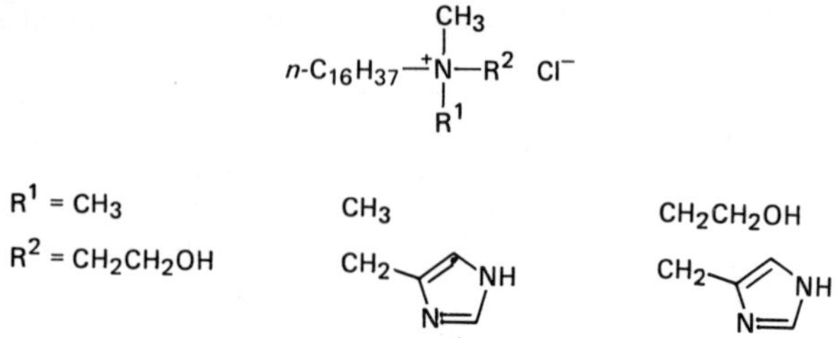

$$n\text{-}C_{16}H_{37}-\overset{\overset{\displaystyle CH_3}{|}}{\underset{\underset{\displaystyle R^1}{|}}{N}}{}^{+}-R^2 \quad Cl^-$$

R¹ = CH₃
R² = CH₂CH₂OH

(35)

CH₃

CH₂ imidazole

(36)

CH₂CH₂OH

CH₂ imidazole

(37)

are effective deacylating agents. The rapid formation of a new covalency by nucleophilic attack on the acyl group was demonstrated by observing a 'burst' of p-nitrophenoxide ion when 38, in excess, is allowed to react with the surfactant (Scheme 22). The final product (39) is formed by O-acylation, which suggests that the imidazole group accepts a proton from the hydroxyl group which attacks the acyl centre. However, the reaction is actually stepwise and an acyl imidazole is initially formed, which in a second slow step acylates the hydroxyl group to form the relatively unreactive ester 39.

The evidence that an acyl imidazole is formed in the rate-limiting step of the reaction is consistent with the observation that potentially bifunctional nucleophilic surfactants such as 37 are poorer deacylating agents than the corresponding imidazole derivatives 36, simply because the density of reactive imidazole groups at the micellar surface is greater with 36 than with 37.

Comicelles of CTABr and the functional surfactant 40 are also excellent catalysts of the deacylation of p-nitrophenyl acetate[196]. Reaction between excess

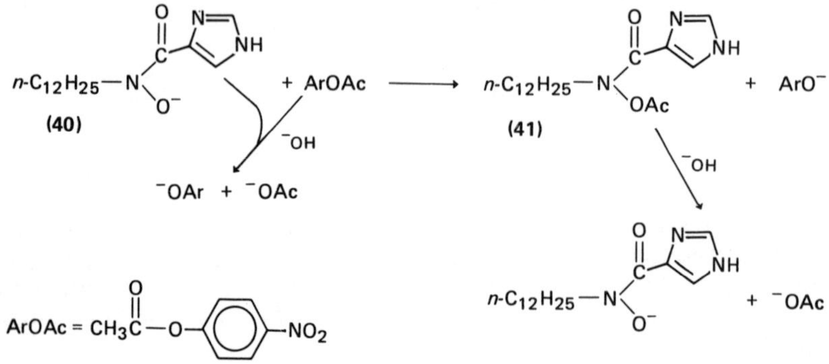

SCHEME 23.

p-nitrophenyl acetate and **40** gives a rapid burst of p-nitrophenoxide ion followed by a slower hydrolysis of the acyl intermediate, **41**. It is suggested that at pH 8.1 the reaction occurs by initial acylation of the hydroxamate function followed by slower deacylation, while at higher pH there is a contribution from a direct reaction to products involving hydroxide ion (Scheme 23), although even then the stepwise reaction predominates.

Even though the postulated direct reaction of hydroxide ion makes only a minor contribution, the surfactant which contains both the hydroxamate and imidazole functions is a slightly better catalyst than the related hydroxamate, **42**. To this extent the evidence for bifunctional catalysis is stronger in this case than in those discussed earlier.

$$CH_3(CH_2)_{11} - \overset{\underset{\displaystyle |}{OH}}{N} - \overset{\overset{\displaystyle O}{\|}}{C} - \bigcirc$$

(42)

The available evidence suggests that it is no easier to observe bifunctional catalysis in micelles than in other model systems. Catalysis with concerted bond-making and -breaking probably imposes geometrical constraints at the reaction centre(s) and upon the positions of the reacting groups which are entropically very unfavourable.

XIII. CONCLUSIONS, CONNECTIONS AND CONJECTURES

The pseudo-phase model of micellar catalysis appears to be applicable, at least qualitatively, to reactions in aqueous surfactant solutions, provided that one takes into account the possibility of induced micellization or the formation of sub-micellar aggregates in the presence of hydrophobic solutes or reactants. A major source of the rate enhancements of bimolecular reactions by both functional and non-functional micelles is the concentration of reactants at the water—micelle interface. However, the significance of quantitative comparisons of second-order rate constants in water and the micellar pseudo-phase is obscured because the comparisons inevitably depend on the choice of concentration units and hence on the assumed volume element in which the micellar reaction takes place.

In principle the concentrations of reactant in the micellar pseudo-phase can be estimated by direct measurement, but there are often serious experimental and theoretical problems, especially for nucleophilic anions generated by acid dissociation. It is difficult to estimate directly the concentration of hydrophilic ions in the micellar pseudo-phase, except in the cases where ion-sensitive electrodes can be used, e.g. for hydrogen or bromide ions. However, the ion-exchange model accounts, at least qualitatively, for the rate—surfactant and rate—counterion profiles of bimolecular reactions between hydrophilic ions and organic substrates.

Although we believe that concentration of reactants at the micellar surface is of key importance we cannot neglect the properties of this surface as a submicroscopic solvent. There is considerable evidence that the micellar surface has a polarity similar to those of protein surfaces, which is less than that of water, but similar to that of ethanol. If the micelle surface acts primarily as a reaction medium of lower polarity than water, then micelles should speed unimolecular reactions such as anionic decarboxylations and the decomposition of monoaryl phosphates and sulphates, but slow S_N1 reactions. This submicroscopic medium effect should also reduce the second-order rate constants in the micellar phase for reactions between non-ionic reactants. All these rate effects have been observed.

Added electrolytes strongly influence micellar catalysis, both by excluding ionic reagents from the micelle, and by changing the properties of the surface. To date this second effect has only been unambiguously observed with unimolecular reactions[92], but it should be possible to alter the rates of bimolecular reactions in the micelle in the same way.

Even if completely general models for micelle-catalysed reactions are never developed, it should be possible to make the experimental controls sufficiently precise to permit meaningful comparisons between various systems, and thus separate the concentration and medium effects of micelles on second- and higher-order reactions. Once this is accomplished, reaction kinetics becomes a powerful probe of micelle structure, sensitive to changes in both the medium and the relative concentrations of competitively binding species.

The relationships between micellar catalysts and other catalytic species such as small non-micellar aggregates and phase-transfer catalysts remain unexplored. Micellar catalysis is a surface phenomenon whereas phase-transfer catalysis[197,198] appears to depend on the volume of the organic phase, even though both systems are similar in that the catalysts speed reactions involving basic or nucleophilic ions and both use hydrophobic cations as catalysts. In very dilute solutions at concentrations near to the cmc, there is a grey area in which it is very difficult to distinguish between catalysis by micelles and by small non-micellar aggregates. In both systems catalysis can originate from concentration of the reactants into a small volume element, but there is no general agreement on the extent to which this is the major source of the rate increase in non-micellar systems.

On a purely practical note it is difficult in preparative chemistry to take advantage of the large rate enhancements or the product specificities sometimes provided by micelles, in part because surfactants are generally high molecular weight compounds whose presence complicates product isolation. But micellized surfactants are often excellent inhibitors and it is possible to use them to eliminate undesired reactions; for example, to stabilize drugs in aqueous solution[35], and to control rates of polymerization[199].

Micelle-catalysed reactions and micellar solutions in general will also be of continuing interest as models for other systems. At first glance the spontaneous formation of micelles appears contrary to the second law of thermodynamics; as though a Maxwell's demon had created ordered aggregates out of a randomly organized surfactant solution. This apparent contradiction is resolved when one recognizes the unique properties of water as a solvent, with micelle formation resulting in an increase in entropy of the whole system and therefore permitting the increased order on the submicroscopic level. The driving force for the reduction in the hydrocarbon—water interfacial area is the decrease in the amount of ordered water structure around the individual hydrocarbon chains. This unique property of aqueous solutions depends on the three-dimensional hydrogen-bonding properties of the water molecule, and is also responsible, in part, for the stability of bilayers[1] and the spontaneous refolding of denatured proteins[200].

Micelle formation is one of the simplest examples of the spontaneous formation of non-covalently bound higher ordered structures from simple molecules in a homogeneous solution. The formation of any higher ordered structure means that new properties are created and consequently new information can be transmitted. For example, micelle formation permits a dramatic increase in the solubility of solutes which are sparingly soluble in water or the development of modest specificity and selectivity for reactions and reactants where none existed before. This spontaneous formation of interfaces has obvious parallels to, and might provide simple models for, theories of the origin of life[201], while a micelle's catalytic and solubility properties can be related to the fact that most living processes occur at interfaces and not in bulk solution[1].

Micelles and miceller catalysed reactions are in some respects experimentally better model systems for enzymes and membranes than are soluble polyelectrolytes, resins or monolayers. Unlike the situation with monolayers, clean uniform surfaces are readily obtained in micellar solutions and reactions are easily followed. Unlike the situation with soluble polyelectrolytes or resins, the composition of the surface can be easily modified. For example, bifunctional micellar interfaces can be created simply by mixing two functional surfactants, and in addition the nature of the interface can be changed by varying the relative surfactant concentrations.

One drawback of micellar surfaces and soluble polyelectrolytes as compared to monolayers and resins is that the concentrations of various species at the surface cannot always be measured as easily because micelles and polyelectrolytes are not separate phases which can be mechanically separated. Another problem in the derivation of general treatments of micellar catalysis is that solutes may significantly perturb micellar structure. To this extent micelles may present more structural complications than polyelectrolytes (and resins) in which the covalent skeleton gives some rigidity to the overall structure. The differences may prove to be more apparent than real, however, because polyelectrolytes (and resins) are conformationally mobile and are therefore also perturbed to some extent by added solutes.

The binding of the substrate to a micelle, or to an enzyme, leads to an increase of the local concentration of the reactants, which manifests itself as a more favourable entropy of activation of a bimolecular reaction*. The importance of entropy terms in the rate enhancements of bimolecular reactions has been cogently developed in recent reviews[9,202,203].

Another possibility is that binding between the substrate and the micelle may bring the reactive portion of the substrate into a region in which there are unfavourable initial-state interactions which are relieved in forming the transition state. For example, a micellar bound anion may be partially or completely desolvated, although the evidence on this point is uncertain.

Much of the work on functional micelles was based on the hope of modelling the active sites of enzymes, and several examples of catalysis by functional micelles have rate enhancements with magnitudes in the range of those found for enzymic catalysis. It is sobering to remember, however, that in each one of these cases the origins of the catalytic activity have not been clearly determined and that micelles, unlike enzymes, are unspecific catalysts.

Finally, the interaction of ionic and non-ionic solutes with surfactants and each other at micelle surfaces may serve as simple models for the much more complex and specific interactions which occur at membrane and protein interfaces.

XIV. ACKNOWLEDGMENTS

Some of our work described in this chapter has been supported by the National Science Foundation and the Arthritis, Digestive and Metabolic Diseases Institute of the United States Public Health Service, and this support is gratefully acknowledged. We thank workers for provided unpublished results cited here. The manuscript was carefully typed by Alyson Posella, to whom we are most grateful.

*This comment assumes that the second-order rate constant is calculated in terms of the total concentration of the reactants averaged over the whole solution volume.

XV. REFERENCES

1. M. K. Jain, *The Bimolecular Lipid Membrane*, Van Nostrand Reinhold Co., New York, 1972.
2. C. A. Bunton, *Prog. Solid State Chem.*, 8, 239 (1973).
3. E. H. Cordes and R. B. Dunlap, *Accs. Chem. Res.*, 2, 329 (1969).
4. E. H. Cordes and C. Gitler, *Prog. Bioorg. Chem.*, 2, 1 (1973).
5. E. H. Cordes (Ed.), *Reaction Kinetics in Micelles*, Plenum Press, New York, 1973.
6. L. S. Romsted, *Ph.D. Thesis*, Indiana University, 1975.
7. C. A. Bunton in 'Micellar Reactions', *Applications of Biomedical Systems in Chemistry*, Part II (Ed. J. B. Jones), John Wiley and Sons, New York, 1976, p. 731.
8. I. V. Berezin, K. Martinek and A. K. Yatsimirski, *Russ. Chem. Rev.*, 42, 787 (1973).
9. W. P. Jencks in *Advances in Enzymology* (Ed. A. Meister), John Wiley and Sons, New York, 1975, p. 219.
10. H. Morawetz, *Adv. in Catalysis*, 20, 341 (1969).
11. J. H. Fendler and E. J. Fendler, *Catalysis in Micellar and Macromolecular Systems*, Academic Press, New York, 1975.
12. K. Shirahama, *Bull. Chem. Soc. Japan*, 48, 2673 (1975).
13. S. J. Dougherty and J. C. Berg, *J. Coll. Interface Sci.*, 49, 135 (1974).
14. K. Martinek, A. K. Yatsimirski, A. V. Levashov and I. V. Berezin in *Micellization, Solubilization and Microemulsions* (Ed. K. L. Mittal), Vol. 2, Plenum Press, New York, 1977, p. 489.
15. L. S. Romsted in *Micellization, Solubilization and Microemulsions* (Ed. K. L. Mittal), Vol. 2, Plenum Press, New York, 1977, p. 509.
16. W. P. Jencks, *Catalysis in Chemistry and Enzymology*, McGraw-Hill, New York, 1969.
17. D. Piszkiewicz, *Kinetics of Chemical and Enzyme Catalyzed Reactions*, Oxford University Press, New York, 1977.
18. J. March, *Advanced Organic Chemistry*, 2nd Ed., McGraw-Hill, New York, 1977.
19. T. C. Bruice and S. Benkovic, *Bioorganic Mechanisms*, Vol. 1, Benjamin, New York, 1966, p. 1.
20. K. Martinek, A. P. Osipov, A. K. Yatsimirski and I. V. Berezin, *Tetrahedron*, 31, 709 (1975).
21. D. Oakenfull, *J. Chem. Soc., Perkin II*, 1006 (1973).
22. L. R. Romsted and E. H. Cordes, *J. Amer. Chem. Soc.*, 90, 4404 (1968).
23. M. T. A. Behme, J. G. Fullington, R. Noel and E. H. Cordes, *J. Amer. Chem. Soc.*, 87, 266 (1965).
24. M. Chevion, J. Katzhendler and S. Sarel, *Israel J. Chem.*, 10, 975 (1972).
25. F. M. Menger and M. J. McCreery, *J. Amer. Chem. Soc.*, 96, 121 (1974).
26. K. Shinoda, T. Nakagawa, B.-I. Tamamushi and T. Isemura, *Colloidal Surfactants*, Academic Press, New York, 1963.
27. G. C. Kresheck in *Water: A Comprehensive Treatise* (Ed. F. Franks), Vol. 4, Plenum Press, New York, 1975, p. 95.
28. M. F. Emerson and A. Holtzer, *J. Phys. Chem.*, 71, 3320 (1967).
29. P. Mukerjee and A. Ray, *J. Phys. Chem.*, 70, 2150 (1966).
30. L. R. Fisher and D. G. Oakenfull, *Quart. Rev. Chem. Soc.*, 6, 25 (1977).
31. C. Tanford, *The Hydrophobic Effect*, John Wiley and Sons, New York, 1973.
32. K. L. Mittal (Ed.), *Micellization, Solubilization and Microemulsions*, Vol. 1 and 2, Plenum Press, New York, 1977.
33. P. Mukerjee, *Adv. Colloid Interface Sci.*, 1, 241 (1967).
34. P. Mukerjee and K. J. Mysels, *Critical Micelle Concentrations of Aqueous Surfactant Systems*, National Bureau of Standards, U.S. Government Printing Office, 1970.
35. P. H. Elworthy, A. T. Florence and C. B. MacFarlane, *Solubilization by Surface Active Agents and Its Application in Chemistry and the Biological Sciences*, Chapman and Hall, London, 1968.
36. P. Mukerjee, in Reference 32, Vol. 1, p. 177.
37. C. Tanford, in Reference 32, Vol. 1, p. 119.
38. C. Tanford, *J. Phys. Chem.*, 76, 3020 (1972).

39. C. Tanford, *J. Phys. Chem.*, **78**, 2469 (1974).
40. C. Tanford, *Proc. Nat. Acad. Sci. USA*, **71**, 1811 (1974).
41. J. N. Israelachvili, D. J. Mitchell and B. W. Ninham, *J. Chem. Soc., Faraday Trans. II*, 1525 (1976).
42. F. Reiss-Husson and V. Luzatti, *J. Phys. Chem.*, **68**, 3504 (1964).
43. N. A. Mazer, M. C. Carey and G. B. Benedek, in Reference 32, Vol. 1, p. 359.
44. D. Stigter, *J. Phys. Chem.*, **70**, 1323 (1966).
45. A. Ray, *J. Amer. Chem. Soc.*, **91**, 6511 (1969).
46. A. Ray, *Nature*, **231**, 313 (1971).
47. A. Ray and G. Nemethy, *J. Phys. Chem.*, **75**, 809 (1971).
48. Recently micelle formation was also observed in sulphuric acid with long chain alkyl methyl ethers. Surface tension – ether concentration profiles showed the sharp break characteristic of micelle formation. The micelle forming species was presumed to be the protonated form of the ether. Private communication, F. M. Menger.
49. U. P. Strauss and S. S. Slowata, *J. Phys. Chem.*, **61**, 411 (1957).
50. H. H. Freedman, J. P. Mason and A. I. Medalia, *J. Org. Chem.*, **23**, 76 (1958).
51. T. Rodulfo, J. A. Hamilton and E. H. Cordes, *J. Org. Chem.*, **39**, 2281 (1974).
52. M. E. Shick (Ed.), *Nonionic Surfactants*, Marcel Dekker, New York 1967.
53. K. Shinoda, *Proc. IVth Intern. Cong. Sur. Act. Sub., Brussels*, **2**, 527 (1964).
54. G. Gouy, *Ann. Phys.*, **7**, 129 (1917).
55. D. C. Grahame, *Chem. Rev.*, **41**, 441 (1947).
56. J. Th. G. Overbeek in *Colloid Science* (Ed. H. R. Krugt), Elsevier Publishing Co., Brussels, 1952, p. 115.
57. J. T. Davies and E. K. Rideal, *Interfacial Phenomena*, Academic Press, New York, 1961, p. 85.
58. M. J. Sparnaay, 'The Electrical Double Layer', *The International Encyclopedia of Physical Chemistry and Chemical Physics*, Vol. 4, Topic 14, Pergamon Press, New York, 1972.
59a. D. Stigter, *J. Phys. Chem.*, **68**, 3603 (1964).
59b. D. Stigter, *J. Colloid Interface Sci.*, **23**, 379 (1967).
59c. D. Stigter, *J. Colloid Interface Sci.*, **47**, 473 (1974).
59d. D. Stigter, *J. Phys. Chem.*, **78**, 2480 (1974).
59e. D. Stigter, *J. Phys. Chem.*, **79**, 1008, 1015 (1975).
60. D. Stigter and K. J. Mysels, *J. Phys. Chem.*, **59**, 45 (1955).
61. D. Stigter, in the press.
62. I. S. Scarpa, H. C. Kiefer and I. M. Klotz, *Intra-Science Chem. Rept.*, **8**, 45 (1974).
63. C. G. Overberger and J. C. Salamone, *Acc. Chem. Res.*, **2**, 217 (1969).
64. V. Gold, C. J. Liddiard and G. D. Morgan in *Proton Transfer Reactions* (Ed. E. F. Caldin and V. Gold), Chapman and Hall, London, 1975, p. 409.
65. F. M. Menger and C. E. Portnoy, *J. Amer. Chem. Soc.*, **89**, 4698 (1967).
66. C. A. Bunton, A. Kamego, M. J. Minch and J. L. Wright, *J. Org. Chem.*, **40**, 1321 (1975).
67. C. A. Bunton, A. A. Kamego and P. Ng, *J. Org. Chem.*, **39**, 3469 (1974).
68. C. A. Lapinte and P. Viout, *Tetrahedron Letters*, 1113 (1973).
69. H. R. Mahler and E. H. Cordes, *Biological Chemistry*, 2nd Ed., Harper and Row, New York, 1971.
70. C. A. Bunton and B. Wolfe, *J. Amer. Chem. Soc.*, **95**, 3742 (1973).
71. D. G. Herries, W. Bishop and F. M. Richards, *J. Phys. Chem.*, **68**, 1842 (1964).
72. S. J. Dougherty and J. C. Berg, *J. Colloid Interface Sci.*, **48**, 110 (1974).
73a. C. A. Bunton and L. Robinson, *J. Amer. Chem. Soc.*, **90**, 5972 (1968).
73b. A. K. Yatsimirski, A. P. Osipov, K. Martinek and I. V. Berezin, *Kolloid. Z.*, **37**, 470 (1975).
74. L. Sepulveda, *J. Colloid Interface Sci.*, **46**, 372 (1974).
75. J. W. Larsen and L. J. Magid, *J. Amer. Chem. Soc.*, **96**, 5774 (1974).
76. L. Sepulveda, unpublished results.
77. J. W. Larsen and L. B. Tepley, *J. Org. Chem.*, **41**, 2968 (1976).
78. C. A. Bunton, K. Ohmenzetter and L. Sepulveda, *J. Phys. Chem.*, **81**, 2000 (1977).
79. L. Sepulveda, unpublished results.

80. H. Chaimovich, A. Blanco, L. Chayet, L. M. Costa, P. M. Monteiro, C. A. Bunton and C. Paik, *Tetrahedron*, **31**, 1139 (1975).
81. C. A. Bunton and L. Robinson, *J. Amer. Chem. Soc.*, **92**, 356 (1970).
82. A. K. Yatsimirski, K. Martinek and I. V. Berezin, *Dokl. Akad. Nauk. SSSR, Chem. Sec. (Engl.)*, **194**, 720 (1970).
83. A. K. Yatsimirski, K. Martinek and I. V. Berezin, *Tetrahedron*, **27**, 2855 (1971).
84. A. K. Yatsimirski, Z. A. Strel'tsova, K. Martinek and I. V. Berezin, *Kinetics and Catalysis*, **15**, 354 (1974).
85. K. Martinek, A. K. Yatsimirski, A. P. Osipov and I. V. Berezin, *Tetrahedron*, **29**, 963 (1973).
86. A. P. Osipov, K. Martinek, A. K. Yatsimirski and I. V. Berezin, *Dokl. Akad. Nauk. SSSR, Phys. Chem. Sec. (Engl.)*, **215**, 370 (1974).
87a. K. Martinek, A. V. Levashov and I. V. Berezin, *Tetrahedron Letters*, 1275 (1975).
87b. K. Martinek, A. P. Osipov, A. K. Yatsimirski, V. A. Dadali and I. V. Berezin, *Tetrahedron Letters*, 1279 (1975).
88. A. P. Osipov, K. Martinek, A. K. Yatsimirski and I. V. Berezin, *Izv. Akad. Nauk. SSSR, Ser. Khim.*, 1984 (1974).
89. P. Mukerjee and K. Banerjee, *J. Phys. Chem.*, **68**, 3567 (1964).
90. G. S. Hartley, *Trans. Faraday Soc.*, **30**, 444 (1934).
91. G. S. Hartley, *Quart. Rev., (London)*, **2**, 152 (1948).
92. C. A. Bunton in Reference 5, p. 73; C. A. Bunton, M. J. Minch, J. Hidalgo and L. Sepulveda, *J. Amer. Chem. Soc.*, **95**, 3262 (1973).
93. R. Kunin, *Ion Exchange Resins*, John Wiley and Sons, New York, 1958.
94. C. A. Bunton and L. G. Ionescu, *J. Amer. Chem. Soc.*, **95**, 2912 (1973).
95. C. A. Bunton and B. Wolfe, *J. Amer. Chem. Soc.*, **96**, 7747 (1974).
96. C. A. Bunton and L. Robinson, *J. Phys. Chem.*, **73**, 4237 (1969).
97. C. A. Bunton and L. Robinson, *J. Phys. Chem.*, **74**, 1062 (1970).
98. C. A. Bunton and M. J. Minch, *J. Phys. Chem.*, **78**, 1490 (1974).
99. C. A. Bunton and R. J. Rubin, *Tetrahedron Letters*, 55 (1975).
100. C. A. Bunton and R. J. Rubin, *J. Amer. Chem. Soc.*, **98**, 4236 (1976); C. A. Bunton, L. S. Romsted and H. J. Smith, *J. Org. Chem.*, **43**, 4299 (1978).
101. E. W. Anaker and H. M. Ghose, *J. Amer. Chem. Soc.*, **90**, 3161 (1968).
102. P. Mukerjee, K. J. Mysels and P. Kapuuan, *J. Phys. Chem.*, **71**, 4166 (1967).
103. R. B. Dunlap and E. H. Cordes, *J. Amer. Chem. Soc.*, **90**, 4395 (1968).
104. J. Baumrucker, M. Calzadilla, M. Centeno, G. Lehrman, M. Urdaneta, P. Lindquist, D. Dunham, M. Price, B. Sears and E. H. Cordes, *J. Amer. Chem. Soc.*, **94**, 8164 (1972).
105. R. N. Lindquist and E. H. Cordes, *J. Amer. Chem. Soc.*, **90**, 1269 (1968).
106. M. Plaisance and L. Ter-Minassian-Saraga, *J. Colloid Interface Sci.*, **56**, 33 (1976).
107. J. A. Marinsky (Ed.), *Ion Exchange*, Vol. 2, Marcel Dekker, New York, 1969.
108. J. Inczédy, *Analytical Applications of Ion Exchangers*, Pergamon Press, London, 1966.
109. Y. Marcus and A. S. Kertes, *Ion Exchange and Solvent Extraction of Metal Complexes*, John Wiley and Sons, New York, 1969, Chap. 4.
110. G. Eisenman, *Biophys. J.*, **2** (2), 259 (1962).
111. C. A. Bunton, L. S. Romsted and G. Savelli, *J. Amer. Chem. Soc.*, in press.
112. K. Shirahama, *Bull. Chem. Soc. Japan*, **48**, 2673 (1975).
113. D. Piszkiewicz, *J. Amer. Chem. Soc.*, **98**, 3053 (1976); **99**, 1550 (1977).
114. D. Piszkiewicz, *J. Amer. Chem. Soc.*, **99**, 7695 (1977).
115. C. A. Bunton and L. Robinson, *J. Org. Chem.*, **34**, 773 (1969).
116. C. A. Bunton, L. Robinson, J. Schaak and M. F. Stam, *J. Org. Chem.*, **36**, 2346 (1971).
117. T. Nakagawa in *Nonionic Surfactants* (Ed. M. J. Schick), Marcel Dekker, New York, 1967, Chap. 17, p. 558.
118. H. B. Klevens, *Chem. Rev.*, **47**, 1 (1950).
119. M. F. Emerson and A. Holtzer, *J. Phys. Chem.*, **71**, 3320 (1967).
120. R. B. Dunlap and E. H. Cordes, *J. Phys. Chem.*, **73**, 361 (1969).
121. A. S. C. Lawrence, B. Boffey, A. Bingham and K. Talbot, *Proc. IVth Intern. Congr. Sur. Act. Sub., Brussels*, **2**, 673 (1964).
122. F. Tokiwa and N. Moriyama, *J. Colloid Interface Sci.*, **30**, 338 (1969).

123. G. D. Parfitt and J. A. Wood, *Kolloid-Z., Z. Polymere,* **229**, 55 (1969).
124. J. W. Larsen and L. B. Tepley, *J. Colloid Interface Sci.,* **49**, 113 (1974).
125. J. Piercy, M. N. Jones and G. Ibbotson, *J. Colloid Interface Sci.,* **37**, 165 (1971).
126. B. W. Barry and E. Shotton, *J. Pharm. Pharmac.,* **19**, 785 (1967).
127a. R. Schiffman, M. Chevion, J. Katzhendler, Ch. Rav-Acha and S. Sarel, *J. Org. Chem.,* **42**, 856 (1977).
127b. R. Schiffman, Ch. Rav-Acha, M. Chevion, J. Katzhendler and S. Sarel, *J. Org. Chem.,* **42**, 3279 (1977).
128. C. A. Bunton, S. Diaz, J. M. Hellyer, Y. Ihara and L. G. Ionescu, *J. Org. Chem.,* **40**, 2313 (1975).
129. C. A. Bunton, A. A. Kamego, M. J. Minch and J. L. Wright, *J. Org. Chem.,* **40**, 1321 (1975).
130. J. L. Kurz, *J. Phys. Chem.,* **66**, 2239 (1962); V. A. Motsavage and H. B. Kostenbauder, *J. Colloid Sci.,* **18**, 603 (1963).
131. H. Nogami and Y. Kunakubo, *Chem. Pharm. Bull. Tokyo,* **11**, 943 (1963).
132. F. M. Menger and C. E. Portnoy, *J. Amer. Chem. Soc.,* **90**, 1875 (1968).
133. C. A. Bunton and M. McAneny, *J. Org. Chem.,* **41**, 36 (1976).
134. F. M. Menger, *J. Amer. Chem. Soc.,* **94**, 909 (1974).
135. F. M. Menger, J. A. Donahue and R. F. Williams, *J. Amer. Chem. Soc.,* **95**, 286 (1973).
136. F. M. Menger and A. C. Vitale, *J. Amer. Chem. Soc.,* **95**, 4931 (1973).
137. B. R. Vijayendran, *J. Colloid Interface Sci.,* **60**, 418 (1977).
138. H. Suzuki, *Bull. Chem. Soc. Japan,* **49**, 381 (1976).
139. C. A. Bunton, A. A. Kamego and P. Ng, *J. Org. Chem.,* **39**, 3469 (1974).
140. J. H. Fendler, E. J. Fendler and L. W. Smith, *J. Chem. Soc., Perkin II,* 2097 (1972).
141. R. A. Moss and D. W. Reger, *J. Amer. Chem. Soc.,* **91**, 7539 (1969).
142. R. A. Moss. C. J. Talkowski, D. W. Reger and C. E. Powell, *J. Amer. Chem. Soc.,* **95**, 3215 (1973).
143. K. Okamoto, T. Kinoshita and H. Yoneda, *J. Chem. Soc., Chem. Commun.,* 922 (1975).
144. C. N. Sukenik and R. G. Bergman, *J. Amer. Chem. Soc.,* **98**, 6613 (1976).
145. F. M. Menger, J. V. Rhee and H. K. Rhee, *J. Org. Chem.,* **40**, 3803 (1975).
146. J. M. Brown and J. A. Jenkins, *J. Chem. Soc., Chem. Commun.,* 458 (1976).
147. W. E. Meyers and G. P. Royer, *J. Amer. Chem. Soc.,* **99**, 6141 (1977).
148. S. Diaz, *Ph.D. Thesis,* University of California, Santa Barbara, 1977.
149. K. E. Mittal and P. Mukerjee in Reference 32, p. 1.
150. A. Williams and K. T. Douglas, *Chem. Rev.,* **75**, 627 (1975).
151. W. Tagaki, S. Kobayashi, K. Kunihara, A. Kurashima, Y. Yoshida and Y. Yano, *J. Chem. Soc., Chem. Commun.,* 843 (1976).
152. M. J. Minch. M. Giaccio and R. Wolff, *J. Amer. Chem. Soc.,* **97**, 3766 (1975).
153. C. A. Bunton and Y. Ihara, *J. Org. Chem.,* **42**, 2865 (1977).
154. V. Gani and C. Lapinte, *Tetrahedron Letters,* 2775 (1973).
155. V. Gani, C. Lapinte and P. Viout, *Tetrahedron Letters,* 4435 (1973).
156. T. J. Broxton, L. W. Deady and N. W. Duddy, *Australian J. Chem.,* **31**, 1525 (1978).
157. A. Cipiciani, P. Linda and G. Savelli, unpublished results.
158. K. J. Mollett and C. J. O'Connor, *J. Chem. Soc., Perkin II,* 369 (1976).
159a. C. A. Bunton and M. McAneny, *J. Org. Chem.,* **42**, 475 (1977).
159b. C. A. Bunton, E. J. Fendler, L. Sepulveda and K.-U. Yang, *J. Amer. Chem. Soc.,* **90**, 5512 (1968).
160. T. C. Bruice, J. Katzhendler and L. R. Fedor, *J. Amer. Chem. Soc.,* **90**, 1333 (1968).
161. C. A. Blyth and J. R. Knowles, *J. Amer. Chem. Soc.,* **93**, 3017, 3021 (1971).
162. J. P. Guthrie, *J. Chem. Soc., Chem. Commun.,* 897 (1972).
163. J. P. Guthrie and Y. Ueda, *Can. J. Chem.,* **54**, 2745 (1976).
164. Y. Okahata, R. Ando and T. Kunitake, *J. Amer. Chem. Soc.,* **99**, 3067 (1977).
165. C. Gitler and A. Ochoa-Solano, *J. Amer. Chem. Soc.,* **90**, 5004 (1968).
166. S. H. Yalkowsky and G. Zographi, *J. Pharm. Sci.,* **59**, 798 (1970).
167. S. H. Yalkowsky and G. Zographi, *J. Colloid Interface Sci.,* **34**, 525 (1970).
168. F. Tokiwa, *Adv. Colloid Interface Sci.,* **3**, 389 (1972).
169. P. Heitman, *European J. Biochem.,* **5**, 305 (1968).

170. W. Tagaki, T. Amada, Y. Yamashita and Y. Yano, *J. Chem. Soc., Chem. Commun.*, 1131 (1972).
171. Reference 19, Chap. 3.
172. R. G. Shorenstein, C. S. Pratt, C.-J. Hsu and T. E. Wagner, *J. Amer. Chem. Soc.,* **90**, 6199 (1968).
173. W. Tagaki, M. Chigira, T. Amada and Y. Yano, *J. Chem. Soc., Chem. Commun.*, 219 (1972).
174. P. Heitman, R. Husung-Bulblitz and H. J. Zunft, *Tetrahedron,* **30**, 4137 (1974).
175. J. M. Brown and C. A. Bunton, *J. Chem. Soc., Chem. Commun.*, 969 (1974).
176. R. A. Moss and W. L. Sunshine, *J. Org. Chem.,* **39**, 1083 (1974).
177. C. N. Sukenik, B.-A. Weissman and R. G. Bergman, *J. Amer. Chem. Soc.,* **97**, 445 (1975); C. N. Sukenik and R. G. Bergman, *J. Amer. Chem. Soc.,* **98**, 6613 (1976).
178. K. Okamoto, T. Kinoshita and H. Yoneda, *J. Chem. Soc., Chem. Commun.*, 922 (1975).
179. R. A. Moss, C. J. Talkowski, D. W. Reger and W. L. Sunshine in Reference 5, p. 99.
180. I. Tabushi, Y. Kuroda and S. Kita, *Tetrahedron Letters*, 643 (1974).
181. T. Kunitake, S. Shinkai and Y. Okahata, *Bull. Chem. Soc. Japan,* **49**, 540 (1976).
182. R. Ueoka, M. Kato and K. Ohkubo, *Tetrahedron Letters*, 2163 (1977).
183. M. Chevion, J. Katzhendler and S. Sarel, *Israel J. Chem.,* **10**, 975 (1972).
184. C. A. Bunton, L. Robinson and M. Stam, *J. Amer. Chem. Soc.,* **92**, 7393 (1970).
185. G. Meyer and P. Viout, *Tetrahedron,* **33**, 1959 (1977).
186. C. A. Bunton and S. Diaz, *J. Amer. Chem. Soc.,* **98**, 5663 (1976).
187. C. A. Bunton and C. H. Paik, *J. Org. Chem.,* **41**, 40 (1976).
188. C. A. Bunton and S. Diaz, *J. Org. Chem.,* **41**, 33 (1976).
189. C. A. Bunton, *Pure Appl. Chem.,* **49**, 969 (1977).
190. J. Hine and Wu-S. Li, *J. Amer. Chem. Soc.,* **98**, 3287 (1976).
191. T. H. Fife, *Adv. Phys. Org. Chem.,* **11**, 1 (1975).
192. R. A. Moss, R. C. Nahas and S. Ramaswami, *J. Amer. Chem. Soc.,* **99**, 627 (1977); R. A. Moss, R. C. Nahas, S. Ramaswami and W. J. Sanders, *Tetrahedron Letters*, 3379 (1975).
193. U. Tonellato, *J. Chem. Soc., Perkin II*, 822 (1977).
194. F. G. Bordwell, *Acc. Chem. Res.,* **3**, 281 (1970); **5**, 374 (1972).
195. E. J. Fendler, J. H. Fendler, R. T. Medary and V. A. Woods, *Chem. Commun.*, 1497 (1971); J. H. Fendler, E. J. Fendler, R. T. Medary and V. A. Woods, *J. Amer. Chem. Soc.,* **94**, 7288 (1972).
196. T. Kunitake, Y. Okahata and T. Sakumoto, *J. Amer. Chem. Soc.,* **98**, 7799 (1976).
197. J. E. Gordon and R. E. Kutina, *J. Amer. Chem. Soc.,* **99**, 3903 (1977).
198. M. Makosza, *Pure Appl. Chem.,* **43**, 439 (1975).
199. K. J. Lissant (Ed.) *Emulsions and Emulsion Technology*, Vol. 6, Parts 1 and 2, Marcel Dekker, New York, 1974.
200. A. L. Lehninger, *Biochemistry*, Worth, New York, 1970, p. 753.
201. F. Clark and R. L. M. Synge (Ed.), *The Origin of Life on Earth*, Pergamon Press, New York, 1959.
202. M. I. Page, *Angew. Chem. (Int. Ed. Engl.),* **16**, 449 (1977).
203. T. C. Bruice, *Ann. Rev. Biochem.,* **45**, 331 (1976).

CHAPTER **18**

The chemistry of thio acid derivatives

J. VOSS

Institute for Organic Chemistry and Biochemistry, University of Hamburg, Hamburg, Germany

I.	STRUCTURE AND PHYSICAL PROPERTIES	. 1022
	A. X-Ray and Electron Diffraction; Microwave Spectroscopy . .	. 1022
	B. Dipole Moments	. 1023
	C. Ultraviolet and Visible Spectra	. 1023
	D. Vibrational Spectra	. 1026
	1. Monothio carboxylic acids	. 1029
	2. Thiocarboxylate anions	. 1030
	3. Thiolo esters	. 1030
	4. Thiono esters	. 1030
	5. Dithio carboxylic acids	. 1030
	6. Dithio carboxylic acid esters	. 1031
	E. Nuclear Magnetic Resonance Spectra	. 1033
	1. Proton n.m.r. spectra	. 1033
	2. Carbon-13 n.m.r. spectra	. 1035
	F. Electron Spin Resonance Spectra	. 1035
	G. Mass Spectra	. 1035
II.	SYNTHESES	. 1036
	A. Thio and Dithio Carboxylic Acids and their Salts . . .	. 1036
	B. Thiolo Esters	. 1038
	C. Thiono Esters	. 1040
	D. Dithio Esters	. 1042
	E. Thioacyl Halides and Anhydrides	. 1044
III.	CHEMICAL PROPERTIES OF THIO CARBOXYLIC ACID DERIVATIVES	. 1044
	A. Prototropic Behaviour	. 1044
	B. Nucleophilic Reactions	. 1045
	1. Reduction	. 1045
	2. Reactions with carbanions	. 1047
	3. Solvolysis	. 1048
	C. Electrophilic Reactions	. 1048
	1. Oxidation	. 1048
	2. Alkylation and acylation	. 1049
	D. Non-polar Reactions	. 1051
	E. Formation of Heterocycles	. 1052
IV.	REFERENCES	. 1053

J. Voss

I. STRUCTURE AND PHYSICAL PROPERTIES

A. X-Ray and Electron Diffraction; Microwave Spectroscopy

Since Janssen's review of 1970[1] a number of investigations have revealed the geometric parameters, i.e. bond lengths and angles, of all types of thio acid derivatives. Representative data are given in Table 1.

Generally a completely planar skeleton of the functional group and the Z configuration is observed. This is in agreement with theoretical considerations, and with other spectroscopic data. Deviations from the normal configuration are observed in the gas phase: Methyl thionoformate exhibits a twisting angle of 15.8° between the Me—O—C— and the O—C=S plane[2], and a minor amount of the E form, which is less stable by 2.767 kJ mol^{-1} with respect to the Z form[3], is observed in the microwave spectrum of monothioformic acid[4]. The length of the C—Y single bond (Y = S or O) increases in the order: dithiocarboxylate < monothiocarboxylate < dithio ester < thiolo ester ≈ monothio carboxylic acid < thiophthalic acid anhydride. Only small differences, due to lattice effects, occur in the two C—S distances of metal dithiocarboxylates, which is evidence for the symmetric charge distribution in the anions. The C=S double bonds of the few thiono and dithio esters studied are almost equal in length. The mean value of ca 163 pm indicates a more pronounced double-bond character with respect to thio amides [d(C=S) = 167 pm].

The O=C—S, S=C—O and S=C—S bond angles are significantly larger than the expected value of 120°, whereas the R—Y—C angles increase in the order thiolo (100°) < dithio (105°) < thiono esters (120°).

TABLE 1. Bond length and angles in thio carboxylic acids and derivatives

R^1—CX—YR^2	Bond lengths (pm)			Angles (degrees)		Ref.
	C=X	C—Y	R^1—C	XCY	CYR^2	
H—CO—SH	121.8	176.3	110.0	126.0	111.8	4
	121.0	177.1^a	110.0	122.4^a	114.6	
Me—CO—S$^-$K$^+$	123.1	170.3	152.8	124.5		5
K^{+-}S—CO—CO—S$^-$K$^+$	122.7	171.2	151.6	126.1		6
MeS—CO—CO—S$^-$K$^+$	122.4	168.8	156.2	126.7		7
K$^+$S—CS—CO—S$^-$K^{+b}	126	170	152	126		8
	122	172	151	127		
H—CO—SMe	120			126	100	9
Ph$_2$CH—CO—SCH$_2$CH$_2$NEt$_2$	120.2	178.0	153.8	123.1	99.6	10
EtS—CO—CO—SEt	120.9	174.9	153.3	126.5	99.4	11
K^{+-}S—CO—CO—SMe	121.3	174.3	156.2	124.6	99.8	7
Ar—S—CO—CO—SAr^c	119.5	175.3	154.1	127.7	101.8	12
MeS—CS—CO—SMe	121.7	176.7	153.8	125.2		13
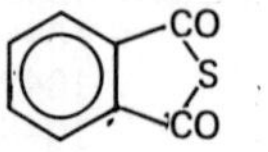	120.0	180.1	148.2	123.1	92.6	14
	119.4	180.1	147.9	124.7		
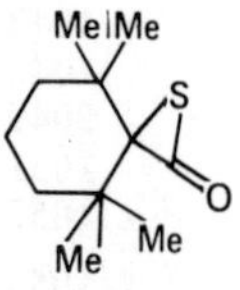	121.3	172.6	146.0	141.4	46.9	15
H—CS—OMe	161.2	136.9	111.4	126.6	115.5	2

TABLE 1 (*continued*)

R¹−CX−YR²	Bond lengths (pm)			Angles (degrees)		Ref.
	C=X	C−Y	R¹−C	XCY	CYR²	
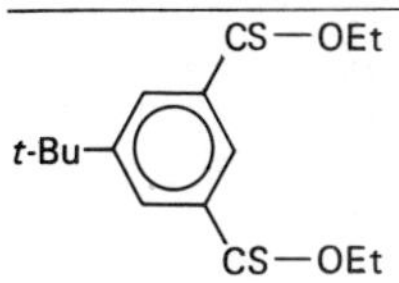 t-Bu〈phenyl〉 CS—OEt (top and bottom)	162.9	132.5	149.1	124.6	120.3	16
	163.3	132.1	148.9	124.8	119.6	
H−CS−S⁻K⁺	164.3	164.3		131.3		17
Me−CS−S⁻K⁺	167.1	167.1	140.5	123.5		18
NC−CS−S⁻NEt₄⁺	167.5	169.3	145.3	128.9		19
Cs⁺⁻O₂C−CS−S⁻Cs⁺	168	168	150.2	128.7		20
K⁺⁻S−CO−CS−S⁻K⁺ᵇ	162	171	152	130		8
	168	169	151	128		
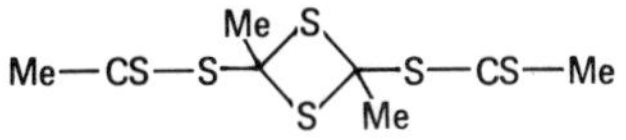 Me—CS—S〈ring〉S—CS—Me	163.1	170.5	149.5	125.4	104.9	21
	160.9	172.2	154.0	125.6	105.2	
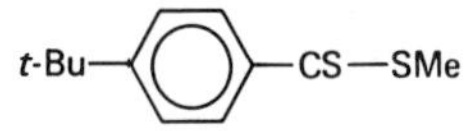 t-Bu〈phenyl〉—CS—SMe	163.0	172.4	148.5	124.2	104.4	22
Ph₃P=C(Ar)−CS−SEtᵈ	169.1	178.0	136.5	121.5	105.2	23
Me−S−CO−CS−SMe	163.1	171.9	153.8	128		13

[a]E configuration, cf. text.
[b]Two independent molecules in the elemental cell.
[c]Ar = 3-chlorophenyl.
[d]Ar = 4-nitrophenyl.

B. Dipole Moments

The dipole moments of all types of thio carboxylic acid derivatives have been studied (Table 2). The results of these measurements are consistent with a planar Z configuration of the molecules, which, as is well known[24], is observed in esters too. If the E configuration is enforced by cyclization, a marked increase of the dipole moment is observed. Polar substituents at the functional group such as trifluoromethyl, acyl and vinyl lead to an increase too, whereas the influence of phenyl rings is not quite straightforward.

Due to the increased atomic radius of sulphur its electronegativity as well as its polarizability is quite different from that of oxygen, and so are the respective bond moments of the C−O, C=O, C−S and C=S fragments. These increments, however, cannot be used to calculate dipole moments of thio acid derivatives by simple vector addition, because there are different mesomeric interactions between them in the various molecules. More reliable seem to be the group moments reported by Exner and coworkers[25] (Table 3).

C. Ultraviolet and Visible Spectra

Although the electronic spectra of some thio carboxylic acid derivatives have been well known for many years, some recent investigations ought to be

J. Voss

TABLE 2. Dipole moments of selected thio carboxylic acid derivatives

Compound	μ(D), in benzene	Reference
HCOSH	1.536^a; $2.868^{a,\,b}$	3
MeCOSH	2.16	26
n-C$_{17}$H$_{35}$COSH	2.14	26
Me$_2$CHCSSH	2.13	27
HCOSMe	1.58^a	9
MeCOSMe	1.43	28
n-C$_{17}$H$_{35}$COSMe	1.37	26
EtCOSEt	1.40	25
CF$_3$COSEt	3.07	29
CF$_3$COSBu-t	3.37	29
CF$_3$COSPh	2.95	29
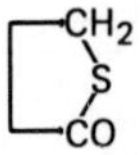	3.83	30
EtOCOCOSEt	1.52	31
EtSCOCOSEt	1.30	31
PhCOSMe	1.70	32
PhCOSEt	1.55	25
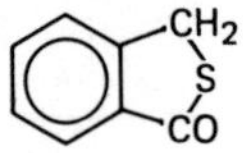	4.31	30
MeCSOEt	2.10	33
	2.22	34
PhCSOEt	2.24	33
	2.60	34
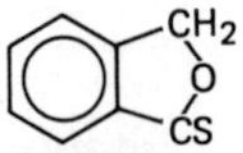	4.87	30
MeCSSME	1.87	35
n-C$_3$H$_7$CSSMe	1.76	28
(PhCH$_2$)$_2$NCPh=CHCSSMe	4.62	36
PhCSSEt	1.74	25
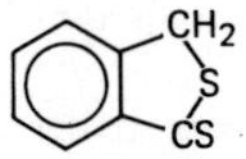	4.54	30
MeCOSCOMe	2.64	37
PhCOSCOPh	3.70	38
	3.88	32
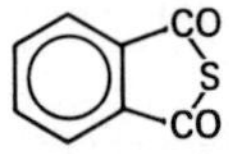	3.84	39

^{a}Determined from the microwave spectrum in the gas phase.
^{b}E configuration.

TABLE 3. Functional-group moments in esters and thio esters, ArX[25]

X	μ_X(D)	Angle with the Ar–C bond (degrees)
–CO–O–	1.89	115
–CO–S–	1.76	127
–CS–O–	2.40	120
–CS–S–	1.78	119

TABLE 4. Ultraviolet–visible spectra of monothio carboxylic acid derivatives

Compound	$n \to \pi^*$ band λ_{max} (nm)	log ϵ	$\pi \to \pi^*$ band λ_{max} (nm)	log ϵ	Solvent[a]	References
HCOSH			223		H_2O	40
MeCOSH	266	1.51	220.5	3.51	EtOH	41
	268	1.60	218.5	3.40	CH	41
HCOS⁻			246	4.93	H_2O	40
MeCOSEt			232	3.71	EtOH	41
			231.5	3.61	CH	41
MeCOSCH=CHBu			254	3.93	EtOH	42
EtOCO–COSEt	330	1.70	273	3.83	iOc	43
EtSCOCOSEt[b]	391/372	1.66/1.72	280	3.82	iOc	43
PhCO–COSEt	410	1.60	267	4.02	iOc	44
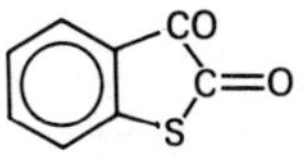 [structure C=O]	406	2.70	302/253	3.41/4.12	Et_2O	39
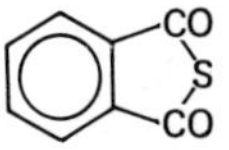 [structure CO S CO]			302/250	3.23/3.99	Et_2O	39
MeOCS(CF$_2$)$_2$CSOMe	399	1.49	243	4.15	iOc	45
EtOCO–CSOEt	399	1.21	253	3.68	iOc	43
EtOCS–CSOEt	526/392	0.68/2.25	255	3.68	iOc	43
MeC(SH)=CHCSOEt	398	3.14	325	4.00	MeOH	46
	415	2.46	338/329	4.04	CH	46
PhC(OH)=CHCSOMe	415	2.50	348	4.31	iOc	47
PhCSOMe	411	2.09	286	4.05	EtOH	48, 49
	417	2.08	287	4.07	CH	48
PhCSOPh	435	1.99	289	4.01	EtOH	48
	441	1.96	288	4.03	CH	48
FCS(CF$_2$)$_2$CSF	428	1.67	294/220	2.23/4.05	iOc	45
t-BuCH$_2$CSCl	480	1.08	304/257	2.24/3.74	CH	50
PhCSCl	518	1.79	317/248	4.08/3.5	MeCN	48
	530	1.82	313/272	4.14/3.5	CH	48

[a]CH = cyclohexane; iOc = isooctane.
[b]S,S-Diethyl dithiooxalate exhibits an additional, long-wavelength band at 425 nm (log ϵ = 0.09), which is assigned to a single–triplet excitation[43].

mentioned, because they are concerned with new types of compounds or are of fundamental interest. Selected data are given in Tables 4–6.

In most cases $n \to \pi^*$ as well as $\pi \to \pi^*$ excitations have been studied, especially in compounds which contain the thiocarbonyl group. Additional bands of different types occur in certain cases. The low-intensity, long-wavelength band of diethyl dithioloxalate at 425 nm has been attributed to a singlet–triplet excitation[43]. Charge transfer (ct) with the solvent causes an intense band at 242 nm (log $\epsilon = 3.25$) in the electronic spectrum of MeCOSH[41,74]. The occurrence of band splittings in the $\pi \to \pi^*$ region of many aromatic dithio esters[48] may well be explained by intramolecular charge transfer between the benzene ring and the functional group, which is further evidenced by the fact, that ortho-substituted derivatives do not exhibit ct bands on account of their twisted configuration[53,65].

The excitation energies for the $n \to \pi^*$ transition decrease in the order RCOSH > RCSOR > RCS$_2^-$ $\approx$ RCSSR > RCS$_2$H, whereas for the $\pi \to \pi^*$ transition energies the order is: RCOSH > RCOSR > RCOS$^-$ $\approx$ RCSOR > RCS$_2$H > RCSSR > RCS$_2^-$. Apparently the excitation of the lone-pair electrons in dithio carboxylic acids is rendered more difficult on ionization.

MO calculations have been successfully applied to the interpretation of the spectra. Reasonable agreement between experimental and theoretical values can be obtained by the PPP method[75,76]. In particular, conjugation of the CS$_2$ group with the benzene ring of aromatic dithio acids and derivatives is reproduced almost quantitatively, as well as the above-mentioned characteristic shift of the $\pi \to \pi^*$ band to lower frequencies going from CS$_2$H to CS$_2^-$. The explicit consideration of 3d orbitals of sulphur is not necessary to reach a simple reproduction of the $\pi \to \pi^*$ bands. Even HMO calculations are suitable for an interpretation of the $n \to \pi^*$ bands[48]. Semiempirical ASMO–SCF–CI calculations show that the 268 nm band of MeCOSH is undoubtedly due to $n \to \pi^*$ excitation. This is experimentally verified by the fact that the first PES band displays vibrational fine structure, which is typical for ionization from n_O electrons while π_S electrons generally show no fine structure[41]. The calculated $n \to \pi^*$ band of MeCOSEt is located at shorter wavelength than that of MeCOSH, and its observation is therefore hindered by the nearby strongly absorbing $\pi \to \pi^*$ transition[41]. The decrease of the first ionization potential (by ca 1 eV $\approx$ 8000 cm^{-1} $\approx$ 125 nm) after substitution of C=O by C=S, which is observed in the PES spectra of MeCOOMe, MeCSOMe, MeCOSMe and MeCSSMe, is in good agreement with CNDO/2 results[77].

Conjugative interaction of the functional groups causes bathochromic shifts of the $n \to \pi^*$ as well as the $\pi \to \pi^*$ bands (Table 4). The substituent effects on the $n \to \pi^*$ transition of aromatic thiono esters[49], and the $n \to \pi^*$ and $\pi \to \pi^*$ transitions of aromatic dithio esters[48] have been discussed in terms of Hammett relationships.

Quite interesting differences occur in the absorption spectra of thio esters R^{1}CYXR^2 (Y and/or X = S) with varying R^2. Substantial bathochromic shifts of the $n \to \pi^*$ bands are observed if R^2 = Me is replaced by R^2 = t-Bu, SiME$_3$, SPh, Ph, (H). This effect is still more pronounced if thioacyl halides and anhydrides are compared with the esters.

D. Vibrational Spectra

Infrared and Raman data of many thio carboxylic acid derivatives have been published in the last decade. Apart from the empirical characterization of compounds the main interest in this field has been focused on the fundamental problem

TABLE 3. Ultraviolet—visible spectra of dithio carboxylic acid derivatives

Compound	n → π* Band		π → π* Band		Solvent[a]	References
	λ_{max} (nm)	log ϵ	λ_{max} (nm)	log ϵ		
HCSSH	510	1.26	295	4.05	Et$_2$O	51
MeCSSH	490	1.24	293	3.83	Et$_2$O	52
t-BuCSSH	481	1.36	300	3.74	Et$_2$O	52
PhCH$_2$CSSH			293	4.23	Et$_2$O	52, 53
Ph(C(OH)=CHCSSH			379		EtOH	54
PhCSSH	518		300		EtOH	48
	526	1.83	333/297	3.66/3.98	Et$_2$O	52, 53
	538	1.85	298	4.00	CH	48
HCSS⁻	386	2.92	331	4.98	H$_2$O	55
MeCSS⁻	446	1.62	333	4.19	H$_2$O	52
t-BuCSS⁻	460	1.54	340	4.10	H$_2$O	52
PhCH$_2$CSS⁻	461	1.65	340	3.90	H$_2$O	52, 53
PhCSS⁻	481	2.15	355	3.70	H$_2$O	52, 53
MeCSSMe	457	1.20	303	4.03	CH	56
n-C$_6$H$_{13}$CSSCH$_2$CO$_2$H	450	1.26	306	3.90	MeOH	57
i-PrCSSPr-i	460	2.08	307	3.68	CH	58
i-PrCSSBu-t	478	1.27	305	4.08	CH	59
c-C$_6$H$_{11}$CSSMe	450	1.24	302	4.08	EtOH	48
	456	1.24	302	4.00	CH	48
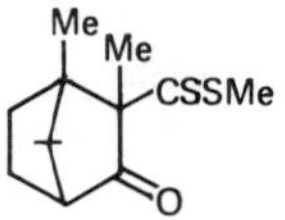 (bicyclic structure: Me, Me, CSSMe, =O)	470	1.23	315	4.02	DI	60
EtSC(NH$_2$)=CHCSSEt	510	1.10	372/325	4.11/3.25	Et$_2$O	61
EtOCOCSSEt	557	1.60	319	3.81	iOc	43
EtSCOCSSMe			328	3.80	CH$_2$Cl$_2$	62
MeSCSCSSMe			362	3.80		63
PhC(OH)=CHCSSMe			381	4.34	EtOH	54
PhCSSMe	498	2.07	330/298	3.8/4.16	EtOH	48
	504	2.11	329/296	3.8/4.12	CH	48
PhCSSBu-t	520	1.97	332/298	3.8/4.08	EtOH	48
	526	2.01	329/296	3.9/4.18	CH	48
PhCSSPh	518	2.12	312	4.30	EtOH	64
	528	2.10	306	4.32	CH	64

TABLE 5. (Continued)

Compound	$n \to \pi^*$ Band		$\pi \to \pi^*$ Band		Solvent[a]	References
	λ_{max}(nm)	log ϵ	λ_{max}(nm)	log ϵ		
4-MeC₆H₄CSSMe	492	2.16	310/245	4.24/3.6	EtOH	48
	501	2.08	307/245	4.20/3.5	CH	48
2-MeC₆H₄CSSMe	481	1.96	310	4.03	EtOH	48
	494	2.00	310	4.02	CH	48
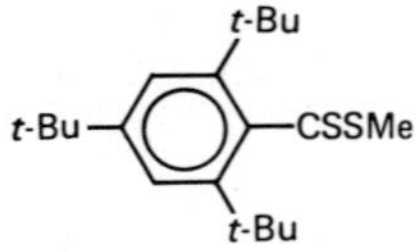 (t-Bu-substituted C₆H₂-CSSMe)	500	1.53	323	3.79	MeOH	65
	500	1.61	325	3.86	CH	65
4-MeOC₆H₄CSSMe	508	2.23	337/290	3.8/4.20	EtOH	48
	511	2.17	335/295	3.7/4.16	CH	48
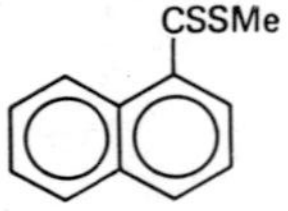 (naphthyl-CSSMe)	488	2.05	308	4.06	EtOH	48
	494	2.11	306	4.02	CH	48
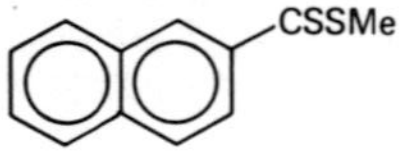 (naphthyl-CSSMe)	495	2.33	319	4.35	EtOH	48
	506	2.30	318	4.44	CH	48
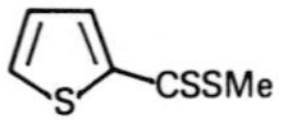 (thienyl-CSSMe)	507	1.98	342	4.20	EtOH	48
	513	1.96	340	4.24	CH	48
PhCSSCOMe	577	2.05	340/310	4.19	CH	66
PhCSSCONHPh	506	2.16	324	4.16	CH₂Cl₂	67
PhCSSCSPh	548	1.36	307	4.08	CH	68
i-PrCSSCSPr-i	517	2.42	351	4.23	CH	59
PhCSSSCSPh	525	1.99	306	4.13		69
4-MeC₆H₄CSSSPh	534		310			70
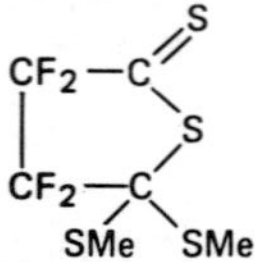 (dithietane structure: S, CF₂-C, S, CF₂-C, SMe, SMe)	513	1.13	320	3.86	iOc	45

TABLE 6. Ultraviolet–visible spectra of Group IV derivatives of thio carboxylic acids (in cyclohexane)

Compound	$n \rightarrow \pi^*$ band		$\pi \rightarrow \pi^*$ band		
	λ_{max}(nm)	log ϵ	λ_{max}(nm)	log ϵ	Reference
PhCOSGeMe$_3$			267/240	3.91/4.11	71
PhCOSSnMe$_3$			271/242	3.91/4.07	71
PhCSOSiMe$_3$	437	1.87	298	3.98	71
MeCSSSiMe$_3$	484	1.08	295	4.03	56
i-PrCSSSiMe$_3$	487	2.08	303	4.07	58
PhCSSSiMe$_3$	531	2.11	298	4.03	58
MeCSSGeMe$_3$	482	1.18	310	4.02	56
PhCSSGeMe$_3$ ·	528	2.04	303	4.12	58
MeCSSSnMe$_3$	474	1.23	317	3.99	56
PhCSSSnMe$_3$	520	1.99	307	4.17	58
PhCSSSnPh$_3$[a]	508	2.18	312	4.31	72
MeCSSPbEt$_3$	476	1.23	330	4.00	56
4-MeC$_6$H$_4$CSSPbPh$_3$	520	2.23	318	4.38	73

[a] In CHCl$_3$.

of the assignment of typical bands in the spectra to specific vibrations of the molecules[78].

1. Monothio carboxylic acids

These normally exhibit the thiolo form 1 rather than the thiono form 2 (cf. Section I. A, B, E). This is especially evident from the occurrence of S—H and C=O

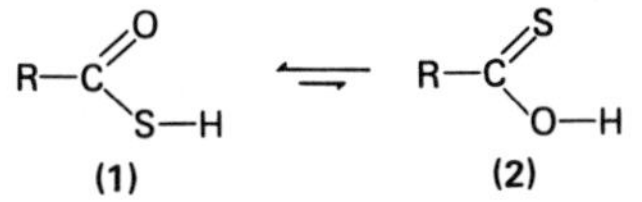

R—C(=O)—S—H ⇌ R—C(=S)—O—H

(1)   (2)

stretching frequencies in the infrared and Raman spectra at 2540–2570 cm^{-1} and 1660–1685 cm^{-1} [40,79-84], which have been unequivocally assigned by normal coordinate analysis[80,81,85] as well as isotope labelling[82,84] [ν(S—D): 1865 cm^{-1} in Me—CO—SD[82]]. However, up to 1% of the hydroxy form 2 has been detected in trichlorothioacetic acid[84] [2, R = CCl$_3$; ν(OH)$_{free}$: 3570 cm^{-1}, ν(OH)$_{ass}$: 3460 cm^{-1}]. Electronegative substituents cause an increase of ν(SH) and ν(CO) (CF$_3$—CO—SH: 2578, 1730 cm^{-1} [79]). The assignment of the C—S stretching is not as straightforward. Most likely bands at 626 cm^{-1} (Me—CO—SH[82,86]), and 712 cm^{-1} (Et—CO—SH[85]) belong to this vibration. In the solid state or solutions of high-concentration dimers of type 3 and 4 are present[81,85,87]. Their stability, which is lower than in the case of carboxylic acids has been studied by quantum-chemical calculations (EHMO and CNDO/2)[80].

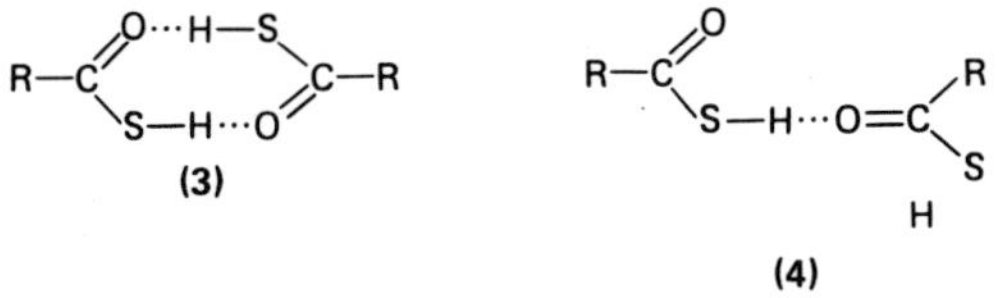

R—C(=O···H—S)(S—H···O)C—R

(3)

R—C(=O)(S—H···O=C(R)(S))

H

(4)

2. Thiocarboxylate anions

These exhibit C$\cdots$O and C$\cdots$S stretching frequencies at 1520 cm^{-1} and 845 cm^{-1}, and a SCO deformation band at 520 cm^{-1} [40,80,88].

3. Thiolo esters

Selected examples of recently studied compounds are given in Table 7. The reasons for the substantial decrease of ν(CO) as compared with normal esters are not fully understood. Probably it is a consequence of the low electronegativity of sulphur, but overlap between the π electrons and 3d orbitals of sulphur may be involved too. ν(C—S) of thiolo esters has been located in the 600—800 cm^{-1} range by most authors.

4. Thiono esters

Marked coupling of the C=S stretching mode to other vibrations of the molecule, and their occurring in the fingerprint region make it difficult to localize C=S bonds in the infrared spectra.

Absorptions near 1200 cm^{-1} have been assigned to the C=S vibration on account of comparison between the infrared and Raman spectra of thiono esters and the corresponding esters. However, normal coordinate calculations on HCSOMe have shown, that the strong band at 1003 cm^{-1} is mainly due to ν(CS), whereas the 1238 cm^{-1} band contains large contributions from the C—O stretching mode[92]. Nevertheless these two absorptions may be conveniently taken as key bands in the vibrational spectra of thiono esters. Table 8 shows some recent examples.

5. Dithio carboxylic acids

Dithio carboxylic acids and their salts exhibit typical bands (Table 9), which are due to the symmetric and antisymmetric stretching vibration of the CS$_2$ group Thorough normal coordinate analyses and isotope studies of Mattes and Stork[97]

TABLE 7. Infrared bands of thiolo esters

Compound	ν(C=O)(cm^{-1})	ν(C—S)(cm^{-1})	Reference
HCOSMe	1660	767	40
MeCOSMe	1715	623	89
MeCOSCH=CHBu	1706		42
MeCOSPh	1713		89
PhCOCOSMe	1670		44
EtOCOCOSEt	1695		31
EtSCOCOSEt	1686	801	31
PhCOSPh	1686		89
PhCOSSiPh$_3$	1695		90
PhCOSGeMe$_3$	1648		71
PhCOSSnMe$_3$	1631		71
R$_2$C—C=O $\underset{S}{\diagdown\diagup}$	1800		15
PhCOSSO$_2$Ph	1690		91

TABLE 8. Infrared bands of thiono esters

Compound	$\nu(C=S)(cm^{-1})$	$\nu(C-O)(cm^{-1})$	Reference
HCSOMe	1003	1238	92
PhCHOHCSOEt	1200		93
MeC(SH)=CHCSOEt	1184		46
PhC(OH)=CHCSOMe	1255/1240		93
EtOCOCSOEt	1284		31
EtOCSCSOEt	1260	1059	94
$R_2NCSCSOEt$	1230	1024	95
PhCSOMe	1230	1058	96
$2\text{-}HOC_6H_4CSOMe$	1230		93
$PhCSOSiMe_3$	1238		71

have shown that especially $\nu_s(CS_2)$ is very strongly coupled with other vibrations of the molecule (cf. Table 9), and that the 848 cm^{-1} band in HCS$_2^-$ is due to ν(CH, out of plane) rather than $\nu_s(CS_2)$[55,99], whereas the latter really appears at 786 cm^{-1}.

Examination of the infrared bands at varying concentrations has indicated that alkane carbodithioic acids are partially associated in the form of hydrogen-bridged dimers[27].

6. Dithio carboxylic acid esters

Strong absorptions near 1200 cm^{-1} and 1000 cm^{-1} are observed in the vibrational spectra (Table 10), which can be used as key bands in the dithio ester series.

TABLE 9. Infrared bands of dithio carboxylic acids and their anions (cm^{-1})

Compound	$\nu(S-H)$	$\nu_{as}(CS_2)$	$\nu_s(SC_2)$			References
MeCSSH	2581	1216				27, 98
EtCSSH	2566	1200–1250				27
i-PrCSSH	$(2505)^a$					
c-C$_3$H$_5$CSSH						27
CF$_3$CSSH	2577	1253	691			98
PhC(OH)=CHCSSH	2490	1070				54
HCS$_2^-$		988	848			55, 99
		982	786			97
MeCS$_2^-$		1141	602			100
		875				56
EtCS$_2^-$		950/925				101
i-PrCS$_2^-$		850				101
$[O_2CCS_2]^{2-}$		1032	766	478^b		88
			27%	59%		
$[SOCCS_2]^{2-}$		1023	836	600	435^b	88
			27%	27%	30%	
PhCS$_2^-$		1020	910	658	435^b	102
			42%	25%	31%	
		1000				103

$^a\nu$(S—H) of the dimer[27].
^{b}Potential energy distribution.

TABLE 10. Infrared bands of dithio esters and related compounds

Compound	ν(C=S)(cm^{-1})	ν(CS—S)(cm^{-1})	Reference
MeCSSMe	1198	870/857/580	35
	1195		96
	1194	862	56
EtCSSMe	1183		96
n-C$_{11}$H$_{23}$CSSCH$_2$CO$_2$H	1224		57
i-PrCSSMe	1205		96
t-BuCSSMe	1101		96
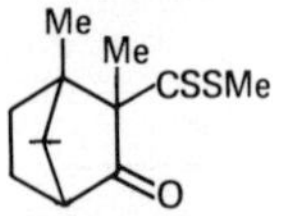 [bicyclic Me Me CSSMe with O structure]	1153/1085		60
H$_2$NCSCH$_2$CSSEt	1060	637/618	61
H$_2$NC(SEt)=CHCSSEt	1256	642	61
PhC(OH)=CHCSSMe	1230		54
EtOCOCSSEt	1259		31
MeSCSCSSMe	1045		63
MeCSSSiMe$_3$	1194/1187	876	56
i-PrCSSSiMe$_3$	1202	848	58
MeCSSGeMe$_3$	1187/1178	873	56
i-PrCSSGeMe$_3$	1198	830	58
MeCSSSnMe$_3$	1184/1172	870	56
i-PrCSSSnMe$_3$	1191	778	58
MeCSSPbEt$_3$	1165/1154	873	56
PhCSSMe	1241/1044		96
PhCSSCH$_2$CH=CH$_2$	1240/1220		103
PhCSSSPh	1245		70
PhCSSSnPh$_3$	1217		72
4-MeC$_6$H$_4$CSSPhPh$_3$	1218		73
PhCSSCOMe	1230		66
PhCSSCONHPh	1250		67
PhCSSCSPh	1245		68
PhCSSSCSPh	1240		69
t-BuCH$_2$CSCl	1230/1215	850	50
PhCSCl	1252/1052		96

An unambiguous assignment is so far not possible. It is, however, likely, and indeed assumed by most authors, that the bands in the 1200 cm^{-1} region should be ascribed to the C=S stretching, although it has been shown by normal coordinate analysis that a reasonable force constant of f(C=S) = 6.06 mdyn/Å yields ν(C=S) = 1100 cm^{-1} in Me—CS—SMe[104]. Intense bands appearing between 850 and 900 cm^{-1} may be due to ν(CS—S) or deformation modes of the dithio ester molecule. This is also true for absorptions near 600 cm^{-1}.

The influence of special structural features on the localization of bands, for instance the decrease of ν(CS) with increasing atomic weight of the metal in trialkylmetal dithioacetates[58], can be seen from the data of Table 10. Anhydrides of dithio carboxylic acids and thioacyl chlorides show the C=S stretching too (Table 10).

E. Nuclear Magnetic Resonance Spectra

1. Proton n.m.r spectra

[1]H-n.m.r. spectroscopy has been extensively used in investigations on thio carboxylic acid derivatives. Some topics of special interest should be mentioned.

The chemical shift, δ(p.p.m.), of the formyl proton increases in the following order: HCO_2H (8.06)[51] $< HCOSH$ (10.18)[40] $< HCS_2H$ (11.6)[51], and HCO_2^- (8.42)[51] $< HCOS^-$ (10.64)[40] $< HCS_2^-$ (12.22)[55]. The thioformyl proton in HCSOMe is less shifted (9.52[92,105]) than the one in HCOSMe (10.12[40]). The chemical shift of α-protons is dependent on the type of compound as well as the substituents. Typical examples are shown in Table 11. The large series of different thio carboxylic acid derivatives studied has enabled Radeglia and coworkers[106] to derive an increment system for the chemical shift of the methylene protons in $R^1CH_2CXYCH_2R^2$ (X, Y = O, S).

Normally, β-oxo- and β-thioxo-thio esters (5) do not exist but are completely tautomerized to the chelates 6. In some cases, however, minor amounts of 5 occur in the equilibrium[107].

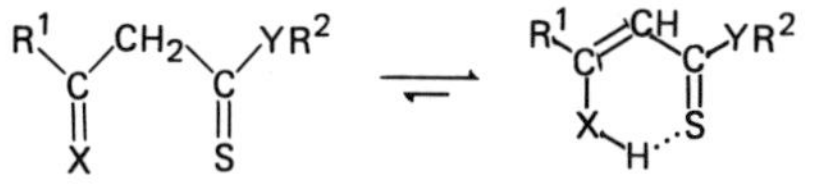

(5)　(6)

R^1	X	YR^2	%5	δ(CH)(p.p.m.)	δ(XH)(p.p.m.)	References
Me	O	OEt	5	5.53	13.57	108
Ph	O	OMe	0	6.27	13.81	47
Ph	O	OEt	19	6.35	14.12	47
Me	S	OEt	0	6.42	8.66	46
Me	O	SMe	6	6.17	14.45	107
Ph	O	SMe	0	6.89	14.93	54, 107, 109

Similarly the thiono ester 7 exists predominantly in the SH form 7b[110], whereas the cyano acetodithioacetate exhibits only the chelate structure 8[111].

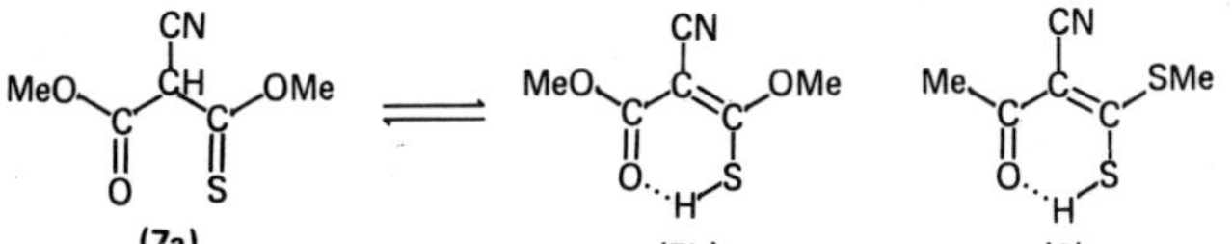

(7a)　(7b)　(8)

δ(CH) = 4.22 p.p.m.　δ(SH) = 14.5 p.p.m.　δ(OH) = 16.08 p.p.m.
(40% in CDCl₃)　(60% in CDCl₃)　(100% in CDCl₃)

The SH proton n.m.r. signals are found at 4.6—4.8 p.p.m. in monothio carboxylic acids[40,82-84,112], and at 5.2—5.9 p.p.m. in dithio carboxylic acids[27,54,107]. From the occurrence of two SH signals in the n.m.r. spectrum of MeCOSH at $-50°C$ and its temperature dependence the existence of E/Z isomers has been deduced and the rotational barriers determined[112]

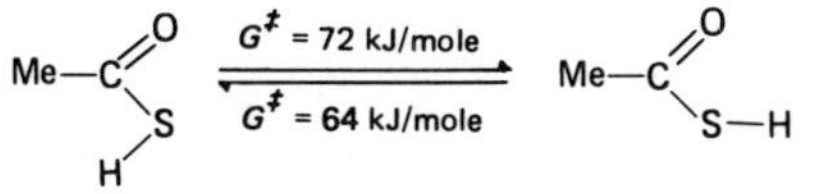

TABLE 11. Chemical shifts of α-protons in the n.m.r. spectra of thio carboxylic acids and their derivatives

Compound	δ(p.p.m.)	Solvent	Reference
CH_3COSEt	2.27	CCl_4	106
EtO_2CCH_2COSEt	3.47	CCl_4	106
$NCCH_2COSEt$	3.62	CCl_4	106
$EtSCOCH_2COSEt$	3.70	CCl_4	106
$PhCH_2COSEt$	3.72	CCl_4	106
$ClCH_2COSH$	4.10	$CDCl_3$	83
$Cl_2CHCOSH$	5.86	$CDCl_3$	83
CH_3CSOEt	2.52	CCl_4	106
EtO_2CCH_2CSOEt	3.69	CCl_4	106
$EtSCOCH_2CSOEt$	3.94	CCl_4	106
$PhCH_2CSOEt$	3.98	CCl_4	106
$EtOCSCH_2CSOEt$	4.11	CCl_4	106
$EtO_2CCH(CN)CSOMe$	4.22[a]	$CDCl_3$	110
$PhCHOHCSOEt$	3.89/4.96	CCl_4	93
t-$BuCH_2CSCl$	3.17	$CFCl_3$	50
CH_3CSSEt	2.80	CCl_4	106
i-$PrCSSPr$-i	3.43	CCl_4	58
EtO_2CCH_2CSSEt	3.89	CCl_4	106
$MeCOCH_2CSSMe$	4.08[b]	CCl_4	107
$EtSCOCH_2CSSEt$	4.16	CCl_4	106
$PhCH_2CSSEt$	4.24	CCl_4	106
$EtSCSCH_2CSSEt$	4.44	CCl_4	106
H_2NCSCH_2CSSEt	4.45	CS_2	61

[a] 40% in equilibrium with 60% of the SH tautomer[110].
[b] 6% in equilibrium with 94% enol[107].

TABLE 12. Chemical shifts, δ(p.p.m.), of alkyl protons in thio esters

R^2	R^1—CO—SR^2	R^1—CS—OR^2	R^1—CS—SR^2
CH_3	2.2–2.5	3.2–4.2	2.5–2.7
References	31, 40, 44, 114, 115	31, 40, 45, 47, 93–95, 105, 110, 115, 116	31, 54, 63, 107, 109, 111, 114, 115, 117, 118
CH_2R	2.8–3.1	4.2–4.7	3.1–3.4
References	31, 44, 61, 106, 115, 119	31, 45, 46, 47, 93–95, 106, 110, 115	31, 61, 106, 115, 118, 120
$CH_2CH=CH_2$	3.6	5.0	4.0
References	119	119	119
CH_2CO_2H			4.1–4.6
References			57
CH_2Ph	4.0–4.3		4.4–4.6
References	31, 121		117, 118
CHR_2	3.5–3.7	5.6–5.7	3.4
References	31, 44	31, 45, 93, 94	58

[1]H-n.m.r. measurements on aliphatic dithio carboxylic acids have shown that association to form dimers SH···S(H) rather than SH···S(C) bridges takes place in these compounds[27]. The range of chemical shifts occurring in various types of thio esters are compiled in Table 12. The protons of O-alkyl groups in the thiono ester series appear at lowest field. Resonance lines due to S-alkyl protons of dithio esters are little shifted with respect to thiol ester alkyl protons.

2. Carbon-13 n.m.r. spectra

Thiolo esters have been systematically studied by Hall and Wemple[113], who have shown that $\delta(C=O)$ is shifted downfield by 15–20 p.p.m. with respect to esters, thus appearing at 190–200 p.p.m. The chemical shift of 194.5 p.p.m. found for MeCOSH has been ascribed to a C=S group[122]. However, the experimental evidence argues strongly against this interpretation, favouring instead a C=O group[113,123,124]. C=S groups really appearing in thio carboxylic acid derivatives show ^{13}C resonance at 215 p.p.m. (t-BuCH$_2$CSCl[50]), 213–229 p.p.m. (R^{1}CSOR^2 [65,108,123,125]), or 225–240 p.p.m. (R^{1}CSSR^2 [21,63,65,123]) in good agreement with the correlation $\delta(C=S) = 1.45 \, \delta(C=O) - 46.5$ p.p.m proposed by Kalinowski and Kessler[124] for the ^{13}C-n.m.r. lines of thio ketones, thio amides, thio ureas, and isothiocyanates, and the increment system of Radeglia and Scheithauer[123].

F. Electron Spin Resonance Spectra

Thio esters are readily reduced to the corresponding radical anions, which can be studied by e.s.r. spectroscopy[43,115,118]. The strong electron-attracting properties of thio and dithio carboxyl groups are reflected in the low (negative) reduction potentials $E_{1/2}$ as determined by polarography (cf. Section III. B.1) as well as the low total spin density remaining in the aromatic ring of thiobenzoate ester radical anions, and the high g-values, i.e. high spin densities at the heavy atom (sulphur) of the functional groups in thiobenzoate and thiooxalate esters (Table 13). No radical anions of thio- or dithioalkanoate esters have been obtained so far.

G. Mass Spectra

Apart from the observation of molecular ions of thio esters, some special fragmentation patterns have been studied in this series. Acyl splitting (A) occurs very frequently in thiolo[40,126-132], thiono[126,129,133] and dithio esters[61,129] as well as in the thioacyl chloride, t-BuCH$_2$CSCl[50]. Alkyl splitting (B) has been observed in β-oxo-[130] and β-thioxothiolo[131], thiono[129,133] and dithio

TABLE 13. g-Values of thio ester radical anions

Compound	X = Y = O	X = O, Y = S	Y = O, X = S	X = Y = S
RY−CX−CX−YR[43]	2.0045	2.0058	2.0101	2.0105[a]
Ph−CX−XR[118]	2.0034		2.0047	2.0071
1,4-RY−CX−C$_6$H$_4$−CS−YR[115]	2.0035	2.0046	2.0073	2.0095
1,3-RY−CX−C$_6$H$_4$−CX−YR[115]	2.0035	2.0038	2.0049	2.0073

[a] RO−CO−CS−SR.

esters[61],[129] too, whereas splitting of the α-substituent (C) has been found in thiolo[128],[130],[131] and dithio esters[61].

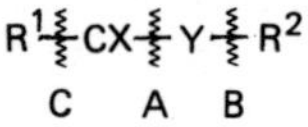

$$R^1 \overset{|}{\underset{|}{\xi}} CX \overset{|}{\underset{|}{\xi}} Y \overset{|}{\underset{|}{\xi}} R^2$$

$$\text{C} \qquad \text{A} \qquad \text{B}$$

Dimethyl tetrathiooxalate is easily split into two halves after electron impact; another interesting fragment ion of this molecule, namely $MeSC \equiv CSME^+$, results from elimination of S_2[63].

The mass spectrum of ethyl thionobenzoate has two peaks due to four centre migration of ethyl from oxygen to sulphur[126],[134], whereas the corresponding migration from sulphur to oxygen in thiolo esters, supposed by Ohno and co-workers[134], could not be confirmed by Bentley and Johnstone[126]. Even in the mass spectra of CF_3COSPh and $PhCOSPh$ rearrangement was not observed, though it should be favoured in these cases[135].

Thiono esters containing hydrogen in the γ-position exhibit McLafferty-type rearrangement in their mass spectra[136]:

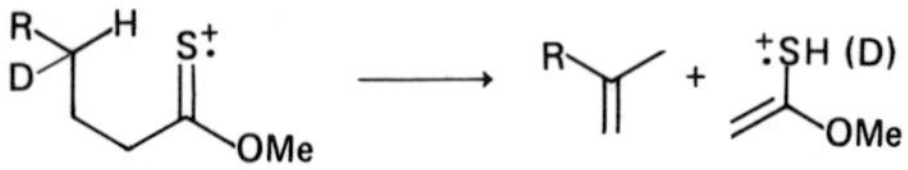

The small isotope effect of only 0.80 may be attributed to either the large size or the lower electronegativity of the sulphur atom compared with the oxygen atom of esters. Thiolo esters show only a very weak McLafferty rearrangement ion[136]. O-Ethyl thioacetyl thioacetate (9) shows a very characteristic behaviour in its mass spectrum. It is fragmented via the dithiolium ions 10 and 11[46].

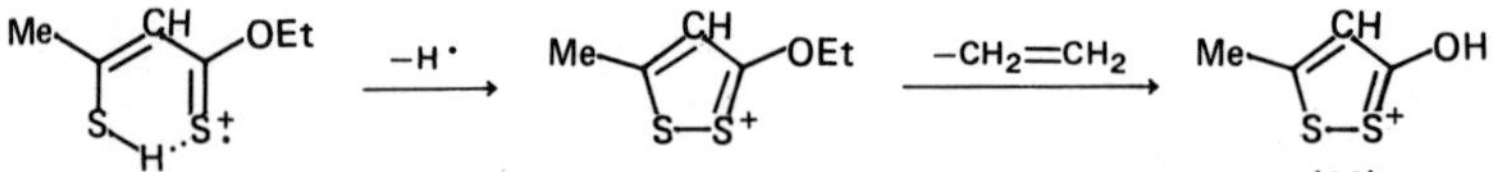

(9)   (10)   (11)

Negative-ion mass spectroscopy of thiono and dithio esters has been studied by Rullkötter and Budzikiewicz[137]. The fragmentation of the molecular anion is accompanied by C to S rearrangement, i.e.:

$$R-CS-SMe^{-} \xrightarrow{-Me^{\cdot}} R-CS-S^{-} \longrightarrow R-S-\bar{C}=S \xrightarrow{-CS} R-S^{-}$$

II. SYNTHESES

A. Thio and Dithio Carboxylic Acids and their Salts

Only few fundamentally new methods for the preparation of *monothio carboxylic acids* have been reported since Janssen's report of 1969[1]. Cleavage of carboxamide anions by CS_2 to form thiocarboxylates[138], and the synthesis of the special compounds $HCOSH$[40], $ClCH_2COSH$[83], $Cl_2CHCOSH$[83], Cl_3CCOSH[84] and $(CF_3)_2CHCOSH$[139] should, however, be mentioned.

Dithio carboxylic acids or their salts have been prepared in many cases in order to obtain the corresponding dithio esters (cf. Section II.D). They may, however, be isolated in the pure state. It has been emphasized by Kato and coworkers[103], that di- and trialkylammonium dithiocarboxylates are readily obtained as stable crystals, which are more useful than metal salts in the purification and preparation of derivatives.

Three types of reactions are convenient for the synthesis of RCS_2H or RCS_2^-:

(1) Reduction of carbon disulphide.
(2) Thiolysis of suitable precursors.
(3) Oxidative sulphuration of compounds of lower oxidation state

$$R-M + CS_2 \longrightarrow R-CS_2^- \; M^+ \xrightarrow{H^+} R-CS_2H \qquad (1)$$

$$M = MgX, \; Li$$

The first reaction involves Grignard or organolithium reagents (equation 1). Although the yields are only moderate in many cases[65,140-142], this method has found widespread application because of its convenience. For examples see the recent reviews by Paquer[143] and Jansons[144], the monographs by McKenzie[145], Duus[146] and Voss and coworkers[147], and the literature cited in Section II.D.

Activated methyl, methylene and methine compounds yield dithiocarboxylates on base-catalysed reaction with CS_2. In the cases shown in equation (2) the free

$$X-\overset{|}{\underset{|}{C}}-H + CS_2 \xrightarrow{base} X-\overset{|}{\underset{|}{C}}-CS_2^- \xrightarrow{H^+} X-\overset{|}{\underset{|}{C}}-CS_2H \qquad (2)$$

$$X = RCO^{54, 107, 111, 148-159}, \; CN^{111, 160}, \; ROCO^{111}, \; CO_2^- \; ^{161}$$

dithio carboxylic acids or their salts are isolated. It is advantageous to prepare the tetraalkylammonium dithiocarboxylates from the methylene compounds by extraction with $Bu_4N^+ \; OH^-$ into CH_2Cl_2 and subsequent reaction with CS_2. Heterocycles such as pyrroles[162] or isocoumarines[163] yield dithio carboxylic acids on reaction with CS_2 too. In certain cases $AlCl_3$ has been used as catalyst in a Friedel–Crafts-type reaction of CS_2 with aromatics[148,149].

The second method, thiolysis, has served recently for the preparation of some dithio carboxylic acids (equations 3–7). Potassium dithioformiate is available from

$$CF_3CN \xrightarrow[25^\circ C]{H_2S} CF_3CSNH_2 \xrightarrow[40^\circ C]{H_2S/HCl} CF_3CS_2H^{98, 150} \qquad (3)$$

$$HCCl_3 \xrightarrow[-3KCl]{2 K_2S} HCS_2^- \; K^+ \xrightarrow{H^+} \text{(12)} \qquad (4)$$

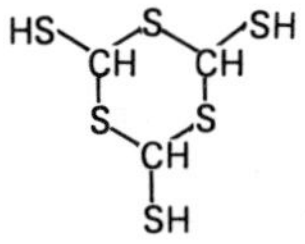

(12)

$$Cl_3CCO_2H + 3 K_2S \longrightarrow K_2(S_2C-CO_2) + 3 KCl + KSH \qquad (5)$$

$$Cl_3CCO_2Ph + 3 H_2S \longrightarrow K_2(S_2C-COS) + 3 KCl + KOPh \qquad (6)$$

$$R^1CS_2R^2 \xrightarrow[-R^2SH]{NaSH} R^1CS_2^- \; Na^+ \xrightarrow{H^+} R^1CS_2H \qquad (7)$$

$CHCl_3$ (equation 4)[55]. Acidification yields amorphous trimeric dithioformic acid (12); the monomer is, however, present in dilute solutions[51]. Unstable cyano and azido dithioformic acids are formed from their sodium salts[151]. Stork and Mattes[152] have obtained potassium di- and tri-thiooxalate by thiolysis of trichloroacetic acid and its phenyl ester, respectively (equations 5 and 6). Dithio esters, which can often be obtained independently in a convenient way (cf. Section II.D), form dithiocarboxylates in good yields on thiolysis (equation 7)[164,165].

Finally, oxidative sulphuration can be achieved by reacting benzyl halides[166-168] or benzaldehydes[169] with elemental sulphur under alkaline conditions (equation 8).

$$ ArCH_2X \xrightarrow[S_8]{NaOCH_3} ArCS_2^-\ Na^+ \xrightarrow{H^+} ArCS_2H \tag{8} $$

B. Thiolo Esters

A Portuguese review on thiol esters appeared in 1972[170]. Thio lactones, on the other hand, were summarized in 1964[171]. The preparation of thiol esters is achieved by either *alkylation* of thio acids and their salts, *acylation* of thiols, or *hydrolysis* of appropriate precursors. Thiol esters of aromatic acids, moreover, are obtained by Friedel–Crafts reaction (equation 9)[172]. α-Thio lactones (13) have been prepared by Schaumann and Behrens[15] (equation 10).

$$ RSCOCl + ArH \xrightarrow{AlCl_3} RSCOAr \tag{9} $$

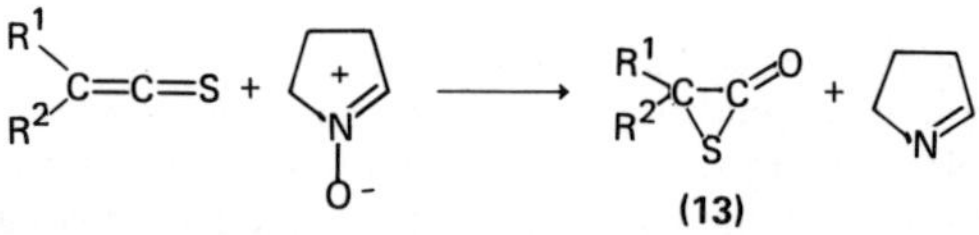

$$ \tag{10} $$

Alkylation of thio acids by alkyl halides is a method of long standing[1]. Alkyl thiosulphates[173] and sulphonates[174] can be used as reagents instead, for instance in the lipid[175], carbohydrate[176] and cephalosporine[177] series. The thioacetate 15 is formed from 14 and Et$_3$NH$^+$ MeCOS^{-178}.

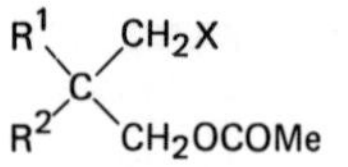

(14) X = $-O\overset{+}{P}(NMe_2)_3ClO_4^-$

(15) X = $-SCOMe$

Activated hydroxy[179], alkoxy[180,181] (equation 11), and carbamate[182] groups are substituted by thio acids too. This is also true for epoxide[183,184].

Thio acids attack olefins yielding thiol esters[185-187]. In many cases this takes place as a radical addition[175,188,189] and is thus achieved photolytically. The mechanism has been studied by Kondo and Tsuda[188]. S-Vinyl thiol esters are obtained by the addition of thio acids to alkynes (equation 12)[42].

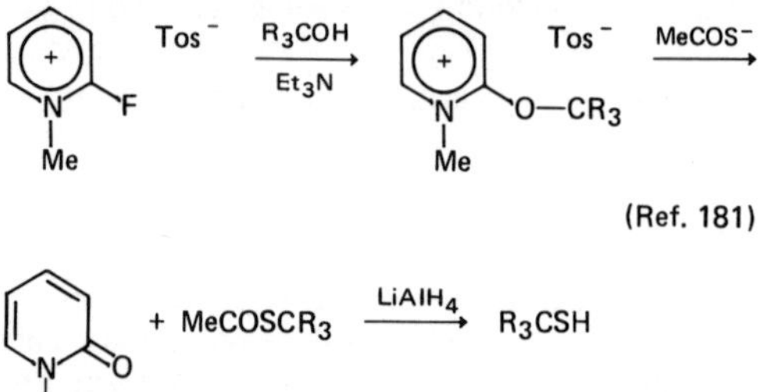

(Ref. 181)   (11)

$$ R^1COSH + HC{\equiv}CR^2 \longrightarrow R^1COSCH{=}CHR^2 \tag{12} $$

Rearrangement of thiono esters (16) to the corresponding thiolo esters (17)[190-197], which can be conceived as an intramolecular alkylation (equation 13),

$$PhCS-OR \longrightarrow PhCO-SR \qquad (13)$$

$$\text{(16a) R = Alkyl} \qquad \text{(17)}$$
$$\text{(16b) R = Aryl}$$

is possible. The alkyl derivatives **16a** need BF_3[191] or $Et_3O^+ BF_4^-$[192] as catalyst, while **16b** is rearranged by heating[190]. Reaction of thio acids with aldehydes[198-201] can yield thiol esters too (equation 14).

$$R^1COSH + CH_2O \longrightarrow R^1COSCH_2OH \longrightarrow R^1COSCH_2X \qquad (14)$$

$$X = Cl, Br, NHR^2$$

Stannyl thiol esters (18) are obtained on reaction of tin hydrides with thio acids[202]. Alkylmetal thiocarboxylates can also be prepared in the conventional way, i.e. from $RCOS^-$ and $ClMMe_3$[91].

$$R^1COSH + R_3^2 SnH \longrightarrow R^1COS-SnR_3^2$$

$$\text{(18)}$$

Acylation of thiols is the second convenient method for the synthesis of thiol esters. Acyl fluorides[203], acyl chlorides[31,44,132,200,204-215], acyl bromides[216], esters[217-221], carboxylic[220,222-225] and phosphoric[226] anhydrides, acyl imidazoles[227] and acyl trialkylammonium fluoroborates[228] have been used as reagents. Compounds of special interest are prepared in this way, i.e. the thiol esters of trihalogeno acetic acids[205], α-oxo carboxylic acids[44], chiral thiols[211] and α-[207] and β-alanin[208] (which can be polymerized), as well as the unsaturated derivatives **19—21**.

$$R^1COSCR^2=CR^3R^4 \qquad R^1COSCH_2CH=CMe_2 \qquad R_2^1C=CHCOSR^2$$

$$\text{(19)}^{120,206,220} \qquad \text{(20)}^{204} \qquad \text{(21)}^{213}$$

Though it is not possible to esterify thiols directly, there are some methods for the preparation of thiol esters from thio carboxylic acids. In these cases an activating (condensating) agent is used (equation 15). Activators are: dicyclohexyl

$$R^1CO_2H + R^2SH \xrightarrow[-H_2O]{\text{activator}} R^1COSR^2 \qquad (15)$$

carbodiimide[229-233], diethoxyphosphoryl cyanide[234], diethoxyphosphoryl azide[234], diphenylphosphoryl chloride[231], triphenylphosphine/dipyridyl disulphide[235-237] and 1-arylsulphonyloxybenzotriazole[231].

Claisen-type rearrangements[238], Pummerer reactions[239] and decarboxylations (equation 16)[240], which yield thiol esters in certain cases, may be taken as acylations too.

$$R^1CO_2^- Na^+ + ClCOSR^2 \xrightarrow[0°C, THF]{-CO_2} R^1COSR^2 + NaCl \qquad (16)$$

Finally, thiol esters are formed on partial *hydrolysis* according to equation (17):

$$R^1CX_2SR^2 \xrightarrow{H_2O} R^1COSR^2 + 2 HX \qquad (17)$$

$$X = Cl^{241,242}, Br^{243,244}, I^{245}, OH^{246,247},$$
$$CN^{247}, SR^{245,246,248}, NR_2^{+115,249,250}$$

C. Thiono Esters

Thiono esters are the least conveniently available types of thio esters because of their decreased stability with respect to the corresponding esters (hydrolysis, oxidation) and thiolo esters (rearrangement), and because it is impossible to attack monothio carboxylic acids at the oxygen atom by normal alkylating agents. Nevertheless, most thiono esters that are needed are now available by any of the following methods.

Reaction of thiocarboxylate anions with Me_3SiCl yields O-trialkylsilyl esters (22) (equation 18)[71,251,252], which are also obtained according to equation (19)[253]. Some O-methyl thiobenzoate is formed on alkylation of thiobenzoic acid with diazomethane, along with a ten-fold amount of S-methyl thiobenzoate[137]. Diazoalkanes also attack thiono esters whereby chain lengthening or branching of the thioacyl group occurs and new thiono esters are formed (equation 20).

$$RCOS^- + Me_3SiCl \longrightarrow RCOSiMe_3 + Cl^- \qquad (18)$$
$$(22)$$

$$RCOCl + (Me_3Si)_2S \longrightarrow 22 \xleftarrow{Et_3N} Me_3SiCl + R^1COS-CSNHR^2 \qquad (19)$$

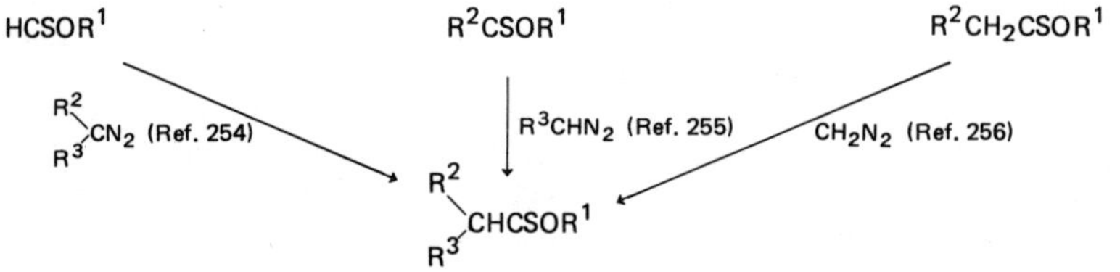

$$HCSOR^1 \qquad\qquad R^2CSOR^1 \qquad\qquad R^2CH_2CSOR^1$$

$$\underset{R^3}{\overset{R^2}{\diagdown}}CN_2 \text{ (Ref. 254)} \qquad \Big\downarrow R^3CHN_2 \text{ (Ref. 255)} \qquad CH_2N_2 \text{ (Ref. 256)} \qquad (20)$$

$$\underset{R^3}{\overset{R^2}{\diagdown}}CHCSOR^1$$

Thionation of esters with P_4S_{10}[115,257-260] or $(EtO)_2PS-SH$[117] has been described in several cases. Sometimes thiolo and dithio esters are formed as by-products (equation 21)[115,117], which cannot always be easily removed. These undesired follow-up reactions, that are due to the long reaction times necessary[257], can be quenched by activating additives such as $NaHCO_3$[258]. Dimeric anisyl thiono-phosphine sulphide has proved to be a valuable thionating agent for esters[413].

$$R^1CO_2R^2 \xrightarrow{P_4S_{10}} R^1CSOR^2 \longrightarrow R^1COSR^2 \xrightarrow{P_4S_{10}} R^1CS_2R^2 \qquad (21)$$

Partial thiolysis of ortho esters yields thiono esters in a convenient way (equation 22)[40,92,261,262]: This method is especially appropriate for the preparation of $HCSOMe$[40,92] and $HCSOEt$[261]. The reaction is catalysed by protons[92,261,262] or Lewis acids ($ZnCl_2$[40,262], $FeCl_3$[262]). Boron sulphide can be used instead of H_2S[133]. $HCSOMe$ is also obtained from the bis-dithiocarbonate 23 by thermolysis (equation 23)[105]:

$$R^1C(OR^2)_3 + H_2S \xrightarrow{catalyst} R^1CSOR^2 + 2 R^2OH \qquad (22)$$

$$HCCl_2OMe + 2 K(EtO-CS_2) \longrightarrow HC(S-CS-OEt)_2OMe \xrightarrow{200^\circ C} HCSOMe \qquad (23)$$
$$(23)$$

Thiolysis of imidates is one of the most powerful tools in the field of thiono ester synthesis. Advantageously, hydrochlorides[263] or acetates[94,264] are used as starting materials instead of the free bases. Nitriles[31,47,93,94,115,137,260,263-272] or amides[31,192,268,272] are useful precursors (equation 24). Thiono esters of oxalic (24)[31,94,264], α, ω-alkanedioic (25)[47,265,266,272], benzenedicarboxylic[115],

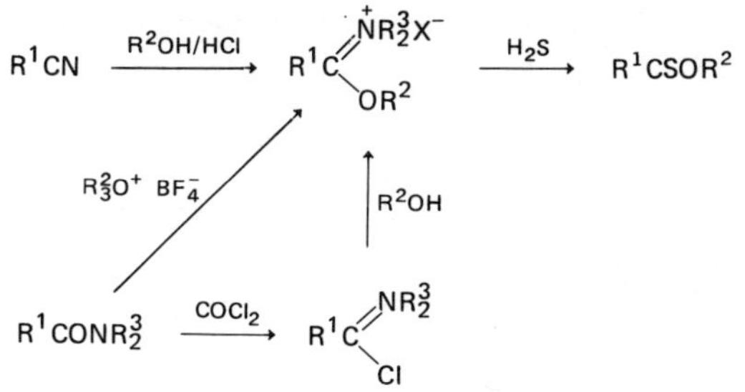

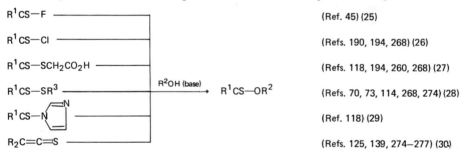

ROCSCXOR   (24a) X = O   (24b) X = S

ROCS(CH$_2$)$_n$CXOR   (25a) X = O, n = 1   (25b) X = S, n = 2–5

hydroxy[93], β-oxo (5/6, X, Y = O)[4,7,93] and β-thioxo (5/6, X = S, Y = 0)[93] carboxylic acids can be prepared in this way. Reaction of 26 with hydrogen sulphide does not yield the α-oxo thiono ester 27, but its reduction product 28[93].

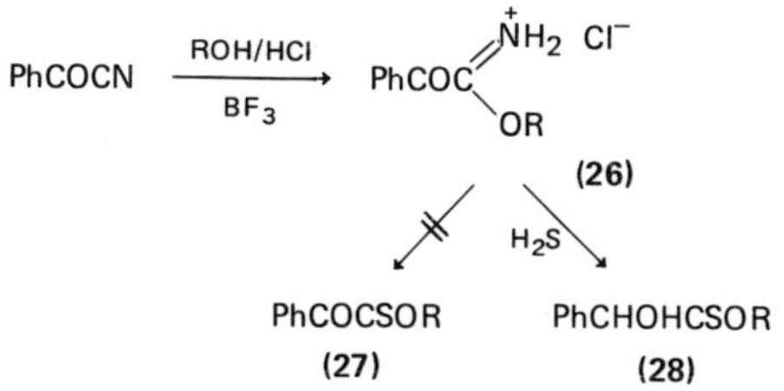

PhCOCN $\xrightarrow[\text{BF}_3]{\text{ROH/HCl}}$ PhCOC$\overset{+}{<}\overset{\text{NH}_2\ \text{Cl}^-}{\underset{\text{OR}}{}}$

(26)

$\cancel{\swarrow}$ $\underset{\text{H}_2\text{S}}{\searrow}$

PhCOCSOR   PhCHOHCSOR

(27)   (28)

S-allyl ketene—*O*, *S*-acetals form thiono esters via Claisen rearrangement[273].
Thioacylation of alcohols and phenols can be achieved by various reagents:

R^{1}CS—F ────────────────────┐   (Ref. 45) (25)

R^{1}CS—Cl ───────────────────┤   (Refs. 190, 194, 268) (26)

R^{1}CS—SCH$_2$CO$_2$H ──────────┤   (Refs. 118, 194, 260, 268) (27)

R^{1}CS—SR^3 ───────────────────┤ $\xrightarrow{\text{R}^2\text{OH (base)}}$ R^{1}CS—OR^2 (Refs. 70, 73, 114, 268, 274) (28)

R^{1}CS—N⟨=N ─────────────────┤   (Ref. 118) (29)

R$_2$C=C=S ───────────────────┘   (Refs. 125, 139, 274–277) (30)

The dithio ester in equation (28) must be activated, i.e. R^1 is an unsaturated residue[114], or R^3 is SPh[70], R^{1}CS$_2$[278], 2,4-(NO$_2$)$_2$ C$_6$H$_3$[268], PbPh$_3$[70] or P(S)(OR)NR$_2$[274]. The thio ketenes involved in equation (30) can be used as stable starting material if R = CF$_3$[118], SiMe$_3$[276]. In other cases they are generated as labile intermediates, for instance by pyrolysis of 1,2,3-thiadiazoles, and the alcohol R^{2}OH is added *in situ*.

Aryloxy thiocarbonyl chlorides yield thiono esters on reaction with activated methylene components (equation 31)[256,279] enamines[231] or aromatic com-

EtO$_2$CCH=C(OSiMe$_3$)OEt $\xrightarrow[\text{–Me}_3\text{SiCl}]{\text{ClCSOAr}}$ (EtO$_2$C)$_2$CHCSOAr $\xrightarrow[\text{–HCl}]{\text{ClCSOAr}}$ (EtO$_2$C)$_2$C(CSOAr)$_2$ (31)

pounds (equation 32)[64]. Xanthates react with methylene phosphoranes in an analogous manner (equation 33)[280].

$$Ar^1H + ClCSOAr^2 \xrightarrow{\text{AlCl}_3} Ar^1CSOAr^2 + HCl \qquad (32)$$

$$Ph_3P{=}CHR^1 + R^2OCS{-}SR^3 \longrightarrow Ph_3P{=}CR^1{-}CSOR^2 \qquad (33)$$

The base-catalysed rearrangement of dithiocarbonates (**29**), accompanied by sulphur extrusion[108], can be used for the preparation of **5/6**, X = 0 (see Section I.E.1).

$$R^1COCH_2SCSOEt \xrightarrow[-S]{\text{NaH}} \textbf{5/6}$$
$$\textbf{(29)}$$

3-Ethoxy-1,2-dithiolium salts are cleaved to enamino thiono esters (**30**)[281]. Thiono esters are also obtained by reaction of elemental sulphur with picoline[282] or pentacarbonyl(methoxyarylcarbene)chromium(0)[116].

$$ArNHCR^1{=}CR^2CSOEt$$
$$\textbf{(30)}$$

D. Dithio Esters

A great number of dithio esters have been prepared by *alkylation of dithiocarboxylates*[54, 60, 62, 65, 109, 111, 114, 118, 137, 140, 141, 143, 144, 153, 156–159, 161, 163, 167, 168, 194, 268, 283–296]. Carboxymethyl dithiocarboxylates $RCS_2CH_2CO_2H$[65,167,194,268,292], including the deuterated derivatives $C_6D_5CS_2CH_2CO_2H$[194] and $4\text{-}t\text{-}BuC_6D_4CS_2CH_2CO_2H$[65], which are important thioacylating reagents, can be obtained in this way. Triphenylplumbyl dithiocarboxylates[73], which can also transfer thioacyl groups, and the analogous R_3Si, R_3Ge and R_3Sn derivatives[56,58,72,73,103] have been prepared from dialkylammonium dithiocarboxylates and triorganyl metal chlorides.

Thiolysis of imidothiolates $R^1C(SR^2){=}NR^3$ or the salts $R^1C(SR^2){=}\overset{+}{N}R_2{}^3$ X^- affords dithiocarboxylates[31, 48, 61, 115, 120, 137, 144, 165, 167, 266, 272, 297–299]. This method has been reviewed by Doyle and Kurzer[292], and by Leon[57].

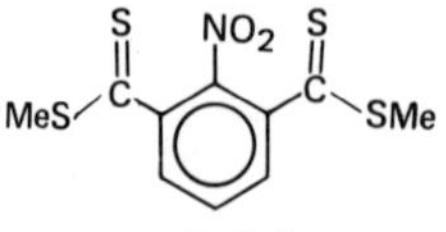

(**31**) (Ref. 259) 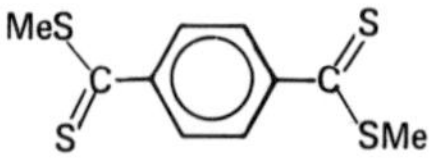   (**32**) (Ref. 115)

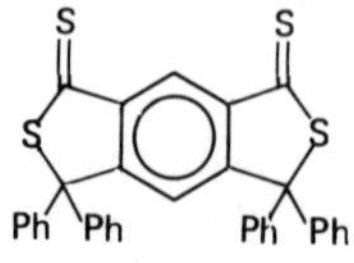

(**33**) (Ref. 300) 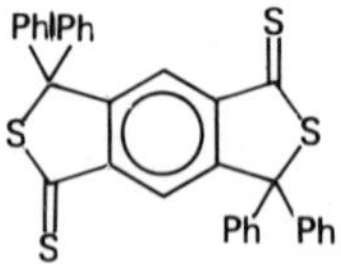   (**34**) (Ref. 300)

It is suitable for the preparation of dialkyl 1,1-dithiooxalates[31], 1,1-dithiomalonates[61,120,266], α,ω-alkane-bis(dithiocarboxylates) $RS_2C(CH_2)_nCS_2R$ ($n = 2-5$)[272], and α-aminodithiocarboxylates[297,299].

Thionation of thiol esters provides dithio esters, e.g. the bis-dithiocarboxylates **31–34**, in a simple fashion[115,117,257–259,300,414].

Thioacylation of thiols can be achieved in a manner analogous to the preparation of thiono esters from alcohols. Carboxymethyl dithiocarboxylates[118,119,194], thioacyl chlorides[50], thioacyl fluorides[45], bis(thioacyl) sulphides[59] and thio ketenes[139,275,301] can be used as reagents.

Propargyl and allenyl ketene mercaptals can be rearranged to dithio esters according to equations (34) and (35)[302]. Ketene dialkyl mercaptals are converted to dithio esters by reaction with lithium (equation 36)[303].

$$\tag{34}$$

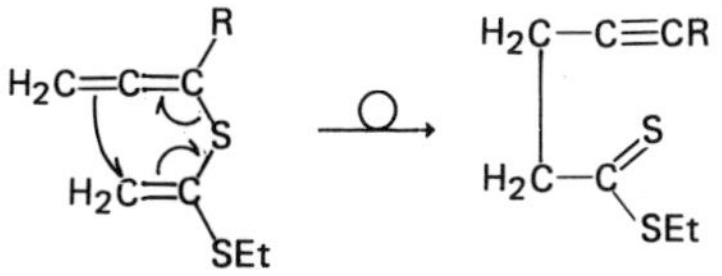

$$\tag{35}$$

$$R^1CH=C(SR^2)_2 \xrightarrow{2\,Li} R^1\bar{C}HCS_2R^2 \xrightarrow{H^+} R^1CH_2CS_2R^2 \tag{36}$$

Aromatic compounds are attacked by chlorodithioformates under Friedel–Crafts conditions[64,118,304,305] yielding aryl and alkyl dithiocarboxylates (equation 37):

$$ArH + ClCS_2R \xrightarrow{AlCl_3} ArCSSR + HCl \tag{37}$$

Dithio esters are also obtained from carbanions and dithiocarbonic acid derivatives (equations 38–40).

$$RMgX + ClCS_2Et \longrightarrow RCS_2Et + MgClX \quad \text{(Ref. 306)} \tag{38}$$

$$Ph_3P=CHR^1 + R^2S-CS_2R^2 \longrightarrow Ph_3P=CR^1-CS_2R^2 \quad \text{(Ref. 280)} \tag{39}$$

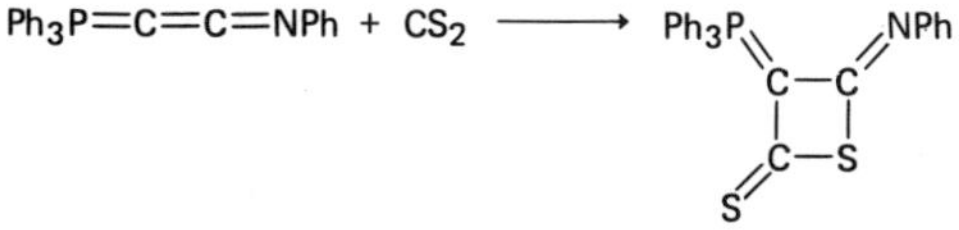

$$Ph_3P=C=C=NPh + CS_2 \longrightarrow$$

$$\text{(Ref. 307)} \tag{40}$$

Dimethyl tetrathiooxalate (**36**), which has been sought after for a long time, has now become available by photolysis of the 1,3-dithiolone-(2) **35** (equation 41)[63].

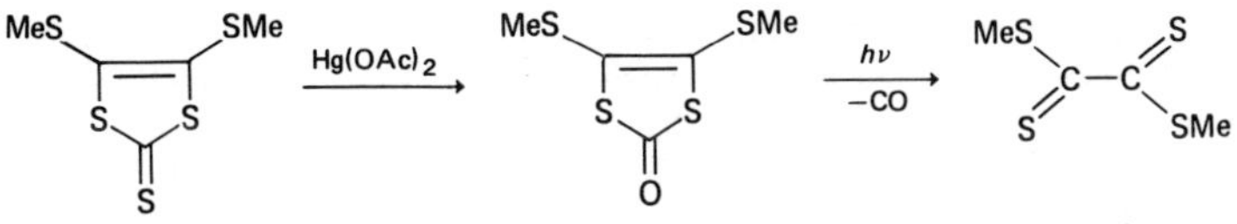

$$\tag{41}$$

3-Methylmercapto-1,2-dithiolium cations (**37**)[36,308], 1,2-dithiolthione-(2)[114], 3-methylmercapto-5-imino-1,2-dithioles[309] and isothiazolethiones-(5)[310] are cleaved by amines to form, for example, the β-aminodithioacrylates **38**.

Finally, dithio esters have been prepared from benzyl sulphides in two steps (equation 42)[311].

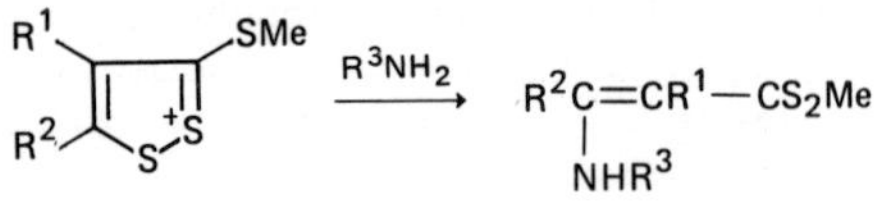

$$R^2C=CR^1-CS_2Me$$
$$\underset{NHR^3}{|}$$

(37)   (38)

$$ArCH_2SR \xrightarrow{Cl_2} ArCHClSR \xrightarrow[NEt_3/DMF]{S_8} ArCS_2R \qquad (42)$$

E. Thioacyl Halides and Anhydrides

Aromatic thioacyl chlorides have been conveniently prepared using phosgene[312] or pyrocatechylphosphorus trichloride[313] as chlorinating agents, or in two steps from dithio esters according to equation (43)[314]:

$$ArCS_2R \xrightarrow{Cl_2} ArCCl_2SCl \xrightarrow{Ph_3P} ArCSCl \qquad (43)$$

Aliphatic thioacyl chlorides, which have long been unknown, have been made available recently. t-BuCH$_2$CSCl (39) and t-BuCHClCSCl can be obtained by addition of hydrogen chloride or chlorine, respectively, to t-BuCH=C=S, generated by flash pyrolysis of 4-t-butyl-1,2,3-thiadiazole or by direct reaction with this heterocycle. The addition of hydrogen chloride to the alkynethiolate t-BuC≡C–S⁻ also gives 39[50]. Addition of thiophosgene to t-BuC(NMe$_2$)=CH$_2$ provides the thioacyl chloride salt (Me$_2$$\overset{+}{N}$=C($t$-Bu)CH$_2$CSCl) Cl⁻[315].

The preparation of α-Oxo thioacyl chlorides according to equations (44)[316] and (45)[317] has been reported. However, mainly trisulphides (RCOCCl$_2$S)$_2$S seem to be formed from methyl ketones[415].

$$RCOCH_3 \xrightarrow{SOCl_2} RCOCHClSCl \xrightarrow{-HCl} RCOCSCl \qquad (44)$$

$$R = t\text{-Bu, Ph, MeO}_2C$$

$$R_2^1NCOCH_2SR^2 \xrightarrow{SO_2Cl_2} R_2^1NCOCCl_2SCl \xrightarrow[-Cl_2]{PPh_3} R_2^1NCOCSCl \qquad (45)$$

Tetrafluorodithiosuccinyl difluoride, FCSCF$_2$CF$_2$CSF, has been obtained from trifluoroiodoethylene and boiling sulphur[45].

The thioacyl bromide (CF$_3$)$_2$CHCSBr has been prepared by Raasch[139] as a purple liquid. It is thus very unlikely that the colourless crystals, described by Barnikow and Gabrio[313], are really thiobenzoyl bromide.

Thiobenzoyl hexafluoroantimonate, (PhCS)⁺ SbF$_6$⁻, is formed from thiobenzoyl chloride and silver hexafluoroantimonate[318].

Acetyl thioacyl sulphides RCS–S–COMe[66] and bis(thioacyl) sulphides ('trithio anhydrides'), RCS–S–CSR, are synthesized from the corresponding aliphatic[53,319] or aromatic[68] dithio carboxylic acids by acetylation or condensation with dicyclohexyl carbodiimide or t-butyl isocyanide.

III. CHEMICAL PROPERTIES OF THIO CARBOXYLIC ACID DERIVATIVES

A. Prototropic Behaviour

The acidity of mono- and dithio carboxylic acids has been discussed in Janssen's review[1] and recently by Jansons[144]. Some pK_a values are given in Table 14. As

TABLE 14. Acid dissociation constants of thio and dithio carboxylic acids

Compound	pK_a	References
HCO—OH	3.75	320
HCO—SH	2.06	320
HCS—SH	0.85	320
MeCO—OH	4.76	1
MeCO—SH	3.35	321
MeCSSH	2.57	321
PhCH$_2$CS—SH	2.05	144, 322
PhCO—OH	4.20	1
PhCO—SH	2.48	1
PhCS—SH	1.92	144, 323
1-C$_{10}$H$_7$CS—SH	1.26	324
2-C$_{10}$H$_7$CS—SH	1.96	324

would be expected from inductive and mesomeric effects the acidity increases in the order $RCO_2H < RCOSH < RCS_2H$.

Thio carboxylic acids[325] and thiolo[325,326] thiono[327,328] and dithio esters[327] exhibit proton acceptor properties too. Thiocarbonyl sulphur protonation and the occurrence of two species, probably E/Z isomers, at low temperature is observed for MeCSOR and $MeCS_2R$[327]. Recently the protonation constants pK_{TH^+} for some aromatic thiono esters in aqueous sulphuric acid have been measured[328]. The relative gas-phase proton affinities have been established as $MeCS_2Me > MeCOSMe > MeCO_2Me$ using ion cyclotron resonance techniques, whereas, due to solvation effects, in solution the basicity of thiolo esters is lower than that of esters[326].

Thio esters exhibit marked CH acidity of their α-protons, which has been discussed by Mayer[329] and is the reason for many typical reactions (cf. Section III). This is especially true for β-oxo and β-thioxo derivatives and has been mentioned in Section I.E.

B. Nucleophilic Reactions

1. Reduction

Thio esters readily take up an electron to form radical anions[43,65,115,118] (cf. Section I.F), which are persistent in aprotic solvents. Selected half-wave potentials for this process are given in Table 15 together with earlier results on the polarography of thio esters in protic media. Not unexpectedly, $-E_{1/2}$ decreases in the order thiol ester > thion ester > dithio ester because of the increased polarizability of thiocarbonyl derivatives and the possibility of d orbital participation in the S-alkyl group.

Controlled potential electroreduction of dithio esters in the presence of electrophiles yields mercaptals. The protons necessary for this reaction stem from the solvent (acetonitrile) or traces of water (equation 46). The stilbene derivative **40** is

$$R^1CS_2R^2 \xrightarrow[R^3X]{2e^-, 2H^+} R^1CH(SR^2)SR^3 + HX (+ R^1C(SR^2){=}C(SR^2)R^1) \qquad (46)$$
$$\text{(40)}$$

TABLE 15. Half-wave potentials for the polarographic reduction of thio esters

Compound	$-E_{1/2}(V)^a$	SSE[b]	Reference
MeCO—SMe	>2.0	A	331
MeCS—OMe	1.57	A	331
EtCS—SMe	1.78	B	330
	1.43	A	331
MeCOCH$_2$CS—OMe	1.43^c	C	332
EtS—COCO—SEt	1.15	B	43
EtO—CSCS—OEt	0.79	B	43
EtO—COCS—SEt	0.87	B	43
PhCO—SMe	2.04	B	115
	1.65	A	331
PhCS—OMe	1.63	B	115
	1.56	A	331
PhCS—SMe	1.34	B	115
	1.11	A	331
PhCS—OPh	1.31	A	64
PhCS—SPh	1.15	B	330
	1.03	A	64

$^a E_{1/2}$ is related to the saturated calomel electrode (SCE), if not otherwise noted.
^{b}SSE (solvent—solute—electrolyte): A = 40% aq. propanol-(2)/ phosphate buffer, B = acetonitrile/Pr$_4$N$^+$ClO$_4^-$, C = acetone/ Et$_4$N$^+$ClO$_4^-$.
^{c}Ag/AgCl reference electrode[332].

formed as by-product[330]. Electroreduction in aqueous solvents as well as the Clemmensen reduction has been extensively studied by Mayer and coworkers[331]. As shown in equation (47) the main products from dithio esters are thio ethers.

$$RCSSMe \xrightarrow[-H_2S]{2 \, 'H_2'} RCH_2SMe \qquad (47)$$

Analogous reduction of thiol esters is efficiently possible only in the aromatic series. Thiono esters are merely hydrolysed to the corresponding esters.

Thiocarbonyl reduction of thion esters to form the corresponding methylene compounds can, however, be achieved by Raney—Ni[269] or hydrogen sulphide[94], whereas hydrogenolysis of the alkyl—oxygen bond occurs with tributyl tin hydride, which reaction implies a very useful method for the conversion of alcohols to hydrocarbons (equation 48)[272]. Toluene is obtained from benzyl thiolacetate and

$$R^1OH \xrightarrow[\quad 2. \, H_2S \quad]{1. \, PhC(NMe_2)^+Cl \, Cl^-} PhCSOR^1 \xrightarrow{Bu_3SnH} R^1H + PhCOSSnBu_3 \qquad (48)$$

lithium in liquid ammonia[333], whereas lithium alanate cleaves thio esters to the corresponding thiols[181]. Semimercaptals (41) are formed from dithio esters and sodium borohydride (equation 49)[334].

$$EtCS_2Me \xrightarrow{NaBH_4} EtCH(SMe)SH \qquad (49)$$
$$(41)$$

2. Reactions with carbanions

Ketones are obtained in good yields in Grignard reactions of the easily available S-(2-pyridyl) thiocarboxylates (equation 50); no carbinols are formed as by-products[335]. This is also true for the reaction of lithium dialkyl cuprates (I) with thiol esters[336].

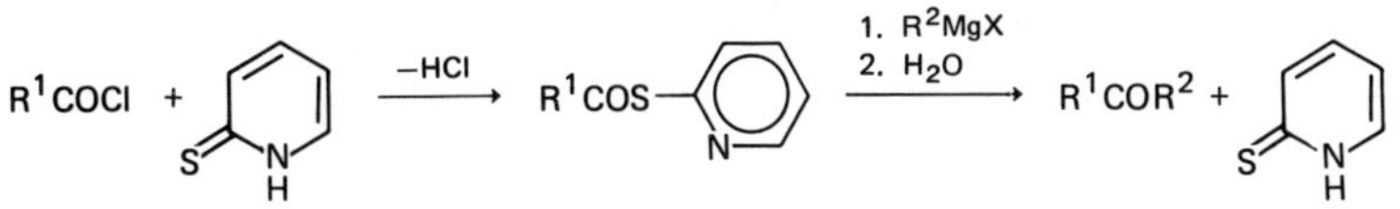

$$R^1COCl + \underset{\substack{\text{(pyridine-2-thione)}}}{\bigcirc} \xrightarrow{-HCl} R^1COS-\bigcirc \xrightarrow[\text{2. } H_2O]{\text{1. } R^2MgX} R^1COR^2 + \bigcirc \qquad (50)$$

Dialkyl bis(thioloxalates) yield α-oxo thiol esters on reaction with Grignard reagents (equation 51)[337].

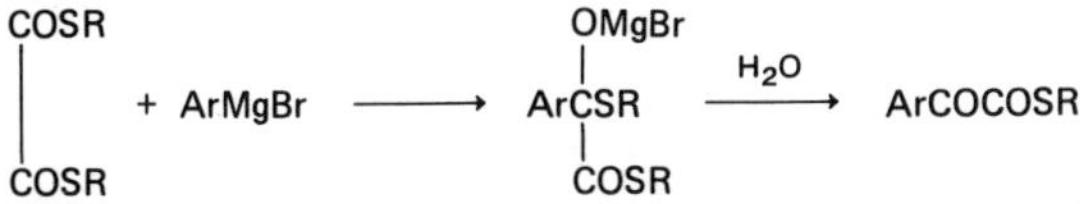

$$\begin{matrix} COSR \\ | \\ | \\ COSR \end{matrix} + ArMgBr \longrightarrow \begin{matrix} OMgBr \\ | \\ ArCSR \\ | \\ COSR \end{matrix} \xrightarrow{H_2O} ArCOCOSR \qquad (51)$$

β-Thioxo ketones are obtained after base-catalysed condensation of thion esters with methyl ketones[338]

Grignard reactions of dithio esters have been reviewed by Paquer[143]. Dithio acetals are the main products, which is indicative of a thiophilic addition of the organometallic reagent to the thiocarbonyl group. Dithio ketals and some stilbene (40) are obtained after methylation of the intermediate. Thiols and thiones are also formed in several cases (equation 52)[339]. The latter reaction path is predominant if

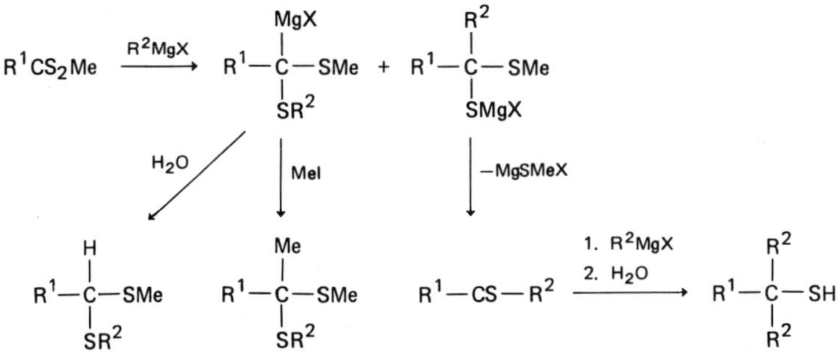

$$R^1CS_2Me \xrightarrow{R^2MgX} \begin{matrix} MgX \\ | \\ R^1-C-SMe \\ | \\ SR^2 \end{matrix} + \begin{matrix} R^2 \\ | \\ R^1-C-SMe \\ | \\ SMgX \end{matrix}$$

$$\swarrow H_2O \qquad \downarrow MeI \qquad \downarrow -MgSMeX \qquad (52)$$

$$\begin{matrix} H \\ | \\ R^1-C-SMe \\ | \\ SR^2 \end{matrix} \qquad \begin{matrix} Me \\ | \\ R^1-C-SMe \\ | \\ SR^2 \end{matrix} \qquad R^1-CS-R^2 \xrightarrow[\text{2. } H_2O]{\text{1. } R^2MgX} \begin{matrix} R^2 \\ | \\ R^1-C-SH \\ | \\ R^2 \end{matrix}$$

unsaturated Grignard reagents ($R^2 = CH_2{=}CHCH_2$, $CH_2{=}CH$, $HC{\equiv}CCH_2$) are used[340,341]. Inversion of allylic chains and direct carbophilic addition rather than initial thiophilic addition followed by [2, 3]sigmatropic shift, which has been postulated in earlier work[147,340], takes place (equation 53)[341]. β-Oxo diothio

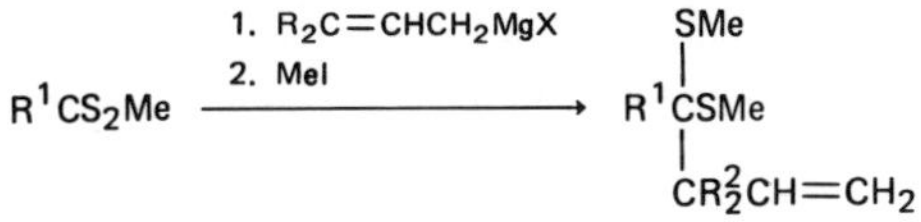

$$R^1CS_2Me \xrightarrow[\text{2. MeI}]{\text{1. } R_2C{=}CHCH_2MgX} \begin{matrix} SMe \\ | \\ R^1CSMe \\ | \\ CR_2^2CH{=}CH_2 \end{matrix} \qquad (53)$$

esters form hydroxyclyclopropenone mercaptals with alkylmagnesium bromides in a stereoselective reaction (equation 54)[342].

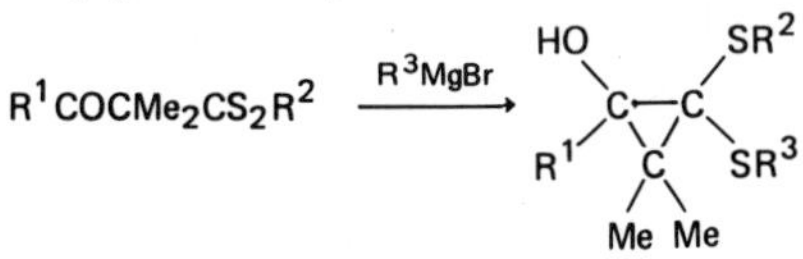

$$R^1COCMe_2CS_2R^2 \xrightarrow{R^3MgBr} \begin{matrix} HO \qquad SR^2 \\ \diagdown \; / \\ C{-}C \\ R^1 \diagup \overset{|}{C} \diagdown SR^3 \\ \diagup \diagdown \\ Me \; Me \end{matrix} \qquad (54)$$

Reaction of phenyl dithiobenzoate with phenyllithium gives bis(phenylthio)-phenylmethane as a product of thiophilic attack (equation 55) together with some stilbene (40)[343].

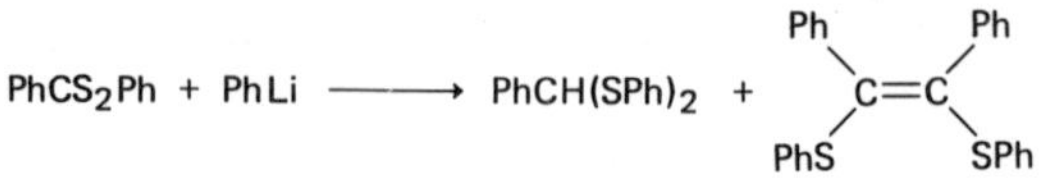

$$PhCS_2Ph + PhLi \longrightarrow PhCH(SPh)_2 + \underset{PhS}{\overset{Ph}{\diagdown}}C{=}C\underset{SPh}{\overset{Ph}{\diagup}} \tag{55}$$

(40)

3. Solvolysis

The reactivity of various nucleophiles towards 2,2,2-trifluoroethyl thiolacetate has been studied systematically by Gregory and Bruice[344]. Rate constants for the alkaline hydrolysis of thiol esters have been reported[29,345]. The alcoholysis of thiol esters (acetyl–CoA) plays an important role in biochemistry[346], which cannot be discussed here. It provides an efficient route to esters, because thiol esters are strong acylating agents[347], which are further activated by metal ions[210,348,349]. The synthesis of macrocyclic lactones[236,350], e.g. in macrolid antibiotics[210,237,350], can be achieved using thiol esters. Alcoholysis and thiolysis of dithio esters have been treated in Sections II.C and II.D. Thiols having functional groups are readily prepared by treating the corresponding thiol esters with 2-amino-ethanethiol[351].

Aminolysis of thiol esters has been intensely studied from a mechanistic point of view[352]. Aminolysis of thiono and dithio esters normally yields thio amides and related compounds. The preparative scope of this reaction is therefore not treated in this chapter. Mechanistic studies have been performed by Tao, Scheithauer and Mayer[353] and by Bruice and Mautner[354].

The unusual thioacyl isothiocyanates ArCSN=C=S are obtained from thioacyl chlorides and sodium thiocyanate[313].

C. Electrophilic Reactions

1. Oxidation

Desulphuration of dithio esters to the corresponding thiol esters and of thion esters to esters can be achieved by various oxidizing agents such as Ag$^+$[115,355], Cu^{2+}[356] Hg^{2+}[357] and KMnO$_4$[358]. Thiono esters may be oxidized to anhydrides by Hg^{2+} as well[359]. The metal-ion promoted (which does not mean 'catalysed') reactions of thio esters have been thoroughly reviewed by Satchell[349]. Ozone yields carboxylic acids and sulphonic acids from thiol esters in acetic acid as solvent[360].

The reactions of phenyl thiolacetate with halogens are shown in equation (56)[361]. Dithio esters yield stable α-bromo derivatives (42), which are useful for

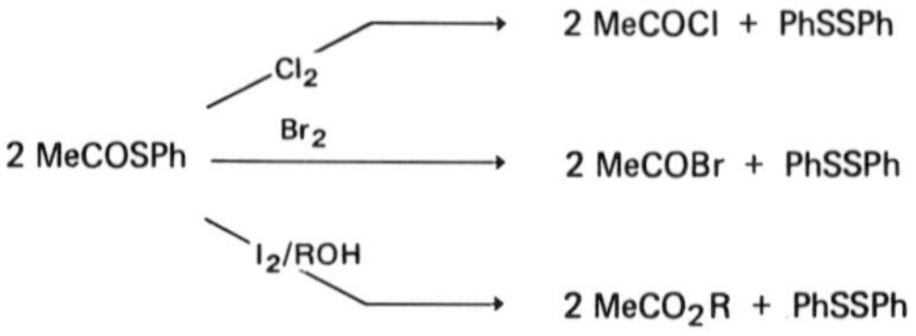

$$2\ MeCOSPh \quad \begin{array}{c} \overset{Cl_2}{\nearrow} \quad 2\ MeCOCl + PhSSPh \\ \xrightarrow{Br_2} \quad 2\ MeCOBr + PhSSPh \\ \underset{I_2/ROH}{\searrow} \quad 2\ MeCO_2R + PhSSPh \end{array} \tag{56}$$

the preparation of highly branched dithio esters according to equation $(57)^{362}$, whereas the disulphides **43** are formed with iodine (equation $58)^{363}$.

$$Me_2CHCS_2Me \xrightarrow{Br_2} Me_2CBrCS_2Me \xrightarrow{MeMgX} Me_3CCS_2Me \tag{57}$$
$$(42)$$

$$RCH_2CS_2Me \xrightarrow[\text{2. } I_2]{\text{1. base}} RCH=\underset{\underset{SMe}{|}}{C}-S-S-\underset{\underset{SMe}{|}}{C}=CHR \tag{58}$$
$$(43)$$

S-Oxides (sulphines) of various types (**44—49**) have been prepared from the corresponding thio carboxylic acid derivatives. Peroxy carboxylic acids such as

$$\underset{\underset{SO}{\|}}{ArCSR}$$
(44) (Refs. 304, 314, 364, 365)

$$\underset{\underset{SO}{\|}}{ArCSOR}$$
(45) (Refs. 304, 365)

$$\underset{\underset{SO}{\|}}{ArCSO_2R}$$
(46) (Refs. 304, 365)

$$\underset{\underset{SO}{\|}}{ArCCl}$$
(47) (Refs. 314, 366)

$$\underset{\underset{SO}{\|}}{ArCN_3}$$
(48) (Ref. 367)

$$\underset{\underset{SO}{\|}}{ArCSCN}$$
(49) (Ref. 366)

m-chloroperbenzoic acid have been mainly used as oxidants, but ozone[364] or chlorine[314] are suitable in certain cases too. **48** and **49** are prepared by nucleophilic substitution of **47** with NaN_3[367] or KSCN[366], whereas **44** (R = Ph) is obtained from **47** and sodium thiophenolate[366].

Phenyl dithiomesitoate yields the thiocarbonyl-S-imide **50** on oxidation with chloramine-T (equation $59)^{368}$.

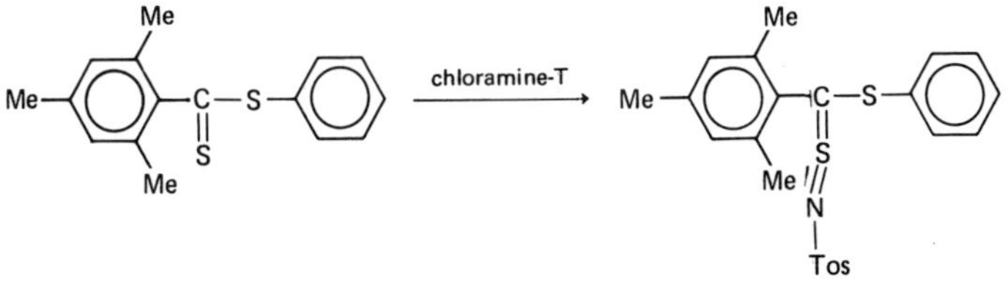

$$(50)$$

The configuration and conformation of **44—47** and **50** has been extensively studied by n.m.r. spectroscopy[304,369], dipole-moment[370] and X-ray diffraction[371] measurements. In many cases stable E and Z isomers of **44—47** and **50** can be cleanly separated because the rotational barrier $\Delta G^{\ddagger}$ for the rotation of the C=S=O group is very high.

2. Alkylation and acylation

Alkylation of thiono and dithio esters with diazoalkanes has already been mentioned in Section II.C, as it provides new thio esters in many cases. Episulphides are the intermediates in this reaction and olefins are formed as typical by-products by sulphur extrusion from the latter (equation $60)^{255}$.

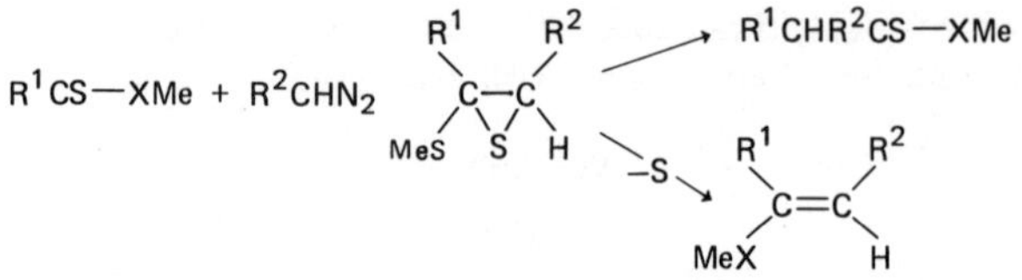

$$R^1CS-XMe + R^2CHN_2 \quad \begin{array}{c} R^1 \quad R^2 \\ \diagdown \diagup \\ C-C \\ \diagup \diagdown \diagdown \\ MeS \quad S \quad H \end{array} \quad \begin{array}{c} \nearrow \quad R^1CHR^2CS-XMe \\ \\ \searrow \quad \begin{array}{c} R^1 \quad R^2 \\ \diagdown \diagup \\ C=C \\ \diagup \diagdown \\ MeX \quad H \end{array} \end{array} \tag{60}$$

Alkylation with alkyl halides occurs at the thiocarbonyl sulphur atom of thio esters and ketene S,S-acetals are formed by deprotonation of the α-position (equation 61)[54, 109, 140, 161, 276, 294, 372]. Ketene S,S-acetals are also formed with dimethyl sulphoxonium methylide. If the substrate dithio ester does not contain α-hydrogen atoms, olefins are the products (equation 62)[373].

$$R_2^1CHCS_2R^2 \xrightarrow[-HX]{R^3X} R_2^1C=C\diagup\diagdown{SR^2}_{SR^3} \tag{61}$$

$$R^1CS_2Me + Me_2SO=CH_2 \longrightarrow R^1C(SMe)=CHR^2 \tag{62}$$
$$R^1 = t\text{-Bu}, R^2 = SMe$$
$$R^1 = Ph, R^2 = H$$

Very strongly alkylating reagents such as FSO_3Me and $ClSO_3Me$ attack methyl thiolbenzoate at the sulphur atom and the sulphonium ion $PhCOSMe_2^+$ is formed[374]. Lithium alkyls abstract protons from the alkyl group of $ArCOSCH_2R^2$ and the rather stable carbanions react with electrophiles to form $ArCOSCHR^1R^2$ (R^2 = Me, $CH_2CH=CH_2$, $SiMe_3$, CHOH–Ph)[375]. Intramolecular alkylation, i.e. rearrangement of thiono esters to the corresponding thiolo esters, takes place by heating[94, 190, 194, 376] or, more efficiently, by catalysis with BF_3/Et_2O[191] or $(Et_3O)^+BF_4^-$[192,193].

Glycidic acid thiol esters, which are easily obtained, rearrange according to equation (63)[377].

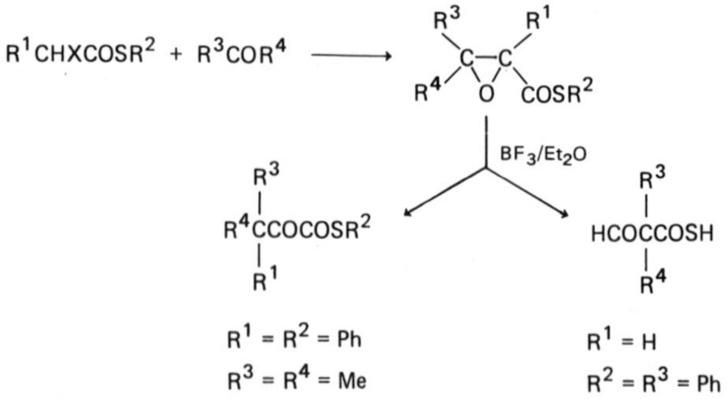

$$R^1CHXCOSR^2 + R^3COR^4 \longrightarrow \begin{array}{c} R^3 \quad R^1 \\ \diagdown \diagup \\ C-C \\ R^4 \diagup \diagdown O \diagdown COSR^2 \end{array} \tag{63}$$

$$\downarrow BF_3/Et_2O$$

$$\swarrow \qquad \searrow$$

$$\begin{array}{c} R^3 \\ | \\ R^4CCOCOSR^2 \\ | \\ R^1 \end{array} \qquad \begin{array}{c} R^3 \\ | \\ HCOCCOSH \\ | \\ R^4 \end{array}$$

$$R^1 = R^2 = Ph \qquad R^1 = H$$
$$R^3 = R^4 = Me \qquad R^2 = R^3 = Ph$$

Metalation in the α-position of thiol esters and subsequent reaction with ketones yields β-hydroxy thiol esters[378].

Base-catalysed self-acylation of ethyl thionoacetate yields ethyl thioacetyl thionoacetate[46,379], whereas dibenzoylmethane is formed from methyl thiolbenzoate and strong bases in a complicated reaction[380]. Activated thiolo[381], thiono[120] or dithio esters[120] are acylated at the α-carbon (equation 64) or thiocarbonyl sulphur atom (equation 65), which is also true for the related sulphenylation reaction (forming disulphides[47]).

$$CH_2(COSEt)_2 \xrightarrow[\text{2. RCOCl}]{\text{1. Mg}} RCOCH(COSEt)_2 \qquad (64)$$

$$EtO_2CCH_2CSXEt \xrightarrow[\text{2. RCOCl}]{\text{1. KOBu-}t} EtO_2CCH=C\begin{array}{c}\diagup SCOR \\ \diagdown XEt\end{array} \qquad (65)$$

$$X = O, S$$

D. Non-polar Reactions

Coyle[382] has reviewed the photochemical behaviour of thiol and thion esters.
Thiol esters give fragmentation products on ultraviolet irradiation according to equation (66). S-Acyl splitting is the main reaction path, while S-alkyl bond

$$R^1COSR^2 \xrightarrow{h\nu} [R^1CO\cdot + \cdot SR^2] \longrightarrow R^2SSR^2 + R^2CHO + R^1COCOR^1 \qquad (66)$$

cleavage gives rise to minor by-products[383]. S-Aryl thiolbenzoates of type **51** are photolytically[384] or thermally[385] $(X = SO_2Me)$ cyclized to the thiaxanthones **52**.

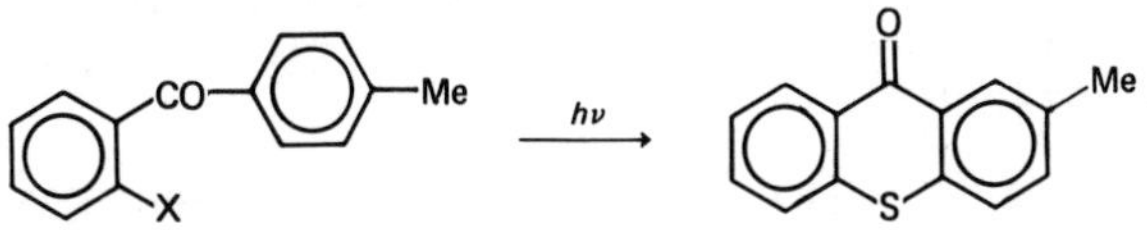

(51) X = Cl, Br, I, SO$_2$Me   (52)

The formation and reactions of polythiol esters, e.g. **53–57**, have been reviewed by Sviridova and Prilezhaeva[386] and recently in a valuable monograph by Goethals[387].

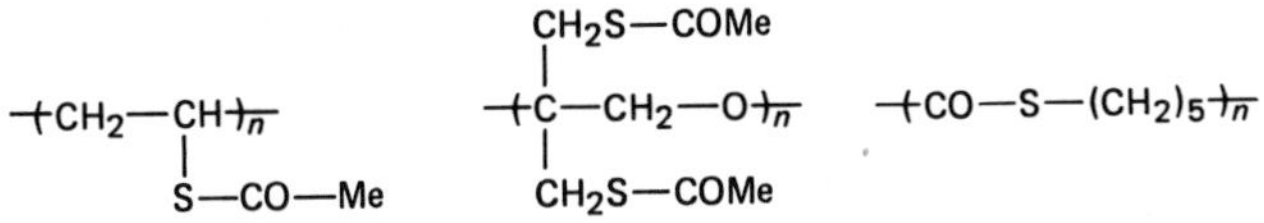

(53)   (54)   (55)

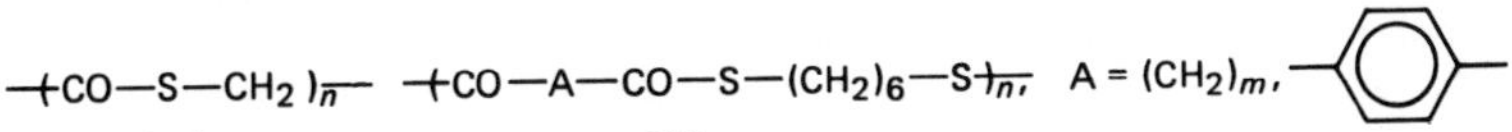

(56)   (57)

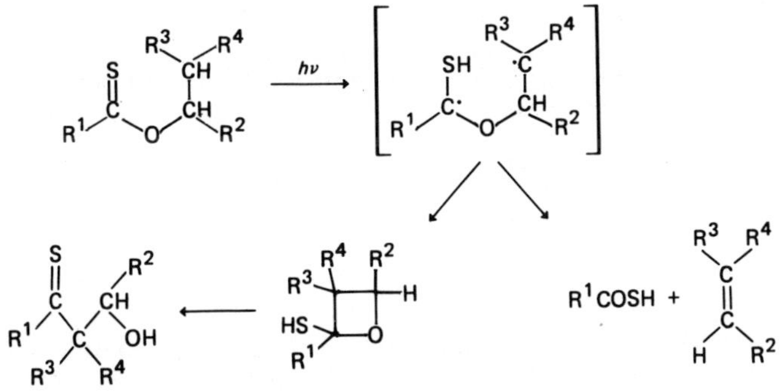

(67)

Norrish Type II photoelimination takes place with thion esters[196, 268, 271, 388] and has been used for the synthesis of olefins (equation 67). On the other hand, thiono—thiolo ester[196] or photo-Fries rearrangement to thioketones via an oxetan[196,389] can occur in suitable cases (equation 67).

O-Ethyl thioalkanoates are photolytically desulphurized and enediol ethers are formed (equation 68)[390].

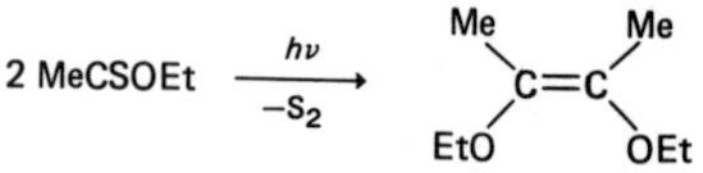

$$2 \text{ MeCSOEt} \xrightarrow[-S_2]{h\nu} \underset{\underset{\text{EtO}}{|}}{\overset{\overset{\text{Me}}{|}}{\text{C}}} = \underset{\underset{\text{OEt}}{|}}{\overset{\overset{\text{Me}}{|}}{\text{C}}} \tag{68}$$

Photolysis in the presence of olefins has been investigated by Ohno and co-workers[391]. Thietans and ketones are formed according to equation (69):

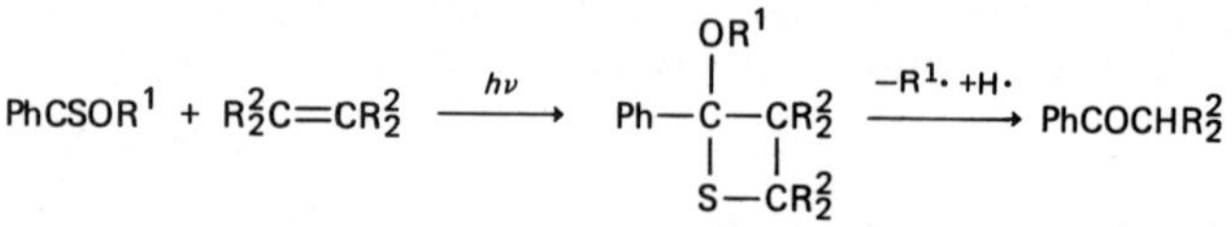

$$\text{PhCSOR}^1 + \text{R}_2^2\text{C}=\text{CR}_2^2 \xrightarrow{h\nu} \underset{\underset{\text{S}-\text{CR}_2^2}{|}}{\overset{\overset{\text{OR}^1}{|}}{\text{Ph}-\text{C}-\text{CR}_2^2}} \xrightarrow{-\text{R}^1, +\text{H}\cdot} \text{PhCOCHR}_2^2 \tag{69}$$

Methyl dithiobenzoate adds 2,3-diphenylazirine on ultraviolet irradiation to yield 2,4,5-triphenyl-5-methylthiothiazoline[392]. Radical-induced cyclizations of alkyl dithioisobutyrate and methyl dithio-Δ^4-pentenoate take place according to equations (70) and (71)[393].

$$\tag{70}$$

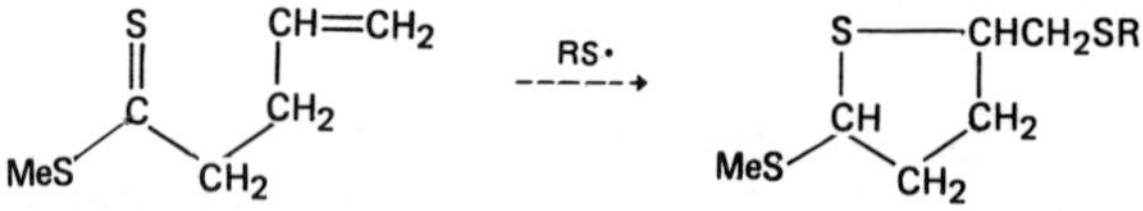

$$\tag{71}$$

E. Formation of Heterocycles

Various heterocycles can be obtained from thio and dithio carboxylic acids and their derivatives. A detailed treatment of this topic is not within the scope of this article. Only a short specification and literature references are therefore given.

Heterocyclic systems without sulphur in the ring, which are available, include pyrrolidinedithione[61], imidazoline[394], oxazoline[95,394], pyrazolone[155], pyrone[395], piperazinedione[207] and quinoxalinone[95,396,397].

In most cases sulphur-containing rings are formed: thiirans (episulphides)[184,255,390], thietan[198,200,391], dithietan[21], thiophene[363,398-400], the bicyclic compounds 58[401] and 59[402], thiapyrane[403], thiapyrone[385,404], 1,3-

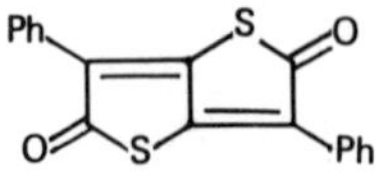

(58)

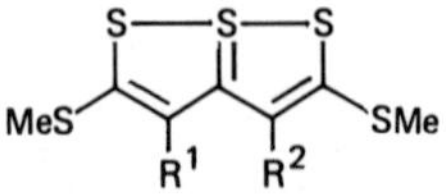

(59)

oxathiolium[400,405] and 1,3-dithiolium[167,168] cations, 1,3-dithioles[168, 255, 294, 393, 406], thiazole[95, 397, 399, 407], isothiazole[281,408], 1,4-oxathian[198],

1,3-dithian[200,294,393], 1,3-thiazine[409], 1,2,3-thiadiazole[255, 263, 266, 410], 1,2,4-thiadiazole[255,411], 1,3,4-thioxazole[114,412].

IV. REFERENCES

1. M. J. Janssen in *The Chemistry of Carboxylic Acids and Esters* (Ed. S. Patai), John Wiley and Sons, London, 1969, Chap. 15, pp. 705–764.
2. J. DeRooij, F. C. Mijlhoff and G. Renes, *J. Mol. Struct.*, **25**, 169 (1975).
3. W. H. Hocking and G. Winnewisser, *Z. Naturforsch.*, **31a**, 995 (1975).
4. W. H. Hocking and G. Winnewisser, *Z. Naturforsch.*, **31a**, 438 (1976); *Chem. Commun.*, 63 (1975).
5. M. M. Borel and M. Ledersert, *Acta Cryst.*, **B30**, 2777 (1974).
6. R. Mattes, W. Meschede and W. Stork, *Chem. Ber.*, **108**, 1 (1975).
7. R. Mattes, W. Meschede and U. Niemer, *Chem. Ber.*, **110**, 2584 (1977).
8. W. Meschede and R. Mattes, *Chem. Ber.*, **109**, 2510 (1976).
9. G. I. L. Jones, D. G. Lister, N. L. Owen, M. C. L. Gerry and P. Palmieri, *J. Mol. Spectrosc.*, **60**, 348 (1976).
10. J. J. Guy and T. A. Hamor, *Acta Cryst.*, **B30**, 2277 (1974).
11. G. Kiel, M. Dräger and U. Reuter, *Chem. Ber.*, **107**, 1483 (1974).
12. M. A. Pelinghelli, A. Tiripicchio and M. Tiripicchio Camellini, *Cryst. Struct. Commun.*, **3**, 159 (1974); *Chem Abstr.*, **80**, 101149 (1974).
13. U. Niemer, K. Mennemann and R. Mattes, *Chem. Ber.*, **111**, 2113 (1978).
14. N. Bresciani Pahor and M. Calligaris, *Acta Cryst.*, **B31**, 2685 (1975).
15. E. Schaumann and U. Behrens, *Angew. Chem.*, **89**, 750 (1977).
16. G. Adiwidjaja and J. Voss, *Chem. Ber.*, **110**, 3792 (1977).
17. R. Engler, G. Kiel and G. Gattow, *Z. Anorg. Allgem. Chem.*, **404**, 71 (1974).
18. M. M. Borel and M. Ledersert, *Z. Anorg. Allgem. Chem.*, **415**, 285 (1975).
19. R. Engler, M. Dräger and G. Gattow, *Z. Anorg. Allgem. Chem.*, **404**, 81 (1974).
20. R. Mattes and W. Meschede, *Chem. Ber.*, **109**, 1832 (1976).
21. M. Mikolajczyk, P. Kiełbasinski, J. H. Barlow and D. H. Russell, *J. Org. Chem.*, **42**, 2345 (1977).
22. G. Adiwidjaja and J. Voss, *J. Chem. Res.*, (S) 256, (M) 2923 (1977).
23. G. Bombieri, E. Forsellini, U. Chiacchio, P. Fiandaca, G. Purello, E. Foresti and R. Graziani, *J. Chem. Soc., Perkin II*, 1404 (1976).
24. G. I. L. Jones and N. L. Owen, *J. Mol. Struct.*, **118**, 1 (1973).
25. O. Exner, V. Jehlička and J. Firl, *Coll. Czech. Chem. Commun.*, **36**, 2936 (1971).
26. Y. Hirabayashi, *Bull. Chem. Soc. Japan*, **38**, 175 (1965).
27. J. M. Beiner, C. G. Andrieu and A. Thuillier, *Compt. Rend.*, **B274**, 407 (1972).
28. H. Lumbroso and P. J. W. Schuijl, *Compt. Rend.*, **C264**, 925 (1967).
29. E. Bock, A. Queen, S. Brownlee, T. A. Nour and M. N. Paddonkow, *Can. J. Chem.*, **52**, 3113 (1974).
30. I. Wallmark, M. H. Krackow, S. H. Chu and H. G. Mautner, *J. Amer. Chem. Soc.*, **92**, 4447 (1970).
31. P. Stäglich, K. Thimm and J. Voss, *Ann. Chem.*, 671 (1974).
32. J. F. Skinner and J. H. Markgraf, *J. Mol. Struct.*, **28**, 177 (1975).
33. H. Lumbroso and P. Reynaud, *Compt. Rend.*, **C262**, 1739 (1966).
34. O. Exner, V. Jehlička and A. Ohno, *Coll. Czech. Chem. Commun.*, **36**, 2157 (1971).
35. K. Herzog, E. Steger, P. Rosmus, S. Scheithauer and R. Mayer, *J. Mol. Struct.*, **3**, 339 (1969).
36. G. LeCostumer and Y. Mollier, *Bull. Soc. Chim. Fr.*, 2958 (1971).
37. P. A. Hopkins and R. J. W. LeFèvre, *J. Chem. Soc., Perkin II*, 338 (1971).
38. O. Exner, P. Dembech, G. Seconi and P. Vivarelli, *J. Chem. Soc., Perkin II*, 1870 (1973).
39. V. Galasso and G. C. Pappalardo, *J. Chem. Soc., Perkin II*, 574 (1976).
40. R. Engler and G. Gattow, *Z. Anorg. Allgem. Chem.*, **388**, 78 (1972).
41. S. Nagata, T. Yamabe and M. Fukui, *J. Phys. Chem.*, **79**, 2335 (1975).
42. J. A. Kampmeier and G. Chen, *J. Amer. Chem. Soc.*, **87**, 2608 (1965).

43. J. Voss, K. Thimm and L. Kistenbrügger, *Tetrahedron*, 33, 259 (1977).
44. K. Thimm, *Dissertation*, Universität Hamburg, 1975.
45. W. J. Middleton, *J. Org. Chem.*, 40, 129 (1975).
46. A. R. Hendrickson and R. L. Martin, *Australian J. Chem.*, 25, 257 (1972).
47. K. Thimm and J. Voss, *Z. Naturforsch.*, 29b, 419 (1974).
48. J. Fabian, S. Scheithauer and R. Mayer, *J. Prakt. Chem.*, 311, 45 (1969).
49. A. Ohno, T. Koizume, Y. Ohnishi and G. Tsuchihashi, *Bull. Chem. Soc. Japan*, 42, 3556 (1969).
50. G. Seybold, *Angew. Chem.*, 87, 710 (1975).
51. R. Engler and G. Gattow, *Z. Anorg. Allgem. Chem.*, 389, 145 (1972).
52. C. Furlani, A. Flamini, A. Sgamellotti, C. Bellitto and O. Piovesana, *J. Chem. Soc., Dalton*, 2404 (1973).
53. C. Furlani and A. Luciani, *Inorg. Chem.*, 7, 1586 (1968).
54. F. C. Larsson and S.-O. Lawesson, *Tetrahedron*, 28, 5341 (1972).
55. R. Engler, G. Gattow and M. Dräger, *Z. Anorg. Allgem. Chem.*, 388, 229 (1972).
56. S. Kato, A. Hori, H. Shiotani, M. Mizuta, N. Hayashi and T. Takakuwa, *J. Organometallic Chem.*, 82, 223 (1974).
57. N. H. Leon, *J. Pharmac. Sci.*, 65, 146 (1976).
58. S. Kato, M. Mizuta and Y. Ishii, *J. Organometallic Chem.*, 55, 121 (1973).
59. S. Kato, T. Takagi, T. Katada and M. Mizuta, *Angew. Chem.*, 89, 820 (1977).
60. A.-M. Lamazouère, J. Sotiropoulos, and P. Bedos, *Compt. Rend.*, C270, 828 (1970).
61. G. Gattow and K. Hanewald, *Chem. Ber.*, 109, 3243 (1976).
62. W. Thiel, H. Viola and R. Mayer, *Z. Chem.*, 17, 366 (1977).
63. T. Kissel, R. Matusch and K. Hartke, *Z. Chem.* 16, 318 (1976).
64. H. Viola, S. Scheithauer and R. Mayer, *Chem. Ber.*, 101, 3517 (1968).
65. C.-P. Klages, *Diplomarbeit*, Universität Hamburg, 1975.
66. S. Kato, K. Sugino, M. Yamada, T. Katada and M. Mizuta, *Angew. Chem.*, 89, 917 (1977).
67. S. Kato, T. Mitani and M. Mizuta, *Bull. Chem. Soc. Japan*, 45, 3653 (1972).
68. S. Kato, T. Katada and M. Mizuta, *Angew. Chem.*, 88, 844 (1976).
69. S. Kato, T. Kato, T. Kataoka and M. Mizuta, *Int. J. Sulfur Chem.*, 8, 437 (1973).
70. T. Katada, S. Tsuji, T. Sugiyama, S. Kato and M. Mizuta, *Chem. Letters*, 441 (1976).
71. S. Kato, W. Akada, A. Mizuta and Y. Ishii, *Bull. Chem. Soc. Japan*, 46, 244 (1973).
72. S. Kato, T. Kato, T. Yamauchi, Y. Shibahashi, E. Kakuda, M. Mizuta and Y. Ishii, *J. Organometallic Chem.*, 76, 215 (1974).
73. T. Katada, S. Kato and M. Mizuta, *Chem. Letters*, 1037 (1975).
74. T. Yamabe, K. Agaki, S. Nagata, H. Kato and K. Fukui, *J. Phys. Chem.*, 80, 611 (1976).
75. M. Bossa, *J. Chem. Soc.(B)*, 1182 (1969).
76. J. Fabian, *Theoret. Chim. Acta*, 12, 200 (1968).
77. A. Flamini, E. Semprini and G. Condorelli, *Chem. Phys. Letters*, 32, 365 (1975).
78. L. J. Bellamy, *The Infrared Spectra of Complex Molecules*, 3rd Ed., Vol. I, Chapman and Hall, London, 1975, Chap. 11.5, p. 214, Chap. 22, p. 394; *Advances in Infrared Group Frequencies*, Vol. II of *The Infrared Spectra of Complex Molecules*, Chapman and Hall, London, 1968, Chap. 5.8.8., p. 170, Chap. 6.6., p. 212.
79. G. A. Crowder, *Appl. Spectrosc.*, 27, 440 (1973).
80. H. S. Randhawa and C. N. R. Rao, *J. Mol. Struct.*, 21, 123 (1974).
81. G. A. Crowder, E. Robertson and K. Potter, *Can. J. Spectrosc.*, 20, 49 (1975); *Chem. Abstr.*, 83, 42488 (1975).
82. H. S. Randhawa, W. Walter and C.-O. Meese, *J. Mol. Struct.*, 37, 187 (1977).
83. H. S. Randhawa and W. Walter, *J. Mol. Struct.*, 38, 89 (1977).
84. H. S. Randhawa, C.-O. Meese and W. Walter, *J. Mol. Struct.* 36, 25 (1977).
85. G. A. Crowder, *J. Mol. Struct.*, 32, 207 (1976).
86. G. A. Crowder, *Appl. Spectrosc.*, 26, 468 (1972).
87. B. Bicca de Alencastro and C. Sandorfy, *Can. J. Chem.*, 51, 1443 (1973).
88. R. Mattes, W. Stork and J. Kahlenberg, *Spectrochim. Acta*, 33A, 643 (1977).
89. A. J. Collings, P. E. Jackson and K. J. Morgan, *J. Chem. Soc.(B)*, 581 (1970).
90. H. Gilman and G. D. Lichtenwalter, *J. Org. Chem.*, 25, 1064 (1960).

91. S. Kato and M. Mizuta, *Bull. Chem. Soc. Japan*, **46**, 860 (1973).
92. P. Stäglich, *Dissertation*, Universität Hamburg, 1974.
93. P. Vinkler, K. Thimm and J. Voss, *Ann. Chem.*, 2083 (1976).
94. K. Hartke and H. Hoppe, *Chem. Ber.*, **107**, 3121 (1974).
95. H. Hoppe and K. Hartke, *Arch. Pharmazie*, **308**, 526 (1975).
96. R. Mayer, E. Schinke, P. Rosmus and S. Scheithauer, *J. Prakt. Chem.*, **312**, 767 (1970).
97. R. Mattes and W. Stork, *Spectrochim. Acta*, **30A**, 1385 (1974).
98. E. Lindner and U. Kunze, *Z. Anorg. Allgem. Chem.*, **383**, 255 (1971); *Chem. Ber.*, **102**, 3347 (1969).
99. H. S. Randhawa and D. H. Sharma, *Z. Physik. Chem.*, **86**, 108 (1973).
100. K. A. Jensen, H. Mygind, P. H. Nielsen and G. Borch, *Acta Chem. Scand.*, **24**, 1492 (1970).
101. S. Kato and M. Mizuta, *Bull. Chem. Soc. Japan*, **45**, 3492 (1972).
102. R. Mattes, W. Stork and I. Pernoll, *Z. Anorg. Allgem. Chem.*, **404**, 97 (1974).
103. S. Kato, T. Mitani and M. Mizuta, *Int. J. Sulfur Chem.*, **8**, 359 (1973).
104. H.-V. Gründler and E. Steger, *Z. Chem.*, **12**, 304 (1972).
105. D. Holsboer and H. Kloosterziel, *Rec. Trav. Chim.*, **91**, 1371 (1972).
106 R. Radeglia, S. Scheithauer and R. Mayer, *Z. Naturforsch.*, **24b**, 283 (1969).
107. M. Saquet and A. Thuillier, *Bull. Soc. Chim. Fr.*, 2841 (1967).
108. A. J. Bridges and G. H. Witham, *J. Chem. Soc., Perkin I*, 1603 (1975).
109. G. Duguay and H. Quiniou, *Bull. Soc. Chim. Fr.*, 637 (1972).
110. K. Hartke and F. Meissner, *Tetrahedron*, **28**, 875 (1972).
111. H. Kolind-Andersen, L. Dalgaard, L. Jensen and S.-O. Lawesson, *Rec. Trav. Chim.*, **92**, 1169 (1973).
112. E. A. Noe, *J. Amer. Chem. Soc.*, **99**, 2803 (1977).
113. C. M. Hall and J. Wemple, *J. Org. Chem.*, **42**, 2118 (1977).
114. J. Maignan and J. Vialle, *Bull. Soc. Chim. Fr.*, 1973 (1973).
115. J. Voss, W. Schmüser and K. Schlapkohl, *J. Chem. Res.*, (S) 144, (M) 1801 (1977).
116. E. O. Fischer and S. Riedmüller, *Chem. Ber.*, **107**, 915 (1974).
117. J. Perregard, B. Pedersen and S.-O. Lawesson, *Acta Chem. Scand.*, **B31**, 460 (1977).
118. J. Voss and K. Schlapkohl, *Tetrahedron*, **31**, 2982 (1975).
119. N. H. Leon and R. S. Asquith, *Tetrahedron*, **26**, 1719 (1970).
120. K. Thimm and J. Voss, *Z. Naturforsch.*, **30b**, 932 (1975).
121. J. M. Purcell and H. Susi, *Appl. Spectrosc.*, **19**, 105 (1965).
122. G. C. Levy and G. L. Nelson, *Carbon-13 Nuclear Magnetic Resonance for Organic Chemists*, Wiley–Interscience, New York, 1972, p. 133.
123. R. Radeglia and S. Scheithauer, *Z. Chem.*, **14**, 20 (1974).
124. H.-O. Kalinowski and H. Kessler, *Angew. Chem.*, **86**, 43 (1974).
125. H. Meier and H. Bühl, *J. Heterocyclic Chem.*, **12**, 605 (1975).
126. T. W. Bentley and R. A. W. Johnstone, *J. Chem. Soc. (B)*, 1804 (1971).
127. W. D. Jamieson and M. E. Peach, *Org. Mass Spectrometry*, **8**, 147 (1974).
128. W. H. McFadden, R. M. Seifert and J. Wasserman, *Anal. Chem.*, **37**, 560 (1965).
129. W. J. McMurray, S. R. Lipsky, R. J. Cushley and H. G. Mautner, *J. Heterocyclic Chem.*, **9**, 1093 (1972).
130. J. H. Bowie, R. G. Cooks, P. Jakobsen, S.-O. Lawesson, and G. Schroll, *Australian J. Chem.*, **20**, 689 (1967).
131. F. Duus, G. Schroll, S.-O. Lawesson, J. H. Bowie and R. G. Cooks, *Arkiv Kemi*, **30**, 347 (1969).
132. J. Martens, K. Praefcke, U. Schulze, H. Schwarz and H. Simon, *Z. Naturforsch.*, **32b**, 657 (1977).
133. J. M. Lalancette and Y. Beauregard, *Tetrahedron Letters*, 5169 (1967).
134. A. Ohno, Y. Ohnishi, T. Koizumi and G. Tsuchihashi, *Tetrahedron Letters*, 4031 (1968).
135. R. A. W. Johnstone, D. W. Payling and A. Prox, *Chem. Commun.*, 826 (1967); R. A. W. Johnstone and D. W. Payling, *Chem. Commun.*, 601 (1968).
136. J. K. McLeod and C. Djerassi, *J. Amer. Chem. Soc.*, **89**, 5182 (1967).
137. J. Rullkötter and H. Budzikiewicz, *Org. Mass Spectrometry*, **11**, 44 (1976).
138. I. Shahak and Y. Sasson, *J. Amer. Chem. Soc.*, **95**, 3440 (1973).

139. M. S. Raasch, *J. Org. Chem.*, 37, 1347 (1972).
140. J. M. Beiner and A. Thuillier, *Compt. Rend.*, C274, 642 (1972).
141. J. Meijer, P. Vermeer and L. Brandsma, *Rec. Trav. Chim.*, 92, 601 (1973).
142. K. A. Jensen and C. Pedersen, *Acta Chem. Scand.*, 15, 1087 (1961).
143. D. Paquer, *Bull. Soc. Chim. Fr.*, 1439 (1975).
144. E. Jansons, *Russ. Chem. Rev.*, 45, 1035 (1976).
145. S. McKenzie, 'Thiocarbonyls, selenocarbonyls and tellurocarbonyls', in *Organic Compounds of Sulphur, Selenium and Tellurium*, Vol. 1 (Ed. D. H. Reid), The Chemical Society, London, 1970, p. 181–247.
146. F. Duus, 'Thiocarbonyl, selenocarbonyl and tellurocarbonyl compounds, in *Organic Compounds of Sulphur, Selenium and Tellurium*, (Ed. D. H. Reid), The Chemical Society, London, Vol. 2, 1973, Chap. 4, p. 200–287; Vol. 3, 1975, Chap. 5, p. 219–321.
147. P. Metzner, D. R. Hogg, W. Walter and J. Voss 'Thiocarbonyl and selenocarbonyl compounds, in *Organic Compounds of Sulphur, Selenium and Tellurium*, Vol. 4 (Ed. D. R. Hogg), The Chemical Society, London, 1977, Chap. 3, p. 124–185.
148. A. Treibs and R. Friess, *Ann. Chem.*, 737, 173 (1970).
149. S. R. Ramadas and P. S. Srinivasan, *Chem. Commun.*, 345 (1972).
150. E. Lindner and H.-G. Karmann, *Angew. Chem.*, 80, 319 (1968).
151. R. Engler and G. Gattow, *Z. Anorg. Allgem. Chem.*, 390, 73 (1972).
152. W. Stork and R. Mattes, *Angew. Chem.*, 87, 452 (1975).
153. T. Takeshima, M. Yokoyama, T. Imamoto, M. Akano and H. Asaba, *J. Org. Chem.*, 34, 730 (1969).
154. T. Takeshima, T. Miyauchi, N. Fukada, S. Koshizawa and M. Muruoka, *J. Chem. Soc., Perkin I*, 1009 (1973).
155. T. Takeshima, N. Fukada, O. Okabe, F. Mineshima and M. Muruoka, *J. Chem. Soc., Perkin I*, 1277 (1975).
156. G. Matolcsy, P. Sohár and B. Bordás, *Chem. Ber.*, 104, 1155 (1971).
157. B. Bordás, P. Sohár, G. Matolcsy and P. Berencsi, *J. Org. Chem.*, 37, 1727 (1972).
158. L. Dalgaard, H. Kolind-Andersen, and S.-O. Lawesson, *Tetrahedron*, 29, 2077 (1973).
159. L. Dalgaard, L. Jensen and S.-O. Lawesson, *Tetrahedron*, 30, 93 (1974).
160. M. Davis, G. Snowling and R. W. Winch, *J. Chem. Soc. (C)*, 124 (1967).
161. D. A. Konen, P. E. Pfeffer and L. S. Silbert, *Tetrahedron*, 32, 2507 (1976).
162. A. Treibs, *Ann. Chem.*, 723, 129 (1969).
163. H. Böhme and F. Ziegler, *Synthesis*, 297 (1973).
164. E. J. Hedgley and H. G. Fletcher, Jr., *J. Org. Chem.*, 30, 1282 (1965).
165. J. Poupaert and A. Bruylants, *Bull. Soc. Chim. Belg.*, 84, 61 (1975).
166. F. Becke and H. Hagen, *German Patent*, 1274121 (1968); *Chem. Abstr.*, 70, 3575 (1969).
167. H. Gotthardt, M. C. Weisshuhn and B. Christl, *Chem. Ber.*, 109, 740 (1976).
168. Y. Ueno, M. Bahry and M. Okawara, *Tetrahedron Letters*, 4607 (1977).
169. G. Mezaraups, L. Kulikova, M. Gertners and E. Jansons, *Khim. Atsetilena Tekhnol. Karbida Kal'tsiya*, 359 (1972); *Chem. Abstr.*, 80, 36894 (1974).
170. H. J. Chaves das Neves, *Rev. Portugues. Farmac.*, 22, 85 (1972).
171. M. G. Lin'kova, N. D. Kuleshova and I. L. Knuyants, *Russ. Chem. Rev.*, 493 (1964).
172. G. O. Olah and P. Schilling, *Ann. Chem.*, 761, 77 (1972).
173. D. L. Klayman and T. S. Woods, *Int. J. Sulfur Chem.*, 8, 5 (1973).
174. D. Daneels, M. Anteunis, L. Van Acker and D. Tavernier, *Tetrahedron*, 31, 327 (1975).
175. F. D. Gunstone, M. G. Hussain and D. M. Smith, *Chem. Phys. Lipids*, 13, 71 (1974); *Chem. Abstr.*, 81, 151447 (1974).
176. M. Chmielewski and R. L. Whistler, *J. Org. Chem.*, 40, 639 (1975).
177. R. Lattrell and G. Lohaus, *Ann. Chem.*, 901 (1974).
178. B. Castro and C. Selve, *Bull. Soc. Chim. Fr.*, 3009 (1974).
179. J. Häusler and U. Schmidt, *Chem. Ber.*, 107, 2804 (1974).
180. M. Imazawa, T. Ueda and T. Ukita, *Chem. Pharm. Bull.*, 23, 604 (1975).
181. K. Hojo, H. Yoshino and T. Mukayama, *Chem. Letters*, 133 (1977).
182. U. Schmidt, A. Perco and E. Oehler, *Chem. Ber.*, 107, 2816 (1974).
183. A. M. Jeffrey, H. J. C. Yeh, D. M. Jerina, R. M. DeMarinis, C. H. Foster, D. E. Piccolo and G. A. Berchtold, *J. Amer. Chem. Soc.*, 96, 6929 (1974).

184. E. Vogel, E. Schmidtbauer and H. J. Altenbach, *Angew. Chem.*, **86**, 818 (1974).
185. L. K. Vinograd, O. D. Shalygina, N. P. Kostyuchenko and N. N. Suvorov, *Khim. Geterosikl. Soedin.*, 1236 (1974); *Chem. Abstr.*, **82**, 53323 (1975).
186. J. A. Marshall, T. F. Schlaf and J. G. Csernansky, *Synth. Commun.*, **5**, 237 (1975).
187. B. M. Lerman, L. I. Umanskaya and G. A. Tolstikov, *Izv. Akad. Nauk SSSR, Ser. Khim.*, 638 (1977); *Chem. Abstr.*, **87**, 39389 (1977).
188. S. Kondo and K. Tsuda, *Makromol. Chem.*, **175**, 2111 (1974).
189. J.-C. Richer and C. Lamarre, *Can. J. Chem.*, **53**, 3005 (1975).
190. Y. Araki and A. Kaji, *Bull. Chem. Soc. Japan*, **43**, 3214 (1970).
191. M. Mori, Y. Ban and T. Oishi, *Int. J. Sulfur Chem. (A)*, 79 (1972).
192. T. Oishi, M. Mori, and Y. Ban, *Tetrahedron Letters*, 1777 (1971).
193. M. Mori, M. Shiozawa, Y. Ban and T. Oishi, *Chem. Pharm. Bull.*, **19**, 2033 (1971).
194. D. H. R. Barton and S. Prabahakar, *J. Chem. Soc., Perkin I*, 781 (1974).
195. K. D. McMichael, *J. Amer. Chem. Soc.*, **89**, 2943 (1967).
196. D. Rungwerth and H. Schwetlick, *Z. Chem.*, **14**, 17 (1974).
197. T. Sheradsky, 'Rearrangements involving thiols', in *The Chemistry of the Thiol Group* (Ed. S. Patai), Part 2, John Wiley and Sons, London, 1974, Chap. 15, p. 698–702.
198. L. I. Gapanovich, M. G. Lin'kova, O. V. Kil'disheva, and I. L. Knuyants, *Izv. Akad. Nauk SSSR, Ser. Khim.*, 152 (1974); *Chem. Abstr.*, **81**, 63570 (1974).
199. N. M. Lysenko, *Latv. PSR. Zinat. Akad. Vestis, Kim. Ser.*, 610 (1974); *Chem. Abstr.*, **82**, 43503 (1975).
200. J. H. Schauble, W. A. van Saun, Jr and J. D. Williams, *J. Org. Chem.*, **39**, 2946 (1974).
201. M. Takagi, S. Goto and T. Matsuda, *J. Chem. Soc., Chem. Commun.*, 92 (1976).
202. W. P. Neumann and J. Schwindt, *Chem. Ber.*, **108**, 1339 (1975).
203. N. Ishikawa and S. Sasaki, *Chem. Letters*, 483 (1977).
204. D. E. A. Rivett, *Tetrahedron Letters*, 1253 (1974).
205. A. M. Kuliev, A. B. Kuliev, G. A. Zeinalova and M. A. Damirov, *Neftekhimiya*, **14**, 137 (1974); *Chem. Abstr.*, **80**, 145681 (1974). A. M. Kuliev, A. B. Kuliev and M. A. Damirov, *Neftekhimiya*, **15**, 470 (1975); *Chem. Abstr.*, **83**, 166614 (1975).
206. R. Couturier, D. Paquer and A. Vibet, *Bull. Soc. Chim., Fr.*, 1670 (1975).
207. Y. Kawabata and M. Kinoshita, *Makromol. Chem.*, **176**, 2797 (1975).
208. Y. Kawabata and M. Kinoshita, *Makromol. Chem.*, **176**, 3243 (1975).
209. J. Korolczuk, M. Daniewski and Z. Mielniczuk, *J. Chromatogr.*, **100**, 165 (1975).
210. S. Masamune, S. Kamata and W. Schilling, *J. Amer. Chem. Soc.*, **97**, 3515 (1975).
211. I. G. Vasi and K. R. Desai, *J. Indian Chem. Soc.*, **52**, 837 (1975).
212. H. Zinner and H. Pinkert, *J. Prakt. Chem.*, **317**, 379 (1975).
213. D. H. Lucast and J. Wemple, *Tetrahedron Letters*, 1103 (1977).
214. R. M. Reynolds, C. Maze and E. Oppenheim, *Mol. Cryst. Liq. Cryst.*, **36**, 41 (1976); *Chem. Abstr.*, **86**, 49316 (1977).
215. N. Johns, *Z. Naturforsch.*, **29c**, 469 (1974).
216. Yu. Gololobov, L. I. Kruglik and V. P. Luk'yanchuk, *Zh. Org. Khim.*, **13**, 515 (1977).
217. Y. Watanabe, S.-I. Shoda and T. Mukayama, *Chem. Letters*, 741 (1976).
218. F. Souto-Bachiller, G. S. Bates and S. Masamune, *J. Chem. Soc., Chem. Commun.*, 719 (1976).
219. T. Mukayama, T. Takeda and K. Atsumi, *Chem. Letters*, 187 (1974).
220. S. Warwel, and B. Ahlfaenger, *Chemikerztg.*, **101**, 103 (1977).
221. E. J. Corey and D. J. Beames, *J. Amer. Chem. Soc.*, **95**, 5829 (1973). R. P. Hatch and S. M. Weinreb, *J. Org. Chem.*, **42**, 3960 (1977).
222. R. Lattrell, *Ann. Chem.*, 1929 (1974).
223. A. V. Fokin, A. F. Kolomiets and T. I. Fedyushina, *Izv. Akad. Nauk SSSR, Ser. Khim.*, 670 (1975); *Chem. Abstr.*, **83**, 27458 (1975).
224. J. C. Jamoulle, *J. Pharm. Belg.*, **29**, 281 (1974); *Chem. Abstr.*, **81**, 105215 (1974).
225. P. G. Nair and C. P. Joshua, *Indian J. Chem.*, **13**, 35 (1975).
226. S. Masamune, S. Kamata, J. Diakur, Y. Sugihara and G. S. Bates, *Can. J. Chem.*, **53**, 3693 (1975).
227. H.-J. Gais, *Angew. Chem.*, **89**, 251 (1977).
228. J. V. Paukstelis and M.-G. Kim, *J. Org. Chem.*, **39**, 1503 (1974).

229. F. Weygand and W. Steglich, *Chem. Ber.*, 93, 2983 (1960).
230. M. V. A. Baig and L. N. Owen, *J. Chem. Soc. (C)*, 540 (1966).
231. K. Horiki, *Synth. Commun.*, 7, 251 (1977).
232. J. R. Grunwell and D. L. Foerst, *Synth. Commun.*, 6, 453 (1976).
233. R. H. White, *Science*, 189, 810 (1975).
234. S. Yamada, Y. Yokoyama and T. Shioiri, *J. Org. Chem.*, 39, 3302 (1974).
235. T. Endo, S. Ikenaga and T. Mukayama, *Bull. Chem. Soc. Japan*, 43, 2632 (1970).
236. E. J. Corey and K. C. Nicolaou, *J. Amer. Chem. Soc.*, 96, 5614 (1974).
237. U. Schmidt, J. Gombos, E. Haslinger and H. Zak, *Chem. Ber.*, 109, 2628 (1976).
238. L. Dalgaard and S.-O. Lawesson, *Acta Chem. Scand.*, B28, 1077 (1974).
239. K. Ogura and G. Tsuchihashi, *J. Amer. Chem. Soc.*, 96, 1960 (1974). S. Iriuchijima, K. Maniwa, and G.-I. Tsuchihashi, *J. Amer. Chem. Soc.*, 97, 596 (1975).
240. R. Gorski, D. J. Dagli, V. A. Patronik and J. Wemple, *Synthesis*, 811 (1974).
241. H. Gotthardt, *Chem. Ber.*, 107, 2544 (1974).
242. H. Allgeier and T. Winkler, *Tetrahedron Letters*, 215 (1976).
243. K. Ogura, S. Furukawa and G. Tsuchihashi, *Bull. Chem. Soc. Japan*, 48, 2219 (1975).
244. H. Saikuchi and J. Matsuo, *Chem. Pharm. Bull.*, 17, 1260 (1969).
245. G. A. Russell and G. J. Mikol, *J. Amer. Chem. Soc.*, 88, 5498 (1966).
246. D. Seebach and R. Bürstinghaus, *Synthesis*, 461 (1975).
247. M. T. Leplawy and A. Redlinski, *Synthesis*, 504 (1975).
248. R. A. Ellison, W. D. Woessner and C. C. Williams, *J. Org. Chem.*, 39, 1430 (1974).
249. S. Cacchi, L. Caglioti and G. Paolucci, *Synthesis*, 120 (1975).
250. O. N. Tolkachev, E. P. Nakova, A. I. Unkolova, G. I. Aranovich and R. P. Evstigneeva, *Zh. Obshch. Khim.*, 44, 410 (1974).
251. H. Ishihara and S. Kato, *Tetrahedron Letters*, 3751 (1972).
252. M. G. Voronkov, R. G. Mirskov, O. S. Ishchenko and S. P. Sitnikova, *Zh. Obshch. Khim.*, 44, 2462 (1974).
253. H. R. Kricheldorf and E. Leppert, *Synthesis*, 435 (1971); *Makromol. Chem.*, 158, 223 (1972).
254. R. Mayer and H. Kröber, *Z. Chem.*, 13, 426 (1973). H. Kröber and R. Mayer, *Int. J. Sulfur Chem.*, 8, 611 (1976).
255. J. M. Beiner, D. Lecadet, D. Paquer, A. Thuillier and J. Vialle, *Bull. Soc. Chim. Fr.*, 1979 (1973). J. M. Beiner, D. Lecadet, D. Paquer and A. Thuillier, *Bull. Soc. Chim. Fr.*, 1983 (1973).
256. G. Barnikow and G. Strickmann, *Z. Chem.*, 8, 335 (1968).
257. C. Trebault and J. Teste, *Bull. Soc. Chim. Fr.*, 2272 (1970). C. Trebault, *Bull. Soc. Chim. Fr.*, 1102 (1971).
258. J. W. Scheeren, P. H. J. Ooms and R. J. F. Nivard, *Synthesis*, 149 (1973).
259. M. S. Chauhan and D. M. McKinnon, *Can. J. Chem.*, 53, 1336 (1975).
260. H. Alper and C. K. Foo, *Inorg. Chem.*, 14, 2928 (1975).
261. R. Mayer and H. Berthold, *Z. Chem.*, 3, 310 (1963).
262. A. Ohno, T. Koizumi and G. Tsuchihashi, *Tetrahedron Letters*, 2083 (1968).
263. U. Schmidt, E. Heymann and K. Kabitzke, *Chem. Ber.*, 96, 1478 (1963).
264. K. Hartke and H. Hoppe, *Chemikerztg.*, 98, 618 (1974).
265. G. Barnikow and G. Strickmann, *Chem. Ber.*, 100, 1428 (1967).
266. S. Scheithauer and R. Mayer, *Chem. Ber.*, 100, 1413 (1967).
267. K. Hartke and G. Gölz *Chem. Ber.*, 106, 2353 (1973).
268. D. H. R. Barton, C. Chavis, M. K. Kaloustian, P. D. Magnus, G. A. Poulton and P. J. West, *J. Chem. Soc., Perkin I*, 1571 (1973). S. Achmatowicz, D. H. R. Barton, P. D. Magnus, G. A. Poulton and P. J. West, *J. Chem. Soc., Perkin I*, 1567 (1973). G. C. Barrett and P. H. Leigh, *FEBS Letters*, 57, 19 (1975). M. Gisin and J. Wirz, *Helv. Chim. Acta*, 58, 1768 (1975).
269. J. Ellis and R. A. Schibeci, *Australian J. Chem.*, 27, 429 (1974).
270. V. I. Cohen, N. Rist and S. Clavel, *Eur. J. Med Chem.–Chim. Ther.*, 10, 134 (1975); *Chem. Abstr.*, 83, 96669 (1975); Eur. J. Med. Chem.–Chim. Ther.*, 10, 140 (1975); *Chem. Abstr.*, 83, 113884 (1975).
271. Y. Ogata, K. Takagi and S. Ihda, *J. Chem. Soc., Perkin I*, 1725 (1975).

272. D. H. R. Barton and S. W. McCombie, *J. Chem. Soc., Perkin I*, 1574 (1975). R. Hoffmann and K. Hartke, *Ann. Chem.*, 1743 (1977).
273. K. Hartke and G. Gölz, *Chem. Ber.*, 107, 566 (1974).
274. L. Legrand and N. Lozac'h, *Bull. Soc. Chim. Fr.*, 1173 (1969).
275. R. Raap, *Can. J. Chem.*, 46, 2251 (1968). G. Seybold and C. Heibl, *Angew Chem.*, 87, 171 (1975); *Chem. Ber.*, 110, 1225 (1977). H. Bühl, B. Seitz and H. Meier, *Tetrahedron*, 33, 449 (1977).
276. S. J. Harris and D. R. M. Walton, *J. Chem. Soc., Chem. Commun.*, 1008 (1976).
277. O. Padwa, A. Au, G. A. Lee and W. Owens, *J. Org. Chem.*, 40, 1142 (1975).
278. K. A. Latif and M. Y. Ali, *Tetrahedron*, 26, 4247 (1970).
279. G. Strickmann and G. Barnikow, *Z. Chem.*, 10, 223 (1970).
280. H. Yoshida, H. Matsuura, T. Ogata and S. Inokawa, *Bull. Chem. Soc. Japan*, 44, 2289 (1971); 48, 2907 (1975).
281. J. Faust, *Z. Chem.*, 15, 478 (1975).
282. R. Mayer, R. Keck and J. D. Eilhauer, *Z. Chem.*, 6, 108 (1966).
283. R. J. S. Beer, P. P. Carr, D. Cartwright, D. Harris and R. A. Slater, *J. Chem. Soc. (C)*, 2490 (1968).
284. F. Clesse and H. Quiniou, *Compt. Rend.*, C268, 637 (1969).
285. Y. Shvo and I. Belsky, *Tetrahedron*, 25, 4649 (1969).
286. A. Previero and J.-F. Pechere, *Biochem. Biophys. Res. Commun.*, 40, 549 (1970).
287. G. Dorange and J. E. Guerchais, *Bull. Soc. Chim. Fr.*, 43 (1971).
288. G. Kobayashi, Y. Matsuda, R. Natuki and Y. Tominaga, *Yakugaku Zasshi*, 91, 1164 (1971). K. Mizuyama, Y. Tominaga, Y. Matsuda and G. Kobayashi, *Yakugaku Zasshi*, 94, 702 (1974); *Chem. Abstr.*, 81, 135915 (1974).
289. I. Shahak and Y. Sasson, *Tetrahedron Letters*, 4207 (1973).
290. D. Laduree, P. Rioult and J. Vialle, *Bull. Soc. Chim. Fr.*, 637 (1973).
291. P. E. Iversen and H. Lund, *Acta Chem. Scand.*, B28, 827 (1974).
292. K. M. Doyle and F. Kurzer, *Chem. Ind.*, 803 (1974).
293. R. Haraoubia, J.-C. Gressier and G. Levesque, *Makromol. Chem.*, 176, 2143 (1975).
294. J. Meijer, P. Vermeer and L. Brandsma, *Rec. Trav. Chim.*, 94, 83 (1975).
295. G. Levesque and J.-C. Gressier, *Bull. Soc. Chim. Fr.*, 1145 (1976). G. Levesque, G. Tabak, F. Outurquin and J.-C. Gressier, *Bull. Soc. Chim. Fr.*, 1156 (1976).
296. S. R. Ramadas and P. S. Srinivasan, *J. Prakt. Chem.*, 319, 169 (1977).
297. H. Hartmann, W. Stapf and J. Heidberg, *Ann. Chem.*, 728, 237 (1969).
298. J. Poupaert, A. Bruylants and P. Crooy, *Synthesis*, 622 (1972).
299. A. Previero, A. Gourdol, J. Derancourt and M. A. Coletti-Previero, *FEBS Letters*, 51, 68 (1975).
300. N. C. Fawcett, P. E. Cassidy and J. C. Lin, *J. Org. Chem.*, 42, 2929 (1977).
301. P. J. W. Schuijl, L. Brandsma and J. F. Arens, *Rec. Trav. Chim.*, 85, 889 (1966).
302. L. Brandsma and D. Schuijl-Laros, *Rec. Trav. Chim.*, 89, 110 (1970).
303. L. Brandsma and P. J. W. Schuijl, *Rec. Trav. Chim.*, 88, 513 (1969).
304. A. Tangerman and B. Zwanenburg, *J. Chem. Soc., Perkin II*, 916 (1975); *Org. Mag. Resonance*, 9, 695 (1977).
305. L. Benati and P. C. Montevecchi, *J. Org. Chem.*, 41, 2639 (1976).
306. H. Viola and R. Mayer, *Z. Chem.*, 16, 354 (1976).
307. H. J. Bestmann and G. Schmid, *Tetrahedron Letters*, 3037 (1977).
308. G. LeCostumer and Y. Mollier, *Bull. Soc. Chim. Fr.*, 3076 (1970); 499 (1971); 3349 (1973). *Compt. Rend.*, C274, 1215 (1972).
309. F. Boberg and W. von Gentzkow, *J. Prakt. Chem.*, 315, 970 (1973).
310. F. Boberg and W. von Gentzkow, *J. Prakt. Chem.*, 315, 965 (1973).
311. W. Thiel, H. Viola and R. Mayer, *Z. Chem.*, 17, 92 (1977).
312. H. Viola and R. Mayer, *Z. Chem.*, 15, 348 (1975).
313. G. Barnikow and T. Gabrio, *Z. Chem.*, 8, 142 (1968).
314. G. E. Veenstra, N. M. Bronold, J. F. M. Smits, A. Tangerman and B. Zwanenburg, *Rec. Trav. Chim.*, 96, 139 (1977).
315. M. Parmentier, J. Galloy, M. Van Meerssche and H. G. Viehe, *Angew. Chem.*, 87, 33 (1975).

1060   J. Voss

316. K. Oka and S. Hara, *Tetrahedron Letters*, 2783 (1976).
317. W. G. Phillips and K. W. Ratts, *J. Org. Chem.*, 37, 1526 (1972).
318. E. Lindner and H.-G. Karmann, *Angew. Chem.*, 80, 567 (1968).
319. S. Kato, A. Hori, T. Takagi and M. Mizuta, *Angew. Chem.*, 89, 820 (1977).
320. R. Engler and G. Gattow, *Z. Anorg. Allgem. Chem.*, 389, 151 (1972).
321. M. A. Bernard, M. M. Borel and G. Dupriez, *Rev. Chim. Miner.*, 12, 181 (1975); *Chem. Abstr.*, 83, 157060 (1975).
322. A. Apsitis and E. Jansons, *Latv. PSR Zinat. Akad. Vestis, Kim. Ser.*, 400 (1968); *Chem. Abstr.*, 70, 43679 (1970).
323. J. Skrivelis, E. Jansons, A. Albeltina and R. Lazdins, *Uch. Zap. Latv. Univ.*, 117, 71 (1970); *Chem. Abstr.*, 77, 118892 (1972).
324. M. Gertners, E. Jansons and V. Jekabsons, *Uch. Zap. Latv. Univ.*, 117, 11 (1970); *Chem. Abstr.*, 77, 118952 (1972).
325. G. A. Olah, A. T. Ku and A. M. White, *J. Org. Chem.*, 34, 1827 (1969).
326. J. R. Grunwell, D. L. Foerst, F. Kaplan and J. Siddigui, *Tetrahedron*, 33, 2781 (1977).
327. G. A. Olah and A. T. Ku, *J. Org. Chem.*, 35, 331 (1970).
328. J. T. Edward, I. Lantos, G. D. Derdall and S. C. Wong, *Can. J. Chem.*, 55, 812 (1977). J. T. Edward, G. D. Derdall and S. C. Wong, *Can. J. Chem.*, 55, 2331 (1977).
329. R. Mayer, *Sulfur Org. Inorg. Chem.*, 3, 325 (1972).
330. L. Kistenbrügger, *Dissertation*, Universität Hamburg, 1978
331. R. Mayer, S. Scheithauer and D. Kunz, *Chem. Ber.*, 99, 1393 (1966).
332. A. M. Bond, A. R. Hendrickson and R. L. Martin, *J. Electrochem. Soc.*, 119, 1325 (1972).
333. J. H. Markgraf, W. M. Hensley and L. I. Shoer, *J. Org. Chem.*, 39, 3168 (1974).
334. J. Meijer and P. Vermeer, *Rec. Trav. Chim.*, 93, 242 (1974).
335. T. Mukayama, M. Araki and H. Takei, *J. Amer. Chem. Soc.*, 95, 4763 (1973). M. Araki, S. Sakata, H. Takei and T. Mukayama, *Bull. Chem. Soc. Japan*, 47, 1777 (1974).
336. R. J. Anderson, C. A. Henrick and L. D. Rosenblum, *J. Amer. Chem. Soc.*, 96, 3654 (1974).
337. I. I. Lapkin, A. S. Rodygin, M. N. Rybakova, L. M. Bykova and M. I. Belonovich, *Zh. Org. Khim.*, 13, 996 (1977).
338. L. Carlsen and F. Duus, *Synthesis*, 256 (1977).
339. L. Leger and M. Saquet, *Compt. Rend.*, C279, 695 (1974). *Bull. Soc. Chim. Fr.*, 657 (1975).
340. L. Leger, M. Saquet, A. Thuillier and S. Julia, *J. Organomet. Chem.*, 96, 313 (1975).
341. S. Masson, M. Saquet and A. Thuillier, *Tetrahedron Letters*, 4179 (1976); *Tetrahedron*, 33, 2949 (1977).
342. J.-L. Burgot, J. Masson and J. Vialle, *Tetrahedron Letters*, 4775 (1976). J. Masson, P. Metzner and J. Vialle, *Tetrahedron*, 33, 3089 (1977).
343. P. Beak and J. W. Worley, *J. Amer. Chem. Soc.*, 94, 597 (1972).
344. M. J. Gregory and T. C. Bruice, *J. Amer. Chem. Soc.*, 89, 2121 (1967).
345. L. S. Prangova and S. I. Stoyanov, *Monatsh. Chem.*, 106, 1045 (1975). J. P. Idoux, P. T. R. Hwang and C. K. Hancock, *J. Org. Chem.*, 38, 4239 (1973). B. Boopsingh and D. P. N. Satchell, *J. Chem. Soc., Perkin II*, 1702 (1972).
346. T. C. Bruice and S. J. Benkovic, *Bioorganic Mechanisms*, Vol. II, W. A. Benjamin Inc., New York, 1966, p. 259.
347. G. Losse, R. Mayer and K. Kuntze, *Z. Chem.*, 7, 104 (1967).
348. C. L. Green, R. P. Houghton and D. A. Phipps, *J. Chem. Soc., Perkin I*, 2623 (1974).
349. D. P. N. Satchell, *Chem. Soc. Reviews*, 6, 345 (1977).
350. T. G. Back, *Tetrahedron*, 33, 3041 (1977). S. Masamune, Y. Hayase, W. Schilling, W. K. Chan and G. S. Bates, *J. Amer. Chem. Soc.*, 99, 6756 (1977).
351. T. Endo, K. Oda and T. Mukayama, *Chem. Letters*, 443 (1974).
352. T. Kömives, A. F. Marton and F. Dutka, *Z. Naturforsch.*, 30b, 138 (1975); *J. Prakt. Chem.*, 318, 248 (1976). B. V. Pachenkov, V. A. Dadali and L. M. Litvinenko, *Zh. Org. Khim.*, 11, 543 (1975). N. M. Oleinik, L. M. Litvinenko, L. P. Kurchenko, N. D. Radchenko and G. K. Geller, *Ukr. Khim. Zh.*, 41, 818 (1975); *Chem. Abstr.*, 84, 4050 (1976). N. M. Oleinik, L. M. Litvinenko, L. P. Kurchenko and N. L. Vasilenko, *Ukr. Khim. Zh.*, 42, 53 (1976); *Chem. Abstr.*, 84, 120735 (1976).

353. N. S. Tao, S. Scheithauer and R. Mayer, *Z. Chem.*, **12**, 133 (1972).
354. P. Y. Bruice and H. G. Mautner, *J. Amer. Chem. Soc.*, **95**, 1582 (1973).
355. E. J. Hedgley and N. H. Leon, *J. Chem. Soc. (C)*, 467 (1970).
356. H. Takahashi, K. Oshima, H. Yamamoto and H. Nozaki, *J. Amer. Chem. Soc.*, **95**, 5803 (1973).
357. D. P. N. Satchell, M. N. White and T. J. Weil, *Chem. Ind.*, 791 (1975). J. Brelivet, P. Appriou and J. Teste, *Bull. Soc. Chim. Fr.*, 1344 (1971).
358. J. Brelivet and J. Teste, *Bull. Soc. Chim. Fr.*, 2289 (1972).
359. J. Ellis, R. D. Frier and R. A. Schibeci, *Australian. J. Chem.*, **24**, 1527 (1971).
360. L. S. Godinho and H. J. Chaves das Neves, *Rev. Port. Farm.*, **23**, 592 (1973); *Chem. Abstr.*, **82**, 43000 (1975).
361. H. Minato, K. Takeda, T. Miura and M. Kobayashi, *Chem. Letters*, 1095 (1977).
362. J. C. Wesdorp, J. Meijer, P. Vermeer, H. J. T. Bos, L. Brandsma and J. F. Arens, *Rec. Trav. Chim.*, **93**, 184 (1974).
363. F. C. V. Larsson, L. Brandsma and S.-O. Lawesson, *Rec. Trav. Chim.*, **93**, 258 (1974).
364. B. Zwanenburg and W. A. J. Janssen, *Synthesis*, 617 (1973).
365. B. Zwanenburg, L. Thijs and J. Strating, *Rec. Trav. Chim.*, **90**, 614 (1971); *Tetrahedron Letters*, 2871 (1968).
366. B. Zwanenburg. L. Thijs and J. Strating, *Rec. Trav. Chim.*, **89**, 687 (1970).
367. A. Holm and L. Carlsen, *Tetrahedron Letters*, 3203 (1973). L. Carlsen, A. Holm, J. P. Snyder, E. Kock and B. Stilkerieg, *Tetrahedron* **33**, 2231 (1977).
368. A. Tangerman and B. Zwanenburg, *Tetrahedron Letters*, 259 (1977).
369. A. Tangerman and B. Zwanenburg, *Tetrahedron Letters*, 5329 (1972); 79 (1973). *J. Chem. Soc., Perkin II*, 532 (1975).
370. A. Tangerman and B. Zwanenburg, *S. Chem. Soc., Perkin II*, 1413 (1974).
371. T. W. Hummelink, *J. Cryst. Mol. Struct.*, **4**, 87 (1974); **4**, 373, (1974); *Cryst. Struct. Common.*, **4**, 441 (1975); **5**, 169 (1976).
372. R. Raap, *Can. J. Chem.*, **46**, 2255 (1968).
373. D. Lecadet, D. Paquer and A. Thuillier, *Compt. Rend.*, C**276**, 875 (1973).
374. H. Minato, T. Miura and M. Kobayashi, *Chem. Letters*, 609 (1977).
375. P. Beak, B. G. McKinnie and D. B. Reitz, *Tetrahedron Letters*, 1839 (1977).
376. P. C. Oele, A. Tinkelenberg and R. Louw, *Tetrahedron Letters*, 2375 (1972).
377. D. J. Dagli and J. Wemple, *J. Org. Chem.*, **39**, 2938 (1974). D. J. Dagli, P.-S. Yu and J. Wemple, *J. Org. Chem.*, **40**, 3173 (1975). D. J. Dagli, R. A. Gorski and J. Wemple, *J. Org. Chem.*, **40**, 1741 (1975).
378. J. Wemple, *Tetrahedron Letters*, 3255 (1975).
379. F. Duus, *Tetrahedron*, **28**, 5923 (1972).
380. P. Beak and R. Farney, *J. Amer. Chem. Soc.*, **95**, 4771 (1973).
381. L. B. Dashkevich and L. V. Konovalova, *Zh. Org. Khim.*, **8**, 2269 (1972).
382. J. D. Coyle, *Chem. Soc. Rev.*, **4**, 523 (1975).
383. Y. Ogata, K. Takagi and Y. Takayanagi, *J. Chem. Soc., Perkin I*, 1244 (1973). J. R. Grunwell, N. A. Marron and S. I. Hanhan, *J. Org. Chem.*, **38**, 1559 (1973). J. Martens and K. Praefcke, *Chem. Ber.*, **107**, 2319 (1974). T. L. Ito and W. P. Weber, *J. Org. Chem.*, **39**, 1691 (1974).
384. G. Buchholz, J. Martens and K. Praefcke, *Synthesis*, 666 (1974).
385. J. Martens and K. Praefcke, *Tetrahedron*, **30**, 2565 (1974).
386. A. V. Sviridova and E. N. Prilezhaeva, *Russ. Chem. Rev.*, **43**, 200 (1974).
387. E. J. Goethals, 'Sulfur-containing polymers', in *Topics in Sulfur Chemistry* Vol. 3, (Ed. A. Senning), G. Thieme Verlag, Stuttgart, 1977, p. 1.
388. S. Achmatowicz, D. H. R. Barton, P. D. Magnus, G. A. Poulton and P. J. West, *Chem. Commun.*, 1014 (1971). D. H. R. Barton, M. Bolton, P. D. Magnus, P. J. West, G. Porter and J. Wirz, *Chem. Commun.*, 632 (1972). D. H. R. Barton, M. Bolton, P. D. Magnus, K. G. Marathe, G. A. Poulton and P. J. West, *J. Chem. Soc., Perkin I*, 1574 (1973).
389. D. H. R. Barton, M. Bolton, P. D. Magnus and P. J. West, *J. Chem. Soc., Perkin I*, 1580 (1973). J. Wirz, *J. Chem Soc., Perkin II*, 1307 (1973).
390. U. Schmidt, K. Kabitzke, J. Boie and C. Osterroth, *Chem. Ber.*, **98**, 3819 (1965). R. Jahn and U. Schmidt, *Chem. Ber.*, **108**, 630 (1975).

1062 J. Voss

391. A. Ohno, *Int. J. Sulfur Chem.*, **6**, 183 (1971). A. Ohno, T. Koizumi and Y. Akasaki, *Tetrahedron Letters*, 4993 (1972); *Bull. Chem. Soc. Japan*, **47**, 319 (1974).
392. A. Padwa, J. Smolanow and S. I. Westmore, *J. Org. Chem.*, **38**, 1333 (1973).
393. G. Levesque, A. Nahjoub and A. Thuillier, *Compt. Rend.*, **C284**, 689 (1977).
394. G. Barnikow and G. Strickmann, *Chem. Ber.*, **100**, 1661 (1967).
395. T. H. Jones and P. J. Kropp, *Tetrahedron Letters*, 3503 (1974).
396. K. Oka and S. Hara, *Heterocycles*, **6**, 941 (1977).
397. K. Thimm and J. Voss, *Z. Naturforsch.*, **30b**, 292 (1975).
398. S. V. Amosova, B. A. Trofimov, N. N. Skatova, O. A. Tarasova, A. G. Trofimova, V. V. Takhistov and M. G. Voronkov, *Dokl. Akad. Nauk SSSR* **215**, 95 (1974); *Chem. Abstr.*, **80**, 145638 (1974). S. I. Zav'yalov, N. A. Rodionova, I. V. Shcherbak, O. V. Dorofeeva, Z. S. Volkova, A. K. Zitsmanis and V. S. Ragovska, *Izv. Akad. Nauk SSSR, Ser. Khim.*, 1895 (1975); *Chem. Abstr.*, **83**, 206165 (1975).
399. K. Hartke and G. Gölz, *Ann Chem.*, 1644 (1973).
400. K. Hirai and T. Ishiba, *Heterocycles*, **3**, 217 (1975).
401. J. Weinstock, J. E. Blank and B. M. Sutton, *J. Org. Chem.*, **39**, 2454 (1974).
402. J.-L. Burgot, *Bull. Soc. Chim. Fr.*, 140 (1974).
403. Y. Tominaga, K. Mizuyama and G. Kobayashi, *Chem. Pharm. Bull.*, **22**, 1670 (1974); *Chem. Abstr.*, **81**, 120390 (1974).
404. L. S. Stanishevskii, I. G. Tishchenko and A. M. Zvonok, *Khim. Geterosikl. Soedin.*, 670 (1975); *Chem. Abstr.*, **83**, 114155 (1975). T. Manimaran, T. K. Thiruvengadam and V. T. Ramakrishnan, *Synthesis*, 739 (1975).
405. K. T. Potts, J. Kane, E. Carnahan, and U. P. Singh, *J. Chem. Soc., Chem. Commun.*, 417 (1975).
406. C. T. Pedersen, *Acta Chem. Scand.*, **28B**, 367 (1974). P. Vermeer, J. Meijer, H. J. T. Bos and L. Brandsma, *Rec. Trav. Chim.*, **93**, 51 (1974).
407. K. Bunge, R. Huisgen, R. Raab and H. J. Sturm, *Chem. Ber.*, **105**, 1307 (1972). G. D. Hartman and L. M. Weinstock, *Synthesis*, 681 (1976).
408. M. Davis, M. C. Dereani, J. L. McVicars and I. J. Morris, *Australian J. Chem.*, **30**, 1815 (1977). J. Faust and R. Mayer, *J. Prakt. Chem.*, **318**, 161 (1976).
409. R. Heymes, G. Amiard and G. Nomine, *Bull. Soc. Chim. Fr.*, 563 (1974).
410. U. Schmidt, *Chem. Ber.*, **100**, 3825 (1967).
411. B. Junge, *Ann. Chem.*, 1961 (1975).
412. R. Huisgen and W. Mack, *Chem. Ber.*, **105**, 2815 (1972). R. Grashey, G. Schroll and M. Weidner, *Chemiker-Ztg.*, **100**, 496 (1976).
413. B. S. Pedersen, S. Scheibye, K. Clausen and S.-O. Lawesson, *Bull. Soc. Chim. Belg.*, **87**, 293 (1978).
414. J. Voss and H. Günther, *Synthesis*, 849 (1978).
415. H. Günther and J. Voss, unpublished results.

CHAPTER **19**

The synthesis of lactones and lactams

JAMES F. WOLFE and MICHAEL A. OGLIARUSO

Virginia Polytechnic Institute and State University, Blacksburg, Virginia 24061, U.S.A.

I.   INTRODUCTION 1064
II. SYNTHESIS OF LACTONES 1065
     A. By Intramolecular Cyclization of Hydroxy Acids, Hydroxy Acid
        Derivatives and Related Compounds 1065
     B. By Intramolecular Cyclization of Unsaturated Acids and Esters . . 1078
        1. Acid-catalysed cyclizations 1078
        2. Photochemical and electrochemical cyclizations . . . 1080
        3. Halolactonization 1081
        4. Intramolecular Diels–Alder reactions 1092
     C. By Acetoacetic Ester and Cyanoacetic Ester Condensations . . 1096
     D. By Aldol Condensations 1098
     E. By Malonic Ester or Malonic Acid Condensation 1102
     F. By Perkin and Stobbe Reactions 1108
     G. By Grignard and Reformatsky Reactions 1110
     H. By Wittig-type Reactions 1115
     I. From α-Anions (Dianions) of Carboxylic Acids 1116
     J. From Lithio Salts of 2-Alkyl-2-oxazolines 1120
     K. By Direct Functionalization of Preformed Lactones . . . 1123
     L. From Ketenes 1125
     M. By Reduction of Anhydrides, Esters and Acids 1138
     N. By Oxidation Reactions 1145
        1. Oxidation of diols 1145
        2. Oxidation of ketones 1168
        3. Oxidation of ethers 1171
        4. Oxidation of olefins 1172
     O. By Carbonylation Reactions 1173
     P. By Cycloaddition of Nitrones to Olefins 1180
     Q. By Rearrangement Reactions 1182
        1. Claisen rearrangements 1182
        2. Carbonium ion rearrangements 1184
        3. Photochemical rearrangements 1186
     R. Lactone Interconversions 1187
     S. Miscellaneous Lactone Syntheses 1189
        1. The Barton reaction 1189
        2. Photolysis of α-diazo esters and amides 1189
        3. Photolysis of 2-alkoxyoxetanes 1191
        4. α-Lactones by photolysis of 1,2-dioxolane-3,5-diones . . 1192
        5. Oxidation of mercaptans, disulphides and related compounds . 1192

6. Addition of diazonium salts to olefins 1192
7. Addition of diethyl dibromomalonate to methyl methacrylate . . 1192
8. Dehydrohalogenation of 2,2-dimethoxy-3-chlorodihydropyrans . . 1193
9. Preparation of homoserine lactone 1193

III. SYNTHESIS OF LACTAMS 1194
A. By Ring-closure Reactions (Chemical) 1195
   1. From amino acids and related compounds 1195
   2. From halo, hydroxy and keto amides 1200
B. By Ring-closure Reactions (Photochemical) 1206
   1. Cyclization of α,β-unsaturated amides 1206
   2. Cyclization of benzanilides 1207
   3. Cyclization of enamides 1207
   4. Cyclization of N-chloroacetyl-β-arylamines . . . 1224
   5. Cyclization of α-diazocarboxamides 1224
   6. Miscellaneous cyclizations 1224
C. By Cycloaddition Reactions 1224
   1. Addition of isocyanates to olefins 1224
   2. From imines 1230
      a. Reaction of imines with ketenes 1230
      b. Reformatsky reaction with imines 1233
      c. Other imine cycloadditions 1265
   3. From nitrones and nitroso compounds 1269
D. By Rearrangements 1272
   1. Ring contractions 1272
      a. Wolff rearrangement 1272
      b. Miscellaneous ring contractions 1273
   2. Ring expansions 1274
      a. Beckmann rearrangement 1274
      b. Schmidt rearrangement 1284
      c. Miscellaneous ring expansions 1285
   3. Claisen rearrangement 1298
E. By Direct Functionalization of Preformed Lactams . . . 1299
F. By Oxidation Reactions 1302
   1. Using halogen 1302
   2. Using chromium or osmium oxides 1304
   3. Using manganese oxides 1305
   4. Using platinum or ruthenium oxides 1306
   5. Via sensitized and unsensitized photooxidation . . . 1309
   6. Via autooxidation 1312
   7. Using miscellaneous reagents 1312
G. Miscellaneous Lactam Syntheses 1314

IV. ACKNOWLEDGMENTS 1315
V. REFERENCES 1315

I. INTRODUCTION

This chapter is devoted to a discussion of recent developments in the synthesis of lactones and lactams, and is meant to supplement our earlier chapter in this volume dealing with the synthesis of carboxylic acids and their acyclic derivatives (Chapter 7).

The primary literature surveyed for this review consists mainly of articles listed in *Chemical Abstracts* from 1966 through mid-1976. In order to treat topics which have not been reviewed before, and to lend continuity and chronological perspective to certain sections, a number of references which appeared prior to 1966 are also included.

Although we have not attempted to make this chapter encyclopaedic, we hope that the numerous lactone and lactam preparations presented in tabular form will be helpful to practitioners of the fine art of organic synthesis in spite of inevitable, but unintentional, omissions.

II. SYNTHESIS OF LACTONES

The first extensive review of lactones covered the synthesis and reactions of β-lactones, and was published in 1954 by Zaugg[1]. A review[2] in 1963, while not concerned with lactones *per se*, discusses many reactions which do give rise to lactones. In 1964 three reviews appeared: the first, by Etienne and Fischer[3], was on the preparation, reactions, etc. of β-lactones; the second, by Rao[4], was on the chemistry of butenolides; and the third, by Ansell and Palmer[5] discussed the cyclization of olefinic acids to ketones and lactones. In 1967 and 1968 three reviews appeared which discussed the synthesis of 2-pyrone[6], the preparation of macrocyclic ketones and lactones from polyacetylenic compounds[7], and the synthesis of substituted lactones, their odour and some transformations[8]. A review in 1972 discussed the preparation, properties and polymerization of β-lactones, ε-caprolactone and lactides[9], and another reported on the preparation, properties and polymerization of hydroxy acids and lactones[10]. The synthesis of α-methylene lactones was reviewed[11] in 1975, while in 1976 Rao reported[12] on recent advances in the chemistry of unsaturated lactones.

Because of the large number and variety of reviews published on all aspects of lactone preparation, this section will mainly be concerned with discussion of newer methods of lactone preparation along with selected recent applications of traditional synthetic methods.

A. By Intramolecular Cyclization of Hydroxy Acids, Hydroxy Acid Derivatives and Related Compounds

Numerous hydroxy acids, hydroxy esters and hydroxylated acid derivatives can be converted to lactones by intramolecular reactions similar to those employed in the synthesis of acyclic esters. Acids containing enolizable carbonyl functions can also serve as useful lactone precursors.

Acid-catalysed cyclization of hydroxy acids comprises a widely used procedure for lactone formation. Examples of intramolecular acid-catalysed condensations yielding γ- and ε-lactones are the reaction of sodium *o*-hydroxymethylbenzoate with concentrated hydrochloric acid, which affords[13] a 67–71% yield of phthalide (equation 1), and cyclization[14] of (R)-(+)-6-hydroxy-4-methylhexanoic acid and

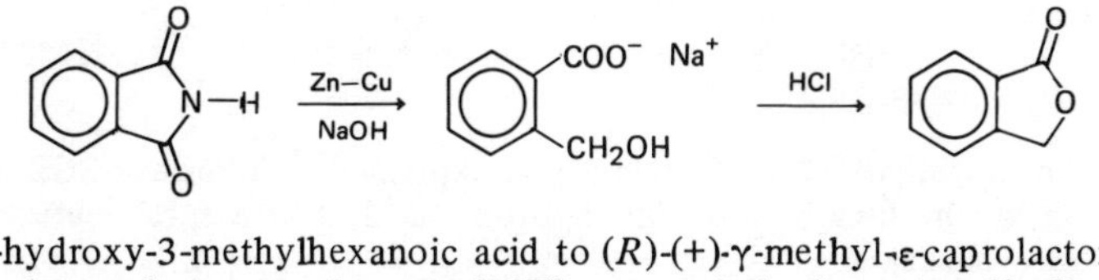

$$ \text{(1)} $$

(R)-(+)-6-hydroxy-3-methylhexanoic acid to (R)-(+)-γ-methyl-ε-caprolactone (35%) and (R)-(−)-β-methyl-ε-caprolactone (59%), respectively (equation 2). Similarly D-gulonic-γ-lactone has been prepared from gulonic acid[15].

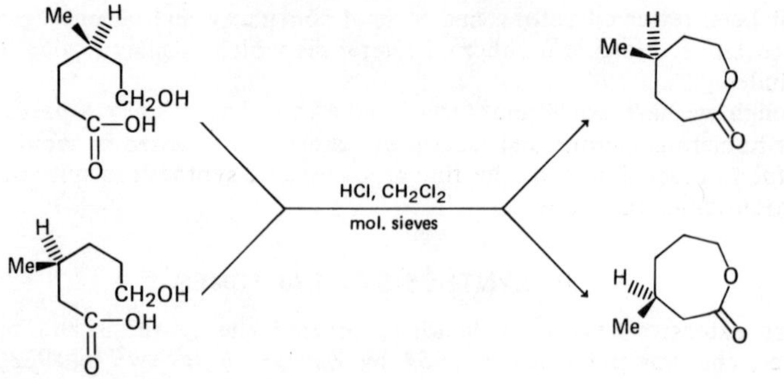

$$(2)$$

The most popular acidic reagent for effecting direct cyclization of hydroxy acids appears to be p-toluenesulphonic acid in a variety of solvents[16-22] (Table 1).

The direct cyclization of β-hydroxy acids with benzenesulphonyl chloride in pyridine at $0-5°C$ (equation 3) has been shown to be a general reaction for the

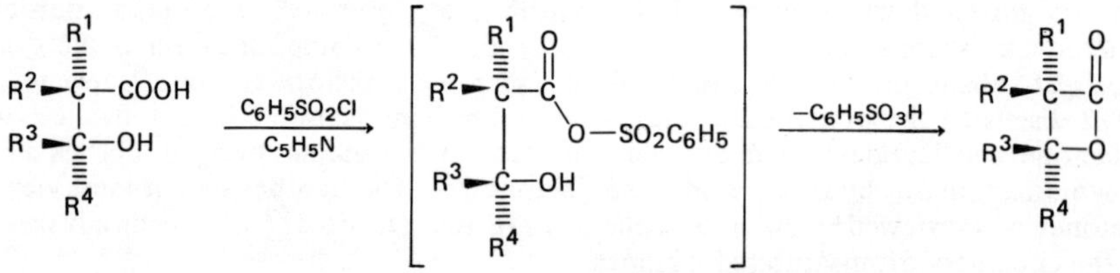

$$(3)$$

formation of tri- and tetra-substituted β-lactones in high yields (Table 2). During these investigations it was observed that hydroxy acids 1, 2 and 3 afforded olefins rather than lactones upon treatment with benzenesulphonyl chloride. Although no explanation was advanced for the absence of lactone formation from 1 or 2, the

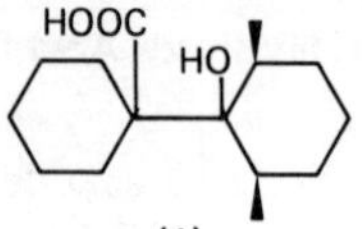

(1)
(Ref. 26)

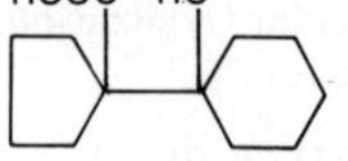

(2)
(Ref. 26)

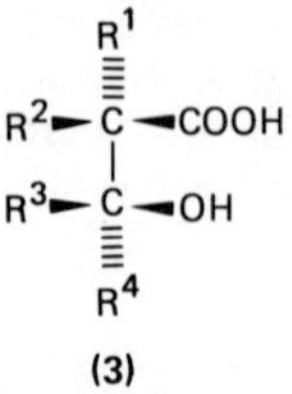

(3)
(Ref. 25)

R^1 = H, H, H, H

R^2 = H, H, H, H

R^3 = Ph, PhCH$_2$, Et, i-Pr

R^4 = Ph, Ph, Ph, Ph

preferred linear dehydration of acids 3 was explained[25] in terms of the absence of substituents at the β-carbon of the hydroxy acid, a structural feature which is essential for cyclization.

Stereoselective cyclizations of hydroxy acids to trisubstituted β-lactones have been reported using methanesulphonyl chloride[27]. For example, the diastereomers

TABLE 1. Cyclization of hydroxy acids to lactones using *p*-toluenesulphonic acid

Hydroxy acid	Conditions	Product	Yield (%)	Reference
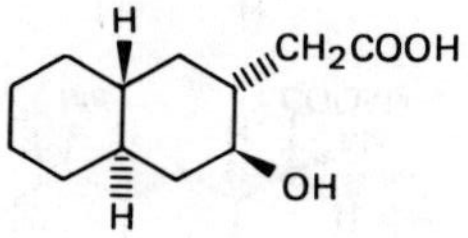	Heat, 10 min.	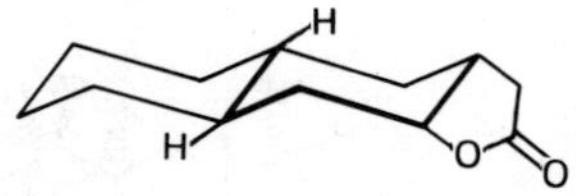	97	16
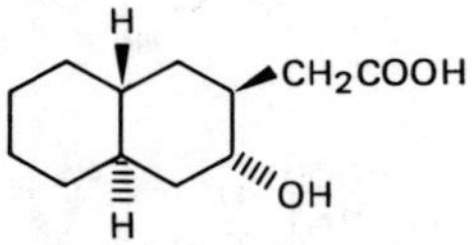	Xylene, heat	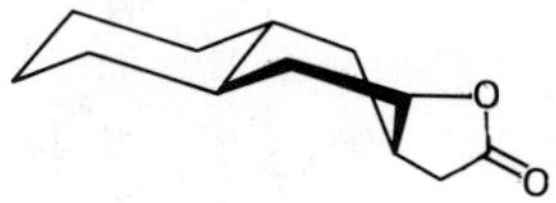	73	16
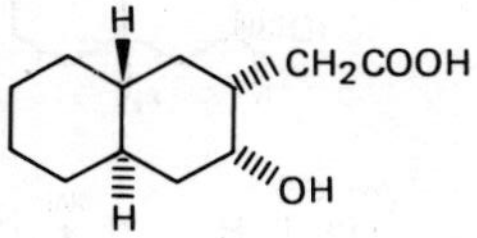	Benzene, heat	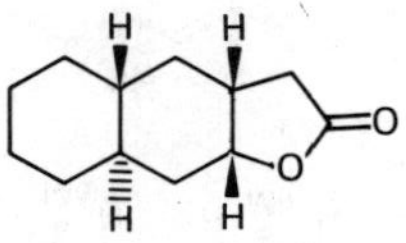	40	16
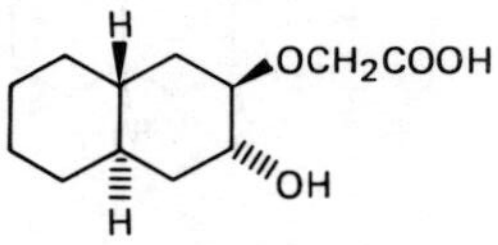	Benzene, heat	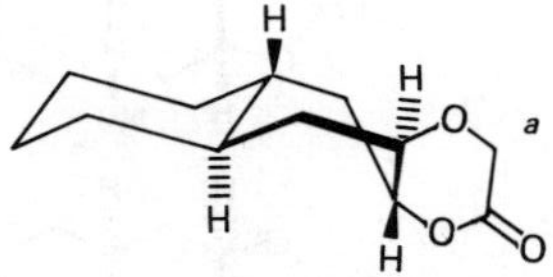	51	16
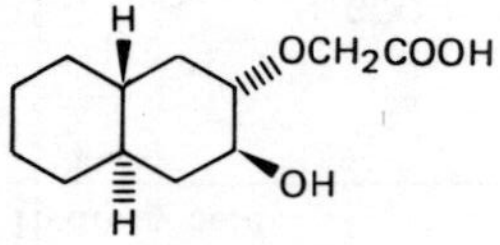	Heat[a]	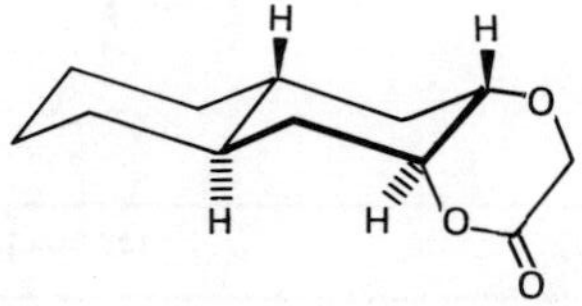	40	16

TABLE 1. (Continued)

Hydroxy acid	Conditions	Product	Yield (%)	Reference
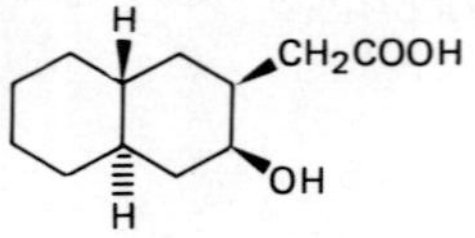	Heat[a]	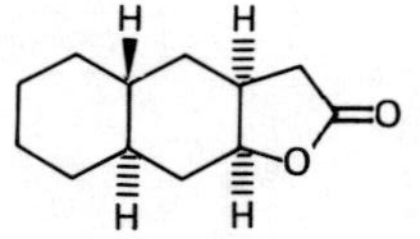	75	16
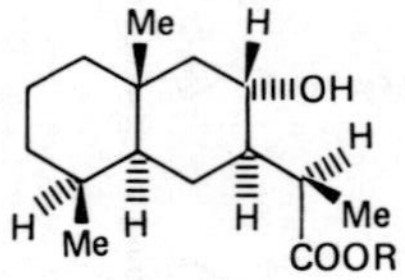	(R = H) Benzene, heat (R = H) Acetic acid, heat (R = Me) Benzene, heat (R = H)[b]	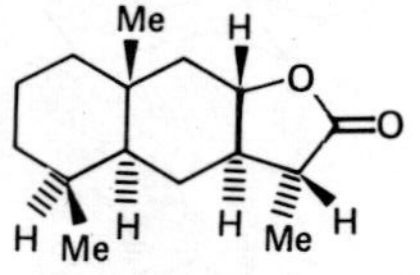	80 88 82 30	17
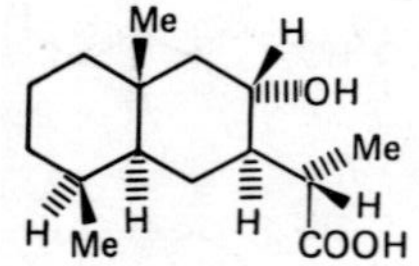	Benzene, acetic acid, heat	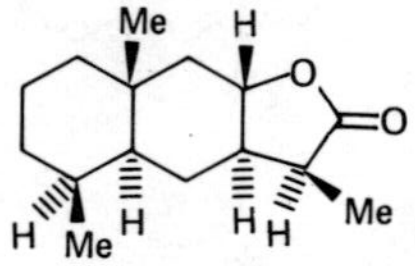	90	17
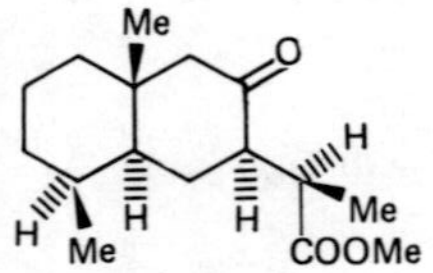	c	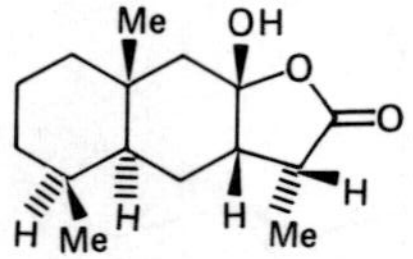	5	17
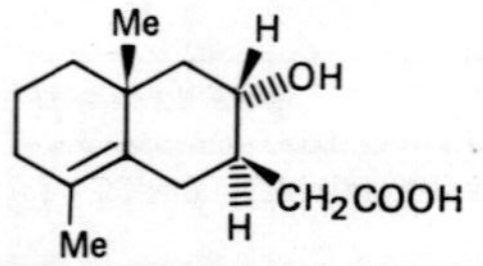	Benzene, heat	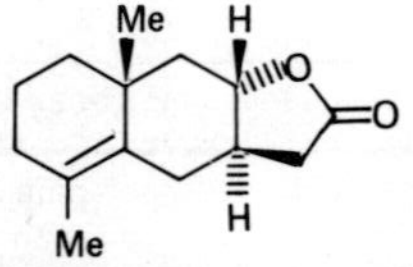	68	19

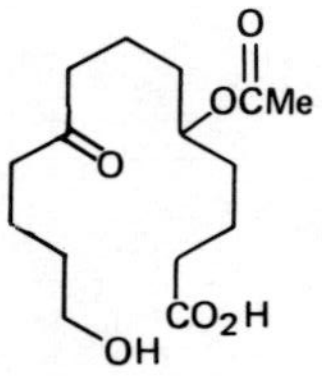

Benzene, heat Et$_3$N[d]

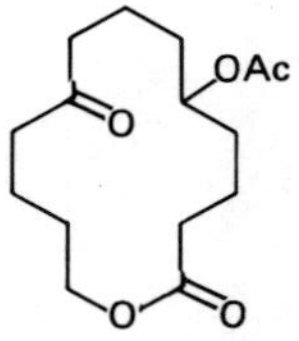

52

21

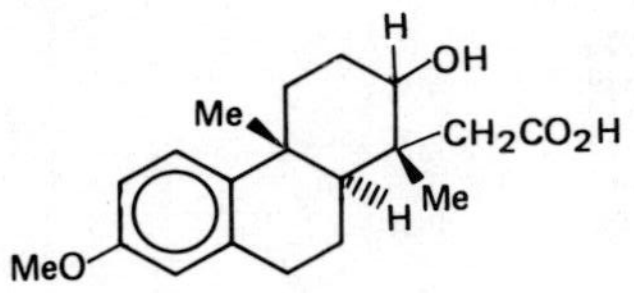

Benzene, heat

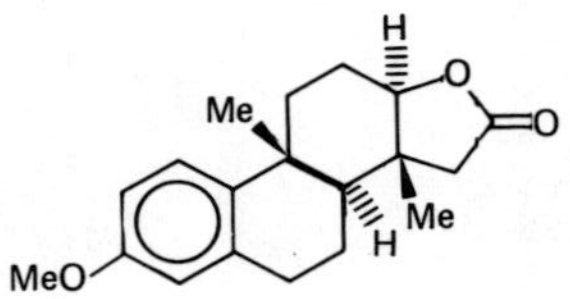

95

22

[a] These products were obtained by heating without p-toluenesulphonic acid.
[b] Using boron trifluoride—etherate in ether without p-toluenesulphonic acid.
[c] Conversion occurred after hydrolysis of the ester, by allowing the mixture to stand at 0°C for 24 hours.
[d] Also prepared in 40% yield by heating a benzene solution containing 1,1-carbonyldiimidazole followed by treatment with a catalytic amount of sodium t-amylate in benzene.

TABLE 2. Preparation of β-lactones from β-hydroxy acids with benzenesulphonyl chloride in pyridine

R^1	R^2	R^3	R^4	Yield (%)	Reference
Me	OMe		—(CH₂)₅—	82	23
Me	OMe		—(CH₂)₄—	83	23
Me	OMe	n-C₃H₇	n-C₃H₇	77	23
Me	OMe	Me	n-C₆H₁₃	45	23
Me	OMe	n-C₆H₁₃	Me	45	23
—CH₂CH=C(Me)CH₂CH₂—		Me	Me	82	24
Me	Me	H	—(CH₂)₂CHMeC₆H₄Me-p	77	24
H	Ph	Ph	Ph	70	25
H	Me	Ph	Ph	37	25
H	t-Bu	Ph	Ph	100	25
Me	H	Me	Ph	87	25
Me	Me	Ph	Ph	95	25
Me	Me	Me	Ph	92	25
Me	PhCH₂	PhCH₂	Ph	85	25
Me	Me	PhCH₂	Ph	30	25
Me	Me	H	Ph	95	25
Me	Me		—(CH₂)₅—	67	25
H	Me	PhCH₂	Ph	93	25
H	Me	Ph	PhCH₂	90	25
—(CH₂)₃—			—(CH₂)₃—	92	26
—(CH₂)₃—			—(CH₂)₅—	90	26
—(CH₂)₄—			—(CH₂)₃—	65	26
—(CH₂)₄—			—(CH₂)₄—	86	26
—(CH₂)₄—			—(CH₂)₅—	80	26
—(CH₂)₅—			—(CH₂)₃—	88	26
—(CH₂)₅—			—(CH₂)₄—	88	26
—(CH₂)₅—			—(CH₂)₅—	94	26
—(CH₂)₆—			—(CH₂)₆—	88	26
—(CH₂)₇—			—(CH₂)₇—	88	26
—(CH₂)₂—CH=CH—(CH₂)₂—			—(CH₂)₅—	77	26
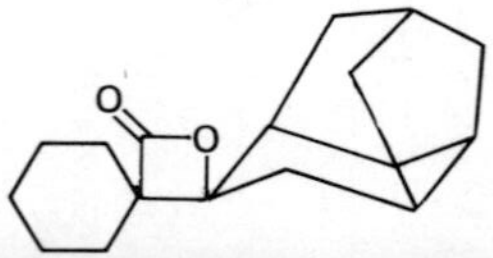				82	26

of α-methyl-α-n-butyl-β-hydroxyheptanoic acid afford the corresponding β-lactones. These lactonizations proceed through formation of intermediate mesyl derivatives, which then undergo internal nucleophilic displacement by the carboxylate group (Scheme 1).

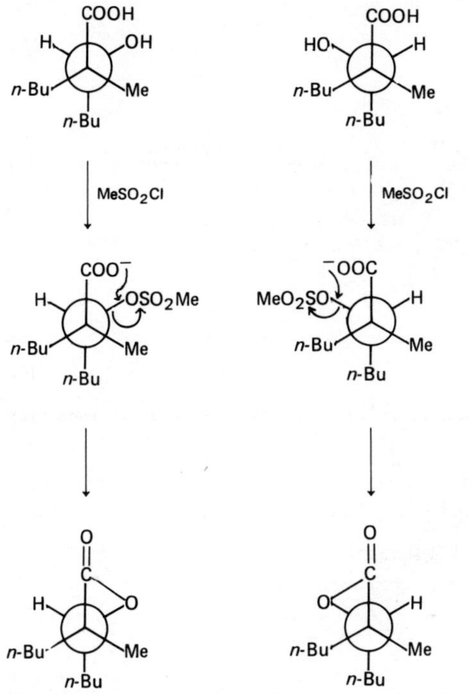

SCHEME 1.

The reaction of hydroxy acids with sodium acetate in acetic anhydride–benzene mixtures is a very effective method of lactonization, which has been used by Woodward and coworkers[28] in the total synthesis of reserpine (equation 4), and by

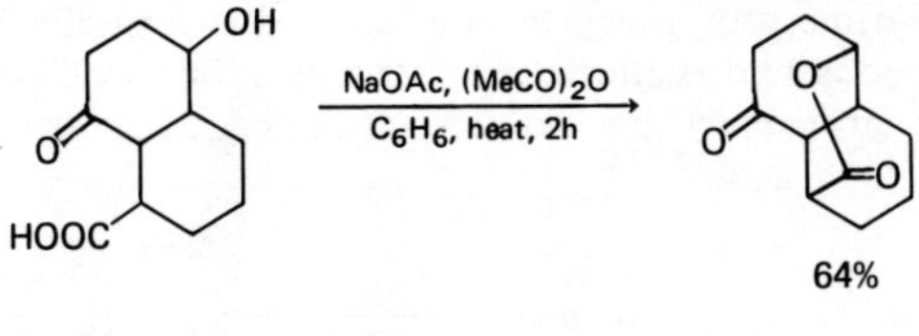

(Ref. 28)   (4)

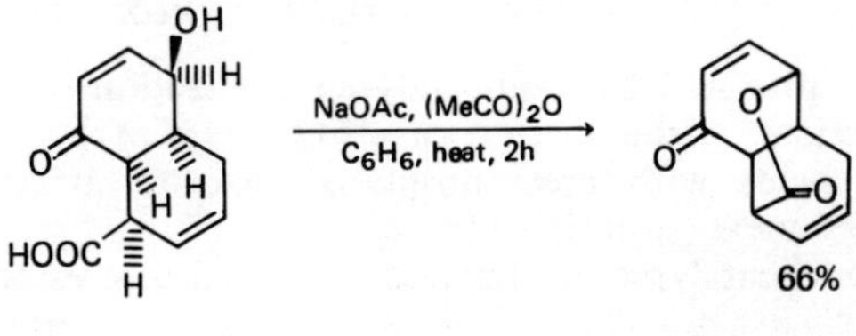

64%

(Ref. 28)   (5)

66%

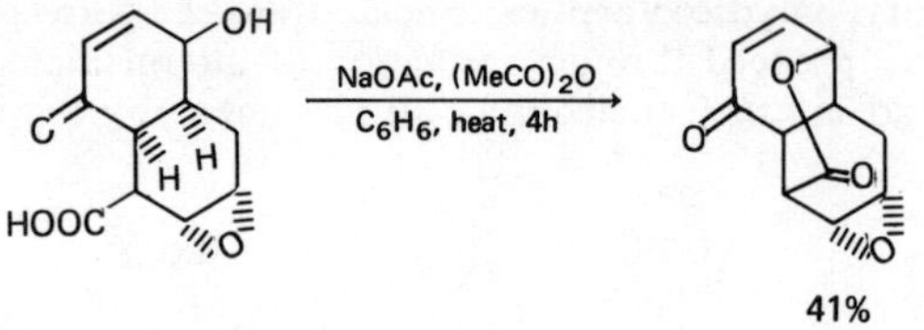

(Ref. 28)   (6)

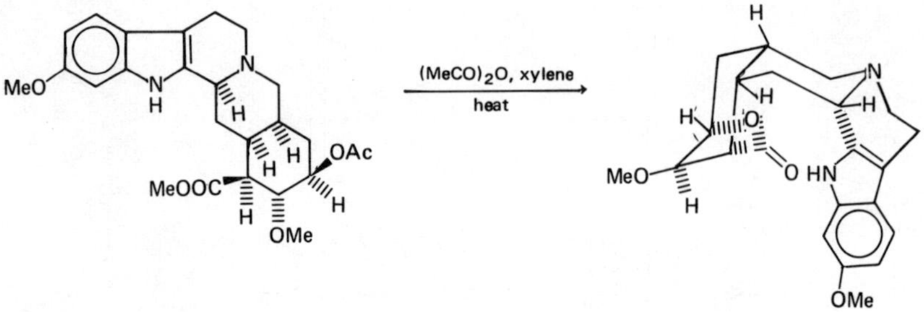

41%

(Ref. 28)   (7)

Meinwald and Frauenglass[29] for the synthesis of various bicyclic lactones (equations 8 and 9).

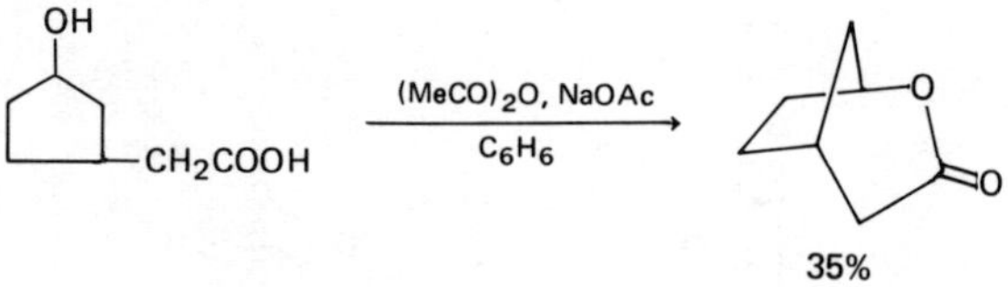

(Ref. 29)   (8)

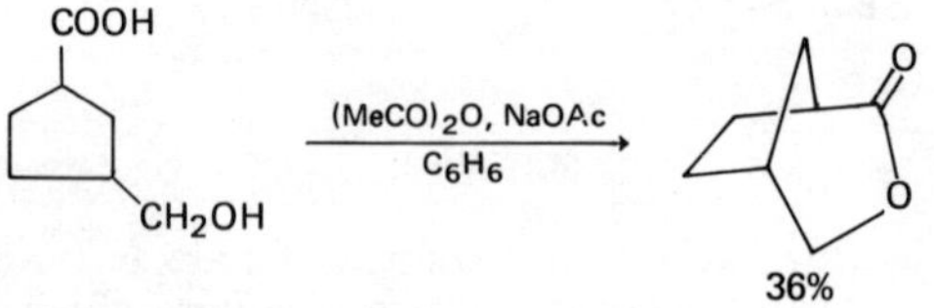

35%

(Ref. 29)   (9)

36%

N,N'-Dicyclohexylcarbodiimide (DCCD) is also an effective reagent for lactone formation from hydroxy acids[16,30-32] as illustrated in Table 3.

Reaction of 3,4-dimethoxyphenylacetic acid with formalin in the presence of acetic acid and aqueous hydrochloric acid affords 6,7-dimethoxy-3-isochromanone (equation 10) in a process which may be regarded as an *in situ* formation and cyclization of a hydroxy acid[33,34]

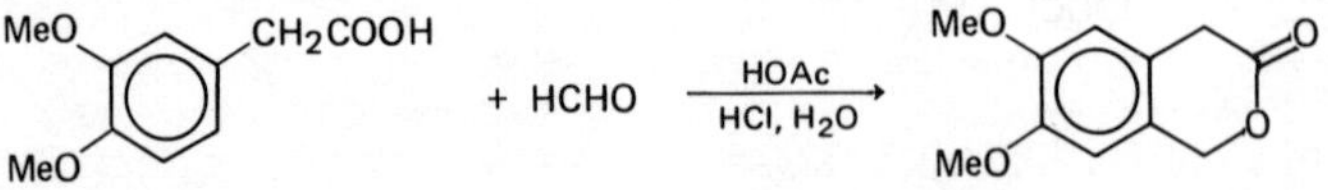

(10)

Lactones can be prepared by acid-catalysed cyclization of hydroxy esters, as shown by the reaction of the methyl or ethyl esters of γ-alkyl-γ-carboethoxy-δ-hydroxyhexanoic acids with metaphosphoric acid to afford[35] the expected δ-lactones in 95–99% yield (equation 11).

Intramolecular acid-catalysed cyclization of γ-hydroxy esters has been found

TABLE 3. Lactonization of hydroxy acids by DCCD

Hydroxy acid	Product	Yield (%)	Reference
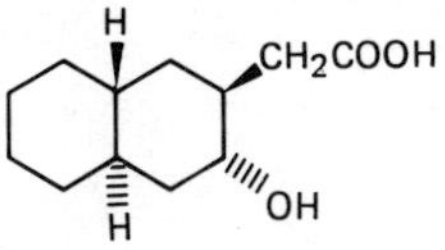		55–86	16
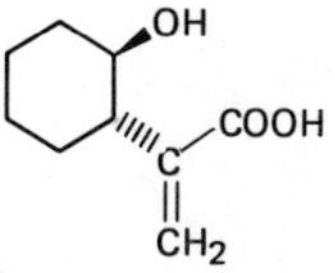	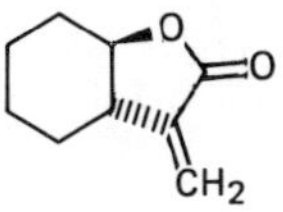	60	30, 31

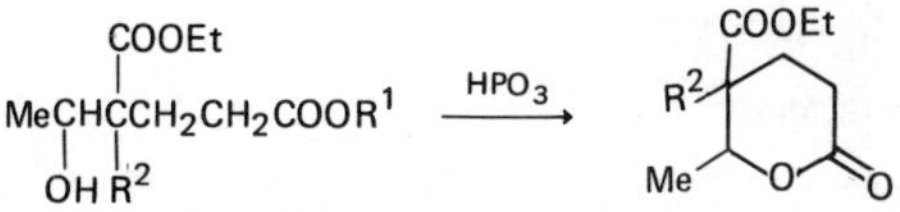

$$\text{R}^1 = \text{Me, Et}$$
$$\text{R}^2 = \text{Et, } n\text{-Pr, } n\text{-Bu, CH}_2\text{CH}_2\text{CH(Me)}_2$$

useful in the preparation of bicyclic lactones. Thus, reaction of diethyl Δ^4-cyclo-hexene-*cis*-1,2-dicarboxylate oxide with dilute aqueous sulphuric acid at 40–50°C gave[36] a 73% yield of diethyl *trans*-4,5-dihydroxycyclohexane-*cis*-1,2-dicarboxylate, which upon partial acid hydrolysis at 80°C afforded the bicyclic lactone shown in equation (12).

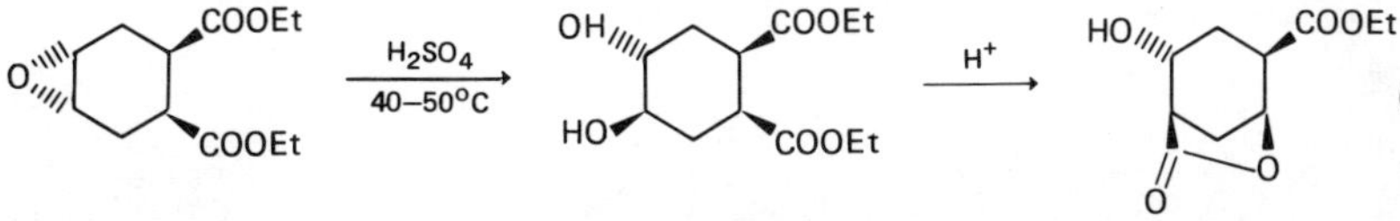

Sulphuric acid-catalysed lactonization of *cis*- and *trans*-N-(carboxymethyl)-4-phenyl-4-ethylpyrrolidin-3-ols, as well as their corresponding methyl and ethyl esters or their 3-acetates, all afforded[37] the bicyclic lactone, 6-phenyl-6-ethyl-1-aza-4-oxabicyclo[3.2.1]octane-3-one (equation 13). The fact that the same lactone was obtained from either the *cis* or *trans* compounds, indicates that the probable mechanism for this transformation involves initial protonation of the $C_{(3)}$-OH or -OR function with subsequent elimination of water or alcohol to create a positive centre at $C_{(3)}$, followed by intramolecular nucleophilic attack by the carbonyl as shown in equation (14).

The use of boron trifluoride—etherate for direct lactonization of hydroxy esters is demonstrated by reaction of methyl 4-hydroxy-6-phenylhex-5-en-1-yne-1-carboxylate to give the lactone of 4-hydroxy-2,2-dimethoxy-6-phenylhex-5-en-1-carboxylic acid, which upon heating at 150°C afforded[38-40] (±)-kawain (4), a constituent of the kawa root (equation 15)[41-44]. Hydroxy esters can also be converted to lactones by means of DCCD (equations 16 and 17)[28,45].

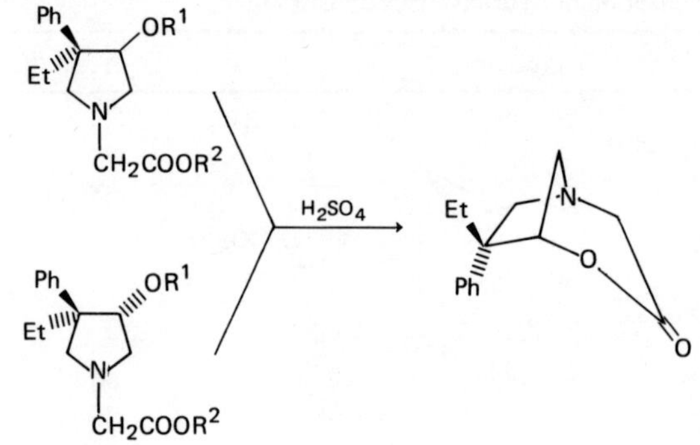

$$R^1 = R^2 = H$$
$$R^1 = H; R^2 = Me$$
$$R^1 = H; R^2 = Et$$
$$R^1 = COMe; R^2 = Me$$

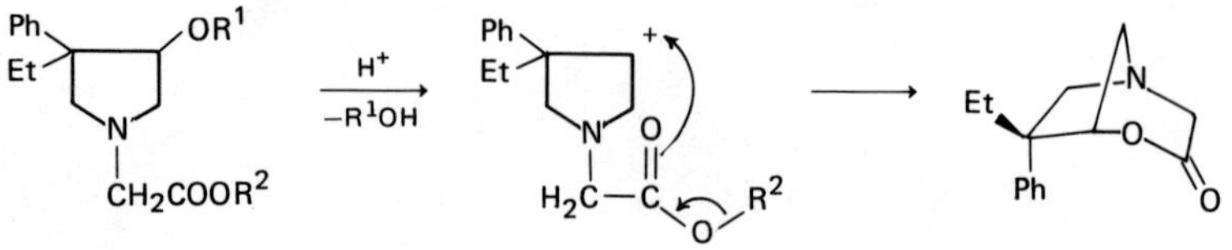

(14)

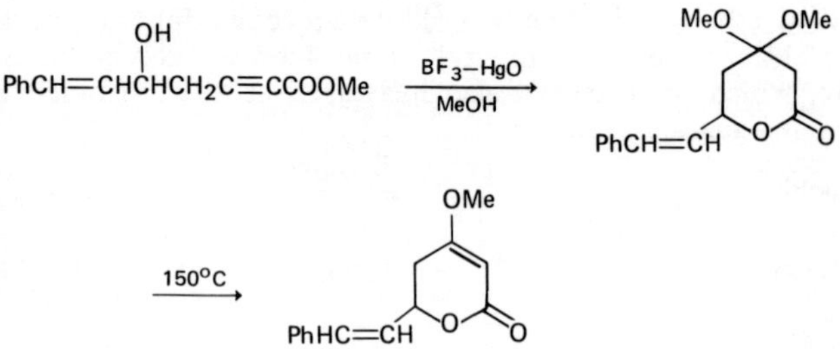

(15)

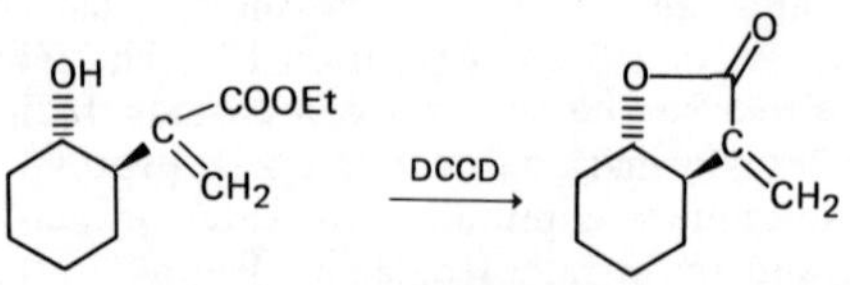

(Ref. 45)   (16)

    In some instances of lactone synthesis from hydroxy esters, the ester function is first saponified, and subsequent acidification leads to lactone formation. Such a procedure has been employed in an alternative synthesis of (±)-kawain[38,46] as shown in equation (18). The synthesis of β-carboxy-γ-tridecyl-γ-butyrolactone is accomplished in a similar fashion (equation 19)[47].

    In general, uncatalysed thermal lactonization of hydroxy esters tends to give significant amounts of polymeric material. For example, distillation of a series of

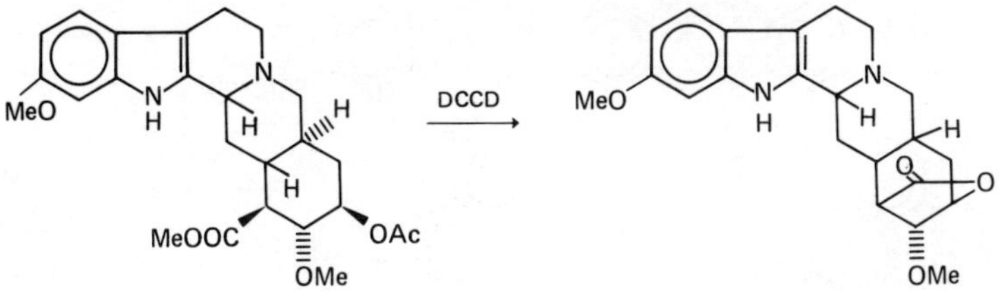

(Ref. 28) (17)

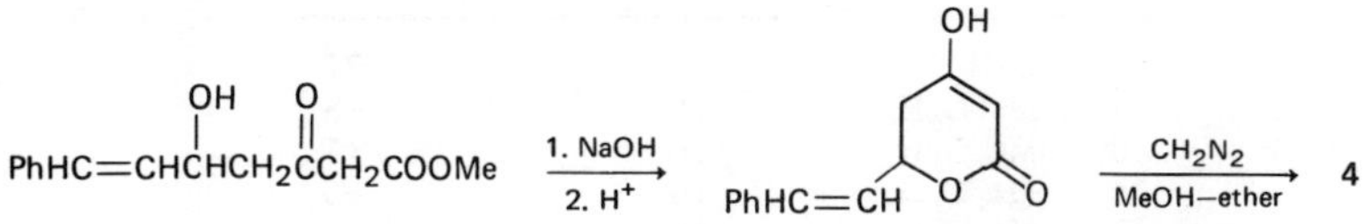

(18)

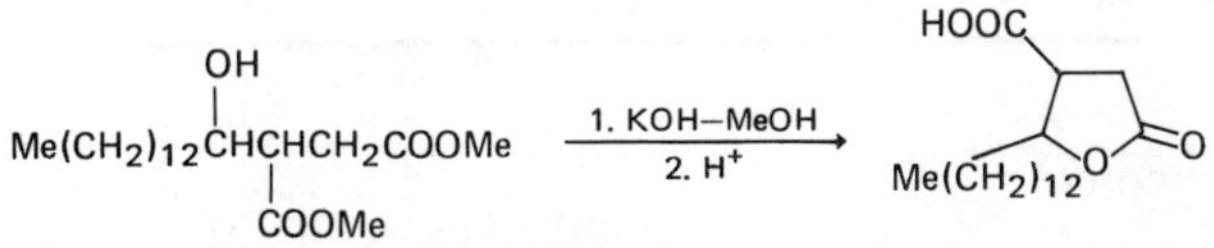

(19)

ethyl α-alkyl-δ-hydroxyhexanoates affords a mixture of unidentified polymers. However, depolymerization of this mixture by distillation in the presence of concentrated sulphuric acid or phosphoric acid produces the corresponding α-alkyl-δ-hydroxyhexanoic acid lactones in good yields (equation 20)[48].

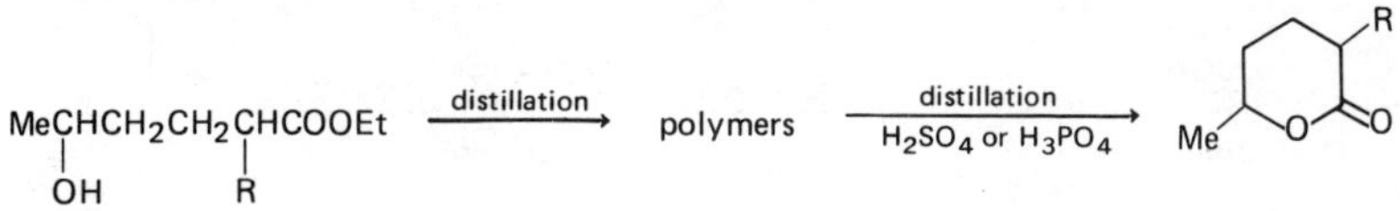

R = Et, Pr, n-Bu, i-Bu, i-amyl (20)

Corey and Nicolaou[49] have recently reported an ingenious method for lactone synthesis in which ω-hydroxy carboxylic acids are first converted to ω-hydroxy-2-pyridinethiol esters, which subsequently undergo facile thermal lactonization (equations 21 and 22). This appears to be one of the most general methods for large-ring lactone synthesis currently available.

Conversion of α|α-dialkyl-β-hydroxy acids to 4-oxo-1,3-dioxanes by reaction with methyl orthopropionate followed by heating these compounds at 150–200°C affords β-lactones in good yields via a proposed concerted mechanism (equation 23)[50].

Reaction of ethyl 1-hydroxymethylcyclopropanecarboxylate with zinc bromide in 48% hydrobromic acid results in cyclopropane ring enlargement to afford α-methylene-γ-butyrolactone in 25% yield (equation 24)[51]. Reactions (25) and (26) provided similar results[51]. This rearrangement has also been observed with cyclopropylmethyl methyl ethers and cyclopropylmethyl bromides (equations 27 and 28)[51].

The conversion of γ- or δ-keto acids to enol lactones is a well-known process, illustrated here by the synthesis[52] of the enol lactone of 1,4,4-trimethylcyclohexan-2-oneacetic acid (equation 29).

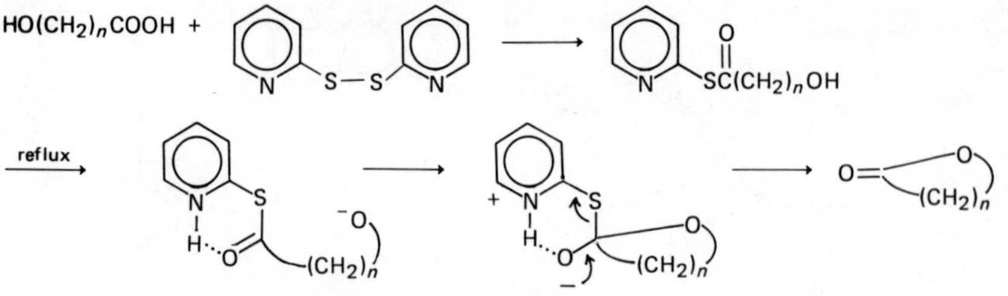

$$\text{HO(CH}_2)_n\text{COOH} + \quad (21)$$

n	Solvent	Reflux time (h)	Isolated yield (%)
5	C_6H_6	10	71
7	$Me_2C_6H_4$	30	8
10	$Me_2C_6H_4$	20	47
11	$Me_2C_6H_4$	10	66
12	$Me_2C_6H_4$	10	68
14	$Me_2C_6H_4$	10	80

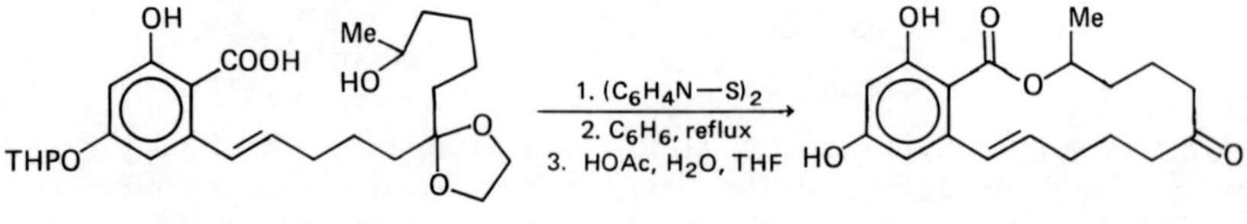

$$\xrightarrow[\begin{array}{c}1.\ (C_6H_4N{-}S)_2\\2.\ C_6H_6,\ \text{reflux}\\3.\ \text{HOAc, }H_2O,\ \text{THF}\end{array}]{} \quad (22)$$

THP = tetrahydropyranyl

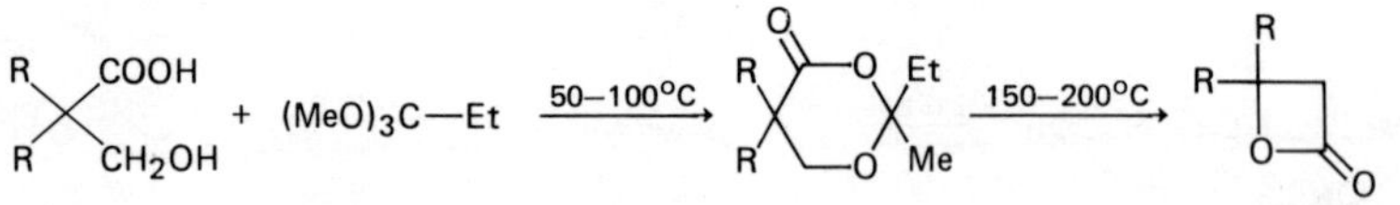

$$(23)$$

R = Me, 87%

R = Et, 84%

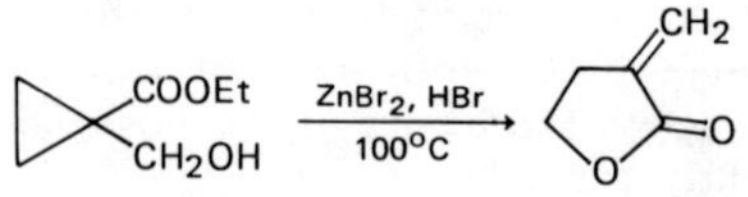

$$\xrightarrow[100^\circ C]{ZnBr_2,\ HBr} \quad (24)$$

$$(25)$$

Treatment of 2-carboethoxymethyl-2-methylcyclohexane-1,3-dione with poly-phosphoric acid (equation 30) results in ring-opening, followed by ring-closure to form lactone **6**, presumably by isomerization of intermediate enol lactone **5**[53]. When the analogous 2-carboethoxymethyl-2-(3-ketopentyl)cyclohexane-1,3-dione is treated under similar conditions, a new mode of cyclization is observed[53], affording fused δ-lactone **7** in 34% yield (equation 31). Formation of unexpected products was also observed when 2,2-di(carboethoxymethyl)cyclohexane-1,3-dione was treated[53] with polyphosphoric acid to afford a 64% yield of dilactone **8** (equa-tion 32). Treatment of 2,2-dimethylcyclohexane-1,3-dione under similar conditions

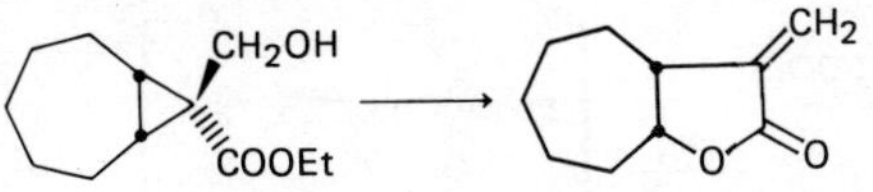

(26)

Reaction	Reaction conditions	Yield (%)
(25)	$ZnBr_2$, 48% HBr, EtOH, 100°C, 6 h	50
(25)	Conc. H_2SO_4, 0°C, 2 h	30
(25)	F_3CSO_4H, C_6H_6	—
(25)	p-MeC$_6$H$_4$SO$_3$H	—
(26)	$ZnBr_2$, 48% HBr, EtOH, 100°, 6 h	43

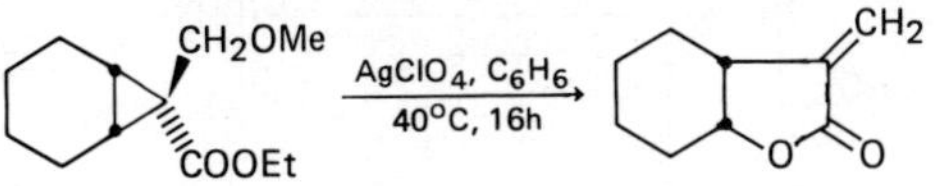

(27)

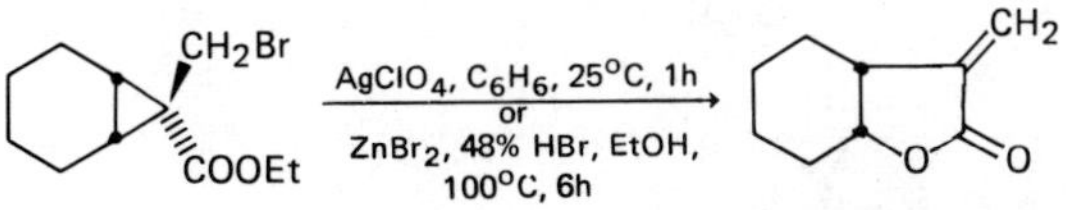

(28)

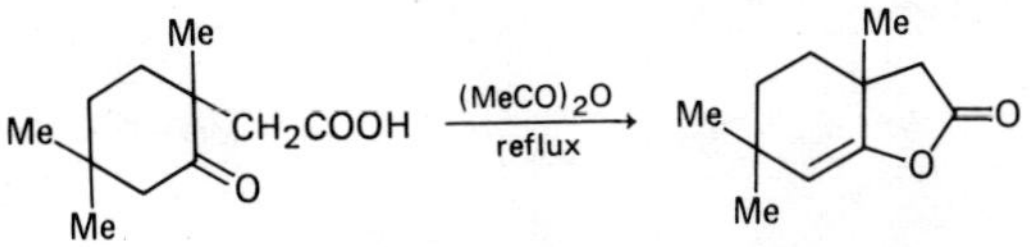

(29)

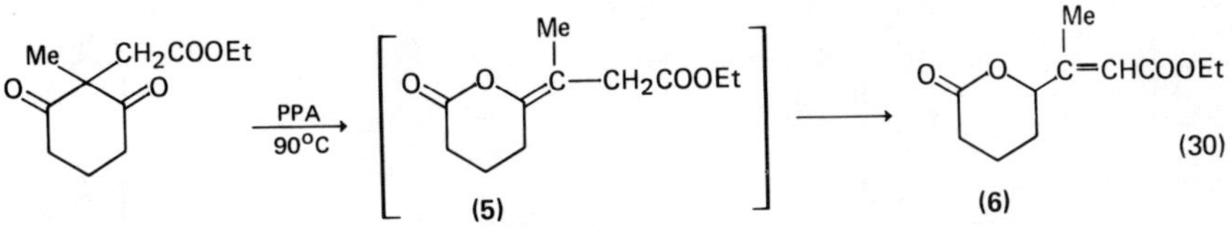

(30)

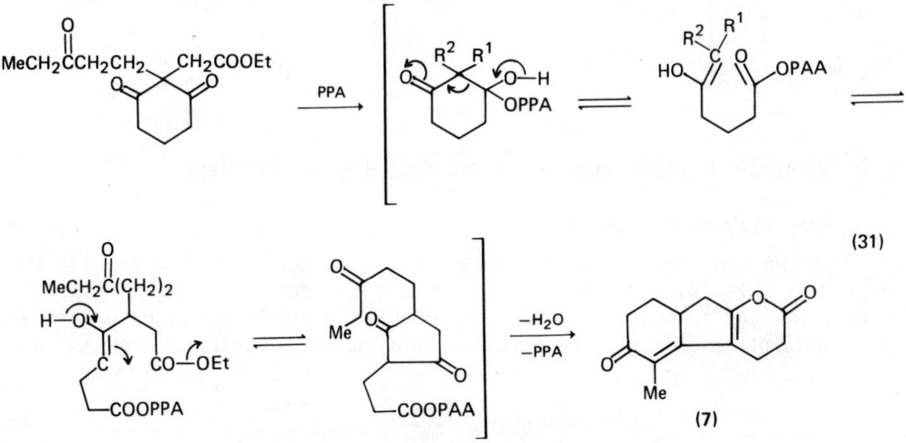

(31)

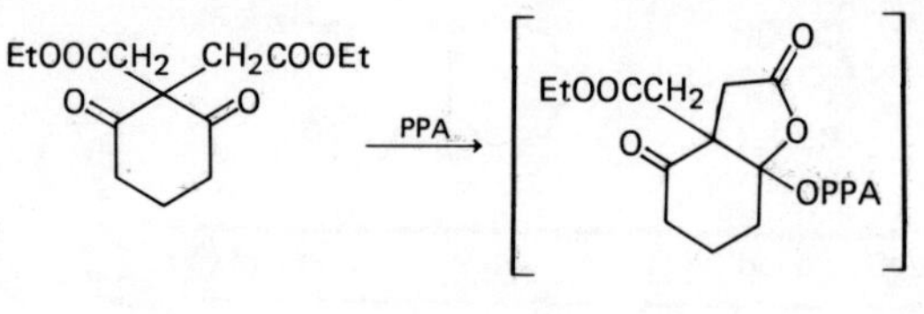

$$(32)$$

afforded no reaction, which was attributed to the deactivating effect of the two methyl groups[53].

Aldehydic acids can be cyclized to enol lactones by treatment with p-toluenesulphonic acid in benzene (equations 33–35)[54].

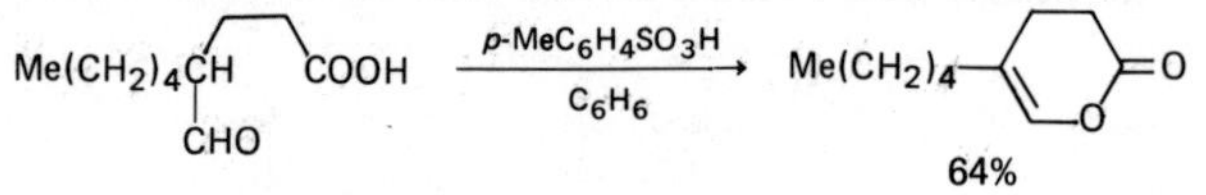

$$(33)$$

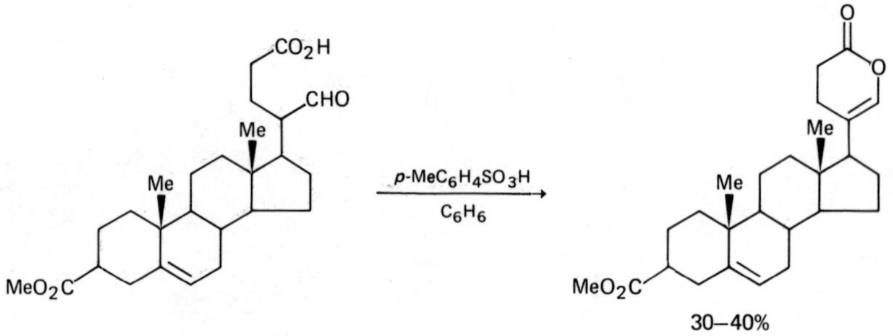

$$(34)$$

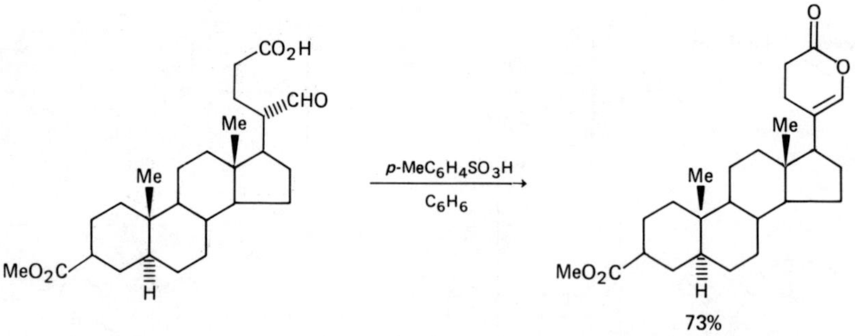

$$(35)$$

B. By Intramolecular Cyclization of Unsaturated Acids and Esters

1. Acid-catalysed cyclizations

Various unsaturated acids and esters have been converted to lactones in the presence of acids[5]. Recent examples include the preparation of 4,4-dimethyl-butyrolactone[55,56] and 4-methyl-4-phenylbutyrolactone[55] by cyclization of 4-methyl-3-pentenoic acid and 4-phenyl-3-pentenoic acid, respectively (equation 36).

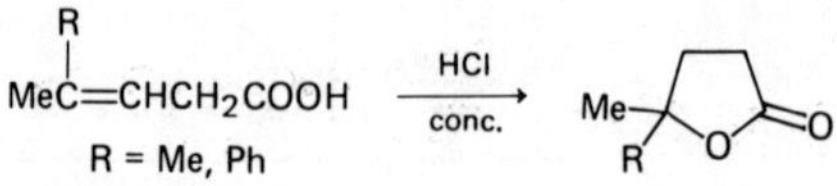

$$(36)$$

Treatment of alkenyl-substituted malonic esters with aqueous acid affords the expected γ-lactones in good yields, while basic hydrolysis produces the γ,δ-unsaturated acid (equation 37)[57].

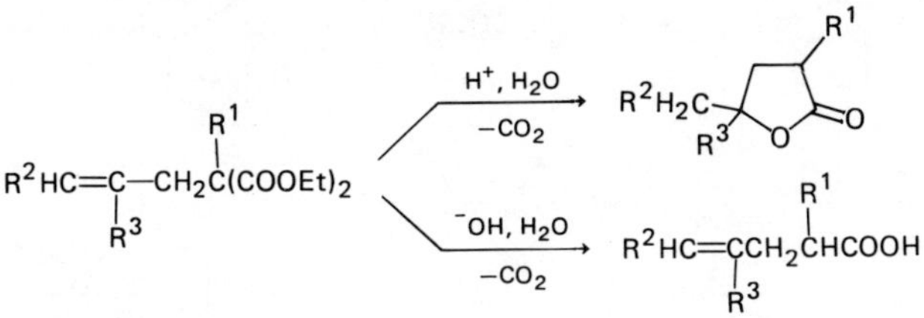

(37)

R[1] = H, n-Bu

R[2] = H, n-Bu, n-C$_5$H$_{11}$

R[3] = H, Me

2-Hydroxy-2,6,6-trimethylcyclohexylideneacetic acid γ-lactone[58-62] has been synthesized[52] by treatment of 2,6,6-trimethylcyclohexene-1-glycolic acid with aqueous sulphuric acid or by simply heating the glycolic acid at 200–220°C (equation 38). Alternatively, this lactone can be prepared[52] by base-catalysed ring closure of 9. Interestingly, treatment of 2,6,6-trimethylcyclohexene-1-glycolic acid with chromic anhydride–pyridine[52] affords 'hydroxyionolactone', which can also be prepared[52] by permanganate oxidation of β-ionone (equation 39).

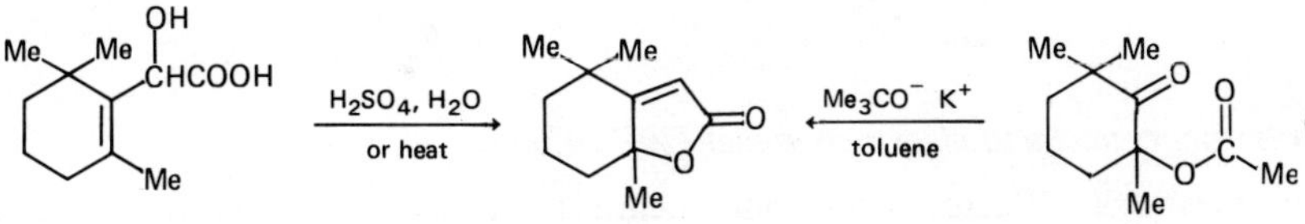

(38)

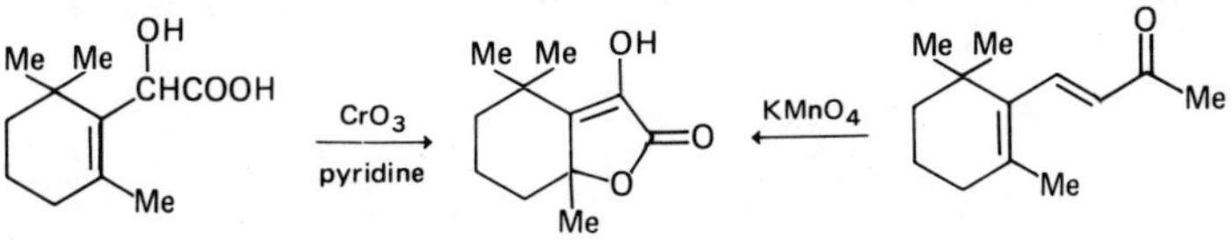

(39)

2,5-Dienoic acids and esters, such as those shown in Table 4, can be converted into α,β-unsaturated δ-lactones upon treatment with 80% sulphuric acid at

TABLE 4. Cyclization of 2,5-dienoic acids and esters using sulphuric acid[63]

Acid or ester	Lactone	Yield (%)
cis-2,5-Hexadienoic acid	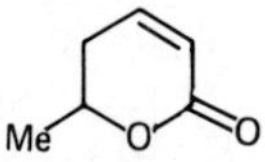	84.6
Methyl cis-2,5-hexadienoate	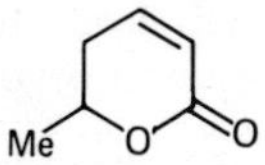	77.0

$0-5°C^{63}$. Carboxylic acids containing multiple unsaturation can undergo rather complex cyclizations[64] in the presence of sulphuric acid as shown in equation (40).

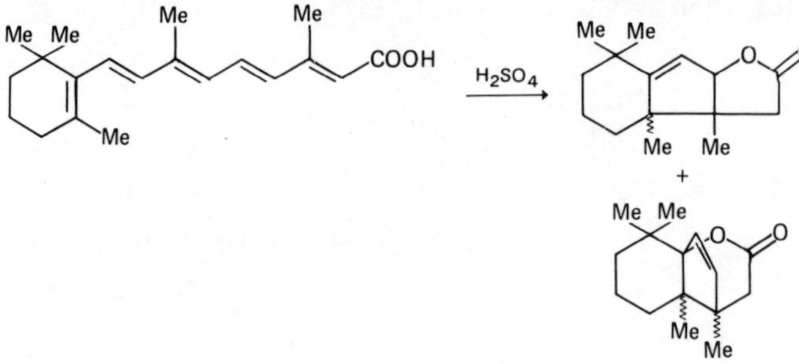

$$\text{(40)}$$

Cyclization of acids and esters containing acetylenic bonds has seen wide application in the preparation of lactones[65-80]. A number of representative examples are presented in Table 5.

o-Phenylbenzoic acids undergo cyclization upon treatment with hydrogen peroxide or chromic anhydride to form lactones in moderate yields (equation 41)[81].

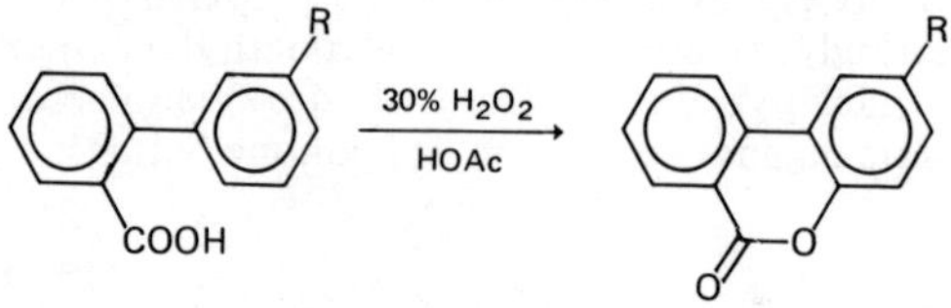

$$\text{(41)}$$

2. Photochemical and electrochemical cyclizations

Preparations of lactones by photochemical cyclization of unsaturated acids or esters have also been reported in the literature[83-89]. Irradiation[82] of a series of α-substituted cinnamic and crotonic acids afforded the corresponding substituted β-lactones (equation 42).

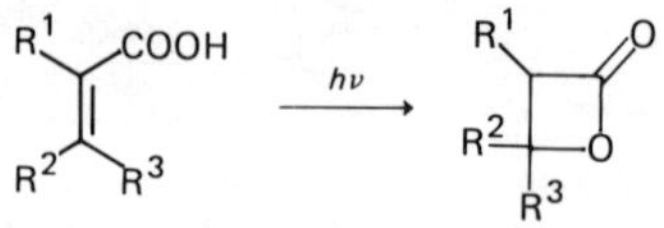

$$\text{(42)}$$

R^1 = Ph, Ph, Me, Ph, Ph, Ph

R^2 = H, Ph, Ph, Me, Ph, p-MeC$_6$H$_4$

R^3 = Ph, H, H, H, Ph, H

γ-Lactones have been prepared by the irradiation-induced addition of alcohols to α,β-unsaturated acids or esters in the presence[83-85] or absence[86,87] of a sensitizer, by the use of ^{60}Co γ-rays[88,89], and by reductive electrochemical addition of acetone[89]. Some γ-lactones prepared by these various methods are listed in Table 6. The mechanism suggested[87] for the photolytic addition in the absence of a sensitizer is shown in equation (43) and involves initial hydrogen abstraction by the excited ester carbonyl, which then leads to α-hydroxyalkyl and allylic radicals. Coupling of the former to the β-carbon of the latter and tautomerization affords a γ-hydroxy ester which then cyclizes.

$$\text{(structure)} \xrightarrow{h\nu} \left[\text{(structure)} \right]^{*} \xrightarrow{R^1R^2CHOH}$$

cis or trans

$$R^1R^2\overset{\cdot}{C}\!-\!OH \; + \; \left[\text{(structure)} \longleftrightarrow \text{(structure)} \right] \xrightarrow{R^1R^2\overset{\cdot}{C}\!-\!OH} \qquad (43)$$

$$\left[\text{(structure)} \right] \longrightarrow \left[\text{(structure)} \right] \xrightarrow{-MeOH} \text{(structure)}$$

3. Halolactonization

The reaction of unsaturated acids with iodine–potassium iodide and bicarbonate in aqueous medium affords iodolactones. This iodolactonization, first reported[90] in 1908, was originally believed to exhibit the following characteristics: (a) α,β-unsaturated acids do not give iodolactones (b) β,γ- as well as γ,δ-unsaturated acids do afford iodolactones, (c) δ,ε acids or acids with the unsaturation further removed from the carboxyl group yield only poorly characterized unsaturated acid iodohydrins and (d) α-keto β,γ-alkenoic acids and α,β,γ,δ-alkenoic acids are exceptional in that no iodolactones are obtained from them. Since α,β-unsaturated acids do not give iodolactones but β,γ-unsaturated acids afford β-iodo-γ-lactones via this procedure, a number of workers[91–93] have used this approach to distinguish α,β-unsaturated acids from β,γ isomers. In a more involved study of this reaction, Van Tamelen and Shamma[94] showed that although β,γ-butenoic acid does not afford any iodolactone even upon long standing, β,γ-pentenoic acid (equation 44)

$$MeCH\!=\!CHCH_2COOH \xrightarrow[\text{NaHCO}_3]{\text{I}_2\text{-KI}} \text{(structure)} \qquad (44)$$

$$\text{(structure)} \xrightarrow[\text{NaHCO}_3]{\text{I}_2\text{-KI}} \text{(structure)} \qquad (45)$$

and Δ^1-cyclohexeneacetic acid (equation 45) rapidly affort the corresponding iodolactones[95]. Van Tamelen further established[94] that although there are two structural possibilities, δ-iodo-γ-lactones and γ-iodo-δ-lactones, for lactones derived

J. F. Wolfe and M. A. Ogliaruso

TABLE 5. Preparation of lactones via intramolecular cyclization of acetylenic acids

Acetylenic acid	Reaction conditions	Product	Yield (%)	Reference
Me—C≡C—C≡C—CH=CH—COOH *cis*	KHCO₃, H₂O	Me—C≡C—C≡C—CH= (butenolide)	55	65–67
PhC≡C—CH=C(COOH)₂	190°C, 10–15 min or AgNO₃	PhHC= (lactone), COOH	85	68
OCH₂—C≡C—C≡C—CH—COOH, CH(OMe)₂	MeOH, AgNO₃	(MeO)₂CH (butenolide), CH₂—CH= with tetrahydropyranyl	80	74
	MeOH, AgNO₃	Me (butenolide), PhHC=	63	76

PhC≡C—CR¹=CR²—COOH
cis

H^+ → **79**

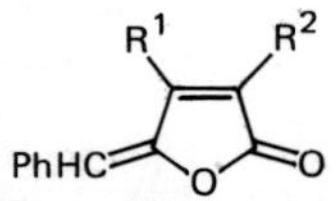

$R^1 = H; R^2 = Ph$ 80
$R^1 = H; R^2 = p\text{-}O_2NC_6H_4$ 85
$R^1 = Me; R^2 = Ph$ 70
$R^1 = Me; R^2 = p\text{-}O_2NC_6H_4$ 75
$R^1 = R^2 = Ph$ 40

PhC≡C—C=CH—COOH
|
Me
cis

50% H_2SO_4, 3h or
$HgSO_4$, dioxane or
acetone → **80**

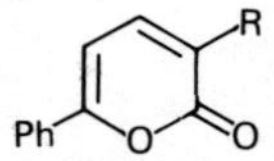

$R = COOH$ 54—80
$R = Ph$ 40—60
$R = \beta\text{-naphthyl}$ 40
$R = p\text{-biphenyl}$ 20
$R = p\text{-MeOC}_6H_4$ 20
$R = m\text{-MeOC}_6H_4$ 20

TABLE 6. γ-Lactones prepared via irradiation-induced and electrochemical addition of alcohols to unsaturated acids or esters

Acid or ester	Alcohol	Method[a]	Product	Yield (%)	Reference
HOOCCH=CHCOOH *cis*	MeC(OH)HR 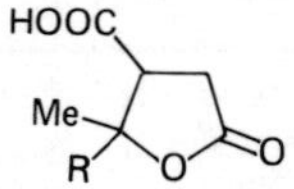				
	R = Me	A		60	83
	R = Et	A		57	83
	R = C_6H_{13}-n	B		40	84
	R = H	B		20	84
MeCH=CHCOOH *trans*	*i*-PrOH	B	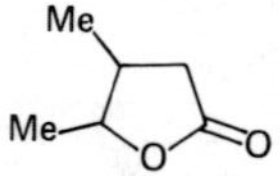	60	84
HOOCC≡CCOOH	*i*-PrOH	C	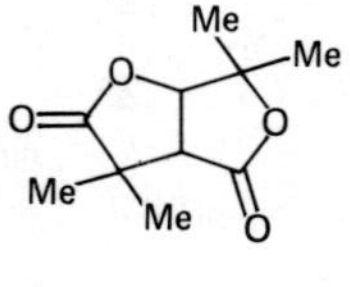	15	85
HC≡CCOOEt	*i*-PrOH	D	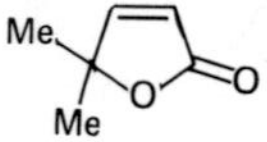	20	86
MeOOCCH=CHCOOMe *cis or trans*	*i*-PrOH	E		from *cis*, 64 from *trans*, 70	87 87

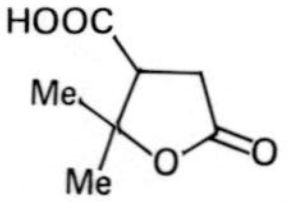

Reagent	Solvent		Product	Yield	Ref.
MeCH=CHCOOMe *trans*	*i*-PrOH	E	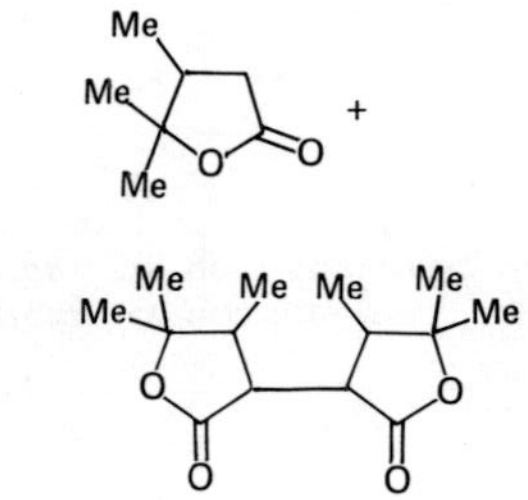 [lactone structures] +	50 + 12	87
MeOOCCH=CHCOOMe *cis* or *trans*	EtOH	E	[bis-lactone structure] *cis + trans*	from *cis* 68 / from *trans* 71	87 / 87
MeCH=CHCOOMe *trans*	EtOH	E	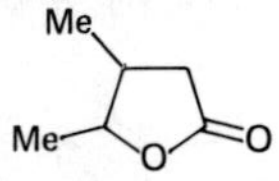 [lactone structure] *cis + trans*	59	87
MeCH=CHCOOEt	MeOH	F / G / H	[lactone structure]	1 / 1 / 1	88 / 88 / 88
MeCH=CHCOOEt	EtOH	F / G / H	[lactone structure] *trans: cis, 4:5*	16 / 4 / 1	88 / 88 / 88

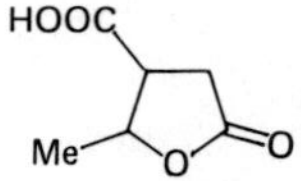

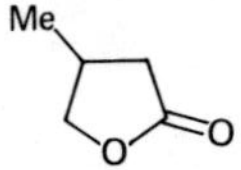

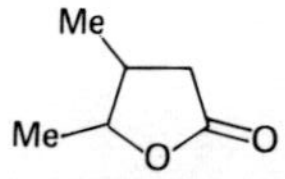

TABLE 6. (Continued)

Acid or ester	Alcohol	Method[a]	Product	Yield (%)	Reference
	i-PrOH	F		54, 44	88, 89
		G		12	88
		H		12	88
		I		30	89
	n-PrOH	F		9	88
			trans: cis, 2:3		
		G	trans: cis, 4:5	4	88
		H	trans: cis, 4:5	5	88
	sec-BuOH	F		18	88
			trans: cis, not determined		
		H	trans: cis, not determined	22	88
	Et₂CHOH				

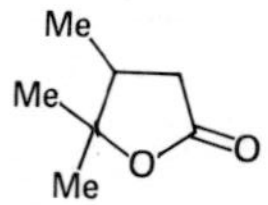

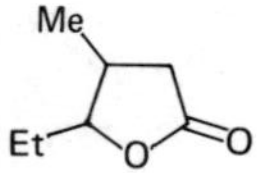

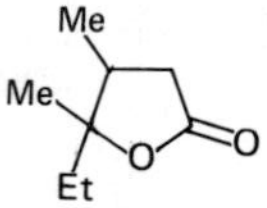

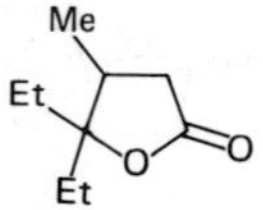

Unsaturated ester	Alcohol	R	Product	Yield (%)	Ref.
H₂C=CHCOOEt	PhCH₂OH	F	Me, Ph lactone (trans: cis, 2:1)	10	88
		H		9	88
	i-PrOH	F	Me, Me lactone	23	88
H₂C=CHCOOH	i-PrOH	F	Me, Me lactone	0	89
		I		95	89
Me₂C=CHCOOEt	i-PrOH	F	Me; Me, Me lactone	41	89
		I		36	89
	i-PrOH	F	Me, Me; Me lactone	29	89
		I		trace	89
EtOOCCH=CHCOOEt *cis or trans*	i-PrOH	F	EtOOC, Me, Me lactone — from *cis*	70	89
		F	from *trans*	22	89
		I	from *cis*	59	89
		I	from *trans*	21	89

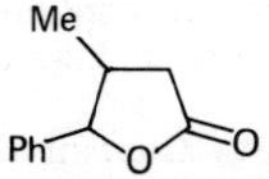

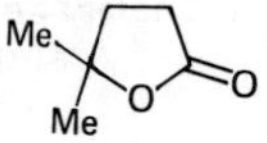

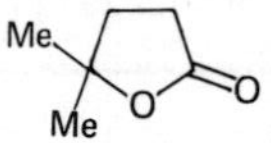

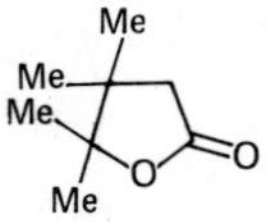

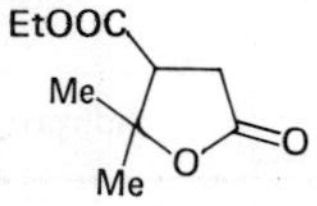

TABLE 6. (Continued)

Acid or ester	Alcohol	Method[a]	Product	Yield (%)	Reference
HOOCCH=CHCOOH *trans*	*i*-PrOH	F	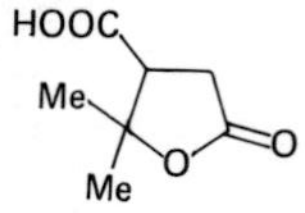	69	89
PhCH=CHCOOEt *trans*	*i*-PrOH	F I		0 0	89 89

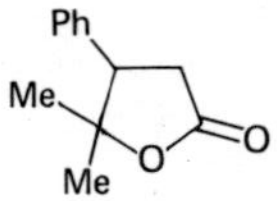

[a] A = Irradiation with an ultraviolet light source for 18 h at 16°C using benzophenone as sensitizer.
B = As above but irradiated for 25 h in the cold.
C = As above but irradiated for 60 h at 35°C.
D = Irradiated using a quartz-contained mercury arc.
E = Irradiated using a 450 W-Hanovia medium-pressure mercury arc.
F = Irradiated using γ-rays in a ⁶⁰Co cavity source at room temperature.
G = Irradiated using a quartz tube for 50 h. with a 500 W high-pressure mercury vapour lamp in the pressure of benzophenone as sensitizer.
H = Irradiated using a Pyrex tube for 72 h. with a 500 W high-pressure mercury vapour lamp in the presence of benzophenone as sensitizer.
I = Electrolysis with acetone, 20% sulphuric acid and water for 1 h with a terminal voltage of 75–95 V in a cylindrical vessel using a mercury pool cathode and a platinum plate anode.

from γ,δ-unsaturated acids, both γ,δ-pentenoic acid (equation 46) and Δ²-cyclohexeneacetic acid (equation 47) give rise to the γ-lactones rather than the

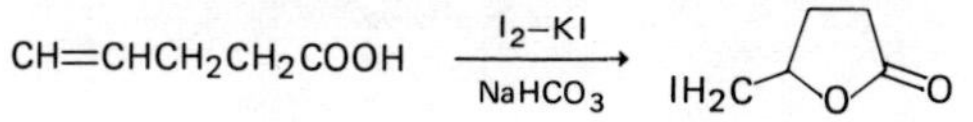

$$CH{=}CHCH_2CH_2COOH \xrightarrow[\text{NaHCO}_3]{\text{I}_2-\text{KI}} \qquad (46)$$

$$\qquad\qquad \xrightarrow[\text{NaHCO}_3]{\text{I}_2-\text{KI}} \qquad (47)$$

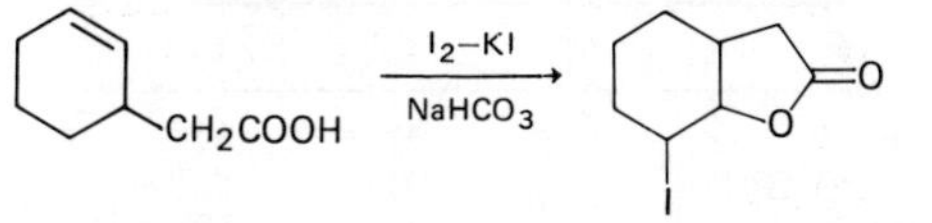

δ isomers originally proposed by Bougault[90]. It was also established[94] that δ,ε-hexenoic acid affords (probably) δ-iodomethyl-δ-valerolactone, again contrary to Bougault's findings, while ε,ζ-heptenoic acid and ω-undecylenic acid led to unstable, poorly defined products.

Halolactonization reactions have also been used to separate[96,97] mixtures of *endo*- and *exo*-norborn-5-enyl acids and *endo*- and *exo*-methylenenorborn-5-enyl acids[98] In the former case[96,97] the *endo* isomer reacts to produce a γ-lactone while the *exo* isomer remains in the aqueous layer as the carboxylate salt; in the latter, both isomers react with bromine in methylene chloride to give lactone products, whereas reaction with iodine in methylene chloride affords the iodolactone from the *endo* isomer only (equation 48)[98]. With the carboxylate salt of the *exo* acid β-lactones are obtained during the bromolactonization but none have been detected during iodolactonization[98].

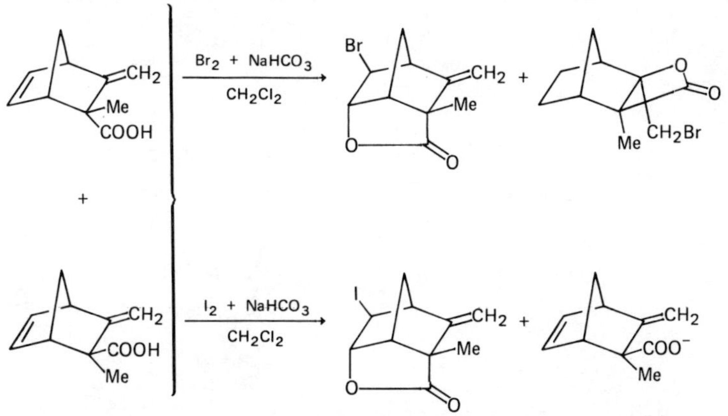

$$\qquad\qquad\qquad\qquad\qquad (48)$$

In order to determine if β-lactone formation was only associated with rigid systems or if conformationally more flexible β,γ-unsaturated acids would also form β-lactones, open-chain β,γ-unsaturated carboxylate salts in aqueous solutions were treated[99] with carbon tetrachloride or methylene chloride solutions containing bromine and were observed to readily cyclize to γ-bromo-β-lactones (equation 49).

In 1972, Barnett and Sohn[100] explained the seeming anomaly that iodolactonization of β,γ-unsaturated carboxylate salts affords β-iodo-γ-lactones, whereas bromolactonization of the same salts affords γ-bromo-β-lactones, and showed that the size of the lactone ring obtained did not depend upon the kind of halogen used, but rather the differences in the experimental procedures used. In iodolactonization conducted under conditions similar to bromolactonization the products obtained

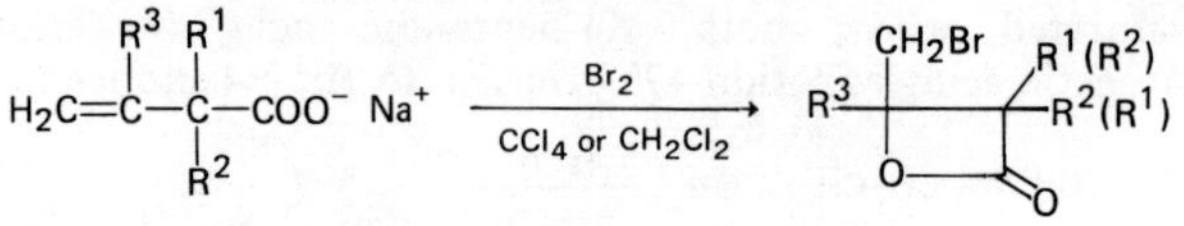

$$\underset{\underset{R^2}{|}}{\overset{R^3 \quad R^1}{\underset{|}{H_2C=C-C-COO^-}}} \; Na^+ \quad \xrightarrow[\text{CCl}_4 \text{ or } \text{CH}_2\text{Cl}_2]{Br_2} \quad R^3 \underset{O}{\overset{CH_2Br}{\underset{\parallel}{\diagdown}}} \overset{R^1(R^2)}{\underset{R^2(R^1)}{\diagup}} \tag{49}$$

R^1	R^2	R^3	Yield (%)
Me	Me	H	83^a
COOEt	Me	Me	88^b
H	H	H	50

^{a}Pure product.
^{b}Mixture of isomers.

are indeed γ-iodo-β-lactones. Thus, if iodolactonization is performed using excess potassium iodide and long reaction times, β,γ-unsaturated acids produce γ-lactones, whereas if short reaction times are used in the absence of potassium iodide, γ-iodo-β-lactones are produced (equation 50). One exception to this rule in stytyl-acetic acid, which is concerted[100] to the γ-lactone regardless of the procedure employed (equation 51).

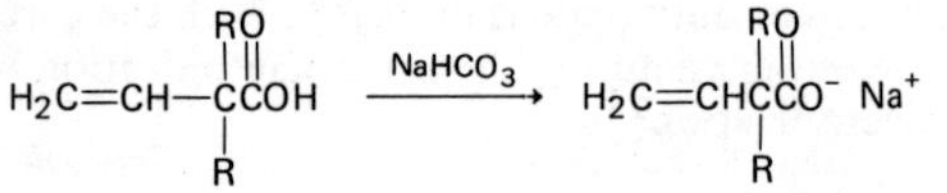

$$\underset{R}{\overset{RO \quad O}{\underset{|}{H_2C=CH-\overset{\parallel}{C}COH}}} \quad \xrightarrow{NaHCO_3} \quad \underset{R}{\overset{RO \quad O}{\underset{|}{H_2C=CHCCO^-}}} Na^+ \tag{50a}$$

$$
\begin{array}{c}
\underset{R}{\overset{RO \quad O}{\underset{|}{H_2C=CHCCO^-}}} Na^+ \\[4pt]
R = Me, H
\end{array}
\qquad
\begin{array}{l}
\xrightarrow[\text{ether}]{I_2} \quad \text{(I—CH}_2 \text{ β-lactone)} \\
\quad R = Me, 70–80\% \\
\quad R = H, \; 20–30\% \\[8pt]
\xrightarrow{\text{excess KI–I}_2} \quad \text{(γ-lactone)} \\
\quad R = Me
\end{array}
\tag{50b}
$$

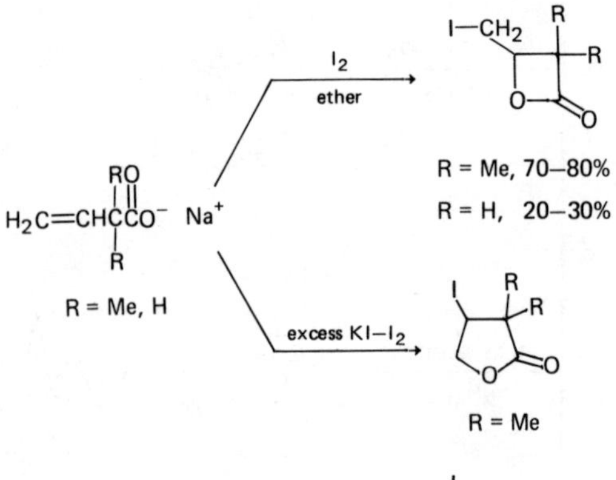

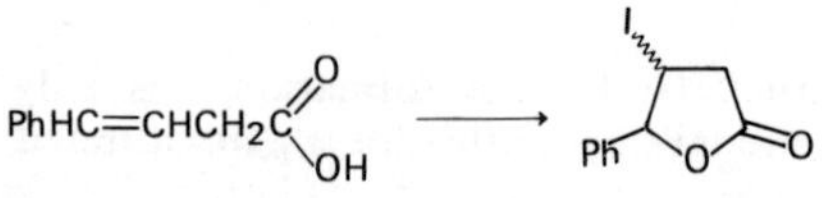

$$PhHC=CHCH_2C\overset{O}{\underset{OH}{\diagdown}} \quad \longrightarrow \quad \text{(iodolactone Ph)} \tag{51}$$

Bromolactonization of Δ^4-cyclohexene-*cis*-1,2-dicarboxylic acid esters has also been reported (equations 52 and 53)[101], while application to this method to linear di- and tetra-carboxylic acid esters affords[102] substituted γ,γ-dilactones (equations 54 and 55).

An interesting application of a bromolactonization-type of reaction in the field of steriod synthesis[6,103] has been reported[104] in the preparation of 5β,14α-bufa-20(22)-enolide using N-bromosuccinimide in carbon tetrachloride (equation 56). Quinone-mediated dehydrogenation[105] of the product in refluxing dioxane con-

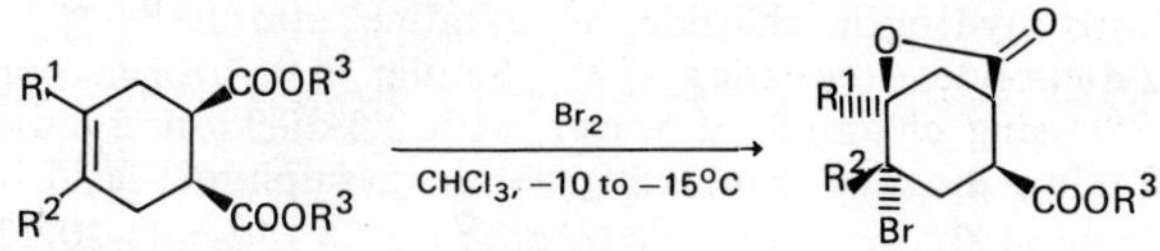

(52)

R^1	R^2	R^3	Yield (%)
Me	H	Me	36
Me	Me	Me	72
Me	Me	Et	75

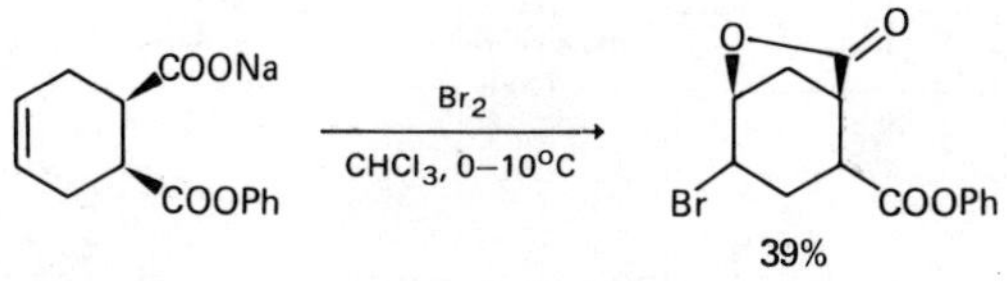

(53)

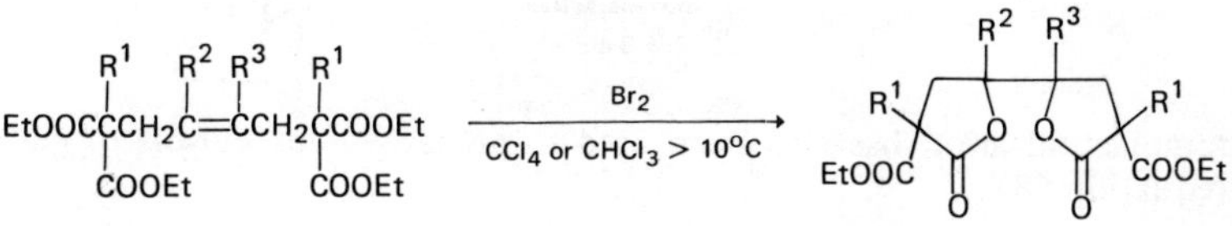

(54)

$R^1 = C_{1-5}$, alkyl or benzyl

$R^2 = $ H, H, Me

$R^3 = $ H, Me, Cl

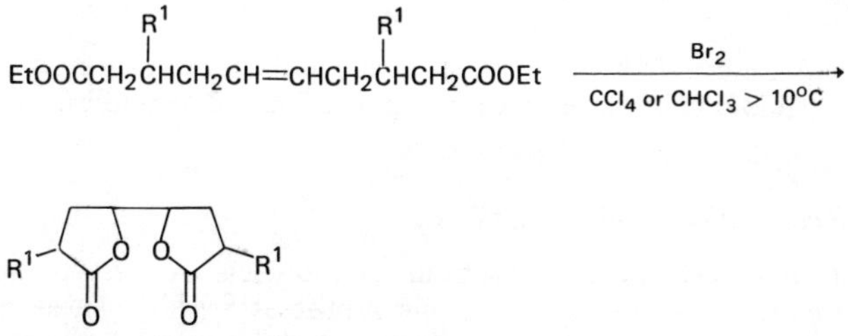

(55)

$R^1 = C_{1-5}$, alkyl or benzyl

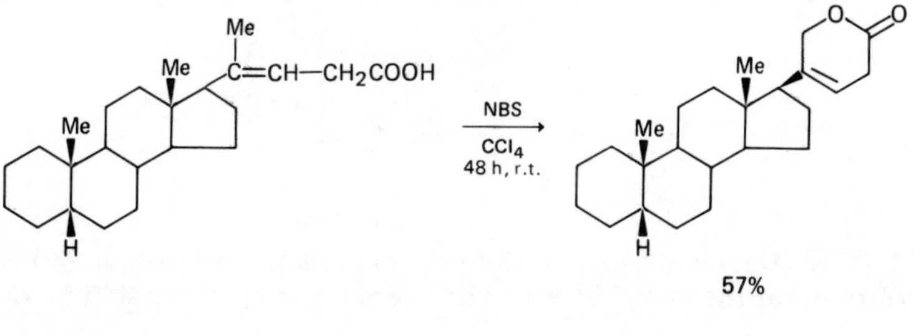

(56)

57%

taining anhydrous hydrogen chloride or pyridine affords[104,106] the isomeric bufa-17(20),22-dienolides (equation 57). Similar dehydrogenation results are obtained[104,106] using chloranil. However, with 2,3-di-chloro-5,6-dicyanoquinone (DDQ) in refluxing dioxane containing p-toluenesulphonic acid, the dehydrogenation is specific at $C_{(21)}$ and $C_{(23)}$ to produce 5β,14α-bufa-20,22-dienolide in quantitative yield.

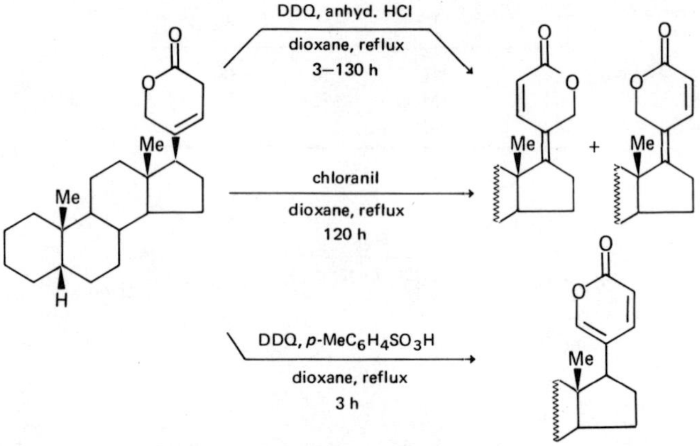

(57)

Bromolactonization has also been used to prepare[107] precursors of gibberellic acid (equation 58).

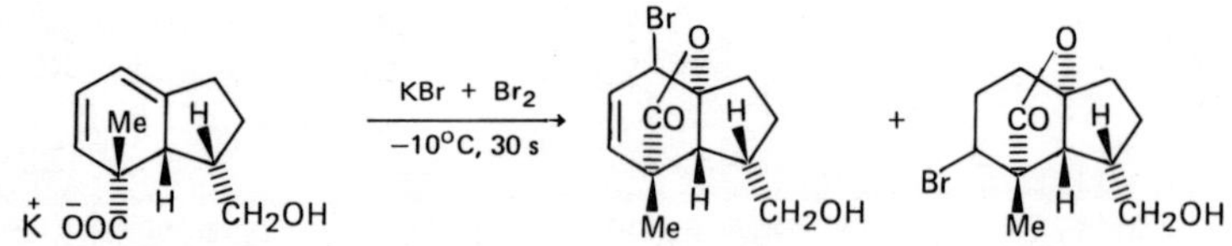

(58)

Arnold and Lindsay[108] have shown that iodolactones can be obtained by the use of cyanogen iodide in place of iodine—potassium iodide and bicarbonate.

4. Intramolecular Diels—Alder reactions

Intramolecular cycloaddition reactions of the Diels—Alder type have been employed in a number of interesting lactone syntheses[109,110]. These reactions may be generalized by viewing them as addition of the dienophilic triple bond of an acetylenic acid ester to a diene function contained in the alkoxy moiety of the ester (equation 59). A number of representative examples of such reactions are given in

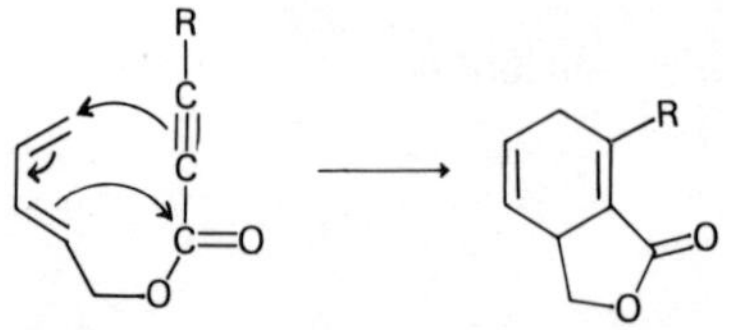

(59)

Table 7. Diels—Alder cyclization of the diene ester shown in equation (60) has been observed to occur thermally[115] via a [1,5] sigmatropic hydrogen shift in which the

TABLE 7. Preparation of lactones by intramolecular Diels–Alder cycloadditions

Acetylenic acid	Reaction conditions	Product	Yield (%)	Reference
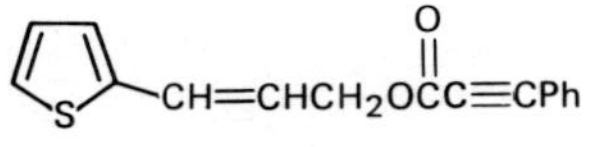 (thiophene)–CH=CHCH$_2$OCC≡CPh, O=; *trans*	(MeCO)$_2$O, reflux, 6 h	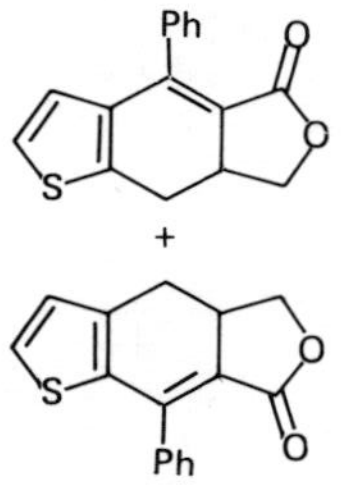 (fused bicyclic lactone with Ph and thiophene) + (isomeric lactone)	24 10	111
PhCH=CR′–C(=O)–O–C(R^2)$_2$–C≡CPh	(MeCO)$_2$O, reflux, 6 h	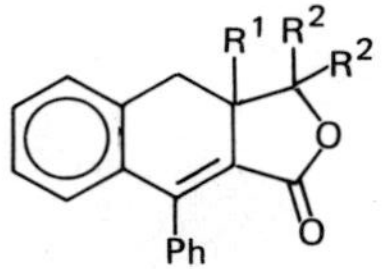 (tricyclic lactone with R^1, R^2, Ph) R^1 = R^2 = H R^1 = H; R^2 = D R^1 = R^2 = D	46 28 30	112
PhC≡CCH$_2$–O–C(=O)–C≡C–Ph	(MeCO)$_2$O, reflux, 5 h	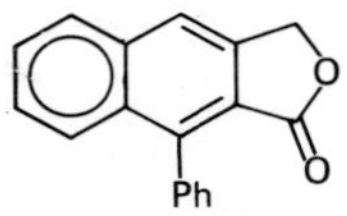 (tricyclic lactone with Ph)	39	112

TABLE 7. (Continued)

Acetylenic acid	Reaction conditions	Product	Yield (%)	Reference
 trans	(MeCO)$_2$O, reflux 6 h or 240°C (0.3 mm)	 (A) + (B) R^1, R^2 = —OCH$_2$O—; R^3 = R^4 = R^5 = OMe R^1, R^2 = —OCH$_2$O—; R^3 = R^4 = OMe; R^5 = H R^1 = R^2 = R^3 = R^4 = OMe; R^5 = H R^1 = R^2 = OMe; R^3 = R^4 = R^5 = H R^1 = R^2 = R^5 = H; R^3 = R^4 = OMe	45(A) + 10(B) 43(A + B) 32 22 43	113, 114

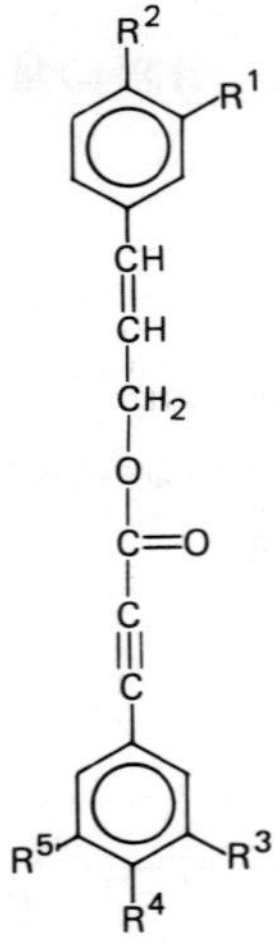

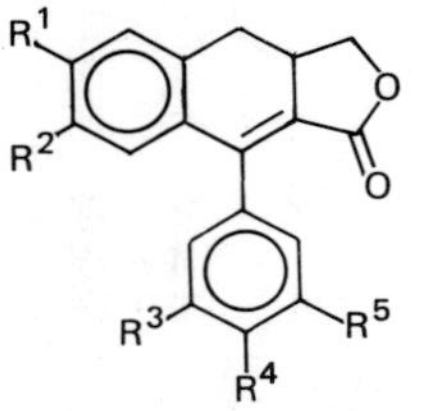

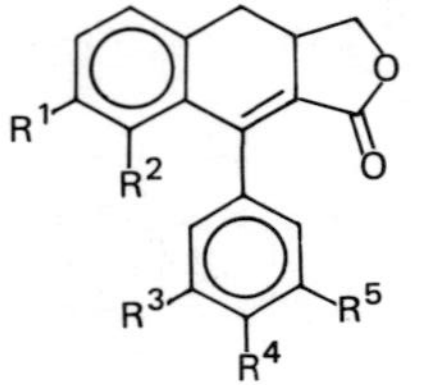

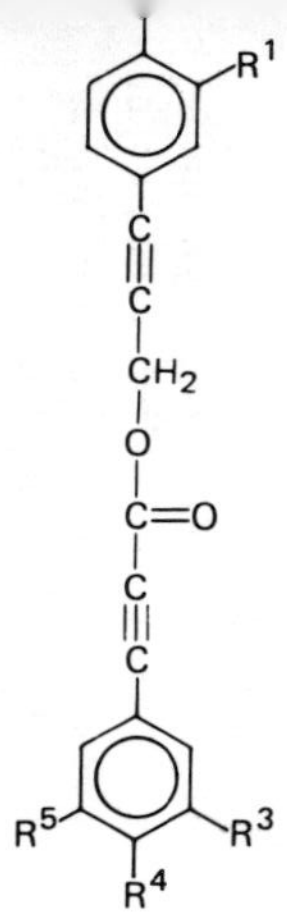

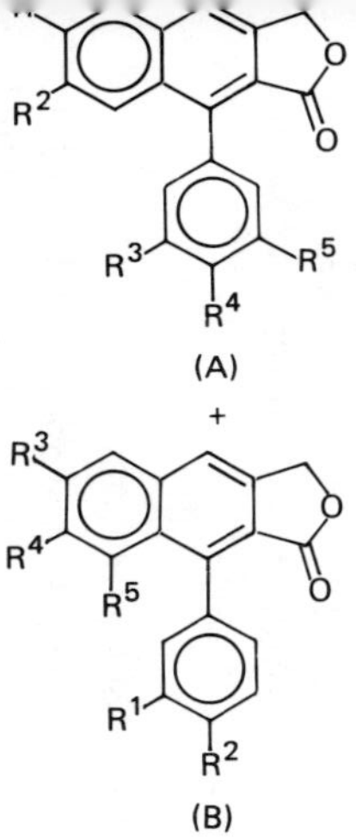

114

$R^1 = R^2 = R^3 = R^4$ OMe; $R^5 = H$

$R^1, R^2 = -OCH_2-O$; $R^3 = R^4 = R^5 = OMe$

$R^1 = R^2 = R^5 = H$; $R^3 = R^4 = OMe$

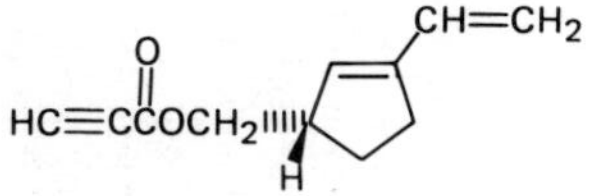

Heat >135°C
or
$(MeCO)_2O$, reflux, 20 h

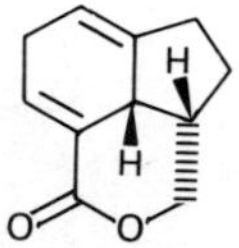

12(B)[a] + 4.5(A)[b]

13(B) + 11(A)[c]

70(A)[d]

70

107

[a] Also prepared by dehydrogenation of dimethyl-α-conidendrin using N-bromosuccinimide.

[b] Also prepared by dehydrogenation of 1-(3,4-dimethoxyphenyl)-3-hydroxymethyl-6,7-dimethoxy-3,4-dihydro-2-naphthoic acid lactone using lead tetraacetate in glacial acetic acid at 80°C.

[c] Also prepared by sublimation of a mixture of podophyllotoxin and 30% Pd–C at 275°C (0.15 mm) for 8 h, followed by cyclization of the sublimate.

[d] Prepared only by hydrogenation of 1-(3,4-dimethoxyphenyl)-3-hydroxymethyl-3,4-dihydro-2-naphthoic acid lactone over 30% Pd–C in p-cymene for 44 h.

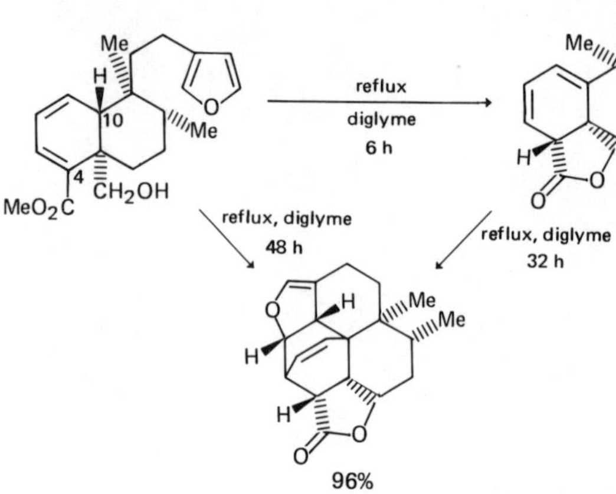

(60)

10β-hydrogen migrates suprafacially to the $C_{(4)}$ position to produce the intermediate lactone. This intermediate then undergoes a Diels—Alder reaction between the disubstituted double bond of the furan (dienophile) and the cyclohexadiene (diene).

C. By Acetoacetic Ester and Cyanoacetic Ester Condensations

Heating ethyl acetoacetate with a trace of sodium bicarbonate affords[116] a 53% yield of dehydroacetic acid (equation 61); however, if ethyl acetoacetate is treated

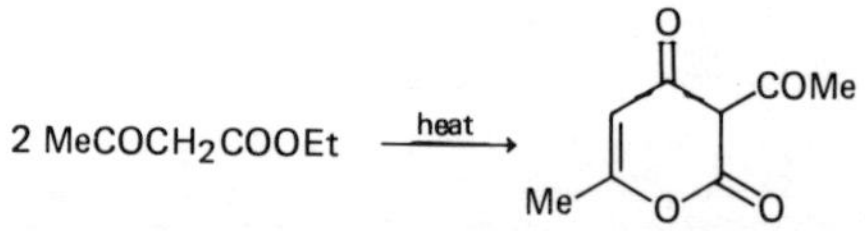

(61)

with concentrated sulphuric acid at room temperature for 5—6 days a mixture of 22—27% of isodehydroacetic acid (4,6-dimethylcoumalic acid) and 27—36% of ethyl isodehydroacetate (ethyl 4,6-dimethylcoumaloate) is obtained (equation 62)[117].

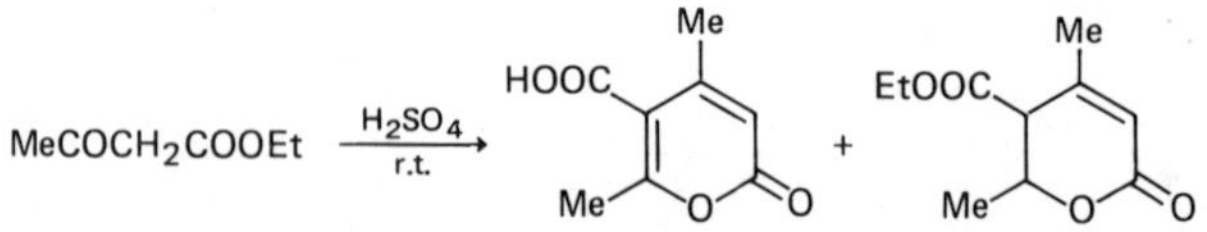

(62)

Condensation of the monocarbanions of acetoacetic ester or ethyl cyanoacetate with a series of α-keto alcohols has been reported[118] to give the corresponding 2-acetyl- and 2-cyano-2-buten-4-olides in 55—96% yield (equations 63 and 64). Alcoholysis[118] of the 2-cyano-3,4,4-trimethyl analogue in the presence of sulphuric acid affords the corresponding ethyl ester (equation 65).

When ethyl acetoacetate is allowed to react[119] with 3-chloro-1,2-epoxypropane (epichlorohydrin) at 45—50°C for 18 hours in the presence of sodium ethoxide, a 61—64% yield of α-acetyl-δ-chloro-γ-valerolactone is obtained (equation 66). A similar report[120] involves the reaction of the carbanion generated from ethyl 2-furoylacetate and propylene oxide, which affords α-furoyl-γ-valerolactone in 54% yield (equation 67).

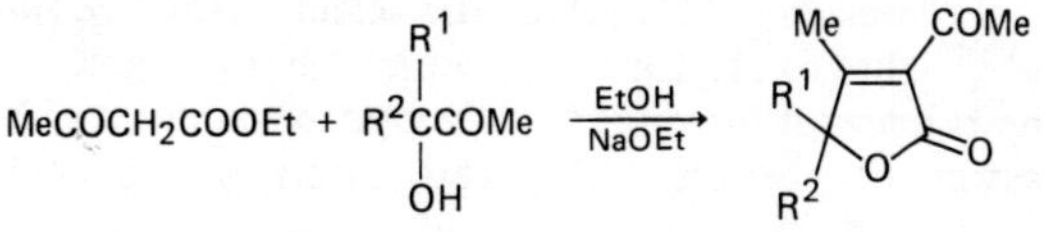

$$\text{MeCOCH}_2\text{COOEt} + \underset{\underset{\text{OH}}{|}}{\overset{\overset{\text{R}^1}{|}}{\text{R}^2\text{CCOMe}}} \xrightarrow[\text{NaOEt}]{\text{EtOH}}$$

(63)

$$\text{R}^1 = \text{Me, Me,} \quad \underset{|}{\overset{|}{(\text{CH}_2)_5}}$$
$$\text{R}^2 = \text{Me, Et,}$$

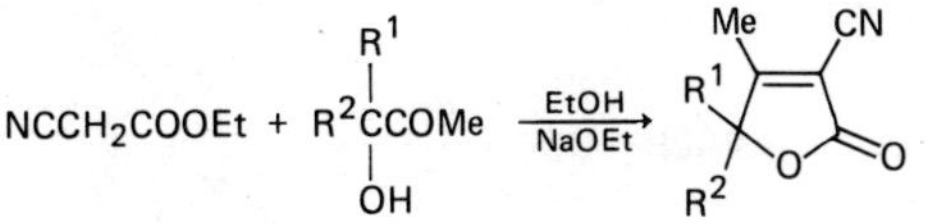

$$\text{NCCH}_2\text{COOEt} + \underset{\underset{\text{OH}}{|}}{\overset{\overset{\text{R}^1}{|}}{\text{R}^2\text{CCOMe}}} \xrightarrow[\text{NaOEt}]{\text{EtOH}}$$

(64)

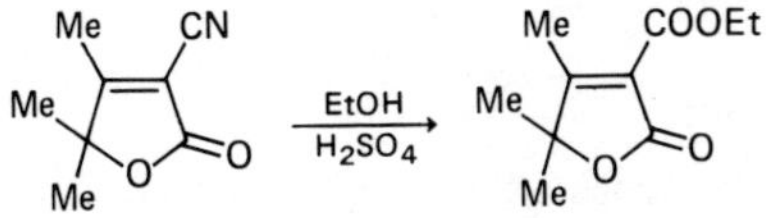

$$\xrightarrow[\text{H}_2\text{SO}_4]{\text{EtOH}}$$

(65)

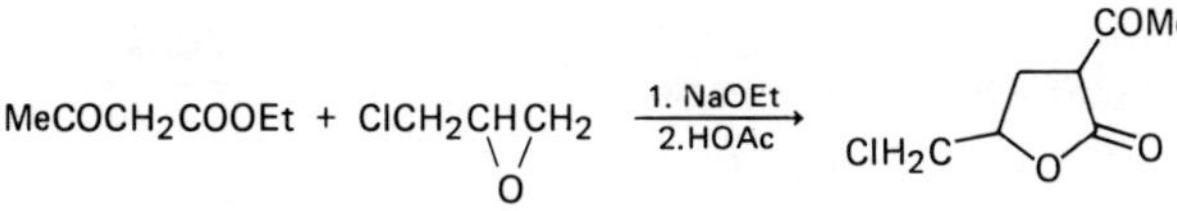

$$\text{MeCOCH}_2\text{COOEt} + \text{ClCH}_2\text{CHCH}_2 \xrightarrow[\text{2.HOAc}]{\text{1. NaOEt}}$$

(66)

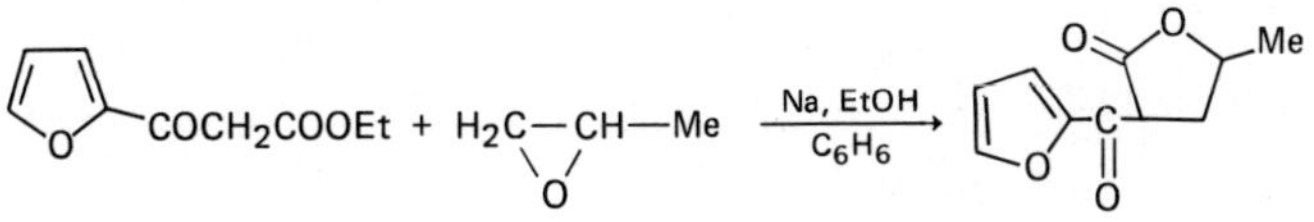

$$\text{COCH}_2\text{COOEt} + \text{H}_2\text{C}-\text{CH}-\text{Me} \xrightarrow[\text{C}_6\text{H}_6]{\text{Na, EtOH}}$$

(67)

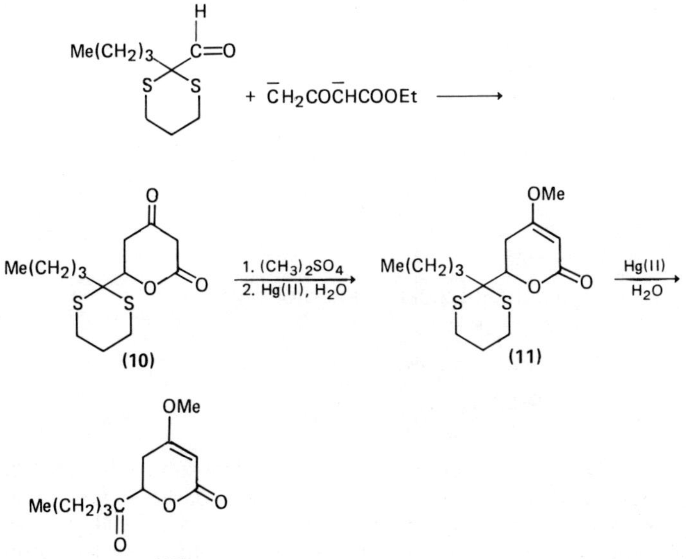

$$\underset{\text{S}\quad\text{S}}{\text{Me(CH}_2)_3}\overset{\overset{\text{H}}{|}}{\text{C}=\text{O}} + \bar{\text{C}}\text{H}_2\text{COCHCOOEt} \longrightarrow$$

$$(\mathbf{10}) \xrightarrow[\text{2. Hg(II), H}_2\text{O}]{\text{1. (CH}_3)_2\text{SO}_4} (\mathbf{11}) \xrightarrow[\text{H}_2\text{O}]{\text{Hg(II)}}$$

(68)

(12)

Reaction of the dianion of acetoacetic ester with 2,2-(propane-1,3-dithio)-hexanal affords[121] the oxolactone **10**, which in turn gives the enol ether **11**. Hydrolysis of the thioacetal provides (±)-didehydropestalotin (**12**).

The condensation of phenols with β-keto esters, β-keto acids or malic acid in the presence of concentrated sulphuric acid affords coumarins, and is known as the von Pechmann reaction[122]. A series of representative preparations of coumarins[125-131] by the von Pechmann reaction are given in Table 8. It may be noted that treatment of malic acid with fuming sulphuric acid in the absence of a phenol affords coumalic acid[123,124].

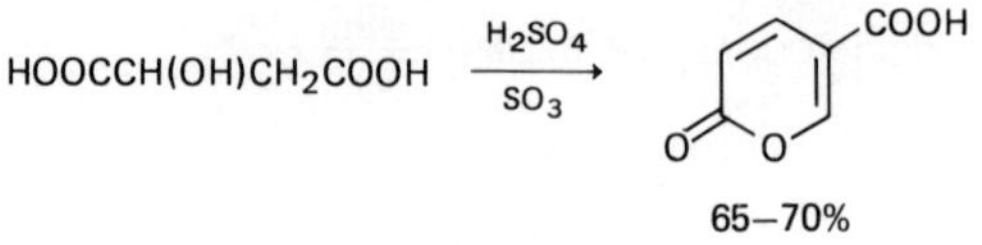

$$\text{HOOCCH(OH)CH}_2\text{COOH} \xrightarrow[\text{SO}_3]{\text{H}_2\text{SO}_4}$$

(69)

65–70%

D. By Aldol Condensations

Base-catalysed aldol condensations of substituted malonic and acetoacetic esters with paraformaldehyde afford good yields of substituted γ-butyrolactones (equations 70–72)[132].

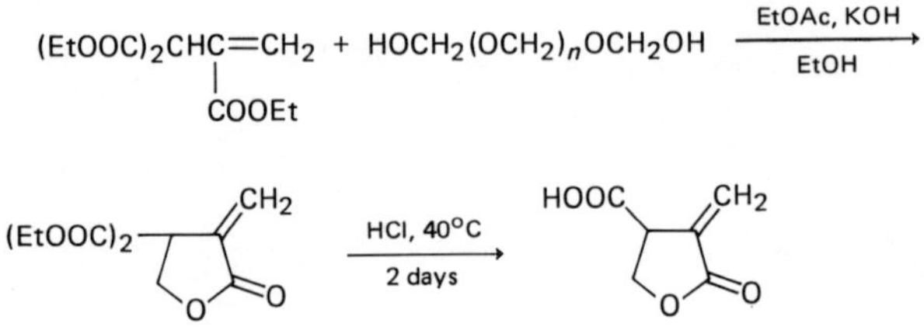

$$(\text{EtOOC})_2\text{CHC}{=}\text{CH}_2 + \text{HOCH}_2(\text{OCH}_2)_n\text{OCH}_2\text{OH} \xrightarrow[\text{EtOH}]{\text{EtOAc, KOH}}$$
$$\text{COOEt}$$

(70)

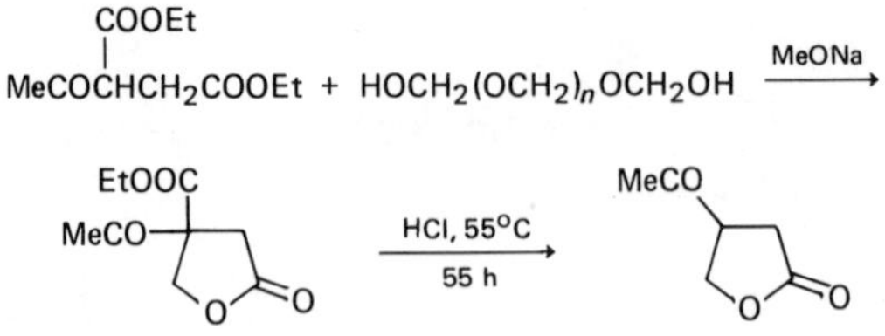

$$\begin{array}{c}\text{COOEt}\\ |\\ \text{MeCOCHCH}_2\text{COOEt}\end{array} + \text{HOCH}_2(\text{OCH}_2)_n\text{OCH}_2\text{OH} \xrightarrow{\text{MeONa}}$$

(71)

$$\begin{array}{c}\text{COOEt}\\ |\\ \text{MeCOCHCHMeCOOEt}\end{array} + \text{HOCH}_2(\text{OCH}_2)_n\text{OCH}_2\text{OH} \xrightarrow{\text{MeONa}}$$

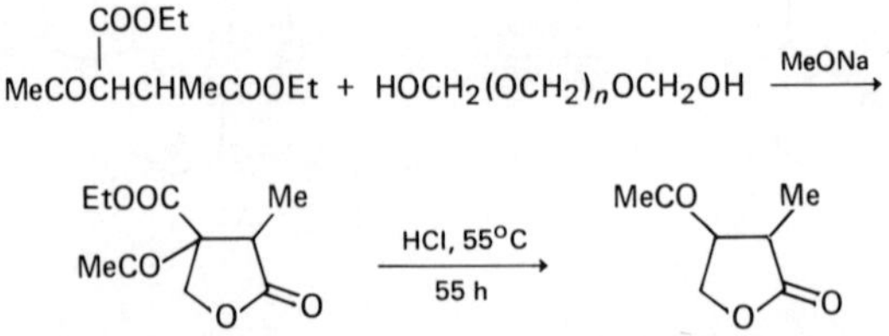

(72)

Although it has been found that steroidal 17-β-hydroxy-16-β-acetic acids[133,134] and 17-β-hydroxy-16-β-propionic acids are easily converted into their respective cis-fused γ- and δ-lactones by simple intramolecular acid-catalysed condensation, the formation of the trans-fused δ-lactones from 17-β-hydroxy-16-α-propionic acids requires a more complex approach[135,136]. The procedure involves base-catalysed

TABLE 8. Synthesis of coumarins by the von Pechmann reaction

Phenol	β-Keto ester (acid)	Product	Yield (%)	Reference
Phenol	Ethyl acetoacetate	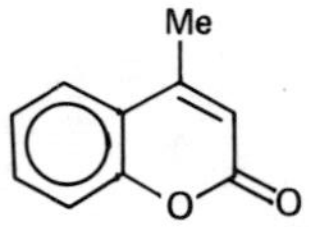	40–55	125
Resorcinol	Ethyl acetoacetate	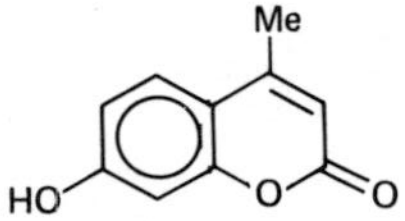	82–90	126
Hydroxyhydroquinone triacetate	Ethyl acetoacetate	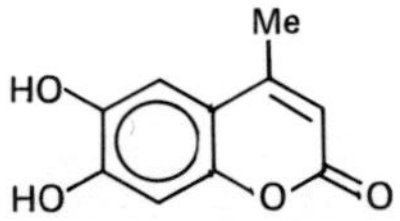	92	127

aldol condensation of 3-β-hydroxy-5-α-androstan-17-one with glyoxylic acid to afford 3-β-hydroxy-17-oxo-5-α-androstan-Δ^{16,α}-acetic acid, the key intermediate in the synthesis. Several additional steps convert this compound into 3-β,17-β-dihydroxy-5-α-androstane-16-α-propionic acid, which upon warming in a solution of acetic anhydride and acetic acid[137,138] afford 3-β-acetoxy-17-β-hydroxy-5-α-androstane-16-α-propionic acid δ-lactone (equation 73). Condensation of 5-α-androstanolone with glyoxylic acid in aqueous methanolic sodium

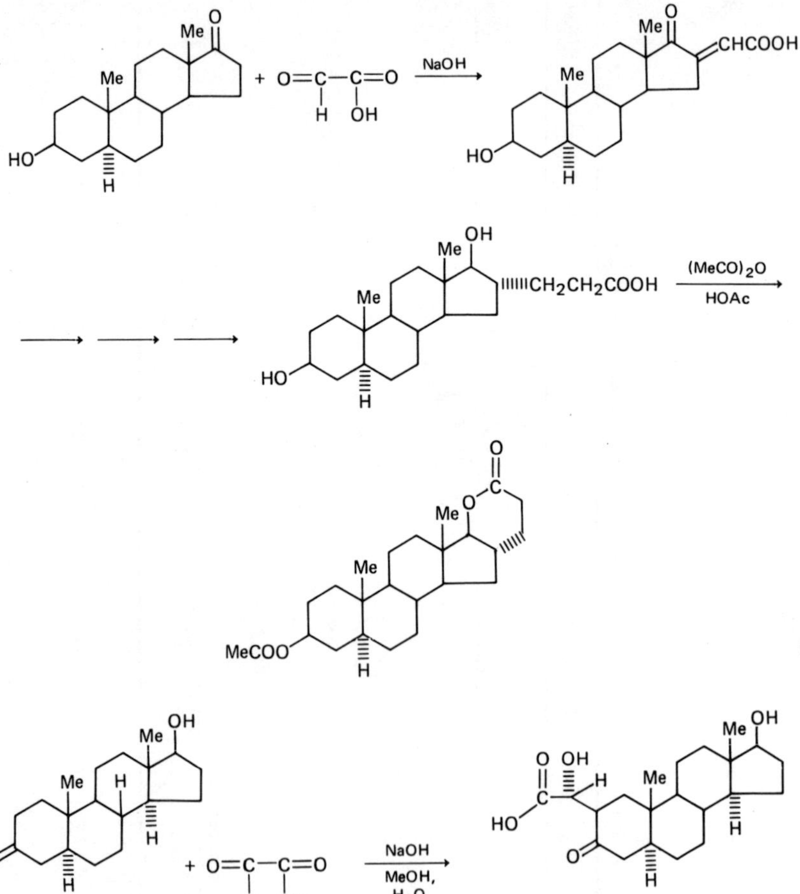

$$(73)$$

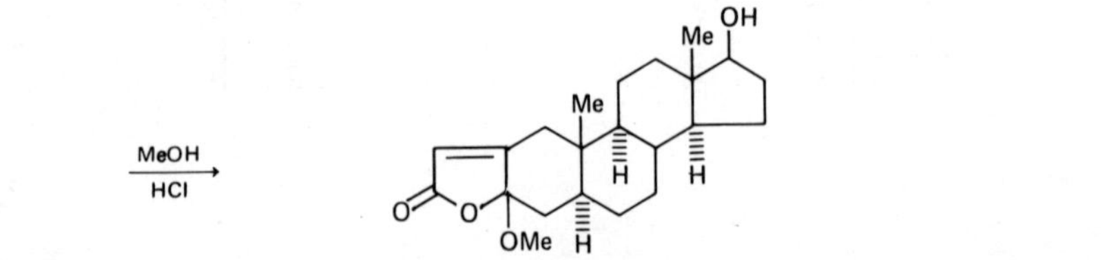

$$(74)$$

hydroxide at room temperature affords[139] 17-β-hydroxy-5-α-androstan-2-α-(α-hydroxyacetic acid)-3-one, which is readily lactonized to 3-ε-methoxy-17-β-hydroxy-5-α-androstan-2-α-(α-hydroxyacetic acid)-3-one-lactol upon treatment with methanolic hydrogen chloride (equation 74). Similarly, 17-α-hydroxy-3-oxo-5-α-androstan-$\Delta^{2,\alpha}$-acetic acid was prepared in 85% yield[139] via condensation of glyoxylic acid with 5-α-androstanolone. Several additional steps converted this product into 3-β,17-β-dihydroxy-5-α-androstan-2-β-acetic acid, which was lactonized upon refluxing with p-toluenesulphonic acid (equation 75).

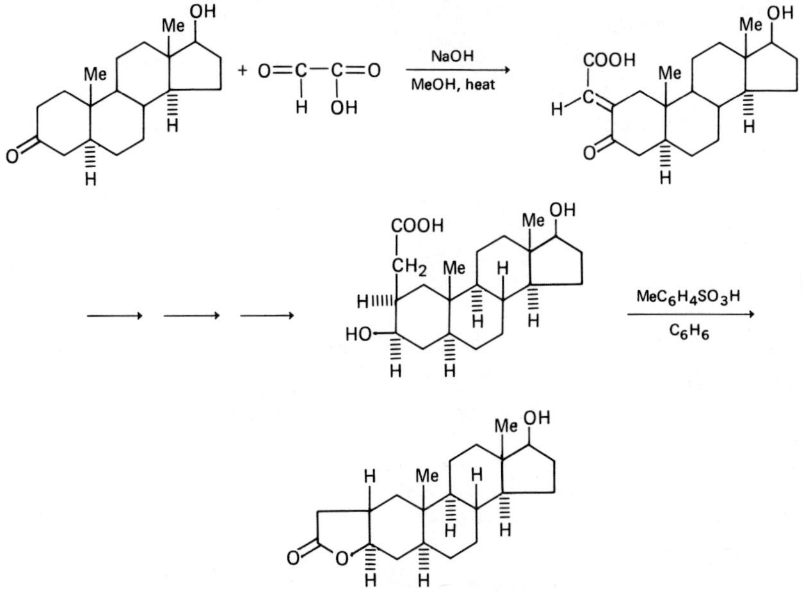

This approach to the preparation of key intermediates in the syntheses of isocardenolides[140], cardenolides[140-142], isobufadienolides[140] and bufadienolides[140] has been investigated.

Intermediates in the total synthesis of fomannosin (**13**), a biologically active metabolite from *Fommes annosus*, have also been prepared[143] via the intramolecular aldol condensation of **14** to form **15**. This product, which contains the formannosane skeleton, appears to be a promising intermediate in the total synthesis of fomannosin.

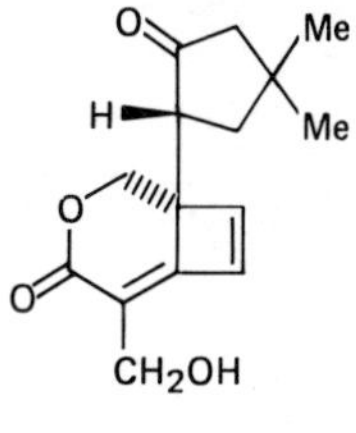

(13)

An interesting example of the use of an intramolecular aldol condensation for construction of α,β-unsaturated butyrolactones may be found in the mercuric sulphate-catalysed hydration of the acetylenic ester of acetoacetic acid (equation 77)[118].

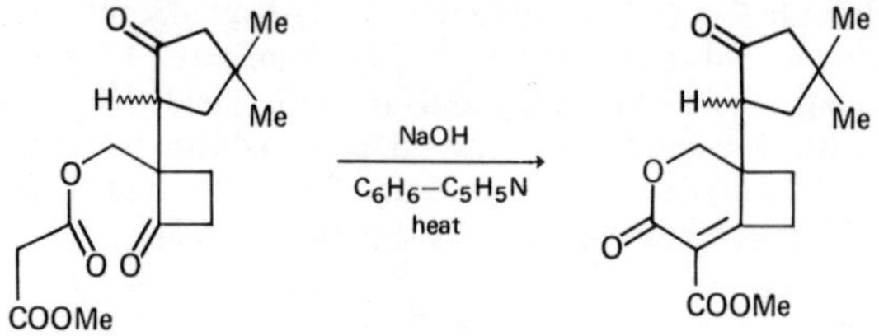

$$(76)$$

(14)   (15)

$$\text{MeCOCH}_2\text{COOCMe}_2\text{C}{\equiv}\text{CH} \xrightarrow[\text{HgSO}_4]{\text{EtOH, H}_2\text{O}} [\text{MeCOCH}_2\text{COOCMe}_2\text{COCH}_3]$$

$$\xrightarrow[\text{-H}_2\text{O}]{\text{aldol}}$$

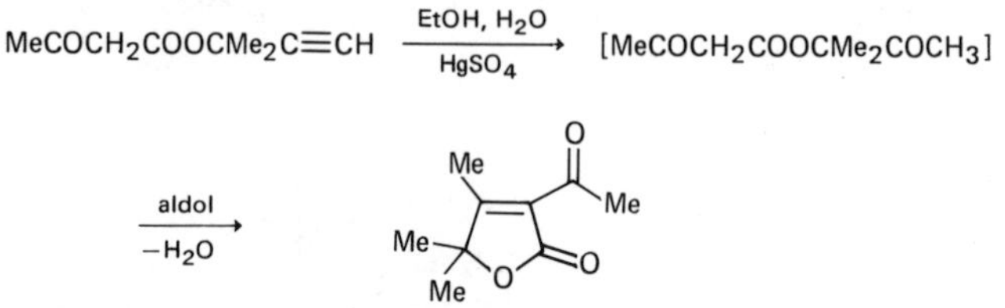

$$(77)$$

E. By Malonic Ester or Malonic Acid Condensation

The condensation of malonic acid or diethyl malonate with o-hydroxy-benzaldehydes or β-alkoxy-α,β-unsaturated aldehydes in piperidine has proved to be a very convenient route to 5,6-fused and 6-substituted-2-pyrones. Using this approach, salicylaldehyde and ethyl malonate were condensed in piperidine—glacial acetic acid solutions to yield[144] a 78–83% conversion to 3-carboethoxycoumarin (ethyl 2-oxo-2H-1-benzopyran-3-carboxylate) (equation 78). The scope of this

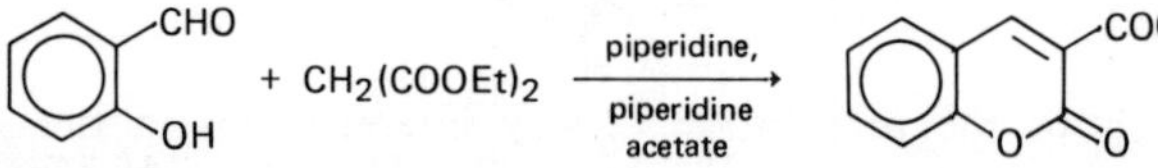

$$(78)$$

method was investigated[145] during the synthesis of several isobufadienolides, and it was found that optimal conditions involve a 1 : 2 : 2 mole ratio of aldehyde, malonic acid and piperidine (or morpholine) in excess pyridine at steam bath temperatures for one hour. Using these conditions, 3-β-acetoxy-20-ethoxy-21-formylpregna-5,20-diene (equation 79), 3-β-acetoxy-20-methoxy-21-formyl-5-α-androst-20-ene (equation 80), 2-formyl-3-methoxy-17-β-acetoxy-5-α-androst-2-ene (equation 81), 3-α-acetoxy-20-methoxy-21-formyl-5-β-pregna-20-ene (equation 82) and 3-α,6-α-diacetoxy-20-methoxy-21-formyl-5-β-pregna-20-ene (equation 83) were converted into 3-β-acetoxy-17-β-(6$'$$\alpha$-pyronyl)-androst-5-ene (54%), 3-β-acetoxy-17-β-(6$'$$\alpha$-pyronyl-5-$\alpha$-androstane (54%), 17-$\beta$-acetoxy-5-$\alpha$-androstano-[2,3-$c$]-2-pyrone (20%), 3-$\alpha$-acetoxy-17-$\beta$-(6$'$$\alpha$-pyronyl)-5-$\beta$-androstane (21%) and 3-$\alpha$,6-$\alpha$-di-

$$(79)$$

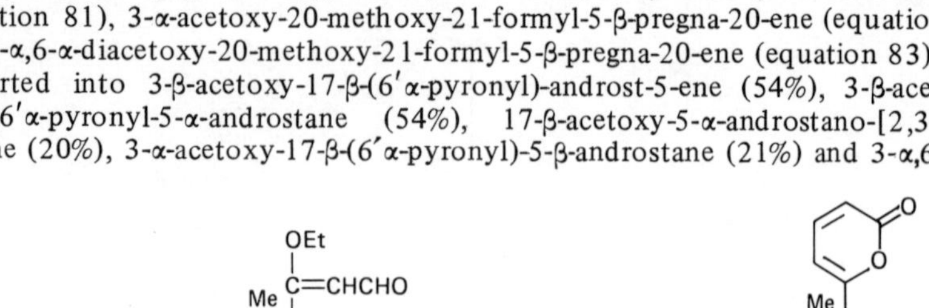

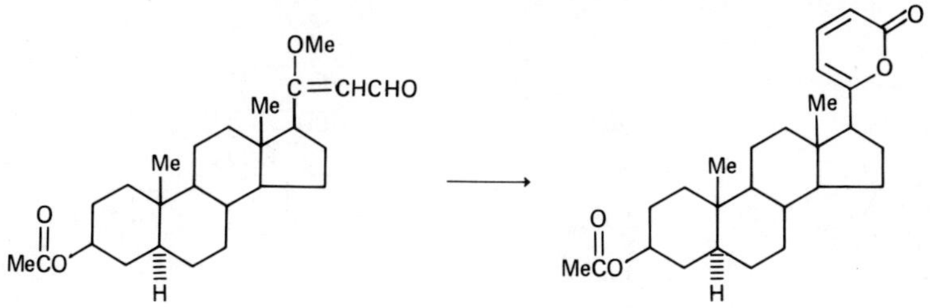

(80)

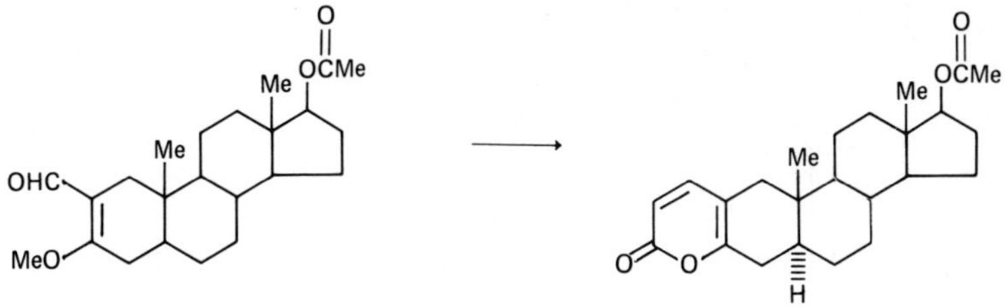

(81)

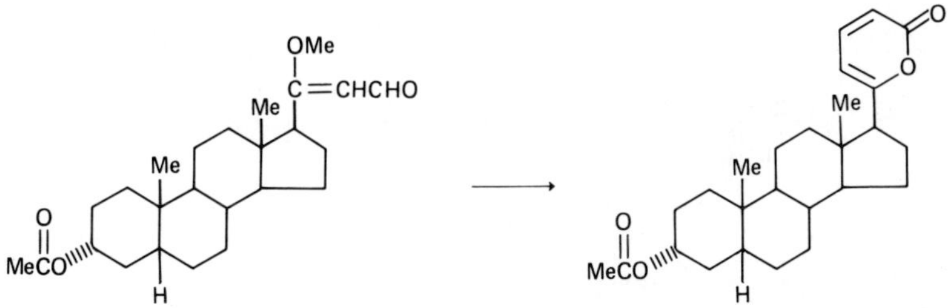

(82)

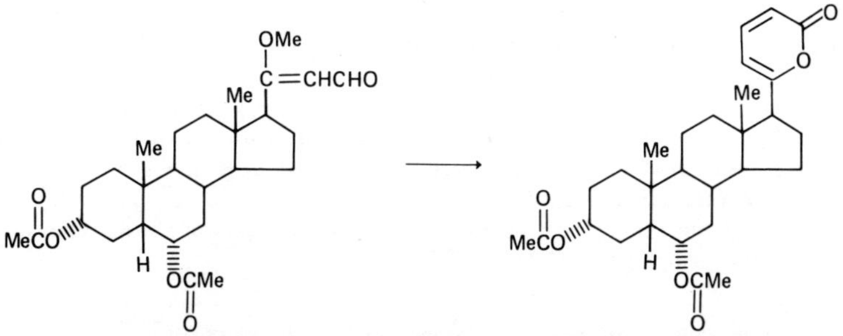

(83)

acetoxy-17-β-(6'α-pyronyl)-5-β-androstane (57%), respectively. The mechanistic pathway proposed[145] for these conversions is shown in equation (84).

Condensation of tertiary α-hydroxy ketones (acyloins) with malonic ester (equation 85) affords unsaturated γ-lactones in good yields (Table 9). The proposal that this reaction occurs via initial transesterification, with subsequent intramolecular condensation of the resulting keto ester, was confirmed by several observations. For example, when weaker bases such as pyridine and triethylamine were used as catalysts it was possible to isolate the intermediate keto esters, which were converted into the unsaturated γ-lactones upon treatment with sodium ethoxide (equation 86).

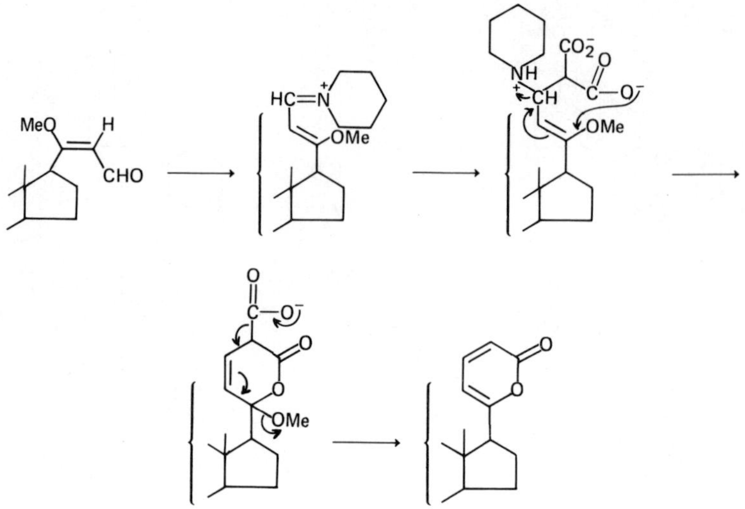

(84)

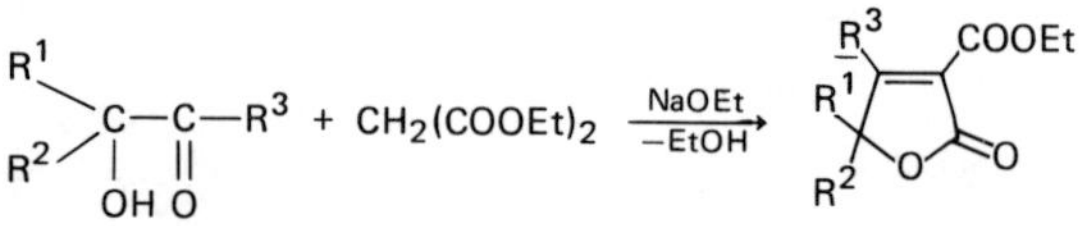

(85)

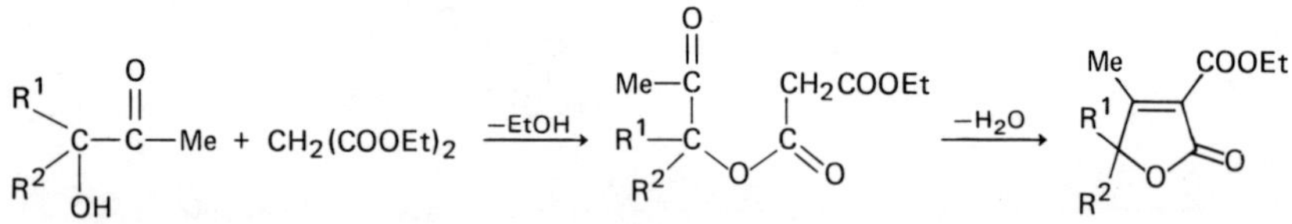

(86)

Condensation of malonic ester anions with epoxides (oxiranes) provides a popular method for the synthesis of lactones. Reaction of diethyl malonate and styrene oxide was originally reported[148] to yield, after hydrolysis and decarboxylation, γ-phenyl-γ-butyrolactone. Other workers have made use of the supposed specificity of this reaction[149-151]. However, DePuy and coworkers[152] reported that this reaction in fact affords a mixture of β-phenyl-γ-butyrolactone (60%) and γ-phenyl-γ-butyrolactone (40%) (equation 87). These results were independently verified by two other groups of workers[153,154].

TABLE 9. Condensations of acyloins with malonic ester to form unsaturated γ-lactones

R^1	R^2	R^3	Yield (%)	Reference
H	Et	Et	—	146
H	Pr	Pr	—	146
H	n-Bu	n-Bu	—	146
H	n-C_5H_{11}	n-C_5H_{11}	—	146
Me	Me	Me	65.5	147
Me	Et	Me	60	147
$-(CH_2)_5-$		Me	61	147

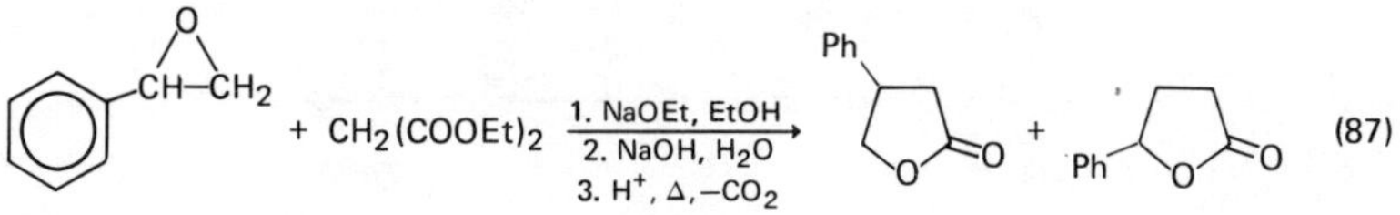

$$
\text{[structure: styrene oxide derivative]} + CH_2(COOEt)_2 \xrightarrow[\substack{2.\ NaOH,\ H_2O \\ 3.\ H^+,\ \Delta,\ -CO_2}]{1.\ NaOEt,\ EtOH} \text{[lactone]} + \text{[Ph lactone]} \quad (87)
$$

Van Tamelen and Bach[47] used the reaction of malonic ester anion with methyl tridecyl glycidate to prepare α,β-dicarbomethoxy-γ-tridecyl-γ-butyrolactone, an important intermediate in the synthesis of d,l-protolichesterinic acid (equation 88).

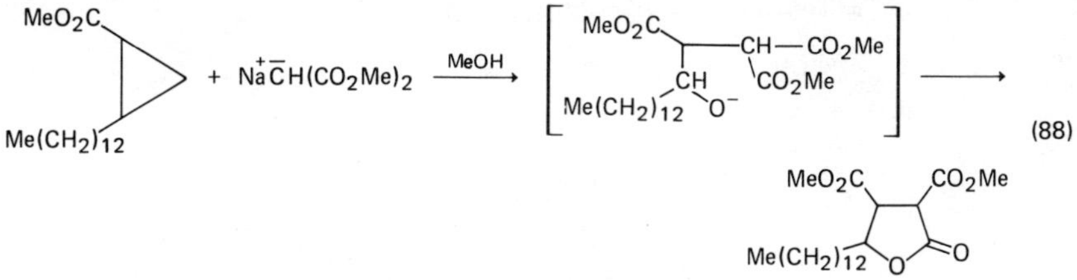

$$
(88)
$$

Dalton and coworkers[155] employed a similar approach to prepare fluorene-9-spiro-4'-(2'-carboxybutyrolactone) (**16**) and fluorene-9-spiro-3'-(2'-ethoxycarbonyl-butyrolactone) (**17**). Thus, condensation of sodium diethylmalonate with fluorene-9-spira-2'-oxiran afforded a 28% yield of **17** and a 20% yield of a diacid, which upon heating under vacuum afforded **16**. Both of these products were also decarb-oxylated to form spiro butyrolactones **18** and **19**. Similarly these workers[155]

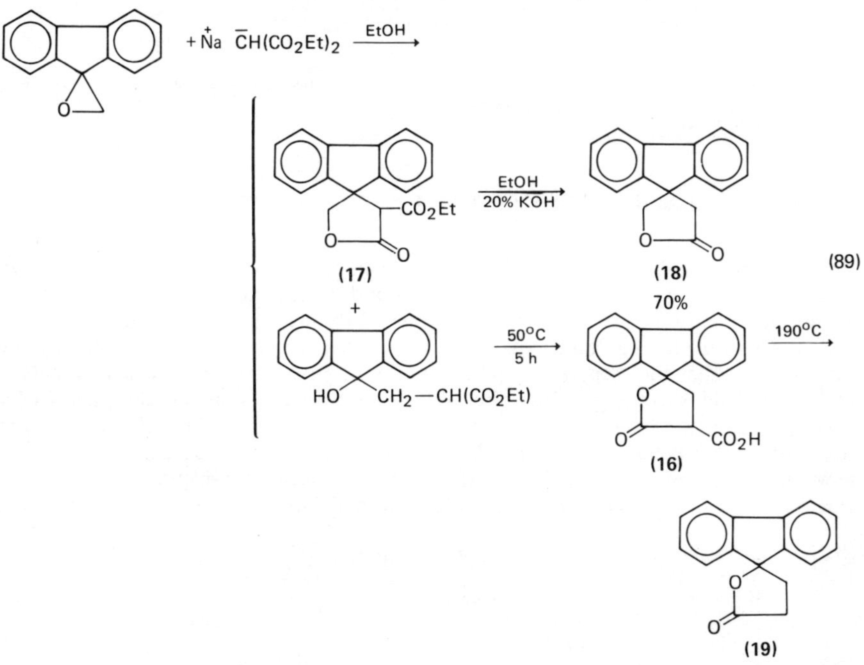

$$
(89)
$$

prepared 2-carboxy-4,4-diphenylbutyrolactone from 2,2-diphenyloxirane, while condensation of sodium diethylmalonate with 2-chloro-1,1-diphenylethanol af-forded the same product (equation 90).

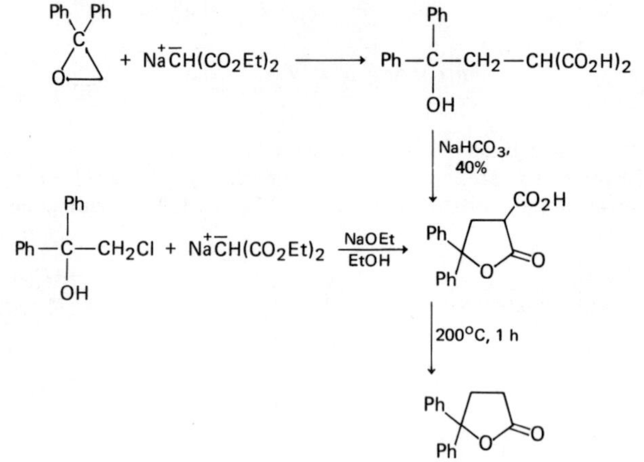

$$\text{(90)}$$

The condensation of 2-chloro-2-methylpropanal with malonic esters in the presence of potassium carbonate to produce γ-butyrolactones (equation 91) has also been studied[156]. At room temperature, in THF, using one equivalent each of

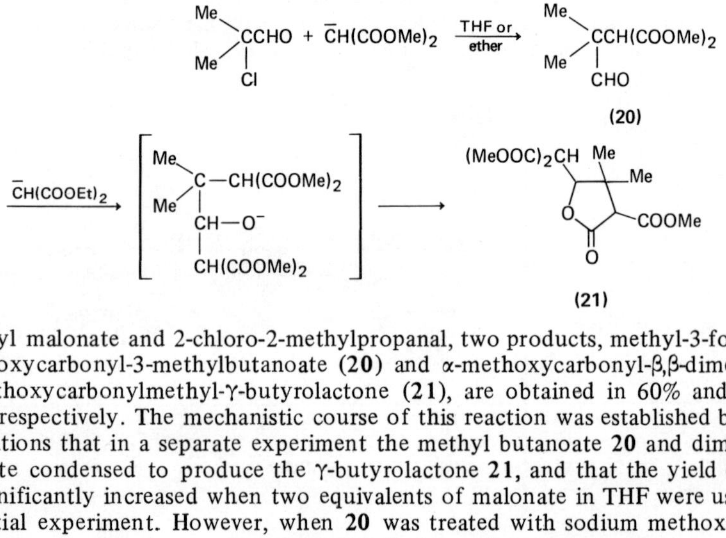

$$\text{(91)}$$

dimethyl malonate and 2-chloro-2-methylpropanal, two products, methyl-3-formyl-2-methoxycarbonyl-3-methylbutanoate (20) and α-methoxycarbonyl-β,β-dimethyl-γ-dimethoxycarbonylmethyl-γ-butyrolactone (21), are obtained in 60% and 26% yields, respectively. The mechanistic course of this reaction was established by the observations that in a separate experiment the methyl butanoate 20 and dimethyl malonate condensed to produce the γ-butyrolactone 21, and that the yield of 21 was significantly increased when two equivalents of malonate in THF were used in the initial experiment. However, when 20 was treated with sodium methoxide, a new lactone, α-methoxycarbonyl-β,β-dimethyl-γ-methoxy-γ-butyrolactone (22), was obtained in 65% yield via intramolecular cyclization (equation 92). Similarly,

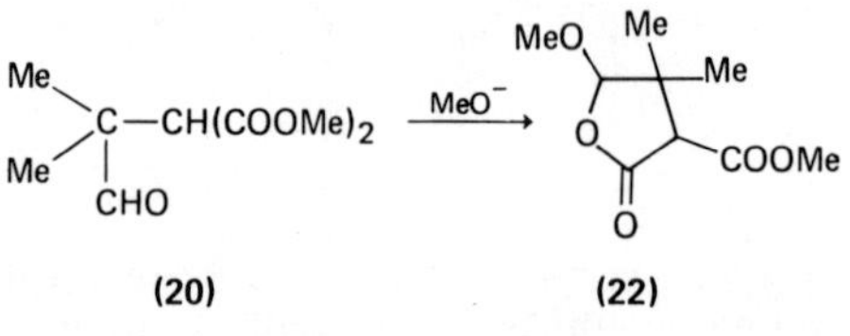

$$\text{(92)}$$

the reaction of 2-chloro-2-methylpropanal with dimethyl malonate in ether containing sodium methoxide (equation 93) also afforded lactone 22, albeit in 20% yield.

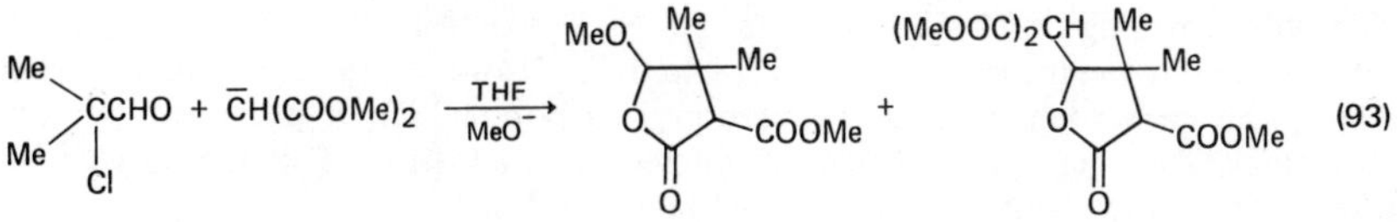

(93)

The major product from this reaction was still **21** (46%). Ester cleavage of **21** to α-carboxy-β,β-dimethyl-γ-carboxymethyl-γ-butyrolactone[156] was effected in 97% yield upon heating with concentrated hydrochloric acid (equation 94) at 70–80° for 24 hrs. Heating this product at 180–200°C for 30 minuted afforded[156] a 98% conversion to β,β-dimethyl-γ-carboxymethyl-γ-butyrolactone.

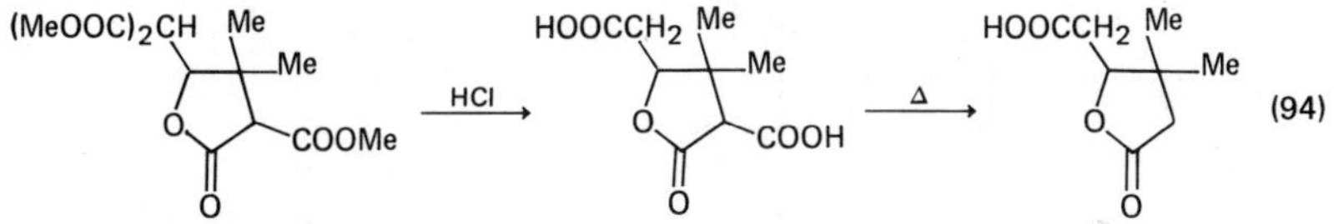

(94)

When 2-chloro-2-methylpropanal is condensed with the methyl or ethyl ester of malonic acid in *aqueous* potassium carbonate[156], α-alkoxycarbonyl-β-dialkoxy-carbonylmethyl-γ,γ-dimethyl-γ-butyrolactones (**23**) are formed in 70–82% yield. This is explained by assuming an epoxide intermediate, which reacts further with malonate as shown in equation (95).

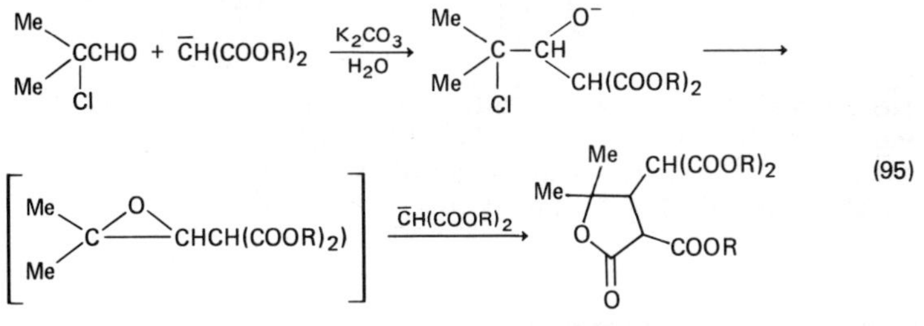

(95)

Hydrolysis of lactone **23** gave the expected diacid (98%) which upon heating afforded terpenylic acid (**24**) (equation 96). These results contradicted a previous

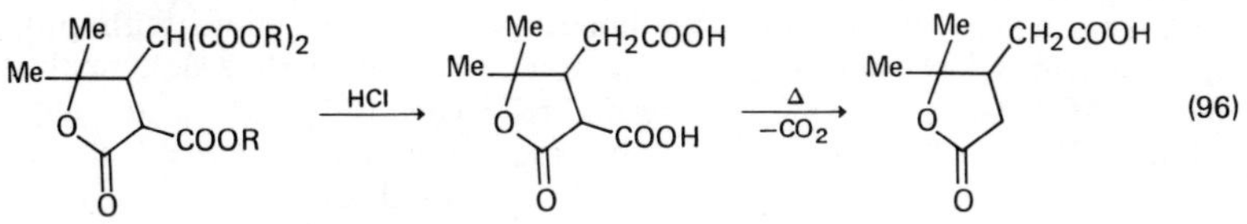

(96)

report[157] that 2-bromo-2-methylpropanal reacted with diethyl sodiomalonate in ethanol to afford α-ethoxycarbonyl-γ,γ-dimethyl-Δ^{α,β}-γ-butenolide. Reinvestigation[156] showed that α-ethoxycarbonyl-β-diethoxycarbonylmethyl-γ,γ-dimethyl-γ-butyrolactone (**23**, R = Et) was indeed formed in 53% yield.

The explanation[156] advanced for the results discussed above maintains that in aprotic solvents such as THF, the carbanion of malonic esters becomes more nucleophilic than in protic solvents and thus attacks the α-carbon of the α-halo-aldehyde forming a C–C bond via an S_N2 reaction. This is followed by an intramolecular cyclization to afford 21 and 22. However, in protic solvents such as water, the carbanion attacks the carbonyl carbon, which is polarized by solvent molecules, forming a C–C bond by nucleophilic addition. This is followed by an intramolecular cyclization to afford the lactones 23.

Malonic ester anion has also been used to obtain *trans*-fused γ-lactones. For example, reaction[158] of sodium diethylmalonate with 3,4-epoxy-1-cyclooctene, 5,6-epoxy-1-cyclooctene and 1,2: 5,6-diepoxycyclooctane, followed by hydrolysis, affords 10-oxo-9-oxabicyclo[6.3.0] undec-2-en-11-carboxylic acid (65%), 10-oxo-9-oxabicyclo[6.3.0] undec-4-en-11-carboxylic acid (70%) and 4,5-epoxy-10-oxo-9-oxabicyclo[6.3.0]-undecan-11-carboxylic acid (60%), respectively (equation 97).

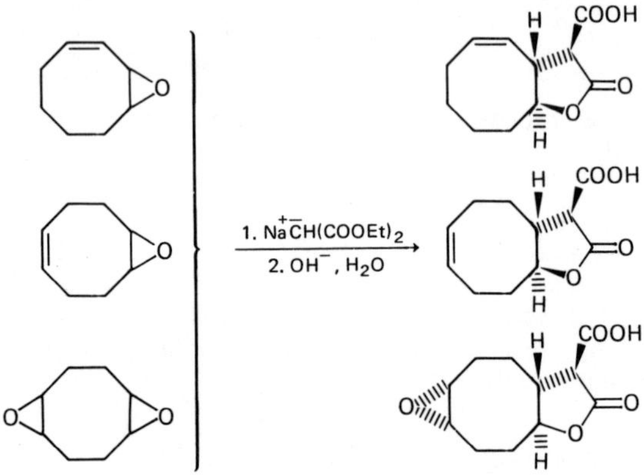

(97)

These acids were converted into their methyl esters by reaction with diazomethane and were decarboxylated to 9-oxobicyclo[6.3.0]undec-2-en-10-one (61%), 9-oxabicyclo[6.3.0]undec-4-en-10-one (90%) and endo-4,5-epoxy-9-oxabicyclo[6.3.0] under-10-one (62%) by heating at 160–180°C.

F. By Perkin and Stobbe Reactions

Although the Perkin reaction[159–162] is not widely used for the direct synthesis of lactones, several applications of this condensation have found some utility in lactone preparation.

A Perkin-type reaction of 2,6-dimethoxy-*p*-benzoquinone with propionic anhydride affords[163] the two products shown in equation (98) along with the propyl diester of 2,6-dimethoxy-*p*-hydroquinone. Using isobutyric acid anhydride, the isobutyl diester of 2,6-dimethoxy-*p*-hydroquinone and the fused lactone (exclusive structure not determined) are formed[163]. To establish the mechanistic course[164] the standard Perkin reaction procedure was modified by using shorter reaction times and lower temperatures. Under these conditions it was possible to isolate the β-lactones 25 and 26, respectively. Transformation of 25 to the mixture of products initially obtained was easily accomplished by heating at 100°C for 48 hours in the presence of sodium propionate and propionic anhydride, β-Lactone 26 could similarly be transformed upon prolonged heating with isobutyric acid anhydride in the presence of sodium isobutyrate, but could not be so transformed upon treatment with acetic acid–sulphuric acid mixtures. Although these and other experiments

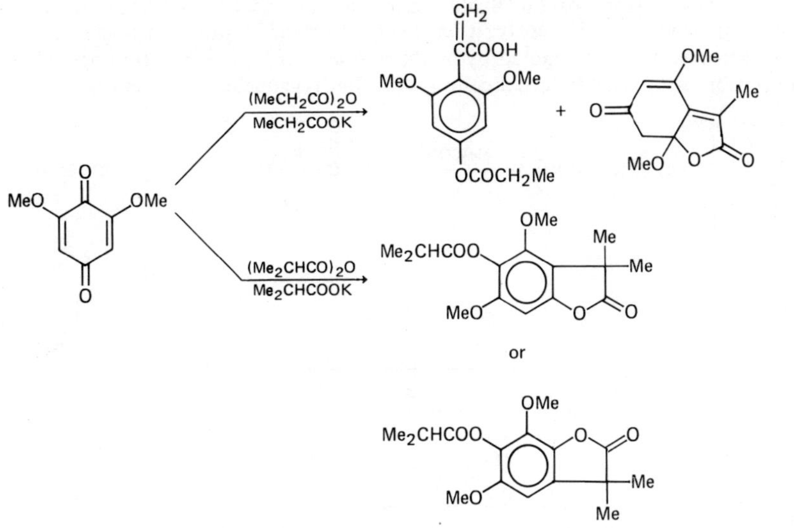

(98)

did not establish with certainty that β-lactones are intermediates in the formation of the observed γ-lactones, they did establish that β-lactones could be formed under Perkin-like reaction conditions.

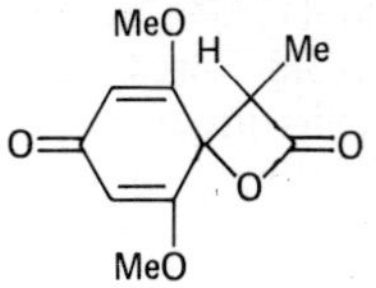

(25) 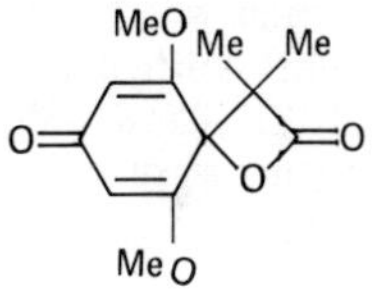   (26)

A variety of α-benzylidene-γ-phenyl-Δ^βⁱγ-butenolides substituted in the aralkyl idene ring with either electron-withdrawing or electron-donating substituents have been prepared[165] by a Perkin-type condensation of 3-benzoylpropionic acid with substituted benzaldehydes in the presence of sodium acetate in acetic anhydride (equation 99).

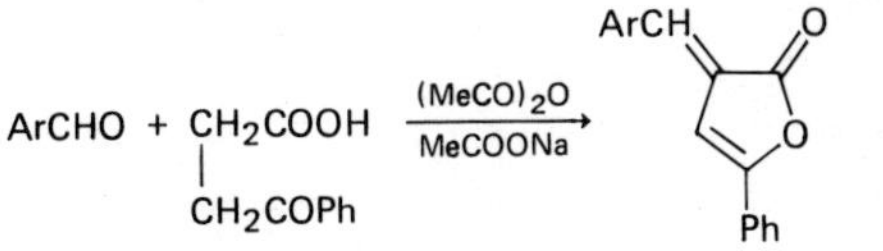

(99)

The preparation[166-169] of aralkylidine- and subsequently arylmethylphthal-ides[168,169], originates with the condensation of phthalic anhydride with arylacetic acids (equation 100).

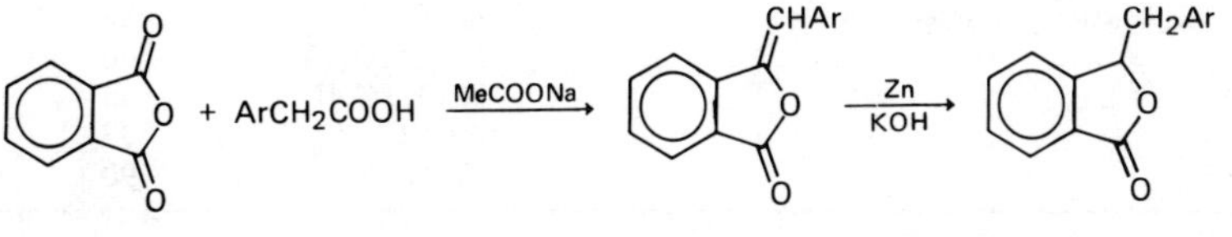

(100)

Various applications of the Stobbe condensation to the synthesis of lactones have been reviewed[170]. Recently, it has been reported[171] that β-carboethoxy-$\Delta^{\beta,\delta}$-δ-valerolactones can be prepared by an intramolecular Stobbe reaction preceded by condensation of tertiary α-keto alcohols with diethyl succinate (equation 101).

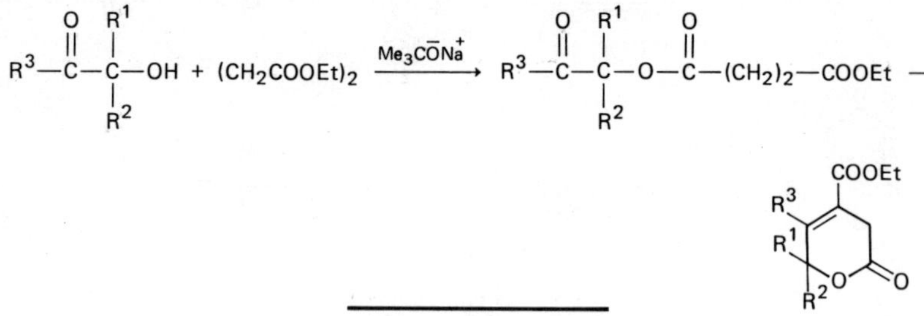

(101)

R^1	R^2	R^3
Me	Me	Me
Me	Me	Me
—(CH$_2$)$_5$—		Me

G. By Grignard and Reformatsky Reactions

During a series of studies[172,173] involving the synthesis of steroids, a general synthesis of δ-lactones was developed. This method[174,175] consists of the reaction of Grignard reagents with glutaraldehyde to afford δ-hydroxyaldehydes in good yields (equation 102). These aldehydes, which exist predominately in cyclic hemi-acetal (δ-lactol) form, were then oxidized to δ-lactones using a variety of reagents as shown in Table 10.

TABLE 10. Synthesis of δ-lactols and δ-lactones by reaction of Grignard reagents (RMgX) with glutaraldehyde[175]

RMgX	Yield of δ-lactol (%)	Oxidizing agent	Yield of δ-lactone (%)
MeCH$_2$MgBr	68.5	Ag$_2$O	50
Me(CH$_2$)$_3$CH$_2$MgBr	–	Ag$_2$O	41
MeCH(CH$_2$)$_2$CH$_2$MgCl 　| 　OCMe$_3$	66	Br$_2$, HOAc	83
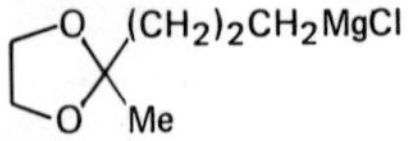	52	Br$_2$, HOAc	77
	64	Br$_2$, HOAc MnO$_2$, C$_6$H$_6$	88 45
	78	Ag$_2$O Na$_2$Cr$_2$O$_7$, HOAc MnO$_2$, C$_6$H$_6$ Ag$_2$CO$_3$, C$_6$H$_5$Me air, MeCOOEt, Pt	86 60 35 33 90

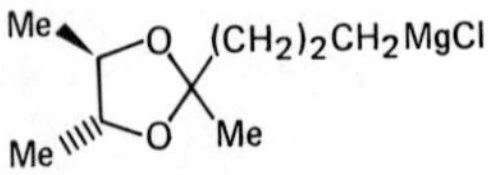

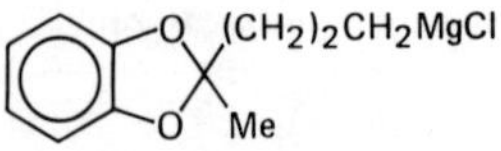

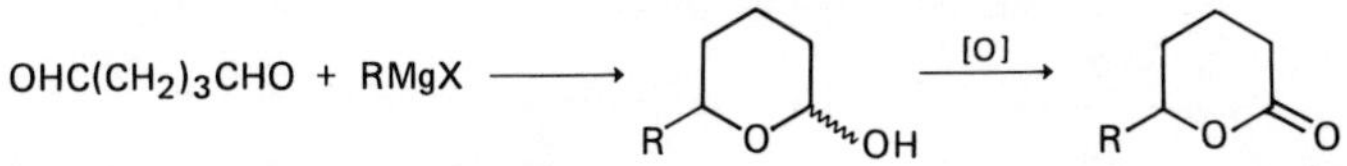

$$OHC(CH_2)_3CHO + RMgX \longrightarrow \underset{R}{} \xrightarrow{[O]} \tag{102}$$

Similarly, addition of methylmagnesium iodide to diethyl acetoglutarate[176] produces a racemic δ-lactone, ethyl terpenylate, which can be easily hydrolysed to terpenylic acid (equation 103).

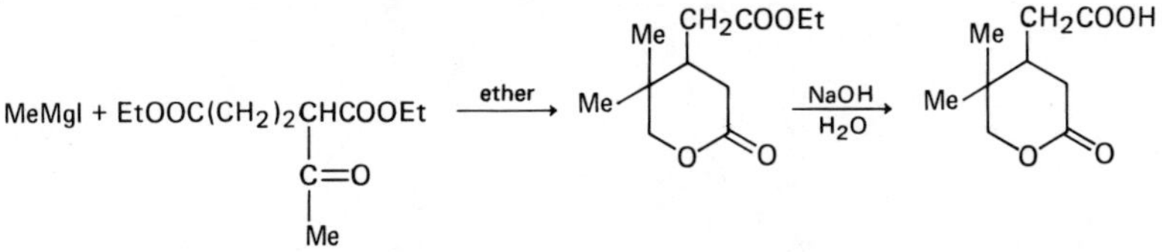

$$MeMgI + EtOOC(CH_2)_2\underset{\underset{Me}{\overset{|}{C=O}}}{\overset{|}{CHCOOEt}} \xrightarrow{ether} \xrightarrow[H_2O]{NaOH} \tag{103}$$

Contrary to previous reports[177,178], it has now been found[179] that addition of the Grignard or Reformatsky reagent formed from ethyl α-bromoisobutyrate to α,β-ethylenic ketones occurs via conjugated addition to produce a mixture of δ-keto

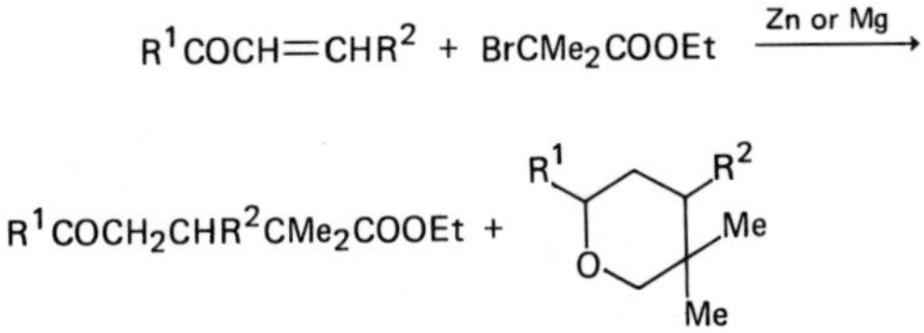

$$R^1COCH{=}CHR^2 + BrCMe_2COOEt \xrightarrow{Zn \text{ or } Mg}$$

$$R^1COCH_2CHR^2CMe_2COOEt + \tag{104}$$

R¹	R²	Method
Et	Ph	Reformatsky
i-Pr	Ph	Reformatsky
Ph	Ph	Grignard
p-MeOC₆H₄	p-MeOC₆H₄	Grignard
Ph	Me	Grignard
n-Pr	Me	Grignard

esters and enolic δ-lactones (equation 104). In addition, several cyclic α,β-unsaturated ketones underwent reaction to afford the δ-lactones shown below:

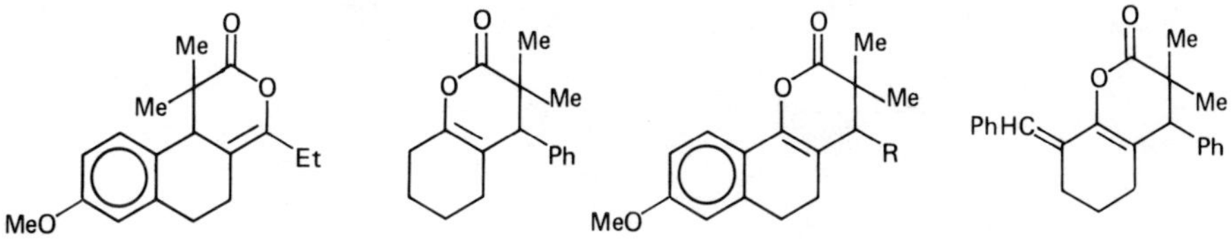

R = Et, Ph

Although these unsaturated ketones underwent smooth conjugate addition, the α,β-unsaturated methyl ketones, 1-acetyl-cyclohex-1-ene, methyl styryl ketone and 3-pentene-2-one, did not undergo conjugated addition with either the Grignard or Reformatsky reagent of ethyl-α-bromoisobutyrate[179].

Using the above approach, the reaction of 16-dehydropregnenolone acetate with

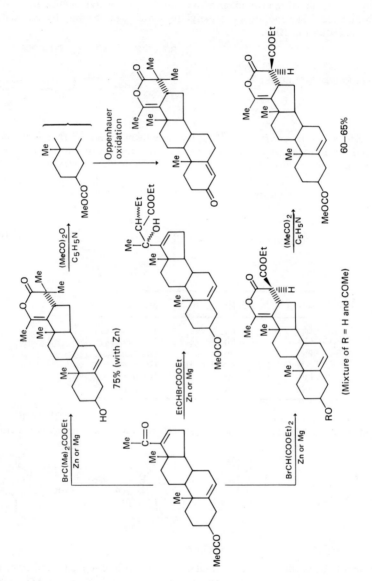

SCHEME 2.

the Grignard or Reformatsky reagents obtained from the ethyl esters of α-bromo-isobutyric, α-bromomalonic and α-bromobutyric acids was investigated[180]. It was found that, although the results depended largely upon the type of α-bromo ester used[181,182], the best yields were obtained with 1 : 6 molar ratio of steroid to Reformatsky reagent, with the Grignard reagent giving less reproducible results. A flow chart listing the reactants used and the products obtained is shown in Scheme 2.

The Reformatsky reagent prepared from diethyl α-methyl-α-bromomalonate has been added[183] to β-acetylenic alcohols to effect the synthesis of various δ-valerol-actones (equation 105).

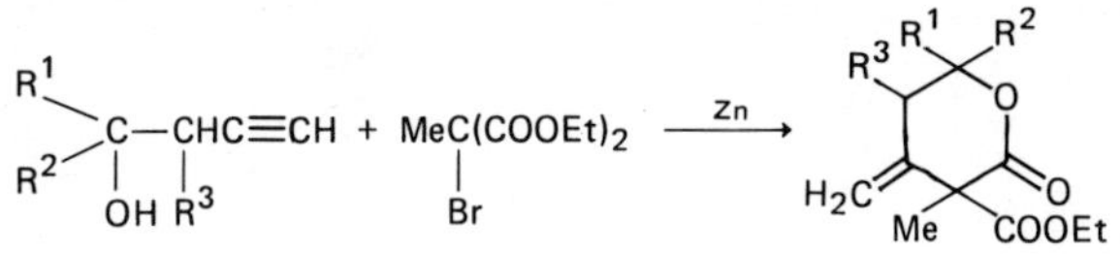

$$
\begin{array}{l}
R^1 \\
\hspace{0.3em}\diagdown C-CHC\equiv CH \ + \ MeC(COOEt)_2 \ \xrightarrow{Zn} \\
R^2 \diagup \ \ \underset{OH}{|} \ \underset{R^3}{|} \hspace{4em} \underset{Br}{|}
\end{array}
\qquad (105)
$$

R^1 = H, H , H , H , Me, Me, Me, Me

R^2 = H, Me, Ph, CHMe$_2$, H , Me, Ph , CHMe$_2$

R^3 = H, H , H , H , Me, Me, Me, Me

Addition of the organozinc reagents derived from α-(bromomethyl) acrylic esters to a variety of aldehydes and ketones in THF affords[184] a single-step synthesis of α-methylene γ-lactones (equation 106). This technique affords good yields (Table

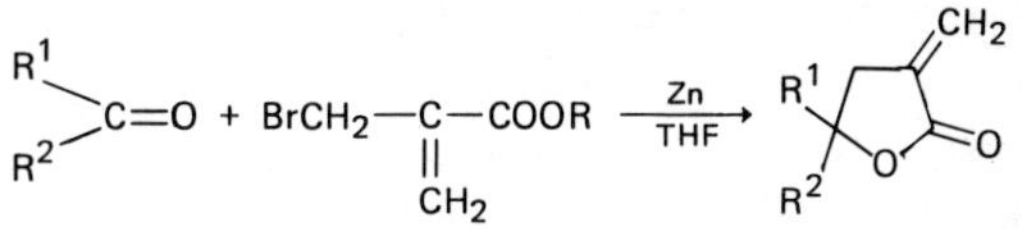

$$
\begin{array}{l}
R^1 \\
\hspace{0.3em}\diagdown C{=}O \ + \ BrCH_2{-}\underset{\underset{CH_2}{\|}}{C}{-}COOR \ \xrightarrow[THF]{Zn} \\
R^2 \diagup
\end{array}
\qquad (106)
$$

11) except in the case of formaldehyde, where a mixture of α-methylenebutyrol-actone and γ-hydroxy-α-methylenebutyric ester is formed in low yields. The analogous reaction of methyl β'-bromotiglate with ketones to produce α-methylene-β-methylbutyrolactones has also been reported (equation 107)[185]. This study also

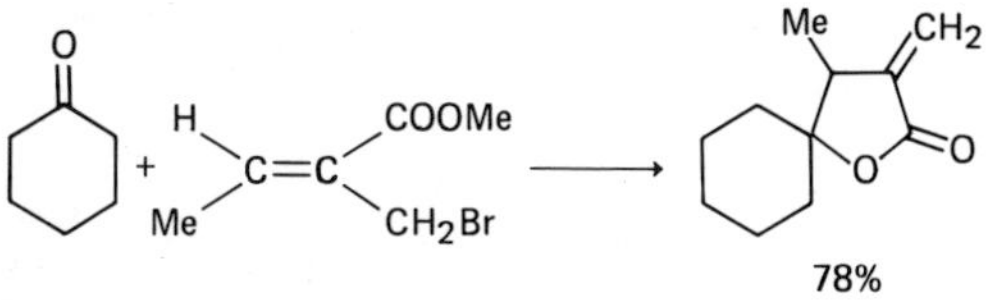

$$
\qquad (107)
$$

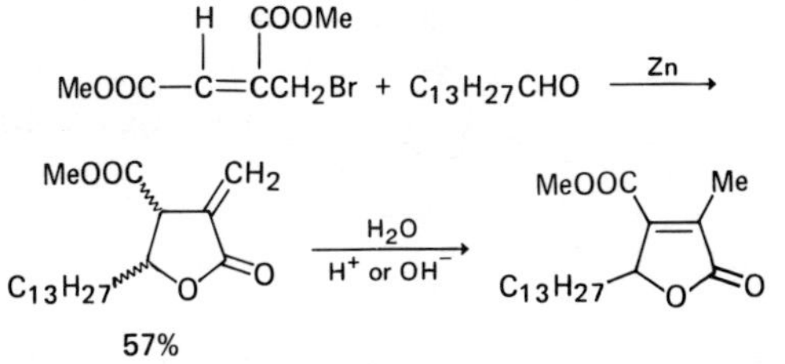

78%

$$
\underset{MeOOC-\underset{|}{C}{=}\underset{|}{C}CH_2Br \ + \ C_{13}H_{27}CHO \ \xrightarrow{Zn}}{\overset{H \quad COOMe}{}}
$$

$$
\xrightarrow[H^+ \text{ or } OH^-]{H_2O}
\qquad (108)
$$

57%

cis : trans = 1 : 1

included the synthesis of both the cis and trans isomers of protolichesterinic ester (equation 108) and trans-protolichesterinic acid itself (equation 109) by reaction of myristic aldehyde with β'-bromomesaconic acid or β'-bromocitraconic anhydride in the presence of zinc.

TABLE 11. Synthesis of α-methylene γ-butyrolactones by reaction of aldehydes and ketones with the zinc reagent derived from α-(bromomethyl)acrylic esters[184]

R^1	R^2	Yield (%)
—(CH₂)₄—		66
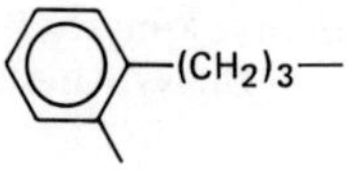 ⟨⟩—(CH₂)₃—		75
Me	Me	42
Ph	Me	78
Ph	Ph	100
Ph	H	100
i-Pr	H	76
PhCH=CH—	Me	92

and

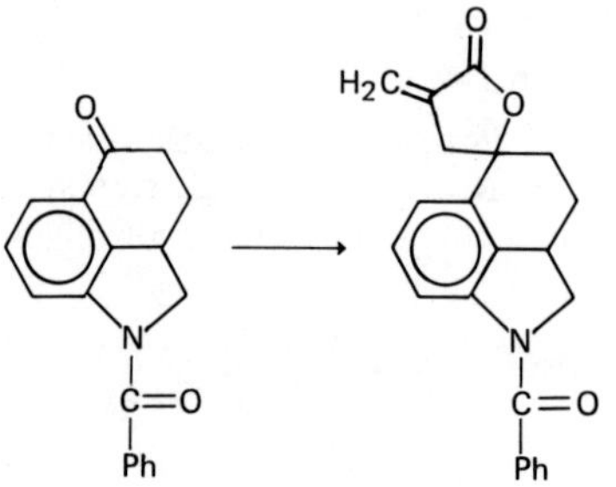

100

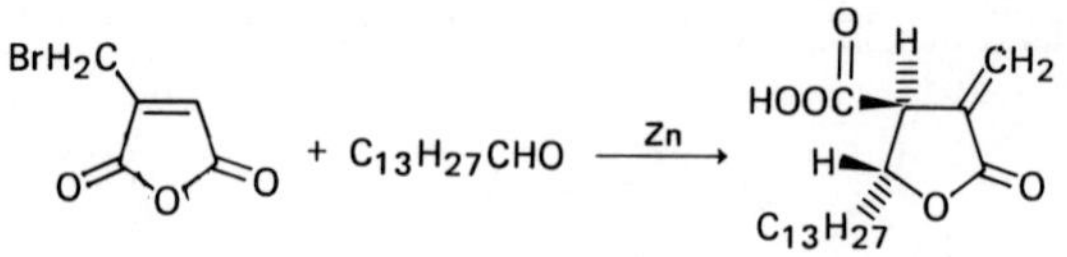

(109)

12%
trans

(110)

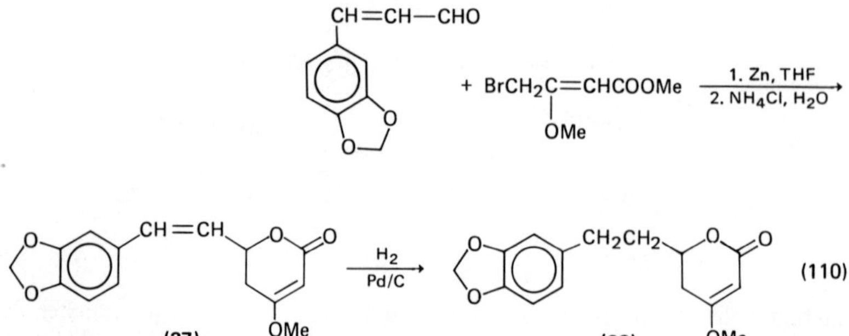

(27)   38%

(28)

The synthesis of d,l-methysiticin (**27**), another lactone constituent of the kawa root[41-44], has been accomplished[186] by a vinylogous Reformatsky-type condensation of 3,4-methylenedioxycinnamaldehyde and methyl γ-bromo-β-methoxycrotonate in THF (equation 110). Catalytic reduction of d,l-methysiticin affords d,l-dihydromethysiticin (**28**).

d,l-Mevalonolactone (**29**) has been synthesized by the Reformatsky condensation[187,188] of methyl or ethyl bromoacetate with either 1,1-dimethoxy-3-oxobutane or 1-acetoxy-3-oxobutane, by the use of ethyl lithioacetate in liquid ammonia[189,190], and by the Reformatsky reaction modification using trimethyl borate[191]. A new synthesis of d,l-mevalonolactone (**29**), which is superior to the methods mentioned above, and which holds considerable promise of generality in lactone preparation[192], consists of condensation of ethyl lithioacetate, with 1-acetoxy-3-oxobutane (equation 111). Hydrolysis of the resulting diester, followed by acidification, affords **29**.

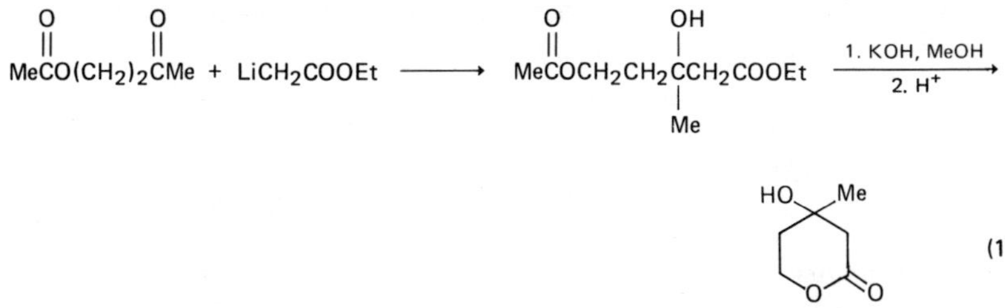

H. By Wittig-type Reactions

Preparation of α-pyrones has been effected via reaction of ethoxycarbonyl-methylenetriphenylphosphorane with a variety of β-diketones (equation 112)[193]

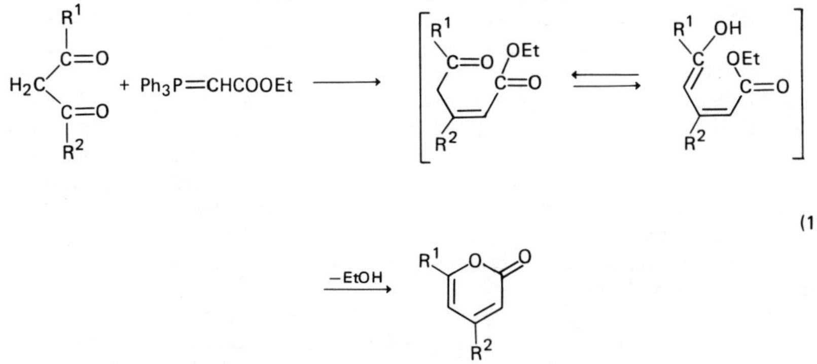

R¹	R²	Yield (%)
Ph	Ph	—
p-MeOC$_6$H$_4$	p-MeOC$_6$H$_4$	—
p-ClC$_6$H$_4$	p-ClC$_6$H$_4$	20
2-thienyl	2-thienyl	17

The mechanism apparently involves initial reaction between the ylide and one of the keto groups of the diketone to form an intermediate keto ester, the enol form of which immediately forms the lactone by ring closure. When 2-benzoylcyclohexanone was used (equation 113)[193] only a 5% yield of 4,5-(tetramethylene)-6-phenyl-2H-pyran-2-one was obtained.

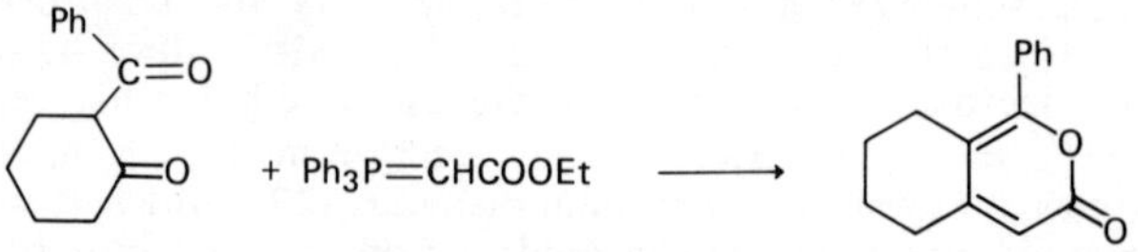

$$+ \; Ph_3P=CHCOOEt \longrightarrow \qquad\qquad\qquad (113)$$

γ-Butyrolactone has been prepared[194] in 62% yield as shown in reaction (114), which, although it is not strictly a Wittig reaction, does involve an intermediate phosphonium salt.

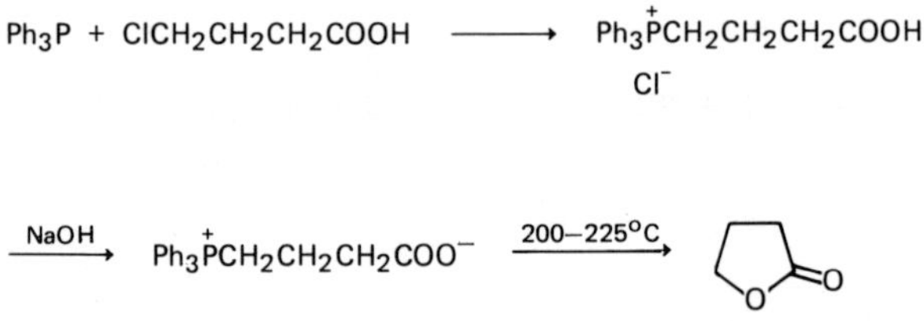

$$Ph_3P \; + \; ClCH_2CH_2CH_2COOH \longrightarrow Ph_3\overset{+}{P}CH_2CH_2CH_2COOH$$
$$Cl^-$$

$$(114)$$

$$\xrightarrow{NaOH} \; Ph_3\overset{+}{P}CH_2CH_2CH_2COO^- \; \xrightarrow{200-225^\circ C} \qquad$$

Wittig reactions which have been employed for functionization of preformed lactones rather than for ring-closure are discussed in Section II.K

I. From α-Anions (Dianions) of Carboxylic Acids

Because of their potential as aldosterone inhibitors, steroidal spiro γ-lactones have been the subject of considerable interest. In 1972 Creger[195] published a method for the preparation of a wide variety of such compounds. This procedure involved dimetalation of several aliphatic carboxylic acids using lithium diisopropylamide (LDA), followed by reaction of the resulting lithio α-anions (dianions) with (17S)-spiro[androst-5-ene-17,2'-oxiran]-3-β-ol (equation 115). Oppenauer oxidation of these spiro lactones afforded the substituted 4',5'-dihydro-(17R)-spiro-[androst-4-ene-17,2'-(3'H)-furan]-3,5'-diones, which upon further oxidation with chloranil in t-butyl alcohol followed by treatment with chloranil in toluene—acetic acid mixtures produced substituted 4',5'-dihydro-(17R)-spiro[androsta-4,6-diene-17,2'-(3'H)-furan]-3,5'-diones (equation 116). Treatment of these products with thiolacetic acid afforded the 7α-thioacetyl derivatives. The parent unsubstituted 4',5'-dihydro-(17R)-spiro[androst-4-ene-17,2'(3'H)-furan]3,5'-dione was not prepared via the oxidation technique discussed above but by the condensation shown in equation (117).

Other conversions reported by Creger[195] include the preparation of 4',5'-dihydro-3β-hydroxy-4'-vinyl-(17R)-spiro[androst-5-ene-17,2'-(3'H)-furan]-5'-one via the reaction of (17S)-spiro[androst-5-ene-17,2'-oxiran]-3β-ol with the crotonic acid anion (equation 118). A large number of steroidal lactones were also synthesized by hydrolysis of various amide or nitrile derivatives as shown in equations (119) and (120).

The generality of the reaction of metalated carboxylic acids with simple and steroidal epoxides is demonstrated by the results summarized in Table 12[195]. A similar approach to the preparation of γ-butyrolactones[196] consists of the reaction

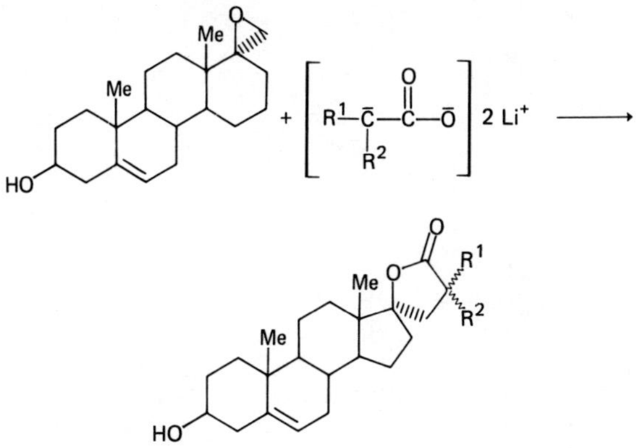

(115)

R¹	R²	Yield (%)
H	H	55
Me	Me	81
H	Me	70
H	Et	87
H	n-Bu	76
H	Ph	75
H	OMe	19

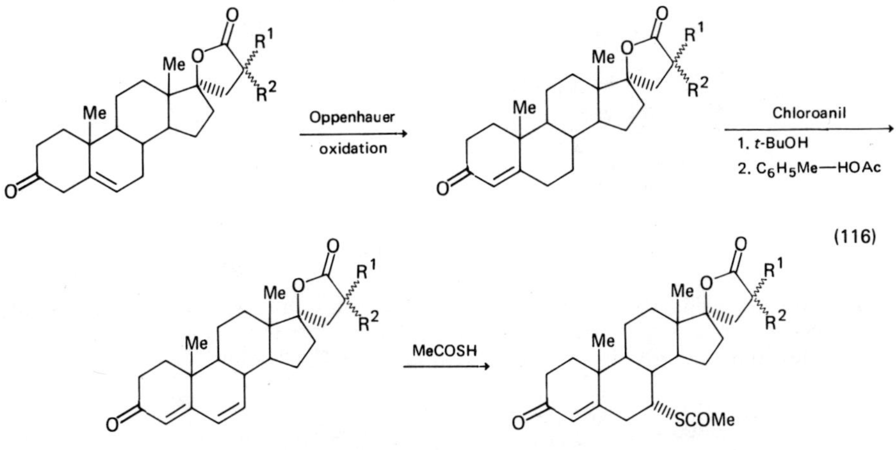

(116)

		Yield (%)		
R¹	R²	Monoene	Diene	Thioacetyl
H	Me	90	85	64
H	Et	84	82	64
Me	Me	77	61	61
H	n-Bu	35	—	—
H	Ph	71	—	—

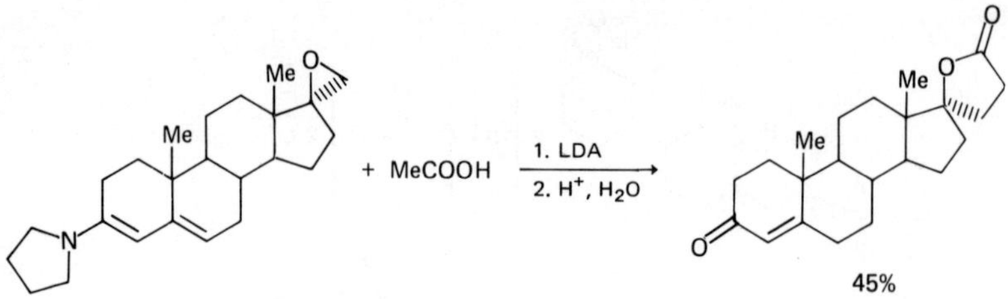

(117)

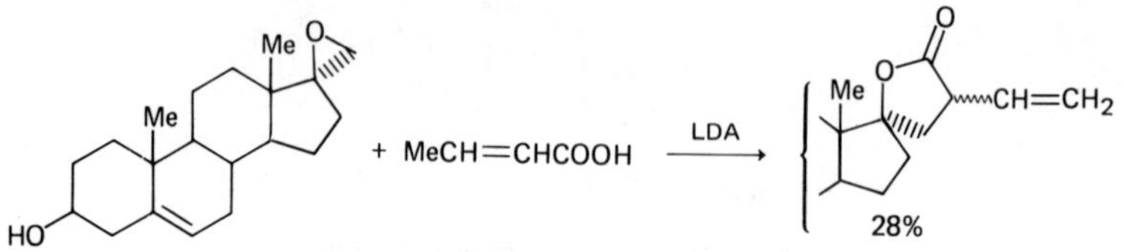

45%

(118)

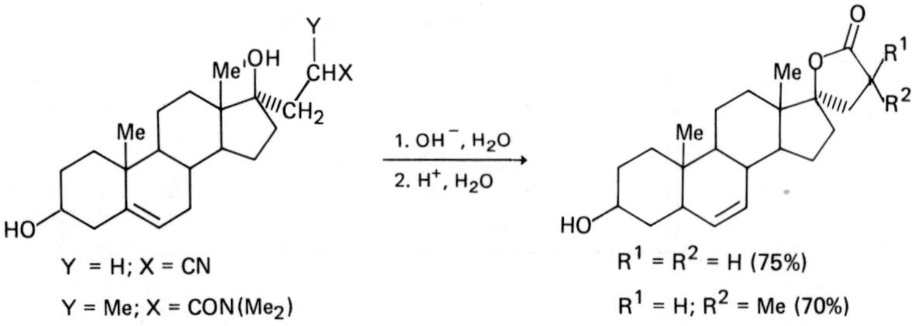

28%

(119)

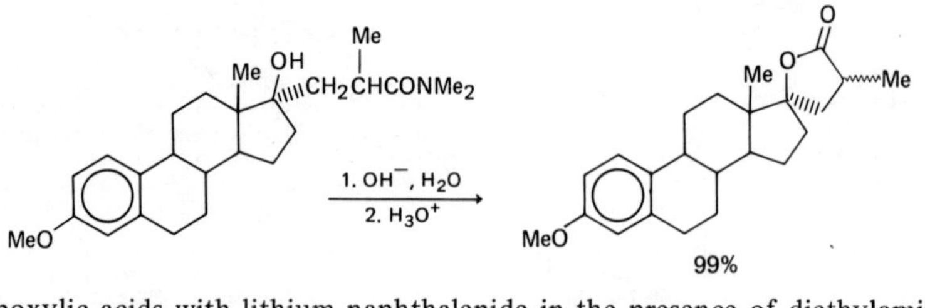

Y = H; X = CN

Y = Me; X = CON(Me$_2$)

R^1 = R^2 = H (75%)

R^1 = H; R^2 = Me (70%)

(120)

99%

of carboxylic acids with lithium naphthalenide in the presence of diethylamine to produce α-anions of the lithium carboxylates, which are then allowed to react with epoxides to afford γ-hydroxy acids (equation 121). Cyclization of these γ-hydroxy

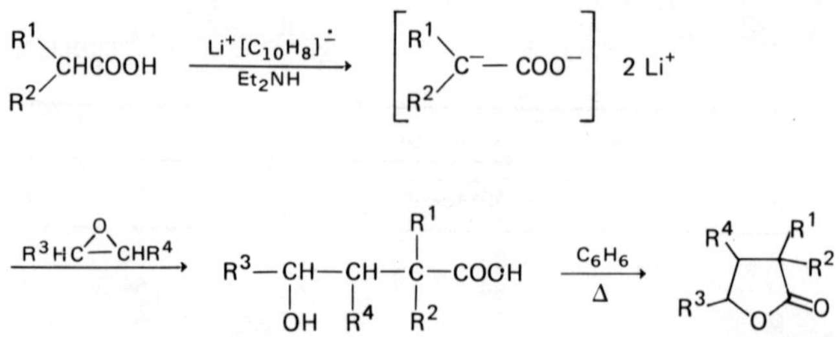

(121)

acids in refluxing benzene provides the lactones shown in Table 13. As may be seen from these results, monosubstituted epoxides react more readily than do disub-

TABLE 12. γ-Lactones by reactions of metalated acids with epoxides

Epoxide	Li⁺ Na⁺ Anion	Product	Yield (%)
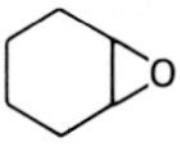	$Me_2\bar{C}COO^-$, then heat in C_6H_5Me	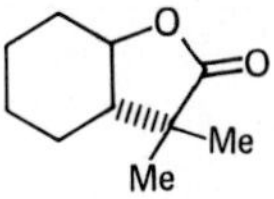	83
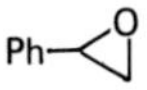 Ph	$Me_2\bar{C}COO^-$, then heat in C_6H_6	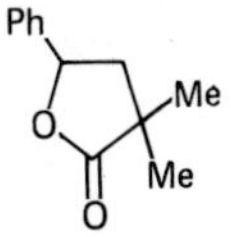	84
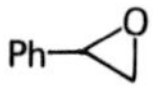 Ph	$Me_3C\bar{C}HCOO^-$	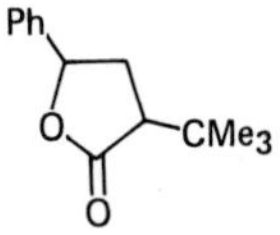	100
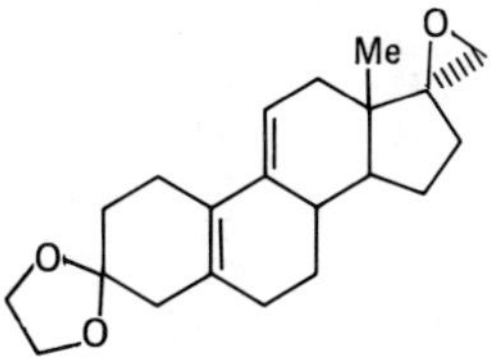	$Me_2\bar{C}COO^-$	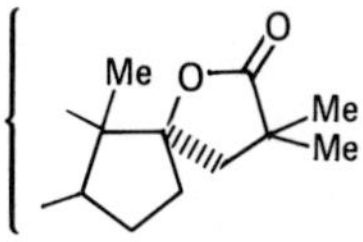	73
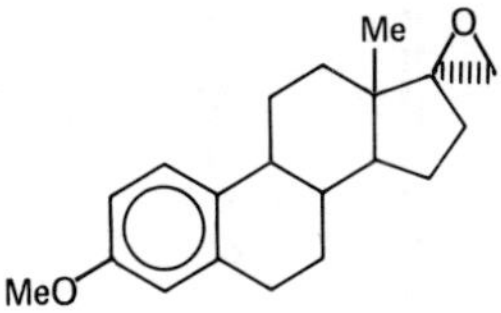	$Me_2\bar{C}COO^-$	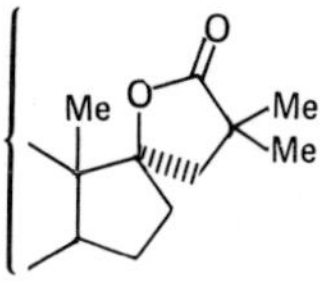	82
As above	$Ph\bar{C}HCOO^-$	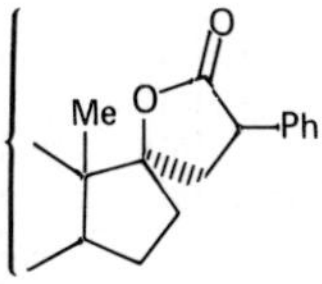	85
As above	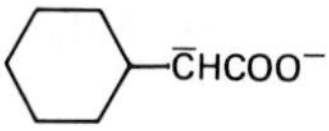 $\bar{C}HCOO^-$	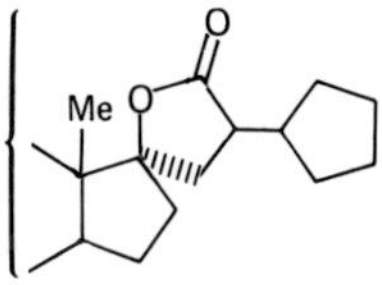	89

stituted epoxides. This reaction difficulty was found to increase to the point where no product was obtained when the di- and trisubstituted epoxides shown in reaction (122) were used.

TABLE 13. γ-Butyrolactones prepared by lithium naphthalenide-promoted reactions of carboxylic acids with epoxides

R^1	R^2	R^3	R^4	Yield (%)
H	H	Me	H	5
H	H	Et	H	22
H	H	Ph	H	31
H	H	$-(CH_2)_4-$		18
Me	H	Me	H	47
Me	H	Et	H	51
Me	H	Ph	H	57
Me	H	$-(CH_2)_4-$		39
Et	H	Me	H	38
Et	H	Et	H	41
Et	H	Ph	H	53
Et	H	$-(CH_2)_4-$		35
Me	Me	Me	H	48
Me	Me	Et	H	73
Me	Me	Ph	H	69
Me	Me	$-(CH_2)_4-$		55
n-Pr	H	Me	H	58
n-Pr	H	Et	H	64
n-Pr	H	Ph	H	69
n-Pr	H	$-(CH_2)_4-$		52
i-Pr	H	Me	H	44
i-Pr	H	Et	H	71
i-Pr	H	Ph	H	53
i-Pr	H	$-(CH_2)_4-$		14
n-Bu	H	Me	H	49
n-Bu	H	Et	H	53
n-Bu	H	Ph	H	66
n-Bu	H	$-(CH_2)_4-$		31
Ph	H	Me	H	52
Ph	H	Et	H	68
Ph	H	Ph	H	55
Ph	H	$-(CH_2)_4-$		23
$Me_2C=CH(CH_2)_2CHMe$	H	Me	H	66
$Me_2C=CH(CH_2)_2CHMe$	H	Et	H	71
$Me_2C=CH(CH_2)_2CHMe$	H	Ph	H	80
$Me_2C=CH(CH_2)_2CHMe$	H	$-(CH_2)_4-$		54
Me	Me	$EtOCH_2$	H	35
Me	Me	i-Pr	H	33
Me	Me	$CH_2=CHCH_2OCH_2$	H	38
Me	Me	n-BuOCH$_2$	H	70
Me	Me	i-BuOCH$_2$	H	81
Me	Me	$PhOCH_2$	H	35
Me	Me	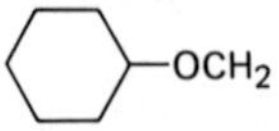 ⬡—OCH$_2$	H	52

J. From Lithio Salts of 2-Alkyl-2-oxazolines

The synthetically versatile[197,198] lithio derivatives of 2,4,4-trimethyl-2-oxazoline and its 2-alkyl homologues have recently been employed[199] in the

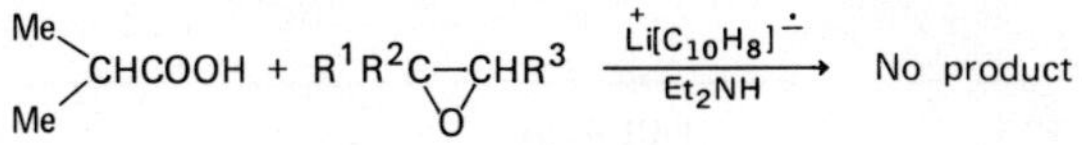

$$\underset{Me}{\overset{Me}{\diagdown}}CHCOOH \ + \ R^1R^2\overset{+}{C}\!\!-\!\!\underset{O}{\overset{}{\diagup\!\!\diagdown}}\!\!CHR^3 \ \xrightarrow[\ Et_2NH\ ]{\ \overset{+}{Li}[C_{10}H_8]^{\cdot\ -}\ } \ \text{No product} \tag{122}$$

R¹	R²	R³
H	—(CH$_2$)$_6$—	
H	—(CH$_2$)$_{10}$—	
Ph	Me	H
CH$_2$=CMe—CH=CH—CH$_2$—	Me	Me
Me	Me	CH$_2$=CMe(CH$_2$)$_2$
Me	Me	—CH$_2$CH(i-Pr)CH$_2$CH$_2$—

preparation of a variety of butyrolactones substituted in the α,β- and/or γ-positions with alkyl groups (equation 123). The procedure involves reaction of lithiated

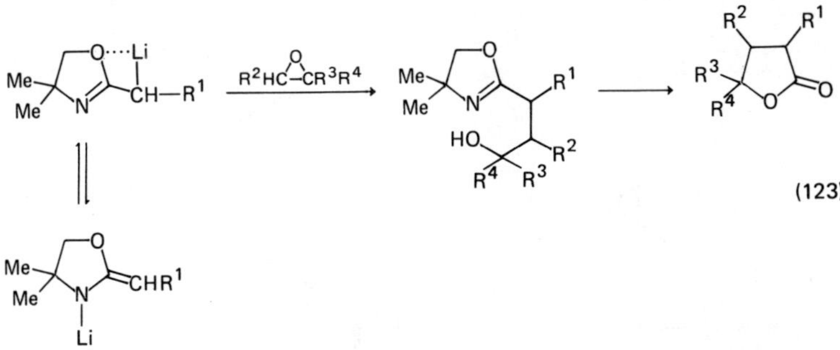

$$\tag{123}$$

oxazolines with an appropriate epoxide. This produces 2-(β-hydroxy-alkyl)oxazolines, which upon hydrolysis with aqueous acid, acidified ethanol or p-toluenesulphonic acid in benzene afford the butyrolactones shown in Table 14.

TABLE 14. γ-Butyrolactones from epoxides and lithio salts of 2-alkyl-2-oxazolines[199]

R¹	Epoxide	Hydrolysis method[a]	Lactone		Overall yield (%)
H	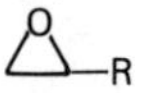	A	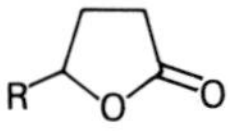	R = H R = Me R = Et	75 72 85
H	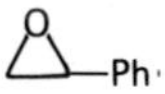	C	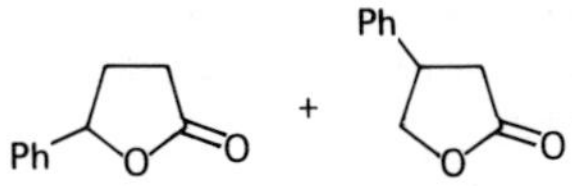 (94%) + (6%)		89
Me		C	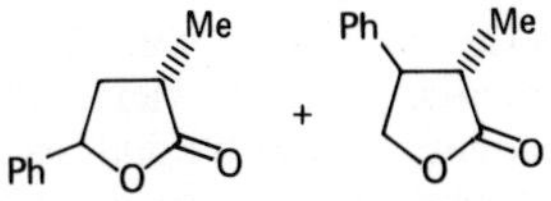 (60%) + (40%)		65

TABLE 14. (Continued)

R^1	Epoxide	Hydrolysis method[a]	Lactone	Overall yield (%)
$Me(CH_2)_4-$	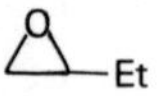	C	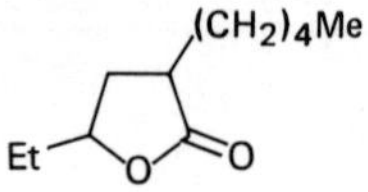	76
H	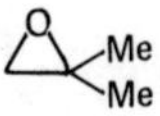	A	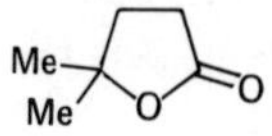	72
H	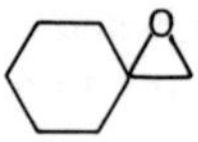	B	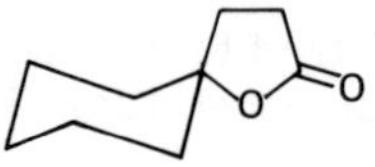	56
$PhCH_2$		C	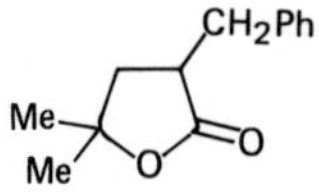	70
H	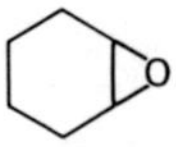	C	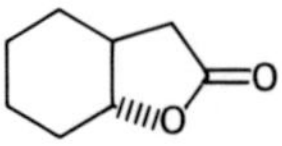	65
H	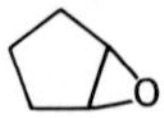	C	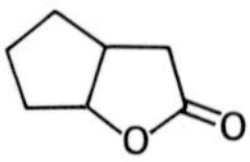	5–6
H	 (cis or trans)	A	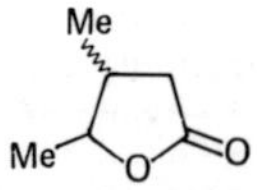 (50:50)	16
Me	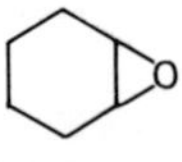	B	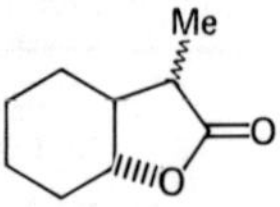	9
H	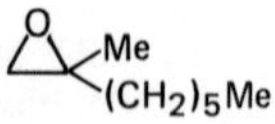	C	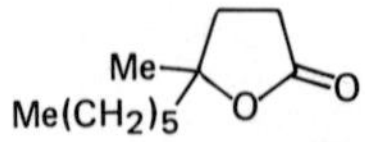	70

[a] Hydrolysis performed in: A = acidic EtOH, B = wet benzene–toluenesulphonic acid, C = acidic aqueous methanol.

It was observed that certain 1,2-disubstituted epoxides, especially those with *trans* substituents, gave low yields of lactones, or in some cases, no product at all.

The oxazoline procedure has also been used in the asymmetric synthesis of 2-substituted γ-butyrolactones as shown in equation (124)[200].

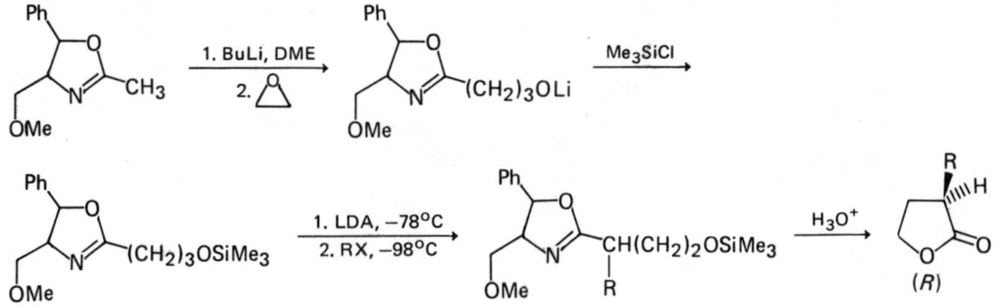

(124)

	Optical	
R	purity (%)	Yield (%)
Me	64.2	58
Et	–	68
n-Pr	73.3	75
Allyl	72.0	60
n-Bu	–	71

K. By Direct Functionalization of Preformed Lactones

The acidity of lactone α-hydrogens permits structural elaboration at the α-position of the lactone nucleus via certain carbanion condensations. One of the earliest examples[201-204] of this type of reaction involved dehydrative aldol condensations of aromatic aldehydes at the α-methylene group of 2,3-dihydrofuran-2-ones. A more recent study[205] of analogous aldol condensations of 2(3*H*)-coumaranone with 2-hydroxybenzaldehydes in the presence of triethylamine revealed that the expected 3-(2-hydroxybenzylidene)2-(3*H*)-coumaranones were produced upon dropwise addition of triethylamine to the reaction mixture at 15°C, while an increase in temperature to 25–40°C during the condensation increased the yield of 3-(2-hydroxyphenyl)coumarins at the expense of the benzylidene products (equation 125). If the temperature were raised to 70°C or if the 3-(2-hydroxybenzylidene)2(3*H*)-coumaranones were treated at 80°C with additional triethylamine, 3-(2-hydroxyphenyl)coumarins resulted via an intramolecular, *in situ* cyclization. Analogous results were obtained[206] in condensations of substituted 2-hydroxybenzaldehydes with γ-aryl-Δ$^{\beta,\gamma}$-butenolides (equation 126).

Several methods for α-alkylation of lactones have appeared in the literature[207-212]. Best results have been obtained by formation of the lactone enolate with a strong base such as LDA, lithium isopropylcyclohexylamide or trityl lithium, followed by treatment of the enolate with an alkyl halide (Table 15). A similar approach[212,213] affords dialkylated products, while attempts to use benzylbromomethyl sulphide as an alkylating agent have failed[214].

Various methods for introducing an α-methylene group into preformed lactones have been discussed in a 1975 review[11] on α-methylene lactones. A procedure[215] which is not discussed in this review involves reactions of an α-phosphono-γ-butyrolactone carbanion with aldehydes, ketones, heterocumulenes and nitrosobenzene to form α-ylidene-γ-butyrolactones (Method A; equation 127). The α-bromo-γ-butyrolactone employed as the starting material for these reactions has also been used in

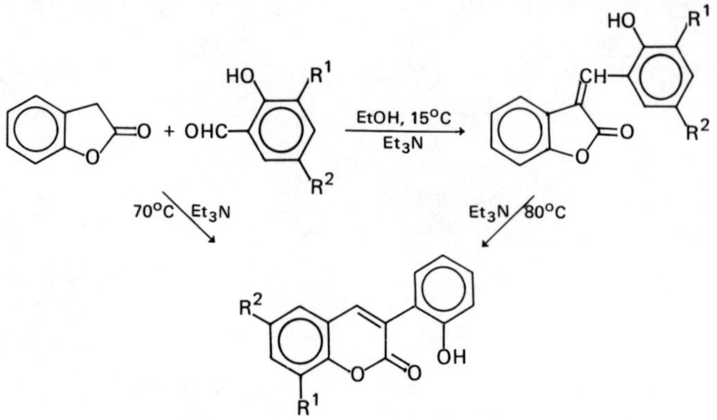

(125)

| R¹ | R² | Yield (%) | |
		3-(2-hydroxybenzylidene)-2(3*H*)-coumaranones	3-(2-Hydroxy-phenyl)coumarins
H	H	76	100
H	Cl	62	89
H	Br	82	96
H	NO₂	62	97
Cl	Cl	91	84
Br	Br	93	81

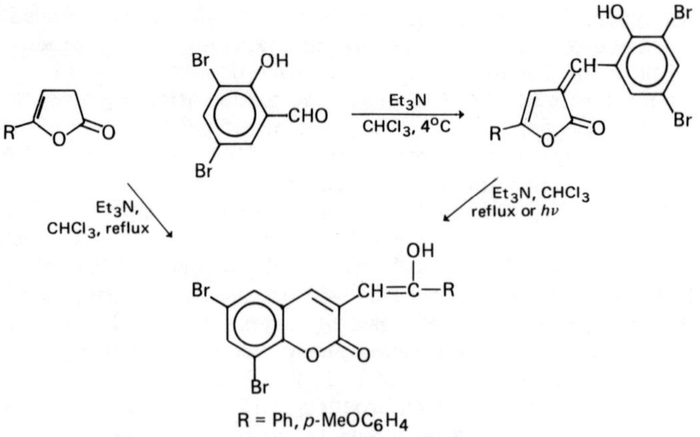

(126)

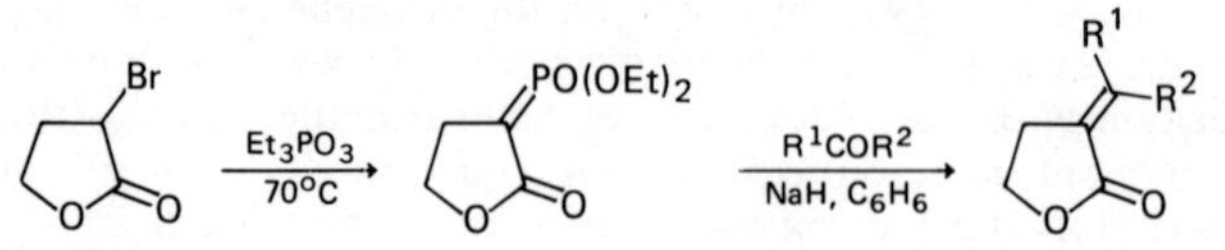

R = Ph, *p*-MeOC₆H₄

(127)

Reformatsky-type reactions to afford similar products (Method B; equation 128)[216]. Results from these two procedures are given in Table 16.

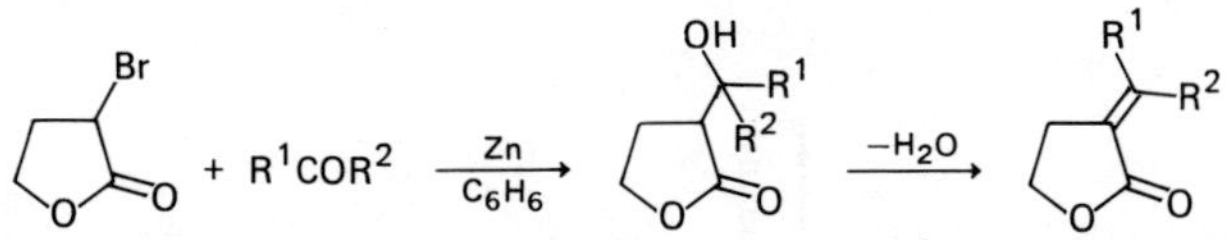

$$\text{(128)}$$

A recent, facile method for the preparation of β-methoxycarbonyl γ-substituted γ-butyrolactones proceeds via generation of the enolate of succinic anhydride in the presence of carboxyl compounds[217]. Thus, addition of a THF solution of 3-phenylpropanal and succinic anhydride at $-78°C$ under argon to a THF solution of lithium 1,1-bis(trimethylsilyl)-3-methyl-1-butoxide afforded, after hydrolysis and treatment with diazomethane, an 80% yield of β-methoxycarbonyl-γ-phenethyl-γ-butyrolactone. The corresponding substituted γ-butyrolactones can be obtained in moderate yields when ketones are used in place of aldehydes in this reaction (equation 129). Methylsuccinic anhydride produces the enolate on the methylene

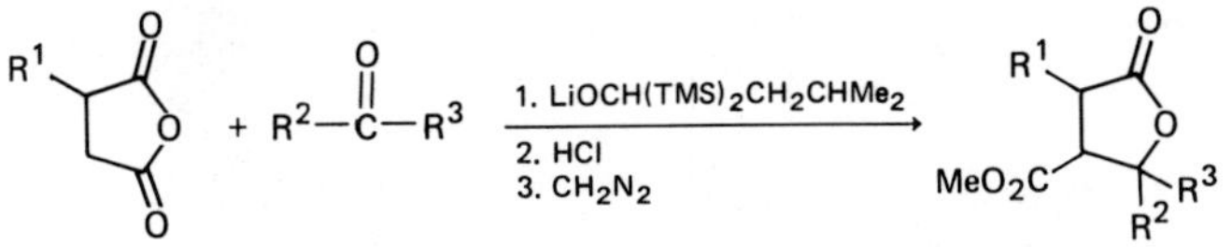

$$\text{(129)}$$

R^1	R^2	R^3	Yield (%)
H	Ph	H	84
H	$PhCH_2CH_2$	H	80
H	$n\text{-PrCH}=CH$	H	82
H	C_5H_{11}	H	78
H	Me_2CH	H	76
H	$-(CH_2)_5-$		51
Me	Ph	H	84
Me	$PhCH_2CH_2$	H	75
Me	$n\text{-PrCH}=CH$	H	85
Me	C_5H_{11}	H	72
Me	Me_2CH	H	70
Me	$-(CH_2)_5-$		57

site and affords the corresponding adduct exclusively, while generation of the enolate from glutaric anhydride does not afford the butyrolactones in yields as high as when succinic anhydride is used.

L. From Ketenes

Simple, as well as substituted ketenes react with aldehydes and ketones via a $(2\pi + 2\pi)$ cycloaddition to afford β-lactones (equation 130).

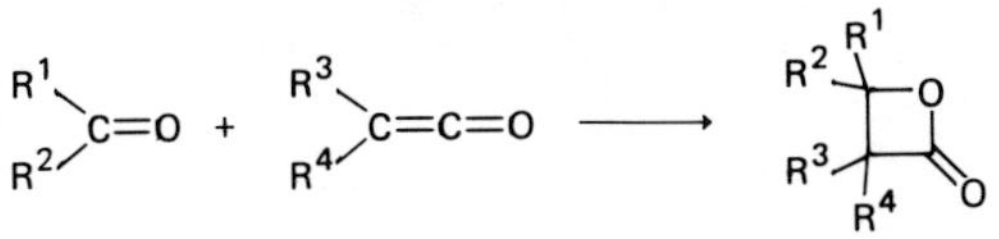

$$\text{(130)}$$

TABLE 15. α-Alkylation of γ-butyrolactone and δ-valerolactone

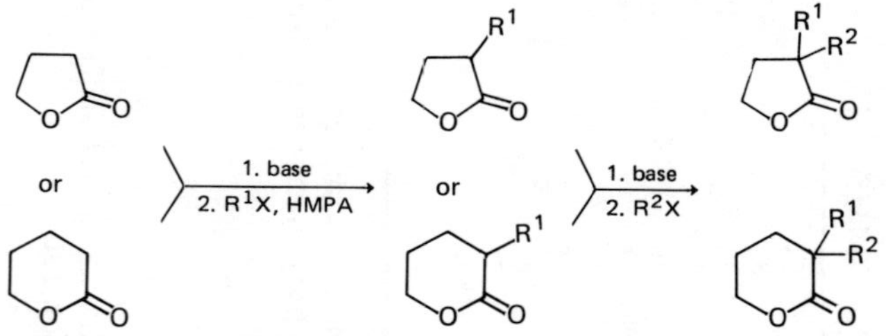

Lactone	Base[a]	R^1	Yield (%)	R^2	Yield (%)	References
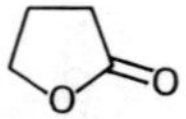	A	Me	56	—	—	211
	A	H$_2$C=CHCH$_2$	74→>90	—	—	211, 212
	A	n-Bu	low	—	—	211
	B	Et	>90	—	—	212
	B	HC≡CCH$_2$	>90	—	—	212
	B	Br(CH$_2$)$_2$CH$_2$	80	—	—	212
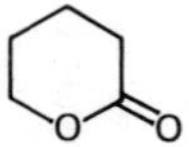	B	Et	>90	—	—	212
	B	H$_2$C=CHCH$_2$	>90	—	—	212
	B	HC≡CCH$_2$	>90	—	—	212

Substrate	Base[a]					Ref.
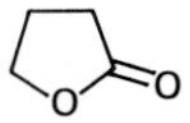 (γ-butyrolactone)	A	Me	80	Me	13	211
	C	Me	—	Me	—	213
	B	Et	—	Et	95	212
	B	Et	—	$H_2C=CHCH_2$	95	212
	B	Et	—	$HC\equiv CCH_2$	95	212
	B	Et	—	$Br(CH_2)_2CH_2$	95	212
(δ-valerolactone)	B	Et	—	Et	95	212
	B	Et	—	$H_2C=CHCH_2$	95	212
	B	Et	—	$HC\equiv CCH_2$	95	212
	B	Et	—	$Br(CH_2)_2CH_2$	95	212

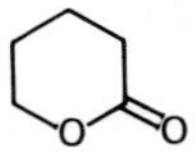

[a] A = Lithium isopropylcyclohexylamide; B = lithium diisopropylamide; C = trityllithium.

TABLE 16. Preparation of α-ylidene-γ-butyrolactones from α-bromo-γ-butyrolactone via Wittig (method A) and Reformatsky (method B) reactions

R^1	R^2	Method	Product	Yield (%)	Reference
H	Ar	A		Ar = Ph　　　　　100 Ar = p-O_2NC_6H$_4$　71 Ar = PhCH=CH　55	215 215 215
H	CCl$_3$	A	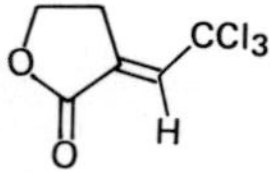	100	215
H	Et or i-Pr	A	 (*cis + trans*)	R = i-Pr　89 R = Et　100	215 215
H	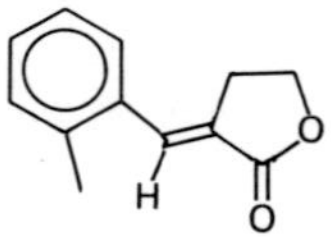	A	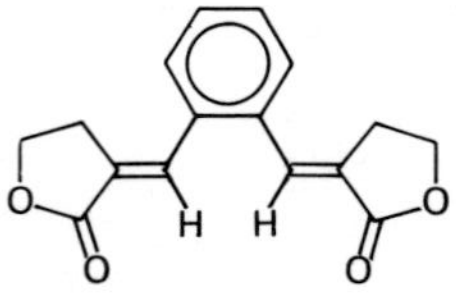 +	47	215

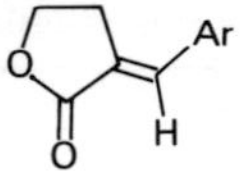

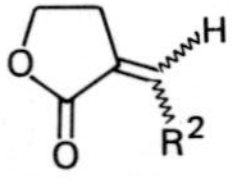

215

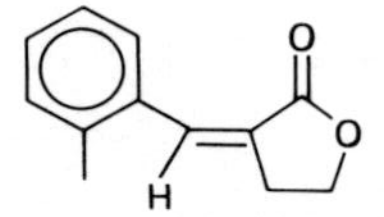

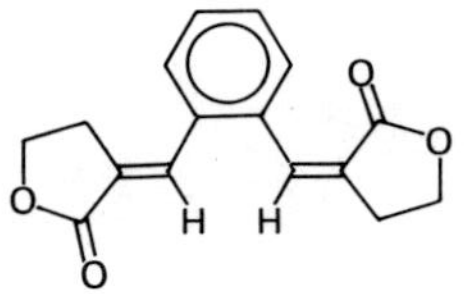

215

215

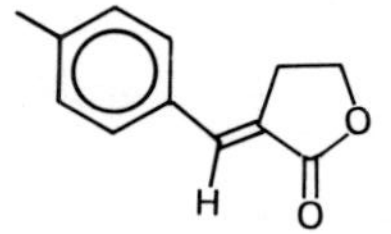

215

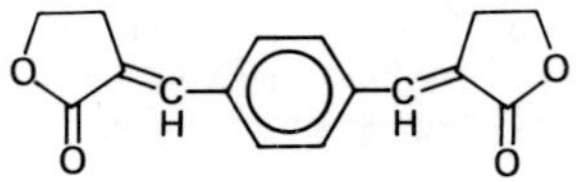

215

215

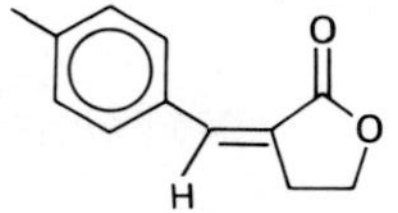

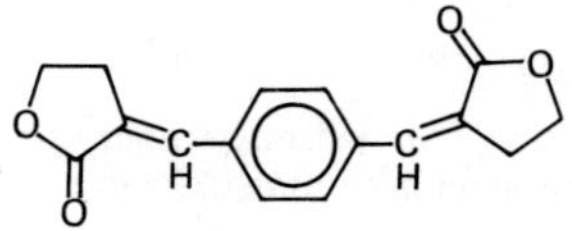

215

7.3

67

4.4

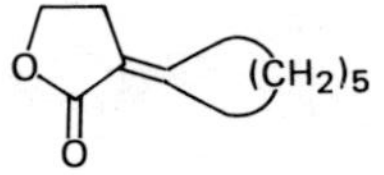

92

91

95

100

$-(CH_2)_5-$

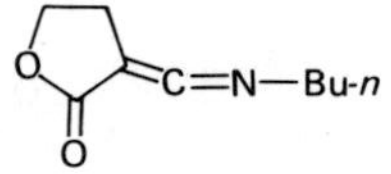

$=N-Bu\text{-}n$

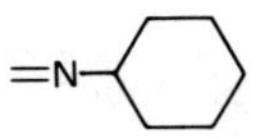

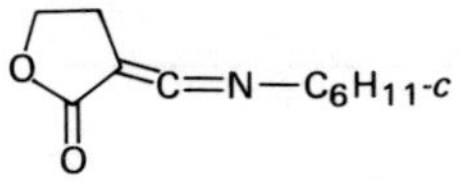

$=N$

$=CPh_2$

A A A A A A A

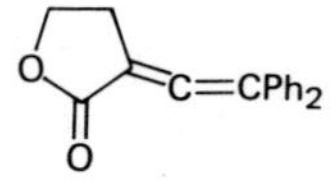

H H H

TABLE 16. (Continued)

R¹	R²	Method	Product	Yield (%)	Reference
Ph–C(=)–Et	Et–C(=)–Ph	A	(structure: 3-ylidene-γ-butyrolactone with Et and Ph)	100	215
	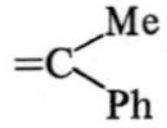 Me–C(=)–Ph	A	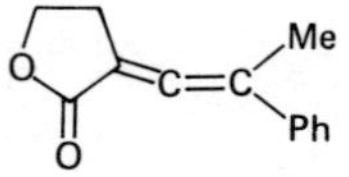 (structure: 3-ylidene-γ-butyrolactone with Me and Ph)	81	215
17α-Methyldihydro-testosterone		A	3-(γ-Butyrolacton-α-ylidene)-17α-methylandrostan-17β-ol + 3-(γ-Butyrolacton-α-yl)-17α-methylandrost-2(or 3)-ene-17β-ol	75 25	216
Cortisone acetate		B	3-(γ-Butyrolacton-α-ylidene)-17α-hydroxy-11-dehydrocorticosterone-21-acetate	—	216

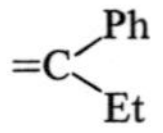

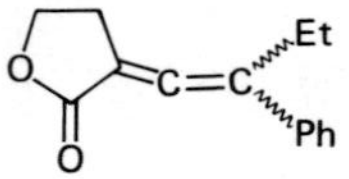

In the preparation of β-butyrolactone, β-propiolactone and β-caprolactone by reaction of the appropriate aldehyde with ketene it was found[218] that boron trifluoride or its etherate complex in THF could be used to increase both the yield and purity of the product. The versatility of both catalysed and uncatalysed reactions of ketenes with aldehydes and ketones may be seen by inspection of Table 17,[219-224], where a representative series of lactone preparations are collected. The first four entries in Table 17 involve γ-lactone formation[219]. Generation of these products is explained[219] by a mechanism involving initial formation of the expected β-lactone, followed by ring-opening, carbonium ion rearrangement and recyclization as shown in equation (131) with methyl t-butyl ketone.

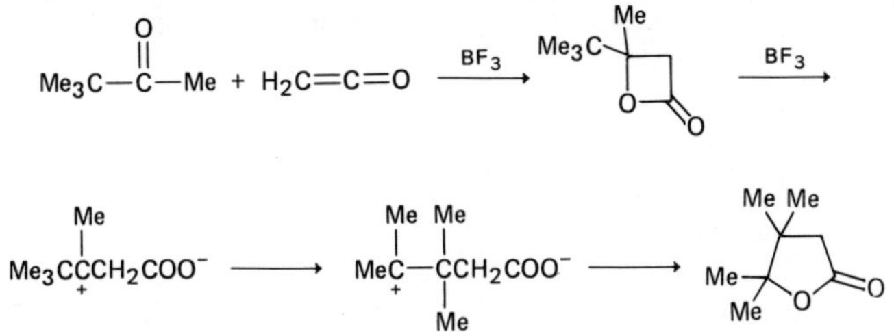

$$
\text{Me}_3\text{C}-\overset{\overset{\text{O}}{\|}}{\text{C}}-\text{Me} + \text{H}_2\text{C}=\text{C}=\text{O} \xrightarrow{\text{BF}_3} \quad \text{[Me}_3\text{C, four-membered lactone]} \xrightarrow{\text{BF}_3}
$$

$$
\underset{+}{\text{Me}_3\overset{}{\text{C}}}\text{CH}_2\text{COO}^- \longrightarrow \overset{\text{Me Me}}{\underset{\text{Me}}{\text{Me}\overset{}{\text{C}}-\underset{+}{\overset{}{\text{C}}}\text{CH}_2\text{COO}^-}} \longrightarrow \text{[Me Me dimethyl γ-lactone]} \tag{131}
$$

Ketenes undergo 1,3-dipolar addition with carbenes derived from diazo ketones to form enol lactones (butenolides)[225] as shown in equation (132) and summarized in Table 18.

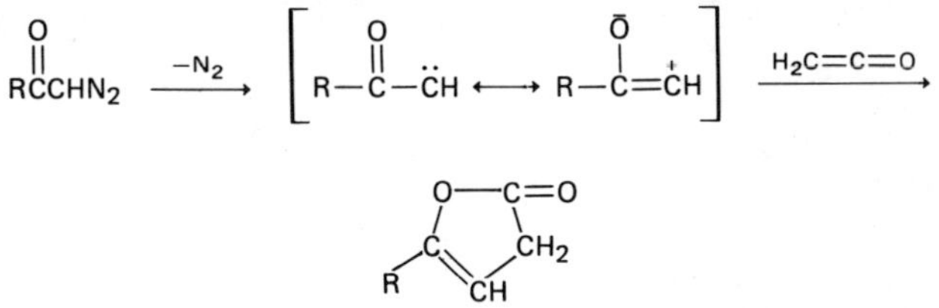

$$
\overset{\overset{\text{O}}{\|}}{\text{RCCHN}_2} \xrightarrow{-\text{N}_2} \left[\overset{\overset{\text{O}}{\|}}{\text{R}-\text{C}-\ddot{\text{C}}\text{H}} \longleftrightarrow \overset{\overset{\bar{\text{O}}}{|}}{\text{R}-\overset{+}{\text{C}}=\text{CH}} \right] \xrightarrow{\text{H}_2\text{C}=\text{C}=\text{O}}
$$

$$
\text{[O—C=O butenolide ring with R, CH, CH}_2\text{]} \tag{132}
$$

An interesting preparation of lactones has been observed during irradiation of several α β-epoxy diazoketones in benzene[226]. The butenolide products obtained are explained in terms of an intermediate epoxy ketene, formed by a Wolff rearrangement, which then undergoes intramolecular cyclization (equation 133).

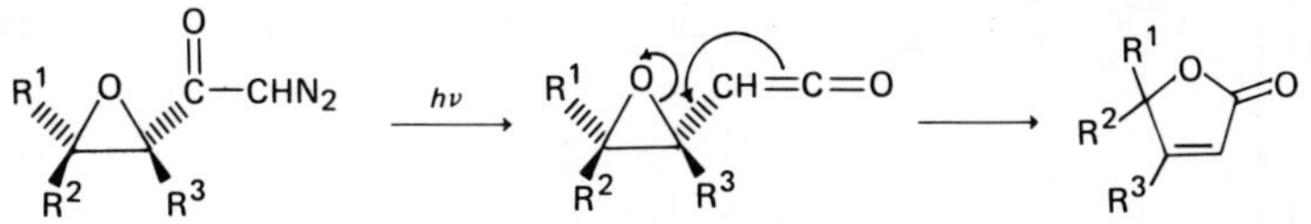

$$
\text{[epoxide with R}^1, \text{R}^2, \text{R}^3, \text{C}-\text{CHN}_2] \xrightarrow{h\nu} \text{[epoxy ketene CH=C=O]} \longrightarrow \text{[butenolide with R}^1, \text{R}^2, \text{R}^3] \tag{133}
$$

R¹	R²	R³	Yield (%)
Ph	Ph	H	90
H	Ph	H	90
H	Ph	Ph	70
Me	Me	H	43

TABLE 17. Preparation of lactones by reaction of aldehydes and ketones with ketene

Aldehyde or ketone	Ketene	Product	Yield (%)	Reference
Me$_2$CHCOMe	H$_2$C=C=O + BF$_3$	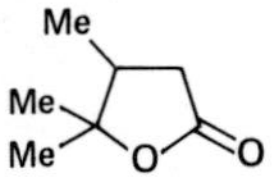	41	219
Me$_3$CCOMe	H$_2$C=C=O + BF$_3$	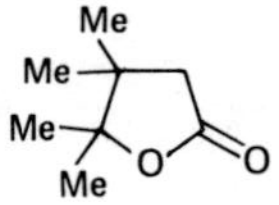	67	219
(CH$_2$)$_5$CHCOMe	H$_2$C=C=O + BF$_3$	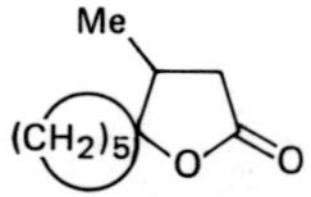	44	219
(CH$_2$)$_5$CCOMe Me	H$_2$C=C=O + BF$_3$	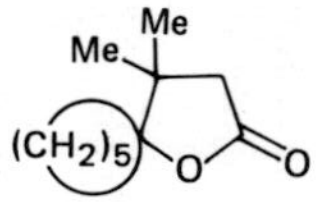	49	219
Cl$_3$CCHO	H$_2$C=C=O[a]	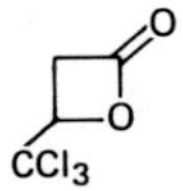	72.2	220, 221
Cl$_3$CCHO	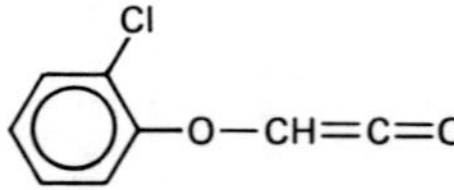	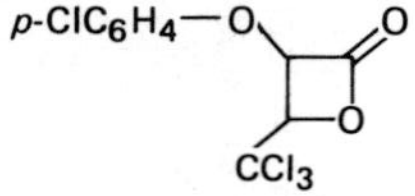	45	221

		Product	Yield (%)	Ref.
Cl_3CCHO	$PhO{-}CH{=}C{=}O$	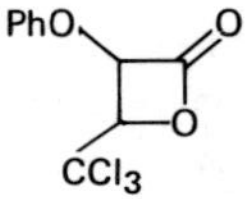 3-PhO-4-CCl_3 β-lactone	36	221
Cl_3CCHO	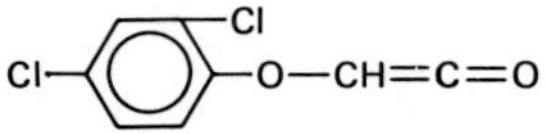 $2,4\text{-}Cl_2C_6H_3{-}O{-}CH{=}C{=}O$	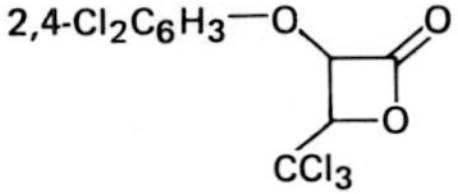 3-(2,4-$Cl_2C_6H_3$O)-4-CCl_3 β-lactone	63	221
Br_3CCHO	$ClHC{=}C{=}O$	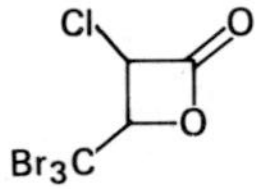 3-Cl-4-CBr_3 β-lactone	11	221
$Cl_3CCOCCl_3$	$H_2C{=}C{=}O$	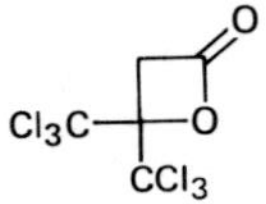 4,4-bis(CCl_3) β-lactone	6	221
F_3CCHO	$H_2C{=}C{=}O$	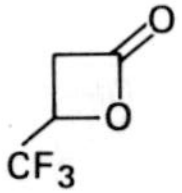 4-CF_3 β-lactone	20	221
Cl_3CCHO	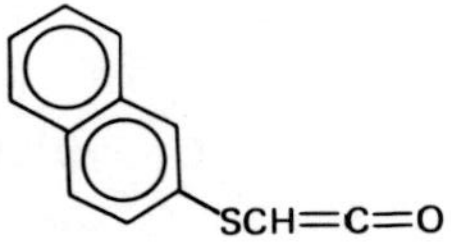 naphthyl-$SCH{=}C{=}O$	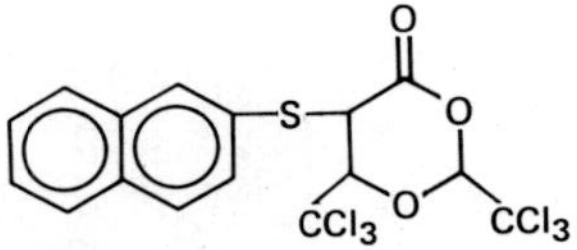 dioxinone, CCl_3/CCl_3	7.8	221
CCl_3CHO	$O{=}C{=}CH(CH_2)_2CH{=}C{=}O$	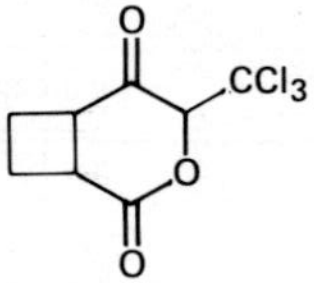 bicyclic CCl_3 lactone	47	221

TABLE 17. (Continued)

Aldehyde or ketone	Ketene	Product	Yield (%)	Reference
Cl_3CCHO	$Cl_2C{=}C{=}O$	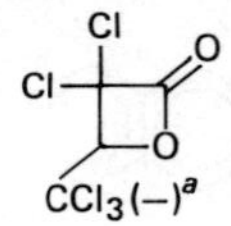 [β-lactone structure with $CCl_3({-})$][a]	39	221
RCHO	$Cl_2C{=}C{=}O$	[β-lactone structure with R]	R = Me 51 R = Me_2Ch 40	222 222
ArCHO	$Cl_2C{=}C{=}O$	[β-lactone structure with Ar]	Ar = Ph 30 Ar = $p{-}ClC_6H_4$ 66	222 222
MeOC—COOEt	$Cl_2C{=}C{=}O$	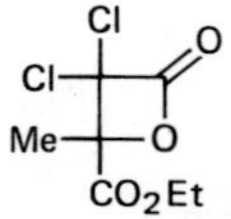 [β-lactone structure with Me, CO_2Et]	33	222
EtOOC—COCOEt	$Cl_2C{=}C{=}O$ 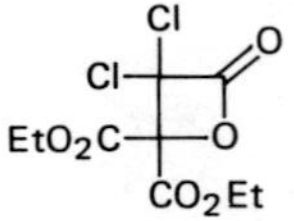	[β-lactone structure with EtO_2C, CO_2Et]	76	222

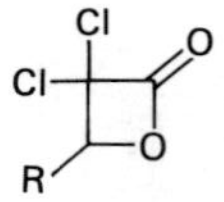

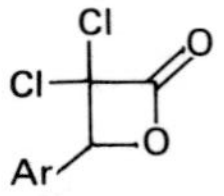

PhOC—COOEt	Cl₂C=C=O	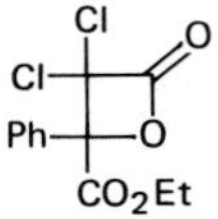	38	222
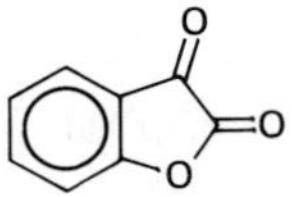	Cl₂C=C=O	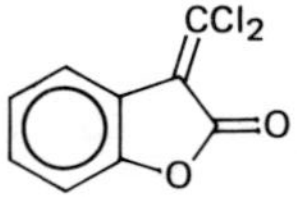	61	222
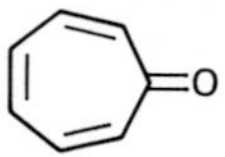	Cl₂C=C=O	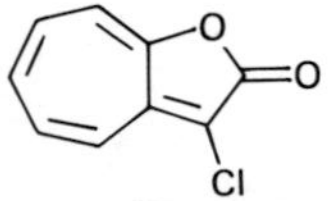	19	224

[a]Ketene prepared from acetyl chloride with *N,N*-dimethyl-α-phenethylamine afforded (−) product, from acetyl chloride with brucine afforded (+) product.

TABLE 18. Preparation of enol lactones by reaction of ketenes with diazoketones[225]

Diazoketone	Ketene	Product	Yield (%)
MeCH$_2$COCHN$_2$	H$_2$C=C=O	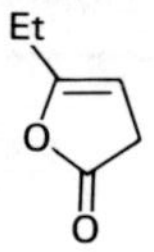	43
PhCOCHN$_2$	H$_2$C=C=O	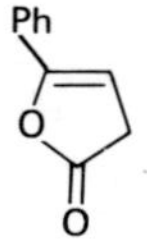	34
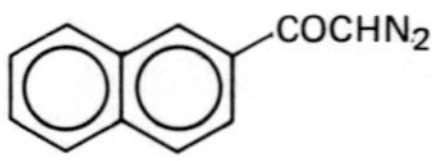	H$_2$C=C=O	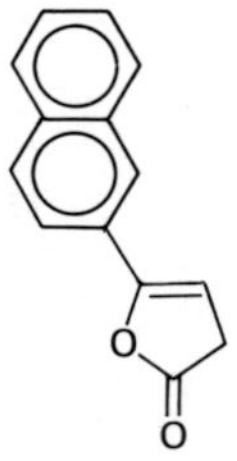	11

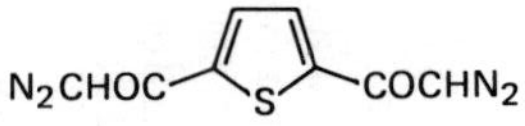

18

29

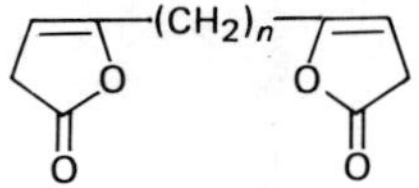

$n = 3$ 40
$n = 4$ 90
$n = 5$ 47
$n = 6$ 32

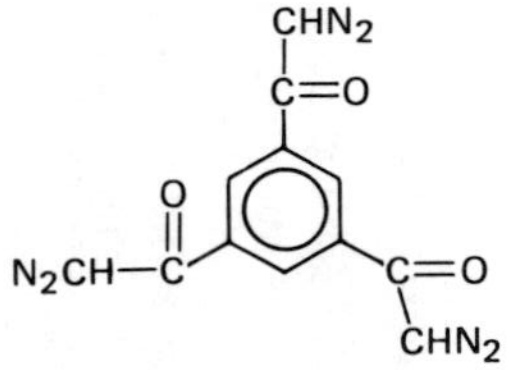

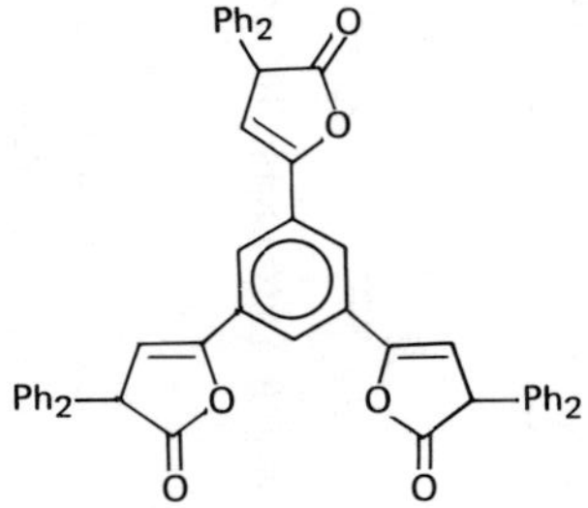

$Ph_2C=C=O$

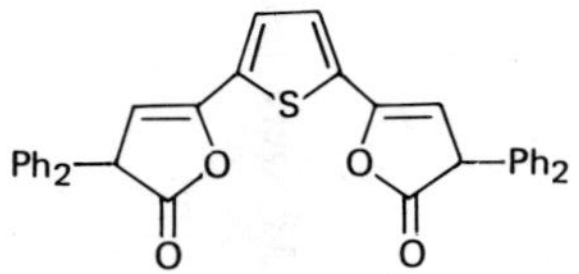

$Ph_2C=C=O$

$H_2C=C=O$

N_2CHOC—$COCHN_2$

N_2CH—C=O ... CHN_2

$N_2CHOC(CH_2)_nCOCHN_2$

Since it is known[227-229] that ozone is an effective epoxidizing agent toward highly hindered alkenes, Wheland and Bartlett[230] treated an emulsion of diphenyl-ketene in ethyl acetate and hexafluoroacetone with ozone at $-78°C$ expecting an α-lactone. Instead they obtained the product shown in equation (134), the structure of which was established by spectroscopy and its alkaline hydrolysis to benzilic

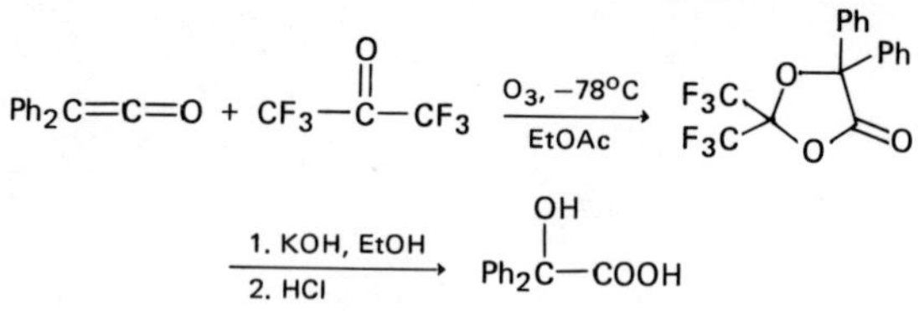

$$Ph_2C=C=O + CF_3-\overset{O}{\underset{\|}{C}}-CF_3 \xrightarrow[\text{EtOAc}]{O_3, -78°C} \quad \text{[structure]} \tag{134}$$

$$\xrightarrow[\text{2. HCl}]{\text{1. KOH, EtOH}} \quad Ph_2\overset{OH}{\underset{|}{C}}-COOH$$

acid. A similar approach was used to prepare di-*t*-butylacetolactone (equation 135)[231]; however, when hexafluoroacetone was added to the chlorotrifluoro-methane (Freon 11) used as the solvent at $-78°C$ and the mixture brought to room

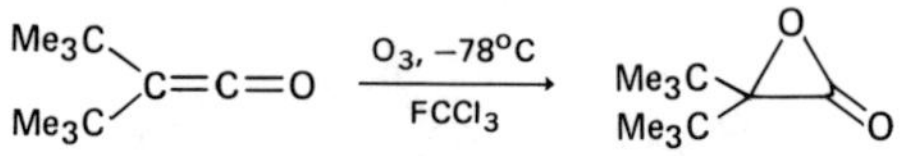

$$\begin{matrix} Me_3C \\ Me_3C \end{matrix}C=C=O \xrightarrow[\text{FCCl}_3]{O_3, -78°C} \quad \text{[structure]} \tag{135}$$

temperature, the two rearrangement products **30** and **31** were isolated upon distillation[230].

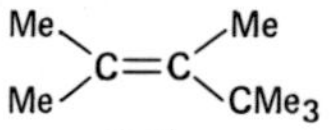

(30)

29%

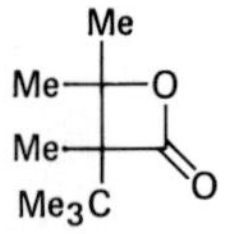

(31)

21%

Although attempts[231] to cause nitrous oxide to react with hydroxyacetylenic compounds in inert solvents have not been very successful, 3-butyn-1-ol did react to afford γ-butyrolactone, presumably via formation and cyclization of the intermediate 2-hydroxyethylketene (equation 136).

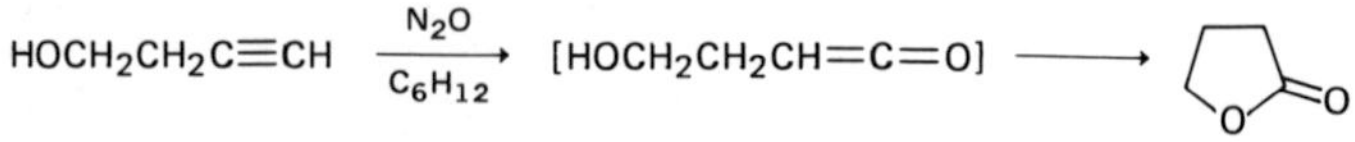

$$HOCH_2CH_2C\equiv CH \xrightarrow[C_6H_{12}]{N_2O} [HOCH_2CH_2CH=C=O] \longrightarrow \quad \text{[structure]} \tag{136}$$

M. By Reduction of Anhydrides, Esters and Acids

Although the first report of the sodium borohydride reduction of an acid anhydride appeared in 1949[232], it was not until 1969 that this method of lactone preparation was thoroughly investigated[233,234]. Since that time, a variety of reagents such as sodium borohydride, lithium aluminium hydride, lithium tri-*t*-butoxyaluminohydride and sodium in ethanol have been used to reduce numerous acid anhydrides to lactones (Table 19)[235-248].

One of the most interesting aspects of this preparative method is the controversy that has developed[236,237,244] concerning which carbonyl group of the anhydride is reduced when one carbonyl function is hindered and the other is relatively free. The majority of unsymmetrical anhydrides undergo reduction at the more hindered carbonyl, irrespective of the reducing agent employed (see first entry in Table 19).

TABLE 19. Preparation of lactones by reduction of acid anhydrides

Anhydride	Reducing agent	Product	Yield (%)	References
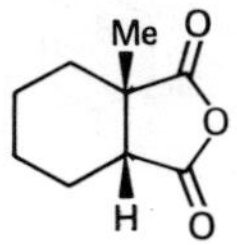	Na + EtOH LiAlH$_4$, THF NaBH$_4$, THF NaBH(OMe)$_3$, THF	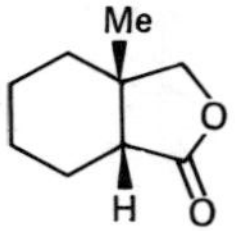	—, 35 78–82 80 78	235, 236 237 236 236
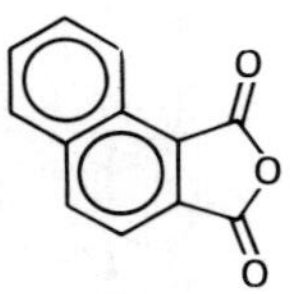	LiAlH$_4$, ether	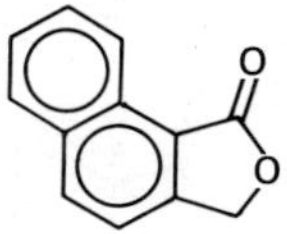	62	238
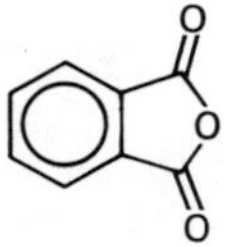	LiAlH$_4$, THF or NaBH$_4$, DMF	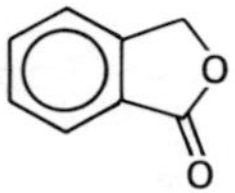	—, 71–73	239, 236
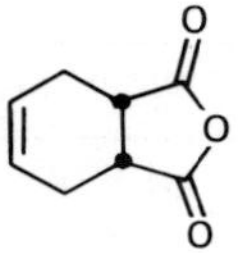	LiAlH$_4$, ether or THF	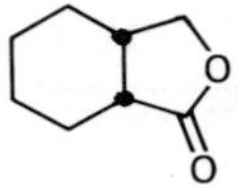	———, 75	239, 237

TABLE 19. (Continued)

Anhydride	Reducing agent	Product	Yield (%)	References
	LiAlH₄, ether THF NaBH₄, THF		– 72.8 76	239 237 236
	LiAlH₄, ether or NaBH₄, THF		––, 65	240, 236
	H₂, Pt or LiAlH₄, THF		–, 89	240, 237
	NaBH₄, in THF NaBH₄, in THF–MeOH		54 40–86	241 248

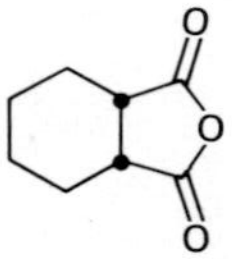

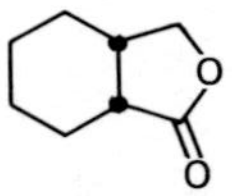

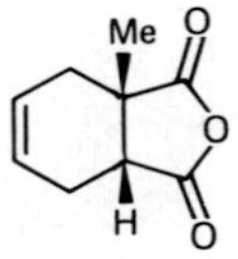

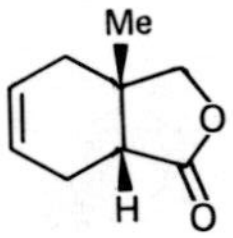

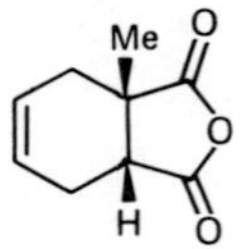

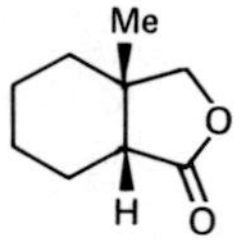

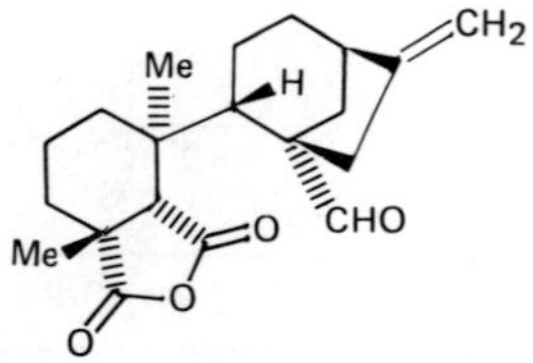

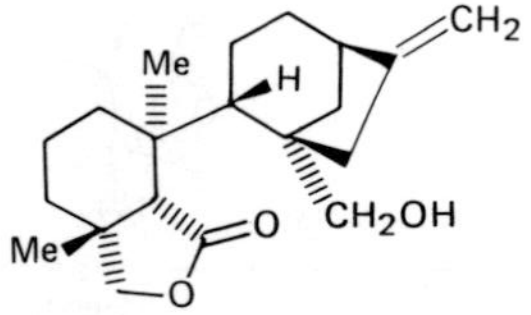

Reagent	Yield	Reference
NaBH₄, in MeOH LiAlH₄ in ether, dioxane	8 60	248 248
LiAlH₄, ether	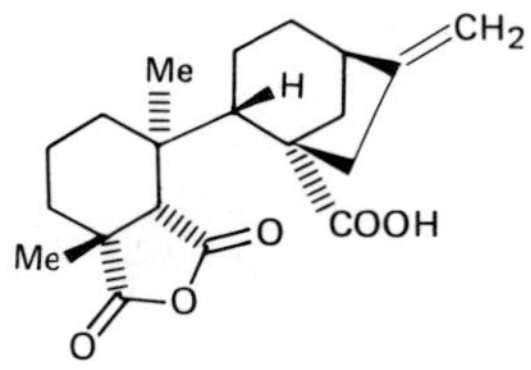 82, 90	241, 248
NaBH₄, i-PrOH	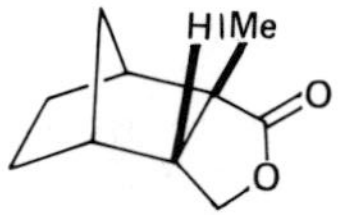 83	242
LiAlH₄, THF	—	243
LiAlH₄, THF	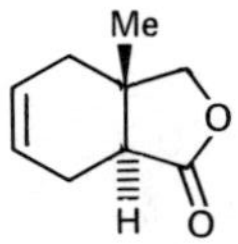 75	237
LiAlH₄, THF	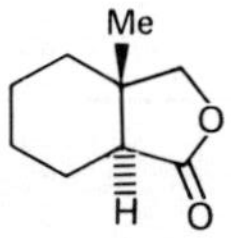 85	237

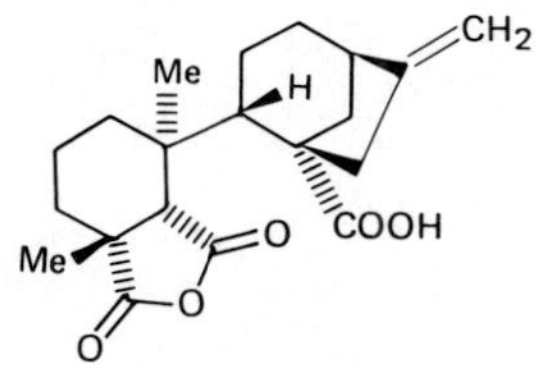

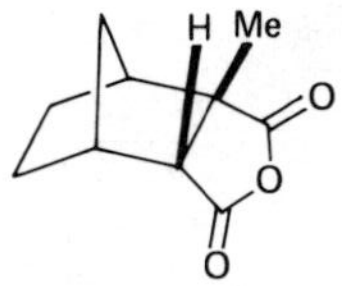

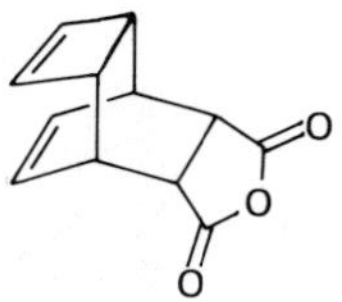

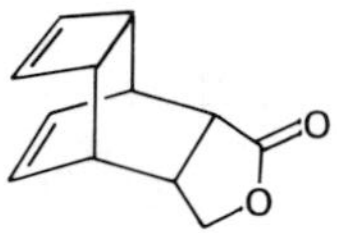

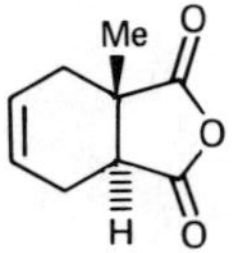

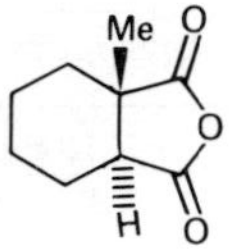

TABLE 19. (Continued)

Anhydride	Reducing agent	Product	Yield (%)	References
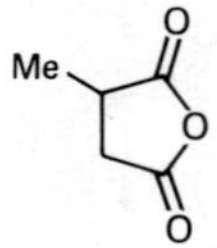	LiAlH$_4$, THF	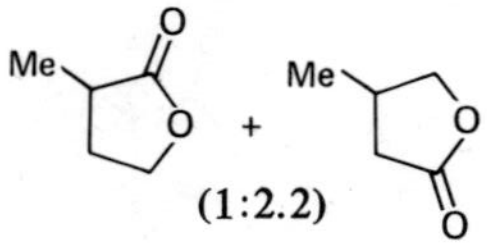 (1:2.2)	69	237
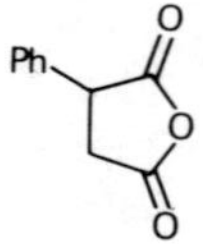	LiAlH$_4$, THF or NaBH$_4$, THF	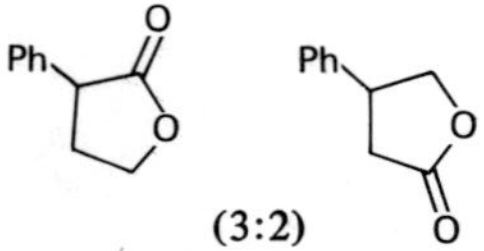 (3:2)	72, 67	237, 236
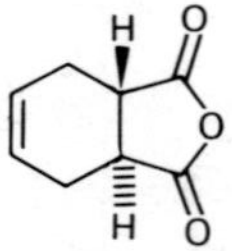	LiAlH$_4$, THF	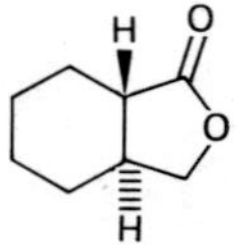	70.4	237
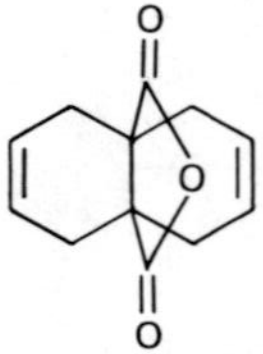	LiAlH$_4$, THF	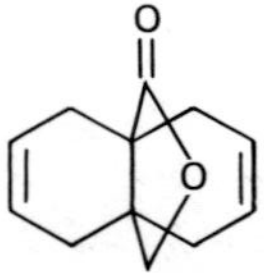	79–83	237
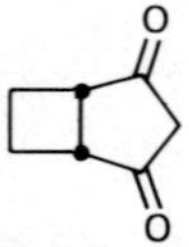	LiAlH$_4$, THF	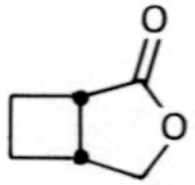	70.5	237

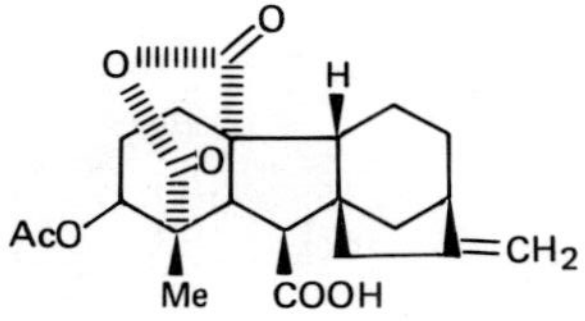

Substrate	Conditions	Product	Yield (%)	Reference
(steroid, AcO, =CH₂, COOH, Me)	LiAlH₄, THF	R = H, major product; R = Ac, minor product	50	244
(thioanhydride)	NaBH₄, C₆H₆–MeOH		80	245
(succinic anhydride)	NaBH₄, THF, 10N HCl in EtOH		51	236
(Me, Me anhydride)	NaBH₄, THF, 6N HCl in H₂O		74	236
(glutaric anhydride)	NaBH₄, THF, 10N HCl in EtOH		67	236
(homophthalic anhydride)	NaBH₄, THF, 10N HCl in EtOH		55	236

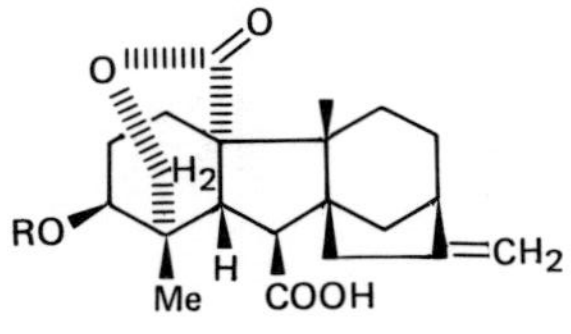

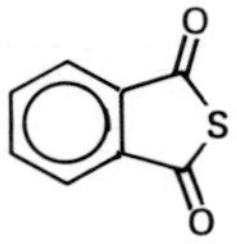

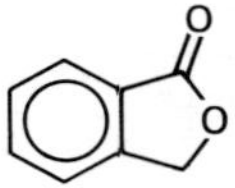

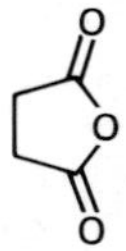

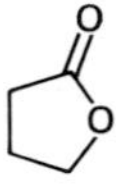

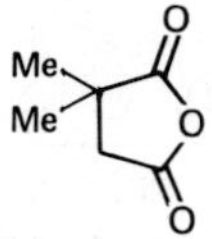

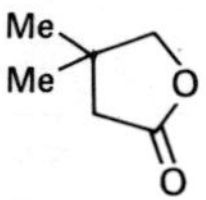

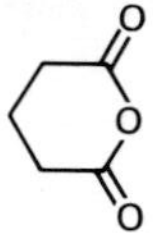

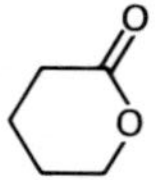

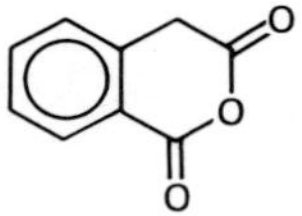

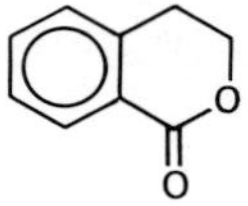

TABLE 19. (Continued)

Anhydride	Reducing agent	Product	Yield (%)	References
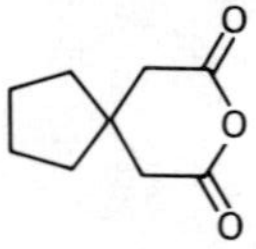	NaBH$_4$, THF 10N HCl in EtOH	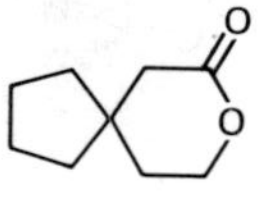	68	236
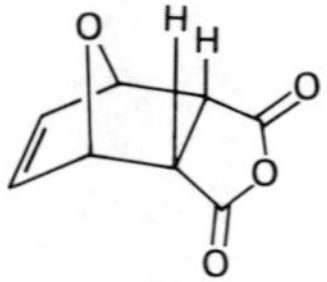	NaBH$_4$, EtOH	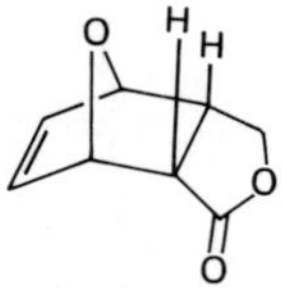	80^a	247

^{a} Heating at 140–150° C (20 Torr) afforded 79% of γ-crotonolactone via a retro Diels–Alder reaction.

Although reactions which have been reported to exhibit the opposite trend are apparently not in question, a uniform explanation for the anomalies is still unavailable.

Lithium aluminium hydride and catalytic[249] reductions of dicarboxylic acids (equation 137)[250,251], and their diesters[249] and monoesters[237,239] have been employed with only modest success for lactone synthesis.

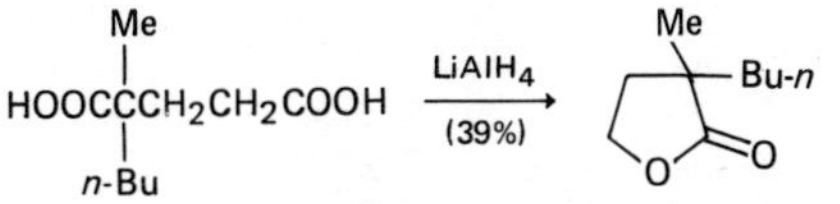

$$\text{HOOCCCH}_2\text{CH}_2\text{COOH} \xrightarrow[\text{(39\%)}]{\text{LiAlH}_4} \qquad \text{(Ref. 250)} \qquad (137)$$

Carboxylic acids and esters containing an aldehyde or ketone carbonyl function at the γ- or δ-position often provide good yields of lactones upon treatment with various reducing agents (Table 20).[252-262]. The choice of reduction conditions is often governed by whether the carboxyl group is free or esterified, for in the latter instances the reducing agent should be capable of reducting the ketone or aldehyde carbonyl without affecting the carboalkoxy function.

N. By Oxidation Reactions

Diols, ketones, ethers, olefins and several other miscellaneous types of compounds can be converted to lactones by oxidative reactions employing a variety of reagents. The following discussion is organized in terms of the type of compound used as starting material.

1. Oxidation of diols

A wide variety of 1,4- and 1,5- diols have been oxidized to lactones by reagents such as copper chromite, chromic acid, manganese dioxide, potassium permanganate and silver carbonate on celite (equation 138)[263-277]. Table 21 contains a representative series of diols along with their lactone oxidation products.

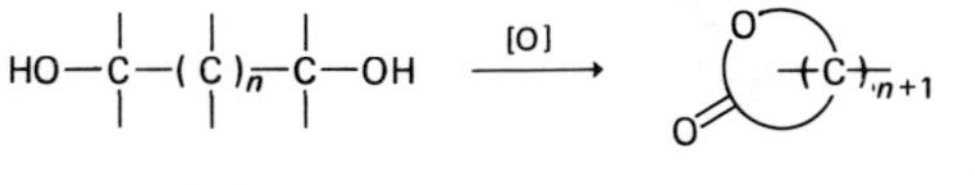

$$\text{HO}-\overset{|}{\text{C}}-(\overset{|}{\text{C}})_{\overline{n}}-\overset{|}{\text{C}}-\text{OH} \xrightarrow{\text{[O]}} \qquad (138)$$

Oxidative cleavage of unsaturated keto diols using lead tetraacetate or sodium periodate has been found to be an effective method for the production of steroidal lactones (equations 139–142)[255,256,278].

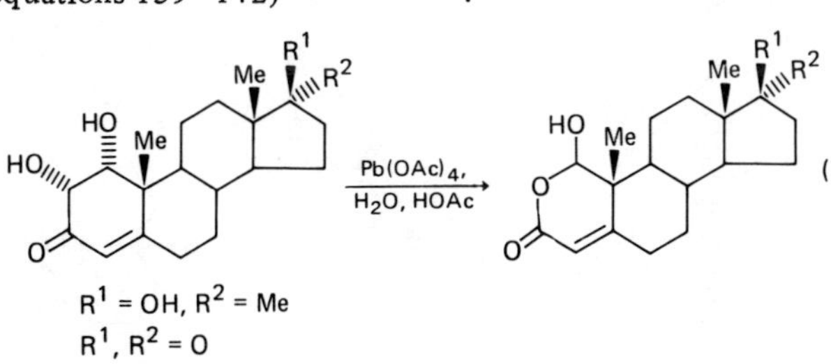

$$\xrightarrow[\text{H}_2\text{O, HOAc}]{\text{Pb(OAc)}_4,} \qquad \text{(Ref. 255)} \qquad (139)$$

R¹ = OH, R² = Me
R¹, R² = O

TABLE 20. Preparation of lactones by reduction of keto and aldehydic acids and esters

Acid or ester	Reducing agent	Product	Yield (%)	References
$ROCCH_2CH_2CH_2COOH$	$Al(i\text{-}PrO)_3$, $i\text{-}PrOH$	$R = n\text{-}C_4H_9$ $R = n\text{-}C_6H_{13}$ $R = n\text{-}C_7H_{15}$	69 64 66	20
$R^1OCCH_2CH_2CH_2COOR^2$	$Al(i\text{-}PrO)_3$, $i\text{-}PrOH$	$R^1 = n\text{-}C_3H_7$ $R^1 = n\text{-}C_5H_{11}$ $R^1 = n\text{-}C_8H_{17}$ $R^1 = n\text{-}C_9H_{19}$ $R^1 = n\text{-}C_4H_9$ $R^1 = n\text{-}C_6H_{13}$	89 69 62 73 87 68	20
	NaBH₄, NaOH			
	NaBH₄, MeOH		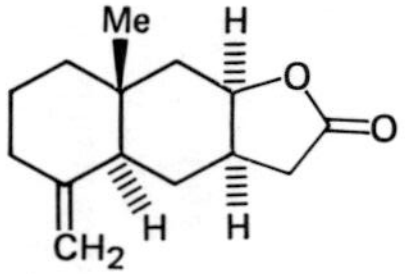 66	253
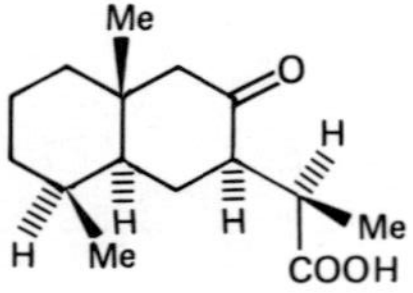	Na, $i\text{-}PrOH$	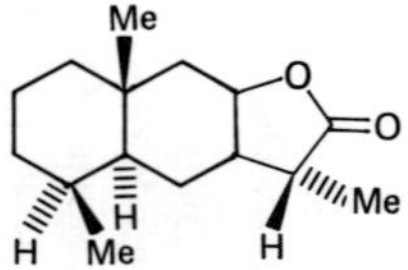	10	17
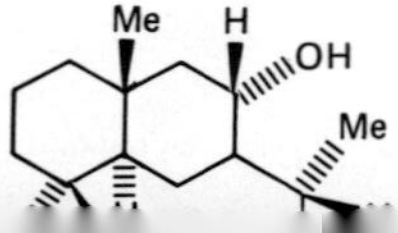	Na, $i\text{-}PrOH$			

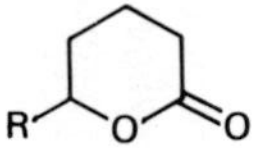

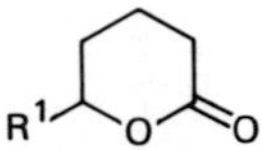

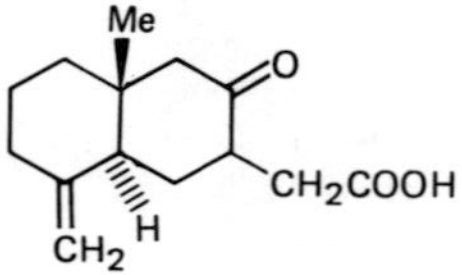

Reactant	Reagent	Product	Yield (%)	Reference
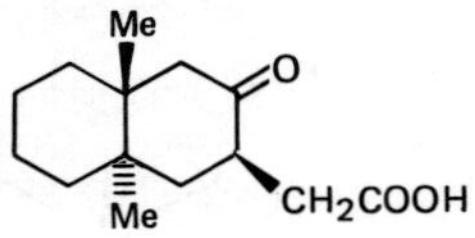 (structure: Me, CH₂COOH, Me, Me)	H₂, PtO₂	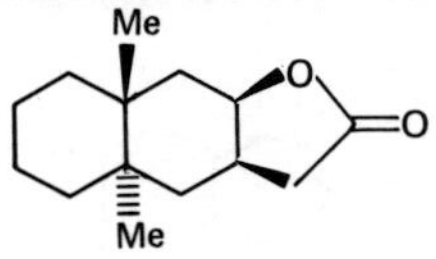 (lactone structure: Me, Me, Me)	73	19, 254
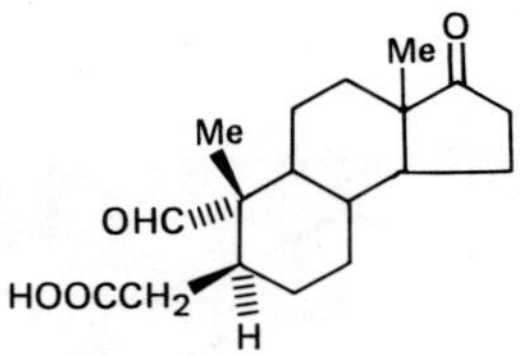 (structure: Me, O, Me, OHC, HOOCCH₂, H)	NaBH₄	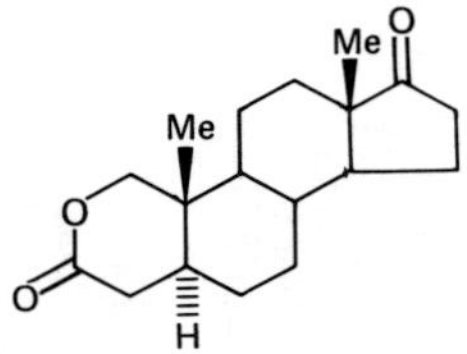 (structure: Me, O, Me, Me)	50–60	255
(steroid structure: Me, C₈H₁₇, Me, HOOC)	NaBH₄[a], NaOH, EtOH Na + EtOH Na + i-PrOH	(steroid lactone: Me, C₈H₁₇, Me, H, O)	34 75[b] 41	257 258 250
(structure: O, CH(COOH)₂)	NaBH₄, NaOH, H₂O	(lactone: O, COOH, H, H)	90	154
(structure: O, O, HOOCCH₂, C₈H₁₇)	H₂, 5% Rh on alumina, then HCl	(bicyclic lactone: O, H, H, C₈H₁₇, H)	>90	261

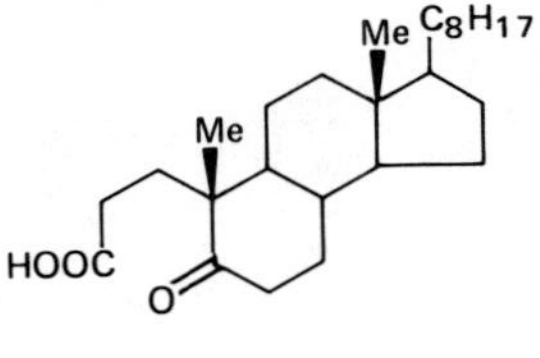

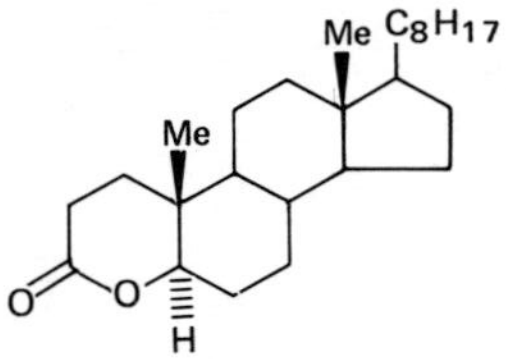

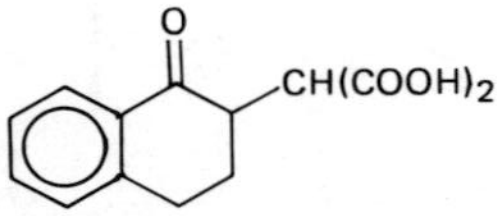

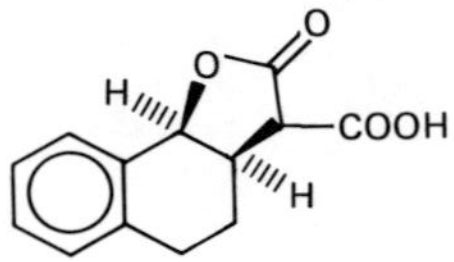

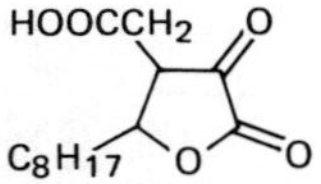

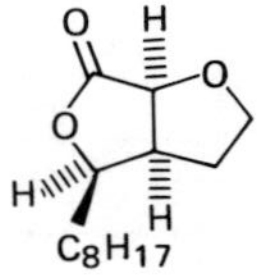

TABLE 20. (Continued)

Acid or ester	Reducing agent	Product	Yield (%)	References
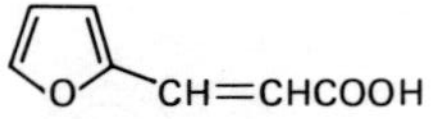 CH=CHCOOH (furan)	Ni(Al)^c, NaOH, H$_2$O	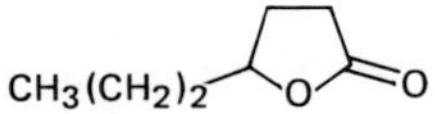 CH$_3$(CH$_2$)$_2$	33–37	262
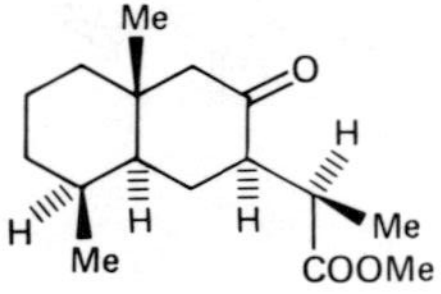	1. 3% NaOMe, MeOH 2. KBH$_4$, MeOH	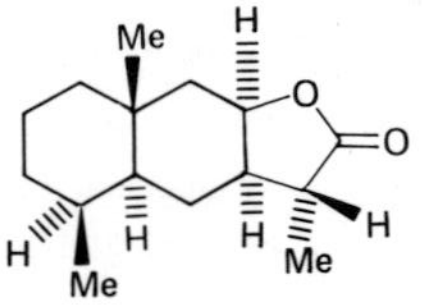	42	17
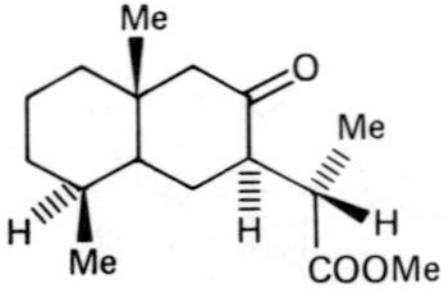	KBH$_4$, MeOH		45	17
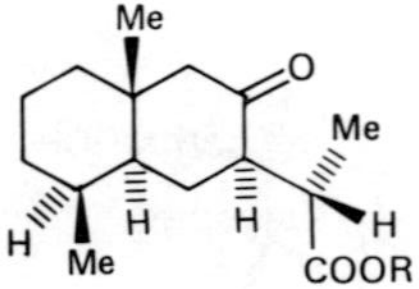	KBH$_4$, MeOH NaHCO$_3$	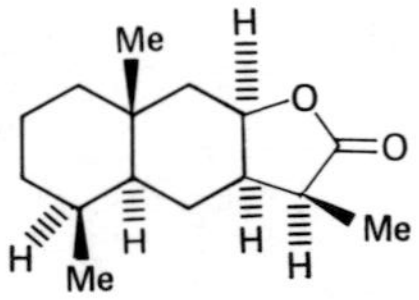	88 (R = H) 45 (R = Me)	17
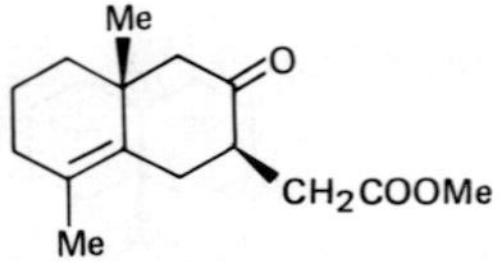	KBH$_4$, MeOH	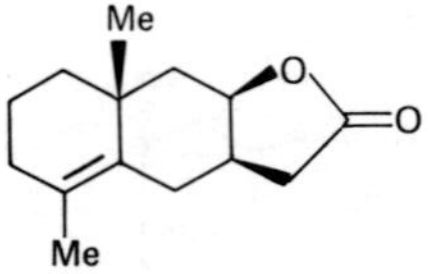	74	19

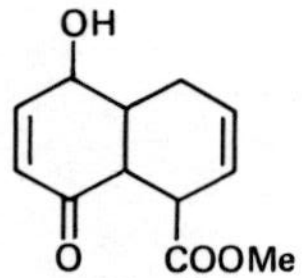

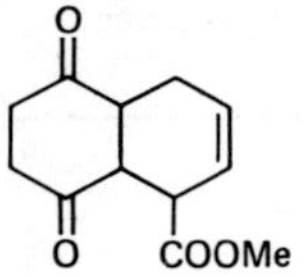

28

22

45

45

65
45

79

50

45

Al(*i*-PrO)₃
i-PrOH

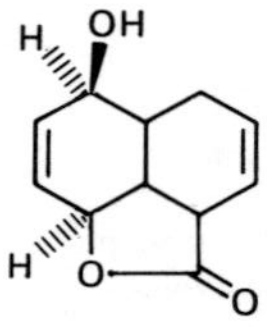

NaBH₄, MeOH

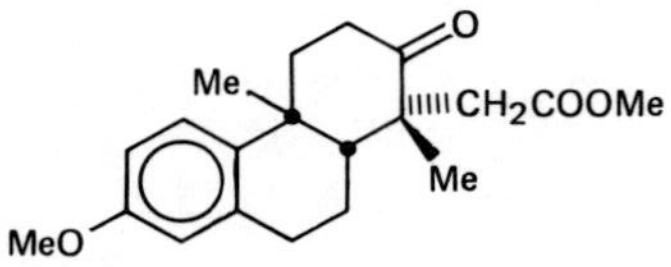

NaBH₄, MeOH 1 h, 0°C

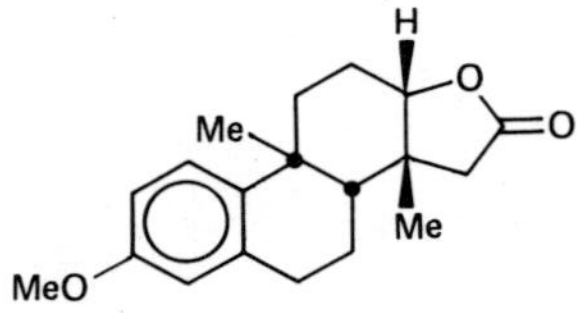

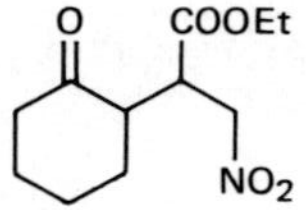

1. NaBH₄, MeOH, 0°C
2. Reflux, 10 h

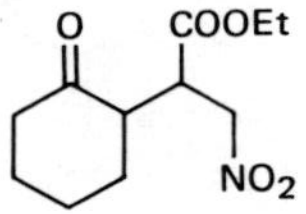

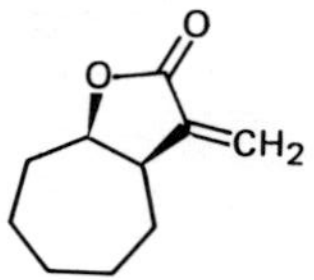

CH₂NO₂

CH₂

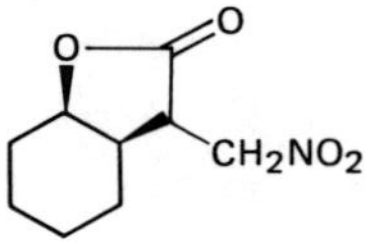

Me

MeO

H

Me

OH

H

H

COOMe

COOMe

OH

CH₂COOMe

Me

Me

MeO

COOEt

NO₂

COOEt

NO₂

TABLE 20. (Continued)

Acid or ester	Reducing agent	Product	Yield (%)	References
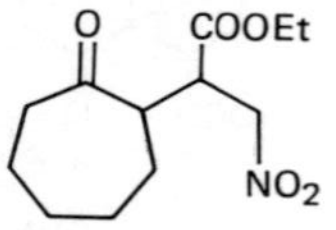	NaBH₄, MeOH	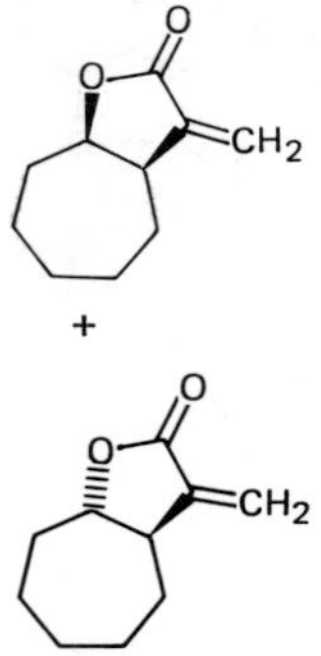	91 (86:14)	45

[a] Reduction with Al(i-PrO)₃ in i-PrOH gave an oily product containing *cis* and *trans* lactones.
[b] Crude product.
[c] Raney nickel–aluminium alloy.

Diol	Oxidizing agent	Product	Yield (%)	References
$HOCH_2CH_2\overset{\displaystyle Me}{\underset{\displaystyle }{CH}}CH_2CH_2OH$	Copper chromite or copper on pumice	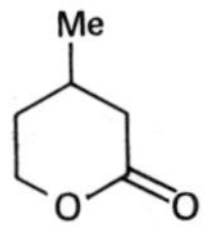	90–95	263, 264
$MeCHCH_2CH_2CH_2OH$ $\overset{\displaystyle }{\underset{\displaystyle OH}{|}}$	Copper chromite	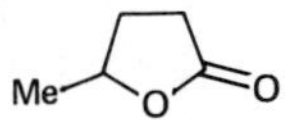	87	265
$HOCH_2CH_2CH_2CH_2CH_2OH$	Copper chromite	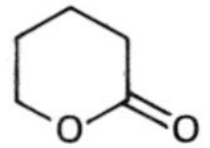	71	266
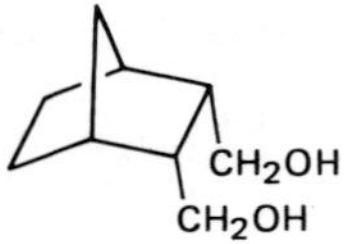		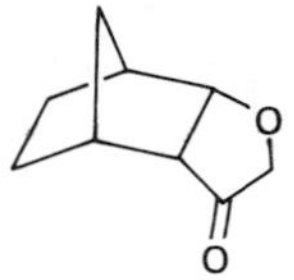		267
	Raney Ni, C_6H_6 KMnO$_4$, NaOH		80 10	
$MeCH_2CH_2CHCHCH_2CH_2OH$ $\overset{\displaystyle }{\underset{\displaystyle OHMe}{|\ \ |}}$	$K_2Cr_2O_7$, AcOH	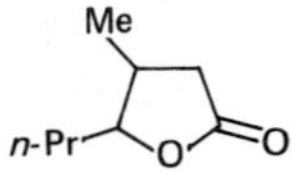	75–80	252
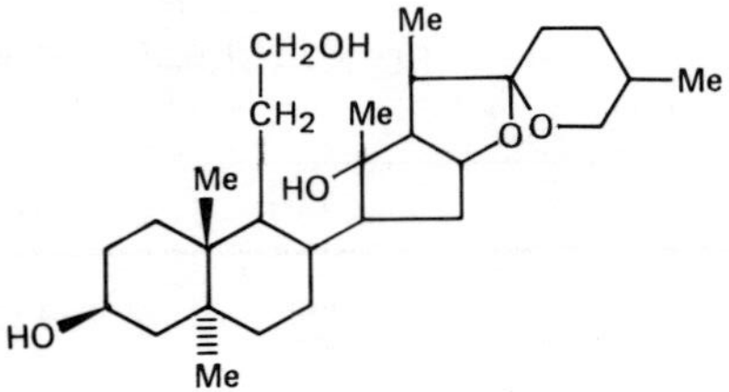	CrO_3, $C_5H_5N^a$	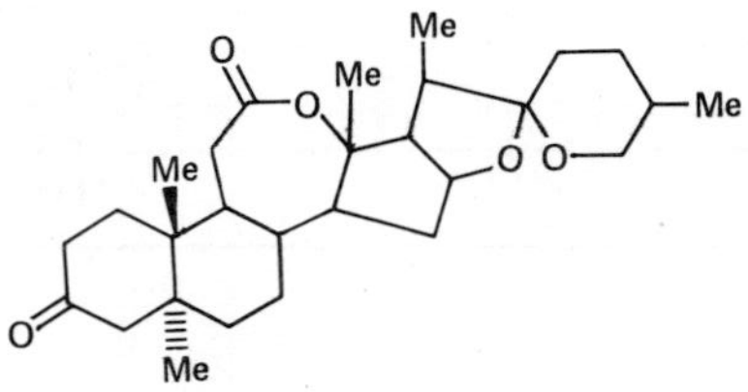	86	268

TABLE 21. (Continued)

Diol	Oxidizing agent	Product	Yield (%)	References
OH $\text{Me(CH}_2)_5\text{C(CH}_2)_2\text{CH}_2\text{OH}$ Me	CrO_3, H_2SO_4, H_2O	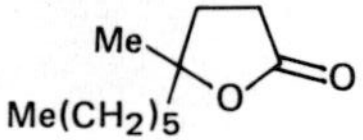	71	270
	$KMnO_4$, H_2O CrO_3, $C_5H_5N^a$ $Na_2Cr_2O_7$, H_2SO_4, H_2O Ag_2CO_3—celite, C_6H_6		71 60 95 95	271 271 271 272
$HOCH_2CH=CHCH_2OH$	CrO_3, $C_5H_5N^a$	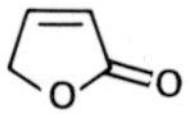	51	271
$HOCH_2CH_2CH_2CH_2OH$	CrO_3, $C_5H_5N^a$	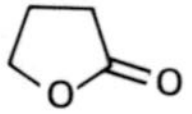	34	271
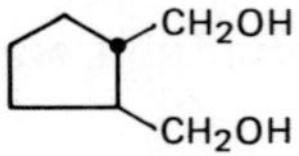	CrO_3, $C_5H_5N^a$	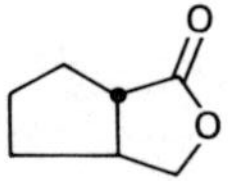	Trace	271
	CrO_3, $C_5H_5N^a$ $Na_2Cr_2O_7$, H_2SO_4, H_2O		Trace 60	271

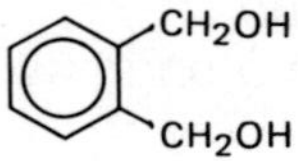

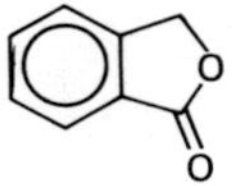

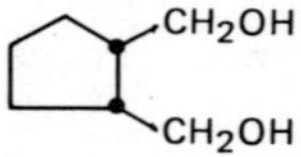

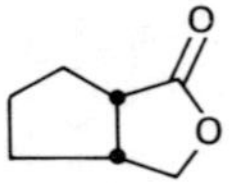

	Conditions		Yield (%)	Refs.
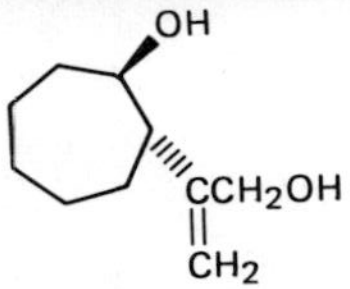	MnO_2, C_6H_6	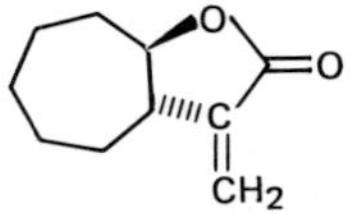	87	30, 31
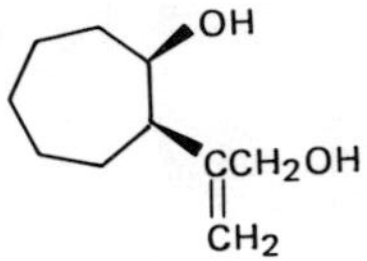	MnO_2, C_6H_6	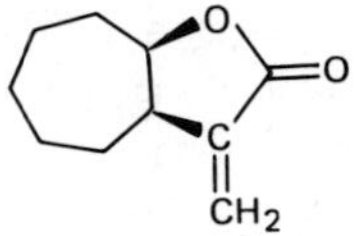	58	31
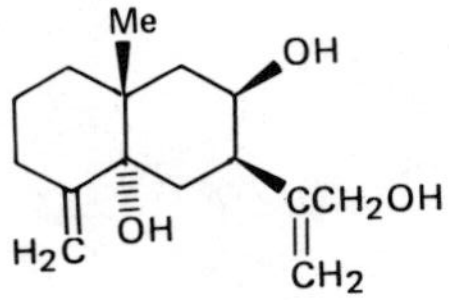	MnO_2, C_6H_6	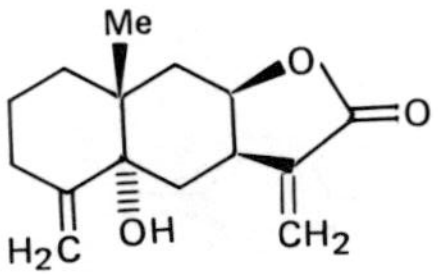	83	19
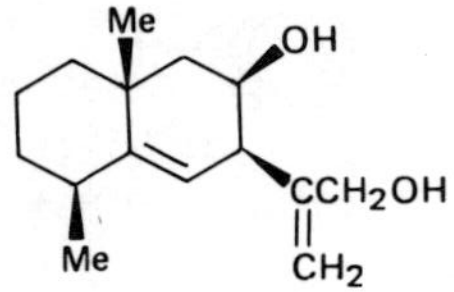	MnO_2, C_6H_6	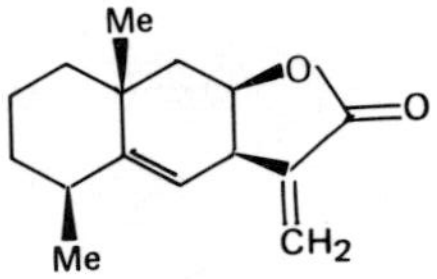	80	19
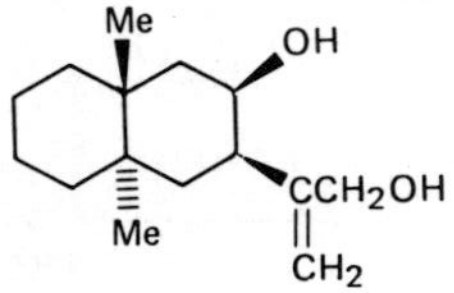	MnO_2, C_6H_6	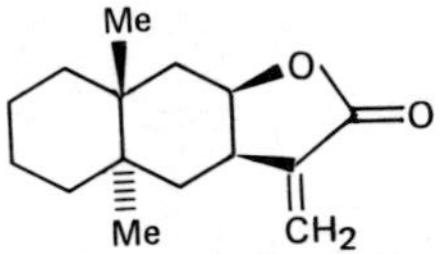	63	254
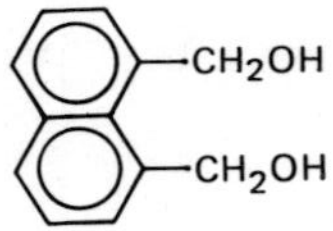	MnO_2, C_6H_6	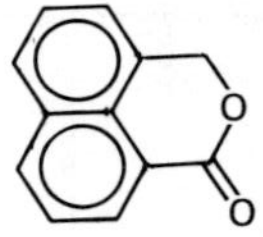	76	273

TABLE 21. (Continued)

Diol	Oxidizing agent	Product	Yield (%)	References
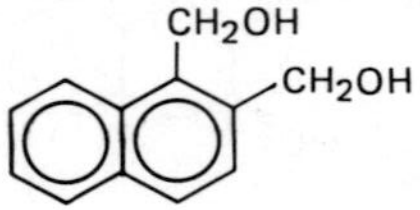	MnO_2, C_6H_6	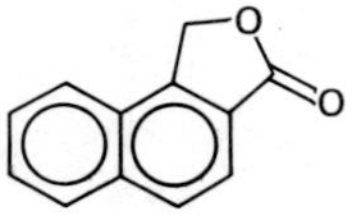	65	273
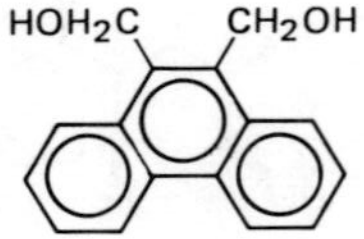	MnO_2, C_6H_6	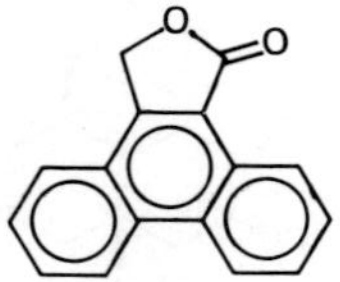	73	273
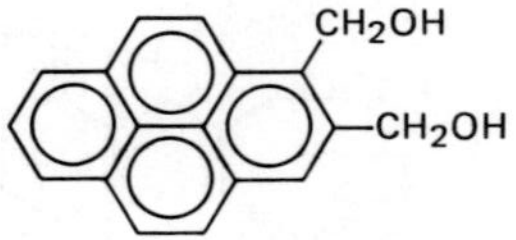	MnO_2, C_6H_6	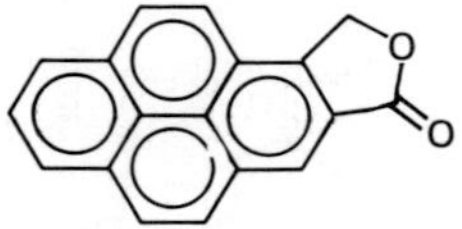	58	273
	CrO_3, $C_5H_5N^a$ (R = Me) MnO_2, MeCN (R = Me) MnO_2, MeCN (R = OH)		55 90 60	274

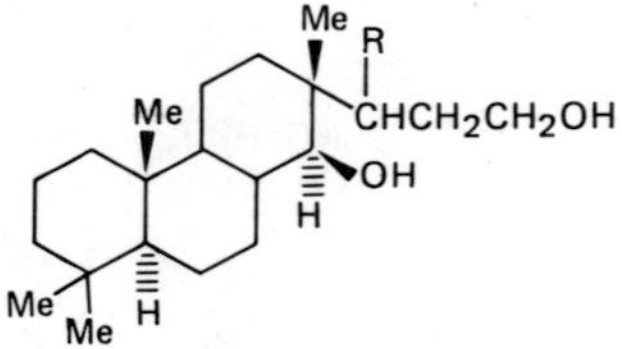

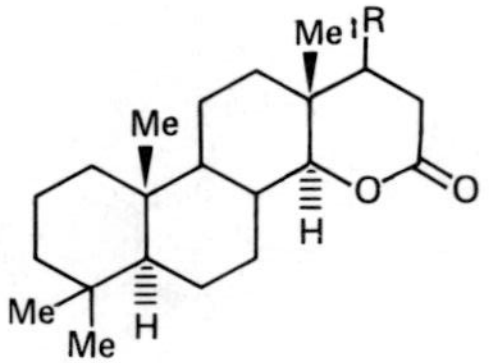

	Reagent		Yield (%)	Reference
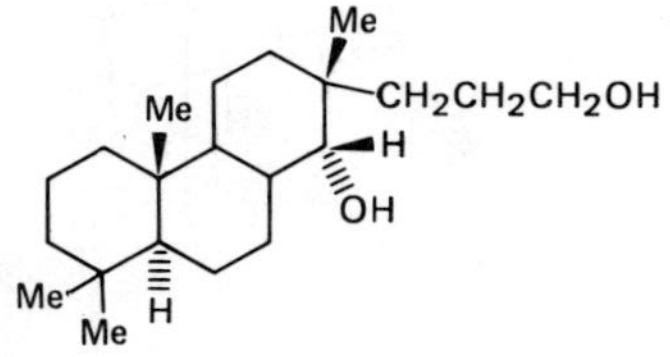 CH$_2$CH$_2$CH$_2$OH	CrO$_3$, C$_5$H$_5$N[a]	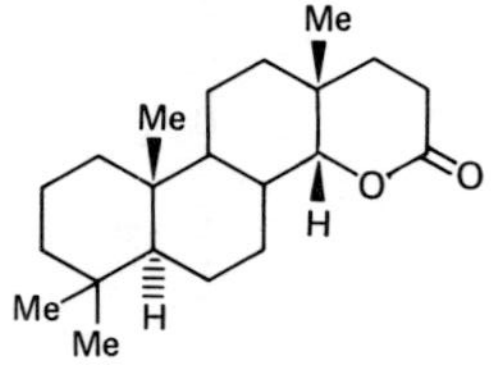	50	274
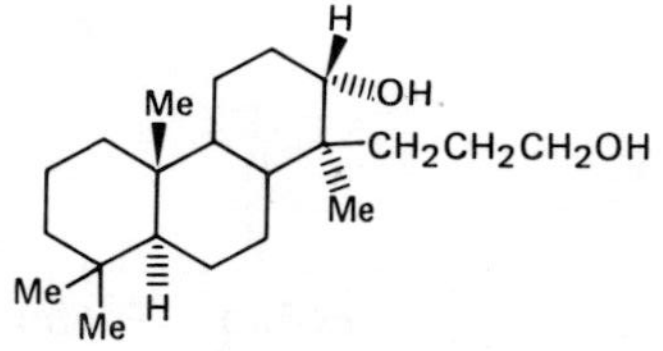 CH$_2$CH$_2$CH$_2$OH	MnO$_2$, C$_6$H$_6$	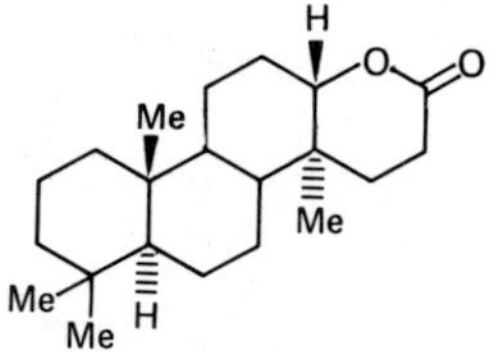	84	276
HOCH$_2$(CH$_2$)$_n$CH$_2$OH	Ag$_2$CO$_3$–celite, C$_6$H$_6$	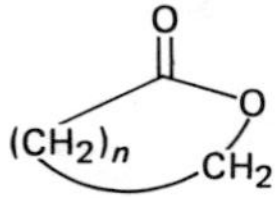	$n = 2$ 52 $n = 3$ 90–94 $n = 4$ 96–100	272, 277
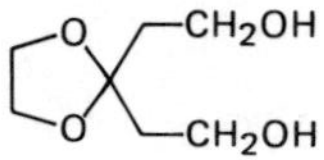 CH$_2$OH / CH$_2$OH	Ag$_2$CO$_3$–celite, C$_6$H$_6$	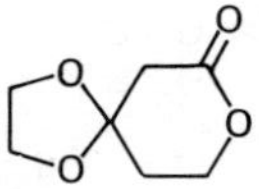	60–79	272, 277
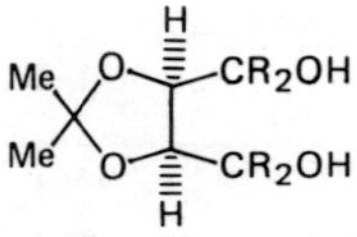 CR$_2$OH / CR$_2$OH R = H, D	Ag$_2$CO$_3$–celite, C$_6$H$_6$	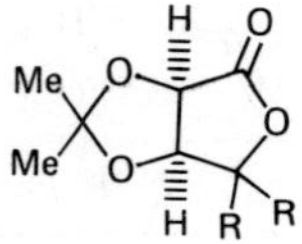	50–65	272, 277

TABLE 21. (Continued)

Diol	Oxidizing agent	Product	Yield (%)	References
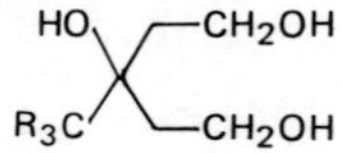	Ag_2CO_3—celite, C_6H_6	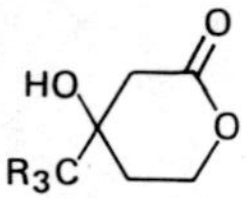	74 —	277
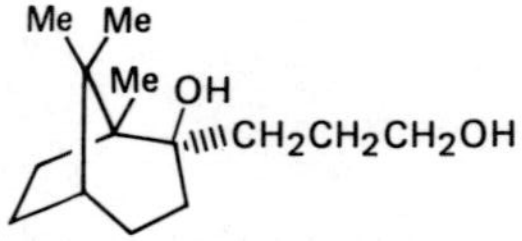	Ag_2CO_3—celite, C_6H_6	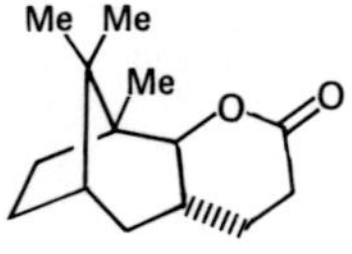 R = H R = D	~100	272
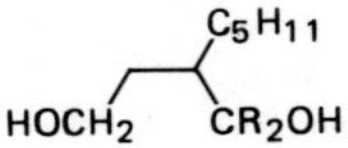	Ag_2CO_3—celite, C_6H_6	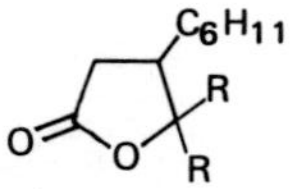 R = H R = D + R = H R = D	60 90 40 10	272
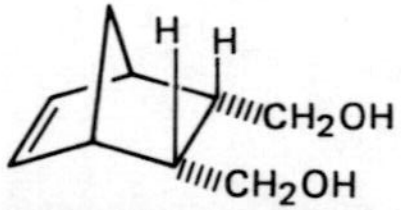	Ag_2CO_3—celite, C_6H_6	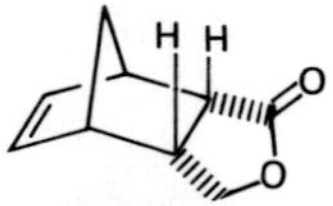	96	272

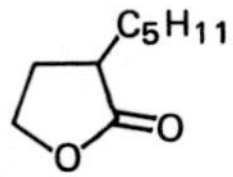

Substrate	Conditions	Product	Yield (%)	Ref.
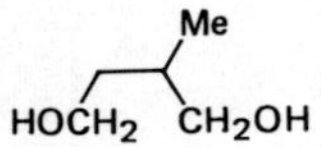 Me / HOCH$_2$—CH$_2$OH (2-methylbutane-1,4-diol)	Ag$_2$CO$_3$—celite, C$_6$H$_6$	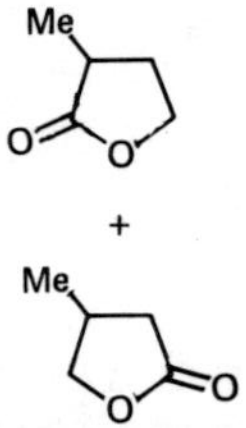 β-methyl-γ-butyrolactone + β-methyl-δ-valerolactone (two lactones, Me)	66	272
CH$_2$= / HOCH$_2$—CH$_2$OH	Ag$_2$CO$_3$—celite, C$_6$H$_6$	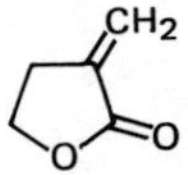 α-methylene lactone (=CH$_2$)	28	272
X(CH$_2$CH$_2$OH)$_2$	Ag$_2$CO$_3$—celite, C$_6$H$_6$	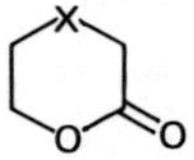 X-containing six-membered lactone	80; X = O 95, X = S 9	272
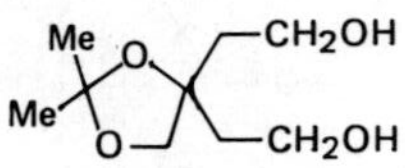 (2,2-dimethyl-1,3-dioxolane bis(CH$_2$OH), Me Me)	Ag$_2$CO$_3$—celite, C$_6$H$_6$	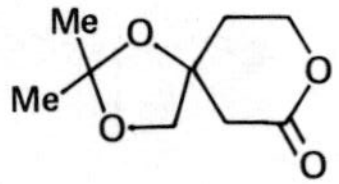 spiro dioxolane-lactone (Me, Me)	77	272
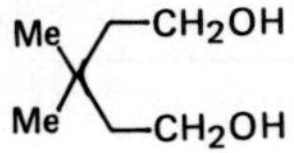 Me—C—Me with two CH$_2$OH	Ag$_2$CO$_3$—celite, C$_6$H$_6$	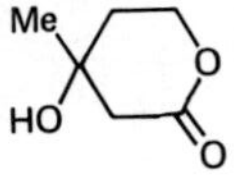 six-membered lactone (Me, HO)	74	272
Me—CHCH$_2$CH$_2$CH$_2$OH / OH	Ag$_2$CO$_3$—celite, C$_6$H$_6$; CHCl$_3$	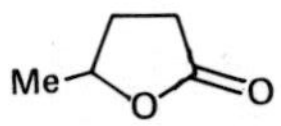 γ-methyl lactone (Me)	41, 72	272

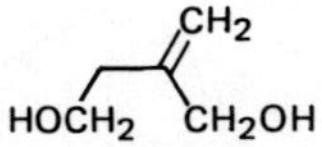

TABLE 21. (Continued)

Diol	Oxidizing agent	Product	Yield (%)	References
				272
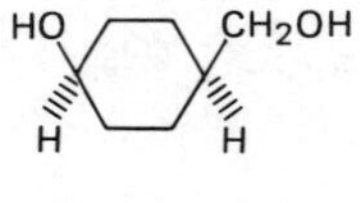	Ag$_2$CO$_3$—celite, C$_6$H$_6$ CHCl$_3$	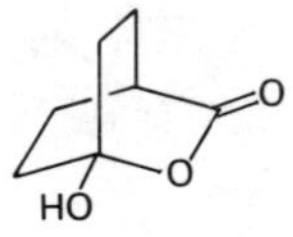	56 14	
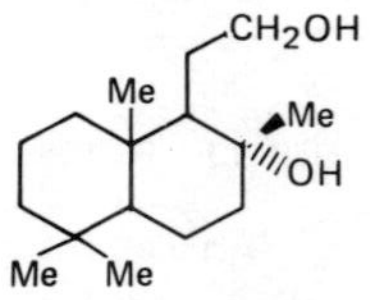	Ag$_2$CO$_3$—celite, C$_6$H$_6$	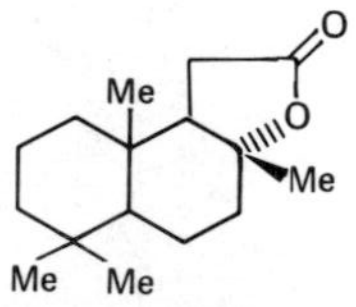	100	272

[a] Chromic anhydride—pyridine complex; see G. I. Poos, G. E. Arth, R. E. Beyler and L. H. Sarett, *J. Amer. Chem. Soc.*, **75**, 427 (1953).

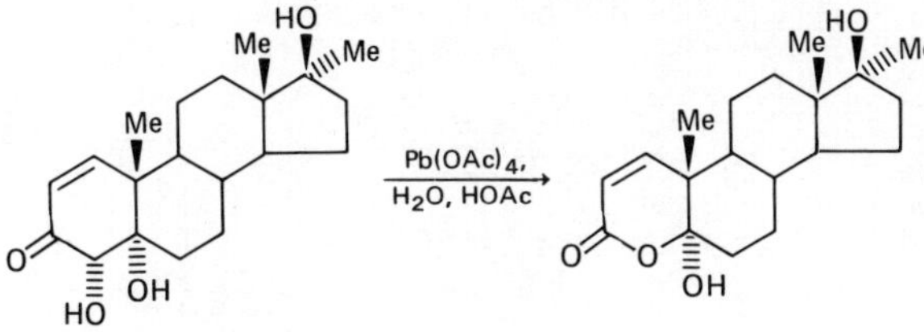

(Ref. 255) (140)

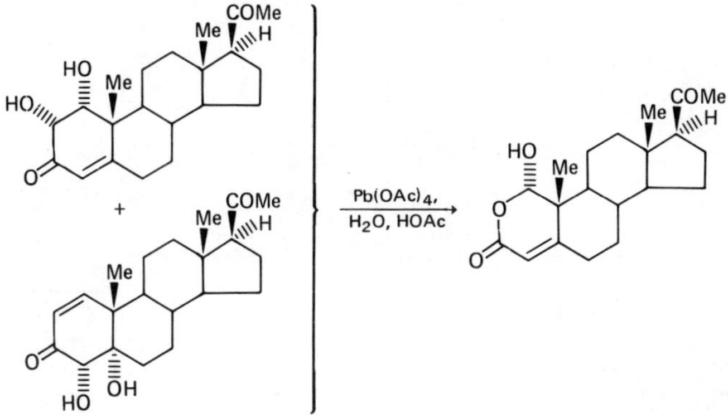

(Ref. 255) (141)

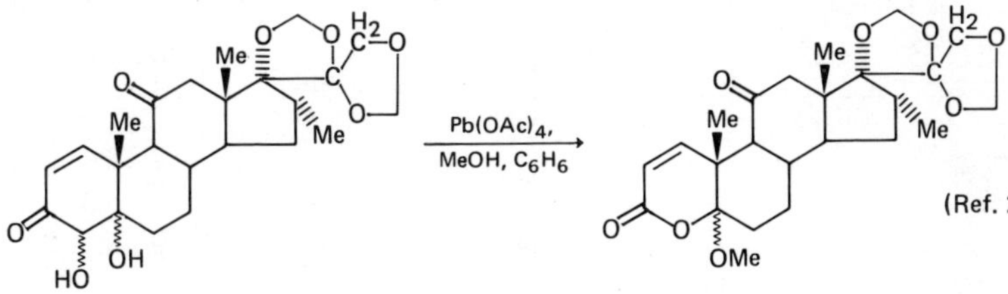

(Ref. 278) (142)

Steroidal δ-hydroxy oximes and lactols derived from the free δ-hydroxy aldehydes can be oxidized to lactones with sodium dichromate[256] or chromic anhydride (equation 143)[279].

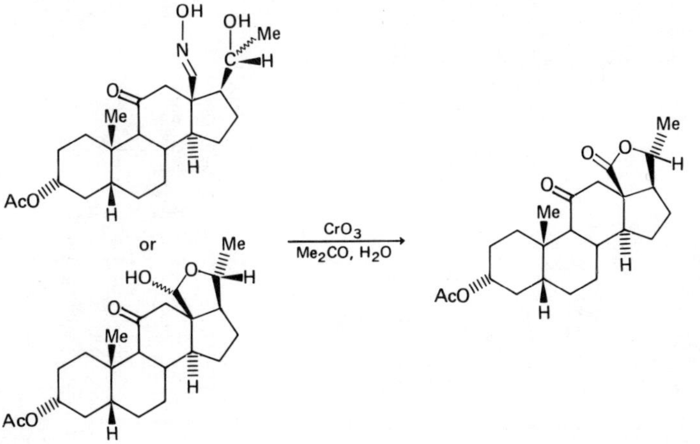

(Ref. 279) (143)

TABLE 22. Preparation of lactones via Baeyer–Villiger oxidation of ketones

Starting material	Reagent	Product	Yield (%)	Reference
	MeCO₃H, 25°C, 80 h.		50	282
	PhCO₃H, CHCl₃, H₂SO₄, 25°C 60 h 12 days 64 h. 30% H₂O₂, HOAc, 12 days	 R = OAc R = OAc R = H R = OAc	18 80 98 90 trace	268
	PhCO₃H, CHCl₃, H₂SO₄, HOAc		90	268

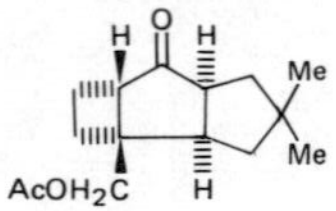

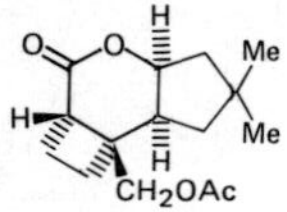

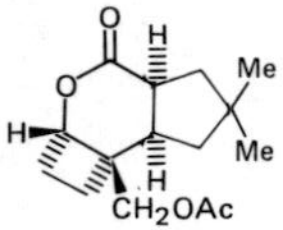

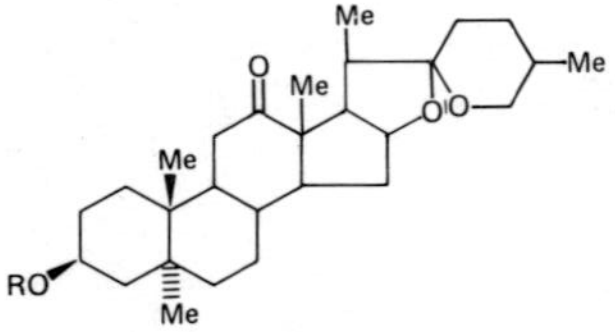

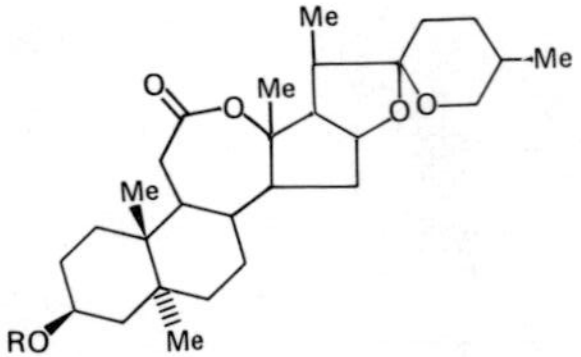

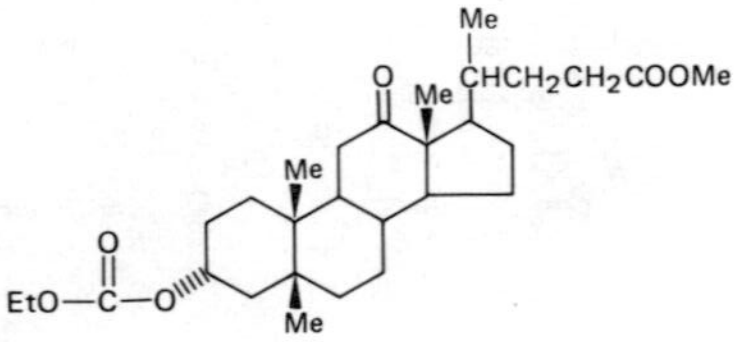

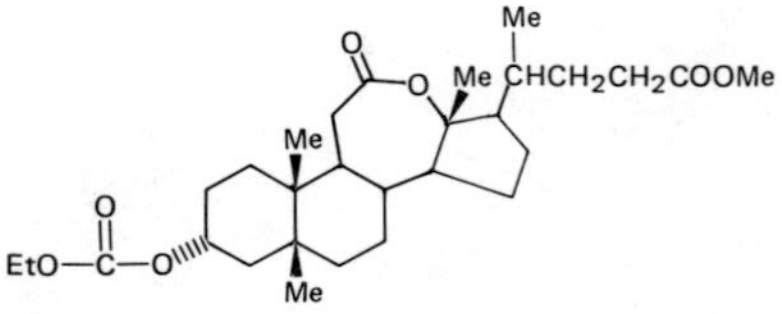

Reagents	Yield	Ref.
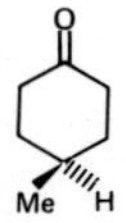 $(CF_3CO)_2O$, 90% H_2O_2	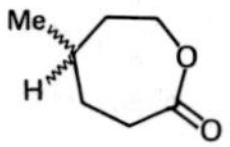 79.2	283
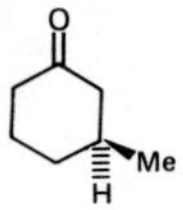 $(CF_3CO)_2O$, 90% H_2O_2	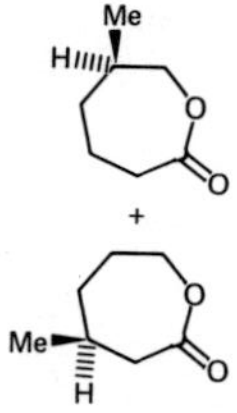 64.3	283
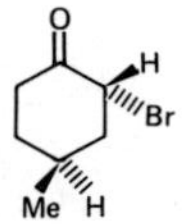 $(CF_3CO)_2O$, 90% H_2O_2, $CHCl_3$	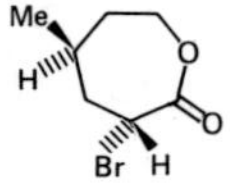 86.7	283
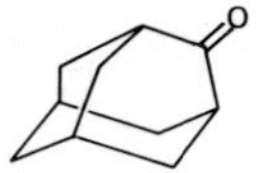 m-$ClC_6H_4CO_3H$ 30% H_2O_2, HOAc	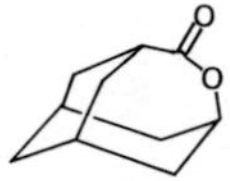 — 90–95	284 285
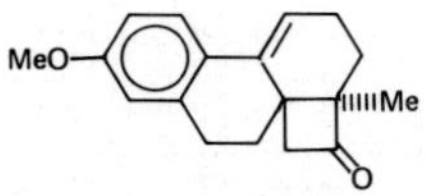 H_2O_2, OH^- or m-$ClC_6H_4CO_3H$, CH_2Cl_2	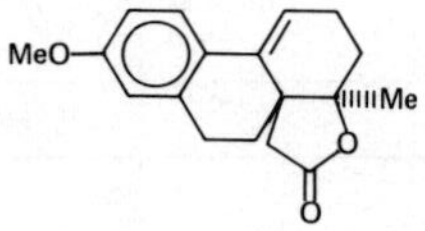 90	286

TABLE 22. (Continued)

Starting material	Reagent	Product	Yield (%)	Reference
	m-ClC$_6$H$_4$CO$_3$H, CH$_2$Cl$_2$			286
	40% MeCO$_3$H, NaOAc, CHCl$_3$	$n = 2$, R = H $n = 2$, R = OMe $n = 1$, R = H	68–72 – 80	289
	40% MeCO$_3$H, 2:3 H$_2$SO$_4$:HOAc MeCO$_3$H, HOAc, NaOAc		56	290
	40% MeCO$_3$H, HOAc, NaOAc		30 82	291
	40% MeCO$_3$H, HOAc, NaOAc		42	291
		(3:2)	35	291

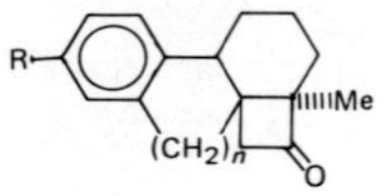

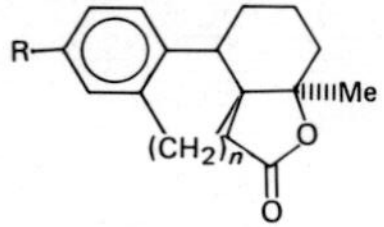

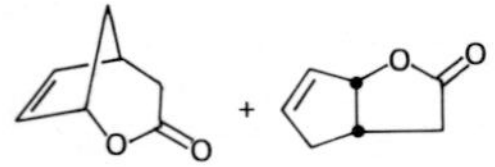

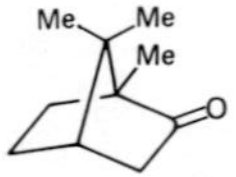

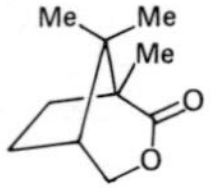

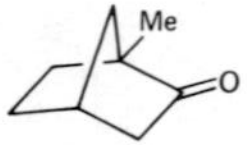

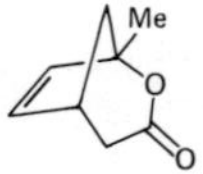

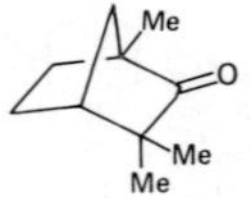

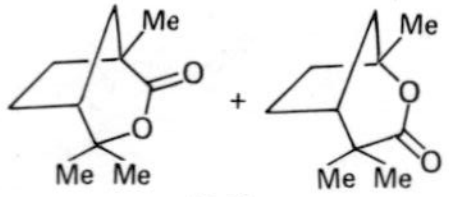

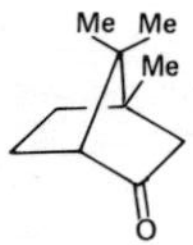

40% MeCO$_3$H, HOAc, NaOAc — 94 — 291

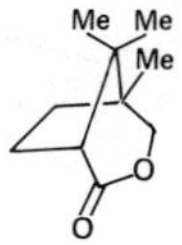

28% MeCO$_3$H, HOAc, NaOAc — 88 — 29
40% MeCO$_3$H, HOAc, H$_2$SO$_4$ — 97

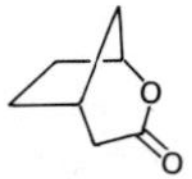

28% MeCO$_3$H, HOAc, NaOAc — 85 — 29

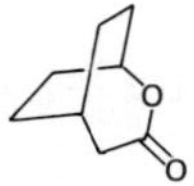

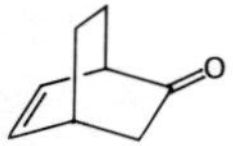

28% MeCO$_3$H, HOAc, NaOAc — 80 — 29

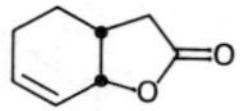

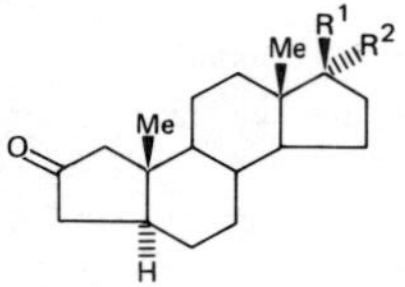

PhCO$_3$H, CHCl$_3$ — 70–90 — 293

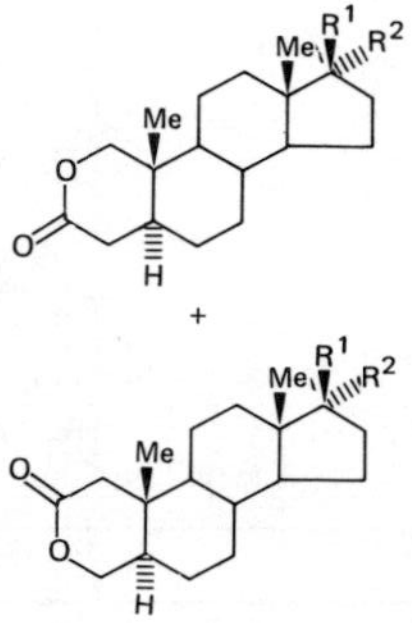

R^1 = C$_8$H$_{17}$, H, OH, OAc, OH
R^2 = H, H, H, H, Me

TABLE 22. (Continued)

Starting material	Reagent	Product	Yield (%)	Reference
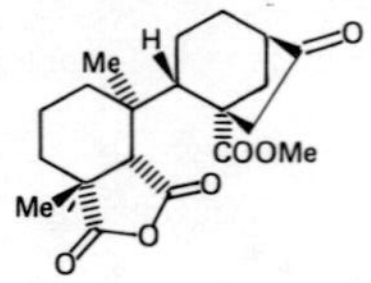 (steroid-type structure, H, Me, Me, COOMe)	$PhCO_3H$, $CHCl_3$, $p\text{-}MeC_6H_4SO_3H$	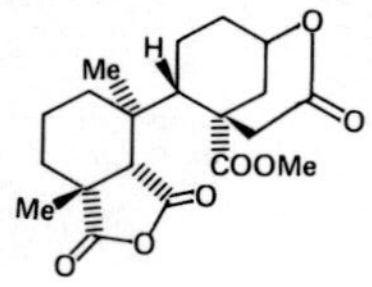 (lactone/anhydride structure, H, Me, Me, COOMe)	~60	248
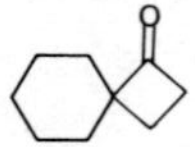 (spiro[3.5] cyclohexane butanone)	30% H_2O_2, MeOH, NaOH HOCl	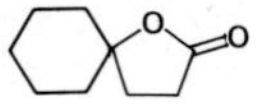 (lactone spiro structure)	82 90	294, 295 295
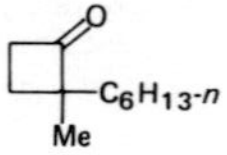 (cyclobutanone with $C_6H_{13}\text{-}n$, Me)	30% H_2O_2, MeOH, NaOH	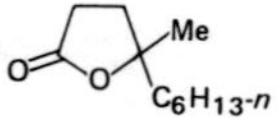 (γ-lactone with Me, $C_6H_{13}\text{-}n$)	100	294, 295
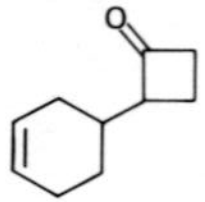 (cyclobutanone with cyclohexenyl)	30% H_2O_2, MeOH, NaOH	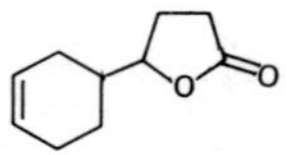 (lactone with cyclohexenyl)	82	294, 295
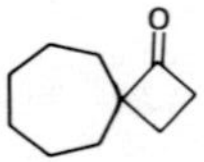 (spiro cycloheptane butanone)	30% H_2O_2, MeOH, NaOH NaOBr	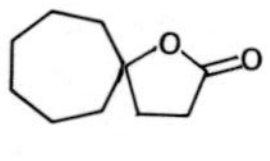 (lactone spiro structure)	94 94	294 295
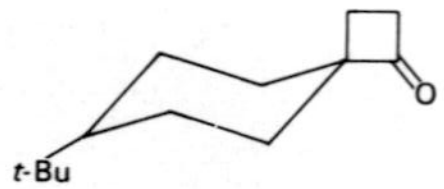 (t-Bu cyclohexyl spiro cyclobutanone)	30% H_2O_2, MeOH, NaOH	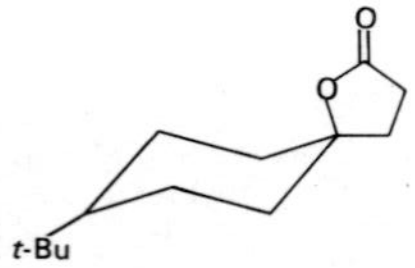 (t-Bu cyclohexyl lactone)	100	294

Substrate	Reagents	Products	Yield (%)	Reference
[structure: spiro cyclobutanone t-Bu cyclohexane], 70:30	30% H₂O₂, MeOH, NaOH	[structures: spiro lactone t-Bu cyclohexane]	100	295
[structure: cyclobutanone]	HOCl, pH 4[a] Me₃CCO₂H	[structure: γ-butyrolactone]	83 22	296 297
[structure: cyclohexanone]	H₂O₂-urea, 85% HCO₂H	[structure: caprolactone]	95	298
[structure: bicyclic ketone]	30% H₂O₂, HOAc	[structure: bicyclic lactone]	85–90	285
[structure: dimethyl bicyclic ketone, Me groups]	30% H₂O₂, HOAc	[structure: dimethyl bicyclic lactone, H, Me groups]	90–95	285
[structure: cage ketone]	30% H₂O₂, HOAc	[structures: two cage lactones]	80–85	285

TABLE 22. (Continued)

Starting material	Reagent	Product	Yield (%)	Reference
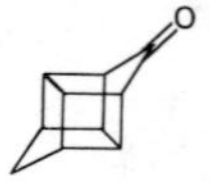	30% H_2O_2, HOAc	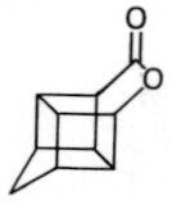	80–85	285
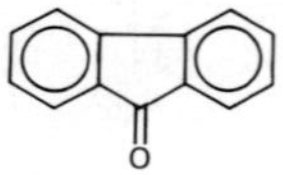	30% H_2O_2, HOAc	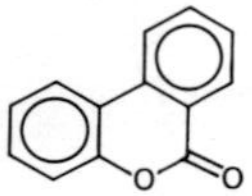	80–85	285
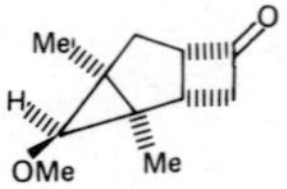	30% H_2O_2, HOAc	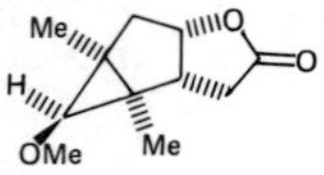	>90	299
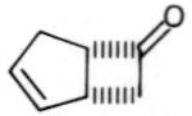	30% H_2O_2, HOAc, H_2O	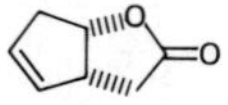	90	299, 300
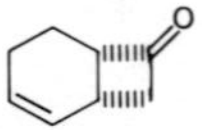	30% H_2O_2, HOAc, H_2O; 30% H_2O_2, MeOH, H_2O, OH^-	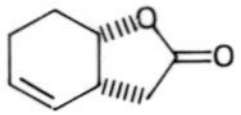	95	303
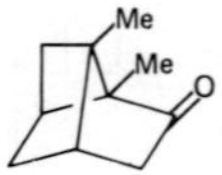	$PhCO_3H$, C_6H_6, p-$MeC_6H_4SO_3H$ or $PhCO_3H$, C_6H_6	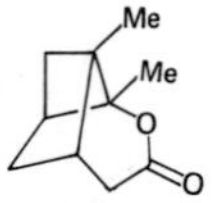 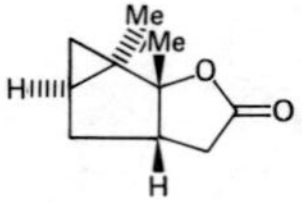 (4:1)	68 78	307

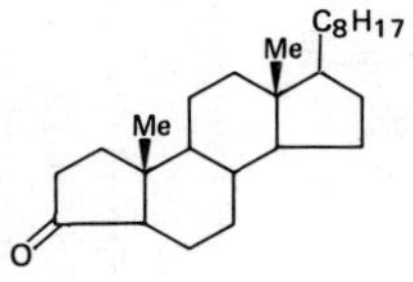

308

90

40

MeCO$_3$H, HOAc, p-MeC$_6$H$_4$SO$_3$H

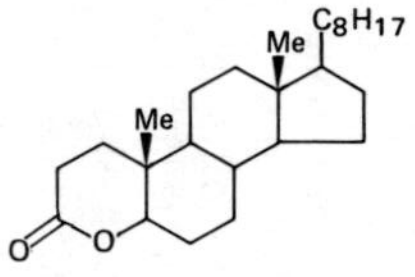

MeCO$_3$H, HOAc, H$_2$SO$_4$

[a] Using this reagent with cyclopentanone and cyclohexanone did not afford any lactone.

2. Oxidation of ketones

The Baeyer–Villiger[280,281] reaction remains the premier oxidative method for the preparation of lactones from cyclic ketones. The mechanism of this reaction has been reviewed in detail[281] and will not be discussed here. Table 22 contains a number of recent examples[282-308].

Oxygen or ozone have been used to convert ketones to lactones. For instance, reaction of cyclopentanone with oxygen in the presence of 1-benzyl-1,4-dihydronicotinamide has been reported[296] to afford a 12% yield of butyrolactone (equation 144). When similar reactions were conducted under nitrogen or in the absence of

$$\text{(144)}$$

the nicotinamide, no lactone was produced. These findings led the authors[296] to conclude that the dihydronicotinamide probably functions as an oxygen carrier, and is converted by oxygen into its hydroperoxide, which then produces the lactone via Baeyer–Villiger oxidation of the ketone.

Various ketones can be oxidized to lactones using potassium t-butoxide and atmospheric oxygen (equations 145–147)[306], Rose Bengal-sensitized photo-oxidation (equation 148)[307] or potassium t-butoxide and oxygen followed by reduction with sodium borohydride (equations 149–151)[308].

$$\text{(145)}$$

$$\text{(146)}$$

84%

$$\text{(147)}$$

Ozonolysis of silyloxyalkenes followed by treatment with sodium borohydride has also been reported[309] to afford lactones (equation 152).

Anodic oxidation of the sodium bisulphite addition products of cyclopentanone and cyclohexanone[310] afford mixtures of γ- and δ-lactones as shown in equations (153) and (154). Since the relative amounts of the lactones obtained by this

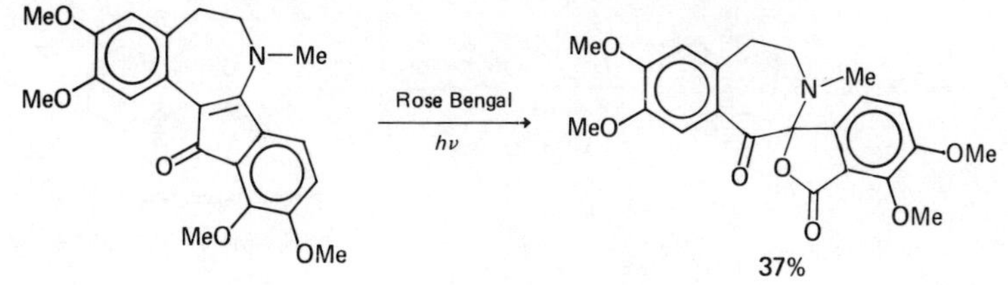
MeO
MeO
N—Me
O
MeO
OMe
Rose Bengal
hν
MeO
MeO
N—Me
O
O
O
OMe
OMe
37%

(148)

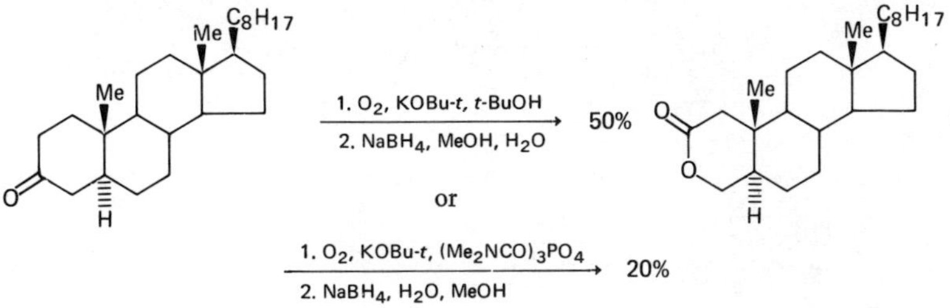
Me
C8H17
Me
O
H
1. O2, KOBu-t, t-BuOH
2. NaBH4, MeOH, H2O
50%
or
1. O2, KOBu-t, (Me2NCO)3PO4
2. NaBH4, H2O, MeOH
20%

(149)

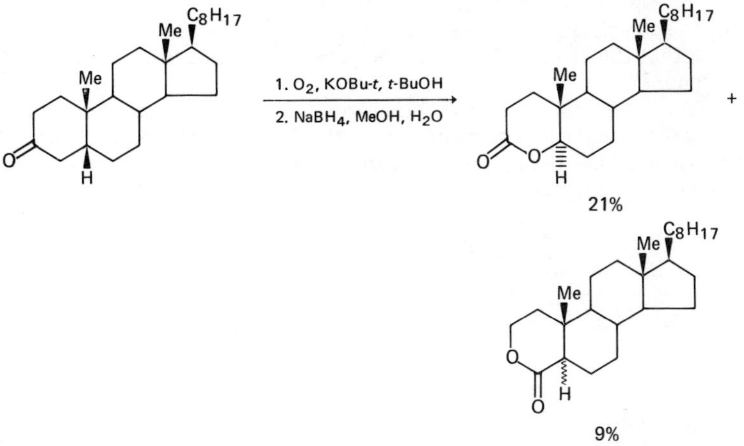
Me
C8H17
Me
O
H
1. O2, KOBu-t, t-BuOH
2. NaBH4, MeOH, H2O
+
21%
9%

(150)

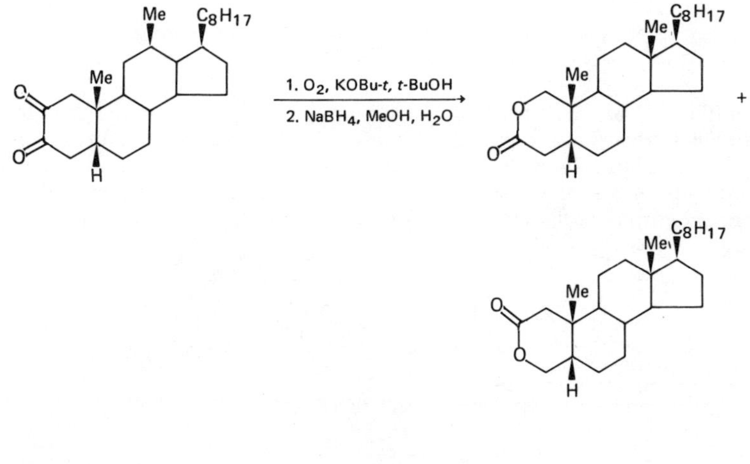
Me
C8H17
Me
O
O
H
1. O2, KOBu-t, t-BuOH
2. NaBH4, MeOH, H2O
+

(151)

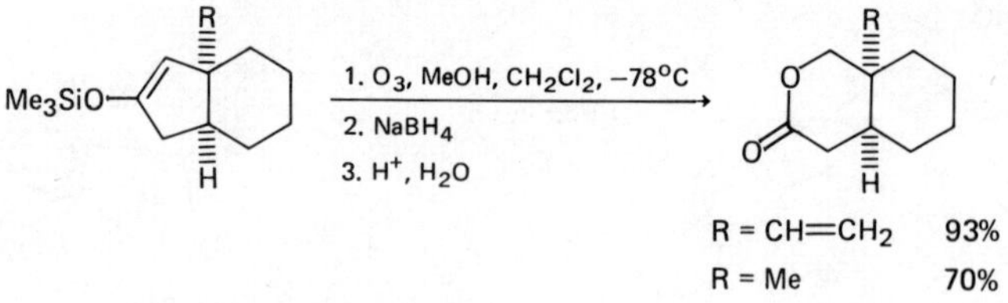

$$\text{(152)}$$

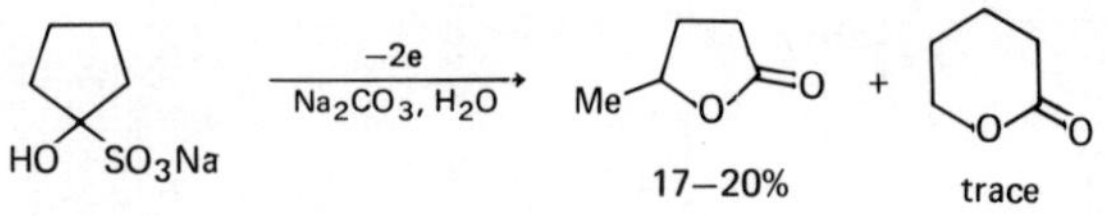

R = CH=CH₂ → R = CH$=$CH$_2$ 93%
R = Me 70%

$$\text{(153)}$$

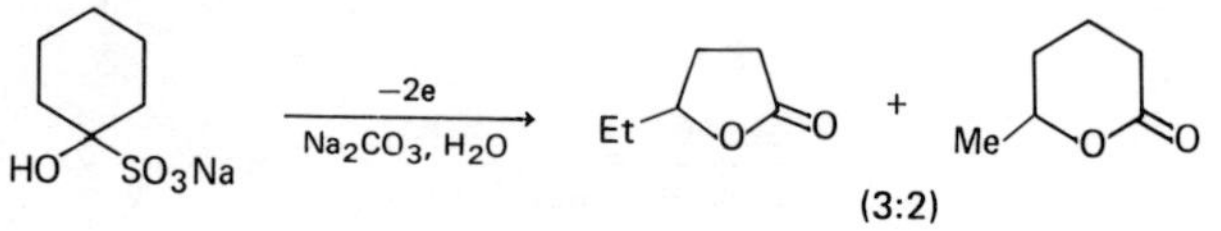

17–20%   trace

$$\text{(154)}$$

(3:2)

method correspond to the relative proportions of the same lactones obtained by acid-catalysed cyclization of 5-hexenoic acid[311,312], the authors find it reasonable to assume that the electrolytic oxidation proceeds via a carbonium ion or oxonium intermediate[310].

Oxidation of 2-adamantanone with ceric ammonium nitrate in aqueous acetonitrile at 60°C has been reported[284] to afford a 73% yield of the corresponding lactone, while similar oxidation of 2-adamantanol gave[284] the same lactone in 50% yield (equation 155).

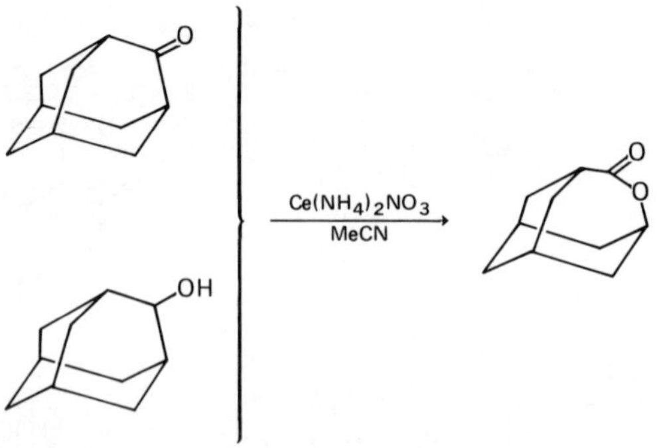

$$\text{(155)}$$

Addition of aqueous methanolic sodium periodate to a crude sample of the hydroxymethylene ketone shown in equation (156) effected[313] a direct conversion to $R(-)$-mevalonolactone, since the acetal group was hydrolysed during isolation of the product. In a similar manner[313] $S(+)$-mevalonolactone was prepared from the analogous hydroxymethylene ketone precursor.

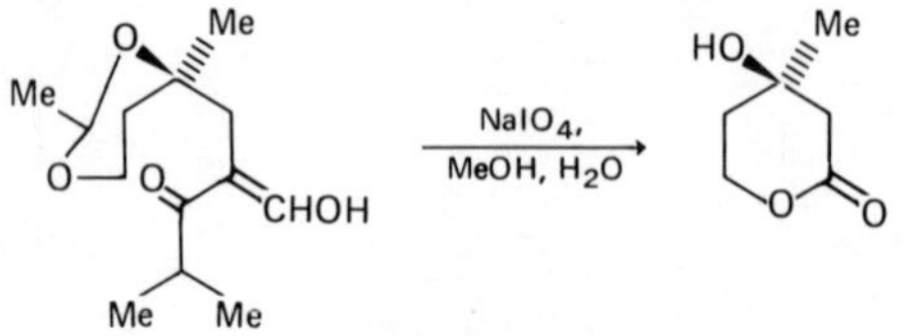

$$\text{(156)}$$

3. Oxidation of ethers

The oxidative conversion of cyclic esters to lactones is not a commonly encountered synthetic procedure; however, it has been found to be useful in several cases, and should not be ignored.

Ruthenium tetroxide has been reported[314] to oxidize tetrahydrofuran to γ-butyrolactone, and tetrahydrofurfuryl alcohol to a coumpound tentatively identified as the corresponding aldehyde lactone. Attempts to convert ethylene oxide to an α-lactone with this reagent were unsuccessful[314].

t-Butyl chromate has been used[315] to obtain spiro lactones from spiroethers. Thus, reaction of 3β-acetoxy-2',3'α-tetrahydrofuran-2',-spiro-17(5-androstene) with t-butyl chromate under standard conditions[316] afforded[315] a 23% yield of 3-(3β-acetoxy-17β-hydroxy-7-oxo-5-androsten-17α-yl) propionic acid lactone (equation 157). Similar results[315] were obtained with the spiro ethers shown in equations (158) and (159).

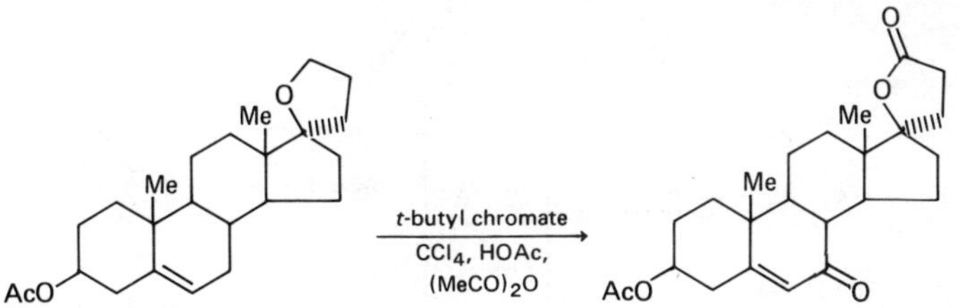

$$(157)$$

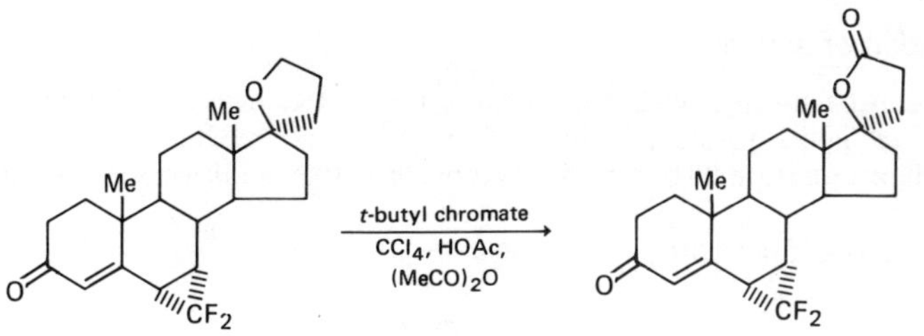

$$(158)$$

40%

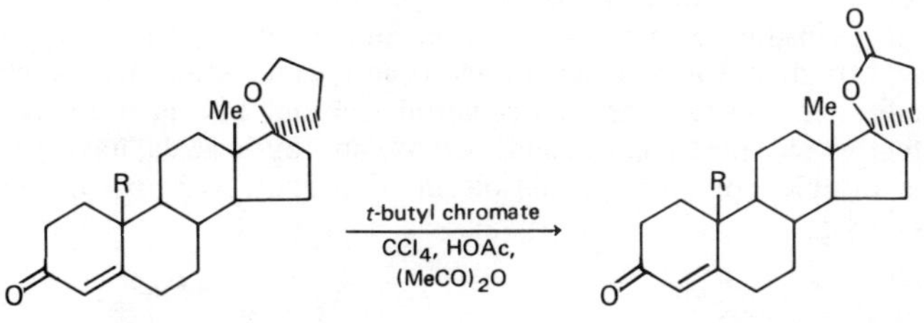

$$(159)$$

R = Me, 48%

R = H, 30%

Photosensitized oxygenation of furan and furan derivatives in the presence of an appropriate sensitizer such as Rose Bengal can be employed for the synthesis of certain butenolides (equations 160 and 161)[317-321].

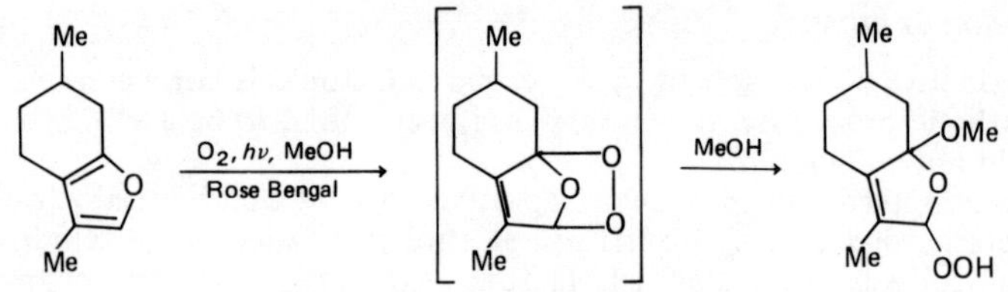

$$O_2, h\nu, MeOH \xrightarrow{\text{Rose Bengal}} \qquad \xrightarrow{MeOH}$$

(Ref. 317, 318)   (160)

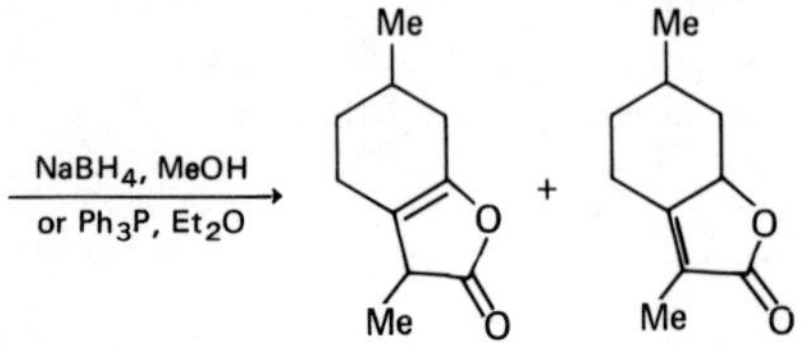

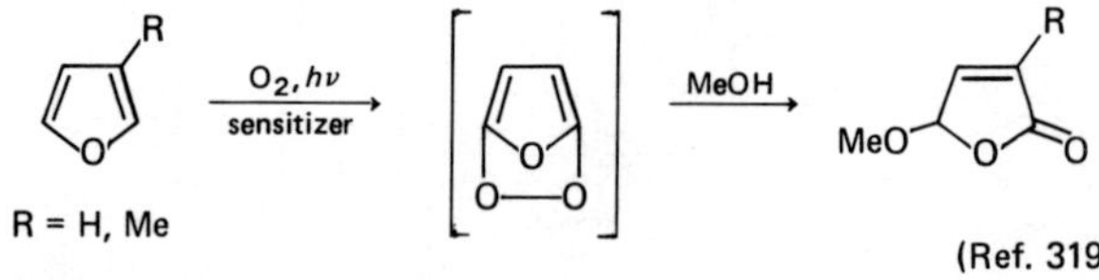

$$\xrightarrow{\text{NaBH}_4, \text{MeOH} \text{ or } \text{Ph}_3\text{P}, \text{Et}_2\text{O}}$$

R = H, Me $\xrightarrow{O_2, h\nu \atop \text{sensitizer}}$ $\xrightarrow{MeOH}$

(Ref. 319–321)   (161)

4. Oxidation of olefins

Oxidation of olefins with excess manganese (III) acetate affords γ-lactones in moderate to good yields (Table 23)[322,323]. The mechanism of this reaction, illustrated in equation (162) with styrene, involves addition of a carboxymethyl

$$PhCH{=}CH_2 + \overset{\cdot}{C}H_2COOH \longrightarrow Ph\overset{\cdot}{C}HCH_2CH_2COOH \xrightarrow{[O]} Ph\overset{+}{C}HCH_2CH_2COOH$$

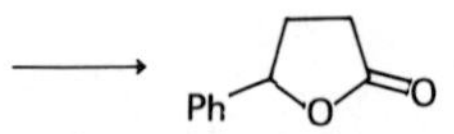

$$\longrightarrow \qquad \qquad (162)$$

radical to the double bond, oxidation of the resulting radical to a carbonium ion, and then ring-closure to form the lactone. Similar results have been observed with manganese dioxide in the presence of acetic anhydride and acetic acid[324].

Manganese (III) acetate, as well as certain cerium and vanadium salts, have been found effective in catalysing the addition of carboxylic acids, having an α-hydrogen across the double bond of various olefins (equation 163) to produce γ-lactones (Table 24)[325].

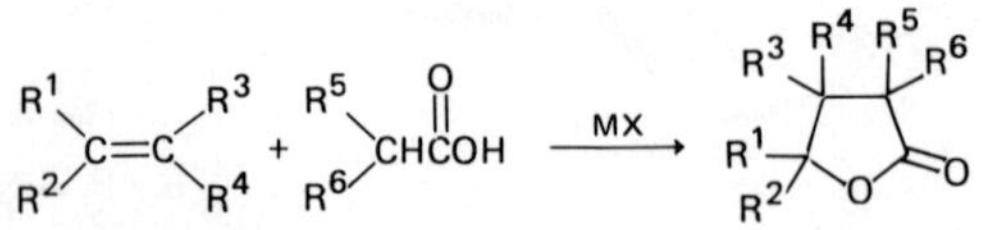

$$\underset{R^2}{\overset{R^1}{>}}C{=}C\underset{R^4}{\overset{R^3}{<}} + \underset{R^6}{\overset{R^5}{>}}CH\overset{O}{\overset{\|}{C}}OH \xrightarrow{MX} \qquad \qquad (163)$$

MX = Mn(OAc)$_3$ · 2H$_2$O, Mn(OAc)$_3$, MnO$_2$, Mn$_2$O$_3$, Ce(OAc)$_4$, Ce(NH$_4$)$_2$(NO$_3$)$_6$, NH$_4$VO$_3$

TABLE 23. Oxidation of olefins to lactones by manganese (III) acetate

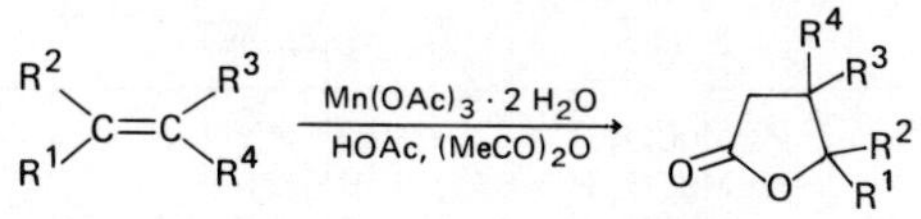

R^1	R^2	R^3	R^4	Yield (%)	References
Ph	H	H	H	75, 60	322, 323
Ph	Me	H	H	83, 74	322, 323
Ph	H	Me	H	21^a, 79	322, 323
PhCH$_2$	H	H	H	16^a	322
Ph	H	H	Ph	20^a	322
Ph	H	Ph	H	16^b	323
Me$_3$C	H	H	H	12^a	322
H	—(CH$_2$)$_4$—		H	10^a	322
n-C$_6$H$_{13}$	H	H	H	74	323
n-Pr	H	n-Pr	H	44^c	323
H	—(CH$_2$)$_6$—		H	62	323

^{a}Yields were not maximized.
^{b}Only one isomer was obtained (presumably *trans*).
^{c}Two isomers in the ratio of 5:1 were obtained.

Oxidation of olefins with lead tetracetate has been shown[326] to produce γ-lactones (equation 164), but yields are generally inferior to those obtained with manganese (III) acetate.

In a rather specialized example of olefin oxidation, *p*-nitroperbenzoic acid has been reported[327] to produce β-lactones from allylallenes (equation 165).

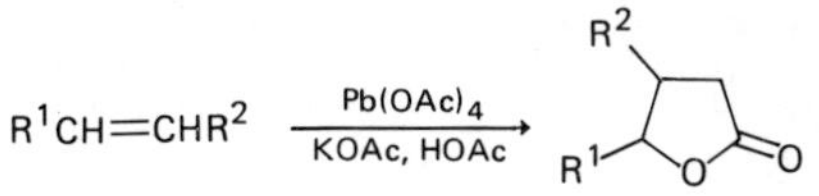

(164)

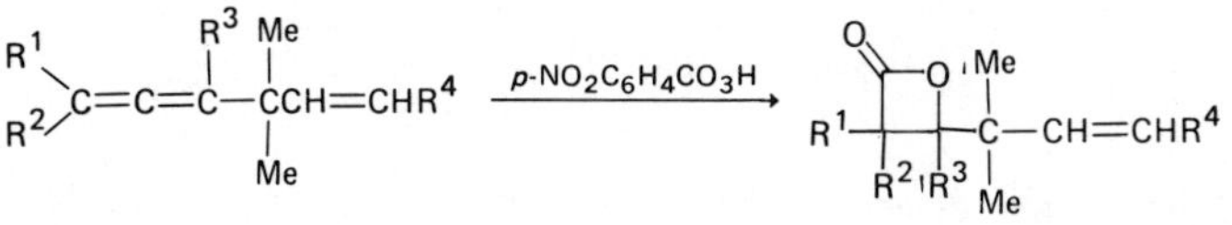

(165)

O. By Carbonylation Reactions

Unsaturated esters undergo carbonylation with carbon monoxide in the presence of hydrogen and dicobalt octacarbonyl to afford lactones (Table 25)[328,329]. These reactions are believed[328] to occur via hydroformylation of the double bond followed by cyclization of the intermediate hydroxy ester under the reaction conditions (equation 166).

Alkenyl and acetylenic alcohols are converted to lactones by carbonylation by nickel tetracarbonyl in the presence of aqueous acid[154,155,330] or dicobalt

TABLE 24. Preparation of γ-lactones by addition of carboxylic acids to olefins[325]

Olefin	Acid	Lactone[a]	Yield (%)
$C_6H_{13}CH=CH_2$	$MeCO_2H$	$R^1 = C_6H_{13}$	74
$PhCH=CH_2$	$MeCO_2H$	$R^1 = Ph$	60
$PhC(Me)=CH_2$	$MeCO_2H$	$R^1 = Ph, R^2 = Me$	74
$Me_2C=CH_2$	$MeCO_2H$	$R^1, R^2 = Me$	30
$Me_3C-CH=CH_2$	$MeCO_2H$	$R^1 = Me_3C$	48
$PrCH=CHPr$ (trans)	$MeCO_2H$	$R^1, R^4 = Pr$	44
$PhCH=CHPh$ (trans)	$MeCO_2H$	$R^1, R^4 = Ph$	16
$PhCH=CHMe$ (trans)	$MeCO_2H$	$R^1 = Ph, R^4 = Me$	79
Cyclooctene	$MeCO_2H$	$R^1, R^3 = -(CH_2)_6-$	62
$PhCH=CHCO_2Me$	$MeCO_2H$	$R^1 = Ph, R^4 = CO_2Me$	45
1,5-Hexadiene	$MeCO_2H$	$R^1 = CH_2=CH(CH_2)_2-$	24
1,7-Octadiene	$MeCO_2H$	$R^1 = CH_2=CH(CH_2)_4-$	26
Butadiene	$MeCO_2H$	$R^1 = CH_2=CH-$	30
Isoprene	$MeCO_2H$	$R^1 = CH_2=C(Me)-$	13
		+	
		$R^1 = CH_2=CH-, R^2 = Me$	37
$Me(CH_2)_4C\equiv CCH_2CH=CH_2$	$MeCO_2$	$R^1 = Me(CH_2)_4C\equiv CCH_2-$	50
$PhCH=CH_2$	$MeCH_2CO_2H$	$R^1 = Ph\ R^5 = Me$	50
$PhCH=CH_2$	$NCCH_2CO_2H$	$R^1 = Ph, R^5 = CN$	41
$C_6H_{13}CH=CH_2$	$NCCH_2CO_2H$	$R^1 = Ph, R^5 = CN$	60
$PhC(Me)=CH_2$	$NCCH_2CO_2H$	$R^1 = Ph, R^2 = Me, R^5 = CN$	43
4-Octene	$NCCH_2CO_2H$	$R^1, R^4 = Pr, R^5 = CN$	49
$PhCH=CHMe$	$NCCH_2CO_2H$	$R^1 = Ph, R^4 = Me, R^5 = CN$	51
Isoprene	$NCCH_2CO_2H$	$R^1 = CH_2=C(Me)-, R^5 = CN +$	5
		$R^1 = CH_2=CH, R^2 = Me, R^5 = CN$	39
$C_6H_{13}CH=CH_2$	$(CH_2CO_2H)_2$	$R^2 = C_6H_{13}, R^5 = CH_2CO_2H$	25

[a]Where not specified R = H.

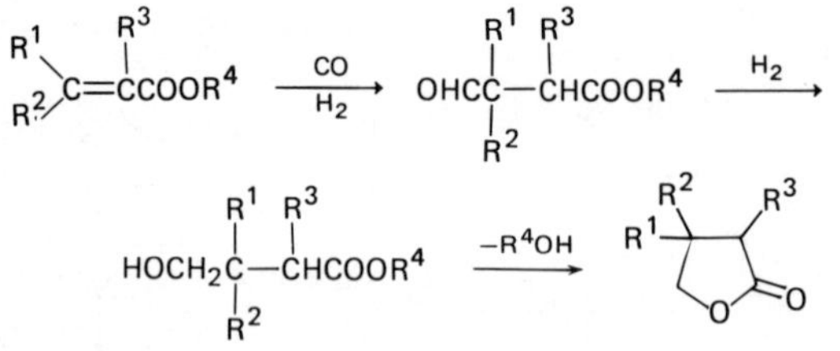

$$\tag{166}$$

octacarbonyl in the presence of carbon monoxide and hydrogen[331] (Table 26). Lactone formation in these cases may be viewed as proceeding by hydrocarboxylation of the unsaturated function with subsequent cyclization of an intermediate hydroxy acid.

Reaction of certain diols and dienes with carbon monoxide or formic acid and a strong mineral acid in the presence of Group IB metal compounds results in Koch–Haaf[332,333] hydrocarboxylation followed by ring-closure to form lactones[334]. As may be seen from equations (167) and (168), these reactions are accompanied by deep-seated carbonium ion rearrangements.

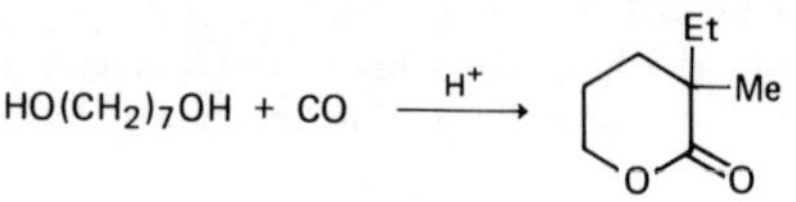

$$HO(CH_2)_7OH + CO \xrightarrow{H^+} \qquad \tag{167}$$

TABLE 25. Preparation of lactones by carbonylation of unsaturated esters with CO, H_2 and $Co_2(CO)_8$ at 200–350°C

Starting material	Product	Yield (%)	Reference
$H_2C=CCO_2R^2$ with R^1 $R^1 = H$, $R^2 = Me$ $R^1 = H$, $R^2 = Et$ $R^1 = R^2 = Me$		69 88 51	328
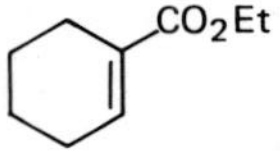 (CO2Et cyclohexene)	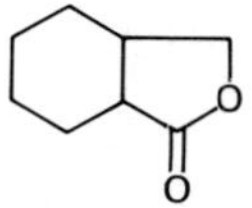	23	328
MeCH=CHCO$_2$R R = Me R = Et		20 + 72 23 + 67	328
MeCH=CMeCO$_2$Et		31	328
		21	
		7	

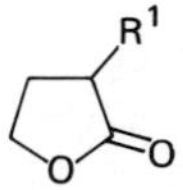

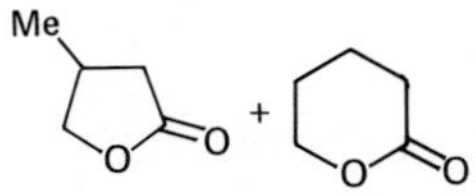

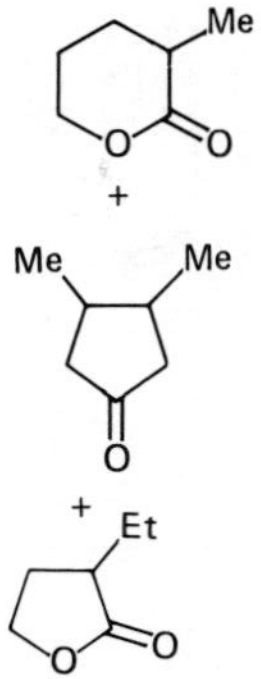

TABLE 25. (Continued)

Starting material	Product	Yield (%)	Reference
CH_2=CHCH$_2$CO$_2$Et	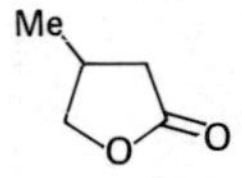	17	328
	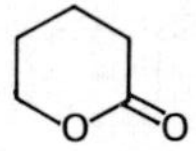	52	
Me$_2$C=CHCO$_2$Et	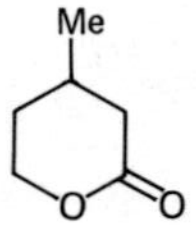	88	328, 329
	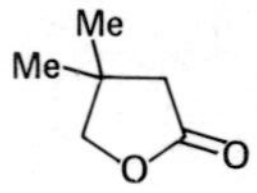	1	
Me$_2$C=CMeCO$_2$Et	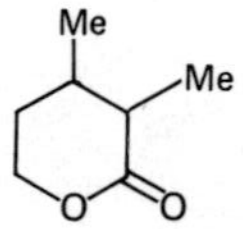	55	328
	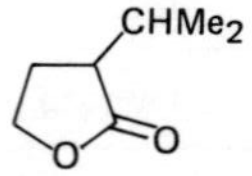	31	

$H_2C{=}CH{-}\overset{\displaystyle Me}{\underset{\displaystyle Me}{C}}{-}CO_2Et$		93 1	328
MeCH=CHCH=CHCO$_2$Et *cis, cis*		49 33	328
PhCH=CHCO$_2$Et		8.5–49	328, 329
EtO$_2$CCH=CHCO$_2$Et *cis* *trans*		47 49	328

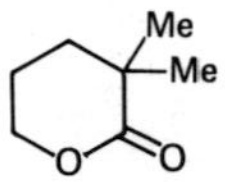

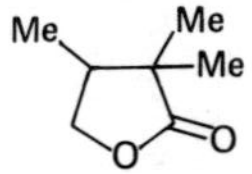

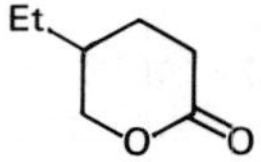

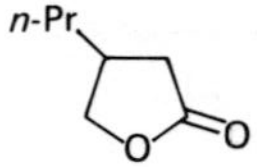

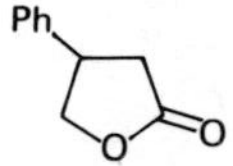

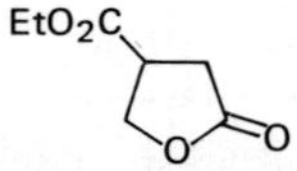

TABLE 26. Preparation of lactones by carbonylation of unsaturated alcohols

Starting material	Reagent[a]	Product	Yield (%)	Reference
$R^1CH=CR^2CH_2OH$	A	$R^1 = H, R^2 = Me$ $R^1 = Me, R^2 = H$	2 2	331
$H_2C=CR^1-\underset{\underset{R^3}{\mid}}{\overset{\overset{R^2}{\mid}}{C}}-CH_2OH$	A A A	$R^1 = H, R^2 = R^3 = Me$ $R^1 = R^2 = R^3 = Me$ $R^1 = H, R^2 = Me, R^3 = Et$	51 + 14 3 + 25 40 + 13	331
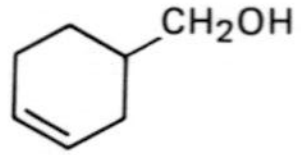	A	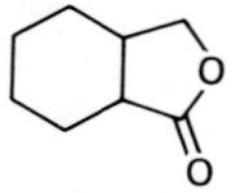	16	331
$H_2C=CHCH_2\underset{\underset{R}{\mid}}{C}HOH$				331

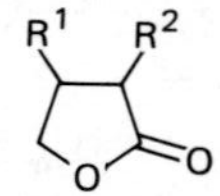

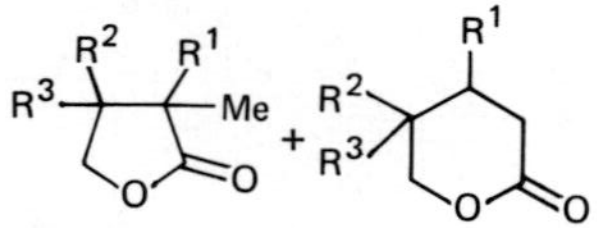

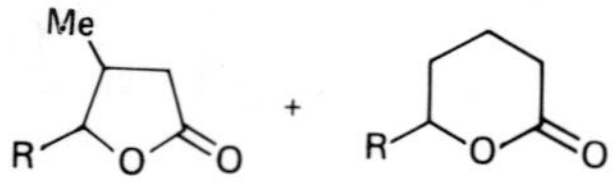

Substrate	Conditions	Substituents	Yield (%)	Reference
$H_2C{=}CHCH_2\overset{\underset{\textstyle Me}{\mid}}{\underset{}{C}}{\overset{\textstyle R}{\mid}}OH$	A A	R = Et R = n-Pr	2 + 73 2 + 0	331
$HOCCH_2C{\equiv}CH$ (with R^1, R^2)	A A A	R = Me R = Et R = i-Bu	10 + 2 29 + 6 10 + 2	330 330 330
	B or C B or C B D	$R^1 = R^2 = H$ $R^1 = H,\ R^2 = Me$ $R^1 = R^2 = Me$ $R^1 = H,\ R^2 = Ph$	23 30–50 10 44	330 330 330 154
$HOCH_2CH_2CH_2C{\equiv}CH$	B	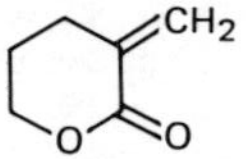	20	330

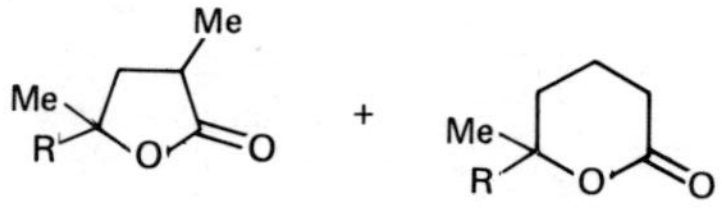

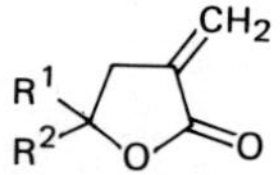

^{a} A = CO, H_2, $Co_2(CO)_8$, 200—350°C; B = $Ni(CO)_4$, HOAc, EtOH, H_2O, 80°C; C = $Ni(CO)_4$, MeOH, HCl; D = $Ni(CO)_4$, HOAc, EtOH, H_2O, hydroquinone.

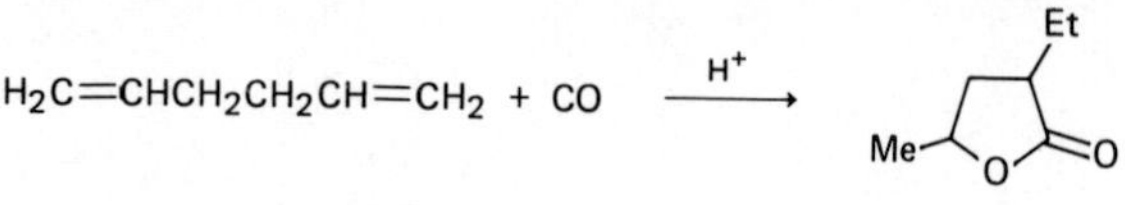

$$\text{H}_2\text{C}\!=\!\text{CHCH}_2\text{CH}_2\text{CH}\!=\!\text{CH}_2 + \text{CO} \xrightarrow{\text{H}^+} \quad \quad \quad \quad \quad \quad \quad (168)$$

$$\text{RCOCl} + \text{HC}\!\equiv\!\text{CH} \xrightarrow[\text{H}_2\text{O, X}^-]{\text{Ni(CO)}_4} \quad \quad \quad \quad \quad (169)$$

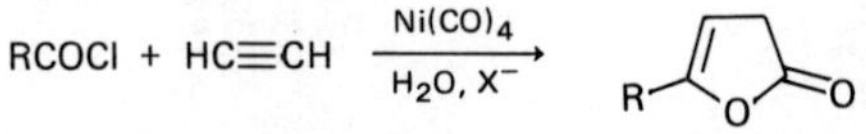

A recent publication[335] describes the synthesis of unsaturated butyrolactones by reaction of acetylenes with acyl chlorides in the presence of nickel tetracarbonyl and halide ion (equation 169).

P. By Cycloaddition of Nitrones to Olefins

An interesting general method for the preparation of γ-lactones from olefins involves initial silver ion-induced addition of N-cyclohexyl-α-chloroaldonitrones to olefins to produce the $(2\pi + 4\pi)$ cycloadduct, which is then treated with base and hydrolysed (equation 170)[336,337].

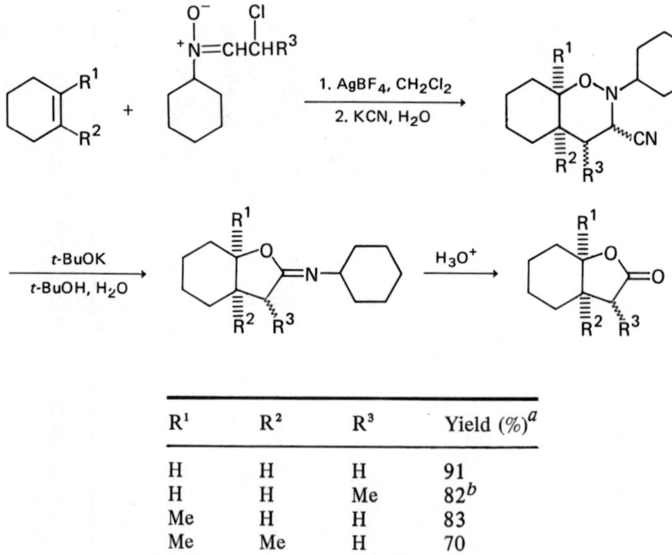

$$ \quad \quad \quad \quad \quad \quad \quad \quad (170)$$

R¹	R²	R³	Yield (%)[a]
H	H	H	91
H	H	Me	82[b]
Me	H	H	83
Me	Me	H	70

[a]Yields reported are only for the hydrolysis (last) step.
[b]Diastereomeric mixture, $\alpha:\beta \approx 4:1$.

Use of the diastereomeric 2-butenes in this reaction (equations 171 and 172)[337] showed the addition to be a stereospecific *cis* process. The reaction may also be performed using N-cyclohexyl-α-chloroethanaldonitrone (equations 173 and 174)[336], N-(t-butyl)-α-chloroethanaldonitrone (equation 175)[336] and N-cyclohexyl-α,β-dichloropropionaldonitrone (equation 176)[337].

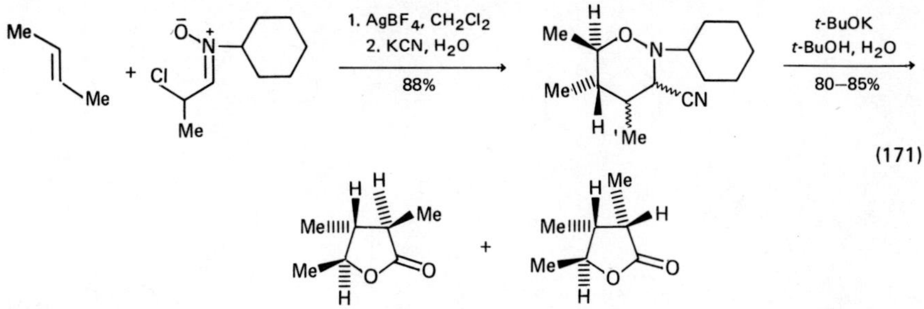

(171)

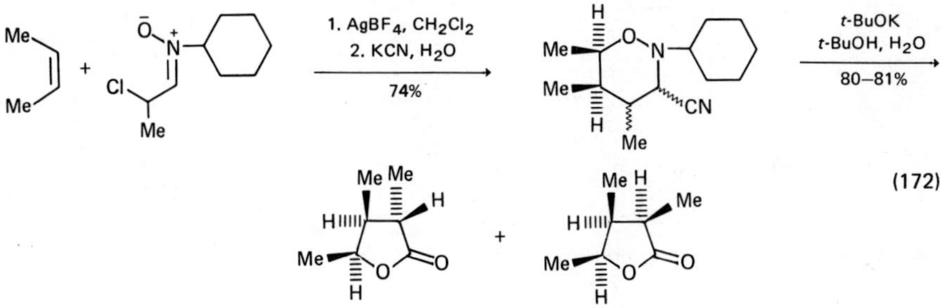

(172)

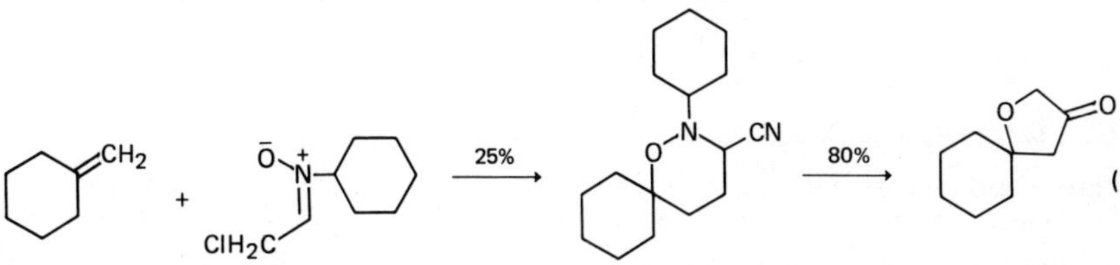

(173)

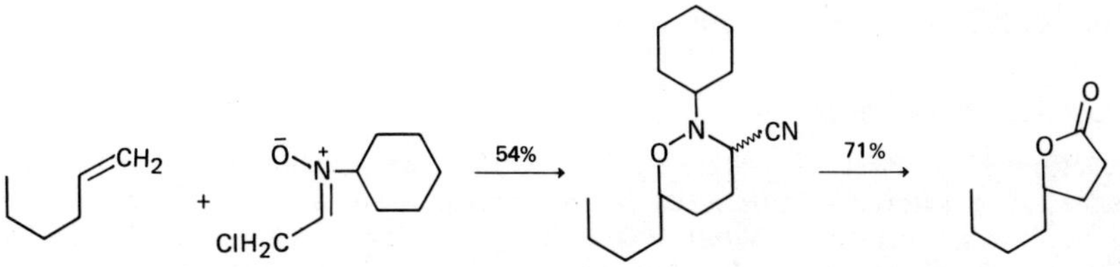

(174)

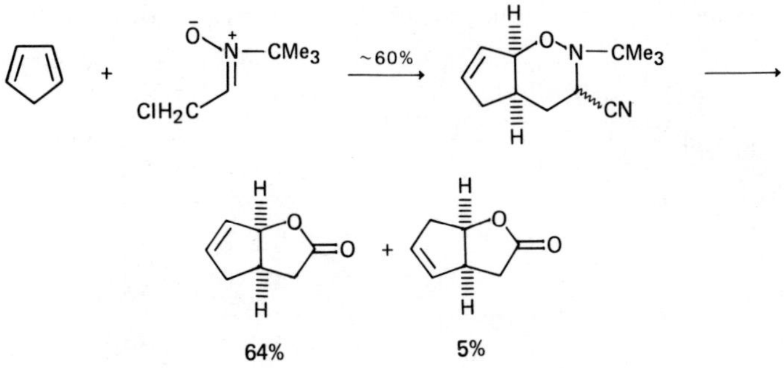

(175)

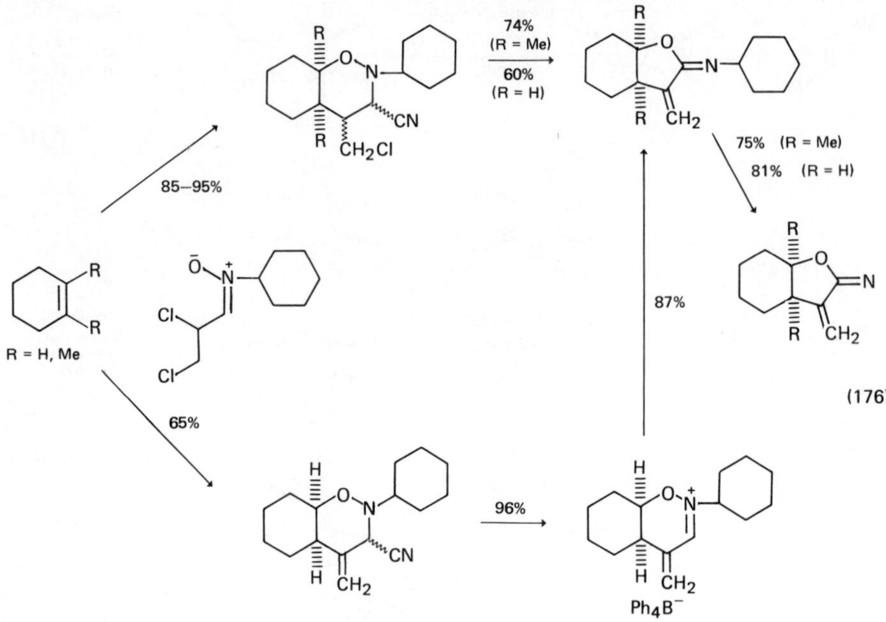

(176)

Q. By Rearrangement Reactions

This section deals with lactone preparations by Claisen, carbonium ion and photochemical rearrangements. The Baeyer–Villiger reaction and certain lactone interconversions, which might also be regarded as rearrangements, are discussed in Sections II. N.2. and II.R, respectively.

1. Claisen rearrangements

Reaction of a series of 2-alkene-1,4-diols with orthocarboxylic esters in the presence of a catalytic amount of hydroquinone or phenol results[338] in the formation of various β-vinyl-γ-butyrolactones via a Claisen rearrangement (Table 27). The proposed mechanism, illustrated in equation (177) involves an exchange of the alkoxy group of the *ortho* ester with the diol, followed by elimination of ethanol to produce a mixed ketene acetal. Rearrangement of this intermediate to a

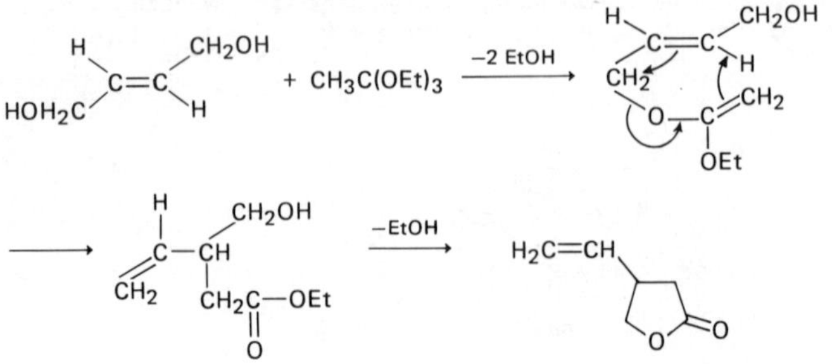

(177)

TABLE 27. γ-Lactones by reaction of *ortho* esters $RCH_2C(OEt)_3$ with unsaturated 1,4-diols[338]

Diol	R	Product	Yield (%)
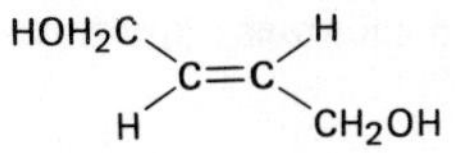 HOH₂C, H / C=C / H, CH₂OH	H	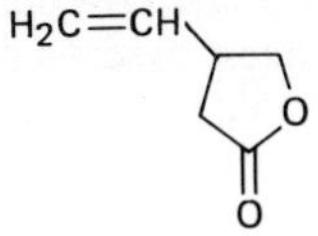 H₂C=CH lactone	89
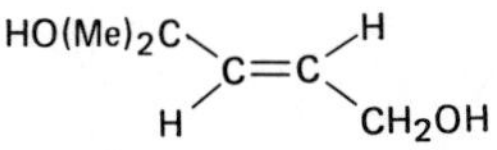 HO(Me)₂C, H / C=C / H, CH₂OH	H	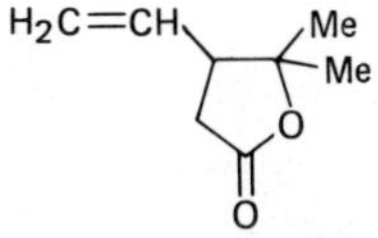 H₂C=CH, Me, Me lactone	91
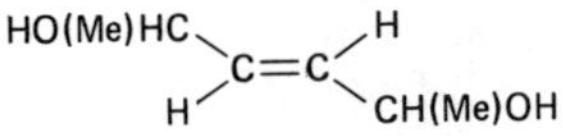 HO(Me)HC, H / C=C / H, CH(Me)OH	H	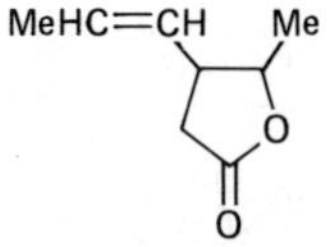 MeHC=CH Me lactone *cis–trans* mixture	52
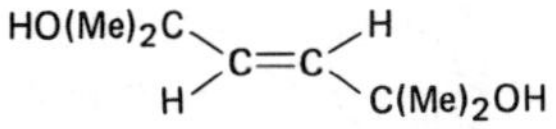 HO(Me)₂C, H / C=C / H, C(Me)₂OH	H	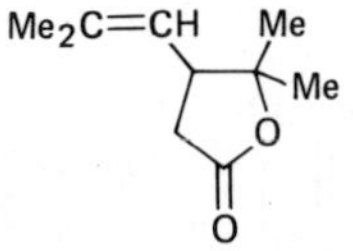 Me₂C=CH Me, Me lactone pyrocin	70
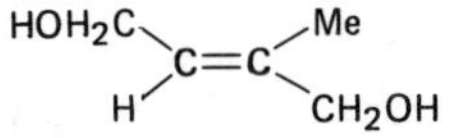 HOH₂C, Me / C=C / H, CH₂OH	H	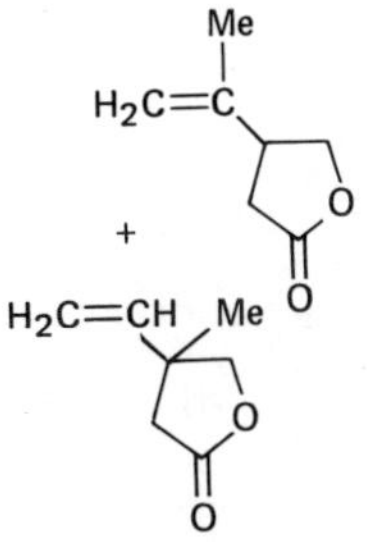 Me, H₂C=C lactone + H₂C=CH Me lactone	81 (ratio 6:4)
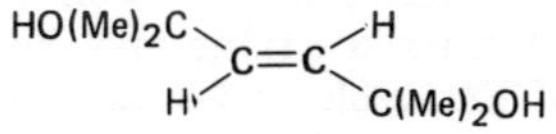 HO(Me)₂C, H / C=C / H, C(Me)₂OH	Me	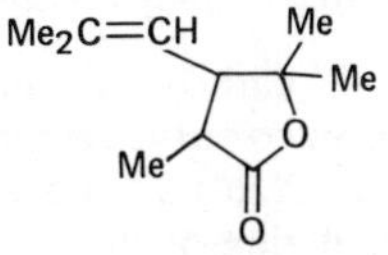 Me₂C=CH Me, Me, Me lactone *cis-trans* mixture	60

β-vinyl-γ-hydroxy carboxylic ester and lactonization under the conditions of the reaction affords the observed lactones. It should be noted that all of the entries in Table 27 are *trans* diols. With substituted *cis*-2-alkene-1,4-diols, γ-lactones were obtained in lower yields. For example, condensation of *cis*-2-butene-1,4-diol with ethyl orthoacetate afforded β-vinyl-γ-butyrolactone in 45% yield, along with 20% of 2-methyl-2-ethoxy-1,3-dioxacyclohept-5-ene. Condensations of allyl alcohols with cyclic orthoesters have also been used to prepare γ- and δ-lactones (equations 178−181)[339].

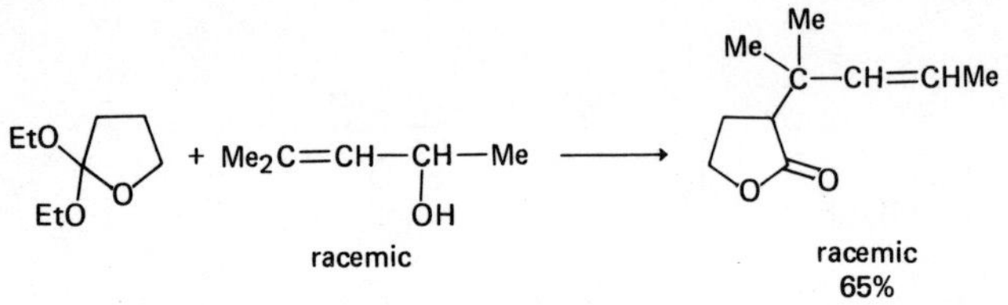

(178)

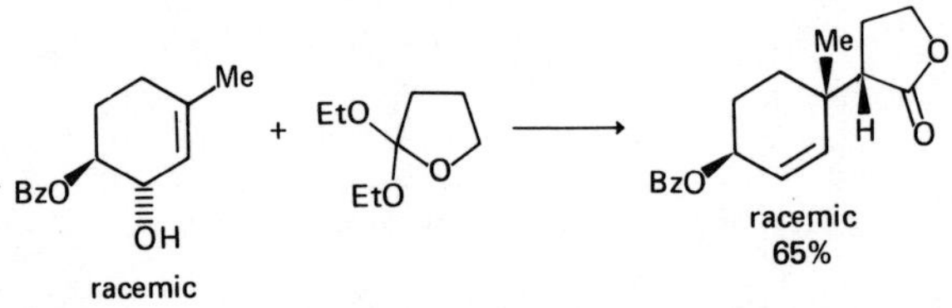

(179)

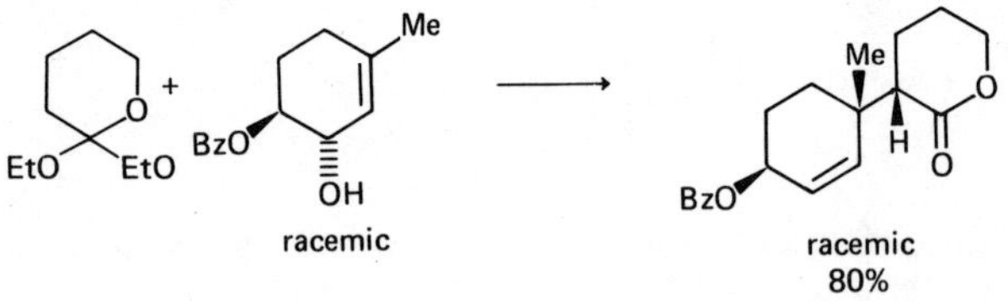

(180)

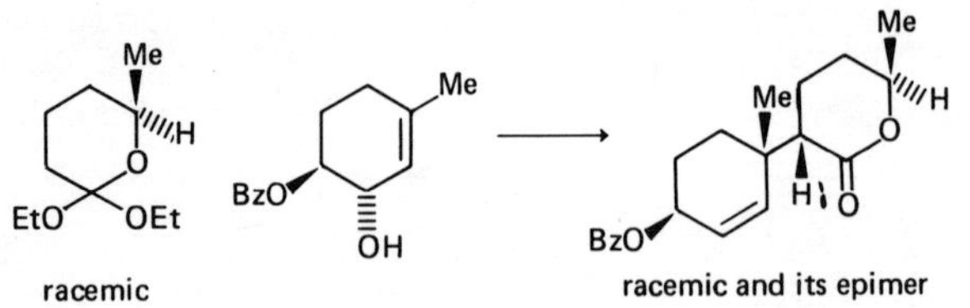

(181)

2. Carbonium ion rearrangements

A number of cyclopropane carboxylic acids undergo acid-catalysed and/or thermal rearrangements to form γ-butyrolactones. The former reactions may be envisioned as occurring via concomitant protonation at the cyclopropyl carbon holding the carboxyl group, and ring-opening to form the most highly substituted carbonium ion, which then interacts with the carboxy group to generate the lactone

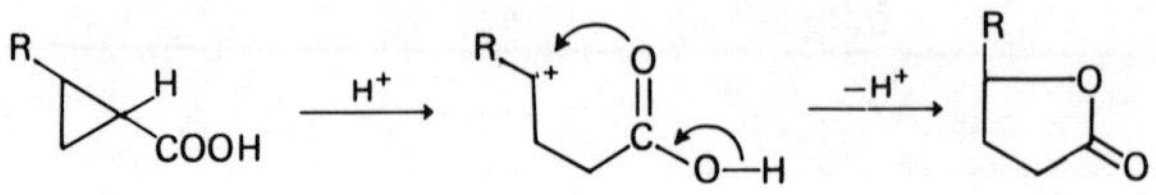

(182)

ring (equation 182). The specific examples given in equations (183)—(185) are representative of this scheme for lactone formation.

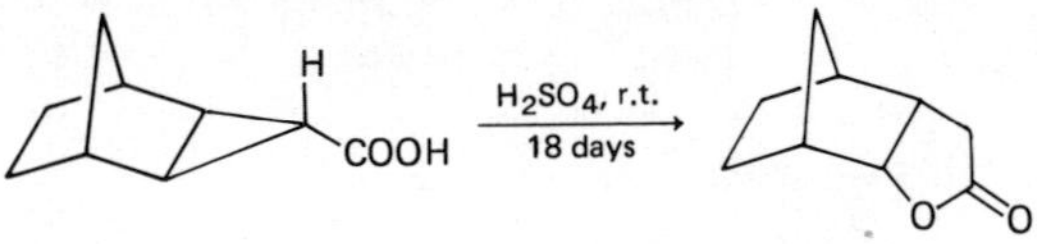

(Ref. 340)   (183)

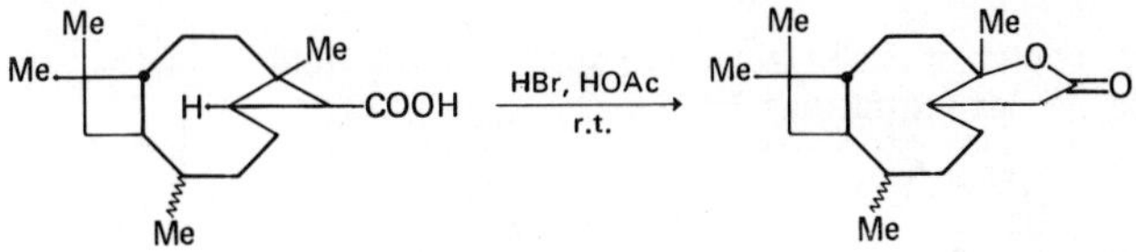

(Ref. 341)   (184)

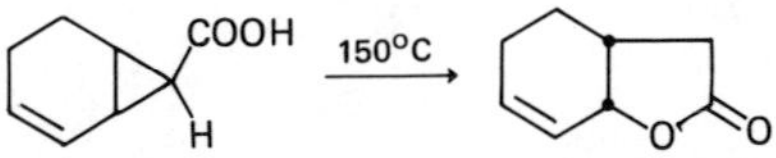

(Ref. 342)   (185)

Certain other monocarboxylic acids containing ring systems which are susceptible to carbonium ion rearrangements can be converted to lactones upon treatment with acid. Thus, both the *endo* and *exo* isomers of (+)-1.5,5-trimethyl-bicyclo[2.1.1]hexane-6-carboxylic acid produce dihydro-β-campholenolactone in 49% yield (equation 186)[343]. The [4.1.0] bicyclic hydroxy ester shown in equation (187) affords an 88% yield of *trans*-fused cycloheptene butyrolactone[344].

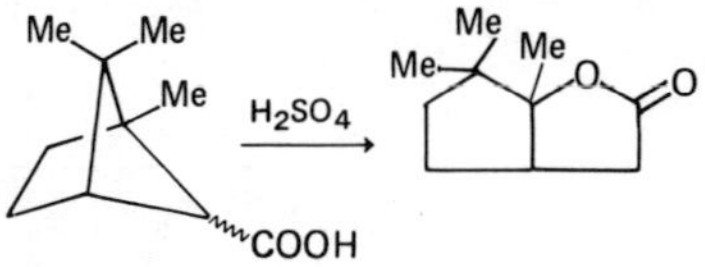

(Ref. 343)   (186)

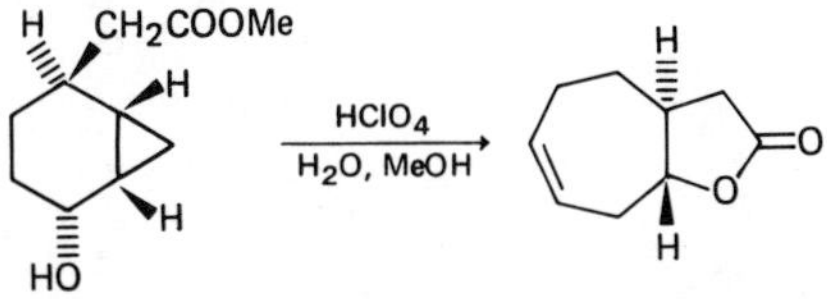

(Ref. 344)   (187)

Cyclopropane-1,1-dicarboxylic acids can serve as useful starting materials for γ-butyrolactones as shown by the reaction of several such acids with deuterated sulphuric acid (equation 188)[345]. The location of the deuterium labels in the final

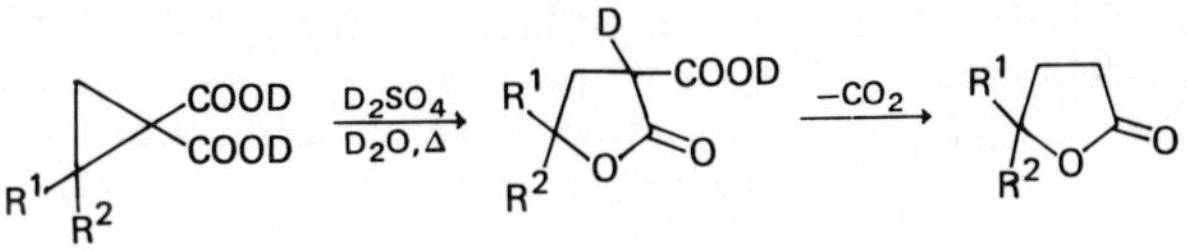

(188)

$$R^1 = R^2 = H$$
$$R^1 = H, R^2 = Me$$
$$R^1 = R^2 = Me$$

products is consistent with operation of a mechanism analogous to that described above for cyclopropanecarboxyclic acids. Thermal decarboxylation of related diacids also affords lactones (equation 189)[342].

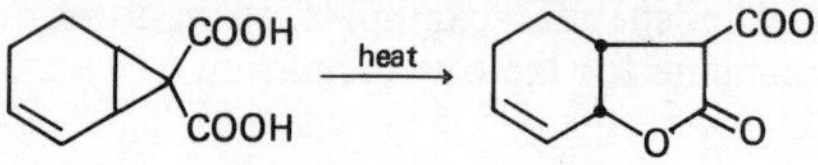

(189)

3. Photochemical rearrangements

Irradiations of β,γ-epoxy cyclic ketones and simple substituted epoxides produce lactones in 35%–65% yields (equations 190–193).

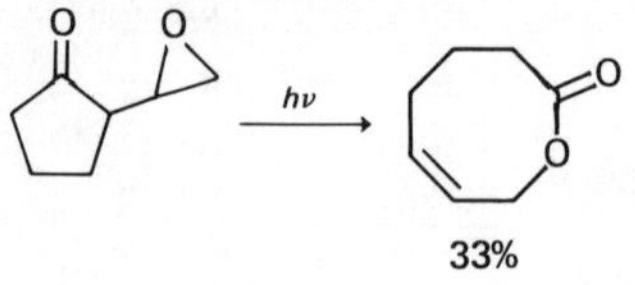

(Ref. 346)   (190)

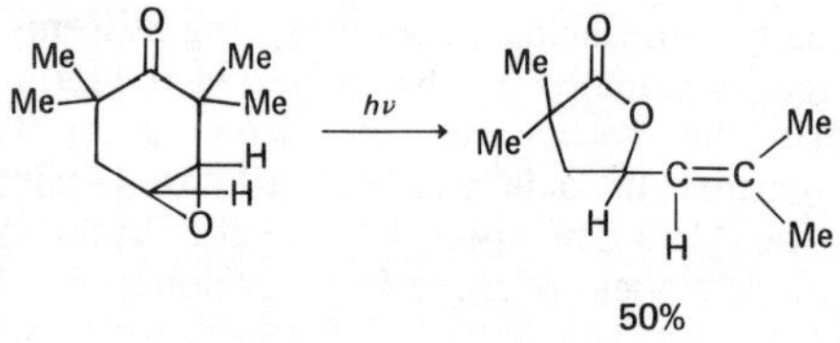

33%

(Ref. 347)   (191)

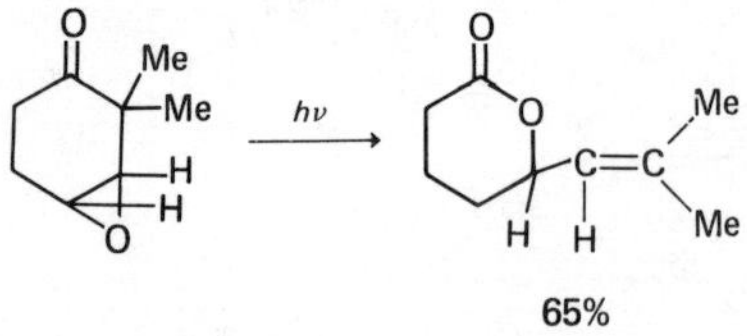

50%

(Ref. 348)   (192)

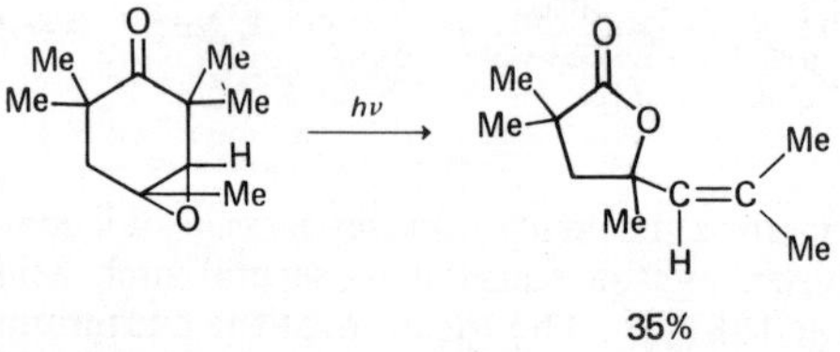

65%

(Ref. 348)   (193)

35%

The photochemical behaviour of the non-enolizable β-diketone, 2,2,5,5-tetra-methyl-1,3-cyclohexanedione, has been studied by several groups of workers[349–352] and all are in essential agreement concerning the products obtained in benzene (equation 194). However, in ethanol or cyclohexane, one group of

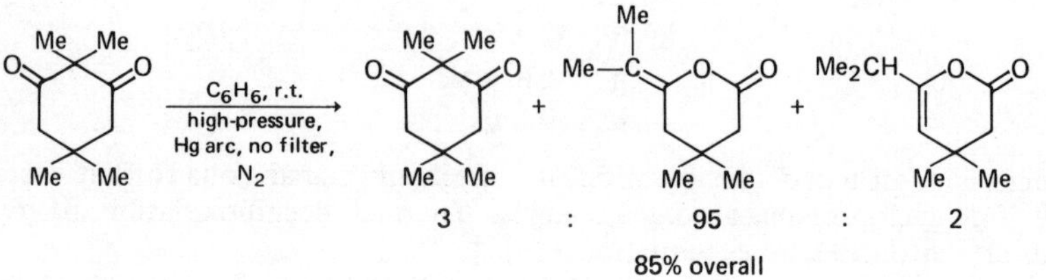

3   :   95   :   2

85% overall

(194)

workers[349] reported a single product, while a second group[352] obtained all the products shown in equation (194).

Interestingly, irradiation of the exocyclic enol lactone, 5-hydroxy-3,3,6-trimethyl-5-heptenoic acid δ-lactone afforded[352] a pseudo-equilibrium mixture (equation 195). Treatment of 2,2-dimethyl-1,3-cyclohexanedione in a similar

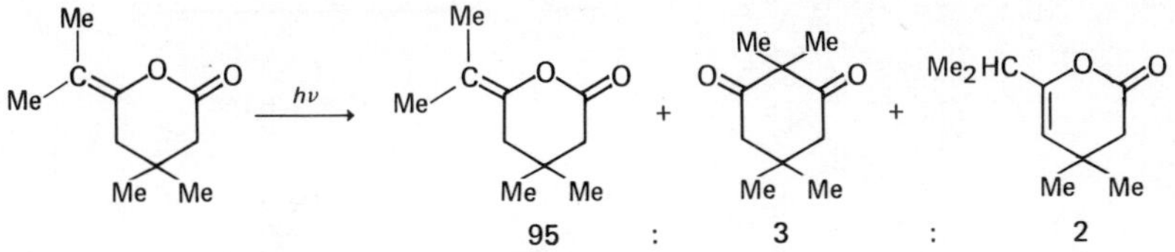

$$95 \quad : \quad 3 \quad : \quad 2 \tag{195}$$

manner afforded[352] exclusively the corresponding enol lactone in 70% yield (equation 196).

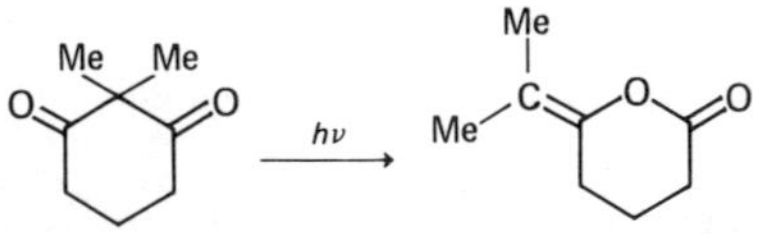

$$\tag{196}$$

R. Lactone Interconversions

Although there are not enough literature reports to permit generalization, the following reactions provide some examples of the synthetic potential of lactone interconversions.

Treatment of d,l-α-campholenic acid lactone with sulphuric acid has been reported[290] to produce the isomeric dihydro-β-campholenolactone (equation 197);

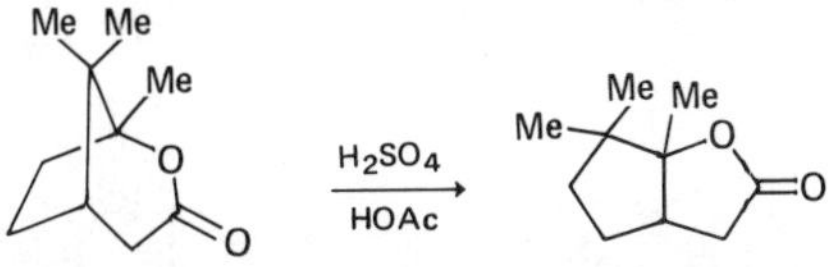

$$\tag{197}$$

however, when the isomeric bicyclic lactone was treated in the same manner no interconversion was observed (equation 198)[291]. This difference in reactivity has

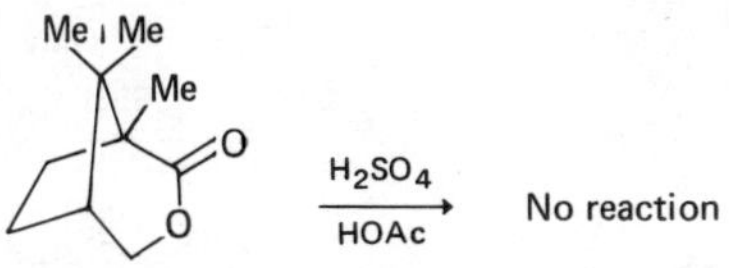

$$\tag{198}$$

been used[291] to obtain analysis of the lactone products obtained from peracetic acid oxidation of camphor (equation 199).

During the elegant synthesis of reserpine, Woodward and coworkers[28] have observed a number of lactone interconversions (equations 200 and 201).

The γ- to δ-lactone interconversion shown in equation (202) has recently[307] been observed during the total synthesis of Rhoeadine alkaloids.

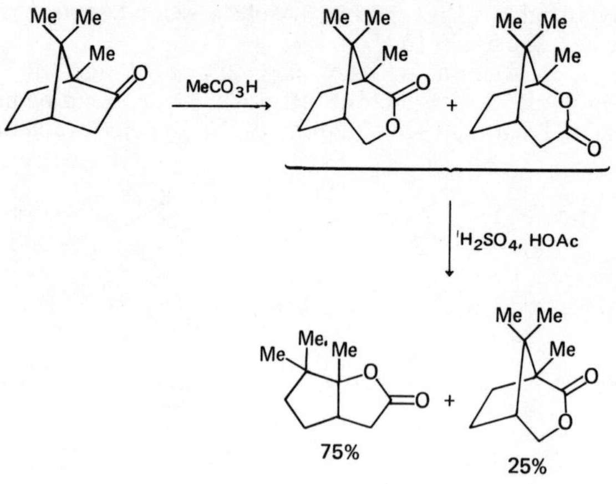

(199)

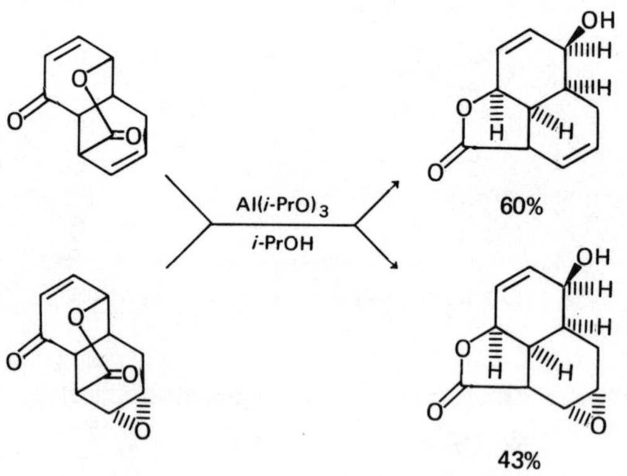

(200)

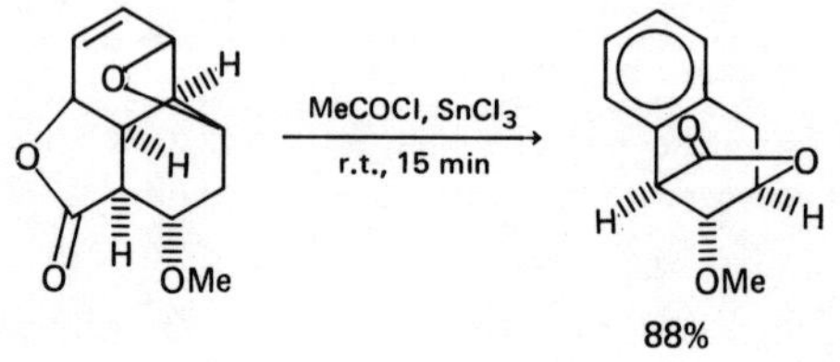

(201)

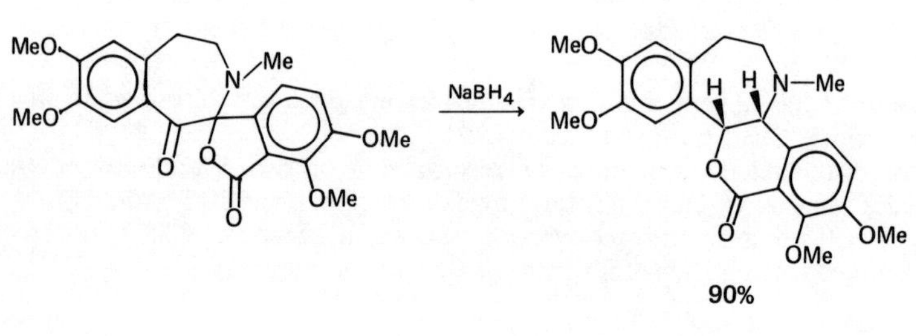

(202)

S. Miscellaneous Lactone Syntheses

The following preparations do not fall conveniently into any of the preceding categories; nevertheless several of them are extremely attractive as general lactone syntheses.

1. *The Barton reaction*

This useful synthesis of lactones[353] consists of reaction of primary or secondary amides with lead tetraacetate or *t*-butyl hypochlorite in the presence of iodine to form *N*-iodo amides, which then undergo a free radical cyclization to lactones when the reaction mixture is photolysed.

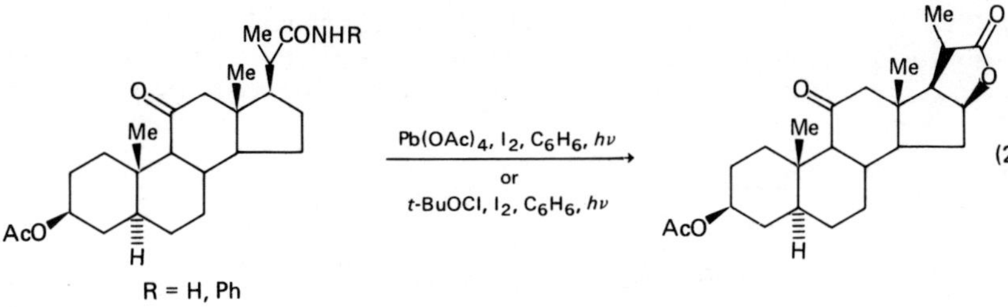
Me CONHR
Me
O
Me
AcO
H
R = H, Ph
Pb(OAc)4, I2, C6H6, hν
or
t-BuOCl, I2, C6H6, hν
Me
O
Me
O
O
Me
AcO
H

(203)

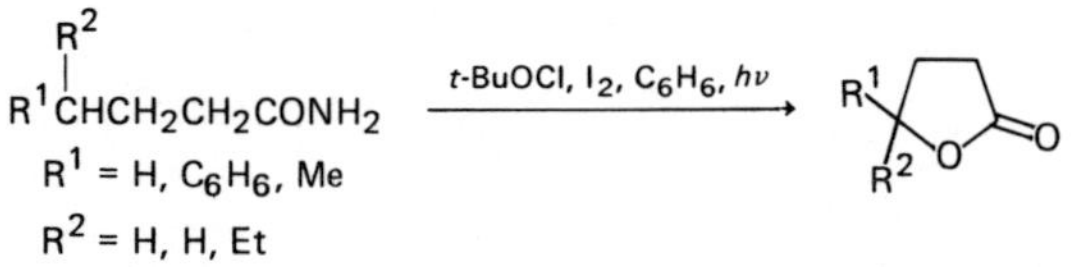
R2
R1CHCH2CH2CONH2
R1 = H, C6H6, Me
R2 = H, H, Et
t-BuOCl, I2, C6H6, hν
R1
R2
O
O

(204)

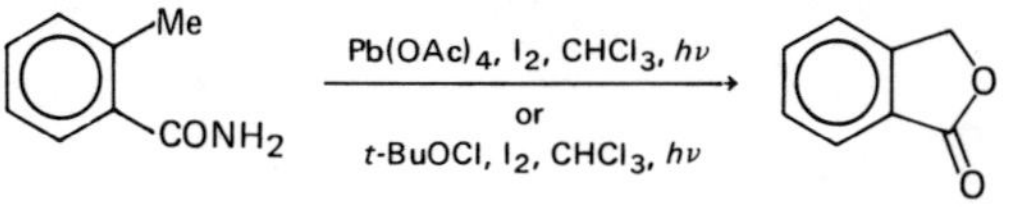
Me
CONH2
Pb(OAc)4, I2, CHCl3, hν
or
t-BuOCl, I2, CHCl3, hν
O
O

(205)

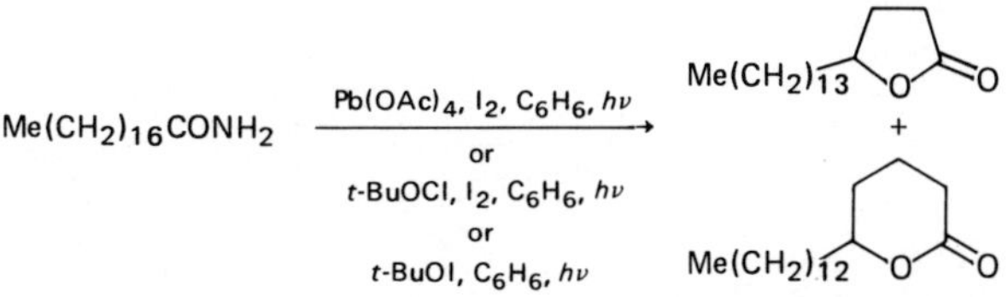
Me(CH2)16CONH2
Pb(OAc)4, I2, C6H6, hν
or
t-BuOCl, I2, C6H6, hν
or
t-BuOI, C6H6, hν
Me(CH2)13
O
O
+
Me(CH2)12
O
O

(206)

In a reaction somewhat related to the Barton reaction, photolysis of *N*-acetyl-3-methyl-3-phenylpropionamide was reported to accord the lactone of 4-phenyl-4-hydroxy-3-methylbutyric acid[354].

2. *Photolysis of α-diazo esters and amides*

Photolysis of certain esters of α-diazo carboxylic acids gives rise to lactones by insertion of the resulting α-carbene into a carbon–hydrogen bond of the alkoxy residue[355]. These reactions are, however, often characterized by low yields. Thus, photolysis of the *t*-butyl esters of diazoacetic acid in cyclohexane affords only a 4% yield of γ,γ-dimethylbutyrolactone (equation 207)[355]. Performing the same

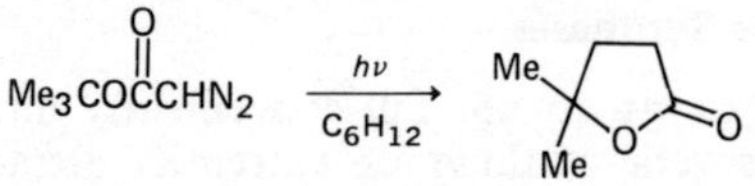

$$Me_3COCCHN_2 \xrightarrow[C_6H_{12}]{h\nu} \qquad (207)$$

reaction on the t-amyl ester of diazoacetic acid[355] affords β,γ,γ-trimethyl- and γ-methyl-γ-ethylbutyrolactone, both in low yields (equation 208). Interestingly,

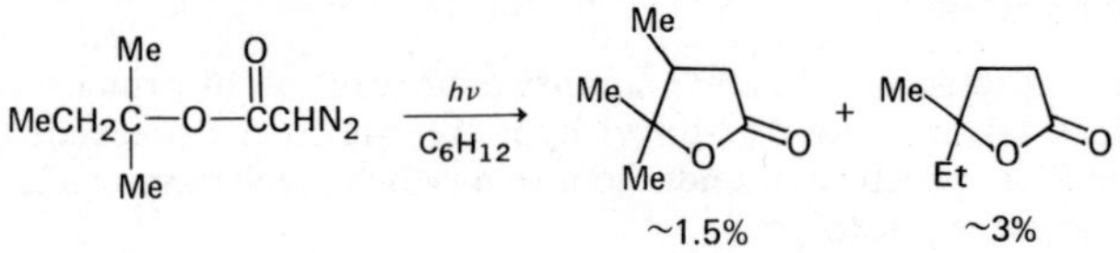

$$MeCH_2\overset{Me}{\underset{Me}{C}}-O-\overset{O}{\overset{\|}{C}}CHN_2 \xrightarrow[C_6H_{12}]{h\nu} \qquad (208)$$

$$\sim 1.5\% \qquad \sim 3\%$$

photolysis[356] of N-[(t-butoxycarbonyl)diazoacetyl]piperidine produced only cis-7-t-butoxycarbonyl-1-azabicyclo[4.2.0]octan-8-one and its *trans* isomer (equation 209), but no γ-lactone. Using N-[(t-butoxycarbonyl)diazoacetyl]pyrro-

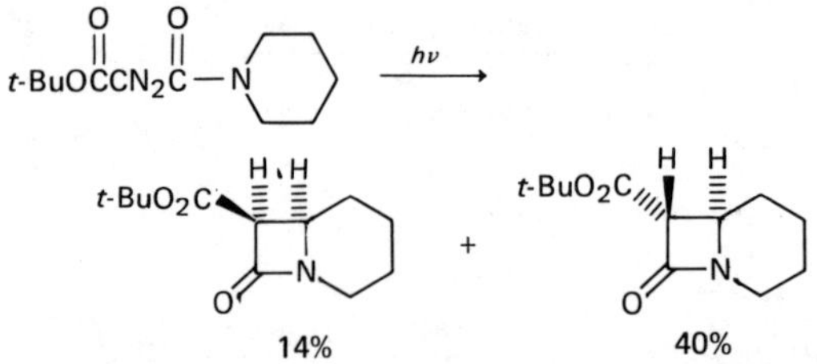

$$14\% \qquad\qquad 40\% \qquad (209)$$

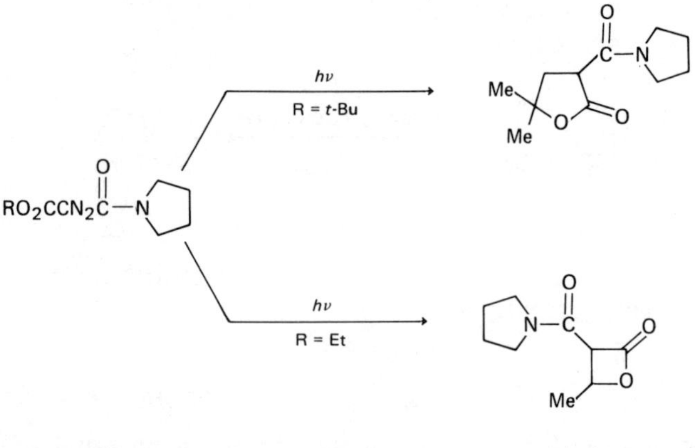

$$(210)$$

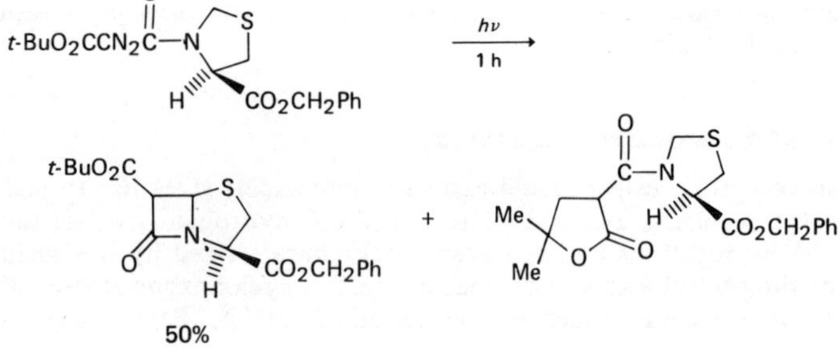

$$(211)$$

$$50\%$$

lidine, only the γ-lactone forms, while from N-[(ethoxycarbonyl)diazoacetyl[pyrrolidine only the β-lactone is obtained (equation 210)[356]. Application of this reaction[356] to N-[(butoxycarbonyl)diazoacetyl]-L-thiazolidine-4-carboxylate substantiated the expectation that the 2-methylene group in the thiazolidine is very susceptible to carbene insertion, since a mixture of β-lactam and its isomeric γ-lactone was obtained (equation 211).

A similar photochemically induced intramolecular insertion has been reported[357] during the photolysis of diethyl diazomalonate with thiobenzophenone in cyclohexane (equation 212).

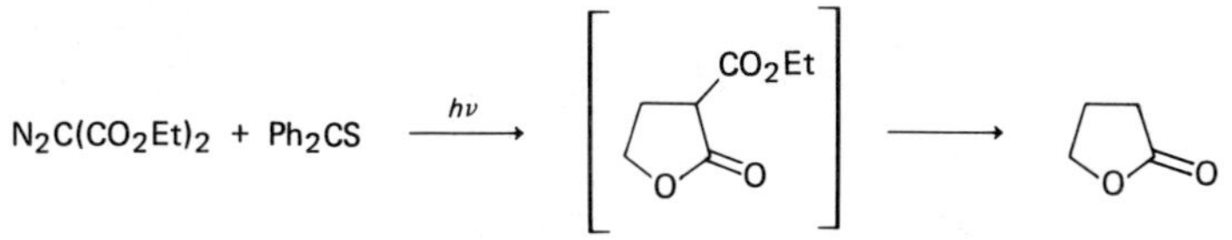

$$N_2C(CO_2Et)_2 + Ph_2CS \xrightarrow{h\nu} \left[\text{(structure with } CO_2Et) \right] \longrightarrow \text{(lactone structure)} \quad (212)$$

3. Photolysis of 2-alkoxyoxetanes

A novel synthesis[358] of tetramethyl-β-propiolactone involves irradiation of an acetonitrile solution of any of the 3,3,4,4-tetramethyloxetanes shown in equation (213) with acetone. This lactone may also be prepared[358] via irradiation, of either

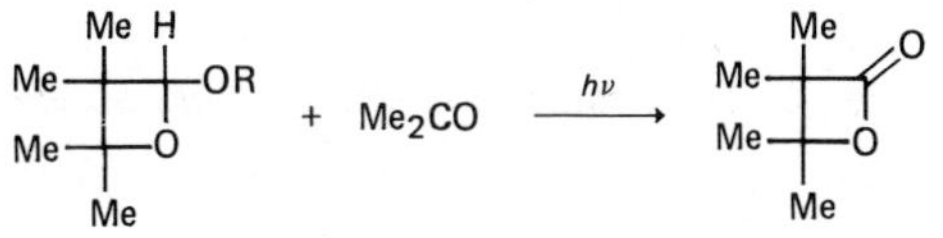

$$\text{(oxetane structure)} + Me_2CO \xrightarrow{h\nu} \text{(lactone structure)} \quad (213)$$

R = —OMe, —OEt, —OPr-n, —OBu-n

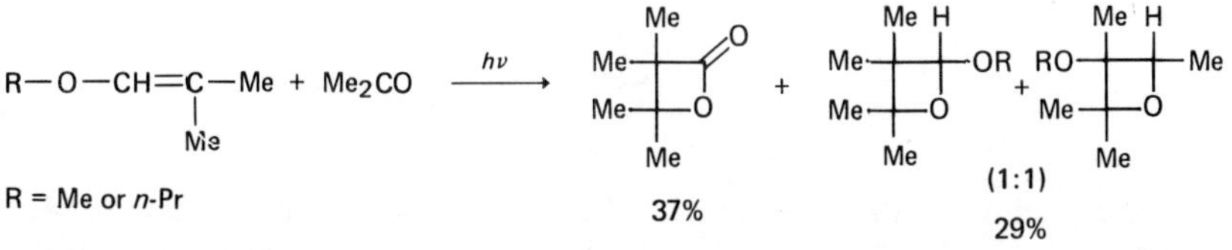

$$R-O-CH=C(Me)-Me + Me_2CO \xrightarrow{h\nu} \text{(lactone)} \underset{37\%}{} + \text{(two structures)} \underset{29\%}{(1:1)} \quad (214)$$

R = Me or n-Pr

methyl or n-propyl β,β-dimethyl vinyl ether with acetone (equation 214). Similar irradiation[358] of acetone with ethyl β,β-diethyl vinyl ether affords α,α-diethyl-β,β-dimethyl-β-propiolactone, which has also been prepared by irradiation of a mixture of isomeric oxetanes with acetone or benzophenone (equation 215). Preparation of

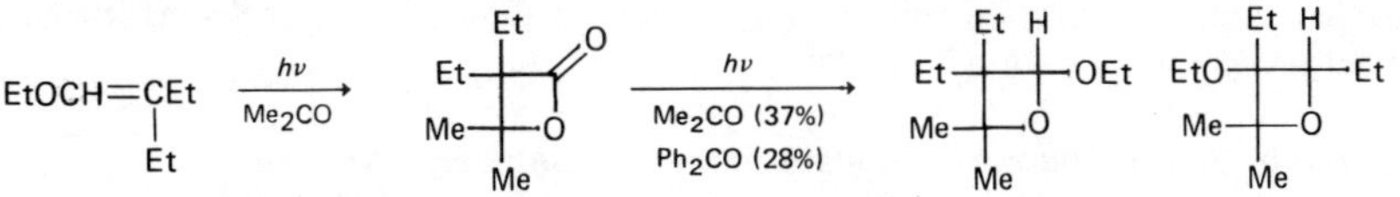

$$EtOCH=CEt(Et) \xrightarrow[Me_2CO]{h\nu} \text{(structure)} \xrightarrow[\substack{Me_2CO (37\%) \\ Ph_2CO (28\%)}]{h\nu} \text{(two structures)} \quad (215)$$

the α,α,β-triethyl-β-propiolactone was accomplished[358] via irradiation of a mixture of the corresponding 2- and 3-methoxyoxetanes with acetone.

4. α-Lactones by photolysis of 1,2-dioxolane-3,5-diones

Methods of preparation of α-lactones are not very common; however, a rather unique, high-yield photochemical synthesis of these elusive compounds via photochemical decarboxylation of 4,4-disubstituted-1,2-dioxolane-3,5-diones has recently been reported[359]. Thus, irradiation of substituted 1,2-dioxalane-3,5-diones as neat liquids at 77 K produces disubstituted α-lactones (equation 216). If the irradiation is performed at room temperature or if the α-lactone is warmed above -100°C a polyester is the only product obtained.

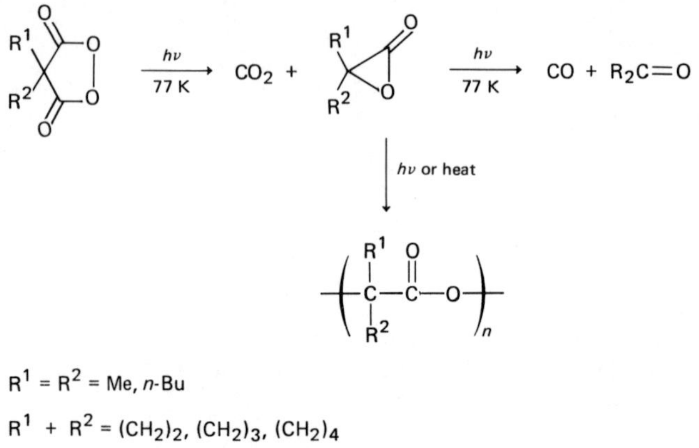

$$R^1 = R^2 = Me, n\text{-}Bu$$

$$R^1 + R^2 = (CH_2)_2, (CH_2)_3, (CH_2)_4$$

5. Oxidation of mercaptans, disulphides and related compounds

When mercaptans and disulphides are treated with an oxidizing agent such as dimethyl sulphoxide under basic conditions in a polar solvent, lactones have been reported[360] as the products. Also prepared were the δ-lactones where $R = n\text{-}C_6H_{13}$, Me, Et and Ph.

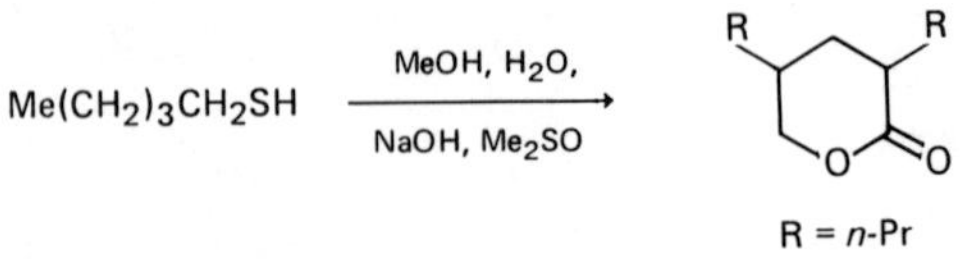

$$Me(CH_2)_3CH_2SH \xrightarrow[\text{NaOH, Me}_2\text{SO}]{\text{MeOH, H}_2\text{O,}} \qquad (217)$$

$$R = n\text{-}Pr$$

The sulphur-donor ligand *ortho*-metalated complexes shown in equation (218) afford lactones upon treatment with 30% hydrogen peroxide or *m*-chloroperbenzoic acid[361].

6. Addition of diazonium salts to olefins

Treatment of olefins with substituted benzenediazonium chlorides in the presence of cuprous chloride and an alkali metal halide affords aryl-substituted butyrolactone esters (equation 219)[362].

7. Addition of diethyl dibromomalonate to methyl methacrylate

Condensation of diethyl dibromomalonate with methyl methacrylate in the presence of iron pentacarbonyl produces the substituted butyrolactone shown in equation (220)[363].

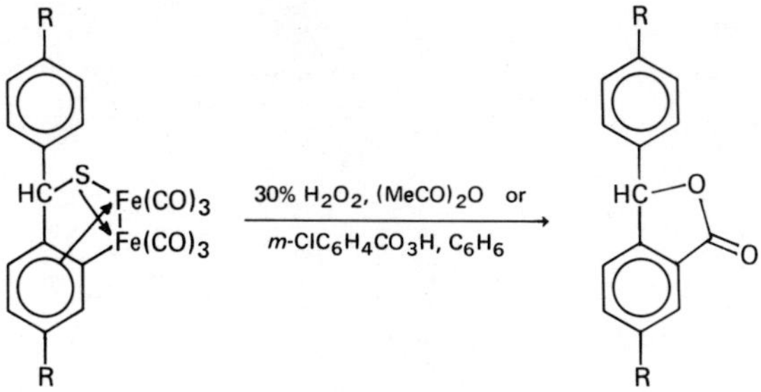

(218)

R	Reagent	Yield (%)
OMe	30% H_2O_2	73
OMe	m-$ClC_6H_4CO_3H$	57
Me	30% H_2O_2	45
H	30% H_2O_2	49

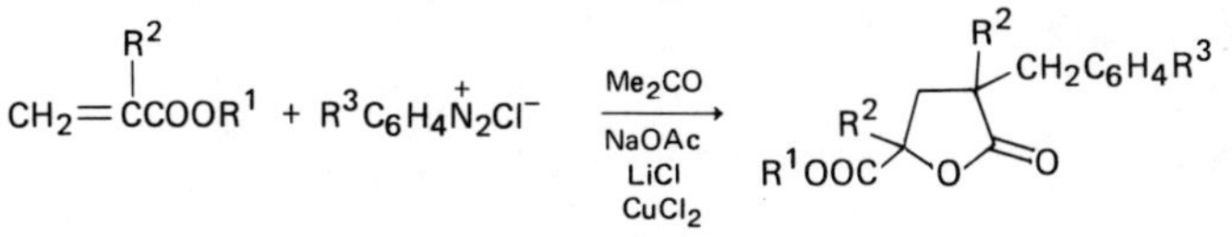

(219)

$R^1 = R^2 = Me; R^3 = p$-Me, p-Cl

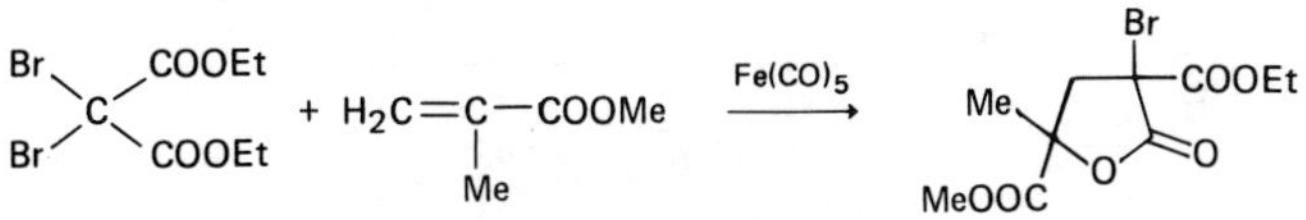

(220)

8. Dehydrohalogenation of 2,2-dimethoxy-3-chlorodihydropyrans

Treatment of a series of substituted 3-chlorodihydropyrans with sodium methoxide in dimethyl sulphoxide or dimethylformamide at room temperature affords the corresponding α-pyrones in good yields (equation 221)[364].

9. Preparation of homoserine lactone

α-Amino-γ-butyrolactone (homoserine lactone), an important intermediate in the synthesis of various amino acids, has been prepared by a two-step sequence in which N-tosyl- or N-benzoylglutamine is converted into N-tosyl- or N-benzoyl-α,γ-diaminobutyric acid with potassium hypobromite, followed by diazotization[365]. A second route involves the reaction of N-acyl methionines with methyl iodide in a mixture of acetic and formic acids to produce their corresponding sulphonium salts, which are then hydrolysed under reflux at pH 6–7 (equation 222). The resulting N-acyl-α-amino-γ-hydroxybutyric acids are then converted into their corresponding lactones using hydrogen chloride[366].

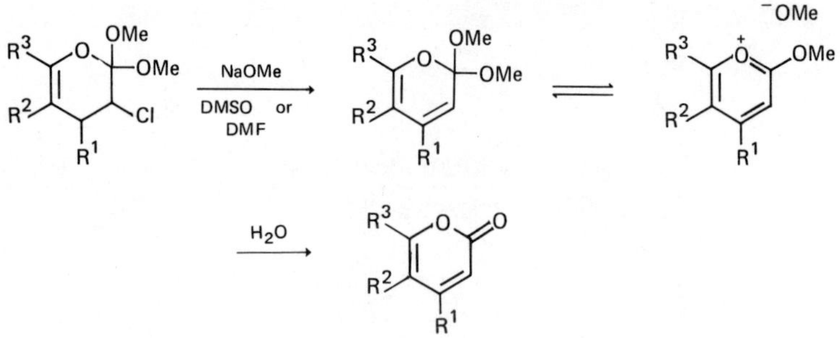

(221)

R^1	R^2	R^3	Yield (%)
Ph	H	H	85
H	H	Ph	64
Ph	H	Ph	72
Et	$-(CH_2)_4-$		78
H	$C_{10}H_{21}$	H	52

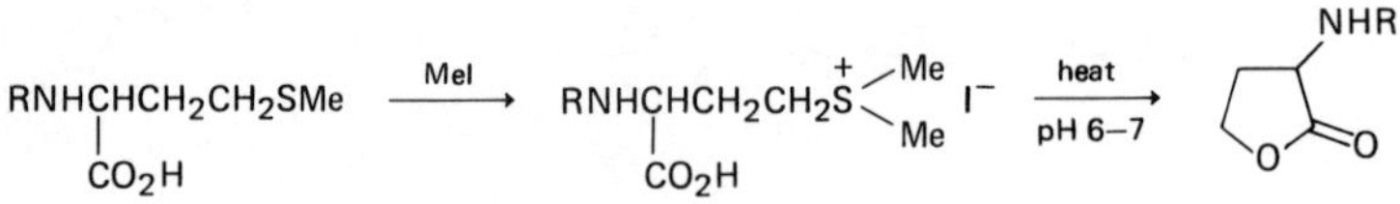

(222)

R	Yield (%)
PhCO	73
p-MeC$_6$H$_4$SO$_2$	92
EtOCO	81
PhCH$_2$OCO	80
Me$_3$COCO	29
O=C\ | O=C/	45

III. SYNTHESIS OF LACTAMS

Information about the synthesis of lactams may be found in numerous review articles, most of which, however, have been limited to the preparation of one particular class of lactam or to the general synthesis of amides.

In 1957 Sheehan and Corey[367] published a review on 'The synthesis of β-lactams'. The synthesis of lactam monomers was reviewed in 1962 by Dachs and Schwartz[368] and by Testa[369]. The synthesis of β-lactams was again reviewed in 1962 by Graf and coworkers[370], while in 1966 a review of the preparation, properties and pharmacology of amides, amino acids and lactams was published by Piovera[371], and in 1967 a discussion of the preparation of β-lactams was published by Muller and Harmer[372].

The first review on 'α-Lactams (aziridinones)' appeared in 1968 from Lengyel and Sheehan[373], while the synthesis of all types of lactams was reviewed first by

Beckwish[374] in 1970 in his chapter on 'Synthesis of amides' for this series, by L'Abbé and Hassner[375] in 1971 in their review of 'New methods for the synthesis of vinyl azides', by Millich and Seshadri[376] in their chapter on lactams in *High Polymers*, by Manhas and Bose[377] in *Chemistry of β-Lactams, Natural and Synthetic*, by Hawkins[378] in his review of 'α-Peroxyamines', and finally, by Mukerjee and Srivastava[379] in a review entitled 'Synthesis of β-lactams'.

A. By Ring-closure Reactions (Chemical)

1. From amino acids and related compounds

Intramolecular reaction of a carboxylic acid or ester function with an appropriately positioned amino group is quite often the method of choice for the synthesis of γ- and δ-lactams. Lactams of smaller and larger ring size are somewhat less frequently synthesized by such procedures, although α-, β- and ε-lactams can be prepared by careful choice of reaction conditions and starting materials. Thermal cyclization of a mixture of *cis*- and *trans*-4-aminocyclohexanecarboxylic acid to produce 3-isoquinuclidone[380] is representative of a typical δ-lactam synthesis (equation 223). Preparation of the γ-lactam, 1,5-dimethyl-2-pyrrolidone[381],

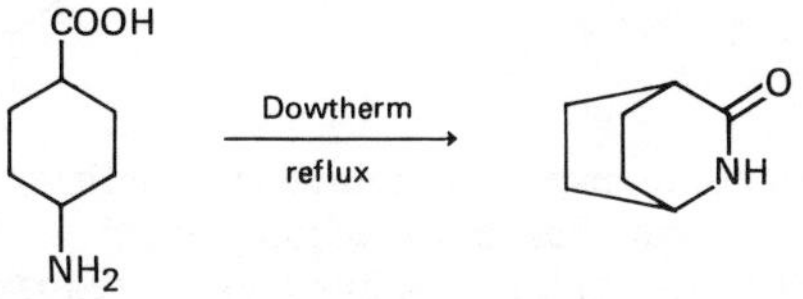

$$\text{(223)}$$

involves a related cyclization of the methylammonium salt of γ-(methylamino) valeric acid (equation 224).

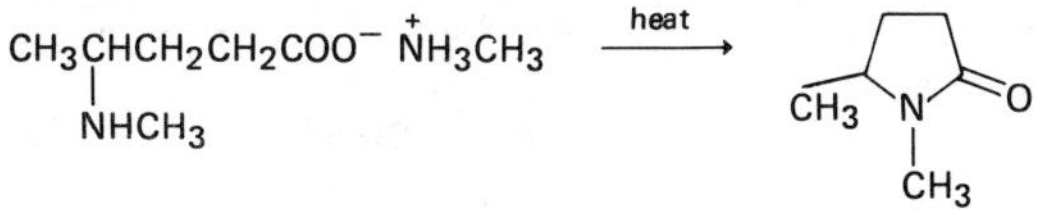

$$\text{(224)}$$

An interesting example[382] of α-lactam (aziridinone) formation involves the synthesis of optically active 3-substituted-1-benzyl-oxycarbonylaziridin-2-ones from *N*-benzyloxycarbonyl L-amino acids by use of phosgene, thionyl chloride or phosphorus oxychloride in THF at −20 to 30°C (equation 225). The cyclization appears to involve initial formation of a mixed anhydride between the *N*-protected amino acid and the dehydrating agent.

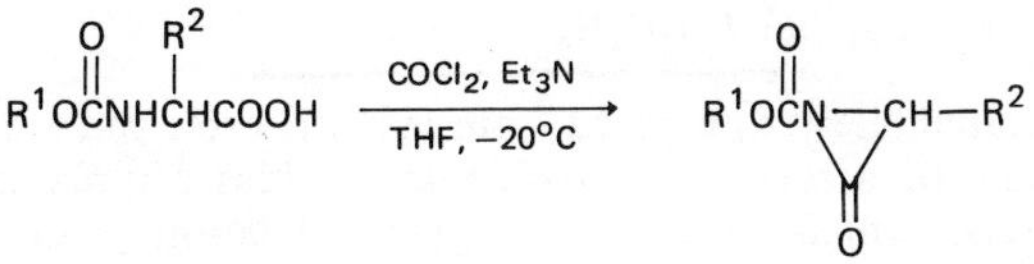

$$\text{(225)}$$

$$R^1 = -CH_2C_6H_5, -CH_2C_6H_4Br\text{-}p, -CH_2C_6H_4Cl\text{-}p,$$

$$R^2 = -CH_2C_6H_5$$

Intramolecular cyclization of amino esters has found numerous applications in lactam synthesis. In some cases the desired cyclizations are accomplished thermally as in the preparations of 5,5-dimethyl-2-pyrrolidone (equation 226)[383], α-(equation 227)[384], β-(equation 228)[385] and δ-methylcaprolactam (equation 229)[385].

$$(CH_3)_2CCH_2CH_2COOMe \quad \xrightarrow[88-96\%]{200°C} \quad \text{(structure)} \qquad (226)$$
|
NH₂

$$H_2N(CH_2)_4CHCOOEt \quad \xrightarrow[44\%]{\text{ethylene glycol} \atop 162-165°C} \quad \text{(structure)} \qquad (227)$$
|
CH₃

$$H_2N(CH_2)_3CHCH_2COOEt \quad \xrightarrow[56\%]{\text{heat}} \quad \text{(structure)} \qquad (228)$$
|
CH₃

$$H_2NCH_2CH(CH_2)_3COOEt \quad \xrightarrow[41\%]{\text{heat}} \quad \text{(structure)} \qquad (229)$$
|
CH₃ (on CH)

Cyclization of dienamino esters, obtained by addition of enamino esters to methyl and ethyl propiolate, has been accomplished at 160–190°C in dipolar aprotic solvents to afford α-pyridones in good yields (equation 230)[386]. Reaction

$$\text{(structure)} + HC\equiv CCOOR^3 \longrightarrow \text{(structure)} \longrightarrow \text{(structure)} \qquad (230)$$
$$R^3 = \text{Me, Et}$$

R¹	R²
H	$CH_2C_6H_5$
Me	H
Me	Me
Ph	H
Ph	Me
$o\text{-}CH_3C_6H_4$	H

of the α,β-unsaturated triester prepared from malonic ester and ethyl pyruvate, with the diethyl acetals of a series of N,N-dimethylamides affords the corresponding dienamino triesters, which in turn undergo cyclization with benzyl amine in refluxing ethanol to afford a series of 1-benzyl-3,4-dicarboethoxy-2(1H)-pyridones (equation 231)[387].

Cyclization of 2-piperidinylacetates to form β-lactams has been effected by means of ethylmagnesium bromide (equation 232)[388]. Yields increase with increasing substitution at the α-carbon of the ester. Similar cyclization of the methyl ester of 3-(methylamino)butyric acid produces[389] N-methyl-β-butyrolactam (equation 233); however, the reaction failed with ethyl 2-pyrrolidinylacetate[388].

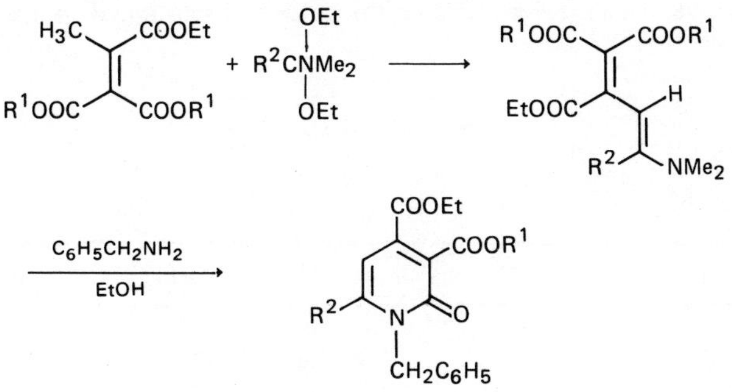

(231)

R^1	R^2
Et	H
Me	H
Me	3-Pyridyl
Me	2-Cyanophenyl
Me	3-Cyano-2-quinolyl

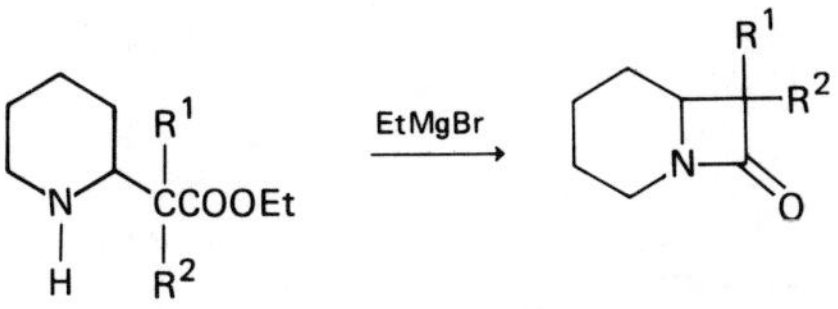

(232)

$$R^1 = Me, \ Me, \ Me, \ Et$$
$$R^2 = H, \ \ Me, \ Et, \ Et$$

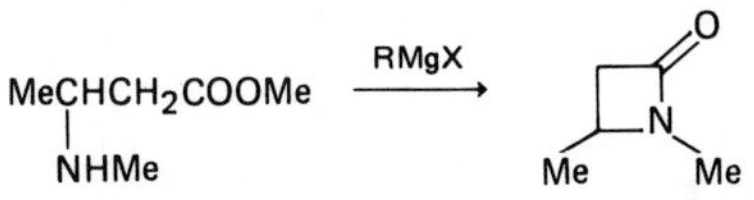

(233)

Listed in Table 28 are various β-aminopropionic acid esters which have been converted to β-lactams by a Grignard reagent. Other examples of β-lactam preparation using this method include the conversion of ethyl 3-phenyl-β-aminopropionate to 4-phenyl-2-azetidinone (equation 234)[401], the conversion of several ethyl

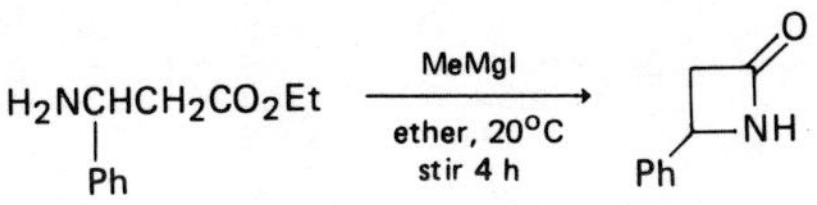

(234)

N-substituted 2-ethyl-2-phenyl-3-aminopropionic acid esters to their corresponding N-substituted 3-ethyl-3-phenyl-2-azetidinones (equation 235)[402], the conversion of methyl 2-substituted 3-phenyl-3-(phenylamino) propionates to a mixture of cis and trans 1-phenyl-3-substituted-4-phenyl-2-azetidinones (equation 236)[403] and the conversion of the methyl, ethyl, isopropyl and benzyl esters of 2-phenyl-3-(benzylamino)propionic acid to 1-benzyl-3-phenyl-2-azetidinone (equation 237)[404].

In connection with a new synthesis of oxindoles[405-407], Gassman and co-

TABLE 28. β-Lactams prepared via the reaction of substituted β-aminopropionic acid esters with a Grignard reagent

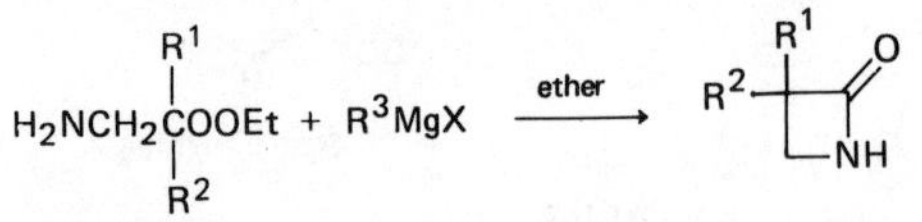

$$\text{H}_2\text{NCH}_2\overset{\overset{\displaystyle R^1}{|}}{\underset{\underset{\displaystyle R^2}{|}}{\text{C}}}\text{COOEt} + R^3\text{MgX} \xrightarrow{\text{ether}} R^2\text{---}\underset{\text{NH}}{\overset{\overset{R^1}{|}}{\square}}\text{O}$$

R¹	R²	R³	Yield (%)	References
H	*n*-Pr	Me	22	390
H	*i*-Bu	Me	67	391
H	*c*-C$_6$H$_{11}$	Me	54	391
H	Ph	Et	–	392
H	*p*-MeC$_6$H$_4$	Me	54	393
H	*p*-MeOC$_6$H$_4$	Me	20	393
H	C$_6$H$_5$CH$_2$	Me	43	391
H	α-naphthyl	Me	49	391
H	*p*-H$_5$C$_6$—C$_6$H$_4$	Me	11	391
Me	Me	Et	80	394
Me	Ph	Et	51	395
CH$_2$OH	Ph	Et	–	396
Et	Et	Me	32	390
Et	Et	Et	92	394
Et	Ph	Me	79	390
Et	Ph	Et	86	394–398
Et	*p*-MeC$_6$H$_4$	Et	88	399
Et	C$_6$H$_5$CH$_2$	Et	64	395
n-Pr	*n*-Pr	Et	91	394
n-Pr	Ph	Et	56	395
i-Pr	Ph	Et	75–79	394, 395
n-Bu	*n*-Bu	Et	99	394
n-Bu	Ph	Et	92	394
Me$_2$N(CH$_2$)$_3$	Ph	Et	32	395
Et$_2$N(CH$_2$)$_2$	Ph	Me	16	399
c-C$_6$H$_{11}$	Ph	Et	80	394
Ph	C$_6$H$_5$CH$_2$	Et	83–87	394
H (as hydrochloride salt)	Ph	Me	52	400

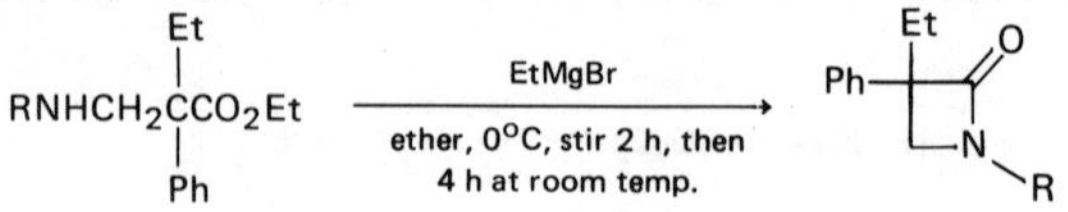

$$\text{RNHCH}_2\overset{\overset{\displaystyle Et}{|}}{\underset{\underset{\displaystyle Ph}{|}}{\text{C}}}\text{CO}_2\text{Et} \xrightarrow[\substack{\text{ether, 0°C, stir 2 h, then} \\ \text{4 h at room temp.}}]{\text{EtMgBr}} \text{Ph---}\underset{\underset{\displaystyle R}{\overset{|}{N}}}{\overset{\overset{Et}{|}}{\square}}\text{O} \qquad (235)$$

R	Yield (%)
n-Pr	23
i-Pr	27
n-Bu	60
C$_6$H$_5$CH$_2$	74

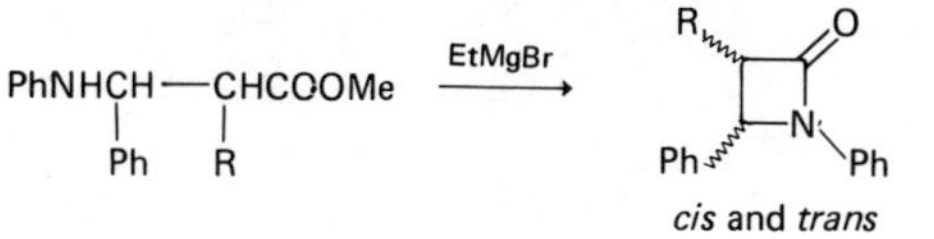

$$PhNHCH\underset{\underset{Ph}{|}}{}-\underset{\underset{R}{|}}{CHCOOMe} \xrightarrow{EtMgBr} \quad \text{(cis and trans)} \tag{236}$$

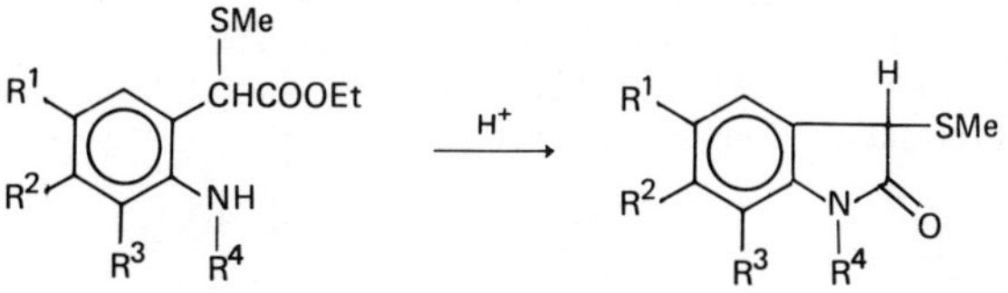

R	Yield (%)
Me	88
i-Pr	95

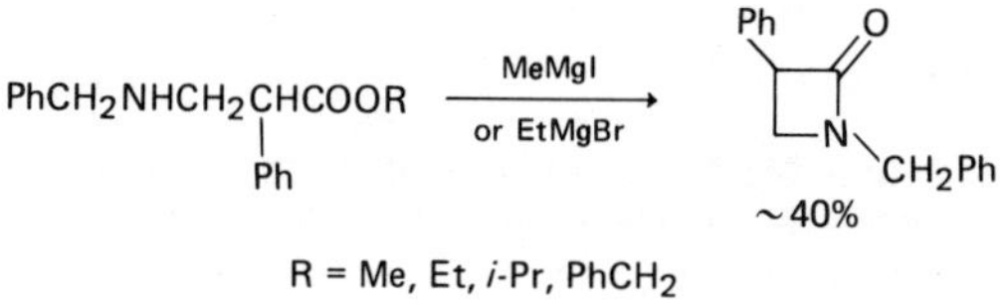

$$PhCH_2NHCH_2\underset{\underset{Ph}{|}}{CHCOOR} \xrightarrow[\text{or EtMgBr}]{MeMgI} \quad \sim40\% \tag{237}$$

R = Me, Et, i-Pr, PhCH$_2$

$$\text{(32)} \xrightarrow{H^+} \text{(33)} \qquad \text{(Refs. 405, 407)} \tag{238}$$

R^1	R^2	R^3	R^4	Yield (%)
Me	H	H	H	34
H	H	Me	H	67
H	H	H	Me	46
NO$_2$	H	H	H	51
H	NO$_2$	H	H	61
COOEt	H	Me	H	66

workers found that amino esters **32** afford 3-methylthiooxindoles **33** upon treatment with dilute acid.

Cyclization of amino esters with 2-pyridone as catalyst[408,409] is quite effective, as illustrated by a recent example (equation 239)[410].

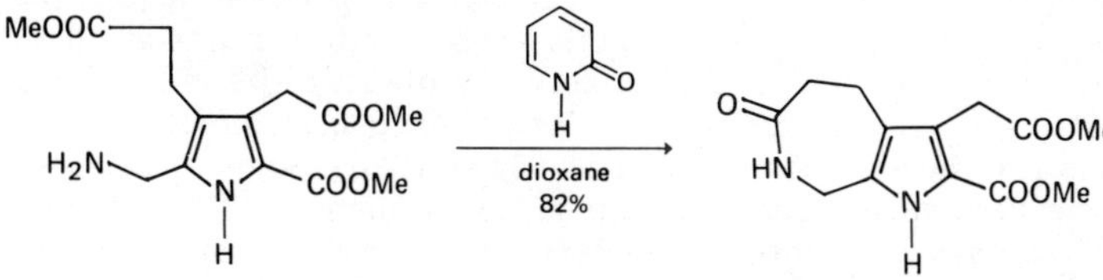

$$\xrightarrow[\substack{\text{dioxane} \\ 82\%}]{} \tag{239}$$

Reductive cyclization of nitro esters such as ethyl 3-carboethoxy-4-nitro-pentanoate[411] can be used for the preparation of γ- and δ-lactams[412,413]. The required nitro esters can often be obtained by Michael addition of a nitroalkane to an appropriate α,β-unsaturated ester (equation 240)[411].

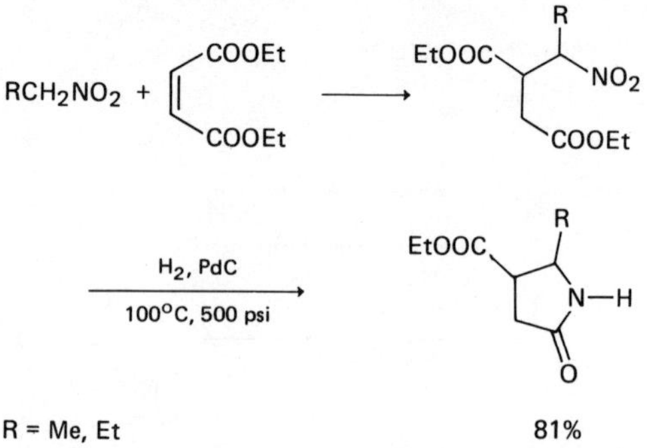

$$RCH_2NO_2 + \quad \begin{matrix} COOEt \\ \parallel \\ COOEt \end{matrix} \quad \longrightarrow \quad \begin{matrix} EtOOC \\ \end{matrix} \begin{matrix} R \\ | \\ \end{matrix} \begin{matrix} NO_2 \\ \\ COOEt \end{matrix}$$

$$\xrightarrow[100^\circ C, 500 \text{ psi}]{H_2, PdC} \quad \begin{matrix} EtOOC \\ \end{matrix} \begin{matrix} R \\ \end{matrix} \begin{matrix} \\ N-H \\ \parallel \\ O \end{matrix} \qquad (240)$$

R = Me, Et   81%

It may be noted that Michael addition of diethyl acetamidomalonate to ethyl acrylate or ethyl crotonate can be accompanied by cyclization of the intermediate adduct to form 2-pyrrolidones[414,415].

In a study of lactam formation from a series of o-aminophenoxyacetamides Cohen and Kirk[416,417] have drawn the conclusion that the mechanism involves simultaneous attack of the aromatic amino function and an external proton donor at the amide carboxyl (equation 241).

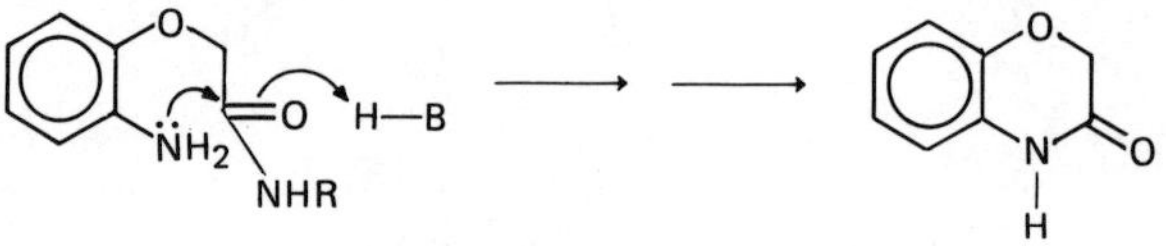

$$\qquad (241)$$

2. From halo, hydroxy and keto amides

Treatment of α-, β-, γ- or δ-halo amides with a suitable basic reagent results in ionization of the amide proton to form a nitrogen anion, which then reacts by intramolecular displacement of halide ion to produce the appropriate lactam (equation 242). The scope and limitations of this method as applied to α-lactam

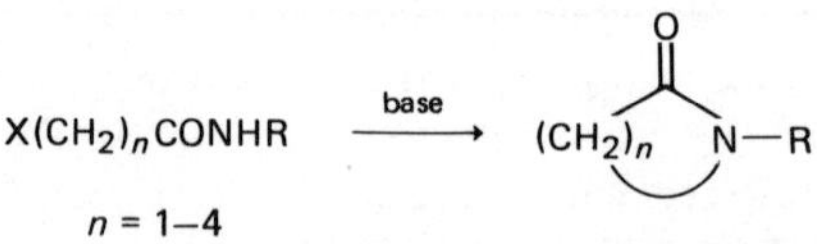

$$X(CH_2)_nCONHR \xrightarrow{\text{base}} \begin{matrix} O \\ \parallel \\ (CH_2)_n \end{matrix} \begin{matrix} \\ N-R \end{matrix} \qquad (242)$$

n = 1—4

synthesis have been discussed[418-424]. Successful preparations require the presence of one or more alkyl or aryl substituents at the α-carbon as well as a bulky N-alkyl group such as t-butyl or 1- or 2-adamantyl. Syntheses of β-, γ- and δ-lactams, but not ε-lactams[425], by cyclizations of prerequisite halo amides are much more general, as may be seen from equations (243)—(250). Some of the basic reagents which have been used include sodium in liquid ammonia[425], sodium hydride in DMSO[425], potassium t-butoxide in DMSO[425] and sodium ethoxide in ethanol[426].

A lactam synthesis first reported by Sheehan and Bose[432], and later exploited by numerous investigators[433-436] consists of intramolecular C-alkylation[435] of N-substituted α-haloacetamides and β-halopropionamides. Alkylation is effected through generation of a carbanion centre in the N-alkyl substituent, where one or preferable both of the substituents R^2 and R^3 shown in the generalized equation

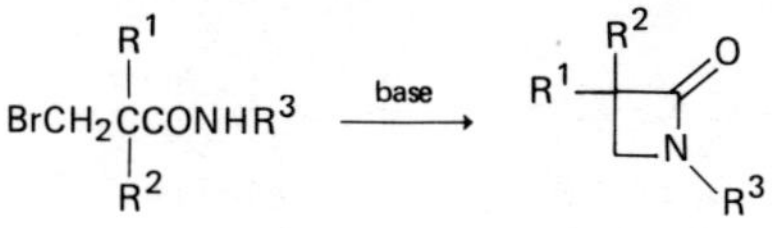

$$\text{BrCH}_2\underset{\underset{R^2}{|}}{\overset{\overset{R^1}{|}}{C}}\text{CONHR}^3 \xrightarrow{\text{base}} \quad (243)$$

R^1	R^2	R^3	Yield (%)	Reference
H	H	Ph	68–95	425
H	H	o-BrC$_6$H$_4$	71	425
H	H	o-FC$_6$H$_4$	90	425
H	H	p-BrC$_6$H$_4$	58	425
Me	Me	p-BrC$_6$H$_4$	55	425
H	H	p-ClC$_6$H$_4$	73	425
H	H	p-IC$_6$H$_4$	80	425
H	H	p-MeOC$_6$H$_4$	50	425
n-Pr	n-Pr	Me	52	427
Me	Ph	H	–	427
Me	Ph	Me	61	427
Me	Ph	C$_6$H$_5$CH$_2$	–	427
Me	Ph	Ph	54	427
Me	Ph	o-O$_2$NC$_6$H$_4$	54	427
Ph	Ph	Me	56	427

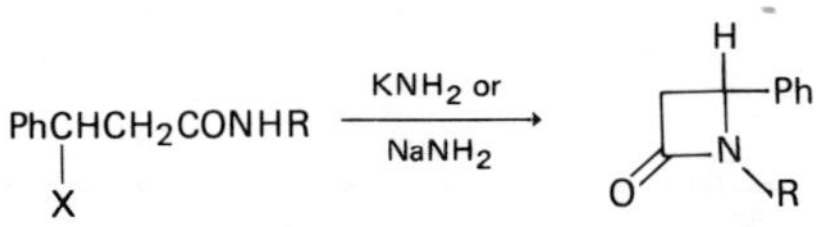

$$\underset{\underset{X}{|}}{\text{PhCHCH}_2}\text{CONHR} \xrightarrow[\text{NaNH}_2]{\text{KNH}_2 \text{ or}} \quad \text{(Ref. 428)} \quad (244)$$

X	R	Yield (%)
Cl	c-C$_6$H$_{11}$	75
Cl	H$_2$NCOCH$_2$	76
Br	Me$_2$CHCH(CO$_2$Et)$_2$	83
Cl	Me$_2$C=C(CO$_2$Et)$_2$	85

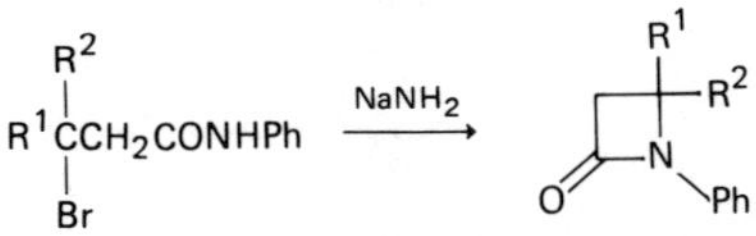

$$\underset{\underset{Br}{|}}{\overset{\overset{R^2}{|}}{R^1}}\text{CCH}_2\text{CONHPh} \xrightarrow{\text{NaNH}_2} \quad \text{(Ref. 429)} \quad (245)$$

R^1	R^2	Yield (%)
H	Me	26
Me	Me	28

(251) are electron-withdrawing functions such as carboalkoxy, phenyl or cyano. In most cases where both R^2 and R^3 are activating groups, triethyl amine[432-438], sodium acetate[439], ethanolic potassium hydroxide[439-441], or basic ion-exchange resins[440] function satisfactorily as the base. With a single activating group sodium hydroxide[443] has proved to be effective. A number of representative examples of this procedure, which have appeared since 1966, are presented in Table 29.

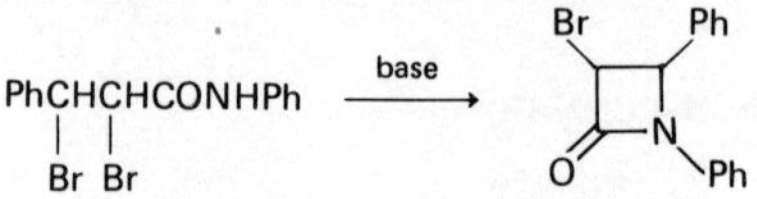

$$\text{PhCHCHCONHPh} \xrightarrow{\text{base}} \text{(lactam)} \quad \text{(Ref. 429)} \qquad (246)$$
$$\underset{\text{Br Br}}{|\ \ |}$$

Base	Yield (%)
NaOH, liq. NH$_3$	96
NaNH$_2$, liq. NH$_3$	78
KNH$_2$, liq. NH$_3$	86
NH$_3$ alone	38

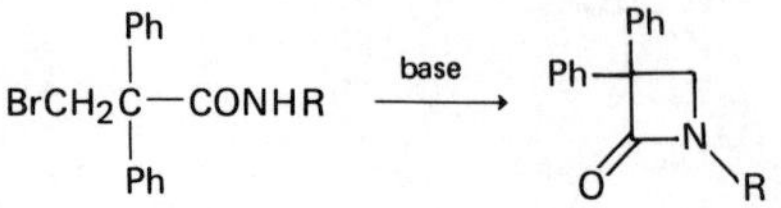

$$\text{BrCH}_2\underset{\overset{|}{\text{Ph}}}{\overset{\overset{\text{Ph}}{|}}{\text{C}}}\text{—CONHR} \xrightarrow{\text{base}} \text{(lactam)} \quad \text{(Ref. 430)} \qquad (247)$$

R	Base	Yield (%)
H	KNH$_2$ in liq. NH$_3$	92
	EtONa in EtOH	90
Ph	KNH$_2$ in liq. NH$_3$	96
	KOH in MeCOEt	91
p-MeOC$_6$H$_4$	EtONa in EtOH	98
p-O$_2$NC$_6$H$_4$	EtONa in EtOH	96
C$_6$H$_5$CH$_2$	NaSH in EtOH	92

TABLE 29. Synthesis of β- and γ-lactams by base-catalysed intramolecular alkylation of N-substituted α-haloacetamides and β-halopropionamides

Starting amide	Base	Lactam	Yield (%)	Reference
ClCH$_2$CONCH (with R^1, Ph, COOEt)	KOH	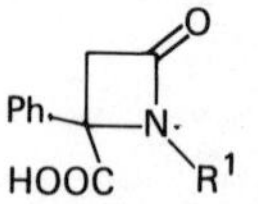 Ph–N–R^1, HOOC		439
		R^1 = Ph	90 (90)^a	
		R^1 = C$_6$H$_4$Cl-p	80	
		R^1 = C$_6$H$_4$Br-p	80	
		R^1 = C$_6$H$_4$Me-p	89 (90)^a	
ClCH$_2$CONCH(COOEt)$_2$ (with R^1)	DMF^b	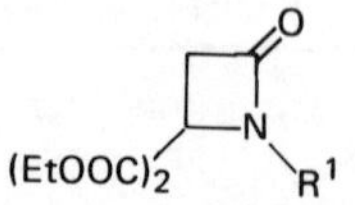 (EtOOC)$_2$–N–R^1		439
		R^1 = Ph	98	
		R^1 = C$_6$H$_4$Cl-p	95–98	
		R^1 = C$_6$H$_4$Br-p	95–97	
		R^1 = C$_6$H$_4$Me-p	95–98	
		R^1 = β-C$_{10}$H$_7$	95–98	

TABLE 29. (Continued)

Starting amide	Base	Lactam	Yield (%)	Reference
$ClCH_2CONCH_2CN$ $\quad\ \ \ \overset{\|}{CH(CH_3)C_6H_5}$	NaH		70	443
$\overset{R^1}{\overset{\|}{Br(CH_2)_2CON-CH(COOEt)_2}}$	KOH			440
		$R^1 = Ph$	85	
		$R^1 = C_6H_4Cl\text{-}p$	90	
		$R^1 = C_6H_4Br\text{-}p$	84	
		$R^1 = C_6H_4Me\text{-}p$	80	
$\overset{R^1}{\overset{\|}{Br(CH_2)_2CONCH}}\overset{-Ph}{\underset{COOEt}{}}$	KOH			440
		$R^1 = Ph$	85	
		$R^1 = C_6H_4Cl\text{-}p$	80	
		$R^1 = C_6H_4Br\text{-}p$	80	
		$R^1 = C_6H_4Me\text{-}p$	80	
$\overset{R^1}{\overset{\|}{Br(CH_2)_2CON-CH}}\overset{-COOH}{\underset{COOEt}{}}$	KOH			440
		$R^1 = Ph$	80	
		$R^1 = C_6H_4Cl\text{-}p$	86	
		$R^1 = C_6H_4Br\text{-}p$	80	
		$R^1 = C_6H_4Me\text{-}p$	80	

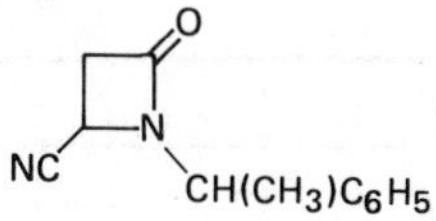

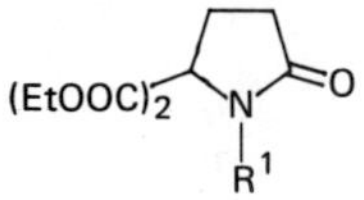

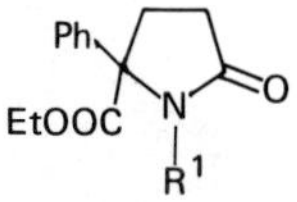

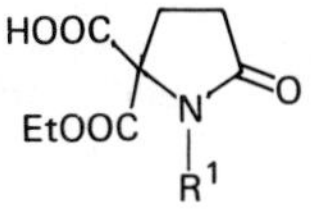

[a] Yield of ethyl ester obtained by heating sodium acetate and starting amide without solvent at 140–145°C.
[b] Reactions carried out in refluxing DMF without added base.

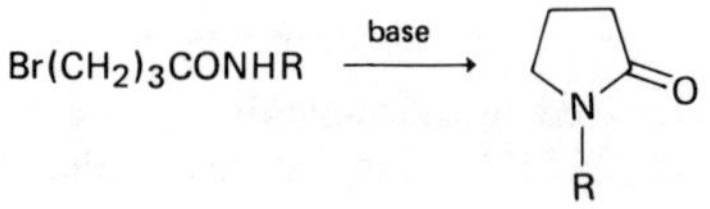

$$Br(CH_2)_3CONHR \xrightarrow{\text{base}} \text{} \quad \text{(Ref. 425)} \qquad (248)$$

R	Yield (%)
Ph	48
$o\text{-}BrC_6H_4$	50
$o\text{-}FC_6H_4$	61
$p\text{-}BrC_6H_4$	67
$p\text{-}ClC_6H_4$	54
$p\text{-}IC_6H_4$	79

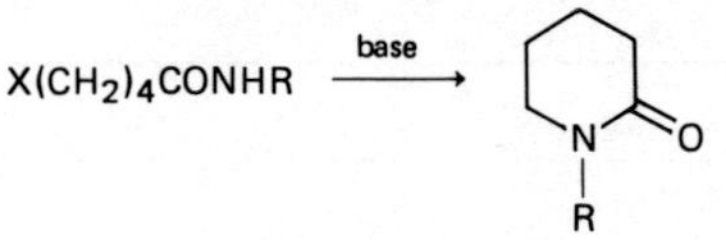

X(CH2)4CONHR
base
N
O
R

(Ref. 425, 426) (249)

X	R	Yield (%)
Br	Ph	61
Cl	H	82
Cl	Me	48
Cl	Et	33
Cl	*n*-Pr	28
Cl	*n*-Bu	36
Cl	*n*-C_8H_{17}	37
Cl	*c*-C_6H_{11}	11
Cl	Ph	96

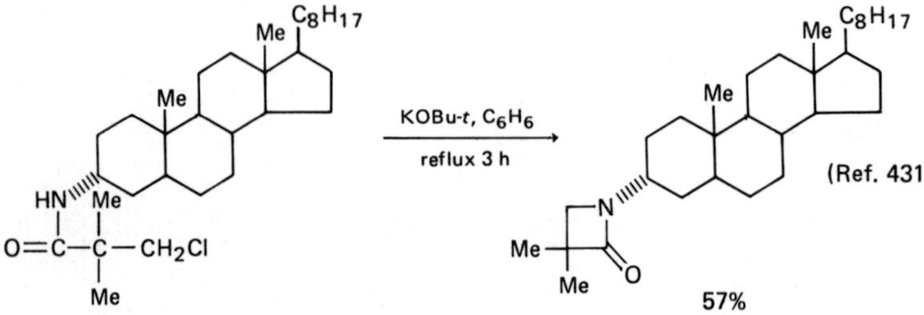

C8H17
Me
Me
HN
Me
O=C—C—CH2Cl
Me
KOBu-t, C6H6
reflux 3 h
N
Me
Me
O
57%

(Ref. 431) (250)

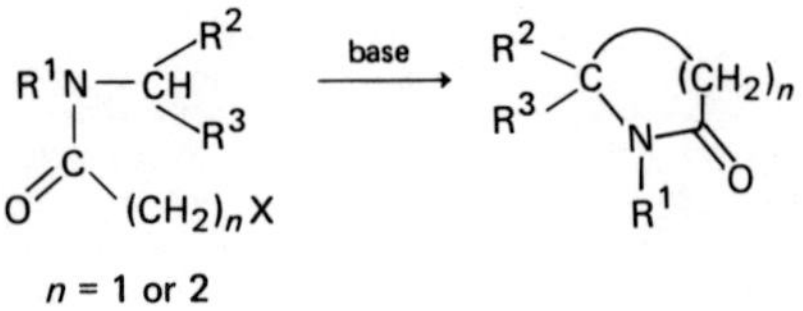

R1N—CH
R2
R3
C
O
(CH2)nX
n = 1 or 2
base
C
(CH2)n
N
R1
O

(251)

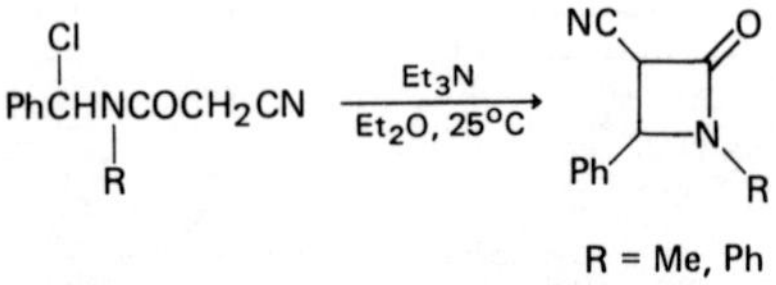

Cl
PhCHNCOCH2CN
R
Et3N
Et2O, 25°C
NC
O
N
Ph
R
R = Me, Ph

(252)

β-Lactams have also been prepared by base-catalysed cyclization of *N*-(α-chlorobenzyl)-β-cyanoamides (equation 252)[444] and by intramolecular Michael addition (equations 253–255)[445].

In addition to the nucleophilic displacements of halide ion shown above, *N*-aryl-α-halo amides can be cyclized via intramolecular Friedel–Crafts alkylation of the *N*-aryl moiety to produce oxindoles, as shown in the synthesis of 3-ethyl-1-methyloxindole from *N*-methyl-α-bromo-*n*-butyranilide (equation 256)[446].

An interesting approach to the cyclization of bromo amides may be seen in the reaction of *N*-(2-bromopropanoyl)aminoacetone with triethyl phosphite to afford an intermediate phosphonate ester, which can then be converted into 2-oxo-3,4-dimethyl-Δ^3-pyrroline via and intramolecular Wittig reaction (equation 257)[447].

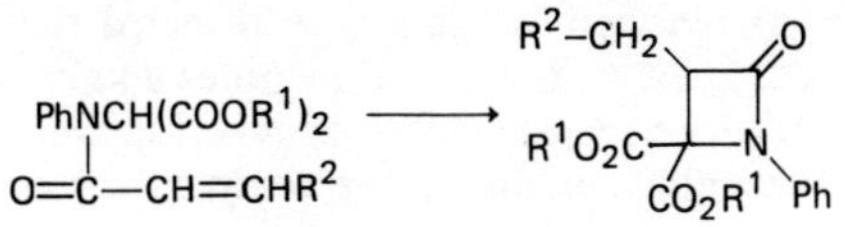

$$
\begin{array}{c}
\text{PhNCH(COOR}^1)_2 \\
\mid \\
\text{O}=\text{C}-\text{CH}=\text{CHR}^2
\end{array}
\longrightarrow
\quad (253)
$$

R^1	R^2	Yield (%)
Et	$p\text{-}O_2NC_6H_4$	94
Me	$p\text{-}O_2NC_6H_4$	74
Et	$o\text{-}O_2NC_6H_4$	70
Et	CO_2Et	–

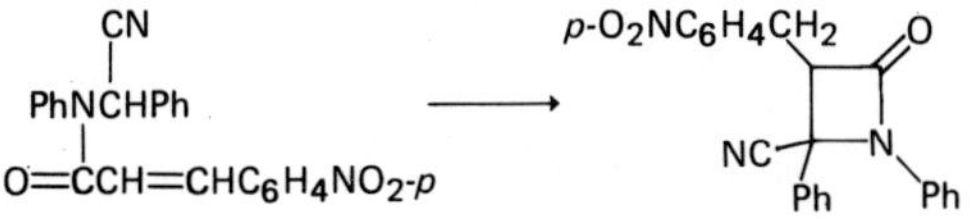

$$\quad (254)$$

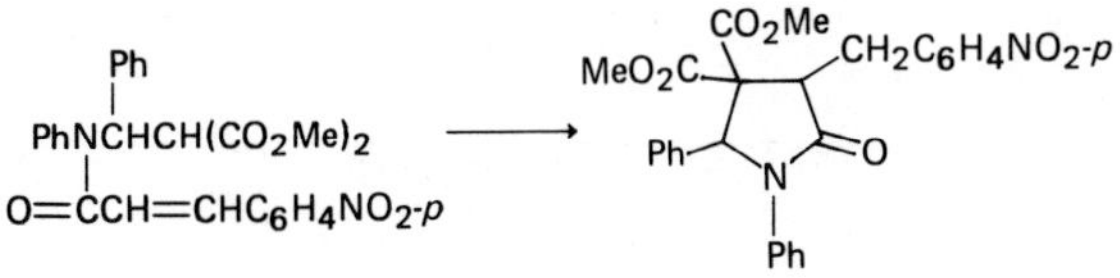

$$\quad (255)$$

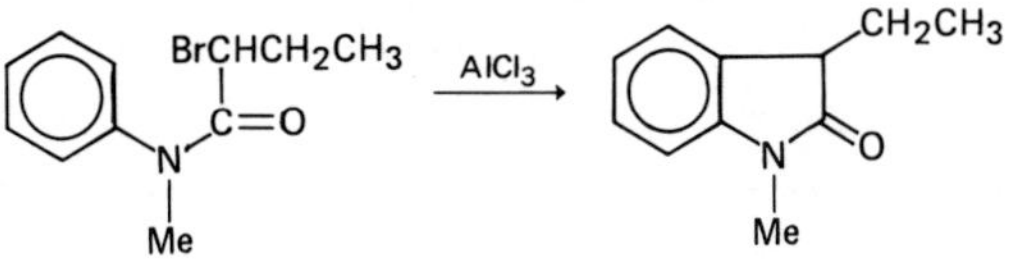

$$\quad (256)$$

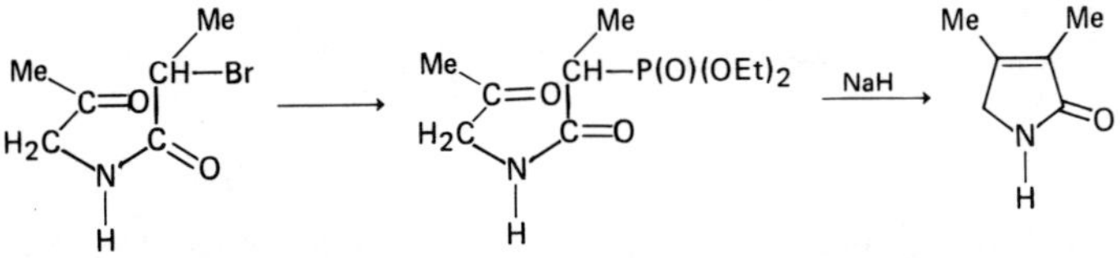

$$\quad (257)$$

γ- and δ-Hydroxy amides obtained from reactions of aldehydes and ketones with the dilithio derivatives of N-substituted benzamides[448] and N-substituted o-toluamides[449] have been cyclized in the presence of cold, concentrated sulphuric acid to form γ- and δ-lactams, respectively (equations 258 and 259)[450-452]. A

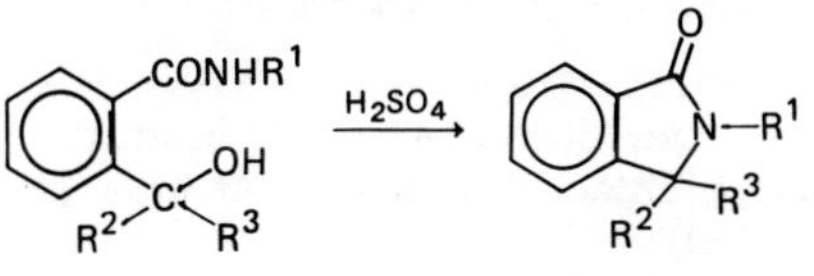

$$\quad (258)$$

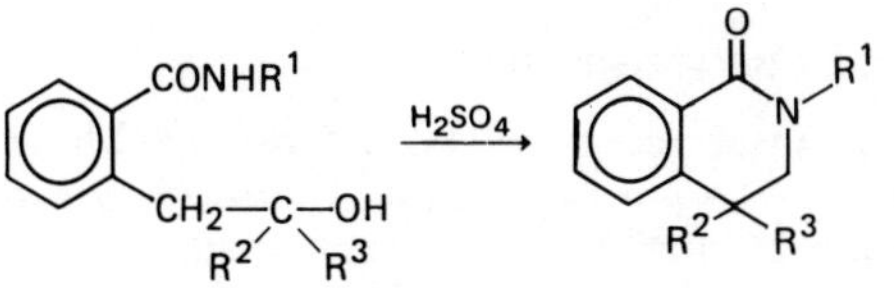

$$\quad (259)$$

mechanistic study[4 5 3] of reactions of this type revealed that in addition to lactam formation, linear dehydration to form olefin amides and cyclodeamination to form δ-lactones also occurred. The major course of reaction was found to be dependent upon the nature of the acidic medium, the temperature and the structure of the hydroxy amide.

A recent patent[4 5 4] claims the preparation of β-lactams by reaction of N-methyldiarylglycolamides with concentrated sulphuric acid in acetic acid (equation 260).

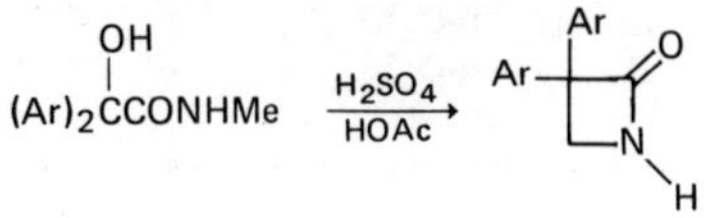

$$(Ar)_2 \overset{\overset{\displaystyle OH}{|}}{C}CONHMe \xrightarrow[\text{HOAc}]{H_2SO_4} \quad (260)$$

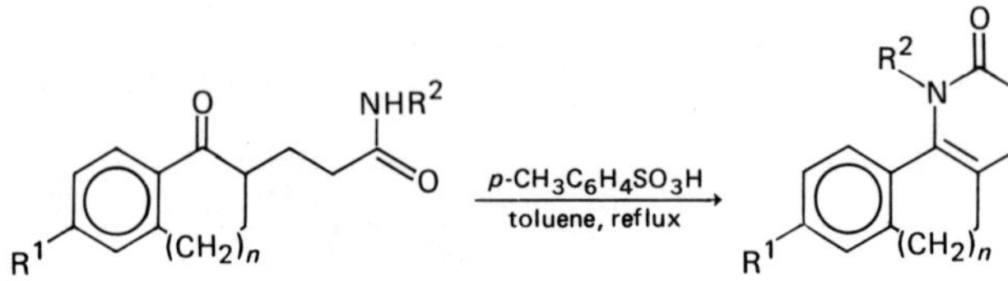

$$(261)$$

$$R^1 = H, R^2 = Me; n = 2$$
$$R^1 = R^2 = H; n = 1$$
$$R^1 = H, R^2 = Me; n = 1$$
$$R^1 = OMe, R^2 = Me; n = 1$$

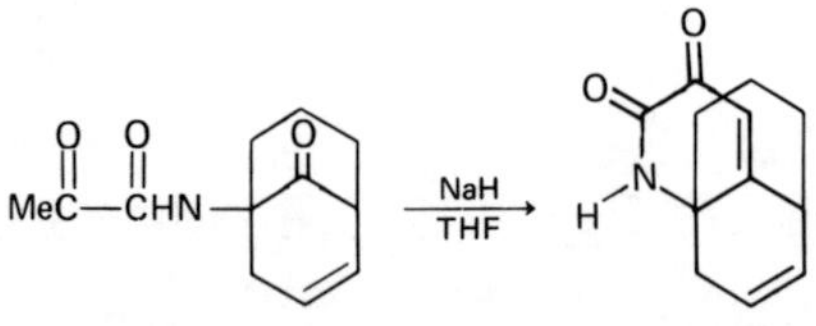

$$MeC\overset{\overset{\displaystyle O}{||}}{}—\overset{\overset{\displaystyle O}{||}}{C}HN— \xrightarrow[\text{THF}]{NaH} \quad (262)$$

(34)   (35)

Acid-catalysed cyclization of a series of δ-keto carboxamides has been found[4 5 5] to afford unsaturated lactams in 80—90% yield (equation 261).

An interesting intramolecular aldol cyclization of α-keto amide 34 afforded the tricyclic lactam 35 (equation 262)[4 5 6].

B. By Ring-closure Reactions (Photochemical)

A large variety of substituted amides have been found to produce lactams upon exposure to ultraviolet and ultraviolet—visible irradiation[4 5 7-4 9 2].

The type of lactam obtained is dependent upon the structural features of the starting amide (Table 30).

1. Cyclization of α,β-unsaturated amides

Irradiation of α,β-unsaturated anilides affords 3,4-dihydrocarbostyrils via ring-closure involving the ortho position of the N-aryl substituent and the β-carbon the acyl moiety (equation 263)[4 5 7-4 6 1]. Unsaturated amides possessing an N-heteroaryl

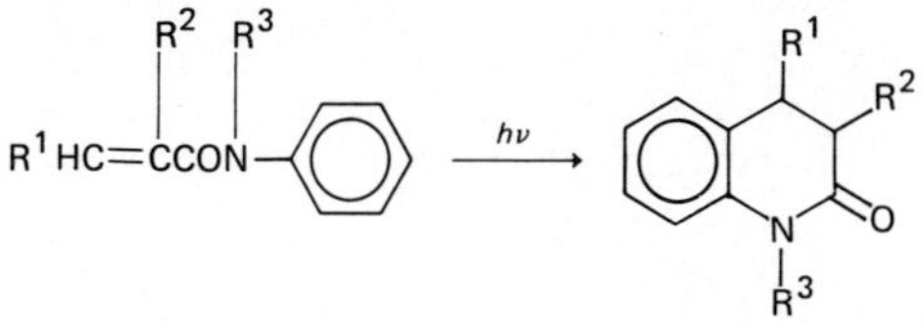

$$R^1HC{=}CCON{-}\bigcirc \quad \xrightarrow{h\nu} \quad \text{(product)} \qquad (263)$$

substituent react similarly (equation 264)[461]. In certain cases where $R^1 = R^2 = Ph$, β-lactam formation can become the major reaction pathway[457,458] (Table 30).

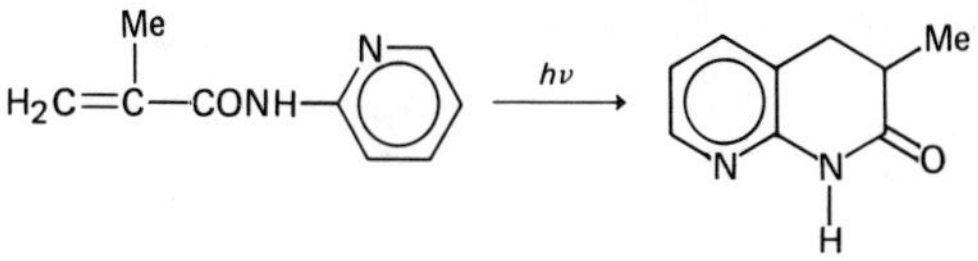

$$H_2C{=}C{-}CONH{-}\bigcirc \quad \xrightarrow{h\nu} \quad \text{(product)} \qquad (264)$$

2. Cyclization of benzanilides

Prolonged irradiation of a benzene solution of benzanilide in the presence of iodine produces phenanthridone in 20% yield (equation 265); however, without

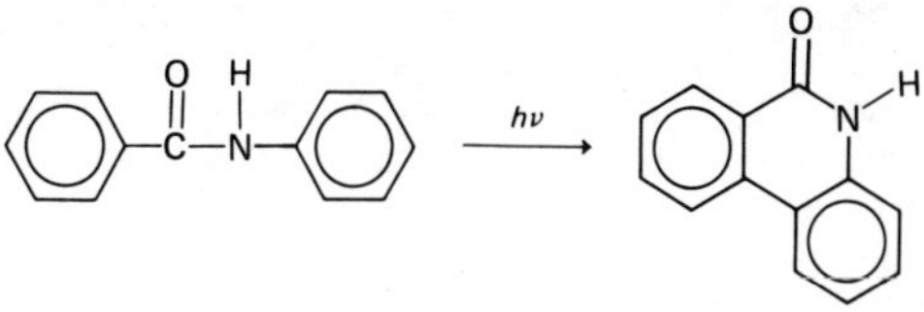

$$\bigcirc{-}\overset{O}{\underset{}{C}}{-}\overset{H}{\underset{}{N}}{-}\bigcirc \quad \xrightarrow{h\nu} \quad \text{(product)} \qquad (265)$$

iodine lactam formation drops to less than 1%[462]. The reaction proceeds more satisfactorily of one or the other of the aromatic residues contains an *ortho* halogen or methoxy group (equation 266)[462–465]. Anilides of thiophene-2-carboxylic acid,

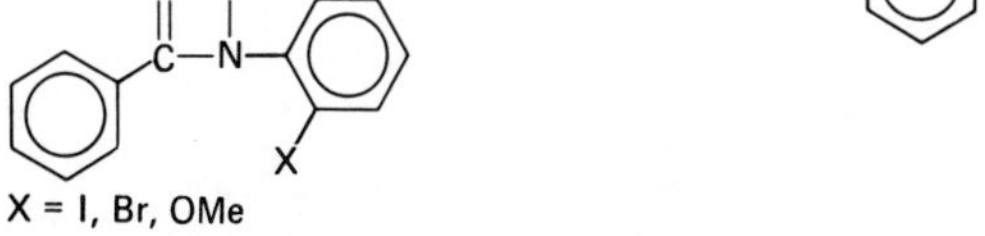

$$\begin{matrix} X = I, Br \\ \\ X = I, Br, OMe \end{matrix} \quad \xrightarrow{h\nu} \quad \text{(product)} \qquad (266)$$

furan-2-carboxylic acid, indole-2-carboxylic acid and indole-3-carboxylic acid participate in similar photoinduced cyclizations (Table 30).

3. Cyclization of enamides

Various enamides of the general type shown in equation (267) have been cyclized in connection with the synthesis of a number of isoquinoline alkaloids[465,466]. Related enamide photocyclizations appear in Table 30.

TABLE 30. Preparation of lactams by intramolecular photocyclization

Amide	Conditions	Product	Yield (%)	Reference
Ph │ PhCH=CCONH₂ *cis*	C₆H₆, 70h		13 3	457, 458
Ph │ PhCH=CCONHPh *cis*	C₆H₆, 23 h		37 2.3 5	457, 458
MeCH=CMeCONHPh *cis*	Ether, HOAc, 6 h	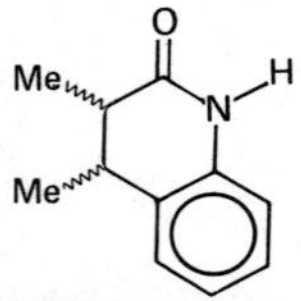	58	459

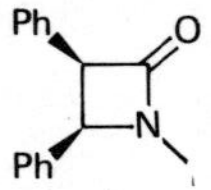

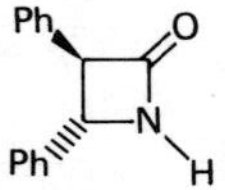

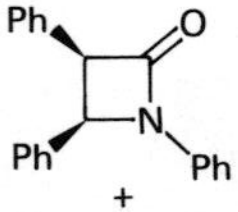

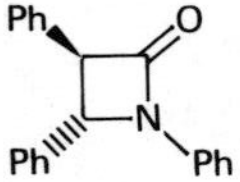

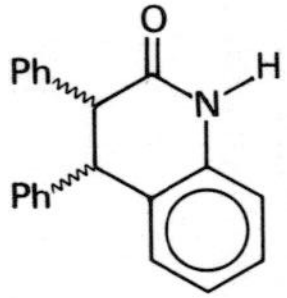

Reactant	Conditions	Product	Yield	Reference
MeCH=CHCONHPh *trans*	Ether, HOAc, 5 days	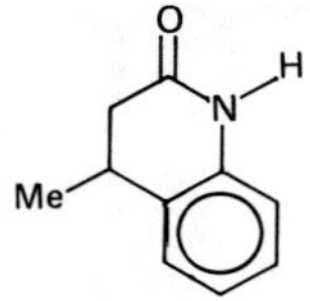 (4-Me-3,4-dihydroquinolin-2(1H)-one)	25	459
H_2C=CMeCONHPh	Ether, HOAc, 9 h	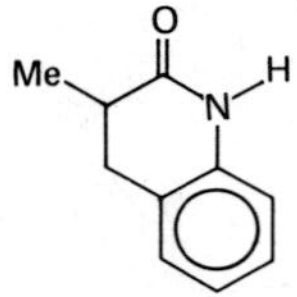 (3-Me-3,4-dihydroquinolin-2(1H)-one)	50 (82ᵃ)	459
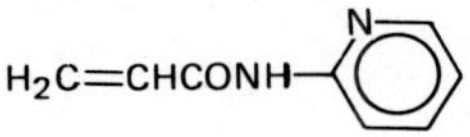 H_2C=CHCONH– (pyridyl)	C_6H_6, HOAc, 3 h	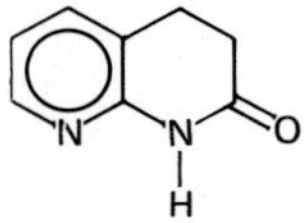 (naphthyridinone)	17	460
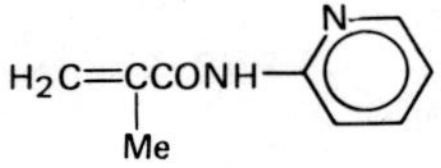 H_2C=CCONH– (Me, pyridyl)	C_6H_6, HOAc, 3 h	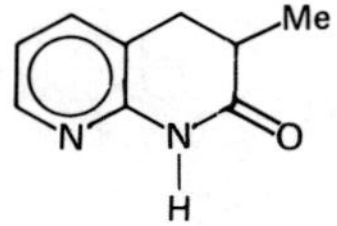 (Me-naphthyridinone)	78	460
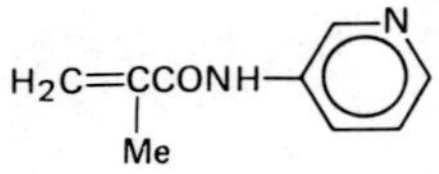 H_2C=CCONH– (Me, pyridyl)	C_6H_6, HOAc, 3 h	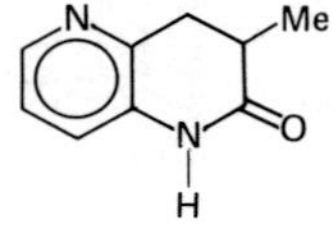 (Me-naphthyridinones) +	53 / 22	460

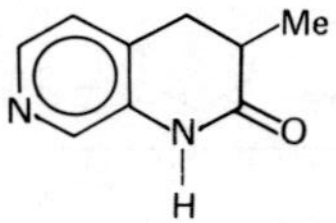

TABLE 30. (Continued)

Amide	Conditions	Product	Yield (%)	Reference
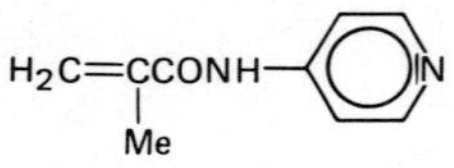	C₆H₆, HOAc, 3 h	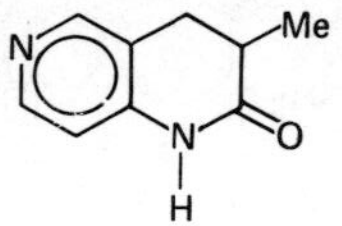	72	460
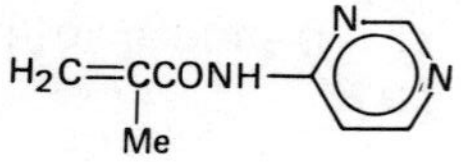	C₆H₆, HOAc, 3 h	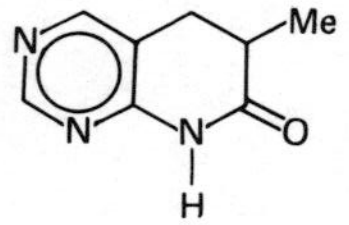	24	460
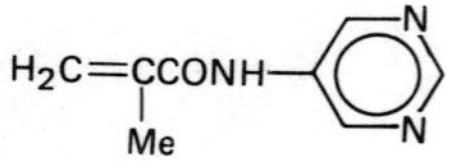	C₆H₆, HOAc, 3 h	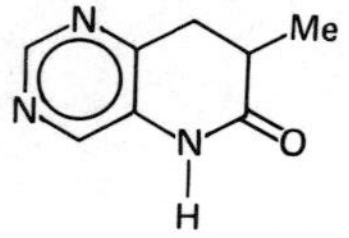	14	460
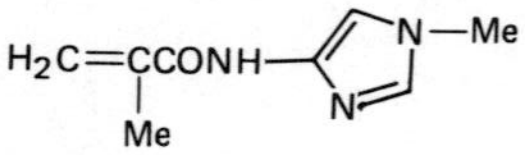	C₆H₆, HOAc, 3 h	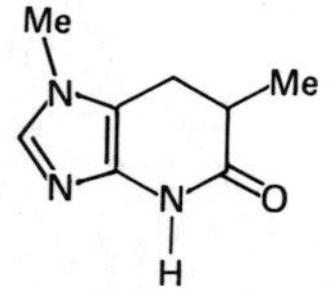	6	460
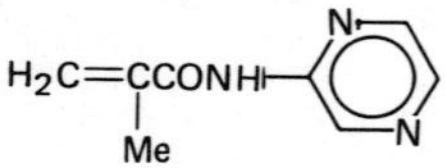	C₆H₆, HOAc, 3 h	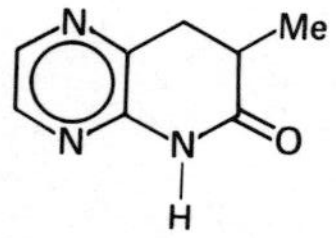	19	460
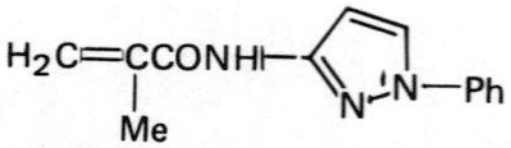	C₆H₆, HOAc, 3 h	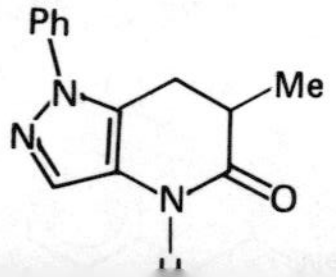	3	460

Reactant	Conditions	Product	Yield (%)	Reference
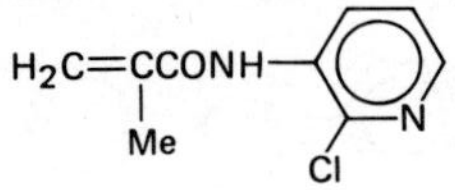 $H_2C=CCONH$– Me (pyridine, Cl)	C_6H_6, HOAc, 3 h	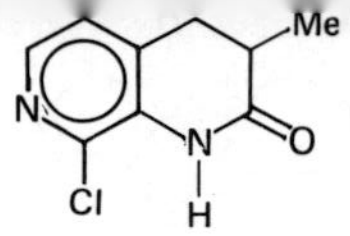 (lactam, Me, Cl, N)	25	460
$R^3CH=CCONR^1Ph$ / R^2	C_6H_6, r.t.	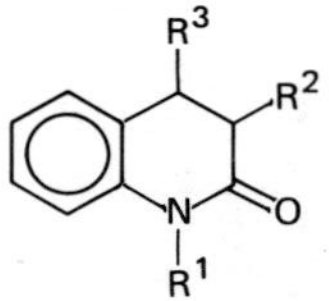 (ring with R^2, R^3, R^1)		461
$R^1 = R^2 = R^3 = H$	150 h		4	
$R^1 = R^2 = Me$, $R^3 = H$	80 h[b]		57	
$R^1 = R^2 = H$, $R^3 = Me$	80 h[b]		61	
PhCONHPh	C_2H_6, I_2, 148 h	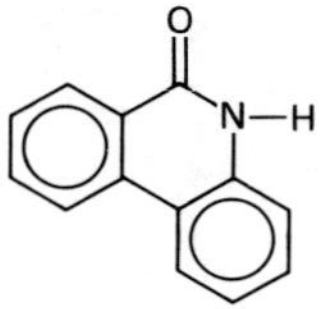 (phenanthridinone)	20	462
(CONHPh, I)	C_6H_6, 126 h	(phenanthridinone)	9	462
	C_6H_6, r.t. 30 h		18[c]	463
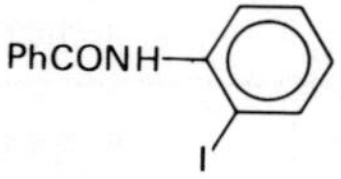 PhCONH– (I)	C_6H_6, 160 h	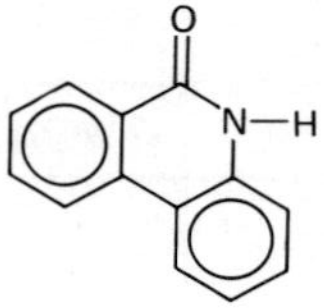 (phenanthridinone)	48	462

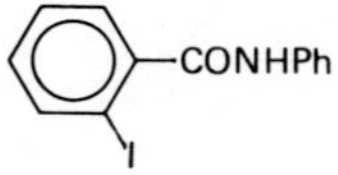

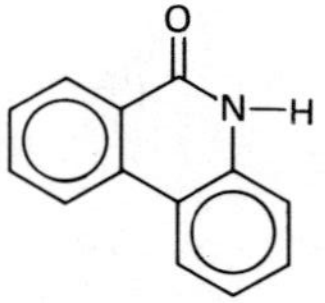

TABLE 30. (Continued)

Amide	Conditions	Product	Yield (%)	Reference
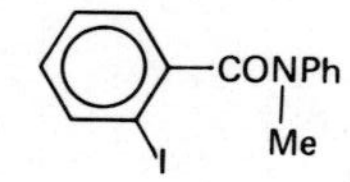	C$_6$H$_6$, 30 h, r.t.	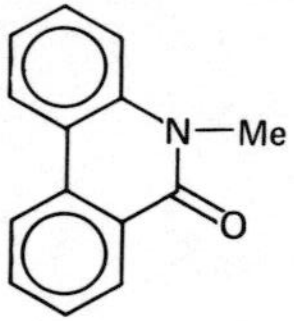	34–36	463
		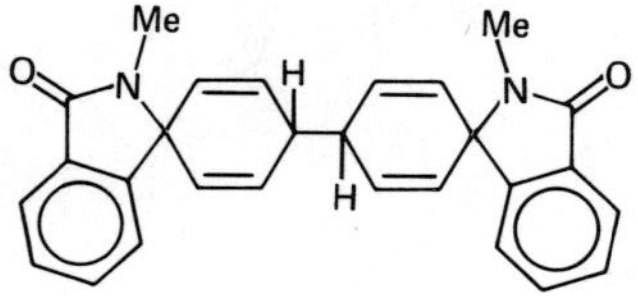	30–36[a]	
		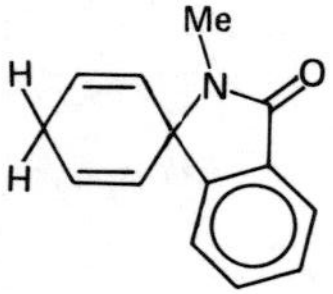	4–5	
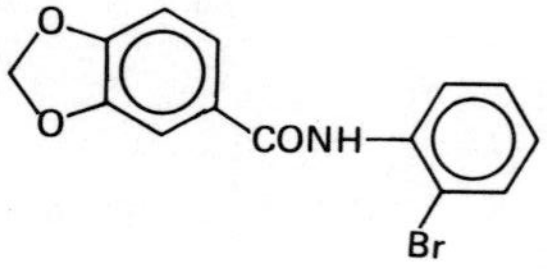	C$_6$H$_6$, MeOH, 24 h	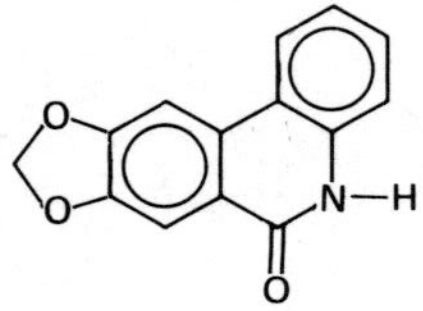	15–20	464
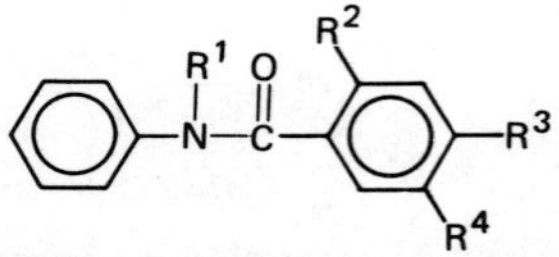		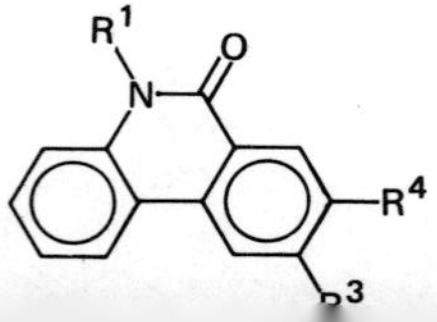		465

$R^1 = R^3 = R^4 = H, R^2 = OMe$
$R^1 = R^3 = R^4 = H, R^2 = O_2CMe$
$R^1 = H, R^2 = R^3 = R^4 = OMe$
$R^1 = Me, R^2 = R^3 = R^4 = OMe$

MeOH — 0
EtOAc, 21 h — 0
EtOAc, 20 h — 55
EtOAc, 12 h — 41

465

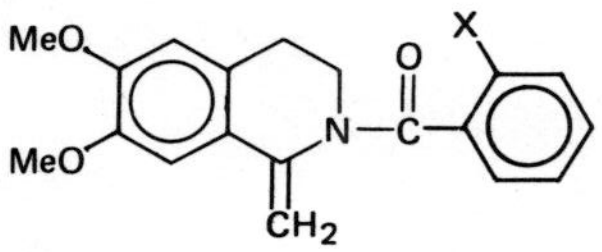

$X = F$
$X = Cl$
$X = Br$
$X = O_2CMe$
$X = SMe$
$X = NO_2$

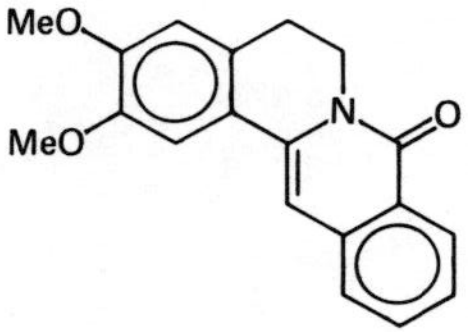

t-BuOH, 2 h

85
50
50
76
55
17

465

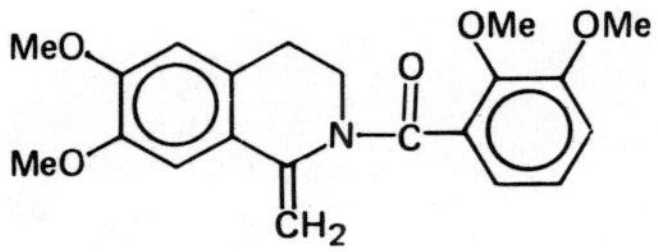

C_6H_6, 4.5 h — 85

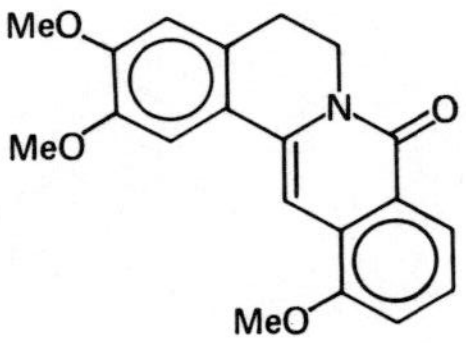

465

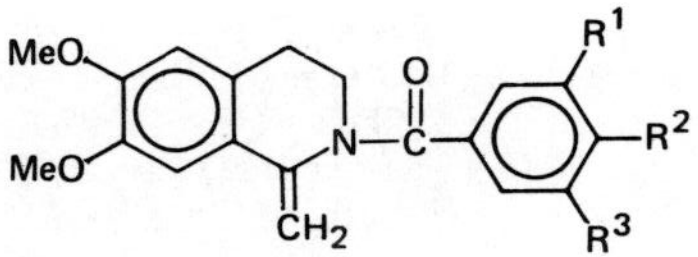

t-BuOH

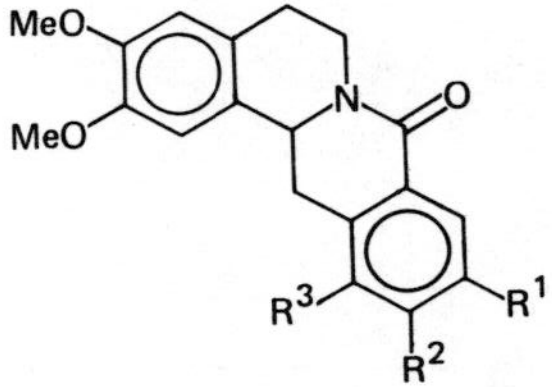

466

TABLE 30. (Continued)

Amide	Conditions	Product	Yield (%)	Reference
$R^1 = R^2 = R^3 = H$	1.5 h		97	
$R^1 = R^2 = OMe, R^3 = H$	2.5 h		94	
$R^1 R^2 = -OCH_2O-, R^3 = OMe$	2.5 h		75	
$R^1 = R^2 = R^3 = OMe$	5 h		70	
$R^1 = OMe, R^2 = O_2CMe, R^3 = H$	12 h		45	
$R^1 = R^3 = H, R^2 = Me$	4.5 h		85	
$R^1 = R^3 = H, R^2 = Cl$	16 h		75	
$R^1 = R^3 = H, R^2 = Ph$	12 h		76	
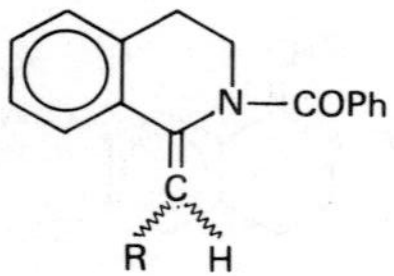 (N—COPh structure, $R = H, Me$)	MeOH, 1–20 h, r.t.	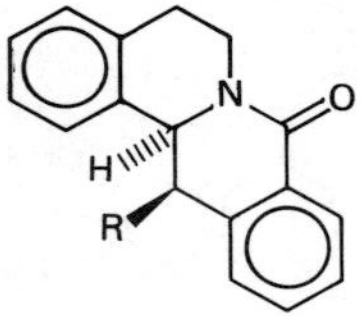	~70	471
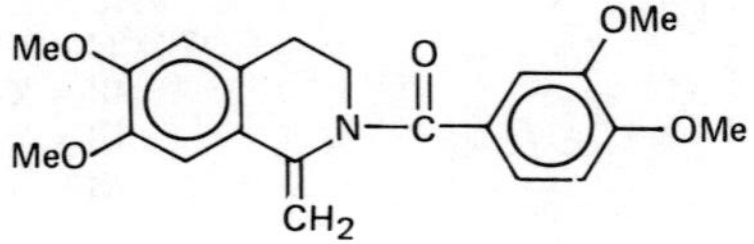 (3,4-dimethoxybenzoyl CH_2 structure)	MeOH, 3 h, r.t.	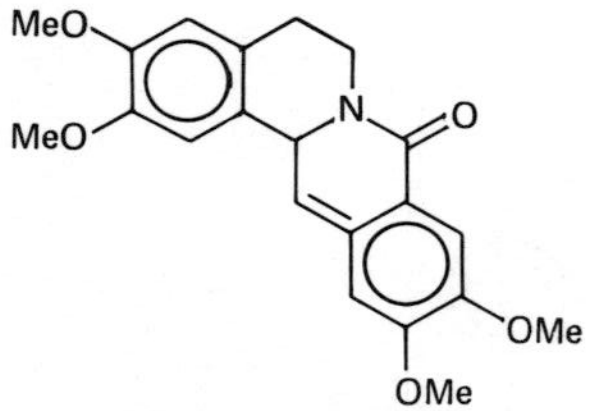	5	471
		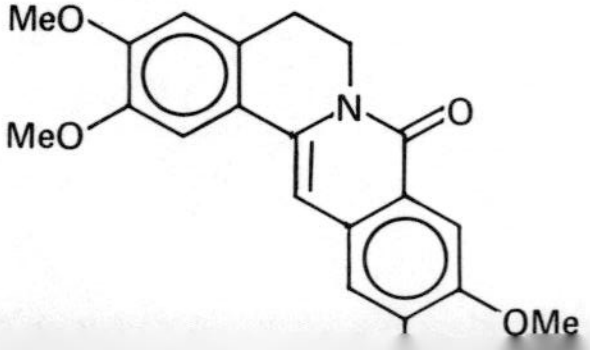	40	

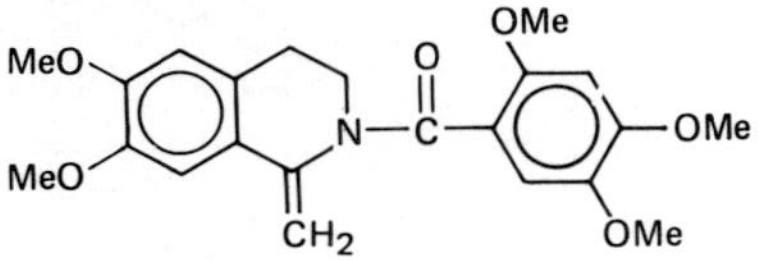

t-BuOH, 2 h — 85 — 465

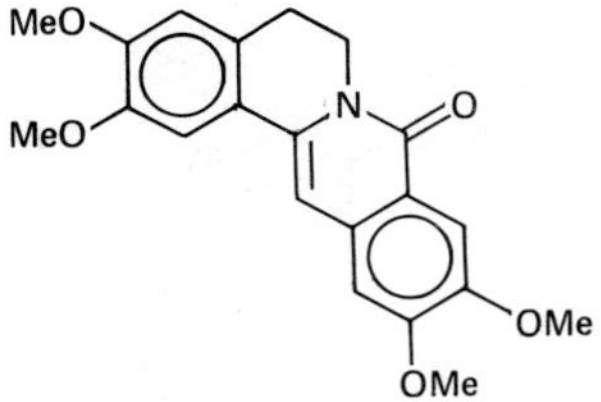

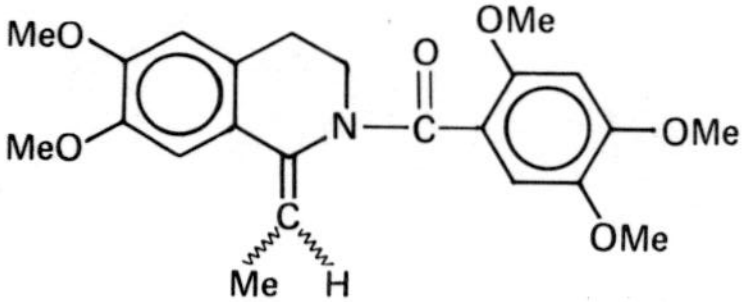

EtOAc, 12 h — 69 — 465

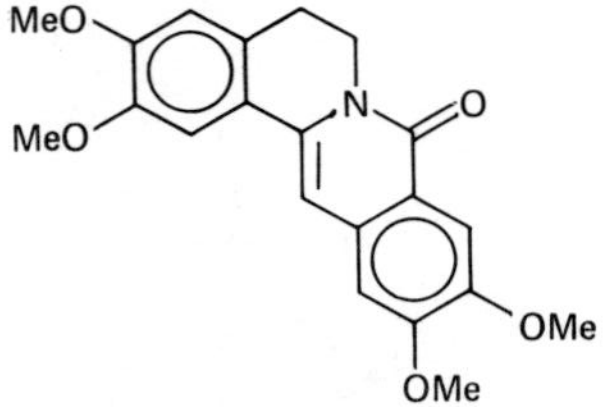

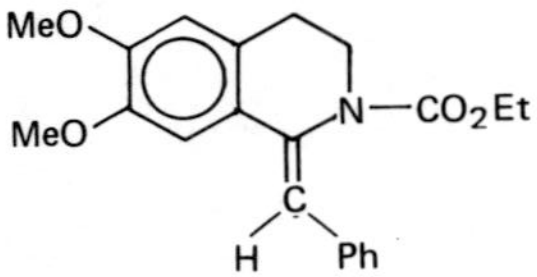

EtOH, 120 h — 64 — 465

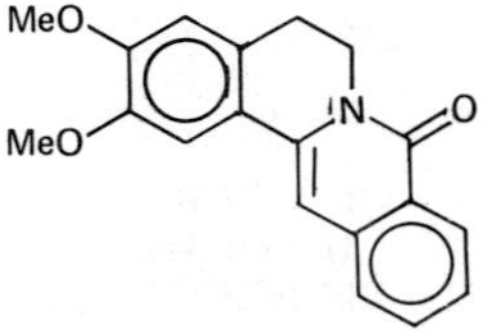

MeOH, 40 h — R = CH$_2$C$_6$H$_5$ 35 / R = Me 15 — 475

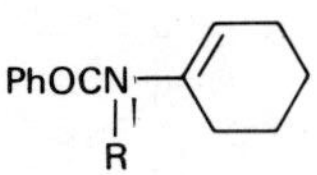

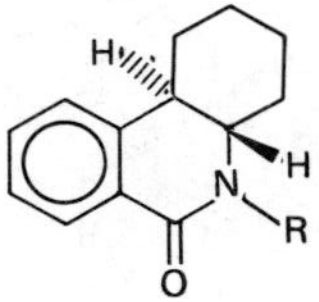

trans

PhOCN — R

TABLE 30. (Continued)

Amide	Conditions	Product	Yield (%)	Reference
	MeOH	$R^1 = CH_2C_6H_5, R^2 = H$ $R^1 = Me, R^2 = H$ $R^1 = CH_2C_6H_5, R^2 = Me$	55 51[e] 20	474
	40 h 40 h 106 h			
	MeOH, r.t., 40 h	$R = CH_2C_6H_5$ $R = Me$	55 51	477
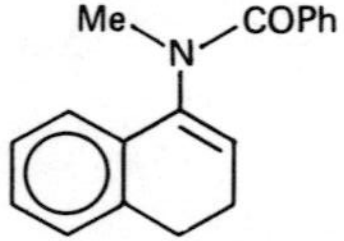	MeOH, I$_2$, 40 h	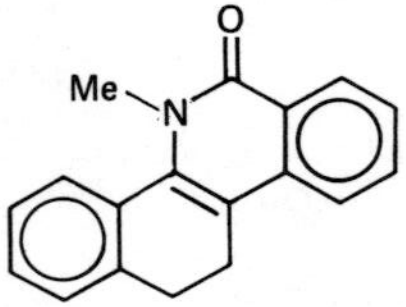	21	474
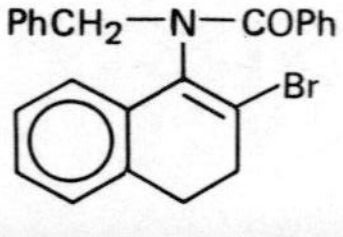	1. MeOH, 15 h 2. KOH, MeOH, reflux 1.5 h	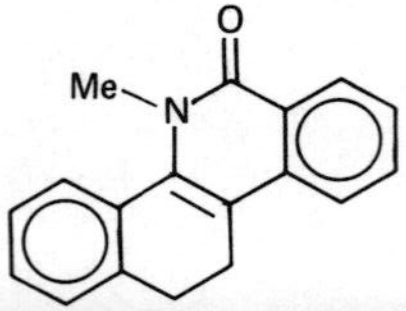	28	474

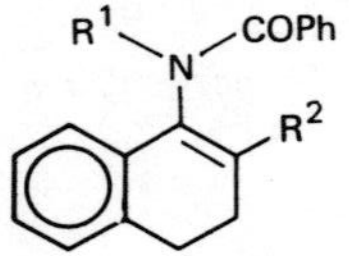

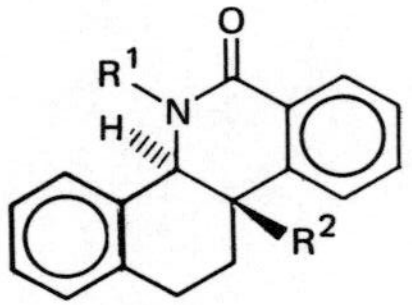

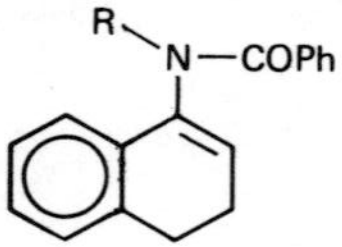

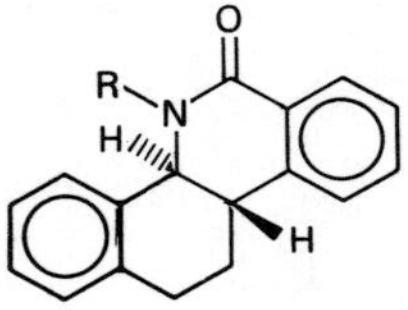

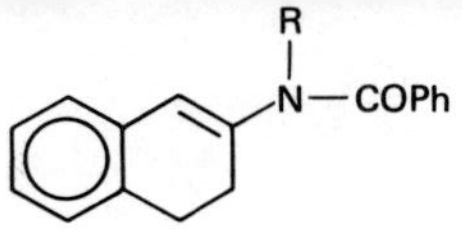

		Conditions	Product	Yield (%)	Ref.
R—N—COPh		Ether, r.t., 24 h		$R = CH_2CH=CH_2$ 71 $R = n\text{-Bu}$ 63 $R = Me$ 40 $R = CH_2C_6H_5$ 47	476
—N—COCH=CH$_2$, CH$_2$Ph		Ether, r.t.	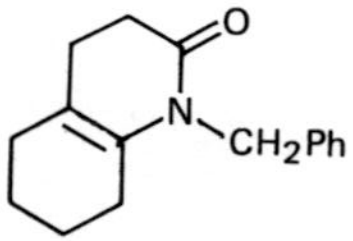	61^f	473
PhCH$_2$—N—COCH=CH$_2$		Ether, r.t.	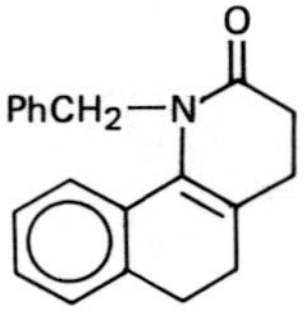	42	473
CH$_2$CH$_2$NHCOCH$_2$X , HO— , X = Cl, X = I		EtOH, H$_2$O, 2 h, 20 min		70, 11	478
CH$_2$CHNHCOCH$_2$Cl, COOH, HO—		H$_2$O, NaOH, (pH 6.5), 45 min		25	478

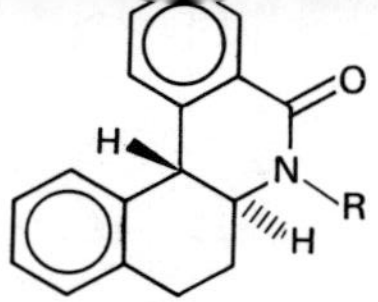

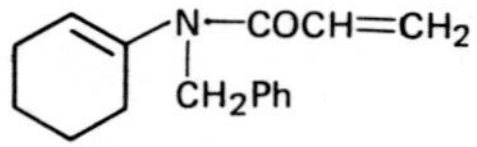

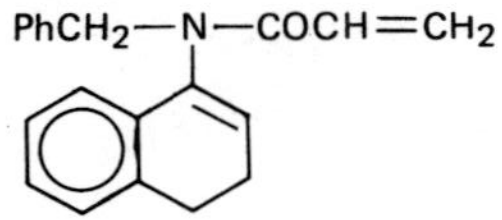

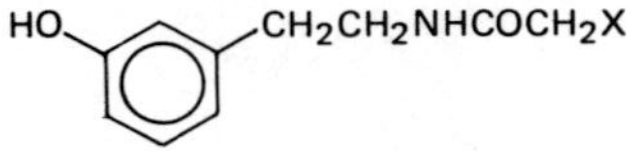

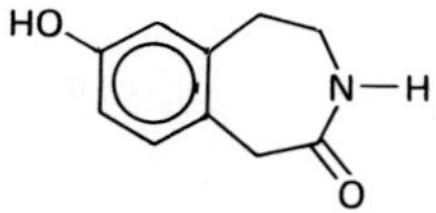

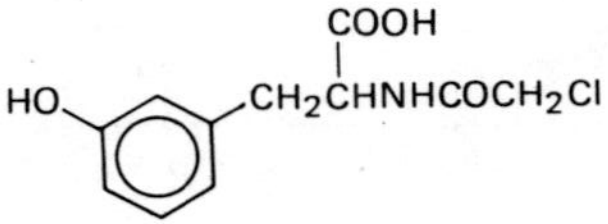

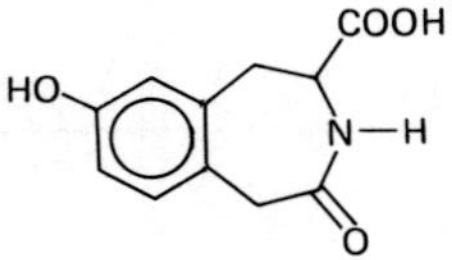

TABLE 30. (Continued)

Amide	Conditions	Product	Yield (%)	Reference
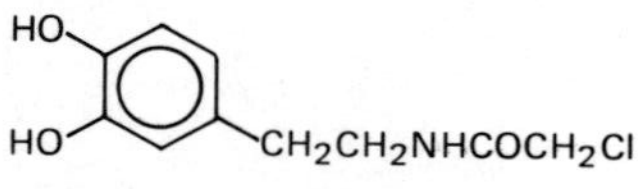	MeOH, (pH 6), 1 h	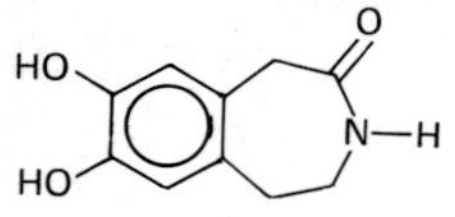	—	478
	EtOH, H$_2$O	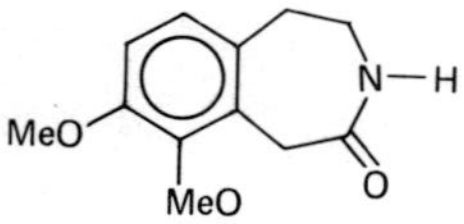	9 (4)[g]	480
		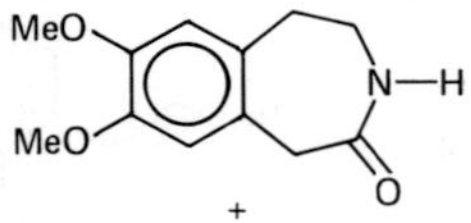	33 (6)[g]	
		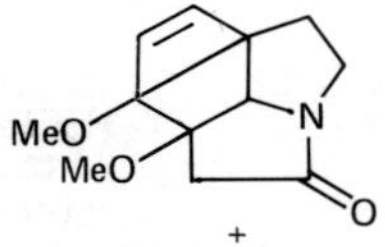	11 (10)[g]	
		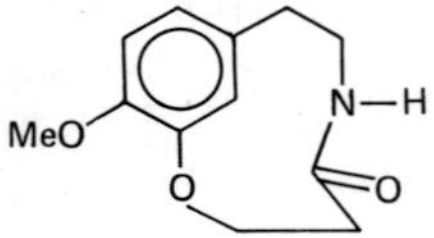	0 (4)[g]	

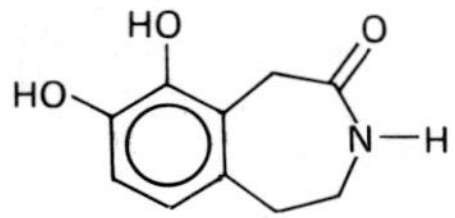

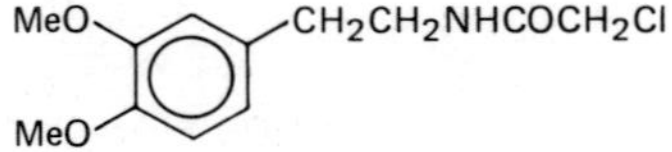

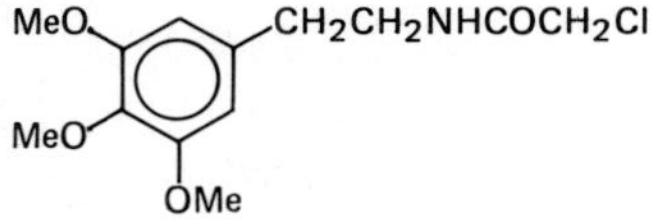

481

10

EtOH, H$_2$O

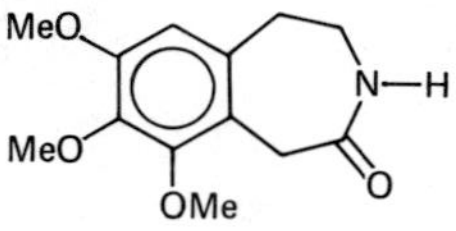

33

47

482

H$_2$O, THF,
NaOAc, 5 h

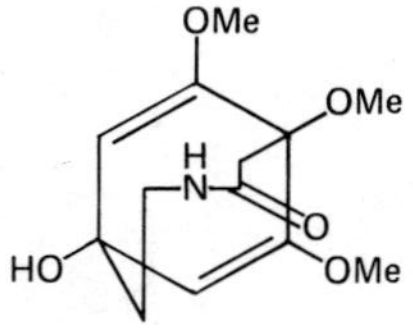

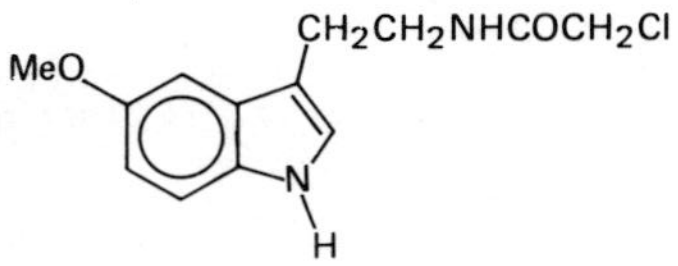

18

H$_2$O, NaBH$_4$, (pH
9.5–10), 30 min

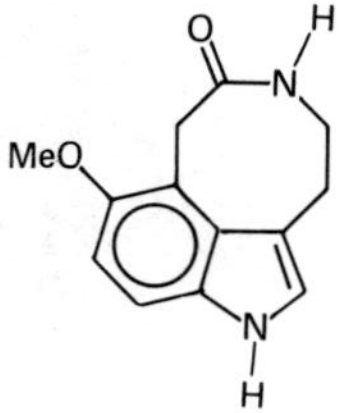

483

CH$_2$CH$_2$NHCOCH$_2$Cl

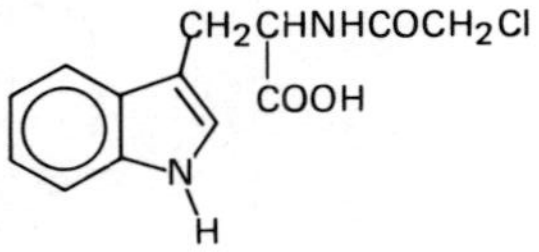

CH$_2$CH$_2$NHCOCH$_2$Cl

CH$_2$CHNHCOCH$_2$Cl
COOH

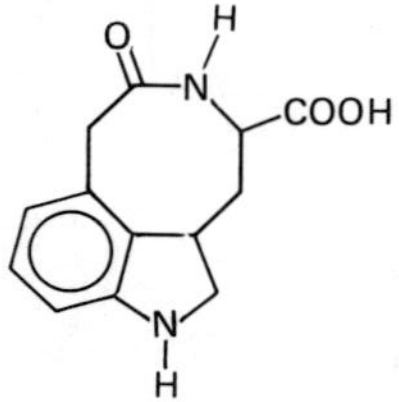

TABLE 30. (Continued)

Amide	Conditions	Product	Yield (%)	Reference
	NaOH, H₂O, 45 min		18	484
	H₂O, 45 min		40	
R = Me, Ph	*hv*		1.2 (R = Me) 2.4 (R = Ph)	
			1.6 (R = Me or Ph)	
		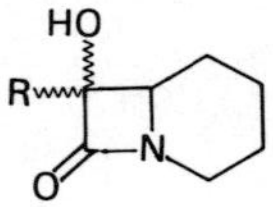	1.2 (R = Me) 1.0 (R = Ph)	
	C₆H₆, N₂, *hv*, 0–10°C	X = S X = SO X = SO₂	11 8–40 70	485

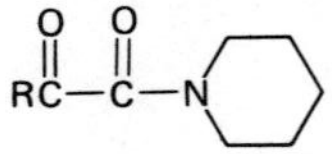

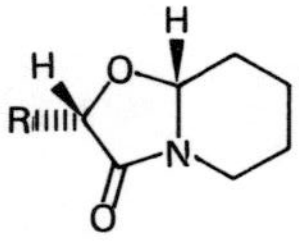

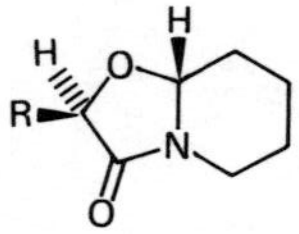

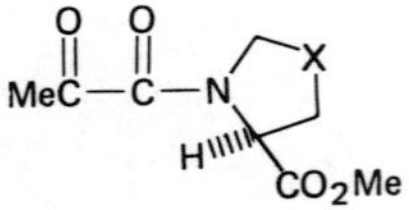

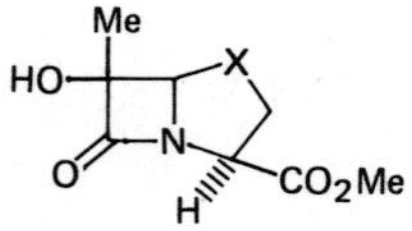

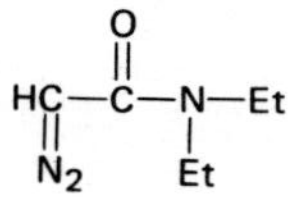 Dioxane	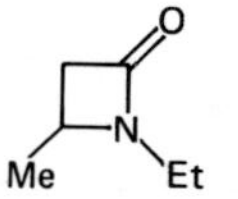 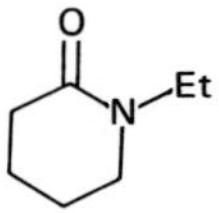 $57\ (43)^h$ $43\ (5)^h$		486
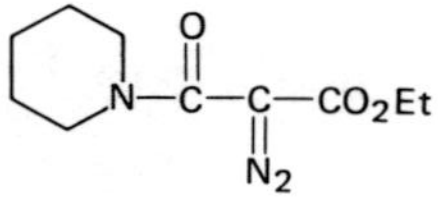 CCl_4, r.t., 2 h	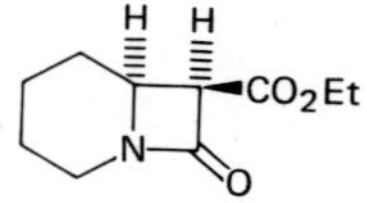 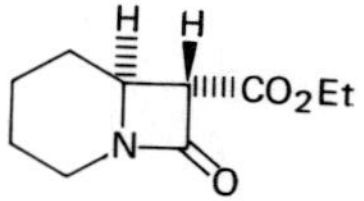 80 (*cis*:*trans*, 1:2)	CO_2Et / CO_2Et	488
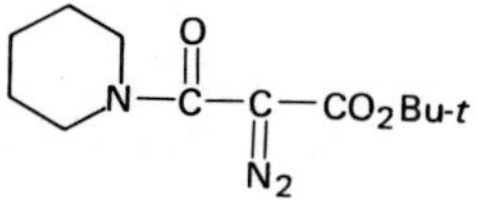 CCl_4, r.t., 1 h	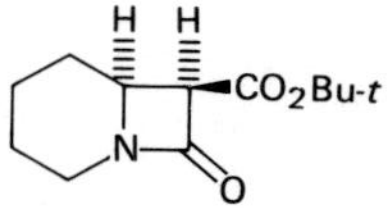 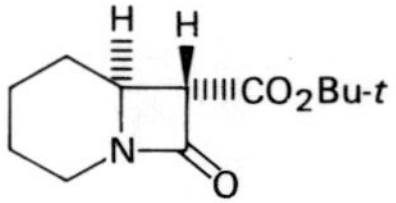 14 40	$CO_2Bu\text{-}t$ / $CO_2Bu\text{-}t$	488
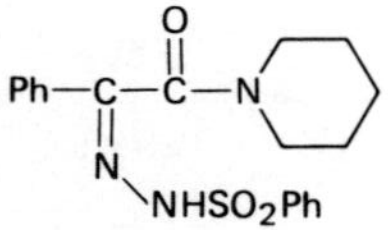 1. NaH, 60°C 2. hv, CH_2Cl_2, 10°C	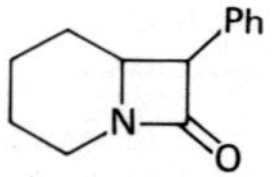 50	Ph	489

Reactant structures:

$HC=C-N-Et$, N_2 , O , Et (diazo amide)

$O=C-C-CO_2Et$, N_2 (piperidine diazo ester)

$O=C-C-CO_2Bu\text{-}t$, N_2 (piperidine diazo ester)

$O=C-C-Ph$, $N=N-NHSO_2Ph$ (piperidine)

TABLE 30. (Continued)

Amide	Conditions	Product	Yield (%)	Reference
$m\text{-RC}_6\text{H}_4\text{C}$—$\overset{\text{O}}{\overset{\|}{\text{C}}}$—$\overset{\overset{\text{CH}_2}{}}{\text{N}}$—$(\text{CH}_2)_n$ (with N_2)	Heat only	Ph-substituted β-lactam with $(\text{CH}_2)_n$		487
$R = H; n = 4$			40	
$R = H; n = 3$			—	
$R = H; n = 5$			—	
$R = \text{NO}_2; n = 4$			32	
$t\text{-BuO}_2\text{C}$—$\overset{\text{O}}{\overset{\|}{\text{C}}}$—$\underset{\text{N}_2}{\text{C}}$—$\text{N}$ (thiazolidine, $\text{CO}_2\text{CH}_2\text{Ph}$)	CCl_4, r.t., 1 h	penam system ($\text{CO}_2\text{Bu-}t$, $\text{CO}_2\text{CH}_2\text{Ph}$)	50	488
$R^1\text{COCH}_2\text{CON}$—$\text{CH}_2R^2$ with CH_2R^3	C_6H_6	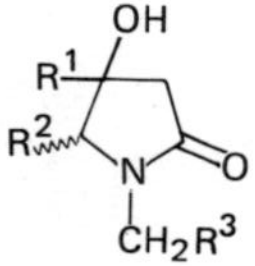 pyrrolidinone (R^1, R^2, OH, CH_2R^3)		467
$R^1 = R^2 = R^3 = \text{Ph}$			80	
$R^1 = \text{Ph}, R^2 = R^3 = \text{H}$			88	
$R^1 = \text{Ph}, R^2 = R^3 = \text{Me}$			60	
$R^1 = \text{Me}, R^2 = R^3 = \text{H}$			73	
$R^1 = R^2 = R^3 = \text{Me}$			76	
$R^1 = \text{Ph}, R^2, R^3 = -\text{CH}_2\text{OCH}_2-$			80	

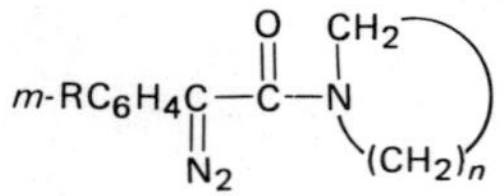

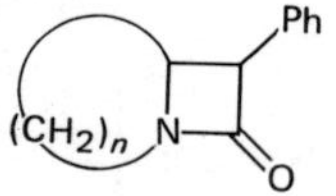

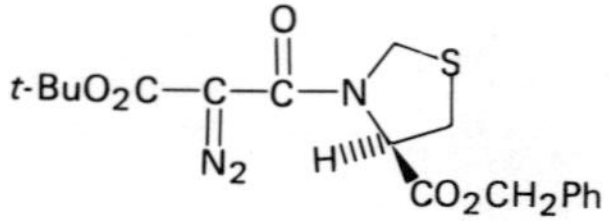

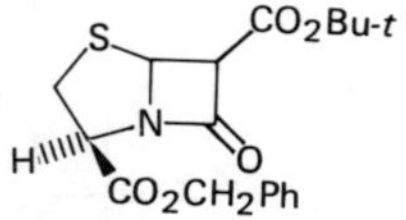

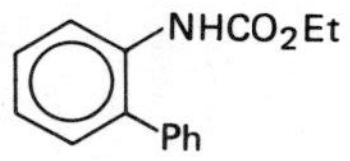

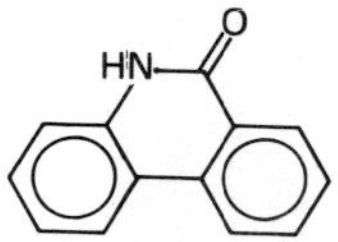

35^i 468

a Yields based upon recovered starting material.
b The effect of solvent was studied in the photocyclizations and the results obtained are shown below:

Solvent	= MeCN,	MeOH,	Me$_2$C=O,	i-PrOH,	n-PrBr,	Et$_2$O,	C$_6$H$_6$,	n-C$_6$H$_{14}$
Yield (%)	= 0,	0,	0,	0,	22,	24,	33,	63
Irrad. time (h)	= 8,	8,	8,	8,	8,	8,	8,	5 min

c This product was also obtained in 35−37% yield by the copper-catalysed decomposition[465] of N-methylbenzanilide-2-diazonium fluoroborate.
d This product was also obtained in 38−40% yield by the copper-catalysed decomposition[465] of N-methylbenzanilide-2-diazonium fluoroborate.
e This product was also prepared by sodium in liquid ammonia reduction of trans-5-benzyl-4b,10b,11,12-tetrahydrobenzo[c]phenanthridin-6-(5H)-one followed by methylation of the resulting trans-4b,10b,11,12-tetrahydrobenzo[c]phenanthridin-6-(5H)-one.
f This product was also prepared by the reaction of benzylamine with methyl 3-(2-cyclohexanone-1)propionate.
g Yield of product when irradiation was performed in THF.
h Yields of product when irradiation was performed in methanol.
i By an analagous reaction (irradiation of ethyl o-biphenylyl carbonate) the corresponding lactone was prepared also in 85% yield.

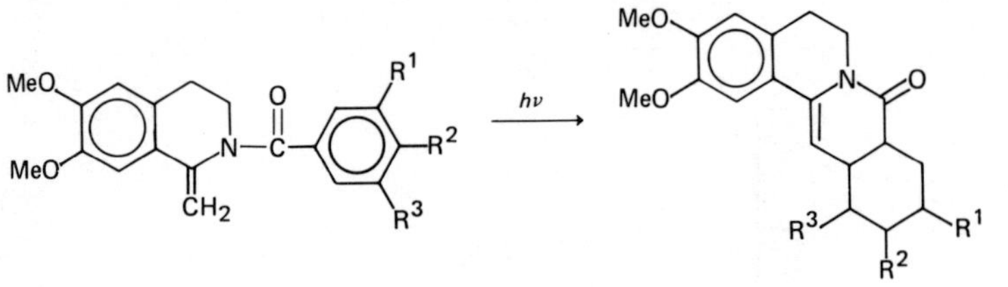

$$(267)$$

4. Cyclization of N-chloroacetyl-β-arylamines

These photocyclizations may be generalized by equation (268). A majority of such reactions have been carried out with N-chloroacetyl-β-phenylethyl amines

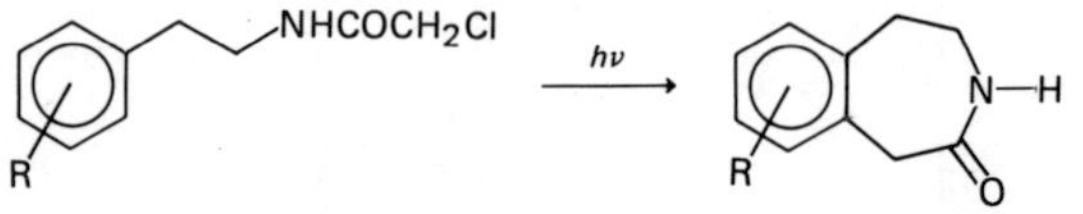

$$(268)$$

containing one or more electron-furnishing groups in the aromatic ring (Table 30). It is interesting to note that the N-chloroacetyl derivatives of the biologically important amines — tryptamine, tyramine, dopamine and normescaline — participate in these cyclizations

5. Cyclization of α-diazocarboxamides

Certain β-lactams have been synthesized by photolysis of α-diazocarboxamides. These reactions proceed by photolytic decomposition of the azo compound to form a carbene intermediate, which then undergoes insertion into a carbon—hydrogen bond of the N-alkyl substituent.

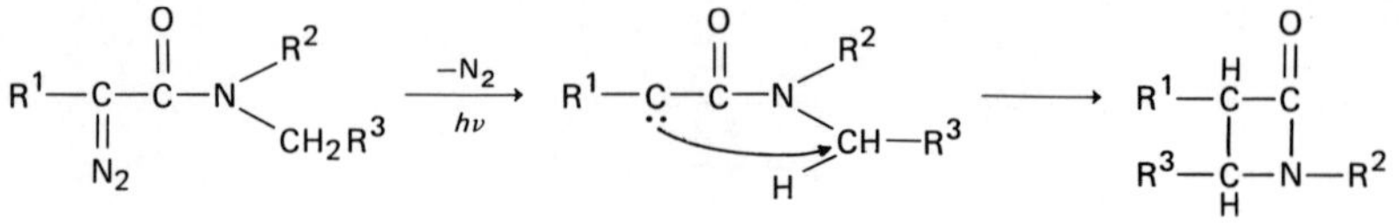

$$(269)$$

6. Miscellaneous cyclizations

The last several entries in Table 30 represent miscellaneous photocyclizations involving β-keto amides[467] and carbamates[468].

C. By Cycloaddition Reactions

1. Addition of isocyanates to olefins

In theory, cycloaddition of isocyanates to olefins should lead directly to β-lactams. In practice, however, simple N-alkyl and N-aryl isocyanates add only to electron-rich olefins such as enamines[493], while successful cycloadditions with simple olefins require the use of an 'activated' isocyanate possessing a strong electron-withdrawing substituent on nitrogen. Since its discovery in 1956[494] chlorosulphonyl isocynate (CSI)[495] has emerged as one of the most widely used isocynate addends for conversion of olefins into β-lactams. The chemistry of CSI has been reviewed[496-498] along with its applications to β-lactam synthesis[367,370,372,377,379,499]. Reaction of CSI with olefins is presumed[499] to involve equilibrium formation of a π complex, which then rearranges to a 1,4-dipolar intermediate having positive charge on the more highly substituted carbon of the original olefin. Combination of the termini of this intermediate completes the stepwise process to form an N-chlorosulphonyl β-lactam (equation 270). In order for the cycloaddition reaction to serve as a viable route to β-lactams,

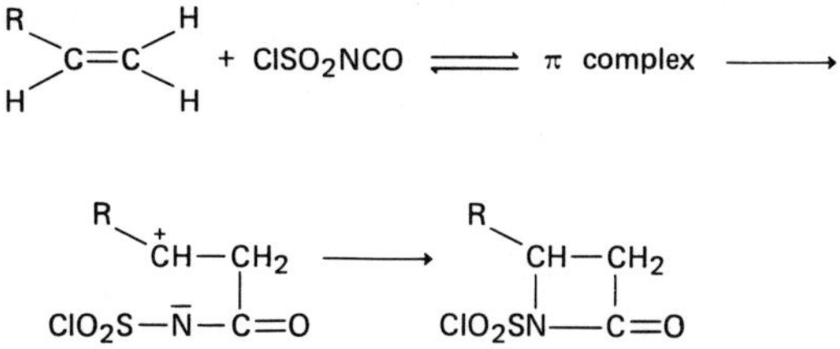

$$
\begin{array}{l}
\overset{R}{\underset{H}{>}}C=C\overset{H}{\underset{H}{<}} \;+\; ClSO_2NCO \;\rightleftharpoons\; \pi \text{ complex} \;\longrightarrow
\end{array}
$$

$$
\underset{ClO_2S-\overset{}{N}-C=O}{\overset{R}{\underset{}{>}}\overset{+}{C}H-CH_2} \longrightarrow \underset{ClO_2SN-C=O}{\overset{R}{\underset{}{>}}CH-CH_2} \tag{270}
$$

the N-chlorosulphonyl group must be removed reductively, preferably by treatment in a suitable organic solvent, with a 25% aqueous sodium sulphite solution[500], or with an aqueous solution of a sulphur oxo acid, or its salt, in the presence of sodium bicarbonate (equation 271)[501,502].

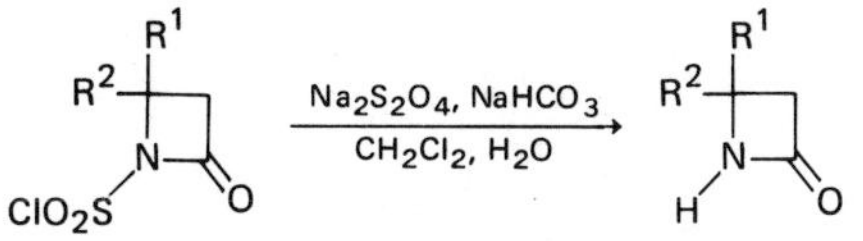

$$
\underset{ClO_2S}{\overset{R^1}{\underset{R^2}{\overset{}{\Box}}}} \xrightarrow[CH_2Cl_2,\ H_2O]{Na_2S_2O_4,\ NaHCO_3} \underset{H}{\overset{R^1}{\underset{R^2}{\overset{}{\Box}}}} \quad \text{(Ref. 501)} \tag{271}
$$

$$R^1 = Ph,\ Ph,\ H,\ Me$$
$$R^2 = H,\ Me,\ Me,\ Me$$

Table 31 contains a number of recent examples. Examination of these reactions reveals that addition of CSI is both a regiospecific and stereospecific reaction. Some regiospecificity is lost with olefins of the type $R^1CH=CHR^2$, where both R^1 and R^2 are simple alkyl groups. Dienes can easily be converted to monoadducts, but

TABLE 31. Synthesis of β-lactams by addition of chlorosulphonyl isocyanate to olefins followed by reduction

Olefin	β-Lactam	Overall yield (%)	Reference
$Me_2C=CH_2$	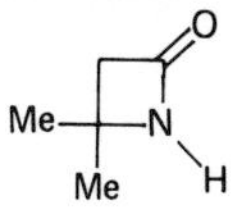 (structure)	51–53	502
MeCH=CHMe *cis*	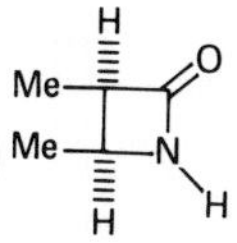 (structure)	85	503

TABLE 31. (Continued)

Olefin	β-Lactam	Overall yield (%)	Reference
MeCH=CHMe *trans*	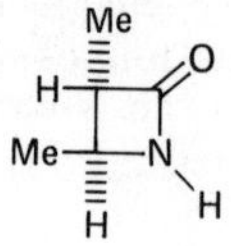	85	503
MeCH=CHPr-*n* *cis*	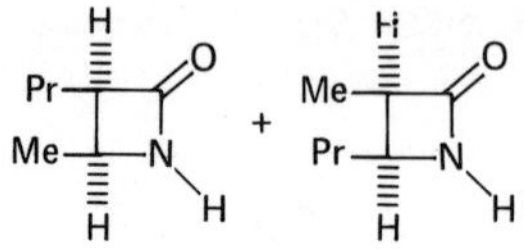 1:3	55	503
MeCH=CHPr-*n* *trans*	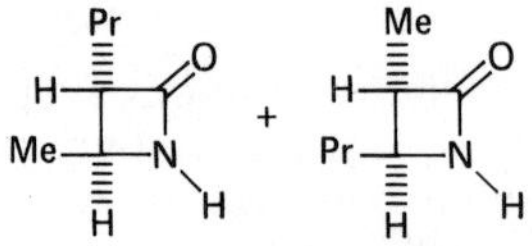 2:3	55	503
Me₂C=CRMe	R = Me R = H	92 98	498, 500
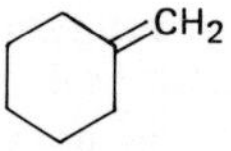	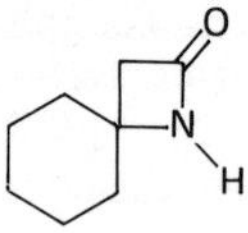	94	500
	n = 3 *n* = 4 *n* = 6	63 57 75	503
	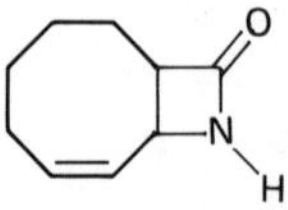	86	500
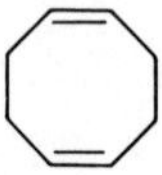	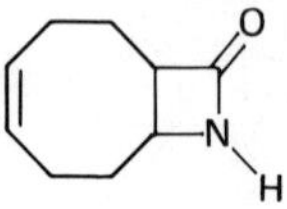	41	503

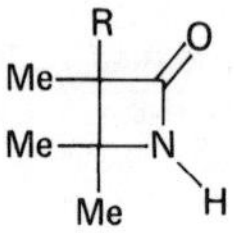

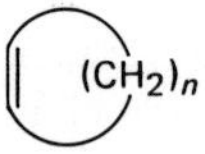

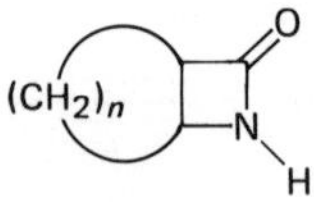

TABLE 31. (Continued)

Olefin	β-Lactam	Overall yield (%)	Reference
	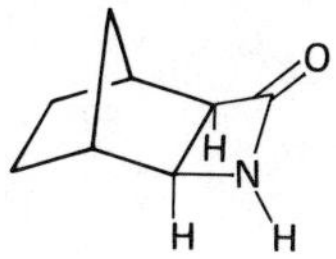	66	499, 500
	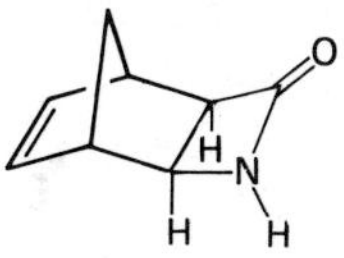	68	499
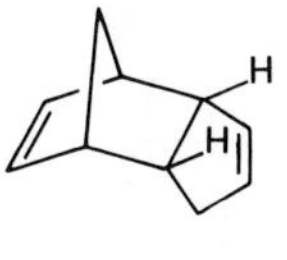	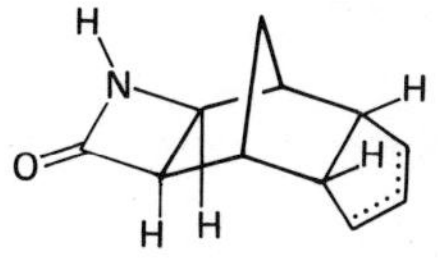	57	499, 503
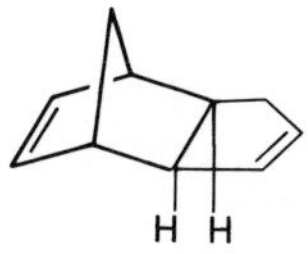	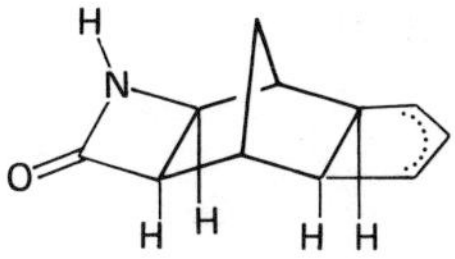	30	499
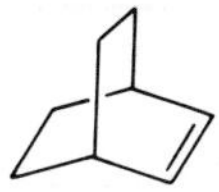	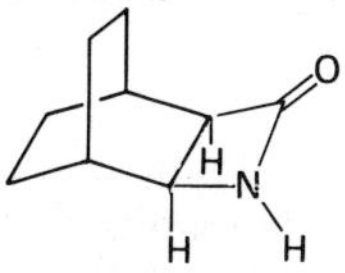	49	499
H$_2$C=CR−CH=CH$_2$	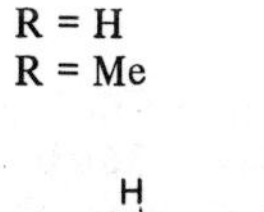	72	500, 504
	R = H	72	
	R = Me	68	
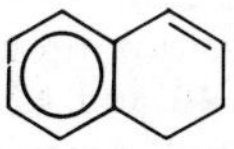	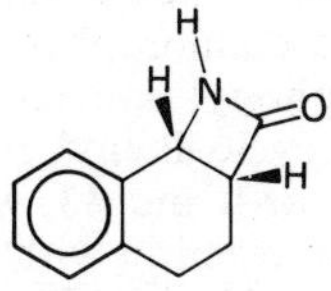	35	505

TABLE 31. (Continued)

Olefin	β-Lactam	Overall yield (%)	Reference
$Me_2C=C=CMe_2$	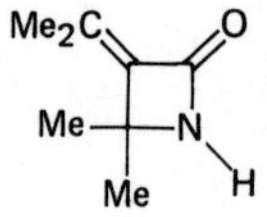	52	506
	+ $CH_2=C(Me)C=CMe_2$ with $CONH_2$	22	
=C=CH_2		26	506
	+ with C=CH_2 and CONH_2	32	
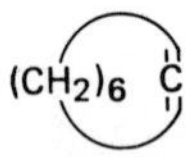 $(CH_2)_6$ C	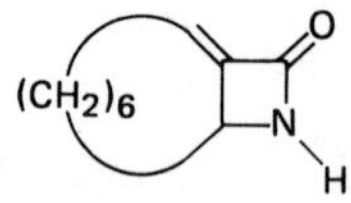 $(CH_2)_6$	36	506

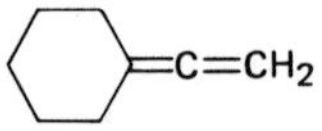

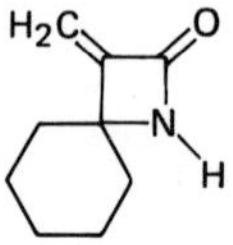

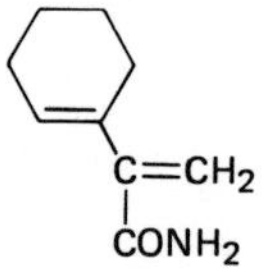

diadducts have not been isolated. Strained double bonds, such as those in a bicyclo [2.2.1] heptene system, tend to react more rapidly than normal unstrained olefins.

As mentioned previously, enamines react with simple isocyanates to afford β-amino-β-lactams (equation 272)[493,507,508].

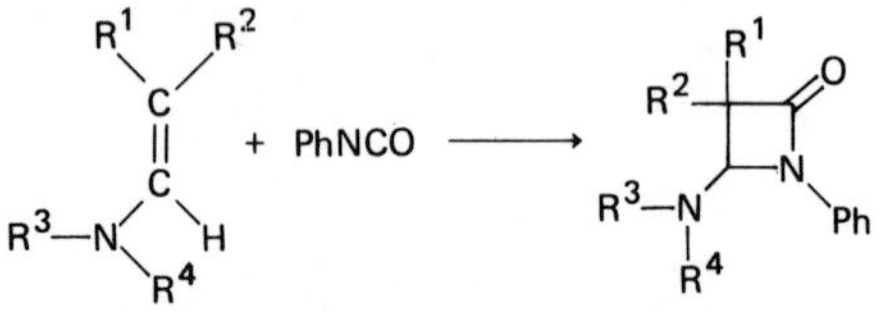

$$(272)$$

Pentahaptocyclopentadienyl dicarbonyl (olefin) iron complexes[509], represented in equation (273) as Fp—olefin complexes **36a,b**, fail to react with either ethyl or phenyl isocyanate, but react with 2.5-dichlorophenyl isocyanate, CSI, p-toluene-sulphonyl isocyanate and methoxysulphonyl isocyanate in a 1,3-addition process to afford butyrolactams **37a–e**[510]. The cycloalkenyl complexes **38, 40** and **42** react similarly[510] to give lactams **39, 41a** or **b** and **43**, while the butynyl complex **44** affords the unsaturated lactam **45** upon reaction with tosyl isocyanate.

Isocyanates of various types undergo $[2\pi + 2\pi]$ cycloaddition with ketenimines to afford β-imino-β-lactams in good yields (equation 278)[511].

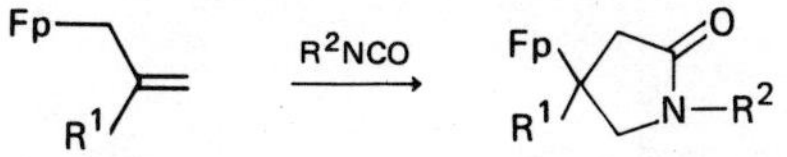

$$\text{(273)}$$

(36)

(a) $R^1 = H$
(b) $R^1 = Me$
$Fp = h^5\text{-}C_5H_5Fe(CO)_2$

(37)

(a) $R^1 = H$, $R^2 = C_6H_3Cl_2\text{-}2,5$
(b) $R^1 = H$, $R^2 = SO_2Cl$
(c) $R^1 = Me$, $R^2 = Ts$
(d) $R^1 = Me$, $R^2 = Ts$
(e) $R^1 = H$, $R^2 = MeOSO_2$

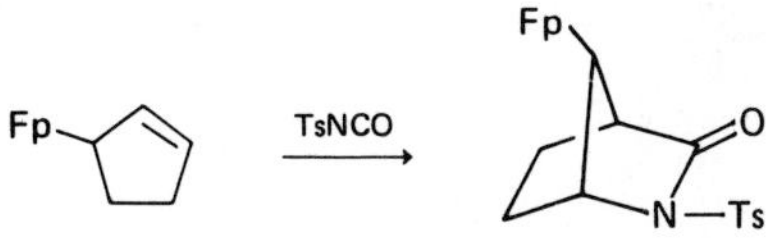

$$\text{(274)}$$

(38)   **(39)**

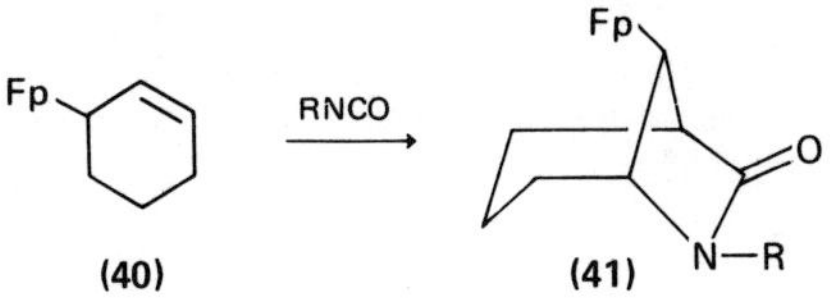

$$\text{(275)}$$

(40)   **(41)**

(a) $R = Ts$; (b) $R = MeOSO_2$

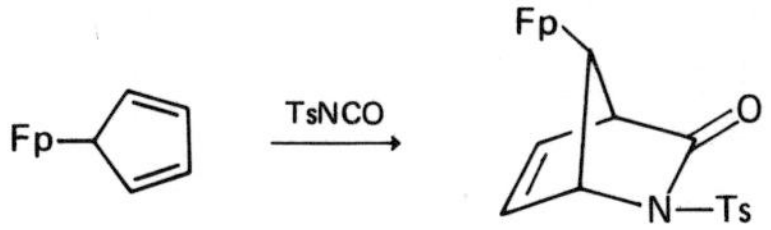

$$\text{(276)}$$

(42)   **(43)**

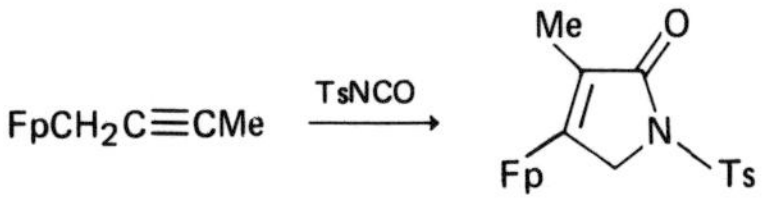

$$\text{(277)}$$

(44)   **(45)**

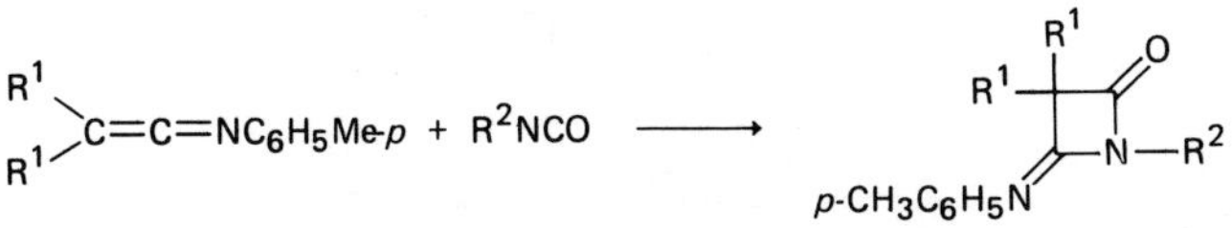

$$\text{(278)}$$

R^1	R^2	Yield (%)
Ph	Ph	83
Ph	$p\text{-MeC}_6H_5$	78
Ph	$p\text{-MeOC}_6H_5$	78
Ph	$p\text{-MeC}_6H_4SO_2$	72
Me	Ph	76
Me	$p\text{-MeC}_6H_4SO_2$	73

Phenyl isocyanate reacts with various acetylenes in the presence of aluminium chloride to afford 3,4-disubstituted carbostyrils (equation 279)[512].
Treatment of o-benzoylbenzaldehyde with aryl isocyanates affords 2,3-disubsti-

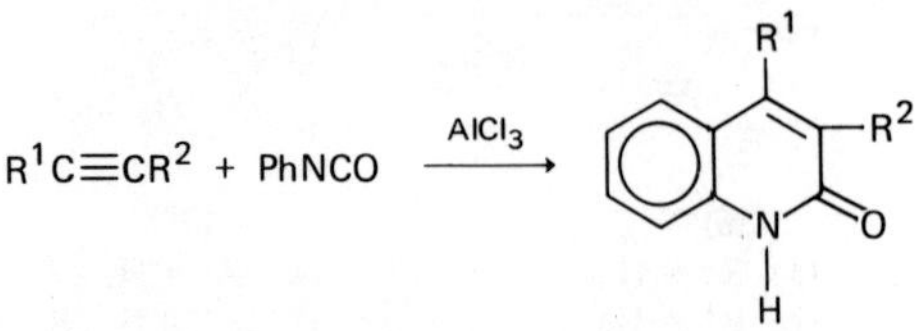

$$R^1C{\equiv}CR^2 + PhNCO \xrightarrow{AlCl_3}$$ (279)

R^1	R^2	Yield (%)
Ph	H	43
Ph	Me_3Si	44
Et	Et	–
n-Bu	Me_3Si	50
Me_3Si	Me_3Si	57
H	Me_3Si	25
Me_3Si	H	89

tuted phthalimidines by a reaction pathway involving intermediate formation of o-benzoylbenzylideneanilines followed by phenyl group migration (equation 280)[513]. Similar results have been observed with aromatic isocyanates and phthalaldehyde (equation 281)[514].

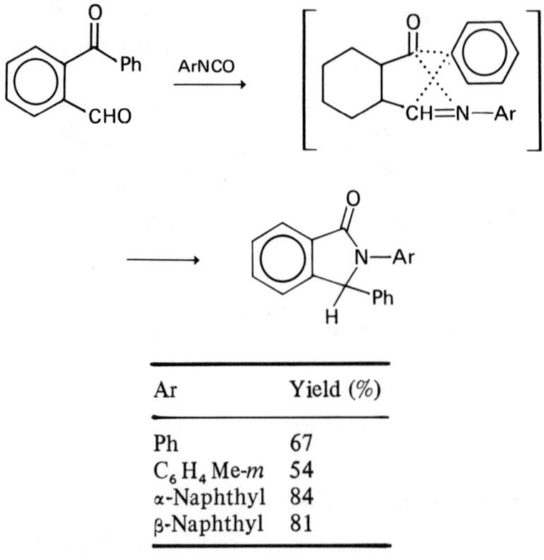

(280)

Ar	Yield (%)
Ph	67
C_6H_4 Me-m	54
α-Naphthyl	84
β-Naphthyl	81

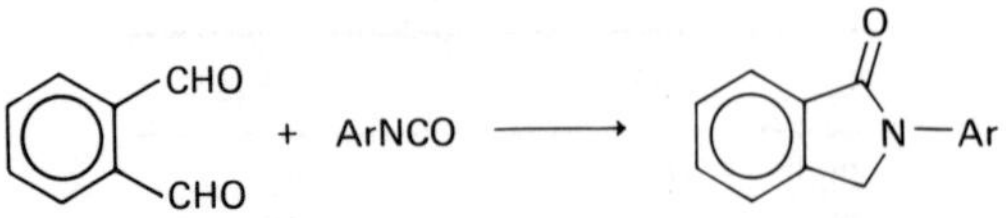

(281)

2. From imines

a. Reaction of imines with ketenes. The most frequently used method for the preparation of lactams involves the reaction of a large variety of imines with ketenes, which are prepared prior to or during the reaction.

In one of the earliest reviews[515] on this method, Staudinger pointed out that the reactivity of ketenes towards benzophenone anil exhibited the following order:

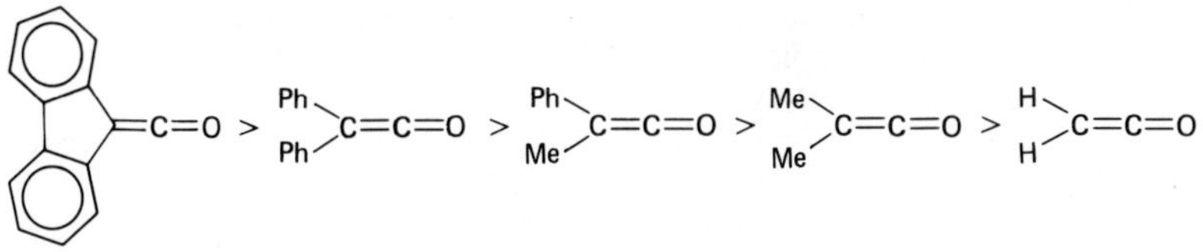

A similar order of ketene reactivity was observed by Brady[516] in a recent investigation of the cycloaddition of ketene itself and fluoro-, chloro-, dibromo-, methylchloro-, phenylchloro-, diphenyl-, phenylethyl-, butylethyl- and dimethylketenes to dicyclohexyl- and diisopropylcarbodiimide.

The mechanism and stereochemistry of the reaction have both been recently elucidated. In 1967, Gomes and Joullie[517] investigated the cycloaddition of ketene to benzylideneaniline in sulphur dioxide as the solvent and obtained the product shown in equation (282) in 52% yield. They concluded from their results, that

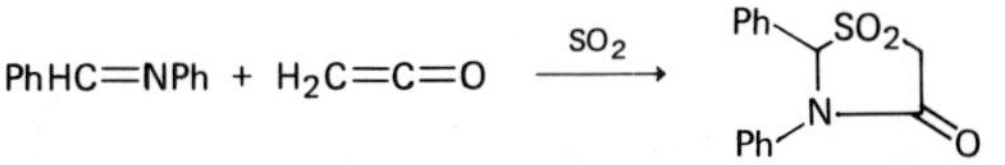

$$PhHC=NPh + H_2C=C=O \xrightarrow{SO_2} \qquad (282)$$

although the cycloaddition may proceed through a concerted mechanism or through the formation and subsequent reaction of a 1,4-dipolar intermediate (equation 283), the latter mechanism appeared more probable. Extension of this

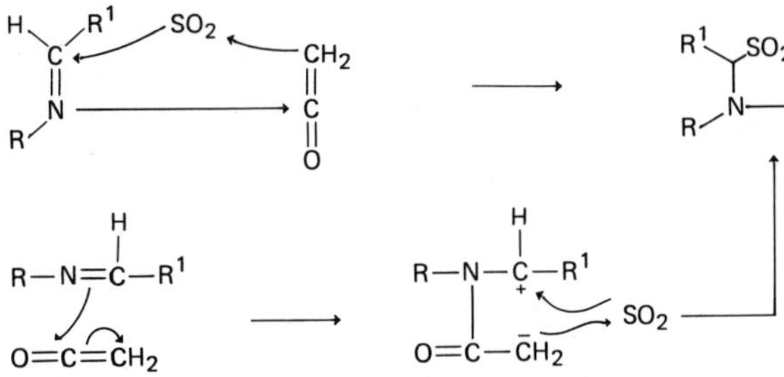

$$(283)$$

mechanism to the reaction of a ketene and an imine in an inert solvent would produce a 1,4-dipolar intermediate as shown in equation (284), which would then cyclize to produce the lactam.

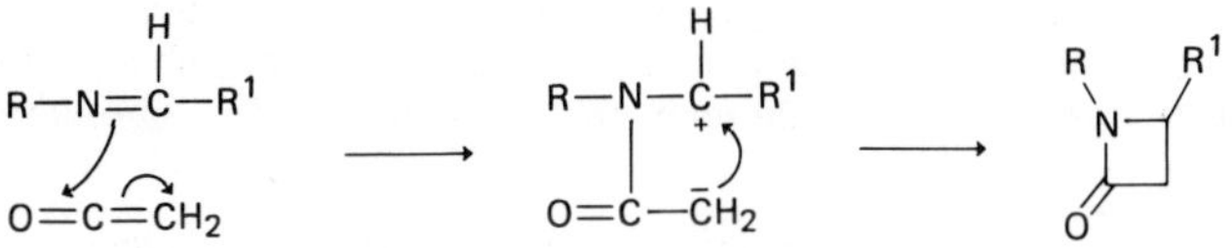

$$(284)$$

In 1968, Luche and Kagan[518] reported that regardless of the method used to generate the ketenes, they added to benzylidene aniline to produce *trans*-β-lactams exclusively (equation 285). This work in conjunction with the study of Sheehan[519] and Bose[520] on the stereochemistry of the β-lactams formed by the reaction of an acid chloride and an imine in the presence of a tertiary amine has produced a controversy in the literature. Based upon the original suggestion of Sheehan[519] that the formation of a ketene from the acid chloride and tertiary amine and subsequent cycloaddition of the ketene to the imine was probably not the pathway

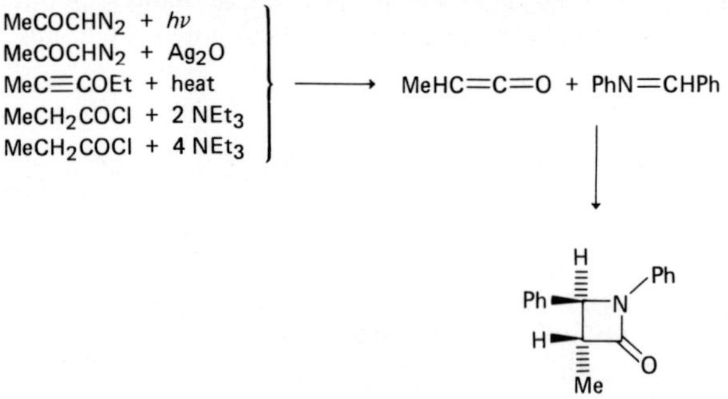

$$
\left.\begin{array}{l}
\text{MeCOCHN}_2 + h\nu \\
\text{MeCOCHN}_2 + \text{Ag}_2\text{O} \\
\text{MeC}{\equiv}\text{COEt} + \text{heat} \\
\text{MeCH}_2\text{COCl} + 2\,\text{NEt}_3 \\
\text{MeCH}_2\text{COCl} + 4\,\text{NEt}_3
\end{array}\right\} \longrightarrow \text{MeHC}{=}\text{C}{=}\text{O} + \text{PhN}{=}\text{CHPh}
$$

(285)

to the β-lactams produced, Bose[520] investigated the initial adduct formed from the reaction of a series of acid chlorides and anils in carbon tetrachloride solution using ^{1}H n.m.r. spectroscopy. He found that the adduct could best be represented by the covalent structure shown in equation (286), and that an equilibrium is established

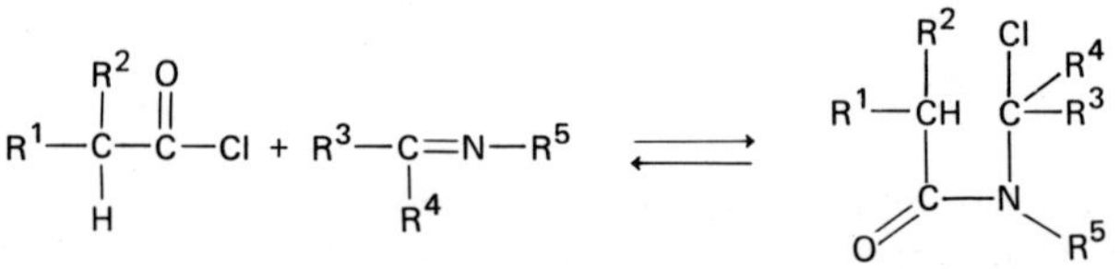

(286)

between the starting materials and the adduct. He further found that in all cases, where a β-lactam was formed both the *cis* and the *trans* isomers were obtained. It was thus concluded, that although the addition of a preformed ketene to an imine produces a *trans*-β-lactam in every case, it appears 'that "the acid chloride reaction" for β-lactam formation by-passes the ketene pathway — at least in those cases where the *cis*-β-lactams are produced'.

Table 32 contains a representative series of lactams produced from the reaction of imines with ketenes[516-594], and although many of the reactions shown do not necessarily involve a ketene intermediate, as can be seen from the discussion presented above, the products obtained are identical with those expected from a formal cycloaddition of a ketene and an imine.

It is interesting to note that the reaction of ketenes with double bonds has also been used to produce diazetidinones when the ketene is allowed to undergo a cycloaddition with an azo compound[515,595-599]. Selected examples of this approach are shown in Table 33. In one instance[597] it has been noted that the diazetidinone obtained by the cycloaddition of diphenylketene and azobenzene dissociates upon heating at 220°C into benzophenone anil, phenyl isocynate and the starting materials, diphenylketene and azobenzene. Recombination of these

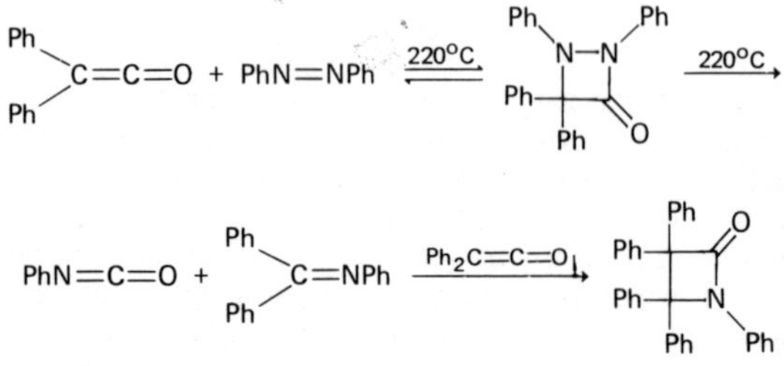

(287)

compounds via a ketene—imine interaction affords 1,3,3,4,4-pentaphenylazetidin-2-one (equation 287)[515].

It has also been reported[600] that irradiation of diphenylacetylene and nitrobenzene for 3 days with a mercury arc lamp affords a 1.8% yield of 1,3,3,4,4-pentaphenylazetidin-2-one. The mechanism proposed involves initial formation of diphenylketene and benzylideneaniline, followed by their subsequent cycloaddition to produce the β-lactam (equations 288 and 289).

$$PhC{\equiv}CPh + PhNO_2 \xrightarrow[\text{ether, N}_2]{h\nu} [PhNO_2]^* + PhC{\equiv}CPh \longrightarrow Ph_2C{=}C{=}O + PhNO \longrightarrow$$

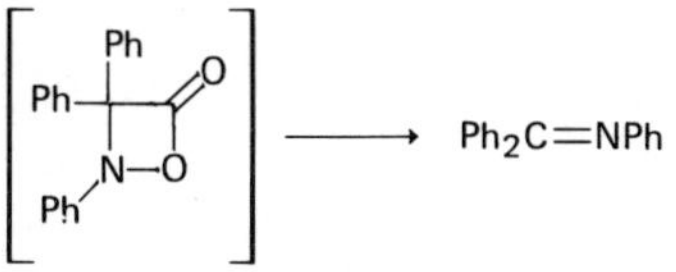

$$\longrightarrow Ph_2C{=}NPh \qquad (288)$$

$$Ph_2C{=}NPh + Ph_2C{=}C{=}O \longrightarrow \qquad (289)$$

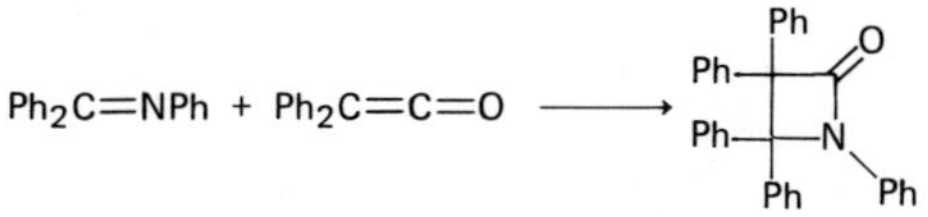

b. Reformatsky reaction with imines.   The main interest in the Reformatsky reaction with imines has not been with their preparative potential, but with their stereochemistry, sinch both *cis* and *trans* isomers may be expected from the addition of a Reformatsky reagent to an anil (equation 290). Studies of this

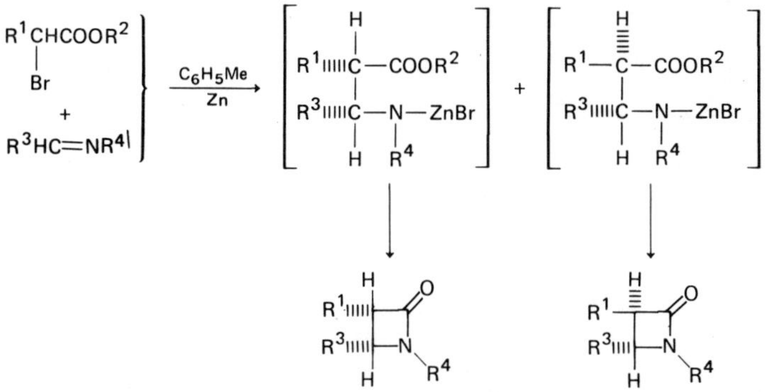

$$(290)$$

reaction using a variety of α-bromo esters have shown[602,603] that as the size of the R^1 group increases the *cis—trans* product ratio decreases, and that the *cis—trans* product ratio is influenced by the solvent (equation 291)[603].

A comparison[519] of the stereochemistry of the Reformatsky reaction with the stereochemistry of the $[2\pi + 2\pi]$ cycloaddition of a ketene and an imine shows the former reaction to yield mixtures of *cis* and *trans* β-lactams, while the latter reactions afford mainly *trans* β-lactam. Also of interest is the observation[605] that a competitive Reformatsky reaction using 1 equivalent of methyl α-bromophenyl acetate and 1 equivalent each of benzylideneaniline and α-deuteriobenzylideneaniline showed a secondary isotope effect of k_H/k_D 0.86 (equation 292), whereas a similar reaction of 1 equivalent of diphenylketene with the same mixture of Schiff bases showed no isotope effect.

In Table 34 are listed β-lactams which have been prepared using a Reformatsky

TABLE 32. Production of lactams by reaction of ketenes with imines

Ketene or ketene precursor	Imine	Conditions	Product	Yield (%)	Reference
$H_2C=C=O$	$i\text{-PrN}=C=CPr\text{-}i$	R.t., 8 h	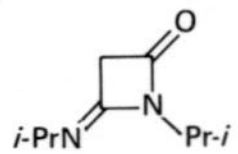	5	516
$H_2C=C=O$	$PhHC=NPh$	SO_2	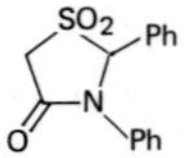	52	517
$H_2C=C=O$	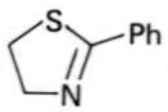	SO_2	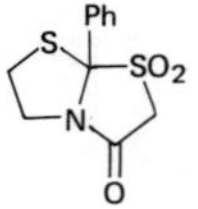	80	517
$H_2C=C=O$	$RC_6H_4CH=NPh$	180–200°C, 1 h	R = o-Me R = m-Me R = p-Me R = o-MeO R = m-MeO R = m-Cl R = p-NMe$_2$	11.5 12 22 39 16 32 62	523
$H_2C=C=O$	$PhHC=N-C_6H_4R$	180–200°C, 1 h	R = o-Me R = m-Me R = p-Me R = o-MeO R = p-MeO R = m-Cl	7 18 13 — 19 34	523

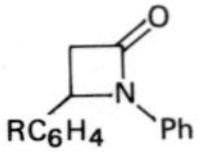

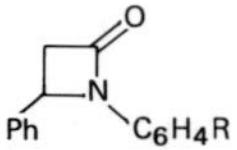

TABLE 32. (Continued)

Ketene or ketene precursor	Imine	Conditions	Product	Yield (%)	Reference
$H_2C=C=O$	PhHC=CHCH=NPh	MeOH, ether, 1 h, reflux	(Ph, N–Ph ring, O)	69	523
MeHC=C=O	PhHC=NPh	MeCOHCN$_2$ + $h\nu$	(Me, Ph, N–Ph β-lactam)	50, 47	518, 525
		MeCOCHN$_2$ + Ag$_2$O		17	518
		MeC≡COEt, heat		30	518
		MeCH$_2$COCl + 2 NEt$_3$		2	518
		MeCH$_2$COCl + 4 NEt$_3$		39	518
MeHC=C=O	Ph$_2$C=NPh	C$_6$H$_6$, N$_2$, $h\nu$, 5 h	(Me, Ph, N–Ph)	48	525
RHC=C=O	i-PrN=C=NPr-i	Hexane, reflux, 2 h C$_6$H$_6$, reflux	R = F	40	516, 526
			R = Cl	20	
RHC=C=O	PhHC=NPh	EtC≡COEt, heat	R = Et	35, 24	518, 525
		EtCH$_2$COCl + 2 NEt$_3$		2	518
		i-PrC≡COEt, heat	R = i-Pr,	78	518
		i-PrCH$_2$COCl + 2.5 NEt$_3$		32	518
		t-BuCH$_2$COCl + 2 NEt$_3$	R = t-Bu,	2	518

TABLE 32. (Continued)

Ketene or ketene precursor	Imine	Conditions	Product	Yield (%)	Reference
$Me_2C=C=O$	PhC=NPh, SMe	EtOAc, 2 days, r.t.	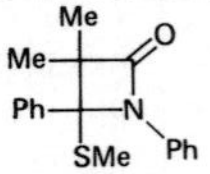	60	528[a]
$Me_2C=C=O$	i-PrN=C=NPr-i	Hexane, reflux, 8 h	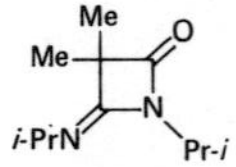	32	516
$R^1R^2C=C=O$		Hexane, reflux, 5 h	$R^1 = Me, R^2 = Cl$ $R^1 = R^2 = Br$	25 59	516
n-BuC=C=O, Et	i-PrN=C=NPr-i	Hexane, reflux, 2 h	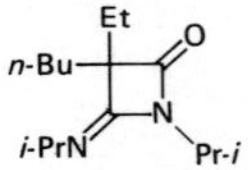	15	516
PhHC=C=O	PhHC=NPh	C_6H_6, N_2, 4 h, 40–50°C C_6H_6, N_2, hv, 5 h PhCOCHN$_2$ + Ag$_2$O PhCOCl + 4 NEt$_3$		35 74 44 6	523 525 518, 532 518, 532
PhHC=C=O	RC$_6$H$_4$CH=NPh	C_6H_6, N_2, 4 h 40–50°C	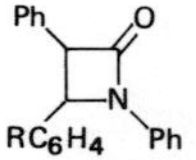		

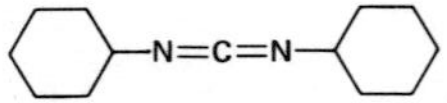

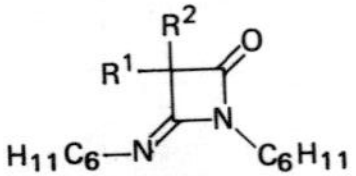

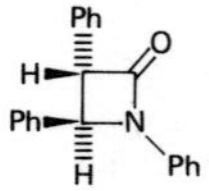

Ketene	Imine	Conditions	Product	R	Yield (%)	Ref.
PhHC=C=O	(2-pyridyl)CH=NPh	C₆H₆, N₂, hv, 5 h	β-lactam (N-Ph, pyridyl)	—	56	525
PhHC=C=O	PhHC=N—C₆H₄R	C₆H₆, N₂, hv, 4 h, 40–50 °C	β-lactam (N-C₆H₄R)	R = o-Me	21	523
				R = m-Me	20	523
				R = p-Me	14	523
				R = o-MeO	25	523
				R = m-MeO	29	523
				R = m-Cl	13.5	523
				R = p-NO₂	—, 78	523, 525
				R = p-NMe₂	90, 79	523, 525
PhHC=C=O	PhHC=N—C₆H₄R	C₆H₆, N₂, hv, 5 h	β-lactam (N-C₆H₄R)	R = o-Me	12	523
				R = m-Me	15	523
				R = p-Me	15	523
				R = o-MeO	5	523
				R = p-MeO	19	523
				R = m-Cl	28	523
				R = p-NMe₂	70	525
PhHC=C=O	Ph₂C=NPh	C₆H₆, N₂, hv, 4–5 h, 40–50 °C	β-lactam (Ph, Ph, N-Ph)		42, 76	523, 525
PhHC=C=O	PhHC=CHCH=NPh	MeOH, reflux, 40–50 °C	dihydropyridinone (N-Ph, Ph)		32	523
PhHC=C=O	PhN=CMe—CMe=NPh	C₆H₆, N₂, hv, 4–5 h, 40–50 °C	dihydropyrimidinone (Me, Me, N-Ph)		47	523

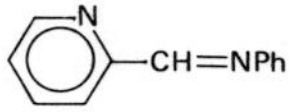

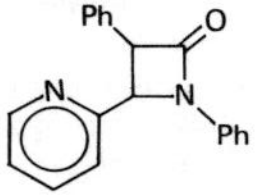

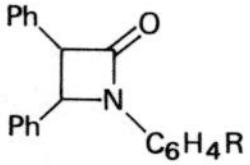

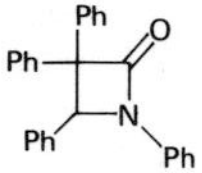

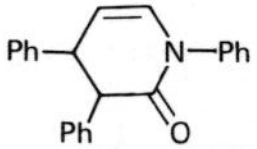

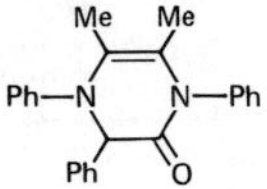

TABLE 32. (Continued)

Ketene or ketene precursor	Imine	Conditions	Product	Yield (%)	Reference
PhHC=C=O	p-O$_2$NC$_6$H$_4$—CH=N—CH$_2$C$_6$H$_5$	C$_6$H$_6$, N$_2$, hv, 5 h	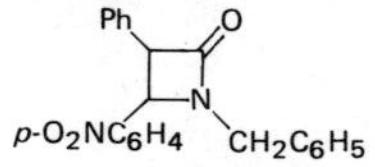 β-lactam: Ph, O, N-CH$_2$C$_6$H$_5$, p-O$_2$NC$_6$H$_4$	65	525
p-RC$_6$H$_4$CH=C=O	PhHC=NPh	C$_6$H$_6$, N$_2$, hv, 5 h	β-lactam: p-RC$_6$H$_4$, O, N-Ph, Ph — R = MeO / R = Cl	65 / 54	525
p-MeOC$_6$H$_4$CH=C=O	PhHC=N—CH$_2$C$_6$H$_5$	C$_6$H$_6$, N$_2$, hv, 5 h	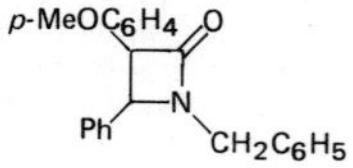 β-lactam: p-MeOC$_6$H$_4$, O, N-CH$_2$C$_6$H$_5$, Ph	36	525
PhRC=C=O	PhHC=NPh	—	β-lactam: Ph, R, O, N-Ph, H; R = Me; cis : $trans$ = 1:4; R = Et; cis : $trans$ = 2:1; R = i-Pr; cis : $trans$ = 9:1		532
PhC=C=O | COOMe	PhHC=NPh	C$_6$H$_6$, N$_2$, hv, 5 h	β-lactam: CO$_2$Me, Ph, O, N-Ph, Ph	14	521, 525
PhC=C=O | COOMe	Ph$_2$C=NPh	C$_6$H$_6$, N$_2$, hv, 5 h	β-lactam: CO$_2$Me, Ph, O, N-Ph, Ph, Ph	35	525

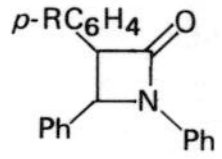

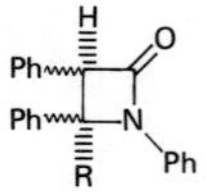

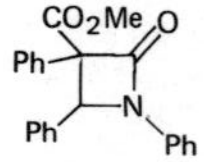

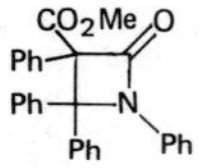

Ketene	Reactant	Conditions	Product	Yield (%)	Reference
PhEtC=O	i-PrN=C=NPr-i	Hexane, 48 h, r.t.	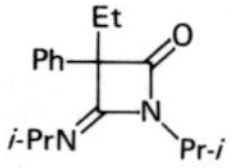 (β-lactam structure)	57	516
PhClC=O	$H_{11}C_6$–N=C=N–C_6H_{11}	Hexane, reflux, 2 h.	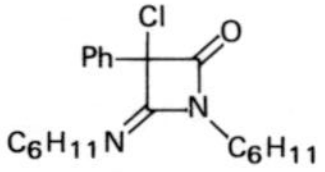 (β-lactam structure)	65	516
Ph_2C=C=O	PhHC=NMe	MeCN, r.t. molar ratio 1:1 C_6H_6, r.t. molar ratio 1:1 C_6H_6, r.t. molar ratio 2:1	(β-lactam structures)	82 71 95	538 538 538
Ph_2C=C=O	PhHC=NMe	MeCN, r.t. molar ratio 2:1	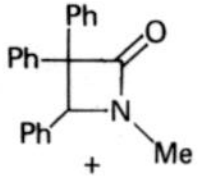 (structure)	19	538
Ph_2C=C=O	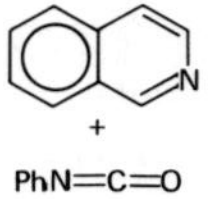 (isoquinoline) + PhN=C=O	R.t.	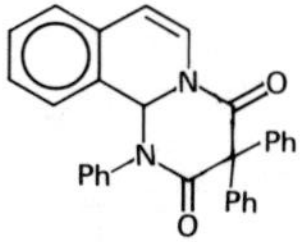 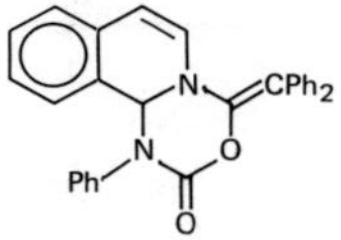 (structures)	81 31	538[b]

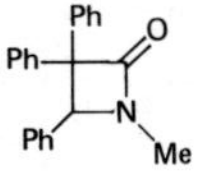

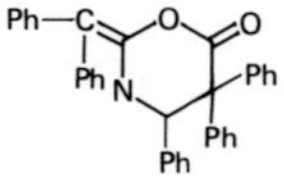

TABLE 32. (Continued)

Ketene or ketene precursor	Imine	Conditions	Product	Yield (%)	Reference
$Ph_2C=C=O$	$PhHC=NPh$	Heat	(Ph, Ph, N–Ph azetidinone)	—	533, 534, 536
		C_6H_6, N_2, $h\nu$, 5 h		71	525
		Ether, stand 1 day		70	535
		C_6H_6, stir 20 min, r.t.		53–65	537
$Ph_2C=C=O$	$RC_6H_4CH=NPh$		$R = OMe$	52	523
		Waterbath, 70–80°C	$R = m\text{-Me}$	63	523
		C_6H_6, 70–80°C, 1 h	$R = p\text{-Me}$	72	523
		Waterbath, 70–80°C	$R = o\text{-MeO}$	30	523
			$R = m\text{-MeO}$	21	523
			$R = m\text{-Cl}$	71	523
		C_6H_6, stir 20 min, r.t.	$R = p\text{-MeO}$	53–65	537
		C_6H_6, 70–80°C, 1 day	$R = p\text{-NMe}_2$	67	523
$Ph_2C=C=O$	$PhHC=N-C_6H_4R$		$R = o\text{-Me}$	72	523
		Waterbath, 70–80°C	$R = m\text{-Me}$	69	523
		Ether, 2 days, r.t.	$R = p\text{-Me}$	21	523
		EtOAc, r.t.	$R = o\text{-MeO}$	55	523
			$R = p\text{-MeO}$	16	523
		C_6H_6, stir 20 min, r.t. Waterbath, 70–80°C	$R = m\text{-Cl}$	53–65	537
		Without solvent, 1 week; in solvent (C_6H_6, ether or EtOAc) 5 h on waterbath; without solvent, melt at 200°C		98	523
			$R = p\text{-NMe}_2$	65	535[c]

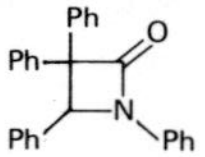

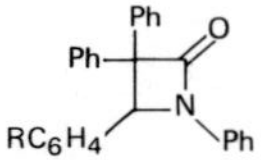

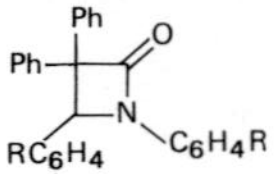

Reactant	Reactant 2	Conditions	Product	Yield	Reference
$Ph_2C=C=O$	$PhN=CMe-MeC=NPh$	EtOAc, reflux, 3 h	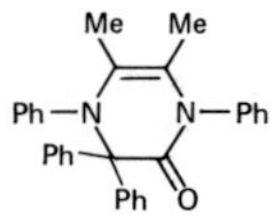 [structure]	61	523
$Ph_2C=C=O$	$PhHC=NCH_2C_6H_5$	C_6H_6, N_2, $h\nu$, 5 h	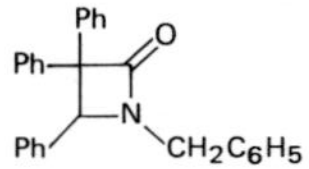 [structure]	46	525
$Ph_2C=C=O$	$Ph_2C=NPh$	C_6H_6, N_2, $h\nu$, 5 h	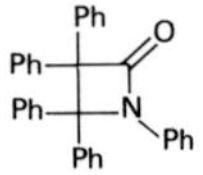 [structure]	72	525[a]
$Ph_2C=C=O$	[thiazoline structure] R = SH; R = NHCOCHPh₂; R = NHCOMe; R = Ph	EtOAc, N_2, stir 20 min; EtOAc, N_2, stir 5 min; EtOAc, C_6H_6, N_2, stir 20 min; C_6H_6, N_2, $h\nu$, 5 h	[structure] R = SCOCHPh₂; R = NHCOCHPh₂; R = NHCOMe; R = Ph	68, —; 54, —; 44; 20, —	523, 531; 523, 531; 523; 525, 531
	R = Me	(molar ratio 2:1); 25°C, 1 week	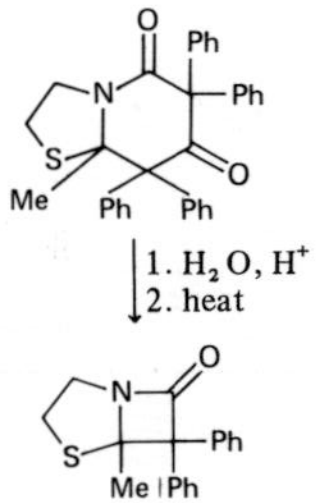 [structure] 1. H_2O, H^+; 2. heat	—, 73	531, 543

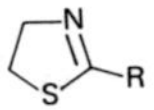

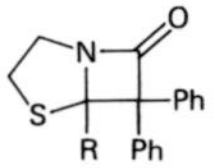

TABLE 32. (Continued)

Ketene or ketene precursor	Imine	Conditions	Product	Yield (%)	Reference
$Ph_2C=C=O$	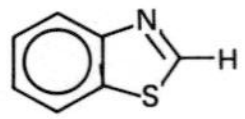	25°C, 1 day (molar ratio 2:1)	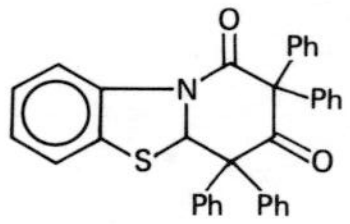	—, 86	531, 543
$Ph_2C=C=O$	$PhC=NR^2$ $\underset{SR^1}{\|}$		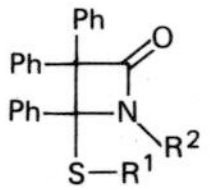		
	$R^1 = Me, R^2 = -CH_2CO_2Me$	C_6H_5Me, 16 h, r.t.		67–69	539
	$R^1 = Me, R^2 = -CH(i\text{-}Pr)CO_2Me$	C_6H_6, reflux, 20 h		72	539
	$R^1 = Me, R^2 = -C(CO_2Me)=CMe_2$	C_6H_6, reflux, 20 h	(2 diastereomers)	61–63	539
	$R^1 = -CH_2CH_2CO_2Me, R^2 = -C(CO_2Me)=CMe_2$	C_6H_6, reflux, overnight		53–64	539
$Ph_2C=C=O$	$R^1SCH=NCHCHMe_2$ $\underset{CO_2Me}{\|}$	C_6H_5Me, reflux 12 h	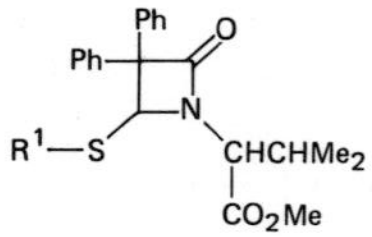		540
			$R^1 = Me$	47^d	
			$R^1 = CH_2C_6H_5$	69^e	
$Ph_2C=C=O$	$Ph_2C=NNHCOPh$	C_6H_5Me, 100°C, 3 h	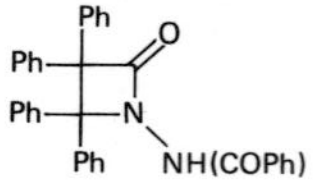	75	541

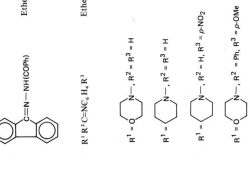

Reagent	Substrate	Conditions	Product	Yield (%)	Reference
$Ph_2C=C=O$	[dibenzo] $C=N-NH(COPh)$	Ether		66	541
$Ph_2C=C=O$	$R^1R^2C=NC_6H_4R^3$	Ether, $h\nu$, 3.5 h			542
	R^1- (morpholine), $R^2=R^3=H$			44–48	
	R^1- (piperidine), $R^2=R^3=H$			79	
	R^1- (piperidine), $R^2=H$, $R^3=p\text{-}NO_2$			75	
	R^1- (morpholine), $R^2=Ph$, $R^3=p\text{-}OMe$			61	
$Ph_2C=C=O$	[benzoxazole]	R.t. 1 week (molar ratio 2:1)		88	543
$Ph_2C=C=O$	$EtOCH=NPh$	R.t. 1 week (molar ratio 2:1)		60	543

TABLE 32. (Continued)

Ketene or ketene precursor	Imine	Conditions	Product	Yield (%)	Reference
$Ph_2C=C=O$	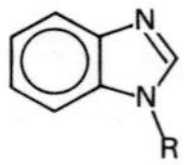 (benzimidazole, R)		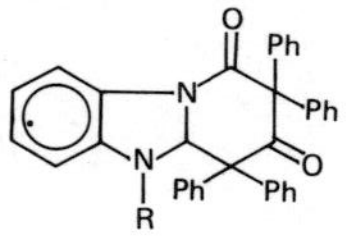		543
	R = H R = —COCHPh₂ R = Me	100°C, 1 h (molar ratio 3:1) 100°C, 1 h (molar ratio 2:1) Ether, r.t., 1 day (molar ratio 2:1)		83 86 85	
$Ph_2C=C=O$	 (imidazole, R)		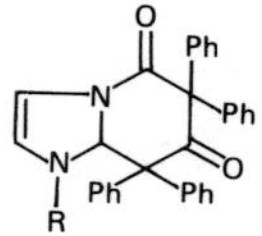		543
	R = Me R = Me R = —COCHPh₂	Ether, r.t., 1 day (molar ratio 1:2) Ether, 100°C, 1 h (molar ratio 3:1) THF, r.t., 1 day (molar ratio 2:1)		64 19 56	
$Ph_2C=C=O$	$RN=C=NR$	C_6H_6, r.t., 2 h	R = C₆H₁₁ R = i-Pr	90 88	516
RCH_2COCl	(cyclohexylidene=NPh)				

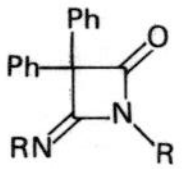

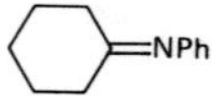

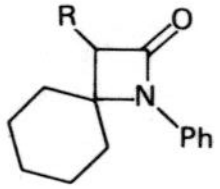

Substrate	Conditions	Product	Stereochem.	Yield (%)	Ref.
RCH₂COCl					
R = Cl	CH₂Cl₂, NEt₃	R = Cl		Trace	545
R = OMe		R = OMe		14	545
R = Ph	CH₂Cl₂, N(Pr-i)₃, 0°C, 3 h stir	R = Ph		—	545, 547
R = OPh	CH₂Cl₂, NEt₃	R = OPh		48	545
R = N₃	CH₂Cl₂, NEt₃, 0°C, 4 h stir	R = N₃		54	545
	CH₂Cl₂, NEt₃			54	547
	CH₂Cl₂, NEt₃	R = (phthalimidyl)		33	545
PhHC=NPh					
R = Cl	CH₂Cl₂, NEt₃, N₂	(β-lactam, N-Ph)	trans	1ᵍ, 62, 20	520, 545
	C₆H₆, NEt₃, 70–75°C,		trans	4	551
	DMF, 180°C		cis : trans = 45:55	—	552
	DMF, NEt₃, 25°C		trans	—	552
	C₆H₆, NEt₃, reflux		trans	60	590
	CH₂Cl₂, NEt₃, N₂		trans	35, 42, 50	520, 545
R = Me			trans	10, 49	520
R = CH=CH₂			trans	34	520
R = CMe₃			trans	50	520, 545
R = OMe			cis : trans = 3:1	85	545
R = CH₂OMe			cis : trans = 1.7:1	0.5	520, 545
R = Ph			trans	20, 59	520, 545
R = OPh			trans	85	520
			cis	(38)	520
			trans	(51)	545
R = MeOC₆H₄			trans	42	545
R = p-O₂NC₆H₄			cis : trans = 3:1	0, 38	520
R = N₃			cis : trans = 1.3:1	40	520
			cis	98	545
R = SCH₂C₆H₅			trans	35—45	548ᵏ
			trans	50	548ᵏ
				40	545

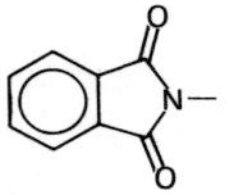

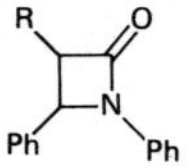

TABLE 32. (Continued)

Ketene or ketene precursor	Imine	Conditions	Product	Yield (%)	Reference
R =		CH$_2$Cl$_2$, NEt$_3$, N$_2$ C$_6$H$_6$, NEt$_3$, r.t. C$_6$H$_6$, NEt$_3$, r.t. stir 1 h	*trans*	30 50 —	545 519 549
			1. EtOH, N$_2$H$_4$, reflux 2 h 2. HCl 3. KOH, H$_2$O ↓ 		371, 401
R 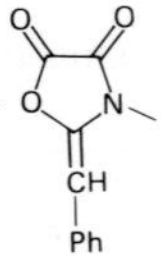 =		CH$_2$Cl$_2$, dioxane, NEt$_3$, r.t., 0.5 h	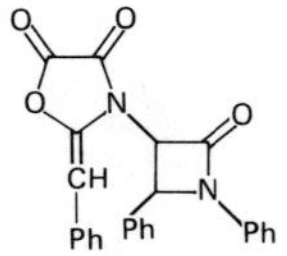	16	580
RCH$_2$COCl	EtOCH=NPh	CH$_2$Cl$_2$, NEt$_3$	*trans* R = Cl R = OMe R = OPh R = SCH$_2$C$_6$H$_5$ R = N$_3$	Trace 18 31 Trace 31	545 545 545 545 545, 548

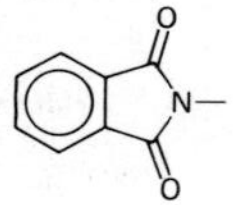

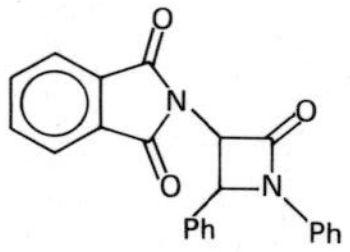

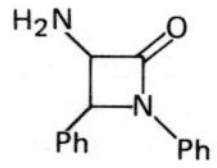

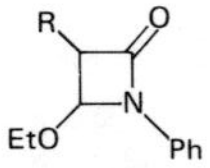

Substrate	Reagent / Conditions	Product	Yield (%)	Ref.
$R = $ (N-CH₂COCl phthalimide)	CH₂Cl₂, NEt₃		42	545, 548
	EtSCH=NPh, C₆H₆, NEt₃, reflux 3 h		31	553
	Ether, NEt₃, r.t., 2 h		33	553

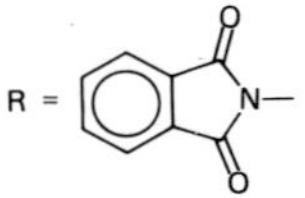

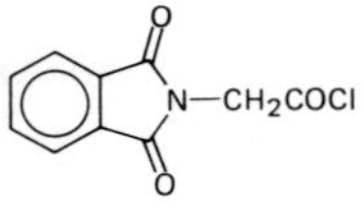

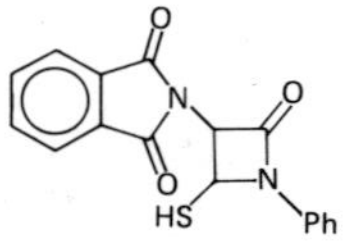

ClCHR¹COOH + R²C₆H₄CH=NC₆H₄R³ → POCl₃, DMF, reflux

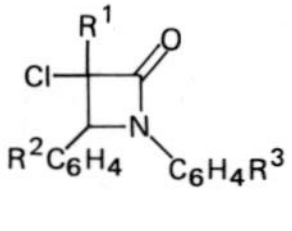

(cis : trans = 53 : 47)

		Conditions	Yield (%)	Ref.
$R^1 = H$	$R^2 = R^3 = H$	2 h	78	550f
	$R^2 = H, R^3 = p\text{-Me}$	2 h	59	552f,h
	$R^2 = H, R^3 = p\text{-MeO}$	3 h	52	550f
	$R^2 = H, R^3 = p\text{-Cl}$	4 h	38	550f
	$R^2 = p\text{-Cl}, R^3 = H$	2.5 h	36	550f
	$R^2 = R^3 = H$	5 h	41	550f
$R^1 = Cl$	$R^2 = H, R^3 = p\text{-Cl}$	POCl₃, C₆H₅Me, reflux 2 h	30	550f
	$R^2 = H, R^3 = p\text{-MeO}$	POCl₃, DMF, reflux 1.5 h	89	550f
	$R^2 = H, R^3 = m\text{-NO}_2$	1.5 h	66	550f
	$R^2 = H, R^3 = p\text{-NO}_2$	140°C, 4 h	44	550f
	$R^2 = H, R^3 = p\text{-Me}$	140°C, 4 h	80	550f
	$R^2 = R^3 = H$	140°C, 6 h	15	550f
	$R^2 = p\text{-Cl}, R^3 = H$	1.5 h	56	550f
$R^1 = Me$	$R^2 = R^3 = H$	4 h	53	550f
	$R^2 = p\text{-Me}, R^3 = H$	1 h	54	550f
		1 h	22	550f

TABLE 32. (Conditions)

Ketene or ketene precursor	Imine	Conditions	Product	Yield (%)	Reference
BrCHCOOH $\mid$ R^1	$R^2C_6H_4CH=NC_6H_4R^3$	POBr$_3$, DMF, reflux	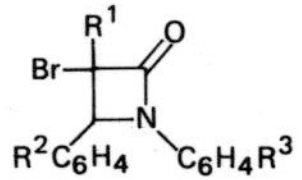		550^f
$R^1 = H$	$R^2 = R^3 = H$	130–140°C, 2 h		53	
	$R^2 = H, R^3 = p\text{-}Me$	150°C, 7 h		47	
	$R^2 = p\text{-}Cl, R^3 = H$	130°C, 5 h		36	
	$R^2 = H, R^3 = p\text{-}MeO$	140°C, 3 h		27	
$R^1 = Br$	$R^2 = R^3 = H$	130–140°C, 2 h		66	
	$R^2 = p\text{-}Cl, R^3 = H$	140°C, 4 h		55	
$R^1 = Me$	$R^2 = R^3 = H$	140°C, 3.5 h		35	
CH$_3$COOH	PhHC=NPh	POCl$_3$, C$_6$H$_5$Me, C$_5$H$_5$N, reflux 1 h	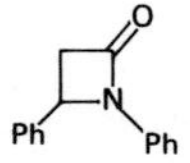	5	550^{f,g}
NC—CHCOOH $\mid$ R^1	$R^2R^3C=NC_6H_4R^4$	POCl$_3$, DMF, C$_6$H$_5$Me, reflux 90 min	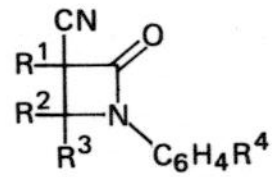		586^f
$R^1 = Me$	$R^2 = H, R^3 = Ph, R^4 = H$			57	
	$R^2 = H, R^3 = Ph, R^4 = o\text{-}Cl$			92	
	$R^2 = H, R^3 = Ph, R^4 = Me$			11	
	$R^2 = H, R^3 = Ph, R^4 = 2,4\text{-dimethyl}$			52	
$R^1 = Et$	$R^2 = H, R^3 = Ph, R^4 = H$			18	
	$R^2 = H, R^3 = Ph, R^4 = 2,4\text{-dimethyl}$			25	
	$R^2 = R^3 = Ph, R^4 = H$			53	
$R^2 = C_6H_5CH_2$	$R^2 = R^3 = Ph, R^4 = p\text{-}MeO$			78	
	$R^2 = H, R^3 = Ph, R^4 = H$			60	
R^1CH_2COCl	$\overset{R^2}{\underset{R^3}{>}}C=NR^4$		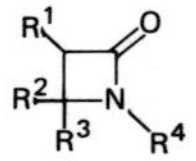		

R¹	R², R³, R⁴	Conditions	cis/trans	Yield (%)	Reference
R¹ = N₃	R², R³ = -(CH₂)₅-, R⁴ = C₆H₁₁	CH₂Cl₂, NEt₃, 0°C, stir 3–4 h		20	547
	R³ = -(CH₂)₆-, R⁴ = Ph			30	547
	R², R³ = -(CH₂)₅-, R⁴ = p-MeOC₆H₄			16	547
	R², R³ = -(CH₂)₅-, R⁴ = p-MeC₆H₄			30	547
	R², R³ = -(CH₂)₅-, R⁴ = p-ClC₆H₄			14	547
	R², R³ = -(CH₂)₅-, R⁴ = o-MeC₆H₄			31	547
	R² = H, R³ = p-O₂NC₆H₄, R⁴ = Ph	Method Aᵏ	cis	35	548
	R² = H, R³ = p-MeOC₆H₄, R⁴ = Ph	Method Aᵏ	cis	25–30	548
		Method Bᵏ	trans	53	548
	R² = H, R³ = p-BrC₆H₄, R⁴ = Ph	Method Aᵏ	cis	30	548
		Method Bᵏ	trans	65	548
	R² = H, R³ = [benzodioxolyl], R⁴ = p-BrC₆H₄	Method Aᵏ	cis	30–35	548
		Method Bᵏ	trans	31	548
	R² = H, R³ = Ph, R⁴ = p-FC₆H₄	Method Aᵏ	cis	23	548
	R² = H, R³ = p-BrC₆H₄, R⁴ = Ph		cis	30	548
	R² = H, R³ = p-FC₆H₄, R⁴ = Ph		cis	19	548
	R² = Me, R² = R³ = R⁴ = Ph	CH₂Cl₂, NEt₃		30	548
	R² = R³ = R⁴ = Ph (Me)			60	548
R¹ = PhCH₂OCOONH	R², R³ = -(CH₂)₂N(CH₂)₂-, R⁴ = p-MeOC₆H₄	CH₂Cl₂, NEt₃, 0°C, stir 3–4 h		68	547
R¹ = OMe	R², R³ = -(CH₂)₅-, R⁴ = Ph	CH₂Cl₂, NEt₃, 0°C, stir 3–4 h		14	547
	R² = Ph, R³ = SMe, R⁴ = Ph	CH₂Cl₂, NEt₃, stir 10 h		90	560, 566
R¹ = Ph	R² = Ph, R³ = SMe, R⁴ = Ph			76	560, 566
R¹ = OPh	R² = Ph, R³ = SMe, R⁴ = Ph			80–90, 64	560, 566
	R² = H, R³ = p-MeOC₆H₄, R⁴ = Ph		cis and trans	32 (cis)	563

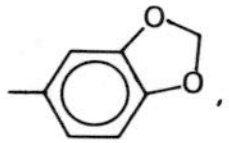

TABLE 32. (Continued)

Ketene or ketene precursor	Imine	Conditions	Product	Yield (%)	Reference
$R^1 = \overset{\mid}{\underset{Pr\text{-}i}{CHCO_2Me}}$					
	$R^2 = Ph, R^3 = CO_2Me,$ $R^4 = p\text{-}MeOC_6H_4$	CH_2Cl_2, NEt_3, N_2, stir overnight	2 isomers	73	564
	$R^2 = Ph, R^3 = SCH_2C_6H_5,$ $R^4 = Ph$	CH_2Cl_2, NEt_3, stir 10 h		71	566
	$R^2 = Ph, R^3 = p\text{-}NO_2C_6H_4CH_2S,$ $R^4 = Ph$			81	566
	$R^2 = H, R^3 = p\text{-}MeOC_6H_4,$ $R^4 = p\text{-}MeC_6H_4$		trans	56	566
$R^1 = SCH_2C_6H_5$ 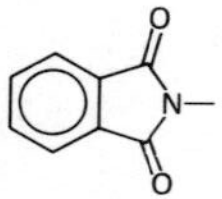	$R^2, R^3 = -(CH_2)_5-, R^4 = Ph$	CH_2Cl_2, NEt_3, 0°C stir 3—4 h		33	547
$R^1 =$ (N-phthalimidyl)	$R^2 = OMe, R^3 = R^4 = Ph$	Ether, NEt_3, 35°C, 4—5 h		50	559
	$R^2 = Ph, R^3 = SMe, R^4 = Ph$	CH_2Cl_2, NEt_3, stir 10 h		69	560, 506
	$R^2 = OEt, R^3 = R^4 = Ph$			55	553
	$R^2 = OCHMe_2, R^3 = R^4 = Ph$			51	553
	$R^2 = SMe, R^3 = R^4 = Ph$			70	553
	$R^2 = H, R^3 = SCH_2C_6H_5,$ $R^4 = CHCHMe_2$	CH_2Cl_2 or C_6H_5Me, NEt_3. reflux 3 h	2 trans isomers	39	540, 556
	$R^2 = H, R^3 = SMe,$ $R^4 = \overset{\mid}{\underset{CO_2Me}{CHCHMe_2}}$	C_6H_5Me, NEt_3, r. t. 2 h	2 trans isomers	40	540
$R^1 = Cl$	$R^2 = H, R^3 = SCH_2C_6H_5,$ $R^4 = \overset{\mid}{\underset{CO_2Me}{CHCHMe_2}}$	CH_2Cl_2 or C_6H_5Me, NEt_3, r. t. 2.5 h	2 trans isomers	45	540, 556
	$R^2 = R^3 = R^4 = Ph$	C_6H_6, NEt_3, 20°C		100	587
	$R^2 = H, R^3 = R^4 = Ph$		trans	70	587
R^1CH_2COCl	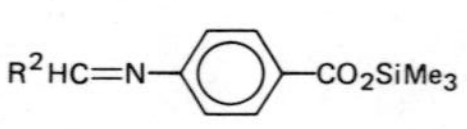 $R^2HC=N$—(p-CO_2SiMe_3 phenyl)	1. CH_2Cl_2, Et_3N, N_2, stir overnight 2. MeOH	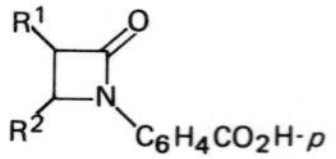 (β-lactam structure, trans)		557

$R^1 = OMe$
$R^1 = OPh$

$R^1 = N_3$

$R^1 CH_2 COCl$

$R^1 = OMe$
$R^1 = OPh$
$R^1 = N_3$

$R^1 =$ (phthalimide)

$R^1 =$ (phthalimide)–N–CH_2COCl

$R^1 = Ph, R^2 = Me, R^3 = Ph$
$R^1 = Ph, R^2 = Et, R^3 = Ph$
$R^1 = Ph, R^2 = Et, R^3 = p\text{-}MeC_6H_4$
$R^1 = p\text{-}MeC_6H_4, R^2 = Et, R^3 = Ph$

$N_3 CH_2 COCl$

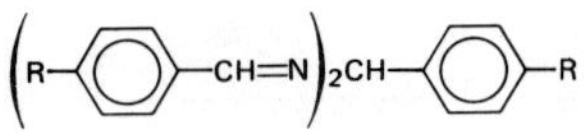

$R^2 = p\text{-}MeOC_6H_4$
$R^2 = p\text{-}MeOC_6H_4$
$R^2 = o\text{-}HOC_6H_4$
$R^2 = p\text{-}Me_2NC_6H_4$
$R^2 = p\text{-}MeOC_6H_4$

Me_3SiO_2C—⟨⟩—$CH{=}NR^2$

$R^2 = p\text{-}MeOC_6H_4$
$R^2 = p\text{-}MeOC_6H_4$
$R^2 = p\text{-}MeOC_6H_4$

$R^2 = p\text{-}MeOC_6H_4$

$R^2 = 3,4\text{-}diMeOC_6H_3CH_2$—
$R^2 = p\text{-}MeOC_6H_4CH_2$—

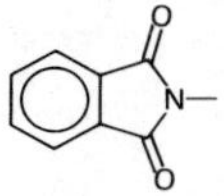

$R^1\overset{NR^2_2}{\underset{}{C}}{=}NR^3$

$(R{=})$ ⟨—$CH{=}$⟩₂CH—⟨⟩—R

1. CH_2Cl_2, Et_3N, N_2, stir overnight
2. MeOH

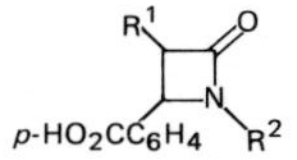

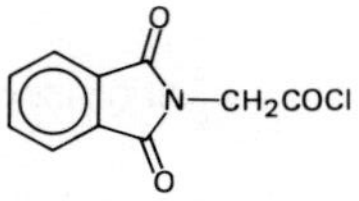

Ether, NEt_3, r.t., 1.5 h

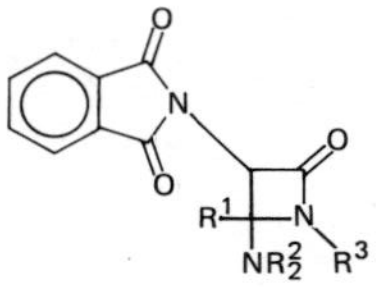

1. CH_2Cl_2, NEt_3, r.t.
2. 10% HCl

$cis:trans = 70:30$	
cis	
$cis:trans = 65:35$	
$cis:trans = 65:35$	
$trans$	

cis

$R = H$
$R = OMe$
$R = Me$

78	
76	
89	
82	
80	557
89	
86	
95	
79	
91	
75	
~100	554ʲ
40	
44	
36	558

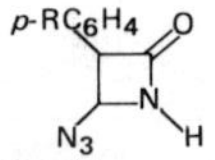

TABLE 32. (Continued)

Ketene or ketene precursor	Imine	Conditions	Product	Yield (%)	Reference
N_3CH_2COCl	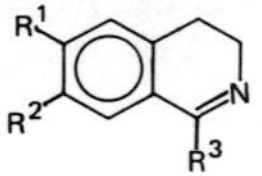	CH_2Cl_2, NEt_3	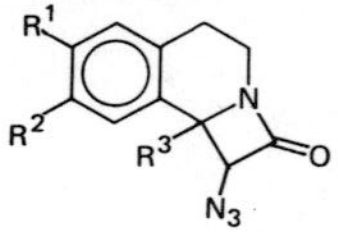		548
$R^1 = R^2 = H$, $R^3 = Ph$				66	
$R^1 = R^2 = H$, $R^3 = p\text{-}O_2NC_6H_4$				77	
$R^1 = R^2 = OMe$, $R^3 = p\text{-}O_2NC_6H_4$				73	
$ClCH_2COCl$	$R^1C_6H_4CH{=}NC_6H_4R^2$	C_6H_6, NEt_3, 70–75°C, 2 h	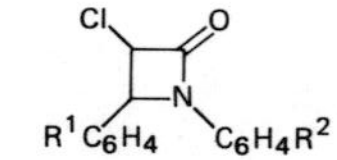 *cis:trans*		551
$R^1 = o\text{-}NO_2$, $R^2 = p\text{-}MeO$			1:1	19	
$R^1 = o\text{-}NO_2$, $R^2 = p\text{-}Cl$			32:68	6	
$R^1 = o\text{-}NO_2$, $R^2 = 2\ 4\text{-diMe}$			22:78	9	
$R^1 = o\text{-}NO_2$, $R^2 = o\text{-}Br$			1:4	2	
$R^2 = o\text{-}NO_2$, $R^2 = H$			44:56	9	
$R^2 = m\text{-}NO_2$, $R^2 = p\text{-}MeO$			0:100	9	
$R^1 = p\text{-}NO_2$, $R^2 = p\text{-}MeO$			0:100	16	
$R^1 = o\text{-}Cl$, $R^2 = p\text{-}MeO$			18:82	30	
$R^1 = o\text{-}Cl$, $R^2 = H$			13:87	28	
$R^1 = p\text{-}Cl$, $R^2 = p\text{-}MeO$			0:100	7	
$R^1 = o\text{-}MeO$, $R^2 = p\text{-}MeO$			1:9	28	
$R^1 = o\text{-}MeO$, $R^2 = H$			0:100	25	
$R^1 = p\text{-}MeO$, $R^2 = p\text{-}MeO$			0:100	65	
$R^1 = o\text{-}Me$, $R^2 = p\text{-}MeO$			1:9	45	
$R^1 = o\text{-}Me$, $R^2 = H$			0:100	20	
$R^1 = o\text{-}(t\text{-Bu})$, $R^2 = p\text{-}MeO$			1:3	10	
$ClCH_2COCl$	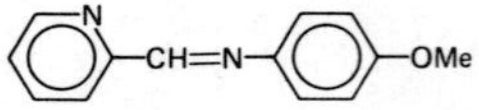	C_6H_6, NEt_3, 70–75°C, 2 h	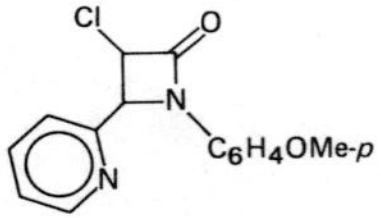 *cis:trans* = 27:73	5	551

Substrate	Reactant	Conditions	Product	cis:trans	Yield (%)	Ref.
MeCH₂COCl	o-O₂NC₆H₄CH=NC₆H₄OMe-p	C₆H₆, NEt₃, 70–75°C, 2 h	[β-lactam: Me, o-O₂NC₆H₄, N–C₆H₄OMe-p] cis:trans = 36:64		25	551
ClCH₂COOH	R¹C₆H₄CH=NC₆H₄R²	POCl₃, DMF, reflux 2 h	[β-lactam: Cl, C₆H₄R², N, R¹C₆H₄]			552^f
	R¹ = o-NO₂, R² = H			1:1	1	
	R¹ = o-NO₂, R² = p-MeO			1:1	1	
	R¹ = p-NO₂, R² = p-MeO			53:47	53	
	R¹ = p-NO₂, R² = H			1:1	40	
	R¹ = o-Cl, R² = H			54:46	38	
	R¹ = p-Cl, R² = p-MeO			48:52	42	
	R¹ = p-Cl, R² = H			1:1	33	
	R¹ = o-MeO, R² = H			1:1	46	
	R¹ = p-MeO, R² = p-MeO			1:1	53	
PhCHClNCOCH₂Cl ($|$ Ph)	[O←P(OR³)₂ / C=N–R¹ on R²C₆H₄]	DMF, 25°C or C₆H₆, reflux	[β-lactam: Cl, Ph, N–Ph] No reaction	cis:trans = 55:45	—	552^i
		DMF, reflux		trans	—	
		DMF, NEt₃, 25°C		trans	—	
		C₆H₆, NEt₃, 25°C				
[phthalimide]N–CH₂COCl		Ether, NEt₃, r.t.	[phthalimidyl β-lactam: p-R²C₆H₄, N–R¹, P(OR³)₂→O]			555

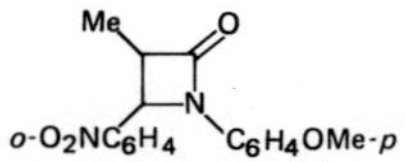

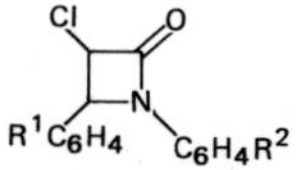

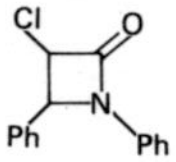

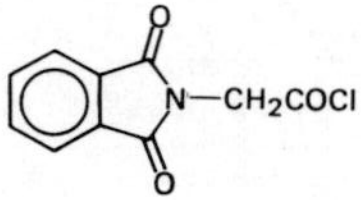

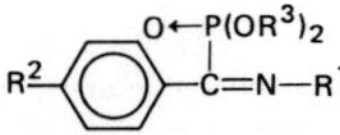

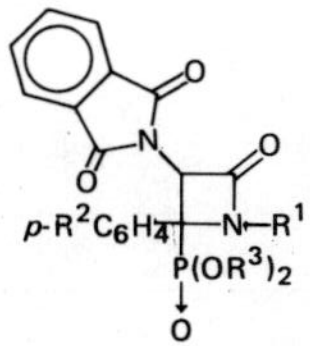

TABLE 32. (Continued)

Ketene or ketene precursor	Imine	Conditions	Product	Yield (%)	Reference
R^1 = Ph, R^2 = H, R^3 = Me				18	
R^3 = Ph, R^2 = H, R^3 = Et				46	
R^2 = Ph, R^2 = Me, R^3 = Me				29	
R^1 = Ph, R^2 = Me, R^3 = Et				32	
R^1 = Ph, R^2 = Cl, R^3 = Me				17	
R^1 = Ph, R^2 = Cl, R^3 = Et				28	
R^1 = Ph, R^2 = Br, R^3 = Me				22	
R^1 = Ph, R^2 = Br, R^3 = Et				27	
R^1 = Ph, R^2 = OMe, R^3 = Me				36	
R^1 = Ph, R^2 = OMe, R^3 = Et				40	
R^1 = Me, R^2 = H, R^3 = Et				24	
R^1CH_2COCl	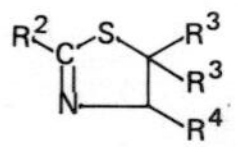	CH_2Cl_2, NEt	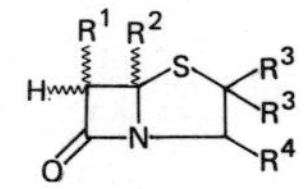		
R^1 = N_3	R^2 = Ph, R^3 = R^4 = H	Reflux 24–26 h.		70	566
	R^3 = Me, R^4 = H			87	566
	R^2 = Ph, R^3 = Me, R^4 = CO_2Me			20–25	574
	R^2 = CO_2Bu-t, R^3 = Me, R^4 = H	Stir overnight		86	568
	R^2 = H, R^3 = Me, R^4 = CO_2Me		*trans*	5–8	572
R^1 = MeO	R^2 = Ph, R^3 = R^4 = H	Reflux 24–26 h	*cis*	90	566
	R^2 = MeS, R^3 = R^4 = H	Stir overnight		73	565
	R^2 = R^3 = H, R^4 = CO_2Et	Reflux 24–26 h		11	566
	R^2 = Ph, R^3 = Me, R^4 = CO_2CHPh$_2$			90	566
	R^2 = p-(PhCH$_2$CO$_2$)C$_6$H$_4$, R^3 = R^4 = H		*cis*	70	566
R^1 = PhO	R^2 = Ph, R^3 = R^4 = H			70	560, 566
	R^2 = p-PhCH$_2$CO$_2$)C$_6$H$_4$, R^3 = R^4 = H			63	566
R^1 = Ph	R^2 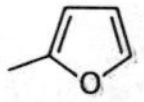 = [furanyl], R^3 = R^4 = H		*cis*	70	566
R^1 = PhCH$_2$OCONH	R^2 = p-O$_2$NC$_6$H$_4$, R^3 = R^4 = H		*trans*	–	575

R¹ = (structure)	Substituents	Conditions	Yield (%)	Reference
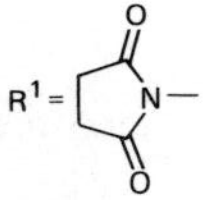 (succinimide)	$R^2 = Ph, R^3 = Me, R^4 = CO_2Me$	Reflux 24–26 h	—	574
		Reflux 6.25 h	13	577, 578
(phthalimide)	$R^2 = Ph, R^3 = R^4 = H$	Reflux 6 h, NEt_3, C_6H_6, reflux 4 h	56	577
			14	577
		Reflux 10 h	16	577
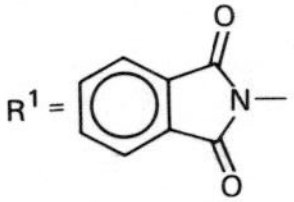	$R^2 = PhCH_2CO_2-, R^3 = R^4 = H$	CH_2Cl_2, NEt_2, reflux 24–26 h	85	569
	$R^2 = Ph, R^3 = Me, R^4 = CO_2Me$	CH_2Cl_2, NEt_3, reflux	—	574
		CH_2Cl_2, NEt_3, reflux 6 h	58.3	582
	$R^2 = Ph, R^3 = Me, R^4 = H$	Ether, NEt_3, reflux 2 h	5	577
	$R^2 = Ph, R^3 = R^4 = H$		40	581
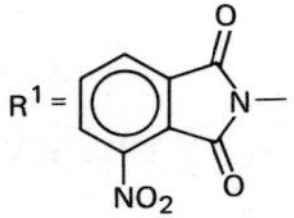 (nitro-phthalimide)	$R^2 = Ph, R^3 = R^4 = H$		17	581
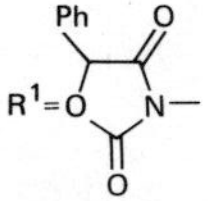 (oxazolidinedione, Ph)	$R^2 = Ph, R^3 = R^4 = H$	CH_2Cl_2, NEt_3, reflux 7 h	28.4	579
		Ether, NEt_3, reflux	poor	579
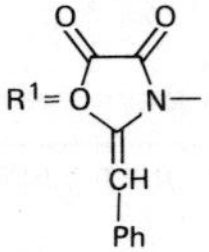 (oxazolidinedione, =CH—Ph)	$R^2 = Ph, R^3 = R^4 = H$	CH_2Cl_2, dioxane, NEt_3, N_2, reflux 6.5 h	45	580

TABLE 32. (Continued)

Ketene or ketene precursor	Imine	Conditions	Product	Yield (%)	Reference
N_3CH_2COCl	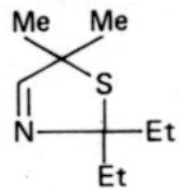	CH_2Cl_2, NEt_3, reflux	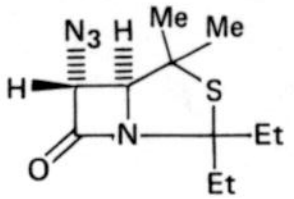	18	570, 573
N_3CH_2COCl	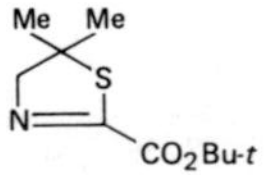	CH_2Cl_2, NEt_3	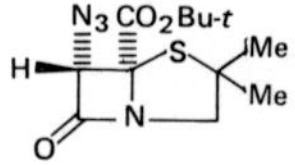	Good	573
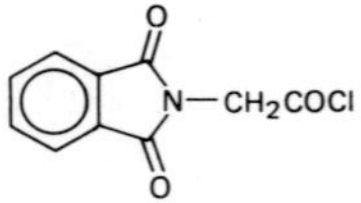	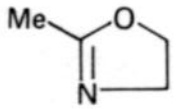	C_6H_6, NEt_3, r.t. 2 h	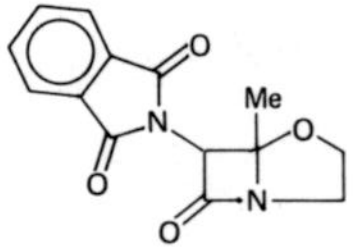	26	583
$ClCH_2COCl$	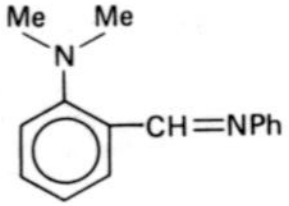	C_6H_6, NEt_3, reflux	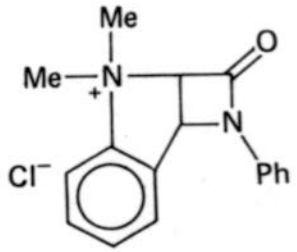	49	591
$ClCH_2COCl$	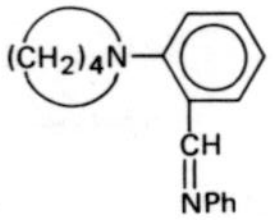	C_6H_6, NEt_3, reflux	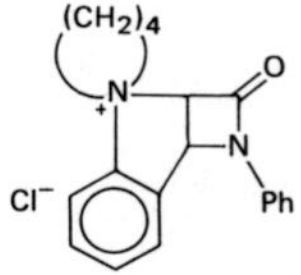	41	591

Reactant	Substituents	Conditions	Product	cis/trans	Yield	Ref.
R¹CH₂COCl + R^2, S, N, R^3, R^4	R^1 = MeO; R^2 = Ph, R^3 = R^4 = H	CH_2Cl_2, NEt_3, reflux, then stir overnight		cis	63	560, 566
	R^1 = PhO; R^2 = Ph, R^3 = R^4 = H			cis	81	560, 566
	R^1 = N₃; R^2 = Ph, R^3 = R^4 = H			cis	23	571
	R^2 = p-$O_2NC_6H_4$, R^3 = R^4 = H			cis	55–60	571, 574
	R^1 = PhCH₂OCONH, R^2 = Ph, R^3 = R^4 = H			trans	50–70	575
	R^1 = PhCH₂OCONH, R^2 = p-$O_2NC_6H_4$, R^3 = R^4 = H			trans	50–70	575
R^1 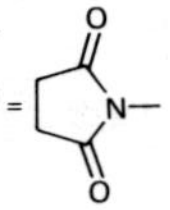 = (succinimide)	R^2 = Ph, R^3 = R^4 = H	C_6H_6, NEt_3, 80°C, 12 h			53	584
R^1 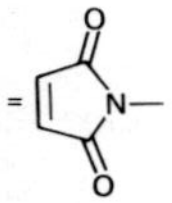 = (maleimide)	R^2 = Ph, R^3 = R^4 = H	C_6H_6, NEt_3, 54°C, 2.5 h			55	584
R^1 = (phthalimide)	R^2 = Ph, R^3 = R^4 = H	C_6H_6, NEt_3, r.t. 2 h; C_6H_6, NEt_3, 80°C, 2.5 h			43; 70.5	583; 584
N_3CH_2COCl + (S, N, Me, CO_2Me)	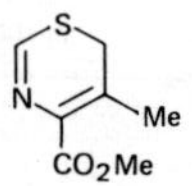	CH_2Cl_2, NEt_3	(N_3, S, N, Me, CO_2Me)		52	567

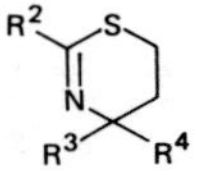

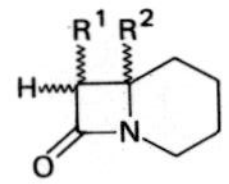

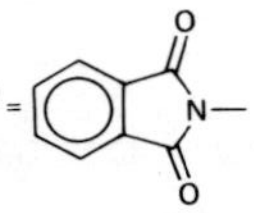

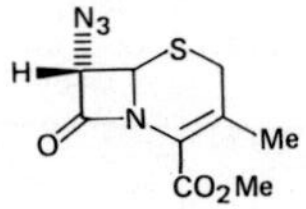

TABLE 32. (Continued)

Ketene or ketene precursor	Imine	Conditions	Product	Yield (%)	Reference
N_3CH_2COCl	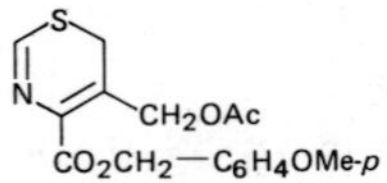	CH_2Cl_2, NEt_3, $-78°C$	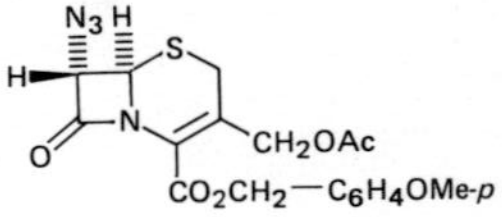	56	567
N_3CH_2COCl	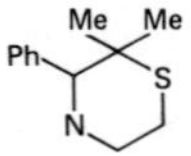	CH_2Cl_2, NEt_3	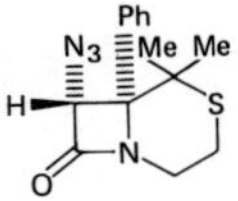	30	571
N_3CH_2COCl	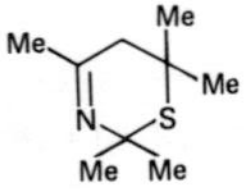	CH_2Cl_2, NEt_3	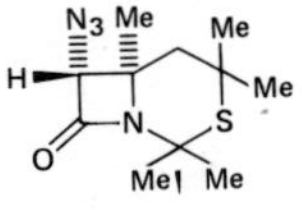	10	571
	 R = Ph R = $C_6H_5CH_2$ R = p-$O_2NC_6H_4$	C_6H_6, NEt_3, reflux 4 h 4 h 3 h, then stir at r.t.		57 25 65	585

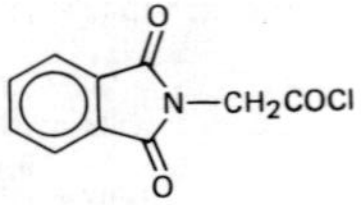

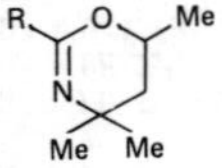

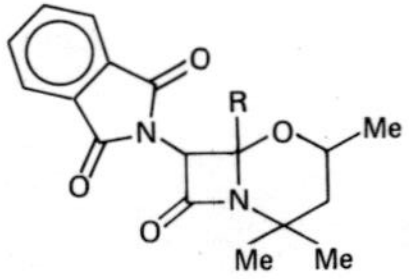

Acyl chloride	Conditions	Substituents	Stereochem.	Yield (%)	Ref.
R¹CH₂COCl	CH₂Cl₂, NEt₃, N₂, r.t.		trans		
R¹ = MeO		R² = Ph; n = 1		68	561, 562
		R² = Ph; n = 2		75	561
		R² = 3,4-(MeO)₂C₆H₃; n = 2		67	561
R¹ = PhO		R² = 3,4-(MeO)₂C₆H₃; n = 1		33	561
		R² = 3-O₂NC₆H₄; n = 1		60	561
		R² = Ph; n = 1		50	561
RCH₂COCl	CH₂Cl₂, NEt₃, N₂, stir overnight, r.t.	R = OMe		25	565
		R = OPh		62	
		R = N₃		31	
$\begin{array}{c} R^1CH_2C(=O)-O \\ R^2C=O \end{array}$	CH₂Cl₂, NEt₃, reflux 1 h, then stir overnight	R³CH=NR⁴			
R¹ = CF₃, OEt, OBu-i; R² = N₃		R³ = R⁴ = Ph	cis and trans	30–70	576
R² = PhO		R³ = p-MeOC₆H₄, R⁴ = p-HO₂CC₆H₄	trans		
		R³ = R⁴ = Ph	cis and trans		
		R³ = p-MeOC₆H₄, R⁴ = CHPh₂	cis		
R¹R²CHCOCl	CH₂Cl₂, NEt₃	R³R⁴C=NR⁵			
R¹ = N₃, R² = Me		R³ = H, R⁴ = R⁵ = Ph		10	548
R¹ = N₃, R² = Et		R³ = H, R⁴ = R⁵ = Ph		9	548
R¹ = N₃, R² = Ph		R³ = H, R⁴ = R⁵ = Ph		5.7	548

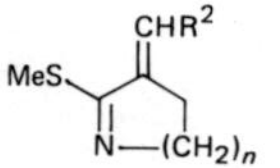

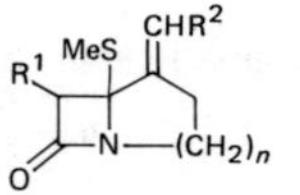

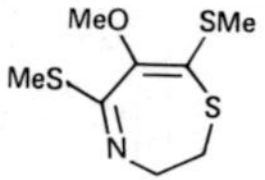

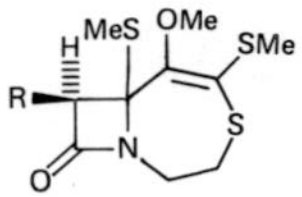

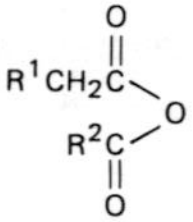

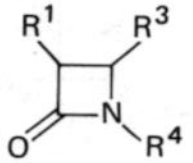

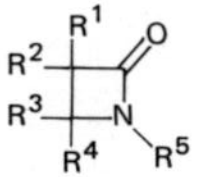

TABLE 32. (Continued)

Ketene or ketene precursor	Imine	Conditions	Product	Yield (%)	Reference
$R^1 = CN$, $R^2 = Me$	$R^3 = H$, $R^4 = R^5 = Ph$	C_6H_6, heat 3 h		53	586
	$R^3 = H$, $R^4 = Ph$, $R^5 = p\text{-MeOPh}$	C_6H_6, heat 3 h		31	586
	$R^3 = R^4 = R^5 = Ph$	C_6H_6, heat 2 h		77	586
	$R^3 = R^4 = Ph$, $R^5 = p\text{-MeOPh}$			82	586
	$R^3 = H$, $R^4 = R^5 = Ph$			49	586
$R^1 = CN$, $R^2 = Ph$	$R^3 = H$, $R^4 = Ph$, $R^5 = p\text{-MeOPh}$	C_6H_6, NEt_3, 20°C		45	586
	$R^3 = R^4 = R^5 = Ph$			100	587
$R^1 = R^2 = Cl$	$R^3 = H$, $R^4 = R^5 = Ph$			100	587*l*
	$R^3 = H$, $R^4 = Ph$, $R^5 = n\text{-Bu}$			70	587
	$R^3 = H$, $R^4 = Ph$, $R^5 = C_6H_{11}$			90	587
$R^1 = Ph$, $R^2 = OAc$	$R^3 = H$, $R^4 = R^5 = Ph$	NEt_3	*trans*	90–95	588
	$R^3 = MeS$, $R^4 = R^5 = Ph$		*cis*	—	588*m*
$\begin{array}{c} R^2 \\ R^1\text{—CHCOCl} \end{array}$	$R^3N{=}C{=}NR^3$		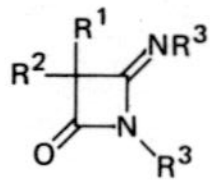		
$R^1 = CN$, $R^2 = Me$	$R^3 = C_6H_{11}$	C_6H_6, 140°, 6 h		88	586
$R^1 R^2 = Cl$	$R^3 = C_6H_{11}$	Cyclohexane, NEt_3, reflux 50 min		88	589
	$R^3 = i\text{-Pr}$	Cyclohexane, NEt_3, reflux 100 min		76	589
$Cl_2CHCOCl$	$PhHC{=}CHCH{=}NR$	C_6H_6, NEt_3	$R = Ph$	45	589
			$R = p\text{-MeC}_6H_4$	67	
			$R = PhHC{=}CHCH{=}N-$	75	

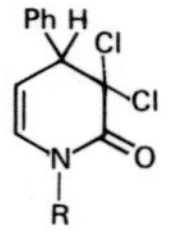

591

R¹R²COCl (Cl)	CH=NR⁴ reagent substituents	conditions	config.	yield
R¹ = R² = Cl	R³ = NO₂, R⁴ = Ph, X = (CH₂)₄	C₆H₆, NEt₃, reflux		75
	R³ = NO₂, R⁴ = Ph, X = (CH₂)₅			85
	R³ = NO₂, R⁴ = Ph, X = (CH₂)₂O(CH₂)₂			49
R¹ = H, R² = Cl	R³ = H, R⁴ = Ph, X = (CH₂)₅		trans	52
	R³ = H, R⁴ = Ph, X = (CH₂)₆		trans	55
	R³ = H, R⁴ = Ph, X = (CH₂)₂O(CH₂)₂		trans	52
R¹ = Cl, R² = Me	R³ = H, R⁴ = Ph, X = Me₂			25
	R³ = H, R⁴ = Ph, X = (CH₂)₄			64
	R³ = H, R⁴ = Ph, X = (CH₂)₅			61
	R³ = H, R⁴ = Ph, X = (CH₂)₆			65
R¹ = Cl, R² = Ph	R³ = H, R⁴ = Ph, X = Me₂	C₆H₆, NEt₃, r.t.		54
	R³ = H, R⁴ = Ph, X = (CH₂)₄	C₆H₆, NEt₃, reflux		10
	R³ = H, R⁴ = Ph, X = (CH₂)₅			63
	R³ = H, R⁴ = Ph, X = (CH₂)₆			75
	R³ = H, R⁴ = Ph, X = (CH₂)₂O(CH₂)₂			32
R¹ = R² = Cl	R³ = H, R⁴ = Ph, X = Me₂			60
	R³ = H, R⁴ = Ph, X = (CH₂)₅			83
R¹ = H, R² = Me	R³ = H, R⁴ = Ph, X = (CH₂)₅			65
R¹ = H, R² = OPh	R³ = H, R⁴ = Ph, X = (CH₂)₅			66
R¹ = H, R² = Cl	R³ = H, R⁴ = C₆H₁₁, X = (CH₂)₅		trans	75
	R³ = H, R⁴ = n-Bu, X = (CH₂)₅		trans	24
R¹ = t-Bu, R² = CN	R³ = H, R⁴ = CH₂C₆H₅, X = (CH₂)₅		trans	44
	R³ = H, R⁴ = Ph, X = (CH₂)₅		trans	66
R¹ = R² = Cl	R³ = H, R⁴ = Ph, X = Me₂		trans	67
	R³ = NO₂, R⁴ = Ph, X = (CH₂)₅	C₆H₆, NEt₃, r.t.		68
				15

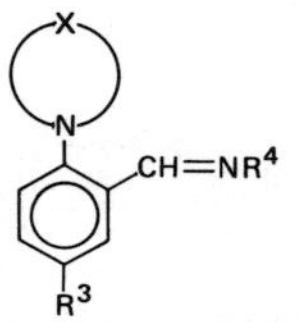

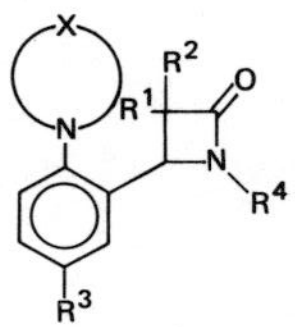

R¹ = H, R² = Ph	R³ = H, R⁴ = Ph, X = Me₂	C₆H₆, NEt₃, reflux 2 h	50

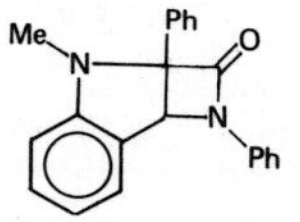

TABLE 32. (Continued)

Ketene or ketene precursor	Imine	Conditions	Product	Yield (%)	Reference
$R^1CH(COR^2)_2$	$R^3R^4C=NR^5$		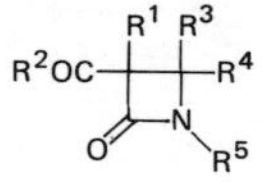		
R^1 = Et, R^2 = Cl	R^3 = H, R^4 = R^5 = Ph	C_6H_6, 120°C, 15 min No solvent, reflux		44.8 44.8	592 592
R^1 = Et, R^2 = OEt	R^3 = H, R^4 = R^5 = Ph	C_6H_6, reflux, 4 h		21.7	592
R^1 = Et, R^2 = Cl	R^3 = H, R^4 = R^5 = Ph	1. C_6H_6, 120°C, 15 min 2. EtOH, reflux 1 h		46.5	592
R^1 = $CH_2C_6H_5$, R^2 = Cl	R^3 = H, R^4 = R^5 = Ph	1. C_6H_6, reflux 4 h 2. EtOH, reflux 1 h or No solvent, 120°C, 15 min 2. EtOH, reflux 1 h	R^1 = $CH_2C_6H_5$ R^2 = OEt R^3 = H R^4 = Ph	70	592
		1. C_6H_6, reflux 1 h 1. C_6H_6, reflux 4 h 2. MeOH, reflux 1 h or No solvent, 110–120°C, 15 min 2. MeOH, reflux 1 h	R^1 = $CH_2C_6H_5$ R^3 = H R^4 = Ph	—	592
R^1 = Ph, R^2 = Cl	R^3 = H, R^4 = R^5 = Ph	C_6H_6, reflux	cis	—	593, 594
	R^3 = H, R^4 = R^5 = Ph	1. C_6H_6, reflux 4 h 2. MeOH, reflux 1 h	R^1 = Ph R^2 = OMe R^3 = H R^4 = Ph	75.8	592
R^1 = i-Pr, R^2 = Cl	R^3 = H, R^4 = Ph, R^5 = 2,4-diMeC_6H_3	C_6H_6, reflux 6 h		72	592

[a] Treatment of this product with Raney Ni in 95% EtOH for 1 h afforded a 30% conversion to 3,3-dimethyl-1,4-diphenylazetidine-2-one.
[b] Structure uncertain. Probable mechanism:

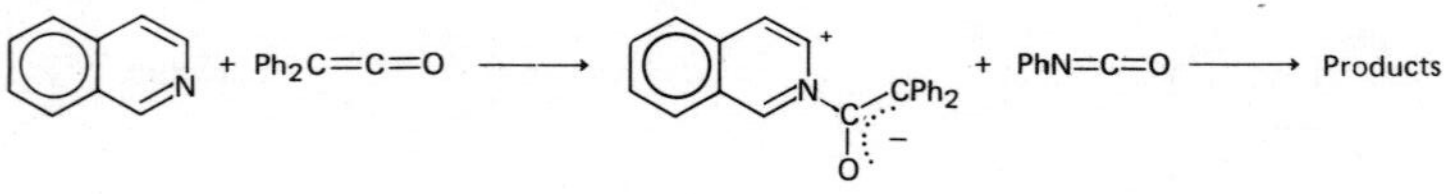

[c]This product was also prepared in this paper by the reaction of 2 moles of diphenylketene and 1 mole of p-nitroso-N,N-dimethylaniline in ether.

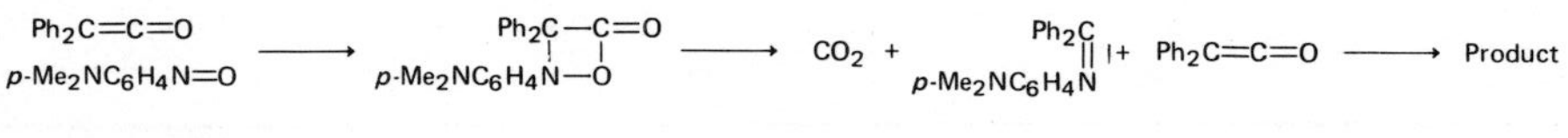

$$Ph_2C=C=O \quad Ph_2C-C=O \longrightarrow CO_2 + \underset{p\text{-Me}_2NC_6H_4N}{\overset{Ph_2C}{\underset{\|\ }{}}}\| + Ph_2C=C=O \longrightarrow \text{Product}$$

$$p\text{-Me}_2NC_6H_4N=O \quad p\text{-Me}_2NC_6H_4N-O$$

[d]Product isolated as a 5:2 mixture of 2 diastereomers.
[e]Product isolated as a 3:1 mixture of 2 diastereomers.
[f]The mechanism for this reaction is believed[550] to be:

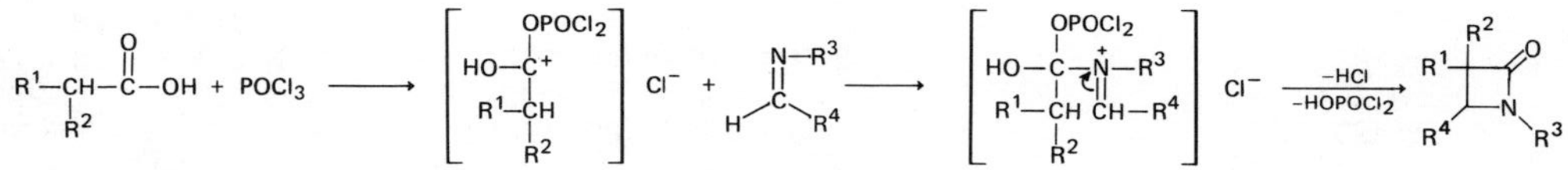

[g]This product is also prepared in this paper by treatment of 1,4-diphenyl-3-bromoazetidin-2-one or 1,4-diphenyl-3,3-dibromazetidin-2-one with zinc in MeOH and liq. NH_3 for 8 h.

[h]Treatment of cis-1,4-diphenyl-3-chloroazetidin-2-one with $POCl_3$ and $ClCH_2COOH$ in DMF for 7 h. afforded 1,4-diphenyl-3-chloroazetidin-2-one in a cis:trans ratio of 53:47. This same ratio of isomers was also obtained by treatment of trans-1,4-diphenyl-3-chloroazetidin-2-one with the same mixture for 22 h. Treatment of the cis isomer as above for only 2 h gave 18% trans, while treatment of the trans isomer for the same length of time afforded no isomerization to the cis isomer.

[i]This reaction illustrates that the lactam product is formed by direct acylation rather than in situ generation of ketene[519,520].

[j]The products from this reaction were not isolated but were cleaved directly by distillation in vacuum to

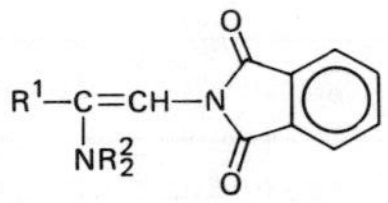

[k]Dropwise addition of a soln. of acid chloride in CH_2Cl_2 to a CH_2Cl_2 soln. of benzylidineaniline and NEt_3 at r.t. was found to produce cis-lactam exclusively (Method A), while addition of NEt_3 to a CH_2Cl_2 solution of acid chloride and benzylidineaniline afforded trans-lactam exclusively (Method B).

[l]Reaction of dichloroacetyl chloride with benzylideneaniline without any NEt_3 afforded $Cl_2CHCONPhCHClPh$, which upon heating to 150°C or refluxing in benzene afforded a 20–30% yield of 1,4-diphenyl-3,3-dichloroazetidin-2-one.

[mm]This product was desulphurized using Raney Ni to afford the previous product with opposite stereochemistry.

TABLE 33. Production of diazetidinones by the reaction of ketenes with azo compounds

Ketene	Azo compound	Product	Conditions	Yield (%)	Reference
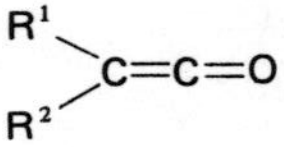 R^1—C=C=O (with R^2)	$R^3N=NR^4$		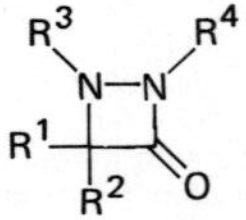		
$R^1 = R^2 = H$	$R^3 = R^4 = Ph$		Hexane, 15°C	—	595
$R^1 = H, R^2 = Ph$	$R^3 = R^4 = Ph$		C_6H_6,	—	515,596
$R^1 = R^2 = Ph$	$R^3 = R^4 = Ph$		100°C, or ether	—	515,596
	(trans)		125–130°C, 42 h, CO_2	25	597
	(cis)		r.t.	92	597
	$R^3 = R^4 = o\text{-MeC}_6H_4$		Ether, $h\nu$, stand overnight	80	597
	$R^3 = R^4 = m\text{-MeC}_6H_4$				
	(cis)			35	597
	(trans)			90	597
	$R^3 = R^4 = C_6H_5CH_2$		C_6H_6, N_2	96	515, 599

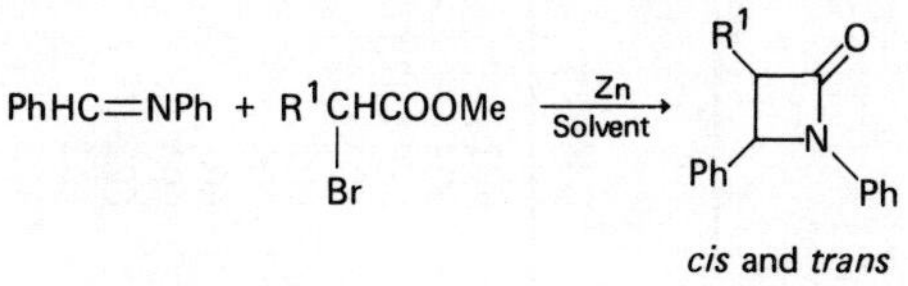

$$\text{PhHC}{=}\text{NPh} + \underset{\underset{\text{Br}}{|}}{\text{R}^1\text{CHCOOMe}} \xrightarrow[\text{Solvent}]{\text{Zn}} \qquad (291)$$

cis and trans

	cis : trans					
Solvent	R^1 = Me	Et	i-Pr	C_6H_{11}	t-Bu	Ph
C_6H_5Me	73:27	64:36	55:45	45:55	25:75	0:100
Et_2O, C_6H_6 (50:50)	–	63:37	80:20	76:24	–	–
THF	80:20	74:26	100:0	100:0	100:0	0:100

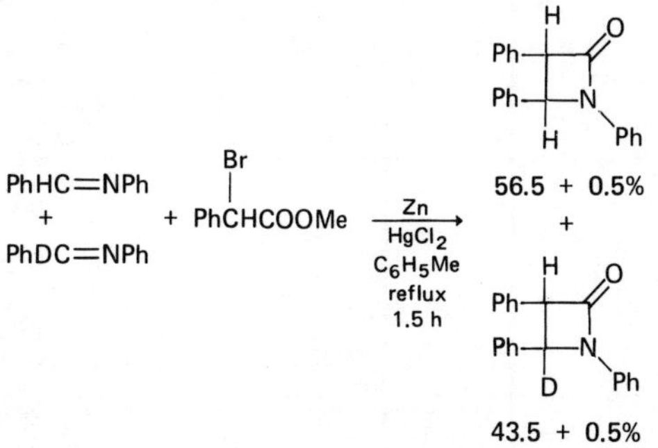

$$\begin{array}{c}\text{PhHC}{=}\text{NPh} \\ + \\ \text{PhDC}{=}\text{NPh}\end{array} + \underset{\underset{\text{Br}}{|}}{\text{PhCHCOOMe}} \xrightarrow[\substack{\text{C}_6\text{H}_5\text{Me} \\ \text{reflux} \\ 1.5\,h}]{\substack{\text{Zn} \\ \text{HgCl}_2}} \quad \begin{array}{c}56.5 + 0.5\% \\ + \\ 43.5 + 0.5\%\end{array} \qquad (292)$$

reaction with imines. One interesting sidelight to these investigations is the study[606] of the time required to epimerize the *cis* β-lactams prepared into their *trans* counterparts. The results obtained[606] are shown in the table below equation (293). In addition, it was also found that in 50 h at 75°C 92% of pure *trans*-1,3,4-triphenylazetidin-2-one was converted into its *cis* epimer.

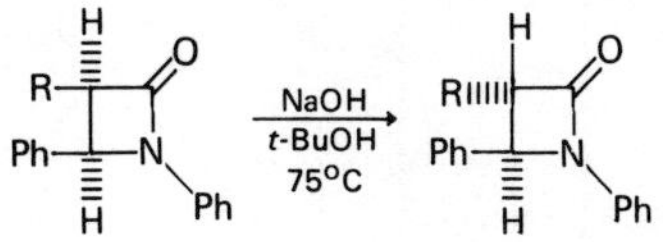

$$\xrightarrow[\substack{t\text{-BuOH} \\ 75°\text{C}}]{\text{NaOH}} \qquad (293)$$

R		% trans			
t(h) = 25	50	75	100	150
Me	42			
i-Pr	55		70	75
C_6H_{11}		80		95
t-Bu 95		98		98

c. *Other imine cycloadditions.* Reaction of anils containing a methyl or methylene group in the α-position of the imine double bond with monosubstituted malonyl chloride has been reported[607] to afford acceptable yields of *n*-aryl-

J. F. Wolfe and M. A. Ogliaruso

TABLE 34. β-Lactam preparation by the reaction of a Reformatsky reagent with an imine

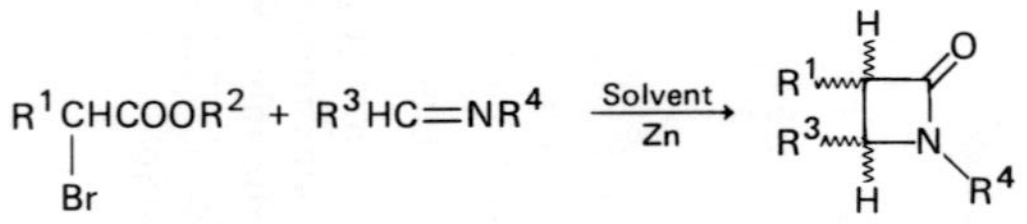

$$R^1CHCOOR^2 + R^3HC{=}NR^4 \xrightarrow[\text{Zn}]{\text{Solvent}}$$

(with Br attached below R^1; product is a β-lactam)

R^1	R^2	R^3	R^4	Solvent	Stereochemistry (cis:trans)	Yield (%)	Reference
H	Et	Ph	Me	C_6H_5Me	—	52	601
H	Et	Ph	Ph	C_6H_5Me	—	56	604
Me	Me	Ph	Ph	C_6H_5Me	73:27	90	518,602,603
				THF	—	94	603
				Et_2O, C_6H_6	—	75	603
Me	Me	Ph	p-BrC_6H_4	THF	—	90	603
Me	Et	Ph	Me	C_6H_5Me	—	81	601
Me	Et	Ph	Ph	C_6H_5Me	—	85	604
Me	Et	Ph	$C_6H_5CH_2$	C_6H_5Me	—	76	601
Me	i-Pr	Ph	Ph	C_6H_5Me	55:45	—	603
Me	t-Bu	Ph	Ph	C_6H_5Me	25:75	—	603
Me	2-(i-Pr)-5-MeC_6H_9	Ph	p-BrC_6H_4	C_6H_5Me	—	80	603
Et	Me	Ph	Ph	C_6H_5Me	64:36	94	518,602,603
				THF	74:26	96	603
				Et_2O, C_6H_6	63:37	72	603
i-Pr	Me	Ph	Ph	C_6H_5Me	55:45	98	518,602,603
				THF	100:0	92	603
				Et_2O, C_6H_6	80:20	98	603

$i\text{-}Pr$	Me	Ph	$p\text{-}MeC_6H_4$	THF	—	95–97	603,606
$i\text{-}Pr$	Me	Ph	$p\text{-}ClC_6H_4$	THF	—	98	603
$i\text{-}Pr$	Me	Ph	$p\text{-}BrC_6H_4$	C_6H_5Me or THF	—	95–98	603,606
$i\text{-}Pr$	Me	$p\text{-}MeOC_6H_4$	Ph	C_6H_5Me	34:66	70	603
$i\text{-}Pr$	$i\text{-}Pr$	Ph	Ph	C_6H_5Me	2:98	93	603
$i\text{-}Pr$	$t\text{-}Bu$	Ph	Ph	C_6H_5Me		—	603
$i\text{-}Pr$	$2\text{-}(i\text{-}Pr)\text{-}5\text{-}MeC_6H_9$	Ph	Ph	C_6H_5Me	45:55	90	518,602,603
C_6H_{11}	Me	Ph	Ph	C_6H_5Me	100:0	92	603
$t\text{-}Bu$	Me	Ph	Ph	THF	76:24	96	603
$t\text{-}Bu$	Me	Ph	Ph	$Et_2O,\ C_6H_6$	25:75	98	518,602,603
$t\text{-}Bu$	$i\text{-}Pr$	Ph	$p\text{-}BrC_6H_4$	C_6H_5Me	100:0	71	603
Ph	Me	Ph	Ph	THF	—	96	603
Ph	Et	Ph	Ph	C_6H_5Me or THF	0:100	98	603
H	$2\text{-}(i\text{-}Pr)\text{-}5\text{-}MeC_6H_3$	Ph	Ph	C_6H_5Me	—	95	518,602,603
$2\text{-}(i\text{-}Pr)\text{-}5\text{-}MeC_6H_3$	Me	Ph	Ph	C_6H_5Me or THF		7	601
$2\text{-}(i\text{-}Pr)\text{-}5\text{-}MeC_6H_3$	$i\text{-}Pr$	Ph	Ph	C_6H_5Me		82	603
Me	Me	Ph	Ph	C_6H_6	38:62	—	603
Me	$i\text{-}Pr$	Ph	Ph	C_6H_5Me	10:90	—	603
$Me\text{--}C(Br)CO_2Et$		Ph	$C_6H_5CH_2$	C_6H_5Me		84	601[a]

[a] Product is:

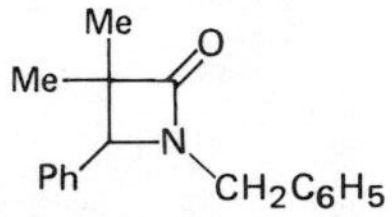

β-lactam: 3,3-dimethyl-4-phenyl-1-($CH_2C_6H_5$)-azetidin-2-one (labels: Me, Me, O, N, $CH_2C_6H_5$, Ph)

4-hydroxy-2-pyridones. The mechanism proposed for this reaction is shown in equation (294).

Initial attack of one acid chloride function on the anil nitrogen affords an imine salt, which then loses two moles of hydrogen chloride consecutively to afford product.

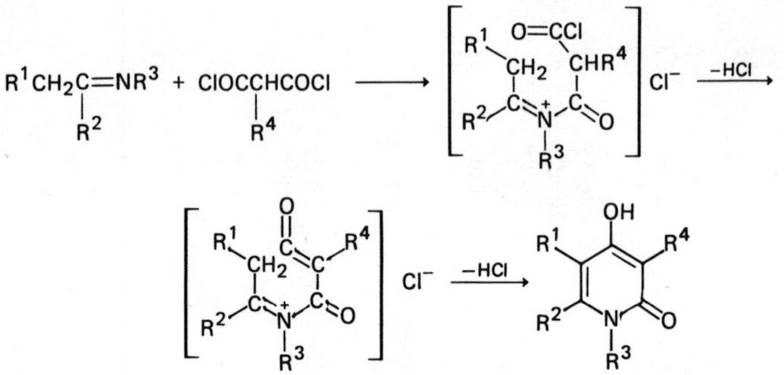

$$(294)$$

R^1	R^2	R^3	R^4	Yield (%)	Conditions
H	Ph	Ph	$CH_2C_6H_5$	42–44	C_6H_5Me, reflux 90 min
				31.2	C_6H_5Me, reflux 30 min
H	Ph	Ph	i-Pr	55.7	C_6H_6, reflux 1 h
H	Ph	p-MeC_6H_4	$CH_2C_6H_5$	81.7	C_6H_6, reflux 45 min
H	Ph	p-MeC_6H_4	n-Bu	42	C_6H_6, reflux 45 min
H	Ph	p-MeC_6H_4	i-Pr	65.9	C_6H_6, reflux 80 min
Me	Ph	Ph	$CH_2C_6H_5$	38.2	C_6H_6, reflux 90 min
Et	Ph	Ph	$CH_2C_6H_5$	39.4	C_6H_6, reflux 2 h
H	s-Bu	Ph	$CH_2C_6H_5$	21	C_6H_6, reflux 80 min

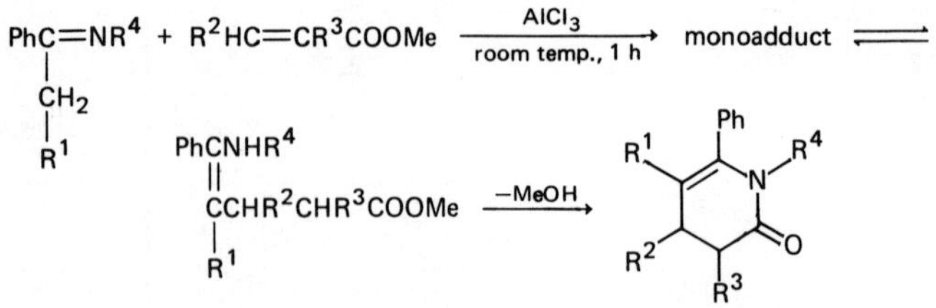

$$(295)$$

R^1	R^2	R^3	R^4	Yield (%)
H	H	H	Ph	85
H	H	Me	Ph	55
Me	H	H	Ph	85
Me	H	Me	Ph	60
H	Ph	H	Ph	30[a]
Me	Ph	H	Ph	25

[a]This yield was obtained at 80°C for 24 hr.; using the conditions shown above (r. t., 1 h) afforded no reaction.

Dihydropyridones have been similarly prepared[608] by the reaction of aromatic ketimines and acrylic esters. The proposed mechanism involves initial formation of an unspecified monoadduct which is proposed to be in equilibrium with the α-substituted anil shown in equation (295). Elimination of methanol affords dihydropyridone via intramolecular condensation. 2-Azetidinylidene ammonium salts (46) afford upon hydrolysis the corresponding 2-azetidinones (47)[609]. The salt 46 is prepared by addition of a *N,N*-dimethyl-1-chloroalkenylamine to a Schiff base. The mechanism proposed (equation 296) involves initial aminoalkenylation of

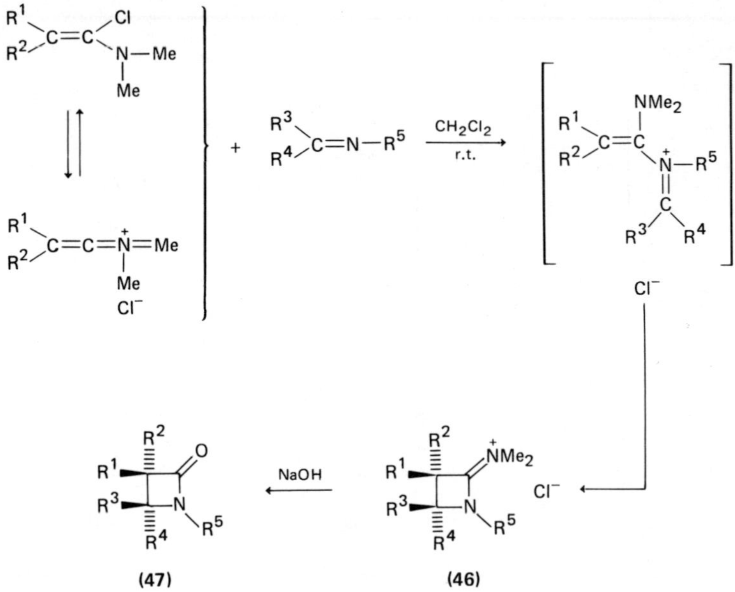

$$(296)$$

					Yield (%)	
R¹	R²	R³	R⁴	R⁵	46	47
Me	Me	Ph	H	Me	74	82
Me	Me	Ph	H	Ph	47	68
Me	Me	C₆H₅CH₂S	H	Me₃C	60	42

the Schiff base to give the intermediate shown, which then cyclizes to afford the salt. Since the intermediate is also in principle available from the reaction of α-chloroalkylideneammonium chloride with Schiff bases, followed by elimination of hydrogen chloride, the authors utilized the reaction of tertiary amides with phosgene followed by reaction with Schiff bases and triethylamine to produce β-lactams as shown in Scheme 3.

3. From nitrones and nitroso compounds

In 1919 Staudinger and Miescher[610] first investigated the reaction of diphenyl-ketene and various nitrones (anil *N*-oxides) and proposed the reaction course shown

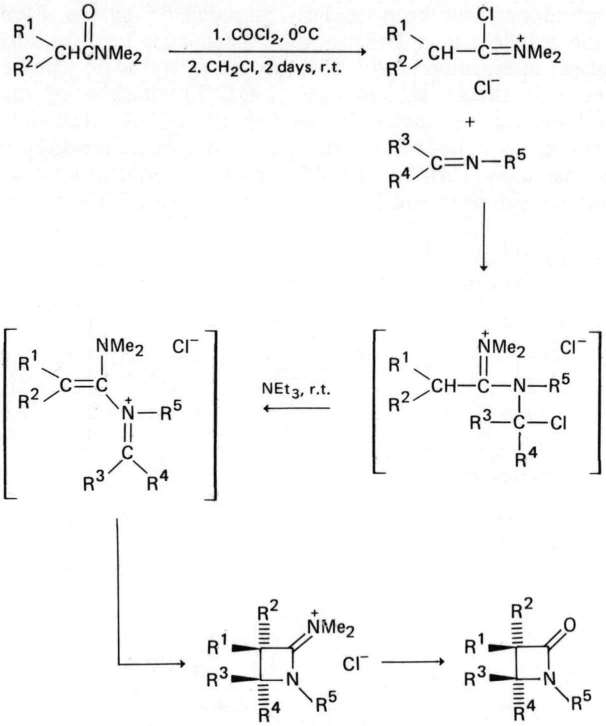

SCHEME 3.

R^1	R^2	R^3	R^4	R^5	Yield (%) 46	Yield (%) 47
Me	Me	Ph	Ph	Ph	65	70
H	CMe$_3$	Ph	H	Ph	80 (*trans*)	0

in equation (297). A similar reaction was investigated in 1938 by Taylor, Owen and Whittaker[611] who proposed the reaction course (298). More recently, however,

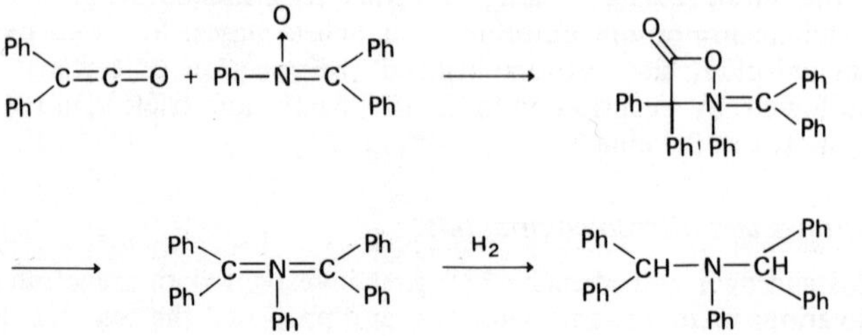

$$(297)$$

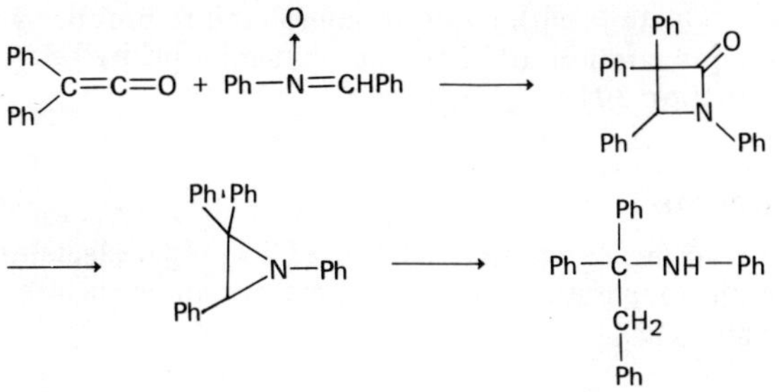

$$\underset{Ph}{\overset{Ph}{>}}C=C=O + Ph-N=CHPh \longrightarrow \text{[β-lactam structure]}$$

(298)

$$\longrightarrow \text{[aziridine structure]} \longrightarrow Ph-\underset{\underset{Ph}{\overset{|}{CH_2}}}{\overset{Ph}{\underset{|}{C}}}-NH-Ph$$

Hassall and Lippman[612] have found that the reaction of diphenylketene and benzylideneaniline oxide affords o-benzylideneaminophenyldiphenylacetic acid, which upon treatment with Adams catalyst in ethyl acetate produces 1-benzyl-3,3-diphenyloxindole and not a β-lactam (equation 299).

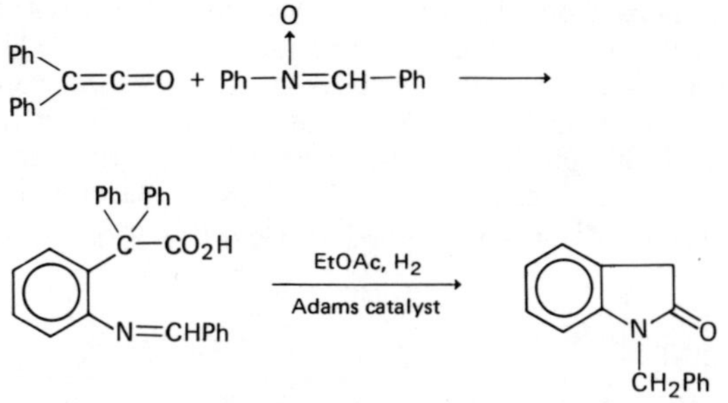

$$\underset{Ph}{\overset{Ph}{>}}C=C=O + Ph-\overset{\overset{O}{\uparrow}}{N}=CH-Ph \longrightarrow$$

(299)

$$\text{[acid structure with } C-CO_2H \text{ and } N=CHPh] \xrightarrow[\text{Adams catalyst}]{\text{EtOAc, } H_2} \text{[oxindole structure, } CH_2Ph]}$$

However, β-lactams have been produced from nitrones by reaction of the nitrones with copper phenylacetylide (equation 300)[613].

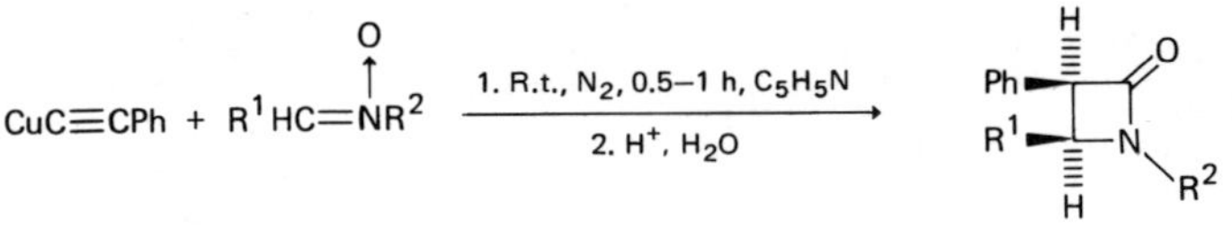

$$CuC\equiv CPh + R^1HC=\overset{\overset{O}{\uparrow}}{N}R^2 \xrightarrow[\text{2. } H^+, H_2O]{\text{1. R.t., } N_2, 0.5-1 \text{ h, } C_5H_5N} \text{[β-lactam structure]}$$

(300)

R^1	R^2	Yield (%)
Ph	Ph	55
Ph	p-ClC$_6$H$_4$	60
o-MeC$_6$H$_4$	Ph	51
o-ClC$_6$H$_4$	Ph	51

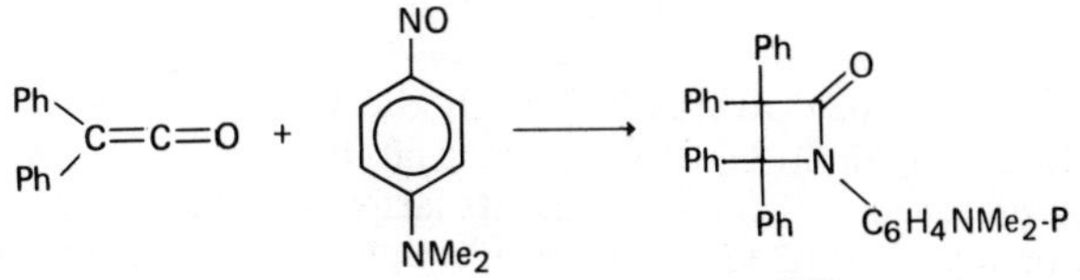

$$\underset{Ph}{\overset{Ph}{>}}C=C=O + \text{[nitroso aromatic with NO and NMe}_2] \longrightarrow \text{[β-lactam structure, } C_6H_4NMe_2\text{-p]}$$

(301)

65%

The reaction of a ketene with a nitroso compound to produce a lactam has been used[614]. in the preparation of 1-(p-dimethylaminophenyl)-3,3,4,4-tetraphenyl-azetidin-2-one (equation 301).

D. By Rearrangements

A number of rearrangements have been used to prepare lactams of varying ring size. In this section, preparations of lactams are presented in terms of the type of rearrangement employed.

1. Ring contractions

a. Wolff rearrangement. By far the most common method for effecting lactam syntheses by ring contraction has been the photolytic Wolff rearrangement. Recently this approach has been studied by Lowe and Ridley[615,616] for the generation of β-lactams from diazopyrrolidinediones. Thus, N-(t-butoxycarbonyl-acetyl)-d,l-alanine ethyl ester and KOBu-t in xylene afforded[615,616] a 60% yield of 5-methylpyrrolidine-2,4-dione, which upon treatment with methane sulphonyl azide in triethylamine produced a 95% yield of 3-diazo-5-methylpyrrolidine-2,4-dione. Photolysis of this product in the presence of t-butyl carbazate afforded a 36% yield of the *cis*-β-lactam and 55% of the *trans*-β-lactam shown in equation (302). Addition of dibenzyl acetylenedicarboxylate to 3-diazo-5-methylpyrrolidine-

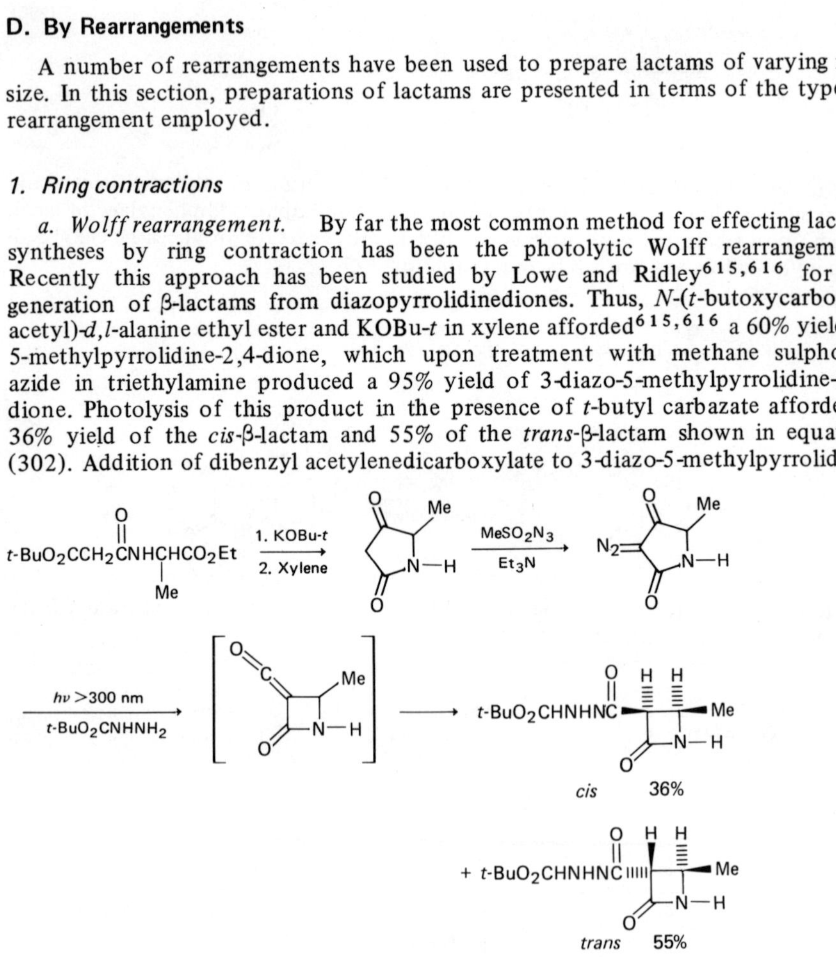

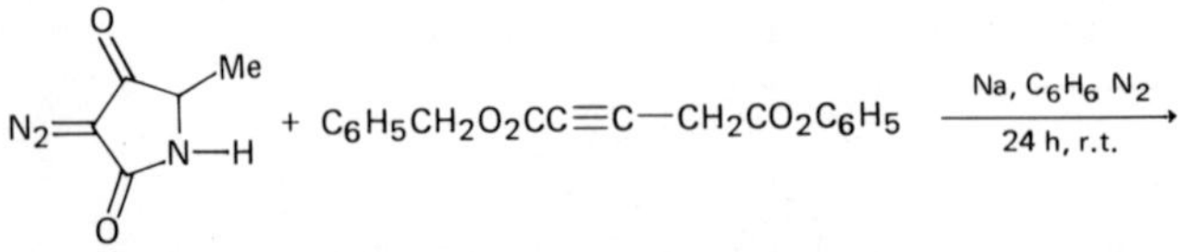

2,4-dione (equation 303) affords [616] both the (E)- and (Z)-dibenzyl (3-diazo-5-methyl-2,4-dioxopyrrolidin-1-yl)fumarates, and irradiation of the (Z) adduct for 0.5 h in the presence of t-butyl carbazate affords both the (E)- and (Z)-trans-β-lactams, dimethyl [cis-3-(3-t-butoxycarbonylcarbazoyl)-2-methyl-4-oxoazetidin-1-yl]maleate. By irradiation for 2 h the (E)-trans-β-lactam is generated exclusively.

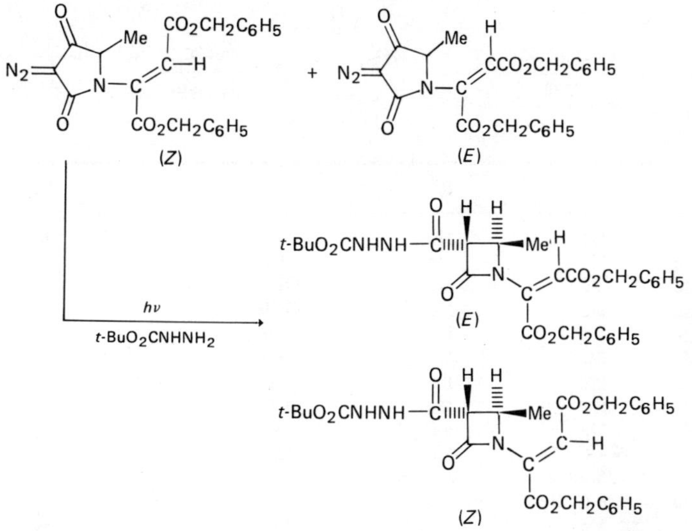

(303)

In a similar manner, irradiation[615,616] of benzyl 6-diazo-5,7-dioxohexahydro-pyrrolidine-3-carboxylate afforded benzyl 7-oxo-6α-[3-(2-phenyl-2-propyloxy-carbonyl)carbazoyl]-5αH-1-azabicyclo-[3.2.0]hexane-2α-carboxylate (equation 304).

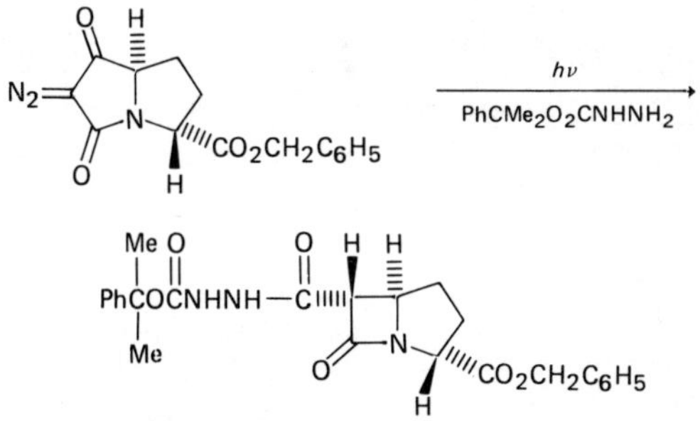

(304)

Using a series of 2,4-pyrrolidinediones ('tetramic acids') as starting materials, Stork and Szajewski[617] demonstrated that the photolytic Wolff rearrangement was a general method for the preparation of carboxy β-lactams (equations 305 and 306).

Ring contraction of 2-acylpyrazolidin-3-ones to afford β-lactams has also been reported[618] to occur upon photolysis (equation 307).

b. Miscellaneous ring contractions. Treatment of substituted α,α-dichloro-succinimides with sodium methoxide in a variety of solvents has been reported[619] to produce both the corresponding ring-opened α-chloroacrylamide and the β-lactam as products, with the proportion of the two products depending upon the nature of the substituents and the solvent used (Scheme 4). Interestingly, if potassium *t*-butoxide is used as the base, the imide again affords the corresponding α-chloroacrylamide (35%) by the mechanism shown, along with a less substituted β-lactam (47%), which arises through a proposed ketene intermediate (equation 308).

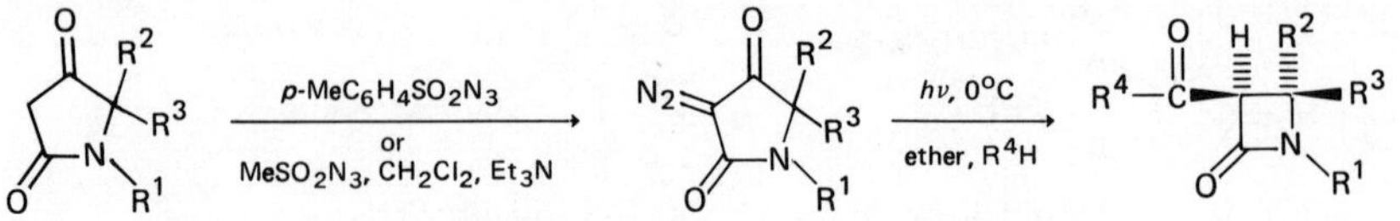

(305)

R^1	R^2	R^3	R^4		Yield (%)
Me	Me	Me	OMe		93
Me	Me	Me	NHNH$_2$		95
Me	Me	Me	OH		98
Me	Me	Me	OCH$_2$CF$_3$		–
Me	Me	Et	OMe ⎫	56:44	64
Me	Et	Me	OMe ⎭		
Me	Me	Et	OH ⎫	58:42	80
Me	Et	Me	OH ⎭		
Me	Me	n-C$_6$H$_{13}$	OMe ⎫	3:1	85–93
Me	n-C$_6$H$_{13}$	Me	OMe ⎭		
Me	Me	n-C$_6$H$_{13}$	NH$_2$ ⎫	3:1	86
Me	n-C$_6$H$_{13}$	Me	NH$_2$ ⎭		
Me	Me	n-C$_6$H$_{13}$	OH ⎫	2:1 or	96
Me	n-C$_6$H$_{13}$	Me	OH ⎭	5:1	80
Me	Me	n-C$_6$H$_{13}$	NHNH$_2$ ⎫	3.5:1	94
Me	n-C$_6$H$_{13}$	Me	NHNH$_2$ ⎭		
H	i-Bu	H	OMe		47
Me	i-Bu	H	OMe		54

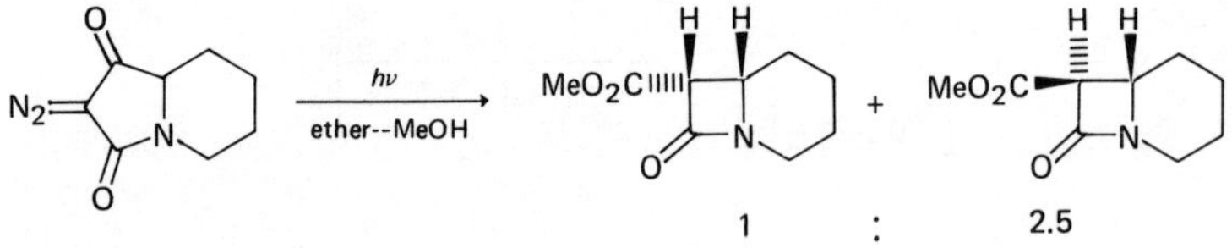

(Ref. 617)

(306)

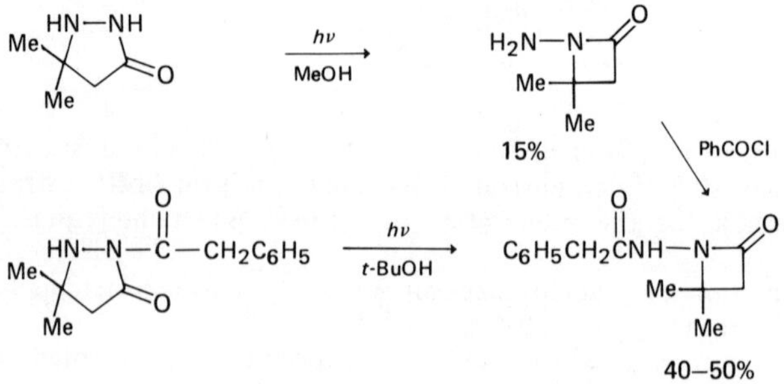

1 : 2.5

(307)

15%

40–50%

2. Ring expansions

a. Beckmann rearrangement. The Beckman rearrangement[620–622] has found extensive use in the preparation of lactams. This reaction generally involves treatment of the oximes of cyclic ketones with H$_2$SO$_4$, PCl$_5$, HCl–HOAc–Ac$_2$O or polyphosphoric acid[622] to convert the hydroxyl group of the oxime into a

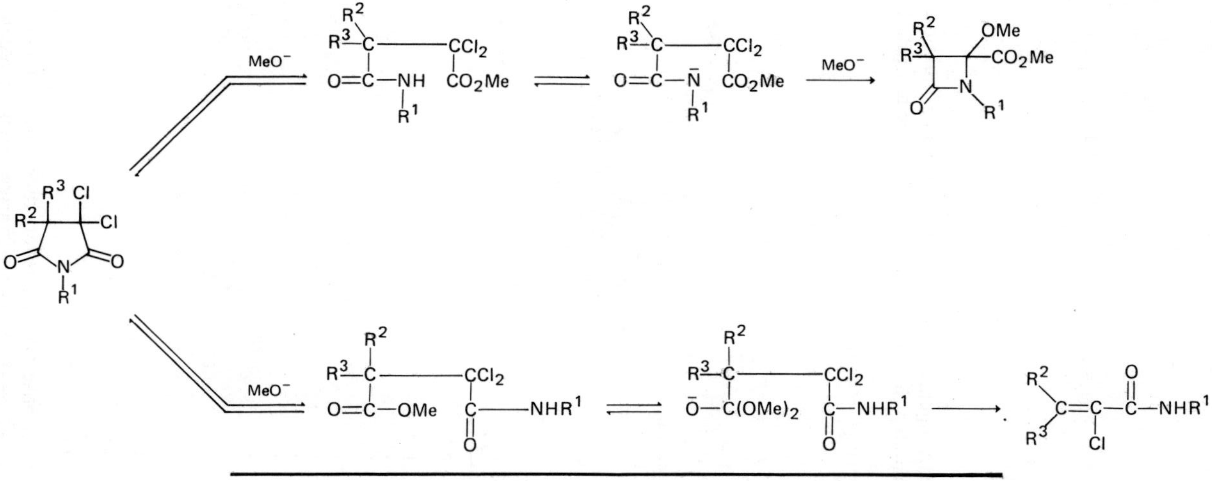

R^1	R^2	R^3	Solvents	Yield (%)	
				Lactam	Amide
Ph	Ph	Me	DMSO	0	100
Ph	Ph	Me	MeOH	0–23	100–77
Ph	Ph	Me	MeOH, DME (50%)	10	90
Ph	Ph	Me	MeOH, dioxane (50%)	25	75
Ph	Ph	Me	MeOH, t-BuOH (50%)	25	75
Ph	Ph	Me	HMPA, MeOH (50%)	5	95
Ph	Ph	Ph	MeOH	20	80
Ph	Me	Me	MeOH	100	0
Ph	$-(CH_2)_4-$	—	MeOH	100	0
Ph	$-(CH_2)_4-$	—	MeOH, DMSO (50%)	90	—

SCHEME 4.

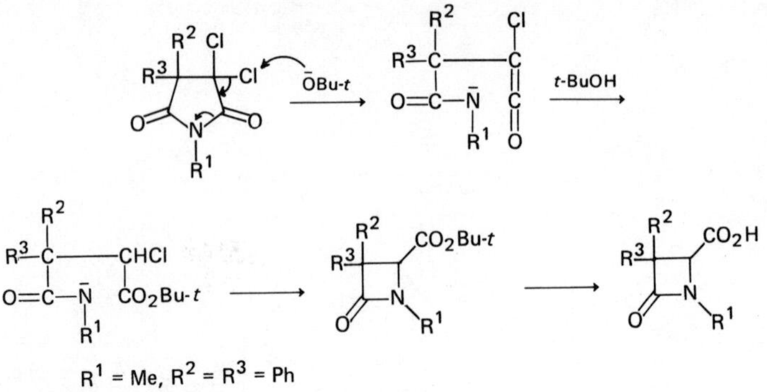

$$(308)$$

$R^1 = Me, R^2 = R^3 = Ph$

better leaving group, followed by rearrangement and tautomerization (equation 309). The group which migrates is normally the one which is *anti* to the hydroxyl

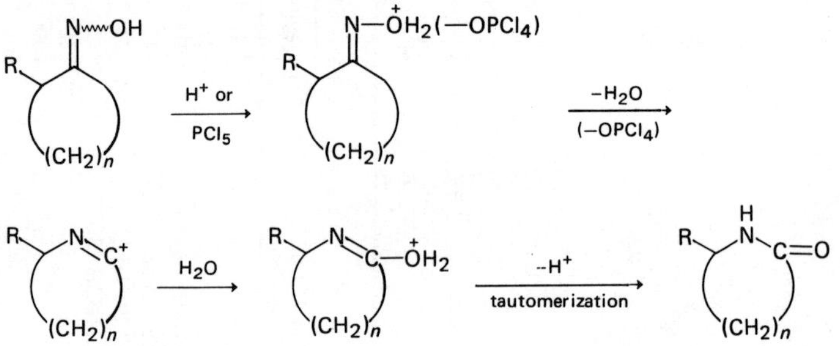

$$(309)$$

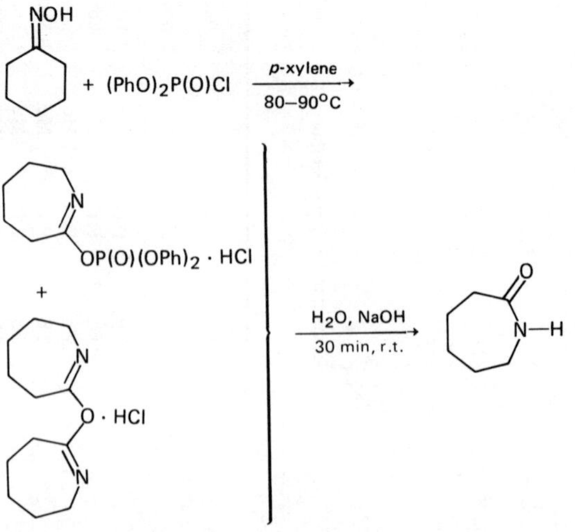

$$(310)$$

group in the oxime. Exceptions have been observed; however, these may involve a *syn* to *anti* isomerization prior to rearrangement. A representative series of recent lactam syntheses via the Beckmann rearrangement[623–642] are compiled in Table 35.

TABLE 35. Preparation of lactams via the Beckmann rearrangement

Oxime	Conditions	Product	Yield (%)	Reference
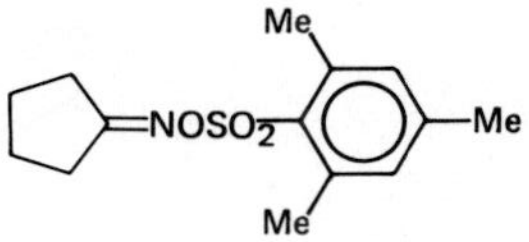	Basic alumina, MeOH	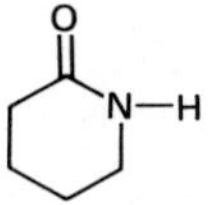	81	624
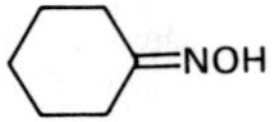	190°C, tetralin, 1 h, boric anhydride	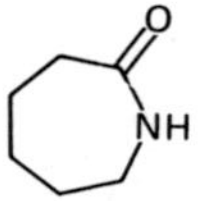	82	627
	1. MeCH, 1 h, 80°C 2. HCl (g)	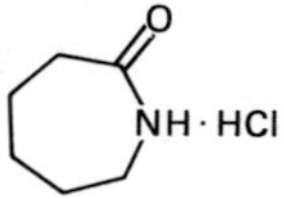	—	628
	AcOH–HCl–MeCN, heat		>90	623
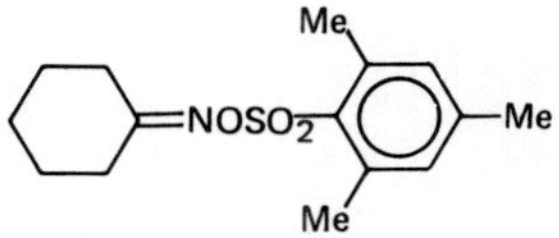	Basic alumina, MeOH	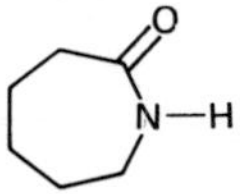	77	624
	1. MeCN, 1 h, 80°C 2. HCl (g)		—	628
	H_2SO_4 or H_3PO_4, 101–103°C		95	625

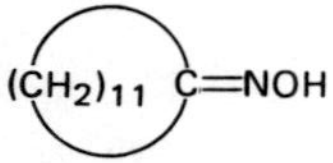

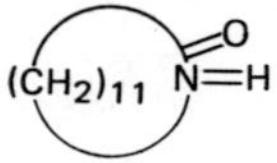

TABLE 35. (Continued)

Oxime	Conditions	Product	Yield (%)	Reference
syn or *anti*	PPA, 148°C, 10 min	from *syn* from *anti*	25 0	631
	1. 19% oleum, 140°C, 1 h 2. MeOH–KOH, 0°C	from *syn* from *anti*	59 59	632
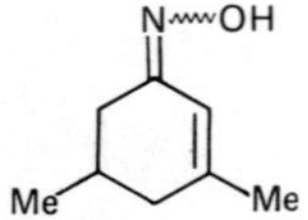 *syn:anti* mixture	PPA, 130°C	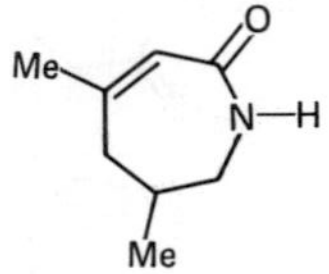	30	633
syn	PCl$_5$, ether PPA, 132–135°C, 10 min PPA		25 21 64	634 636 638

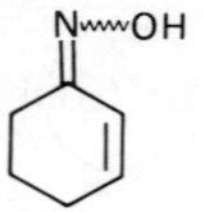

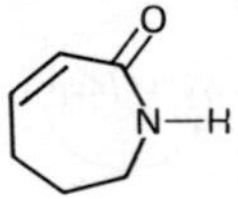

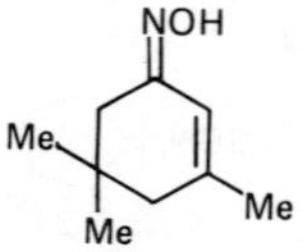

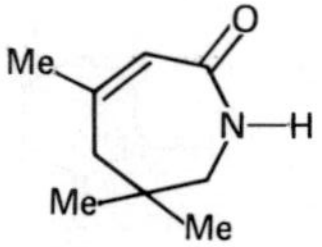

Substrate	Reagents/conditions	Product	Yield (%)	Reference
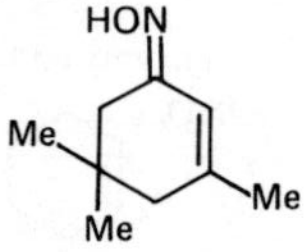 *anti*	PCl$_5$, ether PPA, 132–135°C, 10 min	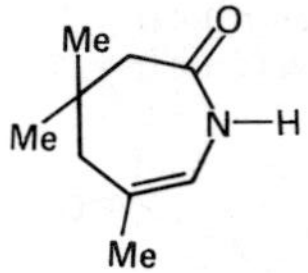	20 72	634 636
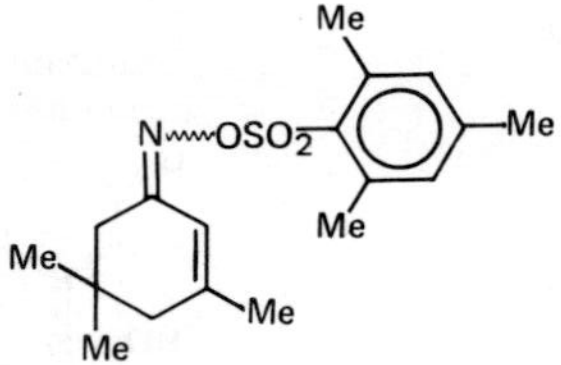	Basic alumina, MeOH	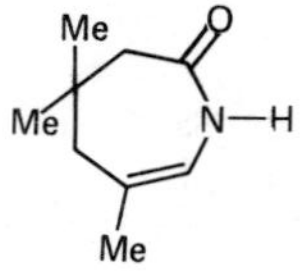	50	624
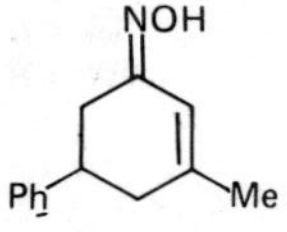 *syn*	PCl$_5$, ether–C$_6$H$_6$	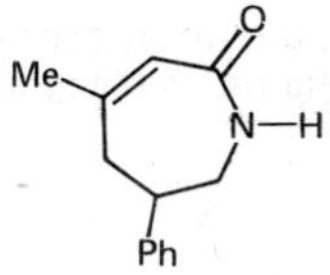	25	634
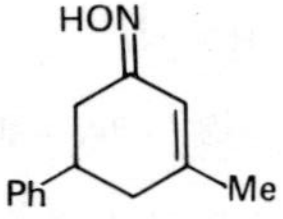 *anti*	PCl$_5$, ether–C$_6$H$_6$	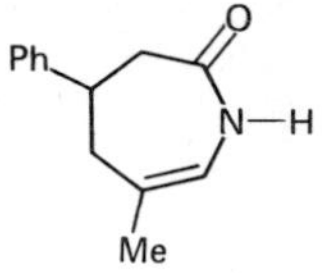	15	634
R = Me R = Et	PPA, 120°C, 1 h	from 2:1 *syn:anti* mixture from *syn* from 4:3 *syn:anti* mixture	66–75 77 88–98	635,637 637 635,637

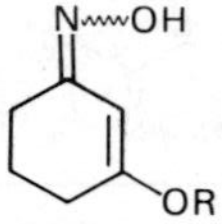

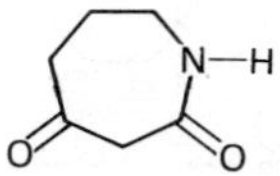

TABLE 35. (Continued)

Oxime	Conditions	Product	Yield (%)	Reference
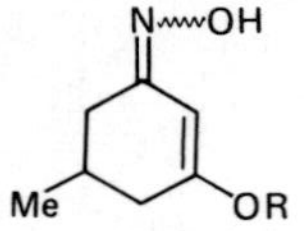 (cyclohexenone oxime, OR; Me; *syn:anti* = 2:1; R = Me or Et)	PPA, 120°C, 1 h	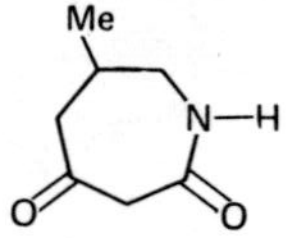 (azepane-2,4-dione, Me, N–H)	78–91	635,637
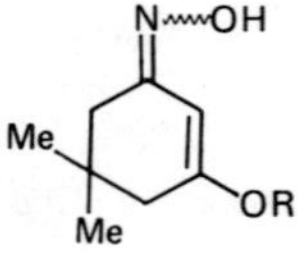 (cyclohexenone oxime, OR; Me, Me; R = Me; R = Et)	PPA, 120°C, 1 h	(azepane-2,4-dione, Me, Me, N–H)	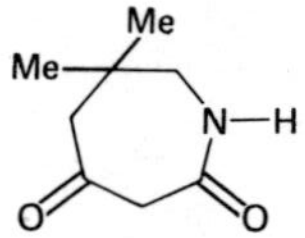 from 5:4 *syn:anti* mixture 78–79; from 3:2 *syn:anti* mixture 81–96	635,637
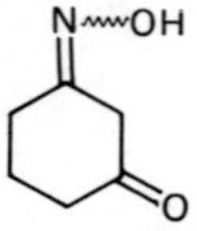 (cyclohexane-1,3-dione oxime, *syn:anti* mixture)	PPA, 120°C, 1 h	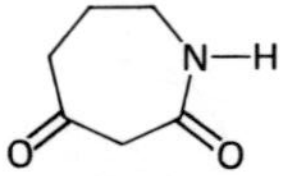 (azepane-2,4-dione, N–H)	30	637
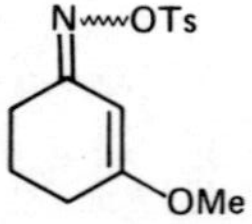 (cyclohexenone oxime, OTs, OMe; *syn* or *anti*)	PPA	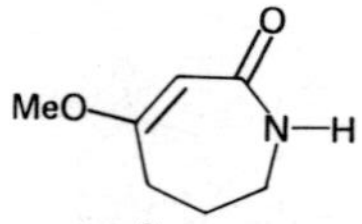 (MeO azepinone, N–H) from *syn*; from *anti*	87; 0	637

PPA, 120°C

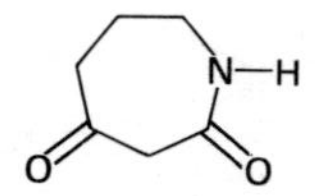

72%

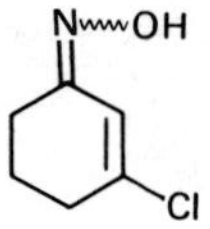

637

26
28

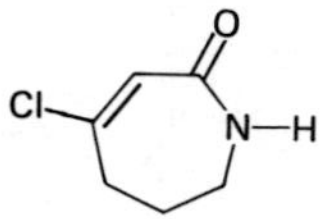

from 4:7 *syn:anti* mixture
from *anti*

PPA, 120°C

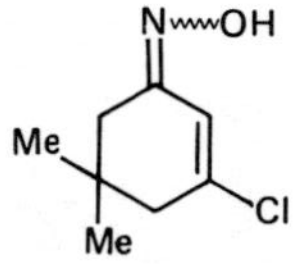

637

34

PPA, 120°

syn:anti = 4:7

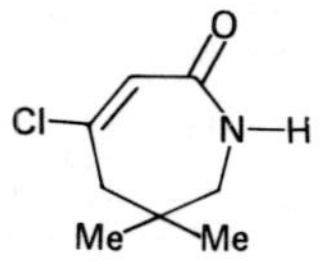

639

—

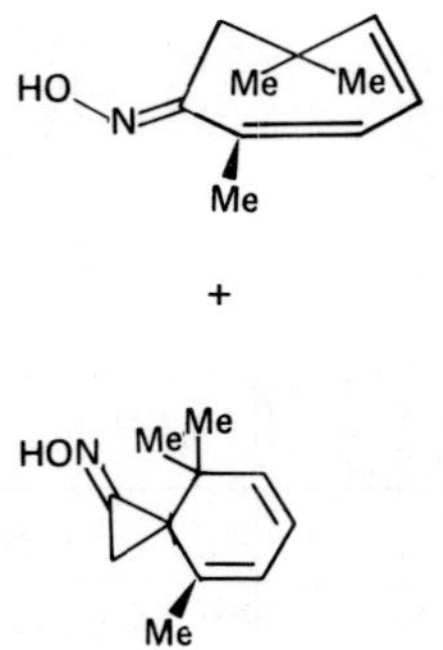

Me$_2$CO, H$_2$O,
p-MeC$_6$H$_4$SO$_2$Cl

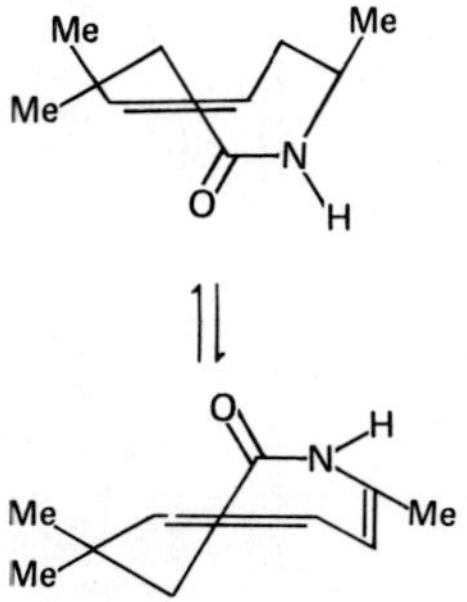

TABLE 35. (Continued)

Oxime	Conditions	Product	Yield (%)	Reference
$Z:E = 3:2$	PPA, 120–130°C, 10 min		91	633
	PPA, 175–180°C, 10 min		93	633
	PPA, 130–135°C, 20 min		52	640
$Z:E = 6:1$	PPA		75	640
	PPA, H_2SO_2	$n = 4$ $n = 5$	35 40	641

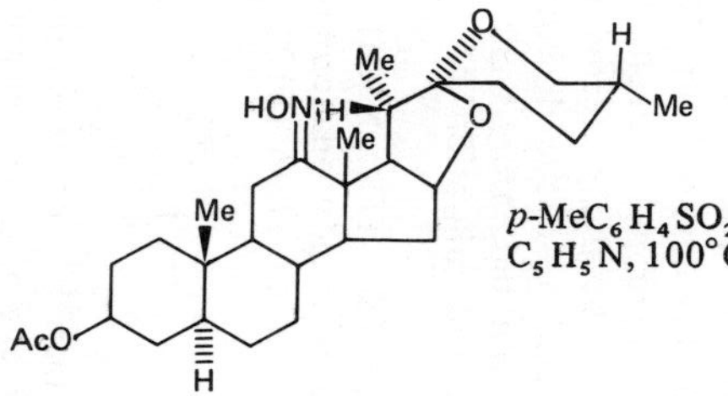

642

642

74

50

p-MeC$_6$H$_4$SO$_2$Cl,
C$_5$H$_5$N, 100°C, 3 h

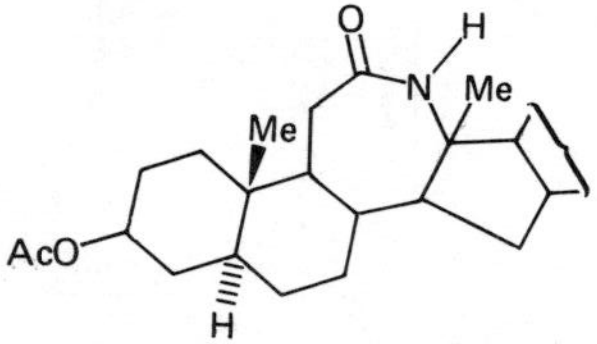

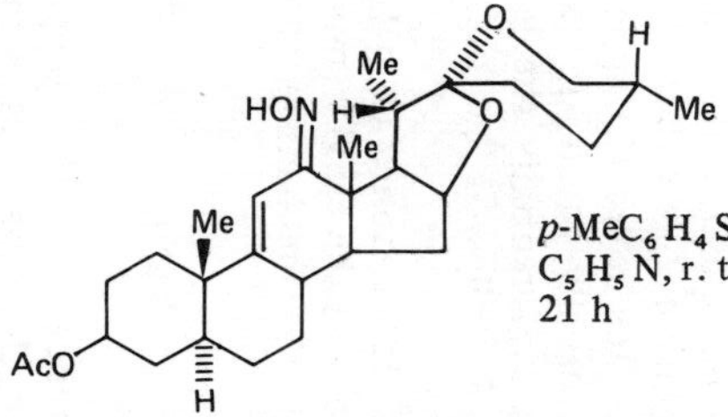

p-MeC$_6$H$_4$SO$_2$Cl,
C$_5$H$_5$N, r. t.,
21 h

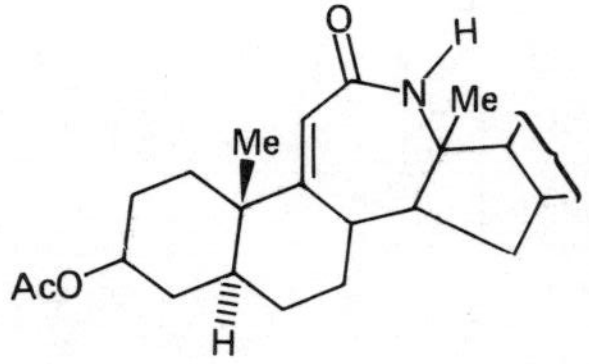

A lactam synthesis which mechanistically resembles a Beckmann rearrangement involves the reaction of cyclohexanone ketoxime in p-xylene with diphenyl chlorophosphite at 80–90°C for 18 hours. This reaction afforded[643] a mixture of the lactim phosphate hydrochloride and bislactim ether hydrochloride shown in equation (310), both of which produced ε-caprolactam upon hydrolysis. Similar preparation of a series of C_4–C_{12} lactams has also been reported[643].

b. *Schmidt rearrangement.*   Among various Schmidt rearrangements[644,645], only the reaction of cyclic ketones with hydrazoic acid gives rise to lactams. The mechanism for this rearrangement is shown in equation (311), and the question of

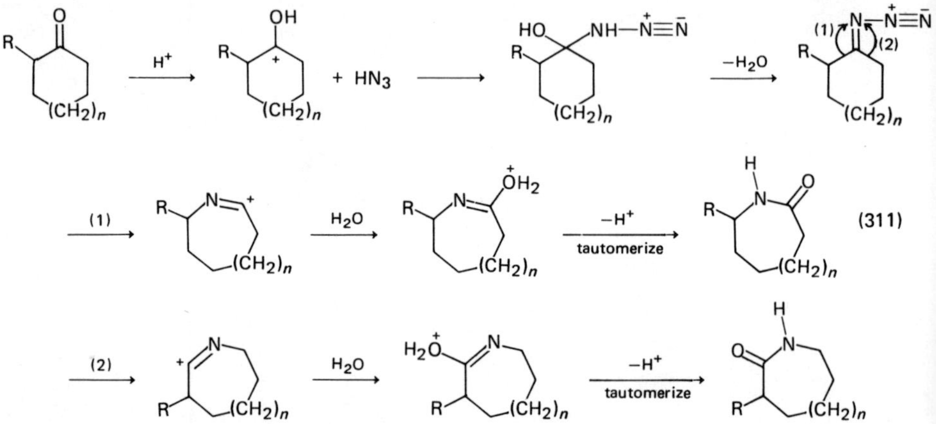

which group migrates during the loss of nitrogen in several different systems has produced errors in the literature and many lively published debates. Table 36[640-656] contains a representative sampling of lactams prepared over the last 25 years by means of the Schmidt rearrangement.

These studies indicate that with saturated cyclic ketones possessing an electron-donating substituent (R = Me, Et, n-Pr, i-Pr, n-Bu, $CH_2 R$, $CHMeC_6 H_5$) in the 2-position, route (1) is the path observed. This route is also observed with some electron-withdrawing substituents such as CN and CO_2 Et. However, when the substituent at position 2 is either Cl or Ph, or if the cyclic ketone contains an α,β-double bond, then route (2) appears to be the preferred reaction path even though mixtures usually result. In the case of cyclic diketones, the azide ion appears to attack preferentially the less hindered, more basic carbonyl function, and this attack is followed by preferential migration of the larger adjacent group[653,654].

Sodium azide in acetone at pH 5.5 ($KH_2 PO_4$ –NaOH buffer) has been reported[657] to convert the ethyl hemiketal of cyclopropanone into γ-butyrolactam in 21% yield via the mechanism shown in equation (312). This reaction was

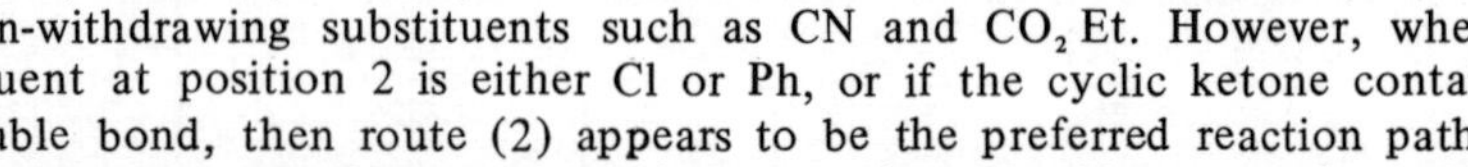

subsequently extended to the preparation[658] of fused-ring β-lactams from 1,1-disubstituted cyclopropanones in the bicyclo[4.1.0] series (equation 313), and made more general by preparation of the corresponding carbinolamines of the cyclic ketones and then affecting the ring enlargement reaction through the

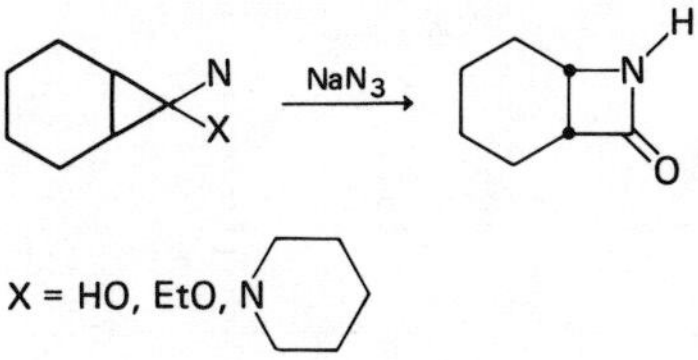

(313)

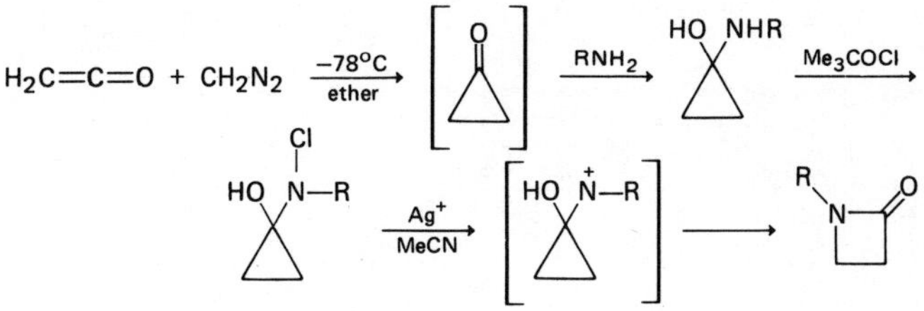

X = HO, EtO, N⟨ ⟩

$$H_2C{=}C{=}O + CH_2N_2 \xrightarrow[\text{ether}]{-78^\circ C} \left[\triangle{=}O \right] \xrightarrow{RNH_2} \text{HO} \ \text{NHR} \xrightarrow{Me_3COCl}$$

(314)

R	Yield (%)
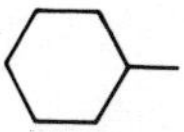 ⟨cyclohexyl⟩	61
Me(CH$_2$)$_3$—	43
MeCH$_2$CHMe	38
Me$_3$C—	52
MeCHCOOEt	65

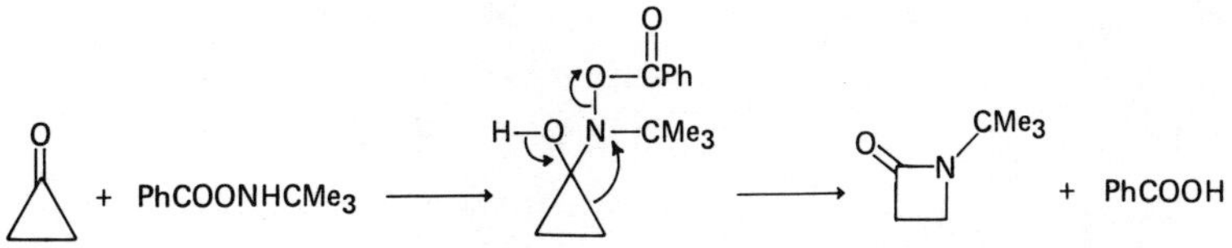

△=O + PhCOONHCMe₃ ⟶ ⟶ + PhCOOH   (315)

nitrenium ion produced from these intermediates (equation 314)[658]. In an effort to extend this method, the same authors[658] investigated leaving groups other than Cl⁻ and N$_2$ in the ring enlargement, and found that the o-benzoyl derivative of N-(t-butyl)hydroxylamine reacted directly with cyclopropanone in ether at −78°C to produce N-(t-butyl)-β-propiolactam in 40% yield (equation 315). It was also found that alkyl hydroxylamines (equation 316)[659] and amino acid esters (equation 317)[660] can be employed in this transformation.

   c. *Miscellaneous ring expansions.*   A novel ring expansion reaction for the preparation of γ-lactams involves the carbonylation of cyclopropylamine using rhodium catalysts (equation 318)[661].

   An interesting disproportionation rearrangement for the preparation of ε-caprolactam involves heating peroxy amines in the presence of a Group I or II element salt in a non-hydrocarbon organic solvent[662]. Thus, 1,1-peroxydicyclohexylamine afforded a 100% conversion to caprolactam and cyclohexanone (equation 319).

   A novel photochemical ring expansion which allows conversion of fused β-lactams to fused bicyclic ring-expanded lactam ethers has also been

TABLE 36. Preparation of lactams via the Schmidt rearrangement

Ketone	Conditions[a]	Product	Yield (%)	Reference
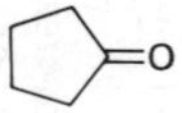 cyclopentanone	A, 3–7°C B, 50°C, 8.5 h	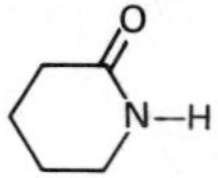 piperidinone	80 83	646 647
2-R-cyclopentanone	A, 3–7°C B, 50°C, 8.5 h	R = Me, Et, Pr, or i-Pr R = Me or Pr	78–94 82–87	646 647
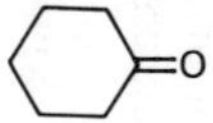 cyclohexanone	B, 50°C, 8.5 h	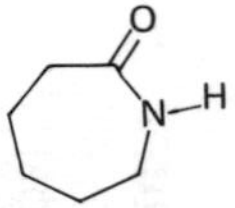 azepanone	89	647
2-R-cyclohexanone	R = Me { A, 3–7°C; B, 50°C, 8.5 h; B, 50°C } R = Et { A, 3–7°C; B, 50°C } R = Pr { A, 3–7°C; B, 50°C, 8.5 h; B, 50°C } R = i-Pr { B, 50°C } R = Bu { B, 50°C }	R-substituted azepanone	74 96 96 84 95 92 95 95 98 94	646 647 640 646 640 646 647 640 640 640

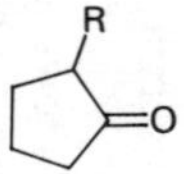

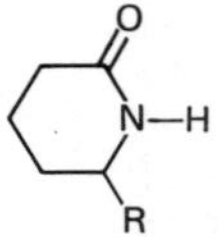

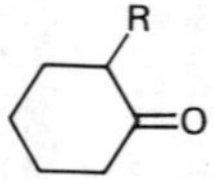

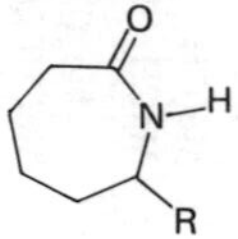

R = CO₂Et
I. = PhCH₂CH₂
R = p-MeOC₆H₄CH₂CH₂
R = o-MeOC₆H₄CH₂CH₂
R = p-MeC₆H₄CH₂CH₂
R = MeCHPh

A, 3–7°C
B, 50°C, 8.5 h
B, 0°, 1.5 h then r.t., 6 h

B, 0°C, 1.5 h then r.t., 6 h

Yield	Ref
80	646
75	647
49	640
40	640
42	640
45	640
37	640

A, 3–7°C
B, 25°C, 8.5 h
B, 35°C, 8.5 h
B, 45°C, 8.5 h
B, 65°C, 8.5 h

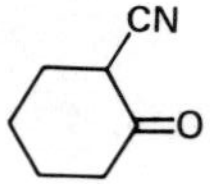

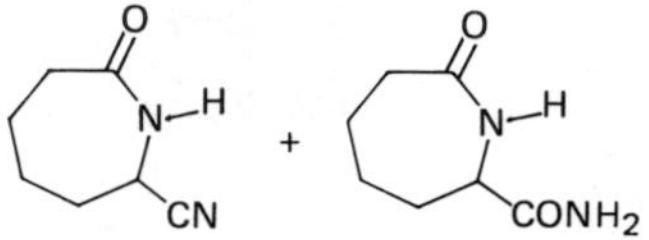

Yield	Ref
70 + 0	646
76 + 8	647
54 + 30	647
37 + 45	647
0 + 83	647

A, 3–7°C
HN₃, CHCl₃, HCl, 30°C
A, EtOH, 40–45°C

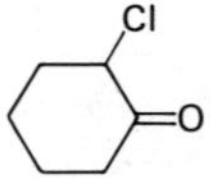

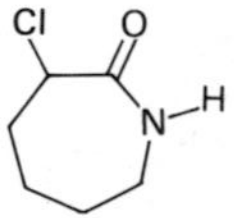

Yield	Ref
9	646
26	646
31	646

A, r.t., 90 min   (1:2)
A, r.t., 90 min   (1:10)
B, r.t., 18 h   (1:3)
B, 55°C

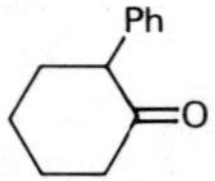

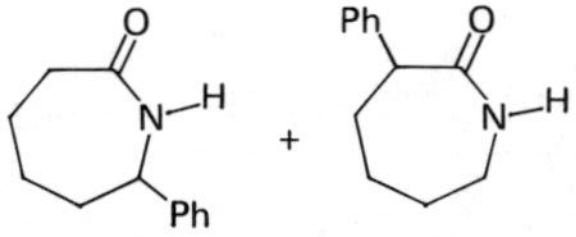

Yield	Ref
67 + 0	641
60	640
55	640
30	640

TABLE 36. (Continued)

Ketone	Conditions[a]	Product	Yield (%)	Reference
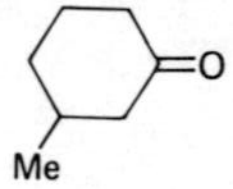	C	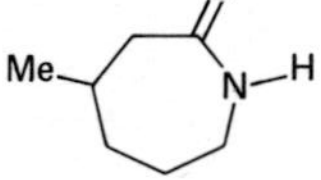 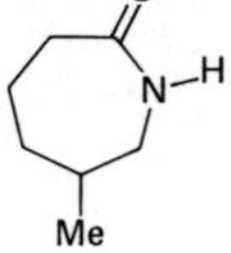	—	640
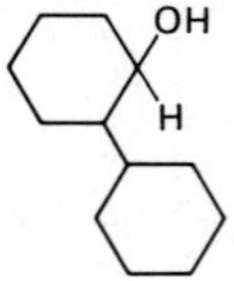	B, r.t.	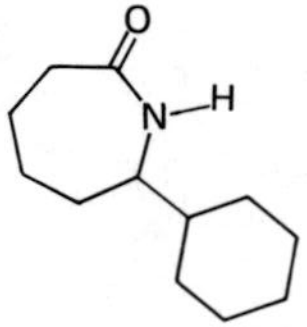	80	640
	B, 50–55°C, 24 h		1.6 1.6 2.0	648

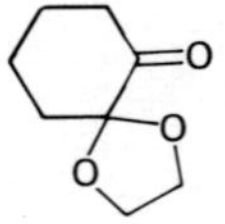

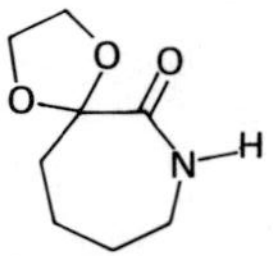

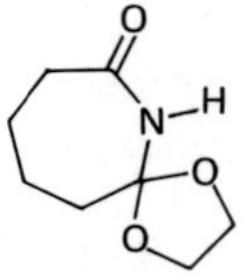

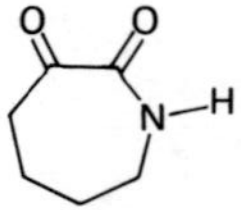

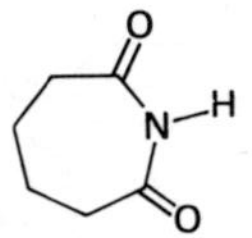

B, 120°C, 2 h

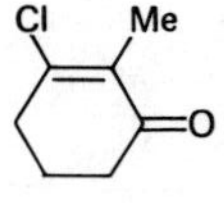

2.2

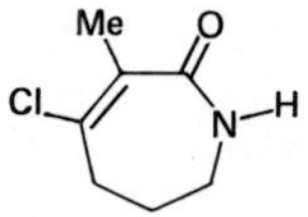

648

B, 0°C, 1.5 h then
55°C, 3.5 h

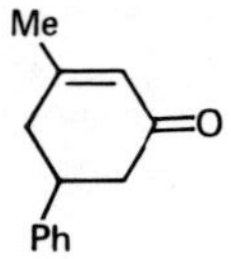

54

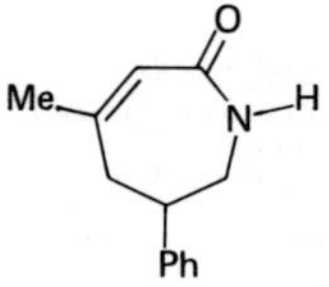

640

30

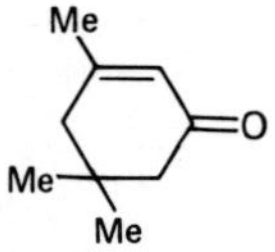

638

D, <35°C

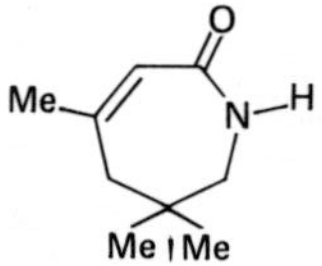

39

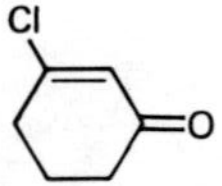

B, 120°C, 1 h

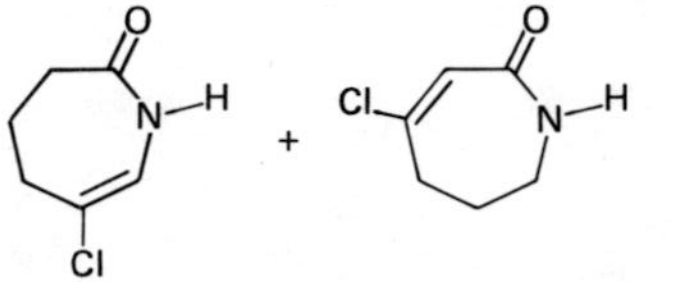

11 + 15

649

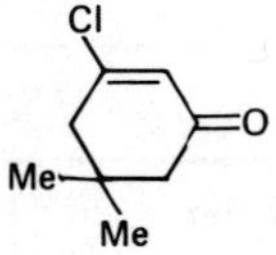

B, 120°C, 2 h

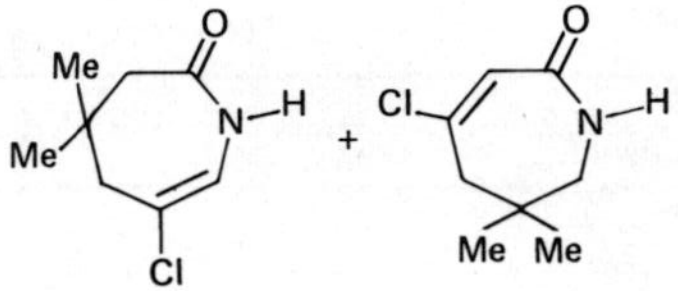

32 + 36

649

TABLE 36. (Continued)

Ketone	Conditions[a]	Product	Yield (%)	Reference
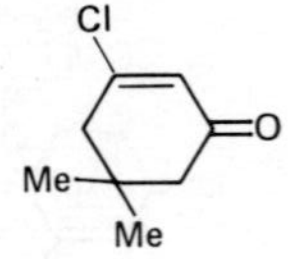	NaN$_3$, MeOH, H$_2$O reflux, 4 h	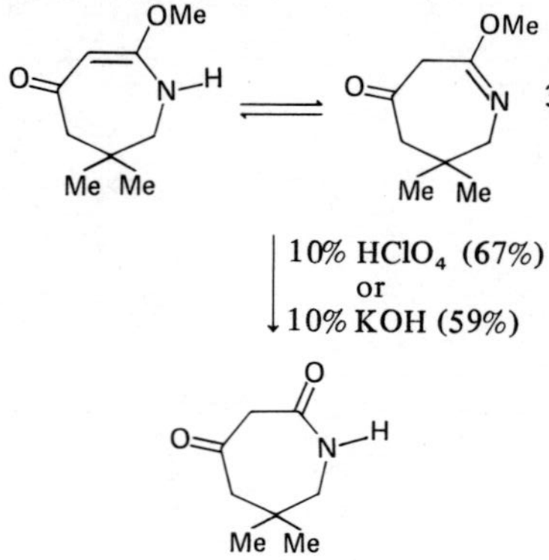	36	650
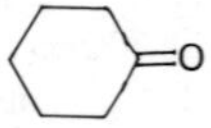	B, 50°C	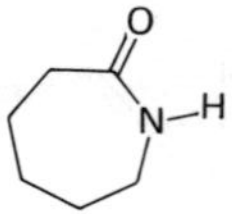	83	647
	B, 50°C	$n = 2$ $n = 3$	90 95	647 647
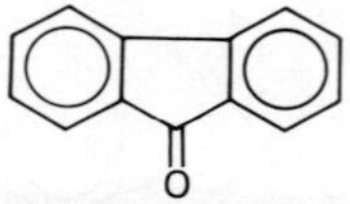	B, 50°C	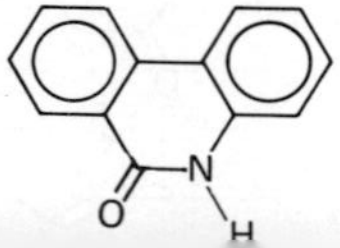	92	647

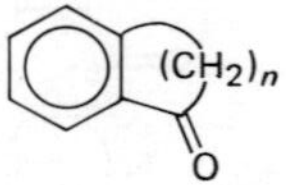

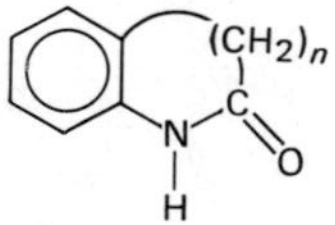

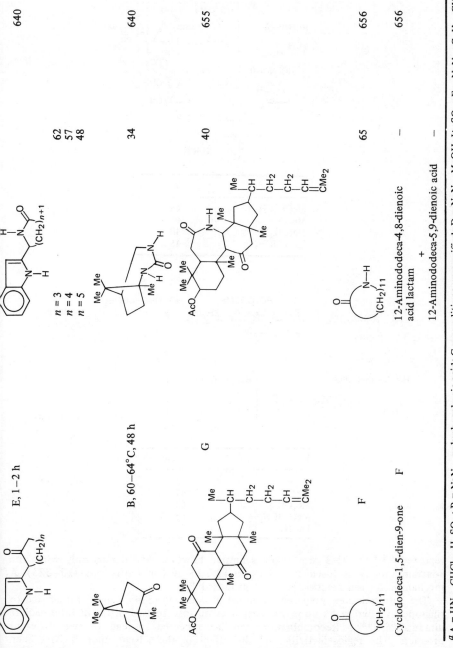

E, 1–2 h

n = 3
n = 4
n = 5

62
57
48

640

B, 60–64°C, 48 h

34

640

G

40

655

F

65

656

Cyclododeca-1,5-dien-9-one   F

12-Aminododeca-4,8-dienoic acid lactam
+
12-Aminododeca-5,9-dienoic acid

—

—

656

^{a} A = HN$_3$, CHCl$_3$, H$_2$SO$_4$; B = NaN$_3$, polyphosphoric acid; C = conditions unspecified; D = NaN$_3$, MeOH, H$_2$SO$_4$; E = NaN$_3$, C$_6$H$_6$, CHCl$_3$, H$_2$SO$_4$; F = NaN$_3$, H$_2$SO$_4$; G = NaN$_3$, CHCl$_3$, H$_2$SO$_4$.

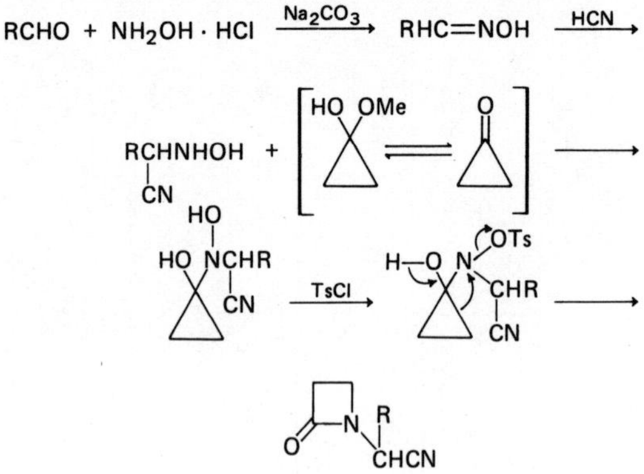

$$RCHO + NH_2OH \cdot HCl \xrightarrow{Na_2CO_3} RHC{=}NOH \xrightarrow{HCN}$$

RCHNHOH + $\left[\begin{array}{c} HO \quad OMe \\ \triangle \end{array} \rightleftharpoons \begin{array}{c} O \\ \triangle \end{array}\right] \longrightarrow$ (316)
|
CN

HO
HO NCHR $\xrightarrow{TsCl}$ H–O N $\overset{OTs}{\underset{CHR}{\triangle}}$ $\longrightarrow$
△ CN CN

O $\triangle$ N $\overset{R}{\underset{CHCN}{|}}$

R	Yield (%)
MeCH$_2$CH$_2$–	45
Me$_3$C–	41
Me(CH$_2$)$_3$–	40
Me$_2$CHCH$_2$–	45

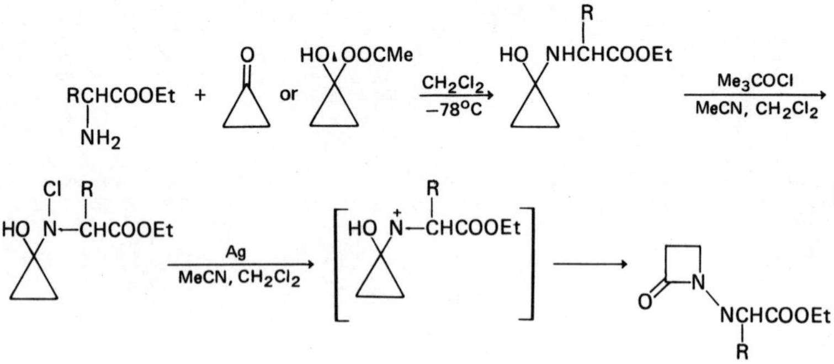

RCHCOOEt + $\overset{O}{\triangle}$ or $\overset{HO \quad OOCMe}{\triangle}$ $\xrightarrow[-78°C]{CH_2Cl_2}$ $\overset{HO \quad NHCHCOOEt}{\underset{\triangle}{}}$ $\xrightarrow{Me_3COCl}{MeCN,\ CH_2Cl_2}$
|
NH$_2$

Cl R
HO N–CHCOOEt $\xrightarrow[MeCN,\ CH_2Cl_2]{Ag}$ $\left[\begin{array}{c} HO \quad \overset{+}{N}-CHCOOEt \\ \triangle \end{array}\right] \longrightarrow$ (317)
△

O $\triangle$ N $\overset{}{\underset{NCHCOOEt}{|}}$
R

R	Yield (%)
H	33
Me	47
Me$_2$CH–	65
Me$_2$CHCH$_2$–	65
PhCH$_2$–	70

reported[663,664]. This conversion is limited however, and occurs only when the β-lactam moiety is fused to a bicyclo [2.2.1] system (equations 320–323). A mechanism for this reaction has been proposed[664].

The spirooxiranes, prepared as shown in equation (324), can be anticipated to ring-open in two ways upon irradiation giving rise to two different intermediate diradicals[665,666]. Recombination can be expected to lead to two different products. The regioselectivity and the effect of the solvent upon it, have been

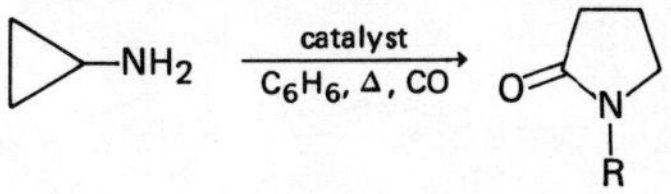

(318)

			Composition of mixture (%)			
T (°C)	CO pressure (atm)	Total yield lactam (%)	$R = (CH_2)_3$	n-Pr	Allyl	H
100[a]	130	10	92	2	5	1
120[a]	150	55	75	19	1	5
140[a]	145	60	62	24	1	12
130[a]	145	22	89	7	2	2
130[a]	150	40	81	16	1	2
130[b]	150	40	28	4	1	67

[a]Catalyst = Rh_6CoO_{16}.
[b]Catalyst = $ClRh(PPh_3)_3$.

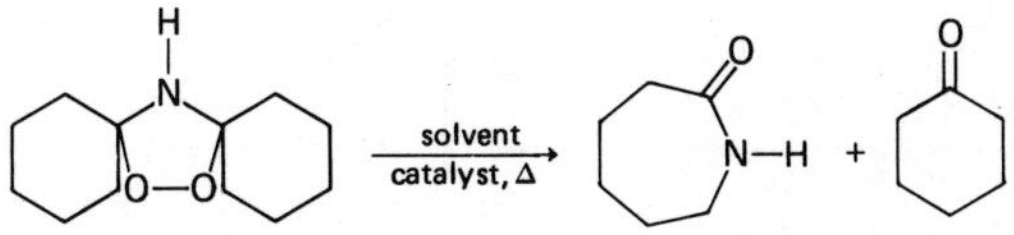

(319)

Solvent = Me_2SO, MeOH, Me_2CO, EtOH, MeCN, or $HO(CH_2)_2OH$

Catalyst = LiBr, $CaCl_2$, NaCl, AgOAc, LiCNS, KF or $SrCl_2$

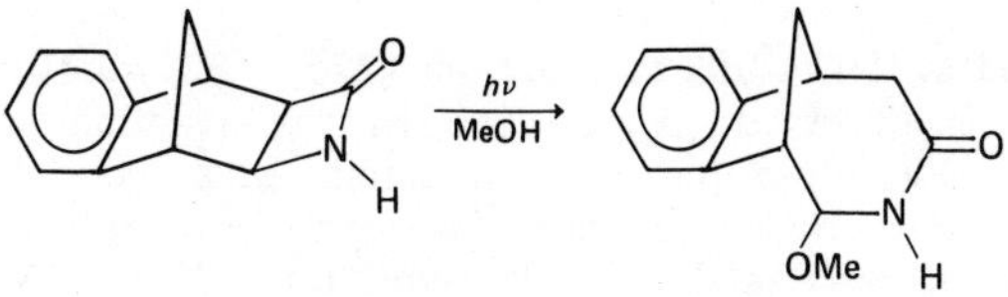

(320)

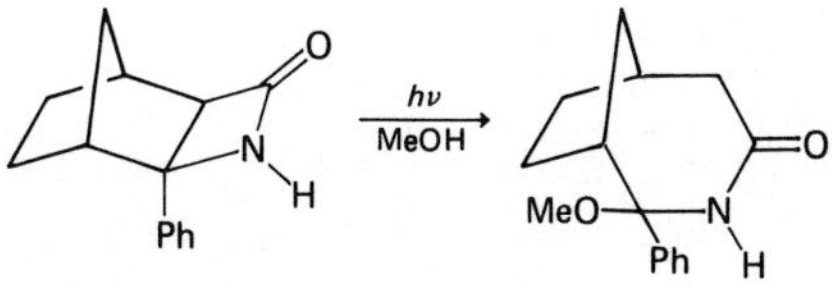

(321)

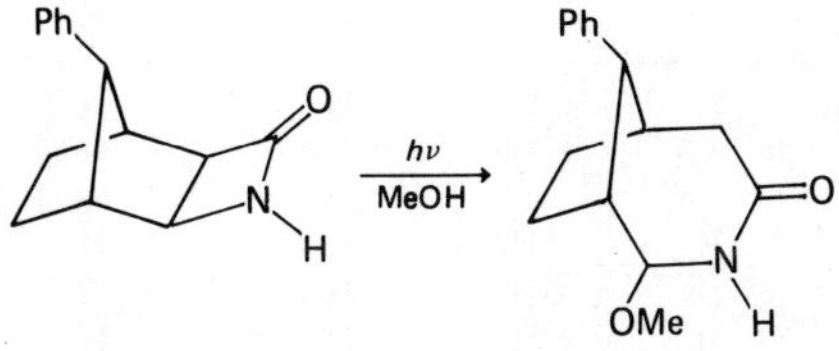

(322)

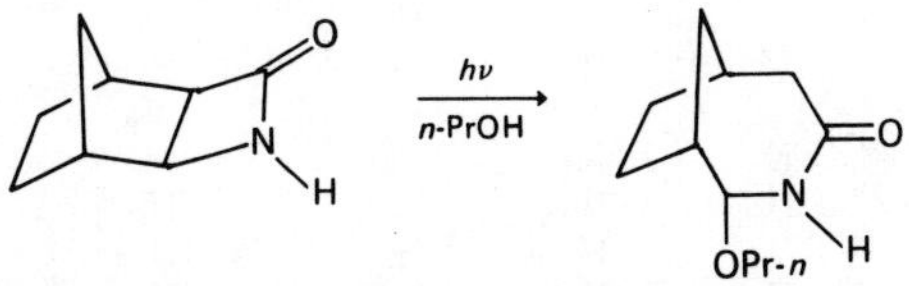

(323)

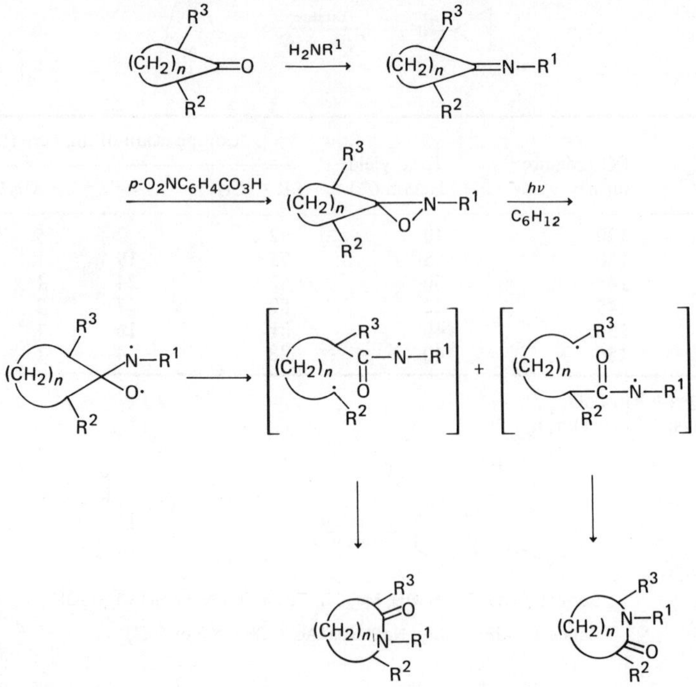

$$(324)$$

investigated, and as the results in equations (325)–(329) indicate, a high degree of regioselectivity is observed. Application of this reaction to the synthesis of N-phenyl- and N-(p-chlorophenyl)caprolactam gave poor results, and failed completely in attempts to prepare N-(p-methoxyphenyl)caprolactam[666]. This reaction can also be performed thermally (equations 330 and 331)[667,668].

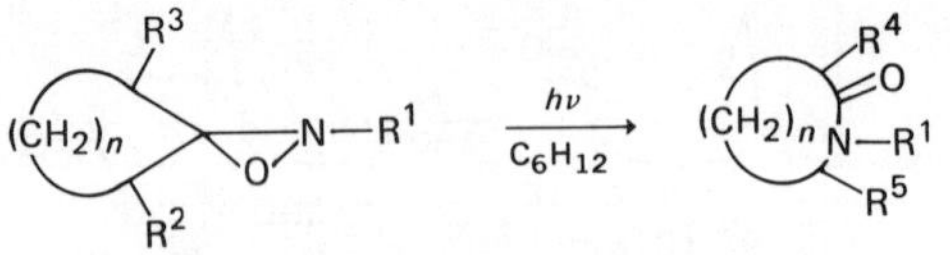

$$(325)$$

n	R^1	R^2	R^3	R^4	R^5		Yield (%)
2	$C_6H_5CH_2$	H	H	H	H		75
2	$C_6H_5CH_2$	Me	H	Me / H	H / Me } 95:5	and	80
3	$C_6H_5CH_2$	H	H	H	H		85
3	C_6H_{13}	H	H	H	H		95
3	Me_2CH	H	H	H	H		85
3	Me	Me	H	Me / H	H / Me } 95:5	and	80
3	$C_6H_5CH_2$	Me	H	Me / H	H / Me } 95:5	and	80
3	$C_6H_5CH_2$	Me	Me	Me	Me		50
4	$C_6H_5CH_2$	H	H	H	H		85
5	$C_6H_5CH_2$	H	H	H	H		85
9	$C_6H_5CH_2$	H	H	H	H		85

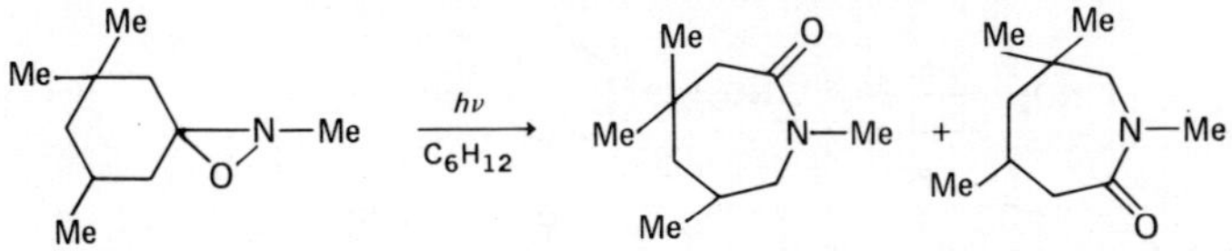

80% (50:50) (Ref. 666) (326)

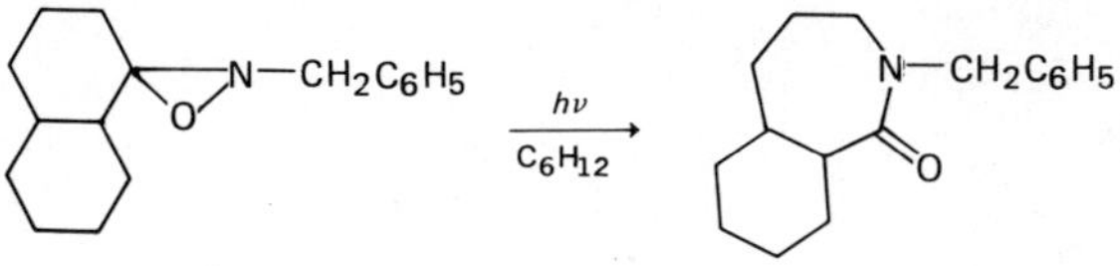

80% (Ref. 665) (327)

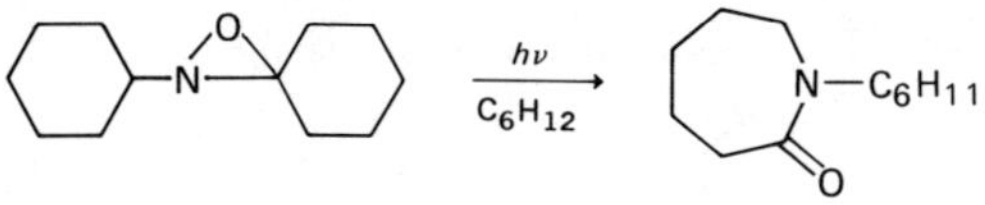

90% (Ref. 648) (328)

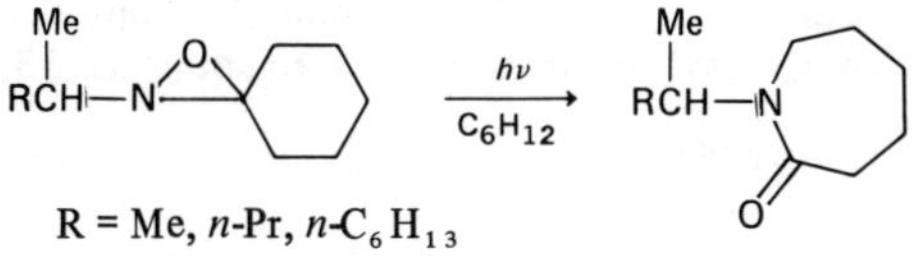

R = Me, n-Pr, n-C$_6$H$_{13}$ (Ref. 648) (329)

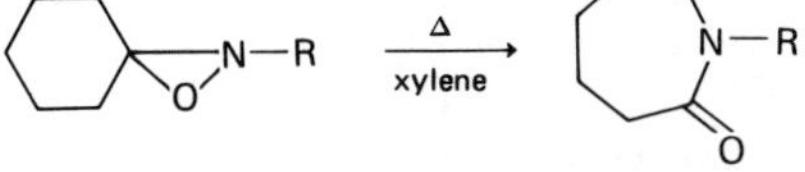

(Ref. 667) (330)

R	Yield (%)
Ph	83
p-ClC$_6$H$_4$	90
c-C$_6$H$_{11}$	85

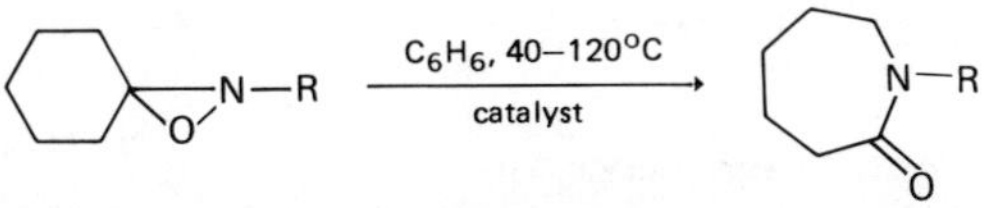

Catalyst = vanadyl bis(acetylacetonate),
V$_2$O$_5$, P$_2$O$_5$, B$_2$O$_3$ or MnO$_3$
R = H, alkyl

(Ref. 668) (331)

The ring expansion of spirooxiranes is also believed to be involved in the production of lactams during irradiation of primary and secondary nitroalkanes in cyclohexane[669]. Thus, irradiation of nitroethane in cyclohexane leads to N-ethyl-caprolactam presumably via a mechanism which involves intermediate formation of a spirooxirane (equation 332). Other primary and secondary nitroalkanes which have also been found[669] to produce lactams are shown in equation (333).

Treatment of cycloalkanecarboxylic acid with nitrosyl pyrosulphuric acid, prepared[670] as shown in Scheme 5, affords the corresponding ring-expanded

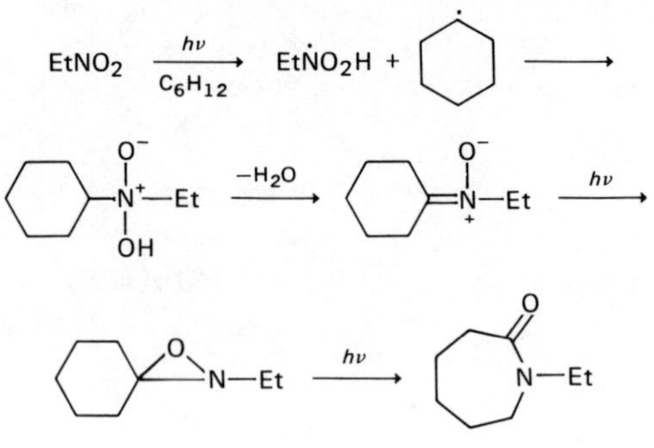

$$\text{EtNO}_2 \xrightarrow[\text{C}_6\text{H}_{12}]{h\nu} \text{Et}\overset{\cdot}{\text{N}}\text{O}_2\text{H} + \overset{\cdot}{\bigcirc} \longrightarrow$$

(332)

$$\text{R}-\text{NO}_2 \xrightarrow[\text{C}_6\text{H}_{12}]{h\nu} \quad \text{N}-\text{C}_6\text{H}_{11} + \quad \text{N}-\text{R}$$

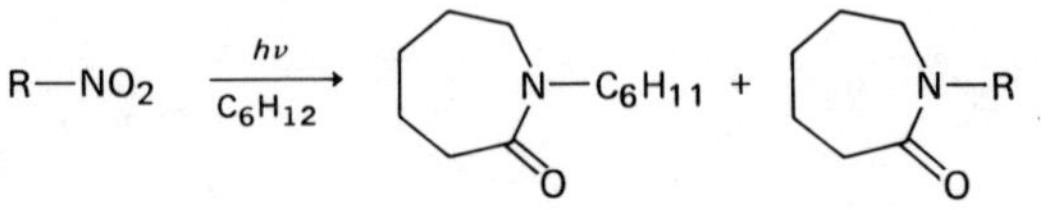

(333)

R = C$_6$H$_5$CH$_2$, C$_6$H$_5$CH$_2$CH$_2$, cyclohexyl, *i*-Pr, *n*-PrCHMe, *n*-C$_6$H$_{13}$CHMe

lactam (equation 334)[670]. Similar results were obtained[671] when the same cyclo-alkanecarboxylic acids were treated with nitrosyl chlorosulphonate (equation 335).

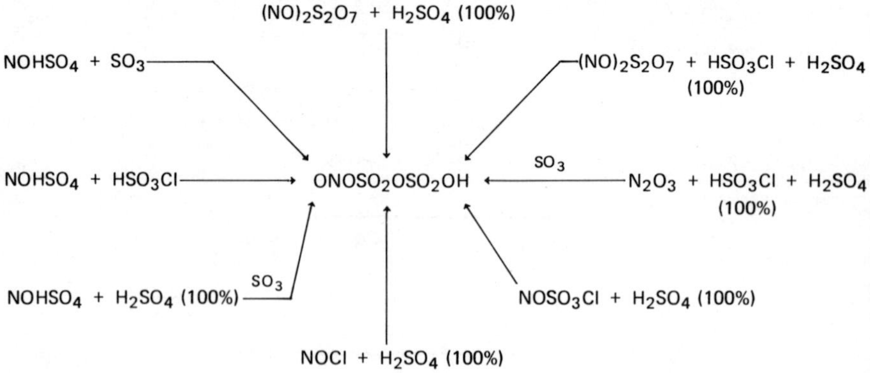

SCHEME 5.

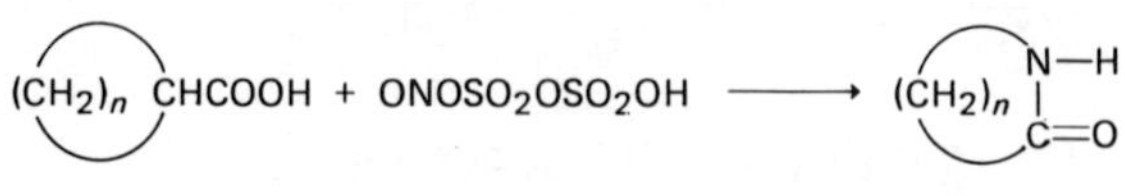

$$(\text{CH}_2)_n\ \text{CHCOOH} + \text{ONOSO}_2\text{OSO}_2\text{OH} \longrightarrow (\text{CH}_2)_n \overset{\text{N}-\text{H}}{\underset{\text{C}=\text{O}}{|}}$$

(334)

n	Yield (%)
4	71
6	87
7	81
11	82

Reaction of the aziridine ring with thionyl chloride has also been reported[672] to afford β-lactams via ring expansion. Thus, reaction of the sodium salt of 1-(*t*-butyl)-

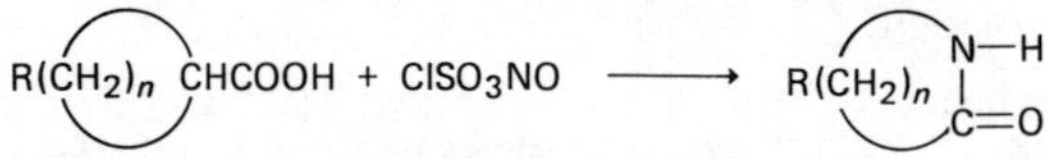

$$R(CH_2)_n \ CHCOOH + ClSO_3NO \longrightarrow R(CH_2)_n \underset{C=O}{\overset{N-H}{|}} \qquad (335)$$

R = H; n = 4, 5, 6, 7, 10, 11

R = 4-Me; n = 5

2-aziridincarboxylic acid with either thionyl chloride and sodium hydride or oxalyl chloride affords 1-(t-butyl-3-chloro-2-azitidinone (equation 336). Similar reaction of the two isomeric 3-methyl-substituted aziridincarboxylates (equation 337) showed this reaction to be stereospecific and led to the conclusion that the rearrangement involved intermediate formation of a mixed anhydride, which ionized to give a novel bicyclic ion which in turn captured Cl⁻ to give the final product (equation 338).

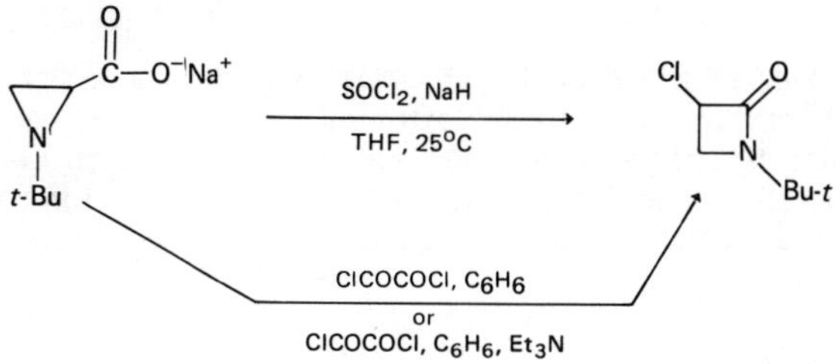

(336)

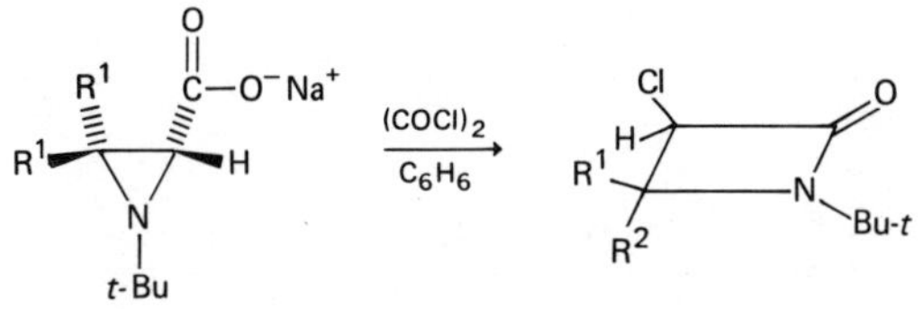

(337)

R^1	R^2	Yield (%)
Me	H	79
H	Me	63

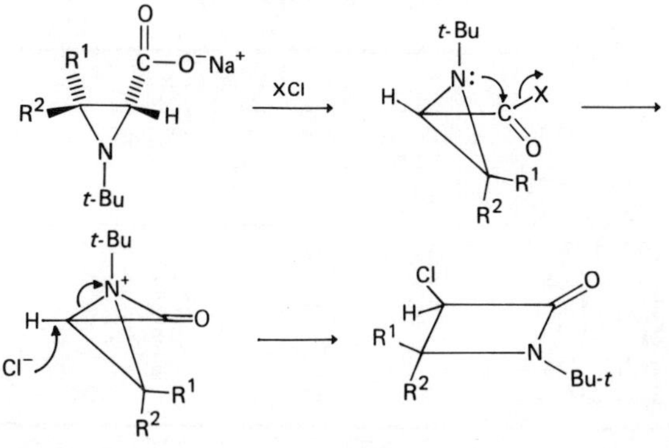

(338)

3. Claisen rearrangement

Thermal treatment of the allyl imidates, 7-allyloxy-3,4,5,6-tetrahydro-2*H*-azepine (48) and (2'*E*)-7-(3'-phenylallyloxy)-3,4,5,6-tetrahydro-2*H*-azepine (49). prepared by extended heating of the methyl imidate 7-methoxy-3,4,5,6-tetrahydro-2*H*-azepine, with excess allyl and cinnamyl alcohol, respectively (equation 339),

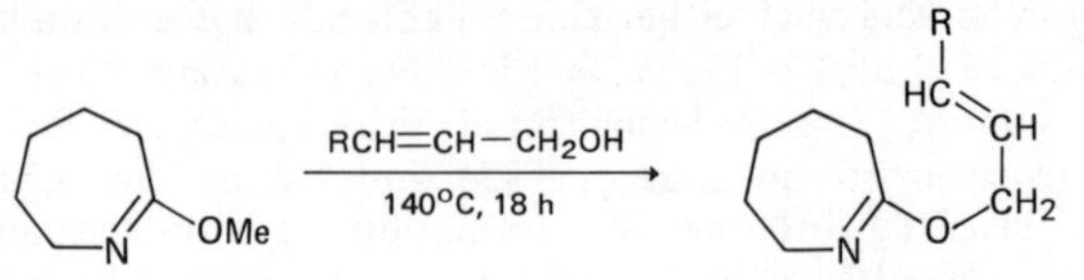

$$RCH{=}CH{-}CH_2OH$$
$$\xrightarrow{\quad 140^\circ C,\ 18\ h \quad}$$

(339)

R = H (48), 49%

R = Ph (49), 61%

affords in both cases the *N*-allyl and *C*-allyl lactams via a sigmatropic Claisen rearrangement[673]. Thus, heating 48 afforded two products, the *C*-allyl lactam (3-allylhexahydro-2*H*-azepin-2-one) (50) and the *N*-allyl lactam (1-allylhexahydro-2*H*-azepin-2-one (51). The *O,N*-ketene acetal shown in equation (340) was postulated as the intermediate, and its formation the rate-determining step, in the

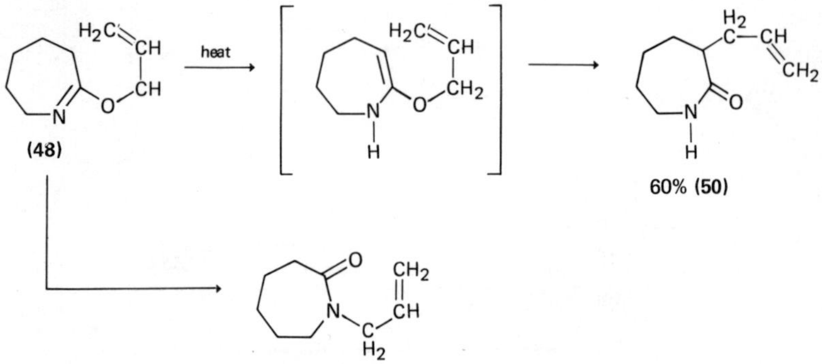

(48)

heat

60% (50)

(340)

(51) 30%

TABLE 37. Effect of temperature on the product distribution for the thermal rearrangement of allyl (48) and cinnamyl (49) imidates

	Yield (%)			
	Products for 48		Products for 49	
T (°C)	50	51	52	53
197–199	32	68	–	–
202.5–205	–	–	95	5
211–213	76	24	–	–
212–214	–	–	78	22
222.5–224.5	–	–	36	64
234–236	69	31	–	–

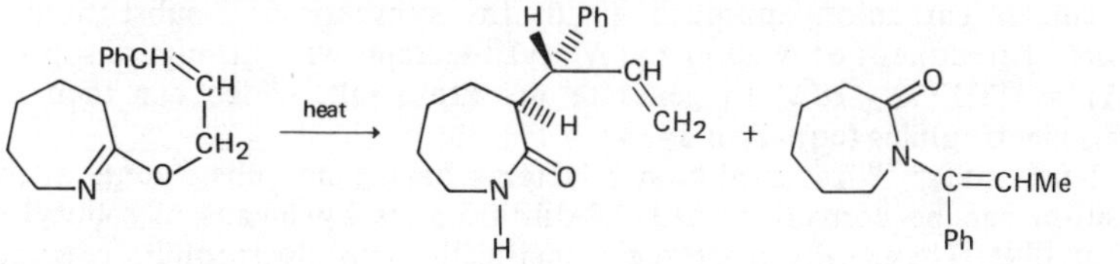

(341)

(49)   (52)   (53)

preparation of **50**. This view was supported by the observation that the yield of **50** was greatly increased by the presence of the bifunctional catalyst 2-pyridone[673]. Similar thermal treatment of the cinnamyl imidate **49**, afforded the *C*-allyl lactam [3-(1′-phenylallyl)hexahydro-2*H*-azepin-2-one] (**52**) and the *N*-propenyl lactam [(*E*)-1-(1′-phenylpropenyl)hexahydro-2*H*-azepin-2-one (**53**). The effect of temperature on the product distribution for the rearrangement of the allyl and cinnamyl imidates was also investigated and the results are given in Table 37[673].

E. By Direct Functionalization of Preformed Lactams

Generation of a carbanion centre adjacent to the lactam carbonyl provides a convenient method for structure elaboration. Gassman and Fox reported[674] that 1-methyl-2-pyrrolidone could be alkylated to afford a series of 3-substituted-1-methyl-2-pyrrolidones (equation 342). Using two molecular equivalents of sodium amide and of methyl iodide afforded 1,3,3-trimethyl-2-pyrrolidone in 45% yield.

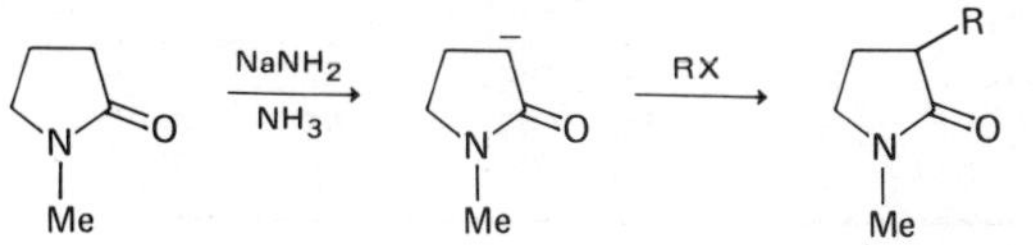

(342)

$$R = CH_3O(CH_2)_3, \ CH_3CH_2, \ (EtO)_2CHCH_2CH_2$$

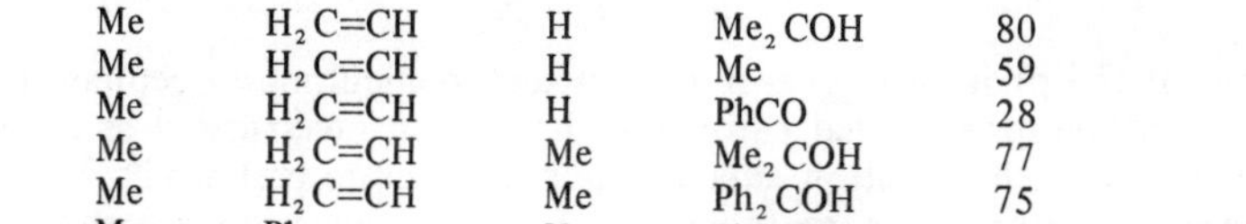

(343)

R^1	R^2	R^3	E	Yield (%)
Me	H$_2$C=CH	H	Me$_2$COH	80
Me	H$_2$C=CH	H	Me	59
Me	H$_2$C=CH	H	PhCO	28
Me	H$_2$C=CH	Me	Me$_2$COH	77
Me	H$_2$C=CH	Me	Ph$_2$COH	75
Me	Ph	H	Ph$_2$COH	50
Ph	Ph	H	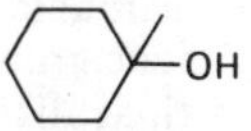 (cyclohexyl)—OH	41
Ph	Ph	H	Me$_2$COH	58
Ph	Ph	H	I	29
Ph	Ph	H	PhCO	61

A similar carbanion approach[675] to the synthesis of 3-substituted β-lactams consists of treatment of N-alkyl and N-aryl β-lactams with lithium diisopropylamide (LDA) in THF at $-78°C$ to generate the lithio salt, which can then react with various electrophiles (equation 343).

A later study[676] revealed that β-lactams having no substituents at the 1- and 3-position can be converted into 1,3-dilithio salts by means of n-butyllithium in THF at $0°C$. These salts react regiospecifically with electrophilic reagents to give 3-substituted β-lactams (equation 344).

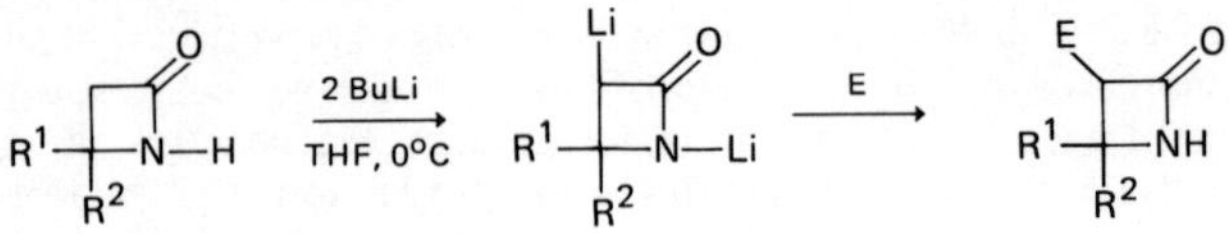

$$(344)$$

R¹	R²	E	Yield (%)
Ph	H	Me₂COH	55
Ph	H	Me	53
Ph	H	n-Bu	66
Ph	H	I	16
H₂C=CH	H	Ph₂COH	88
H₂C=CH	H	[cyclohexanol structure] OH	55
H₂C=CH	H	n-Bu	77
H₂C=CH	H	i-Pr	45
H₂C=CH	Me	Ph₂COH	57
Et	H	Ph₂COH	65
Et	H	n-Bu	65

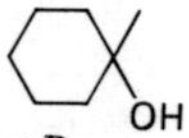

It has been reported[640,648] that attempts to alkylate caprolactam through the dianion intermediate have given a mixture of 1,3-dialkyl and 1-alkyl derivatives. However, lithiation of caprolactim methyl ether with LDA followed by alkylation and hydrolysis of the resulting 3-alkyllactim ether (equation 345) affords a useful alternative[648,677,678] to the dianion method.

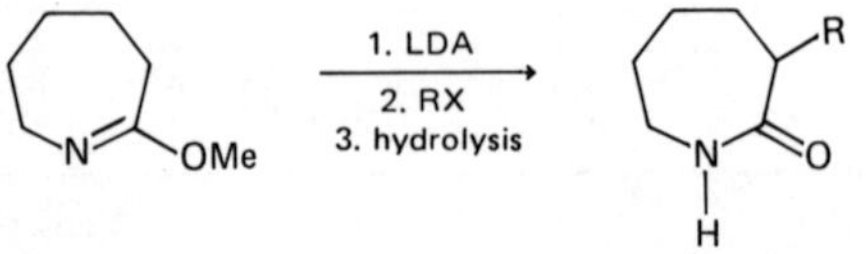

(Ref. 648) $$(345)$$

R = Me, Et, n-Pr

In view of the pharmaceutical importance of penicillin and cephalosporin antibiotics, it is not surprising that carbanions have been investigated as intermediates for substitution at the position adjacent to the β-lactam carbonyl[679]. Among the more successful approaches to the type of functionalization are those involving generation and reactions of carbanions derived from penicillins and cephalosporins containing a 6- or 7-N-arylidene group, which prevents β-elimination of the thiolate ion derived from the fused thiazolidine and dihydrothiazine rings during carbanion formation. The examples given in equations (346)–(349) are typical of this synthetic strategy in the penicillin series.

A new related synthesis of β-lactams[683], involves oxidative decarboxylation of

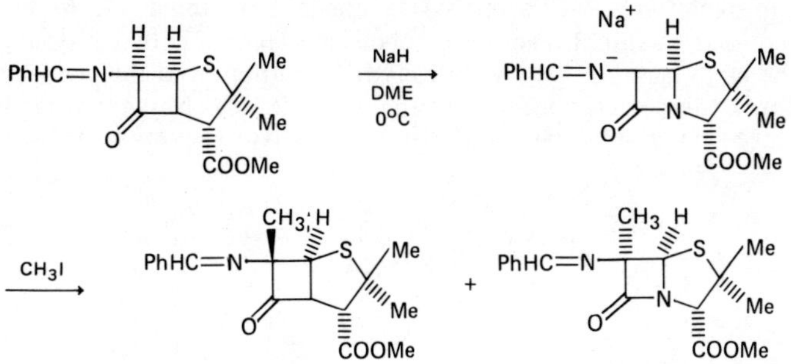

(Ref. 680)   (346)

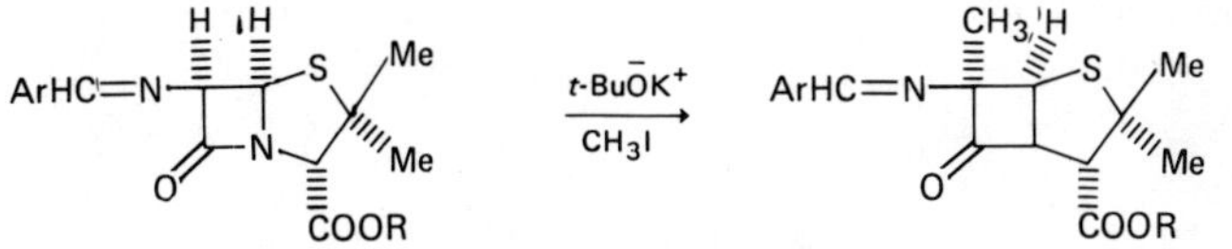

(Ref. 680)   (347)

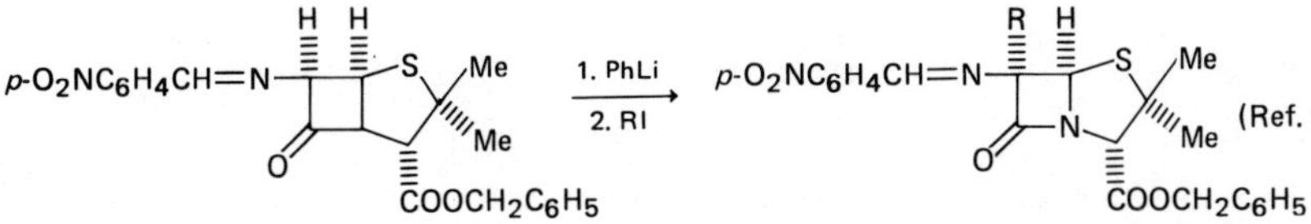

(Ref. 681)

(348)

R = Me, Et

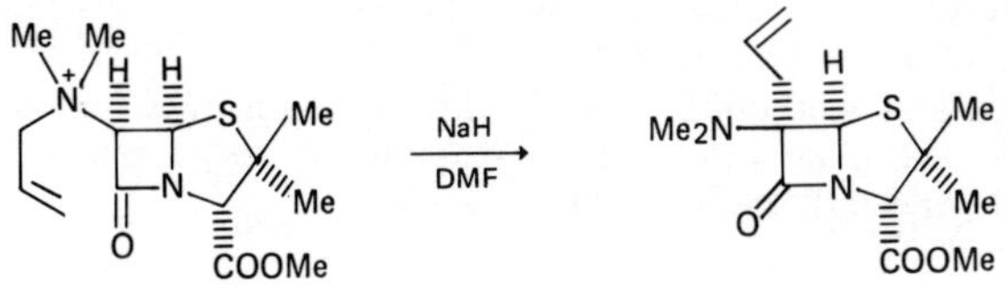

(Ref. 682)   (349)

azetidine 2-carboxylic acids. Oxygenation of the dianion formed from the appropriate acid and two equivalents of LDA in THF and subsequent acidification of the dilithium salt of the resulting hydroperoxy acid leads to decarboxylation and formation of the desired lactam (equation 350).

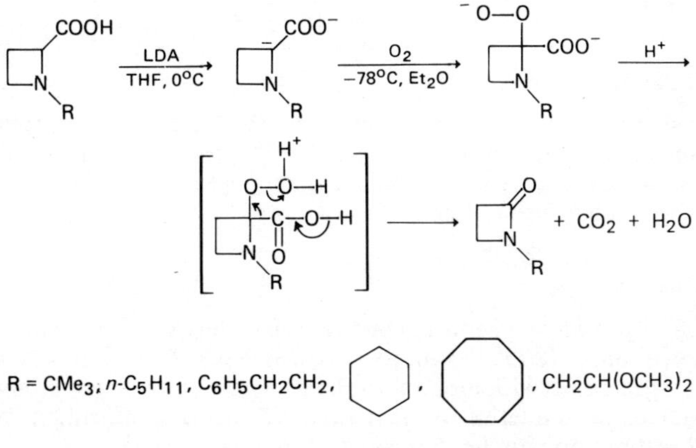

$+ CO_2 + H_2O$ (350)

$R = CMe_3, n\text{-}C_5H_{11}, C_6H_5CH_2CH_2,$, , $CH_2CH(OCH_3)_2$

An interesting route[684] to N-(2-arylethyl) lactams containing 5-, 6- and 7-membered rings consists of initial reaction of O-methyl lactims with a phenacyl halide to form N-phenacyl lactams. Sodium borohydride reduction of the phenacyl carbonyl group affords the corresponding benzylic alcohols, which undergo facile hydrogenolysis to give the desired N-(2-arylethyl) derivatives (equation 351). It

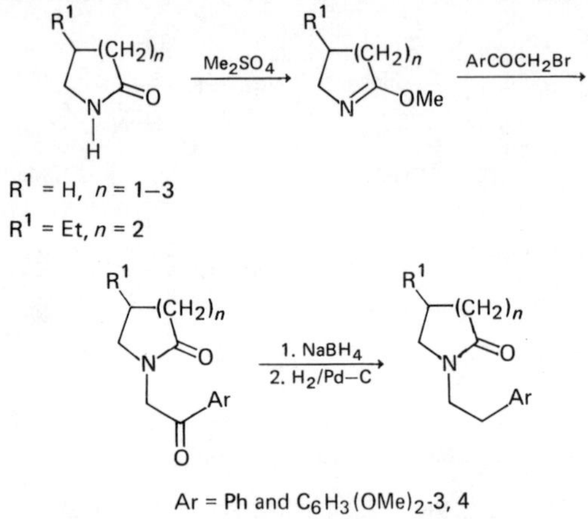

$R^1 = H, n = 1-3$

$R^1 = Et, n = 2$

(351)

Ar = Ph and $C_6H_3(OMe)_2$-3, 4

should be noted that this rather elaborate method of N-alkylation is not necessary with halides that do not undergo facile β-elimination. More routine procedures include reaction of lactams with alkyl halides and sulphates in the presence of sodium hydride[685], reactions with expoxides[686], acetylenes[686] and aldehydes[687], and by thermal rearrangement of allylic lactim ethers[648,688].

A potentially general method[640] for the introduction of alkyl substituents at the 4-position of caprolactam involves reaction of a mixture of Δ^2- and Δ^3-caprolactam with triethylborane (equation 352).

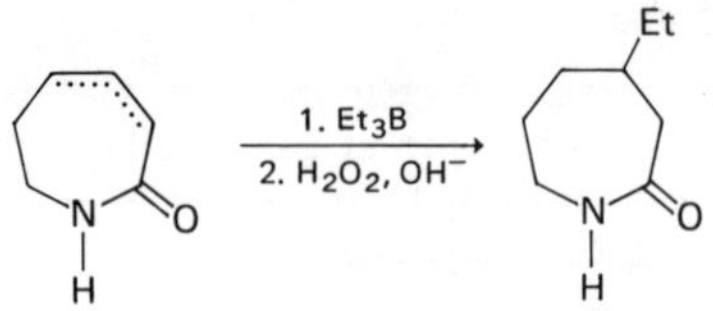

(352)

F. By Oxidation Reactions

The oxidation of nitrogen compounds to lactams using transition metal compounds has been reviewed through 1968[689]. However, in addition to transition metal compounds a variety of other oxidizing agents have been used to convert nitrogen compounds into lactams.

1. Using halogen

The use of bromine under acid conditions to effect the oxidation of nicotine has been studied since 1892[690]. In the original work[690-692] it was reported that treatment of nicotine with bromine in the presence of hydrogen bromide (equation 353) resulted in oxidative bromination of nicotine affording two products identified as dibromocotinine (54) and dibromoticonine (55).

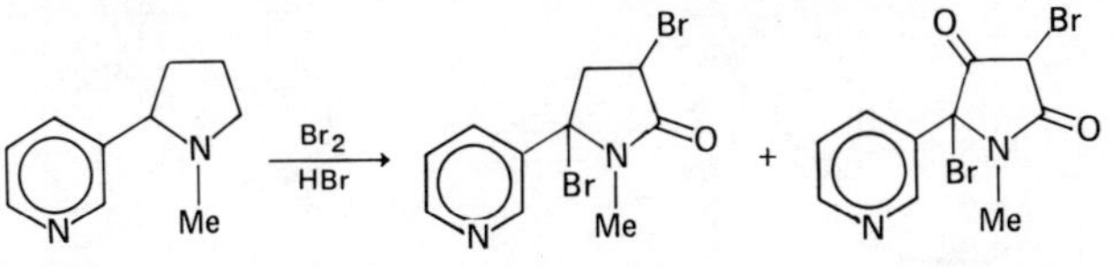

$$(353)$$

(54)   **(55)**

Reinvestigation of the structure of **54** using n.m.r.[693] and mass spectra[694] led to the disclosure that it should be represented as compound **56**. A more recent

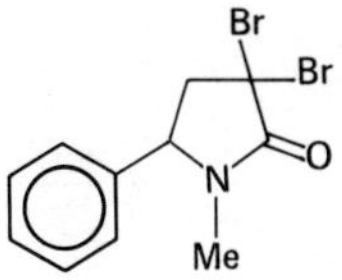

(56)

study[695] on the structure of **55** using chemical and spectral (including ^{13}C-n.m.r.) techniques has established its correct structure to be 3,4-dibromo-5-hydroxy-1-methyl-5-(3-pyridyl)-Δ^3-pyrolin-2-one (**57**).

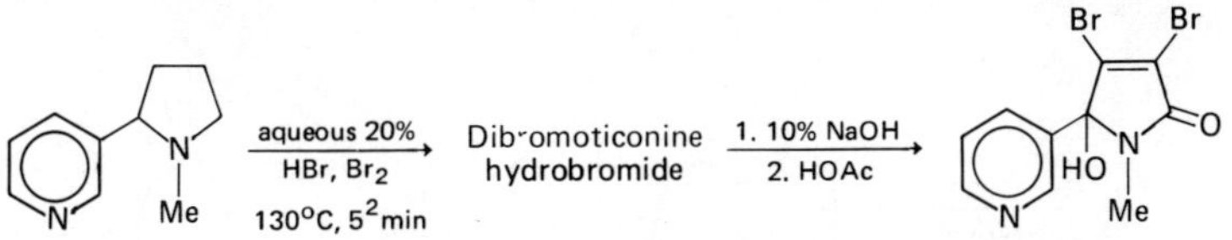

$$(354)$$

(57)

Bromine under basic conditions has been used[696] to oxidize cyclic tertiary amines into lactams (equations 355 and 356). This conversion may be done directly using excess bromine, or via the intermediate formation of the imminium salts, which are easily isolated and convertible into the lactam upon further treatment with additional bromine.

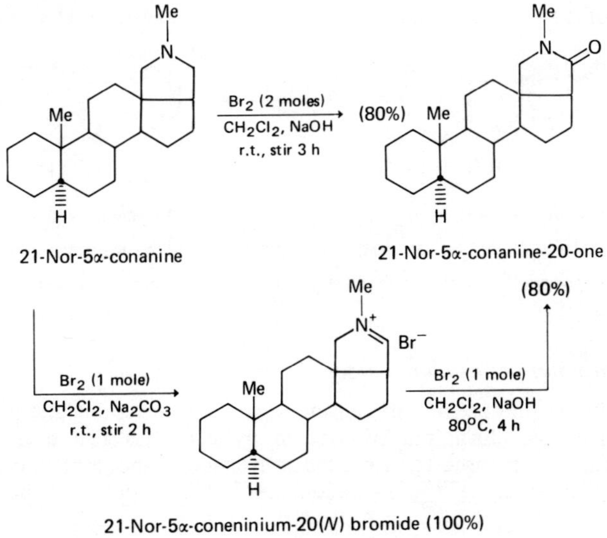

21-Nor-5α-conanine   21-Nor-5α-conanine-20-one

$$(355)$$

21-Nor-5α-coneninium-20(N) bromide (100%)

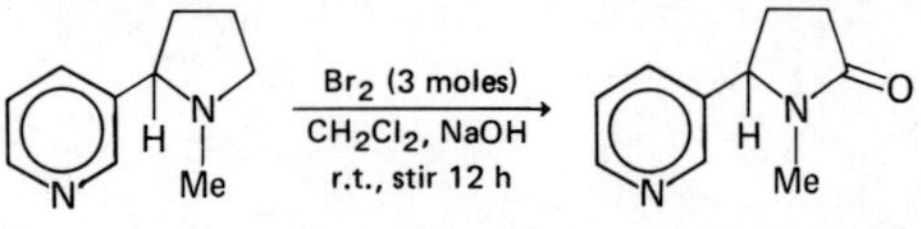
Br2 (3 moles)
CH2Cl2, NaOH
r.t., stir 12 h

Cotinene

(356)

This reaction sequence may also be performed using *N*-bromosuccinimide[695], but slightly different results are obtained if the intermediate iminium salt is further treated with aqueous sodium hydroxide without bromine (equation 357)[695].

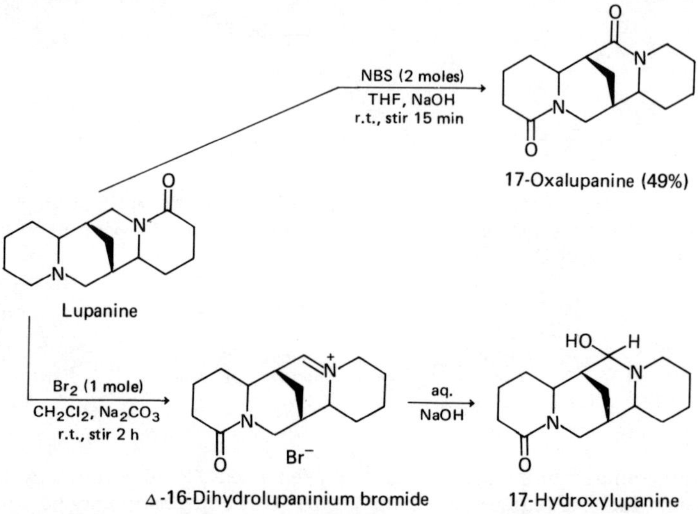
NBS (2 moles)
THF, NaOH
r.t., stir 15 min
17-Oxalupanine (49%)
Lupanine
Br2 (1 mole)
CH2Cl2, Na2CO3
r.t., stir 2 h
Br⁻
aq.
NaOH
Δ-16-Dihydrolupaninium bromide
17-Hydroxylupanine

(357)

Basic solutions of iodine in tetrahydrofuran have been reported to convert ibogamine[697], ibogaine[697] and voacangine[699] to their respective lactams (equation 358). Voacangine lactam has also been prepared[698] by the basic iodine oxidation of dihydrovoacamine followed by acid cleavage of the resulting product.

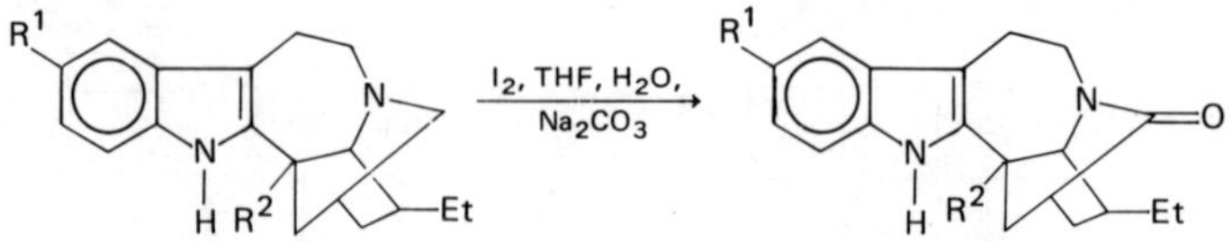
I2, THF, H2O,
Na2CO3

(358)

Ibogamine ($R^1 = R^2 = H$) — Ibogamine lactam

Ibogaine ($R^1 = MeO$, $R^2 = H$) — Ibogaine lactam (89%)

Voacangine ($R^1 = MeO$, $R^2 = COOMe$) — Voacangine lactam (10%)

2. *Using chromium or osmium oxides*

In addition to the use of basic iodine to convert ibogamine and ibogaine to their respective lactams, chromium trioxide in pyridine has also been used[697]. This reagent has also been used to effect the conversion of iboquine (equation 359)[697], iboluteine[697], conanine[700], 3-oxoconanine[700], kopsine and both epimers of

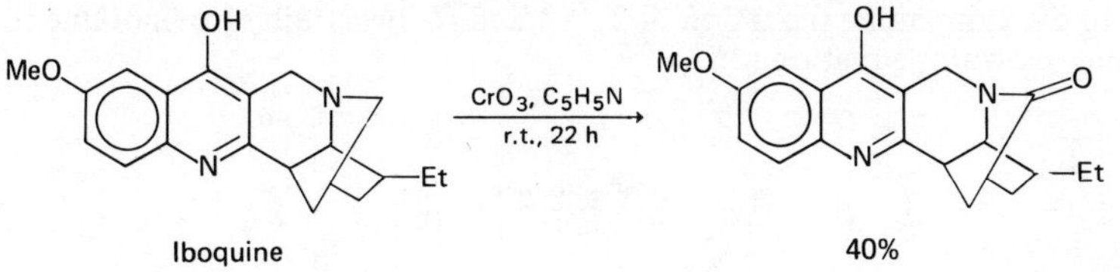

Iboquine   40%   (359)

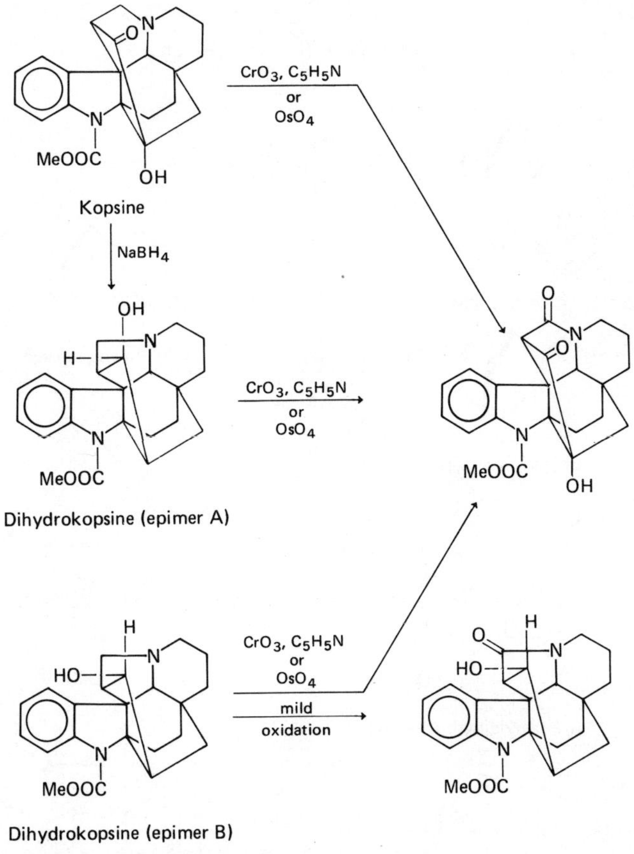

Kopsine

Dihydrokopsine (epimer A)   (360)

Dihydrokopsine (epimer B)

dihydrokopsine (equation 360)[701] into their respective lactams. These latter conversions have also been accomplished using osmium tetroxide[701].

3. Using manganese oxides

Manganese dioxide in acetone has been used to oxidize 4-(3,4-dimethoxyphen-acetyl)- and 4-benzoyl-2-methyl-1,2-dihydroisoquinoline to 4-(3,4-dimethoxyphen-acetyl)-2-methylisocarbostyril and 4-benzoyl-2-methylisocarbostyril, respectively (equation 361)[702], while acetone solutions of potassium permanganate have been used to oxidize dl-lupanine to dl-oxylupanine[703], d-lupanine and 17-hydroxylupanine to d-oxylupanine (equation 362)[704] and N-formyldihydrovindoline to the two lactams shown in equation (363)[705].

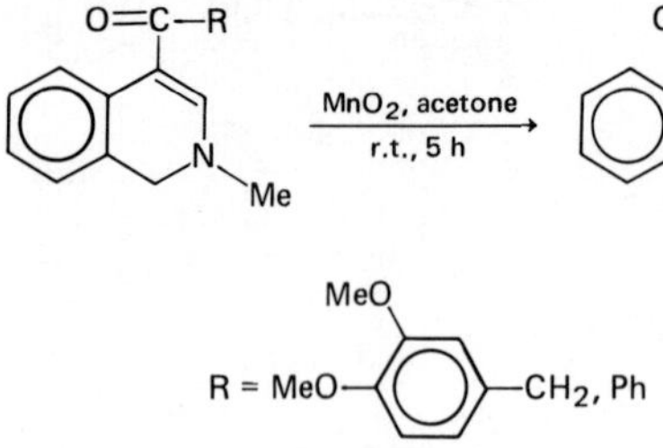

(361)

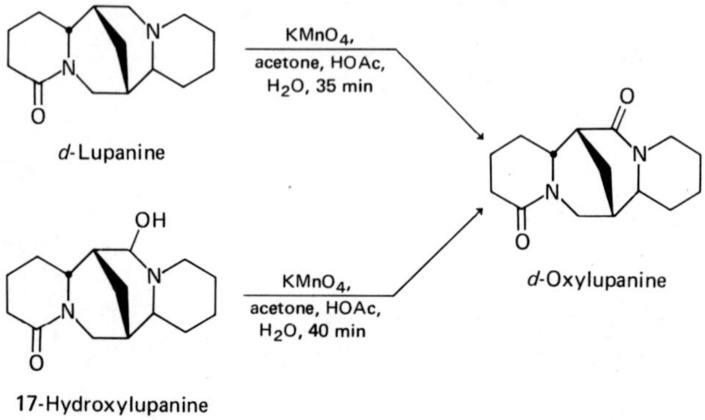

R = MeO—⟨MeO⟩—CH$_2$, Ph

d-Lupanine

17-Hydroxylupanine

d-Oxylupanine

(362)

N-Formyldihydrovindolinine

(25%)

+

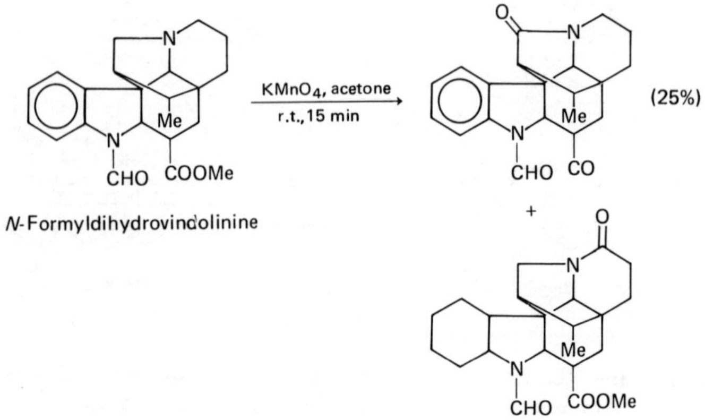

(363)

4. Using platinum or ruthenium oxides

It was originally reported[706] that voacangine, the major alkaloid of *Rejoua aurontiaca* Gaud., was converted into β-hydroxyindolenine by controlled oxidation using platinum and oxygen followed by catalytic reduction. However, a more recent study[699] of this reaction has shown the product to be voacangine lactam

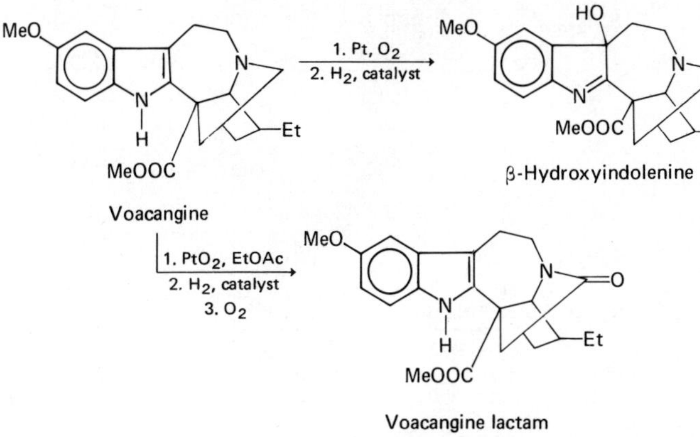

Voacangine

β-Hydroxyindolenine

(364)

Voacangine lactam

(equation 364), identical to the product obtained[698] from the basic iodine oxidation of dihydrovoacamine followed by acid cleavage of the product.

Although unsubstituted amines[707], aziridine[708] and piperidine[708] react with ruthenium tetroxide to produce imides in good yields without oxidation of the nitrogen atom directly, suitable substitution on nitrogen followed by oxidation with ruthenium tetroxide affords β-lactams (equation 365)[708]. This reaction

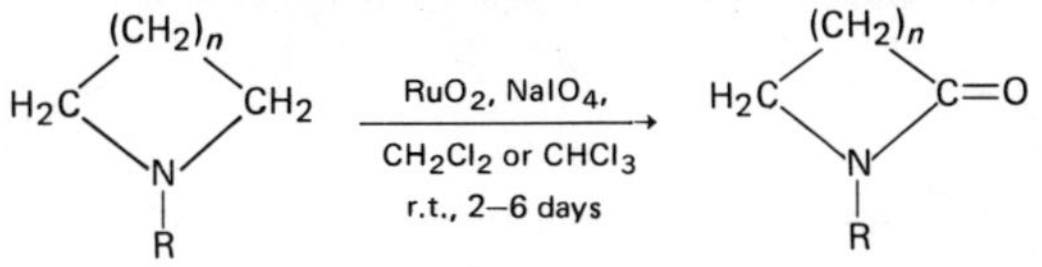

(365)

R	$n = 0$	1	2	3
		Yield (%)		
p-MeC$_6$H$_4$SO$_2$	a	a	46	3–5
MeSO$_2$	a	a	90	85
MeO–C–C– (O O)		22	68	59
HCO		a	34	a
MeCO	a	9–13	45–69	42–60
EtOCO	5	15–33	65	63

a No product could be isolated.

appears more likely to succeed as the ring size increases, and appears to be effected by the electronegativity of the nitrogen substituent. The rate of reaction has also been noted to decrease as the ring size decreases and the electronegativity of the nitrogen substituent increases[708]. By use of the methyloxalyl protecting group it was possible to prepare lactams of varying ring size according to equation (366)[708].

Ruthenium tetroxide has also been reported[709] to oxidize 2-substituted-*N*-acetyl pyrrolidines and piperidines regiospecifically to their corresponding lactams in about 60% yields with retention of absolute configuration (equations 367 and 368).

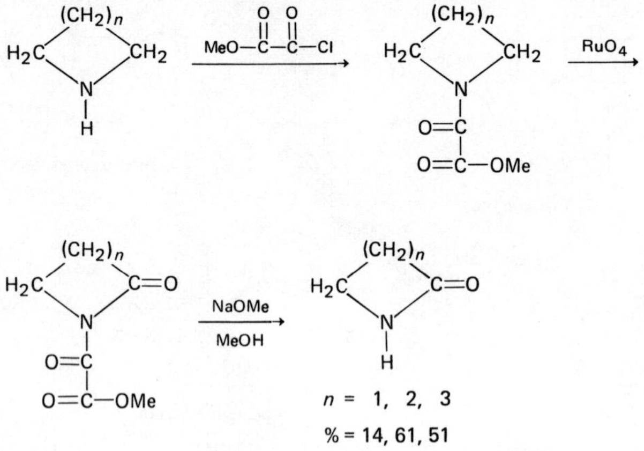

(366)

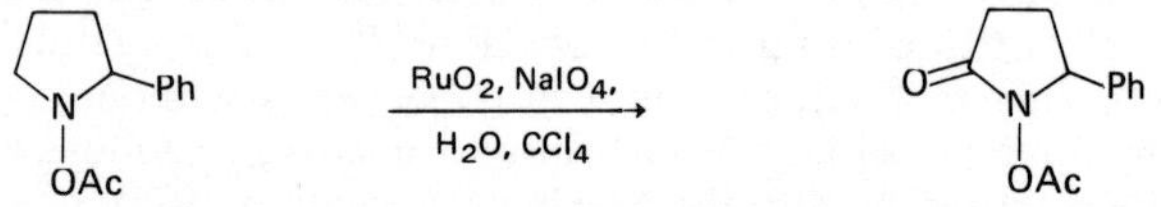

$n = 1, 2, 3$
$\% = 14, 61, 51$

R-(+)-N-Acetyl-2-phenylpyrrolidine   R-(+)-N-Acetyl-5-phenyl-2-pyrrolidinone

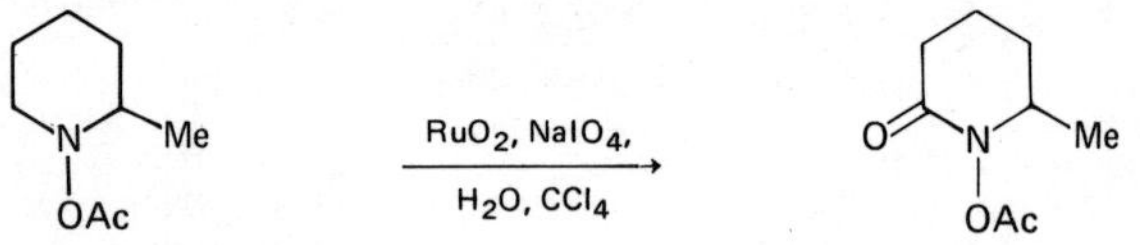

(367)

R-(+)-N-Acetyl-2-methylpiperidine   R-(+)-N-Acetyl-6-methylpiperidone

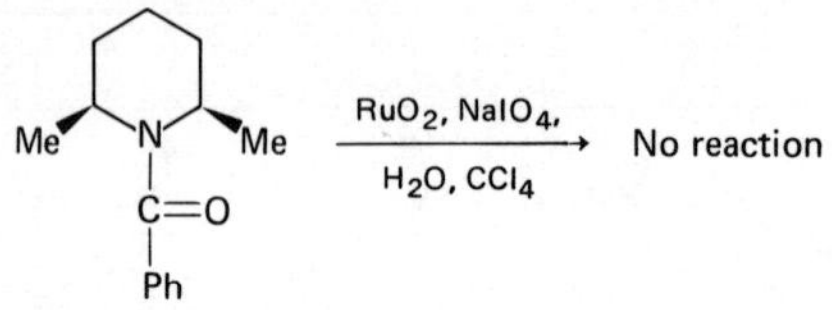

(368)

No reaction   (369)

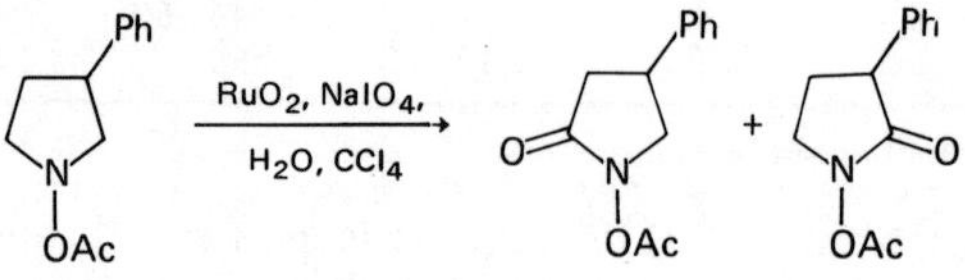

(1:1)

(370)

(1:1)

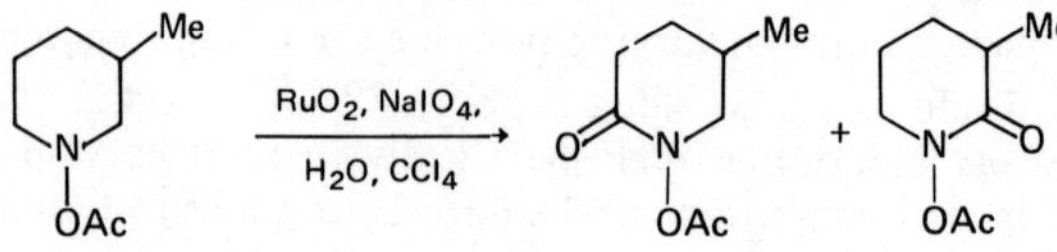

(371)

However, similar oxidation[709] of N-benzoyl-cis-2,6-dimethylpiperidine afforded only recovered starting material (equation 369), while oxidation of similarly 3-substituted N-acylpyrrolidine and piperidine afforded a 1 : 1 mixture of corresponding lactam isomers (equations 370 and 371). Application of this oxidation to N-benzoyl- and n-acetylpiperidine afforded the expected products in good yields (equation 372)[709].

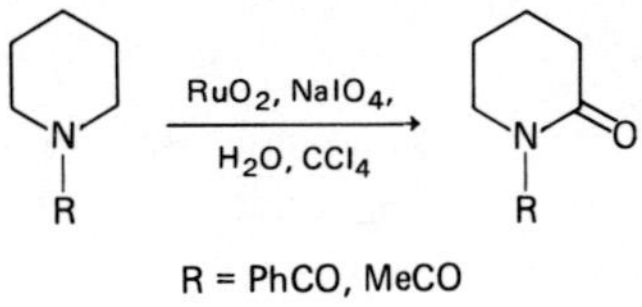

$$\text{(piperidine)} \xrightarrow[\text{H}_2\text{O, CCl}_4]{\text{RuO}_2, \text{NaIO}_4,} \text{(lactam product)} \tag{372}$$

R = PhCO, MeCO

5. Via sensitized and unsensitized photooxidation

Although reaction of ibogaine with ethylmagnesium bromide followed by treatment with oxygen has been reported[699] to produce a 20% yield of iboluteine, benzophenone-sensitized photolysis of this compound affords[699] a 35% yield of ibogaine lactam. Similar treatment[699] of voacangine affords a 5% yield of voacangine lactam, whereas sensitization using Rose Bengal affords a 10% yield of the same product. Rose Bengal-sensitized photooxidation has also been used[710] to effect the conversion shown in equation (373).

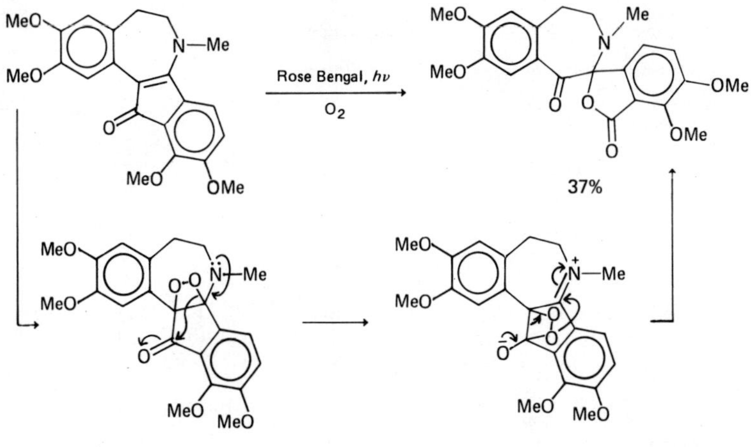

Rose Bengal, hv, O₂

37%

(373)

Conanine

methylene blue, MeOH, hv, O₂

49%

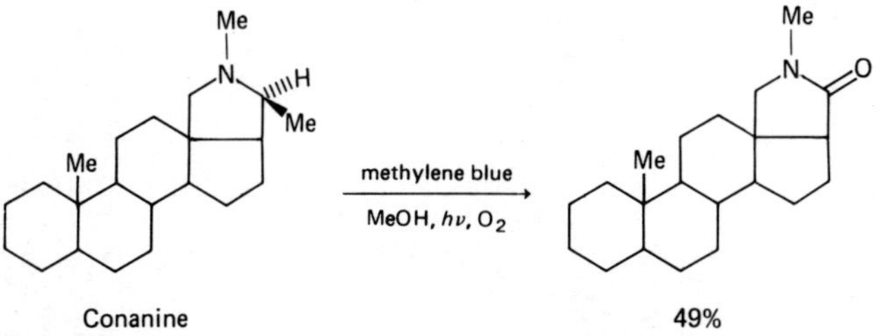

(374)

In addition to benzophenone and Rose Bengal, methylene blue has also been used to sensitize several photooxidations, including the conversion of conanine (equation 374)[711] and sparteine (equation 375)[711] to their corresponding lactams, and lupanine (equation 376)[711] to its corresponding lactam dimer. It has also been employed in the photooxidation of laudanosine (equation 377)[712], a reaction which affords a better yield of product when performed unsensitized[712].

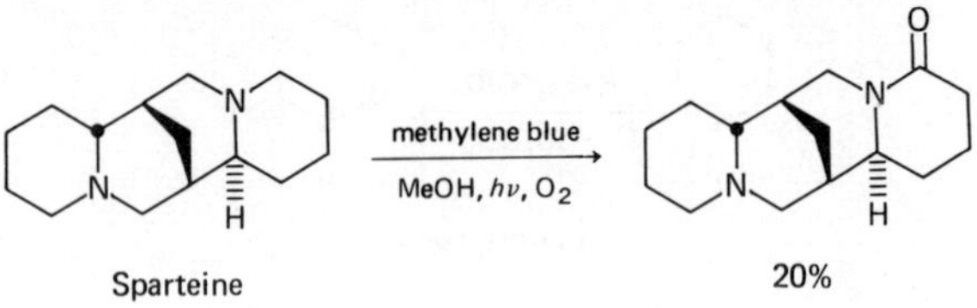

Sparteine — methylene blue / MeOH, $h\nu$, O_2 → 20% (375)

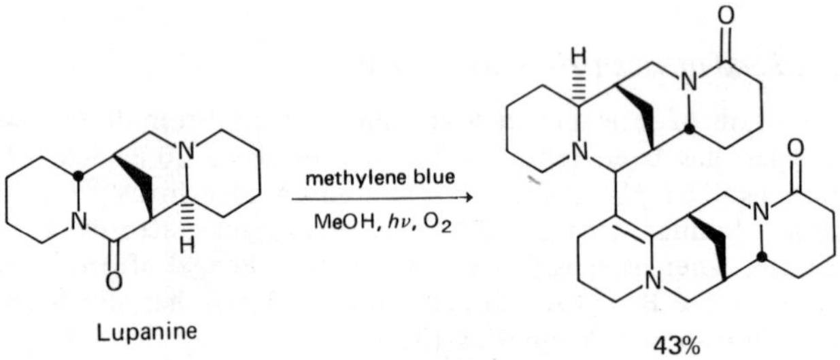

Lupanine — methylene blue / MeOH, $h\nu$, O_2 → 43% (376)

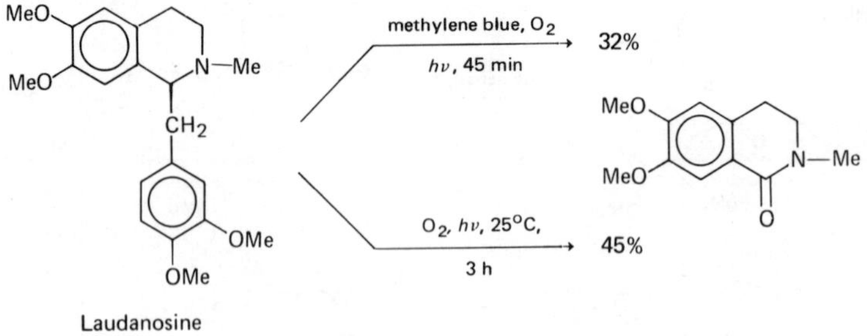

Laudanosine — methylene blue, O_2 / $h\nu$, 45 min → 32%

O_2, $h\nu$, 25°C, 3 h → 45% (377)

Unsensitized photooxidation has also been found to be effective in the production of lactams from a variety[712] of bisbenzylisoquinoline derived alkaloids such as isotetrandine and berbamine (equation 378), tenuipine and micranthine.

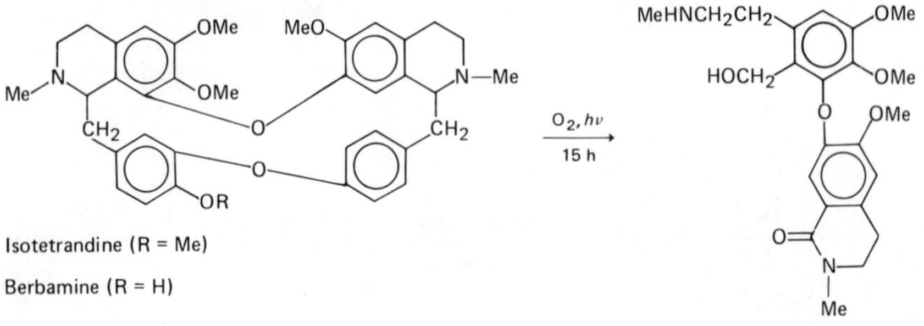

Isotetrandine (R = Me)

Berbamine (R = H)

O_2, $h\nu$ / 15 h → 30% (378)

Photooxidation in the presence of base has been found useful in the conversion of 2'-bromoreticuline to thalifoline (equation 379)[713], and 10-phenyl-9,10-dihydrophenanthridine to N-phenylphenanthridone (equation 380)[714]. This latter conversion has also been accomplished[714] without the use of base via the peroxide dimer followed by cleavage under reflux as shown in equation (381).

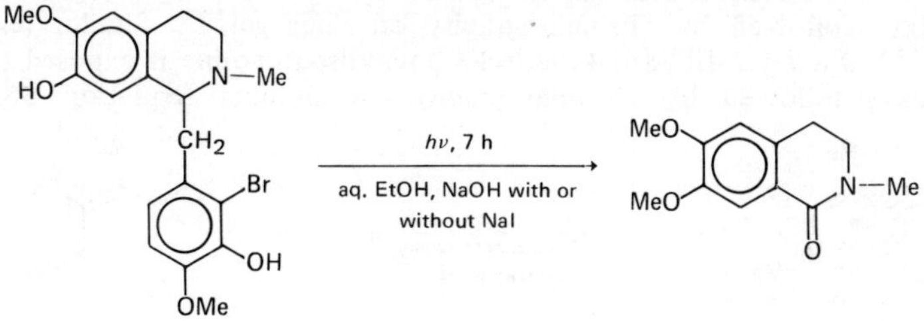

(379)

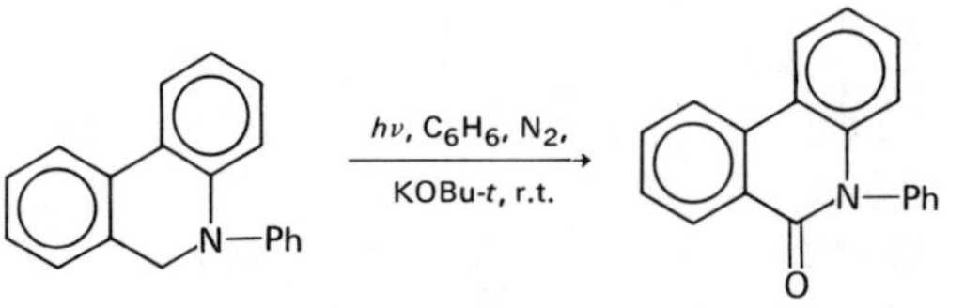

(380)

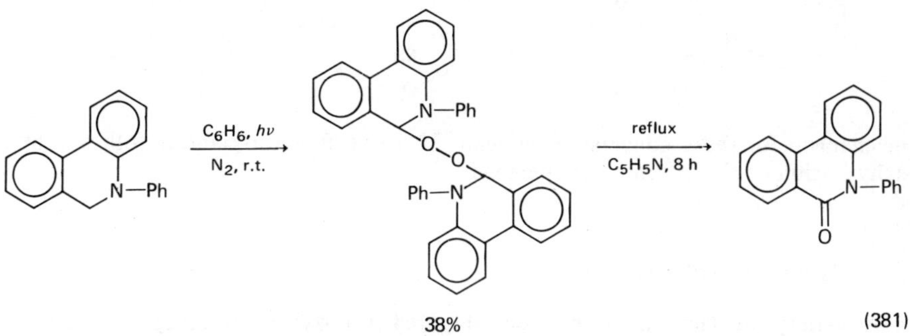

38%

(381)

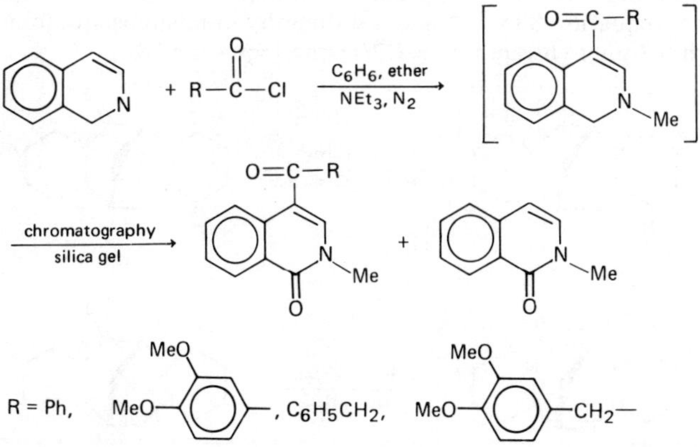

(382)

$$R = Ph, \quad MeO-\!\!\!\left\langle\right\rangle\!\!-\!, \quad C_6H_5CH_2, \quad MeO-\!\!\!\left\langle\right\rangle\!\!-CH_2-$$

6. Via autooxidation

Attempted acylation of 2-methyl-1,2-dihydroisoquinoline using benzoyl,3,4-dimethoxybenzoyl, phenacetyl and 3,4-dimethoxyphencaetyl chlorides has been reported[702] to give acetylated isocarbostyrils in all cases (equation 382). These products arise when the initial reaction products are oxidized by exposure to air for several days followed by chromatography on silica gel[702]. Similar results are obtained[715] when 1,2-dihydro-4-methyl-3-phenylisoquinoline is exposed to air for several days followed by chromatography on alumina (equation 383). The

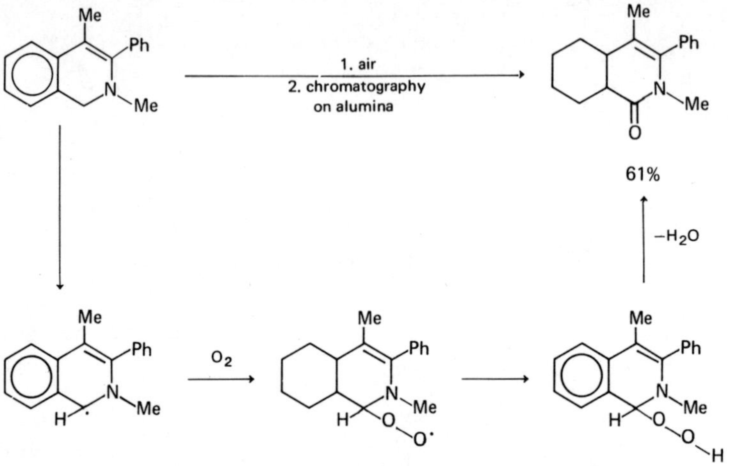

(383)

mechanism for these conversions appears[716] to be an autooxidation followed by a dehydration of the intermediate peroxide.

7. Using miscellaneous reagents

A variety of lactams have been prepared via oxidation using a variety of miscellaneous reagents. For example, potassium hexacyanoferrate has been used to oxidize d-lupanine to d-oxylupainine (equation 384)[704], l-sparteine to l-oxysparteine (equation 385)[704] and 2,4-dimethyl-3-phenylisoquinolinium iodide to 2,4-dimethyl-3-phenylisoquinoline-1(2H)-one (equation 386)[715].

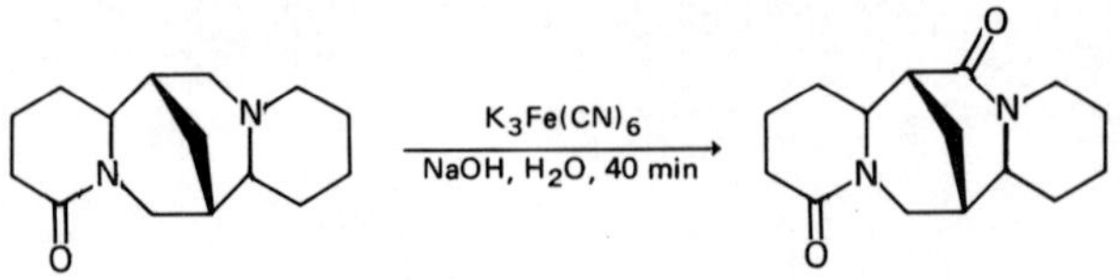

(384)

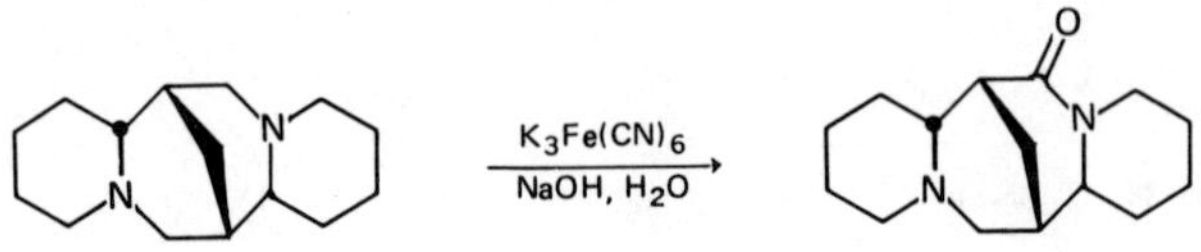

(385)

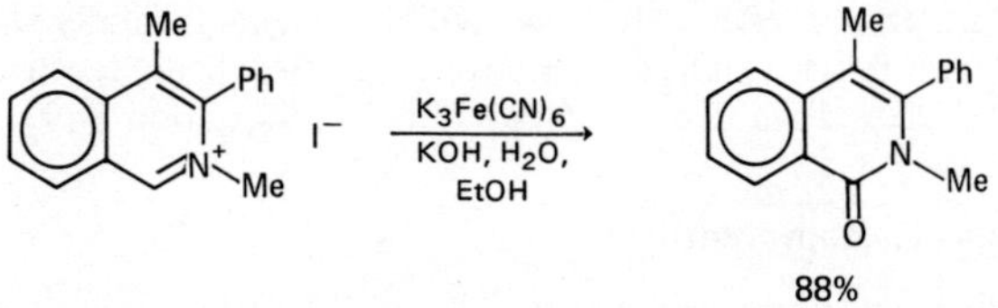

(386)

88%

Wasserman and Tremper[717] have reported that treatment of 1-substituted azetidine-2-carboxylic acid with oxalyl chloride affords the iminium salt shown in equation (387), which upon treatment with *m*-chloroperbenzoic acid in pyridine produces a 70–80% yield of 1-substituted β-lactams. This reaction is reported to be more convenient than the alternative procedure of low-temperature dianion oxygenation reported elsewhere[683] in this review.

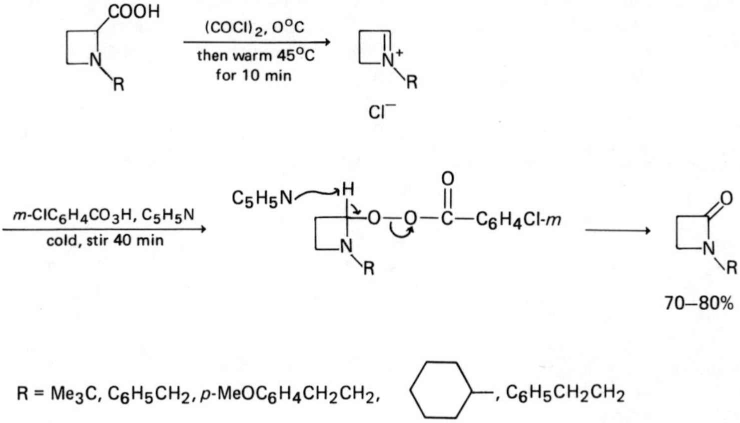

(387)

70–80%

R = Me$_3$C, C$_6$H$_5$CH$_2$, *p*-MeOC$_6$H$_4$CH$_2$CH$_2$, ⬡—, C$_6$H$_5$CH$_2$CH$_2$

Treatment of cyclic amines such as pyrrolidine with a hydroperoxide in the presence of a metal ion catalyst, such as manganic acetylacetonate, cobalt naphthenate or dicyclopentadienyltitanium dichloride affords the corresponding lactam (equation 388)[718].

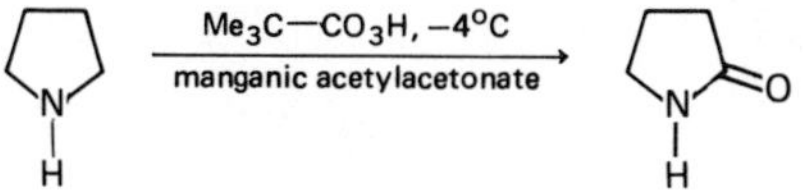

(388)

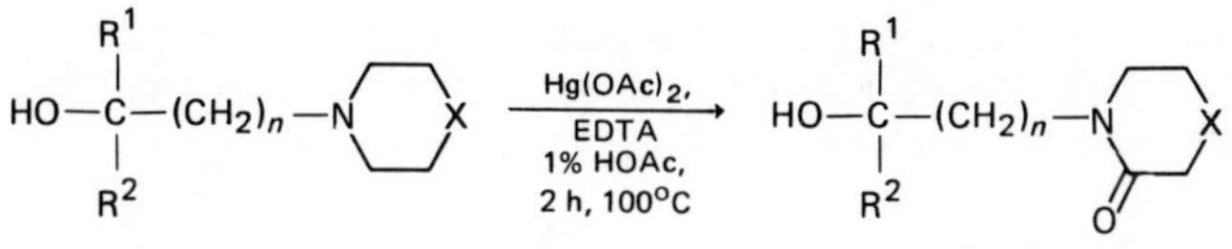

(389)

R^1	R^2	n	X	Yield (%)
H	Ph	2	O	—
H	Ph	2	(CH$_2$)$_2$	71.2
H	PhOCH$_2$	1	(CH$_2$)$_2$	86.2
Ph	Ph	2	CH$_2$	45.0

An interesting preparation of lactams, which appears formally to be an oxidation but which in reality is a dehydrogenation, has also been reported[719] using the Hg(II) salt of ethylenediaminetetraacetate (EDTA) (equation 389).

G. Miscellaneous Lactam Syntheses

The following methods do not qualify for inclusion in one of the foregoing sections, but appear to have sufficient generality to serve as useful, albeit somewhat specialized, synthetic procedures.

Condensations of 4-arylmethylene-2,3-pyrrolidinediones with β-aminocrotonate or with 4-amino-3-penten-2-one result in addition of the nucleophilic vinyl carbon of the enamine to the arylmethylene function, accompanied by cyclization of the amino groups of the addend with the 3-carbonyl group of the pyrrolidinedione. The resulting dihydropyrrolo[3,4,b]pyridin-7-ones can be oxidized by bromine to afford the pyridine-fused δ-lactams shown in equation (390)[720]. When N-phenacylpyridinium bromide was allowed to react with the pyrrolidinediones, the aromatic δ-lactams were formed directly. In the same study[720] it was found that when 4-o-nitrobenzylidene derivatives of 1-substituted 2,3-pyrrolidinediones were treated with tin(II) chloride or with sodium dithionate, reductive cyclization took place to afford 1,2-dihydropyrrolo[3,4,b]quinolin-3-ones (equation 391).

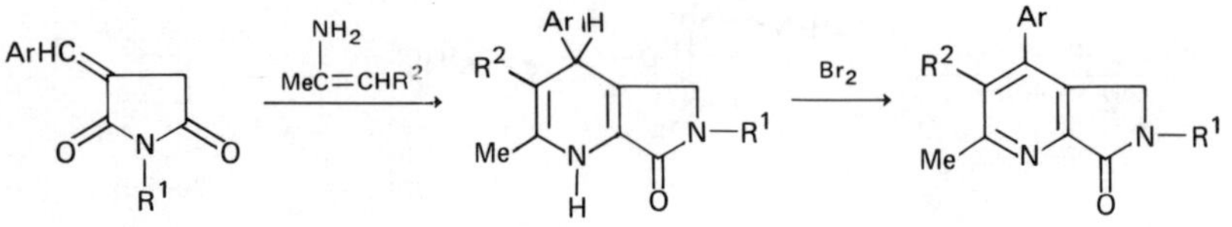

$$R^1 = c\text{-}C_6H_{11}, t\text{-Bu}, C_6H_5CH_2, C_6H_5(CH_2)_2, MeO_2C(CH_2)_2$$
$$R^2 = COOEt, MeCO$$

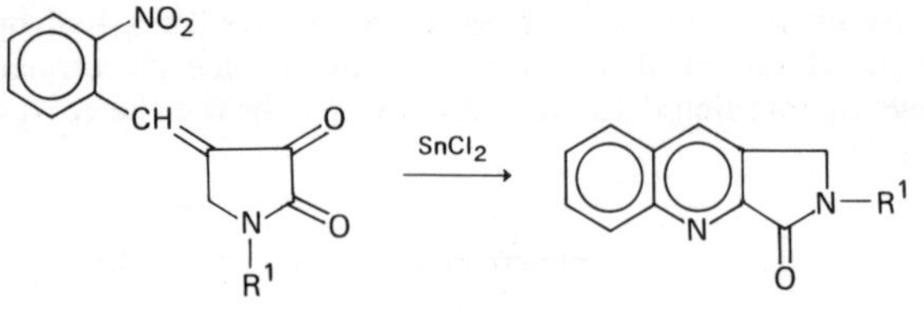

$$R^1 = c\text{-}C_6H_{11}, C_6H_5(CH_2)_2$$

A convenient synthesis of 5-hydroxy- and 5-methoxy-3-pyrrolin-2-ones has been carried out via singlet oxygen addition to an appropriate furan derivative followed by ammonolysis of the resulting pseudo ester (equation 392)[721].

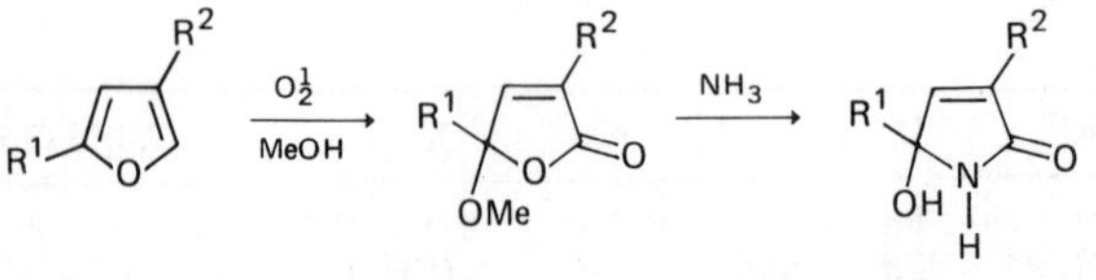

Anils of cycloalkanones have been found[722] to react with oxalyl chloride to afford 1-phenyl-4,5-polymethylene-2,3-pyrrolidinediones (equation 393). When

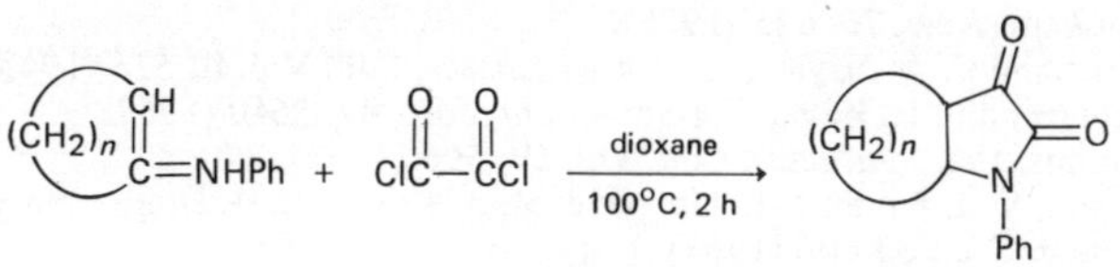

$$n = 3, \quad 4, \quad 5, \quad 6 \tag{393}$$

Yield (%) = 37, 73.8, 75, 84

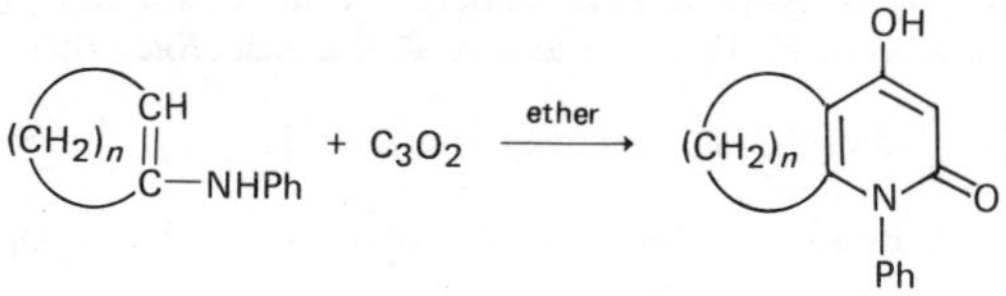

$$n = 3, \ 4, \ 5, \ 6 \tag{394}$$

Yield (%) = 46, 48, 46, 18.7

carbon suboxide is used instead of oxalyl chloride, the same anils afford 4-hydroxy-5,6-polymethylene-2-pyridones (equation 394)[723].

IV. ACKNOWLEDGMENTS

This undertaking was only possible with the dedicated help of our friend and typist, Mrs. Brenda Mills, who typed the entire manuscript and provided all of the structural drawings. Her steadfastness and pleasant nature contributed greatly to the completion of this work, and we are pleased to acknowledge her contributions. Miss Susan Stevens provided us with many hours of help in collecting and organizing the primary literature references used in this review. We are grateful to the Department of Chemistry for providing facilities and financial support during the writing of this chapter. We are also pleased to acknowledge the National Aeronautics and Space Administration (Grants NSG-1064 and NSG-1286) and the National Science Foundation (Grant CHE 74-20520) for support of our research programmes during the writing of this chapter.

V. REFERENCES

1. H. E. Zaugg, *Org. Reactions*, 8, 305 (1954).
2. R. Filler, *Chem. Rev.*, 63, 21 (1963).
3. Y. Etienne and N. Fischer, *The Chemistry of Heterocyclic Compounds* (Ed. A. Weissberger), Vol. 19, Interscience, New York, 1964, p. 729.
4. Y. S. Rao, *Chem. Rev.*, 64, 353 (1964).
5. M. F. Ansell and M. H. Palmer, *Quart. Rev.*, 18, 211 (1964).
6. N. P. Shusherina, N. D. Dmitrieva, E. A. Luk'yanets and R. Y. Levina, *Russ. Chem. Rev.*, 36, 175 (1967).
7. B. Chemielarz, *Tluszcze Srodki Piorace Kosmet*, 12, 21 (1968); *Chem. Abstr.*, 69, 67715e (1968).
8. V. M. Dashunin, R. V. Maeva, G. A. Samatuga and V. N. Belov, *Mezhdunar. Kongr. Efirnym Maslam* (Ed. P. V. Naumenko) 4th ed., Vol. 1, Vses. Nauchno-Issled. Inst. Sint. Nat. Dishistykh Veshchestv, Moscow, 1971, p. 90; *Chem. Abstr.*, 78, 135970x (1973).
9. J. L. Brash and D. J. Lyman, *High Polym.*, 26, 147 (1972).
10. S. G. Cottis, *High Polym.*, 27, 311 (1972).
11. P. A. Grieco, *Synthesis*, 67 (1975).

12. Y. S. Rao, *Chem. Rev.*, 76, 625 (1976).
13. J. H. Gardner and C. A. Naylor, Jr, *Org. Syntheses*, Coll. Vol. II, 526 (1943).
14. C. G. Overberger and H. Kaye, *J. Amer. Chem. Soc.*, 89, 5640 (1967).
15. J. V. Karabinos, *Org. Syntheses*, Coll. Vol. IV, 506 (1963).
16. W. S. Johnson, V. J. Bauer, J. L. Margrave, M. A. Frisch, L. H. Dreger and W. N. Hubbard, *J. Amer. Chem. Soc.*, 83 606 (1961).
17. W. Cocker, L. O. Hopkins, T. B. H. McMurray and M. A. Nisbet, *J. Chem. Soc.*, 4721 (1961).
18. C. Collin-Asselineau, S. Bory and E. Lederer, *Bull. Soc. Chim. Fr.*, 1524 (1955).
19. J. A. Marshall, N. Cohen and A. R. Hochstetler, *J. Amer. Chem. Soc.*, 88, 3408 (1966).
20. G. Lardelli, V. Lamberti, W. T. Weller and A. P. DeJonge, *Rec. Trav. Chim. Pays-Bas*, 86, 481 (1967).
21. J. D. White, S. N. Lodwig, G. L. Trammell and M. P. Fleming, *Tetrahedron Letters*, 3263 (1974).
22. R. E. Ireland, D. A. Evans, D. Glover, G. M. Rubottom and H. Young, *J. Org. Chem.*, 34, 3717 (1969).
23. G. Caron and J. Lessard, *Can. J. Chem.*, 51, 981 (1973).
24. A. P. Krapcho and E. G. E. Jahngen, Jr, *J. Org. Chem.*, 39, 1322 (1974).
25. W. Adam, J. Baeza and J.-C. Liu, *J. Amer. Chem. Soc.*, 94, 2000 (1972).
26. A. P Krapcho and E. G. E. Jahngen, Jr, *J. Org. Chem.*, 39, 1650 (1974).
27. M. V. S. Sultanbawa, *Tetrahedron Letters*, 4569 (1968).
28. R. B. Woodward, F. E. Bader, H. Bickel, A. J. Frey and R. W. Kierstead, *Tetrahedron*, 2, 1 (1958).
29. J. Meinwald and E. Frauenglass, *J. Amer. Chem. Soc.*, 82, 5235 (1960).
30. J. A. Marshall and N. Cohen, *Tetrahedron Letters*, 1997 (1964).
31. J. A. Marshall and N. Cohen, *J. Org. Chem.*, 30, 3475 (1965).
32. V. R. Tadwalkar and A. S. Rao, *Indian J. Chem.*, 9, 1416 (1971).
33. J. Finkelstein and A. Brossi, *J. Heterocyclic Chem.*, 4, 315 (1967).
34. J. Finkelstein and A. Brossi, *Org. Syntheses*, 55, 45 (1976).
35. M. G. Zalinyan, V. S. Arutyunyan and M. T. Dangyan, *Arm. Khim. Zh.*, 26, 827 (1973); *Chem. Abstr.*, 80, 82014w (1974).
36. A. S. Kyazimov, M. M. Movsumzade, A. L. Shabanov and A. A. Babaeva, *Azerb. Khim. Zh.*, 53, (1974); *Chem. Abstr.*, 82, 97725u (1975).
37. A. Hirshfeld, W. Taub and E. Glotter, *Tetrahedron*, 28, 1275 (1972).
38. E. M. P. Fowler and H. B. Henbest, *J. Chem. Soc.*, 3642 (1950).
39. D. Kosterman, *Nature*, 166, 787 (1950).
40. D. Kosterman, *Rec. Trav. Chim.*, 70, 79 (1951).
41. J. Gobley and H. O'Rorke, *J. Pharm. Chim.*, 598 (1860).
42. M. Cuzent, *Compt. Rend.*, 205 (1861).
43. W. Borsche and W. Peitzsch, *Ber.*, 62, 360 (1929).
44. W. Borsche and C. K. Bodenstein, *Ber.*, 62, 2515 (1929) and later papers through 1933.
45. J. W. Patterson and J. E. McMurray, *Chem. Commun.*, 488 (1971).
46. J. B. Brown, H. B. Henbest and E. R. H. Jones, *J. Chem. Soc.*, 3634 (1950).
47. E. E. van Tamelen and S. R. Bach, *J. Amer. Chem. Soc.*, 80, 3079 (1958).
48. O. A. Sarkisyan, A. N. Stepanyan, V. S. Arutyunyan, M. G. Zalinyan and M. T. Dangyan, *Zh. Org. Khim.*, 5, 1648 (1969); *Chem. Abstr.*, 72, 2982g (1970).
49. E. J. Corey and K. C. Nicolaou, *J. Amer. Chem. Soc.*, 96, 5614 (1974).
50. R. C. Blume, *Tetrahedron Letters*, 1047 (1969).
51. P. F. Hudrlik, L. R. Rudnick and S. H. Korzeniowski, *J. Amer. Chem. Soc.*, 95, 6848 (1973).
52. W. C. Bailey, Jr, A. K. Bose, R. M. Ikeda, R. H. Newan, H. V. Secor and C. Varsel, *J. Org. Chem.*, 33, 2819 (1968).
53. A. M. Chalmers and A. J. Baker, *Tetrahedron Letters*, 4529 (1974).
54. G. R. Pettit, D. C. Fessler, K. D. Paull, P. Hofer and J. C. Knight, *J. Org. Chem.*, 35, 1398 (1970).
55. G. S. King and E. S. Waight, *J. Chem. Soc., Perkin Trans. I*, 1499 (1974).

56. E. J. Clarke and R. P. Hildebrand, *J. Inst. Brew.*, **73**, 60 (1967); *Chem. Abstr.*, **67**, 32303a (1967).
57. G. I. Nikishin, M. G. Vinogradov and T. M. Fedorova, *Chem. Commun.*, 693 (1973).
58. Z. Horii, T. Yagami and M. Hanaoka, *Chem. Commun.*, 634 (1966).
59. R. Hodges and A. L. Porte, *Tetrahedron*, **20**, 1463 (1964).
60. T. Wada, *Chem. Pharm. Bull. (Tokyo)*, **12**, 1117 (1964); and **13**, 43 (1965).
61. T. Sakan, S. Isol and S. B. Hyeon, *Tetrahedron Letters*, 1623 (1967).
62. J. Bricout, R. Viani, F. Muggler-Chawan, J. P. Marion, D. Reymond and R. H. Egli, *Helv. Chim. Acta*, **50**, 1517 (1967).
63. G. Agnes and G. P. Chiusoli, *Chim. Ind. (Milan)*, **50**, 194 (1968); *Chem Abstr.*, **69**, 35352t (1968).
64. K. Tsukida, M. Ito and F. Ikeda, *Experientia*, **29**, 1338 (1973).
65. P. K. Christensen, *Acta Chem. Scand.*, **11**, 582 (1957).
66. N. A. Sorensen and K. Stavholt, *Acta Chem. Scand.*, **4**, 1080 (1950).
67. I. Bill, E. R. H. Jones and M. C. Whiting, *J. Chem. Soc.*, 1313 (1958).
68. J. Castaner and J. Pascual, *J. Chem. Soc.*, 3962 (1958).
69. J. Castaner and J. Pascual, *Anales Real. Soc. Espan. Fis. y Quim (Madrid)*, **53B**, 651 (1957); *Chem. Abstr.*, **54**, 3404b (1960).
70. C. Belil, J. Castella, J. Castells, R. Mestres, J. Pascual and F. Settatosa, *Anales Real. Soc. Espan. Fis. y Quim (Madrid)*, **57B**, 617 (1961); *Chem. Abstr.*, **57**, 12455e (1962).
71. C. Belil, J. Pascual and F. Serratosa, *Tetrahedron*, **20**, 2701 (1964).
72. G. I. Nikishin, M. G. Vinogradov and T. M. Fedorova, *Chem. Commun.*, 693 (1973).
73. J. Castells, R. Mestres and J. Pascual, *Anales Real. Soc. Espan. Fis. y Quim (Madrid)*, **60B**, 843 (1964); *Chem Abstr.*, **63**, 11536d (1965).
74. F. Serratosa, *Tetrahedron*, **16**, 185 (1961).
75. J. Bosch, J. Castells and J. Pascual, *Anales Real. Soc. Espan. Fis. y Quim (Madrid)*, **57B**, 469 (1961); *Chem Abstr.*, **56**, 8628d (1962).
76. M. Alguero, J. Bosch, J. Castaner, J. Castella, J. Castells, R. Mestres, J. Pascual and F. Serratosa, *Tetrahedron*, **18**, 1381 (1962).
77. R. H. Wiley, T. H. Crawford and C. E. Staples, *J. Org. Chem.*, **27**, 1535 (1962).
78. R. H. Wiley and C. E. Staples, *J. Org. Chem.*, **28**, 3408 (1963).
79. Y. S. Rao, *Tetrahedron Letters*, 1457 (1975).
80. R. H. Wiley, C. H. Jarboe and F. N. Hayes, *J. Amer. Chem. Soc.*, **79**, 2602 (1957).
81. G. W. Kenner, M. A. Murray and C. M. B. Taylor, *Tetrahedron*, **1**, 259 (1957).
82. O. L. Chapman and W. R. Adams, *J. Amer. Chem. Soc.*, **89**, 4243 (1967).
83. G. O. Schenck, G. Koltzenburg and H. Grossman, *Angew. Chem.*, **69**, 177 (1957).
84. R. Dulou, M. Vilkas and M. Pfau, *Compt. Rend.*, **249**, 429 (1959).
85. M. Pfau, R. Dulou and M. Vilkas, *Compt. Rend.*, **251**, 2188 (1960); **254**, 1817 (1962).
86. G. Buchi and S. H. Feaisheller, *J. Org. Chem.*, **34**, 609 (1969).
87. S. Majeti, *J. Org. Chem.*, **37**, 2914 (1972).
88. M. Tokuda, Y. Kokoyama, T. Taguchi, A. Suzuki and M. Itoh, *J. Org. Chem.*, **37**, 1859 (1972).
89. M. Itoh, T. Taguchi, V. Van Chung, M. Tokuda and A. Suzuki, *J. Org. Chem.*, **37**, 2357 (1972).
90. J. Bougault, *Ann. Chim. Phys.*, **14**, 145 (1908); **15**, 296 (1908).
91. R. P. Linstead and C. J. May, *J. Chem. Soc.*, 2565 (1927).
92. A. W. Schrecker, G. Y. Greenburg and J. L. Hartwell, *J. Amer. Chem. Soc.*, **74**, 5669 (1952).
93. A. W. Schrecker and J. L. Hartwell, *J. Amer. Chem. Soc.*, **74**, 5676 (1952).
94. E. E. van Tamelen and M. Shamma, *J. Amer. Chem. Soc.*, **76**, 2315 (1954).
95. J. Klein, *J. Amer. Chem. Soc.*, **81**, 3611 (1959).
96. S. Beckmann and H. Geiger, *Ber.*, **92**, 2411 (1959).
97. J. A. Berson and A. Remanick, *J. Amer. Chem. Soc.*, **83**, 4947 (1961).
98. W. E. Barnett and J. C. McKenna, *Chem. Commun.*, 551 (1971).
99. W. E. Barnett and J. C. McKenna, *Tetrahedron Letters*, 2595 (1971).
100. W. E. Barnett and W. H. Sohn, *Tetrahedron Letters*, 1777 (1972).

101. M. M. Movsumzade, A. S. Kyazimov, A. L. Shabanov and Z. A. Safarova, *Dokl. Akad. Nauk SSSR*, **30**, 40 (1974); *Chem. Abstr.*, **82**, 111649f (1975).
102. A. A. Akhnazaryan, L. A. Khachatryan and M. T. Dangyan, *U.S.S.R. Patent*, 317,652; *Chem. Abstr.*, **76**, 112917e (1972).
103. S. Sarel, Y. Shalon and Y. Yanuka, *Tetrahedron Letters*, 957, 961 (1969).
104. S. Sarel, Y. Shalon and Y. Yanuka, *Chem. Commun.*, 80 (1970).
105. B. Berkov, L. Cuellar, R. Grezemkovsky, N. V. Avila and A. D. Cross, *Tetrahedron*, **24**, 2851 (1968).
106. S. Sarel, Y. Shalon and Y. Yanuka, *Chem. Commun.*, 81 (1970).
107. E. J. Corey and R. L. Danheiser, *Tetrahedron Letters*, 4477 (1973).
108. R. T. Arnold and K. L. Lindsay, *J. Amer. Chem. Soc.*, **75**, 1048 (1953).
109. K. W. Schulte and K. Baranowsky, *Pharm. Zentr.*, **98**, 403 (1959).
110. K. E. Schutze, J. Reisch and O. Heine, *Arch. Pharm.*, **294**, 234 (1961).
111. L. H. Klemm and K. W. Gopinath, *J. Heterocyclic Chem.*, **2**, 225 (1965).
112. L. H. Klemm, D. H. Lee, K. W. Gopinath and C. E. Klopfenstein, *J. Org. Chem.*, **31**, 2376 (1966).
113. L. H. Klemm and K. W. Gopinath, *Tetrahedron Letters*, 1243 (1963).
114. L. H. Klemm, K. W. Gopinath, D. H. Lee, F. W. Kelly, E. Trod and T. M. McGuire, *Tetrahedron*, **22**, 1797 (1966).
115. E. L. Ghisalferti, P. R. Jefferies and T. G. Payne, *Tetrahedron*, **30**, 3099 (1974).
116. F. Arndt, *Org. Syntheses*, Coll. Vol. III, 231 (1955).
117. N. R. Smith and R. H. Wiley, *Org. Syntheses*, Coll. Vol. IV, 549 (1963).
118. A. A. Avetisyan, Ts. A. Mangasaryan, G. S. Melikyan, M. T. Dangyan and S. G. Matsoyan, *Zh Org. Khim.*, **7**, 962 (1971); *Chem. Abstr.*, **75**, 63047q (1971).
119. G. D. Zuidema, E. van Tamelen and G. Van Zyl, *Org. Syntheses*, Coll. Vol. IV, 10 (1963).
120. D. T. C. Yang and S. W. Pelletier, *Org. Prep. Proced. (Int)*, **7**, 221 (1975).
121. D. Seebach and H. Meyer, *Angew. Chem. (Int. Ed. Engl.)*, **13**, 77 (1974).
122. S. Sethna and R. Phadke, *Org. Reactions*, **7**, 1 (1953).
123. H. Wiley and N. R. Smith, *Org. Syntheses*, Coll. Vol. IV, 201 (1963).
124. H. von Pechmann, *Ann.*, **264**, 272 (1891).
125. E. H. Woodruff, *Org. Syntheses*, Coll. Vol. III, 581 (1955).
126. A. Russell and J. R. Frye, *Org. Syntheses*, Coll. Vol. III, 281 (1955).
127. E. B. Vliet, *Org. Syntheses*, Coll. Vol. I, 360 (1932).
128. U. Kraatz and F. Korte, *Ber.*, **106**, 62 (1973).
129. R. Adams and B. R. Baker, *J. Amer. Chem. Soc.*, **62**, 2405 (1940).
130. G. Powell and T. H. Bembry, *J. Amer. Chem. Soc.*, **62**, 2568 (1940).
131. M. Guyot and C. Mentzer, *Bull. Soc. Chim. Fr.*, 2558 (1965).
132. V. B. Piskov, *Zh. Obsh. Khim.*, **30**, 1390 (1960); *J. Gen. Chem. U.S.S.R.*, **30**, 1421 (1960).
133. P. Kurath and W. Cole, *J. Org. Chem.*, **26**, 1939 (1961).
134. P. Kurath and W. Cole, *J. Org. Chem.*, **26**, 4592 (1961).
135. P. Kurath, W. Cole, J. Tadanier, M. Freifelder, G. R. Stone and E. V. Schuber, *J. Org. Chem.*, **28**, 2189 (1963).
136. M. A. Bielefeld and P. Kurath, *J. Org. Chem.*, **34**, 237 (1969).
137. V. Valcavi and I. L. Sianesi, *Gazz. Chim. Ital.*, **93**, 803 (1963).
138. V. Valcavi, *Gazz. Chim. Ital.*, **93**, 794, 929 (1963).
139. M. Debono, R. M. Molloy and L. E. Patterson, *J. Org. Chem.*, **34**, 3032 (1969).
140. G. R. Pettit, B. Green and G. L. Dunn, *J. Org. Chem.*, **35**, 1367 (1970).
141. H. H. Inhoffen, H. Krösche, K. Radscheit, H. Dettmer and W. Rudolph, *Ann.*, **714**, 8 (1968).
142. H. H. Inhoffen, W. Kreiser and M. Nazir, *Ann.*, **755**, 1, 12 (1972).
143. K. Miyano, J. Ohfune, S. Azuma and T. Matsumoto, *Tetrahedron Letters*, 1545 (1974).
144. E. C. Horning, M. G. Horning and D. A. Dimmig, *Org. Synthese*, Coll. Vol. III, 165 (1955).
145. G. R. Pettit, J. C. Knight and C. L. Heard, *J. Org. Chem.*, **35**, 1393 (1970).
146. A. A. Avetisyan, G. S. Melikyan, M. T. Dangyan and S. G. Matsoyan, *Zh. Org. Khim.*, **8**, 274 (1972); *Chem. Abstr.*, **76**, 139902h (1972).

147. A. A. Avetisyan, G. E. Tatevosyan, Ts. A. Mangasaryan, S. G. Matsoyan and M. T. Dangyan, *Zh. Org. Khim.*, **6**, 962 (1970); *Chem. Abstr.*, **74**, 87430q (1971).
148. R. R. Russell and C. A. Vander Werf, *J. Amer. Chem. Soc.*, **69**, 11 (1947).
149. G. Van Zyl and E. E. van Tamelen, *J. Amer. Chem. Soc.*, **72**, 1357 (1950).
150. S. J. Cristol and R. F. Helmreich, *J. Amer. Chem. Soc.*, **74**, 4083 (1952).
151. E. E. van Tamelen and S. R. Bach, *J. Amer. Chem. Soc.*, **77**, 4683 (1955).
152. C. H. DePuy, F. W. Breitbeil and K. L. Eilers, *J. Org. Chem.*, **29**, 2810 (1964).
153. P. M. G. Bavin, D. P. Hansell and R. G. W. Spickett, *J. Chem. Soc.*, 4535 (1964).
154. L. K. Dalton and B. C. Elmes, *Australian J. Chem.*, **25**, 625 (1972).
155. L. K. Dalton, B. C. Elmes and B. V. Kolczynski, *Australian J. Chem.*, **25**, 633 (1972).
156. A. Takeda, S. Tsubor and Y. Oota, *J. Org. Chem.*, **38**, 4148 (1973).
157. A. Franke and G. Groeger, *Monatsh. Chem.*, **43**, 55 (1922).
158. N. Bensel, H. Marshall and P. Weyerstahl, *Ber.*, **108**, 2697 (1975).
159. J. R. Johnson, *Org. Reactions*, **1**, 210 (1942).
160. H. O. House, *Modern Synthetic Reactions*, 2nd ed., W. A. Benjamin, Menlo Park, California, 1972, Chap. 10.
161. H. E. Carter, *Org. Reactions*, **3**, 198 (1946).
162. E. Baltazzi, *Quart. Rev. (Lond.)*, **9**, 150 (1955).
163. M. Lounasmaa, *Acta Chem. Scand.*, **22**, 70 (1968); **25**, 1849 (1971); **27**, 708 (1973).
164. M. Lounasmaa, *Acta Chem. Scand.*, **26**, 2703 (1972).
165. R. Filler, E. J. Piasek and H. A. Leipold, *Org. Syntheses*, Coll. Vol. V, 80 (1973).
166. R. Weiss, *Org. Syntheses*, Coll. Vol. II. 61 (1943).
167. C. D. Gutsche, E. F. Jason, R. S. Coffey and H. E. Johnson, *J. Amer. Chem. Soc.*, **80**, 5756 (1958).
168. V. I. Bendall and S. S. Dharamski, *J. Chem. Soc., Perkin I*, 2732 (1972).
169. M. Protiva, V. Hnevsova-Seidlova, V. Jirkovsky, L. Novak and Z. J. Vejdelek, *Ceskoslov. Farm.*, **10**, 501 (1962); *Chem. Abstr.*, **57**, 7196f (1962).
170. W. S. Johnson and G. H. Daub, *Org. Reactions*, **6**, 1 (1951).
171. A. A. Avetisyan, K. G. Akopyan and M. T. Dangyan, *Khim. Geterotsikl. Soedin.*, 1604 (1973); *Chem. Abstr.*, **80**, 82006v (1974).
172. G. Saucy and R. Borer, *Helv. Chim. Acta*, **54**, 2121, 2517 (1971).
173. M. Rosenberger, T. P. Fraher and G. Saucy, *Helv. Chim. Acta*, **54**, 2857 (1971).
174. M. Rosenberger, D. Andrews, F. DiMaria, A. J. Duggan and G. Saucy, *Helv. Chim. Acta*, **55**, 249 (1972).
175. P. M. Hardy, A. C. Nicholls and H. N. Rudon, *J. Chem. Soc. (D)*, 565 (1969).
176. R. Sandberg, *Arkiv Kemi*, **16**, 255 (1960).
177. M. Mousseron, M. Mousseron, I. Neyrelles and Y. Beziat, *Bull. Soc. Chim. Fr.*, 1483 (1963).
178. W. R. Vaughan, S. C. Berstein and M. E. Lorber, *J. Org. Chem.*, **30**, 1790 (1965).
179. J. C. Dubois, J. P. Guette and H. B. Kagan, *Bull. Soc. Chim. Fr.*, 3008 (1966).
180. C. Gandolfi, G. Doria, M. Amendola and E. Dradi, *Tetrahedron Letters*, 3923 (1970).
181. E. P. Kohler and G. L. Heritage, *Am. Chem. J.*, **43**, 475 (1911); *Chem Abstr.*, **4**, 2270 (1911).
182. E. P. Kohler and H. Gilman, *J. Amer. Chem. Soc.*, **41**, 683 (1919).
183. M. T. Bertrand, G. Courtois and L. Miginioc, *Compt. Rend., Ser. C*, **280**, 999 (1975); *Chem. Abstr.*, **83**, 96358k (1975).
184. E. Ohler, K. Reininger and U. Schmidt, *Angew. Chem. (Int. Ed. Engl.)*, **9**, 457 (1970).
185. A. Löffler, R. Pratt, J. Pucknat, G. Gelbard and A. S. Dreiding, *Chimia*, **23**, 413 (1969).
186. W. H. Klohs, F. Keller and R. E. Williams, *J. Org. Chem.*, **24**, 1829 (1959).
187. J. W. Cornforth and R. H. Cornforth, 'Natural substances formed biologically from mevalonic acid,' *Biochemical Society Symposium No. 29*, (Ed. T. W. Goodwin), Academic Press, New York, 1970, p. 1
188. W. F. Gray, G. L. Deets and T. Cohen, *J. Org. Chem.*, **33**, 4352 (1968).
189. H. G. Floss, M. Tcheng-Lin, C. Chang, B. Naidov, G. E. Blair, C. I. Abou-Chaar and J. M. Cassady, *J. Amer. Chem. Soc.*, **96**, 1898 (1974).
190. L. Pichat, B. Blagoev and J. C. Hardouin, *Bull. Soc. Chim. Fr.*, 4489 (1968).
191. M. W. Rathke and A. Lindert, *J. Org. Chem.*, **35**, 3966 (1970).

192. R. A. Ellison and P. K. Bhatnagar, *Synthesis*, 719 (1974).
193. A. K. Sorensen and N. A. Klitgaard, *Acta Chem. Scand.*, **24**, 343 (1970).
194. D. B. Denney and L. C. Smith, *J. Org. Chem.*, **27**, 3404 (1962).
195. P. L. Creger, *J. Org. Chem.*, **37**, 1907 (1972).
196. T. Fujita, S. Watanabe and K. Suga. *Australian J. Chem.*, **27**, 2205 (1974).
197. A. I. Meyers, *Heterocycles in Organic Synthesis*, Wiley–Interscience, New York, 1974, Chap. 10.
198. A. I. Meyers and E. D. Mihelick, *Angew. Chem. (Int. Ed. Engl.)*, **15**, 270 (1976).
199. A. I. Meyers, E. D. Mihelick and R. L. Nolen, *J. Org. Chem.*, **39**, 2783 (1974).
200. A. I. Meyers and E. D. Mihelick, *J. Org. Chem.*, **40**, 1186 (1975).
201. J. Thiele, R. Tischbein and E. Lossow, *Ann.*, **319**, 180 (1910).
202. W. F. von Oettingen, *J. Amer. Chem. Soc.*, **52**, 2024 (1930).
203. E. Erlenmeyer and E. Braun, *Ann.*, **333**, 254 (1904).
204. R. Walter and H. Zimmer, *J. Heterocyclic Chem.*, **1**, 205 (1964).
205. R. Walter, H. Zimmer and T. C. Purcell, *J. Org. Chem.*, **31**, 2854 (1966).
206. R. Walter, D. Theodoropoulas and T. C. Purcell, *J. Org. Chem.*, **32**, 1649 (1967).
207. H. Zimmer and J. Rothe, *J. Org. Chem.*, **24**, 28 (1959).
208. W. Reppe *et al.*, *Ann.*, **596**, 158 (1955).
209. P. G. Gassman and B. L. Fox, *J. Org. Chem.*, **31**, 982 (1966).
210. E. Piers, M. B. Geraghty and R. D. Smille, *Chem. Commun.*, 614 (1971).
211. G. H. Posner and G. L. Loomis, *Chem. Commun.*, 892 (1972).
212. J. L. Herrmann and R. H. Schlessinger, *Chem. Commun.*, 711 (1973).
213. A. E. Green, J. C. Muller and G. Ourisson, *Tetrahedron Letters*, 4147 (1971).
214. H. J. Reich and J. H. Renga, *Chem. Commun.*, 135 (1974).
215. T. Hirrami, I. Niki and T. Agawa, *J. Org. Chem.*, **39**, 3236 (1974).
216. R. L. Evans and H. E. Stavely, *U.S. Patent*, 3,248,392; *Chem. Abstr.*, **65**, 2322e (1966).
217. N. Minami and I. Kuwajima, *Tetrahedron Letters*, 1423 (1977).
218. M. Fujii and A. Sudo, *Japanese Patent*, 72–25,065; *Chem. Abstr.*, **77**, 100847q (1972).
219. C. Metzger, D. Borrmann and R. Wegler, *Ber.*, **100**, 1817 (1967).
220. D. Borrmann and R. Wegler, *Ber.*, **100**, 1575 (1967).
221. D. Borrmann and R. Wegler, *Ber.*, **99**, 1245 (1966).
222. D. Borrmann and R. Wegler, *Ber.*, **102**, 64 (1969).
223. A. S. Kende, *Tetrahedron Letters*, 2661 (1967).
224. J. Ciabattoni and H. W. Anderson, *Tetrahedron Letters*, 3377 (1967).
225. W. Ried and H. Mengler, *Ann.*, **678**, 113 (1964).
226. P. M. M. Van Haard, L. Thijs and B. Zwanenburg, *Tetrahedron Letters*, 803 (1975).
227. P. D. Bartlett and M. Stiles, *J. Amer. Chem. Soc.*, **77**, 2806 (1955).
228. P. S. Bailey, *Chem. Rev.*, **58**, 925 (1958).
229. P. S. Bailey and A. G. Lane, *J. Amer. Chem. Soc.*, **89**, 4473 (1967).
230. R. Wheland and P. D. Bartlett, *J. Amer. Chem. Soc.*, **92**, 6057 (1970).
231. G. D. Buckley and W. J. Levy, *J. Chem. Soc.*, 3016 (1951).
232. S. W. Chaikin and W. G. Brown, *J. Amer. Chem. Soc.*, **71**, 122 (1949).
233. W. G. Brown, *Org. Reactions*, **1**, 469 (1951).
234. N. G. Gaylord, *Reduction with Complex Metal Hydrides*, Interscience, New York, 1956, pp. 373–379.
235. R. P. Linstead and A. F. Milledge, *J. Chem. Soc.*, 478 (1936).
236. D. M. Bailey and R. E. Johnson, *J. Org. Chem.*, **35**, 3574 (1970).
237. J. J. Bloomfield and S. L. Lee, *J. Org. Chem.*, **32**, 3919 (1967).
238. F. Weygand, K. G. Kinkel and D. Tietjen, *Ber.*, **83**, 394 (1950).
239. V. Parrini, *Gazz. Chim. Ital.*, **87**, 1147 (1957); *Chem. Abstr.*, **52**, 10001d (1958).
240. R. Granger and H. Techer, *Compt. Rend.*, **250**, 142 (1960).
241. B. E. Cross, R. H. B. Galt and J. R. Hanson, *J. Chem. Soc.*, 5052 (1963).
242. W. R. Vaughan, C. T. Goetschel, M. H. Goodrow and C. L. Warren, *J. Amer. Chem. Soc.*, **85**, 2282 (1963).
243. D. G. Farmun and J. P. Snyder, *Tetrahedron Letters*, 3861 (1965).
244. B. E. Cross and J. C. Stewart, *Tetrahedron Letters*, 3589 (1968).
245. R. H. Schlessinger and I. S. Ponticello, *Chem. Commun.*, 1013 (1969).

246. S. S. G. Sircar, *J. Chem. Soc.*, 898 (1928).
247. S. Takano and K. Ogasawara, *Synthesis*, **42**, (1974).
248. B. E. Cross, R. H. B. Galt and J. R. Hanson, *J. Chem. Soc.*, 5052 (1963).
249. G. Snatzke and G. Zanati, *Ann.*, **684**, 62 (1965).
250. D. S. Noyce and D. B. Denney, *J. Amer. Chem. Soc.*, **72**, 5743 (1950).
251. C. S. Marvel and J. A. Fuller, *J. Amer. Chem. Soc.*, **74**, 1506 (1952).
252. C. Glaret, *Ann. Chim.* [12], 293 (1947).
253. H. Minato and I. Horibe, *Chem. Commun.*, 531 (1965).
254. J. A. Marshall, N. Cohen and F. R. Arenson, *J. Org. Chem.*, **30**, 762 (1965).
255. R. Pappo and C. J. Jung, *Tetrahedron Letters*, 365 (1962).
256. A. L. Nussbaum, F. E. Carlon, E. P. Oliveto, E. Townley, P. Kabasakalian and D. H. R. Barton, *Tetrahedron*, **18**, 373 (1962).
257. J. T. Edward and P. F. Morand, *Can. J. Chem.*, **38**, 1325 (1960).
258. C. C. Bolt, *Rec. Trav. Chim.*, **70**, 940 (1951).
259. T. Tsuda, K. Tanabe, I. Iwai and K. Funakoshi, *J. Amer. Chem. Soc.*, **79**, 5721 (1957).
260. K. J. Divakar, P. P. Sane and A. S. Rao, *Tetrahedron Letters*, 399 (1974).
261. K. Yamada, M. Kato, M. Iyoda and Y. Hirata, *Chem. Commun.*, 499 (1973).
262. E. Schwenk, D. Papa, H. Hankin and H. Ginsberg, *Org. Syntheses*, Coll. Vol. III, 742 (1955); D. Papa, E. Schwenk and H. Ginsberg, *J. Org. Chem.*, **16**, 253 (1951).
263. R. I. Longley, W. S. Emerson and T. C. Shafer, *J. Amer. Chem. Soc.*, **74**, 2012 (1952).
264. R. I. Longley and W. S. Emerson, *Org. Syntheses*, Coll. Vol. IV, 677 (1963).
265. L. P. Kyrides and F. B. Zienty, *J. Amer. Chem. Soc.*, **68**, 1385 (1946).
266. L. E. Schniepp and H. H. Geller, *J. Amer. Chem. Soc.*, **69**, 1545 (1947).
267. J. A. Berson and W. M. Jones, *J. Org. Chem.*, **21**, 1325 (1956).
268. E. S. Rothman, M. E. Wall and C. R. Eddy, *J. Amer. Chem. Soc.*, **76**, 527 (1954).
269. M. F. Murray, B. A. Johnson, R. L. Pederson and A. C. Ott, *J. Amer. Chem. Soc.*, **78**, 981 (1956).
270. P. E. Eaton, G. F. Cooper, R. C. Johnson and R. H. Mueller, *J. Org. Chem.*, **37**, 1947 (1972).
271. V. I. Sternberg and R. J. Perkins, *J. Org. Chem.*, **28**, 323 (1963).
272. M. Fetizon, M. Golfier and J.-M. Louis, *Tetrahedron*, **31**, 171 (1975).
273. S. Hauptmann and A. Blaskovits, *Z. Chem.*, **6**, 466 (1966); *Chem. Abstr.*, **66**, 55262e (1967).
274. P. Johnston, R. C. Sheppard, C. E. Stehr and S. Turner, *J. Chem. Soc. (C)*, 1847 (1966).
275. G. Defaye, M. Fetizon and M. C. Tromeur, *Compt. Rend. Ser. C.*, **265**, 1489 (1967).
276. R. C. Sheppard and S. Turner, *Chem. Commun.*, 682 (1968).
277. M. Fetizon, M. Golfier and J.-M. Louis, *Chem. Commun.*, 1118 (1969).
278. R. Hirschmann, N. G. Steinberg and R. Walker, *J. Amer. Chem. Soc.*, **84**, 1270 (1962).
279. D. H. R. Barton, J. M. Beaton, L. E. Geller and M. M. Pechet, *J. Amer. Chem. Soc.*, **83**, 4076 (1961).
280. C. H. Hassall, *Org. Reactions*, **9**, 73 (1957).
281. P. A. S. Smith in *Molecular Rearrangements*, Vol. 1. (Ed. P. de Mayo), Wiley–Interscience, New York, 1963, pp. 568–591.
282. K. Miyano, J. Ohfune, S. Azuma and T. Matsumoto, *Tetrahedron Letters*, 1545 (1974).
283. C. G. Overberger and H. Kaye, *J. Amer. Chem. Soc.*, **89**, 5640 (1967).
284. P. Soucy, T.-L. Ho and P. Deslongchamps, *Can. J. Chem.*, **50**, 2047 (1972).
285. G. Mehta and P. N. Pandey, *Synthesis*, 404 (1975).
286. V. R. Ghatak and B. Sanyal, *Chem. Commun.*, 876 (1974).
287. H. Levy and R. P. Jacobsen, *J. Biol. Chem.*, **171**, 171 (1947).
288. S. Mori and F. Mukawa, *Bull. Chem. Soc. Japan*, **27**, 479 (1954).
289. J. Meinwald, M. C. Seidel and B. C. Cadoff, *J. Amer. Chem. Soc.*, **80**, 6303 (1958).
290. R. R. Sauers, *J. Amer. Chem. Soc.*, **81**, 925 (1959).
291. R. R. Sauers and G. P. Ahearn, *J. Amer. Chem. Soc.*, **83**, 2759 (1961).
292. S. Hara, N. Matsumoto and M. Takeuchi, *Chem. Ind. (Lond.)*, 2086 (1962).
293. S. Hara, *Chem. Pharm. Bull. (Tokyo)*, **12**, 1531 (1964).
294. M. J. Bogdanowicz, T. Ambelang and B. M. Trost, *Tetrahedron Letters*, 923 (1973).
295. B. M. Trost and M. J. Bogdanowicz, *J. Amer. Chem. Soc.*, **95**, 5321 (1973).

296. J. A. Horton, M. A. Laura, S. M. Kalbag and R. C. Petterson, *J. Org. Chem.*, **34**, 3366 (1969).
297. D. C. Dittmer, R. A. Fouty and J. R. Potoski, *Chem. Ind. (Lond.)*, 152 (1964).
298. K. Kirschke and H. Oberender, *German Patent*, 2,122,598; *Chem. Abstr.*, **76**, 126409c (1972).
299. E. J. Corey, Z. Arnold and J. Hutton, *Tetrahedron Letters*, 307 (1970).
300. P. A. Grieco, *J. Org. Chem.*, **37**, 2363 (1972).
301. N. M. Weinshenker and R. Stephensen, *J. Org. Chem.*, **37**, 3741 (1972).
302. E. J. Corey, N. M. Weinshenker, T. K. Schaaf and W. Huber, *J. Amer. Chem. Soc.*, **91**, 5675 (1969).
303. E. J. Corey and T. Ravindranathan, *Tetrahedron Letters*, 4753 (1961).
304. Y. Tsuda, T. Tanno, A. Ukai and K. Isobe, *Tetrahedron Letters*, 2009 (1971).
305. G. Buchi and I. M. Goldman, *J. Amer. Chem. Soc.*, **79**, 4741 (1957).
306. R. Hanna and G. Ourisson, *Bull. Soc. Chim. Fr.*, **10**, 3742 (1967).
307. K. Orito, R. H. Manske and R. Rodrigo, *J. Amer. Chem. Soc.*, **96**, 1944 (1974).
308. R. Sandmeier and C. Tamm, *Helv. Chim. Acta*, **56**, 2239 (1973).
309. R. D. Clark and C. H. Heathcock, *Tetrahedron Letters*, 1713, 2027 (1974).
310. M. Oyama and M. Ohno, *Tetrahedron Letters*, 5201 (1966).
311. F. Dubois, *Ann.*, **256**, 134 (1890).
312. R. P. Linstead and H. N. Rydon, *J. Chem. Soc.*, 580 (1933).
313. R. H. Cornforth, J. W. Cornforth and G. Popjak, *Tetrahedron*, **18**, 1351 (1962).
314. L. M. Berkowitz and P. N. Rylander, *J. Amer. Chem. Soc.*, **80**, 6682 (1958).
315. G. F. Reynolds, G. H. Rasmusson, L. Birladeanu and G. E. Arth, *Tetrahedron Letters*, 5057 (1970).
316. K. Heusler and A. Wettslein, *Helv. Chim. Acta*, **35**, 284 (1952).
317. C. S. Foote, M. T. Wuesthoff, S. Wexler, I. G. Burstain, R. Denny, G. O. Schenck and K.-H. Schute-Elte, *Tetrahedron*, **23**, 2583 (1967).
318. C. S. Foote, M. T. Wuesthoff and I. G. Burstain, *Tetrahedron*, **23**, 2601 (1967).
319. S. H. Schroeter, R. Appel, R. Brammer and G. O. Schenck, *Ann.*, **692**, 42 (1966).
320. E. Koch and G. O. Schenck, *Ber.*, **99**, 1984 (1966).
321. J. P. van der Merve and C. F. Garbess, *J. South African Inst.*, **17**, 149 (1964); *Chem. Abstr.*, **62**, 9088e (1965).
322. J. B. Bush, Jr and H. Finkbeiner, *J. Amer. Chem. Soc.*, **90**, 5903 (1968).
323. E. I. Heiba, R. M. Dessau and W. J. Koehl, Jr, *J. Amer. Chem. Soc.*, **90**, 5905 (1968).
324. A. Mee, *German Patent*, 1,927,233; *Chem. Abstr.*, **72**, 78456j (1970).
325. E. I. Heiba, R. M. Dessau and W. J. Koehl, Jr, *J. Amer. Chem. Soc.*, **90**, 5905 (1968).
326. E. I. Heiba, R. M. Dessau and W. J. Koehl, Jr, *J. Amer. Chem. Soc.*, **90**, 2706 (1968).
327. J. Grimaldi, M. Malacria and M. Bertrand, *Tetrahedron Letters*, 275 (1974).
328. J. Falbe, M. Huppes and F. Karte, *Ber.*, **97**, 863 (1964).
329. J. Falbe and F. Karte, *Angew. Chem. (Int. Ed Engl.)*, **1**, 657 (1962).
330. E. R. H. Jones, T. Y. Shen. and M. C. Whiting, *J. Chem. Soc.*, 230 (1950).
331. J. Falbe, H.-J. Schulze-Steinen and F. Karte, *Ber.*, **98**, 886 (1965).
332. H. Koch and W. Haaf, *Angew. Chem.*, **70**, 311 (1958).
333. H. Koch and W. Haaf, *Angew. Chem.*, **72**, 628 (1960).
334. Y. Soma and H. Sano, *Japanese Patent*, 61,166 (1974); *Chem. Abstr.*, **81**, 120010f (1974).
335. M. Foa and L. Cassar, *Gazz. Chim. Ital.*, **103**, 805 (1973).
336. T. K. Das Gupta, D. Felix, U. M. Kempe and A. Eschenmoser, *Helv. Chim. Acta*, **55**, 2198 (1972).
337. M. Petrzilka, D. Felix and A. Eschenmoser, *Helv. Chim. Acta*, **56**, 2950 (1973).
338. K. Kondo and F. Mari, *Chem. Letters*, 741 (1974).
339. C. B. Chapleo, P. Hallett, B. Lythgoe and P. W. Wright, *Tetrahedron Letters*, 847 (1974).
340. R. R. Sauers and P. E. Sonnet, *Tetrahedron*, **20**, 1029 (1964).
341. E. W. Wornhoff and V. Dave, *Can. J. Chem.*, **44**, 621 (1966).
342. T. V. Mandelshtam, L. D. Kristol, L. A. Bogdanova and T. N. Ratnikova, *J. Org. Chem. U.S.S.R.*, **4**, 963 (1968).
343. J. Meinwald, A. Lewis and P. G. Gassman, *J. Amer. Chem. Soc.*, **84**, 977 (1962).

344. J. A. Marshall, F. N. Tuller and R. Ellison, *Synthetic Commun.*, **3**, 465 (1973).
345. J. Bus, H. Steinberg and Th. J. DeBoer, *Rec. Trav. Chem. Pays-Bas*, **91**, 657 (1972).
346. R. G. Carlson, J. H.-A. Huber and D. E. Henton, *Chem. Commun.*, 223 (1973).
347. R. K. Murray, Jr and D. L. Goff, *Chem. Commun.*, 881 (1973).
348. R. K. Murray, Jr, T. K. Morgan, Jr, J. A. S. Polley, C. A. Andruskiewicz, Jr and D. L. Goff, *J. Amer. Chem. Soc.*, **97**, 938 (1975).
349. R. C. Cookson, A. G. Edwards, J. Huder and M. Kingsland, *Chem. Commun.*, 98 (1965).
350. H. V. Hostettler, *Tetrahedron Letters*, 1941 (1965).
351. H. Nozaki, Z. Yamaguti and R. Noyari, *Tetrahedron Letters*, 37 (1965).
352. H. Nozaki, Z. Yamaguti, T. Okada, R. Noyari and M. Kawanisi, *Tetrahedron*, **23**, 3993 (1967).
353. D. H. R. Barton, A. L. J. Beckwith and A. Goosen, *J. Chem. Soc.*, 181 (1965).
354. B. Danieli, P. Manitto and G. Russo, *Chim. Ind. (Milan)*, **50**, 553 (1968); *Chem. Abstr.*, **69**, 43375t (1968).
355. W. Kirmse, H. Dietrich and H. W. Bücking, *Tetrahedron Letters*, 1833 (1967).
356. G. Lowe and J. Parker, *Chem. Commun.*, 577 (1971).
357. J. A. Kaufman and S. J. Weininger, *Chem. Commun.*, 593 (1969).
358. S. H. Schroeter, *Tetrahedron Letters*, 1591 (1969).
359. O. L. Chapman, P. W. Wojtkowski, W. Adam, O. Rodriquez and R. Rucktaschel, *J. Amer. Chem. Soc.*, **94**, 1365 (1972).
360. R. A. Dombro, *U.S. Patent*, 3,644,426; *Chem. Abstr.*, **76**, 139953a (1972).
361. H. Alper and W. G. Root, *Chem. Commun.*, 956 (1974).
362. Y. Mori and J. Tsuji, *Japanese Patent*, 73-68,544; *Chem. Abstr.*, **80**, 59724u (1974).
363. T. A. Pudova, F. K. Velichko, L. V. Vinogradova and R. Kh. Freidlina, *Izv. Akad. Nauk SSSR, Ser. Khim.*, 116 (1975); *Chem. Abstr.*, **82**, 111279k (1975).
364. A. Belanger and P. Brassard, *Chem. Commun.*, 863 (1972).
365. K. Jost and J. Rudinger, *Coll. Czech. Chem. Commun.*, **32**, 2485 (1967).
366. H. Sugano and M. Miyoshi, *Bull. Chem. Soc. Japan*, **46**, 669 (1973).
367. J. C. Sheehan and E. J. Corey, *Org. Reactions*, **9**, 388 (1957).
368. K. Dachs and E. Schwartz, *Angew. Chem.*, **1**, 430 (1962).
369. E. Testa, *Farmaco (Pavia) Ed. Sci.*, **17**, (1962); *Chem. Abstr.*, **57**, 9772c (1962).
370. R. Graf, G. Lohaus, K. Börner, E. Schmidt and H. Bestian, *Angew. Chem.*, **1**, 481 (1962).
371. E. Piovera, *Corriere Farm.*, **21**, 512 (1966); *Chem. Abstr.*, **66**, 84282s (1967).
372. L. L. Muller and J. Harmer, *1,2-Cycloaddition Reactions*, Wiley, New York, 1967, p. 173.
373. I. Lengyel and J. C. Sheehan, *Angew. Chem.*, **7**, 25 (1968).
374. A. L. J. Beckwith, *The Chemistry of Amides*, (Ed. J. Zabicky), Interscience, New York, 1970, Chap. 2, p. 73.
375. G. L'Abbé and A. Hassner, *Angew. Chem.*, **10**, 98 (1971).
376. F. Millich and K. V. Seshadri, *High Polymers*, (Ed. H. Mark, C. S. Marvel and H. W. Melville), Vol. 26 (Ed. K. C. Frisch), Wiley–Interscience, New York, 1972, Chap. 3, p. 179.
377. M. S. Manhas and A. K. Bose, *Chemistry of β-Lactams, Natural and Synthetic*, Part I, John Wiley and Sons, New York, 1971, Chap. 1.
378. E. G. E. Hawkins, *Angew. Chem.*, **12**, 783 (1973).
379. A. K. Mukerjee and R. C. Srivastava, *Synthesis*, 327 (1973).
380. W. M. Pearlman, *Org. Syntheses*, Coll. Vol. V, 670 (1973).
381. R. L. Frank, W. R. Schmitz and B. Zeidman, *Org. Syntheses*, Coll. Vol. III, 328 (1955).
382. M. Mityoshi, *Bull. Chem. Soc. Japan*, **46**, 212 (1973).
383. R. B. Moffett, *Org. Syntheses*, Coll. Vol. IV, 357 (1963).
384. P. Cefelin A. Frydrychova, J. Labsky, P. Schmidt and J. Sebenda, *Coll. Czech. Chem. Commun.*, **32**, 2787 (1967).
385. P. Cefelin, J. Labsky and J. Sebenda, *Coll. Czech. Chem. Commun.*, **33**, 1111 (1968).
386. N. Anghelide, C. Draghici and D. Raileanu, *Tetrahedron*, **30**, 623 (1974).
387. R. F. Borch, C. V. Grudzinskas, P. A. Peterson and L. D. Weber, *J. Org. Chem.*, **37**, 1141 (1972).
388. R. H. Earle, Jr, D. T. Hurst and M. Viney, *J. Chem. Soc. (C)*, 2093 (1969).
389. R. Breckpot, *Bull. Soc. Chim. Belges.*, **32**, 412 (1923).
390. E. Testa, L. Fontanella and V. Aresi, *Ann.*, **673**, 60 (1964).

391. E. Testa, A. Bonati, G. Pagani and E. Gatti, *Ann.*, **647**, 92 (1961).
392. E. Testa, F. Fava and L. Fontanella, *Ann.*, **614**, 167 (1958).
393. A. Bonati, G. F. Christiani and E. Testa, *Ann.*, **647**, 83 (1961).
394. E. Testa and L. Fontanella, *Ann.*, **625**, 95 (1959).
395. E. Testa, L. Fontanella, G. F. Christiani and F. Fava, *Ann.*, **614**, 158 (1958).
396. E. Testa and L. Fontanella, *Ann.*, **661**, 187 (1963).
397. L. Fontanella and E. Testa, *Ann.*, **616**, 148 (1958).
398. E. Testa, L. Fontanella and G. F. Cristiani, *Ann.*, **626**, 121 (1959).
399. E. Testa, L. Fontanella and L. Mariani, *Ann.*, **660**, 135 (1962).
400. E. Testa, L. Fontanella, L. Mariani and G. F. Cristiani, *Ann.*, **639**, 157 (1961).
401. E. Testa, L. Fontanella and V. Aresi, *Ann.*, **656**, 114 (1962).
402. G. Cignarella, G. F. Cristiani and E. Testa, *Ann.*, **661**, 181 (1963).
403. J. L. Luche and H. B. Kagan, *Bull. Soc. Chim. Fr.*, 3500 (1969).
404. F. F. Blicke and W. A. Gould, *J. Org. Chem.*, **23**, 1102 (1958).
405. P. G. Gassman and T. J. van Bergen, *J. Amer. Chem. Soc.*, **95**, 2718 (1973).
406. P. G. Gassman, T. J. van Bergen and G. Gruetzmacher, *J. Amer. Chem. Soc.*, **95**, 6508 (1973).
407. P. G. Gassman and T. J. van Bergen, *J. Amer. Chem. Soc.*, **96**, 5508 (1974).
408. H. C. Beyerman and W. M. van den Brink, *Proc. Chem. Soc.*, 226 (1963).
409. H. T. Openshaw and N. Whittaker, *J. Chem. Soc. (C)*, 89 (1969).
410. A. R. Battersby, J. F. Beck and E. McDonald, *J. Chem. Soc., Perkin I*, 160 (1974).
411. K. P. Klein and H. K. Reimschuessel, *J. Polym. Sci., A-1*, **10**, 1987 (1972).
412. K. P. Klein and H. K. Reimschuessel, *J. Polym. Sci., A-1*, **9**, 2717 (1971).
413. H. K. Reimschuessel, K. P. Klein and G. J. Schmitt, *Macromolecules*, **2**, 567 (1969).
414. G. H. Cocalas and W. H. Hartung, *J. Amer. Chem. Soc.*, **79**, 5203 (1957).
415. G. H. Cocalas, S. Avakian and G. J. Martin, *J. Org. Chem.*, **26**, 1313 (1961).
416. K. L. Kirk and L. A. Cohen, *J. Org. Chem.*, **34**, 395 (1969).
417. K. L. Kirk and L. A. Cohen, *J. Amer. Chem. Soc.*, **94**, 8142 (1972).
418. H. E. Baumgarten, R. L. Zey and U. Krolls, *J. Amer. Chem. Soc.*, **83**, 4469 (1961).
419. H. E. Baumgarten, *J. Amer. Chem. Soc.*, **84**, 4975 (1962).
420. H. E. Baumgarten, J. F. Fuerholzer, R. D. Clark and R. D. Thompson, *J. Amer. Chem. Soc.*, **85**, 3303 (1963).
421. J. C. Sheehan and I. Lengyel, *J. Amer. Chem. Soc.*, **86**, 746, 1356 (1964).
422. I. Lengyel and J. C. Sheehan, *Angew. Chem. (Int. Ed. Engl.)*, 7, 25 (1968).
423. E. R. Talaty, C. M. Utermoehlen and L. H. Stekoll, *Synthesis*, 543 (1971).
424. E. R. Talaty, J. P. Madden and L. H. Stekoll, *Angew. Chem. (Int. Ed. Engl.)*, **10**, 753 (1971).
425. M. S. Manhas and S. J. Jeng, *J. Org. Chem.*, **32**, 1246 (1967).
426. H. Wamhoff and F. Karte, *Ber.*, **100**, 2122 (1967).
427. E. Testa, B. J. R. Nicolaus, E. Bellasio and L. Mariani, *Ann.*, **673**, 71 (1964).
428. I. L. Knunyants and N. P. Gambaryan, *Izv. Akad. Nauk SSSR, Otdl. Khim. Nauk*, 1037 (1955).
429. I. L. Knunyants and N. P. Gambaryan, *Izv. Akad. Nauk SSSR, Otdl. Khim. Nauk*, 834 (1957).
430. I. L. Knunyants, E. E. Rytslin and N. P. Gambaryan, *Izv. Akad. Nauk SSSR, Otdl. Khim. Nauk*, 83 (1961).
431. A. K. Bose, B. Anjaneyulu, S. K. Bhattacharya and M. S. Manhas, *Tetrahedron*, **23**, 4769 (1967).
432. J. C. Sheehan and A. K. Bose, *J. Amer. Chem. Soc.*, **72**, 5158 (1950).
433. J. C. Sheehan and A. K. Bose, *J. Amer. Chem. Soc.*, **73**, 1761 (1951).
434. A. K. Bose, B. N. Ghosh-Mazumdar and B. G. Chatterjee, *J. Amer. Chem. Soc.*, **82**, 2382 (1960).
435. A. K. Bose and M. S. Manhas, *J. Org. Chem.*, **27**, 1244 (1962).
436. A. K. Bose, M. S. Manhas and B. N. Ghosh-Mazumdar, *J. Org. Chem.*, **27**, 1458 (1962).
437. B. G. Chatterjee and R. F. Abdulla, *Z. Naturforsch. (B)*, **26**, 395 (1971).
438. B. G. Chatterjee and N. L. Nyss, *Z. Naturforsch. (B)*, **26**, 395 (1971).
439. B. G. Chatterjee and V. V. Rao, *Tetrahedron*, **23**, 487 (1967).

440. B. G. Chatterjee, V. V. Rao, S. K.Roy and H. P. S. Chawla, *Tetrahedron*, **23**, 493 (1967).
441. B. G. Chatterjee, P. N. Moza and S. K. Roy, *J. Org. Chem.*, **28**, 1418 (1963).
442. B. G. Chatterjee, V. V. Rao and B. N. Ghosh-Mazumdar, *J. Org. Chem.*, **30**, 4101 (1965).
443. T. Okawara and K. Harada, *J. Org. Chem.*, **37**, 3286 (1972).
444. H. Bohme, S. Ebel and K. Hartke, *Ber.*, **98**, 1463 (1965).
445. A. K. Bose, M. S. Manhas and R. M. Ramer, *Tetrahedron*, **21**, 449 (1965).
446. M. W. Rutenberg and E. C. Horning, *Org. Syntheses*, Coll. Vol. IV, 620 (1963).
447. H. Plieninger and A. Muller, *Synthesis*, 586 (1970).
448. W. H. Puterbaugh and C. R. Hauser, *J. Org. Chem.*, **29**, 853 (1964).
449. R. L. Vaux, W. H. Puterbaugh and C. R. Hauser, *J. Org. Chem.*, **29**, 3514 (1964).
450. I. T. Barnish, C. -L. Mao, R. L. Gay and C. R. Hauser, *Chem. Commun.*, 564 (1968).
451. E. M. Levi, C. -L. Mao and C. R. Hauser, *Can. J. Chem.*, **47**, 3671 (1969).
452. C. -L. Mao, I. T. Barnish and C. R. Hauser, *J. Heterocyclic Chem.*, **6**, 83 (1969).
453. C. -L. Mao and C. R. Hauser, *J. Org. Chem.*, **35**, 3704 (1970).
454. P. A. Petyunin and G. P. Petyunin, *U.S.S.R. Patent*, 371221 (1973); *Chem. Abstr.*, **79**, 31833j (1973).
455. J. F. Bagli and H. Immer, *J. Org. Chem.*, **35**, 3499 (1970).
456. E. W. Colvin, J. Martin, W. Parker, R. A. Raphael, B. Shroot and M. Doyle, *J. Chem. Soc., Perkin I*, 860 (1972).
457. O. L. Chapman and W. R. Adams, *J. Amer. Chem. Soc.*, **89**, 4243 (1967).
458. O. L. Chapman and W. R. Adams, *J. Amer. Chem. Soc.*, **90**, 2333 (1968).
459. P. G. Cleveland and O. L. Chapman, *Chem. Commun.*, 1064 (1967).
460. M. Ogata and H. Matsumoto, *Chem. Phar. Bull.*, **20**, 2264 (1972).
461. Y. Ogata, K. Takagi and I. Ishino, *J. Org. Chem.*, **36**, 3975 (1971).
462. B. S. Thyagarajan, N. Kharasch, H. B. Lewis and W. Wolf, *Chem. Commun.*, 614 (1967).
436. D. H. Hey, G. H. Jones and M. J. Perkins, *J. Chem. Soc. (C)*, 120 (1971).
464. A. Mondon and K. Krohn, *Ber.*, **105**, 3726 (1972).
465. G. R. Lenz, *J. Org. Chem.*, **39**, 2839 (1974).
466. G. R. Lenz, *J. Org. Chem.*, **39**, 2846 (1974).
467. T. Hasegawa and H. Aoyama, *Chem. Commun.*, 743 (1974).
468. N. C. Yang, A. Shani and G. R. Lenz, *J. Amer. Chem. Soc.*, **88**, 5369 (1966).
469. Y. Kanaoka and K. Itoh, *Synthesis*, 36 (1972).
470. E. Winterfeldt and H. J. Altmann, *Angew. Chem. (Int. Ed. Engl.)*, **7**, 466 (1968).
471. I. Ninomiya and T. Naito, *Chem. Commun.*, 137 (1973).
472. S. M. Kupchan, J. L. Moniot, R. M. Kanojia and J. B. O'Brien, *J. Org. Chem.*, **36**, 2413 (1971).
473. I. Ninomiya, T. Naito and S. Higuchi, *Chem. Commun.*, 1662 (1970).
474. I. Ninomiya, T. Naito, T. Kiguchi and T. Mori, *J. Chem. Soc., Perkin I*, 1696 (1973).
475. I. Ninomiya, T. Naito and T. Kiguchi, *Tetrahedron Letters*, 4451 (1970).
476. I. Ninomiya, T. Naito and T. Mori, *Tetrahedron Letters*, 2259 (1969).
477. I Ninomiya, T. Naito and T Mori, *Tetrahedron Letters*, 3643 (1969).
478. O. Yonemitsu, T. Tokuyama, M. Chaykovsky and B. Witkop, *J. Amer. Chem. Soc.*, **90**, 776 (1968).
479. O. Yonemitsu, B. Witkop and I. L. Karle, *J. Amer. Chem. Soc.*, **89**, 1039 (1967).
480. O. Yonemitsu, Y. Okuno, Y. Kanaoka and B. Witkop, *J. Amer. Chem. Soc.*, **92**, 5686 (1970).
481. O. Yonemitsu, H. Nakai, Y. Kanaoka, I. L. Karle and B. Witkop, *J. Amer. Chem. Soc.*, **91**, 4591 (1969).
482. T. Kobayashi, T. F. Spande, H. Aoyagi and B. Witkop, *J. Med. Chem.*, **12**, 636 (1969).
483. O. Yonemitsu, P. Cerutti and B. Witkop, *J. Amer. Chem. Soc.*, **88**, 3941 (1966).
484. B. Akermark, N.-G. Johansson and B. Sjoberg, *Tetrahedron Letters*, 371 (1969).
485. K. R. Henery-Logan and C. G. Chen, *Tetrahedron Letters*, 1103 (1973).
486. R. R. Rando, *J. Amer. Chem. Soc.*, **92**, 6706 (1970).
487. R. H. Earle, Jr, D. T. Hurst and M. Viney, *J. Chem. Soc. (C)*, 2093 (1969).
488. G. Lowe and J. Parker, *Chem. Commun.*, 577 (1971).
489. E. J. Corey and A. M. Felix, *J. Amer. Chem. Soc.*, **87**, 2518 (1965).
490. D. M. Brunwin, G. Lowe and J. Parker, *Chem. Commun.*, 865 (1971).

491. D. M. Brunwin, G. Lowe and J. Parker, *J. Chem. Soc.* (C), 3756 (1971).
492. G. Lowe and M. V. J. Ramsay, *J. Chem. Soc., Perkin I*, 479 (1973).
493. M. Perelman and S. A. Mizsak, *J. Amer. Chem. Soc.*, 84, 4988 (1962).
494. R. Graf, *Ber.*, 89, 1071 (1956).
495. R. Graf, *Org. Syntheses*, Coll. Vol. V, 226 (1973).
496. R. Graf, *Angew. Chem. (Int. Ed. Engl.)*, 7, 172 (1968).
497. H. Ulrich, *Chem. Rev.*, 65, 369 (1965).
498. R. Graf, *Ann.*, 661, 111 (1963).
499. E. J. Moriconi and W. C. Crawford, *J. Org. Chem.*, 33, 370 (1968).
500. T. Durst and M. J. O'Sullivan, *J. Org. Chem.*, 35 2043 (1970).
501. Farbwerke Hoechst A.-G., French Patent, 2016990 (1970); *Chem. Abstr.*, 75, 63164a (1971).
502. R. Graf, *Org. Syntheses*, Coll. Vol. V, 673 (1973).
503. H. Bestian, H. Biener, K. Clauss and H. Heyn, *Ann.*, 718, 94 (1968).
504. E. J. Moriconi and W. C. Meyer, *Tetrahedron Letters*, 3823 (1968).
505. E. J. Moriconi and P. H. Mazzochi, *J. Org. Chem.*, 31, 1372 (1966).
506. E. J. Moriconi and J. F. Kelly, *J. Amer. Chem. Soc.*, 88, 3657 (1966).
507. M. Perelman and S. A. Mizsak, *J. Amer. Chem. Soc.*, 84, 4988 (1962).
508. G. Opitz and J. Koch, *Angew. Chem.*, 75, 167 (1963).
509. W. P. Giering and M. Rosenblum, *J. Amer. Chem. Soc.*, 93, 5299 (1971).
510. W. P. Giering, S. Raghu, M. Rosenblum, A. Cutler, D. Ehntholt and R. W. Fish, *J. Amer. Chem. Soc.*, 94, 8251 (1972).
511. Naser-ud-din, J. Riegl and L. Skattebol, *Chem. Commun.*, 271 (1973).
512. G. Merault, P. Bourgesis and N. Duffaut, *Bull. Soc. Chim. Fr.*, 1949 (1974).
513. I. Yamamoto, S. Yanagi, A. Mamba and H. Gotoh, *J. Org. Chem.*, 39, 3924 (1974).
514. I. Yamamoto, Y. Tabo, H. Gotoh, T. Minami, Y. Ohshiro and T. Agawa, *Tetrahedron Letters*, 2295 (1971).
515. H. Staudinger, *Die Ketene*, F. Enke Verlag, Stuttgart, 1912.
516. W. T. Brady, E. D. Dorsey and F. H. Parry, III, *J. Org. Chem.*, 34 2846 (1969).
517. A. Gomes and M. M. Joullie, *Chem. Commun.*, 935 (1967).
518. J. -L. Luche and H. B. Kagan, *Bull. Soc. Chim. Fr.*, 2450 (1968).
519. J. C. Sheehan and J. J. Ryan, *J. Amer. Chem. Soc.*, 73, 1204 (1951).
520. A. K. Bose, G. Spiegelman and M. S. Manhas, *Tetrahedron Letters*, 3167 (1971).
521. H. Staudinger, *Ber.*, 50, 1035 (1917).
522. J. Berson and W. M. Jones, *J. Amer. Chem. Soc.*, 78, 1625 (1956).
523. R. Pfleger and A. Jäger, 90, 2460 (1957).
524. R. Graf, *Ann.*, 661, 111 (1963).
525. W. Kirmse and L. Horner, *Ber.*, 89, 2759 (1956).
526. W. T. Brady and E. F. Hoff, Jr, *J. Amer. Chem. Soc.*, 90, 6256 (1968).
527. H. Staudinger and H. W. Klever, *Ber.*, 40, 1149 (1907).
528. A. D. Holley and R. W. Holley, *J. Amer. Chem. Soc.*, 73, 3172 (1951).
529. H. Staudinger, H. W. Klever and P. Kober, *Ann.*, 374, 1 (1910).
530. R. N. Pratt, G. A. Taylor and S. A. Proctor, *J. Chem. Soc. (C)*, 1569 (1967).
531. S. A. Ballard, D. S. Melstrom and C. W. Smith, *The Chemistry of Penicillin* (Ed. H. T. Clarke, J. R. Johnson and R. Robinson), Princeton University Press, Princeton, New Jersey, 1949, pp. 977, 984, 991, 992.
532. J. Decazes, J. L. Luche and H. B. Kagan, *Tetrahedron Letters*, 3661, 3665 (1970).
533. H. Staudinger, *Ber.*, 40, 1145 (1907).
534. H. Staudinger, *Ann.*, 356, 51 (1907).
535. H. Staudinger and S. Jelagin, *Ber.*, 44 365 (1911).
536. H. Staudinger, *Ber.*, 40, 1147 (1907).
537. H. B. Kagan and J. L. Luche, *Tetrahedron Letters*, 3093 (1968).
538. R. Huisgen, B. A. Davis and M. Morikawa, *Angew. Chem. (Int. Ed. Engl.)*, 7, 826 (1968).
539. M. D. Bachi and M. Rothfield, *J. Chem. Soc., Perkin I*, 2326 (1972).
540. M. D. Bachi and O. Goldberg, *J. Chem. Soc., Perkin I*, 2332 (1972).
541. E. Fahr, K. Döppert, K. Königsdorfer and F. Scheckenbach, *Tetrahedron*, 24, 1011 (1968).

542. A. K. Bose and I. Kugajevsky, *Tetrahedron*, **23**, 957 (1967).
543. R. D. Kimbrough, Jr, *J. Org. Chem.*, **29**, 1242 (1964).
544. E. Fahr, K. H. Keil, F. Scheckenbach and A. Jung, *Angew. Chem. (Int. Ed. Engl.)*, **3**, 646 (1964).
545. A. K. Bose, Y. H. Chiang and M. S. Manhas, *Tetrahedron Letters*, 4091 (1972).
546. A. K. Bose, C. S. Narayanan and M. S. Manhas, *Chem. Commun.*, 975 (1970).
547. M. S. Manhas, J. S. Chib, Y. H. Chiang and A. K. Bose, *Tetrahedron*, **25**, 4421 (1969).
548. A. K. Bose, B. Anjaneyulu, S. K. Bhattacharya and M. S. Manhas, *Tetrahedron*, **23**, 4769 (1967).
549. S. M. Deshpande and A. K. Mukerjee, *Indian J. Chem.*, **4**, 79 (1966).
550. E. Ziegler, Th. Wimmer and H. Mittelbach, *Monatsh. Chem.*, **99**, 2128 (1968).
551. D. A. Nelson, *Tetrahedron Letters*, 2543 (1971).
552. D. A. Nelson, *J. Org. Chem.*, **37**, 1447 (1972).
553. L. Paul, A Draeger and G Hilgetag, *Ber.*, **99**, 1957 (1966).
554. G. Hilgetag, L. Paul and A. Draeger, *Ber.*, **96**, 1697 (1963).
555. L. Paul and K. Zieloff, *Ber.*, **99**, 1431 (1966).
556. M. D. Bachi and O. Goldberg, *Chem. Commun.*, 319 (1972).
557. A. K. Bose, S. D. Sharma, J. C. Kapur and M. S. Manhas, *Synthesis*, 216 (1973).
558. J. N. Wells, and R. E. Lee, *J. Org. Chem.*, **34**, 1477 (1969).
559. R. Lattrell, *Angew. Chem. (Int. Ed. Engl.)*, **12**, 925 (1973).
560. A. K. Bose, B. Dayal, H. P. S. Chawla and M. S. Manhas, *Tetrahedron Letters*, 2823 (1972).
561. J. L. Fahey, B. C. Lange, J. M. Van der Veen, G. R. Young and A. K. Bose, *J. Chem. Soc., Perkin I*, 1117 (1977).
562. A. K. Bose and J. L. Fahey, *J. Org. Chem.*, **39**, 115 (1974).
563. A. K. Bose, M. Tsai, S. D. Sharma and M. S. Manhas, *Tetrahedron Letters*, 3851 (1973).
564. A. K. Bose, M. Tsai, J. C. Kapur and M. S. Manhas, *Tetrahedron*, **29**, 2355 (1973).
565. A. K. Bose, J. L. Fahey and M. S. Manhas, *J. Heterocyclic Chem.*, **10**, 791 (1973).
566. A. K. Bose, M. S. Manhas, J. S. Chib, H. P. S. Chawla and B. Dayal, *J. Org. Chem.*, **39**, 2877 (1974).
567. R. W. Ratcliffe and B. G. Christensen, *Tetrahedron Letters*, 4649 (1973).
568. A. K. Bose, G. Spiegelman and M. S. Manhas, *J. Chem. Soc. (C)*, 2468 (1971).
569. J. A. Erickson, *Ph.D. Thesis*, M.I.T., 1953, as reported in Reference 568.
570. A. K. Bose, G. Spiegelman and M. S. Manhas, *J. Chem. Soc. (C)*, 188 (1971).
571. A. K. Bose, V. Sudarsanam. B. Anjaneyulu and M. S. Manhas, *Tetrahedron*, **25**, 1191 (1969).
572. A. K. Bose, G. Spiegelman and M. S. Manhas, *J. Amer. Chem. Soc.*, **90**, 4506 (1968).
573. A. K. Bose, G. Spiegelman and M. S. Manhas, *Chem. Commun.*, 321 (1968).
574. A. K. Bose and B. Anjaneyulu, *Chem. Int. (Lond.)*, 903 (1966).
575. A. K. Bose, H. P. S. Chawla, B. Dayal and M. S. Manhas, *Tetrahedron Letters*, 2503 (1973).
576. A. K. Bose, J. C. Kapur, S. D. Sharma and M. S. Manhas, *Tetrahedron Letters*, 2319 (1973).
577. J. C. Sheehan and G. D. Laubach, *J. Amer. Chem. Soc.*, **73**, 4376 (1951).
578. J. C. Sheehan, E. L. Buhle, E. J. Corey, G. D. Laubach and J. J. Ryan, *J. Amer. Chem. Soc.*, **72**, 3828 (1950).
579. J. C. Sheehan and G. D. Laubach, *J. Amer. Chem. Soc.*, **73**, 4752 (1951).
580. J. C. Sheehan and E. J. Corey, *J. Amer. Chem. Soc.*, **73**, 4756 (1951).
581. J. C. Sheehan and J. J. Ryan, *J. Amer. Chem. Soc.*, **73**, 4367 (1951).
582. J. C. Sheehan, H. W. Hill, Jr and E. L. Buhle, *J. Amer. Chem. Soc.*, **73**, 4373 (1951).
583. S. M. Deshpande and A. K. Mukerjee, *J. Chem. Soc. (C)*, 1241 (1966).
584. L. Paul, P. Polczynski and G. Hilgetag, *Ber.*, **100**, 2761 (1967).
585. J. C. Sheehan and M. Dadic, *J. Heterocyclic Chem.*, **5**, 779 (1968).
586. E. Ziegler and Th. Wimmer, *Ber.*, **99**, 130 (1966).
587. F. Duran and L. Ghosez, *Tetrahedron Letters*, 245 (1970).
588. A. K. Bose, B. Lal, B. Dayal and M. S. Manhas, *Tetrahedron Letters*, 2633 (1974).
589. R. Hull, *J. Chem. Soc. (C)*, 1154 (1967).

590. R. W. Ratcliffe and B. G. Christensen, *Tetrahedron Letters*, 4653 (1973).
591. R. L. Bentley and H. Suschitzky, *J. Chem. Soc., Perkin I*, 1725 (1976).
592. E. Ziegler and G. Kleineberg, *Monatsh. Chem.*, **96**, 1296 (1965).
593. A. K. Bose, J. C. Kapur, B. Dayal and M. S. Manhas, *Tetrahedron Letters*, **39**, 3797 (1973).
594. A. K. Bose, J. C. Kapur, B. Dayal and M. S. Manhas, *J. Org. Chem.*, **39**, 312 (1974).
595. G. O. Schenck and N. Engelhord, *Angew. Chem.*, **68**, 71 (1956).
596. L. Horner, E. Spietschka and A. Gross, *Ann.*, **573**, 17 (1951).
597. A. H. Cook and D. G. Jones, *J. Chem. Soc.*, 184 (1941).
598. C. K. Ingold and S. D. Weaver, *J. Chem. Soc.*, 378 (1925).
599. L. Horner and E. Spietschka, *Ber.*, **89**, 2765 (1956).
600. M. L. Scheinbaum. *J. Org. Chem.*, **29**, 2200 (1964).
601. F. F. Blicke and W. A. Gould, *J. Org. Chem.*, **23**, 1102 (1958).
602. H. B. Kagan, J.-J. Basselier and J.-L. Luche, *Tetrahedron Letters*, 941 (1964).
603. J.-L. Luche and H. B. Kagan, *Bull. Soc. Chim. Fr.*, 3500 (1969).
604. H. Gilman and M. Speeter, *J. Amer. Chem. Soc.*, **65**, 2255 (1943).
605. J.-L. Luche and H. B. Kagan, *Bull. Soc. Chim. Fr.*, 1680 (1969).
606. J.-L. Luche, H. B. Kagan, R. Parthasarathy, G. Tsoucaris, C. de Rango and C. Zeliver, *Tetrahedron*, **24**, 1275 (1968).
607. E. Ziegler and G. Kleineberg, *Monatash. Chem.*, **96**, 1360 (1965).
608. V. Gomez Aranda, J. Barluenga and V. Gotor, *Tetrahedron Letters*, 977 (1974).
609. M. De Poortere, J. Marchand-Bryaert and L. Ghosez, *Angew. Chem. (Int. Ed. Engl.)*, **13**, 267 (1974).
610. H. Staudinger and M. Miescher, *Helv. Chim. Acta*, **2**, 564 (1919).
611. T. W. J. Taylor, J. S. Owen and D. Whittaker, *J. Chem. Soc.*, 206 (1938).
612. C. H. Hassall and A. E. Lippman, *J. Chem. Soc.*, 1059 (1953).
613. M. Kinugasa and S. Hashimoto, *Chem. Commun.*, 466 (1972).
614. J. C. Sheehan and A. K. Bose, *J. Amer. Chem. Soc.*, **72**, 5158 (1950).
615. G. Lowe and D. D. Ridley, *Chem. Commun.*, 328 (1973).
616. G. Lowe and D. D. Ridley, *J. Chem. Soc., Perkin I*, 2024 (1973).
617. G. Stork and R. P. Szajewski, *J. Amer. Chem. Soc.*, **96**, 5787 (1974).
618. C. E. Hatch and P. Y. Johnson, *Tetrahedron Letters*, 2719 (1974).
619. M. F. Chasle and A. Foucaud, *Compt. Rend. (C)*, **270**, 1045 (1970).
620. L. G. Donaruma and W. Z. Heldt, *Org. Reactions*, **11**, 1 (1960).
621. P. A. S. Smith in *Molecular Rearrangements*, (Ed P. de Mayo), Part 1, Interscience, New York, 1963, pp. 483–507.
622. F. Uhlig and H. R. Snyder, *Adv. Org. Chem.*, **1**, 35–81 (1960).
623. Tayo Rayon Co., Ltd., *British Patent*, 1188217; *Chem. Abstr.*, **76**, 59482f (1972).
624. Y. Tamura, H. Fujiwara, K. Sumoto, M. Ikeda and Y. Kita, *Synthesis*, 215 (1973).
625. Inventa A.-G., *British Patent*, 1105805; *Chem. Abstr.*, **68**, 95349m (1968).
626. T. Sonoda, M. Kato and S. Wakamatsu, Japanese Patent, 74135985; *Chem. Abstr.*, **82**, 156964w (1975).
627. S. Yura and K. Horiguchi, Japanese Patent, 7503317; *Chem. Abstr.*, **82**, 156979e (1975).
628. N. V. Stamicarbon, *Dutch Patent*, 6607466; *Chem. Abstr.*, **69**, 18634u (1968).
629. C. G. Overberger and H. Jabloner, *J. Amer. Chem. Soc.*, **85**, 3431 (1963).
630. E. Wenkert and B. F. Barnett, *J. Amer. Chem. Soc.*, **82**, 4671 (1966).
631. F. J. Donat and A. L. Nelson, *J. Org. Chem.*, **22**, 1107 (1957).
632. O. D. Strizhakov, E. N. Zil'berman and S. V. Svetozarskii, *Zh. Obsh. Khim.*, **35**, 628 (1965); *Chem. Abstr.*, **63**, 5535a (1965).
633. E. C. Horning, V. L. Stromberg and H. A. Lloyd, *J. Amer. Chem. Soc.*, **74**, 5153 (1952).
634. R. S. Montgomery and G. Dougherty, *J. Org. Chem.*, **17**, 823 (1952).
635. Y. Tamura, Y. Kita, Y. Matsutaka and M. Terashima, *Chem. Ind. (Lond.)*, 1350 (1970).
636. R. H. Mazur, *J. Org. Chem.*, **26**, 1289 (1961).
637. Y. Tamura, Y. Kita and M. Terashima, *Chem. Pharm. Bull.*, 529 (1971).
638. T. H. Koch, M. A. Geigel and C.-C. Tsai, *J. Org. Chem.*, **38**, 1090 (1973).
639. A. Zabza, H. Kuczynski, Z. Chabudzinski and D. Sedzik-Hibner, *Bull. Acad. Pol. Sci., Ser. Sci. Chim.*, **20**, 841 (1972); *Chem. Abstr.*, **78**, 124728y (1973).

640. T. Duong, R. H. Prager, J. M. Tippett, A. D. Ward and D. I. B. Kerr, *Australian J. Chem.*, **29**, 2667 (1976).
641. P. Rosenmund, D. Sauer and W. Trommer, *Ber.*, **103**, 496 (1970).
642. R. H. Mazur, *J. Amer. Chem. Soc.*, **81**, 1454 (1959).
643. J. Takeuchi and F. Iwata, *Japanese Patent*, 7456972; *Chem. Abstr.*, **83**, 59749t (1975).
644. H. Wolff, *Org. Reactions*, **3**, 307 (1946).
645. P. A. S. Smith in *Molecular Rearrangements*, (Ed. P. de Mayo), Part 1, Interscience, New York, 1963, pp. 507–527.
646. H. Shechter and J. C. Kirk, *Amer. Chem. Soc.*, **73**, 3087 (1951).
647. R. T. Conley, *J. Org. Chem.*, **23**, 1330 (1958).
648. T. Duong. R. H. Prager, A. D. Ward and D. I. B. Kerr, *Australian J. Chem.*, **29**, 2651 (1976).
649. Y. Tamura and Y. Kita, *Chem. Pharm. Bull.*, **19**, 1735 (1971).
650. Y. Tamura, Y. Yoshimura and Y. Kita, *Chem. Pharm. Bull.*, **19**, 1068 (1971).
651. K. Folkers, D. Misiti and H. W. Moore, *Tetrahedron Letters*, 1071 (1965).
652. D. Misiti, H. W. Moore and K. Folkers, *Tetrahedron*, **22**, 1201 (1966).
653. R. W. Richards and R. M. Smith, *Tetrahedron Letters*, 2361 (1966).
654. C. R. Bedford, G. Jones and B. R. Webster, *Tetrahedron Letters*, 2367 (1966).
655. C. S. Barnes, D. H. R. Barton, J. S. Fawcett and B. R. Thomas, *J. Chem. Soc.*, 2339 (1952).
656. K. A. Mueller and W. Kirchhof, *German Patent*, 1242621; *Chem. Abstr.*, **67**, 100022k (1967).
657. H. H. Wasserman, R. E. Cochoy and M. S. Baird, *J. Amer. Chem. Soc.*, **91**, 2375 (1969).
658. H. H. Wasserman, H. W. Adickes and O. Espejo de Ochoa, *J. Amer. Chem. Soc.*, **93**, 5586 (1971).
659. H. H. Wasserman E. L. Glazer and M. J. Hearn, *Tetrahedron Letters*, 4855 (1973).
660. H. H. Wasserman and E. L. Glazer, *J. Org. Chem.*, **40**, 1505 (1975).
661. A. F. M. Iqbal, *Tetrahedron Letters*, 3381 (1971).
662. C. W. Capp, K. W. Denbigh, P. J. Durston and B. W. Harris, *U.S. Patent*, 3583982; *Chem. Abstr.*, **75**, 89054q (1971).
663. H. L. Ammon, P. H. Mazzo chi, W. J. Kopecky, Jr, H. J. Tamburin and P. H. Watts, Jr, *J. Amer. Chem. Soc.*, **95**, 1968 (1973).
664. P. H. Mazzocchi, T. Halchak and H. J. Tamburin, *J. Org. Chem.*, **41**, 2808 (1976).
665. E. Oliveros-Desherces, M. Riviere, J. Parello and A. Lattes, *Compt. Rend (C)*, **275**, 581 (1972).
666. E. Oliveros-Desherces, M. Riviere, J. Parello and A. Lattes, *Tetrahedron Letters*, 851 (1975).
667. H. Krimm. *Ber.*, **91**, 1057 (1958).
668. E. Schmitz, H. U. Heyne and S. Schramm, *German Patent*, 2055165; *Chem. Abstr.*, **75**, 152320r (1971).
669. S. T. Reid, J. N. Tucker and E. J. Wilcox, *J. Chem. Soc., Perkin I*, 1359 (1974).
670. G. Ribaldone and A. Nenz, *Chem. Ind. (Milan)*, **49**, 701 (1967); *Chem. Abstr.*, **67**, 116621r (1967).
671. G. Ribaldone, F. Smai and G. Borsotti, U.S. Patent, 3328394; *Chem. Abstr.*, **68**, 59137g (1968).
672. J. Deyrup and S. C. Clough, *J. Amer. Chem. Soc.*, **91**, 4590 (1969).
673. D. St. C. Black, F. W. Eastwood, R. Okraglik, A. J. Poynton, A. M. Wade and C. H. Welker, *Australian J. Chem.*, **25**, 1483 (1972).
674. P. G. Gassman and B. L. Fox, *J. Org. Chem.*, **31**, 982 (1966).
675. T. Durst and M. J. LeBelle, *Can. J. Chem.*, **50**, 3196 (1972).
676. T. Durst, R. Van Den Elzen and R. Legault, *Can. J. Chem.*, **52**, 3206 (1974).
677. B. M. Trost and R. A. Kunz, *J. Org. Chem.*, **39**, 2476 (1974).
678. B. M. Trost and R. A. Kunz, *J; Amer. Chem. Soc.*, **97**, 7152 (1975).
679. See E. W. H. Böhme, H. E. Applegate, J. B. Ewing, P. T. Funke, M. S. Puar and J. E. Dolfini, *J. Org. Chem.*, **38**, 230 (1973) and references cited therein.
680. E. H. W. Böhme, H. E. Applegate, B. Toplitz, J. E. Dolfini and J. Z. Gougoutas, *J. Amer. Chem. Soc.*, **93**, 4324 (1971).

681. R. A. Firestone, N. Schelechow, D. B. R. Johnston and B. G. Christensen, *Tetrahedron Letters*, 375 (1972).
682. G. V. Kaiser, C. W. Ashbrook and J. E. Baldwin, *J. Amer. Chem. Soc.*, **93**, 2342 (1971).
683. H. H. Wasserman and B. H. Lipshutz, *Tetrahedron Letters*, 4613 (1976).
684. H. Shinozaki and M. Tada, *Chem. Ind. (Lond.)* 177 (1975).
685. C. S. Marvel and W. W. Mayer, Jr, *J. Org. Chem.*, **22**, 1065 (1957).
686. W. Ziegenbein and W. Franke, *Ber.*, **90**, 2291 (1957).
687. R. E. Benson and T. L. Cairns, *J. Amer. Chem. Soc.*, **70**, 2115 (1948).
688. D. St. C. Black, F. W. Eastwood, R. Okraglik, A. J. Poynton, A. M. Wade and C. H. Walker, *Australian J. Chem.*, **25**, 1483 (1972).
689. D. G. Lee, 'Oxidation of Oxygen and Nitrogen-containing Functional Groups With Transition Metal Compounds', in *Oxidation*, (Ed. R. L. Augustine), M. Dekker, New York, Vol. 1, 1969, p. 53.
690. A. Pinner, *Ber.*, **25**, 2807 (1892).
691. A. Pinner, *Ber.*, **26**, 292 (1893).
692. A. Pinner, *Arch. Pharm.*, **28**, 378 (1893).
693. L. D. Quin and P. M. Quan, unpublished results, reported in Reference 695.
694. A. M. Duffield, H. Budzikiewicz and C. Djerassi, *J. Amer. Chem. Soc.*, **87**, 2926 (1965).
695. H. McKennis, Jr, E. R. Bowman, L. D. Quin and R. C. Denney, *J. Chem. Soc., Perkin I*, 2046 (1973).
696. A. Picot and X. Lusinchi, *Synthesis*, 109 (1975).
697. M. F. Bartlett, D. F. Dicke and W. I. Taylor, *J. Amer. Chem. Soc.,* **80**, 126 (1958).
698. G. Buchi, R. E. Manning and S. A. Monti, *J. Amer. Chem. Soc.*, **85**, 1893 (1963).
699. G. B. Guise, E. Ritchie and W. C. Taylor, *Australian. J. Chem.*, **18**, 1279 (1965).
700. A. Cave, C. Kan-Fan, P. Potier, J. Le Men and M. M. Janot, *Tetrahedron*, **23**, 4691. (1967).
701. T. R. Govindachari, B. R. Pai, S. Rajappa, N. Viswanathan, W. G. Kump, K. Nagarajan and H. Schmid, *Helv. Chim. Acta*, **45**, 1146 (1962).
702. M. Sainsbury, S. F. Dyke and A. R. Marshall, *Tetrahedron*, **22**, 2445 (1966).
703. G. R. Clemo and G. C. Leitch, *J. Chem. Soc.*, 1811 (1928).
704. O. E. Edwards, F. H. Clarke and B. Douglas, *Can. J. Chem.*, **32**, 235 (1954).
705. C. Djerassi, M. Cereghetti, H. Budzikiewicz, M. M. Janot, M. Plat and J. Le Men, *Helv. Chim. Acta*, **47**, 827 (1964).
706. F. Percheron, *Ann. Chim.*, **4**, 303 (1959).
707. L. M. Berkowitz and P. N. Rylander, *J. Amer. Chem. Soc.*, **80**, 6682 (1958).
708. J. C. Sheehan and R. W. Tulis, *J. Org. Chem.*, **39**, 2264 (1974).
709. N. Tangari and V. Tortorella, *Chem. Commun.*, 71 (1975).
710. K. Orito, R. H. Manske and R. Rodrigo, *J. Amer. Chem. Soc.*, **96**, 1944 (1974).
711. D. Herlem, Y. Hubert-Brierre, F. Khuong-Huu and R. Goutarel, *Tetrahedron*, **29**, 2195 (1973).
712. I. R. C. Bick, J. B. Brenner and P. Wiriyachitra, *Tetrahedron Letters*, 4795 (1971).
713. T. Kametani, H. Nemoto, T. Nakano, S. Shibuya and K. Fukumoto, *Chem. Ind. (Lond.)*, **28**, 788 (1971).
714. E. Höft, A. Rieche and H. Schultze, *Ann.*, **697**, 181 (1966).
715. J. R. Brooks and D. W. Harcourt, *J. Chem. Soc., Perkin Trans. I*, 2588 (1973).
716. A. G. Davies, *Organic Peroxides,* Butterworths, London, 1961, pp. 28–30.
717. H. H. Wasserman and A. W. Tremper, *Tetrahedron Letters*, 1449 (1977).
718. J. E. McKeon and D. J. Trecker, *U. S. Patent 3634346; Chem. Abstr.*, **76**, 72397b (1972).
719. H. Moehrle and R. Engelsing, *Arch. Phar.*, **303**, 1 (1970).
720. R. Madhav, *J. Chem. Soc., Perkin I*, 2108 (1974).
721. F. Farina, M. V. Martin and M. C. Paredes, *Synthesis*, 167 (1973).
722. E. Ziegler, F. Hradetzky and M. Eder, *Monatsh. Chem.*, **97**, 1391 (1966).
723. E. Ziegler, F. Hradetzky and M. Eder, *Monatsh. Chem.*, **97**, 1394 (1966).

Author Index

This author index is designed to enable the reader to locate an author's name and work with the aid of the reference numbers appearing in the text. The page numbers are printed in normal type in ascending numerical order, followed by the reference numbers in parentheses. The numbers in *italics* refer to the pages on which the references are actually listed.

Aaron, C. 108, 109 (268), *119*, 280 (168), *465*
Aasen, A. 225 (114), *258*
Abdulla, R. F. 369 (761), *477*, 1201 (437), *1324*
Abou-Chaar, C. I. 1115 (189), *1319*
Abrahamsson, S. 225 (122), *259*
Abraitys, V. Y. 656 (32), *747*
Abramovich, V. B. 303 (339), *469*
Abramovitch, B. 416 (1009), 417 (1038), *482, 483*
Abramovitch, R. A. 454 (1375), *490*
Achenbach, A. 663, 668 (53), *748*
Achenbach, H. 161 (183), *173*, 633 (214), *640*
Acheson, R. M. 403 (907), *480*
Achiba, Y. 204 (129, 130), *212*
Achiwa, K. 102 (205), *117*, 290 (244), 314 (426), *466, 470*, 611 (81), *637*
Achmatowicz, S. 1040–1042 (268), 1052 (268, 388), *1058, 1061*
Achwal, W. B. 619 (130, 131), *638*
Ackermann, H. 566, 571 (265), *592*
Ackmann, R. G. 373 (778, 779), *477*
Aczel, T. 140 (87), *171*
Adam, M. 234 (229), *261*
Adam, W. 497 (34), *527*, 620 (148), *639, 649 (14), 655 (29), 661 (29, 38), 691, 700 (29), *747, 748*, 909 (238), *914, 1066, 1070 (25), 1192 (359), *1316, 1323*
Adamowicz, H. 594 (374), *598*
Adams, A. C. 424 (1114), *484*
Adams, B. L. 903 (212), *914*
Adams, G. E. 759 (23), 763, 764 (51), 775, 776 (94), *778, 779*
Adams, J. T. 431 (1167), *485*
Adams, R. 1098 (129), *1318*
Adams, R. N. 782 (4), *821*
Adams, W. R. 165, 168 (206), *173*, 672 (66), *748*, 1080 (82), 1206–1208 (457, 458), *1317, 1325*
Adamson, A. W. 642 (1), 647 (9), *747*
Adduci, J. M. 437 (1233), *487*
Adelman, R. L. 251 (415a), *265*

Adhikary, P. 566 (248), *591*
Adickes, H. W. 1284, 1285 (658), *1329*
Adiwidjaja, G. 1023 (16, 22), *1053*
Adkins, H. 272 (22), 307 (372, 373), *462, 469*
Ady, E. 26, 42, 44 (51), *56*
Afanaseva, A. 302 (334), *468*
Affsprung, H. E. 235 (238, 244), *261*
Afzal, M. 408 (919), *480*
Agaki, K. 1026 (74), *1054*
Agarunov, M. J. 61, 62 (27), *66*
Agawa, T. 1123, 1128–1130 (215), 1230 (514), *1320, 1326*
Agnes, G. 1079, 1080 (63), *1317*
Agosta, W. C. 380 (819), *478*, 725 (191a, b), *752*
Agoutin, M. 798 (67), *822*
Ahearn, G. P. 1162, 1163, 1168, 1187 (291), *1321*
Ahlf, J. 247, 251 (386), *264*
Ahlfaenger, B. 1039 (220), *1057*
Ahnehchi, Y. 507 (92), *528*
Ahuja, S. C. 237 (266), *262*
Aihara, A. 65 (46), *66*
Ainsworth, C. 281, 282 (178), *465*
Airey, J. 735 (209a), *753*
Aizawa, S. 538 (61), *588*
Akada, W. 1029–1031, 1040 (71), *1054*
Akalaev, A. N. 562, 566 (216), *591*
Akano, M. 1037, 1042 (153), *1056*
Akasaki, Y. 1052 (391), *1062*
Akermark, B. 1206, 1220 (484), *1325*
Akhlamova, L. N. 618 (125), *638*
Akhnazaryan, A. A. 1090 (102), *1318*
Aki, O. 812 (138), *824*
Akopyan, K. G. 1110 (171), *1319*
Aksenova, V. P. 869 (53), *910*
Alagona, G. 28 (64), 45, 46 (116), 47 (119), 48 (64), *56, 57*, 251 (422), *265*
Al Assal, L. 504 (81), *528*
Albeltina, A. 1045 (323), *1060*
Albers-Schönberg, G. 603 (17), *636*
Alberts, A. H. 309 (386), *469*
Albertson, N. F. 297 (307), *468*
Albery, W. J. 799 (69), *822*

Albright, J. D.   566 (255), *591*
Alder, E.   300 (327), *468*
Alder, K.   415 (977), *482,* 512 (124, 126), *529*
Alekhin, S. P.   61, 62 (28), *66*
Aleksandrov, Yu. I.   60, 61 (17), *65*
Alencastro, R. V.   254 (453), *266*
Alexander, E. R.   881 (104), 882, 883 (105), *912*
Alexander, J. E.   733, 734 (205), *753*
Alexander, R. G.   131 (37), *170*
Alger, R. S.   769 (75), *779*
Alguero, M.   1080, 1082 (76), *1317*
Ali, A.   100 (170), *117*
Ali, M. Y.   1041 (278), *1059*
Ali, S. M.   508 (100), *528*
Allen, A. O.   774 (92), *779*
Allen, C. F. H.   271 (16), 282 (193), 297 (298), *462, 465, 468*
Allen, G.   235 (246), *261*
Allen, G. R.   273 (66), *463*
Allen, J. C.   289 (229, 230), *466*
Allen, L. C.   214 (2e, i), 218, 234 (2e), *256*
Allen, L. E.   843–845, 848 (72, 73), 852 (73), *857*
Allen, M. J.   782, 818 (8b), 819 (165, 166), *821, 824*
Allen, R. E.   870 (56), *911*
Allen, R. J. P.   874 (75), 876 (75, 85), *911*
Allgeier, H.   1039 (242), *1058*
Allinger, N. L.   381 (823), *478*
Allison, D. A.   250 (405), *265*
Allmand, A. J.   672, 677 (59a), *748*
Allphin, N. L.   318 (454), *471*
Almenningen, A.   22, 23 (27), *56,* 222 (46), *257*
Almlöf, I.   232 (210), *260*
Almlof, J.   32 (91), *57*
Almqvist, A.   111 (310), *120*
Almy, J.   452 (1357), *489*
Aloisi, G. G.   676, 685 (98d), *749*
Alpatova, N. M.   806 (113), *823*
Alper, H.   416 (1030, 1031), *483,* 1040, 1041 (260), *1058,* 1192 (361), *1323*
Alston, W. B.   270 (13), *462*
Altenbach, H. J.   1038, 1052 (184), *1057*
Altman, J.   381 (820), *478*
Altmann, H. J.   1206 (470), *1325*
Altnau, G.   147, 157 (113), *171*
Al'tshuler, R. A.   596 (377), *598*
Alvarez, F. S.   415 (974), *482*
Amada, T.   1001 (170), 1003 (173), *1020*
Amano, T.   202 (116), *211*
Ambelang, T.   1164, 1168 (294), *1321*
Ambler, A. P.   608 (44), *636*
Amendola, M.   1113 (180), *1319*

Ames, S. R.   367 (746), *477*
Amiard, G.   1053 (409), *1062*
Amidon, G. L.   221 (39), *257*
Amin, H. B.   891 (150, 152), 892 (152), 893 (150, 171), 905, 906 (152), *912 913*
Amin, J. H.   415 (972), *482*
Amit, B.   273 (53), 274 (87), *463*
Ammon, H. L.   245 (357), *264,* 1292 (663), *1329*
Amosova, S. V.   1052 (398), *1062*
Amy, J. W.   133 (51), *170*
Anaker, E. W.   978 (101), *1018*
Anand, Nitaya   566, 570, 571 (274), *592*
Anand, R. D.   88, 89 (139), *116*
Anatol, J.   454 (1371), *489*
Anbar, M.   772 (83), *779*
Anderson, A. G. Jr.   440 (1262), *487*
Anderson, C. M.   164 (203), *173*
Anderson, D. R.   392 (866), *479*
Anderson, G. H.   353 (661), *475*
Anderson, H. W.   1131, 1135 (224), *1320*
Anderson, J.   808 (129), *824*
Anderson, J. C.   652, 653 (23, 25), *747*
Anderson, J. D.   807 (119, 120), 820 (119), *823*
Anderson, K.   31 (84), *57*
Anderson, N. H.   768 (71), *779*
Anderson, R. J.   430 (1160), *485,* 847 (80), *857,* 1047 (336), *1060*
Anderson, T. H.   769 (75), *779*
Ando, R.   995, 1002, 1007 (164), *1019*
Ando, S.   327 (449), *472*
Andreev, N.   302 (334), *468*
Andrews, C. E.   867 (38), *910*
Andrews, D.   1110 (174), *1319*
Andrews, L. J.   175 (2), 180 (43), *209, 210*
Andrews, P. R.   247 (380), 248 (390), *264*
Andriaenssens, G. J.   230 (185), *260*
Andrianova, G. M.   539 (79), *588*
Andrieu, C. G.   1024, 1031, 1033, 1035 (27), *1053*
Andruskiewicz, C. A. Jr.   1186 (348), *1323*
Angell, C. L.   620 (137), *638*
Angelo, B.   280 (164, 165, 167, 169, 170), 281 (164, 180), *465*
Angeloni, L.   228 (155, 156), *259*
Anger, V.   603 (13), 631 (193), *636, 640*
Anghelide, N.   1196 (386), *1323*
Angus, H. J. F.   205, 206 (133), *212*
Angyal, S. J.   346 (634), *475*
Anisimova, O. S.   536 (35), 562 (217), 566 (271), 570, 571 (35, 217), 582 (271, 340), 583 (271, 342), *587, 591–593*
Anjaneyulu, B.   1204 (431), 1232 (548, 571, 574), 1245–1247, 1249, 1252 (548), 1254, 1255 (574), 1257 (571, 574), 1258 (571), 1259 (548), *1324, 1327*

Ankli, P.   525 (218), *531*
Anner, G.   299 (321), *468*
Anschütz, R.   630 (190), *639*
Ansell, M. F.   134 (54), *170,* 298 (310),
    373 (781), 438 (1239), *468, 478, 487,*
    501 (64a, b), *527,* 1065, 1078 (5), *1315*
Anteunis, M.   1038 (174), *1056*
Anthoney, M. E.   63 (39) (39), *66*
Antonenko, N. S.   241 (320, 322), 251,
    252 (320), *263,* 632 (209), *640*
Antoniades, E. P.   272 (40), *462*
Antonietta, M.   246 (370), *264*
Aoyagi, H.   1206, 1219 (482), *1325*
Aoyama, H.   1206, 1222, 1224 (467), *1325*
Appel, R.   1171, 1172 (319), *1322*
Applegate, H. E.   364 (721), *476,* 1300
    (679), 1301 (680), *1329*
Appleton, D. C.   700, 735 (142), 736,
    737 (142, 213), *750, 753*
Appleyard, G. D.   444 (1300), *488*
Appriou, P.   1048 (357), *1061*
Apsimon, J. W.   332 (553), *473*
Apsitis, A.   1045 (322), *1060*
Arabamovitch, S.   768 (66), *779*
Arad, Y.   801 (83), *823*
Araki, M.   1047 (335), *1060*
Araki, Y.   1039, 1041, 1050 (190), *1057*
Aranovich, G. I.   1039 (250), *1058*
Archer, S.   452 (1359), *489*
Ard, J. S.   246 (375), *264*
Ardis, A. E.   874 (77), *911*
Arens, J. F.   442 (1277), *488,* 1043 (301),
    1049 (362), *1059, 1061*
Arenson, F. R.   1145, 1147, 1153 (254),
    1321
Aresi, V.   1197 (401), 1198 (390), 1246
    (401), *1323, 1324*
Argyropoulos, N.   272 (49), *463*
Arigoni, D.   345 (630), *475*
Arison, B. H.   603 (17), *636*
Arkell, A.   298 (318), *468*
Armitage, I. M.   252 (428), *265*
Armstrong, G. P.   869 (51), *910*
Armstrong, M. D.   274 (75), *463*
Armstrong, R. A.   104 (246), *118*
Arndt, F.   1096 (116), *1318*
Arndt, F. G.   603 (18), *636*
Arnett, E. M.   218 (19), *257*
Arnold, D. R.   656 (32), 661, 662 (44a, b),
    747, 748
Arnold, P. R.   225 (116), *259*
Arnold, R. T.   360 (702), 361 (712), 391
    (865), *476, 479,* 860 (1), 882 (106),
    909, 912, 1092 (108), *1318*
Arnold, Z.   536 (38, 41), 540, 543, 545,
    554, 556, 557, 559, 562 (38), 566 (38,
    268), 569 (38), *587, 592,* 1166, 1168
    (299), *1322*

Arnone, A.   455 (1383), *490*
Arsenijevic, V.   414 (969), *481*
Arshadi, M. R.   60–62 (19), *66*
Arters, A. A.   417 (1036), *483*
Arth, G. E.   1171 (315), *1322*
Arthur, P.   411 (928), *481*
Aruna, Padam C.   566, 570, 571 (274),
    592
Aruna, V.   566, 570, 571 (274), *592*
Arutyunyan, V. S.   1072 (35), 1075 (48),
    1316
Asaba, H.   1037, 1042 (153), *1056*
Asahara, T.   800 (79), *823*
Asahi, Y.   812 (138), *824*
Asai, Y.   551 (160), *590*
Asano, Y.   238, 239 (288), *262*
Aschenbrenner, H.   793 (55), *822*
Ashbrook, C. W.   416 (1014), *482,* 1301
    (682), *1330*
Ashcroft, S. J.   900 (199), *913*
Asher, J. D.   621 (151), *639*
Asinger, F.   338 (595), 343 (624), 369
    (754), *474, 477*
Aspden, J.   868 (40), *910*
Asplund, L.   248 (395), *264*
Asquith, R. S.   1034, 1043 (119), *1055*
Assarsson, P.   252 (425), *265*
Aston, J. G.   355 (683, 684), 359 (699),
    476
Astrup, A. T.   415 (971), *482*
Atavin, A.   458 (1403), *490*
Atlani, P.   346 (639), *475*
Atlanti, P.   54 (124), *58*
Atsumi, K.   1039 (219), *1057*
Atwell, G. T.   415 (978), *482*
Au, A.   1041 (277), *1059*
Aulchenko, I. S.   330 (521), *472*
Aurivillius, B.   111 (312), *120*
Auschütz, R.   865 (81), *911*
Ausloos, P.   661, 662, 666(41), 671, 672,
    677 (92d, e), 687 (108, 110), 688 (108,
    110, 116), 709 (116), 713 (108, 110,
    116), 714 (175a), *748–751,* 884 (123),
    912
Auterhoff, H.   102 (198), *117*
Auwers, K.   630 (191), *639*
Avaca, L. A.   816, 820 (150), *824*
Avakian, S.   1200 (415), *1324*
Averill, S. J.   874 (77), *911*
Avetisyan, A. A.   1096, 1101 (118), 1104
    (146, 147), 1110 (171), *1318, 1319*
Avigad, G.   88, 91, 92 (127), *116*
Avila, N. V.   1090 (105), *1318*
Avrutskaya, I. A.   804 (100, 101), *823*
Axelrod, L. R.   368 (748), *477*
Ayengar, N. K. N.   739 (219), *753*
Ayscough, P. B.   759, 760 (26), 769, 770
    (77), 773 (86), *778, 779*

Azima, A. 223 (71), *258*
Ažman, A. 24 (58), *56*, 232 (208), *260*
Azuma, S. 1101 (143), 1160, 1168 (282), *1318, 1321*
Azzaro, M. 161 (184), *173*

Baba, H. 31 (87), *57*, 607 (38), *636*
Babad, E. 381 (820), *478*
Babaeva, A. A. 1073 (36), *1316*
Babayan, A. T. 402 (904), *480*
Babiewskij, K. K. 566 (264), *592*
Baccolini, G. 449 (1331), *489*
Bach, E. 102 (207), *117*
Bach, S. R. 1074 (47), 1104 (151), 1105 (47), *1316, 1319*
Bacha, J. D. 508 (102), *528*
Bachelet, J. P. 361 (714), *476*
Bachi, M. D. 1232 (539, 540, 556), 1242 (539, 540), 1250 (540, 556), *1326, 1327*
Bachmann, W. E. 380 (815), *478*
Back, T. G. 1048 (350), *1060*
Baclawk, L. M. 431 (1174), *486*
Bacon, E. 744 (226), *753*
Badash, A. 458 (1405), *490*
Bader, A. R. 500 (56), *527*
Bader, F. E. 1071–1073, 1075, 1149, 1187 (28), *1316*
Bader, R. F. W. 895 (182), *913*
Badger, R. M. 218 (16), *256*
Badilescu, I. I. 375, 379 (798), *478*
Baer, T. A. 286 (216), *466*
Baeza, J. 909 (238), *914*, 1066, 1070 (25), *1316*
Bagchi, P. 298 (317), 311, 314 (399), *468, 470*
Bagger, S. 103 (224), *118*
Bagley, F. D. 890, 898 (148), *912*
Bagli, J. 395 (879), *480*
Bagli, J. F. 1206 (455), *1325*
Bahry, M. 1038, 1042, 1052 (168), *1056*
Baig, M. V. A. 1039 (230), *1058*
Bailey, A. S. 279 (135), *464*
Bailey, D. M. 297 (307), *468*, 501 (62), *527*, 1138–1140, 1142–1144 (236), *1320*
Bailey, P. S. 338 (586, 592, 600), 399 (602, 606), 340 (617), *473, 474*, 1138 (228, 229), *1320*
Bailey, W. C. Jr. 1075, 1079 (52), *1316*
Bailey, W. J. 870 (54, 55), 879 (97), 882 (107), 883, 884 (115), 886 (115, 133), 892 (162), 903, 906 (213a), *910–914*
Baillie, T. B. 819 (162), *824*
Baird, M. C. 827 (32, 37), 828 (37), 835 (37, 46–48, 53, 54), 837 (32, 47, 48, 54), 838 (48), 839, 840 (47, 48, 54), 841 (32, 66), 842 (54), 843 (69), 845 (69), 852 (66, 69), *856, 857*

Baird, M. S. 1284 (657), *1329*
Baird, N. C. 30 (77), *57*
Baitinger, W. E. 133 (51), *170*
Baizer, M. M. 415 (994), *482*, 782 (1), 785 (21), 791 (50), 803 (92), 807 (1, 118–121), 808 (123–125), 809 (123–125), 809 (92, 130, 131), 810 (92), 820 (119, 172), *821–824*
Bak, B. 28, 29 (63), *56*
Bak, T. A. 236, 237 (259), *262*
Baker, A. J. 1076, 1078 (53), *1316*
Baker, B. R. 1098 (129), *1318*
Baker, E. B. 416 (1025), *483*
Baker, J. G. (26), *56*
Baker, J. T. 270 (12), *462*
Baker, J. W. 328 (508), *472*
Baker, R. F. W. 861, 863 (8), *910*
Baker, R. H. 417 (1035), *483*
Balachandran, K. S. 303 (345), *469*
Balakrishnan, T. D. 787 (30), *821*
Balasubramanian, K. 374, 375 (785), *478*
Balcerski, J. S. 101 (172), *117*
Bald, R. W. 561, 563 (196), *590*
Baldwin, J. E. 393 (870), *479*, 1301 (682), *1330*
Baldwin, S. W. 901 (205), *914*
Baldwin, W. H. 894 (177), *913*
Balenovic, K. 380 (818), *478*
Bales, B. L. 758 (11), 761 (37), *778*
Ballantine, J. A. 125, 138 (9), 166 (211), *169, 173*
Ballard, D. A. 386 (857), *479*
Ballard, S. A. 1232, 1241, 1242 (531), *1326*
Ballester, M. 277 (122), *464*
Ballou, C. E. 310 (392), *470*
Baltazzi, E. 1108 (162), *1319*
Balzani, V. 642 (2), *747*
Bamford, C. H. 864 (21), *910*
Bamkole, T. O. 889 (144), *912*
Ban, Y. 412 (947), *481*, 581 (337, 338), 582 (337), 583 (338, 341), *593*, 1039 (191–193), 1040 (192), 1050 (191–193), *1057*
Banerjee, K. 966, 972 (89), *1018*
Bangert, K. F. 536 (20), *587*
Banhidai, B. 457 (1395), *490*
Banhitt, E. H. 499 (47), *527*, 908 (237), *914*
Banoub, J. 557 (188), 558 (188, 189), *590*
Banthorpe, D. V. 888 (139), *912*
Baran, G. 84 (93), *115*
Baranowsky, K. 1092 (109), *1318*
Barb, W. G. 176, 180 (13), *209*
Barbour, A. K. 318, 320, 325, 327 (452), *471*
Barclay, R. 879 (97), *911*

Barco, A. 309 (388), *469*
Bard, A. J. 739 (218), *753,* 782 (14), 803 (88, 89, 93), 808 (88, 89, 126, 127), 820 (171), *821, 823, 824*
Bardi, G. 61 (22), *66*
Barili, P. L. 310 (390), *470*
Bark, L. S. 631 (197, 198), *640*
Bark, S. M. 631 (197), *640*
Barlow, J. H. 1023, 1035, 1052 (21), *1053*
Barltrop, J. A. 208 (152), *212,* 709 (152), 710, 715 (164), 718 (152), *751*
Barluenga, J. 1269 (608), *1328*
Barner, R. 392 (867), *479*
Barnes, C. S. 164 (203), *173,* 1284, 1291 (655), *1329*
Barnes, D. S. 64 (41), *66*
Barnes, E. E. 229 (166), *260*
Barnes, K. K. 782 (2), *821*
Barnes, R. P. 293 (272), *467*
Barnett, B. F. 1276 (630), *1328*
Barnett, B. L. 745 (228b), *753*
Barnett, W. E. 494 (13), 498 (39), *526, 527,* 1089 (98–100), 1090 (100), *1317*
Barnikow, G. 1040 (256, 265), 1041 (256, 279), 1044, 1048 (313), 1052 (394) *1058, 1059, 1062*
Barnish, I. T. 1205 (450, 452), *1325*
Baron, D. 230, 235 (176), *260*
Baron, F. A. 329 (517), *472*
Barr, N. F. 774 (87), *779*
Barr, P. A. 415 (983), *482*
Barradas, R. G. 789 (39), *822*
Barrett, G. C. 102 (191, 206), *117,* 1040–1042, 1052 (268), *1058*
Barroeta, N. 895 (183), *913*
Barrow, G. M. 620, 623 (147), *639*
Barrow, K. D. 610 (70), *637*
Barry, B. W. 984 (126), *1019*
Barry, R. 31 (84), *57*
Barstow, L. E. 443 (1288), *488*
Barth, G. 88 (111, 112), 89 (111), 103, 104 (241), 109 (112), *116, 118*
Bartle, K. D. 183 (58), *210*
Bartlett, L. 104 (247), *118*
Bartlett, M. F. 1304 (697), *1330*
Bartlett, P. A. 272 (33), *462*
Bartlett, P. D. 292 (260, 261), 320 (469), 353 (657), *467, 471, 475,* 498 (38), 499 (52, 53), *527,* 909 (239), *914,* 1138 (227, 230), *1320*
Bartok, M. 348 (642), *475*
Bartok, W. 328 (514), *472*
Bartoletti, I. 292 (255), *467*
Barton, D. 673 (74), *749*
Barton, D. H. R. 272 (52), 275 (92), 310 (395), 369, 370 (756, 757), 447 (1321), 454 (1370), *463, 470, 477, 488, 489,*

524 (205), *530,* 652, 653 (21, 22), *747,* 826 (14), *855,* 882 (111), *912,* 1039 (194), 1040 (268, 272), 1041 (194, 268), 1042 (194, 268, 272), 1043 (194), 1046 (272), 1050 (194), 1052 (268, 388, 389), *1057–1059, 1061,* 1145 (256), 1159 (256, 279), 1189 (353), 1284, 1291 (655), *1321, 1323, 1329*
Barton, D. M. 361 (711), *476*
Barton, T. J. 517 (157b), *529*
Bartoniček, B. 761 (36), *778*
Bartram, K. 340 (614), *474*
Barua, A. B. 312 (405), *470*
Basch, H. 20, 26 (12), 29, 30 (12, 71), *55, 56,* 70, 72, 101 (15), *114,* 229 (167), *260*
Bashir, N. 523 (202), *530*
Bashkirov, A. N. 303 (339), 305 (352), *469*
Bashou, C. 571 (301), *592*
Basselier, J.-J. 689, 700 (119), *750,* 1233, 1266, 1267 (602), *1328*
Bassler, G. C. 612 (90), *637*
Bastiansen, O. 22, 23 (27), *56,* 222 (46), *257*
Basu, S. 181 (51), *210*
Batalin, G. I. 223 (69), *258*
Batcho, A. 571 (288), *592*
Bate, P. 631 (198), *640*
Bately, M. 196 (85), *211*
Bateman, J. H. 901 (206), *914*
Bateman, L. 403 (909), *480*
Bates, G. S. 1039 (218, 226), 1048 (350), *1057, 1060*
Bates, W. W. 248 (391), *264*
Batten, P. L. 369, 370 (756, 757), *477*
Battenberg, E. 517 (157a), *529*
Battersby, A. R. 1199 (410), *1324*
Battiste, M. 505 (85), *528*
Baubouy, R. 313 (412), *470*
Bauer, L. 279 (142), *464*
Bauer, S. H. 218 (16), *256*
Bauer, V. J. 1066–1068, 1072, 1073 (16), *1316*
Baugh, P. J. 761 (33), *778*
Baum, M. E. 338 (593), *474*
Baumann, P. 506 (89), *528,* 712 (180a–c), 719, 725 (180a–c, 184), 727 (180b, c), 745 (180a–c), *751, 752*
Baumgarten, H. E. 167 (215, 218), 168 (218), *174,* 518 (168–170), 520 (170, 176), *530,* 870 (56), *911,* 1200 (418–420), *1324*
Baumgarten, W. 299 (322), *468*
Baumrucker, J. 979 (104), *1018*
Baur, R. 553, 560, 567, 573, 574, 576–578 (172), *590*
Bavin, P. M. G. 1104 (153), *1319*

Baxendale, J. H.　759 (23), *778*
Baydar, A. E.　503, 504 (76), *528*
Bayer, E.　103, 104 (241), *118*
Bažant, V.　916 (16, 17), 918 (31), 919 (16, 17), *943*
Beak, P.　1048 (343), 1050 (375, 380), *1060, 1061*
Beal, D. A.　348 (644, 645), *475*
Beal, G. D.　271 (15), *462*
Beal, P. F.　382 (828), *478*
Beames, D. J.　1039 (221), *1057*
Beare, S. D.　109 (288), *119*
Beaton, J. M.　1159 (279), *1321*
Beauchamp, J. L.　30 (76), *57*
Beaucort, J.-P.　426 (1129), *485*
Beauregard, Y.　1035, 1040 (133), *1055*
Beaute, C.　246 (367), *264*
Bebb, R. L.　293 (271), *467*
Becher, D.　156 (156), *172*
Beck, F.　433 (1191), *486*, 782 (7), 792 (54), 793 (54, 55), 800 (74, 75), 818 (161), *821, 822, 824*
Beck, G.　545, 563, 566, 567 (127), 576, 578, 580 (326), *589, 593*
Beck, J. F.　1199 (410), *1324*
Beck, M.　305 (354), *469*
Becke, F.　450 (1335), *489*, 1038 (166), *1056*
Becker, E.　493 (7), *526*
Becker, E. D.　252 (431), *265*
Becker, E. I.　297 (299), *468*
Becker, F. J.　517 (160), *529*
Becker, H.-D.　328 (512), *472*
Beckmann, S.　1089 (96), *1317*
Beckwith, A. L. J.　442 (1283), *488*, 1189 (353), 1195 (374), *1323*
Bedford, C. R.　1284 (654), *1329*
Bedos, P.　1027, 1032, 1042 (60), *1054*
Beecham, A. F.　78 (61, 62), 79, 83, 84 (61–63), 85, 86 (99, 100), 87 (100, 106, 107), *115*, 495 (18), *526*, 626 (171–173), *639*
Beek, H. C. A. van　672, 674, 678, 682 (82), *749*
Beer, R. J. S.　1042 (283), *1059*
Beer, W.　239 (294), *262*
Beeson, J. H.　519 (174), *530*
Behme, M. T. A.　950 (23), *1016*
Behr, J. P.　315 (430), *470*
Behrens, U.　1022, 1030, 1038 (15), *1053*
Beier, G.　233 (216), *261*
Beiner, J. M.　1024, 1031, 1033, 1035 (27), 1037 (140), 1040 (255), 1042 (140), 1049 (255), 1050 (140), 1052, 1053 (255), *1053, 1056, 1058*
Beisler, J. A.　514 (136), *529*
Beisswenger, A.　496 (27), *527*
Bekkum, H. van 88 (132), *116*, 419 (1071), *483*

Belanger, A.　1193 (364), *1323*
Belikov, V. M.　566 (264), *592*
Belil, C.　1080 (70, 71), *1317*
Bell, I.　310, 314 (389), *470*
Bell, S.　70 (28), *114*
Bellamy, L. J.　88 (129), *116*, 223 (67), 241 (315), 250 (406), 255 (473), *258, 263, 265, 266*, 604 (20, 21), 605 (31), 607 (37), 608 (42, 45, 46), 609 (60), 619 (135, 136), 620 (138, 139, 143, 144), 631 (201–203), 632 (201), *636–640*, 1029 (78), *1054*
Bellasio, E.　1201 (427), *1324*
Bellavista, N. C.　447 (1315), *488*
Bellet, J.　22(35), *56*
Belletire, J.　735 (209a), *753*
Belletire, J. L.　735 (208), *753*
Bellitto, C.　1027 (52), *1054*
Bellot, E. M.　222, 224 (49), *257*
Bellus, D.　649 (11), 652, 657 (11, 20), 694 (20), 733 (11, 20), *747*
Belobrov, V. M.　238 (275), *262*
Belonovich, M. I.　1047 (337), *1060*
Belov, V. N.　806 (115), *823*, 1065 (8), *1315*
Belsky, I.　1042 (285), *1059*
Belyaeva, O. Ya.　596, 597 (386), *599*
Bembry, T. H.　1098 (130), *1318*
Benati, L.　1043 (305), *1059*
Ben-Bassat, J. M.　731 (199), *752*
Bendall, V. I.　1109 (168), *1319*
Bender, H. J.　223 (73), *258*
Bender, M. L.　622 (159), *639*
Benedek, G. B.　952 (43), *1017*
Benesi, H. A.　178 (30), *210*
Benetti, S.　309 (388), *469*
Benezra, S. A.　127, 128 (21), 140 (81), *170, 171*
Benham, J.　380 (819), *478*
Benjamin, L. E.　566 (259), *592*
Benkeser, R. A.　294 (281), *467*, 815 (149), *824*, 883 (116), 884, 885 (128), 886 (116), *912*
Benkovic, S.　948 (19), 1002 (171), *1016, 1020*
Benkovic, S. J.　1048 (346), *1060*
Bennett, G. B.　596, 597 (387), *599*
Bennett, J. E. 757 (2), *778*
Bennett, M. A.　450 (1339), *489*, 826 (22), *856*
Bennett, R. N.　874 (76), *911*
Bennington, F.　596, 597 (385), *599*, 861 (5), *909*
Benoit, F.　127 (18), 128 (25, 27), 130 (18, 33, 34), 131 (35), 158 (35, 168), *169, 170, 173*, 615 (106), *638*
Benoit, F. M.　142 (93), *171*
Bensel, N.　1108 (158), *1319*

Benson, R. E.   517 (159), *529, 537* (51), *587,* 1302 (687), *1330*
Benson, S. W.   59 (4), 60 (8–10), 61 (4, 8–10), 62 (10), 63ʹ(8, 10), 65 (8), *65*
Bentley, F. F.   605, 606 (34), 607 (40), 608 (40, 47), 632 (205, 206), *636, 640*
Bentley, R. L.   1232, 1256, 1261 (591), *1328*
Bentley, T. J.   735 (209a), *753*
Bentley, T. W.   164 (198, 202), *173,* 1035, 1036 (126), *1055*
Benz, W.   135, 136 (61), *170*
Ben-Zvi, Z.   296 (295), *468*
Beránek, L.   916. 919 (16, 17), *943*
Berchtold, G. A.   1038 (183), *1056*
Bercovici, T.   733–735 (204c), *753*
Berencsi, P.   1037, 1042 (157), *1056*
Bereza, S.   498 (41), *527*
Berezin, I. V.   947 (8, 14), 948 (20), 962 (14, 73b), 964 (8, 14, 20, 73b, 82–86, 87a, b, 88), 965 (20, 73b, 83), 966 (14, 20, 86), 967 (83, 88), 968 (73b), 978 (14), 990 (83), 1004 (87b), 1008, 1010 (87a), *1016–1018*
Berg, F.   416 (1013), *482*
Berg, J. C.   947 (13), 962, 980 (72), 983 (13), *1016, 1017*
Berg, V.   223 (72), *258*
Bergbreiter, D. E.   286 (217), *466*
Bergelson, L.   278, 422 (125), *464*
Bergen, T. J. van   (166), *751,* 1197 (405–407), 1199 (405, 407), *1324*
Berger, A.   279 (144), *464*
Berger, S. A.   177 (26), *210*
Bergman, R. G.   989 (144), 1006 (177), *1019, 1020*
Bergmann, E. D.   276 (105), 282, 429 (192), *464, 465,* 828 (42, 43), 829–831 (42), 832, 833 (42, 43), 834, 835, 837, 841, 842, 845 (42), *856*
Bergmann, F.   277 (116), 341 (621b, c), *464, 474,* 603 (4), *635*
Bergmann, G.   276 (104), *464*
Bergthaller, P.   415 (979), *482*
Berkov, B.   1090 (105), *1318*
Berkowitz, L. M.   304 (347), *469,* 1171 (314), 1307 (707), *1322, 1330*
Berlin, A. A.   270 (14), *462*
Berlin, K. P.   273 (60), *463*
Berliner, E.   277, 296 (115), *464*
Berlman, I. B.   647, 698 (10), *747*
Bernard, E.   199 (102), *211*
Bernard, M. A.   1045 (321), *1060*
Bernassau, J.-M.   333 (563), *473*
Bernath, T.   772 (84), *779*
Bernecker, R. R.   138 (71), *171*
Bernitt, D. L.   23 (39), *56*
Bernstein, H. J.   235 (231), *261*

Beroza, M.   333 (560), *473*
Berson, J.   1232 (522), *1326*
Berson, J. A.   1089 (97), 1145, 1151 (267), *1317, 1321*
Berstein, S. C.   1111 (178), *1319*
Bertain, J.   411 (936), *481*
Berthod, H.   26 (54), 45 (113), *56, 57,* 254 (463), *266*
Berthold, H.   1040 (261), *1058*
Berti, G.   310 (390), *470*
Bertie, J. E.   743 (222), *753*
Bertini, F.   455 (1381), *490*
Bertinotti, F.   225 (120), *259*
Bertoluzza, A.   246 (370), *264*
Bertrand, M.   1173 (327), *1322*
Bertrand, M. T.   1113 (183), *1319*
Bessonova, N. N.   249 (402), 253 (445), 255 (467), *265, 266*
Bestian, H.   509 (107), *528,* 1194 (370), 1225 (370, 503), 1226, 1227 (503), *1323, 1326*
Bestmann, H. J.   295 (287), 380 (817), 422 (1100, 1104), 423 (1100), 437 (1230), *467, 478, 484, 487,* 1043 (307), *1059*
Bettoni, G.   102 (201, 202), 103 (201), *117*
Betz, W.   418 (1056), *483,* 536 (17–19), 554 (17), *587*
Bewick, A.   816, 820 (150), *824*
Beychok, S.   102 (184), 105 (184, 253), *117, 118*
Beyer, R. D.   295 (283), *467*
Beyerlin, H. P.   536 (30), 537 (30, 57), 538, 544 (57), 550 (158), 551 (57), 559 (57, 158), 581, 586 (57), *587, 590*
Beyerman, H. C.   887 (136), *912,* 1199 (408), *1324*
Beynon, J. H.   127 (15, 16), 130, 131 (16), 133 (48, 51), 134, 137, 138 (48), 141 (90, 91), 142 (15, 16), 143 (94), 158 (16), *169–171*
Beysens, D.   234 (229), *261*
Bezard, A. von   353 (662), *475*
Beziat, Y.   1111 (177), *1319*
Bezpalova, Z.   101 (179), *117*
Bezuglyy, V. D.   787 (31), *822*
Bhat, H. B.   326, 368, 369 (497), *472*
Bhat, S. N.   241 (325), *263*
Bhatia, K.   774–776 (88), *779*
Bhatnagar, P. K.   427 (1144), *485,* 1115 (192), *1320*
Bhattacharya, S. K.   1204 (431), 1232, 1245–1247, 1249, 1252, 1259 (548), *1324, 1327*
Bhatti, M. ud Din   798, 799 (68), *822*
Bhide, R. S.   294 (276), *467*
Bicca de Alencastro, B.   1029 (87), *1054*

Bick, I. R. C. 1310 (712), *1330*
Bickel, H. 846 (78), *857*, 1071–1073,
  1075, 1149, 1187 (28), *1316*
Biderman, M. 151 (127), *172*
Biehl, E. R. 375, 379 (797), *478*
Bielefeld, M. A. 1098 (136), *1318*
Biellmann, J. F. 156 (158), *172*, 674
  (78), *749*
Bielski, B. H. J. 774 (91, 92), *779*
Biemann, K. 127 (22), 135, 136 (61), 166
  (213), *170, 173*
Biener, H. 1225–1227 (503), *1326*
Bienvenüe-Götz, E. 494 (12), *526*
Bigley, D. B. 131 (37), *170*, 861 (3,
  6, 9–11), 862 (6, 9), 865 (24),
  884 (125), 892 (156), 894 (125,
  176), 899 (125), *909, 910, 912,
  913*
Bill, I. 1080, 1082 (67), *1317*
Billet, D. 277 (117), *464*
Billett, E. H. 408 (917, 918), *480*
Billups, C. 109 (276), *119*
Billups, W. E. 353 (659), *475*
Bingham, A. 984, 987 (121), *1018*
Bingham, R. 431 (1173), *485*
Binkley, R. W. 745 (229), *753*
Binkley, W. W. 745 (229), *753*
Biondi, L. 102 (193), *117*
Birch, A. J. 846, 849 (76), *857*
Birch, S. F. (541), *473*
Bird, C. N. 870 (54), *910*
Bird, C. W. 290 (238, 239), *466*
Birdsall, B. 252 (429), *265*
Birkenmeyer, R. D. 453 (1363), *489*
Birladeanu, L. 1171 (315), *1322*
Birshtein, T. M. 83, 84 (87), *115*
Bischoff, C. 450 (1342), *489*
Bishop, C. A. 884, 886 (124), 904 (214),
  912, 914
Bishop, W. 962, 998 (71), *1017*
Bittlerer, K. 291 (246), *467*
Biz, J. R. van der 887 (135), *912*
Bjorkstam, G. L. 230 (185), *260*
Black, D. R. 138 (73), *171*
Black, D. St. C. 493 (1b), *526*, 1298,
  1299(673), 1302 (688), *1329, 1330*
Black, M. H. 866 (30), *910*
Blackburn, G. M. 493 (1b), 496 (22),
  526, 527
Blades, A. T. 899 (197), *913*
Blagoev, B. 280 (152), *465*, 1115 (190),
  1319
Bláha, K. 76, 83 (50), 101 (178, 179),
  114, 117
Blair, G. E. 1115 (189), *1319*
Blake, C. C. F. 245 (355), *264*
Blake, P. G. 864 (18–20, 22), 865
  (25–27), 868 (22), *910*

Blanchard, H. S. 305 (357), 328 (503),
  469, 472
Blanchard, K. R. 61 (21), *66*
Blanco, A. 963 (80), *1018*
Blank, J. E. 1052 (401), *1062*
Blaser, B. 920 (35), 935 (55, 72), *943,
  944*
Blaskovits, A. 1145, 1153, 1154 (273),
  1321
Blatchford, J. K. 436 (1211), *486*
Blatchly, J. M. 905 (224), *914*
Blau, S. E. 861 (4), *909*
Blazejawicz, L. 305 (350), *469*
Bleckmann, P. 226 (143), *259*
Bleikher, Y. I. 442 (1279), *488*
Blicke, F. F. 280 (151), 371 (766), *465,
  477*, 1197 (404), 1266, 1267 (601),
  1324, 1328
Blinc, R. 217 (6), 230 (184), 231 (200,
  201), *256, 260*
Block, H. S. 365 (732), *477*
Blomquist, A. T. 887 (134), 888 (141,
  142), 901 (141), 903, 906 (213a), *912,
  914*
Bloomfield, J. J. 431 (1168), *485*, 806
  (114), *823*, 1138–1142, 1145 (237),
  1320
Blos, I. 164 (193), *173*
Blossey, E. C. 411 (935), *481*
Blout, E. R. 275 (89), *463*
Blum, J. 827 (38), 828 (38, 42, 43), 829,
  830 (38, 42), 831 (42), 832, 833 (42,
  43), 834 (42, 52), 835 (38, 42, 52),
  837, 841 (42), 842 (42, 52), 845 (42),
  856
Blume, E. 287 (225), *466*
Blume, R. C. 1075 (50), *1316*
Blumenthal, T. 159 (176), 162 (188),
  173
Blunck, F. H. 881 (103), *912*
Blyth, C. A. 995, 1002 (161), *1019*
Boag, G. W. 775, 776 (94), *779*
Boar, R. B. 310 (395), 447 (1321), *470,
  488*
Boatright, L. G. 303 (338), *468*
Bobbitt, J. M. 368 (747), 418 (1060),
  477, 483
Boberg, F. 610, 621 (67), *637*, 1043 (309,
  310), *1059*
Bochwic, B. 302 (333), *468*
Bock, E. 1024, 1048 (29), *1053*
Bock, P. L. 839 (63), *856*
Bocz, A. K. 536 (34), *587*
Bodansky, M. 443 (1285), *488*
Boden, H. 319 (456), *471*, 524 (214),
  531
Bodenstein, C. K. 1073, 1115 (44), *1316*
Bodrikov, I. V. 449 (1333), *489*

Boehner, R. S.   867 (38), *910*
Boer, H.   799 (70), *822*
Boer, Th. J. de   250 (411), *265*
Boerboom, A. J. H.   125 (11), *169*
Boer-Terpstra, T.   525 (221), *531*
Boeseken, J.   338 (583), *473*
Boeyens, J. C. A.   184 (68, 69), 185 (69), *210*
Boffey, B.   984, 987 (121), *1018*
Bogavac, M.   414 (969), *481*
Bogdanova, L. A.   1185 (342), *1322*
Bogdanowicz, M. J.   375, 379 (802), *478*, 1164 (294, 295), 1165 (295), 1168 (294, 295), *1321*
Bogdansky, S.   103 (228), *118*
Boggs, L. E.   137 (67), *170*
Bogri, T.   395 (879), *480*
Bohlmann, F.   147 (113), 154 (143), 156 (159), 157 (113, 162), 158 (165), *171–173*, 312 (409), 340 (614), *470, 474*
Böhme, E. W. H.   1300 (679), 1301 (680), *1329*
Böhme, H.   534 (2), 542 (102, 103), 546 (2), *586, 588*, 1037, 1042 (163), *1056*, 1204 (444), *1325*
Boie, J.   1052 (390), *1061*
Boire, B. A.   710, 718 (160), *751*
Bolan, E. N.   411 (927), *481*
Bolard, J.   103, 104 (238), *118*
Bolsman, T. A. B. M.   687, 743 (104), *749*
Bolt, C. C.   1145, 1147 (258), *1321*
Bolton, I. J.   390, 391 (864), 452 (1352), *479, 489*, 583 (345), *593*
Bolton, M.   1052 (388), *1061*
Bombieri, G.   1023 (23), *1053*
Bonaccorsi, R.   40 (105), *57*
Bonati, A.   1198 (391, 393), *1324*
Bond, A. M.   1046 (332), *1060*
Bondarenko, A. V.   (506), *472*
Bonner, O. D.   252 (424), *265*
Bonnet, P.-H.   312 (409), *470*
Bonnett, R.   539 (69), *588*
Bonniol, A.   103 (223, 239), *118*
Bonora, G. M.   100 (171), 101 (172), *117*
Boopsingh, B.   1048 (345), *1060*
Boorman, E. J.   494 (9), *526*
Borch, G.   1031 (100), *1055*
Borch, R. F.   421 (1087), *484*, 536, 566, 571 (25), *587*, 1196 (387), *1323*
Borchert, A. E.   896, 902 (185), *913*
Bordás, B.   1037, 1042 (156, 157), *1056*
Borden, W. T.   743 (223b), *753*
Bordignon, E.   676, 685 (98d), *749*
Bordwell, F. G.   417 (1035), 452 (1357), *483, 489*, 883 (112), *912*, 1011 (194), *1020*
Borel, E.   318 (448), *471*

Borel, M. M.   1022 (5), 1023 (18), 1045 (321), *1053, 1060*
Borén, H. B.   105 (257–259), 106 (258, 259), *119*
Borer, R.   1110 (172), *1319*
Borgeois, P.   454 (1379), *490*
Borhani, R.   158 (169), *173*
Borisov, S. B.   63 (40), *66*
Borkowski, R.   884 (123), *912*
Bormann, D.   517 (161), *529*
Börner, K.   1194, 1225 (370), *1323*
Borner, P.   535–537, 554, 559, 572 (10), *586*
Bornowskii, H.   551 (164), *590*
Bornstein, J.   411 (925), *481*
Borodina, V. G.   603 (10), *636*
Borovnikov, V. P.   594 (370, 371), 595 (370), 596, 597 (371), *598*
Borowitz, I. J.   395 (882), *480*
Borrell, P.   672, 677, 688, 709, 712, 718, 719 (61), *748*
Borrmann, D.   1131 (219–222), 1132 (219–221), 1133 (221), 1134 (221, 222), 1135 (222), *1320*
Borsche, W.   1073, 1115 (43, 44), *1316*
Borsotti, G.   1296 (671), *1329*
Bory, S.   1066 (18), *1316*
Bos, H. J. T.   442 (1277), *488*, 1049 (362), 1052 (406), *1061, 1062*
Bosch, G. van den   442 (1277), *488*
Bosch, J.   1080 (75, 76), 1082 (76), *1317*
Boschetto, D. J.   839 (63), *856*
Boschung, A. F.   687, 743 (104), *749*
Bose, A. K.   155 (152), 168 (221, 224), *172, 174*, 412 (956), *481*, 521 (181), 523 (201), 524 (181), *530*, 614 (98), *638*, 1075, 1079 (52), 1195 (377), 1200, 1201 (432–436), 1204 (431, 445), 1225 (377), 1231 (520), 1232 (520, 542, 545–548, 557, 560–566, 568, 570–576, 588, 593, 594), 1243 (542), 1245 (520, 545, 547, 548), 1246, 1247 (545, 548), 1249 (547, 548, 560, 563, 566), 1250 (547, 557, 560, 564, 566), 1251 (557), 1252 (548), 1254 (560, 565, 566, 568, 572, 574, 575), 1255 (574), 1256 (570, 573), 1257 (560, 566, 571, 574, 575), 1258 (571), 1259 (548, 561, 562, 565, 576), 1260 (588), 1262 (593, 594), 1263 (520), 1272 (614), *1316, 1323– 1328*
Bosi, P.   22, 25 (36), *56*, 232 (212), *261*
Bosler, M.   386 (844), *479*
Bossa, M.   1026 (75), *1054*
Bosshard, H. H.   539 (73, 74), *588*
Bosshard, H. R.   279 (144), *464*
Bosworth, N.   351, 352 (651), *475*
Botor, B.   420 (1077), *484*

Botsch, H. J.   556, 566, 567 (179), 570, 571 (291), *590, 592*
Bott, K.   276 (111), *464*
Bottorff, E. M.   283 (201), *466*
Bougault, J.   1081, 1089 (90), *1317*
Bougeard, D.   226 (143), *259*
Bounoit, A.   313 (416), *470*
Bourgesis, P.   1229 (512), *1326*
Bourn, A. J. R.   247 (381), *264*
Bournay, J.   226 (138), 227 (146), 228 (159), *259*
Bourne, E. J.   411 (939), *481*
Bourns, A. N.   895 (182), *913*
Bousquet, E. W.   358 (694), *476*
Boussinesq, J.   400 (899), *480*
Boutagy, J.   278 (126), 422 (126, 1092), *464, 484*
Boutamine, N.   571 (302), *593*
Bovey, F. A.   99 (168), *117*
Bowen, M. J.   745 (231), *753*
Bowen, R. O.   774 (91), *779*
Bower, F. A.   870 (56), *911*
Bowers, P. R.   673, 683, 685 (85b), *749*
Bowie, J. H.   123 (2), 132 (39, 40), 133 (44, 45), 144 (100), 146 (106), 147 (111), 148 (106), 149 (121), 150 (39, 122), 151 (40, 132–135), 159 (44, 175, 176), 160 (177), 162 (188–191), 163, 164 (191), 168 (222), *169–174*, 1035, 1036 (130, 131), *1055*
Bowles, A. J.   147 (109), *171*
Bowman, E. R.   1303, 1304 (695), *1330*
Bowman, R. E.   272 (43, 47, 48), 279 (138), *462–464*
Box, H. C.   757 (7), *778*
Boyd, G. V.   503 (76), 504 (76, 77), 525 (77), *528*
Boyer, L.   234 (229), *261*
Brace, N. O.   879 (96), *911*
Brachel, H. v.   542 (100, 101), *588*
Brackman, W.   176 (8), *209*
Brade, H.   516 (149), *529*
Bradley, R. J.   596, 597 (385), *599*
Bradshaw, J. S.   445 (1302), 454 (1374), *488, 490*, (137), *750*
Brady, S. F.   72 (27), *114*
Brady, W. T.   396 (884), *480*, 1231 (516), 1232 (516, 526), 1234 (516), 1235 (516, 526), 1236, 1239, 1244 (516), *1326*
Brainard, R.   709, 715 (154, 161), *751*
Brammer, R.   1171, 1172 (319), *1322*
Branca, S. J.   725 (190), *752*
Brandänge, S.   103, 104 (242, 243), *118*
Brandsma, L.   1037 (141), 1042 (141, 294), 1043 (301–303), 1049 (362, 363), 1050 (294), 1052 (294, 363, 406), 1053 (294), *1056, 1059, 1061, 1062*

Brannen, G. G.   318 (443), *471*
Brasch, J. W.   222 (51), 228 (147), *257, 259*
Brasen, W. R.   280 (149), *465*
Brash, J. L.   496 (29b), *527*, 1065 (9), *1315*
Brassard, P.   1193 (364), *1323*
Bratos, S.   216 (4), 220 (32, 33), *256, 257*
Bratož, S.   214, 218 (2d), 226 (140), *256, 259*
Braun, E.   1123 (203), *1320*
Brechbühler, H.   413 (963), *481*, 536 (26), 543, 557, 561 (26, 109), 566, 584 (26), *587, 589*
Breckpot, R.   1196 (389), *1323*
Bredereck, H.   536 (12, 13, 28–30, 43), 537 (30, 57), 538 (13, 28, 57, 62, 63), 539 (64, 72), 542 (108), 543 (28, 43), 544 (57, 64, 115, 116), 545 (28, 43, 122, 123, 125–129), 546 (62–64), 547 (131–133, 138, 139), 548 (28, 43, 108, 125, 126, 132, 133), 549 (64, 115, 116, 146, 147, 149), 550 (115, 158), 551 (57), 553 (108, 125, 129), 554 (13, 28, 43, 129, 175), 555 (116, 177), 556 (13, 43, 108, 175, 179), 557 (28, 64, 181), 558 (28, 108, 149), 559 (57, 158), 561 (43), 562 (13, 128, 149, 200, 201, 204, 212), 563 (127), 566 (13, 127, 128, 175, 179, 212, 269, 276, 278), 567 (43, 127, 128, 139, 175, 177, 179, 212, 276, 278, 279), 569 (115, 116, 149), 570 (200, 201, 276, 283, 291), 571 (131, 133, 139, 200, 201, 276, 279, 283, 285, 291, 295–298, 303, 305–309), 572 (129, 312), 574 (125, 149, 323, 324), 575 (64, 324), 576 (181, 326), 577 (43, 146, 147, 328, 331), 578 (147, 181, 326), 580 (312, 326, 332, 333), 581 (57), 586 (57, 131, 133, 295, 296, 368), 597 (146, 147, 181), *586–594*
Bredereck, H. J.   534, 553 (3), 586 (368), *586, 594*
Bredereck, K.   539 (72), *588*
Breen, L.   274 (77), *463*
Breger, I. A.   759 (15), 762 (41), *778*
Breitbeil, F. W.   1104 (152), *1319*
Brelivet, J.   1048 (357, 358), *1061*
Brendle, T.   545 (128, 129), 553, 554 (129), 562 (128, 214, 215), 566, 567 (128, 215), 572 (129), 577 (214), 578 (215), *589, 591*
Brennecke, L.   564 (239), *591*
Brenner, A.   888 (138), *912*
Brenner, J. B.   1310 (712), *1330*
Brenniger, W.   374 (783), *478*
Brent, D. A.   909 (241), *914*
Bresciani Pahor, N.   1022 (14), *1053*

Breslow, R.   505 (85), *528*
Breuer, E.   614 (101), *638*
Brewster, J. H.   412 (944), *481*
Brickmann, J.   26, 42, 44 (51), *56*
Bricout, J.   1079 (62), *1317*
Bridges, A. J.   1033, 1035, 1042 (108), *1055*
Bridson, J. N.   435 (1206), *486*
Briegleb, G.   175 (1), 176 (6), 195 (78), 196 (86), 198 (96, 97), 200, 201 (106–108), 202 (105–108, 114, 115), *209, 211*
Briggs, W. S.   888, 899, 900 (143), *912*
Brill, W. F.   270 (12), *462*
Brimelow, H. C.   310, 313 (394), *470*
Brink, M. van der   878 (89), *911*
Brink, W. M. van den   1199 (408), *1324*
Britcher, S. F.   299 (322), *468*
Brittain, E. F. H.   131 (36), 147 (109), *170, 171*
Broaddus, C. D.   294 (280), *467*
Brocklehurst, B.   700, 735–737, (142), *750*
Brocksom, T. J.   429 (1157), *485*
Brockway, L. O.   22 (28), *56*
Brode, G. L.   496 (29a), *527*
Broline, B. M.   451 (1346), *489*
Bronold, N. M.   1044, 1049 (314), *1059*
Brook, P. R.   397 (885, 886), *480*
Brookhart, T.   727, 728 (192b), *752*
Brooks, J. R.   1312 (715), *1330*
Brooks, L. A.   283 (200), *466*
Brossi, A.   1072 (33, 34), *1316*
Brouwer, K. B.   863 (16), *910*
Brown, A. L.   873 (68), *911*
Brown, B. R.   863 (13), *910*
Brown, C. A.   431 (1170), *485*
Brown, C. E.   223 (71), *258*
Brown, C. J.   225, 240 (117), *259*
Brown, C. P.   236 (249), *261*
Brown, D. F.   893, 894 (170), *913*
Brown, E. V.   319 (464, 465), 452 (1358), *471, 489*
Brown, G. B.   273 (62), *463*
Brown, G. M.   225 (121), *259*
Brown, H. C.   240 (312), *263,* 341 (621a), 355 (680), 434 (1195–1202), 435 (1205), 453 (1368), *474, 475, 486, 489,* 825 (5), *855,* 904 (216), *914*
Brown, J. B.   1074 (46), *1316*
Brown, J. M.   989 (146), 1005 (175), *1019, 1020*
Brown, J. N.   581 (336), *593*
Brown, K. H.   446 (1308), *488*
Brown, M.   586 (367), *594*
Brown, M. H.   540 (83, 84), 544, 549 (84), 551 (83), 569, 586 (84), *588*
Brown, O. R.   797 (66), 798 (68), 816 (151, 158), 817 (158), *822, 824*

Brown, R. F. C.   869 (48), 874 (74), *910, 911*
Brown, T. L.   88 (123), *116*
Brown, W. G.   298 (311), *468,* 1138 (232, 233), *1320*
Brownlee, S.   1024, 1048 (29), *1053*
Broxton, T. J.   904 (218), *914,* 992 (156), *1019*
Brubaker, M. M.   892 (158), *913*
Bruce, B. D.   235 (230), *261*
Bruce, G. T.   904 (218), *914*
Brück, B.   566 (247), *591*
Bruckenstein, S.   238 (284), *262,* 799 (69), *822*
Brügel, W.   793 (55), *822*
Bruggink, A.   374 (784), *478*
Bruice, P. Y.   1048 (354), *1061*
Bruice, T. C.   619 (129), *638,* 948 (19), 995 (160), 1002 (171), 1015 (203), *1016, 1019, 1020,* 1048 (344, 346), *1060*
Brundage, D. R. G.   673, 682, 683, 685 (76c), *749*
Bruno, J. J.   619 (129), *638*
Brunwin, D. M.   1206 (490, 491), *1325, 1326*
Brutschy, F. J.   323 (485), 325 (490), *471*
Bruylants, A.   179 (38), 180 (44, 49), 181 (38, 44, 49, 52, 53), 183 (38), 202 (49, 53, 123), 203 (49, 53), *210, 212,* 1038 (165), 1042 (165, 298), *1056, 1059*
Bryce-Smith, D.   205 (133, 135, 137–140), 206 (133, 143–147), 207 (151), *212,* 438 (1237), *487,* 642 (4), *747*
Bryson, T. A.   566, 571 (263), *592*
Brzhezitskaya, L. M.   328 (507), *472*
Buben, H. Ia.   761 (38), *778*
Bubl, E. C.   367 (746), *477*
Buchanan, G. L.   375 (804), *478*
Buchardt, O.   738 (215), *753*
Büchel, K. H.   406 (913, 914), *480,* 502 (68), 526 (223), *528, 531,* 536 (34), *587*
Buchholz, G.   1051 (384), *1061*
Büchi, G.   395 (878, 881), 452 (1355), *480, 489,* 585 (360), *594,* 709, 712 (156), *751,* 888, 905 (140), *912,* 1080, 1084 (86), 1168 (305), 1304, 1307 (698), *1317, 1322, 1330*
Büchi, H.   413 (963), *481,* 536 (26), 543, 557, 561 (26, 109), 566, 584 (26), *587, 589*
Buchwald, S. L.   595 (376), *598*
Buck, A.   331 (530), *472*
Buck, J. S.   365 (728), *476*
Buck, K. T.   745 (228c), *753*
Bücking, H. W.   1189, 1190 (355), *1323*
Buckingham, A. D.   222 (61), *257*

Buckingham, D. A. 103 (212), *118*
Buckles, R. 360 (702), *476*
Buckley, G. D. 335 (574), *473,* 1138 (231), *1320*
Bucourt, R. 79, 80, 84 (65), *115,* 624 (167), *639*
Budanova, L. I. 582 (340), 596 (377), *593, 598*
Buddrus, J. 432 (1186), *486*
Budzikiewicz, H. 123 (1), 135 (57), 153 (140), 160 (178), 164, 168 (195), 169 (225), *169, 170, 172–174,* 517 (155), *529,* 613 (95), 614 (95, 96), 616 (108– 110), 617 (113), 626, 627 (175), 628 (183), 633 (211), *638–640,*1036, 1040, 1042 (137), *1055,* 1303 (694), 1306 (705), *1330*
Budzinski, E. E. 757 (7), *778*
Bueding, E. 422, 423 (1102), *484*
Buehler, C. A. 270 (1), 398 (890), 418 (1057), 437 (1218), 442 (1281), *462, 480, 483, 486, 488*
Buenker, R. J. 25, 26, 30 (47), *56*
Bühl, H. 1035 (125), 1041 (125, 275), 1042 (275), *1055, 1059*
Buhle, E. L. 1232, 1255 (578, 582), *1327*
Bühner, A. 536 (32), *587*
Bukač, Z. 868 (45), *910*
Bukarevich, J. V. 108 (267), *119*
Bulgarovskaya, I. V. 194 (75), *211*
Bulina, V. M. 278 (130), *464*
Bull, B. A. 358 (692), *476*
Bull, D. C. 736, 737 (213), *753*
Bulmer, J. T. 235 (234), *261*
Bumgardner, C. L. 883 (120), *912*
Bunge, K. 1052 (407), *1062*
Bunnenberg, E. 88 (111, 112), 89 (111), 102 (207), 103, 104 (241), 109 (112), *116–118*
Bunton, C. A. 275 (97, 99), 365, 367 (739), *464, 477,* 902 (210), *914,* 947 (2, 7), 956 (7), 957 (2), 958 (7, 66), 959 (66, 67, 70), 962 (73a), 963 (70, 78, 80, 81), 964, 965 (73a), 969 (7, 92), 970 (92), 971 (70, 94, 95), 972 (96– 98), 976 (99, 100), 977 (2, 7, 92), 980 (70), 981 (70, 111), 984 (115, 116), 985 (128, 129, 133), 988 (92, 139), 990 (139), 991 (153), 993 (159a, b), 1005 (175), 1008 (94, 184, 186, 187), 1009 (159a, 188), 1010 (94, 184), 1011 (189), 1014 (92), *1016–1020*
Bunzl, K. 238 (273), *262*
Buravchuk, Yu. P. 241 (323), *263*
Burckhardt, V. 353 (669), 355 (675, 676), *475*
Burden, A. G. 250 (404), *265*
Burdett, K. A. 731, 732 (200a, b), *752*

Bureiko, S. F. 238 (285), *262*
Burgess, K. A. 355 (684), *476*
Burgh Norfolk, S. de 897, 898 (191), *913*
Burgot, J.-L. 1047 (342), 1052 (402), *1060, 1062*
Burgstahler, A. W. 493 (8), *526*
Burke, S. 505 (83), *528*
Burkett, A. R. 281 (175), *465*
Burks, R. E. Jr. 272 (22), *462*
Burlingame, A. L. 123 (3), 148 (115), 154 (144), *169, 172,* 507 (92), *528,* 627 (177), *639*
Burr, G. O. 71 (22), *114*
Burr, J. G. 759 (22), *778*
Burridge, C. 299 (319), *468*
Burrous, M. L. 884, 885 (128), *912*
Bursey, J. M. 154 (144), *172*
Bursey, J. T. 130 (32), *170*
Bursey, M. B. 127, 128 (21), *170*
Bursey, M. M. 130 (32), 140 (81), *170, 171*
Burstain, I. G. 1171, 1172 (317, 318), *1322*
Bürstinghaus, R. 1039 (246), *1058*
Burtner, R. R. 294 (274), *467*
Burton, D. J. 440 (1261), *487*
Burton, M. 671, 672 (92c), *749*
Burton, V. L. 762 (40, 41), *778*
Burwasser, H. 661, 662 (43), *748*
Burwell, R. L. Jr. 373 (774), *477*
Bus, J. 1185 (345), *1323*
Buse, C. T. 375, 379 (800), *478*
Bush, C. A. 73, 99 (37), 102 (185), 105 (185, 256), 109 (278), *114, 117, 119*
Bush, J. B. Jr. 1172, 1173 (322), *1322*
Buss, J. H. 60, 61 (9), *65*
Buswell, R. J. 411 (929), *481*
Busz, J. 534 (1), *586*
Butcher, M. 869 (48), *910*
Butler, A. R. 442 (1284), *488*
Butler, K. 440 (1260), *487*
Butt, G. L. 904 (218), *914*
Buttery, R. G. 137 (67), *170*
Buu, N. T. 869 (52), *910*
Buu-Höi, N. P. 177 (25), *210*
Buurmans, H. M. A. 126 (13), *169*
Buxton, G. V. 766, 767 (63), *779*
Buxton, M. W. 318, 320, 325, 327 (452), *471*
Buysch, H. J. 288 (226, 227), *466*
Buza, D. 594 (374), 597 (394), *598, 599 599*
Buzas, A. 412 (951), *481*
Buzzese, T. 274 (83), *463*
Byerley, J. J. 457 (1400), *490*
Bykova, L. M. 1047 (337), *1060*
Byron, D. J. 273 (59), *463*

Cabassi, F.   228 (150), *259*
Cabiddu, S.   565 (243), *591*
Caccamese, S.   517 (150), *529*
Cacchi, S.   340, 436 (618), 449 (1331), *474, 489,* 1039 (249), *1058*
Cadby, P. A.   437 (1225), *487,* 676, 743 (221a), *753*
Cadoff, B. C.   1162, 1168 (289), *1321*
Cadogan, J. I. G.   289 (229, 230), *466*
Caglioti, L.   340, 436 (618), 444 (1293), *474, 488,* 1039 (249), *1058*
Cahn, R. S.   622 (158), *639*
Cain, B. J.   415 (978), *482*
Cainelli, G.   281 (186, 190, 191), 345 (630), 415 (1004), *465, 475, 482*
Cairns, T. L.   517 (159), *529,* 537 (51), *587,* 1302 (687), *1330*
Calas, R.   454 (1379), *490*
Calder, G. V.   499 (49), *527,* 691, 701 (125c), 703, 738 (144), *750, 751*
Caldin, E. F.   218 (14), *256*
Caldwell, D. J.   71 (29), *114*
Caldwell, J. R.   909 (240), *914*
Calligaris, M.   1022 (14), *1053*
Callingham, R. H.   339 (603), *474*
Callot, H. J.   674 (78), *749*
Calvert, J. G.   612 (91), *637,* 643 (6), 714 (175b), *747, 751*
Calzadilla, M.   979 (104), *1018*
Cambie, R. C.   494 (14), *526*
Campaigne, E.   315 (428), *470*
Campbell, H.   365 (729), *476*
Campbell, J. R.   326, 335 (496), 432 (1181), *472, 486*
Campbell, W. J.   331, 341 (527), *472*
Candlin, J. P.   826 (25), *856*
Cane, D. E.   457 (1397), *490*
Cann, J. R.   101 (183), *117*
Cannon, J. G.   447 (1317), *488*
Cantrell, T. S.   719 (181), *751*
Canziani, P.   632 (208), *640*
Căpăt, M.   603, 631 (14), *636*
Capella, P.   161 (180), *173*
Capobianco, G.   814, 816, 817 (142), *824*
Capp, C. W.   1285 (662), *1329*
Caprioli, R. M.   133 (51), 141 (90, 91), 143 (94), *170, 171*
Caputo, J. A.   326 (495), *472*
Carassiti, V.   642 (2), *747*
Cardillo, G.   281 (186, 190, 191), *465*
Carey, M. C.   952 (43), *1017*
Carl, T. P.   221 (43), *257*
Carlock, J. T.   454 (1374), *490*
Carlon, F. E.   1145, 1159 (256), *1321*
Carlsen, L.   1047 (338), 1049 (367), *1060, 1061*
Carlson, R. G.   731, 733 (201a), *752,* 901 (206), *914,* 1186 (346), *1323*

Carmack, M.   343 (623), 369 (752), *474, 477*
Carnahan, E.   1052 (405), *1062*
Carnahan, J. E.   586 (367), *594*
Carney, R. L.   286 (216), *466*
Caron, G.   280, 281 (171), *465,* 1070 (23), *1316*
Carothers, W. H.   496 (28), *527*
Carpenter, B. K.   584 (356), *594*
Carr, P. P.   1042 (283), *1059*
Carrol, F. I.   80 (67), *115*
Carroll, F. I.   335 (576), *473*
Carroll, J. T.   447 (1318), *488*
Carroll, M. F.   500 (56), *527*
Carruthers, C. R.   104 (246), *118*
Carson, A. S.   61 (20), 63 (39), *66*
Carson, B. B.   274 (74), *463*
Carter, H. E.   1108 (161), *1319*
Cartwright, D.   1042 (283), *1059*
Cartwright, D. C.   225 (129), *259*
Carty, B. J.   312 (406), *470*
Caruthers, L. M.   232, 243 (214), *261*
Casellato, F.   177 (24), 184 (66, 67), *209, 210*
Casey, J. P.   98, 99 (165), *117*
Casida, J. E.   656, 699 (31), *747*
Cason, J.   271 (17), 274 (80), 432 (1178), *462, 463, 486*
Caspi, E.   326 (497), 364 (723), 368, 369 (497), *472, 476,* 846 (77), *857*
Cassady, J. M.   507 (92, 93), *528,* 627 (177), *639,* 1115 (189), *1319*
Cassan, J.   161 (184), *173*
Cassar, L.   290 (240), 291 (251), 292 (254), *466, 467,* 1180 (335), *1322*
Cassidy, P. E.   1042 (300), *1059*
Castaner, J.   1080 (68, 69, 76), 1082 (68, 76), *1317*
Castelfranchi, G.   884 (129), *912*
Castella, J.   1080 (70, 76), 1082 (76), *1317*
Castells, J.   1080 (70, 73, 75, 76), 1082 (76), *1317*
Castle, J. E.   334 (570), *473*
Castro, B.   437 (1228, 1229), 443 (1228), *487,* 1038 (178), *1056*
Casu, B.   177 (24), 184 (67), *209, 210*
Casy, A. F.   610 (69), *637*
Cate, W. E.   398 (890), *480*
Cava, M. P.   162 (185), *173,* 382 (824), *478,* 633 (212), *640,* 874 (73), *911*
Cavalieri, L.   510 (109), *528*
Cavanaugh, J. R.   246 (374), *264*
Cave, A.   1304 (700), *1330*
Cavé, A.   452 (1361), *489*
Cavell, R. G.   250 (405), *265*
Cayen, C.   431 (1173), *485*
Caze, C.   180, 183, 203 (48), *210*

Cazee, C.   127, 143 (20), *170*
Cefelin, P.   1196 (384, 385), *1323*
Centeno, M.   979 (104), *1018*
Čerček, B.   760, 768 (30), *778*
Cereghetti, M.   1306 (705), *1330*
Cerfontain, H.   664, 666, 708, 738 (57a, b), *748*
Cernansky, B.   757 (8), *778*
Çerutti, P.   1206, 1219 (483), *1325*
Červinka, O.   80, 81 (69), *115*
Cha, Sungman   609 (63), *637*
Chabanel, M.   235 (242), *261*
Chabudzinski, Z.   1276, 1281 (639), *1328*
Chafetz, H.   362 (717), *476*
Chaikin, S. W.   1138 (232), *1320*
Chaimovich, H.   963 (80), *1018*
Chakrovarti, K. K.   280, 281 (166), *465*
Challis, B. C.   442 (1284), *488*
Chalmers, A. M.   1076, 1078 (53), *1316*
Chamberlain, G. A.   661, 662, 665 (45, 46), *748*
Chambers, R. D.   354 (672), *475*
Chambers, W. J.   280 (148, 149), *465*
Chan, A. S. Y.   332 (553), *473*
Chan, A. W. K.   655–657 (30), *747*
Chan, T. H.   444 (1294, 1295), *488*
Chan, W. K.   1048 (350), *1060*
Chang, C.   1115 (189), *1319*
Chang, C. W. J.   143 (94), *171*
Chang, C. Y.   247 (382, 388), *264*
Chang, F.   272 (27), *462*
Chang, S. P. de   896, 897 (192), *913*
Chang, Y.   340 (617), *474*
Chantooni, M. K.   223 (75, 76), 240 (75, 305), *258, 263*
Chapleo, C. B.   1184 (339), *1322*
Chapman, C.   805 (106), *823*
Chapman, D.   222 (65), *258*
Chapman, O. L.   165, 168 (206), *173*, 497 (34), 499 (49), 512 (122), *527, 529,* 620 (148), *639,* 655 (29), 658 (35), 661 (29, 37, 40), 672 (66), 691 (29, 37, 125a–e), 700 (29), 701 (125a–e), 702 (125b), 703 (144), 738 (144), *747, 748, 750, 751,* 1080 (82), 1192 (359), 1206 (457–459), 1207 (457, 458), 1208 (457–459), 1209 (459), *1317, 1323, 1325*
Chappell, G. S.   417 (1049), *483*
Chasle, M. F.   1273 (619), *1328*
Chastel, R.   62 (33), *66*
Chastrette, M.   358 (691), *476*
Chatterjee, B. G.   1200 (434), 1201 (434, 437–442), 1202 (439), 1203 (440), *1324, 1325*
Chattopadhyaya, J. B.   451 (1345), *489*
Chau, F.   673 (74), *749*

Chauhan, M. S.   1040, 1042 (259), *1058*
Chauhan, S. M. S.   418 (1055), *483*
Chaula, O. P.   768 (72), *779*
Chaux, R.   313 (415), *470*
Chaves das Neves, H. J.   1038 (170), 1048 (360), *1056, 1061*
Chavis, C.   1040–1042, 1052 (268), *1058*
Chawla, H. P. S.   1201, 1203 (440), 1232 (560, 566, 575), 1249, 1250 (560, 566), 1254, 1257 (560, 566, 575), *1325, 1327*
Chayet, L.   963 (80), *1018*
Chaykovsky, M.   416 (1011), *482,* 1206, 1217, 1218 (478), *1325*
Cheburkov, Y. A.   498 (40), *527*
Chemielarz, B.   1065 (7), *1315*
Chen, C. G.   1206, 1220 (485), *1325*
Chen, C. H.   349 (648), 351 (649), *475*
Chen, C. Y. S.   517 (152), *529*
Chen, E. C.   196, 197 (89), *211*
Chen, G.   1025, 1030, 1038 (42), *1053*
Chen, L.   112 (316), *120*
Chen, P. H.   166 (212), *173,* 495 (21), *527,* 627 (181), *639*
Chen, R. H. K.   296 (288), *467*
Cheng, S.-Y.   313 (419), *470*
Cherluck, R. M.   448 (1323), *489*
Cheronis, N. D.   602 (3), 603 (19), 618 (126), 629 (188), *635, 636, 638, 639*
Cherton, J.-C.   689, 700 (119), *750*
Chessick, J. P.   255 (475), *266*
Cheung, H. T.   275 (89), *463*
Chevion, M.   950 (24), 985 (127a, b), 1008 (127a, b, 183), 1009 (127a), *1016, 1019, 1020*
Chiacchio, U.   1023 (23), *1053*
Chiang, Y. H.   1232, 1245 (545, 547), 1246, 1247 (545), 1249, 1250 (547), *1327*
Chib, J. S.   1232 (547, 566), 1245 (547), 1249, 1250 (547, 566), 1254, 1257 (566), *1327*
Chiba, K.   932 (46), *943*
Chiba, T.   231 (199), *260*
Chick, F.   498 (44), *527*
Chidester, C. G.   282 (195), *465*
Chieh, P. C.   244 (349), *264*
Chigira, M.   1003 (173), *1020*
Chihara, H.   231 (203, 205), *260*
Child, W. C.   307 (372, 373), *469*
Childs, W. V.   803, 808 (88), *823*
Chimiak, A.   102 (203), *117,* 420 (1077), *484*
Chin, A.   676, 743 (221b), *753*
Chinn, L. J.   332 (539), *473*
Chirinko, J. M. Jr.   280 (160), *465*
Chiron, R.   254 (448), *265,* 524 (212), *531*
Chiusoli, G. P.   290 (240, 241, 243), 291 (250), *466, 467,* 1079, 1080 (63), *1317*

Chizhov, O. S. 138 (75, 76), 148 (76), 171
Chladek, S. 557 (183), 558 (183, 186, 187), 562 (183), 563 (227), 590, 591
Chmielewski, M. 1038 (176), 1056
Chodak, J. B. 195 (76), 211
Choi, Young Sang 252 (424), 265
Choloniewska, I. 231 (202), 260
Choo, K. Y. 740 (220b), 753
Chottard, G. 103, 104 (238), 118
Chow, A. W. 415 (970), 481
Chow, Y. L. 272 (52), 463, 652, 653 (21, 22), 747
Chowdhury, M. 181 (51), 210
Christeleit, W. 103 (245), 118
Christensen, B. G. 1232 (567, 590), 1245 (590), 1257, 1258 (567), 1301 (681), 1327, 1328, 1330
Christensen, D. H. 28, 29 (63), 56
Christensen, P. K. 1080, 1082 (65), 1317
Christian, S. D. 235 (238, 244, 245, 247), 236 (248b), 250 (408), 261, 265
Christiani, G. F. 1197 (402), 1198 (393, 395, 398, 400), 1324
Christiansen, R. G. 272 (45), 463
Christie, P. 457 (1394), 490
Christl, B. 1038, 1042, 1052 (167), 1056
Christoffersen, R. E. 28, 29, 31 (65), 32 (89, 90, 92), 33 (92), 45 (115), 56, 57
Christy, K. J. 389 (863), 479
Chruma, J. L. 415 (994), 482
Chu, S. H. 1024 (30), 1053
Chuang, C. K. 318 (446), 471
Chuchani, G. 896, 897 (192), 913
Chudek, J. A. 179 (40), 210
Chulanovskii, V. M. 214 (1c), 256
Ciabattoni, J. 1131, 1135 (224), 1320
Ciardelli, F. 69, 73 (6), 113, 495 (17), 526
Cignarella, G. 1197 (402), 1324
Ciotti, C. J. Jr. 412 (944), 481
Cipiciani, A. 992 (157), 1019
Citron, J. D. 416 (1026), 483
Ciuhandu, C. 603, 631 (14), 636
Clague, A. D. H. 230 (176), 235 (176, 231), 260, 261
Claisen, L. 307 (369), 469
Clapp, L. B. 609 (63), 637
Clarc-Lewis, J. W. 110 (299), 119
Clark, F. 1015 (201), 1020
Clark, G. 826 (13), 855
Clark, J. H. 237 (270), 262
Clark, L. W. 866 (32), 910
Clark, M. 354 (672), 475
Clark, R. D. 167 (215), 174, 335 (575), 362, 363 (720), 473, 476, 498 (46), 518, 520 (170), 527, 530, 1168 (309), 1200 (420), 1322, 1324

Clarke, E. J. 1078 (56), 1317
Clarke, F. H. 1306, 1312 (704), 1330
Clarke, H. T. 273 (64), 318 (451), 319 (461), 463, 471, 521 (183), 530
Clarke, J. T. 355 (684), 476
Clarke, T. G. 304, 315 (346), 317 (442), 469, 471
Clausen, K. 1040 (413), 1062
Clauss, G. 409 (922), 480
Clauss, K. 1225–1227 (503), 1326
Claussen, K. 368 (749), 477
Clavel, S. 1040 (270), 1058
Clayton, R. B. 847 (80), 857
Cleland, J. H. 304 (349), 469
Clemens, D. H. 540, 541 (88), 545–547 (88, 124), 548, 559 (88), 572, 573 (88, 124), 588, 589
Clemens, K. E. 541 (89), 584 (356), 588, 594
Clement, D. A. 849 (86), 857
Clementi, E. 22, 25 (36), 42 (108), 56, 57, 219 (21, 25), 232 (212), 233 (21), 257, 261
Clementi, S. 906 (230), 914
Clements, J. B. 872 (66), 911
Clemo, G. R. 1305 (703), 1330
Clennan, E. L. 511 (118b), 529
Clesse, F. 1042 (284), 1059
Cleveland, P. G. 1206, 1208, 1209 (459), 1325
Close, V. 419 (1070), 483
Closs, G. L. 740 (220a), 753
Closson, R. D. 280 (172), 441 (1273), 465, 488
Closson, W. D. 69 (12), 71 (12, 26), 72 (12, 26, 27), 76 (12), 114, 622 (160), 626 (160, 174), 639
Clough, S. C. 1296 (672), 1329
Clusius, K. 672, 677, 678 (60a), 748
Clutterbuck, P. W. 366–368 (742), 477
Cobb, D. B. 132, 151 (40), 170
Coburn, W. C. 241 (321), 263
Cocalas, G. H. 1200 (414, 415), 1324
Cochoy, R. E. 1284 (657), 1329
Cochran, W. 224 (87), 258
Cocker, W. 1066, 1068, 1146, 1148 (17), 1316
Cockerill, A. F. 609 (59), 637
Coe, P. L. 318, 320, 325, 327 (452), 471
Coffey, R. S. 1109 (167), 1319
Coffin, R. L. 731, 733 (201a), 752
Coffman, D. D. 586 (367), 594
Cohane, B. 234 (227), 261
Cohen, E. 333 (561, 562), 473
Cohen, H. 283 (202), 420 (1078), 466, 484
Cohen, L. A. 1200 (416, 417), 1324

Cohen, N.   1066, 1068 (19), 1072, 1073 (30, 31), 1145 (254), 1147 (19, 254), 1148 (19), 1153 (19, 30, 31, 254), *1316, 1321*

Cohen, S. G.   685 (102a, b), *749, 772* (84), *779*

Cohen, T.   287 (220), *466,* 1115 (188), *1319*

Cohen, V. I.   1040 (270), *1058*

Cohen-Addad, C.   245 (354), *264*

Colangelo, J. P.   711, 713, 714 (171), *751*

Cole, W.   1098 (133–135), *1318*

Coleman, G. H.   790 (47), *822*

Coleman, J. P.   810, 811 (135), *824*

Coletti-Previero, M. A.   1042 (299), *1059*

Colin, G.   419 (1069), *483*

Collet, A.   109 (284), 110 (284, 296), *119*

Colleter, J. C.   255 (478), *266*

Collier, G.   250 (404), *265*

Collin-Asselineau, C.   1066 (18), *1316*

Collings, A. J.   1030 (89), *1054*

Collins, C. H.   883, 884, 886, 887 (119), *912*

Collins, C. J.   386, 387 (849), *479*

Collman, J. P.   292, 456 (257), *467*

Colman, G. H.   273 (61), *463*

Cologne, J.   901 (205), *914*

Colomina, M.   60 (15, 16), 61 (29–31), 62 (16, 29, 30), 63 (15), *65, 66*

Colonna, A.   447 (1315), *488*

Colter, A. K.   202 (120, 121), *212*

Colvin, E. W.   272 (38), *462*, 1206 (456), *1325*

Colwell, W. T.   457 (1394), *490*

Compere, E. L.   276 (107), *464*

Compernolle, F. C.   163, 164 (192), *173*

Comstock, D. A.   774 (91), *779*

Conant, J. B.   406 (915), *480*

Condorelli, G.   1026 (77), *1054*

Conley, R. T.   1284, 1286, 1287, 1290 (647), *1329*

Connelly, S. A.   414 (1000), *482*

Connett, B. E.   205 (139), 207 (151), *212*

Connett, J. E.   62 (34), *66*

Conrad   916 (2), *942*

Conrad, M. E.   19, 21, 22, 25, 26, 28–30 (8), *55*

Consonni, P.   523 (197), *530*

Contento, M.   281 (190, 191), *465*

Conway, B. E.   782 (10), *821*

Conway, W. P.   428 (1154), *485*

Cook, A. H.   1232, 1264 (597), *1328*

Cook, E. W.   819 (164), *824*

Cook, P. F. E.   411 (943), *481*

Cooke, H. G. Jr.   271 (21), *462*

Cooke, P. W.   183 (60), *210*

Cooks, R. G.   132 (38, 39), 144 (38), 147 (111), 150 (39, 122), 162–164 (191), *170–173,* 1035, 1036 (130, 131), *1055*

Cooksey, A. R.   904 (218), *914*

Cookson, J. P.   419 (1064), *483*

Cookson, R. C.   91, 92 (151), *116,* 904 (215), *914,* 1186, 1187 (349), *1323*

Cooper, C. S.   294 (212), *466*

Cooper, G. F.   304 (348), *469,* 1145, 1152 (270), *1321*

Cooper, G. H.   318 (449), *471*

Cooper, R. D. G.   524 (206), *530*

Cope, A. C.   279 (145), 417 (1034), *464, 483,* 883 (120), 901 (204), *912, 914*

Coppens, P.   219 (24), *257*

Coppens, R.   224 (97), *258*

Coppi, G.   274 (83), *463*

Coquelin, J.   440 (1258), *487*

Corbin, J. L.   127, 128, 142, 158 (17), *169*

Corbin, V. L.   430 (1160), *485*

Cordes, E. H.   947 (3–5), 950 (22, 23), 953 (51), 956 (4, 5), 959 (3, 69), 965, 971 (22), 972 (3, 22), 974 (22), 977 (3, 22), 978 (3, 4, 22, 103), 979 (22, 104, 105), 984 (120), *1016–1018*

Corey, E. J.   278 (128, 129), 290 (244), 291 (252, 253), 296 (288), 311 (401), 314 (401, 426), 416 (1007), 422 (1099), 456 (252, 253), 457 (1397), *464, 466, 467, 470, 482, 484, 490,* 514 (138, 139), 515 (142), *529,* 701 (143), *751,* 1039 (221, 236), 1048 (236), *1057, 1058,* 1075 (49), 1092, 1095 (107), 1166 (299, 303), 1168 (299, 302, 303), 1194 (367), 1206, 1221 (489), 1225 (367), 1232 (578, 580), 1246 (580), 1255 (578, 580), *1316, 1318, 1322, 1323, 1325, 1327*

Corey, H. S. Jr.   278 (127), *464*

Corey, M. D.   348 (645), *475*

Cornforth, J. W.   366 (743), *477,* 1115 (187), 1170 (313), *1319, 1322*

Cornforth, R. H.   366 (743), *477,* 1115 (187), 1170 (313), *1319, 1322*

Cornilov, M.   536, 540, 543, 545, 554, 556, 557, 559, 562, 566, 569 (38), *587*

Cornut, J. C.   226 (137), *259*

Corral, R. A.   418 (1061), *483*

Corrodi, H.   369 (755), *477*

Corse, J. W.   138 (73), *171*

Corson, B. B.   333 (566), *473*

Costa, L. M.   963 (80), *1018*

Costain, C. C.   27, 28 (59), *56,* 233, 235 (218), *261*

Costanzo, S. J.   287 (218), *466*

Costin, R. C.   536, 583, 584 (27), *587*

Costisella, B.   279 (132), *464,* 557, 559, 561, 563 (180), 564 (235–239), 565 (235), 572 (180, 317), 573 (317), 576 (327), 596 (383, 384), *590, 591, 593, 599*

Cota, D. J.   453 (1366), *489*
Cotter, J. L.   144 (99), 155, 156 (151), 164 (201, 204, 205), 165 (204), *171– 173*, 614, 615 (97), *638*
Cotterell, G.   430 (1160), *485*
Cottis, S. G.   1065 (10), *1315*
Cotton, F. A.   841 (67), *857*
Coulombeau, C.   611 (79), *637*
Counsell, J. F.   62 (34), *66*
Coupry, C.   230, 235 (176), *260*
Courtois, G.   1113 (183), *1319*
Courtois, J. E.   365 (736, 737), *477*
Courtot, P.   663 (54), *748*
Cousins, P. R.   102 (191), *117*
Coutrot, P.   421 (1089, 1090), *484*
Coutts, R. T.   164 (194), *173*
Couturier, R.   1039 (206), *1057*
Covitz, F. H.   782, 818 (5b), *821*
Cowan, I.   293 (273), *467*
Cowan, J. C.   271 (21), *462*
Cowdrey, W. A.   496 (31), *527*
Cowley, S. W.   140 (84), *171*
Cox, A.   272 (52), *463*, 652, 653 (21, 22), *747*
Cox, E. G.   222 (54), *257*
Cox, G. R.   430 (1160), *485*
Cox, G. W.   224 (96), *258*
Cox, J. C. Jr.   359 (695), *476*
Cox, J. D.   60–65 (6), *65*
Cox, N.   83 (83), *115*
Cox, R. E.   123 (3), *169*
Coyle, J. D.   710 (164), 711 (173), 715 (164), 721, 724 (187), 738 (214a), *751–753*, 1051 (382), *1061*
Crabbé, P.   69, 73 (1–3), 102, 103 (201), 108 (1, 3), *113, 117*, 611 (75, 78, 80), *637*
Craig, J. C.   88 (137), 89 (137, 147), 92, 94 (152), 95 (158), 99 (167), 109 (147, 281), *116, 117, 119*, 369 (760), *477*, 629 (200), *640*
Craig, L. C.   516 (147), *529*, 820 (169), *824*
Craig, W. G.   332 (553), *473*
Cram, D. J.   359 (701), 376 (807, 808), 402 (903), *476, 478, 480*
Crandall, J. K.   709 (158), *751*
Crank, G.   542 (99), *588*
Craven, B. M.   224 (96), *258*
Crawford, R. J.   294 (280), *467*
Crawford, T. H.   1080 (77), *1317*
Crawford, W. C.   1225, 1227 (499), *1326*
Creger, P. L.   280 (154, 156–159), 281 (154, 159), *465*, 500 (59), *527*, 1116 (195), *1320*
Cregge, R. J.   427 (1140), *485*
Cresson, P.   536, 537 (40), 583, 584 (40, 352), *587, 594*

Criegee, R.   330 (522), 340 (615), *472, 474*
Cristensen, H.   294 (277), *467*
Cristol, S. J.   375, 377 (794), *478*, 1104 (150), *1319*
Crombie, L.   331, 341 (525), *472*
Crooy, P.   1042 (298), *1059*
Cross, A. D.   1090 (105), *1318*
Cross, B. E.   1138 (241, 244, 248), 1140, 1141 (241, 248), 1143 (244), 1164 (248), *1320, 1321*
Cross, J. T. D.   826 (16), *855*, 867 (35), 901 (202), *910, 914*
Cross, P. C.   16 (4), *55*
Crouse, D. N.   456, 457 (1390), *490*
Crowder, G. A.   254 (459), *266*, 1029 (79, 81, 85, 86), *1054*
Crowfoot, D.   524 (209), *530*
Crowley, K. J.   719, 722 (179), *751*
Crozby, D. G.   673, 682 (73), *749*
Cruickshank, F. R.   60, 61, 63, 65 (8), *65*
Crump, R. A.   180 (47), *210*
Csernansky, J. G.   1038 (186), *1057*
Csizmadia, I. G.   6 (1), 20 (11), 22 (37), 23 (37, 42), 24 (37), 26 (37, 42), 27 (56), 28 (70), 29, 30 (70, 73), 36 (102), 39 (103), 48 (121), 49 (123), *55–58*
Cuellar, L.   1090 (105), *1318*
Culbertson, T. P.   583 (348), *593*
Cullis, C. F.   321 (471), *471*
Cunneen, J. I.   107 (263), *119*
Cunneen, J. T.   403 (909), *480*
Cunor, C. H.   903, 906 (213a), *914*
Curl, R. F. Jr.   21 (24), 22 (24, 33), 27 (55), *56*
Curnutte, B.   223 (66), *258*
Currand, J.   775, 776 (94), *779*
Currie, M.   223 (82), 224 (84), *258*
Curtin, D. Y.   234 (223), *261*, 882, 899 (108), *912*
Curtis, E. A.   448 (1323), *489*
Curtis, R. F.   125, 138 (9), *169*, 376 (809), *478*
Curtiss, L. A.   19 (6), *55*
Cushley, R. J.   1035, 1036 (129), *1055*
Cushman, M.   452 (1355), *489*, 585 (360), *594*
Cusic, J. W.   294 (274), *467*
Cutler, A.   1228 (510), *1326*
Cuvigny, I.   280 (153), *465*
Cuvigny, T.   414 (1001), 456, 457 (1391), *482, 490*, 547 (134), *589*
Cuzent, M.   1073, 1115 (42), *1316*
Czajkowski, R.   566 (266), *592*
Czekalla, J.   176, 180, 181, 183 (17), 195 (78), 196 (17), 198 (17, 96–99), 202 (105), *209, 211*

Dacanski, J. 499 (49), *527*
Dachs, K. 1194 (268), *1323*
Dack, R. J. 202 (120, 121), *212*
Dadali, V. A. 964, 1004 (87b), *1018,*
1048 (352), *1060*
Dadic, M. 1232, 1258 (585), *1327*
Dagli, D. J. 1039 (240), 1050 (377),
1058, 1061
Dahlbom, R. 102 (207), 103 (211), *117,*
118
Dahlmann, J. 314 (420), *470*
Dahlqvist, K. I. 31 (85), *57*
Dahn, H. 109 (277), *119*
Dailey, B. P. 22, 23 (32), *56*
Dalgaard, L. 132, 151 (40), *170,* 1033,
1034 (111), 1037 (111, 158, 159),
1039 (238), 1042 (111, 158, 159),
1055, 1056, 1058
Dallas, J. L. 230 (192), *260*
Dalton, J. C. 711, 716, 744 (168a, b),
(177), *751*
Dalton, L. K. 1104 (154), 1105 (155),
1147 (154), 1173 (154, 155), 1179
(154), *1319*
Dalton, L. R. 230 (189), *260*
Daly, N. J. 876 (86), 893, 894 (168),
911, 913
Dam, M. van 510 (113), *528*
Damirov, M. A. 1039 (205), *1057*
Dandegaonker, S. H. 91, 92 (151), *116*
Daneels, D. 1038 (174), *1056*
Danen, W. C. 744 (227), *753*
D'Angelo, J. 302 (335), *468*
D'Angelo, P. F. 661, 662 (44b), *748*
Dangyan, M. T. 1072 (35), 1075 (48),
1090 (102), 1096, 1101 (118), 1104
(146, 147), 1110 (171), *1316, 1318,*
1319
Danheiser, R. L. 1092, 1095 (107), *1318*
Danieli, B. 1189 (354), *1323*
Daniels, T. C. 273 (58), *463*
Daniewski, M. 1039 (209), *1057*
Danno, S. 432 (1183), *486*
Dardoize, F. 457 (1393), *490*
Darlow, S. F. 224 (87), 239 (295), *258,*
262
Das, K. G. 155 (152, 154), 156 (154),
172, 614 (98), *638*
Das Gupta, T. K. 1180 (336), *1322*
Dashkevich, L. B. 1050 (381), *1061*
Dashunin, V. M. 1065 (8), *1315*
Daub, G. H. 277 (118), *464,* 1110 (170),
1319
Daub, J. 418 (1056), *483,* 536 (17–19),
554 (17), 566 (244), *587, 591*
Dauben, W. G. 386 (842), *479,* 706
(146b), 729 (195b), 730 (146b), 731,
733 (146b, 201b), 737 (195b), *751, 752*

Dave, H. R. 458 (1409), *490*
Dave, V. 436 (1214), *486,* 1185 (341),
1322
Daviaud, G. 430 (1163–1165), *485*
David, M. P. 905 (223), *914*
Davidkov, K. 225 (114), *258*
Davidson, D. 868 (39), *910*
Davidson, E. R. 29, 30 (74), *57,* 743
(223b), *753*
Davidson, R. S. 204 (127), *212,* 673
(76a–c), 676 (90), 682 (76a–c, 96a–c),
683 (76a–c), 685 (76a–c, 96c), 686,
687, 743 (90), *749*
Davidson, V. L. 332 (544), *473*
Davies, A. G. 353 (654), *475,* 496, 499
(23), *527,* 1312 (716), *1330*
Davies, B. A. 522 (190), *530*
Davies, D. C. 375, 377 (792, 795), 378
(795), *478*
Davies, D. R. 245 (359), *264*
Davies, G. L. O. 176, 177 (15), *209,* 609
(59), *637*
Davies, H. S. 888 (139), *912*
Davies, J. E. 335 (577), *473,* 516 (144),
529
Davies, J. T. 955 (57), *1017*
Davies, L. S. 409 (921), *480*
Davies, M. 239 (298), 247, 248 (384),
262, 264, 606 (36), 609 (61), *636, 637*
Davies, M. M. 221 (42), 238 (280), 246
(378), *257, 262, 264*
Davis, A. G. 417 (1039), *483*
Davis, B. 144 (97), *171*
Davis, B. A. 1232, 1239 (538), *1326*
Davis, C. B. 460 (1413), *490*
Davis, G. T. 452 (1362), *489*
Davis, J. C. 235 (232), *261*
Davis, M. 1037 (160), 1052 (408), *1056,*
1062
Davis, R. H. 315 (429), *470*
Dawson, D. J. 451 (1350, 1351), *489,*
845 (75), *857*
Day, A. R. 307 (371), *469*
Day, E. A. 166, 167 (208), *173,* 626,
627 (176), *639*
Day, R. O. 520 (176), *530*
Day, V. W. 520 (176), *530*
Dayal, B. 1232 (560, 566, 575, 588, 593,
594), 1249, 1250 (566, 575), 1254, 1257
(560, 566, 575), 1260 (588), 1262 (593,
594), *1327, 1328*
Dayan, J. E. 303 (341), *469*
Deady, L. W. 904 (218), *914,* 992 (156),
1019
Dean, F. H. 415 (972), *482*
Dean, F. M. 493, 503 (4), *526*
Dean, P. D. G. 272 (29), *462*
Deb, K. K. 235 (232), *261*

DeBernardo, S. 102 (194), *117,* 566 (266), *592*
DeBoer, Th. J. 1185 (345), *1323*
Debono, M. 369, 371 (758), *477,* 1100, 1101 (139), *1318*
Dec, J. 271 (21), *462*
Decazes, J. 1232, 1236, 1238 (532), *1326*
Decius, J. C. 16 (4), *55*
Dee, P. L. 322 (479), *471*
Deets, G. L. 1115 (188), *1319*
Defaye, G. 1145 (275), *1321*
Degani, Y. 609 (57), *637*
Dega-Szafran, Z. 228 (160), *259*
Degering, E. F. 303 (338), *468*
DeGraff, J. W. M. 419 (1071), *483*
Deguay, G. 566 (261), *592*
Deguchi, Y. 675, 682 (88), *749*
Dehm, D. 506 (86, 87), *528,* 653, 654, 657, 658 (27), 727 (192a, b), 728 (192b), *747, 752*
Dehmlow, E. V. 872 (67), *911*
Dehn, W. M. 366 (741), 386 (846, 857), *477, 479*
Deitch, J. 427 (1147), *485*
De Jong, S. 109, 110 (285), *119*
DeJonge, A. P. 1066, 1146 (20), *1316*
De Jongh, D. C. 123 (3), 162 (185), *169, 173,* 633 (212), *640,* 874 (73), 909 (241), *911, 914*
De Jongh, R. O. 680 (95), *749*
De Kruif, C. G. 61 (24), *66*
De La Cruz, D. 691, 701, 702 (125b), *750*
Delafontaine, J. 63 (37), *66*
Delahay, P. 782 (15), *821*
Delaplane, R. G. 224 (98), *258*
Del Bene, J. E. 19 (8), 20 (16, 17), 21 (8, 16, 17), 22 (8), 25 (8, 16, 17), 26, 28 (8), 29, 30 (8, 16), 31 (17), 45 (117, 118), 46, 47 (118), *55, 57,* 233 (217), 250 (409), *261, 265*
Deldalle, A. 22 (35), *56*
Dell'Erba, C. 110 (308), *120*
Deloche, B. 234 (227), *261*
Delsarte, J. 101 (174), *117*
Delvigne, J. P. 419 (1069), *483*
DeMarinis, R. M. 1038 (183), *1056*
DeMartino, G. 156 (160), *172*
Demas, J. N. 647 (9), *747*
Dembech, P. 1024 (38), *1053*
Dement'eva, L. A. 246 (379), 248 (392), 249 (392, 401), *264, 265*
Demers, J. P. 839 (63), *856*
Demerseman, P. 361 (714), *476*
Demoulin, D. 25, 26 (49), *56*
Dem'yanovich, V. M. 108 (270), *119*
Denbigh, K. W. 1285 (662), *1329*
DeNiro, J. 721, 725 (186), *752*

Denisov, G. S. 238 (276, 285), *262*
Denne, W. A. 245 (362), *264*
Denney, D. B. 1116 (194), 1145, 1147 (250), *1320, 1321*
Denney, R. C. 1303, 1304 (695), *1330*
Denning, G. 887 (134), *912*
Denny, R. 1171, 1172 (317), *1322*
Deno, N. C. 353 (659), 373 (775), *475, 477*
Denzel, T. 295 (287), *467*
Deperasińska, I. 197 (91), *211*
De Poortere, M. 1269 (609), *1328*
Depree, D. O. 280 (172, 173), *465*
De Puy, C. H. 883 (114, 117, 119), 884 (119, 124), 885 (131), 886 (117, 119, 124), 887 (119), 896 (190), 899 (117), 904 (214), *912–914,* 1104 (152), *1319*
Derancourt, J. 1042 (299), *1059*
Deranleau, D. A. 178 (29), *210*
Derbyshire, W. 230 (187), *260*
Derdall, G. D. 1045 (328), *1060*
Dereani, M. C. 1052 (408), *1062*
Derenberg, M. 375, 377 (792), *478*
Derevitskaya, V. A. 345, 414 (632), *475*
Derge, K. 617 (114), *638*
Derible, P. 734 (207), *753*
Derissen, J. C. 232 (215a), *261*
Derissen, J. L. 222 (45, 47, 48), *257*
DeRooij, J. 1022 (2), *1053*
Derrick, P. J. 123 (3), *169*
Desai, K. R. 1039 (211), *1057*
Deshayes, H. 389 (862), 416 (1019), *479, 482,* 692, 693 (131), *750*
Deshpande, R. 206 (146), *212*
Deshpande, S. M. 1232 (549, 583), 1246 (549), 1256, 1257 (583), *1327*
Deslongchamps, P. 54 (124), *58,* 309 (385), 346 (638, 639), 358 (689), *469, 475, 476,* 1161, 1168, 1170 (284), *1321*
Dessau, R. M. 501 (61), *527,* 1172 (323, 325), 1173 (323, 326), 1174 (325), *1322*
Dessy, R. E. 330 (520), *472*
Detoni, S. 224 (89), 228 (156), 240 (89), *258, 259*
Detroparesky, M. L. P. 437 (1232), *487*
Dettmer, H. 1101 (141), *1318*
Detzer, N. 674 (84), *749*
Deuel, H. 318 (448), *471*
Deutsch, J. 145, 146 (103), 164 (197), *171, 173*
Dev, S. 335 (578), 339 (578, 610), 361 (578), *473, 474*
De Villepin, J. 224 (92), 225 (131), *258, 259*
Devon, T. K. 507 (91), *528*
Devore, J. A. 64, 65 (47), *66*

Dewar, M. J. S.　311, 325 (400), *470,*
　864 (21), *910*
De Wolfe, R.　534, 535, 553 (6), *586*
Dey, B. B.　786 (27), *821*
Deyrup, J.　1296 (672), *1329*
Dharamski, S. S.　1109 (168), *1319*
Diakur, J.　1039 (226), *1057*
Diamanti, J.　431 (1173), *485*
Diamond, M. J.　166, 167 (208), *173,*
　626, 627 (176), *639*
Diamond, S. E.　450 (1338), *489*
Diaper, D. G. M.　270 (9), 338 (597), 423
　(1108), *462, 474, 484*
Diasse, P. A.　297 (304), *468*
Diaz, S.　985 (128), 989, 1005 (148),
　1008 (186), 1009 (188), *1019, 1020*
DiBello, C.　84 (92), 102 (200), *115, 117*
Dicarlo, E. N.　233 (220), *261*
Dickel, D. F.　1304 (697), *1330*
Dickinson, H. R.　105 (256), *119*
Diebert, C. E.　894 (178), *913*
Diels, O.　512 (124), *529*
Dieterich, D. A.　234 (223), *261*
Diethelm, H.　281 (179), *465*
Dietrich, H.　1189, 1190 (355), *1323*
Dietsche, I. J.　428 (1154), *485*
Digenis, G. A.　417 (1047), 446 (1309),
　483, 488
Diggle, W. M.　631 (192), *639*
Dijkink, J.　525 (221, 222), *531*
Dillard, R. D.　525 (219), *531*
Dillon, C. S.　800 (78), *823*
Dimant, E.　341 (621b, c), *474*
DiMaria, F.　1110 (174), *1319*
Dimmig, D. A.　1102 (144), *1318*
Dine-Hart, R. A.　164 (201, 204, 205),
　165 (204), *173*
Ding, L. K.　522 (194), *530*
Dinh-Nguhên, Ng　136 (66), *170*
Dinkeldein, U.　536, 549 (15), *587*
Dirkx, I. P.　89, 95 (145), *116*
Disselnkotter, H.　446 (1307), *488*
Ditchfield, R.　19 (9), 20, 21, 25 (16, 17),
　29 (16), 30 (16, 78–80), 31 (17), *55,
　57*
Dittmann, W.　400 (898), *480*
Dittmer, D. C.　1165, 1168 (297), *1322*
Ditzat'eva, L. N.　60, 63 (14), *65*
Divakar, K. J.　508 (101), *528,* 1145
　(260), *1321*
Diversi, P.　291 (245), *467*
Dix, R.　511 (119), *529*
Dixon, J. A.　359 (698), *476*
Djerassi, C.　73 (40), 82 (80, 81), 85, 87
　(81), 88 (111, 112, 114), 89 (111), 102
　(186–188, 207), 103, 104 (241), 107
　(264), 108 (267), 109 (81, 112), *114–
　119,* 123 (1), 128 (26), 135 (56–58),

138, 142 (74), 144 (56), 151 (56, 131),
　152 (139), 153 (139, 140), 156 (156,
　160), 160 (178), 164, 168 (195), 169
　(225), *169–174,* 382 (826), *478,* 517
　(155), *529,* 611 (76), 613 (95), 614
　(95, 96), 616 (108–110), 617 (113),
　623 (161), 626, 627 (175), 628 (183),
　633 (211), *637–640,* 888, 899, 900
　(143), *912,* 1036 (136), *1055,* 1303
　(694), 1306 (705), *1330*
Dmitrieva, N. D.　510 (111), *528,* 1065,
　1090 (6), *1315*
Dobbs, A. J.　768 (71), *779*
Dobkin, J.　673, 683, 685 (85a), *749*
Dobrogowska, L.　223 (78), *258*
Dodds, H. L. H.　496 (22), 527
Dodge, R. A.　274 (74), *463*
Dods, R. E.　240 (308), *263*
Dodson, R. M.　361 (712), *476,* 860 (1),
　882 (106), *909, 912*
Doering, W. von E.　353 (664), 386 (852),
　475, 479, 709, 719, 722 (155), *751*
Doi, Y.　281 (188), 428 (1152), *465, 485*
Dolak, L. A.　453 (1363), *489*
Dolfini, J. E.　1300 (679), 1301 (680),
　1329
Dolgit, I. E.　436 (1212), *486*
Dollish, F. R.　605, 606 (34), 607 (40),
　608 (40, 47), 632 (205, 206), *636, 640*
Dombro, R. A.　1192 (360), *1323*
Dombrowski, D.　248 (394), *264*
Donahue, J. A.　986 (135), *1019*
Donaruma, L. G.　385, 450 (837), *479,*
　1274 (620), *1328*
Donat, F. J.　1276, 1278 (631), *1328*
Donckt, E. V.　238 (279), *262*
Donelson, D. M.　566, 571 (263), *592*
Donohue, I. D.　255 (475), *266*
Doppert, K.　1232, 1242, 1243 (541),
　1326
Dorange, G.　1042 (287), *1059*
Dorfman, L. M.　763, 764 (51), *779*
Doria, G.　1113 (180), *1319*
Dormoy, J. R.　437 (1228, 1229), 443
　(1228), *487*
Dornaver, H.　422 (1104), *484*
Dorofeeva, O. V.　1052 (398), *1062*
Dorokhov, V. A.　541 (90), *588*
Dorsey, E. D.　1231, 1232, 1234–1236,
　1239, 1244 (516), *1326*
Dorsky, J.　359 (699), *476*
Dorst, W.　680 (95), *749*
Dorval, C.　251 (414–416), *265*
Doss, S. H.　88, 89, 91 (136), *116*
Doty, P.　73, 99, 109 (38), *114*
Dougherty, G.　1276, 1278, 1279 (634),
　1328
Dougherty, Ph. L.　761 (34), *778*

Dougherty, S. J. 947 (13), 962, 980 (72), 983 (13), *1016, 1017*
Dougill, M. W. 222 (54), *257*
Douglas, B. 1306, 1312 (704), *1330*
Douglas, K. T. 990 (150), *1019*
Doumaux, A. R. 460 (1415), *490*
Douraghi-Kadeh, K. 437 (1219), *486*
Dowbenkie, R. 276 (109), *464*
Dowling, J. M. 27, 28 (59), *56*
Downing, D. T. 332 (546), *473*
Downward, M. J. 12(2), *55*
Doyle, K. M. 1042 (292), *1059*
Doyle, M. 1206 (456), *1325*
Dozen, Y. 924 (39), 936, 937 (59), 938 (59, 65, 67), *943, 944*
Dradi, E. 1113 (180), *1319*
Draeger, A. 1232 (553, 554), 1247, 1250 (553), 1251 (554), *1327*
Draganić, I. 763–765 (49), *779*
Draganic, I. G. 760 (28, 31), 761 (31, 32), 767 (28, 64), *778, 779*
Draganić, Z. D. 767 (64), *779*
Dräger, M. 1022 (11), 1023 (19), 1027, 1031, 1033, 1037 (55), *1053, 1054*
Draghici, C. 1196 (386), *1323*
Drago, R. S. 250 (410), 251 (419), *265*
Drakenburg, T. 28 (69), 31 (82, 85, 86), *56, 57*
Drakesmith, F. G. 791, 792 (51), *822*
Dreger, L. H. 1066–1068, 1072, 1073 (16), *1316*
Dreickey, E. 663, 668 (53), *748*
Dreiding, A. S. 1113 (185), *1319*
Drewes, S. E. 415 (973), *482*
Dreyfus, M. 42 (107), 44 (107, 112), 45 (112), *57*, 241, 242 (327, 328), 244 (328), *263*
Driesch, A. 322 (479), *471*
Drizina, I. A. 562 (210), *591*
Druckrey, E. 161 (183), *173*, 633 (214), *640*
Dryuk, V. G. 338 (589), *474*
Dubeck, M. 338 (596), *474*
Dubin, J. C. 901 (205), *914*
Dubini, M. 291 (250), *467*
Dubois, F. 1170 (311), *1322*
Dubois, J. C. 1111 (179), *1319*
Dubois, J. E. 358 (691), *476*, 494 (12), *526*
Dubovik, G. Yu. 799 (72), *822*
Duchamp, D. J. 282 (195), *465*
Duddy, N. W. 992 (156), *1019*
Duffaut, N. 1229 (512), *1326*
Duffey, D. C. 895 (180), *913*
Duffield, A. M. 152, 153 (139), 156 (160), 164, 168 (195), 169 (225), *172–174*, 517 (155), *529*, 616 (109, 110), *638*, 1303 (694), *1330*

Duggan, A. J. 1110 (174), *1319*
Duguay, G. 1033, 1034, 1042, 1050 (109), *1055*
Duijneveldt, F. B. van 232 (215a), *261*
Duke, A. J. 397 (885, 886), *480*
Duke, J. M. 236 (254), *261*
Dull, D. 108, 109 (268), *119*, 280 (168), *465*
Dulou, R. 1080, 1084 (84, 85), *1317*
Dulova, V. I. 254 (452), *266*
Dunet, A. 816, 817 (153, 154), *824*
Dunham, D. 979 (104), *1018*
Dunham, N. A. 835, 837, 839, 840 (54), *856*
Dunitz, J. D. 88 (126), *116*, 244 (341–343), 245 (341, 342), 246 (368, 369), *263, 264*
Dunlap, R. B. 947, 959, 972, 977 (3), 978 (3, 103), 984 (120), *1016, 1018*
Dunlop, R. B. 566, 571 (263), *592*
Dunlop, R. S. 224 (83), *258*
Dunn, G. L. 1101 (140), *1318*
Dunn, M. S. 279 (143), *464*
Dunnavant, W. R. 280 (150), 426 (1135, 1137), *465, 485*
Dunstan, D. R. 101 (177), *117*
Duong, T. 1276, 1282 (640), 1284 (648), 1286, 1287 (640), 1288, 1289 (640, 648), 1291 (640), 1295 (648), 1300 (640, 648), 1302 (648), *1329*
Dupire, S. 176 (16), 202 (123), *209, 212*
Dupriez, G. 1045 (321), *1060*
Dupuy, A. E. 167, 168 (219), *174*, 613, 617 (111), *638*
Dupuy, A. E. Jr. 520 (175), *530*, 669 (58), *748*
Duran, F. 1232, 1250, 1260 (587), *1327*
Durand, D. 457 (1398), *490*
Durbeck, J. 406 (914), *480*
Durham, J. E. 805, 806 (112), *823*
Durham, L. J. 271 (17), *462*
Durig, J. R. 620 (145, 146), *639*
Durrani, A. A. 294 (278), *467*
Durst, H. 425 (1125), *485*
Durst, H. D. 414 (1000), *482*
Durst, T. 1225–1227 (500), 1300 (675, 676), *1326, 1329*
Durston, P. J. 1285 (662), *1329*
Dutka, F. 1048 (352), *1060*
Duus, F. 1035, 1036 (131), 1037 (146), 1047 (338), 1050 (379), *1055, 1056, 1060, 1061*
Dvořák, V. 661 (39), *748*
Dwivedi, P. C. 218 (17), 251 (421), *256, 265*
Dyer, E. 893 (166), *913*
Dyke, S. F. 1305, 1312 (702), *1330*
Dytham, R. A. 373 (777, 779, 780), *477, 478*

Dziegielewski, J. 777 (97–100), *780*
Dziomko, V. M. 631 (195, 196), *640*
Dzizenko, A. K. 238 (277), *262*

Eadon, G. 382 (826), *478,* 744 (226), *753*
Earle, R. H. Jr. 1196 (388), 1206, 1222 (487), *1323, 1325*
Eastham, J. F. 386 (843, 849), 387 (849, 858), *479*
Easton, N. R. 525 (219), *531*
Eastwood, F. W. 274 (77), *463,* 542 (99), 585 (363), *588, 594,* 1298, 1299 (673), 1302 (688), *1329, 1330*
Eaton, P. E. 304 (348), 381 (821), *469, 478,* 1145, 1152 (270), *1321*
Ebara, N. 934 (49), *943*
Ebel, S. 1204 (444), *1325*
Eberhardt, R. 320 (466), *471*
Eberius, K. W. 826 (12), *855*
Eberson, L. 240 (302), *263,* 275 (98), 432 (1179), 433 (1194), *464, 486,* 782 (16a), 805, 806 (111), 808 (129), 810, 811 (135), *821, 823, 824*
Ebine, S. 513 (129), *529*
Ebisawa, H. 936 (61), *944*
Eck, J. C. 385 (840), *479*
Eda, B. 757 (3, 5), 758 (5), *778*
Eddy, C. R. 1145, 1151, 1160 (268), *1321*
Edelmann, N. K. 555, 566–568 (178), *590*
Eder, M. 1314 (722), 1315 (723), *1330*
Edge, D. J. 768 (71), *779*
Edington, R. A. 939 (69), *944*
Edward, J. T. 416 (1030), *483,* 1045 (328), *1060,* 1145, 1147 (257), *1321*
Edwards, A. G. 1186, 1187 (349), *1323*
Edwards, B. E. 417 (1033), *483*
Edwards, O. E. 1306, 1312 (704), *1330*
Edwards, R. L. 632, 634 (207), *640*
Edwards, T. T. 332, 368 (556), *473*
Effenberger, F. 536 (12, 29, 30), 537 (30, 57), 538 (57, 62, 63), 544 (57), 545 (122, 123, 128, 129), 546 (62, 63), 547 (138, 139), 550 (158), 551 (57), 553, 554 (129), 555 (177), 556 (179), 559 (57, 158), 562 (128, 200, 201, 204), 566 (128, 179), 567 (128, 139, 177, 179), 570 (200, 201, 283, 291), 571 (139, 200, 201, 283, 291, 295, 297, 303–309), 572 (129), 581 (57), 586 (57, 295, 368), *586–590, 592–594*
Efimov, E. A. 794 (59), *822*
Egan, R. 331 (530), *472*
Egawa, K. 198 (93), *211*
Ege, G. 403 (910), *480,* 539, 557, 572 (76), *588*

Egger, B. 395 (878), *480*
Egglestone, D. 835, 837 (47, 48), 838 (48), 839, 840 (47, 48), *856*
Egli, R. H. 1079 (62), *1317*
Eglinton, G. 435 (1209), 437 (1221), *486,* 883 (118), *912*
Egnell, C. 412 (951), *481*
Ehntholt, D. 1228 (510), *1326*
Eiden, F. 566, 569 (256, 257), 596 (390), *591, 592, 599*
Eigenmann, H. K. 60–63 (10), *65*
Eijck, B. P. van 225 (124), *259*
Eilerman, R. G. 380 (819), *478*
Eilers, K. L. 1104 (152), *1319*
Eilhauer, J. D. 1042 (282), *1059*
Eilingsfeld, H. 539 (67, 70, 71), 550 (67, 157), 557, 559 (157), 569 (67), *588, 590*
Eimutis, E. 177 (23), *209*
Eimutis, E. C. 196, 198, 199 (79), *211*
Einhorn, A. 497 (36), *527,* 908 (235), *914*
Eirich, F. R. 252 (425), *265*
Eisenbraun, E. J. 361 (708), 411 (928), *476, 481*
Eisenman, G. 980, 981 (110), *1018*
Eisenthal, K. B. 202 (111–113), *211*
Eistert, B. 380 (813, 814, 816), *478*
Eizenberg, L. 691, 699, 711, 715 (126), *750*
Ekel, V. A. 787 (31, 33), *822*
El-Abadelah, M. M. 72, 88, 93 (25), *114,* 609 (64), *637*
Elam, E. K. 498 (46), *527*
Elam, E. U. 440 (1264), *487*
El-Berins, R. 539 (65), *588*
Elberling, J. A. 398 (888), *480*
Elek, A. 609 (62b), *637*
Elia, V. J. 167 (215), *174,* 520 (176), *530*
Eliel, E. L. 274 (73), *463*
Elkins, D. 236 (253), *261*
Ellington, P. S. 432 (1185), *486*
Ellis, A. F. 355 (681), *476*
Ellis, B. 255 (479, 481), *266*
Ellis, B. A. 303 (340), *469*
Ellis, J. 1040, 1046 (269), 1048 (359), *1058, 1061*
Ellis, P. D. 566, 571 (263), *592*
Ellis, R. 628 (186), *639*
Ellison, R. 1185 (344), *1323*
Ellison, R. A. 427 (1144), *485,* 1039 (248), *1058,* 1115 (192), *1320*
Ellison, R. D. 224 (88), 225 (126), *258, 259*
Elmer, O. C. 860 (1), *909*
Elmes, B. C. 1104 (154), 1105 (155), 1147 (154), 1173 (154, 155), 1179 (154), *1319*

Elnatanov, Yu. I. 615 (103), *638*
El-Sayed, M. A. 202 (112), *211*
Elsinger, F. 272 (26), *462*
Elston, C. F. 273 (69), *463*
Elworthy, P. H. 952, 984, 989, 1014 (35), *1016*
Elving, P. J. 800 (81), *823*
Emergy, A. 849, 850, 854 (88), *857*
Emerson, M. F. 951–953 (28), 984 (119), *1016, 1018*
Emerson, W. S. 327 (500), 411 (930), *472, 481,* 1145, 1151 (263, 264), *1321*
Emery, E. M. 140 (88), *171*
Emery, T. 275 (95), *463*
Emmert, D. E. 282 (195), *465*
Emmons, W. D. 353 (660), 354 (671, 674), 423 (1093), *475, 484,* 540, 541 (88), 545–547 (88, 124), 548, 559 (88), 572, 573 (88, 124), *588, 589*
Emovon, E. U. 889 (144), 892, 894 (163), 897 (193), 902 (208), *912–914*
Emsley, J. 237 (270, 271), 238 (272), *262*
Endle, R. 509 (106), *528*
Endo, T. 1039 (235), 1048 (351), *1058, 1060*
Endres, L. S. 167 (215), *174*
Engelhardt, H. 238 (273), *262*
Engelhord, N. 1232, 1264 (595), *1328*
Engellbrecht, A. 232, 235, 236 (213), *261*
Engelsing, R. 1314 (719), *1330*
Englard, S. 88, 91, 92 (127), *116*
Engler, E. M. 61 (21), *66*
Engler, R. 1023 (17, 19), 1025 (40), 1027 (51, 55), 1029, 1030 (40), 1031 (55), 1033 (40, 51, 55), 1034–1036 (40), 1037 (51, 55, 151), 1040 (40), 1045 (320), *1053, 1054, 1056, 1060*
Engler, V. R. 254 (460), *266*
English, A. D. 852 (95), *857*
English, J. Jr. 303 (341), *469*
Engoyan, A. P. 562 (220), *591*
Enomoto, K. 936 (61), *944*
Enyo, H. 437 (1227), *487*
Epley, T. D. 251 (419), *265*
Epling, G. A. 672, 678, 679 (91), 739 (219), *749, 753*
Erickson, J. A. 1232, 1255 (569), *1327*
Erickson, J. G. 549, 557 (150), *589*
Erickson, K. L. 422 (1098), *484*
Erickson, R. E. 338 (600), 346 (637), *474, 475*
Erisheva, A. I. 442 (1279), *488*
Erlbach, E. 514 (134), *529*
Erlenmeyer, E. 908 (236), *914,* 1123 (203), *1320*
Érman, M. B. 850, 851 (90), *857*

Erman, W. F. 294 (280), *467*
Ernst, H. 230 (180), *260*
Ertl, H. 743 (225), *753*
Erusalimchik, I. G. 794 (59), *822*
Eschenmoser, A. 272 (26), 413 (963), 451 (1349), *462, 481, 489,* 517 (161), *529,* 536 (26), 543, 557, 561 (26, 109), 566 (26), 583 (343, 350), 584 (26, 343), *587, 589, 593,* 1180 (336, 337), *1322*
Esipov, G. V. 442 (1279), *488*
Eskelson, C. D. 127, 143 (20), *170*
Espejo de Ochoa, O. 1284, 1285 (658), *1329*
Esser, F. 586 (365), *594*
Esterle, J. 511 (117), *529*
Esteve, R. M. 861, 863 (8), *910*
Etemadi, A. H. 148 (119), *172*
Etienne, A. 440 (1258), *487*
Etienne, Y. 496, 497 (24), *527,* 1065 (3), *1315*
Evans, D. 297 (300), *468,* 843, 844 (68), *857*
Evans, D. A. 166 (213), *173,* 447 (1316), *488,* 1066, 1069, 1149 (22), *1316*
Evans, D. D. 416 (1024), *483*
Evans, D. E. 416 (1024), *483*
Evans, D. H. 808 (128), *824*
Evans, D. L. 189 (70, 71), 190 (72), *210, 211*
Evans, E. R. 896 (186), *913*
Evans, H. B. 240 (306), *263*
Evans, J. C. 220 (34), *257*
Evans, R. L. 1125, 1130 (216), *1320*
Evans, T. W. 366 (741), 386 (846), *477, 479*
Evans, W. L. 417 (1040), *483*
Evstigneeva, R. P. 278 (130), *464,* 1039 (250), *1058*
Ewing, J. B. 1300 (679), *1329*
Exarchou, A. 676, 743 (221a), *753*
Exner, O. 1023 (25), 1024 (25, 34, 38), *1053*
Eyring, H. 71 (29), 89, 103, 104 (146), *114, 116*

Fabian, J. 1025 (48), 1026 (48, 76), 1027, 1028, 1042 (48), *1054*
Factor, R. E. 736 (211b), *753*
Fadéeva, V. A. 449 (1328), *489*
Faganeli, J. 562, 571 (211), *591*
Fagnano, C. 246 (370), *264*
Fahey, J. L. 1232 (561, 562, 565), 1254 (565), 1259 (561, 562, 565), *1327*
Fahr, E. 1232 (541, 544), 1242, 1243 (541), *1326, 1327*
Fahrenholtz, K. E. 561 (197), *590*
Fairweather, R. B. 124, 125 (7), *169*

Falbe, J.   1173 (328, 329), 1174 (331), 1175 (328), 1176, 1177 (328, 329), 1178, 1179 (331), *1322*
Falcetta, J.   79 (64), 84 (90), 101 (64), *115*
Farberov, M. I.   314 (427), (506), *470, 472*
Farid, S.   208 (153), *212*
Farina, F.   1314 (721), *1330*
Farina, J. S.   508 (99), *528*
Farkas, J.   459 (1410), *490*
Farkas, L.   672, 677 (59b), *748*
Farmun, D. G.   1138, 1141 (243), *1320*
Farney, R.   1050 (380), *1061*
Farnia, G.   814, 816, 817 (142), *824*
Farragher, A. L.   196 (87), *211*
Farrar, M. W.   360 (703), *476*
Farrer, H. N.   225 (128), *259*
Fasman, G. D.   77, 101 (55), *114*
Fateley, W. G.   27 (57), *56,* 605, 606 (34), 607 (40), 608 (40, 47), 632 (205, 206), *636, 640*
Fatiadi, A. J.   300 (325), 313 (414), 319, 332, 365 (325), *468, 470*
Fatturoso, E.   544 (119), *589*
Faucitano, A.   762, 763 (42), *778*
Faucitano Martinotti, F.   762, 763 (42), *778*
Faust, J.   1042 (281), 1052 (281, 408), *1059, 1062*
Fava, F.   1198 (392, 395), *1324*
Favero, P.   (26), *56*
Fawcett, J. S.   1284, 1291 (655), *1329*
Fawcett, N. C.   1042 (300), *1059*
Feairheller, S. H.   709, 712 (156), *751*
Feairheller, W. R.   605 (32), *636*
Feaisheller, S. H.   1080, 1084 (86), *1317*
Feast, W. J.   163, 164 (192), *173*
Fedor, L. R.   995 (160), *1019*
Fedoronko, M.   804 (97), *823*
Fedorov, B. P.   536 (39, 44), 538 (44), 539 (78, 79), 543 (44, 111), 544, 548 (44), 549 (39), 558 (44), 561 (44, 111), 562, 567, 569 (44), 577 (39), *587– 589*
Fedorova, T. M.   1079 (57), 1080 (72), *1317*
Fedorova, T. N.   279 (147), *465*
Fedorova, V. A.   63 (38), *66*
Fedyushina, T. I.   1039 (223), *1057*
Feeney, J.   252 (429), *265*
Feenstra, T.   222 (57), *257*
Fehlhaber, H. W.   159 (172), *173*
Feichtinger, A.   325 (489), *471*
Feigl, F.   603 (13), 631 (193), *636, 640*
Feil, V. J.   140 (86), *171*
Feinauer, R.   535 (9), *586*
Felix, A. M.   1206, 1221 (489), *1325*
Felix, D.   451 (1349), *489,* 583 (343, 350), 585 (343), *593,* 1180 (336, 337), *1322*

Felner, J.   517 (161), *529*
Fendler, E. J.   947–949, 951, 953, 958, 959, 986, 988 (11), 989 (140), 990 (11), 993 (159b), 1000 (11), 1012 (195), *1016, 1019, 1020*
Fendler, J. H.   947–949, 951, 953, 958, 959, 986, 988 (11), 989 (140), 990, 1000 (11), 1012 (195), *1016, 1019, 1020*
Fenoglio, D. J.   88 (121, 122), *116*
Fenselau, C.   138, 142 (74), *171*
Fenske, D.   230 (180), *260*
Fenwick, R. G.   166 (211), *173*
Ferdorov, Yu. G.   194 (75), *211*
Fergie, R. A.   869 (48), *910*
Ferguson, L. N.   321, 322 (478), *471*
Ferraris, M.   291 (250), *467*
Ferris, J. P.   863 (17), *910*
Ferrito, V.   166 (211), *173*
Ferronato, E. L.   21, 22 (25), *56*
Ferstandig, L. L.   176 (21), 181, 182, 196 (54), *209, 210*
Fessenden, R. W.   763 (44), 768 (68, 72– 74), 774 (90), *778, 779*
Fessler, D. C.   1078 (54), *1316*
Fetizon, M.   333 (563), *473,* 1145 (272, 275, 277), 1152 (272), 1155, 1156 (272, 277), 1157, 1158 (272), *1321*
Feugeas, C.   536 (36), 543 (36, 112), 561 (112), *587, 589*
Feurtill, G. I.   272 (34), *462*
Fiandaca, P.   1023 (23), *1053*
Ficher, F.   496 (27), *527*
Fichter, F.   793 (57), *822*
Fichter, Fr.   782 (18), *821*
Ficini, J.   302 (335), *468,* 500 (57), *527*
Field, F. H.   152 (136), *172*
Fielden, E. M.   772 (83), *779*
Fields, E. K.   162 (186, 187), *173,* 338 (591), 339 (608, 609), *474,* 675 (89), *749,* 874 (71), *911*
Fields, E. S.   451 (1344), *489*
Fieser, L.   298 (316), *468*
Fieser, L. F.   298 (315), *468,* 503 (70), *528,* 789 (41), *822*
Fieser, M.   503 (70), *528,* 789 (41), *822*
Fife, T. H.   1011 (191), *1020*
Filatova, L. S.   594, 595 (370), *598*
Filbey, A. H.   441 (1273), *488*
Filipakis, S. E.   244 (347), *263*
Filira, F.   84 (92), 102 (200), *115, 117*
Filler, R.   503 (73), 504 (80), *528,* 1065 (2), 1109 (165), *1315, 1319*
Finch, N.   415 (989), *482*
Finckenor, L. E.   310 (396), *470*
Finkbeiner, H.   1172, 1173 (322), *1322*
Finkbeiner, H. L.   296 (290), *467*
Finkelstein, J.   1072 (33, 34), *1316*

Finklestein, M. 782 (8c), *821*
Finnegan, R. A. 653, 688, 693, 694 (26), (138), *747, 750*
Finocchiaro, P. 610 (66), *637*
Fioshin, M. Ya. 782 (6), 787 (31), 801 (6), 804 (100, 101), 807 (6), *821–823*
Firestone, R. A. 1301 (681), *1330*
Firl, J. 1023, 1024 (25), *1053*
Fischer, A. 673 (71a), *748*
Fischer, E. 386 (844), *479*, 733–735 (204c), *753*
Fischer, E. O. 1034, 1042 (116), *1055*
Fischer, G. 254 (446), *265*
Fischer, H. 649 (14), *747*
Fischer, H. O. L. 310 (392), *470*
Fischer, M. 518 (162, 163), 522 (191), *529, 530*
Fischer, N. 496, 497 (24), *527,* 1065 (3), *1315*
Fischer, P. 566 (244), *591*
Fischli, A. 517 (161), *529*
Fish, R. W. 1228 (510), *1326*
Fisher, G. J. 871 (61), *911*
Fisher, L. R. 951 (30), *1016*
Fisher, R. R. 566, 571 (263), *592*
Fittig, R. 493 (3), *526*
Flamini, A. 1026 (77), 1027 (52), *1054*
Flechtner, T. W. 725 (189), *752*
Fleck, T. 255 (472), *266*
Fleig, H. 450 (1335), *489*
Fleischauer, P. D. 642 (1), *747*
Fleming, I. 408 (917, 918), *480*
Fleming, M. P. 1066, 1069 (21), *1316*
Flemming, W. 348 (643), *475*
Fletcher, H. G. Jr. 1038 (164), *1056*
Fletcher, S. 816, 817 (158), *824*
Fletcher, V. R. 397 (887), *480*
Flett, M. St. C. 607 (41), *636*
Flippen, J. L. 720, 722 (185e), *752*
Flitsch, W. 524 (213), *531*
Flock, F. H. 415 (977), *482*
Flood, E. 232 (209), *260*
Florence, A. T. 952, 984, 989, 1014 (35), *1016*
Florian, W. 535, 538, 542–544, 550, 551, 553, 554, 557, 559, 562, 566, 568, 572, 573, 581, 583, 586 (11), *586*
Floss, H. G. 1115 (189), *1319*
Flynn, E. H. 524 (203), *530*
Foa, M. 291 (251), 292 (254), *467,* 1180 (335), *1322*
Foerst, D. L. 1039 (232), 1045 (326), *1058, 1060*
Foglia, T. A. 415 (983), *482*
Foglizzo, R. 226 (136), *259*
Föhlisch, B. 566, 569 (252), *591*
Fokin, A. V. 1039 (223), *1057*
Foley, G. E. 507 (97), *528*

Folkers, K. 1284 (651, 652), *1329*
Fones, W. S. 415 (1005), *482*
Fonken, G. S. 414 (967), *481,* 901 (205), *914*
Fontana, A. 109 (275), *119*
Fontanella, L. 523 (197), *530,* 1197 (401), 1198 (390, 392, 394–400), 1246 (401), *1323, 1324*
Foo, C. K. 1040, 1041 (260), *1058*
Foot, S. 566, 569 (267), *592*
Foote, C. S. 1171, 1172 (317, 318), *1322*
Ford, M. E. 284, 285 (211), 415 (998), *466, 482*
Ford, P. C. 642 (3a), *747*
Fordham, W. D. 272 (43, 48), *462, 463*
Ford-Moore, A. H. 386 (856), *479*
Forel, M.-T. 608 (51–53), *636*
Foresti, E. 1023 (23), *1053*
Forman, R. L. 866 (33), 876 (85), *910, 911*
Formosinho, S. J. 724 (188), *752*
Forrest, J. G. 104 (246), *118*
Forsellini, E. 1023 (23), *1053*
Forsen, S. 28 (69), 31 (85, 86), *56, 57*
Forsén, S. 224, 240 (91), *258*
Forsyth, D. A. 904 (218), *914*
Fortunato, B. 632 (208), *640*
Fossen, R. Y. van 874 (73), *911*
Foster, C. H. 1038 (183), *1056*
Foster, D. J. 294 (281), *467*
Foster, G. 417 (1043), *483*
Foster, R. 175 (3), 178 (31, 32), 179 (34, 36, 39, 40), 182, 192 (39), *209, 210*
Foucaud, A. 1273 (619), *1328*
Fougerousse, A. 317 (440), 403 (905), *471, 480*
Fouquey, C. 287 (219), *466*
Fournier, H. 301 (332), *468*
Fouty, R. A. 1165, 1168 (297), *1322*
Fowler, E. M. P. 1073, 1074 (38), *1316*
Fox, B. L. 456, 457 (1386), *490,* 1123 (209), 1299 (674), *1320, 1329*
Fox, J. E. 609 (64), *637*
Fraenkel, G. 241 (314), *263,* 788 (36), *822*
Fraher, T. P. 1110 (173), *1319*
France, W. G. 819 (164), *824*
Franchuk, I. F. 240 (309), *263*
Francis, J. 854 (97), *857*
Frank, C. E. 294 (282), *467*
Frank, R. L. 1195 (381), *1323*
Franke, A. 1107 (157), *1319*
Franke, W. 1302 (686), *1330*
Franklin, E. C. 385, 450 (836), *479*
Franklin, J. L. 35 (101), *57,* 195 (77), *211*
Frankoski, S. P. 618 (119), *638*
Franzen, J. S. 246 (373), 247 (385), *264*

Fraser-Reid, B.　312 (406, 407), *470,* 846, 848 (79), *857*
Frauenglass, E.　1072, 1163 (29), *1316*
Fredga, A.　88, 91 (138), 92, 94 (152), 99 (167), 102 (188), 109 (138, 293−295), 110 (293−295), *116, 117, 119*
Freeburger, M. E.　270 (13), *462*
Freedman, H. H.　953 (50), *1017*
Freeman, J. P.　274 (73), *463*
Freeman, P. K.　375, 377 (794), *478*
Fréhel, D.　54 (124), *58,* 346 (639), *475*
Freiberg, L. A.　381 (823), *478*
Freidlin, G. N.　241 (320, 322), 251, 252 (320), *263*
Frĕidlina, R. Kh.　276 (110), *464,* 1192 (363), *1323*
Freifelder, M.　1098 (135), *1318*
Frejd, T.　110, 111 (305), *120*
Fremery, M. I.　338 (591), 339 (608), *474*
Freon, P.　412 (951), *481*
Freund, H. G.　757 (7), *778*
Frey, A. J.　1071−1073, 1075, 1149, 1187 (28), *1316*
Frey, H. O.　539, 557, 572 (76), *588*
Frič, I.　76, 83 (50), 101 (178, 179), *114, 117*
Fried, I.　782 (11), *821*
Fried, J.　355 (679), *475,* 510 (110), *528*
Friedel, R. A.　133, 134, 136 (47), *170*
Friedman, G. F.　456 (1384), *490*
Friedman, L.　165, 166 (207), *173,* 318 (453), *471,* 874 (72), *911*
Friedrich, K.　323 (483), *471*
Friedrichs, G.　805 (107), *823*
Friend, J. P.　22, 23 (32), *56*
Frier, R. D.　1048 (359), *1061*
Fries, R. W.　834, 835, 837 (50), 844 (71), *856, 857*
Friess, R.　1037 (148), *1056*
Friess, S. L.　353 (668), *475*
Frisch, M. A.　1066−1068, 1072, 1073 (16), *1316*
Froehlich, P. M.　254 (462), *266*
Froemsdorf, D. H.　883 (117, 119), 884 (119), 886 (117, 119), 887 (119), 899 (117), *912*
Fröhlich, I.　458 (1402), *490,* 547 (130), *589*
Frolov, Yu. G.　237 (264), *262*
Fruwert, J.　248 (394), *264*
Fry, A.　151 (132), *172*
Fry, A. J.　782 (8a), *821*
Frydrychova, A.　1196 (384), *1323*
Frye, J. R.　1098, 1099 (126), *1318*
Fryer, R. I.　566 (258, 259), 571 (258), *592*
Fuchs, B.　664, 668 (56a, b), 669, 671 (56b), 708 (56a, b), 730, 731 (56b), *748*

Fuchs, G. I.　226, 235 (133), *259*
Fuchs, O.　535−537, 554, 559, 572 (10), *586*
Fuchs, R.　326 (495), *472,* 513 (127, 128), *529*
Fueki, K.　770 (78), *779*
Fuerholzer, J. F.　1200 (420), *1324*
Fuerholzer, J. J.　518 (170), 519 (173), 520 (170, 178), *530*
Fugiwara, Y.　432 (1183), *486*
Fuglevik, W. J.　241 (318), 251 (418), *263, 265*
Fujewara, K.　594 (373), *598*
Fujii, M.　1131 (218), *1320*
Fujii, Y.　103 (218, 219), *118,* 237 (263), *262*
Fujimori, F.　674 (83), *749*
Fujioka, O.　31 (87), *57,* 607 (38), *636*
Fujishima, S.　924 (39), *943*
Fujita, K.　274 (76), *463*
Fujita, S.　545 (120), *589,* 691, 699, 735 (123), *750*
Fujita, T.　280 (161−163), 281 (161, 162), *465,* 1116 (196), *1320*
Fujita, Y.　675, 682 (88), *749*
Fujiwara, H.　237 (268), *262,* 451 (1348), *489,* 1276, 1277, 1279 (624), *1328*
Fujiyoshi, K.　280, 281 (161, 162), *465*
Fuka, K.　229 (172), *260*
Fukada, N.　1037 (154, 155), 1052 (155), *1056*
Fukuda, M.　565, 566 (240), *591*
Fukui, K.　711 (167), *751,* 1026 (74), *1054*
Fukui, M.　1025, 1026 (41), *1053*
Fukumoto, K.　1311 (713), *1330*
Fukunaga, T.　298 (318), *468*
Fukuoka, S.　455 (1380), *490,* 571 (300), *592*
Fukushima, K.　226 (141), *259*
Fukuyama, K.　224 (103), *258,* 745 (230b), *753*
Fukuyama, T.　243 (339), *263*
Fukuyama, Y.　294 (275), *467*
Fuller, J. A.　1145 (251), *1321*
Fuller, N. A.　275 (97), *464*
Fullington, J. G.　950 (23), *1016*
Fullmer, D. G.　699, 700 (141), 746 (234), *750, 753*
Funakoshi, K.　1145 (259), *1321*
Fung, D. P. C.　417 (1044), *483,* 894 (173), *913*
Funk, G. L.　629 (199), *640*
Funke, B.　534, 557 (8), 559 (192), 562 (8, 212), 566, 567 (212), 568 (8), *586, 590, 591*
Funke, P. T.　155 (152), *172,* 614 (98), *638,* 1300 (679), *1329*

Funke, W. 540, 547 (82), *588*
Furlani, C. 1026 (53), 1027 (52, 53), *1054*
Furuhata, K. 81 (70–72), *115*
Furukawa, S. 287 (223), *466,* 1039 (243), *1058*
Furuya, Y. 928 (44), *943*
Furuyama, S. 916 (25, 26), 917 (25), 929 (25), 934 (49), *943*
Fusco, S. 896 (188), *913*
Fuson, R. C. 273 (63), 358 (692), 403 (911), *463, 476, 480*
Fuson, R. Y. van 633 (212), *640*
Fyfe, C. A. 179 (34), *210*

Gabbott, R. E. 892 (156), *913*
Gabeleva, L. P. 439 (1245), *487*
Gabrio, T, 1044, 1048 (313), *1059*
Gacek, M. 88 (141, 142), 89 (142), 90 (141), 94 (141, 142), 102, 109 (210), *116, 118*
Gadret, M. 255 (478), *266*
Gaede, D. 617 (117), *638*
Gaffield, W. 88 (115, 134), 92, 95 (115), 102 (208), *116, 117*
Gage, J. C. 631 (192), *639*
Gagnaire, D. 800 (73), *822*
Gais, H.-J. 1039 (227), *1057*
Gal, O. 758 (13), 760 (31), 761 (31–33), *778*
Galasso, V. 1024, 1025 (39), *1053*
Gale, L. H. 757 (2), *778*
Galetto, W. G. 88, 92, 95 (115), *116*
Gallagher, B. S. 620 (140), *639*
Gallagher, G. 422 (1097), *484*
Gallagher, K. J. 224 (95), *258*
Galli, R. 455 (1381, 1383), *490*
Gallivan, J. B. 661, 663, 666, 668 (51), *748*
Gallo, G. G. 610 (68), *637*
Galloy, J. 1044 (315), *1059*
Galochkina, G. V. 631 (196), *640*
Galt, R. H. B. 1138, 1140, 1141 (241, 248), 1164 (248), *1320, 1321*
Gambaryan, N. P. 1201 (428, 429), 1202 (429, 430), *1324*
Gamble, A. A. 140 (85), 154 (146), 155 (146, 148), *171, 172*
Gamstedt, E. 109, 110 (295), *119*
Gan, I. 414 (1002), *482*
Gandolfi, C. 571 (310, 311), *593,* 1113 (180), *1319*
Ganem, B. 417 (1048), *483,* 499 (48), *527*
Ganern, B. E. 311, 314 (401), *470*
Gani, V. 991 (154, 155), *1019*
Gano, J. E. 691, 699 (126), 710 (162a–c), 711 (126), 714 (162a–c), 715 (126, 162a–c), *750, 751*

Ganter, C. 179 (35), *210*
Ganz, E. St. C. 805 (103), *823*
Gapanovich, L. I. 1039, 1052 (198), *1057*
Garbarino, G. 110 (308), *120*
Garbess, C. F. 1171, 1172 (321), *1322*
Garcia, E. E. 566 (258, 259, 262), 569 (262), 571 (258), *592*
Gardini, G. P. 455 (1383), *490*
Gardner, D. 874 (74), *911*
Gardner, J. H. 320 (468), *471,* 1065 (13), *1316*
Garegg, P. J. 105 (257–259), 106 (258, 259), *119*
Gariana, P. 415 (997), *482*
Garland, F. 248 (398), *264*
Garrett, A. B. 411 (923), *481*
Garrison, W. E. Jr. 433 (1193), *486*
Garwood, R. F. 298 (310), *468*
Gassman, P. G. 375 (789, 790), 376 (790), 377 (789, 790), 456, 457 (1386), *478, 490,* 1123 (209), 1185 (343), 1197 (405–407), 1199 (405, 407), 1299 (674), *1320, 1322, 1324, 1329*
Gatenbeck, S. 376 (810), *478*
Gatow, G. 254 (460), *266*
Gatsonis, C. 690, 701, 702 (121a), *750*
Gatti, E. 1198 (391), *1324*
Gatti, G. 88 (130), *116*
Gattow, G. 1023 (17, 19), 1025 (40), 1027 (51, 55, 61), 1029, 1030 (40), 1031 (55), 1032 (61), 1033 (40, 51, 55), 1034–1036 (40, 61), 1037 (51, 55, 151), 1040 (40), 1042 (61), 1045 (320), 1052 (61), *1053, 1054, 1056, 1060*
Gaudemar, M. 423 (1109), 457 (1393), *484, 490*
Gaufres, R. 235 (235), *261*
Gaur, J. N. 306 (361), *469*
Gay, B. 863 (16), *910*
Gay, R. L. 1205 (450), *1325*
Gaylord, N. G. 1138 (234), *1320*
Gebreyesus, T. 128 (26), *170*
Gedymin, V. V. 385 (835), *479*
Geer, R. D. 332 (538), *473*
Geffken, O. 443 (1287), *488*
Geffrey, G. A. 19 (10), *55*
Geigel, M. A. 1276, 1278, 1289 (638), *1328*
Geiger, B. 571 (298), *592*
Geiger, H. 1089 (96), *1317*
Geiger, R. E. 83, 84 (86), *115,* 612 (219), *640*
Geipel, H. 351 (650), *475*
Geiseler, G. 248 (394), *264*
Geisenfelder, H. 223 (70), *258*
Geissman, T. A. 85, 86 (95), *115,* 306 (365), *469,* 507 (95), *528,* 625 (169), *639*

Gelbard, G. 1113 (185), *1319*
Geller, G. K. 1048 (352), *1060*
Geller, H. H. 1145, 1151 (262), *1321*
Geller, L. E. 1159 (279), *1321*
Gellert, H. G. 825 (6), *855*
Gempeler, H. 369 (755), *477*
Genêt, J.-P. 500 (57), *527*
Gensler, W. J. 270 (5), *462*
Genson, D. W. 32, 33 (92), *57*
Gent, P. A. 848 (83), *857*
Gentil, M. 554 (174), *590*
Gentzkow, W. von 1043 (309, 310), *1059*
George, M. V. 303 (345), *469*
George, W. O. 88 (128), *116,* 147 (109), *171*
Gerace, M. J. 743 (225), *753*
Geraghty, M. B. 1123 (210), *1320*
Gerhart, G. 536 (23), *587*
Geribaldi, S. 161 (184), *173*
Gerkin, R. M. 422 (1103), *484*
Gerlach, H. 419 (1073), *484*
German, A. L. 420 (1075), *484*
Gerold, C. 340 (612), *474*
Gerry, M. C. L. 1022, 1024 (9), *1053*
Gerschute, D. 297 (304), *468*
Gerson, T. 394 (876), *479*
Gertners, M. 1038 (169), 1045 (324), *1056, 1060*
Gertsev, V. V. 420 (1079), *484*
Gerwin, M. 255 (472), *266*
Geschwend-Steen, K. 583 (350), *593*
Gething, B. 178 (11), *209,* 321, 325 (472), *471*
Getoff, N. 766–768 (62), *779*
Ghatak, V. R. 1161, 1162, 1168 (286), *1321*
Ghisalferti, E. L. 1092 (115), *1318*
Ghislandi, V. 91, 95 (155), *116*
Ghosal, M. 298 (317), 311, 314 (399), *468, 470*
Ghose, H. M. 978 (101), *1018*
Ghosez, L. 1232, 1250, 1260 (587), 1269 (609), *1327, 1328*
Ghosh, M. C. 312 (405), *470*
Ghosh-Mazumdar, B. N. 168 (224), *174,* 1200 (434, 436), 1201 (434, 436, 442), *1324, 1325*
Giaccio, M. 990 (152), *1019*
Giacomello, G. 225 (120), *259*
Giannini, D. D. 252 (428), *265*
Gibbons, C. 673 (74), *749*
Gibbons, E. G. 448 (1324), *489*
Gibbs, D. E. 109 (278), *119*
Gibby, M. G. 231 (193), *260*
Gibson, K. 103 (224), *118*
Gibson, T. W. 711 (170), 745 (228b), *751, 753*
Giering, W. P. 690, 701, 702 (121b), *750,* 1228 (509, 510), *1326*

Gigg, R. 848 (83), *857*
Gigli, R. 61 (22), *66*
Gilardi, R. D. 86 (103), *115*
Gil-Av., E. 146 (104, 105), *171*
Gilbert, A. 205 (135, 139), 206 (144–146), 207 (151), *212*
Gilbert, H. 874 (77), *911*
Gilbert, J. R. 140 (85), 154 (146), 155 (146–148), *171, 172,* 615 (104), *638*
Gilchrist, T. L. 523 (202), *530*
Gilde, H. G. 810, 811 (135), *824*
Gilderson, P. W. 899 (197), *913*
Gilgen, P. 706 (147), *751*
Gill, M. 632, 634 (207), *640*
Giller, S. A. 562 (210), *591*
Gillet, I. E. 807 (122), *823*
Gillier-Pandraud, H. 246 (367), *264*
Gillies, D. G. 247 (381), *264*
Gilman, H. 293 (270), 302 (336), 318 (443), 403 (908), *467, 468, 471, 480,* 1030 (90), *1054,* 1113 (182), 1266 (604), *1319, 1328*
Gilman, N. W. 311, 314 (401), 422 (1099), 453 (1364), *470, 484, 489*
Gilmore, W. F. 396 (883), *480*
Gilpin, J. A. 152, 153 (138), *172,* 613, 614 (94), *638*
Ginsberg, H. 1145, 1148 (262), *1321*
Ginsburg, D. 276 (105), 282 (192), 381 (820), 429 (192), *464, 465, 478,* 731 (199), *752*
Ginzburg, I. M. 249 (402, 403), 253 (445), 254 (450), 255 (467), *265, 266*
Gioia, B. 610 (68), *637*
Giordano, C. 293 (265), *467*
Girard, J.-P. 400 (899), *480*
Girelli, A. 177 (24), 184 (66, 67), *209, 210*
Girijavallabhan, M. 275 (92), *463*
Gisin, M. 1040–1042, 1052 (268), *1058*
Gitler, C. 947, 956, 978 (4), 999 (165), *1016, 1019*
Givens, R. S. 689 (117, 120, 124, 145), 690 (120, 124, 130, 135), 691 (120, 124, 128, 130, 135), 692 (130, 135), 694 (117, 120, 124), 695 (124), 696 (120, 124, 128), 697 (128, 130, 135), 698 (124, 128), 699 (128, 130, 135), 700 (124), 705 (124, 135), 706 (117, 120, 124, 145), 731 (117, 145, 201a), 733 (145, 201a, 202, 203b), 735 (120, 124, 128, 130, 135), 736, 737 (130, 135, 213), *750–753*
Gizler, M. 84 (93), *115*
Gladysheva, T. K. 509 (108), *528*
Glaret, C. 1145, 1151 (252), *1321*
Glasoe, A. P. 236 (254), *261*
Glass, M. A. W. 82, 85, 87, 109 (81), 110, 111 (309), *115, 120*

Glazer, E. L.   1285 (659, 660), *1329*
Glazunov, V. P.   228 (153, 163), 238
    (277), *259, 262*
Gleich, J.   151 (127, 128), *172*
Glier, R.   198 (96), *211*
Gloede, J.   274 (79), 351 (650), *463, 475,*
    534 (5), 557, 559 (180), 561 (5, 180),
    563 (180), 564, 565 (5), 572 (180),
    576 (5), *586, 590*
Glogger, I.   516 (149), *529*
Glotter, E.   80 (68), *115*, 1073 (37), *1316*
Glover, D.   1066, 1069, 1149 (22), *1316*
Glushkov, R. G.   517 (158), *529*, 536 (35),
    537 (52–56), 538 (56), 562 (216–221),
    566 (216, 221, 270–273), 570, 571
    (35, 217), 581 (339), 582 (271, 339,
    340), 583 (271, 342), 594 (369), 596
    (377, 379–381, 386), 597 (379, 381,
    386), *587, 591–593, 598, 599*
Glyde, E.   892 (159), 905 (220–222, 225),
    906 (159, 222, 226), 907 (159, 221,
    222, 226), *913, 914*
Glynn, G. A.   426 (1128), *485*
Gnauk, Th.   564 (239), *591*
Go, T.   928 (44), *943*
Gobley, J.   1073, 1115 (41), *1316*
Godbole, E. V.   138 (70), *171*
Goddard, W. A. III   28, 29 (66), 30 (66,
    76), *56, 57*
Goddu, R. F.   603, 619, 631 (5), *635*
Godinho, L. S.   1048 (360), *1061*
Goeders, C. N.   884, 886 (124), *912*
Goel, O. P.   440 (1263), *487*
Goethals, E. J.   1051 (387), *1061*
Goetschel, C. T.   1138, 1141 (242), *1320*
Goff, D. L.   1186 (347, 348), *1323*
Goheen, D. W.   394 (875), *479*
Gohlke, R. S.   127, 128, 140, 143 (14),
    169
Göknel, E.   549 (147), 577 (147, 328),
    578, 597 (147), *589, 593*
Gold, B. I.   520 (176), *530*
Gold, H.   517 (157a), *529*, 540 (86, 87),
    541, 562 (86), *588*
Gold, P.   744 (226), *753*
Gold, V.   275 (101), *464*, 957 (64), *1017*
Goldberg, N. L.   415 (992), *482*
Goldberg, O.   1232 (540, 556), 1242
    (540), 1250 (540, 556), *1326, 1327*
Golden, D. M.   60, 61 (8, 10), 62 (10), 63
    (8, 10), 65 (8), *65*
Golding, B. T.   594 (372), *598*
Goldman, I. M.   1168 (305), *1322*
Goldschmidt, B. M.   71, 72 (26), *114,*
    626 (174), *639*
Goldschmidt, C. R.   204 (128), *212*
Goldsmith, D.   156 (156), *172*
Goldstein, A.   888 (142), *912*

Goldstein, I. P.   238 (281), *262*
Goldstein, J. H.   240 (306), *263*
Goldstein, M.   512 (125), *529*
Golechek, A. A.   (41, 42), *210*
Golfier, M.   1145 (272, 277), 1152 (272),
    1155, 1156 (272, 277), 1157, 1158
    (272), *1321*
Golič, L.   224 (89, 104, 105), 240 (89),
    258
Gollnick, K.   642 (5), *747*
Gollogly, J. R.   103 (230), *118*
Gololobov, Yu.   1039 (216), *1057*
Golson, J.   167, 168 (219), *174*
Golson, T. H.   167, 168 (219), *174*, 613,
    617 (111), *638, 669* (58), *748*
Gölz, G.   1040 (267), 1041 (273), 1052
    (399), *1058, 1059, 1062*
Gombos, J.   1039, 1048 (237), *1058*
Gomes, A.   1231, 1232, 1234 (517),
    1326
Gomez Aranda, V.   1269 (608), *1328*
Gompper, R.   547 (131–133, 138), 548
    (132, 133), 571 (131, 133, 296, 298),
    586 (131, 133, 296), *589, 592*
Goode, G. C.   196 (88), *211*
Goodin, R. D.   802, 803, 813, 814 (85),
    823
Goodman, C.   759 (15), *778*
Goodman, I.   869 (53), *910*, 939 (69),
    944
Goodman, M.   79 (64), 84 (90), 101 (64),
    115
Goodrow, M. H.   1138, 1141 (242), *1320*
Goodson, T.   416 (1014), *482*
Goodwin, T. H.   231 (195), *260*
Goosen, A.   1189 (353), *1323*
Gopal, H.   335, 336, 340 (579), *473*
Gopichand, Y.   280, 281 (166), *465*
Gopinath, K. W.   1093 (111, 112), 1094
    (113, 114), 1095 (114), *1318*
Gorbunova, T. V.   223 (69), *258*
Gorden, R. Jr.   671, 672, 677 (92e), *749*
Gordini, G. P.   455 (1381), *490*
Gordon, A. J.   335, 336, 340 (579), *473*
Gordon, A. S.   894 (175), *913*
Gordon, J.   224 (93), *258*
Gordon, J. E.   1014 (197), *1020*
Gordon, M. S.   20, 21 (19), *55*
Gordon, S.   763 (46, 50), 764 (54), 765
    (46, 50), 767 (54), *778, 779*
Gore, J.   313 (412), *470*
Gore, P. H.   297 (303), *468*
Gorin, E.   671, 672, 677 (92b), *749*
Gorkom, M. van   88, 91 (144), *116*
Gorodetsky, M.   653, 657, 659 (34a, b),
    733, 734 (204b), *747, 752*
Gorodetsky, M. F.   733, 734 (204a), *752*
Gorres, B. T.   224 (106), *258*

Gorski, R. 1039 (240), *1058*
Gorski, R. A. 1050 (377), *1061*
Gorvin, T. C. 230 (187), *260*
Goswami, K. 312 (405), *470*
Goto, S. 1039 (201), *1057*
Gotoh, H. 1230 (513, 514), *1326*
Gotor, V. 1267 (608), *1328*
Gott, P. 509 (105), 511 (116b), *528, 529*
Gott, P. G. 440 (1264), *487*
Gottarelli, G. 88, 108 (116, 117), *116*
Gotthardt, H. 1038 (167), 1039 (241), 1042, 1052 (167), *1056, 1058*
Gottikh, B. P. 510 (115), *529*
Gough, S. R. 238 (282), *262*
Gougoutas, J. Z. 440 (1256), *487,* 1301 (680), *1329*
Gould, W. A. 1197 (404), 1266, 1267 (601), *1324, 1328*
Goulden, J. D. S. 148 (118), *172,* 236 (253a), *261,* 633 (215), *640*
Gourcy, J. G. 812 (139), *824*
Gourdol, A. 1042 (299), *1059*
Goutarel, R. 1310 (711), *1330*
Gouy, G. 955 (54), *1017*
Govindachari, T. R. 1305 (701), *1330*
Govrilova, T. F. 330 (521), *472*
Goya, S. 596 (391), *599*
Graaf, B. van de 126 (13), *169,* 495 (20), *527*
Graaff, G. B. R. de 331 (523), *472*
Graefe, J. 458 (1402), *490,* 547 (130), *589*
Graf, G. 422, 423 (1100), *484*
Graf, R. 522 (192), *530,* 1194 (370), 1224 (494, 495), 1225 (370, 496, 498, 502), 1226 (498), 1232 (524), *1323, 1326*
Graff, Y. 254 (448), *265,* 524 (212), *531*
Graham, L. L. 247 (382, 388), *264*
Grahame, D. C. 955 (55), *1017*
Graig, B. M. 332 (542), *473*
Gramstad, T. 241 (317, 318), 250 (413), 255 (466), *263, 265, 266*
Gramstad, T. R. 251 (418), *265*
Grand, A. 245 (354), *264*
Granger, R. 400 (899), *480,* 1138, 1140 (240), *1320*
Granik, E. M. 581, 582 (339), *593*
Granik, V. G. 517 (158), *529,* 536 (35), 537 (52, 53, 55), 562 (216–221), 566 (216, 221, 270–273, 275), 570, 571 (35, 217), 581 (339), 582 (271, 339, 340), 583 (271, 342), 594 (369), 596 (377, 379–381, 386), 597 (379, 381, 386), *587, 591–593, 598, 599*
Grant, B. 450 (1338), *489*
Grant, D. M. 252 (428), *265*
Granzano, M. L. 595 (375), *598*

Grashey, R. 1053 (412), *1062*
Grassberger, M. A. 566, 567 (260), *592*
Grasselli, P. 281 (191), *465*
Gravshenko, A. I. 632 (209), *640*
Gray, G. W. 273 (59), *463*
Gray, R. W. 235 (244), *261*
Gray, W. F. 1115 (188), *1319*
Grazia, C. G. de 75 (44), *114*
Graziani, R. 1023 (23), *1053*
Graziano, M. L. 544 (118, 119), *589*
Grebow, P. E. 101 (180), *117*
Greeley, R. H. 415 (995), *482*
Green, A. E. 1123, 1127 (213), *1320*
Green, B. 1101 (140), *1318*
Green, C. L. 1048 (348), *1060*
Green, J. H. S. 88 (128), *116*
Green, M. 418 (1058), *483*
Green, M. L. H. 826 (10), 854 (96), *855, 857*
Greenberg, A. 497 (33), *527,* 620 (149), *639*
Greenburg, G. Y. 1081 (92), *1317*
Greenburg, R. B. 355 (683, 684), *476*
Greene, A. E. 507 (98), *528,* 690, 704, 705 (122), *750*
Greene, R. N. 392 (869), *479*
Greene, R. S. 332 (546), *473*
Greenfeld, D. 673 (70a, b), *748*
Greenwood, F. L. 879 (91, 92), *911*
Gregory, M. J. 1048 (344), *1060*
Grenie, Y. 226 (137), *259*
Grenon, B. J. 426 (1128), *485*
Gresham, T. L. 496 (26), 499 (50), *527*
Gresham, W. F. 328 (509), *472*
Gressier, J.-C. 1042 (293, 295), *1059*
Grezemkovsky, R. 1090 (105), *1318*
Griebenow, W. 566, 567 (276, 278), 570, 571 (276, 289), *592*
Grieco, P. A. 296 (293), *468,* 505 (83), 507 (96), *528,* 1065, 1123 (11), 1166, 1168 (300), *1315, 1322*
Griesbaum, K. 338 (588), *474*
Grieshaber, P. 536, 538, 543, 545, 548, 554, 557, 558 (28), *587*
Griffin, R. N. 861, 863 (8), *910*
Griffith, W. P. 828 (45), *856*
Griffiths, D. M. L. 239 (298), *262*
Griffiths, P. J. F. 255 (479, 481), *266*
Grifone, L. 21, 22 (25), *56*
Grigat, E. 292 (260), 459 (1411), *467, 490*
Grigor'ev, A. B. 596 (380, 386), 597 (386), *599*
Grimaldi, J. 1173 (327), *1322*
Grimm, H. 348 (643), *475*
Grimmer, G. 332, 354 (550), *473*
Grob, C. A. 298 (313), *468,* 525 (218), *531*

Groeger, G. 1107 (157), *1319*
Groff, T. M. 147 (112), *171*
Gromova, E. V. 804 (100, 101), *823*
Gronka, J. 206 (146), *212*
Gronkiewicz, M. 199 (101), 202 (118), *211*
Grønneberg, T. 102, 109 (209), *117*
Gronowitz, S. 106 (260), 107 (262), 109 (286), 110, 111 (305), *119, 120*
Gronse, P. M. 447 (1319), *488*
Gronwell, S. 417 (1050), *483*
Gronzenbach, H. V. 728 (194), *752*
Grosjean, M. 69, 73, 108 (4), *113*, 611 (72, 74), *637*
Gross, A. 279 (141), 382 (829), *464, 478*, 1232, 1264 (596), *1328*
Gross, H. 274 (79), 279 (132), 297 (305), 351 (650), *463, 464, 468, 475*, 534 (5), 551 (164), 552 (170), 561 (5), 564 (5, 235–239), 565 (5, 235), 572, 573 (317), 576 (5, 235, 327) 596 (383, 384), *586, 590, 591, 593, 599*
Gross, M. L. 904 (218), *914*
Gross, P. 536, 537 (14), 559, 574 (193), *587, 590*
Gross, P. M. 222 (60), *257*
Grossman, H. 1080, 1084 (83), *1317*
Grossweiner, L. I. 672 (64, 75b, c, 93), 673 (93), 678 (64, 75b, c, 93), 680, 690 (75b, c), 694, 696 (75c), 699, 735, 736 (75b, c), *748, 749*
Groszos, S. J. 498 (45), *527*
Grouth, P. 225 (114), *258*
Grovenstein, E. Jr. 205 (134), *212*
Grubb, P. W. 208 (152), *212*
Grudzinskas, C. V. 536, 566, 571 (25), *587*, 1196 (387), *1323*
Gruetzmacher, G. 1197 (406), *1324*
Grummitt, O. 297 (299), 331 (530), 417 (1036), *468, 472, 483*, 879 (90), *911*
Gründler, H.-V. 1032 (104), *1055*
Grundy, J. 415 (981), *482*
Grunwald, E. 241 (321), *263*, 496 (32), 497 (37), *527*
Grunwell, J. R. 1039 (232), 1045 (326), 1051 (383), *1058, 1060, 1061*
Grushka, E. 414 (1000), *482*
Grzejszczak, S. 596 (384), *599*
Gschwend, H. W. 286 (215), *466*
Guarnaccia, R. 109 (280), *119*
Guarnieri, A. G. 21, 22 (25), *56*
Gubanova, L. A. 594, 596, 597 (371), *598*
Guerchais, J. E. 1042 (287), *1059*
Guerrieri, F. 290 (240), 291 (250), *466, 467*
Guette, J. P. 111 (179), *1319*
Guha, M. 424 (1115), *484*

Guha, P. G. 272 (23), *462*
Guilleme, J. 235 (233, 242), 236 (233), *261*
Guire, R. F. 232, 243 (214), *261*
Guise, G. B. 1304, 1306, 1309 (699), *1330*
Gul'tyai, V. P. 799 (72), *822*
Gulyas, E. 72, 88, 89, 93 (24), 103 (235), *114, 118*
Gundel, L. 709, 717 (153), *751*
Gunsalus, I. C. 493 (6), *526*
Gunstone, F. D. 270 (3, 6), 333 (559), *462, 473*, 1038 (175), *1056*
Günther, D. 509 (107), *528*
Günther, H. 1042 (414), 1044 (415), *1062*
Gupta, Abha 251 (421), *265*
Gupta, M. P. 224 (102), *258*
Gupta, N. P. 224 (102), *258*
Gupta, P. D. 176 (18), *209*
Gupta, S. K. 445 (1301), *488*
Gurianova, E. N. 238 (281), *262*
Gurke, D. 241, 250, 251 (319), *263*
Guryeva, L. I. 254 (452), *266*
Gusakova, G. V. 238 (276), *262*
Gusevich, L. S. 365 (727), *476*
Gus'kova, T. A. 562, 570, 571 (217), *591*
Gutbrod, H. D. 536 (31), 559, 574 (193), *587, 590*
Guthke, H. 800 (75), *822*
Guthrie, J. P. 554, 557 (173), *590*, 995 (162, 163), 1002 (162), *1019*
Gutsche, C. D. 655–657 (30), 658 (36), *747*, 1109 (167), *1319*
Guttenplan, J. 685 (102a), *749*
Guy, J. J. 1022 (10), *1053*
Guyard, M. 816 (155), *824*
Guyot, M. 1098 (131), *1318*
Guzzi, U. 571 (310, 311), *593*
Gwinn, W. D. 21 (20), 25 (48), *55, 56*

Ha, T.-K. 22, 25, 29, 30 (22), *56*
Haaf, W. 292 (262, 263), 293 (267), *467*, 1174 (332, 333), *1322*
Haag, W. O. 884 (122), *912*
Haage, K. 825 (6), *855*
Haak, P. J. van der 250 (411), *265*
Haake, P. 894 (178), *913*
Haarer, D. 198 (94, 95), 201, 202 (109), *211*
Haase, L. 564, 565, 576 (235), *591*
Habich, A. 392 (867), *479*
Hachiya, K. 82, 83 (76), *115*
Haddon, W. F. 672, 686, 687 (63d), *748*
Hadži, D. 24 (58), *56*, 214 (1a), 216 (4), 218 (13), 220 (32), 223 (81), 224 (89, 94, 104), 225 (115), 226 (140, 142), 228 (149–151), 155–159, 161, 162),

230 (184), 231 (158, 198), 237 (13, 269), 240 (89, 307), 255 (474), *256–260, 262, 263, 266*
Haeberlen, U.   230 (178, 179), 231 (194), *260*
Haefele, L. R.   450 (1336), *489*
Häfliger, O.   240 (312), *263,* 526 (224), *531*
Hafner, K.   536 (20), *587*
Hagedorn, I.   547 (135, 136), 586 (136), *589*
Hagemeier, L. D.   167 (215), *174*
Hagemeyer, H. J. Jr.   419 (1065), *483,* 498 (42), *527*
Hagen, H.   566, 567 (254), *591,* 1038 (166), *1056*
Hagendorn, J. A.   419 (1071), *483*
Hager, F. D.   279 (133), *464*
Hagiwara, D.   412 (957), *481*
Hagler, A. T.   243 (336, 337), *263*
Hagmann, D. L.   323 (485), *471*
Haidukewych, D.   285 (213), 443 (1286), *466, 488*
Haines, R. A.   103 (221, 225, 237), *118*
Haisa, M.   224 (103), 245 (360), *258, 264*
Hajek, M.   450 (1334), *489,* 562 (200, 204), 570, 571 (200, 283), *590, 592*
Hajiev, S. N.   61, 62 (27), *66*
Håkansson, R.   88, 89, 94 (142), 109 (286, 293–295), 110 (293–295, 302–307), 111 (303, 305–307, 310), *116, 119, 120*
Halchak, T.   1292 (664), *1329*
Halcour, K.   343 (624), 369 (754), *474, 477*
Hale, D.   177 (23), *209*
Hale, W. F.   883, 884 (115), 886 (115, 133), *912*
Hall, C. M.   1035 (113), *1055*
Hall, D. R.   594 (372), *598*
Hall, D. T.   304 (349), *469*
Hall, H. K.   619, 620 (133), *638*
Hall, H. T.   826 (13), *855*
Hall, N. M.   415 (970), *481*
Hall, W. L.   882, 899 (109), *912*
Hallam, H. E.   605 (23, 35), 606 (35, 36), 608 (35, 49, 50), *636*
Hallas, G.   411 (937), 415 (980), *481, 482*
Hallcher, R. C.   803, 809, 810 (92), *823*
Hallett, P.   1184 (339), *1322*
Hallock, S.   236 (254), *261*
Halpern, B.   102 (201, 202), 103 (201), 109 (281), *117, 119,* 414 (1002, 1003), 419 (1070), *482, 483*
Halpern, J.   837 (58), *856*
Halsall, T. G.   355 (678), *475*
Halver, J. E.   275 (88), *463*

Halverson, F.   661, 663, 666, 668 (51), *748*
Hamada, T.   415 (986), *482*
Hamdan, A.   286 (215), *466*
Hamilton, D. J.   418 (1059), *483*
Hamilton, J. A.   953 (51), *1017*
Hamilton, W. C.   214 (1d), 221, 233 (41), 244 (352), 245 (361), *256, 257, 264*
Hamilton, W. S.   63 (44), *66*
Hamlin, K. E.   375 (805), 450 (1340), *478, 489*
Hammann, W. C.   429 (1159), *485*
Hammerum, S.   154 (145), 155 (150), *172*
Hammes, G. G.   248 (396), *264*
Hammick, D. Ll.   178 (31), *210*
Hammond, G. S.   205, 206 (136), *212,* 642 (5), 649, 659 (13), (176a), *747, 751,* 883, 884, 886, 887 (119), *912*
Hammond, P. R.   177, 178 (12), *209*
Hamon, D. P. G.   299 (319), *468*
Hamor, T. A.   1022 (10), *1053*
Hampson, N. A.   304, 315 (346), *469*
Hamrin, K.   248 (395), *264*
Hanack, M.   902 (207), *914*
Hanaoka, M.   1079 (58), *1317*
Hancock, C. K.   1048 (345), *1060*
Hanessian, S.   557, 558 (185), *590*
Hanewald, K.   1027, 1032, 1034–1036, 1042, 1052 (61), *1054*
Haney, M. A.   35 (101), *57*
Hanhan, S. I.   1051 (383), *1061*
Hankin, H.   1145, 1148 (262), *1321*
Hankins, D.   242, 252 (333), *263*
Hanna, J.   504 (82), *528*
Hanna, M. W.   758 (11), 761 (37), *778*
Hanna, R.   1168 (306), *1322*
Hanrahan, E. S.   235 (230, 248), *261*
Hansch, C.   236 (253), *261*
Hänsel, R.   627 (179), *639*
Hansell, D. P.   1104 (153), *1319*
Hansen, H. J.   706 (147), *751*
Hansen, J. F.   284 (212), *466*
Hansen, J. G.   102 (198), *117*
Hansen, R. T.   346 (637), *475*
Hanson, G. H.   896 (184), *913*
Hanson, J. R.   1138, 1140, 1141 (241, 248), 1164 (248), *1320, 1321*
Hanson, S. W.   408 (918), *480*
Hansson, B.   103 (211), *118*
Hanuise, J.   439 (1248), *487*
Hanzlik, R. P.   847 (80), *857*
Happ, G. P.   123, 124 (4), *169*
Hara, S.   355 (677), *475,* 585 (362), *594,* 1044 (316), 1052 (396), *1060, 1062,* 1163 (293), 1168 (292, 293), *1321*
Hara, T.   (505), *472*
Harada, K.   1201, 1203 (443), *1325*

Harada, N. 73 (34), 111 (314), 112 (34, 315, 316, 318), *114, 120*
Harada, Y. 674 (83), *749*
Haraoubia, R. 1042 (293), *1059*
Harcourt, D. W. 1312 (715), *1330*
Hardegger, E. 313 (418), 369 (755), *470, 477*
Harden, R. C. 609 (59), *637*
Hardham, W. M. 205, 206 (136), *212*
Harding, L. B. 28, 29 (66), 30 (66, 76), *56, 57*
Hardouin, J. C. 1115 (190), *1319*
Hardy, P. M. 1110 (175), *1319*
Hare, C. R. 103 (227, 228), *118*
Hargreaves, G. 309 (383), *469*
Hargreaves, M. K. 88, 89 (139), *116, 458 (1409), *490*
Harigaya, S. 78 (60), *115*, 622 (157), *639*
Harkins, J. 346 (637), *475*
Harmer, J. 1194, 1225 (372), *1323*
Harmon, R. E. 445 (1301), *488*
Harner, R. S. 236 (248a), *261*
Harnung, S. E. 98 (161), *117*
Harpool, R. D. 242, 246, 247, 252 (332), *263*
Harrington, K. J. 585 (363), *594*
Harris, B. W. 1285 (662), *1329*
Harris, D. 1042 (283), *1059*
Harris, H. E. 110 (300), *119*
Harris, M. M. 110 (300), *119*
Harris, S. A. 274 (74), *463*
Harris, S. J. 1041, 1050 (276), *1059*
Harris, T. M. 424 (1118), *484*
Harrison, A. G. 35 (98), *57*, 136 (62), 138 (69), 142 (93), *170, 171*, 890 (149), *912*
Harrison, H. R. 411 (928), *481*
Harrison, I. T. 346 (635), *475*
Harrison, J. A. 788 (37), 789 (40), 791 (37), 794 (40), 797 (66), 816, 817 (158), *822, 824*
Harrison, J. M. 397 (885, 886), *480*
Harrison, K. 682 (96b), *749*
Harrison, M. C. 20 (11), *55*
Harrison, R. G. 390, 391 (864), 452 (1352), *479, 489*, 583 (345), *593*
Harrison, S. 346 (635), *475*
Harrisson, R. J. 313 (417), *470*
Hart, D. W. 839 (65), *856*
Hart, E. J. 763 (46, 50), 764 (54), 765 (46, 50), 767 (54), 772 (83), *778, 779*
Hart, P. 743 (223b), *753*
Hartan, H.-G. 743 (224b), *753*
Hartke, K. 1027 (63), 1031 (94, 95), 1032 (63), 1033 (110), 1034 (63, 94, 95, 110), 1035, 1036 (63), 1040 (94, 264, 267, 272), 1041 (273), 1042 (272),

1043 (63), 1046 (94, 272), 1050 (94), 1052 (95, 399), *1054, 1055, 1058, 1059, 1062*, 1204 (444), *1325*
Hartley, G. S. 966 (90, 91), *1018*
Hartman, G. D. 1052 (407), *1062*
Hartman, K. P. 23 (39), *56*
Hartman, M. E. 436 (1213), *486*
Hartman, W. 415 (976), *482*
Hartman, W. W. 318 (451), *471*
Hartmann, A. 248 (393), *264*
Hartmann, H. 1042 (297), *1059*
Hartter, P. 99 (166), *117*
Hartung, H. 422, 423 (1100), *484*
Hartung, W. H. 1200 (414), *1324*
Hartwell, J. L. 1081 (92, 93), *1317*
Hartzell, S. L. 423 (1107), *484*
Hasan, S. K. 326 335 (496), *472*
Hase, K. 102 (204), *117*
Hasegawa, T. 1206, 1222, 1224 (467), *1325*
Hasek, R. 509 (105), 511 (116b), *528, 529*
Hasek, R. H. 440 (1264), *487*, 498 (46), *527*
Hashimoto, S. 1271 (613), *1328*
Haslinger, E. 1039, 1048 (237), *1058*
Hassall, C. H. 352–354 (652), 376 (809), *475, 478*, 514 (135), *529*, 1168 (280), 1271 (612), *1321, 1328*
Hasse, L. 534, 561, 564, 565, 576 (5), *586*
Hassner, A. 397 (887), *480, 522* (192), *530*, 1195 (375), *1323*
Hastings, S. H. 195 (77), *211*
Hata, G. 345 (633), 425 (1126), *475, 485*
Hata, N. 745 (232), *753*
Hata, T. 438 (1236), *487*
Hatada, M. 123, 124 (5), *169*
Hatanaka, Y. 720, 722 (185g, i), *752*
Hatch, C. E. 1273 (618), *1328*
Hatch, R. P. 1039 (221), *1057*
Hatfield, L. D. 524 (206), *530*
Hathaway, C. E. 223 (66), *258*
Hatz, E. 413 (963), *481*, 536 (26), 543, 557, 561 (26, 109), 566, 584 (26), *587, 589*
Haug, E. 534, 557, 562, 568 (8), *586*
Haug, P. 69, 71, 72, 76 (12), *114*, 622, 626 (160), *639*
Haugen, G. R. 60, 61, 63, 65 (8), *65*
Hauptmann, S. 1145, 1153, 1154 (273), *1321*
Haurie, M. 222, 226 (64), *257*
Hauser, C. F. 439 (1243, 1244), *487*
Hauser, C. R. 157 (164), *173*, 280 (148– 150), 316 (437), 320 (470), 416 (1009), 417 (1038), 424 (1118), 426 (1132– 1137), 431 (1167), *465, 470, 471, 482–485*, 1205 (448–452), 1206 (453), *1325*

Hausermann, M.   309 (381), 469
Hausigk, D.   375(791, 796), 377 (791),
    378 (791, 796), 379 (791), 406 (914),
    478, 480
Häusler, J.   1038 (179), 1056
Hauthal, G.   574 (320), 593
Havinga, E.   680 (95), 749
Hawk, J. P.   348 (641), 475
Hawkins, C. J.   99 (150), 103 (213, 229,
    230), 109 (150), 116, 118
Hawkins, E. G. E.   315 (433), 353 (653),
    470, 475, 1195 (378), 1323
Haworth, R. D.   503 (74), 528
Hawthorne, M. F.   353 (660), 475
Hay, A. S.   305 (357), 328 (503), 469,
    472
Hayakawa, A.   562, 569 (224), 591
Hayakawa, S.   275 (94), 463
Hayase, Y.   1048 (350), 1060
Hayashi, H.   202 (117), 211
Hayashi, I.   421 (1086), 484
Hayashi, K.   204 (132), 212
Hayashi, N.   1027, 1029, 1031, 1032,
    1042 (56), 1054
Hayashi, S.   234 (225), 261
Hayashi, T.   790 (49), 822
Hayes, E. F.   23, 25 (43), 56
Hayes, F. N.   1080, 1083 (80), 1317
Haynes, L. J.   166 (210), 173
Haynes, P.   458 (1401), 490
Haynes, W. M.   411 (928), 481
Hayon, E.   672, 677, 685, 686 (81a, b),
    749, 763 (45), 766 (58, 59), 767 (58,
    59, 65), 768 (69), (61), 778, 779
Hayward, R. C.   494 (14), 526
Hazdra, J. J.   883 (116), 884, 885 (128),
    886 (116), 912
Hazelrigg, M. J. Jr.   808 (127), 824
Heard, C. L.   1102, 1103 (145), 1318
Hearn, M. J.   1285 (659), 1329
Hearn, M. T. W.   437 (1225), 487
Heathcock, C. H.   362, 363 (720), 476,
    729 (196b), 752, 1168 (309), 1322
Heatherington, K.   504, 525 (77), 528
Heaton, C. D.   176 (21), 181, 182, 196
    (54), 209, 210
Hecht, S. M.   413 (960), 481
Heck, R. F.   292 (255, 256), 456 (1385),
    467, 490, 826 (11), 855
Hedaya, E.   661, 662 (44b), 748
Hederich, V.   572 (314), 593
Hedgley, E. J.   1038 (164), 1048 (355),
    1056, 1061
Heerthes, P. M.   672, 674, 678, 682 (82),
    749, 884 (121), 912
Heffron, P. J.   109, 110 (282), 119
Hegedus, L. S.   291, 456 (252), 467
Hehdi Nafissi, M.   537 (49), 587

Hehemann, D. G.   745 (229), 753
Hehneberg, G.   225 (122), 259
Hehre, W. J.   19 (6, 9), 20 (14, 15, 18),
    21 (15, 18), 23 (44), 24 (18, 44), 25
    (14, 18), 27, 28 (44), 29 (14, 15), 30
    (18, 78), 31 (44), 55−57, 239 (295a),
    262
Heiba, E. I.   501 (61), 527, 1172 (323,
    325), 1173 (323, 326), 1174 (325),
    1322
Heibl, C.   1041, 1043 (275), 1059
Heidberg, J.   1042 (297), 1059
Heikens, D.   420 (1075), 484
Heilmann, S. M.   445 (1303), 488
Heimgartner, H.   706 (147), 751
Heimsch, R. A.   327 (500), 472
Heindl, L.   316 (439), 471
Heine, H. W.   493 (7), 526
Heine, O.   1092 (110), 1318
Heink, W.   230 (180), 260
Heinsohn, G. E.   810 (133), 824
Heise, H.   547 (138), 589
Heisig, G. B.   279 (137), 464
Heitman, P.   1001 (169), 1004 (174),
    1020
Heitmiller, R. F.   493 (6), 526
Heldt, W. Z.   385, 450 (837), 479, 1274
    (620), 1328
Hell, J. W.   496 (28), 527
Hellberg, L. H.   88 (133), 116
Hellmann, H.   276 (111), 464
Hellyer, J. M.   985 (128), 1019
Helmreich, R. F.   1104 (150), 1319
Hemingway, J. C.   507 (93), 528
Hemingway, R. J.   507 (94), 528
Hems, M. A.   205 (140), 212
Henbest, H. B.   452 (1360), 489, 1073
    (38), 1074 (38, 46), 1316
Henderson, W. A. Jr.   661 (47, 48, 50),
    662 (47), 663 (50), 668 (47, 48, 50),
    669 (47, 48), 708 (47), 748
Hendrickson, A. R.   1025, 1031, 1033,
    1034, 1036 (46), 1046 (332), 1050
    (46), 1054, 1060
Hendrickson, J. B.   412, 444 (950), 481
Henery-Logan, K. R.   522 (196), 530,
    1206, 1220 (485), 1325
Henessian, S.   557 (188), 558 (188, 189),
    590
Henklein, P.   562 (203), 590
Hennion, G. F.   322 (479), 358 (690),
    471, 476
Hennis, H. E.   415 (996), 482
Henrick, C. A.   430 (1160), 485, 1047
    (336), 1060
Henry, D. W.   457 (1394), 490
Henscheid, L. G.   745 (228c), 753
Hensley, W. M.   1046 (333), 1060

Henson, D. B. 252 (430), *265*
Henton, D. E. 1186 (346), *1323*
Hentzschel, P. 204 (131), *212*
Herald, D. L. Jr. 310, 314 (391), *470*
Herbert, P. 276 (103), *464*
Herbstein, F. H. 183 (63, 64), 184 (68, 69), 185 (69), 189 (64), *210*
Heritage, G. L. 1113 (181), *1319*
Herlem, D. 1310 (711), *1330*
Hermann, J. L. 501 (65), *527*
Hermann, R. B. 381 (823), *478*
Herndon, W. C. 895 (180), *913*
Herndon, W. S. 892 (161), *913*
Herre, W. 198 (96, 97), 200, 201 (108), 202 (105, 108), *211*
Herrick, E. C. 417 (1034), *483*
Herries, D. G. 962, 998 (71), *1017*
Herriott, A. W. 321, 334 (474), *471*
Hermann, J. L. 427 (1140, 1142), *485*, 1123, 1126, 1127 (212), *1320*
Hershberg, E. B. 303 (337), *468*
Hershberg, E. E. B. 340 (612), *474*
Herslöf, B. 106 (260, 261), 107 (261, 262), *119*
Hertz, H. G. 217 (5), 223 (72–74), 226 (134), *256, 258, 259*
Herz, J. E. 443 (1290), *488*
Herzberg, G. 26 (52), *56*
Herzog, K. 1024, 1032 (35), *1053*
Hesse, G. 310 (393), *470*
Heusler, K. 1171 (316), *1322*
Heweston, G. M. 893, 894 (168), *913*
Hewitt, J. J. 870 (55), 892 (162), *910, 913*
Hey, D. G. 432 (1185), *486*
Hey, D. H. 289 (229, 230), *466*, 1206, 1207, 1211, 1212 (463), *1325*
Heyer, E. W. 392 (869), *479*
Heymann, E. 1040, 1053 (263), *1058*
Heymes, R. 1053 (409), *1062*
Heyn, H. 1225–1227 (503), *1326*
Heyn, M. P. 112 (317), *120*
Heyne, H. U. 1295 (668), *1329*
Heyns, K. 305 (350, 353–355), *469*
Heyrovsky, J. 782 (16b), *821*
Heywood, D. L. 348 (640), 353 (663, 665), *475*
Hickinbottom, W. J. 298 (310), *468*
Hidaka, J. 103 (214, 216, 220, 236), *118*
Hidalgo, J. 969, 970, 977, 988, 1014 (92), *1018*
Higa, T. 440 (1257), *487*
Higashi, F. 412 (948, 949), 443 (948, 949, 1289), *481, 488*
Higgins, C. E. 894 (177), *913*
Higuchi, M. 562, 569 (224), 571 (293), 596, 597 (382), *591, 592, 599*

Higuchi, S. 336 (580), *473*, 1206, 1217 (473), *1325*
Hildebrand, J. H. 178 (30), 203 (124), *210, 212*
Hildebrand, R. 562 (213), *591*
Hildebrand, R. P. 1078 (56), *1317*
Hilgetag, G. 372 (771), *477*, 1232 (553, 554, 584), 1247, 1250 (553), 1251 (554), 1257 (584), *1327*
Hilgetag, K.-P. 351 (650), *475*
Hill, E. A. 904 (218), *914*
Hill, H. W. Jr. 1232, 1255 (582), *1327*
Hill, J. T. 878 (99), *911*
Hill, K. A. 424 (1116), *484*
Hill, R. K. 583 (346), *593*
Hillenbrand, E. F. Jr. 602 (2), 629 (187), *635, 639*
Hillenbrand, G. 147, 157 (113), *171*
Hillis, J. 854 (97), *857*
Hine, J. 1011 (190), *1020*
Hines, L. F. 839 (64), *856*
Hinshelwood, C. 864 (18), *910*
Hinton, J. F. 242, 246, 247 (332), 252 (332, 426), 253 (426), *263, 265*
Hintze, R. E. 642 (3a), *747*
Hirabayashi, Y. 1024 (26), *1053*
Hirai, K. 1052 (400), *1062*
Hiramatsu, T. 513 (130), *529*
Hirata, Y. 1145, 1147 (261), *1321*
Hiratami, K. 544 (117), 572, 573 (316), *589, 593*
Hirayama, M. 802, 803 (86a), 814, 816 (144), *823, 824*
Hirayma, T. 314, 353 (424), *470*
Hirofuji, S. 239 (292), *262*
Hiroi, K. 296 (293), *468*
Hiroka, H. 663, 708 (55), *748*
Hirose, K. 238 (283), *262*
Hiroshi, M. 82, 83 (79), *115*
Hirota, E. 28 (62), *56*
Hirota, K. 123, 124 (5), *169*, 457 (1399), *490*
Hirota, M. 239 (291–293), 296, 297), *262*
Hirota, T. 571 (300, 301), 597 (392, 393), *592, 599*
Hirotsu, K. 224 (100, 109), 225 (110, 112, 113), *258*
Hirowatari, N. 295 (284), *467*
Hirrami, T. 1123, 1128–1130 (215), *1320*
Hirsch, K. A. 538, 546 (63), 555, 567 (177), *588, 590*
Hirschmann, R. 1145, 1159 (278), *1321*
Hirshfeld, A. 1073 (37), *1316*
Hirth, C. G. 156 (158), *172*
Hisano, T. 445 (1304), *488*
Hisatsune, I. C. 23 (39), *56*

Hitchcock, P. B. 826 (30), *856*
Hittenhausen-Gelderblom, R. 160 (179), *173*
Hixon, S. S. 706, 730, 731, 733 (146c), 736 (211a, b), *751, 753*
Hnevsova-Seidlova, V. 1109 (169), *1319*
Ho, A. C. 151 (132, 134), *172*
Ho, C. M. 627 (178), *639*
Ho, D. S. 275 (88), *463*
Ho, H. C. 275 (91), *463*
Ho. T.-L. 272 (35, 36), 275 (91, 93), 306 (360), 309 (382, 385), 358 (689), 437 (1231), 447 (1314), *462, 463, 469, 476, 487, 488,* 1161, 1168, 1170 (284), *1321*
Hobbs, C. F. 548 (144, 145), 554, 557, 572 (145), *589*
Hobbs, M. E. 222 (60), 248 (391), *257, 264*
Hobrock, B. W. 130 (32), *170*
Hochstein, F. A. 298 (311), *468*
Hochstetler, A. R. 1066, 1068, 1147, 1148, 1153 (19), *1316*
Hochstrasser, U. 395 (881), *480*
Hocker, J. 424 (1117), *484,* 550 (154, 156), 557 (182), *589, 590*
Hocking, W. H. 1022 (3, 4), 1024 (3), *1053*
Hodge, P. 375, 377 (792, 795), 378 (795), 439 (1241), *478, 487*
Hodges, R. 1079 (59), *1317*
Hodgkin, D. C. 524 (210), *531*
Hodler, M. 812 (139), *824*
Hodson, D. 413 (961), 437 (1220), *481, 486*
Hoerger, E. 386 (842), *479*
Hofacker, G. L. 220, 226, 228 (28), *257*
Hofer, P. 1078 (54), *1316*
Hofer, R. M. 583 (348), *593*
Hoff, E. F. Jr. 1232, 1235 (526), *1326*
Hoffman, C. 391 (865), *479*
Hoffman, M. Z. 766, 768 (60), *779*
Hoffman, R. 697 (139), *750*
Hoffman, T. D. 376 (807, 808), *478*
Hoffman, W. A. III 412 (956), *481*
Hoffmann, H. 536 (28, 43), 538 (28), 542 (108), 543 (28, 43), 545 (28, 43, 125), 548 (28, 43, 108, 125), 553 (108, 125), 554 (28, 43), 556 (43, 108), 557 (28), 558 (28, 108), 561, 567 (43), 574 (125), 577 (43, 329), 578 (329), *587, 589, 593*
Hoffmann, H. M. R. 541 (89), 584 (356–359), *588, 594*
Hoffmann, R. 1040, 1042, 1046 (272), *1059*
Hoffmann, R. W. 597, 598 (395), *599*
Hofle, G. 417 (1041, 1042), *483*

Hofmann, A. 562, 570 (200, 201), 571 (200, 201, 295), 586 (295), *590, 592*
Hofmann, A. W. 571 (297), *592*
Hofmann, J. Ch. 787 (34), *822*
Hofmann, J. E. 328 (516), *472*
Höft, E. 1311 (714), *1330*
Högfeldt, E. 237 (261), *262*
Hogg, D. R. 1037, 1047 (147), *1056*
Hohn, E. G. 98 (160), *116*
Hojo, K. 1038, 1046 (181), *1056*
Hojo, M. 916, 919, 928 (11), 938 (64), *942, 944*
Holbert, G. W. 499 (48), *527*
Holder, D. 332, 368 (556), *473*
Hole, K. J. 864 (20), *910*
Holeci, I. 323 (481), *471*
Holland, G. 395 (880), *480*
Holley, A. D. 1232, 1236 (528), *1326*
Holley, R. W. 1232, 1236 (528), *1326*
Hollosi, M. 101 (175), *117*
Holm, A. 1049 (367), *1061*
Holm, R. T. 314, 353 (423), *470*
Holman, R. T. 147 (112), *171*
Holmes, J. L. 127 (18), 128 (25), 129 (28, 29), 130 (18, 28, 29, 33, 34), 131 (35), 158 (35, 167, 168), *169, 170, 173,* 615 (106, 107), *638*
Holmstead, R. L. 656 (31), 699 (31, 141), 700 (141), 746 (234), *747, 750, 753*
Holroyd, R. A. 756 (1), *778*
Holsboer, D. 1033, 1034, 1040 (105), *1055*
Holt, G. 413 (961), 437 (1220), *481, 486*
Holtz, H. D. 328 (504), *472*
Holtzer, A. 951–953 (28), 984 (119), *1016, 1018*
Holy, A. 557 (183, 184), 558 (183, 184, 186), 561 (196), 562 (183, 184), 563 (196, 227, 229, 232), 566 (268), *590–592*
Holy, N. 806 (116), *823*
Holzwarth, G. 73, 99, 109 (38), *114*
Homberg, O. 294 (282), *467*
Homer, R. B. 251 (417), *265*
Homes, H. L. 279 (145), 358 (693), *464, 476*
Hong, N. D. 561, 563 (196), *590*
Honkanen, E. 166, 167 (209), *173,* 495 (19), *526*
Hood, F. P. 99 (168), *117*
Hooker, T. M. 109, 110 (292), *119*
Hooker, T. M. Jr. 101 (180), 109 (279), *117, 119*
Hoover, J. R. E. 415 (970), *481*
Hooz, J. 435 (1204, 1206, 1207), *486*
Hope, A. 102 (202), *117*

Hopff, H.   281 (179), 323 (482), *465, 471*
Hopfgarten, F.   111 (313), *120*
Hopkins, L. O.   1066, 1068, 1146, 1148, (17), *1316*
Hopkins, P. A.   1024 (37), *1053*
Hopkinson, A. C.   23, 26 (42), 27 (56), 36 (102), 39 (103), 48 (121), 49 (123), *56–58*
Hoppe, B.   692, 703 (133), *750*
Hoppe, H.   1031, 1034 (94, 95), 1040 (94, 264), 1046, 1050 (94), 1052 (95), *1055, 1058*
Hoppe, I.   457 (1396), *490*
Hopps, H. B.   110, 111 (309), *120*
Hordegger, E.   360 (705), *476*
Hori, A.   1027, 1029, 1031, 1032, 1042 (56), 1044 (319), *1054, 1060*
Horibe, I.   1145, 1146 (253), *1321*
Horiguchi, K.   1276, 1277 (627), *1328*
Horii, Z.   1079 (58), *1317*
Horiki, K.   1039, 1041 (231), *1058*
Horn, P.   536, 538, 543 (28), 544 (115, 116), 545, 548 (28, 125), 549 (115, 116), 550 (115), 553 (125), 554 (28), 555 (116), 557, 558 (28), 569 (115, 116), 574 (125, 322), *587, 589, 593*
Hornberger, C. S. Jr.   493 (6), *526*
Horner, L.   279 (141), 382 (829), *464, 478,* 603 (12), *636,* 805 (102), 813 (140), 817 (102, 159, 160), *823, 824,* 1232 (525, 596, 599), 1235–1238, 1240, 1241 (525), 1264 (596, 599), *1326, 1328*
Horning, E. C.   338 (594), *474,* 1102 (144), 1204 (446), 1276, 1278, 1282 (633), *1318, 1325, 1328*
Horning, M. G.   1102 (144), *1318*
Hornish, R. E.   312 (404), *470*
Horowitz, A.   203 (125), *212*
Horspool, W. M.   653 (24), *747*
Horstmann, Chr.   76 (48), *114*
Horton, D.   341 (620), *474*
Horton, J. A.   1165, 1168 (296), *1322*
Horwitz, J.   109 (276), *119*
Horwitz, J. D.   563 (225), *591*
Hoshino, M.   594 (373), *598*
Hosking, J. R.   331 (523), *472*
Hospital, M.   222 (44), 244 (350), *257, 264*
Hoster, D. P.   902 (211), *914*
Hostettler, H. V.   1186 (350), *1323*
Hott, L.   437 (1230), *487*
Houda, M.   916, 919 (17), *943*
Houghton, R. P.   1048 (348), *1060*
Houk, K. N.   706, 730, 731, 733 (146a), *751*
Houriet, R.   134 (53), *170*

House, H. O.   276, 277 (114), 279 (145, 146), 283 (199), 296 (114), (297), 396 (883), 421 (1082), 431 (1169), *464–466, 468, 480, 484, 485,* 494, (10), *526,* 803 (91), *823,* 1108 (160), *1319*
Housty, J.   222 (44), 244 (350), *257, 264*
Houtman, J. P. W.   884 (121), *912*
Hove, M.   236 (254), *261*
Hover, H.   330 (522), *472*
Howard, J. C.   100 (170), *117*
Howe, I.   135 (59, 60), 137 (59), 143 (96), 155 (149), *170–172*
Howlet, G. J.   248 (390), *264*
Howton, D. R.   315 (429), *470,* 759 (17, 21), 761 (35), 762 (35, 43), 763 (35), *778*
Hoy, D. J.   868 (42), *910*
Hoyte, O. P. A.   237 (271), 238 (272), *262*
Hradetzky, F.   1314 (722), 1315 (723), *1330*
Hrdlovic, P.   652, 657, 694, 733 (20), *747*
Hreble, J. P.   396 (884), *480*
Hribar, J. D.   909 (241), *914*
Hrubesch, A.   792, 793 (54), *822*
Hruby, V. J.   443 (1288), *488*
Hsu, C.-J.   1003 (172), *1020*
Hu, A. T.   61, 62 (26), 64, 65 (48), *66*
Huang, B.-S.   691, 701 (125h,i), *750*
Huang, C. C.   416 (1031), *483*
Huang, E.-S.   507 (95), *528*
Huang, F.   834–838, 842 (51), *856*
Huang, F. J.   843, 844, 849 (70), *857*
Huang, P. C.   72 (27), *114*
Huang, W. H.   680 (69), *748*
Hub, L.   80, 81 (69), *115*
Hubbard, W. N.   1066–1068, 1072, 1073 (16), *1316*
Huber, B.   674 (84), *749*
Huber, J. H.-A.   1186 (346), *1323*
Huber, W.   1168 (302), *1322*
Hubert, J. C.   525 (220), *531*
Hubert-Brierre, Y.   1310 (711), *1330*
Huche, M.   583, 584 (352), 598 (397), *594, 599*
Huckin, S. N.   425 (1120–1122, 1124), *484, 485*
Huder, J.   1186, 1187 (349), *1323*
Hudlicky, T.   461 (1418), *490*
Hudrlik, A. M.   433 (1189), *486*
Hudrlik, P. F.   433 (1189), *486,* 1075 (51), *1316*
Hudson, B. E.   416 (1009), *482*
Hudson, C. S.   621 (153, 154), *639*
Huebner, C. F.   367 (746), *477*

Huffaker, J. E.   386, 387 (849), *479*
Huffman, K. R.   661, 663, 666, 668 (51), *748*
Hughes, D. O.   244 (348), 253 (440), *263, 265*
Hughes, E. D.   496 (31), *527*
Hühle, E.   369 (753), *477*
Huisgen, R.   495, 496, 514 (16), 516 (148, 149), 517 (156), 522 (190), *526, 529, 530,* 619 (134), *638,* 1052 (407), 1053 (412), *1062,* 1232, 1239 (538), *1326*
Hulbert, P. B.   422, 423 (1102), *484*
Huler, E.   243 (336), *263*
Hull, D. C.   419 (1065), *483*
Hull, R.   1232, 1260 (589), *1327*
Hullot, P.   456, 457 (1391), *490*
Hume, E.   735 (209a), *753*
Hummel, C. F.   526 (225), *531*
Hummelink, T. W.   1049 (371), *1061*
Humphrey, J. S. Jr.   652, 657 (33), *747*
Huneck, S.   88 (118), *116*
Hünig, S.   288 (226, 227), 374 (783), *466, 478,* 537, 538, 556 (59), *587*
Hunsdiecker, H.   514 (134), *529*
Hunt, D. F.   603 (16), *636*
Hunt, H. D.   29 (75), *57*
Hunt, J. D.   343 (622), *474*
Hunt, K.   397 (886), *480*
Hunt, M.   868 (43), *910*
Hunter, J. D.   331 (524), *472*
Hunter, T. L.   146 (107), *171*
Huntress, E. H.   303 (337), 411 (925), *468, 481*
Huppes, M.   1173, 1175–1177 (328), *1322*
Hurd, C. D.   881 (103), *912*
Hurst, D. T.   1196 (388), 1206, 1222 (487), *1323, 1325*
Hurwic, J.   554 (174), *590*
Huse, G.   183 (59, 60), *210*
Hussain, M. G.   1038 (175), *1056*
Hussain, S. A. M. T.   501 (63), *527,* 619 (132), *638*
Hussey, G. E.   711, 713, 714 (171), *751*
Husung-Bulblitz, R.   1004 (174), *1020*
Hutchins, R. O.   416 (1018), *482*
Hutton, J.   1166, 1168 (299), *1322*
Huyskens, P.   235 (243), *261*
Hvidt, A.   247 (387), *264*
Hwang, P. T. R.   1048 (345), *1060*
Hyatt, J. A.   710, 715 (163a,b), *751*
Hyde, J. L.   499 (51), *527*
Hyde, P.   209 (154), *212*
Hyde, S. M.   771 (81, 82), *779*
Hyeon, S. B.   1079 (61), *1317*

Iball, J.   179, 182, 192 (39), *210*

Ibbotson, G.   984 (125), *1019*
Ibers, J. A.   214 (1d), 216 (3), 224 (98), *256, 258*
Ibl, N.   787 (34), *822*
Ichihara, A.   406 (916), *480*
Ichikawa, M.   445 (1304), *488*
Ichikawa, Y.   182, 184 (56), *210*
Ide, J.   850, 851 (92), *857*
Ide, W. S.   365 (728), *476*
Idoux, J. P.   1048 (345), *1060*
Igano, K.   303 (344), *469*
Igarashi, K.   102 (204), *117*
Ihara, Y.   985 (128), 991 (153), *1019*
Ihda, S.   1040, 1052 (271), *1058*
Ihrig, P. J.   146 (107), *171*
Ihrman, K. G.   441 (1273), *488*
Ikada, T.   812 (138), *824*
Ikeda, F.   1080 (64), *1317*
Ikeda, M.   451 (1348), *489,* 1276, 1277, 1279 (624), *1328*
Ikeda, R. M.   166 (212), *173,* 495 (21), *527,* 627 (181), *639,* 1075, 1079 (52), *1316*
Ikenaga, S.   1039 (235), *1058*
Iley, D. E.   846, 848 (79), *857*
Ilmet, I.   177 (26), 182 (55), 184 (65), *210*
Il'yasov, A. V.   802 (86b,c), *823*
Imagawa, T.   514 (133), *529*
Imai, H.   515 (143), *529*
Imai, J.   338 (590), *474*
Imamoto, T.   1037, 1042 (153), *1056*
Imazawa, M.   1038 (180), *1056*
Immer, H.   1206 (455), *1325*
Imoto, E.   437 (1227), 448 (1325), *487, 489*
Imoto, M.   413 (959), *481*
Inaba, A.   231 (205), *260*
Inczédy, J.   980 (108), *1018*
Inesi, A.   812 (136, 137), *824*
Ingham, R. K.   318 (443), *471*
Ingold, C. K.   339 (604), *474,* 496 (30, 31), *527,* 622 (158), *639,* 1232 (598), *1328*
Inhoffen, H. H.   340 (614), *474,* 1101 (141, 142), *1318*
Inokawa, S.   1042, 1043 (280), *1059*
Inomata, K.   539 (66), *588*
Inoue, G.   439 (1246), *487*
Inoue, I.   444 (1299), *488*
Inoue, T.   504 (82), *528,* 758 (10), *778*
Inouye, Y.   109 (289), *119*
Inventa, A.-G.   1276, 1277 (625), *1328*
Iogansen, A. V.   228 (152), 246 (371, 379), 248 (392), 249 (392, 401), *259, 264, 265*

Ionescu, L. G.   971 (94), 985 (128), 1008, 1010 (94), *1018, 1019*
Ipaktschi, J.   706, 730, 731, 733 (146b), *751*
Iqbal, A. F. M.   1285 (661), *1329*
Ireland, R. E.   272 (45), 388 (861), 416 (1011), 451 (1350, 1351), *463, 479, 482, 489,* 845 (75), 847 (81), *857,* 1066, 1069, 1149 (22), *1316*
Irie, M.   204 (132), *212*
Irie, S.   204 (132), *212*
Irie, T.   238, 239 (288), *262*
Iriuchijima, S.   400 (900, 901), 401 (900), *480,* 1039 (239), *1058*
Irko, A.   35 (98), *57*
Irwin, W. J.   522 (194), *530*
Isaacs, N. S.   131, 158 (35), *170,* 521 (187), 522 (193), *530,* 615 (106), *638*
Isaacson, A. D.   238 (290), *262*
Isagulyants, V. I.   348 (642), *475*
Isemura, T.   951, 952, 954, 984 (26), *1016*
Ishchenko, O. S.   1040 (252). *1058*
Ishiba, T.   1052 (400), *1062*
Ishibe, N.   511 (118a), *529*
Ishihara, H.   1040 (251), *1058*
Ishihara, T.   289 (236), *466*
Ishii, Y.   551 (160–162), *590,* 1027 (58), 1029 (58, 71, 72), 1030, 1031 (71), 1032 (58, 72), 1034 (58), 1040 (71), 1042 (58, 72), *1054*
Ishikawa, K.   102 (205), *117,* 611 (81), 637
Ishikawa, N.   1039 (203), *1057*
Ishino, I.   1206, 1207, 1211 (461), *1325*
Ishino, Y.   597 (393), *599*
Isobe, K.   1168 (304), *1322*
Isol, S.   1079 (61), *1317*
Israelachvili, J. N.   952 (41), *1017*
Issacs, R.   415 (988), *482*
Issleib, K.   564 (234), 565 (234, 241), *591*
Ito, A.   133 (42), *170*
Ito, H.   336 (580), *473*
Ito, I.   296 (289), *467*
Ito, M.   229 (168), *260,* 1080 (64), *1317*
Ito, N.   406 (916), *480*
Ito, T.   620 (137), *638*
Ito, T. L.   1051 (383), *1061*
Ito, Y.   372 (768), 415 (993), 457 (1399), *477, 482, 490*
Itoh, K.   243 (337b), *263,* 551 (160), *590,* 1206 (469), *1325*
Itoh, M.   412 (957), 430 (1162), *481, 485,* 1080 (88, 89), 1085 (88), 1086, 1087 (88, 89), 1088 (89), *1317*
Ittel, S. D.   852 (95), *857*
Itzchaki, J.   381 (820), *478*

Ivanov, D.   280 (152), *465*
Ivanov, L. L.   278 (130), *464*
Ivanov, O. V.   631 (195, 196), *640*
Ivanov, S. A.   237 (264), *262*
Ivanov, V. B.   735, 736 (210), *753*
Ivanov, V. L.   735, 736 (210), *753*
Ivanova, I. A.   536 (39, 44), 538 (44), 543 (44, 110, 111), 544, 548 (44), 549 (39), 558 (44), 561 (44, 110, 111), 562, 567, 569 (44), 577 (39), *587, 589*
Iversen, P. E.   795 (63), 797 (64, 65), 804 (65), 815 (63–65), 816 (63, 64), *822,* 1042 (291), *1059*
Ives, J. L.   596 (388), *599*
Iwai, I.   1145 (259), *1321*
Iwasaki, M.   217 (8), *256,* 757 (3–6), 758 (5, 10), *778*
Iwata, F.   1284 (643), *1329*
Iwata, S.   42, 44 (109), *57,* 200–202 (110), 205, 206 (141), *211, 212,* 214 (2f), 219 (22), 232 (211), *256, 257, 260*
Iyer, K. N.   345 (631), *475*
Iyoda, J.   291 (249), *467*
Iyoda, M.   1145, 1147 (261), *1321*
Izawa, T.   417 (1052), *483*
Izawa, Y.   289 (236), *466,* 675, 682 (88), *749*

Jabloner, H.   1276 (629), *1328*
Jacklin, A. G.   331, 341 (525), *472*
Jackson, E. L.   365 (735), *477*
Jackson, G. E.   864 (19, 22), 868 (22), *910*
Jackson, M. B.   168 (222), *174*
Jackson, P. E.   1030 (89), *1054*
Jacob, J.   332, 354 (550), *473*
Jacobs, T. L.   88 (133), *116,* 340 (616), 406 (912), *474, 480*
Jacobsen, R. J.   222 (51), 228 (147), *257, 259*
Jacobsen, R. P.   1168 (287), *1321*
Jacobson, H. A.   569 (282), *592*
Jacobson, M.   333 (560), *473*
Jacobson, N.   328 (515), *472*
Jacobson, R. A.   224 (106), *258*
Jacot-Guillarmod, A.   417 (1053), *483*
Jacques, J.   109 (284), 110 (284, 296), *119,* 287 (219), *466*
Jacquier, R.   393, 452 (871), *479*
Jacquignon, P.   177 (25), *210*
Jacquot, C.   440 (1258), *487*
Jadot, J.   341 (526), *472*
Jadzyn, J.   517 (151), *529*
Jaeger, D.   108, 109 (268), *119,* 280 (168), *465*
Jaeger, D. A.   688 (129), 690, 691 (132), 699, 700 (129, 132), 735, 736 (132), 746 (132, 233), *750, 753*

Jaeger, P. 800 (75), *822*
Jaffe, H. H. 239 (299), *262*
Jaffé, H. H. 69, 71, 73 (11), *114*
Jafri, J. A. 743 (223a), *753*
Jäger, A. 1232, 1234–1237, 1240, 1241 (523), *1326*
Jahn, R. 1052 (390), *1061*
Jahngen, E. G. E. Jr. 281 (176, 177, 182), 328 (182), *465,* 1066 (26), 1070 (24, 26), *1316*
Jahnke, D. 825 (6), *855*
Jain, M. K. 947, 1014, 1015 (1), *1016*
Jakobsen, P. 150 (122), *172,* 1035, 1036 (130), *1055*
Jakubetz, W. 232 (213), 233 (216), 235, 236 (213), *261*
Jambresic, I. 380 (818), *478*
James, B. G. 415 (981), *482*
James, B. R. 826 (15, 26), *855*
James, K. 346 (634), *475*
James, M. N. G. 222, 239 (58), *257*
Jamieson, W. D. 1035 (127), *1055*
Jamoulle, J. C. 1039 (224), *1057*
Janda, M. 297 (306), *468*
Jannke, P. J. 300 (330), *468*
Janoschek, R. 30, 44 (81), *57,* 243 (334), *263*
Janot, M. M. 452 (1361), *489,* 1304 (700), 1306 (705), *1330*
Jansen, J. E. 496 (26), *527*
Jansons, E. 1037 (144), 1038 (169), 1042, 1044 (144), 1045 (144, 322–324), *1056, 1060*
Janssen, M. J. 255 (480), *266,* 1022, 1036, 1038, 1044, 1045 (1), *1053*
Janssen, W. A. J. 1049 (364), *1061*
Janzen, E. G. 328 (512), *472*
Jarboe, C. H. 1080, 1083 (80), *1317*
Jardine, F. H. 826 (21, 27), 827 (27), 850 (89), 853 (27), *856, 857*
Jaselskis, B. 306 (363), *469*
Jason, E. F. 1109 (167), *1319*
Jaus, H. 551, 552, 555, 557, 559, 562, 568, 584 (159), *590*
Javaheripour, H. (136), *750*
Jaworska, S. 84 (93), *115*
Jaworski, T. 513 (131), *529,* 896 (189), *913*
Jefferies, P. R. 1092 (115), *1318*
Jefferson, E. G. 275 (101), *464*
Jefford, C. W. 676 (221a), 687 (104), 743 (104, 221a), *749, 753*
Jeffrey, A. M. 1038 (183), *1056*
Jeffrey, G. A. 222 (54), 240 (301a), *257, 262*
Jeger, O. 345 (630), *475*
Jehlička, V. 1023 (25), 1024 (25, 34), *1053*

Jekabsons, V. 1045 (324), *1060*
Jelagin, S. 521 (182b), *530,* 1232, 1240 (535), *1326*
Jencks, W. P. 947 (9), 948, 949, 998, 1010 (16), 1015 (9), *1016*
Jeney de Boresjenö, N. L. T. R. M. von 628 (185), *639*
Jeng, S. 523 (201), *530*
Jeng, S. J. 523 (198), *530,* 1200, 1201, 1203, 1204 (425), *1324*
Jenkins, D. M. 359 (699), *476*
Jenkins, H. D. B. 218 (12), *256*
Jenkins, J. A. 989 (146), *1019*
Jenko, B. 596, 597 (378), *598*
Jennings, J. P. 73 (42), 75 (42, 43), 88, 91 (138), 95 (157), 109 (138), *114, 116,* 623 (163), 624 (163, 164), *639*
Jenny, E. J. 539 (73), *588*
Jensen, B. L. 445 (1301), *488*
Jensen, H. H. 241 (329), 251 (423), *263, 264*
Jensen, K. A. 1031 (100), 1037 (142), *1055, 1056*
Jensen, L. 1033, 1034 (111), 1037, 1042 (111, 159), *1055, 1056*
Jensen, L. H. 245 (357), *264*
Jentsch, W. 539 (77), *588*
Jentschura, U. 222 (63), 229 (173), 230 (63, 177), 231 (197), *257, 260*
Jerina, D. M. 1038 (183), *1056*
Jernow, J. L. 395 (880), *480*
Jeskey, H. 296 (294), *468*
Jesson, J. P. 826, 827, 837 (31), 852 (95), *856, 857*
Jew, S. 495 (15), *526*
Jezierski, A. 777 (99), *780*
Jeżowska-Trzebiatowska, B. 777 (97–99), *780*
Jimenez, P. 61, 62 (30), *66*
Jirkovsky, V. 1109 (169), *1319*
Job, B. E. 127, 130, 131, 142, 158 (16), *169*
Joesten, M. D. 214 (1f), 218 (11), 234 (1f), 250 (410), *256, 265*
Johanson, G. A. 123 (3), *169*
Johansson, A. 29 (72), 45 (72, 114), 46, 47 (72), *56, 57,* 242, 250–252 (331), *263*
Johansson, L. 103 (234), *118*
Johansson, N.-G. 1206, 1220 (484), *1325*
John, J. P. 374, 375 (785), *478*
Johns, N. 1039 (215), *1057*
Johnsen, R. H. 759 (24), *778*
Johnson, B. A. 1145 (269), *1321*
Johnson, C. D. 251 (417), *265*
Johnson, C. K. 225 (126), *259*
Johnson, E. A. (541), *473*
Johnson, F. 296 (292), *468*

Johnson, H. B.   282 (193), *465*
Johnson, H. E.   1109 (167), *1319*
Johnson, H. L.   790 (47), *822*
Johnson, J. B.   629 (199), *640*
Johnson, J. R.   276 (113), 421 (1080), *464, 484,* 521 (183), *530,* 1108 (159), *1319*
Johnson, M. R.   707 (149), *751*
Johnson, P. Y.   1273 (618), *1328*
Johnson, R. C.   304 (348), *469,* 1145, 1152 (270), *1321*
Johnson, R. E.   297 (307), *468,* 501 (62), *527,* 1138–1140, 1142–1144 (236), *1320*
Johnson, T.   731, 732 (200b), *752*
Johnson, W. C. Jr.   70 (14, 16), 95 (16), 99 (14), *114*
Johnson, W. H.   61, 62 (25), *66*
Johnson, W. S.   272 (33, 45), 277 (118, 119), 414 (967), *462–464, 481,* 901 (205), *914,* 1066–1068, 1072, 1073 (16), 1110 (170), *1316, 1319*
Johnston, D. B. R.   1301 (681), *1330*
Johnston, G. A. R.   493 (1b), *526*
Johnston, P.   1145, 1154, 1155 (274), *1321*
Johnstone, R. A. W.   159 (173, 174), 164 (173, 174, 198, 199, 202), *173,* 614 (99, 100), *638,* 1035 (126), 1036 (126, 135), *1055*
Jolly, P. W.   825 (8), *855*
Jolly, W. L.   894 (179), *913*
Jonas, V.   31 (84), *57*
Joncich, M. J.   349 (646), *475*
Jones, A. R.   759 (18–20), *778*
Jones, C. M.   605, 606 (35), 608 (35, 49, 50), *636*
Jones, D. A. K.   893, 894 (165, 169, 170), *913*
Jones, D. G.   1232, 1264 (597), *1328*
Jones, D. W.   183 (58), *210,* 730, 731 (198), *752*
Jones, E.   874 (75, 76), 876 (75), *911*
Jones, E. G.   136 (62), *170*
Jones, E. P.   332 (544), *473*
Jones, E. R. H.   310, 314 (389), 323 (484), 435 (1209), 437 (1221), *470, 471, 486,* 1074 (46), 1080, 1082 (67), 1173, 1179 (330), *1316, 1317, 1322*
Jones, F.   225 (116), *259*
Jones, G.   409 (921), 421 (1081), *480, 484,* 1284 (654), *1329*
Jones, G. H.   1206, 1207, 1211, 1212 (463), *1325*
Jones, G. I. L.   88 (120), *116,* 1022 (9), 1023 (24), 1024 (9), *1053*
Jones, J. I.   916 (12), *942*
Jones, J. K. N.   848 (84), *957*

Jones, K. M.   103 (240), *118*
Jones, M. N.   984 (125), *1019*
Jones, P.   246 (378), *264*
Jones, P. R.   287 (218), *466*
Jones, R. E.   541 (94), *588*
Jones, R. G.   279 (136), 319 (462), *464, 471*
Jones, R. L.   247 (383), 253, 254 (444), *264, 265,* 310, 313 (394), *470*
Jones, R. N.   620 (137, 140), *638, 639*
Jones, T. H.   1052 (395), *1062*
Jones, W. A.   333 (560), *473,* 863 (15), *910*
Jones, W. M.   1145, 1151 (267), 1232 (522), *1321, 1326*
Jönsson, L.   432 (1179), *486*
Jönsson, P. G.   221 (41), 222 (53), 223 (80), 233 (41), *257, 258*
Jordan, E.   316 (437), *470*
Jordan, E. F. Jr.   411 (932), 420 (1074), *481, 484*
Jorgensen, E. C.   95 (156), *116*
Jorgensen, W. L.   837 (60), *856*
Jorgenson, M. J.   422 (1096), *484,* 709 (150, 153, 157, 159), 717 (150, 153, 157), 718 (157), 729 (196a,b), *751, 752*
Joris, L.   241, 250, 251 (319), *263*
Josan, J. S.   585 (363), *594*
Joschek, H.-I.   672, 673, 678 (93), *749*
Joschek, H.-J.   672, 678 (64), *748*
Jose, C. I.   240 (313), *263*
Josephson, S.   103, 104 (242, 243), *118*
Joshua, C. P.   240 (302), *262,* 1039 (225), *1057*
Josien, M. L.   225 (131), *259*
Josimović, Lj.   759 (25), 760 (25, 28), 761 (32), 767 (28), *778*
Jost, K.   1193 (365), *1323*
Joullie, M. M.   1231, 1232, 1234 (517), *1326*
Juenge, E. C.   348 (644, 645), *475*
Julia, S.   1047 (340), *1060*
Julliard, M.   745 (228a), *753*
Junek, H.   541 (95), *588*
Jung, A.   1232 (544), *1327*
Jung, C. J.   1145, 1147, 1159 (255), *1321*
Jung, G.   99 (162, 166), *117*
Junge, B.   1053 (411), *1062*
Junjappa, J.   418 (1055), *483*
Jurgens, W.   805 (109), *823*
Jurist, A. E.   386 (853), *479*
Juszkiewicz, A.   237 (267), *262*
Jutz, C.   567 (280), *592*

Kabalka, G. W.   355 (680), 434 (1196, 1200), *475, 486*
Kabasakalian, P.   1145, 1159 (256), *1321*

Kabat, E. A.　102 (184), 105 (184, 253), 117, 118
Kabitzke, K.　1040 (263), 1052 (390), 1053 (263), 1058, 1061
Kabuss, S.　541 (91), 542 (91, 104–107), 548 (91), 573 (91, 104–107), 588
Kadaba, P. K.　411 (938), 440 (1259), 481, 487
Kadentsev, V. I.　138 (75, 76), 148 (76), 171
Kadorkina, G. K.　615 (103), 638
Kafka, T. M.　501 (64b), 527
Kagan, H. B.　411 (936), 481, 521 (189), 530, 1111 (179), 1197 (403), 1231 (518), 1232 (518, 532, 537), 1233 (602, 603, 605), 1235 (518), 1236 (518, 532), 1238 (532), 1240 (537), 1265 (606), 1266 (518, 602, 603), 1267 (518, 602, 603, 606), 1319, 1324, 1326, 1328
Kagan, J.　730, 731 (197), 752
Kagarise, R. E.　71 (19), 114, 235 (239), 261
Kahlenberg, J.　1030, 1031 (88), 1054
Kaimal, T. N. B.　332 (551), 473
Kairaitis, D. A.　826 (16), 855, 894 (174), 913
Kaiser, E. M.　416 (1022), 450 (1341), 460 (1412), 483, 489, 490
Kaiser, E. W.　403 (911), 480
Kaiser, G. V.　416 (1014), 482, 1301 (682), 1330
Kaji, A.　1039, 1041, 1050 (190), 1057
Kajtár, M.　78 (58), 101 (175), 110 (297), 114, 117, 119
Kako, M.　237 (262), 262
Kakuda, E.　1029, 1032, 1042 (72), 1054
Kala, S.　155 (153), 156 (155), 172
Kalbag, S. M.　1165, 1168 (296), 1322
Kaleciński, J.　777 (97), 780
Kalecińska, E.　777 (97), 780
Kalimin, V. N.　385 (835), 479
Kalina, L. I.　240 (309), 263
Kalinowski, H.-O.　1035 (124), 1055
Kalm, M. J.　271 (16), 462
Kalman, J. R.　432 (1180, 1181), 486
Kalmus, A.　277 (116), 464
Kaloustian, M. K.　1040–1042, 1052 (268), 1058
Kalvoda, J.　299 (321), 468
Kalvoda, L.　459 (1410), 490
Kamai, Gil'm　414 (965), 481
Kamata, K.　284 (206, 207, 211), 285 (207, 211), 466
Kamata, S.　515 (141), 529, 1039 (210, 226), 1048 (210), 1057
Kambe, S.　421 (1086), 484
Kamber, B.　345 (630), 475

Kamego, A.　958, 959 (66), 1017
Kamego, A. A.　959 (67), 985 (129), 988, 990 (139), 1017, 1019
Kamei, H.　28 (67), 56, 253 (433), 265
Kamenar, B.　182 (62), 210
Kametani, T.　1311 (713), 1330
Kamm, O.　271 (15), 307 (375), 318 (450), 462, 469, 471
Kamm, W. F.　307 (375), 469
Kampmeier, J. A.　1025, 1030, 1038 (42), 1053
Kan, K.　565, 566 (240), 591
Kanakam, R.　800 (82), 823
Kanaoka, Y.　412 (947), 415 (986), 481, 482, 674 (80), 720 (185a–i), 722 (80, 185a–i), 723 (185a,f), 724 (185b,d,h), 745 (230b, 235), 749, 752, 753, 1206 (469, 480, 481), 1218 (480), 1219 (481), 1325
Kanda, Y.　202 (116), 211
Kandel, R. J.　661, 662 (42), 748
Kane, J.　1052 (405), 1062
Kane, S. S.　341 (621a), 474
Kaneda, A.　103 (233), 118
Kaneko, R.　312 (411), 470
Kanematsu, K.　513 (130), 529
Kan-Fan, C.　452 (1361), 489, 1304 (700), 1330
Kaniecke, T. J.　424 (1113), 484
Kanojia, R. M.　1206 (472), 1325
Kanters, J. A.　88 (124), 116, 222 (56, 57), 225 (124), 234 (224), 257, 259, 261
Kantlehner, W.　534 (8), 536 (13, 28, 37 42), 538 (13, 28), 539 (42, 64), 540 (42), 543 (28), 544 (64), 545 (28), 546 (64), 548 (28), 549 (64, 149), 553 (37), 554 (13, 28), 555 (176), 556 (13), 557 (8, 28, 42, 64), 558 (28, 149), 559 (42, 192, 193), 562 (8, 13, 149), 566 (13, 254), 567 (254), 568 (8), 569 (149), 573 (318), 574 (149, 193), 575 (64), 577 (37), 581 (334, 335), 586–591, 593
Kantor, S. W.　320 (470), 471
Kapassakalidis, J.　570, 571 (290), 592
Kapaun, G.　574 (323, 324), 575 (324), 593
Kaplan, F.　1045 (326), 1060
Kaplan, L.　416 (1028), 483
Kaplan, M. L.　692, 706 (134), 750
Kaplan, S.　231 (193), 260
Kappeler, H.　272 (42), 462
Kaptein, R.　683 (99a,b), 749
Kapur, J. C.　1232 (557, 564, 576, 593, 594), 1250 (557, 564), 1251 (557), 1259 (576), 1262 (593, 594), 1327, 1328

Kapusainski, J.   302 (333), *468*
Kapuuan, P.   978 (102), *1018*
Karabatsos, G. J.   88 (121, 122), *116,*
    136 (64), *170*
Karabinos, J. V.   1065 (15), *1316*
Karakhanov, R. A.   348 (642), *475*
Kargin, Yu. M.   802 (86b,c), *823*
Karim, A.   507 (94), *528*
Kariyone, K.   446 (1313), *488*
Karl, N.   201, 202 (109), *211*
Karle, I. L.   22 (31), *56,* 86 (102, 103),
    115, 720, 722 (185e), *752,* 1206 (479,
    481), 1219 (481), *1325*
Karle, J.   22 (28, 31), *56,* 86 (102), *115*
Karlsohn, L.   248 (395), *264*
Karmann, H.-G.   1037 (150), 1044 (318),
    1056, 1060
Karnojitzky, V.   365 (730), *477*
Karo, W.   270 (2), 437 (1217), 442
    (1282), 458 (1408), *462, 486, 488, 490*
Karpati, A.   161 (182), *173*
Karrer, P.   340 (613), *474*
Karte, F.   1173 (328, 329), 1174 (331),
    1175 (328), 1176, 1177 (328, 329),
    1178, 1179 (331), 1200, 1204 (426),
    1322, 1324
Karten, M.   868 (39), *910*
Karvonen, P.   166, 167 (209), *173,*
    495 (19), *526*
Karwe, M. V.   620 (142), *639*
Kasafirek, E.   810 (134), *824*
Kasai, P. H.   865 (28), *910*
Kasai, Y.   444 (1292), *488*
Kashdan, D. S.   281, 328 (182), *465*
Kashino, S.   224 (103), 245 (360), *258,*
    264
Kasiwag, I.   386 (848), *479*
Kasler, F.   609, 621 (56), *637*
Kasyanov, V. V.   328 (507), *472*
Kasymhodjaev, P. S.   223 (68), *258*
Katada, T.   1027 (59), 1028, 1032 (66,
    68, 70), 1041 (70), 1043 (59), 1044
    (59, 66, 68), *1054*
Kataka, T.   1029, 1032, 1041, 1042 (73),
    1054
Kataoka, T.   1028, 1032 (69), *1054*
Kathpal, H. B.   30 (77), *57*
Katin, Yu. A.   64 (43), *66*
Katner, A. S.   375 (788), *478*
Kato, H.   1026 (74), *1054*
Kato, M.   1007 (182), *1020,* 1145, 1147
    (261), 1276 (626), *1321, 1328*
Kato, S.   936 (61), *944,* 1027 (56, 58, 59),
    1028 (66–70), 1029 (56, 58, 71–73),
    1030 (71, 91), 1031 (56, 71, 101, 103),
    1032 (56, 58, 66–70, 72, 73, 103),
    1034 (58), 1036 (103), 1039 (91), 1040
    (71, 251), 1041 (70, 73), 1042 (56, 58,

72, 73, 103), 1043 (59), 1044 (59, 66,
    68, 319), *1054, 1055, 1058, 1060*
Kato, T.   537 (50), *587,* 1028 (69), 1029
    (72), 1032 (69, 72), 1042 (72), *1054*
Katon, I. E.   221 (43), *257*
Katon, J. E.   605 (32), *636*
Katritzky, A. R.   525 (216), *531,* 608
    (44), *636*
Katsumi, M.   102 (199), *117*
Katura, T.   82, 83 (79), *115*
Katz, J. L.   243 (338), *263*
Katz, R.   300 (326), 372 (772), *468, 477*
Katzenellenbogen, J. A.   291 (253), 389
    (863), 422 (1099), 456 (253), *467,*
    479, 484
Katzhendler, J.   950 (24), 985 (127a, b),
    995 (160), 1008 (127a, b, 183), 1009
    (127a), *1016, 1019, 1020*
Katzin, L. I.   72, 88, 89, 93 (24), 103
    (222, 235), *114, 118*
Kaučič, V.   224 (105), *258*
Kaufman, J. A.   1191 (357), *1323*
Kaurov, O.   101 (179), *117*
Kawabata, Y.   1039 (207, 208), 1052
    (207), *1057*
Kawada, G.   786, 787 (25), *821*
Kawai, M.   102 (199), *117*
Kawanishi, M.   674, 722 (80), *749*
Kawanisi, M.   514 (133), *529,* 733, 734
    (206), *753,* 1186, 1187 (352), *1323*
Kawano, Y.   502 (67), *528*
Kawanuera, M.   596, 597 (382), *599*
Kawasaki, H.   327 (499), *472*
Kawashima, Y.   294 (275), *467*
Kawazoe, Y.   896, 902 (185), *913*
Kay, M. I.   225 (123), *259*
Kaye, H.   75 (46), *114,* 332 (557), *473,*
    1065 (14), 1161, 1168 (283), *1316,*
    1321
Kazama, Y.   271 (20), *462*
Kazaryan, S. A.   412, 443 (949), *481*
Kazlauskas, R.   314, 355 (422), *470*
Kebarle, P.   34, 35 (95), *57,* 138 (70),
    171, 671, 672, 677 (92f), *749,* 890
    (149), *912*
Keck, H.   547 (131, 132, 138), 548
    (132), 571, 586 (131), *589*
Keck, R.   1042 (282), *1059*
Kedziora, P.   517 (151), *529*
Keefer, L.   102 (208), *117*
Keefer, R. M.   175 (2), 180 (43), *209,*
    210
Keil, K. H.   1232 (544), *1327*
Kekka, S.   245 (360), *264*
Kekulé, A.   534 (1), *586*
Kelkar, G. R.   620 (142), *639*
Keller, F.   1115 (186), *1319*
Keller, K. T.   386 (847), *479*

Keller, L. 22, 25, 29, 30 (22), *56*
Keller, R. 25 (48), *56*
Keller, R. T. 901 (205), *914*
Keller-Schierlein, W. 514 (140), *529*
Kelley, R. B. 275 (90), *463*
Kellogg, M. S. 731, 733 (201b), *752*
Kellogg, R. M. (166), *751*
Kellom, D. B. 882, 899 (108), *912*
Kellve, P. 248 (395), *264*
Kelly, F. M. 899 (195), *913*
Kelly, F. W. 861 (7), 885 (130), 889 (145), 891 (153), 892 (153, 164), *910, 912, 913,* 1094, 1095 (114), *1318*
Kelly, J. F. 168 (223), *174,* 617 (112), *638,* 1228, 1250 (506), *1326*
Kelly, J. P. 131 (36), *170*
Kelsey, J. E. 507 (93), *528,* 627 (177), *639*
Kempe, U. M. 1180 (336), *1322*
Kempf, J. 230 (179), 231 (194), *260*
Kendall, E. C. 339 (611), *474*
Kendall, P. M. 284 (210), *466*
Kende, A. S. 380 (819), 393, 452 (873), *478, 479,* 735 (208, 209a,b), *753,* 1131 (223), *1320*
Kenne, L. 105, 106 (258, 259), *119*
Kenner, G. W. 1080 (81), *1317*
Kenyon, J. 417 (1039), *483,* 496, 499 (23), *527*
Kepner, R. E. 493 (5), *526*
Kerekes, I. 439 (1251, 1253), *487*
Kerfanto, M. 549 (151), *589*
Kern, C. W. 47 (120), *57*
Kerr, D. I. B. 1276, 1282 (640), 1284 (648), 1286, 1287 (640), 1288, 1289 (640, 648), 1291 (640), 1295 (648), 1300 (640, 648), 1302 (648), *1329*
Kerr, J. A. 868 (43), *910*
Kertes, A. S. 980 (109), *1018*
Kertesz, M. 232 (208), *260*
Kessler, H. 1035 (124), *1055*
Kessler, H.-J. 596, 597 (389), *599*
Kessler, Yu. M. 252 (427), *265*
Keszthelyi, C. P. 803, 808 (88), *823*
Ketterman, K. J. 295 (283), *467*
Keubler, N. A. 20, 26, 29, 30 (12), *55*
Khachatryan, L. A. 1090 (102), *1318*
Khafizov, Kh. 615 (103), *638*
Khaleeluddin, K. 902 (210), *914*
Khalil, M. H. 383 (832), *479*
Khan, B. T. 364 (723), *476*
Khan, N. A. 338 (599), *474*
Khan, N. J. 458 (1405), *490*
Kharasch, M. S. 341 (621a), 448 (1322), *474, 489*
Kharasch, N. 1206, 1207, 1211 (462), *1325*
Kharkov, S. N. 869 (53), *910*

Khcheyan, Kh. E. 920, 932 (34), *943*
Kheifets, L. Ya. 787 (33), *822*
Kheifits, L. A. 330 (521), *472*
Khettskhaim, A. 562 (210), *591*
Khitum, V. V. 237 (265), *262*
Kholof, A. A. 423 (1094), *484*
Khrustaleva, K. A. 60, 61 (17), *65*
Khuong-Huu, F. 1310 (711), *1330*
Kiber, C. J. 411 (931), *481*
Kidric, J. 226 (142), *259*
Kidwai, A. R. 458 (1405), *490*
Kieboom, A. P. G. 126 (13), *169*
Kieczykowski, G. R. 427 (1142), *485*
Kiefer, H. 460 (1414), *490*
Kiefer, H. C. 957 (62), *1017*
Kiel, G. 1022 (11), 1023 (17), *1053*
Kiełbasinski, P. 1023, 1035, 1052 (21), *1053*
Kielczewski, M. A. 153 (140), *172,* 613, 614 (95), *638*
Kienitz, H. 916, 917, 928 (15), *942*
Kienitz, L. 534 (8), 552 (165), 557 (8, 165), 559 (165), 562, 568 (8, 165), 584 (165), *586, 590*
Kienzle, F. 343 (622), 395 (880), *474, 480*
Kierstead, R. W. 561 (197), *590,* 1071–1073, 1075, 1149, 1187 (28), *1316*
Kiguchi, T. 1206 (474, 475), 1215 (475), 1216 (474), *1325*
Kihara, K. 299 (323), *468*
Kikuchi, T. 521 (184), *530*
Kikuchi, Y. 804 (99), *823*
Kilburn, E. E. 566, 571 (251), *591*
Kil'disheva, O. V. 1039, 1052 (198), *1057*
Kim, H. 25 (48), *56*
Kim, K. 422 (1098), *484*
Kim, M.-G. 1039 (228), *1057*
Kim, M. K. 103 (233), *118*
Kimble, B. J. 123 (3), *169*
Kimbrough, R. D. Jr. 1232, 1241–1244 (543), *1327*
Kimura, K. 204 (129, 130), *212,* 290 (237), *466*
Kindler, K. 819 (163), *824*
King, A. B. 133 (46), *170*
King, B. M. 166 (213), *173*
King, C. 870 (55), *910*
King, C. G. 774 (87), *779*
King, G. 372 (770), *477*
King, G. S. 1078 (55), *1316*
King, G. S. D. 224 (108), *258*
King, L. C. 361 (710, 711), *476*
King, R. W. 883 (114, 117), 886 (117), 896 (190), 899 (117), *912, 913*
Kingsland, M. 1186, 1187 (349), *1323*

Kingston, D. G. I.   130 (32), 135, 137 (59), 139 (79), *170, 171*
Kingston, D. H.   711 (173), *751*
Kinkel, K. G.   1138, 1139 (238), *1320*
Kinnick, M. D.   278 (131), *464*
Kinoshita, M.   428 (1155), *485,* 1039 (207, 208), 1052 (207), *1057*
Kinoshita, T.   989 (143), 1006 (178), *1019, 1020*
Kinsel, E.   281 (174), *465*
Kintner, R. R.   826 (18), *855*
Kinugasa, M.   1271 (613), *1328*
Kipp, E. B.   103 (221), *118*
Kirby, G. W.   272 (52), *463,* 652, 653 (21, 22), *747*
Kirby, J. A.   891 (151), *912*
Kirchhof, W.   353 (667), 400 (898), *475, 480,* 1284, 1291 (656), *1329*
Kirchhoff, K.   610, 621 (67), *637*
Kirino, K. Y.   774 (89), *779*
Kirk, D. N.   69, 80 (9), 104 (247), *113, 118*
Kirk, J. C.   1284, 1286, 1287 (646), *1329*
Kirk, K. L.   1200 (416, 417), *1324*
Kirkien-Konasiewicz, A.   166 (210), *173*
Kirmse, W.   375 (786), *478,* 1189, 1190 (355), 1232, 1235–1238, 1240, 1241 (525), *1323, 1326*
Kirsanov, A. V.   439 (1252), *487*
Kirschke, K.   1165, 1168 (298), *1322*
Kirson, I.   80 (68), *115*
Kirst, H. A.   291, 456 (253), *467,* 514 (139), *529*
Kiseleva, R. N.   237 (265), *262*
Kiser, R. W.   123 (3), *169*
Kissel, T.   1027, 1032, 1034–1036, 1043 (63), *1054*
Kistenbrügger, L.   1025–1027, 1035, 1045 (43), 1046 (43, 330), *1054, 1060*
Kita, E. J. Jr.   414 (1000), *482*
Kita, S.   1007 (180), *1020*
Kita, Y.   451 (1348), *489,* 1276 (624, 635, 637), 1277 (624), 1279 (624, 635, 637), 1280 (635, 637), 1281 (637), 1284 (649, 650), 1289 (649), 1290 (650), *1328, 1329*
Kitahara, M.   314, 353 (424), *470*
Kitano, M.   243 (339), *263*
Kitzing, R.   161 (183), *173,* 633 (214), *640,* 663, 668 (52, 53), *748*
Kiyohara, Y.   551 (160), *590*
Kiyomoto, A.   78 (60), *115,* 622 (157), *639*
Kjaer, A.   102 (207), *117*
Klages, C.-P.   1026, 1028, 1035, 1037, 1042, 1045 (65), *1054*

Klausner, Y. S.   443 (1285), *488*
Klayman, D. L.   1038 (173), *1056*
Kleiderer, E. C.   311 (398), *470*
Klein, E.   332 (552), *473*
Klein, J.   282 (196), *465,* 1081 (95), *1317*
Klein, K. P.   1199 (411–413), *1324*
Kleineberg, G.   1232, 1262 (592), 1265 (607), *1328*
Kleiner, H. J.   586 (365), *594*
Klemm, K.   547 (131, 132), 548 (132), 571, 586 (131), *589*
Klemm, L. H.   272 (40), *462,* 1093 (111, 112), 1094 (113, 114), 1095 (114), *1318*
Klevens, H. B.   71 (22, 23), *114,* 984 (118), *1018*
Klever, H. W.   498 (43), *527,* 1232 (527, 529), *1326*
Klimov, E. M.   345, 414 (632), *475*
Klingenberg, J. J.   359 (700), *476*
Klitgaard, N. A.   1115, 1116 (193), *1320*
Kloetzel, M. C.   437 (1216), *486*
Klohs, W. H.   1115 (186), *1319*
Kloosterziel, H.   1033, 1034, 1040 (105), *1055*
Klopfenstein, C. E.   1093 (112), *1318*
Klotz, I. M.   957 (62), *1017*
Klotz, J. M.   247 (385), *264*
Kluiber, R. W.   414 (967), *481*
Klyne, W.   69 (9), 71 (21), 73 (40, 42), 75 (42–45, 47), 76 (21), 80 (9), 82 (80), 84 (21), 88 (116, 138), 91 (138), 95 (21, 157), 108 (116), 109 (138, 277), *113–116, 119,* 612 (82, 83, 87, 89), 621 (152), 622 (155, 156), 623 (161–163), 624 (155, 163, 164), 626 (170), *637, 639*
Knappe, W. R.   675 (86), *749*
Knaus, G.   284 (207–211), 285 (207, 211), *466*
Knebler, N. A.   229 (167), *260*
Knecht, L. A.   603 (11), *636*
Kneen, G.   730, 731 (198), *752*
Kneisley, J. W.   403 (911), *480*
Knifton, J.   291 (247, 248), *467*
Knifton, J. F.   441 (1274), *488*
Knight, J. C.   1078 (54), 1102, 1103 (145), *1316, 1318*
Knof, H.   133, 159 (43), *170*
Knott, E. B.   541 (93), *588*
Knowles, J. R.   995, 1002 (161), *1019*
Knowpfel, H.-P.   369 (755), *477*
Knox, J. R.   507 (93), *528*
Knox, L. H.   320 (469), *471*
Knudsen, R. D.   445 (1302), *488*
Knunyants, I. L.   498 (40), *527,* 807 (117), *823,* 1038 (171), 1039, 1052 (198), *1056, 1057,* 1201 (428, 429), 1202 (429, 430), *1324*

Knutson, D. 653, 688, 693, 694 (26), (138), *747, 750*
Kobayashi, G. 1042 (288), 1052 (403), *1059, 1062*
Kobayashi, M. 1048 (361), 1050 (374), *1061*
Kobayashi, S. 457 (1399), *490,* 990 (151), *1019*
Kobayashi, T. 180 (46), 205, 206 (141), *210, 212,* 1206, 1219 (482), *1325*
Kobe, K. A. 786, 787 (26), *821*
Kobelt, M. 526 (224), *531*
Kober, P. 1232 (529), *1326*
Kobilarov, N. 228 (151), *259*
Kobler, H. 584 (355), *594*
Koch, E. 1171, 1172 (320), *1322*
Koch, H. 292 (262, 263), 293 (267), *467,* 1174 (332, 333), *1322*
Koch, J. 1228 (508), *1326*
Koch, T. H. 721, 725 (186), *752,* 1276, 1278, 1289 (638), *1328*
Koch, W. (763), *477*
Kocharyan, S. T. 402 (904), *480*
Kochenour, W. L. 233 (217), *261*
Kocheshkov, K. A. 238 (281), *262*
Kochetkov, N. K. 345, 414 (632), *475,* 510 (115), *529*
Kochevar, I. 685 (101a), *749*
Kochi, J. K. 372 (769), *477,* 508 (102), *528*
Kochloefl, K. 916, 919 (16, 17), *943*
Kochman, M. 102 (203), *117*
Kočíand, K. 918 (29), *943*
Kock, E. 1049 (367), *1061*
Kodama, H. 271 (20), *462*
Kodama, T. 193 (73), *211*
Kodera, K. 123, 124 (6), *169*
Koehl, W. J. Jr. 1172 (323, 325), 1173 (323, 326), 1174 (325), *1322*
Koelsch, C. F. 322 (477), *471*
Koetzle, T. F. 225 (111), *258*
Kofron, W. G. 431 (1174), *486*
Kogan, G. A. 605 (24), *636*
Kogan, V. A. 241 (323), *263*
Kogura, T. 445 (1305), *488*
Kohler, E. P. 1113 (181, 182), *1319*
Kohler, F. 235 (245), *261*
Kohlrausch, K. W. F. 605 (33), 632 (204), *636, 640*
Kohlschütter, U. 230 (178, 179), *260*
Kohnle, J. F. 458 (1401), *490*
Koine, N. 103 (220), *118*
Koizumi, J. 151 (130), *172*
Koizumi, T. 1025, 1026 (49), 1036 (134), 1040 (262), 1052 (391), *1054, 1055, 1058, 1062*
Kojima, I. 237 (262), *262*

Kokoyama, Y. 1080, 1085–1087 (88), *1317*
Kolc, J. 661 (39), *748*
Kolchina, N. A. 603 (10), *636*
Kolczynski, B. V. 1105, 1173 (155), *1319*
Koleske, J. V. 496 (29a), *527*
Kolesov, V. P. 60 (13, 14), 61, 62 (28), 63 (14, 40), *65, 66*
Kolewa, S. 422, 423 (1100), *484*
Kolind-Andersen, H. 1033, 1034 (111), 1037, 1042 (111, 158), *1055, 1056*
Kolleck, M. 368 (749), *477*
Koller, J. 24 (58), *56,* 232 (208), *260*
Kollman, P. 29, 45–47 (72), *56,* 214, 218, 234 (2e), 242, 250–252 (331), *256, 263*
Kollman, P. A. 45 (114), *57*
Kolodny, E. H. 105 (255), *118*
Kolomiets, A. F. 1039 (223), *1057*
Kolomnikov, I. S. 850, 851 (90), *857*
Kolsaker, P. 430 (1166), *485,* 628 (182), *639*
Kolthoff, I. M. 223 (75, 76), 240 (75, 305), *258, 263*
Koltzenburg, G. (85), *779,* 1080, 1084 (83), *1317*
Koma, Y. 868, 869 (46), *910*
Komarov, E. V. 237 (265), *262*
Komeno, T. 107 (264, 266), *119*
Kömives, T. 1048 (352), *1060*
Komoto, R. G. 292, 456 (257), *467*
Konaka, R. 303 (343), *469*
Kondo, K. 312 (403), *470,* 1182, 1183 (338), *1322*
Kondo, M. 254 (447), *265*
Kondo, S. 1038 (188), *1057*
Konen, D. A. 281 (185), *465,* 1037, 1042, 1050 (161), *1056*
Königsdorfer, K. 1232, 1242, 1243 (541), *1326*
Konkol, T. K. 863 (16), *910*
Konno, T. 82, 83 (77–79), *115,* 517 (153, 154), *529*
Kono, D. M. 440 (1262), *487*
Konovalova, L. V. 1050 (381), *1061*
Koope, L. 372 (771), *477*
Kooyman, E. C. 878 (89), 884, 886 (126), 887 (135), 889 (144), 891 (126), 894 (144), 897, 898 (126, 144), 901 (144), *911, 912*
Kopecky, W. J. Jr. 1292 (663), *1329*
Kopp, L. 184 (65), *210*
Koppel, G. A. 278 (131), 421 (1088), *464, 484*
Koppes, W. M. 440 (1261), *487*
Koreeda, M. 112 (318), *120*
Koren, B. 570, 571 (294), *592*

Korickii, A. T.   761 (38), *778*
Korkut, S.   673, 682, 683, 685 (76a), *749*
Korobitsyna, I. K.   365 (727), *476*
Korolczuk, J.   1039 (209), *1057*
Korte, F.   399 (896), 406 (913, 914), *480*, 502 (68, 69), 526 (223), *528, 531*, 536 (34), *587*, 1098 (128), *1318*
Korth, J.   414 (1002), *482*
Kortzeborn, R. N.   28, 29 (63), *56*
Korver, O.   80, 84 (66), 88, 91 (113, 144), 92, 94 (113), 109 (283, 285), 110 (283, 285, 298), *115, 116, 119*, 624, 625 (168), *639*
Korver, P. K.   250 (411), *265*
Korzeniowski, S. H.   1075 (51), *1316*
Kosbahn, W.   108 (273), *119*
Koshel, G. N.   314 (427), *470*
Koshelev, K. K.   231 (204), *260*
Koshizawa, S.   1037 (154), *1056*
Koski, W. S.   802, 814 (87), *823*
Kosley, R. W. Jr.   583, 584 (349, 351), 595 (376), *593, 594, 598*
Kosma, S.   442 (1280), *488*
Kosower, E. M.   72 (27), *114*, 202 (119), *211*
Kosswig, K.   353 (667), *475*
Kost, A. N.   325 (493), *472*
Kost, V. N.   276 (110), *464*
Kostenbauder, H. B.   985 (130), *1019*
Kosterman, D.   1073 (39, 40), *1316*
Kostyanovsky, R. G.   615 (103), *638*
Kostyuchenko, N. P.   566 (272), *592*, 1038 (185), *1057*
Kotano, M.   28 (61), *56*
Koteswaram, P.   225 (130), *259*
Koto, S.   411 (926), *481*
Kovac, F.   570, 571 (294), *592*
Kovacic, P.   903 (212), *914*
Kovaleva, A. S.   278 (130), *464*
Koyama, K.   720, 722 (185a,c,e), 723 (185a), *752*
Koyama, M.   411 (926), *481*
Koyama, T.   571 (300, 301), 597 (392, 393), *592, 599*
Koyama, Y.   243 (340), *263*
Kozarich, J. W.   413 (960), *481*
Kozempel, M. F.   629 (200), *640*
Koztowski, H.   777 (99), *780*
Kozyuberda, A. I.   439 (1245), *487*
Kraatz, A.   567 (280), *592*
Kraatz, U.   399 (896), *480*, 1098 (128), *1318*
Krackow, M. H.   1024 (30), *1053*
Kraemer, R.   503 (75), *528*
Kraeutler, B.   739 (218), *753*
Kraevskii, A. A.   373 (782), *478*
Kraft, M.   130 (31), *170*

Kramer, D. J.   901 (203), *914*
Kramer, K. E.   353 (659), *475*
Krantz, A.   691 (125g–i), 692 (133), 701 (125g–i), 703 (133), *750*
Krapcho, A. P.   281 (176, 177, 182), 328 (182), 426 (1127, 1128), 431 (1173), *465, 485*, 1066 (26), 1070 (24, 26), *1316*
Krasiejko, I.   332 (554), *473*
Krasnec, L.   273 (67), *463*
Krasnov, E. P.   869 (53), *910*
Krasuski, W.   597 (394), *599*
Kratochwill, A.   223 (74), *258*
Kraus, M.   916, 919 (16, 17), *943*
Kraus, M. A.   449 (1326), *489*
Kraus, S.   834, 835, 842 (52), *856*
Kraus, W.   344, 345 (629), *475*
Krause, J. G.   415 (1006), *482*
Krauss, H. J.   709, 712, 719, 722 (151), *751*
Kreienbühl, P.   828, 829 (39), *856*
Kreiser, W.   1101 (142), *1318*
Krem, H.   332 (553), *473*
Kremenskaya, I. N.   631 (195), *640*
Kresheck, G. C.   951, 953, 954, 963 (27), *1016*
Kretschmar, H. J.   572, 576, 580 (315), *593*
Kribota, I.   317, 353 (441), *471*
Kricheldorf, H. R.   441 (1265), *487, 1040* (253), *1058*
Krimen, L. I.   453 (1366), *489*
Krimm, H.   516 (145), *529*, 1295 (667), *1329*
Kringstad, R.   628 (184), *639*
Krisher, L. C.   24 (46), *56*
Krishtalik, L. I.   806 (113), *823*
Kristol, L. D.   1185 (342), *1322*
Kröber, H.   1040 (254), *1058*
Krohn, J.   537, 553 (48), *587*
Krohn, K.   1206, 1207, 1212 (464), *1325*
Kröhnke, F.   536 (24), *587*
Krolls, U.   518 (168), *530*, 1200 (418), *1324*
Kroon, J.   88 (124), *116*, 234 (224), *261*
Kroon, Y.   225 (124), *259*
Kröper, H.   493 (1a), 524 (211), *526, 531*
Kropp, P. J.   709, 712, 719, 722 (151), *751*, 1052 (395), *1062*
Krösche, H.   1101 (141), *1318*
Krubsack, A. J.   440 (1257), *487*
Krueger, P. J.   253 (441), *265*
Kruger, P. E. J.   300 (328), *468*
Kruglik, L. I.   1039 (216), *1057*
Kruk, C.   250 (411), *265*
Krukle, T. I.   562 (210), *591*
Krull, I. S.   661, 662 (44a, b), *748*
Krupička, J.   716 (178), *751*

Kruzhalov, B. D. 920, 932 (34), *943*
Ku, A. T. 1045 (325, 327), *1060*
Kuatani, K. 31 (87), *57*
Kubelka, V. 169 (226), *174*
Kubler, D. G. 315 (432), *470*
Kubo, K. 81 (71), *115*
Kubo, Y. 720, 722 (185j), 745 (230a,b), *752, 753*
Kubota, T. 175 (5), *209,* 217 (9), 229 (175), *256, 260*
Kuča, L. 237 (261), *262*
Kucherov, V. F. 138 (75, 76), 148 (76), *171,* 270, 437 (11), *462*
Kuchitsk, K. 243 (339), *263*
Kuchitsu, K. 28 (61), *56*
Kuczynski, H. 1276, 1281 (639), *1328*
Kudlacek, V. 323 (481), *471*
Kudryashov, L. I. 510 (115), *529*
Kuebler, N. A. 29, 30 (71), *56,* 70, 72, 101 (15), *114*
Kuehne, M. E. 274 (85), *463*
Kugajevsky, I. 168 (221), *174,* 1232, 1243 (542), *1327*
Kugler, F. 369 (755), *477*
Kuhl, R. 367 (744), *477*
Kuhle, E. 343 (625), *474*
Kuhn, A. T. 782 (9), *821*
Kuhn, S. J. 416 (1012, 1025), 439 (1255), *482, 483, 487*
Kuhn, W. F. 495 (21), *527,* 627 (181), *639*
Kuhu, W. F. 166 (212), *173*
Kuivila, H. G. 416 (1027), *483*
Kukhar, V. P. 273 (65), 451 (1347), *463, 489*
Kukhtin, V. A. 414 (965), *481*
Kukolev, V. P. 850, 851 (90), *857*
Kukolich, S. G. 230 (186), *260*
Kuksis, A. 270 (9), 423 (1108), *462, 484*
Kuleshova, N. D. 1038 (171), *1056*
Kulevsky, N. 254 (462), *266*
Kuliev, A. B. 1039 (205), *1057*
Kuliev, A. M. 1039 (205), *1057*
Kulikova, L. 1038 (169), *1056*
Kulpinski, M. S. 307 (376), *469*
Kumakura, S. 193 (73), *211*
Kumamoto, T. 438 (1238), *487*
Kumler, W. D. 609 (54, 55), 613 (54), *637*
Kump, W. G. 1305 (701), *1330*
Kunakubo, Y. 985 (131), *1019*
Kunerth, D. C. 415 (984, 985), *482*
Kunieda, N. 428 (1155), *485*
Kunihara, K. 990 (151), *1019*
Kunin, R. 969, 979 (93), *1018*
Kunitake, T. 995, 1002 (164), 1007 (164, 181), 1012 (196), *1019, 1020*

Kuntze, K. 1048 (347), *1060*
Kunz, D. 1046 (331), *1060*
Kunz, R. A. 425 (1123), *485,* 1300 (677, 678), *1329*
Kunze, U. 1031, 1037 (98), *1055*
Kuo, Y.-N. 281, 282 (178), *465*
Kupchan, S. M. 507 (92–94), *528,* 627 (177), *639,* 1206 (472), *1325*
Kurashima, A. 990 (151), *1019*
Kuratani, K. 607 (38), *636*
Kurath, P. 1098 (133–136), *1318*
Kurchenko, L. P. 1048 (352), *1060*
Kurdyumov, C. 631 (196), *640*
Kureck, J. T. 453 (1368), *489*
Kuri, Z. 770 (78), *779*
Kurihara, T. 514 (137), *529*
Kuriyama, K. 85–87 (96), 107 (264, 266), *115, 119*
Kurkchi, G. A. 246 (379), 248 (392), 249 (392, 401), *264, 265*
Kurland, R. J. 28, 29 (60), *56*
Kurn, N. 279 (144), *464*
Kuroda, Y. 1007 (180), *1020*
Kuroiwa, S. 800 (79), *823*
Kurtz, P. 446 (1307), *488*
Kuryatov, N. S. 537 (55), 562 (219), 566 (270, 271), 581 (339), 582 (271, 339), 583 (271), *587, 591–593*
Kurz, J. L. 985 (130), *1019*
Kurzer, F. 437 (1219), *486,* 1042 (292), *1059*
Kuta, J. 782 (16b), *821*
Kutepow, N. V. 291 (246), *467*
Kutina, R. E. 1014 (197), *1020*
Kutlukova, U. S. 618 (121), *638*
Kutowy, O. 789 (39), *822*
Kuwajima, I. 281 (188), 428 (1152), *465, 485,* 1125 (217), *1320*
Kuwano, H. 245 (363), *264*
Kuzmin, M. G. 735, 736 (210), *753*
Kuznetsova, N. A. 226, 235 (133), *259*
Kuzovkin, V. A. 582 (340), 596 (377), *593, 598*
Kvick, A. 32 (91), *57,* 225 (111), *258*
Kwart, H. 896 (186), 899 (196, 198), 900 (198), 902 (211), *913, 914*
Kwei, G. H. 22 (33), *56*
Kwiatkowski, S. 513 (131), *529*
Kwie, W. 340 (617), *474*
Kwiram, A. L. 230 (188, 189), *260, 758* (9, 12), *778*
Kwok, S. 395 (880), *480*
Kwong, P. T. Y. 143 (95), *171*
Kyazimov, A. S. 1073 (36), 1090 (101), *1316, 1318*
Kyriacou, D. 635 (217), *640*
Kyrides, L. P. 1145, 1151 (265), *1321*
Kyung, Jai Ho 609 (63), *637*

Laakso, P. V.   896, 902 (185), *913*
L'Abbé, G.   1195 (375), *1323*
Labinger, J. A.   839 (65), *856*
Lablache-Combier, A.   204 (126), *212*
Labreux, C. C.   54 (124), *58*
Labsky, J.   1196 (384, 385), *1323*
Lacey, M. J.   127, 128 (19), *169*
Lacey, R. N.   504 (79), *528*
Lachmann, B.   550 (153), *589*
LaCroix, D. E.   628 (186), *639*
Ladbury, J. W.   321 (471), *471*
Ladell, J.   245 (353), *264*
Lader, H.   316 (439), *471*
Ladner, K. H.   252, 253 (426), *265*
Laduree, D.   1042 (290), *1059*
Laffitte, M.   62 (33), 63 (37), *66*
Lagowski, J. M.   525 (216), *531*
Lai, T. F.   225 (119), *259*
Laing, M.   240 (301), *262*
Lake, R. F.   223 (67), *258*
Lakhanpal, M. L.   237 (266), *262*
Lakshminarayana, G.   332 (551), *473*
Lal, B.   412 (956), *481,* 1232, 1260 (588), *1327*
Lala, L. K.   416 (1016), *482*
Lalancette, J. M.   1035, 1040 (133), *1055*
Lamanna, A.   91, 95 (155), *116*
Lamarre, C.   1038 (189), *1057*
Lamazouère, A.-M.   1027, 1032, 1042 (60), *1054*
Lamb, J.   236 (256), *262*
Lamberti, V.   1066, 1146 (20), *1316*
Lamy, Fr.   317 (440), *471*
Lander, G. D.   534, 539 (4), *586*
Landis, P. S.   883 (112), *912*
Landolt, R. G.   392 (868, 869), *479*
Lane, A. G.   1138 (229), *1320*
Lane, J. F.   493 (7), *526*
Lang, G.   571, 574 (292), *592*
Lange, B. C.   1232, 1259 (561), *1327*
Langenbeck, W.   303 (342), *469*
Langer, A. W.   294 (279), *467*
Lansbury, P. T.   276 (112), *464*
Lantos, I.   1045 (328), *1060*
Lapinte, C.   991 (154, 155), *1019*
Lapinte, C. A.   959 (68), *1017*
Lapkin, I. I.   1047 (337), *1060*
LaPlanche, L. A.   31 (83), *57*
Larcheveque, M.   456, 457 (1391), *490*
Lardelli, G.   1066, 1146 (20), *1316*
Large, R.   133, 159 (43), 147 (108), *170, 171*
Larock, R. C.   414 (999), 433 (1190), *482, 486*
Laroff, G. P.   774 (90), *779*
Laronde, R. T.   460 (1413), *490*
Larsen, E.   103 (240), *118*

Larsen, J. W.   962 (75), 963 (77), 980 (75, 77), 984 (124), *1017, 1019*
Larson, D.   25, 31 (50), *56*
Larson, G.   224 (107), *258*
Larsson, F. C.   1027, 1031–1034, 1037, 1042, 1050 (54), *1054*
Larsson, F. C. V.   1049, 1052 (363), *1061*
Larsson, I. M. T.   728 (194), *752*
Lascombe, J.   608 (52), *636*
Lasia, A.   814, 816, 817 (143), *824*
Lassau, C.   457 (1398), *490*
Lasse, A.   307 (368), *469*
Lassegues, J. C.   226 (137), *259*
Lassen, H.   247 (387), *264*
Lastomirsky, R. P.   353 (659), *475*
Latajka, Z.   229 (172a), *260*
Lathan, W. A.   19 (6), 23, 24, 27, 28, 31 (44), 42, 44 (109), *55–57,* 214 (2f), 239 (295a), *256, 262*
Latif, K. A.   1041 (278), *1059*
Latif, N.   686 (103), *749*
Lattes, A.   433 (1188), *486,* 1292 (665, 666), 1294 (666), 1295 (665, 666), *1329*
Lattrell, R.   1038 (177), 1039 (222), *1056, 1057,* 1232, 1250 (559), *1327*
Lau, K. F.   230 (182), *260*
Laubach, G. D.   1232, 1255 (577–579), *1327*
Lauer, R. F.   429 (1156), *485*
Laura, M. A.   1165, 1168 (296), *1322*
Lauransan, J.   249, 254 (400), *265*
Laurence, J.   880 (100), *911*
Laureni, J.   691, 701 (125h,i), *750*
Laurent, A.   454 (1372), *490*
Laurie, V. W.   21 (20), *55*
Lauterbur, P. C.   238 (287), *262*
Lautie, A.   225 (131), *259*
Lauwers, W.   147 (110), *171*
Lavie, D.   80 (68), *115,* 276 (105), *464*
Laviron, E.   795 (62), *822*
Lawesson, S.-O.   132 (39, 40), 144 (100), 146 (106), 147 (111), 148 (106), 149 (121), 150 (39, 122), 151 (40), 162–164 (191), *170–173,* 417 (1050), *483,* 1027, 1031, 1032 (54), 1033 (54, 111), 1034 (54, 111, 117), 1035, 1036 (130, 131), 1037 (54, 111, 158, 159), 1039 (238), 1040 (117, 413), 1042 (54, 111, 117, 158, 159), 1049 (363), 1050 (54), 1052 (363), *1054–1056, 1058, 1061, 1062*
Lawrance, G. A.   99, 109 (150), *116*
Lawrence, A. S. C.   984, 987 (121), *1018*
Lawson, D. N.   827 (34, 37), 828 (37), 835 (37, 46), 844 (37), *856*
Laye, P. G.   61 (20), 63 (39), *66*
Layton, R. B.   435 (1207), *486*

Lazarini, F.　224 (104), *258*
Lazarus, H.　507 (97), *528*
Lazdins, R.　1045 (323), *1060*
Lazniewski, M.　246 (377), *264*
Leandri, G.　110 (308), *120*
Leary, R. E.　885 (131), *912*
Leavall, K. H.　877, 878 (88), *911*
Leaver, I. H.　672, 677, 719 (68), *748*
Lebedeva, N. D.　64 (43), *66*
LeBel, N. A.　448 (1323), *489*
LeBelle, M. J.　1300 (675), *1329*
Le Berre, A.　440 (1258), *487*
Leblanc, N. F.　603, 619, 631 (5), *635*
Leblanc, O. H. Jr.　21 (20), *55*
Lecadet, D.　1040, 1049 (255), 1050 (373), 1052, 1053 (255), *1058, 1061*
Leclerq, B.　869 (53), *910*
LeCostumer, G.　1024 (36), 1043 (36, 308), *1053, 1059*
Led, J. J.　28, 29 (63), *56*
Lederer, E.　1066 (18), *1316*
Lederer, M.　340 (615), *474*
Ledersert, M.　1022 (5), 1023 (18), *1053*
Lednicer, D.　282 (195), 426 (1135), *465, 485*
Ledwith, A.　209 (154, 155), *212*
Lee, C. M.　609, 613 (54), *637*
Lee, C. S.　427 (1140), *485*
Lee, D. A.　62 (34), *66*
Lee, D. G.　304 (349), *469,* 1302 (689), *1330*
Lee, D. H.　1093 (112), 1094, 1095 (114), *1318*
Lee, G. A.　506 (86), *528,* 653, 654, 657, 658 (27), *747,* 1041 (277), *1059*
Lee, H. H.　323 (484), *471*
Lee, J. B.　304, 315 (346), 317 (442), 353 (655), 439 (1240), *469, 471, 475, 487*
Lee, K.-H.　507 (95), *528*
Lee, R. E.　1232, 1251 (558), *1327*
Lee, S. L.　1138–1142, 1145 (237), *1320*
Lee, S.-Y. C.　92, 94 (152), 99 (167), *116, 117*
Lee, Y. H.　240 (310), *263*
Leebrick, J. R.　294 (282), *467*
Leedy, D. W.　814, 816, 818 (141), *824*
Leermakers, P. A.　672, 686, 687, 740 (62, 63a–c), *748*
Leeuwen, H. B.　887 (136), *912*
LeFèvre, R. J. W.　1024 (37), *1053*
Leffler, J.　353 (656), *475*
Leftin, J. H.　146 (104, 105), *171*
Le Gall, L.　249, 254 (400), *265*
Legault, R.　1300 (676), *1329*
Leger, L.　1047 (339, 340), *1060*
Legge, D. I.　321, 322 (480), *471*
Légrádi, L.　631 (194), *640*

Legrand, L.　1041 (274), *1059*
Legrand, M.　69 (4), 71 (30), 73 (4), 79, 80, 84 (65), 88, 89, 95 (30), 101 (176), 108 (4), 109 (287), *113–115, 117, 119,* 611 (72, 74), 624 (167), *637, 639*
Legris, C.　421 (1089, 1090), *484*
Le Guillanton, G.　364 (726), *476*
Lehmann, G.　372 (771), *477*
Lehmann, M. S.　219 (24), *257*
Lehmkuhl, H.　825 (6), *855*
Lehn, J. M.　315 (430), *470*
Lehnert, W.　421 (1083–1085), *484*
Lehninger, A. L.　1014 (200), *1020*
Lehrman, G.　979 (104), *1018*
Leigh, P. H.　1040–1042, 1052 (268), *1058*
Leimgruber, W.　566 (266), 571 (288), *592*
Leipold, H. A.　503 (73), *528,* 1109 (165), *1319*
Leiserowitz, L.　88 (125), *116,* 244 (347), 245 (358), *263, 264*
Leitch, G. C.　1305 (703), *1330*
Leitch, L. C.　125, 126 (10), *169*
Lejeune, P.　164 (196), *173*
Lemal, D. M.　743 (225), *753*
Le Men, J.　452 (1361), *489,* 1304 (700), 1306 (705), *1330*
Lemieux, R. V.　332 (531, 532), *472*
Le Narvor, A.　249, 254 (400), *265*
Lengyel, I.　167 (216, 217), 168 (217), *174,* 518 (167, 171), 519 (172, 173), 520 (177), *530,* 1194 (373), 1200 (421, 422), *1323, 1324*
Lenin, A. H.　415 (992), *482*
Lennon, B. S.　901 (201), *913*
Lenz, G. R.　1206, 1207 (465, 466, 468), 1212 (465), 1213 (465, 466), 1215 (465), 1223 (465, 468), 1224 (468), *1325*
Leo, A.　236 (253), *261*
Leon, N. H.　151 (129), *172,* 1027, 1032 (57), 1034 (57, 119), 1042 (57), 1043 (119), 1048 (355), *1054, 1055, 1061*
L'Eplattenier, F. A.　541 (95), *588*
Leplawy, M. T.　1039 (247), *1058*
Leppert, E.　1040 (253), *1058*
Lerman, B. M.　1038 (187), *1057*
Lerner, L. M.　461 (1419), *490*
Lerner, R. G.　22, 23 (32), *56*
Lesman, T.　164 (197), *173*
Lessard, J.　280, 281 (171), *465,* 1070 (23), *1316*
Lesslie, T. E.　411 (925), *481*
LeSeur, W. M.　315 (428), *470*
Leung, D.　690–692, 697, 699, 705, 735–737 (135), *750*

Levashov, A. V.   947, 962 (14), 964 (14, 87a), 966, 978 (14), 1008, 1010 (87a), *1016, 1018*
Levene, P. A.   430 (1171), *485*
Levesque, G.   1042 (293, 295), 1052, 1053 (393), *1059, 1062*
Levi, E. M.   157 (164), *173,* 1205 (451), *1325*
Levi, N.   690–692, 697, 699, 705, 735 (135), 736 (135, 212a,b), 737 (135), *750, 753*
Levin, V. V.   223 (68), *258*
Levin, Ya. A.   802 (86b,c), *823*
Levina, R. Y.   508 (104), 509 (108), 510 (111), *528,* 1065, 1090 (6), *1315*
Levine, R.   360 (703, 704), *476*
Levine, S. D.   297 (304), *468*
Levinson, A. S.   272 (28), *462*
Levitt, T. E.   444 (1298), *488*
Levy, A. B.   435 (1205), *486*
Levy, E. C.   146 (104), *171*
Levy, E. F.   406 (912), *480*
Levy, G. C.   234 (228), *261,* 1035 (122), *1055*
Levy, H.   1168 (287), *1321*
Levy, H. A.   224 (88), 225 (126), *258, 259*
Levy, M.   801 (83), *823*
Levy, W. J.   335 (574), *473,* 1138 (231), *1320*
Lewin, N.   728 (193b), *752*
Lewis, A.   1185 (343), *1322*
Lewis, E. S.   877 (88), 878 (88, 99), 892 (161), 895 (180), *911, 913*
Lewis, G. S.   416 (1024), *483*
Lewis, H. B.   1206, 1207, 1211 (462), *1325*
Li, Wu-S.   1011 (190), *1020*
Liang, H. T.   909 (239), *914*
Liang, T. H.   499 (52), *527*
Liberles, A.   497 (33), *527,* 620 (149), *639*
Libman, J.   654, 657 (28), 675, 680, 682 (87), 735 (28), *747, 749*
Librando, V.   517 (150), *529*
Lichtel, K. E.   547 (135, 136), 586 (136), *589*
Lichtenwalter, G. D.   1030 (90), *1054*
Lichtin, N. N.   768 (69), *779*
Liddiard, C. J.   957 (64), *1017*
Lide, D. R. Jr.   23 (40), *56*
Liebeskind, L.   425 (1125), *485*
Liedhegener, A.   424 (1117), *484*
Liehr, J. G.   148 (117), 150 (125), 152 (137), *172*
Lien, M. H.   27 (56), 49 (123), *56, 58*
Lienemann, A.   541, 542, 548, 573 (91), *588*

Lifschitz, J.   103 (244), *118*
Lifson, S.   77 (54), *114,* 243 (336, 337), *263*
Lightner, D. A.   107 (264), *119*
Lijinsky, W.   102 (208), *117*
Liler, M.   253 (435), *265*
Lilga, K. T.   757 (7), *778*
Lilie, J.   768 (74), *779*
Lillis, V.   736, 737 (213), *753*
Liminga, R.   222 (53), *257*
Lin, C. Y.   691, 701 (125g), *750*
Lin, I.   307 (371), *469*
Lin, J. C.   1042 (300), *1059*
Lin, L. C.   408, 409 (920), *480*
Lin, L.-S.   711 (169), *751*
Lind, C. D.   272 (40), *462*
Linda, P.   992 (157), *1019*
Linden, G. B.   406 (912), *480*
Linder, P.   219 (23), *257*
Lindert, A.   428 (1149, 1151), *485,* 1115 (191), *1319*
Lindgren, I.   224 (92), *258*
Lindley, P. F.   503, 504 (76), *528*
Lindman, B.   226 (135), *259*
Lindner, D. L.   826, 827, 837 (31), *856*
Lindner, E.   1031 (98), 1037 (98, 150), 1044 (318), *1055, 1056, 1060*
Lindow, D. F.   874 (72), *911*
Lindquist, P.   979 (104), *1018*
Lindquist, R. N.   979 (105), *1018*
Lindsay, A. S.   296 (294), *468*
Lindsay, J. K.   426 (1134), *485*
Lindsay, K. L.   1092 (108), *1318*
Lindsey, A. S.   916 (12), *942*
Ling, Ch.   235 (244), *261*
Ling, H. G.   902 (211), *914*
Linke, S.   435 (1204), *486*
Lin'kova, M. G.   1038 (171), 1039, 1052 (198), *1056, 1057*
Linn, C. B.   454 (1377), *490*
Linnell, R. H.   687, 688, 701 (106), *750*
Linnett, R. H.   214 (1e), *256*
Linstead, P.   373 (778), *477*
Linstead, R. P.   372 (773), *477,* 494 (9), *526,* 860 (2), *909,* 1081 (91), 1138, 1139 (235), 1170 (312), *1317, 1320, 1322*
Linstrumelle, G.   430 (1161), *485*
Liogonkii, B. I.   270 (14), *462*
Lion, H.   195 (84), *211*
Lippert, E.   217 (7), 222 (63), 229 (173), 230 (63, 177), 231 (197), *256, 257, 260*
Lippincott, E. R.   220 (35), *257*
Lippman, A. E.   439 (1247), *487,* 1271 (612), *1328*
Lipschitz, A.   603 (6), *636*

Lipshutz, B. H.　281, 328 (184), 456, 457 (1392), *465, 490,* 523 (199), *530,* 1300, 1313 (683), *1330*
Lipsky, S. R.　1035, 1036 (129), *1055*
Liquori, A. M.　225 (120), *259*
Lischewski, M.　564 (234), 565 (234, 241), *591*
Lisle, J. B.　19 (6), *55*
Lissant, K. J.　1014 (199), *1020*
Lister, D. G.　1022, 1024 (9), *1053*
Listowsky, I.　88, 91, 92 (127), *116*
Litman, B. J.　77, 101 (52), *114,* 612 (84), *637*
Litt, A. D.　685 (102b), *749*
Litvinenko, L. M.　1048 (352), *1060*
Liu, J.　733 (203b), *752*
Liu, J. C.　909 (238), *914*
Liu, J.-C.　1066, 1070 (25), *1316*
Liu, J. H.-S.　733 (202), *752*
Lloyd, D.　562, 567 (222), *591*
Lloyd, H. A.　1276, 1278, 1282 (633), *1328*
Lloyd, K. O.　102, 105 (184), *117*
Löbering, H. G.　567 (280), *592*
Locatelli, P.　762, 763 (42), *778*
Lochow, C. F.　850, 851, 853 (91), *857*
Lock, C. J. L.　835, 837–840 (48), *856*
Locock, R. A.　164 (194), *173*
Lodder, G.　706, 730, 731, 733 (146b), *751*
Loder, D. J.　909 (240), *914*
Loder, J. W.　369 (760), *477*
Lodge, J. E.　205 (137), *212*
Lodwig, S. N.　1066, 1069 (21), *1316*
Loeffler, J. T.　299 (322), *468*
Loening, K. L.　411 (923), *481*
Löffler, A.　1113 (185), *1319*
Loginova, A. A.　254 (450), *266*
Loginova, N. F.　810, 811 (132), *824*
Lohaus, G.　1038 (177), *1056,* 1194, 1225 (370), *1323*
Lohse, F.　313 (418), *470*
Lombana, L.　896, 897 (192), *913*
Lombardo, P.　874, 875 (78), *911*
Long, F. A.　133 (46), 138 (71), 165, 166 (207), *170, 171, 173*
Long, G. L.　733, 734 (205), *753*
Long, J.　35 (99, 100), *57*
Long, J. P.　415 (996), *482*
Long, L. Jr.　338 (585), *473*
Long, R. E.　244 (351), *264*
Longley, R. I.　1145, 1151 (263, 264), *1321*
Longley, R. I. Jr.　411 (930), *481*
Longosz, E. J.　871 (62, 64), *911*
Longstaff, P. A.　826 (22), *856*
Loo, H.-M. van de　344, 345 (629), *475*

Loomis, G. L.　1123, 1126, 1127 (211), *1320*
Lopes, A.　672, 678, 679 (91), 739 (219), *749, 753*
Lorber, M. E.　1111 (178), *1319*
Lord, R. C.　246 (372), *264*
Lorenzi, G. P.　109 (280), *119*
Lorenzo, A. di　634 (216), *640*
Lorkowski, H. J.　453 (1365), *489*
Losse, G.　1048 (347), *1060*
Lossing, F. P.　671, 672, 677 (92f), *749,* 890 (149), *912*
Lossow, E.　1123 (201), *1320*
Loucheux, C.　180, 183, 203 (48), *210*
Loudon, A. G.　166 (210), *173*
Louis, J.-M.　1145 (272, 277), 1152 (272), 1155, 1156 (272, 277), 1157, 1158 (272), *1321*
Lounasmaa, M.　1108 (163, 164), *1319*
Louw, R.　875 (80, 83), 878 (98), 889 (144), 892 (157), 894 (144), 896 (157), 897, 898, 901 (144), 909 (243), *911–914,* 1050 (376), *1061*
Loveridge, E. L.　445 (1302), *488,* (137), *750*
Lovey, A. J.　426 (1127), *485*
Lowe, B. M.　240 (311), *263*
Lowe, G.　523 (200), *530,* 1190, 1191 (356), 1206 (488, 490–492), 1221, 1222 (488), 1272, 1273 (615, 616), *1323, 1325, 1326, 1328*
Løwenstein, H.　247 (387), *264*
Lowrance, W. W.　412 (946), *481*
Lowry, T. M.　611 (71), *637*
Lowy, A.　786, 787 (24), *821*
Lozac'h, N.　1041 (274), *1059*
Lucas, G. B.　354 (671), *475*
Lucast, D. H.　1039 (213), *1057*
Luche, J. L.　411 (936), *481,* 521 (189), *530,* 1197 (403), 1232 (532, 537), 1236, 1238 (532), 1240 (537), *1324, 1326*
Luche, J.-L.　1231, 1232 (518), 1233 (602, 603, 605), 1235, 1236 (518), 1365 (606), 1266 (518, 602, 603), 1267 (518, 602, 603, 606), *1326, 1328*
Luciani, A.　1026, 1027 (53), *1054*
Lucke, E.　374 (783), *478*
Lucker, G. D.　790 (44), *822*
Luft, W.　566, 569 (256, 257), *591, 592*
Lui, Y. H.　229 (170), *260*
Luisi, P. L.　109 (280), *119*
Lukač, S.　759, 760 (27), *778*
Luknitskii, F. I.　412, 444 (953), *481*
Luk'yanchuk, V. P.　1039 (216), *1057*
Luk'yanets, E. A.　510 (111), *528,* 1065, 1090 (6), *1315*
Lum, K. K.　879, 884 (94), 891 (151), 892, 897, 901 (94), *911, 912*

Lumb, J. T.   375–377 (790), *478*
Lumbroso, H.   1024 (28, 33), *1053*
Lumbroso-Bader, N.   230, 235 (176), *260*
Lumpkin, H. E.   140 (87), *171*
Lund, H.   794 (61), 795 (63), 797 (65),
    798 (61), 804 (65), 815 (63, 65, 148),
    816 (63), *822, 824,* 1042 (291), *1059*
Lundgren, G.   240 (310), *263*
Lupton, E. C. Jr.   250 (412), *265*
Lurie, M.   561 (197), *590*
Lusinchi, X.   1303 (696), *1330*
Lutskii, A. E.   241 (320, 322), 251, 252
    (320), *263*
Luttringhaus, A.   620 (141), *639*
Lutz, E. F.   338 (598), *474*
Luzatti, V.   952 (42), *1017*
Lyman, D. J.   496 (29b), *527,* 1065 (9),
    1315
Lynn, W. H.   332, 368 (556), *473*
Lyons, L. E.   196 (85), *211*
Lysenko, N. M.   1039 (199), *1057*
Lythgoe, B.   390, 391 (864), 452 (1352),
    479, 489, 583 (345), *593,* 1184 (339),
    1322
Lyushin, M. M.   328 (507), *472*

Ma, J. C. N.   609 (62a), *637*
Ma, T. S.   602 (3), 603 (19), 618 (126,
    128), 629 (188), *635, 636, 638, 639*
Mabrouk, A. F.   147 (114), *172*
Macchia, B.   310 (390), *470*
Macchia, F.   310 (390), *470*
Maccoll, A.   166 (210), *173,* 868 (40),
    869 (50), 870 (57), 886 (132), 891
    (155), 892 (160), 897 (155, 193),
    910–913
Macdonald, A. L.   223 (81), 224 (85),
    258
MacDonald, C. G.   127, 128 (19), *169*
MacFarlane, C. B.   952, 984, 989, 1014
    (35), *1016*
MacFarlane, P. H.   676, 685 (98c), *749*
Mach, K.   759, 760 (26), *778*
Machida, M.   515 (143), *529,* 745 (230b),
    753
Machiguchi, T.   743 (222), *753*
Machkovskii, A. A.   228 (153), *259*
Machleidt, H.   333 (561, 562), *473*
Maciel, G. E.   230 (190–192), *260*
Mack, W.   1053 (412), *1062*
Mackinnon, H. M.   866 (33), 875 (82),
    910, 911
MacLeod, J. K.   135 (56, 58), 138 (77),
    144, 151 (56), *170, 171*
MacMillan, J. H.   908 (234), *914*
Madden, J. P.   1200 (424), *1324*
Maddox, H.   244 (351), *264*
Madhav, R.   1314 (720), *1330*

Madhaven, V.   768 (69), *779*
Madigan, D. M.   731, 732 (200b–e), 733
    (200c), *752*
Madison, V.   101 (182), *117*
Madsen, J. Ø.   132 (39, 40), 150 (39),
    151 (40), *170*
Madsen, P.   146 (106), 147 (111), 148
    (106), *171*
Maekawa, J.   938 (64), *944*
Maeva, R. V.   1065 (8), *1315*
Maeyama, J.   571 (300), *592*
Mager, K. J.   198 (98, 99), 200–202
    (106), *211*
Magerlein, B. G.   277 (121), *464*
Maggiolo, A.   339 (605), *474*
Magid, L. J.   962, 980 (75), *1017*
Magnus, P. D.   351, 352 (651), 454
    (1370), *475, 489,* 1040–1042 (268),
    1052 (268, 388, 389), *1058, 1061*
Magnusson, R.   300 (327), *468*
Mague, J. T.   827 (32, 37), 828 (37),
    835 (37, 46), 837, 841 (32), 844 (37),
    856
Mah, H.   539 (65), *588*
Mahler, H. R.   959 (69), *1017*
Mahmoud, M. M.   503, 504 (76), *528*
Mai, V. T.   161, 163 (181), *173,* 633
    (213), *640*
Maier, G.   512 (123), *529,* 743 (224a,b),
    753
Maier, T.   534 (8), 555 (176), 557 (8),
    559 (190), 562, 568 (8), 575, 576
    (190), *586, 590*
Maier, W.   236 (255), *261*
Maignan, J.   1034, 1041–1043, 1053
    (114), *1055*
Maigret, B.   241, 242 (327), *263*
Maigret, M.   42, 44 (107), *57*
Mainen, E. L.   270 (13), *462*
Mairanovskii, S. G.   782 (6), 799 (72),
    801, 807 (6), *821, 822*
Mairanovskii, V. G.   810, 811 (132), *824*
Maitlis, P. M.   825 (8), 826 (12), *855*
Majer, J. R.   160 (178), *173*
Majeti, S.   710 (165), 711 (170), 712
    (165), 745 (228b), *751, 753,* 1080,
    1084, 1085 (87), *1317*
Makanishi, K.   251 (420), *265*
Make, S.   438 (1235), *487*
Makosza, M.   1014 (198), *1020*
Maksyutin, Yu. K.   231 (204), *260*
Malacria, M.   1173 (327), *1322*
Malament, D. S.   736 (212a,b), *753*
Malani, C.   333 (568), *473*
Malaprade, L.   365 (733, 734), *477*
Malaspina, L.   61 (22), *66*
Malaval, A.   54 (124), *58,* 346 (639),
    475

Malecki, J. 517 (151), *529*
Málek, J. 450 (1334), *489,* 918 (29), *943*
Mali, M. 231 (201), *260*
Mali, R. S. 294 (276), *467*
Malik, Z. A. 438 (1234), *487*
Malinina, E. S. 631 (196), *640*
Mallan, J. M. 293 (271), *467*
Mallion, R. B. 183 (58), *210*
Malmstrom, L. 376 (810), *478*
Maloy, J. T. 803, 808 (88), *823*
Malte, A. M. (765), *477*
Mamaev, V. P. 594 (370, 371), 595 (370), 596, 597 (371), *598*
Mamba, A. 1230 (513), *1326*
Manchand, P. S. 272 (39), *462*
Mandal, H. G. 237 (266), *262*
Mandelbaum, A. 145 (103), 146 (103–105), 161 (182), 164 (200), *171, 173*
Mandelshtam, T. V. 1185 (342), *1322*
Maneschalchi, F. 415 (1004), *482*
Mangane, M. 177 (25), *210*
Mangasaryan, Ts. A. 1096, 1101 (118), 1104 (147), *1318, 1319*
Manhas, M. S. 168 (224), *174,* 412 (956), *481,* 521 (181), 523 (198, 201), 524 (181), *530,* 1195 (377), 1200, 1201 (425, 435, 436), 1203 (425), 1204 (425, 431, 445), 1225 (377), 1231 (520), 1232 (520, 545–548, 557, 560, 563–566, 568, 570–573, 575, 576, 588, 593, 594), 1245 (520, 545, 547, 548), 1246, 1247 (545, 548), 1249 (547, 548, 560, 563, 566), 1250 (547, 557, 560, 564, 566), 1251 (557), 1252 (548), 1254 (560, 565, 566, 568, 572, 575), 1256 (570, 573), 1257 (560, 566, 571, 575), 1258 (571), 1259 (548, 565, 576), 1260 (588), 1262 (593, 594), 1263 (520), *1323–1328*
Manimaran, T. 1052 (404), *1062*
Manion, M. 904 (218), *914*
Manitto, P. 1189 (354), *1323*
Maniwa, K. 400 (900, 901), 401 (900), *480,* 1039 (239), *1058*
Manly, D. 896 (188), *913*
Mann, C. K. 782 (2), *821*
Mann, F. G. 274 (70), *463*
Manning, D. J. 148 (118), *172*
Manning, R. E. 1304, 1307 (698), *1330*
Manske, R. H. 1166, 1168, 1187 (307), 1309 (710), *1322, 1330*
Mansson, M. 60 (18), 61 (26), 62 (26, 36), 66
Mansukhani, R. 596, 597 (387), *599*
Mantecon, R. E. 443 (1290), *488*
Mao, C. L. 157 (164), *173*

Mao, C.-L. 1205 (450–452), 1206 (453), *1325*
Maquestian, A. 164 (196), *173*
Marathe, K. G. 1052 (388), *1061*
Marazii-Uberti, E. 274 (83), *463*
March, J. 450 (1343), *489,* 948 (18), *1016*
Marchand, A. P. 311, 325 (400), *470*
Marchand-Bryaert, J. 1269 (609), *1328*
Marchenko, N. B. 536, 570, 571 (35), 582 (340), 596 (377, 379–381), 597 (379, 381), *587, 593, 598, 599*
Marchese, F. T. 19, 21, 22, 25, 26, 28–30 (8), *55*
Marchessault, R. H. 101 (173), *117*
Marcus, Y. 980 (109), *1018*
Maréchal, Y. 220 (28, 30), 222 (30), 226 (28, 138), 227 (145, 146), 228 (28), *257, 259*
Mares, J. R. 319 (459), *471*
Margerum, J. 672, 679, 680 (67b), *748*
Margerum, J. D. 672, 679 (65, 67a,c), 680 (67a,c), *748*
Margrave, J. L. 1066–1068, 1072, 1073 (16), *1316*
Mari, F. 1182, 1183 (338), *1322*
Maria, H. J. 25, 31 (50), *56*
Mariani, L. 1198 (399, 400), 1201 (427), *1324*
Mariano, P. S. 706, 730, 731, 733 (146c), *751*
Mariella, R. P. 446 (1308), *488*
Marino, G. 906 (230), *914*
Marinsky, J. A. 980 (107), *1018*
Marion, J. P. 1079 (62), *1317*
Mariono, J. P. 508 (99), *528*
Maritani, I. 432 (1183), *486*
Mark, L. H. 504 (80), *528*
Markel, G. 422 (1101), *484*
Marker, R. E. 357 (687), *476*
Markgraf, J. H. 1024 (32), 1046 (333), *1053, 1060*
Märkl, G. 513 (127, 128), *529*
Markley, K. J. 270 (8), *462*
Marković, V. 760 (29), 763 (29, 48), 765 (48), *778, 779*
Marković, V. M. 766–768 (62), *779*
Markovskii, L. N. 439 (1252, 1254), *487*
Markstein, J. 422 (1098), *484*
Markushina, I. A. 447 (1320), *488*
Marliogen, E. 557, 558 (185), *590*
Maron, L. 105, 106 (258), *119*
Marrero, R. 337 (581), *473*
Marron, N. A. 1051 (383), *1061*
Marschall, H. 508 (103), *528*
Marsh, R. E. 225 (119, 121), *259*
Marshall, A. R. 1305, 1312 (702), *1330*
Marshall, H. 1108 (158), *1319*

Marshall, J. A. 355 (682), 375, 379 (799–801), *476, 478,* 712 (174), *751,* 1038 (186), *1057,* 1066, 1068 (19), 1072, 1073 (30, 31), 1145 (254), 1147 (19, 254), 1148 (19), 1153 (19, 30, 31, 254), 1185 (344), *1316, 1321, 1323*
Marstok, K. M. 253 (439), *265*
Martell, A. E. 103 (233), *118,* 325 (494), *472*
Martens, J. 1035, 1039 (132), 1051 (383–385), 1052 (385), *1055, 1061*
Martensson, O. 232 (210), *260*
Martin, G. J. 1200 (415), *1324*
Martin, J. 296 (292), *468,* 509 (105), 511 (116b), *528, 529,* 1206 (456), *1325*
Martin, J. C. 348 (641), *475*
Martin, L. L. 224 (90), *258*
Martin, M. V. 1314 (721), *1330*
Martin, R. B. 98, 99 (165), 103 (231), *117, 118*
Martin, R. L. 35, 40, 41 (97), *57,* 1025, 1031, 1033, 1034, 1036 (46), 1046 (332), 1050 (46), *1054, 1060*
Martin, S. F. 502, 505, 506 (66), *528*
Martinek, K. 947 (8, 14), 948 (20), 962 (14, 73b), 964 (8, 14, 20, 73b, 82–86, 87a,b, 88), 965 (20, 73b, 83), 966 (14, 20, 86), 967 (83, 88), 968 (73b), 978 (14), 990 (83), 1004 (87b), 1008, 1010 (87a), *1016–1018*
Marton, A. F. 1048 (352), *1060*
Marton, D. 576 (325), *593*
Marton, J. W. Jr. 293 (270), *467*
Maruyama, H. 443 (1291), *488*
Maruyama, K. 720, 722 (185j), 745 (230a,b), *752, 753*
Maruyama, M. 507 (93), *528*
Marvel, C. S. 271 (21), 279 (133, 139, 140), 322 (476), 385 (840), *462, 464, 471, 479,* 879 (93, 96), *911,* 1145 (251), 1302 (685), *1321, 1330*
Marvell, E. N. 349 (646), 392 (866), *475, 479*
Marzocchi, M. P. 228 (155, 156), *259*
Masamune, S. 312 (403), *470,* 515 (141), *529,* 743 (222), *753,* 1039 (210, 218, 226), 1048 (210, 350), *1057, 1060*
Mashkovsky, A. A. 228 (163), 238 (277), *259, 262*
Maslem, E. N. 524 (210), *531*
Mason, J. P. 953 (50), *1017*
Mason, M. S. 275 (88), *463*
Mason, R. 244 (346), *263,* 826 (30), *856*
Mason, R. B. 596,(597 (387), *599*
Mason, S. F. 611 (73), *637*
Masson, J. 1047 (342), *1060*

Masson, S. 1047 (341), *1060*
Massy, M. 430 (1164), *485*
Massy-Barbot, M. 430 (1165), *485*
Mastagli, P. 349 (647), *475*
Masuda, T. 439 (1246), *487*
Masuhara, H. 197 (92), *211*
Masuno, M. 800 (79), *823*
Masure, D. 389 (862), *479,* 525 (217), *531*
Mataga, N. 175 (5), 197 (92), 198 (93), *209, 211,* 217 (9), 229 (175), *256, 260*
Mathar, W. 158 (165), *173*
Matheson, M. S. 763, 765 (46), *778*
Mathieson, A. McL. 85 (101), 104 (251), *115, 118*
Mathieson, A. R. 236 (249), *261*
Matiskella, J. D. 446 (1306), *488*
Matolcsy, G. 1037, 1042 (156, 157), *1056*
Matsen, F. A. 195 (77), *211*
Matsnev, V. V. 442 (1279), *488*
Matsoyan, S. G. 1096, 1101 (118), 1104 (146, 147), *1318, 1319*
Matsuda, T. 1039 (201), *1057*
Matsuda, Y. 1042 (288), *1059*
Matsueda, R. 443 (1291), *488*
Matsui, K. 312 (403), *470*
Matsui, M. 419 (1072), *484*
Matsui, T. 336 (580), *473*
Matsumato, T. 406 (916), *480*
Matsummura, N. 448 (1325), *489*
Matsumoto, H. 1206, 1209–1211 (460), *1325*
Matsumoto, K. 133 (42), *170,* 422 (1091), *484*
Matsumoto, N. 355 (677), *475,* 1168 (292), *1321*
Matsumoto, S. 202 (117), *211,* 571 (301), *592,* 804 (98), *823*
Matsumoto, T. 1101 (143), 1160, 1168 (282), *1318, 1321*
Matsumura, Y. (505), *472*
Matsuo, J. 1039 (244), *1058*
Matsuo, T. 182, 184 (56), 196 (80), *210, 211,* 720 (182, 183), 722 (182), 724, 725 (183), *752*
Matsuoka, N. 103 (216), *118*
Matsuoka, T. 318 (447), *471*
Matsushige, T. (85), *779*
Matsutaka, Y. 1276, 1279, 1280 (635), *1328*
Matsuura, H. 1042, 1043 (280), *1059*
Matsuura, T. 299 (323), 362 (719), *468, 476*
Matsuzaki, K. 412 (947), *481*
Mattes, K. 499 (49), *527,* 703, 738 (144), *751*

Mattes, R. 1022 (6–8, 13), 1023 (8, 13, 20), 1030 (88, 97), 1031 (88, 97, 102), 1037 (152), *1053–1056*
Matthews, A. D. 318 (450), *471*
Matthews, J. S. 419 (1064), *483*
Matthews, W. S. 371 (765), *477*
Mattox, V. R. 339 (611), *474*
Mattson, G. W. 280 (173), *465*
Matusch, R. 1027, 1032, 1034–1036, 1043 (63), *1054*
Matuszewski, B. 690 (130, 135), 691 (128, 130, 135), 692 (130, 135), 696 (128), 697 (128, 130, 135), 698 (128), 699 (128, 130, 135), 705 (135), 735 (128, 130, 135), 736, 737 (130, 135), *750*
Mautner, H. G. 1024 (30), 1035, 1036 (129), 1048 (354), *1053, 1055, 1061*
Mauz, D. 417 (1051), *483*
May, C. J. 1081 (91), *1317*
May, J. A. 786, 787 (26), *821*
May, R. W. 861, 862 (6), *910*
Mayer, C. F. 709 (158), *751*
Mayer, R. 1024 (35), 1025, 1026 (48), 1027 (48, 62, 64), 1028 (48), 1031 (96), 1032 (35, 96), 1033, 1034 (106), 1040 (254, 261, 266), 1042 (48, 62, 64, 266, 282), 1043 (64, 306, 311), 1044 (312), 1045 (329), 1046 (64, 331), 1048 (347, 353), 1052 (408), 1053 (266), *1053–1055, 1058–1062*
Mayer, W. W. Jr. 1302 (685), *1330*
Maze, C. 1039 (214), *1057*
Mazengo, R. Z. 110 (300), *119*
Mazer, N. A. 952 (43), *1017*
Mazur, R. H. 1276 (636, 642), 1278, 1279 (636), 1283 (642), *1328, 1329*
Mazur, Y. 653 (34a–c), 654 (28), 657 (28, 34a–c), 659 (34a–c), 733, 734 (204a–c), 735 (28, 204c), *747, 752, 753*
Mazzocchi, P. H. 745 (231), *753, 1227 (505), 1292 (663, 664), *1326, 1329*
McAdam, A. 224 (84), *258*
McAfee, E. R. 224 (106), *258*
McAlpine, I. M. 896 (187), *913*
McAneny, M. 985 (133), 993, 1009 (159a), *1019*
McCall, J. M. 432 (1182), *486*
McCalley, R. C. 230 (188), *260, 758 (9, 12), *778*
McCallum, K. S. 353 (660), *475*
McCarville, M. E. 25, 31 (50), *56*
McClellan, A. L. 43 (110), *57,* 214 (1b, 2g), 234 (1b), *256*
McCloskey, A. L. 414 (967), *481*
McCloskey, C. M. 273 (61), *463*
McCloskey, J. A. 150 (124), 152 (137), *172*

McClure, J. D. 354 (673), *475*
McCombie, S. W. 1040, 1042, 1046 (272), *1059*
McConnel, J. F. 225 (118), *259*
McCormick, J. P. 311 (402), *470*
McCormick, J. R. D. 278 (127), *464*
McCoy, L. L. 240 (304), *263*
McCreery, M. J. 950 (25), *1016*
McDanie, D. H. 240 (312), *263*
McDermott, F. A. 411 (924), *481*
McDonald, E. 1199 (410), *1324*
McDonald, R. N. 338 (584), *473*
McEachern, D. M. 61 (23), *66*
McElvain, S. M. 319 (463), 418 (1054), *471, 483*
McEvoy, F. J. 273 (66), *463*
McEwen, C. N. 603 (16), *636*
McFadden, W. H. 136 (64), 137 (67), 138 (73), 150 (126), 166, 167 (208), *170–173,* 626, 627 (176), *639,* 1035, 1036 (128), *1055*
McFarlane, P. H. 676, 685 (97), *749*
McGhie, J. F. 297 (300), 369, 370 (756, 757), *468, 477*
McGlashan, M. L. 59 (1), *65*
McGlynn, S. P. 25, 31 (50), *56,* 229 (170), *260*
McGregor, A. 386, 387 (850), *479*
McGregor, D. J. 511 (121b), *529*
McGregor, D. N. 524 (207), *530*
McGuckin, W. F. 339 (611), *474*
McGuire, R. F. 243, 247 (335), *263*
McGuire, T. M. 1094, 1095 (114), *1318*
McIntosh, A. V. (529), *472*
McIntosh, C. L. 499 (49), 512 (122), *527, 529,* 658 (35), 661 (40), 691, 701 (125a, c–e), 703, 738 (144), *747, 748, 750, 751*
McIntosh, D. F. 16 (5), *55*
McIntyre, J. E. 306 (359), 327 (359, 501), 328 (501), *469, 472*
McIver, J. W. Jr. 22 (23), *56*
McKelvey, J. 29, 45–47 (72), *56,* 242, 250–252 (331), *263*
McKenna, J. 700, 735 (142), 736, 737 (142, 213), *750, 753*
McKenna, J. C. 494 (13), *526,* 1089 (98, 99), *1317*
McKenna, J. M. 700, 735 (142), 736, 737 (142, 213), *750, 753*
McKennis, H. Jr. 1303, 1304 (695), *1330*
McKenzie, S. 1037 (145), *1056*
McKeon, J. E. 1313 (718), *1330*
McKillop, A. 343 (622, 626–628), 344 (627, 628), 370 (762), 374 (784), 415 (998), 416 (1023), *474, 477, 478, 482, 483,* 825 (4), *855*
McKinnie, B. G. 1050 (375), *1061*

McKinnon, D. M.   1040, 1042 (259), *1058*
McLachlan, R. D.   253 (443), *265*
McLafferty, F. W.   124 (7), 125 (7, 8), 127, 128 (14), 133, 136 (50), 140, 143 (14), *169, 170,* 672, 686, 687 (63d), *748*
McLauchlan, K. A.   673, 683, 685 (85b), *749*
McLaughlin, K. C.   319 (462), *471*
McLay, G. W.   416 (1023), *483*
McLean, A. F.   871 (61), *911*
McLeese, S. F. de C.   358 (690), *476*
McLeod, D.   865 (28), *910*
McLeod, G. L.   360 (706), *476*
McLeod, I. J.   271 (17), *462*
McLeod, J. K.   1036 (136), *1055*
McMahan, D. G.   520 (176), *530*
McManus, S. P.   447 (1319), *488*
McMichael, K. D.   1039 (195), *1057*
McMillan, A.   274 (75), *463*
McMurray, J. E.   1073, 1074, 1149, 1150 (45), *1316*
McMurray, T. B. H.   1066, 1068, 1146, 1148 (17), *1316*
McMurray, W. J.   1035, 1036 (129), *1055*
McMurry, J. E.   272 (30), 426 (1130), *462, 485*
McNab, M.   562, 567 (222), *591*
McNelis, E.   398 (892), *480,* 936, 937 (57), 938 (57, 68), *943, 944*
McNiven, N. L.   887 (136), *912*
McOmie, J. F. W.   275 (100), *464,* 874 (74), 905 (223), *911, 914*
McPartlin, M.   826 (30), *856*
McVicars, J. L.   1052 (408), *1062*
McVie, G. J.   826 (17), *855*
McWeeny, R.   6 (1), *55*
McWhirter, M.   361 (711), *476*
Mead, T. C.   362 (716, 717), *476*
Mead, W. L.   131 (36), *170*
Meader, A. L., Jr.   382 (827), *478*
Meakin, P. Z.   826, 827, 837 (31), *856*
Meakins, G. D.   432 (1185), *486*
Mecca, T. G.   103 (227, 228), *118*
Mechoulam, R.   296 (295), *468*
Meck, R.   80 (67), *115*
Mecke, R.   608 (43), 620 (141), *636, 639*
Medalia, A. I.   953 (50), *1017*
Medary, R. T.   1012 (195), *1020*
Medete, A.   454 (1371), *489*
Medley, E. E.   866 (31), *910*
Medvedev, B. A.   562 (219), *591*
Mee, A.   333 (564), *473,* 1172 (324), *1322*
Meer, R. van der   420 (1075), *484*
Meerwein, H.   517 (157a), *529,* 535–537 (10, 11), 538 (11), 541 (92), 542–544, 550, 551, 553 (11), 554 (10, 11), 557

(11), 559 (10, 11), 562, 566, 568 (11), 572 (10, 11, 314), 573, 581, 583, 586 (11), *586, 588, 593*
Meese, C. O.   254 (454), *266*
Meese, C.-O.   1029, 1033 (82, 84), 1036 (84), *1054*
Megerle, K.   497 (33), *527,* 620 (149), *639*
Meguro, H.   82, 83 (76–78), 84 (89), 85, 87 (97), *115,* 517 (153, 154), *529*
Mehl, J.   42 (108), *57,* 219, 233 (21), *257*
Mehta, G.   415 (987), *482,* 1161, 1165, 1166, 1168 (285), *1321*
Mehta, S. R.   294 (276), *467*
Meier, H.   1035 (125), 1041 (125, 275), 1043 (275), *1055, 1059*
Meiggs, T. O.   672, 678, 680, 690 (75a–c), 694, 696 (75c), 699 (75b, c), 735, 736 (75a–c), *749*
Meijer, J.   1037 (141), 1042 (141, 294), 1046 (334), 1049 (362), 1050 (294), 1052 (294, 406), 1053 (294), *1056, 1059–1062*
Meikle, D.   297 (304), *468*
Meindl, H.   566, 571 (265), *592*
Meinwald, J.   382 (825), *478,* 1072 (29), 1162 (289), 1163 (29), 1168 (289), 1185 (343), *1316, 1321, 1322*
Meisels, A.   335 (577), *473*
Meissner, F.   1033, 1034 (110), *1055*
Meites, L.   782 (17), *821*
Mekhtiev, S. D.   328 (507), *472*
Mekrykova, T. V.   618 (120), *638*
Melchior, G. H.   673 (71b), *748*
Melchiore, J. J.   325 (492), *472*
Melikyan, G. S.   1096, 1101 (118), 1104 (146), *1318*
Melnick, A. M.   235 (238), *261*
Melnick, B.   687, 743 (104), *749*
Mel'nik, S. Ya.   810 ,811 (132), *824*
Meloan, C. E.   617 (117), *638*
Mels, S. J.   815 (149), *824*
Melstrom, D. S.   302 (336), *468,* 1232, 1241, 1242 (531), *1326*
Melton, C. E.   132, 133 (41), *170*
Melvin, L. S.   514 (138), *529*
Mende, Y.   235, 236 (241), *261*
Menger, F. M.   950 (25), 953 (48), 957, 977 (65), 985 (65, 132), 986 (134–136), 989 (145), 990 (132), *1016, 1017, 1019*
Mengler, H.   1131, 1136 (225), *1320*
Mennemann, K.   1022, 1023 (13), *1053*
Menschutkin, N.   880 (102), *912*
Mentzer, C.   1098 (131), *1318*
Merault, G.   454 (1379), *490,* 1229 (512), *1326*

Merten, R.   542 (100), 550 (154, 156),
    557 (182), *588–590*
Merve, J. P. van der   1171, 1172 (321),
    1322
Merz, A.   276 (108), *464*
Merzoni, S.   290 (243), 291 (250), *466,
    467*
Meschede, W.   1022 (6–8), 1023 (8,
    20), *1053*
Meslin, J. C.   562, 570, 571 (208, 209),
    591
Messe, C. O.   254 (461), *266*
Mestres, R.   1080 (70, 73, 76), 1082
    (76), *1317*
Metayer, C.   566 (261), *592*
Metcalfe, T. P.   310, 313 (394), *470*
Meth-Cohn, O.   676, 685 (98b), *749*
Mettler, C.   784, 786, 787 (19), 790 (43,
    48), 792, 793 (53), 794 (58), 805
    (108), *821–823*
Metzger, C.   1131, 1132 (219), *1320*
Metzner, P.   1037 (147), 1047 (147, 342),
    1056, 1060
Meyer, G.   1008 (185), *1020*
Meyer, G. M.   431 (1171), *485*
Meyer, H.   1098 (121), *1318*
Meyer, H. J.   369 (755), *477*
Meyer, J. W.   649, 659 (13), *747*
Meyer, K.-O.   176, 180, 181, 183, 196,
    198 (17), *209*
Meyer, W.   233 (216), *261*
Meyer, W. C.   1227 (504), *1326*
Meyer, W. L.   272 (28), *462*
Meyers, A. I.   283 (203–205), 284
    (204–211), 285 (207, 211, 213), 286
    (214), 443 (1286), *466, 488, 500 (60),
    527,* 1120 (197–199), 1121 (199),
    1122 (200), *1320*
Meyers, C. Y.   (765), *477*
Meyers, W. E.   989 (147), *1019*
Meyerson, S.   125, 126 (10), 127, 128
    (17), 136 (64), 142 (17), 146 (107),
    158 (17), 162 (186, 187), *169–171,
    173,* 675 (89), *749,* 874 (71), *911*
Meyer zu Reckendorf, Z.   415 (982), *482*
Mezaraups, G.   1038 (169), *1056*
Michael, B. D.   775, 776 (94), *779*
Michalski, J.   273 (54), *463*
Michejda, C. J.   871 (63), *911*
Michelet, D.   800 (76), *822*
Michelot, D.   430 (1161), *485*
Michelsen, P.   107 (262), *119*
Michl, J.   661 (39), *748*
Michurin, A. A.   449 (1333), *489*
Mićić, O. I.   759 (25), 760 (25, 29, 30),
    763 (29, 48, 49), 764 (49), 765 (48,
    49), 768 (30), *778, 779*
Middlemiss, K. M.   232 (215), *261*

Middleton, W. J.   873 (70), *911,* 1025,
    1028, 1034, 1041, 1043, 1044 (45),
    1054
Midland, M. M.   435 (1205), *486*
Midorikawa, H.   421 (1086), *484*
Mielniczuk, Z.   1039 (209), *1057*
Mier, J. D.   420 (1078), *484*
Miescher, K.   376 (811), *478*
Miescher, M.   1269 (610), *1328*
Miesse, C.   297 (299), *468*
Miginiac, P.   430 (1163–1165), *485*
Miginioc, L.   1113 (183), *1319*
Migita, Y.   720, 722 (185b–d, f), 723
    (185f), 724 (185b, d), *752*
Mihelich, E. D.   283 (204, 205), 284
    (204–206), 285 (213), 286 (214), *466,
    500 (60), 527,* 1120 (198, 199), 1121
    (199), 1122 (200), *1320*
Mijlhoff, F. C.   1022 (2), *1053*
Mikawa, Y.   222 (51), 228 (147), *257,
    259*
Mikhailov, B. M.   541 (90), *588*
Mikhailov, V. S.   799 (72), *822*
Mikina, V. D.   60, 61 (17), *65*
Mikita, T.   123, 124 (6), *169*
Mikol, G. J.   (764), *477,* 1039 (245), *1058*
Mikołajczyk, M.   596 (384), *599,* 1023,
    1035, 1052 (21), *1053*
Milano, M.   414 (1000), *482*
Milas, N. A.   316 (435), *470*
Miles, D.   71 (29), *114*
Miles, D. H.   272 (37), *462*
Milewich, L.   368 (748), *477*
Miljamick, P. C.   332 (543), *473*
Millard, B. J.   164 (199), 167 (214), *173,
    174,* 627 (180), 633 (215), *639, 640*
Milledge, A. F.   1138, 1139 (235), *1320*
Miller, D. P.   30 (79, 80), *57*
Miller, Don P.   230 (192), *260*
Miller, E. L.   386, 387 (851), *479*
Miller, F. A.   27 (57), *56*
Miller, F. F.   874 (77), *911*
Miller, J. M.   127, 157 (23), *170*
Miller, J. W.   603 (15), *636*
Miller, L. J.   672, 679, 680 (67c), *748*
Miller, L. L.   454 (1373), *490*
Miller, M. A.   381 (823), *478*
Miller, N. C.   863 (17), *910*
Miller, R. B.   426 (1131), *485*
Miller, R. E.   240 (308), *263*
Miller, R. F.   21, 22 (24), *56*
Miller, R. G.   850, 851, 853 (91), *857*
Miller, R. J.   740 (220a), *753*
Miller, S. I.   672, 678, 680, 690 (75a–c),
    694, 696 (75c), 699 (75b, c), 735,
    736 (75a–c), *749*
Millich, F.   518 (166), *530,* 1195 (376),
    1323

Millington, D. S.   164 (199), *173*
Mills, B. E.   35, 40, 41 (97), *57*
Mills, J. E.   735 (209b), *753*
Mills, K. J.   770, 771 (79), *779*
Mills, R. W.   299 (320), *468*
Milne, H. B.   275 (88), *463*
Milyakov, B. N.   449 (1328), *489, 868* (41), *910*
Mimakata, K.   217 (8), *256*
Mina, J.   294 (275), *467*
Minakata, K.   757 (6), *778*
Minami, N.   1125 (217), *1320*
Minami, T.   1230 (514), *1326*
Minato, H.   1048 (361), 1050 (374), *1061*, 1145, 1146 (253), *1321*
Minato, I.   290 (237), *466*
Minch, M. J.   958, 959 (66), 969, 970 (92), 972 (98), 977 (92), 985 (129), 988 (92), 990 (152), 1014 (92), *1017–1019*
Minden, D. L. von   167, 168 (218), *174*
Mineshima, F.   1037, 1052 (155), *1056*
Minisci, F.   455 (1381–1383), *490*
Mirone, P.   632 (208), *640*
Mirri, A. M.   22 (34), *56*
Mirrington, R. N.   272 (34), *462*
Mirsch, M.   297 (305), *468*
Mirskov, R. G.   1040 (252), *1058*
Mirskova, A. N.   458 (1403), *490*
Mirviss, S. B.   418 (1054), *483*
Mishutin, A. I.   252 (427), *265*
Misiti, D.   1284 (651, 652), *1329*
Mislow, K.   73 (41), 82, 85 (81), 87 (81, 108), 109 (81), 110 (108, 309), 111 (309), *114, 115, 120*
Misra, R.   335, 339, 361 (578), *473*
Mitani, T.   1028 (67), 1031 (103), 1032 (67, 103), 1036, 1042 (103), *1054, 1055*
Mitchell, D. J.   952 (41), *1017*
Mitchell, D. L.   338 (597), *474*
Mitchell, J. Jr.   618 (127), *638*
Mitchell, L. C.   338 (596), *474*
Mitchell, M. J.   162 (185), *173, 633* (212), *640,* 874 (73), *911*
Mitera, J.   169 (226), *174*
Mitra, S. S.   223 (71), *258*
Mitsuhashi, H.   846 (77), *857*
Mitsui, T.   314, 353 (424), *470*
Mitsunobu, O.   412 (955), *481, 514* (137), *529*
Mitta, A. E. A.   437 (1232), *487*
Mittal, J. P.   672, 677, 685, 686 (81a, b), *749*
Mittal, K. E.   989 (149), *1019*
Mittal, K. L.   951, 988 (32), *1016*
Mittal, L. J.   672, 677, 685, 686 (81a, b), *749*
Mittelbach, H.   1232, 1247, 1248, 1263

(550), *1327*
Mittin, Y. V.   443 (1289), *488*
Mityoshi, M.   1195 (382), *1323*
Miukin, J. A.   225 (125), *259*
Miura, T.   1048 (361), 1050 (374), *1061*
Mix, K.   310 (393), *470*
Miyake, A.   345 (633), 425 (1126), *475, 485*
Miyano, K.   1101 (143), 1160, 1168 (282), *1318, 1321*
Miyauchi, T.   1037 (154), *1056*
Miyaura, N.   430 (1162), *485*
Miyazaki, K.   896 (186), *913*
Miyazaki, T.   770 (78), *779*
Miyazawa, T.   23 (38), *56,* 605 (22, 28, 30), 607 (39), *636*
Miyoshi, M.   422 (1091), 449 (1327), *484, 489,* 1193 (366), *1323*
Mizoguchi, T.   674 (80), 720 (185c, d, h), 722 (80, 185c, d, h), 724 (185d, h), *749, 752*
Mizsak, S. A.   1224 (493), 1228 (493, 507), *1326*
Mizushima, S.   31 (87), *57,* 605 (28), 607 (38), *636*
Mizuta, A.   1029–1031, 1040 (71), *1054*
Mizuta, M.   1027 (56, 58, 59), 1028 (66–70), 1029 (56, 58, 72, 73), 1030 (91), 1031 (56, 101, 103), 1032 (56, 58, 66–70, 72, 73, 103), 1034 (58), 1036 (103), 1039 (91), 1041 (70, 73), 1042 (56, 58, 72, 73, 103), 1043 (59), 1044 (59, 66, 68, 319), *1054, 1055 1060*
Mizuyama, K.   1042 (288), 1052 (403), *1059, 1062*
Mo, T.   86 (104), *115*
Mock, W. L.   436 (1213), *486*
Modest, E. J.   507 (97), *528*
Moehrle, H.   306 (362), 319 (460), 371 (767), *469, 471, 477,* 1314 (719), *1330*
Moelwyn-Hughes, K. A.   246 (378), *264*
Moersch, G. W.   281 (175, 183), 328 (183), *465*
Moetz, J.   297 (304), *468*
Moffett, R. B.   (529), *472,* 1195 (383), *1323*
Moffitt, W.   73 (35, 40), 99 (35), *114,* 623 (161), *639*
Mohan, S.   458 (1404), *490*
Moiermeier, L. F.   294 (282), *467*
Moiseikina, N. F.   449 (1329), *489*
Moisio, T.   166, 167 (209), *173,* 495 (19), *526*
Mokrzan, J.   246 (377), *264*
Molenaar-Langeveld, T. A.   250 (411), *265*
Moles, A.   826 (23), 827 (33), *856*
Molin, Iu. H.   761 (38), *778*

Møllendal, H.　253 (439), *265*
Mollett, K. J.　993 (158), *1019*
Mollier, Y.　1024 (36), 1043 (36, 308), *1053, 1059*
Molloy, R. M.　369, 371 (758), *477,* 1100, 1101 (139), *1318*
Momany, F. A.　100 (170), *117,* 232 (214), 243 (214, 335), 247 (335), *261, 263*
Mondal, M. A. S.　420 (1075), *484*
Mondelli, G.　291 (250), *467*
Mondodoev, C. T.　806 (115), *823*
Mondon, A.　1206, 1207, 1212 (464), *1325*
Moniot, J. L.　1206 (472), *1325*
Moniz, W. B.　359 (698), *476*
Monk, C. B.　225 (129), *259*
Monson, R. S.　451 (1346), *489*
Montaudo, G.　610 (66), *637*
Monteil, R. L.　503, 504 (76), *528*
Monteiro, P. M.　963 (80), *1018*
Montevecchi, P. C.　1043 (305), *1059*
Montgomery, R. S.　1276, 1278, 1279 (634), *1328*
Montgomery, W. C.　416 (1015), *482*
Montgudo, G.　517 (150), *529*
Monti, L.　310 (390), *470*
Monti, S. A.　1304, 1307 (698), *1330*
Montzka, T. A.　446 (1306), *488*
Monzeglio, P.　800 (73), *822*
Mooberry, J. B.　583 (344), *593*
Moon, S.　331, 341 (527), *472*
Moore, B.　369 (760), *477*
Moore, D. R.　502, 505, 506 (66), *528*
Moore, G. G.　441 (1266), *487*
Moore, H. W.　1284 (651, 652), *1329*
Moore, J. A.　521 (180), *530*
Moore, L. L.　283 (201), *466*
Morand, P. F.　1145, 1147 (257), *1321*
Morawetz, H.　947, 951, 957 (10), *1016*
Morawitz, H.　198 (95), *211*
Morck, H.　536 (22), *587*
Moreau, C.　346 (638, 639), *475*
Moreau, J.-L.　457 (1393), *490*
Morf, R.　340 (613), *474*
Morgan, G. D.　255 (479), *266,* 957 (64), *1017*
Morgan, K. J.　904 (218), *914,* 1030 (89), *1054*
Morgan, T. K. Jr.　1186 (348), *1323*
Mori, K.　419 (1072), *484*
Mori, M.　581, 583 (338), *593,* 1039 (191–193), 1040 (192), 1050 (191–193), *1057*
Mori, N.　238, 239 (288), *262*
Mori, S.　1168 (288), *1321*
Mori, T.　369 (751), *477,* 1206 (474, 476, 477), 1216 (474, 477), 1217 (476), *1325*

Mori, Y.　1192 (362), *1323*
Moriarty, R. M.　676 (221b), 687 (104), 743 (104, 221b), *749, 753*
Moriconi, E. J.　168 (223), *174,* 526 (225), *531,* 617 (112), *638,* 1225 (499), 1227 (499, 504, 505), 1228, 1250 (506), *1326*
Morikawa, M.　522 (190), *530,* 916, 919, 928 (11), 938 (64), *942, 944,* 1232, 1239 (538), *1326*
Morimoto, A.　812 (138), *824*
Morin, R. D.　596, 597 (385), *599,* 861 (5), *909*
Morishima, I.　253 (438), *265*
Morita, H.　229 (172, 174), 233 (221), *260, 261*
Moritani, I.　361 (707), *476*
Moriwake, T.　277 (120), *464*
Moriyama, N.　984 (122), *1018*
Morley, J. R.　304, 315 (346), *469*
Morokuma, K.　42, 44 (109), *57,* 219 (22), 232 (211), 238 (290), *257, 260, 262*
Morokuma, M. S.　214 (2f), *256*
Morongin, E.　565 (243), *591*
Morozov, L. A.　305 (352), *469*
Morozova, I. D.　802 (86b, c), *823*
Morris, I. J.　1052 (408), *1062*
Morris, L. J.　333 (559), *473*
Morrison, H.　709, 715 (154, 161), *751*
Morrison, W. H. III　295 (285), *467*
Morrissey, A. C.　620 (146), *639*
Morrow, C. J.　536, 583, 584 (27), *587*
Morrow, D. F.　583 (348), *593*
Morschel, H.　572 (314), *593*
Morton, J. R.　685 (100), *749*
Moscowitz, A.　73 (40), 82, 85, 87 (81), 99 (164), 109 (81), *114, 115, 117,* 611 (77), 623 (161), *637, 639*
Mose, W. P.　88 (109, 110), *116*
Moser, W.　866 (34), *910*
Moshentseva, L. V.　325 (493), *472*
Mosher, H. S.　88 (112), 108 (268), 109 (112, 268), *116, 119,* 280 (168), *465*
Mosher, W. A.　359 (695), *476*
Moskowitz, J. W.　242, 252 (333), *263*
Moss, R. A.　989 (141, 142), 1006 (141, 142, 176, 179), 1011, 1012 (192), *1019, 1020*
Motkowska, B.　572, 573 (317), 596 (384), *593, 599*
Motsavage, V. A.　985 (130), *1019*
Motzfeldt, T.　22, 23 (27), *56,* 222 (46), *257*
Moubasher, R.　686 (103), *749*
Moulton, G. C.　757 (8), *778*
Mourev, C.　313 (415), *470*
Moussebois, C.　617 (115), *638*

Mousseron, M.   1111 (177), *1319*
Movsumzade, M. M.   1073 (36), 1090 (101), *1316, 1318*
Moye, A. J.   328 (513), *472*
Moyer, H. R.   325 (492), *472*
Moyle, M.   313 (417), *470*
Moza, P. N.   1201 (441), *1325*
Mozai, T.   597 (393), *599*
Mracec, M.   603, 631 (14), *636*
Muck, D. L.   814, 816, 818 (141), *824*
Mucke, J.   307 (368), *469*
Mudrak, A.   881 (104), 882, 883 (105), *912*
Mueller, K. A.   1284, 1291 (656), *1329*
Mueller, R. H.   304 (348), 388 (861), *469, 479,* 1145, 1152 (270), *1321*
Muffler, H.   545, 553, 554, 572 (129), *589*
Muggler-Chawan, F.   1079 (62), *1317*
Mühlstädt, M.   458 (1402), *490,* 547 (130), *589*
Mukaiyama, T.   413 (958), 417 (1052), 438 (1236, 1238), 443 (1291), 444 (958, 1297), *481, 483, 487, 488,* 504 (82), *528,* 536 (45, 46), 538 (61), 539 (45, 66), 543 (45), 562, 567 (46), *587, 588,* 868 (44, 46), 869 (46, 47), *910*
Mukana, D.   180, 181, 202, 203 (49), *210*
Mukawa, F.   1168 (288), *1321*
Mukayama, T.   1038 (181), 1039 (217, 219, 235), 1046 (181), 1047 (335), 1048 (351), *1056–1058, 1060*
Mukerjee, A. K.   521 (186), *530,* 1195, 1225 (379), 1232 (549, 583), 1246 (549), 1256, 1257 (583), *1323, 1327*
Mukerjee, H.   446 (1311), *488*
Mukerjee, P.   951 (29, 33, 34), 952 (36), 954 (33), 966, 972 (89), 978 (102), 987 (34), 989 (149), 990 (29), *1016, 1018, 1019*
Mukherjee, R.   458 (1406), *490*
Muller, A.   1204 (447), *1325*
Muller, C. T.   493 (5), *526*
Muller, J. C.   1123, 1127 (213), *1320*
Muller, J.-C.   507 (98), *528,* 690, 704, 705 (122), *750*
Muller, L. L.   1194, 1225 (372), *1323*
Müller, N.   236 (251), *261*
Müller-Westerhoff, U.   536, 554 (16), *587*
Mulliken, R. S.   175 (4), 176 (9), 179 (37), *209, 210*
Munch-Peterson, J.   282 (194), *465*
Mungall, W.   103 (227, 228), *118*
Munslow, W. D.   744 (227), *753*
Munson, J. W.   603 (8), *636*
Munson, M. S. B.   152 (136), *172*
Murai, K.   361 (712), *476*
Murakami, M.   273 (55), *463*

Murakami, S.   583 (341), *593*
Murakami, T.   932 (46), *943*
Muramoto, A.   398 (894), *480,* 916, 919, 928 (10), *942*
Murase, I.   372 (768), 415 (991, 993), *477, 482*
Murata, K.   902 (209), *914*
Murawski, J.   867 (36), 871 (59), *910, 911*
Murphy, M. T.   873 (69), *911*
Murphy, R. C.   148, 149 (120), *172*
Murray, M. A.   1080 (81), *1317*
Murray, M. F.   1145 (269), *1321*
Murray, R. D. H.   299 (320), *468*
Murray, R. K., Jr.   1186 (347, 348), *1323*
Murray, R. W.   338 (587), *473*
Murray-Rust, P.   104 (246), *118*
Murrell, J. N.   197 (90), *211*
Murthy, A. S. N.   214 (2a, h), 231 (206), 241 (324–326), 254 (451), *256, 260, 263, 266*
Muruoka, M.   1037 (154, 155), 1052 (155), *1056*
Musser, J. H.   426 (1130), *485*
Mustafin, I. S.   177 (22), *209*
Muszkat, K. A.   673, 683, 685 (85a), *749*
Muto, H.   758 (10), *778*
Muxfeldt, H.   583 (344), *593*
Muzychenko, L. A.   804 (101), *823*
Mygind, H.   1031 (100), *1055*
Mysels, K. J.   951 (34), 955 (60), 978 (102), 987 (34), *1016–1018*

Nace, H. R.   883 (113), 896 (188), *912, 913*
Nafissi-V, M. M.   167, 168 (217), *174*
Nagai, M.   581, 582 (337), *593*
Nagai, U.   102 (199), *117*
Nagakura, S.   31 (87), *57,* 200, 201 (110), 202 (110, 117), 205, 206 (141), *211, 212,* 229 (171, 172, 174), 233 (221), *260, 261,* 607 (38), *636*
Nagamatsu, T.   563, 566 (231), 571 (293), *591, 592*
Nagarajan, K.   1305 (701), *1330*
Nagarayan, G.   250 (407), *265*
Nagasawa, C.   720, 722, 724 (185h), *752*
Nagasawa, H. T.   398 (888), *480*
Nagata, M.   820 (168), *824*
Nagata, S.   1025 (41), 1026 (41, 74), *1053, 1054*
Nagoshi, K.   123, 124 (5), *169*
Nagpal, K.   328 (513), *472*
Nagra, S. S.   869 (50), 870 (57), *910, 911*

Nagy, J. B.　176 (16), 179 (38), 180 (44, 49), 181 (38, 44, 49, 52, 53), 183 (38), 195 (84), 202 (49, 53, 122, 123), 203 (49, 53), *209–212*

Nagy, O. B.　176 (16), 179 (38), 180 (44, 49), 181 (38, 44, 49, 52, 53), 183 (38), 195 (84), 202 (49, 53, 122, 123), 203 (49, 53), *209–212*

Nahas, R. C.　1011, 1012 (192), *1020*

Nahjoub, A.　1052, 1053 (393), *1062*

Nahlovska, Z.　222 (55), *257*

Nahlovsky, B.　222 (55), *257*

Nahringbauer, I.　222 (52), 224 (107), *257, 258*

Naidov, B.　1115 (189), *1319*

Nair, P. G.　1039 (225), *1057*

Naito, T.　1206 (471, 473–477), 1214 (471), 1215 (475), 1216 (474, 477), 1217 (473, 476), *1325*

Nakagawa, A.　515 (143), *529*

Nakagawa, K.　303 (343, 344), *469*

Nakagawa, T.　951, 952, 954 (26), 984 (26, 117), *1016, 1018*

Nakahara, A.　103 (232), *118*

Nakai, H.　674 (80), 720 (185b–d, h), 722 (80, 185b–d, h), 724 (185b, d, h), *749, 752*, 1206, 1219 (481), *1325*

Nakai, M.　372 (768), *477*

Nakai, T.　536 (47), 544 (117), 572, 573 (316), *587, 589, 593*

Nakajima, K.　916, 920 (21), *943*

Nakajima, Y.　514 (137), *529*, 769 (76), *779*

Nakakimura, H.　581, 583 (338), *593*

Nakamura, N.　231 (203, 205), *260*, 850, 851 (92), *857*

Nakana, J.　800 (79), *823*

Nakanishi, H.　229 (174), *260*

Nakanishi, K.　73 (34), 111 (314), 112 (34, 315, 316, 318), *114, 120*

Nakanishi, S.　440 (1260), *487*

Nakano, T.　336 (580), *473,* 1311 (713), *1330*

Nakao, M.　896, 902 (185), *913*

Nakao, Y.　103 (232), *118*

Nakashima, N.　198 (93), *211*

Nakata, H.　140 (82, 83), 144 (101), *171*

Nakata, T.　303 (343), *469*

Nakaya, J.　804 (95), *823*

Nakaya, T.　413 (959), *481*

Nakayama, J.　594 (373), *598*

Nakayama, Y.　182, 184 (56), *210*

Nakina, S.　875 (80a), *911*

Nakova, E. P.　1039 (250), *1058*

Naletova, G. P.　176 (19), 179 (41, 42), *209, 210*

Nambu, H.　434 (1197–1199, 1201), *486*

Nanteuil, M. de　349 (647), *475*

Naoi, Y.　336 (580), *473*

Narain, N. K.　745 (231), *753*

Narayanan, C. N.　345 (631), *475*

Narayanan, C. R.　104 (248–250), *118*

Narayanan, C. S.　1232 (546), *1327*

Narula, A. S.　339 (610), *474*

Naser-ud-Din　810, 811 (135), *824,* 1229 (511), *1326*

Nasielski, J.　238 (279), *262*

Nasipura, D.　424 (1115), *484*

Naskret-Barciszewska, M. Z.　228 (160), *259*

Natarajan, K.　790 (46), *822*

Nathan, W. S.　328 (508), *472*

Nations, R. G.　498 (46), *527*

Natta, F. J. van　496 (28), *527*

Natterstad, J. J.　230 (191), *260*

Natuki, R.　1042 (288), *1059*

Naumov, Y. A.　449 (1329), *489*

Nawa, H.　318 (447), *471*

Nawojska, J.　777 (97), *780*

Nayar, M. S. B.　155, 156 (154), *172*

Naylor, C. A. Jr.　320 (468), *471*, 1065 (13), *1316*

Nazaki, H.　271 (20), *462*

Nazer, M. Z.　72, 88, 93 (25), *114*

Nazir, M.　1101 (142), *1318*

Neadle, D. J.　676, 685 (98a), *749*

Nechiporenko, Ye. K.　787 (31), *822*

Neckers, D. C.　411 (935), *481*, (136), *750*

Needham, L. L.　498 (39), *527*

Needles, H. L.　456, 457 (1387), *490*

Neelakantan, S.　411 (941), 412 (952), *481*

Nefedov, O. M.　436 (1212), *486*

Neff, B. L.　230 (181), *260*

Negishi, A.　312 (403), *470*

Negishi, E.　435 (1208), *486*

Neidlein, R.　542 (102), *588*

Neiswender, D. D. Jr.　359 (698), *476*

Nekrasov, A. S.　303 (339), *469*

Nelke, J. M.　806 (114), *823*

Nelles, J.　917, 919 (28), *943*

Nelsen, S. F.　803 (90), 814, 816 (145), *823, 824*

Nelson, A. L.　1276, 1278 (631), *1328*

Nelson, D. A.　522 (195), *530,* 1232, 1245 (551, 552), 1247 (552), 1252 (551), 1253 (551, 552), *1327*

Nelson, G. L.　1035 (122), *1055*

Nelson, J. D.　445 (1301), *488*

Nelson, P.　444 (1298), *488*

Nemec, M.　297 (306), *468*

Nemethy, G.　953 (47), *1017*

Nemoto, H.　1311 (713), *1330*

Nenz, A.　1295 (670), *1329*

Nesmeyanov, A. N.　276 (110), *464*

Neta, P. 763 (44, 47, 52), 764 (52), 765 (56), 766, 767 (58, 59), (61), 768 (68, 70, 73), 775, 776 (95, 96), *778, 779*
Neubauer, D. 291 (246), *467*
Neubauer, G. 550, 557, 559 (157), *590*
Neubert, P. 614 (102), *638*
Neujmin, H. 671, 672 (92a), *749*
Neumann, H. 585 (361), *594,* 805, 817 (102), *823*
Neumann, W. P. 1039 (202), *1057*
Neuray, M. 341 (526), *472*
Newallis, P. E. 874, 875 (78), *911*
Newan, R. H. 1075, 1079 (52), *1316*
Newkirk, J. D. 359 (699), *476*
Newman, E. R. 878 (99), *911*
Newman, L. G. 179 (35), *210*
Newman, M. S. 272 (24), 273 (56), 274 (81), 277 (121), 298 (318), 318 (444), 319 (456), 330 (520), 338 (599), 349 (648), 351 (649), 358 (693), 383 (828), 411 (923), 415 (1005), 416 (1016), 437 (1223), *462–464, 468, 471, 472, 474–476, 478, 481, 482, 487,* 524 (214), *531*
Newmann, R. C. 31 (84), *57*
Newport, G. L. 721, 724 (187), *752*
Newton, A. S. 759 (16), 771 (80), *778, 779*
Newton, M. D. 22 (23), *56,* 240 (301a), *262,* 743 (223a), *753*
Neyrelles, I. 1111 (177), *1319*
Neywick, C. V. 691, 696–699 (128), 733 (203a, b), 735 (128), *750, 752*
Ng, P. 959 (67), 988, 990 (139), *1017, 1019*
Ng, T. L. 70 (28), *114*
Nguyen-Dinh-Nguyen 136 (66), *170*
Nibbering, N. M. M. 125 (11), 138 (78), 160 (179), *169, 171, 173*
Nichol, L. W. 248 (390), *264*
Nicholas, K. M. 839 (62), *856*
Nicholas, L. 882 (107), 903, 906 (213a), *912, 914*
Nicholls, A. C. 1110 (175), *1319*
Nichols, V. N. 415 (988), *482*
Nicholson, C. 240 (301), *262*
Nicol, C. H. 714 (175b), *751*
Nicolaou, K. C. 514 (138, 140), 515 (142), *529,* 1039, 1048 (236), *1058,* 1075 (49), *1316*
Nicolaus, B. J. R. 1201 (427), *1324*
Nicolet, B. H. 386 (853), *479*
Niedemeyer, A. D. 355 (679), *475*
Nielands, J. B. 275 (95), *463*
Nielsen, C. J. 28 (62), *56*
Nielsen, E. B. 71, 102 (20), *114,* 612, 613, 632 (92), *637*
Nielsen, P. H. 1031 (100), *1055*

Nielsen, S. O. 766–768 (62), *779*
Nielsen, T. 412 (945), *481*
Niemer, U. 1022 (7, 13), 1023 (13), *1053*
Niessen, W. von 42 (108), *57,* 219, 233 (21), *257*
Niimi, H. 551 (161), *590*
Niki, I. 1123, 1128–1130 (215), *1320*
Nikishin, G. I. 279 (147), 289 (231, 232), *465, 466,* 1079 (57), 1080 (72), *1317*
Nikles, E. 360 (705), *476*
Nilsson, W. 316 (438), *471*
Ninham, B. W. 952 (41), *1017*
Ninomiya, I. 1206 (471, 473–477), 1214 (471), 1215 (475), 1216 (474, 477), 1217 (473, 476), *1325*
Nisato, D. 102 (193), *117*
Nisbet, M. A. 1066, 1068, 1146, 1148 (17), *1316*
Nishi, A. 336 (580), *473*
Nishi, M. 562, 565, 569 (207), *590*
Nishihara, A. 317, 353 (441), *471*
Nishihara, H. 101 (181), *117*
Nishihara, K. 101 (181), *117*
Nishikawa, S. 236, 237 (258), *262*
Nishimura, A. 902 (209), *914*
Nishimura, S. 290 (237), 296 (296), *466, 468*
Nishimura, T. 318 (445), *471*
Nishinaga, A. 362 (719), *476*
Nitta, K. 338 (590), *474*
Nivard, R. J. F. 540, 547, 548, 554, 578, 586 (81), *588,* 1040, 1042 (258), *1058*
Niwa, T. 237 (260), *262*
Nixon, J. F. 849 (86), *857*
Niznik, G. E. 295 (285), *467*
Nnadi, J. C. 673 (70a, b), *748*
Nobe, H. 792, 793 (54), *822*
Nobis, J. F. 294 (281), *467*
Noda, S. 757, 758 (5), *778*
Noe, E. A. 1033 (112), *1055*
Noel, R. 950 (23), *1016*
Nogami, H. 985 (131), *1019*
Nohara, A. 309 (387), *469*
Nohe, H. 793 (55, 56), *822*
Nohira, H. 536, 539, 543 (45), *587,* 868 (44), *910*
Nojima, M. 292 (258), 439 (1251, 1253), *467, 487*
Nokami, J. 428 (1155), *485*
Nolde, C. 162–164 (191), *173*
Nolen, R. L. 283, 284 (205), *466,* 500 (60), *527,* 1120, 1121 (199), *1320*
Nomine, G. 1053 (409), *1062*
Nord, F. F. 307 (376, 377), *469*
Nordén, B. 73 (33), 103 (226), 110, 111 (303), *114, 118, 120*
Nordmann, H. G. 536 (24), *587*
Norell, J. R. 453 (1367), *489*

Noritsina, M. V.　447 (1320), *488*
Norman, R. O. C.　341 (528), *472,* 768
　(71), *779*
Normant, H.　280 (153, 167), 456, 457
　(1391), *465, 490,* 547 (134), *589*
Normant, J. F.　389 (862), 416 (1019),
　479, 482
Normt, H.　414 (1001), *482*
Noro, Y.　770 (78), *779*
Norris, W. P.　545 (121), *589,* 894 (175),
　913
Norrish, R. G. W.　672, 677, 688, 709,
　712, 718, 719 (61), *748*
North, B.　690, 701, 702 (121b), *750*
Nosworthy Peto, J. M.　770, 771 (79), *779*
Notani, J.　412 (957), *481*
Notari, R. E.　603 (8, 9), *636*
Nour, T. A.　1024, 1048 (29), *1053*
Novak, A.　218 (15), 222 (64), 224 (92,
　94), 226 (64, 136), 228 (154), 240
　(307), *256–259, 263*
Novák, J. J. K.　82 (82), *115*
Novak, L.　1109 (169), *1319*
Novikov, V. T.　804 (100, 101), *823*
Novasad, J.　932 (48), *943*
Novasad, Z.　918 (29), *943*
Noyari, R.　1186 (351, 352), 1187
　(352), *1323*
Noyce, D. S.　499 (47), *527,* 904 (218),
　908 (237), *914,* 1145, 1147 (250),
　1321
Noyes, W. A.　642 (5), *747*
Noyes, W. A. Jr.　687, 688, 701 (106),
　750
Noyori, R.　278 (128), *464,* 545 (120),
　589, 733, 734 (206), *753*
Nozaki, H.　369 (751), 423 (1105, 1106),
　477, 484, 545 (120), 585 (362), *589,*
　594, 733, 734 (206), *753,* 1048 (356),
　1061, 1186 (351, 352), 1187 (352),
　1323
Nozaki, J.　691, 699, 735 (123), *750*
Nukada, N.　248 (389), *264*
Nunson, B.　35 (100), *57*
Nurullaev, H. G.　61, 62 (27), *66*
Nussbaum, A. L.　1145, 1159 (256),
　1321
Nussey, B.　151 (133, 135), *172*
Nützel, K.　825 (2, 3), 837 (3), *855*
Nyi, K.　381 (821), *478*
Nyman, C. J.　843–845, 852 (69), *857*
Nyquist, R. A.　253 (442, 443), *265*
Nyss, N. L.　1201 (438), *1324*

Oakenfull, D.　949, 976, 1002 (21), *1016*
Oakenfull, D. G.　951 (30), *1016*
Oberender, H.　1165, 1168 (298), *1322*
Obigin, Y. N.　289 (234), *466*

Obradović, M.　226 (142), *259*
O'Brien, J. B.　1206 (472), *1325*
Ocampo, J.　819 (166), *824*
Occolowitz, J. L.　128 (24), *170*
Ochiai, M.　581, 582 (337), *593,* 770
　(78), *779,* 812 (138), *824*
Ochoa-Solano, A.　999 (165), *1019*
Ockman, T.　274 (77), *463*
O'Connor, C. J.　993 (158), *1019*
O'Connor, G. L.　883 (113), *912*
Oda, K.　745 (230b), *753,* 1048 (351),
　1060
Oda, O.　850, 851 (92, 93), *857*
Oda, R.　274 (76), *463,* 562, 565, 569
　(207), *590*
Odaira, Y.　290 (237), 296 (296), *466,*
　468, 711 (167), *751*
Odani, M.　511 (118a), *529*
Odinokov, S. E.　228 (152, 153, 163),
　238 (277), *259, 262*
Odinokov, V. N.　339 (601), *474*
Odinokova, A. I.　339 (601), *474*
Oehler, E.　1038 (182), *1056*
Oehler, R.　411 (929), *481*
Oehlschlager, A. C.　849, 850, 854 (88),
　857
Oehme, G.　239 (294, 300), *262*
Oele, P. C.　875 (80, 83), 892, 896 (157),
　911, 913, 1050 (376), *1061*
Oettingen, W. F. von　1123 (202), *1320*
Oettle, W. F.　689 (117, 120, 124, 145),
　690, 691 (120, 124), 694 (117, 120,
　124), 695 (124), 696 (120, 124), 698,
　700, 705 (124), 706 (117, 120, 124,
　145), 731 (117, 145, 201a), 733 (145,
　201a), 735 (120, 124), *750–752*
Oettmeier, W.　334 (571), *473*
Offen, H. W.　199 (103), *211*
Ogandyhanyan, S. M.　402 (904), *480*
Ogasawara, K.　505 (84), *528,* 1138, 1144
　(247), *1321*
Ogata, M.　1206, 1209–1211 (460), *1325*
Ogata, T.　1042, 1043 (280), *1059*
Ogata, Y.　289 (236), 398 (894), *466,*
　480, 916 (10, 11, 19–21), 919 (10,
　11), 920 (19–21), 924 (19, 20), 928
　(10, 11, 19, 44), 938 (64), *942–944,*
　1040 (271), 1051 (383), 1052 (271),
　1058, 1061, 1206, 1207, 1211 (461),
　1325
Ogibin, Y. N.　289 (231, 232), *466*
Ogilvie, J.　318 (455), 319 (458), *471*
Ogiso, H.　333 (558), *473*
Ogiwara, H.　720, 722, 724 (185d, h), *752*
O'Gorran, J. M.　22 (29), *56*
Ogura, H.　81 (70–72), *115*
Ogura, K.　287 (221–223), *466,* 1039
　(239, 243), *1058*

Ohashi, M. 111 (314), *120*, 153 (140), *172*, 613, 614 (95), *638*
Ohashi, Y. 108, 109 (268), *119*, 280 (168), *465*
Ohfune, J. 1101 (143), 1160, 1168 (282), *1318, 1321*
Ohga, K. 720 (182, 183), 722 (182), 724, 725 (183), *752*
Ohkubo, K. 1007 (182), *1020*
Ohkura, K. 224 (103), *258*
Ohler, E. 1113, 1114 (184), *1319*
Ohloff, G. 313 (413), *470*, 887 (137), *912*
Ohlson, R. 106 (260), *119*
Ohmenzetter, K. 963 (78), *1017*
Ohmori, S. 571 (300, 301), 597 (392, 393), *592, 599*
Ohnishi, T. 151 (130), *172*
Ohnishi, Y. 1025, 1026 (49), 1036 (134), *1054, 1055*
Ohno, A. 151 (130), *172*, 1024 (34), 1025, 1026 (49), 1036 (134), 1040 (262), 1052 (391), *1053–1055, 1058, 1062*
Ohno, K. 826 (20), 827 (36), 828, 829 (40, 41), 831 (40), 834 (40, 41, 55), 835 (40, 41, 49), 837 (41, 56), 838, 839, 841 (41), 843 (41, 74, 82), 844 (36, 41, 74), 845 (36, 41), 847, 848 (41, 74), 852 (41), *855–857*
Ohno, M. 1168, 1170 (310), *1322*
Ohshiro, Y. 1230 (514), *1326*
Oine, T. 444 (1299), *488*, 506 (86), *528*, 653, 654, 657, 658 (27), *747*
Oishi, T. 581 (337, 338), 582 (337), 583 (338, 341), *593*, 1039 (191–193), 1040 (192), 1050 (191–193), *1057*
Ojima, I. 445 (1305), *488*
Ojtkowski, P. W. 655, 661, 691, 700 (29), *747*
Oka, K. 1044 (316), 1052 (396), *1060, 1062*
Okabe, K. 416 (1021), *483*
Okabe, O. 1037, 1052 (155), *1056*
Okada, T. 733, 734 (206), *753*, 1186, 1187 (352), *1323*
Okahata, Y. 995, 1002 (164), 1007 (164, 181), 1012 (196), *1019, 1020*
Okamoto, K. 989 (143), 1006 (178), *1019, 1020*
Okamoto, Y. 565, 566 (240), *591*
Okamura, K. 101 (173), *117*
Okano, M. 562, 565, 569 (207), *590*
Okawara, M. 536 (47), 544 (117), 572, 573 (316), *587, 589, 593*, 1038, 1042, 1052 (168), *1056*
Okawara, T. 1201, 1203 (443), *1325*
Okaya, Y. 225 (123), *259*

Oki, M. 239 (291–293, 296, 297), *262*
Okonnischnikova, G. 436 (1212), *486*
Okraglik, R. 1298, 1299 (673), 1302 (688), *1329, 1330*
Okuda, T. 78 (60), *115*, 622 (157), *639*
Okuno, Y. 1206, 1218 (480), *1325*
Olah, G. A. 274 (86), 296 (301), 416 (1012, 1025), 439 (1251, 1253, 1255), 454 (1376), *463, 468, 482, 483, 487, 490*, 828, 829 (39), *856*, 1045 (325, 327), *1060*
Olah, G. O. 1038 (172), *1056*
Olah, J. A. 274 (86), 296 (301), *463, 468*
Olbricht, T. 375 (786), *478*
Oldenziel, O. H. 343, 344 (627, 628), *474*
Oldham, A. R. 826 (25), *856*
Oldham, W. J. (541), *473*
Olechowski, J. R. 334 (573), *473*
Oleinik, B. N. 60, 61 (17), *65*
Oleinik, N. M. 1048 (352), *1060*
Olesen, J. A. 721, 725 (186), *752*
Oliveira, P. A. M. de 454 (1378), *490*
Oliveros-Desherces, E. 1292 (665, 666), 1294 (666), 1295 (665, 666), *1329*
Oliveto, E. P. 310 (396), 340 (612), *470, 474*, 1145, 1159 (256), *1321*
Ollis, W. D. 501 (63), *527*, 619 (132), *638*
Olovsson, I. 223 (80), *258*
Olschwang, D. 536 (36), 543 (36, 112, 113), 554 (174), 561 (112, 113), *587, 589, 590*
Olson, A. R. 499 (51), *527*
Omura, S. 515 (143), *529*
O'Neal, H. E. 60, 61 63 (8), 64 (47), 65 (8, 47), *65, 66*
O'Neill, W. A. 327, 328 (510), *472*
Onesta, R. 884 (129), *912*
Ong, E. C. 98 (161), *117*
Ong, J. 392 (866), *479*
Ono, I. 745 (232), *753*
Ono, K. 441 (1271, 1272), *487*
Ono, S. 790 (49), 804 (95), *822, 823*
Ono, Y. 123, 124 (6), *169*
Onopchenko, A. 328 (502), *472*
Ooms, P. H. J. 1040, 1042 (258), *1058*
Oonk, H. A. J. 61 (24), *66*
Oota, Y. 1106, 1107 (156), *1319*
Openshaw, H. T. 446 (1312), *488*, 1199 (409), *1324*
Opitz, G. 511 (116a), *529*, 1228 (508), *1326*
Oppenauer, R. V. 417 (1045), *483*
Oppenheim, A. 880 (101), *912*
Oppenheim, E. 1039 (214), *1057*
Oppenheimer, E. 828–835, 837, 841, 842, 845 (42), *856*

Orazi, O. O. 418 (1061), *483*
Orchin, M. 69, 71, 73 (11), *114,* 436 (1211), *486*
Qrel, B. 224 (89), 228 (150, 155–159, 161), 231 (158), 240 (89), *258, 259*
Orenski, P. J. 71, 72 (26), *114,* 626 (174), *639*
Orfanos, V. 536 (20), *587*
Orgel, L. E. 179 (37), *210*
Ori, M. 854 (97), *857*
Oriel, P. 77 (53), *114*
Orito, K. 1166, 1168, 1187 (307), 1309 (710), *1322, 1330*
Orloff, M. K. 661, 663, 666, 668 (51), *748*
Orlov, I. G. 231 (204), *260*
O'Rorke, H. 1073, 1115 (41), *1316*
Oroshnik, W. (96), *823*
Orr, G. 499 (49), *527,* 691, 701 (125c), 703, 738 (144), *750, 751*
Ort, M. R. 785 (21), 808 (123, 124), *821, 824*
Orttmann, H. 341 (621d), *474*
Orville-Thomas, W. J. 218 (17), 231 (202), 251 (421), *256, 260, 265*
Orzech, C. E. 135 (64), *170*
Osborn, J. A. 826 (21, 27), 827 (27, 32, 34, 37), 828 (37), 835 (37, 46), 837 (32, 59), 840 (59), 841 (32), 843 (68), 844 (37, 68), 853 (27), *856, 857*
Oshima, K. 423 (1105), *484,* 1048 (356), *1061*
Osintseva, L. V. 176 (19), *209*
Osipov, A. P. 948 (20), 962 (73b), 964 (20, 73b, 85, 86, 87b, 88), 965 (20, 73b), 966 (20, 86), 967 (88), 968 (73b), 1004 (87b), *1016–1018*
Osipov, O. A. 241 (323), *263*
Oster, H. 323 (483), *471*
Østerberg, O. 236, 237 (259), 248 (397), *262, 264*
Osterroth, C. 1052 (390), *1061*
Ostlund, N. S. 22 (23), *56*
Ostromuisslenskii, I. I. 879 (95), *911*
O'Sullivan, M. J. 1225–1227 (500), *1326*
Otsuyi, Y. 448 (1325), *489*
Ott, A. C. 1145 (269), *1321*
Ott, D. 495, 496, 514 (16), *526*
Ott, H. 619 (134), *638*
Ottersen, T. 241 (329, 330), 251 (423), 253 (330), *263, 265*
Ottnad, M. 99 (162, 166), *117*
Otto, E. 82 (75), *115*
Ottolenghi, M. 204 (128), *212*
Oude-Alink, B. A. M. 655–657 (30), 658 (36), *747*
Ourisson, G. 507 (98), *528,* 689 (118), 690 (122), 704, 705 (118, 122), *750,*

1123, 1127 (213), 1168 (306), *1320, 1322*
Outurquin, F. 1042 (295), *1059*
Ovchinnikova, R. A. 618 (123), *638*
Overbeek, J. Th. G. 955 (56), *1017*
Overberger, C. G. 75 (46), 85, 107 (94), *114, 115,* 332 (557), *473,* 896, 902 (185), *913,* 957 (63), *1017,* 1065 (14), 1161, 1168 (283), 1276 (629), *1316, 1321, 1328*
Overman, L. E. 452 (1356), *489*
Oversby, J. P. 759, 760 (26), 769, 770 (77), 773 (86), *778, 779*
Owen, G. R. 416 (1008), *482*
Owen, J. S. 1270 (611), *1328*
Owen, L. N. 1039 (230), *1058*
Owen, N. L. 88 (120), *116,* 1022 (9), 1023 (24), 1024 (9), *1053*
Owens, W. 653, 654, 657, 658 (27), *747,* 1041 (277), *1059*
Owsley, D. C. 806 (114), *823*
Oyama, M. 1168, 1170 (310), *1322*
Ozaki, Y. 691, 699, 735 (123), *750*

Paabo, M. 238 (280), *262*
Pabon, H. J. J. 136 (65), *170*
Pacansky, J. 512 (122), *529,* 691, 701 (125a–d), 702 (125b), 703, 738 (144), *750, 751*
Pace, R. J. 223 (67), 250 (406), *258, 265*
Pachenkov, B. V. 1048 (352), *1060*
Pachler, K. G. R. 300 (329), *468*
Pacifici, J. G. 710, 715 (163a, b), *751*
Paddonkow, M. N. 1024, 1048 (29), *1053*
Padmasani, R. 411 (941), 412 (952), *481*
Padwa, A. 506 (86, 87), *528,* 653, 654, 657, 658 (27), 706, 707 (148), 727 (192a, b), 728 (192b), 738 (214b), *747, 751–753,* 1052 (392), *1062*
Padwa, O. 1041 (277), *1059*
Paessler, P. 450 (1335), *489*
Paetzold, P. I. 442 (1280), *488*
Pagani, G. 1198 (391), *1324*
Pagano, J. S. 507 (95), *528*
Paganov, A. 272 (50), *463*
Page, F. M. 196 (87, 88), *211*
Page, G. A. 329 (518), *472*
Page, M. I. 1015 (202), *1020*
Pai, B. R. 1305 (701), *1330*
Paik, C. 963 (80), *1018*
Paik, C. H. 1008 (187), *1020*
Pailer, M. 415 (979), *482,* 627 (179), *639*
Pailthorpe, D. 88 (128), *116*
Pakhomov, V. P. 537 (55), 562 (219), 566 (270, 271), 581 (339), 582 (271, 339), 583 (271), *587, 591–593*
Pal, P. R. 446 (1311), *488*
Pala, G. 274 (83), *463*

Paldus, J.   12 (2), *55*
Palla, G.   455 (1383), *490*
Palm, V. A.   766 (57), *779*
Palmer, M. H.   501 (64a), *527,* 826 (17),
    855, 1065, 1078 (5), *1315*
Palmer, P. J.   416 (1024), *483*
Palmieri, P.   1022, 1024 (9), *1053*
Pandell, A. J.   305 (358), *469*
Pandey, P. N.   1161, 1165, 1166, 1168
    (285), *1321*
Pandey, R. C.   335, 339, 361 (578), *473*
Pannier, R.   453 (1365), *489*
Paolucci, G.   1039 (249), *1058*
Papa, D.   274 (82), *463,* 1145, 1148
    (262), *1321*
Papathanasopoulos, N.   507 (97), *528*
Papina, T. S.   60 (13), *65*
Pappalardo, G. C.   1024, 1025 (39), *1053*
Pappas, S. P.   733, 734 (205), *753*
Pappo, R.   282, 429 (192), *465,* 1145,
    1147, 1159 (255), *1321*
Paquer, D.   1037 (143), 1039 (206), 1040
    (255), 1042, 1047 (143), 1049 (255),
    1050 (373), 1052, 1053 (255),
    1056–1058, 1061
Paquette, L. A.   517 (157b), *529*
Paraskewas, S.   329 (519), 450 (1337),
    472, 489
Paredes, M. C.   1314 (721), *1330*
Parekh, M. M.   339 (604), *474*
Parello, J.   89 (148), *116,* 1292 (665,
    666), 1294 (666), 1295 (665, 666),
    1329
Parfitt, G. D.   984 (123), *1019*
Parham, W. E.   416 (1015), *482*
Parish, E. J.   272 (37), *462*
Parish, R. C.   411 (942), *481*
Parker, C. Λ.   198 (100), *211*
Parker, J.   1190, 1191 (356), 1206 (488,
    490, 491), 1221, 1222 (488), *1323,*
    1325, 1326
Parker, J. R.   632 (210), *640*
Parker, K. A.   448 (1324), *489,* 583, 584
    (349, 351), 595 (376), *593, 594, 598*
Parker, R. G.   167, 168 (218), *174*
Parker, W.   1206 (456), *1325*
Parmentier, M.   1044 (315), *1059*
Parrini, V.   1138–1140, 1145 (239), *1320*
Parris, C. L.   454 (1369), *489*
Parrish, R. C.   417 (1046), *483*
Parry, D. R.   376 (809), *478*
Parry, F. H. III   1231, 1232, 1234–1236,
    1239, 1244 (516), *1326*
Parshin, V. A.   596 (377), *598*
Parsons, A. E.   605, 609 (29), *636*
Parsons, G. H.   772 (84), *779*
Parthasarathy, R.   1265, 1267 (606), *1328*
Parthington, P.   252 (429), *265*

Partyka, R. A.   446 (1306), *488*
Pascual, J.   1080 (68–71, 73, 75, 76),
    1082 (68, 76), *1317*
Pashinnik, V. E.   439 (1252, 1254), *487*
Pasternak, R. A.   245 (359), *264*
Pasternak, V. I.   451 (1347), *489*
Patai, S.   69 (8), *113,* 341 (621b, c), *474,*
    826 (19), *855*
Patchornik, A.   273 (53), 274 (87), *463,*
    609 (57), *637*
Pathy, M. S. V.   800 (82), *823*
Patin, H.   826 (14), *855*
Patnaik, D.   246 (378), *264*
Patrick, C. R.   178 (11), *209,* 321, 325
    (472), *471*
Patronik, V. A.   1039 (240), *1058*
Pattenden, G.   415 (981), *482*
Patterson, J. W.   1073, 1074, 1149, 1150
    (45), *1316*
Patterson, L. E.   1100, 1101 (139), *1318*
Pattison, F. L. M.   298 (314), 415 (972),
    468, 482
Patton, J. W.   936 (60), *943*
Patton, W.   419 (1070), *483*
Patumtevapibal, S.   709 (159), *751*
Paukstelis, J. V.   1039 (228), *1057*
Paul, H.   415 (990), *482*
Paul, I. C.   224 (90), 234 (223), *258,*
    261
Paul, L.   1232 (553–555, 584), 1247,
    1250 (553), 1251 (554), 1253 (555),
    1257 (584), *1327*
Paull, K. D.   1078 (54), *1316*
Paulsen, H.   305 (353), *469*
Paulson, P. L.   653 (24), *747*
Pavlichenko, V. F.   806 (113), *823*
Pavlichev, A. F.   920, 932 (34), *943*
Pavlik, J. W.   511 (118b), *529*
Pavlov, S.   414 (969), *481*
Pawelka, Z.   238 (274), *262*
Pawlak, M.   313 (413), 395 (881), *470,*
    480
Pawlak, Z.   223 (77, 78), *258*
Payling, D. W.   159 (173, 174), 164 (173,
    174, 198), *173,* 614 (99, 100), *638,*
    1036 (135), *1055*
Payne, G. B.   364 (724), 400 (897), *476,*
    480
Payne, T. G.   1092 (115), *1318*
Paytin, B. M.   537, 538 (56), *587*
Peace, P. W.   436 (1215), *486*
Peach, M. E.   1035 (127), *1055*
Pearl, I. A.   307 (379, 380), 314 (380,
    425), *469, 470*
Pearlman, W. M.   1195 (380), *1323*
Pearson, D. E.   270 (1), 293 (273), 418
    (1057), 437 (1218), 442 (1281), *462,*
    467, 483, 486, 488

Pearson, D. W. 494 (12), *526*
Pearson, H. 252 (428), *265*
Pearson, J. M. 209 (155), *212*
Pechere, J.-F. 1042 (286), *1059*
Pechet, M. M. 1159 (279), *1321*
Pechmann, H. von 510 (112), 511 (120), *528, 529,* 1098 (124), *1318*
Pedersen, B. 1034, 1040, 1042 (117), *1055*
Pedersen, B. F. 224 (86), *258*
Pedersen, B. S. 1040 (413), *1062*
Pedersen, C. 1037 (142), *1056*
Pedersen, C. T. 1052 (406), *1062*
Pederson, C. J. 272 (25), *462*
Pederson, K. J. 863 (14), *910*
Pederson, R. L. 1145 (269), *1321*
Peel, T. E. 870 (58), *911*
Peerdeman, A. F. 88 (124), *116*
Peitzsch, W. 1073, 1115 (43), *1316*
Pelah, Z. 153 (140), *172,* 613, 614 (95), *638*
Pelinghelli, M. A. 1022 (12), *1053*
Pelizza, F. 184 (67), *210*
Pelletier, S. W. 310, 314 (391), 333 (558), *470, 473,* 1096 (120), *1318*
Pelter, A. 444 (1298), *488,* 566, 569 (267), *592*
Penfold, B. R. 255 (476), *266*
Peng, C. T. 273 (58), *463*
Penturelli, C. D. 236 (248a), *261*
Pentz, C. A. 602 (2), 629 (187), *635, 639*
Peover, M. E. 195 (81–83), *211*
Percheron, F. 1306 (706), *1330*
Perciaccante, V. 84 (90), *115*
Perco, A. 1038 (182), *1056*
Pereira, J. F. 235 (245), *261*
Pereira, W. 419 (1070), *483*
Pereira, W. E. Jr. 88, 89 (137), 109 (281), *116, 119*
Perelman, M. 1224 (493), 1228 (493, 507), *1326*
Perepelkova, T. I. 238 (281), *262*
Peresleni, E. M. 562 (220), *591*
Pereyre, M. 419 (1069), *483*
Perez-Ossorio, R. 61, 62 (29, 30), *66*
Perie, J. J. 433 (1188), *486*
Perkampus, H. H. 71, 72 (17), *114,* 613 (93), *637*
Perkins, M. J. 1206, 1207, 1211, 1212 (463), *1325*
Perkins, R. J. 1145, 1152 (271), *1321*
Perlin, A. S. 365 (738), *477*
Perlstein, J. 420 (1076), *484*
Pernoll, I. 1031 (102), *1055*
Perold, G. W. 300 (328, 329), *468,* 689, 704, 705 (118), *750*
Perotti, A. 762, 763 (42), *778*

Perregard, J. 1034, 1040, 1042 (117), *1055*
Perricaudet, M. 23, 24, 27, 28, 31 (45), 40 (106), *56, 57*
Perry, D. H. 275 (100), *464*
Perry, S. G. 275 (97, 99), *464*
Pershin, G. N. 562, 570, 571 (217), *591*
Persianova, I. V. 562 (220), 566 (273, 275), 581, 582 (339), *591–593*
Person, W. B. 175 (4), 178 (28), *209, 210*
Persson, N. O. 226 (135), *259*
Pesaro, M. 517 (161), *529*
Pete, J.-P. 692, 693 (131), *750*
Peter, A. 369 (759), *477*
Peter, F. A. 765 (56), *779*
Peter, H. 846 (78), *857*
Peter-Katalinic, J. 403 (906), *480*
Petersen, C. S. 244 (344, 345), *263*
Petersen, J. D. 642 (3a), *747*
Petersen, J. W. 386 (842), *479*
Petersen, R. D. 154 (143), *172*
Peterson, D. A. 536, 566, 571 (25), *587*
Peterson, D. L. 73 (39), *114*
Peterson, M. R. 16 (5), 22–24, 26 (37), 27 (56), 49 (123), *55, 56, 58*
Peterson, P. A. 1196 (387), *1323*
Petković, Lj. 761 (32, 36), *778*
Petragani, N. 429 (1157), *485*
Petraitis, J. J. 595 (376), *598*
Petrakovich, V. E. 618 (125), *638*
Petric, A. 570, 571 (294), *592*
Petrov, A. D. 289 (231–233), *466*
Petrovich, J. P. 807 (120), 808 (123–125), *823, 824*
Petrusis, C. T. 672, 679, 680 (67a), *748*
Petrzilka, M. 1180 (337), *1322*
Petterson, R. C. 1165, 1168 (296), *1322*
Pettit, G. R. 1078 (54), 1101 (140), 1102, 1103 (145), *1316, 1318*
Petyunin, G. P. 1206 (454), *1325*
Petyunin, P. A. 1206 (454), *1325*
Petz, W. 566 (245), 598 (396), *591, 599*
Peyerimhoff, S. D. 25 (47), 26 (47, 53), 30 (47), *56*
Pfau, M. 282 (198), *466,* 745 (228a), *753,* 1080, 1084 (84, 85), *1317*
Pfeffer, P. E. 280 (155, 160), 281 (174, 185, 189), 415 (983), *465, 482,* 1037, 1042, 1050 (161), *1056*
Pfeiffer, P. 103 (245), *118,* 176 (10), *209,* 224 (93), *258*
Pfeiffer, R. 88 (133), *116*
Pfeil, E. 517 (157a), *529*
Pfister, G. 847 (81), *857*
Pfleger, R. 1232, 1234–1237, 1240, 1241 (523), *1326*
Pföhler, P. 828, 837, 841, 842 (44), *856*
Phadke, R. 1098 (122), *1318*

Philip, A. 335 (576), 431 (1172), *473, 485*
Philips, K. D. 563 (225), *591*
Phillips, B. 348 (64), 353 (663, 665, 666), *475*
Phillips, D. D. 496 (25b), *527*
Phillips, G. O. 761 (33), 777 (101), *778, 780*
Phillips, W. G. 1044 (317), *1060*
Philpott, M. R. 198 (95), *211*
Phipps, D. A. 1048 (348), *1060*
Piantadosi, C. 507 (95), *528*
Piasek, E. J. 503 (73), 504 (80), *528,* 1109 (165), *1319*
Piatak, D. M. 326, 368, 369 (497), *472*
Piccolo, D. E. 1038 (183), *1056*
Pichat, L. 1115 (190), *1319*
Pichat, J. 426 (1129), *485*
Picker, D. 321, 334 (474), *471*
Pickett, D. J. 800 (77), *822*
Pickholtz, Y. 834, 835, 842 (52), *856*
Pickles, C. K. 183 (58), *210*
Pickworth Glusker, J. 225 (125), *259*
Picot, A. 1303 (696), *1330*
Picot, F. 89 (148), *116*
Piechucki, C. 273 (54), 423 (1095), *463, 484*
Piekarski, S. 500 (55), *527*
Pielipp, L. 596 (390), *599*
Piercy, J. 984 (125), *1019*
Piers, E. 1123 (210), *1320*
Piette, L. H. 28 (68), *56*
Piffori, G. 523 (197), *530*
Pilcher, G. 60–63 (6), 64 (6, 41), 65 (6), *65, 66*
Pilette, Y. P. 182, 196, 204 (57), *210*
Pilgrim, W. R. 674 (78), *749*
Pilotti, Å. 105, 106 (259), *119*
Pimentel, C. C. 43 (110), *57*
Pimentel, G. C. 214 (1b, 2g), 234 (1b), *256*
Pincock, R. E. 292 (260, 261), *467*
Pines, A. 231 (193), *260*
Pines, H. 884 (122), *912*
Pinhey, J. T. 432 (1180, 1181), *486*
Pinkert, H. 1039 (212), *1057*
Pinkerton, J. M. M. 236 (256), *262*
Pinkey, J. 314, 355 (422), *470*
Pinner, A. 1302 (690–692), *1330*
Pintar, M. 231 (200), *260*
Piovera, E. 1194, 1246 (371), *1323*
Piovesana, O. 1027 (52), *1054*
Pirc, V. 562 (206), *590*
Piretti, M. 161 (180), *173*
Pirkle, W. H. 109 (288), *119*
Pirson, D. 235 (243), *261*
Pisanenko, N. P. 273 (65), *463*
Piskala, A. 570, 571 (284), *592*

Piskov, V. B. 1098 (132), *1318*
Piszkiewicz, D. 948, 959 (17), 983 (113, 114), 998 (17), *1016, 1018*
Pittman, C. U. 447 (1319), *488*
Pitts, J. N. 612 (91), *637*
Pitts, J. N. Jr. 642 (5), 643 (6), 651, 655 (16–18), 659, 660 (16), 704 (16, 17), *747*
Pitts, W. 83 (84), *115*
Pitzer, K. S. 23 (38), *56*
Place, B. D. 581 (336), *593*
Plaisance, M. 979 (106), *1018*
Plamondon, J. 435 (1203), *486*
Plat, M. 1306 (705), *1330*
Platt, J. R. 71 (22, 23), 73 (32), *114*
Platthaus, D. 247, 251 (386), *264*
Plekhanov, V. G. 615 (103), *638*
Plieninger, H. 403 (910), *480, 539 (65), 588,* 1204 (447), *1325*
Pobiner, H. 357 (686), *476*
Podurovskaya, O. M. 618 (125), *638*
Poeth, T. 287 (220), *466*
Pogonowski, C. S. 505 (83), *528*
Pogorelyi, V. K. 254 (456–458), *266*
Pohl, H. A. 222 (60), *257*
Poindexter, G. S. 415 (988), *482*
Polanc, S. 562, 571 (205, 211), *590, 591*
Poland, D. 44 (111), *57*
Polczynski, P. 415 (990), *482,* 1232, 1257 (584), *1327*
Poleshchuk, O. Kh. 231 (204), *260*
Polgar, N. 279 (135), 333 (567, 568), *464, 473*
Polievktov, M. K. 537 (52, 53), 596 (380, 386), 597 (386), *587, 599*
Pollak, A. 562 (206), *590*
Polley, J. A. S. 1186 (348), *1323*
Pollini, G. P. 309 (388), *469*
Pollitt, R. J. 676, 685 (98a), *749*
Poloni, M. 444 (1293), *488*
Polonski, T. 90, 94, 95 (149), 102 (203), *116, 117*
Polya, J. G. 603 (7), *636*
Pommer, H. 340 (614), *474*
Pong, R. G. S. 691, 701 (125f, h, i), *750*
Ponomarev, A. A. 447 (1320), *488*
Ponticello, I. S. 1138, 1143 (245), *1320*
Popjak, G. 331 (524), 366 (743), *472, 477,* 1170 (313), *1322*
Pople, J. A. 19 (6, 9, 10), 20 (14–19), 21 (15–19), 22 (23), 23 (44), 24 (18, 44), 25 (14, 16–18), 27, 28 (44), 29 (14–16), 30 (16, 18, 78–80), 31 (17, 44), *55–57,* 250 (409), *265*
Pople, J. S. 239 (295a), *262*
Popov, E. M. 605 (24), *636*
Popov, S. 382 (826), *478*

Porkert, H.   572 (312), 580 (312, 332, 333), *593*
Porro, T. J.   246 (372), *264*
Port, G. N. J.   32 (93, 94), 33 (94), *57*
Porte, A. L.   1079 (59), *1317*
Portella, C.   692, 693 (131), *750*
Porter, C. W.   609 (55), *637*
Porter, G.   672, 678 (60b), *748,* 1052 (388), *1061*
Porter, J. W. G.   274 (70), *463*
Porter, Q. N.   144 (102), 158 (102, 166), 160 (102), *171, 173*
Portier, P.   452 (1361), *489*
Portnoy, C. E.   957, 977 (65), 985 (65, 132), 990 (132), *1017, 1019*
Posner, G. H.   429 (1158), *485,* 1123, 1126, 1127 (211), *1320*
Posposil, J. A.   891 (151), *912*
Post, B.   222 (50), 243 (338), 245 (353), *257, 263, 264*
Posthumus, M. A.   125 (11), *169*
Potapov, V. M.   108 (270), *119*
Potashnik, R.   204 (128), *212*
Potekhin, V. M.   362 (715), *476*
Potgieter, D. J. J.   628 (185), *639*
Potier, P.   1304 (700), *1330*
Potoski, J. R.   1165, 1168 (297), *1322*
Potter, E.   615 (104), *638*
Potter, K.   1029 (81), *1054*
Potter, N. H.   373 (775), *477*
Potterson, A. L.   225 (125), *259*
Pottle, M.   225 (127), *259*
Potts, K. T.   1052 (405), *1062*
Poulton, G. A.   1040–1042 (268), 1052 (268, 388), *1058, 1061*
Poupaert, J.   1038 (165), 1042 (165, 298), *1056, 1059*
Povlain, E.   221 (40), *257*
Powell, C. E.   989, 1006 (142), *1019*
Powell, D. L.   241 (316), *263*
Powell, G.   1098 (130), *1318*
Powell, H. M.   183 (59, 60), *210*
Powell, R. G.   103, 104 (243), *118*
Powell, S. G.   303 (337), *468*
Power, D. M.   777 (101), *780*
Poynton, A. J.   1298, 1299 (673), 1302 (688), *1329, 1330*
Pozemka, M.   323 (481), *471*
Poziomek, E. J.   868 (42), *910*
Prabahakar, S.   1039, 1041–1043, 1050 (194), *1057*
Praefcke, K.   1035, 1039 (132), 1051 (383–385), 1052 (385), *1055, 1061*
Prager, R. H.   1276, 1282 (640), 1284 (648), 1286, 1287 (640); 1288, 1289 (640, 648), 1291 (640), 1295 (648), 1300 (640, 648), 1302 (648), *1329*
Prangova, L. S.   1048 (345), *1060*

Pratt, C. S.   1003 (172), *1020*
Pratt, K. F.   218 (12), *256*
Pratt, R.   1113 (185), *1319*
Pratt, R. N.   1232 (530), *1326*
Pratt, Y. T.   361 (713), *476*
Prausnitz, J. M.   203 (124), *212*
Precht, H.   880 (101), *912*
Preckel, M.   375, 379 (803), *478, 566,* 567 (277), *592*
Prelog, V.   526 (224), *531,* 622 (158), *639*
Premović, P.   758 (13), *778*
Premru, L.   225 (115), *259*
Preobrazhenskii, N. A.   373 (782), *478*
Preston, J.   416 (1008), *482*
Previero, A.   1042 (286, 299), *1059*
Price, A. H.   180 (47), *210,* 238 (282), *262*
Price, C. C.   346 (636), *475*
Price, G. G.   272 (31), *462*
Price, H. C.   71 (18), 109 (18, 290), *114, 119*
Price, M.   979 (104), *1018*
Price, M. J.   418 (1059), *483*
Prichard, W. W.   359 (697), 417 (1032), 442 (1278), *476, 483, 488*
Prileschajew, N.   337 (582), *473*
Prilezhaeva, E. N.   1051 (386), *1061*
Prince, R. H.   850 (87), *857*
Principe, P. A.   297 (304), *468*
Prinzbach, H.   161 (183), *173,* 633 (214), *640,* 663, 668 (52, 53), *748*
Pritchard, H.   865 (26), *910*
Pritchard, J. G.   458 (1409), *490*
Prochorow, J.   199 (101, 102), *211*
Proctor, S. A.   1232 (530), *1326*
Profft, E.   517 (160), *529*
Prokipcak, J. M.   417 (1044), *483*
Prosen, E. J.   61, 62 (25), *66*
Proskurovskaya, I. V.   799 (72), *822*
Proskuryakov, V. A.   362 (715), *476*
Prosvirnova, L. N.   618 (124), *638*
Protiva, M.   1109 (169), *1319*
Protopopova, T. V.   547 (140), 572 (313), *589, 593*
Prout, C. K.   104 (246), *118,* 183 (61, 62), *210*
Proverb, R. J.   908 (234), *914*
Prox, A.   159, 164 (173), *173,* 614 (99), *638,* 1036 (135), *1055*
Prudnikov, A. I.   305 (352), *469*
Pryor, W. A.   324 (488), *471*
Przhiyalgovskaya, N. M.   787 (33), 806 (115), *822, 823*
Puar, M. S.   1300 (679), *1329*
Pucknat, J.   1113 (185), *1319*
Pudova, T. A.   1192 (363), *1323*

Puglisi, V. J. 803 (89), 808 (89, 126), *823, 824*
Pulay, P. 16, 19 (3), *55*
Pullin, J. A. 248 (399), *264*
Pullman, A. 23, 24, 27, 28, 31 (45), 32 (93, 94), 33 (94), 40 (104–106), 42 (107), 44 (107, 112, 113), 45 (112, 113, 116), 46 (116), 47 (119), *56, 57,* 241, 242 (327, 328), 244 (328), 251 (422), 254 (463), *263, 265, 266*
Pullman, B. 26 (54), *56*
Pura, J. L. 585 (363), *594*
Purcell, J. M. 246 (374), *264,* 1034 (121), *1055*
Purcell, K. F. 254 (464), *266*
Purcell, T. A. 272 (38), *462*
Purcell, T. C. 1123 (205, 206), *1320*
Purdum, W. R. 273 (60), *463*
Purello, G. 1023 (23), *1053*
Puskas, I. 162 (187), *173,* 332, 368 (556), *473*
Pustinger, J. V. 605 (32), *636*
Put, J. 163, 164 (192), *173*
Puterbaugh, W. H. 426 (1132, 1133), *485,* 1205 (448, 449), *1325*
Putten, F. H. van 875 (79), *911*
Pyatnoka, Y. B. 278 (130), *464*
Pye, E. L. 253 (434), *265*
Pysh, E. S. 101 (172), *117*

Quadrifoglio, F. 100 (169), *117*
Quan, P. M. 1303 (693), *1330*
Quast, H. 550, 552 (166), *590*
Queen, A. 1024, 1048 (29), *1053*
Quin, L. D. 1303 (693, 695), 1304 (695), *1330*
Quiniou, H. 562 (208, 209), 566 (261), 570, 571 (208, 209), *591, 592,* 1033, 1034 (109), 1042 (109, 284), 1050 (109), *1055, 1059*
Quinkert, G. 369 (750), *477*

Raab, R. 1052 (407), *1062*
Raab, R. E. 222 (61), *257*
Raaen, V. F. 386, 387 (849), *479*
Raap, R. 1041, 1043 (275), 1050 (372), *1059, 1061*
Raasch, M. S. 334 (570), *473,* 1036, 1041, 1043, 1044 (139), *1056*
Rabani, J. 763, 765 (46), 768 (66), *778, 779*
Raber, D. J. 415 (997), *482*
Rabjohn, N. 273 (63), 314 (421), *463, 470*
Raciszewski, Z. 180 (45), 206 (148), *210, 212*
Rackham, D. M. 609 (59), *637*
Radak, B. 758 (13), 761 (36), *778*

Radchenko, N. D. 1048 (352), *1060*
Radeglia, R. 1033, 1034 (106), 1035 (123), *1055*
Radom, L. 19 (10), 20, 21 (15, 18), 23 (44), 24 (18, 44), 25 (18), 27, 28 (44), 29 (15), 30 (18, 78), 31 (44), *55–57,* 239 (295a), *262*
Radscheit, K. 1101 (141), *1318*
Rae, I. R. 274 (77), *463*
Raecke, B. 398 (889, 891), *480,* 916 (4–7), 917, 918 (4), 920 (33, 35), 932 (4), 935 (4, 55), 939 (4, 72), *942–944*
Raffelson, H. 280 (151), *465*
Raghu, S. 1228 (510), *1326*
Ragovska, V. S. 1052 (398), *1062*
Raha, C. 416 (1010), *482*
Rahman, M. B. 88, 93, 94, 107 (143), *116*
Raileanu, D. 1196 (386), *1323*
Rainer, G. 571 (303, 306), *593*
Raistrick, H. 367 (745), *477*
Rajappa, S. 1305 (701), *1330*
Rajasekharan Pillai, V. N. 240 (302), *262*
Rajnvajn, J. 235 (240), 237 (269), *261, 262*
Rakhmankulov, D. L. 348 (642), *475*
Rakoff, H. 147 (112), *171*
Rakoutz, M. 791 (52), *822*
Rakshys, J. W. 241, 250, 251 (319), *263*
Ramachandran, V. 454 (1373), *490*
Ramadas, S. R. 1037 (149), 1042 (296), *1056, 1059*
Ramakaeva, R. F. 787 (31, 33), *822*
Ramakrishnan, V. T. 1052 (404), *1062*
Ramana, D. V. 155 (153), 156 (155), *172*
Ramanzade, Z. M. 328 (507), *472*
Rama Rao, A. V. 451 (1345), *489*
Ramaswami, S. 1011, 1012 (192), *1020*
Rambacher, P. 438 (1235), *487*
Ramdas, P. K. 240 (302), *262*
Ramer, R. M. 1204 (445), *1325*
Ramirez, O. 908 (233), *914*
Ramirez, R. S. 437 (1233), *487*
Rampazzo, L. 812 (136, 137), *824*
Ramsay, B. G. 384 (834), *479*
Ramsay, C. C. R. 144 (102), 158 (102, 166), 160 (102), *171, 173*
Ramsay, G. C. 672, 677, 719 (68), *748*
Ramsay, M. V. J. 1206 (492), *1326*
Ranade, A. C. 294 (276), *467*
Randal, E. W. 247 (381), *264*
Randhawa, H. S. 251 (421), 254 (454, 455, 461), *265, 266,* 1029 (80, 82–84), 1030 (80), 1031 (99), 1033 (82–84), 1034 (83), 1036 (83, 84), *1054, 1055*

Rando, R. R. 709, 719, 722 (155), *751,* 1206, 1221 (486), *1325*

Rango, C. de 1265, 1267 (606), *1328*

Rao, A. S. 110 (299), *119,* 508 (101), *528,* 1072 (32), 1145 (260), *1316, 1321*

Rao, B. D. N. 254 (451), *266*

Rao, C. N. R. 214 (2a, h), 218 (17), 231 (206), 241 (324–326), 251 (421), 254 (451, 455), *256, 260, 263, 265, 266,* 1029, 1030 (80), *1054*

Rao, D. V. 205 (134), *212*

Rao, K. G. 241 (326), *263*

Rao, K. R. K. 240 (313), *263*

Rao, R. N. 417 (1033), *483*

Rao, S. P. 306 (361), *469*

Rao, V. V. 1201 (439, 440, 442), 1202 (439), 1203 (440), *1324, 1325*

Rao, Y. S. 503 (71), *528,* 1065 (4, 12), 1080, 1083 (79), *1315–1317*

Rapala, R. T. 322 (475), *471*

Rapaport, H. 316 (438), *471*

Raphael, R. A. 272 (38), 299 (320), *462, 468,* 1206 (456), *1325*

Rappe, C. 361 (709), *476*

Rappe, G. 394 (877), *480*

Rappoport, H. 536, 583, 584 (27), *587*

Rappoport, Z. 203 (125), *212,* 432 (1177), *486*

Rascher, W. 368 (749), *477*

Rashba, P. M. 182 (55), *210*

Rasmussen, J. K. 522 (192), *530*

Rasmussen, J. R. 839 (63), *856*

Rasmusson, G. H. 1171 (315), *1322*

Raspin, K. A. 850 (87), *857*

Rassat, A. 611 (79), *637*

Rassing, J. 236 (252, 259), 237 (259), 248 (397, 398), *261, 262, 264*

Ratajczak, H. 218 (17), 229 (172a), 231 (202), 251 (421), *256, 260, 265*

Ratchford, W. P. 376 (812), *478*

Ratcliff, M. A. 866 (31), *910*

Ratcliffe, R. W. 1232 (567, 590), 1245 (590), 1257, 1258 (567), *1327, 1328*

Rathke, M. 427 (1141), 434 (1200), *485, 486*

Rathke, M. W. 355 (680), 423 (1107, 1110), 427 (1138, 1139, 1143, 1146–1148), 428 (1149, 1151), 434 (1196, 1199), *475, 484–486,* 1115 (191), *1319*

Rathman, T. L. 376 (806), *478*

Ratner, M. 220, 226, 228 (28), *257*

Ratnikova, T. N. 1185 (342), *1322*

Ratts, K. W. 1044 (317), *1060*

Ratuský, J. 398 (893, 895), *480,* 916 (9, 13, 14, 18, 22–24, 27), 917 (18, 23, 24, 30), 918 (23), 919 (18, 22–24), 920

(18, 23, 32, 36, 37), 921 (18, 24, 30), 923 (38), 924 (24, 36, 37), 926 (41), 927 (14, 18, 23, 24, 42), 928 (13, 14, 23, 24, 43), 929 (14, 18, 30, 45), 932 (47, 48), 933 (47), 934 (47, 51–53), 935 (47, 53, 56), 936 (38, 58, 62, 63), 937 (63), 938 (47, 66), 939 (27, 62, 70, 73, 74), 941 (27, 73), 942 (27), *942–944*

Rausser, R. 340 (612), *474*

Rav-Acha, Ch. 985, 1008 (127a, b), 1009 (127a), *1019*

Ravens, D. A. S. 306, 327 (359), *469*

Ravindranathan, T. 416 (1008), *482,* 1166, 1168 (303), *1322*

Rawlinson, D. J. 432 (1175, 1176), *486*

Ray, A. 951 (29), 953 (45–47), 990 (29), *1016, 1017*

Ray, F. E. 319 (457), *471*

Ray, N. H. 298 (312), *468*

Raymond, G. 102 (195, 196), *117*

Read, J. 887 (136), *912*

Reboul, O. 389 (862), *479*

Rebsdat, S. 536, 538, 543, 545, 548 (28), 549 (146), 554 (28), 557 (28, 181), 558 (28), 576 (181), 577 (146, 331), 578 (181), 581 (334), 597 (146, 181), *587, 589, 590, 593*

Recca, A. 517 (150), *529*

Records, R. 102 (207), 107 (264), *117, 119*

Redlinsky, A. 1039 (247), *1058*

Redwood, A. M. 274 (77), *463*

Reed, R. 88 (133), *116*

Reed, R. I. 129, 130 (30), 158 (169), *170, 173*

Reef, L. J. 493 (6), *526*

Rees, B. 48 (122), *58*

Reese, C. B. 272 (44), 416 (1008), *462, 482,* 652, 653 (23, 25), *747*

Reeve, L. 672, 677 (59a), *748*

Reeve, W. 276 (106), *464*

Reeves, L. W. 222 (62), *257*

Reeves, P. C. 375, 379 (797), *478*

Regan, M. T. 837–840 (57), *856*

Regan, T. R. 834–838, 842 (51), *856*

Reger, D. W. 989 (141, 142), 1006 (141, 142, 179), *1019, 1020*

Regitz, M. 381 (822), 424 (1117), *478, 484,* 550 (155), *589*

Rehberg, C. E. 419 (1067), *483*

Rehn, H. 570, 571 (291), *592*

Reich, H. J. 1123 (214), *1320*

Reichstein, J. 320 (466), 333 (565), *471, 473*

Reichstein, T. 353 (669), 355 (675, 676), *475*

Reid, E. B. 498 (45), *527*

Reid, E. E. 279 (134), *464*
Reid, S. T. 1295 (669), *1329*
Reid, W. K. 129, 130 (30), *170*
Reiffen, M. 597, 598 (395), *599*
Reiffer, S. 281 (181), *465*
Reimer, M. 103 (237), *118*
Reimschuessel, H. K. 1199 (411–413), *1324*
Reinertshofer, J. 517 (156), *529*
Reinhart, J. 805 (103), *823*
Reinheckel, H. 825 (6), *855*
Reinhoudt, D. N. 495 (20), *527*
Reininger, K. 1113, 1114 (184), *1319*
Reinmuth, O. 448 (1322), *489*
Reis, H. 291 (246), *467*
Reisch, J. 1092 (110), *1318*
Reishakhrit, L. S. 618 (122–124), *638*
Reiss-Husson, F. 952 (42), *1017*
Reitsema, R. H. 318 (454), *471*
Reitz, D. B. 1050 (375), *1061*
Relles, H. M. 439 (1242), *487*
Ramanick, A. 1089 (97), *1317*
Rempel, G. L. 457 (1400), *490*
Rempfer, H. 547 (131, 132), 548 (132), 571, 586 (131), *589*
Renes, G. 1022 (2), *1053*
Renfrow, W. B. 272 (41), *462*
Renga, J. H. 1123 (214), *1320*
Renwick, J. D. 88 (118, 119), 108 (119), *116*
Reppe, W. 290 (242), *466,* 496 (25a), *527,* 1123 (208), *1320*
Reppert, R. E. 688, 709, 713 (116), *750*
Resemann, W. 571 (308), *593*
Resink, J. J. 138 (78), *171*
Reuben, J. 231 (196), *260*
Reubke, K. J. 255 (469), *266*
Reuter, F. 366–368 (742), *477*
Reuter, U. 1022 (11), *1053*
Rey-Lafon, M. 608 (51–53), *636*
Reymond, D. 1079 (62), *1317*
Reynaud, P. 1024 (33), *1053*
Reynolds, D. D. 417 (1040), *483*
Reynolds, G. F. 1171 (315), *1322*
Reynolds, R. M. 1039 (214), *1057*
Reyntjens, D. 255 (465), *266*
Rhee, H. K. 989 (145), *1019*
Rhee, J. V. 989 (145), *1019*
Rhoads, S. J. 388 (860), *479*
Rhodes, R. E. 826 (24), *856*
Ribaldone, G. 1295 (670), 1296 (670, 671), *1329*
Rice, F. O. 873 (69), *911*
Richard, R. L. 318 (449), *471*
Richards, F. M. 962, 998 (71), *1017*
Richards, R. W. 1284 (653), *1329*
Richardson, D. 709, 715 (154), *751*

Richardson, F. S. 83 (83–85), 84 (85), 92 (153), 95 (153, 159), *115, 116*
Richardson, G. 439 (1241), *487*
Richardson, W. H. 358 (688), *476*
Richer, J.-C. 1038 (189), *1057*
Richman, J. E. 427 (1140), *485, 552* (171), *590*
Richter, B. 615 (105), *638*
Richter, M. 303 (342), *469*
Richter, R. 562 (202), *590*
Richter, W. 312 (408), 452 (1353), *470, 489,* 536 (33), *587*
Richter, W. J. 148 (115, 117), 150 (125), 154 (144), *172*
Richter, W. von 915 (1), *942*
Rickborn, B. 422 (1103), *484*
Rickert, H. 512 (126), *529*
Rideal, E. K. 955 (57), *1017*
Ridley, D. D. 1272, 1273 (615, 616), *1328*
Rieche, A. 339 (607), *474,* 1311 (714), *1330*
Ried, W. 409 (922), *480,* 503 (75), *528,* 1131, 1136 (225), *1320*
Riedel, O. 916, 917, 928 (15), *942*
Rieder, W. 369 (755), *477*
Riedmüller, S. 1034, 1042 (116), *1055*
Riegel, B. (529), *472*
Rieger, P. H. 814 (147), *824*
Riegl, J. 1229 (511), *1326*
Riehl, J. J. 317 (440), 403 (905), *471, 480*
Rieke, R. 423 (1112), *484*
Rieke, R. D. 728 (193a), *752*
Riese, H. 569 (281), *592*
Rievescke, G. Jr. 319 (457), *471*
Rif, I. I. 362 (715), *476*
Rifi, M. R. 782, 818 (5b), *821*
Rigaudy, J. 734 (207), *753*
Rimpler, M. 99 (162), *117*
Rincon, M. 895 (183), *913*
Ringer, B. J. 61, 62 (26), *66*
Rio, G. 525 (217), *531*
Rioult, P. 1042 (290), *1059*
Riphagen, B. G. 415 (973), *482*
Ripperger, H. 76 (48), 102 (189), *114, 117*
Rist, N. 1040 (270), *1058*
Ritchie, E. 1304, 1306, 1309 (699), *1330*
Ritchie, P. D. 866 (33), 873 (68), 874 (75, 76), 875 (82), 876 (75, 85), *910, 911*
Ritter, J. J. 424 (1113), *484*
Riva, A. 674 (79), *749*
Rivett, D. E. A. 1039 (204), *1057*
Riviere, M. 1292 (665, 666), 1294 (666), 1295 (665, 666), *1329*
Rivilis, F. Sh. 598 (398), *599*

Rizzo, V. 109 (280), *119*
Rø, G. 245 (356, 365), *264*
Robb, M. A. 12 (2), 28 (70), 29, 30 (70, 73), *55–57*
Robba, M. 571 (302), *593*
Robbins, M. D. 274 (78), *463*
Roberts, J. D. 179 (35), *210*, 252 (428), 253 (436), *265*, 610 (65), *637*
Roberts, J. L. 494 (14), *526*
Roberts, R. M. 392 (867–869), 423 (1094), *479, 484*
Roberts, S. M. 508 (100), *528*
Roberts, W. 272 (32), *462*
Robertson, E. 1029 (81), *1054*
Robertson, G. R. 299 (324), *468*
Robertson, I. M. 231 (195), *260*
Robertson, J. M. 193 (74), *211*
Robertson, P. M. 787 (34), *822*
Robey, R. L. 343, 344 (627, 628), *474*
Robin, M. B. 20, 26 (12), 29, 30 (12, 71), *55, 56*, 70, 72, 101 (15), *114*, 229 (167), *260*
Robin, R. 816 (155), *824*
Robinson, C. 777 (101), *780*
Robinson, C. H. 310 (396), 422, 423 (1102), *470, 484*
Robinson, L. 962 (73a), 963 (81), 964, 965 (73a), 972 (96, 97), 984 (115, 116), 1008, 1010 (184), *1017, 1018, 1020*
Robinson, R. 279 (135), 353 (658), *464, 475*, 521 (183), *530*
Robinson, W. T. 189 (70, 71), 190 (72), *210, 211*
Robson, J. H. 406 (912), *480*
Robson, R. 208 (152), *212*
Rochling, H. 406 (914), *480*
Roda, G. 674 (79), *749*
Rode, B. M. 232 (213), 233 (216), 235, 236 (213), *261*
Rodewald, P. G. 501 (61), *527*
Rodger, M. N. 883 (118), *912*
Rodgers, A. S. 60, 61, 63, 65 (8), *65*
Rodionova, N. A. 1052 (398), *1062*
Rodionow, W. M. 805 (104), *823*
Rodrigo, R. 1166, 1168, 1187 (307), 1309 (710), *1322, 1330*
Rodrigues, R. 429 (1157), *485*
Rodriguez, H. R. 315 (431), *470*
Rodriguez, O. 497 (34), *527*, 620 (148), *639*, 655, 661, 691, 700 (29), *747*, 1192 (359), *1323*
Rodulfo, T. 953 (51), *1017*
Rodygin, A. S. 1047 (337), *1060*
Roebke, H. 355 (682), 375, 379 (799), *476, 478*
Roebuck, D. S. P. 328 (511), *472*
Roehr, J. 542 (103), *588*

Roelofsen, D. P. 419 (1071), *483*
Roelofsen, G. 222 (56, 57), 234 (224), *257, 261*
Roeske, R. 414 (968), *481*
Roff, G. A. 176, 177 (15), *209*
Rogasch, P. E. 255 (473), *266*
Roger, R. 386, 387 (850), *479*
Rogers, J. B. 440 (1256), *487*
Rogers, M. T. 31 (83), *57*
Rogers, T. G. 460, 461 (1416), *490*
Rogers-Low, B. W. 524 (209), *530*
Rogić, M. M. 434 (1197–1200, 1202), *486*
Rogit, H. M. 434 (1196), *486*
Rohrs, E. J. 415 (976), *482*
Rohwedder, W. K. 147 (114), *172*
Rojahn, W. 332 (552), *473*
Rol, N. C. 126 (12), *169*
Rollefson, G. K. 687, 688 (105a), *750*
Roller, R. S. 652, 657 (33), *747*
Rolston, J. H. 292 (261), *467*
Roman, S. R. 422 (1099), *484*
Romanin, A. 814, 816, 817 (142), *824*
Romanowska, K. 251 (421), *265*
Romsted, L. R. 950, 965, 971, 972, 974, 977–979 (22), *1016*
Romsted, L. S. 947 (6, 15), 951, 956 (6), 969 (6, 15), 970, 971, 973, 975 (6), 976 (6, 15, 100), 977 (6, 15), 978–980 (6), 981 (111), *1016, 1018*
Ronchi, A. U. 281 (191), *465*
Ronchi, A. W. 281 (186), *465*
Rondestvedt, C. S. Jr. 441 (1267–1269), *487*
Roof, A. A. M. 664, 666, 708 (57a–c), 738 (57a–c, 217), 739, 743, 746 (217), *748, 753*
Root, W. G. 1192 (361), *1323*
Roothaan, C. C. J. 6 (1), *55*
Ropp, G. A. 132, 133 (41), *170*
Ros, P. 23, 26, 39 (41), *56*
Rosado, O. 458 (1407), *490*
Rose, J. G. 447 (1317), *488*
Rose, P. I. 236 (251), *261*
Rosen, P. 395 (880), *480*
Rosenberg, A. 109 (291), *119*
Rosenberg, H. M. 177 (23), 196, 198, 199 (79), *209, 211*
Rosenberg, H. R. 320 (466), *471*
Rosenberger, M. 1110 (173, 174), *1319*
Rosenblatt, D. H. 452 (1362), *489*
Rosenblum, L. D. 1047 (336), *1060*
Rosenblum, M. 690, 701, 702 (121a, b), *750*, 839 (62), *856*, 1228 (509, 510), *1326*
Rosenfeld, D. D. 328 (514, 516), *472*
Rosenfelder, W. J. 882 (111), *912*
Rosenfield, J. S. 99 (164), *117*

Rosenman, H.   828, 832, 833 (43), *856*
Rosenmund, P.   1276, 1282, 1287 (641), *1329*
Rosenthal, D.   355 (679), *475*
Rosini, G.   444 (1293), 449 (1331), *488, 489*
Rosmus, P.   1024 (35), 1031 (96), 1032 (35, 96), *1053, 1055*
Rosowsky, A.   507 (97), *528*
Ross, R. A.   868 (40), *910*
Ross, S. D.   782 (8c), *821*
Ross, W. A.   297 (300), *468*
Rosser, C. A.   61 (20), *66*
Rossi, J.-C.   400 (899), *480*
Rossi, R.   291 (245), *467*
Rosso, P. D.   731, 732 (200b), *752*
Rossotti, F. J. C.   104 (246), *118*, 225 (128), *259*
Rostock, K.   422 (1104), *484*
Roth, R.   691, 701, 702 (125b), *750*
Rothe, J.   506 (88), *528*, 1123 (207), *1320*
Rothenberg, S.   23, 25 (43), 29, 45−47 (72), *56*, 242, 250−252 (331), *263*
Rotherstein, B.   500 (58), *527*
Rothfield, M.   1232, 1242 (539), *1326*
Rothman, E. S.   270 (7), 420 (1076), 441 (1266), *462, 484, 487*, 1145, 1151, 1160 (268), *1321*
Rothstein, E.   826 (17), *855*
Rotschild, W. G.   233 (219), 243 (337a), *261, 263*
Rouault, G. F.   333, 334, 339 (569), *473*
Roux, M. V.   60 (15, 16), 61 (29, 31), 62 (16, 29), 63 (15), *65, 66*
Rowe, J. E.   275 (96), *464*
Roy, A. K.   759, 760 (26), *778*
Roy, S. K.   89 (147), 95 (158), 109 (147), *116*, 1201 (440, 441), 1203 (440), *1325*
Royal, J. K.   687, 688 (105a), *750*
Royals, E. E.   884 (127), *912*
Royer, G. P.   989 (147), *1019*
Royer, R.   361 (714), *476*
Royston, G. C.   334 (573), *473*
Rozenberg, M. Sh.   246 (371), *264*
Rozenberger, H. J.   230 (183), *260*
Rubin, R. J.   976 (99, 100), *1018*
Rubottom, G. M.   337 (581), *473*, 1066, 1069, 1149 (22), *1316*
Ruch, J. E.   629 (189), *639*
Rucktäschel, R.   497 (34), *527*, 620 (148), *639*, 655, 661, 691, 700 (29), *747*, 1192 (359), *1323*
Rudd, E. J.   782 (8c), *821*
Rüdiger, W.   164 (193), *173*

Rudinger, J.   101 (178), *117*, 1193 (365), *1323*
Rudloff, E. von   332 (531−537, 547−549, 555), *472, 473*
Rudnick, L. R.   1075 (51), *1316*
Rudolph, W.   1101 (141), *1318*
Rufer, C.   596, 597 (389), *599*
Ruggieri, P. de   571 (310, 311), *593*
Ruhemann, S.   510 (114), *528*
Ruhoff, J. R.   279 (134), 311 (397), *464, 470*
Rullkötter, J.   1036, 1040, 1042 (137), *1055*
Rumin, R.   663 (54), *748*
Rummens, F. H. A.   879 (98), *911*
Rummert, G.   340 (614), *474*
Rundle, R. E.   222 (59), *257*
Runge, W.   108 (273, 274), *119*, 148 (117), *172*
Rungwerth, D.   1039, 1052 (196), *1057*
Rupp, R.   84 (90), *115*
Ruppert, J. F.   423 (1111), *484*
Rusche, J.   297 (305), *468*, 551 (164), *590*
Rusoff, I.   71 (23), *114*
Rusoff, I. I.   71 (22), *114*
Russell, A.   1098, 1099 (126), *1318*
Russell, D. H.   1023, 1035, 1052 (21), *1053*
Russell, D. W.   676, 685 (97, 98c), *749*
Russell, G. A.   328 (512, 513), (764), *472, 477*, 803 (94), *823*, 1039 (245), *1058*
Russell, R. R.   1104 (148), *1319*
Russo, G.   1189 (354), *1323*
Rutenberg, M. W.   1204 (446), *1325*
Ruter, J.   381 (822), *478*
Rutherford, K. C.   894 (173), *913*
Rutherford, K. G.   417 (1044), *483*
Rutledge, P. S.   494 (14), *526*
Rutledge, T. F.   433 (1187), *486*
Rütz, W.   559−561 (191), *590*
Ruzicka, L.   331 (523), *472*, 526 (224), *531*
Ruzo, L. O,   656, 699 (31), *747*
Ryan, J. A.   32 (88), *57*
Ryan, J. J.   1231 (519), 1232 (519, 578, 581), 1233, 1246 (519), 1255 (578, 581), 1263 (519), *1326, 1327*
Ryan, M. D.   808 (128), *824*
Ryang, M.   455 (1380), *490*
Rybakova, M. N.   1047 (337), *1060*
Rydon, H. N.   1110 (175), 1170 (312), *1319, 1322*
Rye, R. T. B.   870 (58), *911*
Ryhage, R.   133 (49), 134 (49, 55), *170*

Rylander, P. N.   304 (347), *469,* 499
(53), *527,* 825 (7), *855,* 1171 (314),
1307 (707), *1322, 1330*
Rytslin, E. E.   1202 (430), *1324*
Ryvolova, A.   816 (156), *824*
Ryvolova-Kejharova, A.   816, 817 (157),
824

Sabbah, R.   62 (33), 63 (37), *66*
Sabin, J. R.   219 (23), *257*
Sabine, T. M.   224 (96, 97), *258*
Sabol, M. A.   815 (149), *824*
Sacco, A.   826 (23), 827 (33), *856*
Sackur, O.   394 (874), *479*
Saegusa, T.   307 (374), 372 (768), 415
(991, 993), 457 (1399), *469, 477, 482,
490*
Saenger, W.   440 (1256), *487*
Safarova, Z. A.   1090 (101), *1318*
Saffhill, R.   272 (44), *462*
Safiullin, G. S.   439 (1245), *487*
Said, A.   419 (1068), *483*
Saigo, K.   413, 444 (958), *481*
Saikuchi, H.   1039 (244), *1058*
Sainsbury, M.   1305, 1312 (702), *1330*
Saito, H.   248 (389), *264*
Sakabe, N.   416 (1021), *483*
Sakai, K.   336 (580), *473,* 850, 851 (92,
93), *857*
Sakai, S.   551 (160, 162), *590*
Sakakibara, T.   290 (237), 296 (296),
466, 468
Sakamoto, K.   916, 920, 924 (19, 20),
928 (19), *942*
Sakan, T.   1079 (61), *1317*
Sakata, S.   1047 (335), *1060*
Saki, S.   551 (161), *590*
Sakota, N.   101 (181), 103 (220), *117,
118*
Saksena, B. D.   71 (19), *114*
Sakumoto, T.   1012 (196), *1020*
Sakuragi, H.   745 (232), *753*
Sakurai, B.   805 (105), 816, 817 (152),
820 (170), *823, 824*
Sakurai, H.   273 (55), *463,* 565, 566
(240), *591*
Salame, L. W. F.   417 (1039), *483*
Salamone, J. C.   957 (63), *1017*
Salbaum, H.   295 (287), *467*
Salem, L.   729 (195a, b), 737 (195b),
752, 837 (60), *856*
Salmon-Legagneur, F.   270 (10), *462*
Salomone, R. A.   168 (223), *174,* 617
(112), *638*
Salvadori, P.   69, 73 (6), 99 (163), *113,
117,* 495 (17), *526*
Salzmann, T. N.   428 (1153), *485*
Sam, D. J.   321, 334 (473), *471*

Samatuga, G. A.   1065 (8), *1315*
Samek, Z.   621 (150), *639*
Sammes, P. G.   275 (92), *463,* 524 (208),
530, 602, 603 (1), *635*
Sample, S.   156 (156), *172*
Samson, C.   22 (35), *56*
Sanabia, J. A. de   649 (14), *747*
Sandberg, R.   417 (1050), *483,* 1111
(176), *1319*
Sandborn, L. T.   358 (694), *476*
Sandel, V. R.   651, 680, 688, 694, 695,
699, 700, 735 (15), *747*
Sandeman, I.   71, 72 (17), *114*
Sanders, W. J.   1011, 1012 (192), *1020*
Sandhu, J. S.   458 (1404), *490*
Sandler, S. R.   270 (2), 415 (971), 416
(1013), 437 (1217), 442 (1282), 458
(1408), *462, 482, 486, 488, 490*
Sandmeier, R.   1167, 1168 (308), *1322*
Sandner, M. R.   649 (12), *747*
Sandorfy, C.   214 (1g), 254 (453), *256,
266,* 1029 (87), *1054*
Sandoval, O.   61 (23), *66*
Sandström, J.   108 (269), *119,* 255
(466), *266*
Sane, P. P.   508 (101), *528,* 1145 (260),
1321
Sankaran, D. K.   272 (23), *462*
Sanno, Y.   309 (387), *469*
Sano, H.   291 (249), 292 (259), 293
(268), *467,* 1174 (334), *1322*
Sano, S.   336 (580), *473*
Sano, T.   235, 236 (241), 237 (260), *261,
262*
Santis, V. de   895 (183), *913*
Santry, D. P.   232 (215), *261*
Sanyal, B.   1161, 1162, 1168 (286), *1321*
Saquet, M.   1033, 1034, 1037 (107), 1047
(339–341), *1055, 1060*
Saraf, S. D.   438 (1234), 439 (1249),
487
Sarel, S.   274 (81), 300 (326), 372 (772),
463, 468, 477, 614 (101), *638,* 950
(24), 985 (127a, b), 1008 (127a, b,
183), 1009 (127a), *1016, 1019, 1020,*
1090 (103, 104), 1092 (104, 106), *1318*
Sarett, L. H.   353 (670), *475*
Sargeson, A. M.   103 (212), *118*
Sarkisyan, O. A.   1075 (48), *1316*
Sarma, M. R.   104 (248, 249), *118*
Särnstrand, C.   111 (312, 313), *120*
Sasada, Y.   245 (364), *264*
Sasaki, S.   1039 (203), *1057*
Sasaki, T.   513 (130), *529*
Sasse, H. J.   535–537, 554, 559, 572
(10), *586*
Sasson, Y.   1036 (138), 1042 (289), *1055,
1059*

Sastry, K. S. 797 (66), *822*
Satchell, D. P. N. 1048 (345, 349, 357), *1060, 1061*
Sato, M. 513 (129), *529*
Sato, S. 769 (76), *779*
Sato, T. 273 (57), *463,* 504 (82), *528,* 902 (209), *914*
Sato, Y. 674 (80), 720 (185b–d, h), 722 (80, 185b–d, h), 724 (185b, d, h), *749, 752*
Satoh, Y. 571 (301), *592*
Saucy, G. 1110 (172–174), *1319*
Sauer, D. 1276, 1282, 1287 (641), *1329*
Sauer, J. C. 419 (1066), *483*
Sauers, R. R. 514 (136), *529,* 626 (171), *639,* 1162 (290, 291), 1163 (291), 1168 (290, 291), 1185 (340), 1187 (290, 291), *1321, 1322*
Saumagne, P. 249, 254 (400), *265*
Saun, W. A. van Jr. 1039, 1052, 1053 (200), *1057*
Saunders, K. J. 147 (108), *171*
Saunders, R. A. 133, 134, 137, 138 (48), *170*
Saunders, W. H. 871 (63), *911*
Saunders, W. H. Jr. 711, 716, 744 (168a, b), *751*
Saur, H. 566, 567, 571 (249), *591*
Saus, A. 369 (754), *477*
Sauter, R. 571 (307, 309), *593*
Sauve, D. M. 294 (281), *467*
Sauvetre, R. 389 (862), *479*
Savelli, G. 981 (111), 992 (157), *1018, 1019*
Savelyev, V. A. 220 (26), *257*
Savignac, P. 414 (1001), *482*
Savtome, K. 301 (331), *468*
Sawada, S. 583 (346), *593*
Sawant, B. M. 104 (250), *118*
Saxby, M. J. 158 (170, 171), *173*
Sayrac, T. 743 (224b), *753*
Saytzeff, A. 493 (2), *526*
Sbrana, G. 228 (155, 156), *259*
Scala, A. A. 711, 713, 714 (171), *751*
Scanlon, B. 304, 315 (346), *469*
Scarpa, I. S. 957 (62), *1017*
Scarpati, R. 544 (118, 119), 595 (375), *589, 598*
Scatchard, G. 179 (33), *210*
Schaad, L. J. 214 (1f), 218 (11), 234 (1f), *256*
Schaaf, T. K. 278 (129), *464,* 1168 (302), *1322*
Schaafsma, K. 672, 674, 678, 682 (82), *749*
Schaak, J. 984 (116), *1018*
Schach von Wittenau, M. 888, 905 (140), *912*

Schade, G. 887 (137), *912*
Schaden, G. 627 (179), *639*
Schaeffer, J. J. 431 (1168), *485*
Schaeffer, J. R. 306 (364), *469*
Schäfer, H. 787 (35), *822*
Schäfer, U. 512 (123), *529*
Schäfer, W. 343 (624, 625), 369 (753, 754), *474, 477*
Schäffer, W. 614 (102), *638*
Schaffner, K. 706 (146d), 728 (194), 730, 731, 733 (146d), *751, 752*
Schank, K. 272 (46), *463*
Schanzer, W. 672, 677, 678 (60a), *748*
Scharf, G. 664, 668 (56a, b), 669, 671 (56b), 708 (56a, b), 730, 731 (56b), *748*
Scharp, J. 277 (123), *464*
Scharrer, R. P. F. 872 (64a), *911*
Schaub, F. 430 (1160), *485*
Schauble, J. H. 1039, 1052, 1053 (200), *1057*
Schaumann, E. 255 (468, 469), *266,* 1022, 1030, 1038 (15), *1053*
Scheben, J. A. 294 (282), *467*
Schechter, H. 888, 905 (140), *912*
Scheckenbach, F. 1232 (541, 544), 1242, 1243 (541), *1326, 1327*
Scheel, D. 566 (248), *591*
Scheele, W. 248 (393), *264*
Scheer, J. C. 884, 886, 891, 897, 898 (126), *912*
Scheeren, J. W. 540, 547, 548, 554, 578, 586 (81), *588,* 875 (84), *911,* 1040, 1042 (258), *1058*
Scheffer, J. R. 710, 718 (160), 728 (193a), *751, 752*
Scheibye, S. 1040 (413), *1062*
Scheinbaum, M. L. 1233 (600), *1328*
Scheiner, S. 47 (120), *57*
Scheithauer, S. 1024 (35), 1025, 1026 (48), 1027 (48, 64), 1028 (48), 1031 (96), 1032 (35, 96), 1033, 1034 (106), 1035 (123), 1040 (266), 1042 (48, 64, 266), 1043 (64), 1046 (64, 331), 1048 (353), 1053 (266), *1053–1055, 1058, 1060, 1061*
Schelechow, N. 1301 (681), *1330*
Schellenberger, A. 239 (294, 300), 254 (446), *262, 265*
Schellman, J. 101 (182), *117*
Schellman, J. A. 69 (7), 71 (20), 76 (51), 77 (52–54), 83, 99 (51), 101 (52), 102 (20), 109, 110 (292), 113 (51), *113, 114, 119,* 612 (84, 92), 613, 632 (92), *637*
Schenck, G. O. 206 (142), *212,* 1080, 1084 (83), 1171, 1172 (317, 319, 320), 1232, 1264 (595), *1317, 1322, 1328*

Schenck, H. U.　545, 548 (126), *589*
Scheraga, H. A.　44 (111), *57,* 100 (170), *117,* 225 (127), 232 (214), 243 (214, 335), 247 (335), *259, 261, 263*
Scherrer, J. R.　228 (148), *259*
Schibeci, R. A.　1040, 1046 (269), 1048 (359), *1058, 1061*
Schied, D.　574 (320), *593*
Schiffman, R.　985, 1008 (127a, b), 1009 (127a), *1019*
Schiller, J. C.　195 (77), *211*
Schilling, P.　1038 (172), *1056*
Schilling, W.　515 (141), *529,* 1039 (210), 1048 (210, 350), *1057, 1060*
Schindel, W. G.　292 (261), *467*
Schindlbauer, H.　449 (1330), *489*
Schinke, E.　1031, 1032 (96), *1055*
Schinz, H.　888 (138), *912*
Schirp, H.　916 (6), 920 (35), 935 (55), 939 (72), *942–944*
Schisla, R. M.　429 (1159), *485*
Schissel, P. O.　661, 662 (44b), *748*
Schittenhelm, D.　306 (362), *469*
Schiys, H.　398 (891), *480*
Schlack, P.　518 (165), *530,* 536 (33), *587*
Schlaf, T. F.　1038 (186), *1057*
Schlapkohl, K.　1034, 1035 (115, 118), 1039, 1040 (115), 1041 (118), 1042 (115, 118), 1043 (118), 1045 (115, 118), 1046, 1048 (115), *1055*
Schlenk, H.　394 (876), *479*
Schlessinger, R. H.　427 (1140, 1142), *485,* 501 (65), *527,* 1123, 1126, 1127 (212), 1138, 1143 (245), *1320*
Schleyer, P. v. R.　61 (21), *66,* 241, 250, 251 (319), *263*
Schlittler, E.　415 (989), *482*
Schlitze, G. R.　610, 621 (67), *637*
Schlueng, R. W.　439 (1242), *487*
Schmeltz, I.　415 (983), *482*
Schmetzer, J.　566 (244), *591*
Schmid, G.　1043 (307), *1059*
Schmid, H.　392 (867), 403 (906), *479, 480,* 706 (147), *751,* 1305 (701), *1330*
Schmid, H. J.　298 (313), *468*
Schmid, W.　364 (723), *476*
Schmidt, E.　1194, 1225 (370), *1323*
Schmidt, E. A.　541 (89), 584 (356–359), *588, 594*
Schmidt, G. M.　88 (125), *116*
Schmidt, G. M. J.　244 (347), 245 (358), *263, 264*
Schmidt, P.　1196 (384), *1323*
Schmidt, R. F.　874 (77), *911*
Schmidt, S. P.　738 (216), *753*
Schmidt, T.　369 (755), *477*
Schmidt, U.　1038 (179, 182), 1039

(237), 1040 (263), 1048 (237), 1052 (390), 1053 (263, 410), *1056, 1058, 1061, 1062,* 1113, 1114 (184), *1319*
Schmidtbauer, E.　1038, 1052 (184), *1057*
Schmiegel, J. L.　108, 109 (268), *119,* 280 (168), *465*
Schmiegel, K. K.　901 (205), *914*
Schmillen, A.　198 (98, 99), *211*
Schmitt, E.　550, 552 (166), *590*
Schmitt, G. J.　1199 (413), *1324*
Schmitz, E.　1295 (668), *1329*
Schmitz, W. R.　1195 (381), *1323*
Schmüser, W.　1034, 1035, 1039, 1040, 1042, 1045, 1046, 1048 (115), *1055*
Schnakenberg, G. H. F.　493 (6), *526*
Schneider, H.-J.　902 (207), *914*
Schneider, R. S.　583 (344), *593*
Schneider, W. G.　28 (68), *56,* 222 (62), *257*
Schneider, W. P.　277 (119), *464*
Schneider-Berlöhr, H.　902 (207), *914*
Schniepp, L. E.　1145, 1151 (266), *1321*
Schnoes, H. K.　507 (92), *528,* 627 (177), *639*
Schoch, S.　164 (193), *173*
Schoemaker, H. E.　525 (221), *531*
Scholler, D.　692, 693 (131), *750*
Schöllkopf, U.　287 (224, 225), 413 (964), 457 (1395, 1396), *466, 481, 490,* 536 (23), *587*
Schomaker, V.　22 (29), *56*
Schön, N.　535–538, 542–544, 550, 551, 554, 557, 559, 562 (11), 565 (242), 566 (11, 242), 568, 572, 573, 581, 583, 586 (11), *586, 591*
Schönberg, A.　292 (255), 386 (847), 456 (1385), *467, 479, 490,* 686 (103), *749*
Schöneshöfer, M.　774, 776 (93), *779*
Schoone, J. C.　88 (124), *116*
Schöpf, C.　367 (744), *477*
Schors, A.　305 (356), *469*
Schosser, H. P.　571 (303), *593*
Schössler, W.　550 (155), *589*
Schrader, B.　226 (143), *259*
Schrader, H.　916 (3), *942*
Schramm, S.　1295 (668), *1329*
Schrauzer, G. N.　825 (9), *855*
Schreiber, J.　272 (26), 413 (963), *462, 481,* 536 (26), 543, 557, 561 (26, 109), 566, 584 (26), *587, 589*
Schreiber, K.　76 (48), *114*
Schreiber, W. L.　725 (191b), *752*
Schrier, E. E.　225 (127), *259*
Schriesheim, A.　328 (514–516), 357 (686), *472, 476*
Schroder, E.　450 (1342), *489*

Schröder, R. 220 (35), *257,* 287 (224, 225), 413 (964), *466, 481*

Schrodt, H. 535–537, 554, 559, 572 (10), *586*

Schroeter, S. H. 1171, 1172 (319), 1191 (358), *1322, 1323*

Schroll, G. 144 (100), 146 (106), 147 (111), 148 (106), 149 (121), 150 (122), *171, 172,* 1035, 1036 (130, 131), 1053 (412), *1055, 1062*

Schryver, F. C. de 163, 164 (192), *173,* 511 (121a), *529*

Schuber, E. V. 1098 (135), *1318*

Schubert, B. 452 (1353), *489*

Schubert, W. M. 826 (18), *855*

Schuh, H. G. v. 547, 548, 571, 586 (133), *589*

Schuijl, P. J. W. 1024 (28), 1043 (301, 303), *1053, 1059*

Schuijl-Laros, D. 1043 (302), *1059*

Schuizer, A. W. 871 (61), *911*

Schulenberg, J. W. 452 (1359), *489*

Schuler, M. A. 774–776 (88), *779*

Schuler, R. H. 763 (44, 52), 764 (52), 774 (88–90), 775, 776 (88, 95, 96), *778, 779*

Schulte, K. W. 1092 (109), *1318*

Schulte-Frohlinde, D. (85), *779*

Schultze, H. 1311 (714), *1330*

Schulz, J. G. D. 328 (502), *472*

Schulze, K. E. 1092 (110), *1318*

Schulze, U. 1035, 1039 (132), *1055*

Schulze-Steinen, H.-J. 1174, 1178, 1179 (331), *1322*

Schunk, E. 358 (691), *476*

Schuster, D. I. 679, 697 (94), *749*

Schuster, G. B. 738 (216), *753*

Schuster, H. 200, 201 (108), 202 (108, 114, 115), *211*

Schuster, P. 218 (20), 231 (207), 233 (207, 216), 238 (289), *257, 260–262*

Schute-Elte, K.-H. 1171, 1172 (317), *1322*

Schütt, H. 939 (72), *944*

Schuzer, R. 272 (42), *462*

Schwab, G. M. 869 (49), *910*

Schwab, P. A. 338 (584), *473*

Schwang, H. 82, 85, 87 (73), *115*

Schwartz, A. L. 461 (1419), *490*

Schwartz, E. 1194 (368), *1323*

Schwartz, J. 839 (65), *856*

Schwartz, M. E. 23, 25 (43), *56*

Schwartz, R. N. 758 (11), 761 (37), *778*

Schwartzman, S. M. 412, 444 (950), *481*

Schwarz, H. 144 (98), 147 (113), 154 (143), 156 (159), 157 (113, 162, 163), 158 (165), *171–173,* 615 (105), *638,* 1035, 1039 (132), *1055*

Schwarz, K. 596, 597 (389), *599*

Schwedova, I. B. 436 (1212), *486*

Schweizer, D. 536, 538, 554, 556, 562, 566 (13), 577 (330), *586, 593*

Schweizer, E. 571 (305), *593*

Schweizer, E. E. 883 (120), *912*

Schwenk, E. 274 (82), *463,* 1145, 1148 (262), *1321*

Schwetlick, H. 1039, 1052 (196), *1057*

Schwindt, J. 1039 (202), *1057*

Schwörer, F. 766–768 (62), *779*

Scoffham, K. 789, 794 (40), *822*

Scola, D. A. 418 (1060), *483*

Scopes, P. M. 71 (21), 73 (42), 75 (42–45), 76, 84 (21), 88 (109, 110, 116–119, 138, 143), 91 (138), 93, 94 (143), 95 (21, 157), 101 (177), 104 (247), 107 (143), 108 (116, 117, 119, 271), 109 (138, 277), *114, 116–119,* 612 (82, 83, 87, 89), 621 (152), 622 (155), 623 (162, 163), 624 (155, 163, 164), 626 (170), *637, 639*

Scott, A. I. 69, 71 (10), 87 (105), *113, 115,* 507 (91), *528*

Scott, R. L. 203 (124), *212*

Scott, W. M. 127, 143 (20), *170*

Scrimgeour, S. N. 179, 182, 192 (39), *210*

Scrocco, E. 28 (64), 40 (105), 45, 46 (116), 48 (64), *56, 57,* 251 (422), *265*

Scroggins, M. W. 603 (15), *636*

Seagebarth, E. 24 (46), *56*

Sealy, R. C. 673, 683, 685 (85b), *749*

Seamans, R. E. 440 (1263), *487*

Searby, G. M. 234 (229), *261*

Searles, S. 620, 623 (147), *639, 875* (80a), *911*

Sears, B. 979 (104), *1018*

Šebenda, J. 518 (164), *530,* 868 (45), *910,* 1196 (384, 385), *1323*

Seconi, G. 1024 (38), *1053*

Secor, H. V. 1075, 1079 (52), *1316*

Sedlatschek, H. 176 (20), *209*

Sedzik-Hibner, D. 1276, 1281 (639), *1328*

Seebach, D. 456, 457 (1390), *490,* 1039 (246), *1058,* 1098 (121), *1318*

Seefelder, M. 539 (67, 70, 71), 549 (148), 550 (67, 157), 555 (148), 557 (148, 157), 559 (157), 561, 562 (148, 157), 569 (67), *588–590*

Seeliger, W. 293 (266), *467*

Seeman, J. I. 731, 733 (201b), *752*

Segur, J. B. 271 (15), *462*

Segwick, R. D. 759 (23), *778*

Sehested, K. 766–768 (62), *779*

Sehon, A. H. 866 (29, 30), *910*

Seibert, W. E. 839 (65), *856*

Seidel, M.   566 (250), *591*
Seidel, M. C.   566, 571 (251), *591,* 1162, 1168 (289), *1321*
Seifert, R. M.   150 (126), *172,* 1035, 1036 (128), *1055*
Seitz, B.   1041, 1043 (275), *1059*
Seitz, D. E.   375, 379 (800, 801), *478*
Seitz, G.   536 (22), *587*
Šek, B.   562, 571 (205), *590*
Seka, R.   176 (20), *209*
Seki, K.   312 (411), *470,* 674 (83), *749*
Sekine, T.   804 (99), *823*
Selke, E.   147 (114), *172*
Sellers, R. M.   766, 767 (63), *779*
Selman, J.   387 (858), *479*
Selman, S.   386 (843), *479*
Selve, C.   1038 (178), *1056*
Semenov, A. A.   598 (398), *599*
Semmelhack, M. F.   810 (133), *824, 826* (13), *855*
Semprini, E.   1026 (77), *1054*
Senda, K.-I.   711 (167), *751*
Senko, M. E.   222 (50), *257*
Senning, A.   569 (282), *592*
Senter, G.   908 (236), *914*
Sepulveda, L.   962 (74, 76), 963 (78, 79), 969 (76, 92), 970, 977, 988 (92), 993 (159b), 1014 (92), *1017–1019*
Sergievski, V. V.   237 (264), *262*
Serokhvostova, V. E.   596, 597 (386), *599*
Serota, S.   332 (540), 420 (1076), 441 (1266), *473, 484, 487*
Serratosa, F.   1080 (70, 71, 74, 76), 1082 (74, 76), *1317*
Serum, J. W.   147 (110), 156 (157), *171, 172*
Seshadri, K. V.   1195 (376), *1323*
Seshadri, T. R.   411 (941), 412 (952), *481*
Seshari, K. V.   518 (166), *530*
Sethi, P. S.   458 (1404), *490*
Sethna, S.   1098 (122), *1318*
Setínek, K.   916 (17), 918 (31), 919 (17), *943*
Setser, D. W.   744 (227), *753*
Setterquist, R. A.   870 (56), *911*
Setton, R.   411 (936), *481*
Severin, T.   566 (247, 248), *591*
Seybold, G.   1025, 1032, 1034, 1035 (50), 1041 (275), 1043 (50, 275), 1044 (50), *1054, 1059*
Seyferth, D.   513 (132), *529*
Seymour, D.   406 (912), *480*
Sgamellotti, A.   1027 (52), *1054*
Shabanov, A. L.   1073 (36), 1090 (101), *1316, 1318*
Shachidayatov, Ch.   138, 148 (76), *171*
Shacklett, C. D.   386 (855), *479*

Shaefer, F. C.   449 (1332), *489*
Shafer, T. C.   327 (500), *472,* 1145, 1151 (263), *1321*
Shah, A.   415 (1006), *482*
Shahak, I.   1036 (138), 1042 (289), *1055, 1059*
Shakhidayatov, Kh.   138 (75), *171*
Shalon, Y.   1090 (103, 104), 1092 (104, 106), *1318*
Shalygina, O. D.   1038 (185), *1057*
Shambhu, M .B.   417 (1047), 446 (1309), *483, 488*
Shamma, M.   315 (431), *470,* 494 (11), *526,* 1081, 1089 (94), *1317*
Shams El Din, A. M.   790 (45), *822*
Shani, A.   408, 409 (920), *480,* 1206, 1223, 1224 (468), *1325*
Shanker, G.   619 (130, 131), *638*
Shannon, J. S.   127, 128 (19), *169*
Shapiro, R. H.   140 (80), 141 (89–92), 143 (94), 155 (80), 156 (157), *171, 172*
Shapiro, S. H.   283 (200), *466*
Sharkey, A. G.   133, 134, 136 (47), *170*
Sharkey, W. H.   873 (70), *911*
Sharma, D. H.   1031 (99), *1055*
Sharma, S. D.   1232 (557, 563, 576), 1249 (563), 1250, 1251 (557), 1259 (576), *1327*
Sharma, S. K.   306 (361), *469*
Sharpless, K. B.   429 (1156), *485,* 847 (80), *857*
Sharvit, J.   164 (200), *173,* 614 (101), *638*
Shastlivtseva, N. V.   179 (41, 42), *210*
Shaver, F. W.   496 (26), *527*
Shaw, B. L.   435 (1209), 437 (1221), *486*
Shaw, J. E.   415 (984, 985), *482*
Shaw, K. N. F.   274 (75), *463*
Shaw, R.   60 (5, 8), 61, 63, 65 (8), *65*
Shays, D. B.   386, 387 (851), *479*
Shchakina, G. G.   868 (41), *910*
Shchegol, S. S.   328 (507), *472*
Shcherbak, I. V.   1052 (398), *1062*
Shealer, S. E.   208 (153), *212*
Shebab, A. H.   504 (81), *528*
Shechter, H.   1284, 1286, 1287 (646), *1329*
Sheehan, J. C.   167, 168 (217), *174,* 518 (167, 171), 519 (172–174), 520 (177), 522 (196), *530,* 537 (49), *587,* 691 (127), *750,* 1194 (367, 373), 1200 (421, 422, 432, 433), 1201 (432, 433), 1225 (367), 1231 (519), 1232 (519, 577–582, 585), 1233 (519), 1246 (519, 580), 1255 (577–582), 1258 (585), 1263 (519), 1272 (614), 1307 (708), *1323, 1324, 1326–1328, 1330*

Sheehan, J. T. 542 (98), *588*
Sheehan, M. 359 (701), *476*
Sheinker, Yu. 596, 597 (379), *599*
Sheinker, Yu. N. 562 (218, 220), 566 (272, 273, 275), *591, 592*
Sheldon, R. A. 372 (769), *477*
Sheldrick, G. 503 (74), *528*
Shemyakin, M. M. 278, 422 (125), *464*
Shen, C.-M. 437 (1224), *487*
Shen, T. Y. 1173, 1179 (330), *1322*
Shenton, F. L. 731, 732 (200a), *752*
Shephard, B. R. 372 (773), *477*
Shephard, I. S. 373 (781), *478*
Shepherd, R. G. 566 (255), *591*
Sheppard, N. 220 (27), 226 (140), 254 (449), *257, 259, 266*
Sheppard, R. V. 75 (45), *114*, 1145 (274, 276), 1154 (274), 1155 (274, 276), *1321*
Sheppard, W. A. 895 (181), *913*
Sheradsky, T. 1039 (197), *1057*
Sherry, A. D. 218 (18), 254 (464), *257, 266*
Sherry, J. J. 415 (984), *482*
Shevtchenko, V. I. 273 (65), *463*
Shibahashi, Y. 1029, 1032, 1042 (72), *1054*
Shibanov, V. V. 63 (38), *66*
Shibuya, S. 1311 (713), *1330*
Shick, M. E. 954 (52), *1017*
Shids, S. 769 (76), *779*
Shigemitsu, Y. 711 (167), *751*
Shillady, D. D. 83, 84 (85), *115*
Shimada, A. 224 (99–101, 109), 225 (110, 112, 113), *258*
Shimada, E. 413, 444 (958), *481*
Shimanouchi, T. 226 (144), 243 (337b, 340), 255 (471), *259, 263, 266*, 605 (28), *636*
Shimizu, K. 800 (79), *823*
Shimizu, S. 245 (360), *264*
Shimizu, Y. 846 (77), *857*
Shimoji, K. 423 (1105, 1106), *484*
Shimura, Y. 103 (214, 216, 220, 236), *118*
Shiner, V. J. 275 (97), *464*
Shiner, V. J. Jr. 365, 367 (739), *477*
Shingu, H. 938 (67), *944*
Shinkai, S. 1007 (181), *1020*
Shinoda, K. 951, 952 (26), 954 (26, 53), 984 (26), *1016, 1017*
Shinohara, Y. 571 (301), *592*
Shinozaki, H. 1302 (684), *1330*
Shinozaki, K. 812 (138), *824*
Shioiri, T. 444 (1292), *488*, 1039 (234), *1058*
Shiona, M. 444 (1297), *488*
Shiotani, H. 1027, 1029, 1031, 1032,

1042 (56), *1054*
Shiozawa, M. 1039, 1050 (193), *1057*
Shipman, L. L. 28, 29, 31 (65), 32 (89, 90), 45 (115), *56, 57*
Shirahama, H. 278 (129), *464*
Shirahama, K. 947 (12), 983 (112), *1016, 1018*
Shirai, H. 273 (57), *463*
Shirk, J. S. 691, 701 (125f), *750*
Shirley, D. A. 35, 40, 41 (97), *57*
Shivers, J. C. 416 (1009), *482*
Shkurina, T. N. 608, 609 (48), *636*
Shoda, S.-I. 1039 (217), *1057*
Shoer, L. I. 1046 (333), *1060*
Shoesmith, D. W. 788 (37), 789 (39), 791 (37), *822*
Shoji, K. 445 (1304), *488*
Shoppee, C. W. 328 (508), 339 (604), *472, 474*
Shorenstein, R. G. 1003 (172), *1020*
Short, G. D. 198 (100), *211*
Shorter, J. 250 (404), *265*, 309 (384), *469*
Shorygin, P. P. 608, 609 (48), *636*
Shostakovskii, M. F. 608, 609 (48), *636*
Shotton, E. 984 (126), *1019*
Showell, J. S. 270 (7), *462*
Showler, A. J. 411 (943), *481*
Shrecker, A. W. 1081 (92, 93), *1317*
Shriner, R. L. 311 (398), *470*
Shrivastava, G. P. 233, 235 (218), *261*
Shroot, B. 1206 (456), *1325*
Shropshire, E. Y. 540, 541, 545–548, 559, 572, 573 (88), *588*
Shubart, R. 283 (202), *466*
Shultz, J. L. 133, 134, 136 (47), *170*
Shumkov, V. G. 237 (265), *262*
Shurpach, V. I. 238 (275), *262*
Shurvell, N. F. 235 (234), *261*
Shusherina, N. P. 508 (104), 509 (108), 510 (111), *528, 1065, 1090 (6), *1315*
Shuster, P. 214 (1g), *256*
Shvo, Y. 1042 (285), *1059*
Sianesi, I. L. 1100 (137), *1318*
Sicher, J. 88 (131), *116*, 716 (178), *751*
Siddall, J. B. 430 (1160), *485*
Siddall, T. H. III 253 (432, 434), *265*
Siddigui, J. 1045 (326), *1060*
Sidel'kovskaya, F. P. 608, 609 (48), *636*
Siegbahn, H. 248 (395), *264*
Siegbahn, K. 248 (395), *264*
Siegel, S. 253 (437), *265*
Siemion, I. Z. 84 (93), 111 (311), *115, 120, 777 (97, 98), *780*
Siggia, S. 618 (119), *638*
Sigimori, A. 761 (39), *778*
Signor, A. 102 (192, 193), *117, 676, 685 (98d), *749*

Siládi, J.   918 (29), *943*
Silbert, L. S.   270 (7), 280 (155, 160), 281 (174, 185, 189), 415 (983), *462, 465, 482,* 1037, 1042, 1050 (161), *1056*
Silhavy, P.   450 (1334), *489*
Silverstein, R. M.   612 (90), *637*
Sim, G. A.   231 (195), *260,* 621 (151), *639*
Simanouti, T.   31 (87), *57,* 607 (38), *636*
Simchen, G.   534 (7), 536 (12, 28, 29, 43), 538 (28, 60), 539 (64), 542 (108), 543 (28, 43), 544 (64, 115, 116), 545 (28, 43, 122, 123, 125–127), 546 (64), 548 (28, 43, 108, 125, 126), 549 (64, 115, 116, 146, 147, 149), 550 (115), 553 (108, 125), 554 (28, 43, 175), 555 (116), 556 (43, 108, 175), 557 (28, 64, 181), 558 (28, 108, 149), 559 (60), 561 (43), 562 (149, 212), 563 (127), 566 (127, 175, 212, 246, 269, 276, 278), 567 (43, 127, 175, 212, 276, 278, 279), 569 (115, 116, 149), 570 (276), 571 (246, 276, 279), 572 (312), 574 (125, 149, 323, 324), 575 (64, 324), 576 (181, 326), 577 (43, 146, 147, 328, 331), 578 (147, 181, 326), 580 (312, 326, 332, 333), 581 (334, 335), 586 (60), 597 (146, 147, 181), *586–593*
Simes, J. J.   314, 355 (422), *470*
Simić, M.   763 (45), 766 (58–60), 767 (58, 59, 65), 768 (60), (61), *778, 779*
Simmonds, P. G.   866 (31), *910*
Simmons, H. E.   321, 334 (473), *471,* 552 (171), *590*
Simon, C.   793 (57), *822*
Simon, E.   110, 111 (309), *120*
Simon, H.   1035, 1039 (132), *1055*
Simon, M. S.   440 (1256), *487*
Simonaitis, R.   651, 655 (16–18), 659, 660 (16), 704 (16, 17), *747*
Simons, J. P.   207 (150), *212,* 644 (8), *747*
Simpson, R. B.   225 (132), *259*
Simpson, W. R. J.   596, 597 (387), *599*
Simpson, W. T.   29 (75), *57,* 73 (39), *114,* 229 (166), *260*
Sina, A.   686 (103), *749*
Singer, A. W.   319 (463), *471*
Singer, G. M.   454 (1375), *490*
Singer, L. A.   292 (261), *467*
Singer, R. J.   603 (12), *636*
Singer, R.-J.   813 (140), 817 (159), *824*
Singh, B.   447, 449 (1318), *488*
Singh, S.   110 (300, 301), *119*
Singh, U. P.   1052 (405), *1062*

Sinha, B.   298 (317), 311, 314 (399), *468, 470*
Sinke, G. C.   60 (7), 61, 62 (26), 63 (7), 64, 65 (48), *65, 66*
Sioda, R. E.   802, 814 (87), *823*
Šipoš, F.   88 (131), *116*
Sircar, S. S. G.   1138 (246), *1321*
Sisido, K.   271 (20), *462*
Sitnikova, S. P.   1040 (252), *1058*
Sivertsen, B. K.   86 (104), *115*
Sixina, F. L. J.   884, 886, 891, 897, 898 (126), *912*
Sixma, F. L. J.   89, 95 (145), *116*
Sjöberg, B.   88, 91 (138), 102 (188, 190, 207), 103 (190, 211), 108 (190), 109 (138), *116–118,* 1206, 1220 (484), *1325*
Sjöberg, S.   88, 91 (113, 138), 92, 94 (113), 109 (138), *116*
Skapski, A. C.   827 (35), *856*
Skarzyński, M.   609 (58), *637*
Skatova, N. N.   1052 (398), *1062*
Skattebol, L.   1229 (511), *1326*
Skell, P. S.   882, 899 (109), *912*
Skinner, J. F.   1024 (32), *1053*
Sklar, N.   222 (50), *257*
Sklarz, B.   365 (740), *477*
Skoldinov, A. P.   547 (140), 571, 572 (313), *589, 593*
Skramstad, J.   110, 111 (305), *120*
Skrivelis, J.   1045 (323), *1060*
Skuratov, S. M.   61, 62 (28), *66*
Slater, R. A.   1042 (283), *1059*
Slaugh, L. H.   458 (1401), *490*
Slavutskaya, G. M.   60 (13, 14), 61, 62 (28), 63 (14, 40), *65, 66*
Slooff, G.   338 (583), *473*
Slopianka, M.   452 (1353), 456, 457 (1389), *489, 490*
Slowata, S. S.   953 (49), *1017*
Slutsky, J.   899 (196, 198), 900 (198), *913*
Slutsky, L. J.   238 (286), *262*
Smai, F.   1296 (671), *1329*
Small, R. W. H.   244 (348), 245 (355, 362), 253 (440), *263–265*
Small, V. R. Jr.   417 (1048), *483*
Smart, B. W.   279 (143), *464*
Smelyanskaya, E. M.   194 (75), *211*
Smentowski, F. J.   328 (512), *472*
Smerkolj, R.   218, 237 (13), *256*
Smerz, O.   571, 586 (296), *592*
Smille, R. D.   1123 (210), *1320*
Smirnov, V. A.   782, 801, 807 (6), *821*
Smissman, E. E.   417 (1049), *483*
Smit, P. H.   232 (215a), *261*
Smith, A. A.   103 (225), *118*

Smith, A. B. III 380 (819), 384 (833), *478, 479,* 725 (190, 191a, b), *752*
Smith, B. F. 426 (1131), *485*
Smith, C. 501 (63), *527,* 619 (132), *638*
Smith, C. W. 314, 353 (423), 364 (724), 400 (897), *470, 476, 480,* 1232, 1241, 1242 (531), *1326*
Smith, D. G. 240 (311), *263*
Smith, D. H. 627 (177), *639*
Smith, D. M. 1038 (175), *1056*
Smith, D. W. 253 (441), *265*
Smith, E. 880 (100), *911*
Smith, F. X. 711 (169), *751*
Smith, G. 367 (745), *477*
Smith, G. G. 140 (84), *171,* 861 (4, 7), 879 (94), 882 (106), 884 (94), 885 (130), 889 (145), 890 (146–148), 891 (146, 147, 151, 153), 892 (94, 153, 164), 893, 894 (165, 169, 170), 896 (146), 897 (94), 898 (148, 194), 899 (195), 901 (94, 203), 904 (146, 147), 905 (146), 908 (234), *909–914*
Smith, H. A. 386 (855), *479*
Smith, H. J. 976 (100), *1018*
Smith, H. Q. 340 (612), *474*
Smith, J. G. 127, 157 (23), *170,* 353 (661), *475*
Smith, J. L. 603 (17), *636*
Smith, J. S. 125 (8), *169*
Smith, L. C. 1116 (194), *1320*
Smith, L. H. 353 (658), *475*
Smith, L. I. 333, 334, 339 (569), 359 (697), *473, 476*
Smith, L. W. 989 (140), *1019*
Smith, M. J. 700, 735–737 (142), *750*
Smith, N. R. 1096 (117), 1098 (123), *1318*
Smith, P. A. S. 385, 450 (838), *479,* 1168 (281), 1274 (621), 1284 (645), *1321, 1328, 1329*
Smith, R. J. D. 620 (137), *638*
Smith, R. M. 1284 (653), *1329*
Smith, W. 333 (567), *473*
Smith, W. H. 803 (93), *823*
Smith, W. T. 360 (706), *476*
Smithers, R. 541 (89), *588*
Smits, J. F. M. 1044, 1049 (314), *1059*
Smolanow, J. 1052 (392), *1062*
Smolarsky, M. 279 (144), *464*
Smolders, R. R. 439 (1248), *487*
Smolyansky, A. L. 238 (276), *262*
Smoylan, Z. S. 439 (1245), *487*
Smrt, J. 557, 558, 562 (183), *590*
Smutný, J. 918 (29), *943*
Snatzke, F. 78 (58), 80, 81 (69), 82, 85, 87 (74), 107 (265), *114, 115, 119*
Snatzke, G. 69 (5), 72 (25), 73 (5), 76 (48), 78 (57, 58), 80 (68, 69), 81 (69),

82 (73–75), 85 (73, 74, 98), 87 (73, 74), 88 (25, 136), 89, 91 (136), 93 (25), 101 (175), 107 (265), 110 (297), *113–117, 119,* 1145 (249), *1321*
Snatzke, G. N. 609 (64), *637*
Sneeden, R. P. A. 293 (269), *467*
Snow, J. T. 435 (1203), *486*
Snow, J. W. 109 (279), *119*
Snowling, G. 1037 (160), *1056*
Snyder, H. R. 273 (69), 283 (200), 365 (728), 385 (839), *463, 466, 476, 479,* 541 (94), *588,* 1274 (622), *1328*
Snyder, J. P. 1049 (367), *1061,* 1138, 1141 (243), *1320*
Snyder, P. A. 70, 95 (16), *114*
Snyman, J. A. 183, 189 (64), *210*
Sobczyk, L. 214 (1h), 238 (273, 274), *256, 262*
Sobti, A. 80 (67), *115*
Sochneva, E. O. 596 (379, 380), 597 (379), *599*
Soest, T. C. van 109, 110 (285), *119*
Sohár, P. 1037, 1042 (156, 157), *1056*
Sohn, W. H. 1089, 1090 (100), *1317*
Sokodowaska, T. 420 (1077), *484*
Sokol, P. E. 297 (302), *468*
Sokolov, N. D. 214 (1c), 220 (26), *256, 257*
Sokolov, V. P. 241 (323), *263*
Soldan, F. 534, 546 (2), *586*
Solly, R. 874 (74), *911*
Solov'eva, L. D. 108 (270), *119*
Soloway, S. 603 (6), *636*
Soma, Y. 1174 (334), *1322*
Soman, R. 583 (346), *593*
Somerville, L. F. 297 (298), *468*
Son, M. O. 936 (60), *943*
Sondheimer, F. 312 (410), *470*
Sonnet, P. E. 1185 (340), *1322*
Sonneveld, W. 134 (52), 136 (65), 138 (72), *170, 171*
Sonoda, A. 361 (707), *476*
Sonoda, T. 1276 (626), *1328*
Sorensen, A. K. 1115, 1116 (193), *1320*
Sørensen, C. S. 103 (224), *118*
Sorensen, G. O. 28 (62), *56*
Sorensen, N. A. 1080, 1082 (66), *1317*
Šorm, F. 398 (893, 895), 459 (1410), *480, 490,* 799 (71), *822,* 916 (13, 14, 16, 22, 23), 917, 918 (23), 919 (16, 22, 23), 927 (14, 23, 42), 928 (13, 14, 23), 929 (14), 932 (48), 936 (58), *942, 943*
Sørum, H. 245 (356, 365), *264*
Sosnovsky, G. 275 (102), 432 (1175, 1176), *464, 486*
Sotgin, F. 565 (243), *591*

Sotiropoulos, J. 1027, 1032, 1042 (60), *1054*

Sotnikova, N. N. 802 (86b, c), *823*

Soucy, P. 309 (385), 358 (689), *469, 476,* 1161, 1168, 1170 (284), *1321*

Souma, Y. 291 (249), 292 (259), 293 (268), *467*

Sousa, L. R. 707 (149), *751*

Southwick, P. L. 289 (235), *466*

Souto-Bachiller, F. 1039 (218), *1057*

Souto-Bachiller, F. A. 743 (222), *753*

Spaargaren, K. 250 (411), *265*

Spande, T. F. 1206, 1219 (482), *1325*

Spangler, R. J. 745 (228c), *753*

Sparnaay, M. J. 955 (58), *1017*

Sparrow, D. R. 75 (44), *114*

Spasskaya, R. I. 449 (1328), *489,* 868 (41), *910*

Spassow, A. 416 (1017), *482*

Speaker, T. J. 300 (330), *468*

Speakman, J. C. 223 (79, 81), 224 (83–85), 233 (222), *258, 261*

Speckamp, W. N. 525 (220–222), *531*

Speers, L. 353 (664), *475*

Speeter, M. 1266 (604), *1328*

Speh, P. 534 (8), 536, 539, 540 (42), 557 (8, 42), 559 (42), 562, 568 (8), *586, 587*

Spencer, J. N. 236 (248a), *261*

Spencer, T. A. 901 (205), *914*

Sperber, N. 274 (82), *463*

Spickett, R. G. W. 1104 (153), *1319*

Spiegelman, G. 1231 (520), 1232 (520, 568, 570, 572, 573), 1245 (520), 1254 (568, 572), 1256 (570, 573), 1263 (520), *1326, 1327*

Spielman, M. A. 343 (623), 369 (752), *474, 477*

Spiess, H. W. 230 (179), 231 (194), *260*

Spietschka, E. 382 (829), *478,* 1232, 1264 (596, 599), *1328*

Spille, J. 535–537, 554, 559, 572 (10), *586*

Spillman, L. J. 359 (697), *476*

Spinelli, D. 110 (308), *120*

Spiteller, G. 130 (31), 134 (53), 157 (161), *170, 172*

Spiteller-Friedmann, M. 134 (53), *170*

Spitzer, W. A. 731, 733 (201b), *752*

Spitzner, E. B. 566, 567, 571 (253), *591*

Spivey, H. O. 248 (396), *264*

Splitter, J. 879 (90), *911*

Spoerri, P. E. (96), *823*

Spotswood, T. M. 168 (222), *174,* 610 (70), *637*

Spotswood, T. McL. 180 (50), *210*

Sprecher, M. 654, 657, 735 (28), *747*

Sprenger, H. E. 536 (21), *587*

Spry, D. O. 524 (206), *530*

Srinivasan, P. S. 1037 (149), 1042 (296), *1056, 1059*

Srinivasan, R. 651 (19), *747*

Srinivasan, T. K. K. 104 (249), *118*

Srivastava, R. C. 521 (186), *530,* 1195, 1225 (379), *1323*

Srivastava, R. D. 176 (18), *209*

Srogi, J. 297 (306), *468*

Staab, H. A. 412, 444 (954), *481,* 867 (37), *910*

Stace, A. J. 155 (147), *172*

Stacey, M. 411 (939), *481*

Stachel, H. D. 551 (163), 552 (168), *590*

Stäglich, P. 1024 (31), 1030, 1031 (31, 92), 1032 (31), 1033 (92), 1034, 1039 (31), 1040 (31, 92), 1042 (31), *1053, 1055*

Staley, R. H. 30 (76), *57*

Stam, M. 1008, 1010 (184), *1020*

Stam, M. F. 984 (116), *1018*

Stamicarbon, N. V. 1276, 1277 (628), *1328*

Stanbury, P. 522 (193), *530*

Stanishevskii, L. S. 1052 (404), *1062*

Stanovnik, B. 562 (205, 206, 211, 223), 569 (223), 570 (294), 571 (205, 211, 294), 596, 597 (378), *590–592, 598*

Stapf, W. 1042 (297), *1059*

Staples, C. E. 1080 (77, 78), *1317*

Stapp, P. R. 307 (378), *469*

Starcher, P. S. 353 (666), *475*

Starks, C. 334 (572), *473*

Starks, C. M. 271 (18), *462*

Starting, J. 281 (181), *465*

Stasiewicz, M. 904 (218), *914*

Staudinger, H. 386 (845), *479,* 498 (41, 43), 509 (106), 521 (182a, b, 188), 524 (188), *527, 528, 530,* 1230 (515), 1232 (515, 521, 527, 529, 533–536), 1233 (515), 1238 (521), 1240 (533–536), 1264 (515), 1269 (610), *1326, 1328*

Staunton, J. 361 (708), *476*

Stavely, H. E. 1125, 1130 (216), *1320*

Stavholt, K. 1080, 1082 (66), *1317*

Stavrovskaya, A. V. 547 (140), 571, 572 (313), *589, 593*

Steacie, E. W. R. 671, 672, 677 (92d), *749*

Steahly, G. W. 516 (146), *529*

Stearns, J. A. 417 (1036), *483*

Steele, W. V. 61 (20), *66*

Steen, K. 451 (1349), *489,* 583, 584 (343), *593*

Steenbeckeliers, G. 22 (35), *56*

Steger, E. 1024 (35), 1032 (35, 104), *1053, 1055*

Steglich, W.   334 (571), 417 (1041, 1042), *473, 483,* 1039 (229), *1058*
Stehr, C. E.   1145, 1154, 1155 (274), *1321*
Steiger, M.   333 (565), *473*
Steiger, R. E.   274 (71, 72), 282 (197), *463, 466*
Stein, W.   916 (6), 920 (35), 939 (72), *942–944*
Steinberg, H.   1185 (345), *1323*
Steinberg, N. G.   1145, 1159 (278), *1321*
Steinberger, N.   788 (36), *822*
Steiner, K.   369 (755), *477*
Steiner, P. R.   673 (76a, b), 682 (76a, b, 96a–c), 683 (76a, b), 685 (76a, b, 96c), *749*
Steinmetz, R.   206 (142), *212*
Stekoll, L. H.   1200 (423, 424), *1324*
Stelakatos, G. C.   272 (49, 50), *463*
Stemple, N. R.   225 (123), *259*
Stenhagen, E.   133 (49), 134 (49, 55), *170*
Stenkamp, L. Z.   29, 30 (74), *57*
Stepanyan, A. N.   1075 (48), *1316*
Stephan, H.   805 (106), *823*
Stephens, J. R.   360 (704), *476*
Stephens, R.   318, 320, 325, 327 (452), *471*
Stephens, R. E.   246 (373), *264*
Stephensen, R.   1168 (301), *1322*
Stepišnik, J.   231 (198), *260*
Sterling, C. J. M.   444 (1300), *488*
Stermitz, F. R.   680 (69), *748*
Stern, R. L.   411 (927), *481*
Sternberg, V. I.   411 (933), *481,* 1145, 1152 (271), *1321*
Sternhell, S.   432 (1180, 1181), *486*
Stetter, H.   271 (19), *462*
Stevens, C. L.   418 (1054), *483*
Stevens, E. D.   219 (24), *257*
Stevens, K. L.   136 (64), *170*
Stevens, R. E.   306 (364), *469*
Stevens, Th. C.   235 (247), *261*
Stevens, T. L.   236 (248b), *261*
Stevens, W.   875 (84), *911*
Stewart, D. W.   123, 124 (4), *169*
Stewart, F. D.   874 (77), *911*
Stewart, J. C.   1138, 1143 (244), *1320*
Stewart, R.   298, 332 (308), *468*
Stewart, R. C.   276 (112), *464*
Stewart, W. E.   252 (426), 253 (426, 432, 434), *265*
Stewart, W. W.   521 (185), *530*
Stibor, I.   297 (306), *468*
Stigter, D.   952 (44), 955 (59a, 60), 956 (59a–e, 61), *1017*
Stiles, M.   296 (290), *467,* 1138 (227), *1320*

Stilkerieg, B.   1049 (367), *1061*
Stille, J. K.   270 (13), *462,* 834, 835 (50, 51), 836 (51), 837 (50, 51, 57), 838 (51, 57), 839 (57, 64), 840 (57), 842 (51), *856*
Stiller, E. T.   297 (304), *468*
Stillinger, F. H.   242, 252 (333), *263*
Stilz, W.   540, 547 (80), *588*
Stimson, V. R.   826 (16), *855,* 867 (35), 894 (174), 901 (201, 202), *910, 913, 914*
St Jean, T.   129, 130 (28), *170*
Stock, L. M.   411 (942), 417 (1046), *481, 483,* 904 (216), *914*
Stöcklin, W.   85, 86 (95), *115,* 625 (169), *639*
Stoddart, J. F.   501 (63), *527,* 619 (132), *638*
Stodola, F. H.   279 (137), *464*
Stokes, P. D.   816 (151), *824*
Stolle, W. T.   711, 713, 714 (171), *751*
Stoltenberg, J.   360 (702), *476*
Stone, A. J.   615 (104), *638*
Stone, A. L.   105 (252, 254, 255), *118*
Stone, G. R.   1098 (135), *1318*
Stone, R. H.   886 (132), *912*
Stoneberg, R. L.   416 (1027), *483*
Stoodley, R. J.   384 (834), *479,* 524 (204), *530*
Stopp, G.   535–538, 542–544, 550, 551, 553, 554, 557, 559, 562, 566, 568, 572, 573, 581, 583, 586 (11), *586*
Storesuno, H. J.   430 (1166), *485*
Stork, G.   335 (577), 395 (882), *473, 480,* 1273 (617), *1328*
Stork, W.   1022 (6), 1030 (88, 97), 1031 (88, 97, 102), 1037 (152), *1053–1056*
Storm, P. O.   771 (80), *779*
Stothers, J. B.   298 (314), *468*
Stotter, P. L.   312 (404), *470*
Stoughton, R. W.   333 (566), *473*
Stoyanov, S. I.   1048 (345), *1060*
Stoyanovich, F. M.   536 (39, 44), 538 (44), 539 (78, 79), 543 (44, 110, 111), 544, 548 (44), 549 (39), 558 (44), 561 (44, 110, 111), 562, 567, 569 (44), 577 (39), *587–589*
St. Pancescu   306 (367), *469*
Strachan, E.   672, 678 (60b), *748*
Straley, J. M.   424 (1114), *484*
Strand, T. G.   222 (55), *257*
Stratford, M. J. W.   452 (1360), *489*
Strating, J.   277 (123), 295 (286), 309 (386), *464, 467, 469,* 1049 (365, 366), *1061*
Straub, B.   109 (280), *119*
Strauss, U. P.   953 (49), *1017*
Strege, P. E.   428 (1154), *485*

Streicher, R. 917, 919 (28), *943*
Streith, J. 701 (143), *751*
Strel'tsova, Z. A. 964 (84), *1018*
Strichler, P. 88 (126), *116*
Strickland, E. H. 109 (276), *119*
Strickland, R. 83, 84 (85), *115*
Strickland, R. W. 92 (153), 95 (153, 159), *116*
Strickmann, G. 1040 (256, 265), 1041 (256, 279), 1052 (394), *1058, 1059, 1062*
Stridh, G. 60 (12), *65*
Strizhakov, O. D. 1276, 1278 (632), *1328*
Strocchi, A. 161 (180), *173*
Strogova, O. A. 414 (965), *481*
Stroh, H. 571 (299), *592*
Ströhmeier, W. 828, 837, 841, 842 (44), *856*
Strohschein, R. J. 596, 597 (387), *599*
Strom, K. A. 894 (179), *913*
Stromberg, V. L. 1276, 1278, 1282 (633), *1328*
Strotter, P. L. 424 (1116), *484*
Stroud, M. M. A. 862 (12), *910*
Strouse, C. E. 195 (76), *211*
Struve, W. S. 380 (815), *478*
Studebaker, J. F. 199 (103), *211*
Stull, D. R. 60, 63 (7), *65*
Stumpf, W. 353 (667), 400 (898), *475, 480*
Sturm, J. J. 1052 (407), *1062*
Styskin, E. L. (506), *472*
Su, C. J. 101 (173), *117*
Subramanian, G. S. 787 (29, 30), 790 (46), *821, 822*
Sucrow, W. 312 (408), 452 (1353, 1354), 456, 457 (1388, 1389), *470, 489, 490,* 583 (353, 354), *594*
Sudarsanam, V. 1232, 1257, 1258 (571), *1327*
Sudo, A. 1131 (218), *1320*
Sueda, N. 514 (133), *529*
Suga, K. 280 (161–163), 281 (161, 162), 359 (696), *465, 476,* 1116 (196), *1320*
Suga, T. 299 (323), 332 (548, 549), *468, 473*
Sugano, H. 1193 (366), *1323*
Sugasawa, S. 804 (98), *823*
Suggs, J. W. 416 (1007), *482*
Sugihara, J. M. 140 (86), *171*
Sugihara, Y. 1039 (226), *1057*
Sugimori, A. 674 (83), *749*
Sugino, K. 1028, 1032, 1044 (66), *1054*
Sugisaki, R. 28 (62), *56*
Sugita, J. 303 (344), *469*
Sugiyama, T. 674 (83), *749,* 1028, 1032, 1041 (70), *1054*

Sukenik, C. N. 989 (144), 1006 (177), *1019, 1020*
Sukhoruskin, A. G. 566 (270), *592*
Sullivan, D. F. 423 (1107), 427 (1146, 1148), *484, 485*
Sullivan, R. E. 123 (3), *169*
Sulston, J. E. 272 (44), *462*
Sultanbawa, M. V. S. 1066 (27), *1316*
Sulzbacher, M. 276 (104), *464*
Sumoto, K. 451 (1348), *489,* 1276, 1277, 1279 (624), *1328*
Sunami, M. 511 (118a), *529*
Sundberg, R. J. 711 (169), *751*
Sung, M. 506 (89), *528*
Sung, M.-T. 712, 719, 725, 745 (180a), *751*
Sunner, S. 60 (18), *66*
Sunners, B. 28 (68), *56*
Sunshine, W. L. 1006 (176, 179), *1020*
Suschitzky, H. 1232, 1256, 1261 (591), *1328*
Susi, H. 214 (2c), *256,* 1034 (121), *1055*
Sussman, S. 411 (934), *481*
Susuki, T. 441 (1275, 1276), *488*
Sutcliffe, B. T. 6 (1), 20 (11), *55*
Sutcliffe, L. H. 309 (383), *469*
Suter, H. 792 (54), 793 (54–56), *822*
Suter, U. 109 (280), *119*
Sutherland, G. B. B. M. 221 (42), *257*
Sutton, B. M. 1052 (401), *1062*
Sutton, L. E. 19, 26, 29 (7), *55*
Suvorov, N. N. 1038 (185), *1057*
Suzi, H. 228 (148), 246 (374, 375), *259, 264*
Suzuki, A. 293 (264), 430 (1162), *467, 485,* 1080 (88, 89), 1085 (88), 1086, 1087 (88, 89), 1088 (89), *1317*
Suzuki, H. 987 (138), *1019*
Suzuki, I. 255 (471), *266,* 605 (25–27), *636*
Suzuki, M. 226 (144), *259,* 312 (411), 422 (1091), *470, 484*
Suzuki, N. 675, 682 (88), *749*
Suzuki. T. 102 (204), *117*
Šváb, V. 918 (29), *943*
Svateeva, L. A. 618 (125), *638*
Svensson, A. 110, 111 (307), *120*
Svensson, C. 111 (312, 313), *120*
Svensson, S. 105 (257–259), 106 (258, 259), *119*
Svetozarskii, S. V. 1276, 1278 (632), *1328*
Sviridova, A. V. 1051 (386), *1061*
Swahn, C. G. 105, 106 (259), *119*
Swain, C. G. 250 (412), *265,* 861, 863 (8), *910*
Swain, G. 503 (72), *528*
Swallow, A. J. 759 (14), 768 (67), *778, 779*

Swamer, F. W.   431 (1167), *485*
Swaminathan, S.   374, 375 (785), *478*
Swan, G. A.   375, 377 (793), *478*
Swann, B. P.   343 (626–628), 344 (627, 628), 370 (762), *474, 477*
Swann, S. Jr.   411 (929), 433 (1193), *481, 486,* 782 (5a), 790 (44), 818 (5a), 820 (167), *821, 822, 824*
Sweeny, J. G.   583 (347), *593*
Sweet, A. J.   319 (458), *471*
Swensen, W. E.   278 (127), *464*
Swenson, C. A.   252 (430), *265,* 517 (152), *529*
Swenton, J. S.   731, 732 (200a–e), 733 (200c), *752*
Swern, D.   270 (4, 7), 411 (932), 420 (1074, 1076), *462, 481, 484*
Sychev, O. F.   231 (204), *260*
Synerholm, M. G.   416 (1029), *483*
Synge, R. L. M.   1015 (201), *1020*
Szafran, M.   228 (160), *259*
Szajewski, R. P.   1273 (617), *1328*
Szarek, W. A.   848 (84), *857*
Szczesniak, M. M.   251 (421), *265*
Szmuszkovicz, J.   298 (316), 341 (621b), 414 (966), *468, 474, 481,* 524 (215), *531,* 901 (205), *914*
Szpovar, Z.   223 (78), *258*
Szutka, A.   763, 765 (50), *779*
Szwarc, M.   867 (36), 871 (59), *910, 911*

Tabak, D.   454 (1378), *490*
Tabak, G.   1042 (295), *1059*
Tabo, Y.   1230 (514), *1326*
Tabor, W. J.   24 (46), *56*
Tabushi, I.   274 (76), *463,* 1007 (180), *1020*
Tada, M.   1302 (684), *1330*
Tada, R.   386 (841), *479*
Tadanier, J.   1098 (135), *1318*
Taddia, R.   309 (388), *469*
Tadwalkar, V. R.   1072 (32), *1316*
Tafel, J.   805 (107, 109, 110), 819 (162), *823, 824*
Taft, R. W.   241, 250, 251 (319), *263,* (55), *779*
Tagaki, W.   990 (151), 1001 (170), 1003 (173), *1019, 1020*
Tagiri, A.   82, 83 (76), 84 (89), *115*
Taguchi, H.   423 (1105, 1106), *484,* 585 (362), *594*
Taguchi, T.   896, 902 (185), *913,* 1080 (88, 89), 1085 (88), 1086, 1087 (88, 89), 1088 (89), *1317*
Taillefer, R.   54 (124), *58*
Taits, S. Z.   799 (72), *822*
Tajima, K.   438 (1236), *487*
Takada, A.   537 (50), *587*

Takada, S.   416 (1021), *483*
Takada, Y.   439 (1246), *487*
Takadate, A.   596 (391), *599*
Takagi, K.   1040 (271), 1051 (383), 1052 (271), *1058, 1061,* 1206, 1207, 1211 (461), *1325*
Takagi, M.   1039 (201), *1057*
Takagi, T.   1027, 1043 (59), 1044 (59, 319), *1054, 1060*
Takahashi, H.   1048 (356), *1061*
Takahashi, K.   277 (120), 345 (633), 425 (1126), *464, 475, 485*
Takahashi, N.   85, 87 (97), *115*
Takahashi, T.   804 (99), *823*
Takahashi, Y.   293 (264), 336 (580), *467, 473*
Takakaki, Y.   245 (364), *264*
Takakuwa, T.   1027, 1029, 1031, 1032, 1042 (56), *1054*
Takami, Y.   296 (289), *467*
Takano, S.   1138, 1144 (247), *1321*
Takaya, H.   545 (120), *589*
Takayanagi, H.   81 (70–72), *115,* 275 (94), *463*
Takayanagi, Y.   1051 (383), *1061*
Takebe, N.   457 (1400), *490*
Takeda, A.   277 (120), *464,* 1106, 1107 (156), *1319*
Takeda, K.   107 (264, 266), *119,* 1048 (361), *1061*
Takeda, T.   1039 (219), *1057*
Takei, H.   868 (44, 46), 869 (46, 47), *910,* 1047 (335), *1060*
Takei, S.   502 (67), *528*
Takei, W. J.   245 (366), *264*
Takeshima, T.   1037 (153–155), 1042 (153), 1052 (155), *1056*
Takeuchi, F.   674 (83), *749*
Takeuchi, J.   1284 (643), *1329*
Takeuchi, M.   355 (677), *475,* 1168 (299), *1321*
Takeuchi, T.   133 (42), *170*
Takhistov, V. V.   1052 (398), *1062*
Takigawa, T.   419 (1072), *484*
Takumaru, K.   745 (232), *753*
Takusagawa, F.   224 (99–101, 109), 225 (110–113), *258*
Talaty, E. R.   167, 168 (219, 220), *174,* 520 (175), *530,* 613, 617 (111), *638,* 669 (58), *748,* 1200 (423, 424), *1324*
Talbot, K.   984, 987 (121), *1018*
Talkowski, C. J.   989 (142), 1006 (142, 179), *1019, 1020*
Talley, C. P.   617 (118), *638*
Tam, S. Y.   312 (407), *470*
Tamamushi, B.-I.   951, 952, 954, 984 (26), *1016*
Tamburin, H. J.   1292 (663, 664), *1329*

Tamelen, E. E. van 311 (402), *470*, 494
    (11), *526*, 847 (80), *857*, 1074 (47),
    1081, 1089 (94), 1096 (119), 1104
    (149, 151), 1105 (47), *1316–1319*
Tamimoto, S. 562, 565, 569 (207), *590*
Tamm, C. 1167, 1168 (308), *1322*
Tamres, M. 620, 623 (147), *639*
Tamura, C. 245 (363), *264*
Tamura, Y. 451 (1348), *489*, 1276 (624,
    635, 637), 1277 (624), 1279 (624,
    635, 637), 1280 (635, 637), 1281 (637),
    1284 (649, 650), 1289 (649), 1290
    (650), *1328, 1329*
Tan, A. T. C. H. 866 (29), *910*
Tanabe, K. 1145 (259), *1321*
Tanaka, F. S. 673 (77), *749*
Tanaka, J. 73 (31), *114*, 199 (104),
    200–202 (110), *211*, 229 (169), *260*
Tanaka, K. 446 (1313), *488*
Tanaka, M. 237 (262, 263), 238 (283),
    262
Tanaka, T. 596 (391), *599*
Tanaka, Y. 248 (389), *264*
Tanako, S. 505 (84), *528*
Tanford, C. 951 (31), 952 (37–40),
    1016, 1017
Tang, C.-S. 673, 682 (73), *749*
Tangari, N. 1307, 1309 (709), *1330*
Tangerman, A. 1043 (304), 1044 (314),
    1049 (304, 314, 368–370), *1059, 1061*
Tani, J. 444 (1299), *488*
Tanizawa, K. 412 (947), *481*
Tannenbaum, H. P. 135, 137 (59), 139
    (79), *170, 171*
Tanno, T. 1168 (304), *1322*
Tao, N. S. 1048 (353), *1061*
Tapia, O. 221 (40), *257*
Tarasov, B. P. 249 (403), *265*
Tarasova, O. A. 1052 (398), *1062*
Tarbell, D. S. 329 (518), 388 (859), *472,
    479*, 498 (38), *527*, 871 (62–64), 872
    (64a), *911*
Tardew, P. L. 603 (7), *636*
Tardivel, R. 454 (1372), *490*
Tarker, M. F. 339 (603), *474*
Taschner, E. 420 (1077), *484*
Tatematsu, A. 140 (82, 83), 144 (101),
    171
Tatevosyan, G. E. 1104 (147), *1319*
Tatlow, J. C. 178 (11), *209*, 318, 320
    (452), 321 (472), 325 (452, 472), 327
    (452), 411 (939), *471, 481*
Tatsumi, T. 292 (258), *467*
Tatsumoto, N. 235 (241), 236 (241,
    257, 258), 237 (258, 260), *261, 262*
Taub, W. 1073 (37), *1316*
Taube, A. 614 (101), *638*
Taube, H. 450 (1338), *489*

Taussig, P. R. 888, 901 (141), *912*
Tavernier, D. 1038 (174), *1056*
Tayaka, T. 437 (1227), *487*
Taylor, C. M. B. 1080 (81), *1317*
Taylor, E. C. 343 (622, 626–628), 344
    (627, 628), 370 (762), 416 (1023), *474,
    477, 483*, 825 (4), *855*
Taylor, E. R. 273 (64), 319 (461), *463,
    471*
Taylor, G. A. 316 (436), *470*, 1232
    (530), *1326*
Taylor, H. A. 661, 662 (42, 43), *748*
Taylor, H. S. 671, 672, 677 (92b), *749*
Taylor, J. D. 461 (1417, 1418), *490*
Taylor, J. W. 205 (134), *212*
Taylor, K. G. 415 (988), *482*
Taylor, M. D. 235 (236), *261*
Taylor, P. J. 865 (23), *910*
Taylor, P. S. 700, 735–737 (142), *750*
Taylor, R. 862 (12), 872 (65), 882 (110),
    885 (127a), 890 (146–148), 891 (146,
    147, 150, 151a, 154), 892 (152, 159),
    893 (150, 165, 171, 172), 894 (154,
    165, 172), 895 (65), 896 (146), 897
    (191), 898 (148, 191), 899 (110, 172),
    900 (110, 172, 200), 903 (127a, 213),
    904 (146, 147, 217), 905 (146, 151a,
    152, 219–225), 906 (152, 159, 213,
    222, 226–230), 907 (159, 219, 221,
    222, 226, 228, 231, 232), *910–914*
Taylor, T. W. J. 1270 (611), *1328*
Taylor, W. C. 1304, 1306, 1309 (699),
    1330
Taylor, W. I. 1304 (697), *1330*
Tcheng-Lin, M. 1115 (189), *1319*
Tchoubar, B. 393 (872), 394 (874), 452
    (872), *479*
Techer, H. 1138, 1140 (240), *1320*
Tedder, J. M. 411 (939, 940), *481*
Teitelbaum, C. 800 (81), *823*
Temin, S. C. 338 (593), *474*
Temple, D. L. 283, 284 (205), 285 (213),
    466
Temple, R. D. 364 (725), *476*
Templeman, M. B. 235 (236), *261*
Ten Brink, R. E. 432 (1182), *486*
Tengi, J. P. 566 (266), *592*
Tepley, L. B. 963, 980 (77), 984 (124),
    1017, 1019
Teplý, J. 759, 760 (25, 27), *778*
Teranishi, A. Y. 429 (1156), *485*
Terashima, M. 1276, 1279, 1280 (635,
    637), 1281 (637), *1328*
Terashima, S. 278 (129), *464*, 495 (15),
    526
Terem, B. 812 (139), *824*
Terenin, A. 671, 672 (92a), *749*
Terentsev, P. B. 325 (493), *472*

Ter-Minassian-Saraga, L.   979 (106), *1018*
Terpstra, D.   234 (228), *261*
Tesh, K. S.   786, 787 (24), *821*
Testa, E.   523 (197), *530,* 1194 (369),
    1197 (401, 402), 1198 (390–400),
    1201 (427), 1246 (401), *1323, 1324*
Teste, J.   1040, 1042 (257), 1048 (357,
    358), *1058, 1061*
Thacher, A. F.   422 (1096), *484*
Thackeray, S.   700, 735–737 (142), *750*
Thalmann, A.   419 (1073), *484*
Thayer, G.   512 (125), *529*
Theilig, G.   547, 548, 571, 586 (133), *589*
Theobald, D. W.   355 (678), *475*
Theodoropoulas, D.   1123 (206), *1320*
Thiel, W.   1027, 1042 (62), 1043 (311),
    1054, 1059
Thiele, J.   1123 (201), *1320*
Thiemann, W.   88 (140), 93, 94 (154),
    116
Thijs, L.   383 (831), *479,* 1049 (365,
    366), *1061,* 1131 (226), *1320*
Thimm, K.   1024 (31), 1025 (43, 44, 47),
    1026, 1027 (43), 1030 (31, 44), 1031
    (31, 93), 1032 (31), 1033 (47), 1034
    (31, 44, 47, 93, 120), 1035 (43), 1039
    (31, 44, 120), 1040 (31, 47, 93), 1041
    (47, 93), 1042 (31, 120), 1045, 1046
    (43), 1050 (47, 120), 1052 (397),
    1053–1055, 1062
Thiruvengadam, T. K.   1052 (404), *1062*
Thomas, A. F.   148 (116), *172*
Thomas, B. R.   1284, 1291 (655), *1329*
Thomas, C. B.   341 (528), *472*
Thomas, D. K.   247, 248 (384), *264,* 609
    (61), *637*
Thomas, J. K.   763 (46, 50), 764 (53,
    54), 765 (46, 50), 767 (54), *778, 779*
Thomas, J. O.   32 (91), *57*
Thomas, P. J.   891, 897 (155), *913*
Thomas, R.   225 (111), *258,* 278 (126),
    422 (126, 1092), *464, 484*
Thomas, R. K.   218, 229 (10), *256*
Thomas, R. N.   88, 93, 94, 107 (143), 109
    (277), *116, 119*
Thomason, S. C.   315 (432), *470*
Thompson, H. W.   214 (1a), *256*
Thompson, L. R.   415 (996), *482*
Thompson, R. D.   518, 520 (170), *530,*
    1200 (420), *1324*
Thomson, J. B.   135 (58), *170*
Thorne, M. P.   893 (167, 172), 894, 899
    (172), 900 (167, 172, 199), *913*
Thornton, E. R.   161, 163 (181), *173,*
    633 (213), *640*
Thorstensen, J. H.   416 (1018),
    482
Thosi, N.   246 (367), *264*

Thuillier, A.   1024, 1031 (27), 1033 (27,
    107), 1034 (107), 1035 (27), 1037
    (107, 140), 1040 (255), 1042 (140),
    1047 (340, 341), 1049 (255), 1050
    (140, 373), 1052, 1053 (255, 393),
    1053, 1055, 1056, 1058, 1060–1062
Thurman, J. C.   861 (9–11), 862 (9),
    865 (24), *910*
Thyagarajan, B. S.   1206, 1207, 1211
    (462), *1325*
Tiberi, R.   310 (396), *470*
Tichonova, N. A.   566 (264), *592*
Tichý, M.   88 (131), *116*
Tietjen, D.   1138, 1139 (238), *1320*
Tikhonov, V. P.   226, 235 (133), *259*
Tille, H.   565 (241), *591*
Tillett, J. G.   140 (85), 154 (146), 155
    (146, 148), *171, 172*
Timmons, C. J.   71, 72 (17), *114*
Tinkelberg, A.   889 (144), 892 (157),
    894 (144), 896 (157), 897, 898, 901
    (144), *912, 913*
Tinkelenberg, A.   1050 (376), *1061*
Tinoco, I.   70, 83 (13), *114*
Tinoco, I. Jr.   70 (14), 73 (36, 37), 99
    (14, 36, 37), *114*
Tinoco, J.   332 (543), *473*
Tippett, J. M.   1276, 1282, 1286–1289,
    1291, 1300 (640), *1329*
Tiripicchio, A.   1022 (12), *1053*
Tiripicchio Camellini, M.   1022 (12), *1053*
Tirouflet, J.   795 (62), 816 (155), *822,
    824*
Tischbein, R.   1123 (201), *1320*
Tishchenko, I. G.   1052 (404), *1062*
Tishchenko, W.   307 (370), *469*
Tisler, M.   562 (205, 211, 223), 569
    (223), 570 (294), 571 (205, 211, 294),
    596, 597 (378), *590–592, 598*
Tissier, C.   798 (67), *822*
Titov, E. V.   238 (275), *262*
Tobias, C. W.   782 (15), *821*
Todd, A. R.   503 (72), *528*
Todd, D.   325 (494), *472*
Todd, J. F. J.   131 (37), *170*
Toder, B. H.   725 (190), *752*
Togue, M. W.   437 (1223), *487*
Toh, S. H.   904 (218), *914*
Tohadze, K. G.   238 (285), *262*
Tojo, T.   362 (719), *476*
Tokiwa, F.   984 (122), 1000 (168), *1018,
    1019*
Tokizawa, M.   868 (44), 869 (47), *910*
Tokoroyama, T.   294 (275), *467*
Tokuda, M.   1080 (88, 89), 1085 (88),
    1086, 1087 (88, 89), 1088 (89), *1317*
Tokura, N.   292 (258), 368 (841), 416
    (1020), *467, 479, 483*

Tokuyama, T. 1206, 1217, 1218 (478), *1325*

Toland, W. G. 176 (21), 181, 182, 196 (54), *209, 210*

Toland, W. G. Jr. 323 (485, 486), 324 (486, 487), 325 (490, 491), *471*

Tolgyesi, W. S. 416 (1025), *483*

Tolkachev, O. N. 1039 (250), *1058*

Tolman, C. A. 826, 827, 837 (31), 852 (94, 95), *856, 857*

Tolstikov, G. A. 339 (601), *474,* 1038 (187), *1057*

Tom, G. M. 450 (1338), *489*

Tomasi, J. 28 (64), 40 (105), 45, 46 (116), 47 (119), 48 (64), *56, 57,* 251 (422), *265*

Tomer, K. B. 128 (26), 140 (80), 141 (89–92), 143 (94), 151 (127, 128, 131), 155 (80, 150), *170–172*

Tomilov, A. P. 782, 801 (6), 806 (113), 807 (6), *821, 823*

Tominaga, Y. 1042 (288), 1052 (403), *1059, 1062*

Tominoga, M. 419 (1072), *484*

Tomita, N. 293 (264), *467*

Tomlin, J. E. 297 (300), *468*

Tomlinson, A. D. 865 (25–27), *910*

Tomlinson, G. E. 223 (66), *258*

Tomoto, T. 413 (959), *481*

Tonellato, U. 1011, 1012 (193), *1020*

Tonhara, H. 251 (420), *265*

Toniolo, C. 79 (64), 84 (90, 92), 88, 89 (135), 100 (171), 101 (64, 172), 102 (192, 193, 197, 200), 109 (275), *115–117, 119*

Toome, V. 102 (194–196), *117*

Toplitz, B. 1301 (680), *1329*

Toppet, S. 511 (121a), *529*

Topsom, R. D. 904 (218), *914*

Töregård, B. 106 (260), *119*

Torii, S. 277 (120), *464*

Torikai, A. 770 (78), *779*

Toriyama, K. 757 (3–5), 758 (5), *778*

Toropov, A. P. 618 (121), *638*

Torri, G. 161 (184), *173*

Tortorella, V. 102 (201, 202), 103 (201), *117,* 1307, 1309 (709), *1330*

Toru, T. 515 (142), *529*

Torzo, F. 814, 816, 817 (142), *824*

Tossi, R. 246 (370), *264*

Toubiana, R. 627 (178), *639*

Tovzin, A. M. 427 (1145), *485*

Towne, J. C. 127, 143 (20), *170*

Townley, E. 1145, 1159 (256), *1321*

Toyoda, K. 238 (278), *262*

Traficante, D. D. 230 (190), *260*

Trammell, G. L. 1066, 1069 (21), *1316*

Trampuž, C. 226 (142), *259*

Traut, H. 567, 571 (279), *592*

Trautner, K. 406 (914), *480*

Traverso, J. J. 542 (97), *588*

Trebault, C. 1040, 1042 (257), *1058*

Treboganov, A. D. 373 (782), *478*

Trebra, R. L. von 382, 383 (830), *479*

Trecker, D. J. 460 (1415), 490, 649 (12), *747,* 1313 (718), *1330*

Treiber, H. J. 547, 567, 571 (139), *589*

Treibs, A. 1037 (148, 162), *1056*

Treibs, W. 341 (621d), *474*

Tremper, A. W. 1313 (717), *1330*

Triem, H. 369 (754), *477*

Trimitsis, G. B. 295 (283), 461 (1418), *467, 490*

Tritschler, W. 541 (91), 542 (91, 104–107), 548 (91), 573 (91, 104–107), *588*

Trod, E. 1094, 1095 (114), *1318*

Trofimov, B. A. 1052 (398), *1062*

Trofimova, A. G. 1052 (398), *1062*

Tromeur, M. C. 1145 (275), *1321*

Trommer, W. 1276, 1282, 1287 (641), *1329*

Tronchet, J. M. J. 341 (620), *474*

Tronor, B. V. 176 (19), *209*

Trontelj, Z. 231 (201), *260*

Troparevsky, A. 437 (1232), *487*

Trost, B. M. 375, 379 (802, 803), 425 (1123), 428 (1153, 1154), *478, 485,* 566, 567 (277), *592,* 1164 (294, 295), 1165 (295), 1168 (294, 295), 1300 (677, 678), *1321, 1329*

Trotman-Dickenson, A. F. 868 (43), *910*

Trotter, J. 61 (32), *66,* 244 (349), *264*

Troughton, P. G. H. 827 (35), *856*

Trublood, K. N. 244 (351), *264*

Truesdale, E. A. 692, 706 (134), *750*

Trumpler, G. 790 (45), *822*

Truter, M. R. 255 (477), *266*

Trzmielewska, H. 635 (218), *640*

Tsai, C.-C. 1276, 1278, 1289 (638), *1328*

Tsai, M. 1232 (563, 564), 1249 (563), 1250 (564), *1327*

Tsangaris, J. M. 103 (231), *118*

Tsatsaronis, G. 571 (304), *593*

Tsay, Y.-g. 735 (209b), *753*

Tschesche, R. 333 (561, 562), *473*

Tsou, K. C. 415 (971), *482*

Tsoucaris, G. 1265, 1267 (606), *1328*

Tsuboi, M. 31 (87), *57,* 246 (376), 255 (471), *264, 266,* 607 (38), *636*

Tsubor, S. 1106, 1107 (156), *1319*

Tsuchida, M. 398 (894), *480,* 916, 919, 928 (10), *942*

Tsuchihashi, G. 151 (130), *172,* 287 (221–223), 400 (900, 901), 401 (900), *466, 480,* 1025, 1026 (49), 1036 (134), 1039 (239, 243), 1040 (262), *1054, 1055, 1058*

Tsuchihashi, G.-I. 1039 (239), *1058*
Tsuchiya, T. 902 (209), *914*
Tsuda, K. 1038 (188), *1057*
Tsuda, M. 251 (420), *265*
Tsuda, T. 1145 (259), *1321*
Tsuda, Y. 1168 (304), *1322*
Tsuji, J. 275 (94), 441 (1271, 1272, 1275, 1276), *463, 487, 488,* 826 (20), 827 (36), 828, 829 (40, 41), 831 (40), 834 (40, 41, 55), 835 (40, 41, 49), 837 (41, 56), 838, 839, 841 (41), 843 (41, 74, 82), 844 (36, 41), 845 (36, 41, 74), 847, 848 (41, 74), 852 (41), *855–857,* 1192 (362), *1323*
Tsuji, S. 1028, 1032, 1041 (70), *1054*
Tsujino, N. 197 (92), *211*
Tsukida, K. 1080 (64), *1317*
Tsunetsugu, J. 513 (129), *529*
Tsutsui, M. 854 (97), *857*
Tsutsumi, S. 455 (1380), *490*
Tsuzuki, Y. 238, 239 (288), *262*
Tucker, E. E. 217 (7), 250 (408), *256, 265*
Tucker, J. N. 1295 (669), *1329*
Tucker, M. P. 676, 743 (221b), *753*
Tucker, S. H. 365 (729), *476*
Tuley, W. F. 322 (476), *471*
Tulis, R. W. 1307 (708), *1330*
Tuller, F. N. 1185 (344), *1323*
Tulloch, A. P. 332 (542), *473*
Tumolo, A. L. 339 (605), 346 (636), *474, 475*
Tumolo, M. 339 (605), *474*
Tuncay, A. 295 (283), *467*
Tunemoto, D. 312 (403), *470*
Turk, J. 156 (157), *172*
Turnbull, H. C. 874 (77), *911*
Turner, G. 835, 837–840 (48), *856*
Turner, H. S. 916 (12), *942*
Turner, L. M. 411 (935), *481*
Turner, R. B. 339 (611), *474*
Turner, S. 75 (45), *114,* 1145 (274, 276), 1154 (274), 1155 (274, 276), *1321*
Turner, S. R. 209 (155), *212*
Turrion, C. 61 (29–31), 62 (29, 30), *66*
Turrion, C. J. 60 (15, 16), 62 (16), 63 (15), *65*
Turro, N. 729, 737 (195b), *752*
Turro, N. J. 206 (149), *212,* 672, 686, 687 (63d), 698 (140), (177), *748, 750, 751*
Tur'yan, Ya. I. 618 (120), *638*
Tutsch, R. 223 (72), 226 (134), *258, 259*
Tuzimura, K. 82, 83 (76–78), 84 (89), 85, 87 (97), 102 (204), *115, 117,* 517 (153, 154), *529*

Tykva, R. 916–920 (23), 923 (38), 927 (23, 42), 928 (23), 936 (38, 58, 62), 939 (62), *943, 944*
Tyman, J. H. P. 294 (278), *467*
Tysee, D. A. 809 (130, 131), *824*
Tyulakov, V. I. 439 (1245), *487*

Ubbelohde, A. R. 224 (95), *258*
Uccella, N. 155 (149), *172*
Uchibayashi, M. 318 (447), *471*
Uchida, I. 85–87 (96), *115*
Udupa, H. V. K. 786 (27, 28), 787 (29, 30), 790 (46), 800 (82), *821–823*
Udupa, K. S. 787 (29, 30), 790 (46), *821, 822*
Ueda, T. 537 (50), *587,* 1038 (180), *1056*
Ueda, Y. 995 (163), *1019*
Uefuji, T. 101 (181), *117*
Ueki, M. 443 (1291), *488*
Ueno, Y. 1038, 1042, 1052 (168), *1056*
Ueoka, R. 1007 (182), *1020*
Ueshima, T. 307 (374), *469*
Uff, B. C. 353 (655), *475*
Ugi, I. 441 (1270), *487*
Ugo, R. 825 (9), 826 (23), 827 (33), *855, 856*
Uhlig, F. 385 (839), *479,* 1274 (622), *1328*
Uhm, S. J. 423 (1112), *484*
Ukai, A. 1168 (304), *1322*
Ukita, T. 1038 (180), *1056*
Uliss, D. B. 167 (216, 217), 168 (217), *174*
Ullman, E. F. 712, 719, 725 (180a–c), 727 (180b, c), 745 (180a–c), *751*
Ullman, F. 506 (89), *528*
Ulrich, H. 539 (68), 562 (202), *588, 590,* 1225 (497), *1326*
Umani-Ronchi, A. 281 (190), *465*
Umanskaya, L. I. 1038 (187), *1057*
Umemura, J. 234 (225, 226), *261*
Umen, M. J. 803 (91), *823*
Umetani, T. 309 (387), *469*
Umezawa, K. 691 (127), *750*
Undheim, K. 88 (141, 142), 89 (142), 90 (141), 94 (141, 142), 102 (186, 187, 209, 210), 109 (209, 210), *116–118*
Unge, T. 109, 110 (294), *119*
Unkolova, A. I. 1039 (250), *1058*
Unrau, A. M. 849, 850, 854 (88), *857*
Untereker, D. F. 238 (284), *262*
Upham, R. A. 603 (16), *636*
Urban, R. S. 386 (852), *479*
Urdaneta, M. 979 (104), *1018*
Urry, D. W. 71 (29), 77 (56), 84 (91), 89 (146), 100 (169), 101 (56, 91), 103, 104 (146), *114–117*

Usui, M.   413, 444 (958), *481*
Utermoehlen, C. M.   167, 168 (220), *174,*
   520 (175), *530,* 1200 (423), *1324*
Utley, J. H. P.   810, 811 (135), 812 (139),
   824
Uyeo, S.   521 (184), *530*

Vaalburg, W.   295 (286), *467*
Vacek, K.   759, 760 (27), *778*
Vahlensieck, H. J.   202 (105), *211*
Vairamani, M.   155 (153), 156 (155), *172*
Vala, M.   229 (169), *260*
Valcavi, V.   1100 (137, 138), *1318*
Valkanas, G.   323 (482), *471*
Vallely, F.   826 (17), *855*
Vallén, S.   103, 104 (242, 243), *118*
Van Acker, L.   1038 (174), *1056*
Van Boom, J. H.   416 (1008), *482*
Van Chung, V.   1080, 1086–1088 (89),
   1317
Vandenbranden, S.   617 (115), *638*
Van Den Elzen, R.   1300 (676), *1329*
Van der Berghe, J.   382, 383 (830), *479*
Van der Steen, D.   136 (65), *170*
Van der Veen, J. M.   1232, 1259 (561),
   1327
Vanderwalle, M.   147 (110), *171*
Van der Werf, C. A.   1104 (148), *1319*
Vanderzee, C. E.   60 (18), *66*
VanFossen, R. Y.   162 (185), *173*
Van Haard, P. M. M.   383 (831), *479,*
   1131 (226), *1320*
Van Hecke, P. E.   230 (181), *260*
Van Heyningen, E. M.   272 (51), 416
   (1014), *463, 482,* 561 (198), *590*
Van Meerbeck, M.   511 (121a), *529*
Van Meerssche, M.   1044 (315), *1059*
Van Raalte, D.   35 (98), *57,* 138 (69),
   170
Van Steenis, J.   884 (121), *912*
Van Volkenburgh, R.   334 (573), *473*
Van Zyl, G.   1096 (119), 1104 (149),
   1318, 1319
Varbroggen, H.   413 (962), *481*
Varsel, C.   1075, 1079 (52), *1316*
Vartires, I.   803 (93), *823*
Vas, S.   306 (363), *469*
Vasi, I. G.   1039 (211), *1057*
Vasileff, R. T.   272 (51), *463,* 561 (198),
   590
Vasilenko, N. L.   1048 (352), *1060*
Vasil'eva, E. I.   276 (110), *464*
Vasil'eva, T. T.   276 (110), *464*
Vaska, L.   826 (24), *856*
Vasol'eva, T. F.   65 (45), *66*
Vassilev, G.   281 (187), *465*
Vasyanini, G. I.   449 (1333), *489*
Vaughan, R. W.   230 (182), *260*

Vaughan, W. R.   394 (875), *479,* 1111
   (178), 1138, 1141 (242), *1319, 1320*
Vaughn, H. L.   274 (78), *463*
Vaux, R. L.   1205 (449), *1325*
Vecchi, C.   177 (24), 184 (66), *209, 210*
Vecchi, M.   (141), *172*
Vecher, R.   234 (229), *261*
Vedejs, E.   428 (1150), *485*
Veenstra, G. E.   1044, 1049 (314), *1059*
Veierov, D.   733–735 (204c), *753*
Veillard, A.   48 (122), *58*
Vejdelek, Z. J.   1109 (169), *1319*
Velasquez, O.   908 (233), *914*
Velichko, F. K.   1192 (363), *1323*
Velluz, L.   69, 73, 108 (4), *113,* 611 (72,
   74), *637*
Vendrova, O. E.   108 (270), *119*
Venema, A.   138 (78), 160 (179), *171,*
   173
Venkateswarlu, A.   278 (129), *464*
Venkateswarlu, P.   254 (451), *266*
Venturello, C.   290 (243), *466*
Verbit, L.   71 (18), 109 (18, 282, 289,
   290), 110 (282, 299), *114, 119*
Vercek, B.   562, 571 (205), *590*
Verdin, D.   771 (81, 82), *779*
Verdol, J. A.   903, 906 (213a), *914*
Verkade, P. E.   88 (132), *116*
Vermeer, P.   1037 (141), 1042 (141,
   294), 1046 (334), 1049 (362), 1050
   (294), 1052 (294, 406), 1053 (294),
   1056, 1059–1062
Vermeeren, H. P. W.   878 (89), *911*
Vermeulen, N. M. J.   628 (185), *639*
Vernon, J. M.   651, 655 (18), *747,* 877
   (87), *911*
Verschoar, H. M.   419 (1071), *483*
Vesheva, L. V.   618 (122–124), *638*
Vesley, G. F.   411 (933), *481,* 672, 686,
   687, 740 (62, 63a–c), *748*
Vetter, W.   154 (142), (141), *172*
Vialle, J.   1034 (114), 1040 (255), 1041
   (114), 1042 (114, 290), 1043 (114),
   1047 (342), 1049, 1052 (255), 1053
   (114, 255), *1055, 1058–1060*
Viani, R.   1079 (62), *1317*
Vibet, A.   1039 (206), *1057*
Vickery, B.   205 (135, 138), *212*
Vickery, G.   206 (143), *212*
Vida, J. A.   461 (1420), *490*
Viehe, H. G.   437 (1226), *487,* 1044
   (315), *1059*
Viehe, R. G.   341 (619), *474*
Vieles, P.   103 (239), *118*
Viennet, R.   71, 88, 89, 95 (30), 101
   (176), 109 (287), *114, 117, 119*
Vigevani, A.   610 (68), *637*
Vijayendran, B. R.   987 (137), *1019*

Vikana, O. 250 (413), *265*
Vilkas, M. 1080, 1084 (84, 85), *1317*
Villani, F. J. 307 (377), *469*
Villepin, J. de 228 (154), *259*
Villieras, J. 389 (862), 421 (1090), *479, 484*
Vilsmaier, E. 422, 423 (1100), *484*
Viney, M. 1196 (388), 1206, 1222 (487), *1323, 1325*
Vinkler, P. 255 (470), *266,* 1031, 1034, 1040, 1041 (93), *1055*
Vinograd, L. K. 1038 (185), *1057*
Vinogradov, M. G. 279 (147), *465,* 1079 (57), 1080 (72), *1317*
Vinogradov, S. N. 214 (1e), *256*
Vinogradova, L. V. 1192 (363), *1323*
Viola, A. 908 (234), *914*
Viola, H. 1027, 1042 (62, 64), 1043 (64, 306, 311), 1044 (312), 1046 (64), *1054, 1059*
Viout, P. 959 (68), 991 (155), 1008 (185), *1017, 1019, 1020*
Vipond, P. M. 70, 95 (16), *114*
Virmani, Vinod 566, 570, 571 (274), *592*
Viswanathan, N. 1305 (701), *1330*
Vitale, A. C. 986 (136), *1019*
Vivarelli, P. 1024 (38), *1053*
Vlasov, G. M. 439 (1245), *487*
Vlasova, T. 582 (340), *593*
Vlasova, T. F. 536 (35), 562 (217, 218), 570, 571 (35, 217), 596 (379–381, 386), 597 (379, 381, 386), *587, 591, 599*
Vliet, E. B. 1098, 1099 (127), *1318*
Voelter, W. 88 (111, 112), 89 (111), 103, 104 (241), 109 (112), *116, 118*
Voevodskii, V. V. 761 (38), *778*
Vofsi, D. 801 (83), *823*
Vogel, E. 508 (103), *528,* 1038, 1052 (184), *1057*
Vogel, H.-H. 288 (228), *466*
Voglet, N. 439 (1248), *487*
Vogt, B. R. 382 (824), *478*
Volke, J. 794 (60), *822*
Volkova, V. 794 (60), *822*
Volkova, Z. S. 1052 (398), *1062*
Vollrath, R. 552 (169), *590*
Volman, D. H. 687, 688 (105b), *750*
Volosov, A. P. 83, 84 (87), *115*
Vol'pin, M. E. 850, 851 (90), *857*
Vora, M. 806 (116), *823*
Vorbrüggen, H. 559, 561 (194, 195), 574 (195, 321), *590, 593*
Vorhees, K. J. 899 (195), *913*
Voronkov, M. G. 1040 (252), 1052 (398), *1058, 1062*
Voss, J. 255 (472), *266,* 1023 (16, 22), 1024 (31), 1025 (43, 47), 1026, 1027 (43), 1030 (31), 1031 (31, 93), 1032

(31), 1033 (47), 1034 (31, 47, 93, 115, 118, 120), 1035 (43, 115, 118), 1037 (147), 1039 (31, 115, 120), 1040 (31, 47, 93, 115), 1041 (47, 93, 118), 1042 (31, 115, 118, 120, 414), 1043 (118), 1044 (415), 1045 (43, 115, 118), 1046 (43, 115), 1047 (147), 1048 (115), 1050 (47, 120), 1052 (397), *1053– 1056, 1060, 1062*
Vries, G. de 305 (356), *469*
Vuitel, L. 417 (1053), *483,* 541 (95), *588*
Vvedenskii, A. A. 65 (45), *66*
Vyazankin, N. S. 807 (117), *823*
V'zerova, I. 365 (727), *476*

Wacek, A. von 353 (662), *475*
Wackerle, L. 441 (1270), *487*
Wacks, M. E. 127, 143 (20), *170*
Wada, S. 236 (250), *261*
Wada, T. 1079 (60), *1317*
Waddell, T. G. 85, 86 (95), *115,* 625 (169), *639*
Waddey, W. E. 273 (68), *463*
Waddington, D. J. 877 (87), *911*
Wade, A. M. 1298, 1299 (673), 1302 (688), *1329, 1330*
Wade, T. 102 (207), *117*
Wadia, M. S. 104 (249), *118*
Wadso, I. 62 (35), 64, 65 (42), *66*
Wadsworth, W. S. Jr. 423 (1093), *484*
Wagenknecht, J. H. 415 (994), *482,* 784 (20), 785 (22), 787 (20), *821*
Wagner, F. 564, 566, 567 (233), 573, 575 (319), *591, 593*
Wagner, P. J. 685 (101a, b, 176a–c), *749, 751*
Wagner, R. B. 356 (685), 357 (687), *476*
Wagner, R. M. 567 (280), *592*
Wagner, T. E. 1003 (172), *1020*
Wagnière, G. H. 83, 84 (86), *115,* 612 (219), *640*
Wahl, G. H. Jr. 110, 111 (309), *120*
Wahl, R. 536, 538, 543 (28), 545, 548 (28, 125), 553 (125), 554 (28, 175), 556 (175), 557, 558 (28), 566, 567 (175), 571 (286), 574 (125, 323), *587, 589, 590, 592, 593*
Waight, E. S. 1078 (55), *1316*
Wakabayashi, N. 451 (1344), *489*
Wakamatsu, S. 1276 (626), *1328*
Wakefield, B. J. 373 (778, 779), *477, 825 (1), *855*
Wal, R. J. van der 332 (545), *473*
Walborsky, H. M. 295 (284, 285), 315 (429), *467, 470,* 843–845, 848 (72, 73), 852 (73), *857*

Walia, J. S.   316 (439), *471*
Walia, P. S.   316 (439), *471*
Walker, C. H.   1302 (688), *1330*
Walker, G. B.   272 (41), *462*
Walker, J. A.   393 (870), *479*
Walker, K. A. M.   846, 849 (76), *857*
Walker, R.   1145, 1159 (278), *1321*
Wall, D. K.   413 (961), 437 (1220), *481, 486*
Wall, M. E.   332 (540), *473,* 1145, 1151, 1160 (268), *1321*
Wallace, J. G.   305 (351), *469*
Wallace, T. J.   328 (515), 329 (517), 357 (686), *472, 476*
Wallach, O.   386 (854), *479*
Walley, A. R.   700, 735−737 (142), *750*
Wallis, S. R.   904 (215), *914*
Wallmark, I.   1024 (30), *1053*
Walsh, A. D.   70 (28), *114*
Walsh, E. J. Jr.   416 (1027), *483*
Walsh, I.   159 (176), *173*
Walsh, R.   60, 61, 63, 65 (8), *65*
Walshaw, K. B.   355 (678), *475*
Walter, R.   1123 (204−206), *1320*
Walter, W.   254 (454, 461), 255 (468−470, 472), *266,* 537, 553 (48), *587,* 1029, 1033 (82−84), 1034 (83), 1036 (83, 84), 1037, 1047 (147), *1054, 1056*
Walther, W.   154 (142), (141), *172*
Walton, D. R. M.   1041, 1050 (276), *1059*
Walton, E.   504 (78), *528*
Walz, H.   516 (148, 149), *529*
Wamhoff, H.   399 (896), 406 (914), *480,* 502 (69), *528,* 1200, 1204 (426), *1324*
Wan, J. K. S.   740 (220b), *753*
Wang, S. Y.   673 (70a, b), *748*
Wansbrough-Jones, O. H.   672, 677 (59b), *748*
Wanzlick, H. W.   550 (152, 153), 586 (152, 365), *589, 594*
Ward, A. D.   275 (96), 437 (1225), *464, 487,* 1276, 1282 (640), 1284 (648), 1286, 1287 (640), 1288, 1289 (640, 648), 1291 (640), 1295 (648), 1300 (640, 648), 1302 (648), *1329*
Ward, A. M.   908 (236), *914*
Ward, D. J.   848 (84), *857*
Ward, T. R.   235 (248), *261*
Wardley, A. A.   178 (31), *210*
Warner, C. D.   450 (1341), *489*
Warner, D.   230 (187), *260*
Warnhoff, E. W.   436 (1214), *486,* 609 (62a), *637*
Warren, C. L.   1138, 1141 (242), *1320*
Warren, J. P.   610 (65), *637*
Warren, P. C.   672, 686, 687, 740 (63c), *748*
Warrener, R. N.   164 (203), *173*

Warring, W. S.   503 (72), *528*
Warwel, S.   1039 (220), *1057*
Wasada, N.   902 (209), *914*
Washüttl, J.   617 (116), *638*
Wasserman, H.   281, 328 (184), *465*
Wasserman, H. H.   436 (1210), 437 (1222), 456, 457 (1392), *486, 490,* 523 (199), *530,* 596 (388), *599,* 1284 (657, 658), 1285 (658−660), 1300 (683), 1313 (683, 717), *1329, 1330*
Wasserman, J.   150 (126), *172,* 1035, 1036 (128), *1055*
Wassiliadou-Mitcheli, N.   415 (982), *482*
Watabe, M.   103 (217), *118*
Watanabe, H.   815 (149), *824*
Watanabe, K.   444 (1297), *488*
Watanabe, N.   251 (420), *265*
Watanabe, S.   280 (161−163), 281 (161, 162), 359 (696), *465, 476,* 1116 (196), *1320*
Watanabe, T.   245 (364), *264*
Watanabe, Y.   571 (301), *592,* 1039 (217), *1057*
Waters, R. M.   451 (1344), *489*
Waters, W. A.   298, 332 (309), *468*
Watkins, A. R.   204 (131), *212*
Watkins, D. A. M.   673, 682 (72), *749*
Watkinson, J. G.   235 (246), *261*
Watson, T. G.   314, 355 (422), *470*
Watt, A. N.   415 (974), *482*
Watter, R. A.   763 (47), *778*
Watts, P. C.   296 (292), *468*
Watts, P. H. Jr.   1292 (663), *1329*
Waugh, J. S.   230 (181), 231 (193), *260*
Wawzonek, S.   445 (1303), *488,* 805, 806 (112), *823*
Weaver, J. C.   230 (181), *260*
Weaver, S. D.   1232 (598), *1328*
Webb, A. D.   493 (5), *526*
Webb, J.   95 (159), *116*
Webb, K. H.   235 (246), *261*
Webb, L. A.   769 (75), *779*
Webb, R. L.   566, 567, 571 (253), *591*
Weber, H.   571 (287), *592*
Weber, L. D.   536, 566, 571 (25), *587,* 1196 (387), *1323*
Weber, W. P.   1051 (383), *1061*
Webster, B. R.   1284 (654), *1329*
Weedon, B. C. L.   372 (773), 373 (776−781), 433 (1192), *477, 478, 486,* 810, 811 (135), *824*
Week, A. P. M. van der   875 (79, 84), *911*
Wegler, R.   343 (625), 369 (753), *474, 477,* 1131 (219−222), 1132 (219−221), 1133 (221), 1134 (221, 222), 1135 (222), *1320*
Wegrzynski, B.   102 (196), *117*
Weidemann, E. G.   221 (38), *257*

Weidinger, H.   539 (67, 70, 71), 550 (67, 157), 557, 559 (157), 569 (67), *588, 590*
Weidler, A.-M.   102 (187), *117*
Weidmann, E.   220 (36), *257*
Weidner, M.   1053 (412), *1062*
Weigang, O. E.   612 (88), *637*
Weigang, O. E. Jr.   76 (49), 98 (160, 161), *114, 116, 117*
Weigele, M.   102 (194), *117,* 566 (266), *592*
Weil, H.   786 (23), 789 (38), *821, 822*
Weil, T. J.   1048 (357), *1061*
Weiler, L.   150 (123), *172,* 425 (1119–1122, 1124), *484, 485*
Weill, G.   101 (174), *117*
Weinberg, N. L.   782 (3), 790 (42), 807, 818 (3), *821, 822*
Weiner, S. A.   803 (94), *823*
Weingarten, H.   444 (1296), *488,* 548 (141–145), 552 (167), 554 (145), 555 (178), 557 (145), 566–568 (178), 572 (145), 585 (364), *589, 590, 594*
Weininger, S. J.   161, 163 (181), *173,* 633 (213), *640,* 1191 (357), *1323*
Weinreb, S. M.   1039 (221), *1057*
Weinsenborn, F. L.   364 (721), *476*
Weinshenker, N. M.   437 (1224), *487,* 1168 (301, 302), *1322*
Weinstein, M.   673, 683, 685 (85a), *749*
Weinstein, S.   146 (104, 105), *171*
Weinstock, J.   1052 (401), *1062*
Weinstock, L. M.   1052 (407), *1062*
Weise, A.   576 (325), *593*
Weise, J. K.   85, 107 (94), *115*
Weiss, D. S.   672, 686, 687 (63d), *748*
Weiss, J. J.   176 (7), *209*
Weiss, K.   182, 196, 204 (57), *210*
Weiss, L. B.   499 (48), *527*
Weiss, R.   48 (122), *58,* 1109 (166), *1319*
Weiss, R. G.   711 (172), *751*
Weiss, U.   108 (272), *119*
Weissbach, R.   817 (160), *824*
Weissberger, A.   411 (931), *481*
Weisshuhn, M. C.   1038, 1042, 1052 (167), *1056*
Weissman, B.-A.   1006 (177), *1020*
Weizmann, C.   276 (104), *464*
Weizmann, M.   341 (621b, c), *474*
Welb, R. L.   422 (1097), *484*
Welber, J. A.   272 (51), *463,* 561 (198), *590*
Welch, S. C.   415 (975), *482*
Welinder, H.   275 (98), *464*
Welker, C. H.   1298, 1299 (673), *1329*
Weller, W. T.   1066, 1146 (20), *1316*
Wellman, K. M.   88 (114), 103 (227, 228), *116, 118*

Wells, C. H. J.   176 (14, 15), 177 (15, 27), 179, 180 (14), 182 (27), *209, 210*
Wells, D.   690, 701, 702 (121b), *750*
Wells, J. N.   1232, 1251 (558), *1327*
Welsh, W.   511 (120), *529*
Weltner, W.   235 (237), *261*
Welzel, P.   82, 85, 87 (73), *115, 159 (172), 173*
Wemple, J.   1035 (113), 1039 (213, 240), 1050 (377, 378), *1055, 1057, 1058, 1061*
Wenkert, E.   566, 567, 571 (253), *591,* 1276 (630), *1328*
Wentland, S. H.   436 (1210), *486*
Wentrup, C.   909 (242), *914*
Wentworth, W. E.   196, 197 (89), *211*
Wepster, R. M.   88 (132), *116*
Werner, D.   507 (94), *528*
Werner, R. L.   248 (399), *264*
Werner-Zamojska, F.   110 (297), *119*
Werstivk, E. S.   412 (945), *481*
Wertheimer, R.   22 (35), *56*
Wesdorp, J. C.   1049 (362), *1061*
Wesigerber, G.   406 (914), *480*
West, G.   727, 728 (192b), *752*
West, P. J.   1040–1042 (268), 1052 (268, 388, 389), *1058, 1061*
West, P. R.   768 (71), *779*
West, R.   241 (316), *263*
Westheimer, F. H.   863 (15), *910*
Westley, J. W.   109 (281), *119*
Westmore, S. I.   1052 (392), *1062*
Weston, A. W.   375 (805), 450 (1340), *478, 489*
Westphal, G.   562 (203), 571 (299), *590, 592*
Westrum, E. F.   60 (7, 11), 63 (7), *65*
Wetmore, D. E.   493 (8), *526*
Wetmore, S. I. Jr.   706, 707 (148), *751*
Wettslein, A.   1171 (316), *1322*
Wetzel, W. H.   890, 891, 896, 904, 905 (146), *912*
Wexler, S.   1171, 1172 (317), *1322*
Weyell, D. J.   416 (1024), *483*
Weyerstahl, P.   508 (103), *528,* 1108 (158), *1319*
Weygand, F.   334 (571), 380 (817), *473, 478,* 1039 (229), *1058,* 1138, 1139 (238), *1320*
Whalley, A. R.   736, 737 (213), *753*
Whalley, W. B.   75 (44), *114,* 375 (787), *478*
Wharton, P. J.   437 (1222), *486*
Wheeler, T. W.   382 (825), *478*
Wheeler, W.   458 (1407), *490*
Wheland, R.   1138 (230), *1320*
Whistler, R. L.   1038 (176), *1056*
White, A. M.   1045 (325), *1060*

White, D. L. 513 (132), *529*
White, D. R. 278 (124), *464*
White, E. H. 274 (84), 419 (1062, 1063), *463, 483*
White, J. D. 423 (1111), *484,* 1066, 1069 (21), *1316*
White, L. (137), *750*
White, M. N. 1048 (357), *1061*
White, R. D. 238 (286), *262*
White, R. H. 1039 (233), *1058*
White, R. W. 354 (674), *475*
White, W. A. 548 (141, 142), 552 (167), *589, 590*
Whitehead, C. W. 542 (96, 97), *588*
Whitehead, W. L. 759 (15), *778*
Whitesides, G. M. 286 (217), *466, 839* (63), *856*
Whitfield, G. F. 134 (54), *170*
Whitfield, R. E. 456, 457 (1387), *490*
Whiting, M. C. 272 (31, 32), 310, 314 (389), 323 (484), 435 (1209), 437 (1221), *462, 470, 471, 486,* 1080, 1082 (67), 1173, 1179 (330), *1317, 1322*
Whitlock, B. J. 296 (291), *467*
Whitlock, H. W. Jr. 296 (291), *467*
Whitmore, F. C. 320 (467), *471*
Whitnack, G. C. 805 (103), *823*
Whittaker, A. G. 253 (437), *265*
Whittaker, D. 902 (210), *914,* 1270 (611), *1328*
Whittaker, N. 446 (1312), *488,* 1199 (409), *1324*
Whitten, J. L. 32 (88), *57*
Whittle, E. 661, 662, 665 (45, 46), *748*
Wibaut, J. P. 799 (70), *822,* 887 (136), *912*
Wiberg, K. B. 332 (538), (763), 417 (1043), *473, 477, 483*
Wiberg, N. 547, 586 (137), *589*
Wick, A. 517 (161), *529*
Wick, A. E. 451 (1349), *489,* 583 (343, 350), 584 (343), *593*
Widdowson, D. A. 310 (395), 447 (1321), *470, 488*
Widmer, E. 369 (755), *477*
Wieland, Th. 111 (311), *120*
Wiemann, J. 246 (367), *264*
Wien, R. G. 673 (77), *749*
Wijnberg, J. B. P. A. 525 (220), *531*
Wijnen, M. H. J. 687 (107, 109, 111–113), 688 (107, 109, 111–115), 693 (114), *750*
Wiklund, E. 110 (303, 304, 306), 111 (303, 306), *120*
Wilcox, E. J. 1295 (669), *1329*
Wilder, R. S. 318 (455), *471*
Wilds, A. L. 382 (827, 830), 383 (830), *478, 479*

Wiley, H. 1098 (123), *1318*
Wiley, J. C. 454 (1377), *490*
Wiley, R. 511 (117), *529*
Wiley, R. H. 273 (68), *463,* 500 (54), *527,* 1080 (77, 78, 80), 1083 (80), 1096 (117), *1317, 1318*
Wilke, G. 825 (8), *855*
Wilkerson, A. K. 195 (76), *211*
Wilkes, J. B. 323 (485), 325 (490, 491), *471*
Wilkinson, G. 826 (21, 27), 827 (27, 32, 34, 37), 828 (37), 835 (37, 46), 837 (32), 841 (32, 67), 843 (68, 69), 844 (37, 68, 69), 845 (69), 849 (85), 850 (89), 852 (69), 853 (27), *856, 857*
Will, F. 495 (21), *527,* 627 (181), *639*
Will, F. III 166 (212), *173*
Willemarte, A. 816, 817 (153, 154), *824*
Willfang, G. 517 (157a), *529*
Willhalm, B. 148 (116), *172*
Williams, A. 622, 624 (155), *639,* 990 (150), *1019*
Williams, A. E. 127, 130, 131 (16), 133, 134, 137, 138 (48), 142, 158 (16), *169, 170*
Williams, B. C. 179, 182, 192 (39), *210*
Williams, B. D. 162 (190), *173*
Williams, C. C. 1039 (248), *1058*
Williams, D. E. 222 (59), *257*
Williams, D. H. 123 (1), 132 (38), 135 (57, 59, 60), 137 (59), 143 (96), 144 (38, 97, 100), 146 (106), 147 (111), 148 (106), 149 (121), 155 (149), 160 (178), *169–173,* 614 (96), 616 (108), 617 (113), 626, 627 (175), 628 (183), 633 (211), *638–640*
Williams, D. L. H. 494 (12), *526*
Williams, G. J. B. 222, 239 (58), *257*
Williams, J. D. 1039, 1052, 1053 (200), *1057*
Williams, J. L. R. 879 (93), *911*
Williams, K. 414 (1003), *482*
Williams, P. H. 354 (673), *475*
Williams, R. E. 1115 (186), *1319*
Williams, R. F. 986 (135), *1019*
Williams, R. L. 88 (129), *116,* 241 (315), *263*
Williams, V. Z. 22 (30), *56*
Williamson, K. L. 253 (436), *265,* 901 (205), *914*
Wills, J. 709, 718 (152), *751*
Wilsmore, N. T. 498 (44), *527*
Wilsmore, N. T. M. 871 (60), *911*
Wilson, C. L. 801 (84), *823*
Wilson, C. V. 432 (1184), *486*
Wilson, E. B. 16 (4), *55,* 222, 224 (49), *257*
Wilson, E. B. Jr. 28, 29 (60), *56*

Wilson, G. L.   127, 157 (23), *170*
Wilson, J. A.   177, 182 (27), *210*
Wilson, J. D.   444 (1296), *488,* 548 (143, 144), *589*
Wilson, J. M.   153 (140), *172,* 613, 614 (95), *638*
Wilson, K. B.   801 (84), *823*
Wilson, R. C.   273 (59), *463*
Wilson, W. C.   306 (366), *469*
Wilt, M. H.   339 (603), *474*
Wimmer, Th.   1232 (550, 586), 1247 (550), 1248 (550, 586), 1260 (586), 1263 (550), *1327*
Wims, A. I.   321, 322 (478), *471*
Winberg, H.   537, 545, 586 (58), *587*
Winberg, H. E.   540 (85), 544, 548, 549 (114), *562* (85), 586 (114, 366, 367), *588, 589, 594*
Winch, R. W.   1037 (160), *1056*
Windholz, T. B.   872 (66), *911*
Winestock, C. H.   382, 383 (830), *479*
Winkelmann, H. D.   547 (135), *589*
Winkler, D.   456, 457 (1389), *490*
Winkler, F. K.   244 (341–343), 245 (341, 342), 246 (368, 369), *263, 264*
Winkler, J.   108 (273, 274), *119*
Winkler, T.   1039 (242), *1058*
Winnacker, E. L.   517 (161), *529*
Winnewisser, G.   1022 (3, 4), 1024 (3), *1053*
Winnik, M. A.   137 (68), 143 (95), *170, 171*
Winstein, S.   406 (912), 432 (1177), *480, 486,* 496 (32), 497 (37), *527*
Winter, R.   505 (85), *528*
Winter, S. R.   292, 456 (257), *467*
Winterfeldt, E.   1206 (470), *1325*
Wiriyachitra, P.   1310 (712), *1330*
Wirz, J.   1040–1042 (268), 1052 (268, 388, 389), *1058, 1061*
Wise, R. M.   273 (56), *463*
Wislicenus, W.   916 (2), *942*
Witham, G. H.   1033, 1035, 1042 (108), *1055*
Witkop, B.   720, 722 (185e), *752,* 1206 (478–483), 1217 (478), 1218 (478, 480), 1219 (481–483), *1325*
Witkowski, A.   220 (29, 30), 222 (30), 226 (139), *257, 259*
Witt, L. C.   63 (44), *66*
Wittbecker, E. L.   357 (687), *476*
Witte, H.   293 (266), *467*
Witte, K.   895 (180), *913*
Woerden, H. F. van   664, 666, 708, 738 (57a, b), *748*
Woessner, W. D.   1039 (248), *1058*
Wogatsuma, S.   336 (580), *473*
Wojcik, M.   220 (31), 226 (139), *257, 259*

Wojciki, A.   839 (61), *856*
Wojtkowski, P. W.   497 (34), *527,* 620 (148), *639,* 1192 (359), *1323*
Wojtowiak, B.   235 (233, 242), 236 (233), *261*
Woldring, M. G.   295 (286), *467*
Wolf, F.   307 (368), *469*
Wolf, H.   78, 80 (59), 84 (59, 88), *115,* 368 (749), *477,* 612 (85, 86), 624 (85, 165, 166), 625 (165), *637, 639*
Wolf, W.   1206, 1207, 1211 (462), *1325*
Wolfbeis, O. S.   541 (95), *588*
Wolfe, B.   959, 963 (70), 971 (70, 95), 980, 981 (70), *1017, 1018*
Wolfe, J. F.   376 (806), 424 (1118), 460 (1416), 461 (1416–1418), *478, 484, 490*
Wolfe, S.   326, 335 (496), *472*
Wolff, G.   150 (124), *172*
Wolff, H.   1284 (644), *1329*
Wolff, M. E.   313 (419), *470*
Wolff, R.   990 (152), *1019*
Wolff, R. E.   150 (124), *172*
Wolff, S.   725 (191b), *752*
Wolfrom, M. L.   368 (747), *477*
Wolfschütz, R.   157 (163), *172*
Wolinsky, J.   416 (1018), *482*
Wollast, P.   439 (1248), *487*
Wolter, H.   916 (3), *942*
Wong, C. L.   103 (229, 230), *118*
Wong, C. M.   275 (91, 93), 437 (1231), *463, 487*
Wong, C. S.   837 (58), *856*
Wong, G. B.   272 (30), *462*
Wong, L. T. L.   444 (1294, 1295), *488*
Wong, N. P.   628 (186), *639*
Wong, R. Y.   415 (975), *482*
Wong, S. C.   1045 (328), *1060*
Woo, E. P.   312 (410), *470*
Wood, H. B. Jr.   338 (594), *474*
Wood, J. A.   984 (123), *1019*
Wood, J. K.   814 (146), *824*
Wood, J. L.   214 (2b), *256*
Wood, L. L.   318 (444), *471*
Wood, N. F.   272 (27), *462*
Woodruff, E. H.   1098, 1099 (125), *1318*
Woodruff, R. A.   416 (1022), *483*
Woods, L.   511 (119), *529*
Woods, R.   800 (80), *823*
Woods, T. S.   1038 (173), *1056*
Woods, V. A.   1012 (195), *1020*
Woodward, G. E.   320 (467), *471*
Woodward, I.   224 (95), *258*
Woodward, J.   193 (74), *211*
Woodward, R. B.   73 (40), *114,* 273 (53), *463,* 623 (161), *639,* 697 (139), *750,* 1071–1073, 1075, 1149, 1187 (28), *1316*

Woody, R. W. 70, 83 (13), *114*
Woolford, R. G. 298 (314), *468*
Woolsey, N. F. 382 (830), 383 (830, 832), *479*
Wordie, J. D. 274 (80), *463*
Worley, J. W. 1048 (343), *1060*
Wornhoff, E. W. 1185 (341), *1322*
Worth, G. T. 19, 21, 22, 25, 26, 28–30 (8), *55*
Wren, C. M. 884 (125), 894 (125, 176), 899 (125), *912, 913*
Wright, C. M. 603, 619, 631 (5), *635*
Wright, G. C. 893 (166), *913*
Wright, I. 416 (1014), *482*
Wright, J. D. 183 (61), *210*
Wright, J. L. 958, 959 (66), 985 (129), *1017, 1019*
Wright, P. W. 1184 (339), *1322*
Wright, W. B. 224 (108), *258*
Wrighton, M. S. 642 (3b), *747*
Wrixon, A. D. 87 (105), *115*
Wu, D. K. 278 (124), *464*
Wu, G.-S. 759 (17, 21), 761–763 (35), *778*
Wubbels, G. 727, 728 (192b), *752*
Wuest, H. 452 (1355), *489*, 585 (360), *594*
Wuesthoff, M. T. 1171, 1172 (317, 318), *1322*
Wulfman, D. S. 436 (1215), *486*
Wunderlich, K. 572 (314), *593*
Wynberg, H. 277 (123), 281 (181), 295 (286), 309 (386), *464, 465, 467, 469*
Wyn-Jones, F. K. 240 (308), *263*

Yagami, T. 1079 (58), *1317*
Yagi, F. 597 (392), *599*
Yahner, J. A. 281, 282 (178), *465*
Yale, H. L. 446 (1310), *488*, 542 (98), *588*
Yalkowsky, S. H. 1000 (166, 167), *1019*
Yamabayashi, T. 675, 682 (88), *749*
Yamabe, T. 1025 (41), 1026 (41, 74), *1053, 1054*
Yamada, K. 1145, 1147 (261), *1321*
Yamada, M. 412 (955), *481*, 1028, 1032, 1044 (66), *1054*
Yamada, S. 444 (1292), *488*, 611 (81), *637*, 1039 (234), *1058*
Yamada, S.-I. 102 (205), *117*
Yamaguchi, T. 536 (45, 46), 538 (61), 539 (45, 66), 543 (45), 562, 567 (46), *587, 588*
Yamaguti, Y. 733, 734 (206), *753*
Yamaguti, Z. 1186 (351, 352), 1187 (352), *1323*
Yamamoto, H. 278 (129), 369 (751), 423 (1105, 1106), *464, 477, 484*, 585 (362), *594*

Yamamoto, I. 338 (590), *474*, 1230 (513, 514), *1326*
Yamamoto, K. 457 (1394), *490*
Yamamoto, T. 231 (205), *260*
Yamamoto, Y. 204 (132), *212*
Yamanaka, C. 198 (93), *211*
Yamanoto, H. 1048 (356), *1061*
Yamashita, J. 936 (61), *944*
Yamashita, Y. 1001 (170), *1020*
Yamato, M. 571 (300, 301), 597 (392, 393), *592, 599*
Yamauchi, O. 103 (232), *118*
Yamauchi, T. 1029, 1032, 1042 (72), *1054*
Yamazaki, N. 412, 443 (948, 949), *481*
Yamazaki, T. 301 (331), *468*, 820 (168), *824*
Yamdagui, R. 34, 35 (95), *57*
Yan, J. F. 243, 247 (335), *263*
Yanagi, S. 1230 (513), *1326*
Yang, D. T. C. 364 (722), *476*, 1096 (120), *1318*
Yang, K.-U. 993 (159b), *1019*
Yang, K.-W. 447 (1317), *488*
Yang, N. C. 408, 409 (920), *480*, 1206, 1223, 1224 (468), *1325*
Yang, S. S. 408, 409 (920), *480*
Yano, Y. 990 (151), 1001 (170), 1003 (173), *1019, 1020*
Yanovskaya, L. A. 138 (75, 76), 148 (76), *171*, 270, 437 (11), *462*
Yanuka, Y. 300 (326), 372 (772), *468, 477*, 1090 (103, 104), 1092 (104, 106), *1318*
Yap, K. S. 800 (77), *822*
Yarchak, M. L. 711, 716, 744 (168a, b), *751*
Yashiro, T. 273 (57), *463*
Yaslovitskii, A. V. 442 (1279), *488*
Yasuda, H. 421 (1086), *484*
Yasui, T. 103 (214, 215), *118*
Yasunaga, T. 235 (241), 236 (241, 258), 237 (258, 260), *261, 262*
Yates, B. L. 898 (194), 908 (233, 234), *913, 914*
Yates, D. H. 731, 732 (200a), *752*
Yates, K. 23, 26 (42), 36 (102), 49 (123), *56–58*
Yatsimirski, A. K. 947 (8, 14), 948 (20), 962 (14, 73b), 964 (8, 14, 20, 73b, 82–86, 87b, 88), 965 (20, 73b, 83), 966 (14, 20, 86), 967 (83, 88), 968 (73b), 978 (14), 990 (83), 1004 (87b), *1016–1018*
Yazawa, H. 446 (1313), *488*
Yeaw, J. S. 274 (74), *463*
Yeh, C. Y. 102, 105 (185), *117*
Yeh, H. J. C. 1038 (183), *1056*

Yeo, A. N. H. 136 (63), *170*
Yeung, H. W. 523 (200), *530*
Yogev, A. 653, 657, 659 (34a, c), *747*
Yogev, Y. 733–735 (204c), *753*
Yokoyama, K. 386 (841), *479*
Yokoyama, M. 1037, 1042 (153), *1056*
Yokoyama, Y. 444 (1292), *488,* 1039 (234), *1058*
Yoneda, F. 562 (224), 563, 566 (231), 569 (224), 571 (293), 596, 597 (382), *591, 592, 599*
Yoneda, H. 103 (219), *118,* 989 (143), 1006 (178), *1019, 1020*
Yoneda, N. 293 (264), *467*
Yonemitsu, O. 412 (947), 415 (986), *481, 482,* 1206 (478–481, 483), 1217 (478), 1218 (478, 480), 1219 (481, 483), *1325*
Yonezawa, T. 253 (438), *265*
Yoon, V. C. 739 (219), *753*
Yorke, M. 416 (1027), *483*
Yoshida, H. 1042, 1043 (280), *1059*
Yoshida, K. 199 (104), *211*
Yoshida, M. 711 (172), *751*
Yoshida, T. 435 (1208), 450 (1339), *486, 489*
Yoshida, Y. 990 (151), *1019*
Yoshida, Z. 180 (46), *210*
Yoshifuji, M. 826 (13), *855*
Yoshikawa, S. 103 (217), *118*
Yoshimura, Y. 1284, 1290 (650), *1329*
Yoshino, H. 1038, 1046 (181), *1056*
Yosida, J. 786, 787 (25), *821*
Young, G. A. R. 375 (804), *478*
Young, G. R. 1232, 1259 (561), *1327*
Young, H. 1066, 1069, 1149 (22), *1316*
Young, J. F. 826 (21, 27), 827, 853 (27), *856*
Young, R. N. 454 (1370), *489*
Young, W. G. 432 (1177), *486*
Youngquist, M. J. 901 (204), *914*
Yu. P.-S. 1050 (377), *1061*
Yukimoto, Y. 513 (130), *529*
Yun, H. H. 460 (1412), *490*
Yura, S. 1276, 1277 (627), *1328*
Yurekli, M. 63 (39), *66*

Zabusova, S. E. 806 (113), *823*
Zabza, A. 777 (98), *780,* 1276, 1281 (639), *1328*
Zagurskaya, L. M. 615 (103), *638*
Zaikin, A. D. 63 (38), *66*
Zak, H. 1039, 1048 (237), *1058*
Zakai, M. 439 (1249), *487*
Zakharkin, L. I. 276 (110), 385 (835), *464, 479*
Zalar, F. V. 375 (789, 790), 376 (790), 377 (789, 790), *478*

Zalinyan, M. G. 1072 (35), 1075 (48), *1316*
Zamorani, A. 674 (79), *749*
Zanati, G. 1145 (249), *1321*
Zapadinski, B. I. 270 (14), *462*
Zappelli, P. 340, 436 (618), *474*
Zatorski, A. 596 (384), *599*
Zatulina, O. S. 618 (122), *638*
Zaugg, H. E. 322 (475), *471,* 497, 499 (35), 521 (179), *527, 530,* 1065 (1), *1315*
Závada, J. 716 (178), *751*
Zav'yalov, S. I. 1052 (398), *1062*
Zbinden, R. 619, 620 (133), *638*
Zdansky, G. 99 (167), *117*
Zeegers-Huyskens, Th. 251 (414–416), 255 (465), *265, 266*
Zehr, R. D. 733, 734 (205), *753*
Zeidler, M. D. 217 (5), *256*
Zeidman, B. 1195 (381), *1323*
Zeinalova, G. A. 1039 (205), *1057*
Zelenskaya, M .G. 608, 609 (48), *636*
Zeliver, C. 1265, 1267 (606), *1328*
Zelsmann, H. R. 227 (145), *259*
Zemlicka, J. 557 (183, 184), 558 (183, 184, 186, 187), 562 (183, 184), 563 (226–230), *590, 591*
Zen, S. 411 (926), *481*
Zeppa, A. 812 (137), *824*
Zerbi, G. 22, 25 (36), *56,* 232 (212), *261*
Zervas, L. 272 (50), *463*
Zetta, L. 88 (130), *116*
Zey, R. L. 518 (168), *530,* 1200 (418), *1324*
Zeyfang, D. 538, 546 (62, 63), 555, 567 (177), *588, 590*
Zheltova, V. N. 605 (24), *636*
Zhemaiduk, L. P. 339 (601), *474*
Zhidkova, A. M. 537 (55), 562 (217–220), 566 (271–273), 570, 571 (217), 582 (271), 583 (271, 342), 594 (369), *587, 591–593, 598*
Zhil'tsova, E. N. 65 (45), *66*
Zia, A. 438 (1234), *487*
Ziegenbein, W. 536 (21), *587,* 1302 (686), *1330*
Ziegler, E. 1232 (550, 586, 592), 1247 (550), 1248 (550, 586), 1260 (586), 1262 (592), 1263 (550), 1265 (607), 1314 (722), 1315 (723), *1327, 1328, 1330*
Ziegler, F. 1037, 1042 (163), *1056*
Ziegler, F. E. 583 (347), *593*
Ziegler, K. 825 (6), *855*
Zieloff, K. 1232, 1253 (555), *1327*
Zienty, F. B. 516 (146), *529,* 1145, 1151 (265), *1321*

Index

Ziessow, D.   231 (197), *260*
Ziffer, H.   108 (272), *119*
Zilberman, E.   449 (1328), *489*
Zilberman, E. N.   868 (41), *910*
Zil'berman, E. N.   1276, 1278 (632), *1328*
Zimmer, H.   506 (88, 90), *528,* 1123 (204, 205, 207), *1320*
Zimmerman, H. E.   402 (902), *480,* 645 (7), 651 (15), 680 (7, 15), 688, 694 (15), 695 (7, 15), 699, 700 (15), 706 (146c), 728 (193a, b), 730, 731, 733 (146c), 735 (7, 15), *747, 751, 752*
Zimmermann, F.   511 (116a), *529*
Zimmermann, H.   223 (70), 230 (179), 231 (194), *258, 260*
Zinkevich, E. P.   373 (782), *478*
Zinner, G.   443 (1287), *488,* 552 (169, 170), *590*
Zinner, H.   1039 (212), *1057*
Ziolkowski, F.   876 (86), 893, 894 (168), *911, 913*
Zirrolli, J. A.   148, 149 (120), *172*
Zitrin, S.   282 (196), *465*
Zitsmanis, A. K.   1052 (398), *1062*
Zlotskii, S. S.   348 (642), *475*
Zographi, G.   1000 (166, 167), *1019*
Zollinger, H.   539 (73, 74), *588*

Zolotarev, B. M.   138 (75, 76), 148 (76), *171*
Zorina, E. F.   458 (1403), *490*
Zsindely, J.   403 (906), *480*
Zubhov, V. A.   83, 84 (87), *115*
Zuidema, G. D.   1096 (119), *1318*
Zuliani, G.   21, 22 (25), *56*
Zuman, P.   816, 817 (157), *824*
Zundel, G.   214 (1g), 221 (37, 38), 228 (164, 165), *256, 257, 259, 260*
Zunft, H. J.   1004 (174), *1020*
Zupan, M.   562 (206), *590*
Zupančič, I.   231 (200), *260*
Zurbach, E. P.   233 (220), *261*
Zvonkova, Z. V.   194 (75), *211*
Zvonok, A. M.   1052 (404), *1062*
Zvorykina, V. C.   805 (104), *823*
Zwanenburg, B.   383 (831), *479,* 1043 (304), 1044 (314), 1049 (304, 314, 364–366, 368–370), *1059, 1061,* 1131 (226), *1320*
Zweifel, G.   435 (1203), *486*
Zweig, A.   661 (47–51), 662 (47, 49), 663 (49–51), 666 (49, 51), 668 (47–51), 669 (47, 48), 708 (47, 49), *748*
Zwiesler, M. L.   281, 328 (183), *465*
Zwolinski, B. J.   226 (141), *259*

Subject Index

Acetal exchange 543
Acetals, oxidation of, in carboxylic acid
    synthesis 344–352
O,N-Acetals 569
O,S-Acetals 561
Acetamide,
    association constants of phenols with
        251
    dimers of 244
    mass spectrum of 152
    pyrolysis of 867, 868
Acetamides, hydrogen bonding in
    243–251, 253–255
Acetanilides, mass spectra of 155, 156
Acetates, pyrolysis of 875, 878, 881, 886
    alumina-catalysed elimination in 900
    reactivity of 880, 881, 892, 894
Acetic acid,
    mass spectra of 124, 133
    pyrolysis of 864, 868
    radiolysis of 759, 760
    self-association of 222, 226, 232
        chemical shift of the proton in 230
        isotropic ^{13}C shifts in 231
        thermodynamics and kinetics 235,
            236
Acetic anhydride,
    pyrolysis of 871
    radiolysis of 773
Acetoacetamide, pyrolysis of 868
Acetoacetanilides, intramolecular hydrogen
    bonding in 254
Acetoacetic acid, pyrolysis of 863, 868
Acetoacetic esters,
    2-alkyl-substituted, cathodic reduction
        of 805
    condensation reactions of 1096–1099
    in synthesis of carboxylic acids 279
    in synthesis of substituted acetate
        esters 424, 425
    pyrolysis of 896
N-Acetoacetylamino acids, chiroptical
    properties of 84, 102
3-Acetoxybicyclo[2.2.2]octene-5,6-
    dicarboxylic anhydride, mass

spectrum of 161, 162
2-(1-Acetoxyethyl)pyridine *N*-oxide,
    rearrangement in pyrolysis of 903
7-Acetoxy-7-methylnorbornane, pyrolysis
    of 901
7-Acetoxy-7-methylnorbornene, pyrolysis
    of 901
N-Acetyl-(*S*)-alanineamide, CD spectrum
    of 84
N-Acetyl-L-alanine-*N'*-methylamide, CD
    spectrum of 99
Acetylamino acid amides 101, 102
N-Acetylamino acid *N'*-methylamides, CD
    spectra of 102
Acetyl chloride, oxidative addition of, to
    RhCl(PPh$_3$)$_2$ 838
Acetylcholine, *ab initio* investigations of
    32, 33
Acetylene, reaction of, with orthoamides
    569
Acetylenecarboxylic esters, mass spectra
    of 148
Acetylenedicarboxylic acid, hydrogen
    bonding in hydrate of 224
Acetyl fluoride, CI calculations for 21
4-*O*-Acetyl-β-D-galactopyranosides,
    chiroptical properties of 106
N-Acetylgalactosamine 105
N-Acetylglucosamine 105
N$^\alpha$-Acetylhistidine 947
    deacylation of *p*-nitrophenyl esters
        by 999
S-Acetyl-(2*R*)-mercaptopropionic acid,
    chiroptical properties of 107
N-Acetylneuraminic acid, chiroptical
    properties of 105
Acetylsalicylic acid, polarized spectra of
    229
Acetylthioacyl sulphides, synthesis of
    1044
Acetylthio chromophore 93
Acid dissociation constants – *see also* pK_a
    of thio carboxylic acids 1044, 1045
Acidity 33–35
    intrinsic, of carboxylic acids 34

Acrylates, reaction of, with dimethylaceta-
     mide acetal   582
Acrylic acid,
     pyrolysis of   866
     reduction of   801
     u.v. spectra of   108
Acrylonitrile, reaction of, with
     dimethylacetamide acetal   582
N-Acylacetidines, 2-substituted, mass
     spectra of   615
Acylamination, in amide synthesis   453
Acylation,
     in acid anhydride synthesis   438
     in amide synthesis   442–449
     in imide synthesis   458–460
     of orthoamides   575–577, 581, 582
     of thio esters   1049–1051
     of thiols   1038, 1039
Acyl azides, reaction of, with amide acetals
     576, 577
N-Acylaziridines, 2-substituted, mass spectra
     of   615
Acyl chlorides,
     CD spectra of   109, 110
     ferric hydroxamate test for   603
     oxidative addition of, to RhCl(PPh$_3$)$_2$
     837
Acyl cyanides,
     O,N-acetals of   569
     O,N-aminals of   569
3-Acyl-sn-glycerols, chiroptical properties
     of   106
N-Acylguanidines   562
Acyl halides,
     ab initio investigations of   20–22
     acylation with, in amide synthesis   445,
     446
     alcoholysis of, in ester synthesis   415,
     416
     aliphatic – see Aliphatic acyl halides
     as proton acceptors   240, 241
     decarbonylation of   827–842
         mechanism of   837–842
         regioselective   837
         stoichiometric   841
     heats of formation of   64, 65
     hydrolysis of, in carboxylic acid
         synthesis   275
     mass spectra of   160
     oxidative addition of, to RhClCO(PPh$_3$)$_2$
     841
     pyrolysis of   867
     synthesis of   438–442
         by carbonylations   441
         from carboxylic acids and anhydrides
         438–440
         from esters   440, 441
         from trihalides   441

N$^\alpha$-Acylhistidines, in mixed micelles   1003
Acylimidazoles   992
N-Acyl imino esters, in amide acetal
     synthesis   541
Acyl isocyanate O,N-acetals   580
Acylium ion   134
Acyllactam rearrangement   526
α-Acyllactams, quantitative analysis by
     n.m.r. spectroscopy   621
Acyllactone rearrangement   502
α-Acyllactones, quantitative analysis by
     n.m.r. spectroscopy   621
Acyloxylation, in ester synthesis   432,
     433
N-Acylpiperidides, mass spectra of   154
Acyl radicals   758
Acyl splitting   1035, 1036
S-Acyl splitting   1051
Acylthio chromophore   104, 107
α-Acylthio lactones, quantitative analysis
     by n.m.r. spectroscopy   621
Acylthio ureas, chiroptical properties of
     102, 103
Adipamide, hydrogen bonding in   244
Adipic anhydride, pyrolysis of   873
Aerosol O.T.[sodium di(2-ethylhexyl)-
     sulphosuccinate in octane], as
     micelle   986
Alanine, CD spectrum of   89, 98
L-Alanine   95
(S,S)-Alaninediketopiperazine, CD spectrum
     of   84
Alanines, CD spectra of   101
(S)-Alaninol, CD spectrum of   89
Alcohol formation, in cathodic reduction,
     of aromatic acids   790–792
     of aromatic amides   817, 818
     of esters   805
     of heterocyclic acids   797–799
Alcohols,
     oxidation of, in carboxylic acid
         synthesis   298–306
     reduction of, to hydrocarbons   1046
     thioacylation of   1041
Alcoholysis,
     of orthoamides   557
     of thiolo esters   1048
Aldehyde formation, in cathodic reduction,
     of amides   814–817
     of aromatic acids   784–789
     of esters   804, 805
     of heterocyclic acids   794–800
Aldehydes,
     aliphatic – see Aliphatic aldehydes
     aryl – see Aryl aldehydes
     decarbonylation of   843–855
         mechanism of   852–855
     oxidation of   453

in carboxylic acid synthesis
    306–317
primary – *see* Primary aldehydes
reaction of, with orthoamides   573,
    574
secondary – *see* Secondary aldehydes
tertiary – *see* Tertiary aldehydes
Aldimines, chiroptical properties of   102,
    103
Aldol condensation, in lactone synthesis
    1098–1102
Aldonic acid, γ-lactones of   84
Aliphatic acid imides, mass spectra of
    162–164
Aliphatic acids,
    cathodic reduction of   800, 801
    mass spectra of   123–126
Aliphatic acyl halides,
    decarbonylation of   834–837, 842
        regioselective   837
    dehydrohalogenation of   834–837
        regioselective   837
Aliphatic aldehydes, decarbonylation of,
    retention of configuration during
    852
Aliphatic amides, mass spectra of
    152–154, 613, 614
Aliphatic amines, reaction of, with
    orthoamides   561, 562
Aliphatic anhydrides, mass spectra of   161,
    162
Aliphatic esters, mass spectra of   133–139
Alkaline-earth metal benzenecarboxylates,
    transcarboxylation of   935
Alkanecarboxylates, long-chain, deacylation
    by long-chain alkylamines   995
n-Alkanecarboxylates, anionic micelles of
    985
Alkanoic acids,
    pyrolysis of   864–867
        Arrhenius parameters for   864
        four-centre processes in   864, 866
        six-centre process in   864
Alkenes, reaction of, with orthoamides
    582, 583
Alkenoic acids – *see also* Unsaturated acids
    pyrolysis of   860–863, 877
β-Alkoxy amides, pyrolysis of   869
Alkoxycarbonyl group rearrangement   403
Alkoxy(dialkylamino)carbenes   581
5-Alkoxy-5-dialkylamino-2,4-diarylimino-
    imidazolines   577, 578
5-Alkoxy-5-dialkylamino-2,4-dithio-
    imidazolines   577, 578
2-Alkoxyoxetanes, photolysis of   1191
Alkyl acetoacetates, mass spectra of   149,
    150
Alkyl acids, conformational analysis of   88

α-Alkyl acids, chiroptical properties of
    88–95
Alkylallenecarboxylic acids, CD spectra of
    108
N-Alkylamides   606, 607
    oxidation of   460
Alkylamine oxides, micellized, pK_a deter-
    minations of   1000
n-Alkylamines, as deacylating agents   1002
Alkylammonium carboxylates, reverse
    micelles of   1012
N-Alkylanilines, in triaminomethane
    synthesis   546
Alkylation,
    in imide synthesis   460
    of dithiocarboxylates   1042
    of orthoamides   572, 573, 581, 582
    of thio acids   1038
    of thio esters   1049, 1050
Alkyl chlorides, decomposition of alkoxy-
    methylene iminium chlorides to
    539
trans-2-Alkylcyclohexyl acetates, pyrolysis
    of, relative rates of elimination in
    897
Alkyl diphenylphosphinates, pyrolysis of
    894
α-Alkyl esters, chiroptical properties of
    88–95
N-Alkylguanidines   562
Alkylhydrazines, reaction of, with ortho-
    amides   561, 562
Alkyl hydroxamates, deacylation by   995
n-Alkylimidazoles, as deacylating agents
    1002
N-Alkylimidazoles, micellar catalysed
    reaction of   966
2-Alkylmalic acids, CD spectra of the
    molybdate complexes of   104
S-Alkylmercaptopropionic acids, chiroptical
    properties of   93
2-Alkyl-2-oxazolines, lithio salts of, in
    lactone synthesis   1120–1123
Alkyl phenoxythioacetates, mass spectra
    of   151
α-Alkylphenylacetic acids, CD spectra of
    109
Alkyl phenylalkanoates, mass spectra of
    138, 139
Alkyl splitting   1035, 1036
S-Alkyl splitting   1051
Alkylsuccinic acids, chiroptical properties
    of   91, 94
D-(+)-Alkylsuccinic acids, CD spectra of 91
N-Alkylsulphonamides, reaction of, with
    orthoesters   542
Alkylsulphonates, in tris(acylamino)-
    methane synthesis   547

2-Alkyltartaric acids, CD spectra of the
    molybdate complexes of   104
S-Alkyl thioacetates, rate spread for
    pyrolysis of   892
Alkylthiomethylene iminium slats, in
    amide thioacetal synthesis   536
Alkylthio nitriles   539
Alkyl thiophenoxyacetates, mass spectra
    of   151
β-Alkylthiopropionic ester, mass spectrum
    of   150
n-Alkyltrimethylammonium bromides, as
    micelles   946, 947, 962, 963,
    965–968, 971, 972, 974, 975
Alkyl-type radicals   757
Alkynes, reaction of, with orthoamides
    582, 583
Allenecarboxylic acids, chiroptical
    properties of   108
Allenecarboxylic esters, mass spectra of
    148
L-Allocystathionine, CD spectra of   99
Allyl benzoate, pyrolysis of   878
Allyl esters, pyrolysis of   877–880
    effect of substituents on the rate of
        rearrangement in   878, 880
    cis β-elimination in   878
Allyl formate, pyrolysis of   877
Allylic acetates, cathodic cleavage of   812,
    813
Allylic alcohols,
    aldehyde tautomers of   849, 850
    decarbonylation of   849, 850, 854
Allylic couplings   621
Allylic oxygen chirality rule   87
Allylic rearrangement   583, 584
Allyl oxalate, pyrolysis of   878
Allyl phenylacetate,
    chemical ionization mass spectrum of
        152
    pyrolysis of   878
Allyl trifluoroacetate, pyrolysis of   878
Aluminium carbide, in transcarboxylations
    917
Aluminium hydride complex, in reduction
    of di(alkylthio)methylene iminium
    salts to amide thioacetals   544
Amide acetals,
    acylation of   575–577
    alcoholysis of   557, 558, 595
    alkylation of   572, 573
    conductivity measurements on   542
    heterocyclic   595
    hydrolysis of   557
    reaction of,
        to form heterocyclic compounds
            569, 570, 596, 597
        with acyl cyanides   544

with carbon acids   566, 569, 596
with carboxylic acids   561
with compounds containing acidic
    NH₂ or NH groups   561–564,
    596
with compounds containing acidic
    PH groups   564, 565, 596
with electrophilic reagents
    572–581
with heterocumulenes   577
with hydrocyanic acid   544
with isocyanates   577, 597, 598
with organometallic compounds
    565
with phenols   559, 560
reactions on the α-methylene groups of
    581–586
synthesis of   535–545, 594, 595
thiolysis of   561
Amide–amide interaction   615
Amide carbonyl group, reduction of, to a
    methylene group   818–820
Amide chloride, reaction of, with alkoxide
    to form amide acetals   534
Amide chromophore   70
Amide–dialkylsulphate adducts,
    in amide acetal synthesis   536
    in amide thioacetal synthesis   536
Amide formation, in cathodic reduction of
    imides   818–820
Amide hemihydrohalides, hydrogen bonding
    in   245, 246
Amide–phosphoroxychloride adducts, in
    amide thioacetal synthesis   539
Amides,
    ab initio investigations of   27–32
    acylation of   546, 547
    alcoholysis of, in ester synthesis   418,
        419
    aliphatic – see Aliphatic amides
    alkylation of   546, 547
    N-alkyl-substituted, pyrolysis of   870
    aromatic – see Aromatic amides,
        Arylamides
    as proton acceptors   250, 251
    as proton donors   248, 249
    cathodic reduction of   813–820
    chiral, enantiomeric purity of   609
    chiroptical properties of   68–70, 72,
        87–95, 99–105, 611, 612
    detection and determination of
        602–618
        chromatographic methods for   617,
            618
        polarographic methods for   618
        titration methods for   618
    N,N-disubstituted, reaction with
        oxonium salts to give iminium

salts 535
heats of formation of 63–65
hydrogen bonding in 241–254
hydrolysis of, in carboxylic acid
    synthesis 274
internal rotation in 252, 253
i.r. spectra 604–608
    of open-chain *trans* N-monosub-
        stituted 607
mass spectra of 152–160, 613–616
n.m.r. spectra of 609, 610
    ^{15}N-n.m.r. 610
of α,β-unsaturated acids, mass spectra
    of 615
primary – *see* Primary amides
pyrolysis of 867–871
    five-centre process in 869
    four-centre process in 868
    reactivity in 881, 892
    six-centre processes in 868, 870,
        871
Raman spectra of 604–608
    of N,N-dialkyl-substituted 607
    of open-chain *trans* N-monosub-
        stituted 607
reaction of,
    with dialkyl sulphates 536
    with hydroxylamine sulphate and
        ferric chloride, kinetics and
        mechanism of 603
    with orthoamides 561, 562
    with tetrakis(dimethylamino)-
        titanium 548
secondary – *see* Secondary amides
spot tests for 603
sterically hindered 607
N-substituted, configuration of 609
synthesis of 442–458
    by acylaminations 453
    by acylations 442–449
    by carboxamidations 454–456
    by cleavages 450
    by condensations 456
    by hydrolysis 449
    by oxidations 452
    by rearrangements 450–452
tertiary – *see* Tertiary amides
unsaturated – *see* Unsaturated amides
u.v. spectra of 612, 613
Amide thioacetals,
    reaction of 558, 562, 565, 567, 569,
        577, 596, 597
    synthesis of 535–545, 594, 595
Amide transitions 73
Amide I vibrations 604, 605
    in primary amides 606
    in secondary amides 607
    in tertiary amides 607

Amide II vibrations 604, 605
    in primary amides 606
    in secondary amides 607
Amide III vibrations 604, 605
    in secondary amides 607
Amide IV vibrations 604, 605
Amide V vibrations 604, 605
Amide VI vibrations 604, 605
Amide–water association 251, 252
Amidines 562
    reaction of, with orthoamides 561, 562
Amidinium salts,
    in orthoamide synthesis 540, 541, 545,
        546
    reaction of, with orthoamides 561,
        562
Aminal esters 535
    reaction of 557, 559–562, 564–566,
        569, 574, 575, 578, 580, 585,
        596
    synthesis of 545–550
O,N-Aminals 569
Aminal thio esters 535
    reaction of 567
    synthesis of 545
Amine formation, in cathodic reduction of
    amides 818–820
Amines,
    aliphatic – *see* Aliphatic amines
    aromatic – *see* Aromatic amines
    micellar catalysed reaction of 976
    oxidation of 371, 452
        in carboxylic acid synthesis 371,
        372
    reaction of, with orthoamides
        561–564
Amino acids,
    ab initio investigations of 32, 33
    chiroptical properties 95–104
        of N-acyl derivatives 101
        of metal complexes 103, 104
        of selenium-containing 99
        of N-substituted derivatives 102,
        103
    cyclizations using 1195–1200
    reaction of, with amide acetals 561
α-Amino acids, chiroptical properties of
    88–90
2-Aminobenzamides, hydrogen bonding in
    255
o-Aminobenzoic acid, intramolecular
    hydrogen bonding in 240
Aminobenzoic acids, intermolecular
    hydrogen bonding in 224
α-Amino-α,α-diphenylacetamides, mass
    spectra of 614
α-Amino esters, chiroptical properties of
    88–90

Aminolysis, of thio esters   1048
2-Amino-3-methylbenzoic acid, hydrogen
      bonding in   225
Amino-1,3,4-oxadiazoles, reaction of, with
      orthoamides   561, 562
o-Aminophenols   562
Aminopurines, reaction of, with ortho-
      amides   561, 562
Aminopyrazines, reaction of, with ortho-
      amides   561, 562
Aminopyridazine-N-oxides, reaction of,
      with orthoamides   561, 562
Aminopyridazines, reaction of, with ortho-
      amides   561, 562
Aminopyridazinium salts, reaction of, with
      orthoamides   561, 562
Aminopyridines, reaction of, with ortho-
      amides   561, 562
Aminopyridinium salts, reaction of, with
      orthoamides   561, 562
Aminopyrimidines, reaction of, with ortho-
      amides   561, 562
L-3-Aminopyrrolid-2-one   77
p-Aminosalicylic acid, hydrogen bonding
      in   225
Amino sugars, study of the structure and
      stereochemistry of   105
Amino-1,2,3,4-tetrazoles, reaction of, with
      orthoamides   561, 562
Aminothiazoles, reaction of, with ortho-
      amides   561, 562
o-Aminothiophenols   562
Amino-S-triazines, reaction of, with ortho-
      amides   561, 562
Amino-1,2,4-triazoles, reaction of, with
      orthoamides   561, 562
Ammonium ribonucleoside-2′,3-phosphate,
      cyclization of   558
Anchimeric assistance, in pyrolysis of
      esters   901
α-Angelica lactone,
      mass spectrum of   166
      synthesis of   503
β-Angelica lactone,
      mass spectrum of   166
      synthesis of   505
Angular methyl group, introduction of,
      into bicyclic compounds   848
Anhydrides,
      acylations with, in amide synthesis   446
      alcoholysis of, in ester synthesis   416, 417
      aliphatic — see Aliphatic anhydrides
      aromatic — see Aromatic anhydrides
      as proton acceptors   240, 241
      cathodic reduction of   801, 802, 805
      complexes of   175–209
      detection and determination of
            628–635
            by reaction with hydroxylamine
                  631
            chromatographic methods for   634
            luminescent method for   631
            polarographic methods for   635
            thermometric method for   631
            titration methods for   629, 630,
                  635
      ferric hydroxamate test for   603
      hydrolysis of, in carboxylic acid
            synthesis   275
      in acyl halide synthesis   438–440
      i.r. spectra of   631, 632
      mass spectra of   161, 162, 633, 634
      n.m.r. spectra of   632
      photodecarbonylation of   661–669,
            738, 739
      photodecarboxylation of   708
      pyrolysis of   871–874
            four-centre processes in   871, 873
            six-centre processes in   872
            three-centre process in   873
      radiolysis of   773
      Raman spectra of   631, 632
      reduction of, in lactone synthesis
            1138–1144
      spot tests for   631
      synthesis of   437, 438
      unsaturated — see Unsaturated
            anhydrides
      u.v. spectra of   632, 633, 670
Anilides, micellar catalysed hydrolysis of
      991, 992
Aniline, micellar catalysed reaction of, with
      2,4-dinitrofluorobenzene   964, 983
Anion-radical formation,
      for amides   813, 814
      for conjugated esters and anhydrides
            801–805
      for imides   814
      for thio esters   1035, 1045
α-Anions,
      of carboxylic acids, in synthesis of
            substituted acids   280–282
      of esters, in synthesis of substituted
            esters   426–429
3-(p-Anisyl)butanoic acid, CD spectrum of
      110
2-(p-Anisyl)propionic acid, CD spectrum of
      110
Anodic dimerization, in ester synthesis
      433
Anthranilic acid, mass spectrum of   127,
      128
Anthraquinone-1-carboxylates, mass spectra
      of   151
Arenes, oxidation of, in carboxylic acid
      synthesis   317–331

Arndt–Eistert rearrangement   380–385
Aromatic acid imides, mass spectra of
        164, 165
Aromatic acids,
    cathodic reduction of   784–794
    mass spectra of   127, 128
    transcarboxylation of salts of   915–942
Aromatic acyl cyanides – see Aroyl
        cyanides
Aromatic acyl halides – see Aroyl halides
Aromatic amides,
    cathodic reduction of   817, 818
    mass spectra of   154–157, 614
Aromatic amines, reaction of, with
        orthoamides   561, 562
Aromatic amino acids, CD spectra of   98
Aromatic anhydrides, mass spectra of   162
Aromatic esters, mass spectra of   139–143
Aroyl chlorides, CD spectra of   109, 110
Aroyl cyanides, decarbonylation of
        828–834
Aroyl halides, decarbonylation of
        828–834
    stoichiometric   842
Arrhenius parameters, for thermal
        decomposition of alkanoic acids
        864
Arylacetyl halides, stoichiometric
        decarbonylation of   842
Aryl aldehydes, decarbonylation of
        843–847
Arylamides, aryl migration in   614
N-Arylamides   607
Arylamino acids, CD spectra of   109
Arylcarnitines   1000
Aryl esters, mass spectra of   139, 140
1-Arylethyl acetates,
    pyrolysis of,
        in determination of electrophilic
                aromatic substituent constants
                904, 905
        linear free-energy correlations
                against σ^+-constants for   890
        relative rates of elimination
                correlated with σ^+-constants
                for   890
1-Arylethyl esters,
    pyrolysis of   890, 891, 904, 905
        Hammett ρ-factors for   891
1-Arylethyl phenylcarbonates, pyrolysis of,
    in determination of electrophilic
        aromatic substituent constants   905,
        906
Arylhydrazines, reaction of, with ortho-
        amides   561, 562
Aryl iodides, preparation of   828
2-Aryl-3-methoxymaleic anhydrides   634
Aryl phosphate dianions   993

Aryl phosphate esters, micellar catalysed
        hydrolysis of   958, 959
Aryl sulphate esters, micellar catalysed
        hydrolysis of   958, 959
Ascorbic acid (vitamin C)   503
    radiolysis of, in aqueous solution
        774–777
Association, in solid amides   243–246
Atomic orbitals   4–9
    overlap matrix of   8
Autooxidation, in lactam synthesis   1312
Axial haloketone rule   82
Aza vinylogues of amidinium salts, in amide
        acetal synthesis   541
Azetes, fragmentation of   512
L-Azetidinecarboxylic acid   95
Azides, reaction of, with amide acetals
        576, 577
Azidinium salts, reaction of, with amide
        acetals   576, 577
Azidoformate, reaction of, with ketene
        acetals   544
Azines, reaction of, with orthoamides
        573–575
Aziridine, in amide acetal synthesis   540
Azlactones, chiroptical properties of   84
Azobenzene-o-carboxylic acids, intra-
        molecular hydrogen bonding in   240
Azomethines,
    chiroptical properties of   102, 103
    reaction of, with orthoamides
        573–575

Baeyer–Villiger reaction   353–355
Baker–Nathan order   907
Barium benzoate, transcarboxylation of
        934, 935
Barium hemimellitate, as transcarboxy-
        lation product   935
Barium isophthalate,
    as transcarboxylation product   935
    transcarboxylation of   934, 935
Barium naphthalenecarboxylates, trans-
        carboxylation of   938
Barium 1,2-naphthalenedicarboxylate, as
        transcarboxylation product,
    of barium α-naphthoate   937
    of barium β-naphthoate   937
Barium 1,8-naphthalenedicarboxylate, as
        transcarboxylation product of
        barium α-naphthoate   937
Barium 2,3-naphthalenedicarboxylate, as
        transcarboxylation product of
        barium β-naphthoate   937
Barium α-naphthoate, transcarboxylation
        of   937
Barium β-naphthoate, transcarboxylation
        of   937

Barium phthalate,
  as transcarboxylation product of barium
    benzoate 935
  transcarboxylation of 935
Barium terephthalate, as transcarboxylation
    product 935
Barium trimellitate, as transcarboxylation
    product 935
Barium trimesate, as transcarboxylation
    product 935
Barton reaction 1189
Bases, use of in oxidation,
  of alcohols 298
  of aldehydes 306–309
  of double and triple bonds 331
  of ketones 355–357
Basic groups, use of micelles to assist
    substrate incorporation in 989
Basicity 33
  of carboxylic acids 35
  of orthoamides 553
Basis sets 9, 10
  contracted AO 9
  double-zeta 9, 10
  minimal (single-zeta) 9
Beckmann rearrangement,
  in amide synthesis 450
  in carboxylic acid synthesis 385, 386
  in lactam synthesis 1274–1284
Benzaldehyde, formation of, in cathodic
    reduction of benzoic acid 784
Benzamide, mass spectrum of 156
Benzamides, intramolecular hydrogen
    bonding in 254
Benzanilides, cyclization of 1207–1223
Benzene,
  as transcarboxylation product,
    in mixed transcarboxylations 940
    of potassium benzoate 921, 922
  reaction of, with maleic anhydride 205
Benzenecarboxylates, transcarboxylation
    of 916–935
  effect of catalysts on 932, 933
  effect of cations on 933–935
  in an atmosphere of ^{14}C-labelled carbon
    dioxide 926–929
  in fused KCNO or KCNS 924–928
  mechanism of 930–932
Benzidine rearrangement, micellar catalysis
    of 976
Benzilic acid rearrangement 386–388
Benzimidazole 965
  micellar catalysed reactions of
    966–969
  salt effects on 968
Benzoates,
  pyrolysis of 874, 876, 878
    Hammett ρ-factors for 893

  relative rates of 891, 892
  substituted, transcarboxylation of 938,
    939
Benzoate sector rule 111–113
Benzofuran, reaction with dimethylmaleic
    anhydride 208
Benzoic acid 24
  cathodic reduction of, to benzaldehyde
    784
  electron spectra of 229
  hydrogen bonding in, thermodynamics
    and kinetics of 235
  mass spectra of 127
  proton distributions in dimers of 231
  u.v. spectra of 72
  vibrational spectra of 226
Benzoic anhydride, pyrolysis of 874
Benzoic carbonic anhydrides, pyrolysis of
    871
Benzopropiolactone 499
Benzoylamino acids, chiroptical properties
    of 102
Benzoyl cyanide 828
Benzoyl derivatives, chiroptical properties
    of 111–113
Benzoyl fluoride,
  mass spectrum of 160
  STO-3G calculations on the planar
    conformation of 21
Benzoylimide, mass spectrum of 164
Benzyl acetates, photosolvolysis of 651,
    694, 746
Benzyl alcohol, formation of, in cathodic
    reduction of benzoates 805
Benzyl alcohols, synthesis of 790
Benzyl anion 678
Benzyl benzoate, mass spectrum of 142,
    143
Benzyldimethylcarbinyl acetate, as possible
    intermediate in pyrolysis of
    neophyl acetate 903
Benzyl ethyl ether, formation of, in
    cathodic reduction of ethyl benzoate
    805
Benzyl halogenides, reaction of, with
    ketene O,N-acetals 581
Benzylic acyl halides, decarbonylation of
    834–837
Benzylidenemalonamide, mass spectrum
    of 158
Benzylidenemalonic esters, mass spectra
    of 144, 145
Benzylidenemalonyl chloride, mass
    spectrum of 160
Benzyl migration, 1,3-suprafacial 697
Benzyl phenylacetate, photodecarboxyla-
    tion of 694, 743
2-Benzylpropene, formation of, in

pyrolysis of neophyl acetate   902
Benzyl radical   678, 699
Benzylthio chromophore   93
Benzyne   703, 738
    synthesis of   499
Bicyclic γ-lactones, mass spectra of   166
Bifunctional catalysis   1011–1013
Bile salts, as micelles   953
Bimolecular reactions,
    inhibition of, by inert electrolytes   949
    micellar catalysis of   949, 959, 960
        kinetic models for   960–964
    micellar inhibition of   957, 976, 977
1,1'-Binaphthyls, chiroptical properties of
    110
Biological molecules, *ab initio* investigations
    of   32, 33
Biphenylcarboxylates, transcarboxylation
    of   938
Biphenylcarboxylic acids, chiroptical
    properties of   110
Bisbut-2-yl carbonate, pyrolysis of   886
2,2-Bis(dialkylamino)-1,3-benzdioxoles,
    reaction of, with Grignard reagents
    565
Bis(dialkylamino)difluoromethanes, in urea
    acetal synthesis   551
α,α-Bis(dialkylamino)nitriles   535
    reaction of   557, 560–562, 567, 573,
        574, 577, 585
    synthesis of   545, 546, 549, 550
Bis(dialkylphosphine)dimethylamino-
    methanes   565
Bis(dimethylamino)dialkylphosphino-
    methanes   565
Biselenienylcarboxylic acids, chiroptical
    properties of   110
Bis(ethylcarbonic)dicarboxylic anhydride,
    pyrolysis of   872
Bisignate curves   80, 98, 106, 113
Bis(thioacyl) sulphides, synthesis of   1044
3,3'-Bithienyl, 2,2'-diformyl derivative of
    CD spectrum of   111
3,3'-Bithienylcarboxylic acids, chiroptical
    properties of   110
3,3'-Bithienyldicarboxylic acids, CD
    spectra of   111
$B_{Ac}2$ mechanism, for deacylation of
    p-nitrophenyl alkanoates   990
Bond-fixation effects   907
Borate complexes, CD spectra of   84
Boric acid thio esters, in amide thioacetal
    synthesis   541
Born–Oppenheimer (fixed nuclei)
    approximation   6, 7
Bornyl acetate, pyrolysis of   902
Boron fluorides, in oxidation of ethers,
    acetals and ketals   344, 345

α-Bromoalkylcarboxylic acids, CD spectra
    of   92, 93
m-Bromobenzaldoxime, micellar catalysed
    acylation of   990
β-Bromocarboxylate ions, decomposition
    of, effect of micelles on   988
β-Bromoethyl benzoate, mass spectrum of
    141
(S)-2-Bromo-4-methylpentanoic acid   93
2-Bromonaphthalene-1,4,5,8-tetracarboxylic
    acid dianhydride, complexes of   177
(S)-2-Bromopropanoic acid, CD spectrum of
    93
Brown σ^+ function   614
Buffer effects, in micellar catalysed
    reactions   950, 959, 966, 971, 972,
        986, 1001
'Burst' kinetics   996, 1008, 1012
But-3-enoic acid, pyrolysis of, concerted
    cyclic mechanism for   861
Butenolides, chirality rules for   83–85
α,β-Butenolides, preparation and properties
    of   505, 506
β,γ-Butenolides, preparation and properties
    of   503, 504
t-Butoxycarbonyl group in oligopeptides
    100, 101
N-t-Butylacetamide,
    hydrogen bonding in   247
    pyrolysis of   870
But-2-yl acetate, pyrolysis of   886
2-Butyl acetate, pyrolysis of   884
    rate of elimination in   890
dl-erythro-3d-2-Butyl acetate, pyrolysis of
    882
dl-threo-3d-2-Butyl acetate, pyrolysis of
    882
n-Butyl acetate, mass spectrum of   138
t-Butyl acetate, pyrolysis of, rate of
    elimination in   894
n-Butyl benzoate, mass spectrum of   142
t-Butyl chloroacetate, pyrolysis of, rate of
    elimination in   894
t-Butyl dichloroacetate, pyrolysis of, rate
    of elimination in   893
2-Butyl halides, pyrolysis of   886
n-Butyl hexanoates, mass spectra of   137
N-n-Butyl-γ-lactam, mass spectrum of   169
N-(t-Butyl)methoxyacetamide, intra-
    molecular hydrogen bonding in
    254
n-Butyl propionate, mass spectrum of   138
t-Butyl thioacetate, pyrolysis of, reactivity
    in   892
Butyric acid, mass spectrum of   124, 125
γ-Butyrolactone   493
    mass spectrum of   166
    photolysis of   651, 659

γ-Butyrolactones,
    configuration assignments for   619
    equilibrium studies on   619

Cadmium benzenecarboxylates, trans-
        carboxylation of   933, 935
Cadmium benzoate, transcarboxylation of
        933
Cadmium naphthalenecarboxylates,
        transcarboxylation of   938
Cadmium salts, as catalysts in trans-
        carboxylations   932, 933
Caesium benzenecarboxylates, trans-
        carboxylation of   934
Caesium naphthalenecarboxylates, trans-
        carboxylation of   938
Caesium terephthalate, as trans-
        carboxylation product   934
Cahn—Ingold—Prelog convention   622
Calcium benzenecarboxylates, trans-
        carboxylation of   935
Calcium carbide, in transcarboxylations   917
Camphorolactams   79
Caprolactam, hydrogen bonding in   248
ε-Caprolactam, hydrogen bonding in   244,
        246
ε-Caprolactones, sector rule for   75
Caprylolactam, hydrogen bonding in   244
Carbamates, pyrolysis of   876, 891, 900
    reactivity of   880
Carbanion reactions, in carboxylic acid
        synthesis   287, 288
Carbanions,
    formation of, in transcarboxylations
        930, 931, 937
    reaction of, with thio esters   1047,
        1048
Carbohydrates,
    acetamides of   104
    acetates of   105, 106
    decarbonylation of   848, 849
Carbon acids,
    cathodic carboxylation of   809
    reaction of, with orthoamides
        566—569
Carbonamides, heterocyclic, mass spectra
        of   614
Carbonate esters,
    decomposition of, by micelles of
        N^α-stearoylhistidine   1003
    pyrolysis of   872, 881, 891, 893
    reactivity of   880, 894
Carbonation, of organometallic reagents,
        in carboxylic acid synthesis
        293—296
Carbon disulphide, reduction of, in
        synthesis of thio carboxylic acids
        1037

Carbonic acid imides   563
Carbonic acid orthoamides, synthesis of
        550—553
Carbonium ions, reaction of, with
        orthoamides   572
Carbon-13 labelling, in mass spectroscopy
        128, 136, 138, 139, 168
Carbon-14 labelling, in transcarboxylations
        926—929, 935, 936
Carbon-13 nuclear magnetic resonance
        spectroscopy, of thio acid
        derivatives   1035
Carbonylation reactions, in lactone
        synthesis   1173—1180
Carbophilic addition   1047
Carboxamidation, in amide synthesis
        454—456
Carboxonium ions   559
    reaction of, with orthoamides   572
Carboxyalkylation, of aromatic compounds,
        in carboxylic acid synthesis   289,
        290
Carboxy—hydroxyalkyl radicals   767
Carboxylate anion   95
Carboxylate cation, formation of, in
        transcarboxylations   917, 928, 930,
        931, 937
Carboxylates, as leaving groups   810
Carboxylate salts,
    acylation of   438
    alkylation of, in ester synthesis   414,
        415
Carboxylate sector rule   95, 98, 99
Carboxyl—carboxyl interactions   129, 615
Carboxylic acids,
    acylations with, in amide synthesis
        442—444
    aliphatic — see Aliphatic acids
    α-anions (dianions) of, in lactone
        synthesis   1116—1120
    aromatic — see Aromatic acids
    cathodic reduction of   782—801
    chiroptical properties   70—73,
        87—105, 108—113
        of allenic   108
        of aromatic   108—113
        of naturally occurring   88, 108
        of α-substituted   87—104
        of α,β-unsaturated   108
    dehydrative coupling of   437
    direct esterification of   411—414
    heats of formation of   60—62
    hydrogen bonding in   221—240
    in acyl halide synthesis   438—440
    keto and aldehydic, reduction of
        1145—1150
    liquid, structure of   222, 223
    mass spectra of   123—133

oxidation of 372
photochemical hydrogen-abstraction
   reactions of 718–722
photodecarboxylation of 671–687,
   739–743
pyrolysis of 860–867
   four-centre processes in 864, 866
   six-centre process in 864
radiolysis of 755–769
reaction of,
   with alkyl *t*-butyl ethers, to yield
      esters 345, 346
   with amide acetals 561
synthesis of 270–411
   by carbonation of organometallic
      reagents 293–296
   by cleavages 373–380
   by condensations 276–288
   by electrophilic substitutions 296
   by free-radical processes 288–290
   by hydrocarboxylations 290–293
   by hydrolysis 271–276
   by oxidations 297–373
   by rearrangements 380–411
unsaturated – *see* Alkenoic acids,
   Unsaturated acids
Carboxylic acid–water systems 225, 226,
   237, 238
Carboxylic dithiocarbamic anhydrides,
   pyrolysis of 872
Carboxylic radicals 757
Carcinogens 629
Cardiac glycosides 503
Catalysis, general acid or base 1008–1010
Catechol dichloromethyl ethers, in urea
   acetal synthesis 551
Catechols, in amide acetal synthesis 539
Cathodic carboxylation, of α,β-unsaturated
   esters 808–810
Cathodic reduction,
   of amides 813–820
   of anhydrides 801, 802, 805
   of carboxylic acids 782–801
   of esters 801–813
   of imides 813, 814, 816–820
   of lactams 813
   of lactones 801, 804, 812
Cephalosporanic acids, cathodic reduction
   of 812
Cephalosporin C 521
Cephalosporin derivatives, p.m.r. spectra of
   610
Cephamycins 603
Ceric ammonium nitrate, in oxidation
   reactions 309, 358
Cetyltrimethylammonium bromide
   (CTABr), as micelle 946, 962, 963,
   965–968, 972, 974, 980, 988,

990–994, 996, 999, 1001, 1003,
   1004, 1007
Cetyltrimethylammonium chloride
   (CTACl), as micelle 988
Charge transfer 94
   with solvent in thio acids 1026
Charge-transfer absorption of EDA
   complexes 195–198
Charge-transfer band, of benzoic acid u.v.
   spectrum 72
Cheddar cheese lactones, determination of
   628
Chelate compounds, of α-amino and
   α-hydroxy acids 103, 104
Chelate formation, between borate ions and
   vicinal *cis* diols 84
Chemical ionization mass spectrometry 603
   of esters 151, 152
Chemical-shift reagents 610
Chemical shifts,
   for thio acid derivatives 1033–1035
      increment system for 1033
   in hydrogen-bonded acids 217,
      229–231
Chiral head groups 1006
Chiral recognition 1006
Chloroacetamide, hydrogen bonding in
   243
2-Chloroacetamide, pyrolysis of 869
Chloroacetates, pyrolysis of 886
   reactivity in 880, 894
N-Chloroacetyl-β-arylamines, cyclization of
   1208–1224
α-Chloroalkylcarboxylic acids, CD spectra
   of 92, 93
p-Chlorobenzaldehyde, reaction of, with
   aminal esters 574
2-Chlorobenzamide, mass spectrum of 156
2-Chlorobenzimidazolium salts, in
   synthesis of cyclic tris(dialkylamino)-
   alkoxymethanes 552
2-Chlorobenzoic acid, cathodic reduction of
   789
Chlorodifluoromethane, in triaminomethane synthesis 546
(*S*)-2-Chloro-3,3-dimethylbutanoic acid, CD
   spectrum of 93
Chlorodi(methylthio)methane, in amide
   acetal synthesis 542
Chlorodiphenoxymethane, in amide acetal
   synthesis 542
Chloroform, in orthoamide synthesis 546,
   547
   as precursor of dichlorocarbene 540
Chloroformamidinium chloride, in synthesis
   of orthocarbamic acid esters 550
Chloroformates, pyrolysis of 892, 894
   reactivity in 880

Chloromethylene iminium salts,
  in amide acetal synthesis   536, 539
  in triaminomethane synthesis   546
α-Chloropropionic acid, chiroptical
      properties of   95
(S)-2-Chloropropanoic acid, CD spectrum
      of   93
Chloro substituent effects, multiple   907
L-(−)-Chlorosuccinic acid, CD spectra of
      91
Chlorotris(triphenylphosphine)rhodium
      (I) [RhCl(PPh₃)₃],
  oxidative addition reactions of,
      with hydrogen   827
      with hydrogen chloride   827
      with methyl iodide   827, 837
      with RCO—H group   852
  reactions of   827
      with acyl halides   827–842
      with aldehydes   843–855
  structure of   826, 837
  synthesis of   826
Choline derivatives, as micelles   991, 994,
      1008
Chromatography,
  in detection and determination,
      of amides   617, 618
      of anhydrides   634
      of lactones   628
  sources of error in   617
Chromium oxides, use of in oxidation,
  in lactam synthesis   1304, 1305
  of alcohols   298–300
  of aldehydes   309, 310
  of arenes   317–319
  of double and triple bonds   331, 332
  of ethers   346
Chromophores,
  inherently dissymmetric   82
  optically active   611
Chymotrypsin   998, 1007
α-Chymotrypsin, chiroptical properties of
      100
Cinnamic acid, mass spectrum of   131
trans-Cinnamyl formate, pyrolysis of   878
Circular dichroism (CD) spectra   68–113
  at low temperatures   80, 88
  of amides   72, 73, 87–95, 99–105,
      611, 612
  of carboxylic acids and esters   70–73,
      87–113
  of lactams   73–85, 517, 611, 612
  of lactones   73–87, 495, 621,
      624–626
Citraconic acid, mass spectrum of   131
Citric acid, hydrogen bonding in   225
Claisen condensation,
  in carboxylic acid synthesis   283

in ester synthesis   431
Claisen rearrangement   583, 584, 598
  in amide synthesis   451
  in carboxylic acid synthesis   388–393
  in lactam synthesis   1298, 1299
  in lactone synthesis   1182–1184
  in monothio ester synthesis   1039,
      1041
α-Cleavage   730, 731, 738
  of aliphatic amides   152, 153
  of carboxylic acids   123
  of esters   134
β-Cleavage, of carboxylic acids   124
γ-Cleavage,
  of aliphatic amides   153
  of carboxylic acids   123
  of esters   134
Cleavage reactions   373–380
  cathodic   810–813
  of ethers   373
  of ketones   373–376, 450
Clemmensen reduction, of dithio esters
      1046
CNDO/2 calculations,
  and dimers of monothio carboxylic acids
      1029
  and electronic spectra of thio esters
      1026
CNDO/CI method   612
  for calculation of rotational strengths
      84
CNDO/MO models, for studying the
      conformational dependence of the
      chiroptical properties of α-substituted
      propionic acids   95
CNDO/S calculations,
  for alkylallenecarboxylic acids   108
  of rotational strengths   84
Cobalt complexes, of amino and hydroxy
      acids   103, 104
Coefficient matrix   8
Comicelles   995, 998, 1000, 1002, 1012
1:1 Complexes, formation of, in deacylation
      of aryl esters   995, 999, 1002–1004
Computer-generated plots   973, 981
Condensation reactions,
  in amide synthesis   456
  in carboxylic acid synthesis   276–288
  in ester synthesis   420–431
  in lactone synthesis   1096–1100
  of thiono esters   1047
Conductivity measurements,
  on micellar solutions   987, 988
  on orthoamides   553
Configuration interaction (CI) wave
      functions   4–6
Conformation, of natural products   882
β-Conformation,

of oligopeptides 100
of polyamino acids 99
of polypeptides 100
Conformational isomerism, of amide group
605
Cooperative catalysis 983
Coordinate analysis, normal 1029, 1030
Copper complexes, of amino and hydroxy
acids 103, 104
Correlation energy 6
Cotton effect 68–113
in amides 72, 73, 87–95, 99–105, 611,
612
in carboxylic acids and esters 70–73,
87, 113
in lactams 73–85, 611, 612
in lactones 73–87, 621–626
Coumalic acid 510
α-Coumaranone, pyrolysis of 909
Coumarin, pyrolysis of 909
Counterions,
binding of, in micellar solutions 955,
959, 970
competition between, for 'sites' on
micellar surface 960
distribution of, in micellar solutions
955
Couplets, in CD spectra 99, 113
Coupling reactions, in carboxylic acid
synthesis 286, 287
Covalent participation 1011
Crotonamide, hydrogen bonding in 245
Crotonic acid, pyrolysis of 866
Crotonic acid anhydride, pyrolysis of 872
Crotonolactone, mass spectrum of 166
Crystalline concentration (zone freezing)
631
C=S stretching mode in thiono esters,
coupling of 1030
$C_{(\alpha)}$ substituents, influence of 78, 82
t-Cumyl chlorides, S_N1 solvolysis of 904
Cupric benzenecarboxylates, transcarboxylation of 935
Cyanoacetamide, hydrogen bonding in
244
Cyanoacetamides, pyrolysis of 868
Cyanoacetic ester, condensation reactions
of 1096–1099
Cyanoformates, pyrolysis of 894
Cyclic anhydrides,
mass spectra of 161, 162
pyrolysis of 873
Cyclization,
in lactam synthesis,
chemical 1195–1206
photochemical 1206–1224
in lactone synthesis 493–495
intramolecular,

of hydroxy acids and derivatives
1065–1078
of unsaturated acids and esters
1078–1096
of olefinic aldehydes 850, 851
Cycloaddition,
in lactam synthesis 1224–1272
in lactone synthesis 1180–1182
Cycloalkene-1,2-dicarboxylic esters, mass
spectra of 145
Cycloalkyl acetates, pyrolysis of 886–888
Cycloalkyl amides, mass spectra of 614
Cyclobutadiene,
as photoproduct,
of β-lactones 743
of α-pyrone 701–703
formation of 512
1,1-Cyclobutanedicarboxylic acid,
hydrogen bonding in 230
Cyclodecyl acetate, pyrolysis of 888, 889
3,5-Cyclohexadiene-1,2-dicarboxylic acid,
synthesis of 792, 793
Cyclohexan-1,2-dicarboxylic acid, mass
spectrum of 130
Cyclohexane, reaction of, with maleic
anhydride 207
Cyclohexene-1,2-dicarboxylic acids, mass
spectra of 130
Cyclohexenol acetate, mass spectrum of
140
Cyclohexyl acetates, cis-2-substituted,
pyrolysis of 882
Cyclohexyl trifluoroacetate, pyrolysis of
894
Cyclo-γ-oligoglutamic acids, CD spectra of
101
Δ^2-Cyclopentenone derivatives, i.r. and
Raman spectra of 620
Cyclopropanecarboxylic acid, pyrolysis of
865
Cyclopropenecarboamide, hydrogen
bonding in 244
Cyclopropylacetic acid, pyrolysis of 865
α,β-Cyclopropyllactones, ring chirality
rules for 81, 82
L-Cystine, CD spectrum of 99

Dansylamino acids, chiroptical properties
of 102
Darzens condensation,
in carboxylic acid synthesis 277
in ester synthesis 421, 422
Deacylation, micellar effects on 946–1015
Deamination, of micellized amines 1006
Decanol, effect on micellar catalysed
reactions 984
S-(+)-5-Decanolide, CD spectrum of 80,
81

Decarbonylation,
   by PdCl$_2$ 848
   by RhCl(PPh$_3$)$_3$ 825–855
   in pyrolysis reactions   866
   of acyl halides   827–842
      mechanism of   837–842
      regioselective   837
   of aldehydes   843–855
      mechanism of   852–855
      olefin elimination as side-reaction
         in   852
      solvent effects on   842
      stoichiometric   834, 841–843
Decarboxylation,
   in synthesis of thiolo esters   1039
   of alkanoic acids   864–867
      mechanism of   864, 865
   of alkenoic acids   860–863
   of β-keto acids   860, 861, 863
Decarboxylation–recarboxylation
      mechanism, intermolecular, ionic,
      for transcarboxylation   941
   of benzenecarboxylates   932
   of heterocyclic carboxylates   939
   of naphthalenedicarboxylates   937
   of polynuclear carboxylates   939
n-Decylguanidinium ion, addition of, to
      micellar catalysed reactions   994
n-Decyltrimethylammonium bromide, as
      micelle   971, 974
Dehydration, of alkanoic acids   864
Dehydrative coupling, of carboxylic acids
      437
Dehydrohalogenation   1193
   of aliphatic acyl halides   834–837
      regioselective   837
Dephosphorylation, micellar catalysis of
      1008
Deprotonation reactions, theory of   33–41
Derivative properties   16
Deslongchamps model   54
Desulphuration, of dithio esters   1048
β-Deuterium isotope effects, in pyrolysis
      of esters   881, 899, 900, 904
Deuterium labelling,
   in formamide acetals   578, 580, 581
   in mass spectrometry   124–128, 133,
      135–138, 140, 142, 168
   in micellar catalysed reactions   1005,
      1010
   in photolysis reactions   702, 713, 744
   in transcarboxylations   929, 930, 934,
      935, 938
   in vibrational spectroscopy of monothio
      carboxylic acids   1029
Diacetamide, hydrogen bonding in   249
1,2-Diacetoxymethylcyclobutane,
      pyrolysis of   903

Diacyl amides   603
Diagonal matrix   8
Dialkoxyalkyltriethylammonium fluoro-
      borates, in amide acetal synthesis
      542
Dialkoxydialkylaminoacetonitriles   576
N,N-Dialkylalkoxymethylene iminium
      salts, in amide acetal synthesis
      535–539
N,N-Dialkylalkylmercaptomethylene
      iminium salts, in amide acetal
      synthesis   535–539
α-Dialkylamino-α-alkoxynitriles,
      reaction of   557, 558, 562, 574, 577,
      584
   synthesis of   535, 539, 544
2-Dialkylamino-4-alkylidene-1,3-dioxolones
      584
1-Dialkylamino-1-arylnitriles   574
2-Dialkylaminobenzo-1,3-dioxole   557
Dialkylaminodialkoxyacetonitriles   576
2-Dialkylaminodioxanes   543
2-Dialkylamino-1,3-dioxanes   557
5-Dialkylamino-2,4-dioxoimidazolines,
      5-substituted   577
2-Dialkylaminodioxolanes   543
2-Dialkylamino-1,3-dioxolanes   557
N,N-Dialkylaminomethane phosphonic
      esters   564
Dialkylammonium dithiocarboxylates,
      synthesis of   1036
Dialkylphosphine oxides, reaction of, with
      dimethylformamide dimethylacetal
      564
Dialkyl sulphates,
   in tris(acylamino)methane synthesis   547
   reaction of,
      with amides   536
      with N-substituted lactams   537
Diamides, mass spectra of   158
1,2-Diaminoethylene   575
N,N-Diaryl amide acetals, preparation of
      542
Diarylcarbodiimides, reaction of, with
      amide acetals   577, 578
1,2-Diarylethyl acetates, pyrolysis of,
      Hammett ρ-factors for   890
1,4-Diaza[2.2.2]bicyclooctane, reaction
      with anhydrides   203
Diazoalkanes, reaction with thio esters
      1040, 1049
α-Diazo amides, photolysis of   1189–1191
α-Diazocarboxamides, cyclization of   1224
α-Diazo esters, photolysis of   1189–1191
Diazomethane, reaction with thiobenzoic
      acid   1040
Diazonium salts, addition of, to olefins
      1192

Dibenzoate chirality rule   112, 113
Dibromoacetonitrile, in orthoamide
    synthesis   549
1,3-Dibromo ketones, in amide acetal
    synthesis   541
Dibromopyromellitic dianhydride,
    complexes of   197
Dibutylacetamide, hydrogen bonding in
    245
Dicarboxylic acids, mass spectra of
    129–131
Dicarboxylic esters, mass spectra of
    143–147
Dichloroacetates, pyrolysis of   886
Dichloroacetic acid, hydrogen bonding in
    230, 237
2,6-Dichlorobenzamide, mass spectrum of
    156
Dichlorocarbene, in amide acetal
    synthesis   540
Dichlorofluoromethane, in triamino-
    methane synthesis   546
Dichloromaleic anhydride, complexes of
    176, 179, 180, 197
Dichloromaleic thioanhydride, complexes
    of   176
Dichlorophthalic anhydride, complexes of
    176, 200
3,6-Dichlorophthalic anhydride, complexes
    of   181, 197
Dicyanophenylmethyl benzoate, pyrolysis
    of, intramolecular electrophilic
    aromatic substitution mechanism
    for   874
Dielectric constants, of lactams   516
Diels–Adler reactions, intramolecular
    1092–1096
Diethyl adipate   807
Diethyl carbonate, pyrolysis of   894
Diethyl dibromomalonate, addition of, to
    methyl methacrylate   1192, 1193
N,N-Diethyldodecanamide, hydrogen
    bonding in   250
Diethyl fumarate,
    anion radical of   803
    cathodic reduction of   803, 808
Diethyl maleate, anion radical of   803
Diethyl malonate, mass spectrum of   144
Diethyl oxalate,
    anion radical of   803
    cathodic reduction of   803, 804
Diethyl phenylmalonate   810
Diethyl succinate, radiolysis of   771
Difluoroacetamide, hydrogen bonding in
    244
Difluoroacetic acid, pyrolysis of   865
Difluorocarbene   865
Difluoromethyl trifluoroacetate   865

α,α-Difluorotrialkylamines, in amide acetal
    synthesis   540
Digitoxigenin   503
Diglycerides, chiroptical properties of
    106, 107
Dihalides, hydrolysis of, in carboxylic acid
    synthesis   276
Dihedral angle   80
1,2-Dihydrophthalic anhydride, mass
    spectrum of   161, 162
Dihydroxyfumaric acid, hydrogen bonding
    in dihydrate of   224
Diisopropylamides, mass spectra of   615
Diketen   500
    synthesis of   498
Diketopiperazines,
    $m_1\mu_2$ mechanism for   83
    quadrant rule for   76, 77
γ-Dilactones, α,β-unsaturated, mass spectra
    of   628
1,2-Dilauryl-3-myristoyl-sn-glycerol,
    chiroptical properties of   107
Dimedon derivatives, chiroptical properties
    of   102, 103
Dimerization   41–45 – see also
    Self-association
    of dithio carboxylic acids   1031, 1035
    of formamide   42, 44, 45
    of formic acid   25, 42, 43, 222, 226
    of monothio carboxylic acids   1029
2,2-Dimethoxy-3-chlorodihydropyrans,
    dehydrohalogenation of   1193
1,1-Dimethoxypyrrolidinomethane
    561
Dimethylacetamide, hydrogen bonding in
    250, 251
N,N-Dimethylacetamide, hydrogen bonding
    in   254
Diemthylaminelithium dimethylamide,
    reaction with N,N-dimethyl-
    chloromethylene iminium chloride
    546
Dimethylaminoalkoxycarbenes   597
Dimethylaminoalkylthiomethane
    phosphonic esters   565
Dimethylaminobis(dialkylphosphinyl)-
    methanes   564
α-N,N-Dimethylamino tertiary amides, CD
    spectra of   94
N,N-Dimethylamylamine   820
N,N-Dimethylanthranilic acid, hydrogen
    bonding in   240
N,N-Dimethylbenzamide, cathodic
    reduction of   813, 819
N,N-Dimethylbenzamides,
    hydrogen bonding in   250
    mass spectra of   156
N,N-Dimethylbenzylamine   819

Dimethyl benzylidenemalonate, mass spectrum of 158

Dimethyl benzylidenemalonates, mass spectra of 144, 145

2,2-Dimethylbut-3-enoic acid, pyrolysis of 860, 861
transition state in 861

4,4-Dimethylbut-2-enoic acid, pyrolysis of 860

N,N-Dimethylcinnamides, hydrogen bonding in 250

Dimethyl cyclobut-3-ene-1,2-dicarboxylates, mass spectra of 146

N,N-Dimethyldithiocarbamic acid esters 577, 578

N,N-Dimethyl-*N*-dodecylglycine, as micelle 958

Dimethylformamide,
hydrogen bonding in 251, 252
negative-ion mass spectrum of 159

Dimethyl fumarates, mass spectra of 146, 147

Dimethyl maleate, cathodic reduction of, in the presence of carbon dioxide 809

Dimethyl maleates, mass spectra of 146, 147

Dimethylmaleic anhydride, reactions of 208

Dimethyl muconates, mass spectra of 146

2,2-Dimethylpent-3-enoic acid, pyrolysis of 861

Dimethyl phenylsuccinate, mass spectrum of 144

2,2-Dimethylstyrene, formation of, in pyrolysis of neophyl acetate 902

2,3-Dimethylsuccinic acid, chiroptical properties of 94

Dimethyl tetrathiooxalate, synthesis of 1043

N,N-Dimethylthioacetamide, hydrogen bonding in 254

N,N-Dimethylurethan, in synthesis of orthocarbamic acid esters 550

N,N-Dimethylvaleramide, cathodic reduction of 820

α,α-Dimorpholineacetonitrile, synthesis of 549

1,2-Dimyristoyl-sn-glycerol, chiroptical properties of 107

2,4-Dinitrochlorobenzene, micellar catalysed decomposition of 1008, 1011

2,4-Dinitrofluorobenzene,
micellar catalysed decomposition of 1008
micellar catalysed reaction with aniline 964, 983

3,6-Dinitronaphthalene-1,8-dicarboxylic anhydride, complexes of 182

Dinitrophenyl phosphate dianion 993

3,5-Dinitrophthalic anhydride, complexes of 176, 181, 197

Diols,
dibenzoate chirality rule for 112, 113
oxidation of, in lactone synthesis 1145, 1151–1159

3,6-Dioxo-1,4-cyclohexadiene-1,2,4,5-tetracarboxylic dianhydride 177

1,2-Dioxolane-3,5-diones, photolysis of 1192

1,2-Dipalmitoyl-3-myristoyl-sn-glycerol, chiroptical properties of 107

dl-erythro-2d-1,2-Diphenylethyl acetate, pyrolysis of 882

dl-threo-2d-1,2-Diphenylethyl acetate, pyrolysis of 882

3,5-Diphenyl-5-hydroxy-2-pyrrolin-4-ones, chiroptical properties of 102

N,N-Diphenyl phenylacetamide, mass spectrum of 155

Diphenyl maleate, pyrolysis of 875

Diphenylphosphinates, pyrolysis of 894

Dipicolinic acid, cathodic reduction of 797, 798

2,6-Dipicolinic acid, hydrogen bonding in monohydrate of 225

Dipole–dipole couplings 73

Dipole moments,
of formamide 13–15, 28, 29
of formic acid 25
of formyl fluoride 21
of lactams 517
of lactones 495, 619, 622
of orthoamides 554
of *S*-oxides 1049
of 2-pyrrolidone 609
of thio acid derivatives 1023, 1024
of thiocarbonyl-*S*-imides 1049

Direct field effects 907

Dismutation, of aminal esters 553

Dissociation, of orthoamides 553

Dissociation energy 33

Dissymmetry factor 79

Disulphide chromophore 99

Disulphides, oxidation of 1192

Disulphonylamines, reaction of, with orthoamides 563

Diterpene carboxylic acids, conformational analysis of 88

1,3-Dithian 1053

Dithietan 1052

Dithiocarbamates, chiroptical properties of 102, 103

Dithiocarbonic acid *O,S*-esters 577, 578

Dithiolcarbonates, formation of, from rearrangement of xanthates 896
Dithiocarboxylates, alkylation of 1042
Dithio carboxylic acids,
    chemical properties of 1044, 1045, 1052
    i.r. spectra of 1030, 1031
    n.m.r. spectra of 1033, 1035
    synthesis of 1036–1038
    u.v. spectra of 1027, 1028
Dithio esters,
    acylation of 1050, 1051
    alkylation of 1049, 1050
    i.r. spectra of 1031, 1032
    mass spectra of 1035, 1036
    n.m.r. spectra of 1034
    oxidation of 1048, 1049
    photolysis of 1052
    proton acceptor properties of 1045
    reaction of, with carbanions 1047, 1048
    reduction of 1045, 1046
    solvolysis of 1048
    synthesis of 1042–1044
    u.v. spectra of 1027, 1028
Dithioformic acid, synthesis of 1037
1,3-Dithioles 1052
1,3-Dithiolium cations 1052
1,3-Dithiolone-(2), photolysis of 1043
Dithioparabanic acid, orthoamides of 549
Dithiopyromellitic dianhydride, complexes of 177
n-Dodecane thiol, as functional micelle 1002
S-(+)-5-Dodecanolide, CD spectrum of 80
N-Dodecanoyl-DL-cysteinates 1001
N-Dodecanoylglycinates, as micelles 1001
N^α-Dodecanoyl-L-histidine, mixed micelles containing 1003
Dodecylamine, in deacylation of p-nitrobenzoyl phosphate dianion 993
Dodecylammonium chloride 955
N^α-Dodecylglycine, mixed micelles containing 1004
p-Dodecyloxybenzenesulphonic acid 981
n-Dodecyltrimethylammonium bromide, as micelle 971, 974, 975, 979
Doebner reaction, in carboxylic acid synthesis 277
Double bonds, oxidation of, in carboxylic acid synthesis 331–344
Double proton transfer 43,44
Dowex 1 979
Drug stabilization, in micellar solutions 989, 1014
'Dynamic' carbanion species, in transcarboxylation,

of benzenecarboxylates 931
of naphthalenecarboxylates 937

$E_{1/2}$, for oxidation of the amide group, effects of structure on 618
EHMO calculations, and dimers of monothio carboxylic acids 1029
Eigenvalues 18
Electrochemical cyclization, in lactone synthesis 1080, 1081, 1084–1088
Electrohydrodimerization (EHD) reaction of α,β-unsaturated esters 806–808
Electrolyte effects,
    in micellar catalysed reactions 952, 968, 972, 977–980, 987, 1000, 1014
    in micellar inhibited reactions 949, 977
Electron affinity 33, 34
    of anhydrides 195–197
π-Electron densities, reactivity and 907
Electron density formalism 7
Electron diffraction spectroscopy, of thio acid derivatives 1022, 1023
Electron–electron repulsion 7
Electronic effects, on direction of elimination in pyrolysis of esters 884, 890–898
Electronic transition energies,
    of formamide 29–31
    of formic acid 25
Electron impact 123, 125, 130
Electron kinetic energy 7
Electron paramagnetic resonance spectroscopy, for micellar catalysed reactions 988
Electron spectroscopy 69–73
    of hydrogen bonding 217, 218, 229
    of thio acid derivatives 1023, 1025–1029
Electron spin resonance spectroscopy, of thio esters 1035
Electrophiles, reaction of orthoamides with 572–581
Electrophilic aromatic substituent constants, determined from pyrolysis of 1-arylethyl acetates 904–907
Electrophilic reactions, of thio esters 1048–1051
Electrophilic substituent effects, in the gas phase 890
Electroreduction, of thio esters, 1045, 1046
Elephantin 507
E1-like transition states, in pyrolysis of esters 881, 883, 886, 897, 898, 904

E$_i$-like transition states, in pyrolysis of
          esters   881, 883, 886, 897, 903, 904
Elimination,
     in orthoamides   584–586
     in pyrolysis of esters containing non-
               vinylic β-hydrogen atoms
               880–907
          as a model for electrophilic
               substitution   904–907
          cis nature of   880–884
          cyclic nature of   881
          direction of   884
          isotope effects on   898–900
          neighbouring-group effects in   900,
               901
          rearrangements and   901–903
          trans   889, 890
     syn   716
ElcB mechanism, for decomposition of
          p-nitrophenyl alkanoates in presence
          of cationic micelles   990
E2 mechanism, for decomposition of
          β-bromocarboxylate ions in cetyl-
          trimethylammonium bromide   988
Enamides, cyclization of   1207–1224
Enamines, in carboxylic acid synthesis
          288
Endocyclic alkenes,
          formation of, in pyrolysis of cycloalkyl
               esters   889
          stability of   889
Ene reaction   504
Energy hypersurface, critical points of   17,
          18
     order of   18
Enthalpy of formation   59–65
     of formyl fluoride   20, 21
     of orthoamides   554
Entropy of activation, in pyrolysis of
          esters   881
Entropy of formation, of orthoamides
          554
EPA (ether:isopentane:ethanol = 5:5:2)
          81, 91
Ephedrine derivatives, as functional micelles
          1008
Episulphides   1052
     as intermediates in alkylation of thio
          esters   1049
Epoxides   585
α,β-Epoxy lactones, ring chirality rules for
          81, 82
E2 reactions, of alkyl halides   1011
Ester chromophore   70
Esters,
     acylations with, in amide synthesis
          446, 447
     aliphatic – *see* Aliphatic esters

aromatic – *see* Aromatic esters, Aryl
          esters
     as proton acceptors   240, 241
     cathodic reduction of   801–813
     chiroptical properties   87–108
          of naturally occurring   108
          of α-substituted   87–104
          of α,β-unsaturated   108
     containing non-vinylic β-hydrogen
          atoms, mechanism of pyrolysis
          of   880–907
     ferric hydroxamate test for   603
     heats of formation of   62, 63
     hydrolysis of,
          in carboxylic acid synthesis
               271–273
          theory of   49–54
     in acyl halide synthesis   440, 441
     keto and aldehydic, reduction of
          1145–1150
     mass spectra of   133–152
     photochemical hydrogen-abstraction
          reactions of   708–718, 744, 745
     photodecarbonylation of   652–661,
          738
     photodecarboxylation of   687–708,
          743
     photorearrangement of   729–735
     pyrolysis of   874–908
          breaking of the β-C–H bond in
               897, 899
          conformational effects in   889
          eclipsing interactions in   889
          effect of electron withdrawal at the
               γ-carbon on the reactivity in
               893
          effect of β-substituents on   896
          five-centre process in   875
          four-centre processes in   874–876
          Hammett ρ-factors for   883, 893,
               896, 900
          isotope effects in   881, 898–900
          mechanism of the elimination in
               880–907
          neighbouring-group effects in   900,
               901
          nucleophilicity of the carbonyl
               oxygen in   893
          nucleophilic substitution in   895
          partial formation of a carbocation
               in   890, 891, 904
          endo:exo product yields   883
          rearrangements in   896, 901–903
          six-centre processes in   875, 877,
               880, 908
          solvent effects in   900
          steric acceleration in   881, 882,
               885, 889, 897, 900–902

steric interactions in 886
σ^0-values for *ortho* substituents in 894
radiolysis of 769–773
synthesis of 411–436
  by condensation reactions 420–431
  by free-radical processes 431–434
  by solvolytic reactions 411–420
  from acetylenes 435
  from diazo esters 436
  from organoboranes 434
thiolysis of 1040
thionation of 1040
unsaturated — *see* Unsaturated esters
Ethane thiol, as functional micelle 1001
Ethanol, effect on micellar catalysed reactions 984
Ether formation, in cathodic reduction, of esters 805
Ethers,
  cleavage reactions of 373
  oxidation of,
    in carboxylic acid synthesis 344–352
    in lactone synthesis 1171, 1172
o-Ethoxybenzamide, mass spectrum of 157
2-Ethoxysuccinic acid, CD spectrum of 91
N-Ethylacetamide, hydrogen bonding in 247
Ethyl acetate, molecular ion of 136
Ethyl acetoacetate, mass spectrum of 150
Ethyl acrylate, cathodic reduction of 807
Ethyl benzoate,
  cathodic reduction of 805
  electron-transfer reduction of, by naphthalene anion 806
Ethyl benzoates, mass spectra of 141
Ethyl carbonic acid anhydride, pyrolysis of 872
Ethyl *trans*-crotonate, pyrolysis of 894
Ethylene dibenzoate, pyrolysis of 876
Ethyl formate, mass spectrum of 137, 138
Ethyl glyoxylate, hemiacetal of 804
2-Ethyl-3-ketohexanoic acid, pyrolysis of 863
Ethyl lactate, CD spectrum of 89
Ethylmethyl carbonate, pyrolysis of 894
Ethyl phenylacetate, cathodic carboxylation of 810
Ethyl 2-thiazolecarboxylate, cathodic reduction of 804
S-Ethyl thiobenzoate, mass spectrum of 151
Ethynyl substituents at the β-carbon of esters, effect on rate of pyrolysis of 899
Eudesman lactone 73, 74, 79
Euparotin 507
Exchange of deuterium by protium, in transcarboxylations of mixtures of benzenecarboxylates labelled and non-labelled by deuterium on the benzene ring 929, 930
Excimer fluorescence 229
Excited-state EDA complexes 204
Excited structures 4
Exciton chirality method 111–113
Exciton coupling 99
Exocyclic alkenes, formation of, from cycloalkyl esters 889
Exponential-Type Functions (ETF) 9, 10

Favorskii rearrangement,
  in amide synthesis 452
  in carboxylic acid synthesis 393–398
Feist's acid 731
Fermi resonance 220, 249, 607
Ferric hydroxamate test 603
  for lactones 619, 628
Fluorene, in ketene aminal synthesis 568
Fluorescence spectroscopy,
  for micellar catalysed reactions 988
  of EDA complexes 198–202
2-Fluoroacetamide, intramolecular hydrogen bonding in 253
Fluorobenzene ion 160
Fluoromaleic anhydride, pyrolysis of 873
Fluoromalonic acid, hydrogen bonding in 234
Force constant matrix 17
Force constants 16, 17
  of formic acid 25
Force-field calculations, on amide group 605
Formamide,
  barrier to rotation about C–N bond of 28
  dimerization of 42, 44, 45
  electronic transition energies of 29–31
  geometry of 27, 28
  hydrogen bonding in 247, 252, 253
    ab initio studies of 241–243
  hydrolysis of 48
  in tris(formylamino)methane synthesis 547
  ionization potentials of 29
  mass spectrum of 152
  protonation of 39–41
  pyrolysis of 869
  reaction of, with tris(dimethylamino)-alkoxymethanes 562
  solvation of 45–47
  variation of dipole moment of, with basis-set size 13–15, 28, 29

Formamides,
N-substituted 618
N,N-substituted, Raman spectra of 607, 608
Formamidines, N,N,N-trisubstituted 564
Formamidinium carbaminates 578
Formamidinium salts, N,N,N',N'-tetra-substituted, in orthoamide synthesis 540, 541, 545, 546
Formanilides, mass spectra of 154, 155
Formate ion, electronic states of 26
Formates,
mass spectra of 137, 138
pyrolysis of 877, 878
reactivity of 880
Formic acid,
barrier to in-plane tautomerization of 24
barrier to rotation of 23, 24
chain formation in 215, 222, 232
deuteron quadrupole coupling constant of 230
dimerization of 42, 43, 222, 226
vibrational frequency shifts on 25
dipole moment of 25
electronic transition energies of 25
cis–trans energy difference of 23, 24
expulsion of, in fragmentation processes 125
force constants of 25
geometrical parameters of 22–24
ionization potentials of 25, 26
molecular ion of, in mass spectra 129
negative-ion mass spectrum of 133
protonation of 35–41
pyrolysis of 864
u.v. spectra of 70
vibrational spectra of 227
Formyl chloride, as intermediate in pyrolysis of formamide 869
Formyl fluoride, ab initio investigations of 20–22
Formyl radical 768
Fragmentations 123–169
involving hydrogen migrations 153, 154
involving rearrangements 124–126, 134–139
simple 123, 124, 133, 134, 140, 141
Free energy, of orthoamides 554
Free-energy differences, between cis and trans isomers of 2,4-disubstituted γ-butyrolactones 619
Free-radical processes,
in carboxylic acid synthesis 288–290
in ester synthesis 431–434
Free radicals, neutral 757
Friedel–Crafts reaction, in synthesis of thio

acid derivatives 1037, 1038, 1043
Fumaramide, hydrogen bonding in 245
Fumaric acid,
cathodic hydrogenation of 800
mass spectra of 131, 133
Functional-group moments, of esters and thio esters 1025
Furan, electrophilic aromatic substituent constants for 904, 906
Furancarboxylic acids, transcarboxylation of 939
Furanedicarboxylic acid, self-association of 222

Gaillardin, mass spectrum of 627
Galactonolactones, colorimetric estimation of 619
Galactose, chiroptical properties of 105
Gangleosides, chiroptical properties of 105
Gas chromatography,
alkali-fusion 617
use of in determination,
of amides 617
of anhydrides 634
of lactones 628
Gaussian-Type Functions (GTF) 9, 10
primitive 10
Gaussian-Type Orbitals (GTO) 9
Gel filtration method,
for measuring the fraction of micellar incorporated nucleophile 962
for measuring substrate binding constants 974
General Restricted Hartree–Fock (GRHF) theory 12
Germaacetic acid, decomposition of 894
Gibberellins 87
Gluconolactones, colorimetric estimation of 619
Glucoronic acid 105
Glucose, chiroptical properties of 105
Glutaconic anhydrides, rearrangement of 511
Glycerides, chiroptical properties of 106, 107
Glycine,
ab initio investigations of 32, 33
electron density distribution for 32
polypeptides of 32
zwitterion of 32
Glycolic acid, intramolecular hydrogen bonding in 240
1,2-Glycols, dibenzoate chirality rule for 112, 113
Glycosaminoglycans, chiroptical properties of 105
Glycosidation, of protected acylated sugars 558

Glycosides, chiroptical properties of   105
Glyoxylic acid, synthesis of   800
Gouy−Chapman theory   955, 956, 969, 978
Grignard reagents,
   in lactone synthesis   1110−1113
   in thio carboxylic acid synthesis   1037
   reaction of,
      with orthoamides   565
      with thio esters   1047
Ground configurations   4, 5
Group additivity   60−65
Guanidines, reaction of, with orthoamides
   561, 562
Guanidinium salts, in orthoamide synthesis
   548

Half-wave potentials, for reduction of thio
   esters   1035, 1046
Halides, in oxidation reactions   349−351
Halo acids, mass spectra of   132
N-Haloacyl amides, reaction of, with
   orthoamides   580
Halo amides, cyclizations using
   1200−1204
α-Halo esters, conformational analysis of 88
Haloform reaction, in oxidation of ketones
   358−361
α-Halogeno acids, chiroptical properties of
   88, 92, 93, 95
α-Halogeno esters, chiroptical properties of
   88−93, 95
N-Halogenosulphonamide   580
Halogens, as oxidizing agents in lactam
   synthesis   1302−1304
Halolactonization   494, 1081, 1089−1092
α-Halopropionanilides, chiroptical
   properties of   93
Hamiltonian   6, 7, 13, 14
   expectation value of   6
Hammett ρ-factors, for pyrolysis of esters
   883, 893, 896, 900
Hammett relationships   1026
Hartree−Fock equations   6−12
Hartree−Fock Limit (HFL)   6
Hartree−Fock problems,
   closed-shell   6−10
   open-shell   10−12
H/D exchange,
   in elucidation of transcarboxylation
      mechanism   934, 935
   in formamide acetals   578, 580, 581
Heat of formation − see Enthalpy of
   formation
Helenalin   507
α-Helix conformation, of polyamino acids
   99
Hemimellitic acid, hydrogen bonding in
   dihydrate of   224

Henkel reaction   398
N-Heptylimidazole, micellar catalysed
   acylation of   966, 967
Herzig−Meyer method   609
Hessian matrix   17, 18
Heterocumulenes   577−580
Heterocycles,
   electrophilic reactivities of   904
   synthesis of, from thio acid derivatives
      1052, 1053
Heterocyclic acids, cathodic reduction of
   794−800
Heterocyclic carboxylates, transcarboxylation of   939
Heterocyclic syntheses, using orthoamides
   569−572
n-Hexadecyltrimethylammonium bromide,
   as micelle   946, 971, 975
Hexafluoroisopropanol (HFIP), as solvent
   95, 99
Hexamethylbenzene, reaction of, with
   maleic anhydride   206
Hexamethylguanidinium chloride   559
N,N,N',N',N'',N''-Hexamethylguanidinium
   chloride, in orthoamide synthesis
   548
Hexamethylguanidinium cyanide   576
Hex-3-endoic acid, pyrolysis of, concerted
   cyclic mechanism for   861
n-Hexyl acetates, mass spectra of   135
L-Histidine, in chelation with transition
   metals   104
Histidine derivatives, as functional micelles
   1005
HMO calculations, and electronic spectra
   of thio acid derivatives   1026
Hofmann rearrangement, of primary amides
   602, 603
Hofmeister series   956, 978
3-Homoadamantyl acetate, pyrolysis of
   903
Homocholine derivatives, functional
   micelles of   1008
Homoconjugate ions, in carboxylic acid
   hydrogen bonding   223, 224, 237,
   238
Homophthalic acid anhydride   197
Homoserine lactone, synthesis of   1193,
   1194
Hückel method   907
Hudson's rule   621
Hydantoins, chiroptical properties of   102
Hydrated electrons   763−769, 772, 775
Hydration numbers, of carboxylic acids
   237
Hydrazobenzene   976
Hydrocarboxylation, in carboxylic acid
   synthesis   290−293

Hydrocyanic acid, reaction with alicyclic
　　amines　549
Hydrogen abstracting reactivity and
　　thermodynamic bond dissociation
　　energies　713, 714
Hydrogen-abstraction reactions　708−727
　　reversibility of　714, 715
γ‑Hydrogen-abstraction reactions　712,
　　713, 725
　　isotope effect for　713
δ‑Hydrogen-abstraction reactions
　　722−724
　　substituent effects on　724
Hydrogenation, of the aromatic nucleus
　　790
Hydrogen bonding　214−254
　　in amides　241−254
　　in carboxylic acids　221−240
　　in thio acids　254
　　in thio amides　254, 255
　　spectroscopy of　216−218, 226−231
　　structural aspects of　214−216
　　symmetry of　228, 231
　　theory of　41−47, 218−221, 231−234,
　　241−243
　　thermodynamics and kinetics of　218,
　　234−238
Hydrogen migration　135, 136, 142, 153,
　　154
Hydrogen radicals　763−768, 772, 775
Hydrogen randomization (scrambling)　136
1,5-Hydrogen shifts, in pyrolysis of esters
　　containing non-vinylic β-hydrogen
　　atoms　881
Hydrolysis,
　　in synthesis of thiolo esters　1038,
　　1039
　　of acyl halides　275
　　of amides　48, 274
　　of anhydrides　275
　　of dihalides　276
　　of esters　49−54, 271
　　of nitriles　271, 449
　　of orthoamides　557
　　of thiolo esters　1048
　　of trihalides　275
　　theory of　47−54
Hydrophobic interactions　225, 226
　　between micelles and solute　951
　　between solutes　952
Hydrophobicity,
　　of nucleophile in micellar reactions
　　949, 998
　　of substrate in micellar reactions　949,
　　950, 973, 974
Hydroxamic acid−iron complex, stability
　　of　603

Hydroxamic acids, as functional micelles
　　1007
Hydroxy acids, intramolecular cyclization
　　of　1065−1078
α-Hydroxy acids, chiroptical properties of
　　88−91
β-Hydroxy acids, reaction of, with amide
　　acetals　585
β-Hydroxyalkenes, pyrolysis of　908
β-Hydroxyalkynes, pyrolysis of　908
Hydroxy amides, cyclizations using　1205,
　　1206
p-Hydroxybenzoic acid, hydrogen bonding
　　in monohydrate of　224
o-Hydroxybenzoic acids, intramolecular
　　hydrogen bonding in　239
α-Hydroxybenzylacetic acid, CD spectrum
　　of　110
Hydroxycarbene, from photolysis of
　　pyruvic acid　686, 740
Hydroxy esters, mass spectra of　148
α-Hydroxy esters, chiroptical properties of
　　88−91
β-Hydroxy esters, pyrolysis of　908
2-Hydroxyethyl acetate, radiolysis of　772
o-(2-Hydroxyethyl)benzamide, mass
　　spectrum of　157
β-Hydroxyethyl groups, in functional
　　micelles　1012
2-(1-Hydroxyethyl)pyridine N-oxide,
　　pyrolysis of　903
β-Hydroxy ketones, pyrolysis of　908
Hydroxy-γ-lactone, mass spectrum of　166
Hydroxylalkyl derivatives, as functional
　　micelles　1008−1011
Hydroxylamine, reaction of, with anhydrides
　　631
α-Hydroxy-β-methylvaleric acid, CD
　　spectrum of　91
α-Hydroxyphenylacetic acid, CD spectrum
　　of　110
Hydroxyproline, CD spectrum of　89
3-Hydroxypyridinium derivatives of amino
　　acids, chiroptical properties of　102
Hyperconjugation　863
　　in pyrolysis of esters　899, 906
Hypochlorite, in oxidation of alcohols　298

Imidates, thiolysis of　1040, 1041
Imidazole anion, in micellar catalysis
　　1004, 1005
Imidazole-2-carboxylic acid, cathodic
　　reduction of　795−797
Imidazole derivatives, as functional micelles
　　995, 1002, 1004, 1012
Imidazole group, acylation of　996
Imidazoline　1052

Imide carbonyl group, reduction of, to a
    methylene group 818–820
Imides,
    cathodic reduction of 813, 814,
        816–820
    ferric hydroxamate test for 603
    hydrogen bonding in 254
    mass spectra of 162–165
    photochemical hydrogen-abstraction
        reactions of 722–724, 745
    photodecarbonylation of 669–671
    reaction of, with orthoamides
        562–564
    synthesis of,
        by acylation reactions 458–460
        by C- and N-alkylations 460
        by oxidation reactions 460, 461
Imidothiolates, thiolysis of 1042
Imidozolinediones, in synthesis of urea
    acetals 551
Imines,
    reaction of, with ketenes 1230–1263
    Reformatsky reaction with 1233,
        1265–1267
Iminium alkylsulphates,
    in amide acetal synthesis 536
    in amide thioacetal synthesis 536
Iminium salts, reduction of, to give amide
    acetals 544
Indenes, reaction of, with dimethylmaleic
    anhydride 208
INDO calculations, on acetolactone and its
    difluoro and dimethyl derivatives
    620
INDO models, for studying the conforma-
    tional dependence of the chiroptical
    properties of α-substituted
    propionic acids 95
INDO/MO model, for calculation of
    rotational strengths 83, 84
Infrared spectroscopy,
    of amides 604–608
    of anhydrides 631, 632
    of hydrogen bonding 216, 217
    of lactams 517, 608, 609
    of lactones 495, 619–621
    of orthoamides 553
    of products of photolysis of α-pyrone
        701, 702
    of thio acid derivatives 1026,
        1029–1032
Inorganic acids, reaction of, with
    orthoamides 572
Integrals,
    coulomb 7
    exchange 7
    molecular 8

two-electron 7
Internal coordinates 19, 20
Internal motion of atoms 17
Iodine–pyridine, in oxidation of ketones
    361
Iodoacetic acid, pyrolysis of 865
Ion cyclotron resonance 162
Ion-exchange constant 970, 978, 980
    for two counterions between micellar
        surface and the aqueous phase
        946
Ion-exchange models, in micellar catalysis
    969, 970, 978–980, 982
Ion-exchange resins 956, 957, 960, 969
Ionization-initiated reactions 756–759
Ionization potentials,
    of carbon 14
    of formamide 29
    of formic acid 25, 26
    of hydrogen 33, 34
Ion–molecule reactions 757
Ion-pair formation from excited EDA
    complexes 204
Ion-selective electrodes 980
Iron pentacarbonyl, reaction of, with
    orthoamides of carbonic acid 566
Isobornyl acetate, pyrolysis of 902
Isobutyramide, mass spectrum of 152
Isocyanates,
    acylations with, in amide synthesis 448
    in orthoamide synthesis 542, 549
    reaction of, with orthoamides 577
Isodehydroacetic acid 510
Isoimides 868
Isoleucine, CD spectrum of 98
Isomerism,
    cis–trans,
        in formic acid H-bonded chains 222
        in products of decarbonylation and
            dehydrohalogenation of
            aliphatic acyl chlorides 834
        in products of pyrolysis of esters
            containing non-vinylic
            β-hydrogen atoms 880
    E/Z, in thio acids 1033
Isonicotinic acid,
    anion radicals of methyl and ethyl
        esters of 803
    electrochemistry of 794, 795, 799,
        803
Isonicotinic amide, cathodic reduction of
    815
Isonitrile rearrangement 399
Isophthalic acid, cathodic reduction of
    793
N-Isopropylacetamides, hydrogen bonding
    in 247

o-Isopropylbenzoic acid, mass spectrum of
  127, 128
1,2-Isopropyliden-3-acyl-sn-glycerols,
  chiroptical properties of 106
1,2-Isopropyliden-3-thioacetyl-sn-glycerol,
  CD spectrum of 107
Isoquinoline, electrophilic aromatic
  substituent constants for 904, 906,
  907
Isothiazole 1052
Isothiocyanates,
  in orthoamide synthesis 549
  reaction of, with amide acetals 577,
    578
Isotope effects,
  in mass spectrometry 135, 142
  in micellar catalysed reactions 1005,
    1010
  in photochemical reactions 681, 713,
    717, 732, 733, 744, 745
  in pyrolysis reactions,
    of carboxylic acids 861
    of esters 881, 898–900, 904
Isotope studies 1030, 1031 – see also
  Carbon-13 labelling, Carbon-14
  labelling, Deuterium labelling,
  Oxygen-18 labelling
Isotropic absorption (u.v.) spectra 69–73

Karl Fischer reagent 603
Kawa lactones, mass spectra of 627
Ketalation 574
Ketals, oxidation of, in carboxylic acid
  synthesis 344–352
Ketene O,N-acetals, synthesis of 538
Ketene aminals 568
Ketene dialkyl mercaptals, reaction of, to
  give dithio esters 1043
Ketenes,
  acylations with, in amide synthesis 447
  alcoholysis of, in ester synthesis 417,
    418
  cycloaddition of, in lactone synthesis
    1125, 1131–1138
  reaction of,
    with azo compounds 1232, 1264
    with imines 1230–1263
β-Keto acids, pyrolysis of 860, 861, 863,
  868
Keto amides, cyclizations using 1206
β-Keto amides, pyrolysis of 868, 896
α-Keto carboxylic acids, photosensitization
  of 687, 742, 743
Keto esters, mass spectra of 149, 150
Ketones,
  cleavage reactions of 373–379, 450
  oxidation of,
    in carboxylic acid synthesis

  353–371
  in lactone synthesis 1160–1170
  reaction of, with orthoamides 573,
    574
Kinetic acidity 555
Kinetic isotope effect 732, 733
  β-deuterium, in pyrolysis of esters 881
Knoevenagel reaction 420, 421
Koopmans' theorem 25, 29

Lactam acetals,
  reaction of 562, 565, 566, 572, 595,
    596
  reactions on the α-methylene groups of
    581–586
  synthesis of 535, 537, 538, 594
Lactams,
  cathodic reduction of 813
  chemical properties of 517, 518
  chiroptical properties of 73–85, 517,
    611, 612
  colorimetric estimation of 604
  determination of, by titration methods
    618
  dielectric constants of 516
  dipole moments of 517
  hydrogen bonding in 244, 246, 248,
    253
  i.r. spectra of 517, 608, 609
  mass spectra of 167–169, 517, 616,
    617
  n.m.r. spectra of 517, 610
  oxidation of 460
  photodecarbonylation of 669–671
  pyrolysis of 909
  radiolysis of 777
  Raman spectra of 608, 609
  N-substituted, in lactam acetal synthesis
    537
  synthesis of 1194–1315
    by cycloadditions 1224–1272
    by direct functionalization of
      preformed lactams
      1299–1302
    by oxidations 1302–1314
    by rearrangements 1271–1299
    by ring-closures (chemical)
      1195–1206
    by ring-closures (photochemical)
      1206–1224
  general methods of 516
  types of 492, 493
  u.v. spectra of 517, 613
α-Lactams,
  mass spectra of 165, 167, 168
  preparation and properties of 518–520
β-Lactams,
  3,3-disubstituted 610

i.r. and Raman spectra of   608
mass spectra of   168
n.m.r. spectra of   610
properties of   523, 524
synthesis of   521–523
γ-Lactams,
   calculation of rotational strengths for
      83, 84
   chirality rules for   78, 79, 81
   influence of $C_{(\alpha)}$ substituents on CD
      spectra of   82
   mass spectra of   168, 169
   preparation and properties of   524, 525
   quadrant rules for   77
δ-Lactams   526
   calculation of rotational strengths for
      83
   ring chirality rules for   81
ε-Lactams   526
   ring chirality rules for   81
Lactates, CD spectra of   104
Lactic acid, chiroptical properties of   91,
   95
(S)-Lactic acid, CD spectrum of   89
Lactim ethers   517
   in lactam acetal synthesis   537
Lactone hydroxamates, kinetics of the
     formation of   619
Lactone interconversions   1187, 1188
γ-Lactone ring,
   conformational mobility of   621
   stereoconfiguration of   621
δ-Lactone ring, conformation of   624
Lactones,
   as proton acceptors   240, 241
   cathodic reduction of   801, 804, 812
   chemical properties of   496
   chiroptical properties   73–87, 495,
      621, 624–626
     of saturated   73–85
     of α,β-unsaturated   85–87
     of β,γ-unsaturated   87
   detection and determination of
     618–628
     chromatographic methods for   628
     colorimetric methods for   619
     titration methods for   619
   dipole moments of   495, 619, 622
   ferric hydroxamate test for   619
   i.r. spectra of   495, 619–621
   mass spectra of   165–167, 495, 496,
     626–628
   n.m.r. spectra of   621
   oxidation of, in carboxylic acid
     synthesis   371, 372
   photochemical hydrogen-abstraction
     reactions of   719–722, 746
   photodecarbonylation of   652–661

photodecarboxylation of   687–708,
   743
photorearrangement of   727–729
pyrolysis of   908, 909
   cis nature of the elimination in   908
radiolysis of   773–777
Raman spectra of   619–621
synthesis of   1065–1194
   by carbonylations   1173–1180
   by condensations   1096–1100
   by cyclizations   1065–1096
   by cycloadditions   1180–1182
   by direct functionalization of
     preformed lactones
     1123–1130
   by Grignard and Reformatsky
     reactions   1100–1115
   by lactone interconversions   1187,
     1188
   by oxidations   1145, 1151–1173,
     1192
   by rearrangements   1182–1187
   by reductions   1138–1150
   by Wittig-type reactions   1115,
     1116
   from α-anions (dianions) of
     carboxylic acids   1116–1120
   from ketenes   1125, 1131–1138
   from lithio salts of 2-alkyl-2-
     oxazolines   1120–1123
   general methods of   493–495
   types of   492, 493
   u.v. spectra of   626
   α,β-unsaturated, five-membered,
     stereochemistry of   621
α-Lactones,
   i.r. and Raman spectra of   620
   synthesis of   496, 497
β-Lactones,
   i.r. and Raman spectra of   620
   mass spectra of   165
   properties of   499, 500
   synthesis of   497, 498
γ-Lactones,
   calculation of rotational strengths for
     83, 84
   chirality rules for   78, 79, 81
   endocyclic conjugated   620
   i.r. and Raman spectra of   620
   mass spectra of   166, 626, 627
   preparation and properties of   500–508
   pyrolysis of   909
   α,β-unsaturated, chirality rules for
     85–87
δ-Lactones,
   calculation of rotational strengths for
     83, 84
   CD spectra of   625

chirality rules for   78, 80, 81
i.r. and Raman spectra of   619, 620
mass spectra of   167, 626, 627
preparation and properties of   508–514
pyrolysis of   909
α,β-unsaturated, chirality rules for
        85–87
ε-Lactones,
    preparation and properties of   514, 515
    β,γ-unsaturated, chirality rule for   87
Langmuir adsorption isotherm   956
L-Lanthionine, CD spectrum of   99
Lead (II) benzenecarboxylates, trans-
    carboxylation of   935
Lemieux–von Rudloff reagent, in oxidation
    of double and triple bonds   332,
    333
Leucine, CD spectrum of   98
Lewis acids, reaction of, with orthoamides
    572
Light scattering measurements, on micellar
    solutions   988
Linderenolide   87
Linear dichroism   73
Liquid chromatography, high-speed, high-
    resolution   617
Liquid crystals   226
Lithium benzenecarboxylates, transcar-
    boxylation of   934
Litium dialkylphosphides, reaction of, with
    dimethylformamide acetal   565
Lithium dimethylamide, in triaminomethane
    synthesis   547
Long-chain acid amides, determination of
    617
Long-chain acids, mass spectra of   126
Long-chain esters, mass spectra of   136,
    137
Luminescent method, for the determination
    of anhydrides   631
Lycoctonamic acid   863
Lysolecithin, as micelle   958

Macrolides   514
Maleic acid,
    cathodic hydrogenation of   800
    intramolecular hydrogen bonding in
        239
    mass spectra of   131, 133
    self-association of   222
Maleic anhydride,
    complexes of   176, 179, 180, 197
    negative-ion mass spectrum of   162
    photochemical reactions of   205–209
    pyrolysis of   873
Maleimide, hydrogen bonding in   249
Malic acid, CD spectrum of   91
Malondiamide, reaction of, with tris-

(dimethylamino)alkoxymethanes,
    562
Malonic acid,
    condensation reactions of, in lactone
        synthesis   1102–1108
    hydrogen bonding in   230
    mass spectrum of   129, 130
Malonic acids, intramolecular hydrogen
    bonding in   240
Malonic esters,
    in carboxylic acid synthesis   279
    in ester synthesis   426
    in lactone synthesis   1102–1108
Mandelic acids,
    chiroptical properties of   109, 110
    thiophene analogues of, CD spectra of
        109
Manganese oxides, use of in oxidation,
    in lactam synthesis   1305, 1306
    of alcohols   301–303
    of aldehydes   310–313
    of arenes   319–321
    of double and triple bonds   333, 334
Mannose, chiroptical properties of   105
Mass-action model of micelle formation
    954
Mass spectrometry,
    of acyl halides   160
    of amides   152–160, 613–616
    of anhydrides   161, 162, 633, 634
    of carboxylic acids   123–133
    of esters   133–152
    of imides   162–165
    of lactams   167–169, 517, 616, 617
    of lactones   165–167, 495, 496,
        626–628
    of thio acid derivatives   1035, 1036
McLafferty rearrangement,
    of carboxylic acids   124–126
    of esters,
        aliphatic   134–136, 139
        keto   150
        thio   151, 1036
$m_1\mu_2$ Mechanism   83
Mellitic trianhydride, complexes of   177,
    184, 197
(−)-Menthyl acetate, pyrolysis of   888
Mercaptans, oxidation of   1192
α-Mercapto acids, chiroptical properties of
    88, 93–95
Mercaptobenzothiazole, hydrogen bonding
    in   255
α-Mercapto esters, chiroptical properties of
    88, 93–95
α-Mercaptopropionic acid, chiroptical
    properties of   95
(R)-2-Mercaptopropionic acid, ORD and
    CD spectra of   93

Mesaconic acid, mass spectrum of 131
Metal complexes, of amino and hydroxy
    acids 103, 104
Metal-ion promoted reactions, of thio
    esters 1048
Metal oxides, use of in oxidation,
    of alcohols 303–305
    of arenes 325, 326
Metaphosphate ion, elimination of, from
    p-nitrobenzoyl phosphate 993
Methacrylic acid, pyrolysis of 866
Methacrylic anhydride, pyrolysis of 873
Methoxybenzoates, mass spectra of 143
o-Methoxybenzoic acid, intramolecular
    hydrogen bonding in 239
p-Methoxycarbonylbenzyl acetate, cathodic
    reduction of 811
Methoxy esters, mass spectra of 148, 149
S-Methoxymethyl acetates, pyrolysis of
    875
S-Methoxymethyl thioacetates, pyrolysis of
    875
1,2-Methoxynaphthoic acid, hydrogen
    bonding in 239
2,3-Methoxynaphthoic acid, hydrogen
    bonding in 239
Methoxyphenylacetic anhydrides, sub-
    stituent effects on 708
N-Methylacetamide,
    $ab\ initio$ investigations of 31
    hydrogen bonding in 243, 246–249,
      251
Methylacetanilides, mass spectra of 156
N-Methylacetanilides, p-substituted,
    micellar catalysed hydrolysis of
    991
Methyl acetate,
    mass spectrum of 134
    radiolysis 770
    of aqueous solutions of 772
N-Methylacetylimide, mass spectrum of
    163
Methyl O-acetyl-(S)-lactate, CD spectrum
    of 84
Methyl (O-d$_3$-acetyl)salicylate, mass
    spectrum of 140
Methyl acrylate, cathodic carboxylation of
    809
2-Methylallyl formate, pyrolysis of 877
N-Methylbenzimidazole 965
Methyl benzoate,
    anion radical of 802
    cathodic reduction of 802, 805
Methyl benzoates, mass spectra of
    139–141, 143
N-Methyl-N-benzyl-o-chlorobenzamide,
    hydrogen bonding in 253
Methyl 3-bromo-3-phenylpropionate, mass

spectrum of 144
2-Methylbut-2-enoic esters, mass spectra of
    148
Methyl butyrate, mass spectrum of 134
Methyl butyrates, γ-substituted, mass
    spectra of 135
α-Methylbutyric acid, CD spectrum of 91
L-(+)-2-Methylbutyric acid, CD spectrum
    of 91
1-Methylcycloalkyl acetates, pyrolysis of
    889
1-Methylcyclohexanol, acetate derived from,
    pyrolysis of 883
N-(1-Methylcyclohexyl)acetamide,
    pyrolysis of 870
$trans$-2-Methylcyclohexyl acetate,
    pyrolysis of 888, 889
1-Methyl-1,4-dihydronicotinamide,
    hydrogen bonding in 245
Methylene dibenzoate, pyrolysis of 876
α-Methylene-γ-lactones, preparation and
    properties of 506–508
α-Methylene-δ-lactones, preparation of
    508
Methyl esters, mass spectra of 133–141,
    143–151
N-Methylformamide, $ab\ initio$ investigations
    of 31
o-Methylformanilide, hydrogen bonding in
    247
Methyl formate,
    $ab\ initio$ investigations of 27
    base-catalysed hydrolysis of 49–54
    mass spectrum of 137, 138
2-Methyl-1-indanyl acetate, pyrolysis of
    881
N-Methyl-N-laurylhydroxamate ion, in
    mixed micelles 1007
Methyl methacrylate, addition of diethyl
    dibromomalonate to 1192, 1193
Methyl N-methylcarbamates, pyrolysis of
    876
Methyl myristate, mass spectrum of 134
Methyl octadecanoate, mass spectrum of
    136
Methyl oleate, radiolysis of 771
Methyl orthobenzoate 983
    micellar caatlysed hydrolysis of 972
    effect of alcohols on 984
Methyl 10-oxoundecanoate, mass spectrum
    of 150
Methyl 4-phenylbutanoates, mass spectra
    of 139
2-Methyl-3-phenyl-γ-butyrolactones 619
Methyl 6-phenylhexanoate, mass spectrum
    of 139
Methyl 3-phenylpropionate, mass spectrum
    of 138

N-Methyl-*N*-phenyltrifluoroacetamides,
  mass spectra of 164
N-Methylphthalimide, cathodic reduction
  of 814
Methyl propionate, mass spectrum of 134
trans-3-(6-Methyl-2-pyridilthio)propenic
  acid, hydrogen bonding in 225
3-Methylpyrrolid-2-one 77, 84
5-Methylpyrrolid-2-one 84
L-5-Methylpyrrolid-2-one 77
N-Methylpyrrolidone 524
  hydrogen bonding in 246
Methylpyrrolid-2-ones 77
  calculation of rotational strengths of
    84
Methyl sorbate, mass spectrum of 147
Methyl substituents, activating effects of
  907
Methylsuccinic anhydride, pyrolysis of
  873
N-Methylsuccinimide, cathodic reduction
  of 820
Methylthiepan-2-ones, chiroptical properties
  of 107
N-Methylthiocarbamoylamino acids,
  chiroptical properties of 102
Methyl thionvalerate, mass spectrum of
  151
N-Methyltrichloroacetamide, hydrogen
  bonding in 248
N-Methyl trifluoroacetanilides, mass
  spectra of 159
Methyl *p*-(trifluoromethyl)benzoate,
  cathodic reduction of 811
Methyl undecanoate, mass spectrum of
  150
Methyl valerate, mass spectrum of 134
Methyl yellow 629
Micellar catalysed reactions 946–1015
  buffer effects in 950, 959, 966, 971,
    972, 986, 1001
  concentration effect in 1013
  effect of non-electrolytes on 949, 984
  effect of nucleophile hydrophobicity on
    949, 998
  effect of substrate hydrophobicity on
    949, 950, 973, 974
  electrolyte effects in 952, 968, 972,
    977–980, 987, 1000, 1014
  inhibited or unaffected by micelles of
    the opposite charge 948
  kinetic models for 960–984
  simple 957–960
  medium effects in 1013
  pseudo-phase model for 954, 957,
    960, 969
  transition-state effects in 958
Micellar inhibited reactions 976, 977,

    984, 992, 1001
  effect of nucleophile hydrophobicity on
    949
  effect of substrate hydrophobicity on
    949
  electrolyte effects in 949, 977
  kinetic models for 960
  simple 957, 959
Micellar solutions, water structure in 1014
Micelles,
  anionic, examples of 952
  as models for enzyme catalysis 948,
    957, 998
  cationic, examples of 952
  chain-length of 950, 956
  critical concentration (cmc) of 946,
    951, 963, 974, 975, 977, 978,
    986, 987, 995, 1014
  degree of ionization (α) of 946, 955,
    956
  fraction of counterions bound to (β)
    946, 970, 978
  functional 949, 988, 995–1011
  non-functional 989–995, 998
  non-ionic 953, 984
  reverse 953, 985, 986, 1012
  rough 955
  shape of 952, 954
  size of 952, 978
  smooth 955
  Stern layer of 954, 956, 963, 969,
    970, 972, 973, 975, 979, 980,
    987, 988, 994, 1007
  structure of 951–957, 988
  submicroscopic reaction medium of
    958
  surface of 955, 969
  zwitterionic 952, 959, 985, 1009
Micelle–water interface, polarity of 958
Michaelis–Menten equation 957
Michael reaction,
  in carboxylic acid synthesis 282
  in ester synthesis 429, 430
Microwave spectroscopy, of thio acid
    derivatives 1022, 1023
Migration 727–735
  1,2-acyl 731
  1,3-acyl 733
  alkyl, in isomerization of triphenyl-
    phosphinerhodium compounds
    852
  double-bond, in products of decarbonyl-
    ation and dehydrohalogenation of
    aliphatic acyl chlorides 834
  of aryl groups 727–729
  of carboxyl groups 727, 733–735
Migratory aptitudes 728
Migratory preferences 729

Minimum-energy pathway 18
Mixed indicators 629
MO calculations, and electronic spectra of thio acid derivatives 1026
Model problems 16
Molecular anions 756
Molecular complexes, of carboxylic acids, hydrogen bonding in 224, 225, 237, 238
Molecular geometry 19
of formamide 27, 28
of formic acid 22–24
Molecular orbital energies 6, 8
Molecular orbitals (MO) 4–6
canonical (CMO) 4
localized (LMO) 4
Molybdate complexes, of α-hydroxy acids 104
Monocarboxyalkyl radicals 767
Monochloroacetic acid, hydrogen bonding in hydrates of 237
α-Monochloroacetic acid, self-association of 222
Monofluoroacetic acid, pyrolysis of 865
Monothio carboxylic acids,
chemical properties of 1044, 1045, 1052
n.m.r. spectra of 1033, 1034
synthesis of 1036
u.v. spectra of 1025
vibrational spectra of 1029
Monothio esters – see Thiolo esters, Thiono esters
Monothioformic acid, microwave spectrum of 1022
Morpholine, in α,α-bis(dialkylamino)-acetonitrile synthesis 549
Morpholine method, for determination of anhydrides 629, 630
Multiple equilibria 178
N^{α}-Myristoylhistidine 947
Myristoylhistidines, deacylation of p-nitrophenyl esters by 999

Nalidixic acid, photodecarboxylation of 674
Naphthalene, as transcarboxylation product 936, 940
Naphthalenecarboxylates, transcarboxylation of 935–938
Naphthalene-1,8-dicarboxylic anhydride 197
Naphthalene-1,4,5,8-tetracarboxylic acid anhydride 177, 197
Naphthalic 1,8-anhydride, complexes of 209
β-Naphthol, reaction of, with aminal esters 561

Negative-ion mass spectrometry,
of amides 159, 160
of anhydrides 162
of carboxylic acids 123, 132, 133
of esters 151, 152
of thio esters 1036
Neighbouring-group effects, in pyrolysis of esters 900, 901
Neopentyl esters, mass spectra of 136
Neophyl acetate, pyrolysis of 902
Newman projection 624
Nickel tetracarbonyl, reaction of, with tetrakis(dimethylamino) methane 566
Nicotinamide, hydrogen bonding in 252
Nicotinic acid, photodecarboxylation of 674
Nitrene insertion 545
Nitric acid, in ketone oxidation 358
Nitriles,
alcoholysis of, in ester synthesis 418, 419
hydrolysis of,
in amide synthesis 449, 450
in carboxylic acid synthesis 273
Nitroacetanilides, mass spectra of 155
p-Nitrobenzaldehyde diethyl acetal, micellar catalysed hydrolysis of 964, 980–982
6-Nitrobenzisoxazole-3-carboxylate ions, micellar catalysed decarboxylation of 958
o-Nitrobenzoic acid, mass spectrum of 128
Nitrobenzoic acids, negative-ion mass spectra of 133
p-Nitrobenzoyl phosphate, micellar catalysed reactions of 993, 994
p-Nitrobenzoyl phosphate dianion, micellar catalysed deacylation of 1010
Nitrogen oxides, use of in oxidation,
of alcohols 303
of aldehydes 313
of arenes 321–323
of double and triple bonds 334, 335
3-Nitronaphthalene-1,8-dicarboxylic acid anhydride 177, 197
Nitrones, cycloaddition of,
in lactam synthesis 1269–1271
in lactone synthesis 1180–1182
o-Nitrophenylacetamide, negative-ion mass spectra of 159
(R)-p-Nitrophenyl-N-acetylphenylalanine, micellar catalysed deacylation of 1005
Nitrophenyl benzoates, negative-ion mass spectra of 151
p-Nitrophenyldiphenyl phosphate 981, 991

micellar inhibited reaction of   984
p-Nitrophenyl *p*-guanidinobenzoate
     hydrochloride, deacylation of, in
     reverse micelles   986
p-Nitrophenyl 2-phenylpropionates, micellar
     catalysed hydrolysis of   1005
3-Nitrophthalic anhydride,
     complexes of   180
     negative-ion mass spectrum of   162
4-Nitrophthalic anhydride,
     complexes of   181, 197
     negative-ion mass spectrum of   162
Nitrophthalic anhydrides, reaction of, with
     nitrogen bases   203
3-Nitro-2-pyridylamino acids, chiroptical
     properties of   102
Nitroso compounds, reaction of, with
     ketenes   1271, 1272
N-Nitroso derivatives of amino acids,
     chiroptical properties of   102, 103
Non-Relativistic Limit (NRL)   6
Norrish Type II photoelimination, of
     thiono esters   1052
Norvalines, CD spectra of   101
Nuclear–electron attraction   7
Nuclear magnetic resonance spectroscopy,
     of amides   609, 610
     of anhydrides   632
     of hydrogen bonding   217, 229
     of lactams   517, 608–610
     of lactones   621
     of orthoamides   553, 554
     of *S*-oxides   1049
     of thio acid derivatives   1033–1035
     of thiocarbonyl-*S*-imides   1049
Nucleophile binding,
     constant for   946, 962, 968
     in micellar catalysed reactions   1001
Nucleophiles, reaction of, with orthoamides
     554–571
Nucleophilic groups, use of micelles to
     assist substrate incorporation in
     989
Nucleophilic reactions,
     of hydroxy moiety of hydroxyalkyl
          surfactants, with halides and with
          carbocations   1008
     of thio esters   1045–1048
'Nylon 3'   518
'Nylon 6' (Perlon)   518

Octadecylammonium benzoate, as micelle
     986
Octadecyltrimethylammonium bromide,
     comicelle of   1001
n-Octadecyltrimethylammonium bromide,
     as micelle   971, 974, 975

Octant rule, in lactones and lactams   73,
     75, 76, 623
Octyloxyamine, in deacylation of
     p-nitrobenzoyl phosphate   994
p-Octyloxybenzyltrimethylammonium
     hydroxide   981
n-Octyltrimethylammonium bromide, as
     micelle   971, 974
Olefinic aldehydes, cyclization of   850,
     851
Olefins,
     addition of diazonium salts to   1192
     as products of decarbonylation reactions
          of aliphatic acyl halides
          834–837, 842
     cycloaddition of isocyanates to
          1224–1230
     cycloaddition of nitrones to
          1180–1182
     elimination of,
          in reactions of acyl halides with
               triphenylphosphinerhodium
               compounds   840, 841
          in reactions of aldehydes with
               triphenylphosphinerhodium
               compounds   852, 853
     isomerization of   834, 854
     oxidation of, in lactone synthesis
          1172, 1173
     synthesis of, from amide acetals   585
Oleic acid, radiolysis of   762, 763
Oligopeptides, chiroptical properties of
     99–102
Operators   13, 14
     expectation values of   13
     one-electron   13
     two-electron   13
Optical characteristics, of radicals produced
     by pulse radiolysis of aqueous
     solutions of acids   766
Optical rotatory dispersion (ORD)
     spectroscopy   68, 69, 104–107
     of amides   611, 612
     of carboxylic acids   93, 109
     of lactams   517, 611, 612
     of lactones   495, 621, 624–626
Optical rotatory power   611
Optimal geometries   19
Orbital exponents   (ζ)   9
d-Orbital participation, in thio acid
     derivatives   1045
Organotin compounds, in synthesis of
     orthocarbamic acid esters   551
Orotic acid, hydrogen bonding in
     dihydrate of   224
Orthoamides,
     physical characteristics and structure of
          553, 554

reactions of   554–586
 on the α-methylene groups of amide
  and lactam acetals   581–586
 with electrophilic reagents
  572–581
 with nucleophilic reagents
  554–571
 reactivity of   556, 557
 synthesis of   535–553
Orthocarbamic acid esters   535
 reaction of   557, 559, 568, 575
 synthesis of   550, 551
Ortho effect   127, 128, 157, 159
Orthoesters,
 in amide acetal synthesis   541, 542
 in triaminomethane synthesis   547, 548
Orthooxalic acid   549
Osmium oxides, as oxidizing agents in
  lactam synthesis   1305
Oxadi-π-methane rearrangements   706,
  731–733
Oxalic acid,
 cathodic reduction of   800
 hydrogen bonding in dihydrate of   230
 mass spectrum of   129
 radiolysis of   760, 761
 self-association of   222
α-Oxalic acid, vibrational spectra of   226
β-Oxalic acid, vibrational spectra of   226
Oxalic acid amide imide thio esters   577,
  578
Oxamide, reaction of, with tris(dimethylamino)alkoxymethanes   562
1,4-Oxathian   1052
1,3-Oxathiolium cations   1052
1,3-Oxazin-6-one   512
Oxazoline   1052
Oxazolines, in carboxylic acid synthesis
  283–286
Oxetans, formation via EDA complexes
  208
Oxidation,
 of acetals   344–352
 of alcohols   298–306
 of aldehydes   306–317, 453
 of *N*-alkylamides   460
 of amines   371, 452
 of arenes   317–331
 of carboxylic acids   372
 of double and triple bonds   331–344
 of ethers   344–352, 1171, 1172
 of ketals   344–353
 of ketones   353–371, 1160–1170
 of lactams   460
 of lactones   371, 372
 of olefins   1172, 1173
 of thio esters   1048, 1049
Oxidation states, of carbon   2–4

Oxidative sulphuration, in synthesis of thio
  carboxylic acids   1037, 1038
S-Oxides,
 configuration and conformation of
  1049
 synthesis of   1049
Oximate ions, as functional micelles   1007
Oxime derivatives, as functional micelles
  1007, 1008
Oxonium salts,
 in tris(acylamino)methane synthesis
  547
 reaction with *N,N*-disubstituted amides
  to give iminium salts   535
α-Oxo thioacyl chlorides, synthesis of
  1044
Oxycelluloses, estimation of   619
Oxygen, use of in oxidation,
 of alcohols   305
 of aldehydes   313
 of arenes   326–331
 of ketones   361, 362
Oxygen-18 labelling,
 in mass spectrometry   124, 141, 143,
  148
 in photolysis of lactones   705
 in photosolvolysis reactions   735
 in study of oxygen interchange reaction
  in photolysis of esters   697,
  700, 746
Ozone, use of in oxidation,
 of aldehydes   313
 of arenes   326–331
 of double and triple bonds   336–341
 of ethers and acetals   346–348
 of ketones   361–365

Pachystermine A   521
Palladium metal, as catalyst in dehydro-
  halogenation of aliphatic acyl
  chlorides   834
Parabanic acid, orthoamides of   549
Partition, between water and benzene of
  carboxylic acids   237
Pelargolactam, hydrogen bonding in   244
Penicillin, i.r. and Raman spectra of   608
Penicillin V   521
Penicillins   603
 p.m.r. spectra of   610
 radiolysis of, in aqueous solution   777
Pentafluorobenzamide, mass spectrum of
  156
Pentafluorobenzoic acid, cathodic reduction
  of   791
Pent-4-enoic acid   861, 865
Pentenolides, chirality rules for   83–85
n-Pentyl acetates, mass spectra of   135

Peptides 602
  reaction of, with amide acetals 561
Peracids, use of in oxidation 346–348
  of acetals 348
  of ketones 353–355
Periodates, use of in oxidation,
  of alcohols 300, 301
  of aldehydes 310
  of arenes 319
  of double and triple bonds 336–341
  of ketones 365–369
Periodic acid, use of in oxidation,
  of alcohols 300, 301
  of aldehydes 310
  of arenes 319
Perkin condensation, in lactone synthesis
    1108, 1109
Perkin reaction, in carboxylic acid
    synthesis 276
Peroxides, use of in oxidation,
  of alcohols 305
  of arenes 326–331
  of dioxanes 348
  of ketones 361–365
Peroxy acids, in oxidation of double and
    triple bonds 336–341
Peroxycarboxyl radicals 767
PES band 1026
pH, micellar effects on 950
Pharmaceutical materials 602
Phase-transfer catalysts 986, 1007, 1014
Phenols,
  reaction of, with orthoamides
    559–561
  thioacylation of 1041
α-Phenoxypropionic acids, CD spectra of
    109
Phenyl acetate, radiolysis of 770
Phenyl acetates, mass spectra of 139, 140
Phenylacetates, pyrolysis of 878
  Hammett ρ-factors for 893
  relative rates of 891, 892
Phenylacetic acid,
  hydrogen bonding in, thermodynamics
    and kinetics of 235
  pyrolysis of 869
Phenylacetic acids,
  chiroptical properties of 108–110
  photodecarboxylation of 678–682
    CIDNP studies of 682
    substituent-effect study of 679
Phenyl acrylate, pyrolysis of 875
Phenylalanine, CD spectrum of 98
ω-Phenylalkanoic acids, mass spectra of
    125, 126
Phenylallenecarboxylic acids, CD spectra of
    108
Phenyl butanoate, mass spectrum of 140

4-Phenylbutanoic acid, mass spectrum of
    125
N-Phenylcarbamates, pyrolysis of,
  Hammett ρ-factors for 893
  relative rates of 891, 892
Phenylcarbonates, pyrolysis of,
  Hammett ρ-factors for 893
  relative rates of 891, 892
4-Phenylcrotonic acid 860
cis-2-Phenylcyclohexanol, xanthate derived
    from, pyrolysis of 883
Phenyldiazomethane 575
2-Phenyl-2,2-diphenylacetamide, pyrolysis
    of 868
o-Phenylenediamine 562
Phenyl ethers 559
6-Phenylhexanoic acid, mass spectrum of
    126
Phenylmercaptoic acid, pyrolysis of 865
Phenyl α-methylacrylate, pyrolysis of
    875
Phenyl nitrobenzoates, negative-ion mass
    spectra of 151
Phenyl (p-nitrophenyl)acetate, negative-ion
    mass spectrum of 151
3-Phenylpropionic acid, mass spectrum of
    125
3-Phenylpropionyl halides, mass spectra of
    160
Phenyl substituents at the β-carbon of esters,
    effect on rate of pyrolysis of 897
Phenylthioacetamide, hydrogen bonding in
    255
Phenylthiohydantoins, chiroptical
    properties of 102, 103
Phenyl valerate, mass spectrum of 140
Phosphines, reaction of, with dimethyl-
    formamide dimethylacetal 564
Phosphonic esters 564, 565
Phosphorescence spectroscopy,
  in micellar catalysed reactions 988
  of EDA complexes 202
Phosphoric acid dialkyl ester chlorides,
    reaction of, with amide acetals
    576, 577
Phosphorus trichloride,
  in tris(acylamino)methane synthesis 547
  reaction of, with amide acetals 576,
    577
Phosphoryl chloride, in tris(acylamino)-
    methane synthesis 547
Photochemical hydrogen abstraction
    708–727
Photochemical reactions,
  in carboxylic acid synthesis 369, 408
  in ketone oxidation 369
  in lactam synthesis 1206–1224,
    1309–1311

in lactone synthesis 1186, 1187, 1189–1191
involving EDA complexes 204–209
of thio esters 1051
Photodecarbonylation 651
of anhydrides 661–669
of esters 652–661
of imides 669–671
of lactams 669–671
of lactones 652–661
Photodecarboxylation 671
of acids 671–687
pH-dependent 677
of anhydrides 708
of esters 687–708
of lactones 687–708
of RXCH$_2$CO$_2$H,
CIDNP studies of 684
reductive 739
Photodeconjugation 717, 718, 722
Photo-Fries rearrangement,
of esters 652, 657, 693, 694, 735
of thiono esters 1052
Photooxidation, in lactam synthesis 1309–1311
Photorearrangements 408–410, 727–735, 1186, 1187
Photosolvolysis 735–738
of benzyl acetates 651, 694
Phthalates,
cathodic reduction of 805
determination of the anhydride content of 635
pyrolysis of 894
Phthalic acid,
cathodic reduction of, to tetrahydro-phthalic acid 793
mass spectrum of 130, 131
Phthalic acids, cathodic reduction of 793
Phthalic anhydride,
anion radical of 802
complexes of 176, 179, 180, 197, 200, 202
determination of 635
mass spectrum of 162
pyrolysis of 874
Phthalic thioanhydride, complexes of 176
Phthalide,
formation of, in cathodic reduction of phthalates and phthalic anhydride 805
pyrolysis of 909
Phthalimide,
cathodic reduction of 814, 816, 817
hydrogen bonding in 249
Phthalimides,
cathodic reduction of 819
mass spectra of 164

Phthalimidines, cathodic reduction of 819
2-Phthalimidobiphenyl, mass spectrum of 164, 165
4-Phthalimidobiphenyl, mass spectrum of 164, 165
N-Phthaloyl derivatives of amino acids, chiroptical properties of 102, 103
Picolinic acid, cathodic reduction of 797–799
Piperazinedione 1052
Piperidine, in α,α-bis(dialkylamino)-acetonitrile synthesis 549
Pivalanilides, mass spectra of 156
pK, of radicals produced by pulse radiolysis of aqueous solutions of acids 766
pK$_a$ – see also Acid dissociation constants
apparent,
effect of electrolyte on 1000
effect of micelles on 959, 972
of m-bromobenzaldoxime 991
of choline 1010
Platinum oxides, as oxidizing agents in lactam synthesis 1306, 1307
Polarography,
of amides 618
of phthalic anhydride 635
of thio esters 1035, 1045, 1046
Polar solvents, in pyrolysis of β-keto acids 863
Poly-L-alanine, CD spectrum of 100
Polyamides,
analysis by alkali-fusion gas chromatography 617
pyrolysis of 869
Polycyclic acyl compounds, decarbonylation of 828
Polycyclic imides, mass spectra of 164
Polyelectrolytes, soluble 957
Polyesters, conformation of, by chiroptical methods 101
Polyglutamic acid, CD spectra of 99
Poly-γ-methylglutamate, CD spectra of 99, 100
Polynitro(2,5-;3,6-;4,5-;2,4,5-)naphthalene-1,8-dicarboxylic acid anhydride, complexes of 177
Polypeptides, chiroptical properties of 99–102
Polyproline helices 99
Polysoaps, as micelles 953
Polythiol esters 1051
Polyvinylpyridinium chloride 979
Positive holes 756
Positive ions, generated as a result of electron impact fragmentations 123–133
Potassium 9-anthracenecarboxylate, as reactant in mixed transcarboxylations 940

Potassium benzenecarboxylates, trans-
    carboxylation of 916−932
  effect of catalysts on 932
  in an atmosphere of ^{14} C- labelled carbon
      dioxide 926−929
  in fused KCNO or KCNS 924−928
  mechanism of 930−932
  time dependence of the content of
      reaction mixtures in 918−924
Potassium benzenedicarboxylates, trans-
    carboxylation of 918−921
  effect of catalysts on 932, 933
  in fused KCNO 924−927
  mechanism of 930, 931
Potassium benzenehexacarboxylates,
    transcarboxylation of mixtures
    containing 924
Potassium benzenepentacarboxylates,
    transcarboxylation of mixtures
    containing 924
Potassium benzenetetracarboxylates,
  as transcarboxylation products, of
      benzenetricarboxylates in fused
      KCNO 926
  transcarboxylation of mixtures
      containing 924
Potassium benzenetricarboxylates, trans-
    carboxylation of mixtures
    containing 923, 924
  in fused KCNO 924−926
Potassium benzoate,
  as reactant in mixed transcarboxylations
      940
  as transcarboxylation product,
    of potassium isophthalate 920,
        921, 926
    of potassium phthalate 918−920,
        925
    of potassium terephthalate in fused
        KCNO 926
  transcarboxylation of 921−923, 938
    in fused KCNO 925, 928
Potassium 2-biphenylcarboxylate,
  as reactant in mixed tanscarboxylations
      940
  transcarboxylation of 938
Potassium 2,2′-biphenyldicarboxylate,
  as reactant in mixed transcarboxylations
      940
  transcarboxylation of 938
Potassium 4,4′-biphenyldicarboxylate, as
    transcarboxylation product 938
Potassium ethyl phthalate, cathodic
    reduction of 805
Potassium 2-furoate, as reactant in mixed
    transcarboxylations 940
Potassium hemimellitate, as trans-
    carboxylation product,

of potassium benzoate 921, 922, 928
of potassium isophthalate 920, 921,
    926
of potassium phthalate 918, 919, 925
of potassium terephthalate in fused
    KCNO 927
Potassium hydrogen maleate,
  charge-transfer bands of 229
  vibrational spectra of 228
Potassium isonicotinate, as reactant in
    mixed transcarboxylations 940
Potassium isophthalate,
  as transcarboxylation product,
    of potassium benzoate 921, 928
    of potassium phthalate 918, 919,
        925
    of potassium terephthalate in fused
        KCNO 927
  transcarboxylation of 920, 921
    in fused KCNO 926
Potassium lauryl sulphate 980
Potassium 2,6-naphthalenedicarboxylate, as
    transcarboxylation product 935,
    936
Potassium naphthalenedicarboxylates,
    transcarboxylation of 935, 936
  in fused KCNO 936
Potassium naphthalenemonocarboxylates,
    as transcarboxylation products 936
Potassium naphthalenetricarboxylates, as
    transcarboxylation products 936
Potassium α-naphthoate, as reactant in
    mixed transcarboxylations 940
Potassium β-naphthoate, as reactant in
    mixed transcarboxylations 940
Potassium naphthoates, transcarboxylation
    of 936
  in fused KCNO 937
Potassium nicotinate, as reactant in mixed
    transcarboxylations 940
Potassium 9-phenanthrenecarboxylate, as
    reactant in mixed transcarboxylations
    940
Potassium phthalate,
  as transcarboxylation product,
    of potassium benzoate 921, 922,
        928
    of potassium isophthalate 920,
        921, 926
    of potassium terephthalate in fused
        KCNO 927
  transcarboxylation of 918, 919
    effect of catalysts on 932, 933
    in fused KCNO 925
    mechanism of 930, 931
Potassium phthalimide, reaction with
    N,N-dimethylchloromethylene
    iminium chloride 546

Potassium picolinate, as reactant in mixed transcarboxylations 940
Potassium pyrazinecarboxylate, as reactant in mixed transcarboxylations 940
Potassium 3,4-pyridinedicarboxylate, as reactant in mixed transcarboxylations 940
Potassium α-pyrrolecarboxylate, as reactant in mixed transcarboxylations 940
Potassium 4-sulphobenzoate 939
Potassium terephthalate,
  as transcarboxylation product,
    of potassium benzoate 921–923, 928
    of potassium isophthalate 920, 921, 926
    of potassium phthalate 918, 919, 925
  transcarboxylation of, in fused KCNO 927
Potassium α-thiophenecarboxylate, as reactant in mixed transcarboxylations 940
Potassium toluates, transcarboxylation of 938
Potassium trimellitate, as transcarboxylation product,
  of potassium benzoate 921, 922, 928
  of potassium isophthalate 920, 921, 926
  of potassium phthalate 918, 919, 925
  of potassium terephthalate in fused KCNO 927
Potassium trimesate, as transcarboxylation product,
  of potassium benzoate 921, 922, 928
  of potassium isophthalate 920, 921, 926
  of potassium phthalate 918, 919, 925
  of potassium terephthalate in fused KCNO 927
Potentiometric titration 1000
PPP method 1026
Primary aldehydes, decarbonylation of 843
Primary amides,
  i.r. and Raman spectra of 604–606
  mass spectra of 613–615
Proline, CD spectrum of 89, 95, 98
L-Proline 95
Propanoates, pyrolysis of, reactivity in 880
Propanoic acid, pyrolysis of 864, 866
Propargyl aldehydes,
  O,N-acetals of 569
  O,N-aminals of 569
Propionamide, mass spectrum of 152
Propionic acid,

hydrogen bonding in 230, 234, 236
  negative-ion mass spectrum of 133
2-Propyl acetate, pyrolysis of, rate of elimination in 890
n-Propyl benzoate, mass spectrum of 142
N-Propyl-3-carbomoylpyridinium iodide 979
n-Propyl formate, mass spectrum of 137, 138
Proteins 602
  fundamental functional group of 2, 27
Proton affinity 33–36
  definition of 33
Protonation reactions, theory of 33–41
Proton dynamics, in hydrogen bonding 231–234
Proton magnetic resonance spectroscopy,
  in quantitative analysis 609
  of thio acid derivatives 1033–1035
Proton-transfer reactions 42–44, 238
Proton tunnelling 233
Proximity effects 128, 133, 151
Pseudo-phase models, in micellar catalysis 954, 957, 960, 969
Pummerer reactions, in synthesis of thiolo esters 1039
Pummerer rearrangement 400, 401
Pyrazincarboxamides, hydrogen bonding in 245
Pyrazinecarboxylic acids, transcarboxylation of 939
Pyrazolone 1052
Pyridine,
  electrophilic aromatic substituent constants for 904, 905, 907
  oximate ions of 991
Pyridine aldehydes, hydration kinetics of 798
Pyridinecarboxylates, transcarboxylation of 939
2,3-Pyridinecarboxylic acid, intramolecular hydrogen bonding in 240
Pyridine acids, intermolecular hydrogen bonding in 224
2,3-Pyridinedicarboxylic anhydride, complexes of 197
Pyridine N-oxide, electrophilic aromatic substituent constants for 904, 906, 907
Pyridine-N-oxide-2-carboxylic acid, intramolecular hydrogen bonding in 240
Pyridines, carboxamido-substituted, separation of, by liquid chromatography 617
2-Pyridone, hydrogen bonding in 248
2-Pyridones, mass spectra of 617
Pyridylacetic acids,
  photodecarboxylation of 680

pyrolysis of 865
Pyrolysis,
of acyl halides 867
of amides 867–871
of anhydrides 871–874
of carboxylic acids 860–867
of esters 874–908
of lactams 909
of lactones 908, 909
of orthoamides 584–586
Pyromellitic dianhydride, complexes of
176, 181, 183–193, 196–198, 201,
202, 204
Pyromellitic dithioanhydride, complexes
of 194
Pyrone 1052
2-Pyrone, pyrolysis of 909
α-Pyrone, photolysis of 701–703
2-Pyrones,
properties of 511–514
synthesis of 510, 511
Pyrrolecarboxylic acids, transcarboxylation
of 939
Pyrrolidides, mass spectra of 154
Pyrrolidine, in α,α-bis(dialkylamino)-
acetonitrile synthesis 549
Pyrrolidine amides, mass spectra of 153,
154
Pyrrolidinedithione, 1052
2-Pyrrolidone, dipole moment of 609
Pyruvamides, intramolecular hydrogen
bonding in 254
Pyruvic acid,
hydrogen bonding in 239
photolysis of 686
CIDNP studies of 740

Quadrant rules 76, 77, 612
Quinoline, electrophilic aromatic
substituent constants for 904, 906,
907
Quinoline aldehydes, oximate ions of 991
Quinolinecarboxylic acids,
anilides of, mass spectra of 614
transcarboxylation of 939
Quinoxalinone 1052

Radical mechanism in pyrolysis of alkanoic
acids 866
Radicals,
acyl 758
alkyl-type 757
anion 801–805, 813, 814, 1035, 1045
benzyl 678, 699
carboxy-hydroxyalkyl 767
carboxylic 757
·CO₂H 677
formyl 768

hydrogen 763–768, 772, 775
monocarboxyalkyl 767
·OH 763, 764, 766, 772, 774, 775
peroxycarboxyl 767
Radiolysis,
of anhydrides 773
of aqueous solutions,
of ascorbic acid 774–777
of carboxylic acids 763–769
of esters 772, 773
of penicillins 777
of carboxylic acids 755–769
of esters 769–773
of lactams 777
of lactones 773–777
of water 763
Raecke process 398
Raman spectroscopy,
of amides 604–608
of anhydrides 631, 632
of hydrogen bonding 217
of lactams 608, 609
of lactones 619–621
of thio acid derivatives 1026,
1029–1032
Random coil arrangement, of polyamino
acids 99
Rate constants,
for reaction of ascorbic acid with H,
e⁻aq and OH radicals 776
for reaction of carboxylic acids with H,
e⁻aq and OH radicals 763–765
Reaction coordinates 18
Rearrangements,
acid-catalysed 403–405
base-catalysed 393–398, 403–408
of dithiocarbonates 1042
carbonium ion 1184–1186
in amide synthesis 450–452
in carboxylic acid synthesis 380–410
in lactam synthesis 1272–1299
in lactone synthesis 1182–1187
in pyrolysis of esters 896, 901–903
involved with fragmentations 124–126,
134–139
photochemical 408–410, 652, 657,
693, 694, 727–725, 1052, 1186,
1187
β,γ-unsaturated ester 732
Reductive elimination reactions 853
of triphenylphosphinerhodium
compounds 839, 840
Reformatsky reaction,
in ester synthesis 423, 424
in lactam synthesis 1233, 1265–1267
in lactone synthesis 1100–1115
[RhCl(C₂H₄)₂]₂ 848
RhCl₂(COCH₂CH₂Ph)(PPh₃)₂, X-ray

crystal structure determination of
838
RhClCO(PPh₃)₂   827, 839, 840
oxidative addition of acyl halides to
841
structure of   837
RhCl₂(COR)CO(PPh₃)₂ compounds   841
RhCl₂(COR)(PPh₃)₂ compounds   838,
841
RhCl(PMePh₂)₃   848
RhCl(PPh₃)₂   826, 837, 838
oxidative addition,
of acetyl chloride to   838
of acyl chlorides to   837
structure of   837
RhCl(PPh₃)₃ — see Chlorotris
(triphenylphosphine)rhodium (I)
RhMeI₂(PPh₃)₂   837
RhRCl₂CO(PPh₃)₂ compounds   839
elimination from,
of olefin and hydrogen chloride
839
of RCl   839
RhRCl₂(PPh₃)₂ compounds   841
RhRHClCO(PPh₃)₂ compounds, reductive
elimination of alkane from   852
D-Ribonolactone, cathodic reduction of
804, 805
Ring chirality   78–82
Ring contractions,
in lactam synthesis   1272–1275
oxidative   399, 400
Ring expansions, in lactam synthesis
1274–1297
Ritter reaction   453, 454
Rotational barriers,
of formamide   28
of formic acid   23, 24
of E/Z isomers of thio acids   1033
Rotational modes of a molecule   17
Rubidium benzenecarboxylates,
transcarboxylation of   934
Rubidium terephthalate, as
transcarboxylation product   934
Ruthenium oxides, use of in oxidation,
in lactam synthesis   1307–1309
of double and triple bonds   335, 336
RXH-acidic compounds, reaction of, with
orthoamides   554–557
RXH₂-acidic compounds, reaction of, with
orthoamides   554–557
Rydberg states   25

Saccharides, chiroptical properties of   105
Salicylic acid,
cathodic reduction of   785–788,
791
hydrogen bonding in   240

Salicylic acids, mass spectra of   127, 128
'Sandwich' complex, as possible
intermediate in transcarboxylations
931, 932
Santonins, mass spectra of   627
Schmidt rearrangement, in lactam synthesis
1284–1291
Secondary aldehydes, decarbonylation of
843
Secondary amides,
N-deuterated   605
i.r. and Raman spectra of   604–607
mass spectra of   152, 613–615
Sector rules   73–78
for lactams   612
for lactones   623, 624, 626
Selenide chromophores   99
Selenium oxides, use of in oxidation,
of aldehydes   314
of arenes   325
Selenolactams, i.r. and Raman spectra of   608
Self-association — see also Dimerization
of amides   243–248
of carboxylic acids, structural aspects of
221–226
Self-Consistent Field (SCF) method   8–10
Semicarbazides, reaction of, with
orthoamides   561, 562
Semiconcerted processes, in pyrolysis of
esters containing non-vinylic
β-hydrogen atoms   880
Semiempirical ASMO–SCF–CI calculations,
and electronic spectra of thio acid
derivatives   1026
Sesquiterpene lactones   621
CD spectra of   625
mass spectra of   627
separation of, by thin-layer
chromatography   628
[2,3] Sigmatropic shift, in reactions of
unsaturated Grignard reagents with
dithio esters   1047
Silver oxides, in aldehyde oxidation   314,
315
Slater-Type Orbitals (STO)   9
SN1 mechanism, for decomposition of
β-bromocarboxylate ions in dilute
aqueous alkali   988
SNi mechanism, in pyrolysis reactions,
of anhydrides   872
of esters   872, 895
-Sn- (stereospecifically numbered)   106,
107
Sodium benzenecarboxylates, transcarboxylation of   934
Sodium 1,4-benzenedisulphonate   939
Sodium benzenesulphonate, thermal
disproportionation of   939

Sodium laurate, as micelle   976, 977
Sodium lauryl sulphate(NaLS), as micelle
  947, 955, 959, 963–966, 972, 976,
  980, 981, 983–985, 987, 992,
  1007
  aggregation number of   956
  density of   956
Sodium naphthalenecarboxylates,
  transcarboxylation of   938
Sodium α-naphthoate, transcarboxylation
  of   937
Sodium β-naphthoate, transcarboxylation of
  937
Sodium saccharin, reaction with *N,N*-
  dimethylchloromethylene iminium
  chloride   546
Sodium terephthalate, as transcarboxylation
  product   934
Sodium trimesate, as transcarboxylation
  product   934
Solubility methods, for measuring the
  fraction of micellar incorporated
  nucleophile   962
Solvation   41
  of formamide   45–47
Solvent competition in complex formation
  183
Solvent effects, in pyrolysis of esters   900
Solvent isotope exchange   745
Solvent structure   953
Solvolytic reactions, in ester synthesis
  411–420
Sorbamide, hydrogen bonding in   244
Spatial functions   5
Specific ion electrodes   960
  for measuring the fraction of micellar
    incorporated nucleophile   962
Spectral methods, for measuring the
  fraction of micellar incorporated
  nucleophile   962
Spin density, in thio esters   1035
Spin-Unrestricted Hartree–Fock (SUHF)
  equations   12
Spirocyclic urea acetals, preparation of
  551
Spot tests,
  for amides   603
  for anhydrides   631
S_N1 reactions,
  in which the stereochemistry is changed
    on micellization   989, 1006
  micellar inhibition of   959
S_N2 reactions, of alkyl halides   1011
S_N1 solvolysis, of *t*-cumyl chlorides   904
Stable structures, theory of   17
Statistical effects, on the direction of
  elimination in pyrolysis of esters
  884, 885

N^α-Stearoylhistidine, as micelle   1003
Stereochemical probes   694
Steric acceleration, in pyrolysis of esters
  881, 882, 885, 889, 890, 897,
  900–902
Steric inhibition of conjugation   862
Stern theory   955, 956
Steroid carboxylic acids, chiroptical
  properties of   88, 108
Steroid ketones, ketalation of   574
Steroid lactones,
  chirality rules for   78
  sector rule for   75
Steroids,
  acetamides of   104, 105
  acetates of   104, 105
Stevens rearrangement   402, 574
trans-Stilbene   882
Stobbe condensation,
  in carboxylic acid synthesis   277
  in lactone synthesis   1100
Stretching vibrations of the CS_2 group
  1030
Strontium benzenecarboxylates, transcar-
  boxylation of   935
Strophantidin   503
Styrylacetic acid   860
α-Substituent, splitting of, in fragmentation
  processes   1036
Substituent pattern,
  influence on Cotton effect of   78, 86
  of the aromatic ring, influence on CD
    spectra   110
Substitution reactions, electrophilic   296
Substrate aggregation   985
Substrate binding constant (K_s)   946, 957,
  958, 961, 962, 973
Succinamide, hydrogen bonding in   245
Succinic acid,
  mass spectrum of   130
  vibrational spectra of   226
Succinic acids, α-substituted, chiroptical
  properties of   91, 92, 94
Succinic anhydride,
  mass spectrum of   161
  pyrolysis of   873
Succinimide,
  hydrogen bonding in   249
  mass spectrum of   163
γ-Sugar lactones   621
δ-Sugar lactones   621
Sulphide chromophore, in sulphur-
  containing amino acids   98, 99
Sulphines – *see* *S*-Oxides
Sulphonamides,
  chiroptical properties of   89, 90
  reaction of, with orthoamides   561,
    562

Sulphonyl azides, reaction of, with amide acetals 576, 577
N-Sulphonylformamidines 580
Sulphur, 3d orbitals of 1026, 1030
Sulphur compounds, use of in oxidation,
  of arenes 323–325
  of ketones 369
Sulphur extrusion 1042, 1049
Sulphuryl chloride (SO_2Cl_2), in
    tris(acylamino)methane synthesis
    547
Sultam derivatives, of amino acids,
    chiroptical properties of 89, 90
Superposition of configurations (SOC)
    wave functions 4
Surface potential, at the Stern layer of
    micelles 955
Surfactants,
  anionic, examples of 952
  cationic, examples of 952
  chain-length of 950
  functional 949, 988, 995–1011
  non-functional 989–995, 998
  non-ionic 953, 984
  zwitterionic 952, 959, 985, 1009
Synartetic assistance, in pyrolysis of esters
  902
Synthesis,
  of acyl halides 438–442
  of amides 442–458
  of anhydrides 437, 438
  of carboxylic acids 270–411
  of esters 411–436
  of imides 458–461
  of lactams 516, 1194–1315
  of lactones 493–495, 1065–1194
  of orthoamides 535–553
  of thio acid derivatives 1036–1044

Tafel rearrangement 805
Taft's equation 765
Taft σ*-values, for ethyl esters 895
Tartaric acid, hydrogen bonding in 225
(+)-Tartaric acid, CD spectrum of 93
Tartronic acid, hydrogen bonding in 225
Terephthalamides, cathodic reduction of
  817
Terephthalic acid, cathodic reduction of
  793
Terpene acids, chiroptical properties of
  108
Tertiary aldehydes, stoichiometric
    decarbonylation of 843
Tertiary amides,
  i.r. spectra of 604, 607
  mass spectra of 153, 613–615
  Raman spectra of 604, 607

Tertiary hydrogen phthalates, pyrolysis of
  894
Tetraalkylammonium dithiocarboxylates,
  synthesis of 1037
Tetraaminoethylenes 547
Tetrabromophthalic anhydride,
  complexes of 176, 181, 197, 201
  mass spectrum of 162
Tetrachlorophthalic anhydride,
  complexes of 176, 180, 195, 196, 202
  mass spectrum of 162
  reaction of, with triethylamine 203
n-Tetradecyltrimethylammonium bromide
  947
  as micelle 971, 975, 978, 979, 988
n-Tetradecyltrimethylammonium chloride,
  as micelle 947, 972, 979
Tetraethyl ethenetetracarboxylate,
  anion radical of 803
  dianion of 803
Tetra-n-hexylammonium benzoate, as
  micelle 986
Tetrahydrophthalic anhydride, mass
  spectrum of 161
Tetrakisaminomethane, pentacyclic 552,
  553
Tetrakis(dialkylamino)methanes 535
  reaction of 566, 568, 585
  synthesis of 552, 553
Tetrakis(dimethylamino)ethylene,
  methanolysis of 545
Tetrakis(dimethylamino)titanium, in
    tris(dimethylamino)alkane synthesis
    548
(R)-(−)-2,2′,5,5′-Tetramethyl-3,3′-
    bithienyl-4,4′-dicarboxylic acid, CD
    spectrum of 111, 112
(R)-(+)-4,4′,5,5′-Tetramethyl-3,3′-
    bithienyl-2,2′-dicarboxylic acid, CD
    spectrum of 111, 112
Tetronic acid, mass spectrum of 166
Thermochemistry 59–65
Thermodynamic acidity 555
Thermodynamic parameters, of
  orthoamides 554
Thermodynamic stability of products,
    effect on direction of elimination in
    pyrolysis of esters 884–890
Thermometric method, for the determina-
    tion of anhydrides 631
1-Thiacyclobutane-3-carboxylic acid-1-
    oxide, hydrogen bonding in 225
1,2,3-Thiadiazole 1053
1,2,4-Thiadiazole 1053
Thiapyrane 1052
2-Thiapyridone, hydrogen bonding in 255
Thiapyrone 1052
1,3-Thiazine 1053

Thiazole 1052
Thiazole-2-carboxylic acid, cathodic
    reduction of 797
Thiazolidine-2-thione, hydrogen bonding in
    255
Thiepan-2-ones, chiroptical properties of
    85, 107
Thietan 1052
Thiirans 1052
Thin-layer chromatography, for detection
    and determination,
    of amides 617
    of lactones 628
Thioacetamide, hydrogen bonding in 255
Thioacetates, pyrolysis of, reactivity in
    881, 892
Thioacetic acid, hydrogen bonding in 231,
    254
3-Thioacetyl-sn-glycerol, CD spectrum of
    107
Thioacyl anhydrides, synthesis of 1044
Thioacylation,
    of alcohols and phenols 1041
    of thiols 1043
Thioacyl halides,
    mass spectra of 1035
    synthesis of 1044
    u.v. spectra of 1025
Thioacyl isothiocyanates 1048
Thio amides,
    N,N-disubstituted, formation of 561
    hydrogen bonding in 254
    mass spectra of 158
    reaction of, with orthoamides 561,
        562
Thioanilides, hydrogen bonding in 255
Thiobenzamide, mass spectrum of 158
Thiobenzoic acid, hydrogen bonding in
    254
Thiobenzoylamino acids, chiroptical
    properties of 102
Thiobenzoyl hexafluoroantimonate,
    synthesis of 1044
Thiocarbonates, pyrolysis of, formation of
    thio esters in 895
Thiocarbonyl-S-imides,
    configuration and conformation of
        1049
    synthesis of 1049
Thiocarboxylate anions, vibrational
    spectra of 1030
Thio carboxylic acids,
    alkylation of 1038
    bond angles of 1022, 1023
    bond lengths of 1022, 1023
    dipole moments of 1023, 1024
    electron diffraction spectra of 1022,
        1023

formation of heterocycles from 1052
hydrogen bonding in 254
mass spectra of 132
microwave spectra of 1022, 1023
n.m.r. spectra of 1033–1035
prototropic behaviour of 1044, 1045
synthesis of 1036–1038
u.v. spectra of 1023, 1025–1028
vibrational spectra of 1029–1031
X-ray diffraction of 1022, 1023
Thio derivatives of amino acids, chiroptical
    properties of 102, 103
Thio esters,
    bond lengths of 1022, 1023
    dipole moments of 1024
    electrophilic reactions of 1048–1051
    e.s.r. spectra of 1035
    functional-group moments of 1025
    i.r. spectra of 1030–1032
    mass spectra of 150, 151, 1035, 1036
    n.m.r. spectra of 1033–1035
    non-polar reactions of 1051, 1052
    nucleophilic reactions of 1045–1048
    prototropic behaviour of 1045
    synthesis of 1038–1044
    u.v. spectra of 1025, 1027, 1028
Thioglycollic acids, mass spectra of 132
Thioglycollic esters, mass spectra of 150
Thiolacetates, chiroptical properties of 107
Thiolactams, i.r. and Raman spectra of 608
α-Thiolactones, synthesis of 1038
ε-Thiolactones, ORD spectra of 85
Thiolcarbamates, formation of, from
    rearrangement of thioncarbamates
    896
Thiolcarbonates, formation of, from
    rearrangement of thioncarbonates
    896
Thiolo esters,
    acylation of 1050, 1051
    alkylation of 1050
    chiroptical properties of 107
    i.r. spectra of 1030
    mass spectra of 1035, 1036
    n.m.r. spectra of 1034, 1035
    oxidation of 1048
    photolysis of 1051
    proton acceptor properties of 1045
    reaction of, with carbanions 1047
    reduction of 1045, 1046
    solvolysis of 1048
    synthesis of 1038, 1039
    thionation of 1042
Thiols,
    acylation of 1038, 1039
    reaction of, with amide–phosphoroxy-
        chloride adducts 539
    thioacylation of 1043

Thiolysis,
  in thio carboxylic acid synthesis 1037
  in thio ester synthesis 1040–1042
  of orthoamides 561
Thionacetates, pyrolysis of, reactivity of
    881, 896
Thionamide chromophore 108
Thionamides, chiroptical properties of
    102, 103
Thionation 1040, 1042
Thioncarbamates, pyrolysis of, rearrange-
    ment in 896
Thioncarbonates, pyrolysis of, rearrange-
    ment in 896
Thiono derivatives, of amino acids,
    chiroptical properties of 102, 103
Thiono esters,
  acylation of 1050, 1051
  alkylation of 1049
  i.r. spectra of 1030, 1031
  mass spectra of 1035, 1036
  n.m.r. spectra of 1033, 1034
  oxidation of 1048
  photolysis of 1051, 1052
  proton acceptor properties of 1045
  reaction of, with carbanions 1047
  rearrangement of 1039
  reduction of 1045, 1046
  solvolysis of 1048
  synthesis of 1040–1042
Thiophene 1052
  electrophilic aromatic substituent
      constants for 904, 906
Thiophenecarboxylic acids, transcarboxy-
    lation of 939
Thiophilic addition 1047
Thiophilic attack 1048
Thio ureas,
  in synthesis of orthocarbamic acid
      esters 550
  reaction of, with orthoamides 561,
      562
1,3,4-Thioxazole 1053
Third-order reactions, micellar catalysis of
    949, 975, 976
Thujyl acetate, pyrolysis of 888
Thymol blue 629
Titration methods, for determination,
  of amides 618
  of anhydrides 629, 630, 635
  of lactams 618
  of lactones 619
o-Toluic acid, mass spectrum of 127, 128
Tolylacetic acid-d, isotope effect of 681
Topology, of the energy surface for the
    tetrahedral intermediate in the
    base-catalysed hydrolysis of methyl
    formate 49, 50, 53

Transamidation 448, 449
Transamination 544, 548
Transcarboxylation 915–942
  effect of catalysts on 932, 933
  effect of cations on 933–935
  intermolecular, ionic mechanism of
      916–940
  ionic character of the liberation of the
      carboxylate group from the
      benzene ring in 917
  mixed 940–942
    as a method for determination of the
        unknown transcarboxylation
        mechanism of one partner
        942
  of mixtures of benzenecarboxylates
      labelled and non-labelled by
      deuterium on the benzene ring,
      exchange of deuterium by
      protium in 929, 930
  simultaneous decarboxylation–
      recarboxylation in 917, 931
Transesterification 419, 420
Transition states,
  in fragmentation processes,
    four-membered 125, 140
    six-membered 125
  in pyrolysis of N-t-butylacetoacetamide,
      eight-membered 868
  in pyrolysis of 2,2-dimethylbut-3-enoic
      acid 861
  in pyrolysis of esters 890
    comparison between t-butyl and
        i-propyl 892
    polarity of 881, 886, 892, 893
    seven-membered 895, 902
    six-membered 877, 878, 883
  in pyrolysis of β-keto acids, six-
      membered 863
  in pyrolysis of phenylacetamides,
      four-membered 869
  in pyrolysis of propanoic acid, five-
      membered 866
  in pyrolysis of trimethylacetic acid,
      seven-membered 867
  theory of 17
  with electron-deficient centres 890
Translational modes of a molecule 17
Transprotonation process, in trans-
    carboxylation 932, 937, 939
Trialkoxyacetonitrile 575
Trialkylammonium dithiocarboxylates,
    synthesis of 1036
Triazolines 544
Tributyl phosphate, pyrolysis of 894
Trichloroacetamide, hydrogen bonding in
    249
Trichloroacetanilides, mass spectra of 155

Trichloroacetic acid, hydrogen bonding in
231, 232
thermodynamics and kinetics of 235,
237
vibrational spectra of 228
Trichloroacetic acid ester, as precursor of
dichlorocarbene in amide acetal
synthesis 540
1,1,1-Trichloroethane, reaction of, with
secondary amines 534
Trichloroisocyanuric acid, in oxidation of
ethers 348
Trichlorothioacetic acid, hydrogen bonding
in 254
Triethylamine, reaction with tetrachlorophthalic anhydride 203
Triethyloxonium fluoroborate, in synthesis
of orthocarbamic acid esters 550
Trifluoroacetamides, mass spectra of 158
Trifluoroacetates, pyrolysis of 886
Trifluoroacetic acid,
electrocatalytic reduction of 800
hydrogen bonding in 230, 232, 235
electron spectra of 229
vibrational spectra of 228
pyrolysis of 865
Trifluoroacetylacetic acid 863
1-Trifluoroacetylindole, hydrolysis of,
micellar effects on 992
3'-Trifluoromethyldiphenylamine-2-
carboxylic acid, hydrogen bonding
in 225
Triglycerides, chiroptical properties of
107
Trihalides,
in acyl halide synthesis 441
in carboxylic acid synthesis 275
Trihalomethanes, in triaminomethane
synthesis 546, 547
α-Trimethylammonio acids, CD spectra of
94
α-N,N,N-Trimethylammonio tertiary
amides, CD spectra of 94, 95
α-Trimethylammonium acids, chiroptical
properties of 88, 90, 91, 94, 95
α-Trimethylammonium esters, chiroptical
properties of 88, 90, 91, 94, 95
Trimethyl-1,1,2-ethanetricarboxylate 809
Trioctylmethylammonium chloride, as
deacylation catalyst 1002, 1007
Tri-n-octylmethylammonium chloride, as
functional micelle 996
Triols, dibenzoate chirality rule for 112,
113
2,2,2-Triphenylacetamide, pyrolysis of
868
Triple bonds, oxidation of, in carboxylic
acid synthesis 331–344

Tris(acylamino)methanes 535
reaction of 562, 567, 571, 578, 586,
597
synthesis of 545–550
Tris(dialkylamino)alkoxymethanes 535
reaction of 568, 576, 584
synthesis of 552
Tris(dialkylamino)methanes 535
reaction of 557, 558, 562, 566, 573,
574, 577, 578
synthesis of 545–550
Tris(diarylamino)methanes 535
Tris(dimethylamino)phenylmethane,
synthesis of 548
Tris(N-methyl-N-phenylamino)methane,
synthesis of 546
'Trithio anhydrides', synthesis of 1044
Tryptophane, CD spectrum of 98
'Twinning' of reactants 995, 1002
Type I cleavage reactions 731, 734, 735,
738
Type II reactions 677, 678, 712–718,
722, 744
Tyrosine, CD spectrum of 98

Ultrafiltration method 980
for measuring the fraction of micellar
incorporated nucleophile 962
Ultraviolet spectroscopy 69–73
of acetic acid 70
of acrylic acid 108
of acylthio derivatives 107
of alanine in hexafluoroisopropanol 70
of amides 612, 613
of amino acids 95, 98
of anhydrides 632, 633, 670
of benzoic acid 72
of carboxylate anions 71
of formic acid 70
of lactams 517, 613
of lactones 626
of oligopeptides 101
of polypeptides 99, 100
of thio acid derivatives 1023,
1025–1029
Unimolecular reactions,
micellar catalysis of 949, 969, 975,
976, 993, 994
pseudo-phase model for 960
salt effects in 977
simple kinetic model for 957–959
micellar inhibition of 959
Unsaturated acids — see also Alkenoic acids
intramolecular cyclization of
1078–1096
mass spectra of 131, 132
α,β-Unsaturated acids, isomerization to
β,γ-unsaturated acids 860

β,γ-Unsaturated acids, pyrolysis of
860–863, 877
substituent effects in 861
Unsaturated amides, mass spectra of 157,
158, 615
α,β-Unsaturated amides, cyclization of
1206–1223
α,β-Unsaturated anhydrides, pyrolysis of
873, 874
Unsaturated esters,
intramolecular cyclization of
1078–1096
mass spectra of 147, 148
Urea, effect of, on micellar catalysed
reactions 984
Urea acetals 535
alcoholysis of 550
reaction of 559, 568, 576, 584
synthesis of 551, 552
Ureas,
reaction of,
with orthoamides 561, 562
with orthoesters 542
N,N,N',N'-tetrasubstituted, in
synthesis of orthocarbamic acid
esters 550

Vacuum ultraviolet CD spectra,
of amino acids 95, 98
of oligopeptides 101
of polypeptides 99, 100
Valence Bond (VB) wave functions 4, 5
δ-Valerolactam, hydrogen bonding in 246
Valine, CD spectrum of 98
g-Values, of thio ester radical anions 1035
Variation theorem 6
Vermeerin, identification of 628
Vermiculine 515
Vernolepin 507
Vernolide, mass spectrum of 627
Vibrational modes of a molecule 17
Vibrational spectroscopy – see also
Infrared spectroscopy, Raman
spectroscopy
of hydrogen-bonded carboxylic acids
226–228
Vilsmeier–Haack intermediates 547
Vinyl benzoate, pyrolysis of 876
cis β-elimination in 876
Vinyl esters, pyrolysis of 876, 877
Vinyl ethers, pyrolysis of 877
Vinylic aldehydes, decarbonylation of
848
retention of configuration during 852

N-Vinyllactams, i.r. and Raman spectra of
608
Vinylogous amide acetals, synthesis of
538
Vinylogous amide chromophore 103
Vinyl substituents at the β-carbon of esters,
effect on rate of pyrolysis of 897
S-Vinyl thiol esters, synthesis of 1038

Water pools 953, 986
Wave functions 4–16
Wavelength effect, in photochemical
reactions 660, 725, 727, 735, 745
Wildfire toxin 521
Willgerodt reaction 369, 452
Wilson's GF matrix method 16
Wittig-type reactions,
in amide synthesis 457
in carboxylic acid synthesis 278
in ester synthesis 422, 423
in lactone synthesis 1115, 1116
Wolff rearrangement,
in carboxylic acid synthesis 380–385
in lactam synthesis 1272, 1273

Xanthates, pyrolysis of 883
reactivity in 896
rearrangement in 896
transition state for 883
X-ray diffraction,
of S-oxides 1049
of thio acid derivatives 1022, 1023
of thiocarbonyl-S-imides 1049

Yukawa–Tsuno version of the Hammett
equation 897

Zearalenone dimethyl ether 515
Zeta potential, in micelles 955
Zinc benzenecarboxylates, transcarboxylation of 935
Zinc naphthalenecarboxylates, transcarboxylation of 938
Zinc salts, as catalysts, in transcarboxylations 932
Zwitterionic form,
of γ-aminobutyric acid 26
of glycine 32
Zwitterions,
as intermediates in pyrolysis,
of alkanoic acids 865, 866
of anhydrides 874
of β-keto acids 863
as micelles 952, 959, 985, 1005